BASIC LABORATORY METHODS FOR BIOTECHNOLOGY

Textbook and Laboratory Reference

Second Edition

Lisa A. Seidman

Program Director, Biotechnology Laboratory Technician Program,
Madison Area Technical College

Cynthia J. Moore

Associate Professor of Biological Sciences,
Illinois State University

Benjamin Cummings
San Francisco Boston New York
Cape Town Hong Kong London Madrid Mexico City
Montreal Munich Paris Singapore Sydney Tokyo Toronto

Editor-in-Chief: Beth Wilbur
Executive Editor: Gary Carlson
Assistant Editor: Kaci Smith
Executive Marketing Manager: Lauren Harp
Executive Managing Editor: Erin Gregg
Managing Editor: Mike Early
Production Supervisor: Shannon Tozier
Production Service, Illustrations, and Text Design: Nesbitt Graphics
Cover Design: Big Wig Design
Manufacturing Buyer: Michael Penne
Director, Image Resource Center: Melinda Patelli
Manager, Rights and Permissions: Zina Arabia
Photo Research: Maureen Spuhler
Text and Cover Printer: Courier, Stoughton
Cover Images: Eric Myer/Taxi/Getty; Comstock Photography/Veer; UpperCut Images/Veer; UpperCut
Images/Veer; Randy Allbritton/Photodisc/Getty Images; Andrew Brookes/CORBIS

ISBN   10-digit: 0-321-57014-6
       13-digit: 978-0-321-57014-7

          Library of Congress Cataloging-in-Publication Data

Seidman, Lisa A.
  Basic laboratory methods for biotechnology : textbook and laboratory reference / Lisa A. Seidman, Cynthia J.
Moore. — 2nd ed.
     p. ; cm.
  Includes bibliographical references and index.
  ISBN-13: 978-0-321-57014-7
  ISBN-10: 0-321-57014-6
  1. Biotechnology—Laboratory manuals. I. Moore, Cynthia J. II. Title.
  [DNLM: 1. Biotechnology—methods. 2. Laboratory Techniques and Procedures.
  TP 248.24 S458b 2009]
  TP248.24.S45 2009
  660.6078—dc22
                              2008034096

2 3 4 5 6 7 8 9 10—CRS—12 11 10 09

Benjamin Cummings
is an imprint of

www.pearsonhighered.com

This book is dedicated to our patient supporters who have survived yet another round of deadlines and angst: My mother, Frances R. Seidman, who is excited that her name will appear in a book in the archives of the Library of Congress and who has been waiting for this book to be finished with bated breath for ever and ever; also to Percy, Ilana, and Steve, who have always been there for us.

TABLE OF CONTENTS

PREFACE TO THE SECOND EDITION

More than ten years have gone by since we began writing the first edition of this textbook. Many things have changed since we began writing—in our lives, in the lives of our students, and in the biotechnology industry. But the fundamentals of good laboratory practice that underlie work in the laboratory have not changed much, and the importance of understanding the basics has not changed at all. Understanding the basics is still critical for anyone who wants to work effectively in the laboratory.

In this edition, we have maintained those features of the text that make it accessible to readers who may have limited backgrounds in science and math. Beginners in biotechnology must master many concepts and an associated vocabulary in order to progress to more advanced scientific reading. Some of the words in this vocabulary will be entirely new, other words may seem familiar, yet they have unexpected meanings in the technical literature. We introduce new concepts and associated vocabulary as clearly and sequentially as possible. We include a thorough glossary (now at the back of the text). There are many examples and problem sets provided that demonstrate and promote analytical reasoning. We include questions to aid students in reviewing the material presented. We have added more figures and photos to this edition to better illustrate concepts, equipment, and procedures, and to enhance our readers' visual understanding of the laboratory landscape. We have also included more stories and case studies to probe related ethical, societal, and human issues.

The most substantial changes in the second edition are in the first eight chapters of the book, which have been revised and expanded to reflect the extraordinary changes that have occurred in biotechnology. Chapters 1 and 2 introduce the biotechnology industry, an enterprise that transforms biological knowledge into widely diverse products; biofuels, biopharmaceuticals, DNA fingerprints, industrial enzymes, and stem cell therapies all fit comfortably within biotechnology's domain. A new chapter, Chapter 3, is devoted to a discussion of the development and production of biopharmaceuticals because of their central role in the growth of the biotechnology industry. Chapters 4 through 8, which cover quality and regulatory affairs, have been expanded and updated, including an introduction to new regulatory initiatives by the Food and Drug Administration. Every biotechnologist needs at least some familiarity with the business and quality aspects of making products and providing services.

The remainder of the textbook, Chapters 9 through 36, is an introduction to fundamental biotechnology laboratory practices. Most of the chapters in the later part of the book required updating and revision to remain current with modifications in equipment and general laboratory practices. Chapter 24, focusing on performing quality assays and tests, was added to provide a more systematic introduction to this important topic, which is seldom explicitly taught. The discussion of laboratory solutions was expanded with a new chapter on culture media for bacterial and mammalian cells. Needless to say, the chapters dealing with computers required substantial revision, particularly Chapter 36 that discusses the Internet. But there, as in the rest of the book, we have tried to emphasize underlying principles of data storage, analysis, and computation; these remain relevant even as the technology advances at a rapid pace.

PREFACE TO THE FIRST EDITION

This is an exciting time to work in biotechnology. The Human Genome Project is generating fundamental genetic information at a breathtaking rate; basic research findings are being applied in medicine, agriculture, and the environment; and a variety of new biotechnology products are moving into production. Behind each of these accomplishments are teams of scientists and technicians whose everyday work makes such achievements possible.

For the past twelve years, we have been working with students who are beginning their careers as technicians and bench scientists in biotechnology laboratories. In order to best assist our students, we, and our colleagues elsewhere in the United States, have explored what entry level biotechnologists do at work and what abilities they need to perform this work. We have been impressed with the complexity and diversity of technical roles and responsibilities, and the importance of the skills that bench workers bring to their jobs. This book emerges partly from our experiences working with students and our explorations into the nature of the laboratory workplace*.

This book also results from our personal experiences in the laboratory. As graduate students we struggled to master the "laboratory lore" that was passed among "post-docs" and graduate students in a not always coherent chain. Some of what is in this book is the systematic introduction to laboratory lore that we wish we had received.

The result of our efforts is not a laboratory manual; this text contains few step-by-step procedures. Nor is it a book about molecular genetics, immunology, or cell culture—there are already many excellent specialized texts and manuals on these topics. This book rather is a textbook/reference manual on basic laboratory methods and the principles that underlie those methods. These basics are important to every biotechnologist, regardless of whether one is cloning DNA or purifying proteins, whether one is working in an academic setting or is employed in a company.

We intend this book to assist students preparing to become biotechnology laboratory professionals, those who already work in the laboratory, and biology students who are learning to operate effectively in the laboratory. Others who may also find this book helpful include high school teachers and their advanced students, and industry trainers. We have endeavored to make this text accessible to beginning college students with a limited science and math background. Some sections, such as the math review in Unit III, could be skipped or skimmed by more experienced readers. At the same time as we tried to make this book practical and accessible, we also endeavored to provide enough background theory so that readers will understand the methods they use and will be prepared to solve the unavoidable problems that arise in any laboratory.

Although we focus on the biotechnology laboratory, the majority of topics we cover are of importance to individuals working in any biology laboratory. A few topics, such as quality regulations and standards, are included because they are important for those working in biotechnology companies. As biotechnology companies mature, their focus shifts from research into commercial production. As this maturation occurs, scientists and technicians often find that they must add terms like "GMP," "ISO 9000," and "quality systems" to their technical vocabulary. This book therefore weaves a conversation about regulations and standards into many chapters.

We are aware that the basic methods in this book (such as how to mix a solution or weigh a sample) are less glamorous than learning how to manipulate DNA, or how to clone a sheep. However, we also know that, in practice, the most sophisticated and remarkable accomplishments of biotechnology are possible only when the most basic laboratory work is done properly.

---

*The results of some of these discussions about the biotechnology workplace are summarized in the National Voluntary Skill Standards Documents in Agricultural Biotechnology and the Biosciences. (FFA, "National Voluntary Occupational Skill Standards: Agricultural Biotechnology Technician," National FFA Foundation, Madison, WI, 1994 and "Gateway to the Future, Skill Standards for the Bioscience Industry," Education Development Center, Newton, MA, Inc., 1995.)

ABOUT THE AUTHORS

Lisa Seidman received her Ph.D. from the University of Wisconsin and has taught for more than 20 years in the Biotechnology Laboratory Technician Program at Madison Area Technical College in Madison, Wisconsin. Cynthia Moore received her Ph.D. in Microbiology from Temple University School of Medicine. She is currently an Associate Professor of Biological Sciences at Illinois State University, where she also serves as Director of Biology Teacher Education.

The authors welcome comments and feedback from readers. You can e-mail Lisa Seidman at lseidman@matcmadison.edu and Cynthia Moore at cjmoor1@ilstu.edu.

ACKNOWLEDGMENTS

As in the first edition, many people have contributed to this book and we appreciate all their help. We thank our students who provided feedback and purpose over the years. We thank our many talented colleagues who used the first edition in their classes, and who provided support, ideas, editing, encouragement, and feedback. Thank you to Jeanette Mowery, Diana Brandner, Rebecca Josvai, Joseph Lowndes, Mary Ellen Kraus, Linnea Fletcher, and Elaine Johnson for their much-appreciated support. We thank the staff at Benjamin Cummings for their skill and hard work, including Gary Carlson, Kaci Smith, and Shannon Tozier. We also very much appreciate the work of the talented production team including the staff at Nesbitt Graphics, Inc. and photo researcher Maureen Spuhler. Many thanks to the expert reviewers who helped make this a much better book: Craig Caldwell, Salt Lake Community College; Michael Fino, MiraCosta College; Todd Freeman, Illinois State University; Collins Jones, Montgomery College; Melanie Lenahan, Raritan Valley Community College; Ying-Tsu Loh, City College of San Francisco; Nancy Magill, Indiana University; Charlotte Mulvihill, Oklahoma City Community College; Traci Nanni-Dimmey, Berdan Institute; Virginia Naumann, St Louis Community College; Trish Phelps, Austin Community College; Rebecca Siepelt, Middle Tennessee State University; Salvatore Sparace, Clemson University; Duncan Walker, Array BioPharma; and Dwayne Zeiler, Bradley University.

While we do not want to endorse the products of any company over any other, we do want to acknowledge and thank the many companies that so graciously provided us with illustrations and technical information.

This material is based in part on work supported by the National Science Foundation Advanced Technology Education Initiative, under grant number 0501520. Any opinions, findings, conclusions, or recommendations expressed in this material are those of the authors and do not necessarily reflect the views of the National Science Foundation.

UNIT I

Introduction to the Biotechnology Workplace

Biotechnology is a sweeping term that is used to refer to many things—scientific discoveries, laboratory techniques, processes, commercial enterprises, and more. Biotechnology is ancient, and yet is as cutting edge as today's latest scientific discoveries.

The discoveries of biological researchers provide the foundation of scientific knowledge in which biotechnology is rooted. Bio*technology,* however, is about **technology,** *the application of knowledge to provide goods and services valued by people.* Unit I provides a brief, panoramic introduction to the science underlying

biotechnology and discusses some of the many products and services that emerge from that science.

Chapter 1 is an overview of the many areas of biological discovery that provide the basis for a wide array of biotechnology products.

Chapter 2 discusses the organization of biotechnology companies and their role in transforming scientific knowledge into products.

Chapter 3 focuses on the development of biopharmaceuticals and other medical products.

BIBLIOGRAPHY FOR UNIT I

There are many good books, journals, and websites on biotechnology. A few examples that discuss topics covered in this unit are listed here, but this list is by no means comprehensive and new resources are appearing constantly. Specific quotes and article references are directly cited in the text.

Books

Clark, David, and Russell, Lonnie. *Molecular Biology Made Simple and Fun.* 3rd ed. St. Louis: Cache River Press, 2005.

International Intellectual Property Section, Office of Trade Negotiations, Department of Foreign Affairs and Trade. *Intellectual Property and Biotechnology: A Training Handbook.* Australia, 2001. (Available online at http://www.dfat.gov.au/ip.)

Kreuzer, Helen, and Massey, Adrianne. *Biology and Biotechnology, Science, Applications, and Issues.* Washington, DC: ASM Press, 2005.

Kreuzer, Helen, and Massey, Adrianne. *Recombinant DNA and Biotechnology: A Guide for Teachers.* Washington, DC: ASM Press, 1996.

Micklos, David A., Freyer, Greg A. with Crotty, David A. *DNA Science, a First Course.* 2nd ed. New York: Cold Spring Harbor Laboratory Press, 2003.

Thieman, William J., and Palladino, Michael A. *Introduction to Biotechnology.* San Francisco: Benjamin Cummings, 2004.

Renneberg, Reinhard. *Biotechnology for Beginners.* Burlington, VA: Academic Press, 2006. (A beautifully illustrated and interesting to read introductory text.)

Watson, James D., Witkowski, Jan, Myers, Richard, and Caudy, Amy. *Recombinant DNA: Genes and Genomics, A Short Course.* 3rd ed. New York: W.H. Freeman and Co., 2007.

Articles

Chapman, Kenneth, Fields, Timothy, and Smith, Barbara. "The Case of the q.c. Unit." *Pharmaceutical Technology* (January 1996): 74–9.

Cookson, Clive, Rennie, John, Soares, Christine, Gardner, Richard, Watson, Tim, Waldmeir, Patti, and Stix, Gary. "The Future of Stem Cells: A Special Report with Financial Times." *Scientific American* (July 2005): A1.

Flamm, Eric L. "How FDA Approved Chymosin: A Case History." *Bio/technology* (1991): 349–51.

Stix, Gary. "Hitting the Genetic Off Switch." *Scientific American*, October 2004, 98–101. (An article about RNAi.)

Stix, Gary. "The Land of Milk and Money." *Scientific American,* November 2005, 102–5. (This article is about the use of animals to produce biopharmaceuticals.)

Harlin, Michael B. "How to Prove When You Made Your Invention." *BioProcess International,* May 2004, 24–8.

National Cancer Institute. "Gene Therapy for Cancer: Questions and Answers." Washington, DC: U.S. National Institutes of Health. http://www.cancer.gov/cancertopics/factsheet/therapy/gene.

Zachary, G. Pascal. "GM Comes to Sub-Saharan Africa (sort of)." Tempe, AZ: Consortium for Science, Policy, and Outcomes at Arizona State University. http://www.cspo.org/ourlibrary/perspectives/pascal.htm.

Websites

BioPharm International. A monthly print magazine that is available online with excellent information about the biopharmaceutical industry. http://www.biopharminternational.com/biopharm.

Periodic *BioPharm International* supplements provide more comprehensive information on selected topics. The following are examples of their supplements:

- *The BioPharm Guide to Biopharmaceutical Development,* 2nd ed. *BioPharm International Journal,* March 2002.

- *The Bio Process: A Primer for Biotech Manufacturing, BioPharm International Journal* 4, sup. 2. (March 2006).

Biotech Work Portal. A website funded by the U.S. Department of Labor, with information about biotechnology careers. Job titles discussed range from librarian to molecular biologist, manufacturing technician, buyer, metrologist, and many more. http://www.biotechwork.org

The Dolan DNA Learning Center. Award-winning website with information, videos, animations, and interactive activities covering molecular biology and biotechnology. This extensive website can be used as an online textbook for high school and college courses. http://www.dnalc.org

National Cancer Institute, NCI. This U.S. federal agency has an informative website with information about cancer and its treatment, including gene therapy. http://www.cancer.gov

United States Patent and Trademark Office. This U.S. federal agency has an informative website with information about patenting. http://www.uspto.gov/web/offices/com/iip/index.htm

CHAPTER 1

The Modern Biotechnology Industry: A Broad Overview

I. DNA, THE HEART OF MODERN BIOTECHNOLOGY

A. Introduction

Humans are naturally curious. Human history from ancient to modern times is punctuated by discoveries and the application of those discoveries to create tools, foods, medicines, dwellings, weapons, art, and the like. *The application of knowledge to make products useful to humans is called* **technology.** The research laboratory is a relatively recent manifestation of human curiosity, a place invented to study nature. In our time, significant discoveries about the intricacies of living systems have emerged from laboratory studies and have been transformed into the products of "biotechnology."

What is biotechnology? In a broad sense, **biotechnology** *refers to the use of organisms, or materials derived from organisms, to make useful products.* In this sense, biotechnology is not new. Thousands of years ago, humans discovered how to use organisms to make products such as wine, cheese, and bread. When people use the term "biotechnology," however, they are usually thinking about the more dazzling products of modern biotechnology, such as cloned sheep, gene therapies, and DNA fingerprints—not bread and cheese.

Modern biotechnology is deeply rooted in basic research from the biology laboratory. Over the years, researchers around the world have explored fundamental questions in biology such as the complexities of cellular function, the mechanisms by which information is passed from generation to generation, how individuals develop from a single, fertilized egg cell, how the complex immune system is coordinated, and many others. This is **basic research,** *research that is performed in order to understand nature.* The modern biotechnology industry has emerged as knowledge from basic biological research is transformed into products. In this modern sense, **biotechnology** *is the transformation of biological knowledge and discovery into useful products.*

It is important for individuals working in the biotechnology industry to have a sense of its breadth and scope. This chapter therefore provides a brief (and somewhat idiosyncratic) overview of the modern biotechnology industry, the products of the biotechnology industry, and the processes by which biological knowledge is transformed into a host of products. We begin this overview with a very brief introduction to the key biological molecules—DNA, RNA, and proteins—because it is not possible to understand biotechnology without some understanding of these molecules. For a more comprehensive introduction to the science and applications of biotechnology, see this unit's bibliography on p. 65.

B. A Brief Overview of Molecular Biology

i. DNA TELLS THE CELL HOW TO MAKE PROTEINS

Biotechnology makes use of powerful tools to manipulate DNA (deoxyribonucleic acid); in fact, the term "biotechnology" is sometimes used to refer to the manipulation of DNA. The image of the DNA double helix has become so pervasive in our culture that it is easy to forget that biologists did not understand the structure and function of DNA until the middle of the twentieth century. At that time, the work of many researchers led to the realization that DNA is the chemical of inheritance. **DNA** *is the genetic substance by which parents pass information to their offspring.* DNA tells the offspring how to grow and develop to form an organism with that individual's unique traits.

How can a molecule of DNA contain and transmit all the information necessary to build an organism? The key to answering this question was the discovery of the molecular structure of DNA. The structure of DNA was elucidated in the 1950s by the research of various scientists including Erwin Chargaff, Rosalind Franklin, Maurice Wilkins, James Watson, and Francis Crick. The work of these scientists culminated in a series of scientific papers, the most famous of which, by Watson and Crick, was published in 1953 in the prestigious journal, *Nature* (Watson, James, and Crick, Francis. "A Structure for

Deoxyribose Nucleic Acid." *Nature* 171 (1953): 737–8.). Watson, Crick, and Wilkins won a Nobel Prize in 1962 for their contributions.

DNA is a linear molecule consisting of four types of *molecular subunits,* called **nucleotides,** connected one after another into long strands. The four types of nucleotide are distinguished from one another because they each contain a different **base.** There are four bases in DNA: adenine, guanine, thymine, and cytosine. In most situations DNA is double-stranded, meaning that two linear strands of DNA associate with one another. Double-stranded DNA twists to form the famous "double helix," see Figure 1.1.

It is the sequence (order) of the four types of nucleotides making up a strand of DNA that contains information. A particular sequence of nucleotides "tells" the cell how to build a particular protein. Therefore, a **gene** *can be considered to be an ordered sequence of nucleotides (thus a stretch of DNA) that contains information that "tells" the cell how to make a particular protein.* The proteins provide structure to the cell and do the work of the cell, hence, by encoding proteins, the DNA "controls" the cell.

Cells contain many *genes that are organized into long DNA macromolecules called* **chromosomes.** In bacteria, the entire collection of genes, the *genome,* lies on a single chromosome. Organisms that are more complex have many more genes that are arranged onto multiple chromosomes, which are sequestered in *a specialized membrane-bound area of the cell, the* **nucleus.** Bacterial cells do not have a nucleus. Bacteria are termed **prokaryotic,** *meaning they lack a nucleus. Cells with a nucleus are termed* **eukaryotic.** The cells of plants, animals, and yeast are eukaryotic.

Human cells each (with a few exceptions) contain two copies of an individual genome, one copy from the mother and another from the father. The human genome consists of about three billion base pairs of DNA where a *base pair* is two nucleotides across from each other, one on each of the two opposing DNA strands, see Figure 1.1c. These three billion base pairs are arranged onto 23 chromosomes. Since most human cells have two copies of the genome, they contain 46 chromosomes, 23 from each parent. These chromosomes, which are inherited from an individual's parents, contain the information required to "build" that individual.

Every cell in a particular individual (with a few exceptions) contains the same genes. Clearly, however, a nerve cell is different than a muscle cell or a skin cell. Different cells have different characteristics because only some of the genes present in that cell are "turned on," that is, make the protein for which they code. *When a gene is "turned on," the cell makes the protein that the gene encodes and we say that the gene coding for that protein is* **expressed.** In muscle cells, for example, genes are expressed that code for proteins required for muscle contraction. Other genes are expressed in nerve cells, still others in skin cells, and yet others in retinal

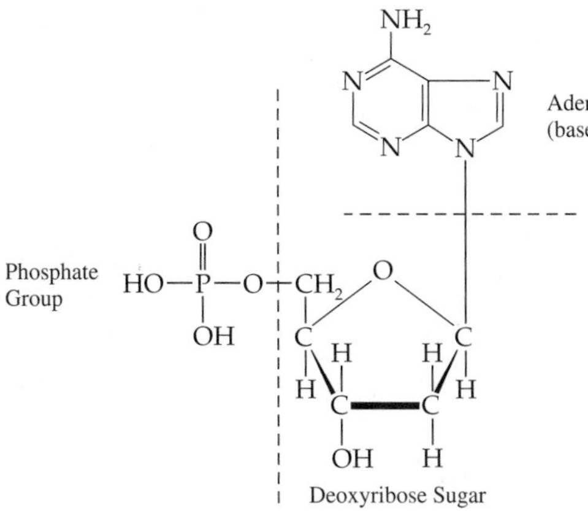

Figure 1.1.   The Structure of DNA.
a. The basic subunit of DNA is the nucleotide, which consists of a sugar, a phosphate group, and a nitrogenous base, in this case, adenine. Each of the four types of DNA nucleotide has a different base but the rest of the nucleotide is the same. **b.** The subunits of DNA are connected by covalent phosphodiester bonds between the phosphate group on one nucleotide and the sugar on the next. The bases extend out to the side. In solution, the phosphate groups lose a H^+ and so have a negative charge. Three nucleotides linked together are shown here. **c.** Chromosomal DNA consists of two strands of DNA held together by hydrogen bonds between complementary bases. A guanine will always pair with a cytosine and an adenine with a thymine. Observe that three hydrogen bonds stabilize each G–C linkage, but only two hydrogen bonds stabilize each A–T linkage. The order in which the four types of nucleotides are arranged along the strands encodes genetic information.

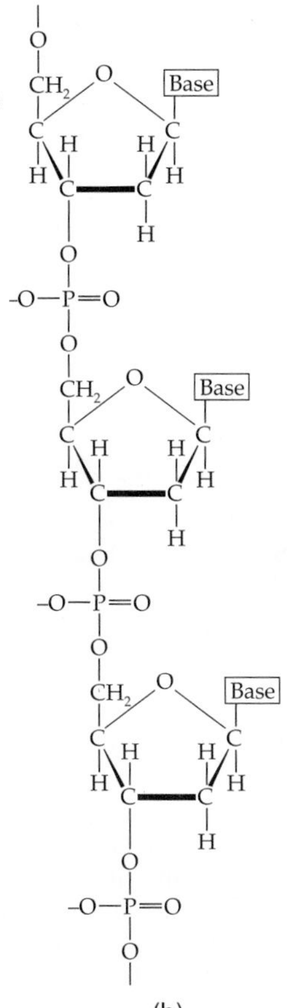

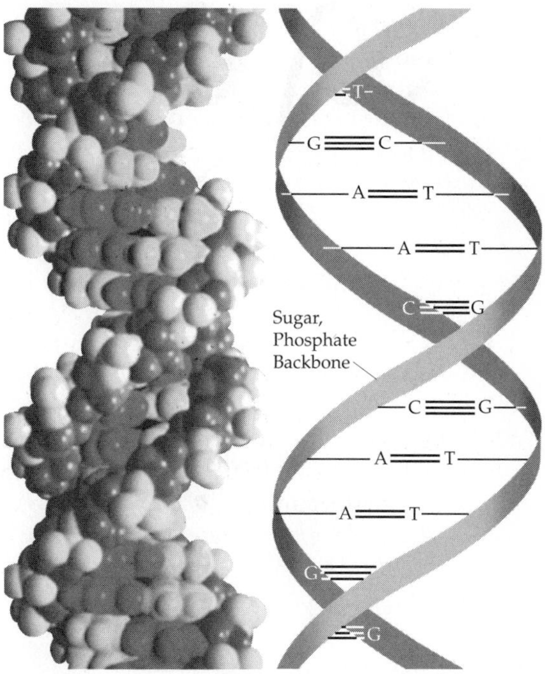

cells. **Differentiation** *is the process in which an embryonic cell, which originally has the capacity to become any type of cell in the body, matures into a particular cell type with a specialized structure and function (e.g., muscle, nerve, skin, retina).* **Regulatory DNA sequences** *act as "switches" to control which genes are turned on at a given time in a given cell.* DNA thus plays its vital role by directing each cell to make the correct proteins required to do the work of that cell.

ii. PROTEINS PERFORM THE WORK OF CELLS

Every cell in an organism has many **proteins** and it is these proteins that give the cell its structure and perform the work of that cell. **Proteins** *are diverse molecules composed of chains of varying numbers of amino acid building blocks that link together, like beads on a chain, and then fold into complex three-dimensional structures.* Each protein has its own structure that results from the amino acids that comprise it; different amino acids arranged in different orders give rise to differently structured proteins. The varied structures of proteins allow them to perform many tasks. Hemoglobin, for example, is a protein found in red blood cells; it has a unique structure that enables it to transport oxy-

gen. Antibodies are proteins that recognize and help the body neutralize foreign invaders (e.g., bacterial or viral pathogens).

iii. THE ASSEMBLY OF PROTEINS

Proteins are assembled from amino acid building blocks by a specialized cellular component, the **ribosome.** DNA conveys information to the ribosome, directing it to stitch together specific amino acids in a specific order to make a particular protein. *Information moves from DNA to ribosomes via an intermediary molecule, called* **messenger ribonucleic acid, mRNA.** When a protein is to be made by the cell, the DNA that encodes the information for how to manufacture that protein is used to **transcribe,** or *synthesize,* mRNA. The mRNA molecules travel to the ribosomes where they direct the assembly of a protein molecule from amino acid subunits. Information thus flows as follows:

$$DNA \rightarrow mRNA \rightarrow protein$$

This pathway by which information flows from DNA via mRNA to code for protein is sometimes termed *The Central Dogma* of biology.

The process in which a protein is manufactured is called **translation,** see Figure 1.2 on p. 6.

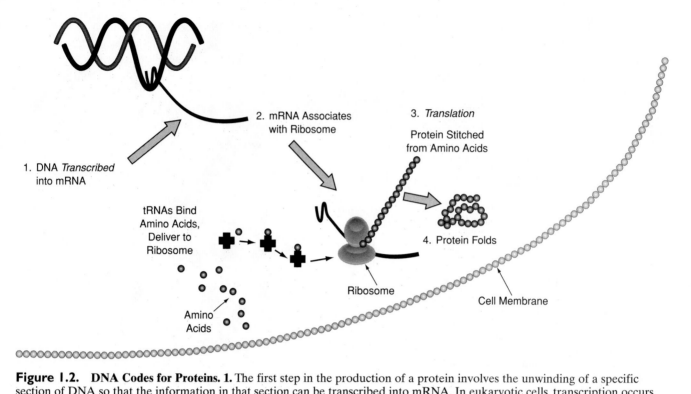

Figure 1.2. DNA Codes for Proteins. 1. The first step in the production of a protein involves the unwinding of a specific section of DNA so that the information in that section can be transcribed into mRNA. In eukaryotic cells, transcription occurs in the membrane-bound nucleus and the resulting mRNA exits the nucleus. **2.** The mRNA associates with a ribosome. **3.** Amino acid subunits are picked up by transporter molecules, called tRNA, and are taken to the ribosome where they are added to the growing protein chain in the order directed by the mRNA. This is called *translation*. **4.** Proteins fold into specific three-dimensional shapes based on their amino acid sequences.

C. Introduction to Recombinant DNA Techniques

i. THE TOOLS OF BIOTECHNOLOGISTS

As scientists have come to understand the structures and functions of DNA, RNA, and proteins, they have also devised tools to manipulate these biological molecules. Tools to manipulate DNA include:

- Enzymes that cut DNA at specific sites.
- Enzymes that ligate (join) DNA strands together.
- Techniques to visualize DNA.
- Techniques to separate DNA fragments from one another.
- Techniques to identify fragments of DNA with specific sequences.
- Techniques that amplify DNA (generate many copies of a specific gene fragment).
- Techniques to determine the nucleotide sequence of a piece of DNA.
- Techniques to synthesize DNA.

The process of studying DNA and other biological molecules is ongoing. As scientists develop ever more sophisticated methods of manipulating biological molecules, they use these methods to probe more deeply into the workings of biological systems. As the functions of biological molecules are better understood, researchers are better able to manipulate these molecules, making new discoveries about the intricacies of life.

One of the most powerful accomplishments of modern biotechnology is the development of tools to manipulate DNA in such a way that genetic information (DNA) from one organism is transferred to another. *When a biologist causes a cell or organism to take up a gene from another organism, we say the cell or organism is* **genetically modified** *or* **genetically engineered,** see Figure 1.3. The related term **recombinant DNA (rDNA)** *refers to DNA that contains sequences of DNA from different sources that were brought together (i.e., re-combined) using the tools of biotechnology.* Under the proper conditions, a genetically modified cell will express (produce) a protein encoded by an introduced gene. The ability of biologists to create genetically modified organisms that make proteins they would not ordinarily produce is so powerful that the term "revolutionary" is often applied to it.

DNA from one organism can be transferred into another in various ways. **Plasmids** *are small, circular molecules of DNA that occur naturally in many types of bacteria and yeast and that exist separately from their chromosomes.* Scientists often ligate a gene of interest into a plasmid using enzymes that are able to stitch together two pieces of DNA. Under proper conditions, plasmids are readily taken up by bacterial cells (and other types of cells as well) and can carry with them a gene of interest. Plasmids are called **vectors** *when they*

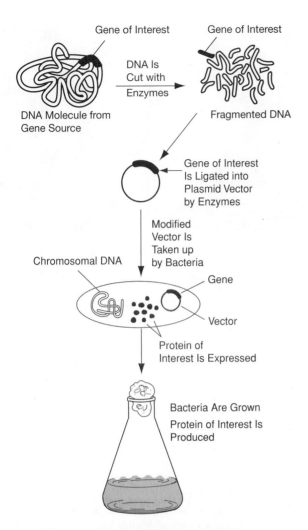

Figure 1.3.   The Genetic Modification of Bacteria. A gene of interest is isolated and inserted into a vector. The vector is taken up by bacteria, which are now said to be *transformed,* or genetically modified. The transformed bacteria can be grown in flasks or larger containers. Under the proper conditions, the bacteria express the protein encoded by the foreign DNA.

carry a desired gene into recipient cells. A gene of interest can also be introduced into the DNA of a virus so that the virus acts as a vector when it infects a host cell. In some cases, DNA can be directly injected into a recipient cell. Electrical current can also be used to induce cells to take up foreign DNA. *When a bacterial cell takes up foreign DNA (e.g., takes up a plasmid vector), it is said to be* **transformed.** *When a eukaryotic cell takes up foreign DNA, it is said to be* **transfected.**

ii. Using Genetically Modified Organisms

By the early 1970s, scientists and entrepreneurs realized that the powerful techniques of manipulating DNA could be utilized to make products of commercial importance; the modern biotechnology industry emerged from this vision. The scientific community, however, was also concerned that this new, powerful technology might create unknown and potentially significant safety issues. Prominent scientists called for a temporary halt

in research—a request without precedent in the scientific community—in order to assess the safety of recombinant DNA methods. A Recombinant DNA Advisory Committee (RAC) was established by the National Institutes of Health (NIH) to study the issues. A larger gathering, the Asilomar Conference, was convened in February, 1975, to discuss the safety of recombinant DNA technology. The conclusion that emerged from this scientific conference was that most rDNA work should continue, but appropriate safeguards in the form of physical and biological containment procedures should be put in place.

Shortly after the Asilomar Conference, the company, Genentech, was founded; Genentech and Cetus Corporation are generally acknowledged to be the first modern biotechnology companies. According to the Genentech website (http://www.gene.com), the company was formed in 1976 by business investor Robert Swanson and scientist Herbert Boyer who saw the commercial potential of the new methods of manipulating DNA. By 1978 Genentech scientists were able to transfer the gene coding for human insulin into bacteria. Insulin is a protein hormone required for proper regulation of sugar levels in blood and cells, and is used to treat diabetes. Type I diabetes is a serious, relatively common disease that occurs when pancreatic cells that normally produce insulin are destroyed. The researchers were also able to induce the bacteria to express the human gene and thus produce human insulin at levels that allowed for commercial production.

Bacteria that contain an introduced gene, such as insulin, can be grown in large quantities in *special vats called* **fermenters.** *The large-scale cultivation of bacteria to produce a product is called* **fermentation.** The bacteria produce the product of interest, which can then be isolated and purified using protein-separation techniques. The bacteria thus become a "factory" to manufacture the protein product of interest. This technology, which takes advantage of a cellular process that nature has optimized over eons, lies at the heart of modern biotechnology.

The production of insulin in bacteria was a scientific accomplishment; bacteria have absolutely no use for insulin and would never produce it without human intervention. Insulin made by recombinant DNA methods was also a medical achievement. Prior to the 1980s, insulin to treat diabetics was purified from the pancreas of animals slaughtered for human consumption. Animal insulin is similar, but not identical to human insulin; therefore some diabetics developed allergies to the drug. In 1982 human insulin made by recombinant DNA technology was approved for use by patients. The production of insulin by bacteria is a classic example of how basic scientific knowledge was transformed into a product that helps millions of diabetic patients—and also confirmed the significant commercial potential of recombinant DNA methods.

Bacteria were the first genetically engineered cells used to make a commercial product. Researchers, however, soon learned to genetically modify other types of cells as well, including those from mammals, yeasts, insects, and plants. **Cultured cells** *are those grown in flasks, dishes, vats or other containers outside a living organism.* (The term "cell culture" most commonly refers to eukaryotic cells, although prokaryotic cells are also grown under "culture" conditions.) *Some cells can be induced to divide indefinitely in culture using particular procedures, and so immortal* **cell lines** *have been estab-* lished. Certain immortal cell lines are commonly used for production purposes. The techniques used to grow eukaryotic cells for production are analogous to those used in bacterial fermentation, though the growth conditions microbes require are somewhat different than those of eukaryotic cells. *The specialized growth chambers used for eukaryotic cells are usually called* **bioreactors** (instead of *fermenters*). Figure 1.4 illustrates the use of genetically modified mammalian cells to produce a protein product.

We have so far talked about introducing foreign DNA into cells and then growing the cells to high densities in fermenters or bioreactors. It is also possible to introduce a gene of interest into whole plants and animals, although this is more complex than manipulating cultured cells. *A plant or animal whose cells are genetically modified is called* **transgenic.** Figure 1.5 shows one of the first transgenic animals. The two animals in this photo are litter mates. The mouse on the left, however, was genetically modified by the introduction of the gene for rat growth hormone. This was accomplished by microinject-

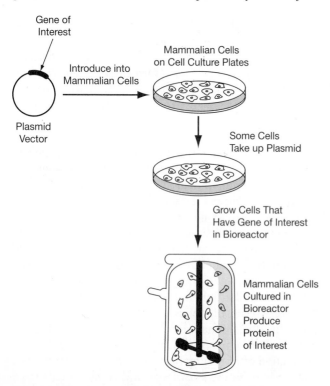

Figure 1.4. The Use of Genetically Modified Cultured Mammalian Cells to Produce a Protein of Interest. Once cells have been transfected with a gene of interest, they can be grown in large quantities in a bioreactor and the protein product can be isolated.

Figure 1.5. The First Transgenic Animal. The gene for rat growth hormone was microinjected into a fertilized mouse egg, which was then implanted into a surrogate mother. The resulting transgenic mouse on the left expressed the gene, resulting in an exceptionally large animal. The mouse on the right is normal size. (Photo courtesy R. L. Brinster/Peter Arnold.)

ing the growth hormone gene into a fertilized mouse egg; thus all the cells in the resulting mouse contained the foreign gene. The rat growth hormone gene was expressed, causing the transgenic animal to be unusually large.

Biologists can produce genetically modified microorganisms and cultured cells, transgenic animals, and transgenic plants. There are many commercial applications that involve genetically modified organisms. One application is to use cultured cells to produce proteins of value, such as insulin. It is also possible to create transgenic plants and animals that have desirable characteristics. For example, transgenic crop plants may have enhanced resistance to disease. Figure 1.6 summarizes some of these points regarding the genetic modifications of organisms.

D. The Genome Projects

A handful of genes had been identified and isolated by the 1980s, each one of which was painstakingly studied by researchers over a period of many years. Some of these genes, like the gene that codes for insulin, became the basis for early biotechnology products. By the 1990s, the pace of gene identification had accelerated from a trickle into a torrent.

The Human Genome Project (HGP), which began formally in 1990 and ended in 2003, contributed to the accelerated pace of gene discovery. The goals of the HGP, as stated on the project's website, were to "determine the complete sequence of the 3 billion DNA subunits (bases) [that comprise the human genome], identify all human genes, and make them accessible for further biological study." As we saw earlier in this chapter, the sequence of DNA nucleotides is of tremendous importance because it encodes the information for creating an individual organism. *The study of entire genomes, their functions and regulation, has come to be called* **genomics.**

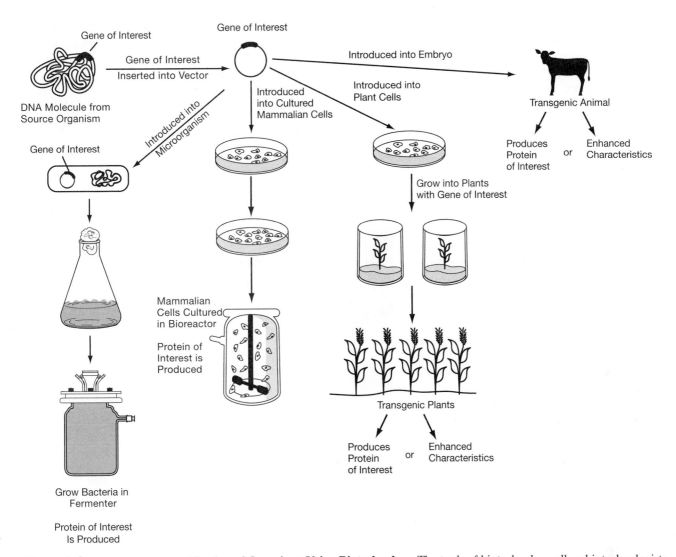

Figure 1.6.   The Genetic Modification of Organisms Using Biotechnology. The tools of biotechnology allow biotechnologists to insert recombinant DNA into various cell types including those of bacteria, cultured mammalian cells, whole plants, and whole animals.

The information gathered by the Human Genome Project plays a key role in shaping the growth of the biotechnology industry today. Medicine is the area where the applications of knowledge from the HGP are most obvious. Scientists and clinicians have learned a great deal about dozens of genetic conditions, such as muscular dystrophies, inherited colon cancer, Alzheimer's disease, and familial breast cancer, see Figure 1.7 on p. 10. Understanding the genetic basis of health and disease is already leading to better diagnosis of illnesses and is also leading to therapies that treat the root causes of disease rather than simply treating symptoms. Sequencing methods continue to improve. It is already possible, and in the near future it may be economical, to sequence the genomes of individual people with the goal of helping individuals lead healthier lives.

The sequencing technologies first used for sequencing human DNA are enabling researchers to efficiently determine the DNA sequences of many other organisms, from bacteria to mice and rice. For example, the genomes of several important pathogenic microorganisms have been sequenced and this information is expected to lead to better diagnosis and treatment of the diseases they cause. Understanding the genomes of other bacteria is expected to help scientists find new methods of harnessing energy from biomass, more effective ways of using bacteria to clean up pollutants in the environment, and better methods for using bacteria in industrial processing.

The sequence of the rice genome was published in 2005 and is helping scientists understand the traits that make rice the most important food source for more than 50% of the population of the earth. This information can be used to improve rice and other crops.

The knowledge acquired from the various genome projects has had a significant impact across the biological sciences. This knowledge has deepened our understanding of the evolutionary relationships among organisms; striking similarities in genes among disparate organisms demonstrate the continuity of life on earth.

The genome projects and other molecular biology endeavors create vast amounts of information. Huge computer databases are required to store and organize all this genomic information. Consider, for example, that the human genome contains three billion base pairs. If each of these base pairs were written as a single letter, this information would fill 200 telephone directories, each 500 pages long. It would take almost a century to recite this information speaking one letter per second for 24 hours each day. And, this is only the genetic sequence for humans! *Biologists, mathematicians, and computer scientists have come together to find ways to store, analyze, and utilize vast amounts of biological information; the field they created is called* **bioinformatics.**

136 Million Base Pairs

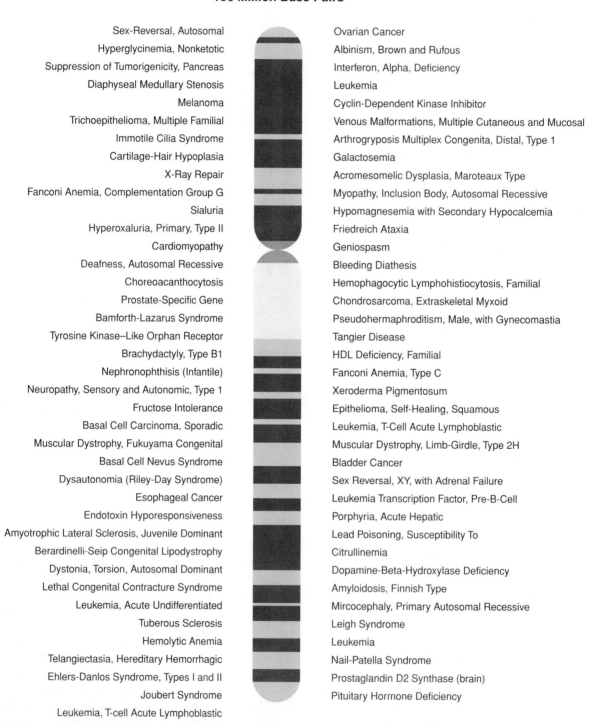

Sex-Reversal, Autosomal	Ovarian Cancer
Hyperglycinemia, Nonketotic	Albinism, Brown and Rufous
Suppression of Tumorigenicity, Pancreas	Interferon, Alpha, Deficiency
Diaphyseal Medullary Stenosis	Leukemia
Melanoma	Cyclin-Dependent Kinase Inhibitor
Trichoepithelioma, Multiple Familial	Venous Malformations, Multiple Cutaneous and Mucosal
Immotile Cilia Syndrome	Arthrogryposis Multiplex Congenita, Distal, Type 1
Cartilage-Hair Hypoplasia	Galactosemia
X-Ray Repair	Acromesomelic Dysplasia, Maroteaux Type
Fanconi Anemia, Complementation Group G	Myopathy, Inclusion Body, Autosomal Recessive
Sialuria	Hypomagnesemia with Secondary Hypocalcemia
Hyperoxaluria, Primary, Type II	Friedreich Ataxia
Cardiomyopathy	Geniospasm
Deafness, Autosomal Recessive	Bleeding Diathesis
Choreoacanthocytosis	Hemophagocytic Lymphohistiocytosis, Familial
Prostate-Specific Gene	Chondrosarcoma, Extraskeletal Myxoid
Bamforth-Lazarus Syndrome	Pseudohermaphroditism, Male, with Gynecomastia
Tyrosine Kinase–Like Orphan Receptor	Tangier Disease
Brachydactyly, Type B1	HDL Deficiency, Familial
Nephronophthisis (Infantile)	Fanconi Anemia, Type C
Neuropathy, Sensory and Autonomic, Type 1	Xeroderma Pigmentosum
Fructose Intolerance	Epithelioma, Self-Healing, Squamous
Basal Cell Carcinoma, Sporadic	Leukemia, T-Cell Acute Lymphoblastic
Muscular Dystrophy, Fukuyama Congenital	Muscular Dystrophy, Limb-Girdle, Type 2H
Basal Cell Nevus Syndrome	Bladder Cancer
Dysautonomia (Riley-Day Syndrome)	Sex Reversal, XY, with Adrenal Failure
Esophageal Cancer	Leukemia Transcription Factor, Pre-B-Cell
Endotoxin Hyporesponsiveness	Porphyria, Acute Hepatic
Amyotrophic Lateral Sclerosis, Juvenile Dominant	Lead Poisoning, Susceptibility To
Berardinelli-Seip Congenital Lipodystrophy	Citrullinemia
Dystonia, Torsion, Autosomal Dominant	Dopamine-Beta-Hydroxylase Deficiency
Lethal Congenital Contracture Syndrome	Amyloidosis, Finnish Type
Leukemia, Acute Undifferentiated	Mircocephaly, Primary Autosomal Recessive
Tuberous Sclerosis	Leigh Syndrome
Hemolytic Anemia	Leukemia
Telangiectasia, Hereditary Hemorrhagic	Nail-Patella Syndrome
Ehlers-Danlos Syndrome, Types I and II	Prostaglandin D2 Synthase (brain)
Joubert Syndrome	Pituitary Hormone Deficiency
Leukemia, T-cell Acute Lymphoblastic	

Figure 1.7. Chromosome 9. This is a graphical representation of one human chromosome, chromosome 9, as elucidated in the Human Genome Project. It shows a number of traits and disorders that are now known to be controlled or partially controlled by genes located along chromosome 9. (Image credit: Genome Management Information System, Oak Ridge National Laboratory, http://genomics.energy.gov.)

Sequencing genes is only the beginning of a long process. As people study individual genes, the information they obtain is used to "annotate" the information in the databases so that stretches of code take on meaning. Genes code for proteins, and it is the proteins that do the work of the cell. One of the tasks of biologists today is discovering more about *how proteins interact with one another and work in living systems, a field that is called* **proteomics.**

II. APPLICATIONS OF RECOMBINANT DNA TECHNOLOGY

A. Biopharmaceuticals

The use of genetically modified cells as "factories" to manufacture therapeutic proteins, like human insulin, was the first, and is still probably the most commercially important modern biotechnology application. **Biopharmaceuticals** *are defined here as therapeutic products, like insulin, that are manufactured using genetically modified organisms as production systems.* A few examples of biopharmaceutical products that are in use by patients are shown in Table 1.1 on p. 12; there are many more in addition to these. We will devote much of this unit to exploring biopharmaceuticals because of their commercial importance, their significance in the development of the biotechnology industry, and their value in alleviating illness.

"Traditional" drugs manufactured by pharmaceutical companies are small molecules that are most often produced by chemical synthesis, for example, aspirin. Chemically synthesized drugs have been manufactured for a long time, are well-characterized, and their manufacture has been optimized by pharmaceutical companies. But some disorders cannot be treated with small molecule drugs. Type I diabetics, for example, require regular administration of insulin, which is a protein. Proteins are large, complex molecules that cannot easily be chemically synthesized on the scale required for their use as a therapeutic. Before the advent of recombinant DNA technology, proteins and other complex biopharmaceuticals had to be purified from natural sources (often plants or animals). For instance, growth hormone, used to treat dwarfism in children, was isolated from the pituitary glands of human cadavers. Isolating natural products from animals and plants has drawbacks. Some children who received growth hormone from human cadavers tragically contracted an otherwise rare, fatal neurological disease, Creutzfeldt-Jakob disease, caused by contaminants in the brain tissue from which the hormone was isolated. Thousands of individuals with hemophilia contracted AIDS because the blood factors they require to control bleeding were isolated from human blood contaminated with the HIV virus. Contamination is always a concern with products isolated from natural sources.

Another problem with isolating drugs from plants and animals is that the drug might not be available in large quantities from its natural source. Interferon, discovered in 1957, is a protein made by animal cells as a defense against viral infection. Interferon is released into the blood and causes the body to mount a response against the attacking viruses. Scientists quickly realized that interferon had the potential to be a therapeutic agent against infectious agents, but only tiny quantities could be isolated from blood, not nearly enough to test or to use clinically. It was not until recombinant DNA technology was developed in the early 1980s that researchers could obtain enough interferon to test the compound in animals and humans. Interferon drugs did not turn out to be as useful in fighting infectious agents as scientists predicted, but interferon has proven efficacious in treating patients with certain cancers and other conditions.

There are so many advantages to producing protein therapeutics using recombinant DNA technology, and the market for these products is so large, that pharmaceutical companies rapidly created biopharmaceutical production facilities and biotechnology companies sprang up to develop new products. There are now many types of biopharmaceuticals on the market including enzymes, vaccines, antibodies, thrombolytics (drugs that dissolve blood clots), blood clotting factors (drugs that help blood to clot), and cytokines (small proteins that act as chemical messengers). A financial report published in 2006 stated that more than $51 billion was invested in discovering and developing biopharmaceuticals in the year 2005 (Burrill and Company, http://www.burrillandco.com/burrill/pr_1141780182).

Some biopharmaceutical agents, such as insulin, human growth factor, and blood clotting factors, are proteins that are missing or deficient in afflicted patients and are therefore administered to treat them. Other biopharmaceuticals, such as vaccines, play other roles in promoting health. Some biopharmaceutical agents, like insulin and human growth hormone, take the place of drug compounds previously isolated from natural sources. Yet other biopharmaceutical products are entirely new and could not exist without recombinant DNA technology; Herceptin, described later in this chapter, is an example. Small, chemically synthesized drugs are still very much in use, and pharmaceutical companies continue to find new ones, but the existence of biopharmaceuticals greatly expands the possibilities for therapeutic agents.

B. Production Systems for Biopharmaceuticals

i. CULTURED CELLS

Genetically modified bacterial cells were the first type of host cells used to manufacture biopharmaceuticals. Bacterial cells are often still used because they require relatively inexpensive growth medium and are easily grown in culture. Yeast cells are also used sometimes to manufacture products and are also relatively inexpensive and

Table 1.1. Examples of Biopharmaceutical Products

Recombinant DNA Products Produced in Bacteria (*E. coli*)

Human Insulin **Approved 1982, Genentech**

Used to treat diabetes. Before insulin from genetically modified bacteria was available, it was usually purified from the pancreas of animals. Insulin derived from genetically modified bacteria provides a more reliable source than animals and is less likely to cause allergic reactions in patients.

Human Growth Hormone **Approved 1985, Genentech**

Used to treat dwarfism. Prior to the introduction of recombinant DNA-derived human growth hormone, dwarfism was treated with hormone purified from pituitary glands collected from human cadavers. Some children who received hormone isolated from human sources eventually died of the neurodegenerative disease, Creutzfeldt-Jakob disease.

Interferon α-2b **Approved 1986 (for treatment of leukemia), Hoffmann-La Roche**

Various types of interferons are used to treat a variety of cancers and viral diseases including: hairy cell leukemia, AIDS-related Kaposi's sarcoma, renal cell carcinoma, and chronic hepatitis B. The potential of interferons in the treatment of disease was recognized in 1957, but it was not available in the amounts and purity required for experimentation and clinical trials until the advent of biotechnology.

Recombinant DNA Products Produced in Yeast Cells

Hepatitis B Vaccine **Approved 1986, Merck & Co., Inc.**

Used to prevent infection with hepatitis B. Hepatitis B is a common and serious viral illness for which there is no known cure. Before the recombinant vaccine was developed, a vaccine was prepared from the plasma of hepatitis-infected humans. This source of vaccine was limited and there were concerns about its purity. The substance required to make the hepatitis B vaccine is now produced by genetically modified yeast, which makes the vaccine more available and reduces the possibility of contamination.

Human Albumin **Approved 2005, Delta Biotechnology Ltd.**

Used as a stabilizer in vaccine manufacture. Human albumin, a blood protein, is routinely used as a stabilizer in vaccine production, and has been shown to help prevent fever in vaccinated children. There are concerns that blood proteins isolated from natural sources (humans or animals) might be contaminated with viruses or prions. Human albumin made using recombinant DNA technology alleviates this concern.

Human Papillomavirus (HPV) Vaccine **Approved 2006, Merck & Co., Inc.**

Developed to prevent infection with several types of HPV. HPV is a virus that is sexually transmitted and that causes genital warts and, in some cases, cervical cancer. This vaccine is therefore considered to be an anticancer vaccine.

Recombinant DNA Products Produced in Cultured Mammalian Cells

Erythropoietin **Approved 1989, Amgen**

Used to treat anemia due to renal failure, AZT treatment (used for AIDS patients), and chemotherapy (in cancer patients). Erythropoietin (EPO) is a glycoprotein, produced in the kidney, which stimulates the production and maturation of red blood cells. EPO occurs naturally in very small quantities in human urine and therefore, prior to its production by recombinant DNA methods, had never been available in sufficient quantity for clinical testing. The conventional treatment for anemia due to renal failure was blood transfusion. Recombinant EPO provided a new approach to the treatment of anemia.

Factor VIII **Approved 1992, Baxter**

Used to treat hemophilia. Before the introduction of recombinant Factor VIII, many hemophilia patients contracted AIDS from Factor VIII derived from infected human plasma.

Tumor Necrosis Factor Blocker **Approved 1998 (for rheumatoid arthritis), Immunex**

Used to treat various types of arthritis and psoriasis. People with an inflammatory disease (e.g., rheumatoid arthritis, ankylosing spondylitis, Crohn's disease, psoriasis) have too much tumor necrosis factor (TNF) in their bodies, which stimulates an inflammatory response. The recombinant DNA drug, Enbrel, reduces the amount of TNF to normal levels.

Monoclonal Antibody Product*

Murine Monoclonal Antibody to CD3 **Approved 1986, Ortho Biotech**

Used to suppress organ rejection by patients receiving kidney transplants. This product is a highly purified antibody that attacks the T cells of patients. T cells are involved in transplant rejection. This antibody has been effective in treating patients who do not respond to conventional anti-rejection treatments.

*See Table 1.2 on p. 18 for more examples of monoclonal antibody products.

simple to grow. But bacteria and yeast are not able to produce all human proteins in an active form. This is because many human proteins are modified by cells after they are assembled from their amino acid subunits, and bacteria and yeast often do not perform these modifications the same way that human cells do. **Glycosylation** *is a common, important type of modification in which complex, specific-branched carbohydrates are attached to a protein,* see Figure 1.8. These branched structures affect how the protein functions when it is administered to a patient. When proteins must be modified in specific ways in order to be functional, bacteria and yeast are generally are not used as host cells and mammalian cells are commonly used instead.

Mammalian cells are relatively fragile and they grow more slowly than bacterial cells. This means mammalian cells are more expensive to grow than bacteria. Mammalian cells require a more complex growth medium than bacteria or yeast. This means that more impurities from the medium must be removed from products when mammalian cells are used in manufacturing than is the case when bacteria or yeast are used. Mammalian cells have the further disadvantage that they are more likely than bacteria or yeast to harbor contaminants that are pathogenic to humans; these contaminants must be removed during processing. Despite all these disadvantages, a survey published in 2004 by the trade magazine, *BioProcess International,* showed that mammalian cells have become the most common type used for biopharmaceutical production (Langer, Eric S. "Manufacturing Capacity Put on Simmer." *Bio Process International,* February 2004.). There are various mammalian cell lines used for biopharmaceutical production, the most common of which is the CHO cell line. CHO cells are the descendents of cells originally isolated from a *C*hinese *h*amster *o*vary, hence, their acronym.

CHO cells were described by Theodore Puck and his colleagues in a 1958 paper (Puck, Theodore T., Cieciura, Steven J., and Robinson, Arthur. "Genetics of Somatic Mammalian Cells III. Long-Term Cultivation of Euploid Cells from Human and Animal Subjects." *Journal of Experimental Medicine* 108 no. 6 (1958): 945–56.). Puck and his colleagues not only described their work, they also deposited some of their CHO cells with the American Type Culture Collection (ATCC), a global, nonprofit organization that stores and distributes biological resources, particularly cells and tissues. Puck's CHO cell line and other cell lines derived from it can be easily purchased from the ATCC.

Puck and his colleagues were probably not thinking about creating a production system for the then nonexistent biopharmaceutical industry when they isolated CHO cells in the 1950s. The purpose of their work was to find improved methods for culturing cells in the laboratory, which, at the time, was difficult and often unsuccessful. It is interesting to note that in order to promote healthy cell growth, Puck's group tried adding serum isolated from the blood of fetal calves to the growth medium. They found that fetal calf serum was very helpful in coaxing cells to grow. Indeed, fetal calf serum is so helpful to cultured cells that its addition became standard in mammalian cell culture. Unfortunately, a few people in Great Britain have died of variant Creutzfeldt-Jakob disease thought to be caused by a pathogen found in cows. This means the pathogen could also be present in fetal calf serum. This concern, and other similar problems relating to animal products, has led the biotechnology industry to make a concerted (and still ongoing) effort to find alternatives to the use of fetal calf serum and other animal-derived materials (see also Chapter 30).

ii. ANIMALS

At this time, the usual production systems for protein products are cultured bacteria, yeast, mammalian, or sometimes insect cells. However, transgenic plants and farm animals are also being developed for production. Since the 1980s, biotechnologists have speculated that farm animals, like goats, sheep, or cows, might be genetically modified so that they would produce a therapeutic protein in their milk. When transgenic mammals are used as production systems, the gene for the therapeutic protein is joined to the "on-switch" for a milk production gene. The resulting DNA is called the *genetic construct*. The genetic construct is sometimes painstakingly injected under a microscope into the nucleus of a fertilized egg from a sheep or other host species. Each embryo that results from a successful microinjection is transferred into a sheep surrogate mother (assuming sheep are the animal of choice) who gives birth to transgenic lambs. Some of these transgenic females will produce the therapeutic protein in their milk. Any transgenic sheep can then be bred in a normal fashion to form a herd of animals that produce the therapeutic protein in their milk.

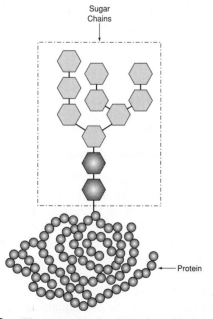

Sugar
Chains

Protein

Figure 1.8.   Glycoproteins Are Proteins with Attached Sugar Chains.

The therapeutic protein is harvested from the milk of transgenic sheep (or other animals) using standard dairy methods. Then, the protein goes through a series of purification steps to ensure that it is pure and uncontaminated by any material from the host animal, see Figure 1.9.

As an alternative to microinjection, the genetic construct can be transfected into host cells (from a goat in this example) that are maintained in a cell culture lab. A single transfected cultured cell can then be fused with a goat egg cell that has had its nucleus removed. The result of the fusion is an embryo with genetic information only from the cultured cell. The embryo that results from the fusion of the transfected cell with an egg cell is implanted into a surrogate mother and the process proceeds as for microinjection.

Animals have advantages as production systems for proteins. Transgenic animals can produce proteins that are folded into their proper shapes and are properly glycosylated. CHO cells will accomplish these tasks, but farm animals can, in principle, provide much larger amounts of high-quality protein that is noninvasively harvested from their milk. Transgenic animal technology has been slow to develop despite its advantages, due to technical difficulties, regulatory issues, and societal concerns about this use of animals. At the time of writing, only one drug made in transgenic animals has been approved for marketing, ATryn, which was approved in 2006 in the European Union. ATryn is antithrombin, which inhibits coagulation of blood. This drug is used for surgical patients who have a genetic disorder and do not make enough antithrombin themselves. ATryn is a product of GTC Biotherapeutics, Inc.

iii. PLANTS

Materials derived from plants have been used as medicines for thousands of years and some modern drugs are still derived from plants. The use, however, of genetically modified plants to manufacture drugs is new. Plants that have been transfected with a genetic construct can be used to produce a protein of interest in their stems, leaves, shoots, or roots.

At the time of writing, plant-based biopharmaceuticals have not entered commercial production, but many companies are testing them. For example, testing is in progress of a drug made in transgenic corn to aid cystic fibrosis patients, a drug made in transgenic tobacco to treat non-Hodgkin's lymphoma, insulin made in safflowers, and vaccines made in potatoes. Transgenic plants, like transgenic animals, offer the potential of high yields while avoiding controversy relating to the use of animals. Also, plants are unlikely to harbor pathogens that are dangerous to human patients. At this time, a major concern with plant-based biopharmaceuticals is the possibility that altered genetic material contained in the pollen grains might escape and fertilize nearby crops or wild plants. This would allow the altered genetic material to spread where it should not and might expose people to unforeseen drug products in foods.

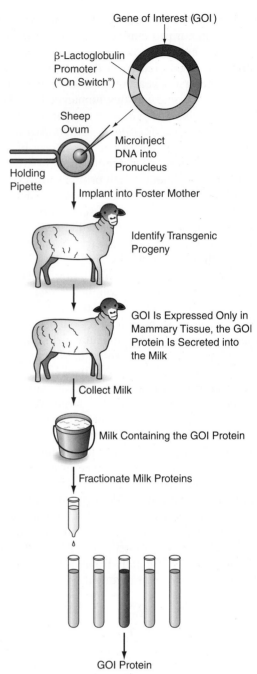

Figure 1.9. Production of a Protein Product Using a Transgenic Farm Animal. In this example, the gene of interest (GOI) is a therapeutic protein that is to be harvested from the milk of a transgenic farm animal.

iv. OTHER PRODUCTION SYSTEMS

Scientists are working on new strategies as all these current issues relating to cell, plant, and animal production systems are being resolved. Some scientists, for example, are attempting to engineer bacterial and yeast cells so they can fold and glycosylate proteins in a manner like human cells. Cultured insect cells are also in occasional use, as are transgenic chickens that produce therapeutic proteins in their eggs. Scientists in some companies are genetically engineering the tiny aquatic duckweed plant, *Lemna,* which can be grown in a sealed vessel inside a production facility. This system reduces the potential for the escape of genetic information to other plants. Other scientists are working on synthesizing proteins by chemical reactions—without using cells at all. It is hoped that eventually cheaper and more efficient drug-production systems will be developed that have no environmental risks and do not raise ethical concerns.

C. Vaccines

Vaccines are sometimes made using recombinant DNA technology. **Vaccines** *are agents that are used to enhance the immune system,* in the most familiar case to protect against infection by a pathogen, such as polio or diphtheria. Vaccines are also being tested as agents for cancer treatment.

The 18th-century English physician, Edward Jenner, is usually credited with devising the first vaccine. Jenner realized that milkmaids who had been infected with cowpox did not get smallpox. Since cowpox infection is mild and smallpox is often fatal, Jenner got the idea of intentionally infecting people with the milder disease to prevent the more severe one. Jenner experimented by injecting a healthy boy, who had never had cowpox or smallpox, with some fluid from the cowpox sore of a milkmaid. When the boy recovered from cowpox, Jenner exposed him to smallpox. The boy did not get sick, and Jenner termed the procedure "vaccination," which is Latin for "pertaining to cows."

HISTORICAL STORY

Transporting a Vaccine to the New World

An interesting side note to the smallpox vaccine story relates to the storage, stability, administration, and transport of the vaccine. An important part of the development of any new therapeutic product is determining how best to package and store it, how to administer it, and how long it is stable. These problems are particularly acute for drugs to be used in countries where there is a scarcity of refrigeration, sterile water for reconstituting drugs, and sterile syringes. There were no refrigerators or pharmaceutical manufacturers in Edward Jenner's time. The Spanish physician, Francisco Xavier Balmis, was ordered by the Spanish King, Charles IV, to sail across the ocean to bring vaccination to the Spanish colonies in the Americas. Because the cowpox fluid was not stable over long periods, Balmis used a chain of orphan boys on the voyages. He would infect one boy with cowpox, wait a week or so, and then infect the next, and so on, as they sailed all the way across the sea. On reaching land, the boys were used to vaccinate other people and then were fostered out to families. Due to the efforts of Balmis and other physicians, vaccination against smallpox spread rapidly across Europe and the Americas.

The story of smallpox vaccination is still relevant today for several reasons. It is interesting from the point of view of understanding the immune system, the nature of vaccines, and issues with storage, administration, and transport. The story also raises interesting ethical questions. Jenner's experi-ments, intentionally exposing a boy to smallpox, and the use of children to store and produce a vaccine, seriously violate the ethical principles applied today, but were within the accepted ethical boundaries of the 1700s and 1800s.

Many other vaccines have been developed since Jenner's time, for example, to prevent polio, diphtheria, and tetanus. Prior to the advent of recombinant DNA technology, most vaccines were made by growing the virus or bacterium that caused the disease, and then either killing the pathogen or damaging it in some way so that it could not cause full-blown illness. The dead or attenuated (weakened) pathogen was then administered to people, eliciting an immune system response that protected the person if they were later exposed to the pathogen. The flu vaccine is still made this way; flu virus is grown each year in millions of fertilized chicken eggs. Vaccines made in the traditional fashion are usually fine, but they may, in a few cases, cause the disease they are intended to prevent.

Recombinant DNA technology provides a new method of vaccine production that avoids the risk of causing the disease it is meant to prevent. The vaccine against hepatitis B, for example, is made by taking a gene that codes for a protein found on the surface of the virus and inserting the gene into host yeast cells. The yeast then manufacture large amounts of the viral protein. The resulting protein is purified and formulated into an injectable vaccine that, when administered, triggers an immune response against the hepatitis B virus. There is no risk that the vaccine will cause the disease it is intended to prevent because only a single viral protein is used to make the vaccine.

Transgenic plants are also being tested as production systems for ingestible vaccines. This method has the major advantages that plant-based vaccines do not require refrigeration or injection. There are, however, various issues to be resolved before plant vaccines are widely marketed.

Yet another new type of biotechnology-based vaccine, DNA vaccines, is predicted to become common in the future. **DNA vaccines** *are made from vectors that have been genetically engineered to include the DNA coding for one or two specific proteins from the infectious agent.* When injected into the cells of a person (or other animal), the DNA is expressed, leading to the synthesis of proteins from the infectious agent. The recipient's immune system responds to the new proteins by mounting a protective immune response. At the time of writing, no DNA vaccines have been approved for human use, but a DNA vaccine against the virus that causes West Nile Disease has been licensed for horses.

D. Monoclonal Antibodies

Monoclonal antibodies (Mabs) are biopharmaceuticals made in a somewhat different way than has been discussed so far. **Antibodies** *are proteins made by the* *immune system that recognize and bind to substances invading the body*—such as bacteria, viruses, and foreign proteins—thus aiding in their destruction. *Substances that trigger the production of antibodies are called* **antigens.** Antibodies are produced by B cells, a type of white blood cell. A particular antibody binds only a particular target antigen, somewhat like a specific key only fitting into a particular lock, see Figure 1.10.

Before the 1970s, antibodies were made by exposing an animal, like a mouse or a rabbit, to the target material, the antigen, of interest. *The animal would mount an immune response against the antigen, producing many antibodies,* **polyclonal antibodies,** *which could then be isolated from the blood serum of the animal.* Polyclonal antibodies are useful but they have two problems. First, they are not homogeneous molecules since an animal will usually produce diverse antibodies directed against various parts of the antigen. Second, polyclonal antibodies cannot be indefinitely obtained; a certain popu-

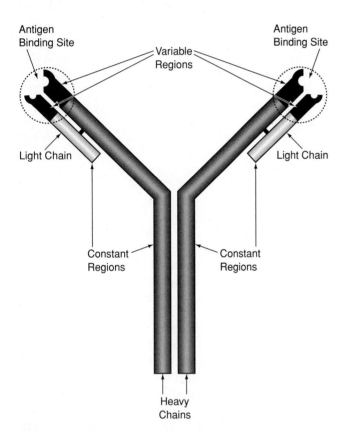

Figure 1.10. Antibody Structure. Antibodies are large proteins that recognize and help neutralize foreign substances. Antibodies have a characteristic Y-shaped structure. They are composed of subunits: two identical heavy chains plus two identical light chains that are linked to each other. Antibodies are all structurally similar except for a variable region at the end of each Y arm. An individual has millions of different antibody populations, each of which has a structurally distinct variable region. *The part of an antigen that an antibody recognizes and binds is called an* **epitope.** Each antibody is able to recognize a specific antigen from among millions of different molecules because each epitope interacts in a highly specific way with the structure of a particular matching variable region.

lation of antibodies can be harvested only from the animal in which it is produced. When the animal dies, the source of antibodies is gone. These limitations have kept polyclonal antibodies from being widely used in medicine. **Monoclonal antibodies** do not have these two limitations; *they are exceptionally homogenous populations of antibodies directed against a specific target, and they can be produced indefinitely in culture.*

Monoclonal antibodies were first described in 1975 by George Köhler and Cesar Milstein (Köhler, George, and Milstein, Cesar. "Continuous Cultures of Fused Cells Secreting Antibody of Predefined Specificity." *Nature* 256 (1975): 495–7.). Monoclonal antibodies are produced by fusing together two cells: an antibody-producing cell from a mouse, and a tumor cell. Suppose, for example, that one wants to make a monoclonal antibody against a specific protein that is thought to be important in making cancer cells divide. One obtains the protein of interest and injects it into a laboratory mouse. The mouse mounts an immune response against the protein, making different types of antibodies that recognize different parts of the antigen. The different types of antibodies are made by different B cells. Each B cell divides many times to form a clone and each clone makes only one type of antibody. The mouse is later sacrificed and its B cells are isolated from its spleen. *These B cells are then fused with mouse myeloma cells to form* **hybridoma cells.** The myeloma cells are immortal; that is, they will divide indefinitely in culture. Each fused hybridoma cell thus has two important qualities: it produces a single, identical type of antibody molecule, as does each clone of B cells, and it will divide indefinitely like the myeloma cells. It is necessary to fuse the B cells with the myeloma cells because the B cells are not immortal; they will eventually stop dividing. The fused cells are diluted in culture to isolate individual hybridoma cells that divide repeatedly to form a uniform clone of cells, all of which make the same antibody; hence, *monoclonal antibodies.*

Monoclonal antibodies were rapidly adopted for use in research laboratories to detect and follow specific proteins in cells and tissues. Their potential use in the clinic was also quickly imagined; people reasoned that monoclonal antibodies could be used to search out and destroy specific pathogens and damaged cells. Monoclonal antibodies were optimistically termed "magic bullets" because they could, in theory, find and bind a very specific target. Monoclonal antibodies, however, were not easily adapted to clinical uses. The first monoclonal antibody was approved for medical use in the United States in 1986 (see Table 1.1) but their general development into useful products was initially slow. One of the biggest obstacles to using monoclonal antibodies in the clinic was that the standard procedure of producing them yields mouse antibodies. Although mouse antibodies are similar to human ones, mouse antibodies are detected as being nonself by the human immune system and are destroyed. Various sophisticated approaches to humanize monoclonal antibodies were eventually devised using recombinant DNA technology. Scientists furthermore developed methods to efficiently manufacture large amounts of humanized monoclonal antibodies using mammalian cells growing in bioreactors, see Figure 1.11.

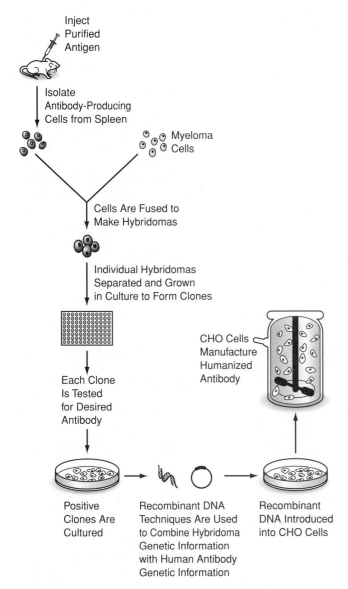

Figure 1.11.   A Method of Manufacturing Humanized Monoclonal Antibodies. Various methods exist to manufacture humanized monoclonal antibodies. Here, hybridomas are first produced. A mouse is inoculated with the antigen of interest, which stimulates the proliferation of cells that produce antibodies against the antigen. These cells are harvested from the mouse's spleen and are then fused with myeloma cells to produce hybridomas. Individual hybridoma cells are transferred to separate wells in a 96-well plate where they are cultured for several days. The original hybridomas divide to form clones of cells. Each clone is tested to see if it is producing the desired antibody. Positive clones are maintained in culture. Using the tools of genetic engineering, the genetic information from a clone that is producing the desired antibody is combined with the genetic information for producing a human antibody. CHO cells are transfected with the recombined DNA. The CHO cells are placed in a bioreactor where they manufacture an antibody that recognizes the antigen of interest (as did the hybridoma cell), yet has most of the structural features of a human antibody.

Improvements in the technologies for humanizing and manufacturing monoclonal antibodies have at last resulted in successful monoclonal antibody therapeutics. Roughly half of the monoclonal antibodies that have been approved for use in patients are used to treat cancer. Others are used to treat inflammatory and autoimmune diseases, transplant rejection, and other disorders. Table 1.2 on p. 18 provides examples of monoclonal antibodies that have been approved for therapeutic use in humans. Observe that each of these monoclonal antibodies works by selectively finding, binding, and inhibiting a specific cellular protein target that is associated with diseased cells.

Monoclonal antibodies are also used in medical diagnosis. Home pregnancy kits, for example, are based on monoclonal antibodies that specifically bind to a hormone that increases during pregnancy. Another example is the use of monoclonal antibodies in tests that detect illicit use of performance enhancing drugs by athletes, as is discussed in a case study in Chapter 24.

The success of monoclonal antibodies has an economic impact, as well as a medical one. An independent market analyst reported "antibody-based treatments, which currently generate global revenues of around $20 billion, would see sales grow 14 percent a year between 2006 and 2012." ("Antibodies Forecast to Stay Fastest-Growing Drugs." *New York Times,* October 11, 2007. http://www.reuters.com/article/ousiv/idusL11261949200 71012.) Another industry analyst reported in 2005 that "the strong growth of the biopharmaceutical industry is driven by monoclonal antibodies." (Werner, Rolf G. "The Development and Production of Biopharmaceuticals: Technological and Economic Success Factors." *BioProcess International,* September 2005: 6–15.)

E. Gene Therapy

Biopharmaceuticals are very valuable to patients and to the companies that produce them. But biopharmaceuticals and other drugs have limitations. Diseases like diabetes and hemophilia can be managed by repeated administration of drugs, but the disease is not cured in this way. Some disorders are not treatable at all with drugs. Ever since recombinant DNA technologies were first developed, scientists have been exploring the possibility of genetically modifying patients to better treat certain disorders. **Gene therapy** *involves replacing a gene that is missing, or correcting the function of a faulty gene, in order to treat or cure an illness.*

The most obvious approach for gene therapy is to treat diseases that are caused when an individual inherits a genetic defect in a single gene. It is straightforward in such cases to imagine treating the patient by administering the properly functioning gene. The first gene therapy trials thus treated genetic disorders caused by single gene defects.

One strategy for gene therapy involves removing cells from the patient, genetically modifying them, and then returning them to the patient, see Figure 1.12 on p. 19. In 1990, Ashanti De Silva became the first patient

Table 1.2. MONOCLONAL ANTIBODY THERAPEUTICS

Mab Name	Trade Name	Disease Treated	Year of Approval for Use in Patients	Target	What Target Is
Rituximab	Rituxan	Non-Hodgkin's lymphoma	1997	CD20	Protein found on the surface of B cells, which grow aberrantly in this form of cancer
Trastuzumab	Herceptin	Breast cancer	1998	Her2	See Case Study p. 19
Palivizumab	Synagis	Human Respiratory Syncytial Virus, RSV (virus associated with infant mortality)	1998	RSV fusion protein	Protein on surface of RSV that is involved in viral insertion into lung cells
Infliximab	Remicade	Rheumatoid arthritis and Crohn's disease (autoimmune diseases)	1999	Tumor necrosis factor (TNF)	Protein that is associated with immune responses
Gemtuzumab Ozogamicin	Mylotarg	Acute myelogenous leukemia (AML)	2000	CD33*	Cell-surface molecule expressed by cancerous blood cells but not found on normal stem cells needed to repopulate the bone marrow
Alemtuzumab	Campath	Chronic lymphocytic leukemia (CLL)	2001	CD52	Molecule found on white blood cells, which grow aberrantly in this form of cancer
Ibritumomab tiuxetan	Zevalin	Non-Hodgkin's lymphoma	2002	CD20*	See Rituximab
Tositumomab	Bexxar	Non-Hodgkin's lymphoma	2003	CD20*	See Rituximab
Omalizumab	Xolair	Asthma	2003	Immunoglobin E	Class of antibody that is produced in excess during allergic reactions
Cetuximab	Erbitux	Colorectal cancer Head and neck cancers	2004 2006	Epidermal growth factor receptor (EGFR)	Cell surface protein that receives growth promoting messages associated with abnormal growth in these cancers
Bevacizumab	Avastin	Colorectal cancer	2004	Vascular endothelial growth factor (VEGF)	Compound that stimulates new blood vessel formation that in cancer is associated with the growth of tumors
Panitumumab	Vectibix	Colorectal cancer	2006	EGFR	See Cetuximab
Ranibizumab	Lucentis	Age-related macular degeneration	2006	Vascular epithelial growth factor A (VEGF-A)	Growth factor that is associated with abnormal blood vessel growth and leakage in the back of diseased eyes
Eculizumab	Soliris	Paroxysmal nocturnal hemoglobinuria (blood disorder)	2007	Complement system protein (C5)	Protein that is associated with the abnormal destruction of red blood cells in diseased patients

*These antibodies are conjugated with a radionuclide or toxic agent that contributes to the destruction of cancer cells.
Sources: The American Cancer Society website provided much of the information for this table. http://www.cancer.org/docroot/ETO/ content/ETO_1_4X_Monoclonal_Antibody_Therapy_Passive_Immunotherapy.asp?sitearea=ETO. Each of the drugs also has its own website that provided information about their modes of action.

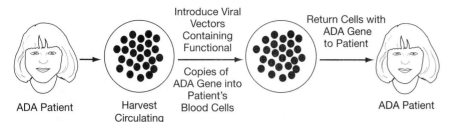

Figure 1.12. Gene Therapy to Treat ADA Deficiency. Blood cells were removed from the patient. Copies of the ADA gene were inserted into the cells, which were then returned to her. The genetically modified cells produced the protein that is missing in the affected child, thereby improving the function of her immune system.

CASE STUDY

The Development of Herceptin to Treat Breast Cancer

The development of the monoclonal antibody drug, Herceptin, demonstrates how scientific knowledge of a disease process can be used to rationally design a focused treatment. Cancer researchers in the 1970s and 1980s learned that cancer cells, which divide over and over again malignantly, often have genetic alterations that drive their abnormal growth. Dr. Dennis Slamon and colleagues at the University of California found that there is a genetic alteration in a specific gene called Her2 in about 25% of women with breast cancer. The Her2 gene codes for the Her2 protein. The job of the Her2 protein is to reside on the surface of cells and act as a receptor to accept signals from growth factors—chemicals from outside the cell that carry growth-regulating orders. In a normal cell, two copies of the Her2 gene are present and the cell makes modest amounts of Her2 protein. Sometimes a mutation (change) occurs in a cell so that the Her2 gene is amplified, resulting in more than two copies of the gene. This amplification results in the production of too much receptor protein. When too much receptor protein is present, the cell binds too much growth factor and divides and multiplies more actively than normal. This mutation is associated with an aggressive form of breast cancer. The Her2 gene is thus a normal gene that causes cancer when it becomes overexpressed.

Scientists reasoned that if they designed a drug that would locate and bind to the Her2 receptor protein, it would block the receptor and prevent the cell from receiving growth signals and dividing malignantly. Monoclonal antibodies can target and bind to a specific protein, in this case the Her2 receptor, so scientists decided to create a monoclonal antibody as the blocking agent. Scientists successfully created a monoclonal antibody, named Herceptin, which recognizes the Her2 receptor protein, see Figure 1.13. Experiments in human volunteers showed that treatment with Herceptin improves survival in women with Her2-positive breast cancer. Herceptin thus became one of the first novel cancer treatments to emerge from basic research into the fundamental mechanisms of cancer cell growth. Herceptin does not target all dividing cells, like standard chemotherapy treatments for cancer, but rather is specific for those expressing the Her2 receptor on their surface. Herceptin is manufactured and marketed by Genentech Corporation.

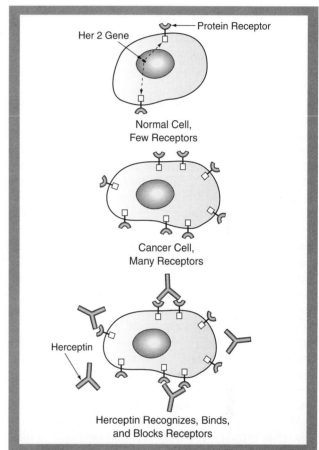

Figure 1.13. Herceptin Blocks Receptor Protein in Cancerous Cells and Slows Disease Progression. In normal breast tissue the Her2 gene causes a receptor protein to be made that accepts growth regulating signals. Some breast cancer cells have more copies of the Her2 gene than normal and therefore produce too much receptor protein leading to an accelerated rate of cell division and proliferation. Herceptin recognizes the receptor protein, binds to it, and thus interrupts the growth signals.

to receive such gene therapy in an approved procedure. She was four years old and suffering from severe combined immunodeficiency (SCID), an inherited condition that caused her to lack a functional gene for an enzyme, adenosine deaminase, which is needed for proper functioning of the immune system. Ashanti was therefore highly vulnerable to infection. Her condition improved after therapy, but she must continue to receive repeated gene therapy treatments because the treated cells do not survive indefinitely. She also takes the enzyme adenosine deaminase as a drug; therefore, no one is certain how well her gene therapy treatment works by itself.

In 2000, French physicians used gene therapy to treat 10 boys with a different form of immunodeficiency disease, X-SCID. Children with this disorder completely lack certain cells that are part of a normal immune system, so they suffer from recurrent and often fatal infections. Unlike the immunodeficiency disease afflicting Ashanti, X-SCID cannot be treated by administering a missing enzyme as a drug. Infants with this disorder are sometimes treated successfully with bone marrow transplants that provide them with normally functioning immune cells. Other times, however, no matching bone marrow donor can be found, as was the case for the children treated in this gene therapy trial.

The X-SCID gene therapy trial employed a different strategy than was used with Ashanti De Silva. The therapeutic gene was carried into the children's own bone marrow stem cells by a type of nonpathogenic virus (stem cells are discussed in more detail below). At the time of writing, nine of these children appear to have been completely cured of the disease without taking supplemental medicine. Unfortunately, three of the ten children developed leukemia and one child died. Researchers found that the leukemia occurred because the viral vector inserted itself into the children's chromosomes near a gene that codes for a protein that causes cancer when it is overexpressed. Researchers at the Salk Institute for Biological Studies treated mice with the same gene therapy used in the X-SCID trial and found that one-third of the animals developed a lymph node cancer later in their lives (Woods, Niels-Bjarne, Bottero, Virginie, Schmidt, Manfred, von Kalle, Christof, and Verma, Inder M. "Therapeutic Gene Causing Lymphoma." *Nature* 440 [2006]: 1123.).

While these first moderately successful gene therapy trials involved children who had inherited defects in a single gene, researchers imagine treating more complex diseases with gene therapy. We saw in the Herceptin case study that cancer can occur when a normal gene becomes altered. Scientists are experimenting with treating cancer patients by administering a gene that codes for a protein that turns off the action of malfunctioning genes in cancer cells.

While gene therapy is, in principle, promising and has been intensely pursued by researchers around the world, it is not a technology that has quickly fulfilled its promise. Gene therapy is beset by technical problems, particularly in delivering and targeting the "good" gene to the right cells. In 1999, 18-year-old Jesse Gelsinger died after receiving an experimental gene therapy treatment. His death was apparently due to a massive immune response triggered by exposure to a viral vector. At the time of writing, the X-SCID experiment is considered to be the most successful gene therapy trial, and the children who developed leukemia have been treated for that disease, but obviously the method is not perfected. This situation may change as technical problems are slowly resolved.

F. Genetic Engineering and Agriculture

Recombinant DNA technology not only has impacted medicine, but also agriculture. Conventional methods of plant and animal breeding have long been used to enhance the characteristics of crops and livestock. In conventional breeding, plants or animals with desirable qualities, for example, plants with large fruit or resistance to insects, have been selectively mated with other plants with desired traits. It is possible, however, to use recombinant DNA technologies to genetically modify plants and animals more quickly, and with better control, than is possible using traditional methods. Conventional breeding does not allow genes from unrelated plant or animal families to combine because the plants or animals do not breed with one another. Also, in conventional breeding there is an uncontrolled mixing of all the genes from both parents. Recombinant DNA methods allow the introduction of only a specific gene of interest into the offspring.

Genetically modified varieties of soybeans, corn, and cotton were introduced into commercial production beginning in 1996. The first trait that was introduced into crop plants using recombinant DNA methods was a feature that makes the plants easier for farmers to grow, not a trait with obvious appeal to consumers. These plants contain an introduced gene that makes them resistant to the herbicide glyphosate (trade name: Roundup). Glyphosate kills plants and is nontoxic to animals. The use of glyphosate in the past was limited because it kills all plants—including the crop. Farmers now plant genetically modified, herbicide-resistant plants so they can use glyphosate to kill weeds without harming their crops.

A second widely adopted genetic modification of plants was the introduction of a gene from the soil bacterium, *Bacillus thuringiensis* (Bt). Plants with the Bt gene produce a protein that is toxic to insects that destroy crops, thus helping protect genetically modified plants from insect damage. This method has the environmental advantage of protecting crops while reducing the use of pesticides.

Genetically modified crop varieties rapidly gained acceptance with farmers in the United States. Survey data from the U.S. Department of Agriculture (USDA) in 2005 indicated that more than 50% of all corn, 79% of all cotton, and 87% of all soybean crops in the United States were genetically modified with an herbicide resistance gene, a Bt gene, or both.

While the majority of genetically modified crops now have herbicide or insect resistance, other genetically modified crop varieties are being field tested and are beginning to enter the marketplace. These new genetically modified crops are expected to provide many benefits to consumers and to society. Crops are being modified with genes that make them resistant to viral and fungal diseases; that make them tolerant of cold, drought, and high salt environments; and that allow them to be grown in marginal soils and climates. Genetically modified plants are being developed with increased nutritional value (e.g., added vitamins, iron, and antioxidants). Fruits and vegetables are being developed that will have a longer shelf life, are seedless, are more flavorful, and so on. Biotechnology methods are similarly being used with livestock to produce animals that are faster growing; produce more meat, milk, eggs, or wool; or have altered nutritional value. The USDA website (http://www.ers.usda .gov) is a good source of updates about the status of agricultural biotechnology in the United States.

While genetic engineering is increasingly being applied to agriculture, particularly in the United States, the use of these technologies in food production is among the more controversial of its applications. One issue relates specifically to the use of Bt crops. Organic farmers have a long tradition of applying the bacterium, *Bacillus thuringiensis*, to their crops. Organic farmers fear that the widespread use of genetically modified Bt crops will result in target pests that develop resistance to the Bt toxin. Bt seed producers and farmers planting genetically modified Bt crops therefore use resistance management strategies to avoid, or at least postpone, the appearance of Bt-resistant pests. Seed producers genetically modify their seeds in such a way that the plants will produce high levels of Bt toxin sufficient to kill all target insects except for a few very rare resistant individuals. Farmers plant small refuges of non-genetically modified crops within their Bt crop fields. The logic behind the refuges is that susceptible, nonresistant insects that come from the refuges will mate with the few Bt-resistant individuals. Their resulting offspring will still be susceptible to the Bt toxin. This combined strategy appears to be successful to date, although scientists are working on new methods to protect crops in anticipation of the appearance of Bt-resistant pests.

Another concern associated with Bt crops is that desirable insects will be harmed along with target pests. Early research, for example, suggested that monarch butterflies might be harmed by eating pollen from Bt corn plants. Later research indicated that monarch butterflies are not at risk, though the general concern remains a topic of discussion.

Other concerns relating to genetically modified foods are that food produced using genetic engineering methods might not be as safe as conventional foods, that animals may be stressed by the introduction of foreign genes, that genes that make plants more hardy may unintentionally be transferred to weeds, and that the introduction of genetically modified organisms into the environment may have unforeseen, adverse effects. There are also economic and political issues that are of concern. These various issues will presumably be resolved as the agricultural biotechnology industry matures. In the meantime, the trend has been toward a gradual introduction of more genetically modified crops in the United States with slower acceptance in Europe and other parts of the world.

G. Other Products of Recombinant DNA Technology

Although pharmaceuticals and modified crops are the most well-known applications of genetic engineering, there are other applications as well. Many industries have been quietly incorporating recombinant DNA technology into their manufacturing for years. For example, rennin is an enzyme used in the manufacture of cheese. Before the advent of genetically modified organisms, rennin was isolated from the fourth stomach of unweaned calves. Rennin is now produced by genetically modified bacteria.

Almost all laundry detergents have enzymes to break down and remove substances that soil clothing. Lipases break down oils, proteases break down proteins, and amylases break down starches. These enzymes are typically manufactured by genetically modified microorganisms. Subtilisin is an example of a protease manufactured by genetically modified bacteria. According to the National Library of Medicine database of ingredients in household products, subtilisin is found in more than 30 common household products, ranging from laundry detergents to contact lens cleaner (http://householdproducts.nlm.nih .gov/cgi-bin/household/brands?tbl=chem&id=2167).

There are other industrial processes that have been enhanced by biotechnology. Traditional methods of paper manufacturing, for example, use harsh chemicals that are released into the environment. Paper manufacturers now can substitute enzymes manufactured by genetically modified bacteria for these chemicals. Another example is genetically modified bacteria that can break down pollutants in contaminated soil and water.

III. OTHER BIOTECHNOLOGIES

A. Cell Therapy, Regenerative Medicine, and Tissue Engineering

So far we have discussed medical and nonmedical applications of recombinant DNA technology, where DNA from one organism in inserted into another to either produce a product or to alter the traits of the recipient cells. There are, however, other methodologies that have arisen as biological research discoveries are transformed into applications and products. These new methodologies are also considered to be biotechnology.

We noted above that drugs are often successful therapies, but they have limitations for treating and curing many disorders. Scientists are therefore exploring the possibility not only of administering genes to patients, but also transferring whole cells and tissues. Methods involving the transfer of cells and tissues into patients are variously called *cell-based therapies, regenerative medicine,* and *tissue engineering.*

The human body naturally can repair itself to a limited extent; for example, cuts, scrapes, and broken bones can heal. More extensive medical repair of damaged body parts is usually limited to transplantation of tissue and organs from donors. There are far more people in need of transplants than there are donors. Tissue engineering has the promise of greatly expanding the options for repairing diseased and damaged tissues and organs.

Tissue engineering *helps the body heal itself through the delivery of therapeutic cells.* These cells may be administered along with biomolecules that turn on healing processes and supporting structures that provide a scaffold to guide new tissue growth. In one scenario, a scaffold and growth factors are implanted at the site of injury inside the patient. Alternatively, tissue may be grown outside the patient and later be transplanted.

In an experimental application of tissue engineering, seven children born with spina bifida and severely shrunken bladders received new bladders grown on a scaffold in the laboratory and seeded with their own cells (Aldhous, Peter. "Lab Grown Bladder Shows Big Promise." *New Scientist,* April 8, 2006.). The researchers noted modest improvement in bladder function in the patients. Examples of tissue engineering products that have been approved for patient use include: Dermagraft, which is used to treat diabetic foot ulcers; Carticel, which is used to repair damaged knee cartilage; and TransCyte, which is a temporary skin substitute for burn victims.

Researchers are also exploring the use of organs harvested from animals in order to meet the immediate demand for organs for transplants. *Transplantation of organs, tissues, or cells from one species to another is called* **xenotransplantation.** Whole organs from animals are not in widespread use but hundreds of thousands of patients have received heart valves from pigs since the procedure was commercially introduced in 1975. There have also been preliminary experiments in which diabetic patients received transplants of encapsulated porcine pancreatic islet cells. The islet cells produce insulin in response to high glucose levels in the recipients. The encapsulation is intended to reduce the chance of rejection by the patient's immune system.

Xenotransplantation raises various concerns. Tissue rejection is a problem when animal tissue is transplanted into humans. There is also the possibility that xenotransplantation might introduce animal diseases into humans. Some people have ethical concerns about the use of animals for this purpose. Researchers are therefore conducting experiments aimed at producing transgenic animals whose organs are not rejected by humans. Regulatory agencies are working to establish safety requirements, for example, requiring that animals used for xenotransplantation be bred in captivity to reduce the possibility of unexpected disease transmission. Ethical concerns continue to be the subject of discussion.

The use of stem cells to treat disease and injury is probably the most well-known of the regenerative medicine technologies, see Figure 1.14. Dr. James Thomson and colleagues at the University of Wisconsin published the first report describing the identification and isolation of stem cells from human embryos in 1998. These embryos were obtained from fertility clinics where they were originally produced to treat infertility and were donated for research with the informed consent of donor couples who no longer wanted them. Once established, an embryonic stem cell line can be maintained indefinitely in

culture, or the cells can be frozen for later use. In the relatively short time since Thomson's first report, stem cells have created a flurry of excitement and have contributed to major discoveries in cell biology. At the same time, they have been the focus of discussion and controversy, even to the extent of dominating political campaigns.

The controversy over the use of human stem cells centers on the fact that they are currently isolated from human embryos. Critics voice concerns about the ethics of destroying human embryos. There is also the fear that stem cell technologies will lead to cloning of adult humans, with its own set of complex ethical issues.

The excitement generated by stem cells centers on their potential to treat common, serious maladies that are presently intractable. To understand the potential of stem cells, consider that every human being begins as a fertilized egg. This single egg cell must divide many times to form a complete individual. If the cells of the embryo, however, simply divided over and over again, no person would form because the cells also need to differentiate into muscles, nerves, skin, bones, arms, legs, brains, eyes, and all the other structures of the body. **Embryonic stem cells** *are cells from an embryo that have the potential to differentiate into any cell type in the body.* The potential of stem cells to become any cell type is the key to their potential as therapeutics. Stem cells, for example, might someday be used to replace pancreatic islet cells in people with Type I diabetes so their bodies would again produce insulin. Stem cells are being explored to treat patients with Parkinson's disease who experience a loss of motor control when brain cells that are supposed to produce dopamine fail to do so. Stem cells are being studied to see if they can repair heart muscle cells damaged in a heart attack, nerve cells

Figure 1.14. A Colony of Cultured Human Embryonic Stem Cells. Cultured stem cells grow in colonies, as shown here. (The stem cell colony covers most of the photo. The individual stem cells are too small to see in this photograph.) Spindle-shaped mouse fibroblast feeder cells are visible on the edges of the photo. The mouse cells are placed on the culture dishes along with the human stem cells in order to provide nutrients and growth factors that promote the growth of the stem cells.

damaged in spinal cord injuries, cells damaged by arthritis, and a vast number of other conditions.

Therapeutic cloning, or **somatic cell nuclear transfer (SCNT),** *is an experimental technology in which stem cells would be created to treat an individual patient,* see Figure 1.15. In SCNT, genetic material would be taken from the cell of a patient and transferred into an enucleated human egg. The result would be a new embryo that would be allowed to undergo just a few cell divisions, at which time the stem cells in the embryo would be harvested. This method could, in theory, provide material for transplantation into a patient that the individual's immune system would not reject. At the time of writing, there is no documented example of successful SCNT with human cells, although in 2007 the first announcement was made of successful SCNT using skin cells and eggs from a primate. While some people look forward to the development of SCNT to treat serious disorders, others believe that no human embryo should ever be intentionally destroyed. The issue of obtaining eggs for the procedure is also problematic, since harvesting eggs from women is difficult and poses some risk to the donors.

Two reports published in late 2007 hold the promise that it will eventually be possible to bypass embryos in the production of therapeutic stem cells. One report was from James Thomson's laboratory, the other from Shinya Yamanaka's laboratory in Japan. These two groups were able to *transfect human skin cells with genes that caused them to look and act like human embryonic stem cells; these cells are called* **induced pluripotent stem cells, iPS.** While this is an important advance, the transfection methods used by these groups are not at this time suitable for medical use because they might cause mutations that might lead to tumors in tissues grown from the cells. Researchers now hope to find new ways to reprogram genes within cells to make them act like stem cells, without inserting new genes into the cells.

Another avenue of research relates to adult stem cells. **Adult stem cells** *are undifferentiated cells found among differentiated cells in a tissue or organ that can differentiate when needed to form the specialized cell types found in that organ.* The use of adult stem cells is far less contentious than that of embryonic stem cells. However, they are difficult to isolate and scientists have had less success coaxing them to form specific cell types than they have had with embryonic stem cells. One use of these cells is already routine; bone marrow transplants have long been used to treat leukemia, certain autoimmune diseases, and other disorders. Bone marrow transplants work because bone marrow contains blood stem cells.

There are not only many societal issues involving stem cells, but technical ones as well. For example, stem cells from any source that are used to treat a disorder must differentiate into the right type of cell within a

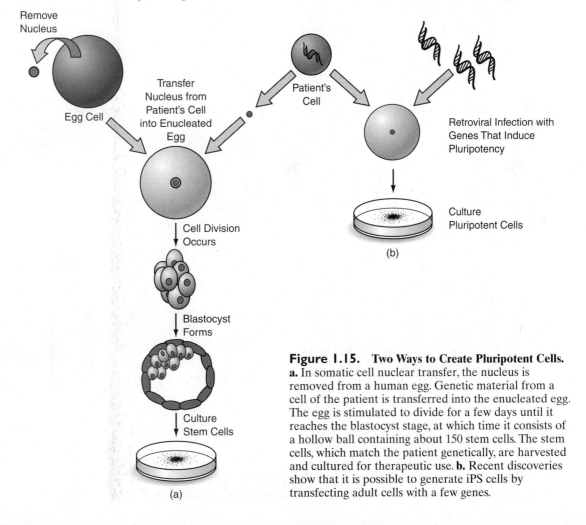

Figure 1.15.   Two Ways to Create Pluripotent Cells.
a. In somatic cell nuclear transfer, the nucleus is removed from a human egg. Genetic material from a cell of the patient is transferred into the enucleated egg. The egg is stimulated to divide for a few days until it reaches the blastocyst stage, at which time it consists of a hollow ball containing about 150 stem cells. The stem cells, which match the patient genetically, are harvested and cultured for therapeutic use. **b.** Recent discoveries show that it is possible to generate iPS cells by transfecting adult cells with a few genes.

patient. Presumably the cells will take cues from the environment in which they are placed, but this is not yet proven. Also, stem cells have the potential to divide indefinitely—a hallmark of cancer cells and a potential obstacle to their use in patients. Another problem with established stem cell lines is that the patient's immune system will reject them (hence the interest in SCNT or in reprogramming the patient's own skin cells).

While stem cells have fomented a very public debate centering on issues relating to embryos, another debate is raging more quietly. This is the debate over patent rights, which determines who can profit if stem cells become commercially successful. Several groups, including the University of Wisconsin, Geron Corporation, which bought patent rights for James Thomson's original discoveries, and various other parties are arguing in court about the stem cell patents. This is not an isolated situation. Commercial biotechnology is about transforming knowledge and ideas into products that are profitable. The issue of who controls the profits from a given discovery is of critical importance and so patent lawyers almost always play a key role in the development of biotechnology products and applications.

B. Cloning

The term "clone" means a copy. When used as a verb, the term *clone* means to make a copy. In biotechnology, a clone can refer to an exact copy of a DNA segment produced using recombinant DNA technology, or one or more cells derived from a single ancestral cell (as discussed previously in the section on monoclonal antibodies), or one or more organisms derived by asexual reproduction that are genetically identical (or nearly identical) to a parent. In this section we discuss the last of these definitions. **Whole animal cloning** *is the production of one or more adult animals that have the same genetic information.*

Cloning occurs naturally in people and in other animals when an early embryo splits, giving rise to identical twins. Scientists working with laboratory and farm animals in the 1980s intentionally created twins by splitting embryos in the laboratory into parts, each of which was then implanted into a surrogate mother. Using this technique, scientists cloned embryos that resulted from a normal mating of two animals, so the clones were identical to one another, but they were not identical to either parent and their genetic traits could not be fully predicted.

While these early cloning activities proceeded with little fanfare, Dolly, the cloned sheep, became an international celebrity. Dolly was not produced by splitting an embryo. She was made by taking the nucleus of an adult sheep's mammary cell and inserting it into another sheep's egg whose nucleus had been removed (somatic cell nuclear transfer), see Figure 1.16. The resulting egg thus contained only the genetic information from the mammary cell. The egg was placed inside a surrogate mother sheep where it developed into a baby sheep, named Dolly. Dolly did not arise from any

mating between two animals and she was genetically equivalent to the adult sheep from whose mammary cell the nucleus was removed. Dolly was born in July, 1996, at the Roslin Institute in Scotland under the direction of Ian Wilmut. She was euthanized in 2003 because she suffered from various health problems, including serious lung disease. Dolly did not live a normal sheep lifespan, but it is unclear whether her health problems were related to the fact that she was cloned.

Scientists have now used somatic cell nuclear transfer to clone and produce adult sheep, cows, goats, pigs, rats, mice, rabbits, cats, horses, and mules. Cloned animals may have an impact on agriculture—a prospect that is not without controversy. (An FDA report on this issue is available at http://www.fda.gov/cvm/CloneRiskAssessment.htm.) At the time of writing there is no documented case of a cloned primate (the group to which humans belong) using somatic cell nuclear transfer. There is, of course, much controversy

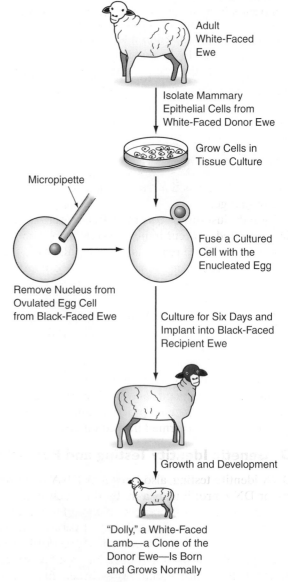

Figure 1.16. The Cloning of Dolly. Somatic cell nuclear transfer was used to create an adult animal.

as to whether the idea should ever be pursued with humans. The possibility of human cloning, both from a technical perspective and an ethical perspective, is likely to be a source of discussion for years to come.

C. RNAi

Biotechnology continues to branch out, sometimes in surprising ways, as biologists make new discoveries about the workings of living systems. An example of this is the discovery of RNA (ribonucleic acid) interference, RNAi. **RNAi** *is a phenomenon in which short RNA molecules inside cells can turn off the activity of specific genes.* The existence of RNAi came as a surprise to the scientific community when it was first discovered in plants and small nematode worms; the existence of such a system of turning off gene expression was never suspected. While the novelty of the discovery created something of a stir in the 1990s, the phenomenon achieved blockbuster status in the scientific community in 2001 when scientists demonstrated that the RNAi phenomenon occurs in mammalian cells. In 2006, two American researchers, Andrew Z. Fire and Craig C. Mello, won the Nobel Prize in Physiology or Medicine for their pioneering work on RNAi in nematode worms.

Biologists were quick to see many practical applications of RNAi. Researchers immediately began using RNAi as a tool to investigate the role of genes and the inner workings of cells. Scientists can, for example, add carefully designed RNA fragments to cells in order to "turn down" the expression of a specific gene, and then observe the effect that the loss of a gene product has on the cell. This allows the researcher to study the normal role of that gene.

By 2004, just three years after RNA interference was demonstrated to occur in human cells, many companies were making RNAi reagents or testing RNAi for use in treating or preventing human disease, see Figure 1.17. The first early RNAi experiments began in humans in 2004. These experiments involved treating macular degeneration, a disease that causes blindness. Other experiments are being conducted on a wide variety of ailments including AIDS, respiratory diseases caused by pathogens, and various inherited diseases. RNAi thus provides an example of how a basic research discovery, originally made in plants and small worms, turns out to have a great number of practical applications and is quickly being transformed into commercial products.

D. Genetic Identity Testing and Forensics

DNA identity testing, also known as DNA fingerprinting or DNA profiling, is a method of identifying individuals based on differences in their DNA. Identity testing uses DNA extracted from tissue (hair, bone, blood, etc.) to distinguish individuals. About 99.9% of every person's DNA is identical, but in a few stretches of DNA there are variations between humans that make each of us distinct from one another (except for identical twins, whose DNA is the same). These variable

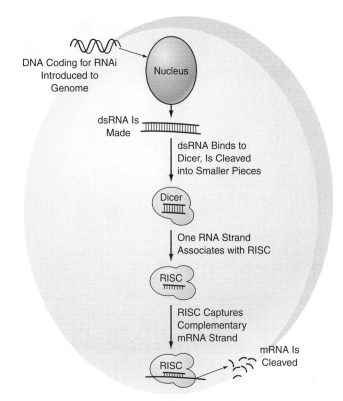

Figure 1.17. The Use of RNAi to Treat Disease. The goal of these experiments is to switch off, or decrease, the expression of a disease-causing gene. A cell is transfected with DNA coding for a specific RNA. The cell makes the RNA, which is double stranded (ds). The dsRNA enters the cytoplasm where it triggers the RNAi process. The dsRNA is first snipped into short pieces by an enzyme called Dicer. The short pieces of dsRNA are incorporated into a protein complex called RISC (RNA-inducing silencing complex), which cleaves and discards one strand of the dsRNA. The remaining strand guides the complex to bind with its complementary mRNA, which codes for the protein associated with the disease. The captured mRNA is then chopped apart so that the protein for which it codes is not made.

regions are used to make a DNA "fingerprint" that is unique for each person. (Keep in mind that the human genome consists of about 3 billion base pairs. A 0.1% variation between individuals could represent 3 million base pair differences in sequences.) The most famous use of this technology is to match DNA from a crime scene with that of a suspect. If the DNA matches, it provides evidence that the individual was at the scene. The idea of DNA fingerprinting is well-known, thanks to the high-profile murder trial of O.J. Simpson and some very successful television crime dramas.

DNA fingerprinting was developed in England in 1984 by Sir Alec Jeffreys, a professor at the University of Leicester, see Figure 1.18 on p. 26. In a brief interview posted by the Australian Broadcasting Corporation, Jeffreys talks about his discovery of DNA fingerprinting (http://www.abc.net.au/science/features/biotech/1980.htm). He says "It was invented purely by accident . . . from a different project in my laboratory looking at how genes evolve, and, in fact properly started with a lump of . . . seal meat which we used to get at a gene which we were interested in studying its

Figure 1.18.   Sir Alec Jeffreys Showing an Example of a DNA Fingerprint. (Photo Courtesy Homer Sykes/Aamy.)

evolution . . . We got that gene, looked at its human counterpart and purely by chance inside that gene was a bit of DNA which was a key to unlock the door on genetic fingerprinting . . . [I]t was absolutely blindingly obvious that what we had was a technology that could be used for identification, establishing family relationships and the like. So for me, it was very much a Eureka, my life literally changed in five minutes flat, in a darkroom, when I pulled out that first DNA fingerprint and saw just what we'd stumbled upon." As with many discoveries, this one was made by accident, but Jeffreys was quick to see the profound implications of his discovery.

DNA fingerprinting is so much more accurate than older blood typing methods of identification that it was quickly adopted around the world for identification purposes. Forensic scientists in the United States now commonly use a fingerprinting method that is a modification of the one pioneered by Jeffreys. (Forensic identity testing is also discussed in a Case Study on pp. 31–32).

DNA identity testing, though it is probably most famous for its use in criminal investigation, is also used to reliably determine paternity and other family relationships. It is used to identify bodies in the case of a disaster. Analysts use DNA fingerprinting in wildlife investigations, for example, to identify whether an item was illegally poached from an endangered species. DNA fingerprinting can be used to detect pathogens in food, air, and water. It can be used to determine the pedigree of a valuable animal, such as a racehorse. DNA fingerprinting methods are therefore powerful and widely applicable to a variety of circumstances.

HISTORICAL STORY

DNA Fingerprinting

Sir Alec Jeffreys's professional life did indeed change with his discovery of DNA fingerprinting. Almost immediately after Jeffreys published his fingerprinting method, he was called upon to help a family embroiled in an immigration dispute. The family, who were citizens of the United Kingdom, claimed that their son had visited Ghana and was returning home to his family after the trip. The immigration authorities would not allow the boy to return to his family, claiming that his passport had been altered and that he was an unrelated imposter. Jeffreys used his DNA fingerprinting technique to conclusively show that the boy was indeed the son of the family in question and, in a happy ending, the boy was reunited with his family ("All Things Considered," National Public Broadcasting Archive. July 15, 2005).

Soon after, Jeffreys was called on to use his method in a tragic crime investigation involving two teenage girls who had been raped and murdered. A man had confessed to the rape and murder of one of the victims, but not the other. The police were certain the murders were related and asked Jeffreys to use his method to analyze semen isolated from the victims and the suspect. The DNA analysis showed that the semen isolated from both girls was from the same man. DNA fingerprinting moreover showed that the man who had confessed was not the perpetrator of either crime. Thus, the very first time DNA fingerprinting was used in a crime scene investigation, it established a suspect's innocence, not his guilt. The police went on to mount a dramatic investigation in which they asked all the men in the local community to submit their blood for DNA testing. More than 5,000 men complied but none were a match for the crime scene DNA. The case would have ended there were it not that someone overheard a man telling his friends that he had been asked by the local baker to give his blood in place of the baker. This conversation was reported to investigators who took a DNA sample from the baker. The baker's DNA was a perfect match for the semen taken from the victims and the baker confessed to both crimes. DNA fingerprinting thus quickly proved its value as a forensics tool.

Dr. Mary-Claire King, a scientist recognized for her pioneering work studying genes that cause familial breast cancer, used DNA fingerprinting in another significant way. Between 1976 and 1983 there was a brutal dictatorship in Argentina. Thousands of men, women, and children were tortured, killed, and "disappeared" during this reign of terror. Some of the women who disappeared were pregnant at the time of their capture and gave birth before their death. Also, some very young children were seized before their parents were killed. The babies were sold or given to military families who illegally adopted and raised them. When the military regime was deposed, grandmothers of these kidnapped children began to search for them. The grandmothers followed leads from school registrars, who saw children arrive with forged papers, and from others who had noticed babies suddenly appear in military families. As the grandmothers found individuals they thought were their grandchildren, they appealed to scientists to use DNA fingerprinting to definitively identify their family members. Mary-Claire King, then a professor at the University of California, went to Argentina and obtained court orders that enabled her to test the DNA of individuals who were thought to have been kidnapped as babies. In this way, about 50 children were eventually matched with their grandmothers and reunited with their birth families.

E. Biofuels

The use of biomass to manufacture biofuels and other valuable products is one of the fastest-growing areas in biotechnology. **Biomass** *is organic material created by recently living organisms, particularly plants, but also including manure, food waste, sewage waste, and such.* Biotechnologists are working on ways to substitute *biomass-derived fuels,* **biofuels,** for fossil fuels (gas, coal, and oil). Biofuels are generating enthusiasm partly because biomass is a renewable resource that can be produced almost anywhere, whereas fossil fuel deposits are rapidly being depleted. Substituting biofuels for fossil fuels is also expected to have a positive environmental impact by reducing the rate of global warming. This is because the use of biomass to provide energy can be, in principal, "carbon neutral." The carbon dioxide released when harvested material is burned exactly replaces carbon dioxide that the plants used in photosynthesis during their lifetimes. The burning of fossil fuels, in contrast, releases carbon dioxide that was previously sequestered in the ground. (The situation is complicated, however, by the fact that farmers use fossil fuel when cultivating crops.)

Ethanol, which is already widely used in E10 gasoline, is one of the first, and currently most important, biofuels. Ethanol is the same alcohol that is present in wine, beer, and liquor. Wine, for example, is made by yeast that break down the sugar, glucose, present in grapes to form ethanol and carbon dioxide. *The metabolic process in which sugar is converted to ethanol by yeast or bacteria is called* **fermentation.** Ethanol for biofuels is also produced by fermentation.

Plant substances that contain a lot of sugar, such as sugar cane and corn kernels, are easily and relatively inexpensively converted to ethanol using a standard fermentation process. The diversion of food crops to produce fuel, however, is obviously problematic. Scientists are therefore looking for ways to produce fuel from nonedible plant materials, such as plant waste (e.g., corn stalks) and grasses grown on nonfarmable land. This is proving to be a difficult technical challenge. Nearly all plant cell walls are stiffened by cellulose, which consists of long chains of glucose. The cellulose in any plant waste material is therefore a potentially abundant source of fermentable biomass. The challenge is that cellulose is extremely resistant to digestion. To solve this problem, scientists are developing new methods to break down cellulose into its glucose subunits. Many of the methods involve enzymes, called *cellulases,* which are isolated from bacteria and fungi. Significant scientific work is required to make the digestion of cellulose cost-efficient enough to operate at a commercial scale, see Figure 1.19 on p. 28.

Biotechnologists are involved in many aspects of the biofuels industry. Scientists, engineers, and process operators already design, build, and operate large-scale plants to convert corn kernels into ethanol, and more ethanol facilities are in construction. Researchers are optimizing the fermentation process for biofuel production by finding the best conditions and the most efficient microbes. Scientists are working to improve the enzymes used to break down cellulose. Yet other projects aim to modify crops to be better sources of biofuels. Genetic engineering tools are being used to help solve these problems. The conversion of biomass to fuels is a rapidly growing area of biotechnology that brings together environmental biology, agriculture, industrial microbiology, genetic engineering, and bioprocessing.

IV. SUMMARY

Biotechnology can be understood as the transformation of knowledge about living systems into useful applications and products. Biotechnology is not a new phenomenon because for thousands of years curious humans have made discoveries that have led to useful products. But in the past 50 or so years, the pace of discovery has exploded and applications of this knowledge are expanding every day. Often, as was the case with DNA identity testing, new knowledge is quickly transformed into widespread application. In the 1980s DNA fingerprinting was discovered, in the 1990s it was developed for use around the world, and by the beginning of the next decade DNA fingerprinting was a fixture of popular culture. Other applications are slower to mature, but are likely to do so in the future.

Modern biotechnology is based on the discovery of methods to manipulate DNA, but the industry clearly incorporates many other methodologies as well. Protein scientists isolate, model, purify, and study proteins. Cell culture biologists and fermentation specialists develop and refine methods for growing cells in culture. Engineers work together with scientists to develop manufacturing and protein purification processes. Methods using RNA to control gene expression are becoming ever more important. There are a host of analytical methods that are used in the laboratory for testing materials at all stages of product discovery and development. Pharmacology, medicine, forensics, agricultural science, environmental science, and manufacturing technology are key parts of the biotechnology industry. Combine all these technical fields with business, ethics, and more, and this becomes the modern biotechnology industry.

We have also glimpsed in this chapter how biotechnology is a commercial enterprise and as such, it is driven by profit and finances, patents, and business concerns. Yet, it is important to remember that biotechnology is also driven by people who study science out of curiosity, and by many people who hope their work will help others in ways unrelated to monetary profit.

PRACTICE PROBLEMS

1. The process illustrated in Figure 1.3 is sometimes called *gene cloning.* Why is this process called *cloning*?

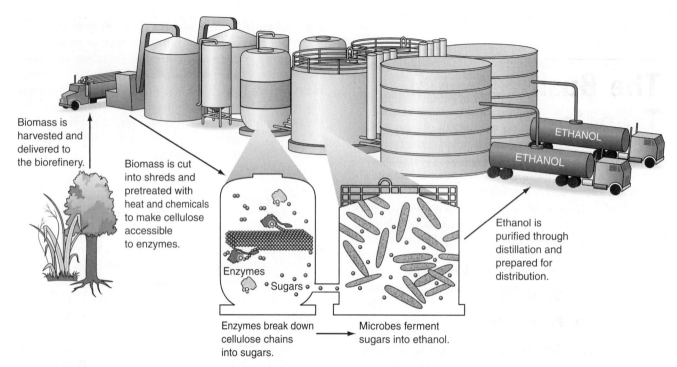

Biomass is harvested and delivered to the biorefinery.

Biomass is cut into shreds and pretreated with heat and chemicals to make cellulose accessible to enzymes.

Enzymes

Sugars

Ethanol is purified through distillation and prepared for distribution.

Enzymes break down cellulose chains into sugars.

Microbes ferment sugars into ethanol.

Figure 1.19.   An Artist's Conception of a Future, Large-Scale, Cellulosic Ethanol Production Facility. (*Source:* Genome Management Information System, Oak Ridge National Laboratory, the U.S. Department of Energy Genome Programs, http://genomics.energy.gov.)

2. The following words are sometimes confused with one another. Define them:

   a. Transfection

   b. Transformation

   c. Translation

   d. Transcription

3. Which of the following biotechnology applications involve transfection of cells, which involve transformation of cells, and which involve neither transfection nor transformation?

   a. Producing crop plants that are resistant to the herbicide glyphosate.

   b. Growing cultured human skin cells to use for tissue engineering (e.g., to treat burn victims).

   c. Making wine.

   d. Manufacturing human epidermal growth factor (a protein) in bacteria.

   e. Harvesting stem cells from an embryo.

4. **a.** What is the role of the antibodies that a person, or other animal, normally produces? (If you are unsure, consult a biology reference.)

   b. Compare and contrast monoclonal antibody therapeutic agents to the antibodies normally produced in the body. Refer to Table 1.2 in your answer.

5. The monoclonal antibody Synagis is used to fight infection by respiratory syncytial virus (RSV). The manufacturers provide an animation of the action of this antibody at their website, http://www.synagis.com/how-synagis-works.aspx.

   a. The general mechanism of action of monoclonal antibody drugs is to recognize a specific protein target, bind to this target, and block the target's action with therapeutic effect. Watch the animation and explain how this mechanism of action applies to Synagis.

   b. Synagis protects babies against a viral agent, but it is not a vaccine, at least not in a conventional sense. What is the difference between Synagis and a conventional vaccine (like the one Jenner invented against smallpox)?

6. Compare and contrast *therapeutic cloning* with the form of cloning used to create the sheep, Dolly.

7. Insulin and other biopharmaceuticals are seldom administered orally. What might happen to these drugs if they were swallowed?

8. The gene therapy treatment used for Ashanti De Silva must be periodically repeated whereas the surviving children who received gene therapy for X-SCID are presumed to be cured. What's the difference between the two treatments?

9. How might the tools of biotechnology be used to make it more practical and cost-effective to manufacture ethanol from cellulose? (There are many answers to this question; be creative.)

CHAPTER 2

The Business of Biotechnology: The Transformation of Knowledge into Products

I. PRODUCT LIFECYCLES IN BIOTECHNOLOGY

A. Overview

As we saw in Chapter 1, the modern biotechnology industry transforms biological knowledge into a variety of products ranging from cancer therapeutics to DNA fingerprints to lens cleaning solutions. The transformation of an idea into a product does not happen all at once; rather, it is a process that usually requires years of effort. It is possible to describe this process as a product's *lifecycle*. The typical biotechnology product is "born" in the research laboratory when a discovery results in an idea for a product. The potential product must then go through a period of **development** *during which the idea is transformed into an actual, workable product.* If the development period is successful, then the product enters **production**—*the final stage in a* *successful lifecycle including sales and expansion and growth of the product.* As we will see later in this section, this product lifecycle is roughly mirrored in the organization of biotechnology companies.

B. Research and Discovery

The biotechnology industry is built on basic scientific discoveries that are usually made by scientists and their students in universities, colleges, research institutes, and medical centers. Sometimes research scientists imagine the potential commercial applications of their work and decide to explore that potential. The initial work to explore commercial applications of a research discovery is conducted in a research laboratory, often still in an academic institution. Early experiments may involve, for example, testing a treatment in cultured cells, testing a compound in animals, using biochemical assays to explore a molecule of significance, using computer simulations of molecules and their interactions, and growing plants in controlled environments. This is also the period when intellectual property is secured (discussed later in this chapter) for the idea, the product, its potential uses, and its processes of manufacturing.

If these early studies yield promising results, then researchers may decide to begin a company to develop the commercial applications of the idea. Many biotechnology companies are thus initially founded by research scientists from universities and other research institutions. Sometimes the discovery and early research that results in a product is conducted by scientists in a company that already exists. Most biotechnology companies have research scientists and technicians whose job is to explore new ideas for products.

C. Development

Product development requires a cycle of rigorous testing, modification, and continued testing of a potential product to optimize its utility for its intended purpose. The development phase of a product's lifecycle is the transition between its discovery in a research laboratory and its use as a commercial product.

The development stage includes determining the properties of the product, specifying the properties the product must have to be effective, and describing how to make the product. The development phase is a time of transition, evolution, and evaluation. As development progresses, the characteristics of the product become established, methods of production become increasingly consistent and systematized, and the specifications for raw materials are codified. It is during this evolutionary period that quality is designed and built into the product.

A manufactured product is initially made in small quantities in the laboratory. A major part of development is the **scale-up** *of the processes used in the laboratory so that larger amounts of the product can be made in a consistent, reproducible, and economically efficient manner.* Depending on the quantities of the product to be sold, scale-up may be a fairly simple process of moving from one size flask to another. In other situations scale-up involves initially making the product in a medium-sized **pilot plant** *where further development*

Bruker AXS designs and manufactures analytical X-ray systems for elemental analysis, materials research and structural investigations.

Covance - a global drug development services company - provides an industry-leading portfolio of preclinical, clinical development and commercial service offerings.

Lucigen Corporation focuses on developing new, much more effective products and technologies for gene cloning and genomics.

Gilson manufactures HPLC systems, high-throughput robotic workstations, pipettes, fraction collectors, SPE systems, detectors, injectors, and much more.

From DNA sequence analysis to microarrays, DNASTAR is a leader in meeting the software needs of molecular biologists for research.

EPICENTRE Biotechnologies is a manufacturer and seller of high-quality molecular biology products.

Monsanto is an agricultural company that applies innovation and technology to help farmers around the world.

Invitrogen provides products and services that support research institutions as well as pharmaceutical and biotechnology companies.

Promega's products help researchers successfully explore gene, protein and cellular interactions.

Figure 2.1. Hundreds of Biotechnology and Associated Companies in Locations Around the World Provide a Wide Variety of Products. A few companies shown here illustrate the diversity of biotechnology companies.

Table 2.1. THE RESPONSIBILITIES OF THE RESEARCH AND DEVELOPMENT UNIT(S) IN A BIOTECHNOLOGY COMPANY

- *Discovering a potential product with commercial value*
  performing scientific research
- *Characterizing and documenting the properties of the product, such as:*
  composition, physical and chemical properties
  strength, potency, or effect of the product
  purity of the product and steps required to avoid contamination
  applications of the product
  safety concerns in the use of the product
- *Establishing and documenting product specifications* (descriptions of properties that every batch of the final product must have to be released for sale)
- *Developing and documenting methods to test the product to be sure it meets its specifications*
- *Developing and documenting processes to make the product*
- *Describing and documenting the features of any cells or organisms required to make the product*
- *Determining and documenting the raw materials required to make the product and establishing specifications to characterize these materials*
- *Describing and documenting equipment and facilities required to make the product*
- *Determining and documenting stability and shelf life of the product*
- *Developing and documenting a plan for production of the product*
- *Scaling-up production*

occurs, and then eventually in full-scale production facilities. Scale-up of an agricultural product may mean moving from the greenhouse, to a small experimental plot, and finally to large fields.

Sometimes ideas are transformed into commercial products in an academic institution or a research institute, but it is often in biotechnology companies that this transformation occurs, see Figure 2.1. Biotechnology companies have organizational units, generally called **research and development (R&D),** *which find ideas for products, perform research and testing to see if the ideas are feasible, and develop promising ideas into actual products.* Larger companies often separate the *research* part of a product's lifecycle from its *development,* and have separate functional units for each. The development of pharmaceutical/biopharmaceutical products is so extensive that multiple teams and more than one company often participate in their development. The general responsibilities of the R&D functional unit(s) are summarized in Table 2.1.

CASE STUDY

Crime Scene Investigation: A Behind-the-Scenes Story

There is a gritty crime scene; enter a freshly coiffured investigator who discovers a bit of blood, rushes it back to the laboratory, and translates the blood stain into the evidence that unlocks the case. Behind the scenes of the popular television crime scene investigation dramas are real scientists who create laboratory tools that are used in actual forensic investigations.

DNA "fingerprinting," one of the most famous of the modern forensic technologies, was developed in 1984 by the scientist, Sir Alec Jeffreys. Although about 99.9% of every person's DNA is identical, there are a few differences between humans. These variations, called polymorphisms, are the basis for DNA typing. Jeffreys found that certain regions of human DNA contain specific sequences of nucleotides that are repeated over and over again, one after the other. He also discovered that the number of times a sequence repeats is polymorphic (has many forms) and so can be used to distinguish the DNA from individuals. Jeffreys used a method called restriction fragment length polymorphism (RFLP) analysis to detect the differences in numbers of repeats in the DNA from different people. Although RFLP analysis was successfully used in criminal investigations in the 1980s and

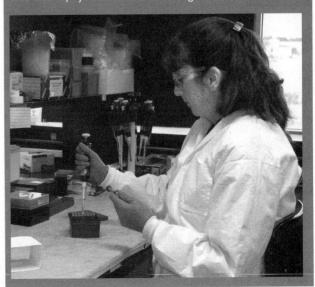

1990s, it has limitations. RFLP analysis requires relatively large amounts of undegraded DNA, more than is often available from a crime scene. RFLP analysis is used to examine only one sequence of DNA at a time, is not easily automated, and is slow. Many researchers in academia and in biotechnology companies sought to improve DNA typing technology after Jeffreys's initial discovery. Scientists quickly adapted other types of polymorphisms in addition to the repeated sequences Jeffreys used, including SNPs and STRs. **Single nucleotide polymorphisms (SNPs),** are places in DNA where a single nucleotide differs from person to person. A **short tandem repeat (STR)** is a type of polymorphism, similar to that used by Jeffreys, but smaller. Scientists also learned to use the **polymerase chain reaction (PCR),** a powerful method for amplifying DNA, to deal with the problem that DNA is often limited in quantity and poor in quality, see Figure 2.2. Furthermore, scientists reasoned that if they could look at many polymorphic DNA locations (loci) at one time, they could quickly and reliably distinguish one individual from another with little chance of two people sharing the same "fingerprint."

Research scientists at Promega Corporation were among the many creative people studying DNA typing. Their challenge was to develop a method to distinguish individuals from one another that is fast, reliable, has virtually no chance of error, and that could be readily used by forensic analysts with different levels of training and possibly different equipment. At the time Promega was beginning their DNA typing project, Dawn Rabbach (shown in the photo shown in the photo on p. 31) was a student in a two-year associate degree biotechnology program, and a part-time entry-level technician at Promega whose job was to prepare laboratory reagents. Dawn was selected to work on the DNA typing team because of her hands-on skills at the lab bench and because she was known to pay careful attention to detail. Dawn and her colleagues worked feverishly to be one of the first companies to develop a commercial method of DNA typing that would meet the stringent requirements of forensics laboratories. During this fast-paced period of research and development, Ms. Rabbach performed many studies, carefully modifying one factor at a time to help determine the best methods of typing. She tested STRs to see if this type of polymorphism would work in an automated system. She tested different methods of labeling DNA and experimented to determine how much DNA should be analyzed in each test to obtain the most accurate results. The efforts of the Promega R&D team led to a patented, automated DNA typing system, which provides a fingerprint from 16 polymorphic loci simultaneously. The team packaged the method into a kit that was tested and accepted by the forensics community. The team's efforts also contributed to the development of the **Combined DNA Index System (CODIS)** a national database of DNA fingerprints from convicted offenders. In the late 1990s, the FBI was creating their database and they needed to select a standard set of polymorphisms that every crime lab in the United States would analyze. They

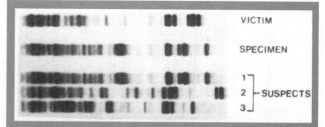

Figure 2.2. DNA Fingerprints Using STRs. DNA was extracted from a victim and three suspects. Their DNA, and also DNA from the crime scene (specimen), was treated with PCR to specifically amplify specific STRs. The amplified DNA was analyzed using electrophoresis (electrophoresis is discussed in Chapter 33) resulting in the patterns of bands shown in this figure. Analysts can compare the band patterns to see if DNA from the crime scene (specimen) matches the pattern of any of the suspects. (Photo courtesy Orchid Cellmark.)

selected for their data base a set of 13 loci, many of which had been developed by the Promega scientific team.

Ms. Rabbach, who is now a Senior Research and Development Scientist at Promega Corporation, continues to transform the ideas of research scientists into commercial products. Her job responsibilities are typical of those in biotechnology company R&D laboratories and include:

- performing laboratory experimentation necessary to develop new molecular biology products
- documenting her work so that her discoveries can be patented and her results are repeatable
- writing **standard operating procedures (SOPs)** that are used by manufacturing personnel when her team's products are ready for commercial production
- developing assays that are used by quality-control analysts when they test the products her team developed
- writing standard operating procedures that are used by quality-control analysts
- troubleshooting problems that arise as customers with different equipment and processes adopt Promega products
- anticipating issues that arise as software and instrumentation change
- traveling to customer laboratories to test methods and equipment
- speaking at forensic conferences

D. Production

If a product successfully makes it through the development phase, then it enters the production phase and is produced, marketed, sold, and used. In most biotechnology companies, a distinct production unit is responsible for these tasks.

The systems used for production in biotechnology companies are diverse. Manufacturing a product may involve growing bacteria in laboratory-sized flasks and isolating product from the cultures. In other situations, manufacturing involves growing bacteria or other types

Figure 2.3. Industrial Bioreactors. Each of these bioreactors (foreground and left side of picture) contains a cell broth that produces proteins that are purified to make biopharmaceuticals. (Photo courtesy of the National Research Council of Canada.)

of cells in fermenters or bioreactors that can be several stories tall, and using industrial-scale equipment to purify products from the cultures, see Figure 2.3. Production may involve isolating biological molecules, growing plant cells in petri dishes, cultivating crops in a field, maintaining laboratory animals, or even keeping farm animals. The details of production thus vary greatly from company to company. Certain functions, however, are generally the responsibility of the production team, regardless of the nature of the product and company, see Table 2.2.

E. Quality Control and Quality Assurance

As a product moves through its lifecycle from development and into production, the tasks of the quality control (QC) and quality assurance (QA) units mature along with it. These are functional units in a company or facility that are responsible for assuring the quality of

the product. You may see the terms *quality control* and *quality assurance* used differently in various sources; in fact, the two terms are sometimes used interchangeably. Here, we define **quality control (QC),** *as the unit that is chiefly responsible for monitoring processes and performing laboratory testing.* A quality-control technician might, for example, test a product in the laboratory to make sure it has the proper attributes before it is released for sale to customers. The quality-control laboratory is discussed in more detail in the last section of this chapter.

Quality assurance (QA) *refers to all the activities, people, and systems that ensure the final quality of products.* The quality assurance unit plays a key role during development and production in assuring that the procedures that are used adhere to the company's requirements. A person who works in quality assurance might, for example, check documentation (paperwork) associated with a product to be certain that it is completed properly, stored in the right place, and is accessible if needed. Another example of a quality-assurance task would be participating in the investigation and correction of a problem, as will be discussed in Chapter 4. The goal of the quality-assurance team is to assure that all the processes in the company come together to produce a quality product. The responsibilities of the quality units are summarized in Table 2.3 on p. 34.

F. Regulatory Affairs

When products are regulated by the government (e.g., as they are with biopharmaceuticals) a **regulatory affairs staff** is needed *to interpret the rules and guidelines of regulatory agencies and to ensure that the company complies with these requirements.* Regulatory affairs specialists work with their counterparts in the regulatory agencies (e.g., the Food and Drug Administration, the Environmental Protection Agency) to make sure that the data, documentation, and forms that the company produces are sufficient to support their product (e.g., pharmaceuticals, genetically modified crops).

Table 2.2. THE RESPONSIBILITIES OF THE PRODUCTION UNIT IN A BIOTECHNOLOGY COMPANY

- *Making the product*
- *Working with large-scale equipment and/or large volume reactions* (not applicable to all biotechnology companies or products)
- *Routine monitoring and control of the environment as required for the product* (e.g., maintaining the proper temperature or sterility requirements)
- *Routine cleaning, calibration, and maintenance of equipment*
- *Following written procedures and performing tasks associated with producing the product*
- *Monitoring processes associated with making the product*
- *Working with automated equipment*
- *Initiating corrective actions if problems arise*
- *Completing forms, labeling, filling in logbooks, and maintaining other required documents*

Table 2.3. THE RESPONSIBILITIES OF THE QUALITY CONTROL AND QUALITY ASSURANCE UNITS IN A BIOTECHNOLOGY COMPANY

QC

- *Monitoring equipment, facilities, environment, personnel, and product*
- *Testing samples of the product and the materials that go into making the product to determine whether they are acceptable*
- *Comparing data to established standards*

QA

- *Reviewing all production procedures in the company*
- *Ensuring that all documents are accurate, complete, secure, and available when needed*
- *Deciding whether or not to approve each batch of product for release to consumers*
- *Reviewing customer complaints*

G. The Lifecycle of a Company

Biotechnology companies usually undergo a maturation process that reflects the lifecycle of their products. Start-up companies are those that, like their product(s), are at the beginning of their lifecycle. The emphasis in a start-up company is on research and development and nearly everyone in the company is likely to be working in this area. Start-up companies typically have only a few employees and a single individual may play a number of roles, perhaps doing research and making business connections. Depending on the nature of the product, the same facilities may be used to perform research, development, and manufacturing of the first batches of product for sale.

The start-up of a company or the beginning of a new product's lifecycle is a time of excitement, change, and unpredictability. Research and development scientists often do not know which ideas will result in a product, which methods will work best, and how much success they will ultimately have in their endeavors. Creativity and a willingness to work with uncertainty characterize most investigators in R&D and in academic research laboratories.

If a product is successful, then the company must evolve to manufacture it. During development, the company likely must expand, more individuals might be hired, and staff will start to specialize in certain functions. Mature companies have a separate production team that works in a relatively predictable environment.

As a company grows along with its product(s), a variety of job opportunities open, see Figure 2.4, and new functional areas are established to support the development, production, marketing, and sales of the product(s) including:

- *Business development,* which identifies, researches from an economic perspective, analyzes, and brings to market new products or collaborations.
- *Marketing and sales,* which is responsible for interacting with customers.
- *Business divisions,* such as accounting and human relations, which keep the organization running.
- *Dispensing,* which puts products that are produced in bulk into individual containers for customer use.
- *Metrology,* which ensures that instruments (such as those used in laboratories and those used to monitor production conditions) operate properly.
- *Clinical Research Management* (for companies whose products are therapeutic agents), which monitors experiments with human subjects to ensure compliance with regulations, guidelines, and company procedures.

The "culture" of a successful company is likely to change as it makes the transition from a small start-up with few employees who fill many roles to a larger company with specialized staff. Everyone in a small company knows everyone else; this is not true of larger companies. Communication becomes less direct as the company grows physically and in numbers of staff, so processes need to be established to ensure good communication. Small companies, if they have a promising product, are

Lab Analyst Wanted

Work for one of the worlds leading Bio-pharmaceutical companies.
Responsibilities:
1. Testing of raw materials, in-process and finished product samples as required
2. Environmental monitoring of classified/controlled areas
3. Monitor water
4. Microbial ID
5. Data trending
6. Control of lab supplies
7. Assist in the preparation of Laboratory Investigation Reports
8. Method Validation Protocol preparation & execution
9. Equipment and Process Validation Protocol execution
Equal Opportunity Employer

Quality Check Inc.

Associate Scientist

Research in the fields of autoimmune diseases and oncology.

You will be required to conduct in-vitro and ex-vivo experiments using various technologies (e.g. multiplexed protein arrays, immunoassays and activity assays) that will assist in biomarker identification and assay development.

You will work closely with scientists and project teams and effectively communicate the study's results.

GenHealth
Applying genetic techniques to improve your health
Equal Opportunity Employer

DNA Today Associate Scientist

The Associate Scientist position is responsible for assisting in the development of fermentation procedures necessary for the licensure of a biologics drug product in the European market and also assisting in the development of a second fermentation process for the U.S. market. This individual is responsible for hands on operations related to development, scale-up and characterization of fermentation and primary recovery processes.

DNA Today is an Equal Opportunity Employer

Research Assistant

Assisting on key R&D projects you will be responsible for conducting a variety of experimental techniques in the search of novel cancer treating therapeutics.

Your duties in this position will include:
- Calibration and maintenance of laboratory equipment
- Quality documentation in accordance to strict regulations
- Preparation of media, sample set up and identification of organisms
- Bioburden and sterility testing
Req. # 06-91578
Equal Opportunity Employer

PINNACLE PROJECTS

Figure 2.4. Newspaper Job Postings Reflect the Many, Diverse Opportunities in the Biotechnology and Medical Products Industry.

often purchased by larger ones. When a company is bought, its staff suddenly acquires new colleagues they have never met, who work in different facilities—perhaps on different continents. As a company grows, and is perhaps bought, decisions will be made by different people and individuals may have less input. Changes that affect individual workers are common as the products of a company move through their lifecycle.

II. INTELLECTUAL PROPERTY AND THE BIOTECHNOLOGY INDUSTRY

A. Intellectual Property and Patents: Overview

We have seen that the biotechnology industry is fueled by scientific discoveries that are transformed into valuable products. The economic value of a company therefore depends not only on its physical assets (e.g., buildings, laboratories, production facilities), but also on its scientific knowledge. The knowledge base of a company, individual, or other entity is a kind of property, called intellectual property. **Intellectual property (IP)** *embodies creations of the mind and intellect.* The success of most biotechnology organizations depends on their intellectual property. It is not enough, however, for an organization to have intellectual property—it must also protect that property. This section introduces the use of patents and trade secrets to protect IP.

Intellectual property includes such familiar items as literary and artistic creations and product trademarks. Intellectual property also encompasses *inventions,* which is the category of most interest to us as biotechnologists. Machines, such as digital cameras or computers, often come to mind when we think of inventions, but the term *invention* can broadly refer to many types of innovations. New therapeutic compounds, diagnostic kits, genetically modified organisms, methods to genetically transform bacteria, and industrial enzymes are examples of biotechnology *inventions.*

Intellectual property is different than more tangible forms of property, like a car or a piece of land. One of the differences is that it is relatively easy to steal intellectual property; one need only copy it. A person might expend considerable efforts devising something new only to have another person copy and profit from it. Governments therefore provide protections for creators of intellectual property. These protections include patents, trade secrets, and copyrights. A **patent** *is a type of intellectual property protection that is an agreement between the government, represented by the Patent Office, and an "inventor" whereby the government gives the inventor the right to exclude others from using an "invention" in certain ways.* In exchange for this right, the inventor must fully disclose the invention to the public. Governments protect intellectual property not only to reward inventors with a means to protect and profit from their work, but

also to promote innovation. To obtain a patent, an inventor must, in effect, teach everyone to make and use the invention. In this way, new innovations spread quickly.

Useful chemicals, enzymes, and drugs are items that are inventions for patent purposes. Processes or methods of accomplishing a task can also be patented as inventions. A process to manufacture a biopharmaceutical agent, a method to manufacture an industrial enzyme, a method for DNA identity testing, a method to insert genetic information into mammalian cells, a method of purifying a biological compound, and a method of administering a therapeutic compound are examples of inventions that are processes. Manufactured items that are parts of other devices, and improvements of previous inventions, are also inventions.

One cannot patent a human being, a naturally occurring organism, a law of nature, natural or physical phenomena, or abstract ideas. It is not possible, for example, to patent a mathematical theorem, nor the discovery of the structure of DNA. One cannot patent an idea or discovery alone. It is the practical application of an idea or a discovery in order to solve a problem or supply a need that makes it patentable.

In the United States, patents are granted by the U.S. Patent and Trademark Office (USPTO). Other countries have similar offices. Patents only apply in the country in which they are granted. It is therefore necessary to receive a patent in every country where one wants protection.

The term of a new patent is commonly 20 years from the date on which the application for the patent was filed. Anyone can use the invention when the patent expires. Generic drugs, for example, are compounds that were originally patented by a pharmaceutical company but their patent protection has expired.

If you own property such as land or a car, you have the right to use or sell it. A patent is different. A patent grants the inventor the right to *stop others* from selling, using, or exporting the invention. A patent can, however, be *licensed,* as explained below.

If one party holds a patent and another party uses that invention without permission, it is called **infringement.** In this case, the patent holder will normally send the other party a letter telling them to stop. The patent holder may offer the offending party a license. A **license** *is an agreement by the patent holder that it will not enforce the right of exclusion against the licensee (the party wishing to use the patented invention).* The licensee pays the patent holder fees and/or a share of revenues. There are many situations where a company wants to manufacture a product that is partly protected by one or more other patents. The company may then license the technologies it needs in order to develop and produce its own product.

There are cases where two parties disagree as to whether one is infringing on the other's patent, in which case the dispute may be decided in court. If the court agrees with the patent holder, then the court might compel the other party to pay damages to the patent holder. This situation is not uncommon in the biotechnology industry and the economic stakes are often high in these disputes.

B. Invention and Biotechnology

The Boyer-Cohen patent, awarded to Stanford University in 1980, was a landmark in the history of modern biotechnology. This patent covered the discovery by Stanley N. Cohen and Herbert W. Boyer of a method to transform one organism with the DNA from another organism (see Chapter 1 for an explanation). Stanford required that any commercial party (business) wanting to use this method of genetic transformation had to obtain a license and pay fees, plus a royalty on the sales of any product developed using the method. Stanford allowed multiple businesses to license the Boyer-Cohen invention and so the method of "genetic engineering" quickly spread around the world.

Biotechnology involves living things and so leads to the issue of whether it is possible to patent an organism. This is not a new question, although it remains a topic of much discussion. In 1873 Louis Pasteur patented a yeast "free from . . . germs or disease." More recently, Ananda Chakrabarty genetically modified a bacterium so that it could degrade crude oil. His attempt to patent this bacterium was originally turned down by a patent examiner, who argued that the law dictated that living things are not patentable. Ultimately however, in 1980, the Supreme Court, in a close 5 to 4 decision, ruled that this bacterium was patentable because it was not found in nature. The Supreme Court's interpretation of U.S. patent law opened the door to patents for other genetically modified organ-

Figure 2.5.   The First Patented Transgenic Animal. The Oncomouse, a genetically altered mouse that gets cancer very easily, was developed by researchers at Harvard University to serve as a model organism in cancer research. In 1988, the Oncomouse became the first transgenic animal to be patented in the United States. The patenting of an animal raises ethical and societal concerns and Oncomouse patents have been repeatedly challenged by activists, particularly in Europe and Canada. (Photo courtesy Science Museum/Science & Society Picture Library.)

isms and cells including plants, nonhuman animals, hybridoma cells, viruses, cell lines, and embryonic stem cells, see Figure 2.5.

Humans cannot be patented, but rulings allow the patent of products isolated from humans, for example, proteins, cell lines, and genes. A scientist, for example, who purifies a type of interferon from blood can patent that molecule. Interferon existing in a person, however, is not patentable. Genes can be patented if their sequence is determined, if the gene was not previously discovered, if it is well-described, and if it has a useful application. Uses of genes include: prenatal screening, prognosis of disease (e.g., cancer progression), diagnosis of disease, treatment of disease, and determination of risk factors for disease. Recent rulings prohibit patenting a gene sequence if its usefulness is unknown.

C. Getting a Patent

Intellectual property is the foundation for biotechnology businesses. It is therefore essential that biotechnologists, particularly those who work in a research or development environment, are familiar with the requirements for obtaining a patent so that their innovations can be protected. This section introduces these requirements.

Obtaining a patent requires submission of a patent application and fees (that can run from hundreds to thousands of dollars). If an employee invents something in the course of their work for an employer, then the employer (e.g., a biotechnology company or university) will generally submit the application and will own that patent if it is awarded. However, inventors have the right to be recognized for their inventions so even if the patent is owned by someone else, the inventor's name(s) is required on the patent application. If an employee creates an invention outside of their scope of work for the company, then the employee would normally submit a patent application on her or his own behalf.

The patent application must describe the innovation in sufficient detail that another skilled person could use the invention described. Patent applications submitted in the United States must contain one or more *claims*. The invention is described in the claims and the scope of protection requested by the applicant is defined by the claims. The scope includes the applications and situations that the applicant wants the patent to cover, somewhat as a deed to land establishes its boundaries.

A patent applicant must sign an oath claiming inventorship and must prove the following in the patent application:

- the innovation is new
- the invention is not obvious
- the invention is useful
- he or she was the first to invent the new item

To prove that an invention is new and that it is not obvious, the invention must not have been previously described or discovered by another before the patent application was filed. The innovation must not be obvious from other people's work. For a gene or protein, for example, this usually means the inventor must learn the chemical structure or sequence of the gene or protein. If the structure or sequence was already known, then this requirement cannot be met. This requirement means that it is important to conduct a patent search to see what other people have already patented. It is now possible to search on-line, full-text patent databases that make it vastly more convenient to conduct a patent search than was possible when only paper copies were available.

The patent application must clearly specify the invention's potential usefulness. A newly discovered protein, for example, might be useful as a drug to treat a particular illness or might have application in an industrial process. A newly discovered gene might be used in a diagnostic kit for diagnosing a disease or as a gene therapy agent.

The fourth requirement, that the applicant is the first to create the new innovation, is very important. There are situations where more than one party applies for a patent on the same invention. Only one party will be awarded a patent on this invention, that is, will have *priority*. The laws of many countries unambiguously assign priority to the party that submits a patent application first. At the time of writing, this is not true in the United States, where the date of invention is used to establish priority.

In the United States, proving a date of invention currently includes proving the date of *conception* of the invention and the date of *reduction to practice*. **Conception** *is defined as the formation, in the mind of the inventor, of the complete invention (as defined in the patent claim).* Conception requires that the inventor knows how to make the invention and how to use it in a practical way. If an invention, for example, is a new compound isolated from blood, then the inventor must know its structure, how to isolate or synthesize it, and how to use it for a practical purpose. Conception occurs in the mind of the inventor but the concept must be disclosed to someone else in order to establish the date of conception. This disclosure must be documented so that a patent examiner or court will be certain that it is valid.

Reduction to practice *is constructing a prototype of the invention or performing a method or process* (as described in the patent claim). Reduction to practice requires establishing that the invention works for its intended purpose.

The date of conception is the relevant date for the purpose of determining priority for a patent if the inventor can prove *diligence*. To prove **diligence** *between conception and reduction to practice, the inventor must prove that he or she never abandoned the invention and worked nearly every day on it (except for vacations and sick days).* If an inventor cannot prove diligence, then the date of reduction to practice is the invention date.

Inventors must have solid documentation and corroboration supporting their ideas and experiments to prove the dates of conception and reduction to practice. Care-

fully maintained laboratory notebooks are thus crucial for patent purposes. The credibility of a laboratory notebook is directly related to the care taken to ensure that it is properly maintained. (Proper documentation is discussed in Chapter 6.) Corroboration is independent evidence from other people that supports the inventor's testimony and documentation. Corroboration comes from witnesses, who have seen one or more aspects of the invention in progress, and from documents read by others. Each page in a laboratory notebook therefore must be witnessed (read), signed, and dated by another person for the purposes of corroboration. Co-workers should make a habit of reading, discussing, and signing one another's notebooks every day or so. It is important that the witnesses understand the invention at the time it was disclosed to them. Other documents, such as e-mails, may be considered to supplement a laboratory notebook.

D. Trade Secrets

Science advances due to the open exchange of knowledge; publication of scientific discoveries is critical to the success of academic research scientists. The patent system similarly promotes exchange of information and knowledge. There are, however, situations where biotechnology companies do not want to share information or materials so they can protect their competitive advantage. **Trade secrets** *are private information or physical materials that give a competitive advantage to the owner.* To qualify as a trade secret, information must be valuable, it must be secret, and it must give the holder a competitive advantage. The formula for making Coca-Cola is a commonly cited example of a trade secret. A biotechnology trade secret might be, for example, a cell line to manufacture a product, or information about how to grow a valuable cell line. Trade secrets are not registered with the government, as are patents. Rather they are actively protected (e.g., with passwords on computers, locks on file cabinets, limited access to the facility).

One of the most important ways to protect trade secrets is by the use of nondisclosure (confidentiality) agreements. A **nondisclosure agreement** *is a contract in which the parties promise to protect the confidentiality of secret information that is disclosed during employment or another type of business transaction.* A person who violates a nondisclosure agreement can be taken to court and sued for damages. It is standard practice for employers to require employees, interns, consultants, vendors, and anyone else involved in a company to sign a nondisclosure agreement. The knowledge of a biotechnology company is critical to its success. It is essential that employees respect confidentiality agreements and not disclose secret information intentionally or through carelessness.

E. Patent Issues

The use of patents to protect inventions is generally viewed as being good both for individuals and societies because the system protects inventors and promotes innovation. The system, however, provides many opportunities for discussion and disagreement.

It is not uncommon for two parties to disagree as to whether an innovation is new or whether it is obvious. One party may have a patent that it believes covers an invention that another party is claiming is new. One party may believe that another party's patent claims are too broad and therefore stifle their innovations. These sorts of disputes often must be resolved by the court system and can be very expensive for both parties. Winning or losing these court battles can have a profound effect on a company's fortunes and future.

Society as a whole has a stake in patent disputes because society wants patents to be broad enough to reward innovation but not so broad as to prevent others from building on previous work. A patent whose claims are too broad may stifle the innovations of others. It is often difficult for a patent examiner to have sufficient knowledge in the complex, technical arena of biotechnology to evaluate whether a patent's claims are too broad.

Some people believe that it is unethical to patent a living organism, a gene, or a protein and hence object to many biotechnology patents. There is also the concern that patenting of genes prevents the use and development of important diagnostic tools and treatments if the patent holder chooses not to develop them or prices them out of the reach of the needy.

Some people object to patents on plants because food is an absolute necessity for life, and patents might cause certain foods to be too expensive for the poor. In the past, traditional breeding of varieties of crop staples that are hardy and high producing was largely the concern of researchers in public, government institutions and universities. The results of their work were freely available. Biotechnology, however, has introduced new commercial interests to agriculture.

These are only a few of the issues relating to IP, a complex, sometimes contentious area that is of vital importance to the biotechnology industry. Every biotechnologist should know something about intellectual property, and many find it provides a stimulating career.

III. THE MANY ROLES OF THE LABORATORY IN THE BIOTECHNOLOGY INDUSTRY

A. What Is a Laboratory?

We have talked about the lifecycle of products and how the organization of biotechnology companies reflects those lifecycles. Laboratory scientists, technicians, and analysts play key roles at all stages of a biotechnology product's lifecycle, from its discovery, through its development, and into the production stage. This section therefore explores the laboratory in a bit more detail.

Research laboratories, which have existed for hundreds of years, *are spaces set aside to study the complex-*

ities of nature in a controlled manner. Observations can be made, and experiments can be performed in the laboratory in which the researcher controls the factors of interest. For example, if researchers are interested in the effect of light on plant growth, they can carefully control the light that plants receive in a laboratory. Outside the laboratory, the researcher has little control over light exposure or many other important factors—such as rain, temperature, and insects—that may affect the plants' growth. The research laboratory is the site of discoveries that root biotechnology; without research, modern biotechnology would not exist.

Many research laboratories are located in academic institutions and research institutes. Important biological research is also conducted in medical centers. Some biotechnology researchers work in laboratories associated with biotechnology companies. In all of these settings, the tasks of biological research, that is, making observations and performing experiments, are similar. The purpose, however, of research in a medical center or a company is usually to find and develop practical applications of knowledge, whereas university research may be "basic"—that is, the pursuit of knowledge for its own sake.

There is another important category of laboratory that can be distinguished from the research laboratory. This is the **testing laboratory,** *a place where analysts test samples.* The product of a testing laboratory is a test result, such as a measurement of the blood glucose level in a sample, a DNA "fingerprint," or a report on pollutants in a lake. Clinical laboratories are a familiar type of testing lab where samples from patients are tested. Forensics laboratories are testing labs where samples from crime scenes are tested. Samples from the environment are tested in environmental laboratories. A quality-control laboratory in a company is a type of testing laboratory where samples of products and raw materials are tested.

What then, makes a place a laboratory? We can say that a **laboratory** *is a workplace whose product is data, information, or knowledge.* This is a reasonable definition with a couple of caveats. First, people in laboratories do produce tangible items such as photographs, antibodies, purified proteins, and printouts. These materials, however, are produced with the purpose of learning more about a system or a sample, answering a research question, or documenting what has been discovered. The tangible materials that emerge from a laboratory are not produced for commercial sale. Another caveat is that biotechnology companies often produce small amounts of products for sale in facilities that are also used for research and development, or that are similar to research laboratories. However, we will consider a laboratory-like facility used for producing a commercial product to actually be a small-scale production facility, rather than a laboratory.

There is one more type of laboratory that is perhaps the most familiar one—the teaching laboratory. A

teaching laboratory *is a space set aside in which students learn about nature.* Often students learn things that other people (like their teachers) already know. Students also practice using the methods that research scientists use when performing experiments. The product of a teaching laboratory is knowledgeable students.

B. Laboratories and the Lifecycle of a Biotechnology Product

i. RESEARCH AND DEVELOPMENT

The discoveries that lead to biotechnology products occur in research laboratories. The efforts of many scientists from many laboratories often interconnect in the discovery of a single product. Chapter 1 discussed, for example, how the scientific discoveries of various scientists over the years led to genetic engineering and how research into the mechanisms of cancer led to the drug Herceptin. Another example is provided in the Gleevec Case Study on p. 52.

Development, the transformation of discoveries into products and applications, is primarily a laboratory-based function. For example, development of a new product made by genetically modified cells might involve experimenting with different host cells to see which is best, optimizing their culture medium to maximize protein expression, and developing optimized purification methods. All these tasks require scientifically skilled laboratory personnel.

ii. QUALITY CONTROL

The quality-control laboratory is a type of testing laboratory that is essential in helping ensure the quality of products through development and during production. The quality-control tests performed in biotechnology companies are often sophisticated and diverse. Quality-control analysts perform:

- environmental monitoring
- tests of raw materials
- tests of in-process samples
- tests of product

Environmental monitoring in this context refers to the air, water, surfaces, and equipment in a facility. QC technicians might, for example, take swabs from a piece of production equipment after it has been cleaned to be certain that no remnants of the last product run remain.

Incoming raw materials are evaluated by quality-control analysts to be sure they meet the requirements of production. In a pharmaceutical company, regulations require that every incoming raw material be quarantined and tested to assure its quality before it is released for production. This means that even if the manufacturer of the raw material has tested the material and provided documentation to that effect, the pharmaceutical company still needs to confirm, with laboratory testing, that the raw material is acceptable.

In a company that does not make pharmaceuticals, the requirements for testing raw materials may be less stringent, but a process still needs to exist to confirm that raw materials are acceptable.

Quality-control analysts may perform *in-process* testing; tests done while a product is in the process of manufacturing to be sure everything is proceeding normally. For example, analysts might check the purity of a material at an intermediate stage in a purification process.

Quality-control analysts run a series of tests on samples from each batch of final product in order to see if that batch is good enough to be released for sale. They compare the results of the tests to **specifications** *that are numerical limits, ranges, or other results that the tests must meet if the product is good.* For example, if the product is a chemical entity, QC analysts might perform a battery of chemical tests to confirm that the batch contains the right compound and that it is pure.

For pharmaceutical products, final product testing is tightly controlled and QC laboratory personnel have a key role in ensuring the quality of the final product. Regulations require that *samples from every batch manufactured for clinical trials or for sale are tested in a variety of ways to be certain that they meet all specifications for the product;* this is called **lot release testing.** The drug will only be released if it meets all the specifications; if it does not, the drug is rejected. The criteria on which to base acceptance or rejection of a drug product are established and justified based on data obtained from material used during development of the product. For a biopharmaceutical product as many as 40 different assays may be used to test the product (*The BioPharm Guide to Biopharmaceutical Development.* Eugene, OR: Advanstar Publishing, 2002: 15.). The R&D unit must optimize every test and prove that all the assays used in QC are effective and contribute to evaluating the quality of the product.

The various tests that are used for testing a product are sometimes categorized as follows.

1. **Tests of general characteristics.** Examples of such tests include evaluating appearance, color, and clarity; and measuring pH, particulate, and moisture content.

2. **Tests of identity.** In this context, *identity* refers to whether a particular substance or substances (e.g., the active ingredient in a drug) is present. For drug products, identity tests should be highly specific for the drug substance and should be based on unique aspects of its molecular structure or other specific properties. (*Identity testing* may refer to the identification of individual people or other organisms. In the present context, identity testing means to identify a particular molecular entity.)

3. **Tests of purity. Purity** *is the relative absence of undesired, extraneous matter in a product.* For drug products, the absence of contaminants is of the utmost importance. There are various methods to test

for purity or impurity. Some tests detect specific contaminants (e.g., the HIV virus) and other tests detect classes of contaminants (e.g., bacteria). Tests for impurities may be qualitative **limit tests,** *which is simply whether or not the impurity is present at levels above a certain detection limit.* Impurity tests may also be quantitative, in which case the amount or concentration of the impurity is measured. In the pharmaceutical industry, a battery of different tests for many potential contaminants is required.

4. **Quantitation/concentration tests.** These assays test the amount or concentration of a substance. For example, many biotechnology products are proteins. The amount of protein in a product, or at an intermediate stage of processing, is frequently determined.

5. **Potency/activity tests. Potency** or **activity** *is the specific ability of the product to produce a desired result.* For a drug, potency refers to the drug's ability to have its desired therapeutic effect on a human or other animal. The same principle applies to many other products; activity refers to the product's ability to perform as desired. For example, a restriction enzyme is intended to cut DNA at specific sites. The activity of a restriction enzyme therefore relates to how much DNA it can specifically cut in a set amount of time.

EXAMPLE

A company develops a recombinant DNA production system in *E. coli* to produce a therapeutic protein. They establish final product specifications and assay methods to ensure that the product meets the specifications. Before a drug lot can be released for sale, it must be tested to be sure that it meets all these specifications. These tests might include the following:

RELEASE SPECIFICATIONS FOR PROTEIN THERAPEUTIC XYZ

Characteristic	Specification	Assay Method
Appearance	Clear, colorless solution	Visual inspection
E. coli DNA	< 0.01 μg/μg product	Southern blot
E. coli Protein	Undetectable	Specific immunoassays
E. coli RNA	Undetectable	Agarose gel electrophoresis
Residual ethanol	< 250 ppm	Gas chromatography
Endotoxin	< 0.1 EU/μg	LAL assay
Sterility	No growth in 14 days	Assay in USP*
Retrovirus	Undetectable	Infectivity assay

*US Pharmacopeia, a compendium of accepted methods for the pharmaceutical industry.

IV. SUMMARY

This chapter focused on the business side of biotechnology; the production of commercial products. We have seen how products go through a lifecycle beginning with their conception in a research laboratory, moving through development, and then into production. The chapter discussed the organization of biotechnology companies into functional units. We have further seen how knowledge is essential to the biotechnology industry and is a valuable form of property that is protected by the company, by laws, and by the courts.

In Chapters 1 and 2, we observed that the work of laboratory personnel is woven throughout the enterprise of biotechnology. The discoveries that drive biotechnology emerge from the research laboratory. The transformation of ideas into products requires development scientists and technicians. Laboratory analysts play many roles in testing experimental samples, testing products, and performing other sorts of analyses.

Chapter 3 continues the discussion begun in Chapter 2 by looking in more detail at a specific type of biotechnology product, biopharmaceuticals. Unit II continues to probe the business of biotechnology by introducing the many issues relating to making products that are of good quality.

PRACTICE PROBLEMS

1. A new, effective anticancer compound is discovered. The compound is unfortunately found in the stems of a rare, slow-growing plant; therefore very little of the compound is available. A small, start-up biotechnology company is formed by a team of research scientists who plan to isolate the gene for this anticancer agent. After several years of dedicated, difficult work, they obtain the gene that codes for the anticancer agent. They are then able to insert the gene into bacteria and are elated to find that the bacteria make the anticancer agent. They develop methods to grow and harvest large quantities of these bacteria and to isolate the product from them. At this point, the company is purchased (for a lot of money) by a pharmaceutical company that takes over the tasks of testing the compound in animals and humans. Eventually the drug is approved and is sold for patients. Over a number of years, the following tasks were performed by staff in the biotechnology and pharmaceutical companies. Label each task as being primarily the job of research and development personnel, production personnel, quality-control technicians, or quality-assurance personnel. The tasks are not necessarily listed in the order in which they would be performed.

   a. Identify the gene that codes for the anticancer agent.

b. Design and develop a new assay to check whether the anticancer compound is present in a sample at a certain level.

c. Devise methods to grow large amounts of the bacteria in such a way that the compound is consistently produced.

d. Develop a system to keep track of all documents.

e. Sterilize the equipment used to produce the drug product.

f. Devise a method to insert the gene into bacterial cells so that the cells make the desired anticancer agent.

g. Perform laboratory tests of the anticancer agent to see that it meets its specifications before it is released for sale to patients.

h. Design purification methods to purify the compound from the bacteria.

i. Monitor pH and temperature levels in the fermenters during production runs.

j. Purify the anticancer product that is to be used by patients.

k. Determine the chemical composition of the anticancer agent.

l. Review all documentation associated with a batch of product to be sure it is correct and complete.

m. Perform laboratory tests of incoming raw materials to be sure they are suitable for use.

2. Suppose you are beginning a small start-up biotechnology company. You have a strain of bacteria that is good at degrading certain industrial by-products and you hope that your bacteria can be used to remediate (clean) contaminated soil. You need to rent a space for your company. What features would you be looking for in your first space? There are many answers to this; you can be creative.

CHAPTER 3

The Lifecycle of Pharmaceutical Products

I. INTRODUCTION: WHAT ARE PHARMACEUTICALS?

Chapter 3 continues the discussion of product lifecycles that was introduced in Chapter 2. This chapter focuses first on pharmaceutical products in general, and then more specifically on one type of pharmaceutical product, biopharmaceuticals. These products warrant their own chapter because of their importance in the biotechnology industry. Pharmaceuticals/biopharmaceuticals are of significant medical value, have a major worldwide economic impact, provide interesting and challenging career opportunities, and are associated with major advances in the biological sciences.

To begin with some terminology, **pharmaceutical products** *are chemical agents with therapeutic activity in the body that are used to treat, correct, or prevent the symptoms of illnesses, injuries, and disorders in humans and other animals.* The familiar word, *drug,* is generally used synonymously with *pharmaceutical* although drugs include not only therapeutic compounds but also agents that are used in the body for nontherapeutic (e.g., "recreational") or harmful purposes.

Biopharmaceutical is often used (as it was in Chapter 1) to refer to a category of pharmaceutical products that are manufactured by genetically modified organisms. These products are generally proteins. The term "biopharmaceutical" is also used more broadly, for example, to refer to any drug manufactured by living cells or organisms (whether or not recombinant DNA techniques are involved), and also to include whole cells or tissues used therapeutically. *Biopharmaceutical* can also refer to any drug that is a large biological molecule; these are usually proteins, but can be RNA or DNA (e.g., DNA for gene therapy). As biotechnologists we are interested in biopharmaceutical agents according to any of these definitions.

It is helpful to contrast large biological molecule drugs with small molecule drugs in order to better understand the characteristics of biopharmaceuticals (however one defines the term). To give a sense of the relative sizes involved, aspirin is an example of a small molecule drug with a molecular weight less than 200. Rituxan is an example of a large biological molecule drug (a monoclonal antibody) with a molecular weight of about 145,000. Insulin, discussed in Chapter 1, is one of the smallest protein drugs with a formula weight of about 6,000, see Figure 3.1.

Small molecule drugs are often chemically synthesized in factories. Large biological molecule drugs cannot at this time be efficiently chemically synthesized, with a few exceptions (e.g., DNA used for gene therapy). Large biological molecule drugs are isolated from organisms or are manufactured in carefully nurtured living systems (cells, plants, or animals). Large biological molecule drugs are more difficult, complex, and expensive to produce than small molecule agents because living systems are required for their production. Moreover, the exact structure and function of a biological molecule is likely to be dependent on the particular living system in which it was produced, which means that ensuring consistency and potency from batch to batch is challenging. Biological molecules are destroyed by heat, so assuring their sterility poses another technical challenge. The heat sensitivity of biological molecules also has implications relating to their stability during shipping and storage.

Although they are difficult and expensive to manufacture, protein therapeutics have the important advantage that they have a relevant, known activity in the body. Proteins are molecules that perform the work of cells. Insulin and growth hormone, for example, are proteins whose work is to carry messages between cells. It is therefore likely that a protein drug administered to play its normal role in the body (e.g., insulin administered to diabetics) will have a positive effect. In contrast, tens of thousands of small chemicals must be screened to find one with a desired effect.

Small and large molecules tend to pose different issues when administered to a patient. Proteins have the problem of being immunogenic. **Immunogenicity** *is the ability of an agent to stimulate an immune reaction.* Immunogenicity can be a problem in drug development in two ways. First, if a drug agent is immunogenic, a patient may have a dangerous adverse (e.g., anaphylactic) reaction to it. Second, a patient may raise antibodies against the protein drug. The antibodies can bind to and inhibit the activity of the drug, thereby rendering it useless. A slightly different issue is that large, biological

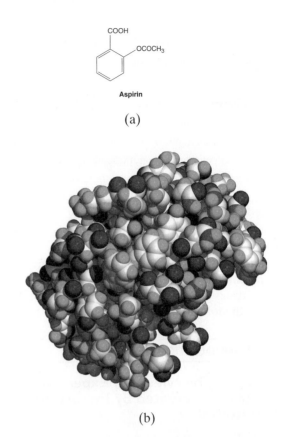

Aspirin

(a)

(b)

Figure 3.1. A Small Molecule Drug Compared to a Protein Drug. a. Aspirin is relatively small. Its formula is $C_9H_8O_4$, meaning it has 21 atoms. **b.** The protein drug, insulin. Each sphere in this picture represents an atom; more than 700 atoms are present. (Image courtesy K. Seddon & T. Evans, QUB Photo Researchers.)

molecules may not be toxic in the sense that a foreign chemical is toxic, but they can be very dangerous if their activity is exaggerated by the immune system once administered to a patient.

Small molecules, in contrast to large ones, are less likely to induce immunological effects. It is not unusual, however, for a small chemical agent to have off-target effects (i.e., it binds to unintended sites in cells with an undesirable effect). As the body breaks down small molecule drugs, metabolites are generated that often have unintended consequences.

Yet another difference between small and large molecule drugs is that biological molecules are almost always destroyed in the gut by digestive enzymes, so they cannot be taken orally. This makes administering these drugs to a patient more difficult than administering small chemical drugs, which are typically taken by mouth. Some of the differences between small molecule and protein pharmaceuticals are summarized in Table 3.1 on p. 44

Despite the important distinctions between small pharmaceutical agents and large biological molecule agents, the basic outline of their lifecycles is the same. For that reason, the first part of this chapter is a discussion of the lifecycle of pharmaceutical products in general. The

Table 3.1. COMPARISON OF SMALL CHEMICAL AND PROTEIN PHARMACEUTICALS

Small Chemical Therapeutic Agents	Protein Therapeutic Agents
Often chemically synthesized	Produced by cells
Relatively inexpensive to manufacture	Expensive to manufacture
Relatively homogenous	Production system may result in heterogeneous molecules
Relatively easily characterized	Complex and expensive to characterize
Typically enter cells and act on intracellular targets	Often act on cell surface receptors
Have less selectivity in their actions; therefore more side effects	Selective action with fewer side effects
Usually administered orally	Are destroyed in digestive tract; cannot be administered orally
Immunogenicity usually not significant problem	Immunogenicity is a major consideration

later part of the chapter turns to topics specific to the development stage of the lifecycle of large, biological molecule drugs that are manufactured using biotechnology methodologies.

II. THE PHARMACEUTICAL LIFECYCLE

A. Lifecycle Overview: An Expensive Process Punctuated by Important Milestones

The lifecycle of modern pharmaceuticals is a long one. Pharmaceuticals begin in research laboratories from which emerges a candidate drug substance, also called the **active pharmaceutical ingredient (API),** *a compound that is thought to have therapeutic activity in the body.* The potential drug then enters a prolonged development period. The goal of the drug-development process is to turn this promising drug candidate into a product for manufacture and sale. While the goal is straightforward, drug development is complex. It is dangerous to administer a poorly designed or improperly manufactured drug to a patient. Potential drugs therefore go through years of testing to ensure their safety and efficacy. The manufacturing processes for pharmaceuticals are similarly honed over years to ensure that they produce a consistent, safe, uncontaminated product.

The government regulates the development and then the manufacturing of pharmaceuticals to guard that they are thoroughly tested and properly manufactured. This oversight is provided by the Food and Drug Administration (FDA) in the United States and similar agencies in other countries. The FDA exerts its authority partly by controlling "gates" at key milestones in the pharmaceutical lifecycle. The interaction of regulatory agencies with the pharmaceutical product lifecycle contributes to their safety—and the challenges they pose to companies that develop them.

Figure 3.2 is an outline of the lifecycle of a pharmaceutical product showing its important milestones. The research and discovery stage results in the identification of a promising possible drug agent, the first milestone. The development phase begins in laboratories and animal testing facilities. Early development is called the **preclinical** *stage because it comes before testing in humans.* If the candidate drug performs well in the preclinical testing, then the company can file with the FDA an **Investigational New Drug Application (IND)** *which requests permission to begin testing the substance in human subjects.* If the FDA grants the application, then another important milestone is passed and the product enters the **clinical development stage,** *where pharmaceutical products are tested in humans.* If the drug is successful in clinical testing, then the company can submit to the FDA a **New Drug Application (NDA)** *which requests permission to begin marketing the drug.* If the FDA approves the product, then a major milestone is passed and the post-approval stage of the product's lifecycle begins. After approval, the company can sell the product. The FDA will continue to interact with the company to help ensure that the product is manufactured consistently and that safety surveillance continues.

The company or institution that is developing the drug is the drug sponsor, but the sponsoring organization is unlikely to perform all the tasks of development. Various development functions are often performed by **contract research organizations (CROs),** *which are outside organizations paid to perform particular tasks.* It is common, for example, for the sponsor to contract with a testing company that has expertise in animal testing or human testing. Biotechnology companies often contract out development tasks ranging from laboratory testing, through animal and human testing, and into manufacturing. Even extremely large pharmaceutical companies often contract with testing companies for specialized services.

Drug development is costly. According to a report from The Tufts Center for the Study of Drug Development (an independent, academic, nonprofit research group affiliated with Tufts University) it takes 10 to 15 years from the time research begins to the time a new pharmaceutical product receives approval from the FDA. They further state that a drug company typically spends $800 million to create a successful product. A major reason why drug development is so costly is that the vast majority of potential drugs fail. The Tufts Cen-

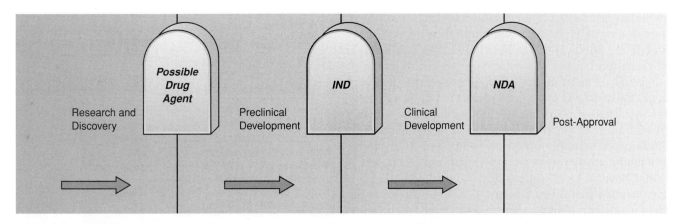

Figure 3.2. **Overview of the Lifecycle of a Pharmaceutical Product.** In the United States, the FDA controls "gates" at critical milestones in the lifecycle. Similar agencies in other countries provide oversight of drug development and manufacture.

ter cites studies showing that of every 5000 compounds tested, only five are promising enough to reach the stage of human testing and only one of these five is eventually approved for patient use, see Figure 3.3. Although some analysts dispute that drug companies spend as much as $800 million to create a new drug, by any estimate the cost of drug development is measured in hundreds of millions of dollars.

B. The Beginning of the Lifecycle: Research and Discovery

There are many ways by which therapeutic compounds are discovered and/or designed; just a few examples will be described here. One method, used since ancient times, is to search for naturally occurring substances, particularly from plants, which have potency for treating illness and injury. The word *drug* is derived from an old Dutch word *droog*, which means *dry*, since in the past, most drugs were made from dried plants. Scientists now are often able to identify and manufacture the active ingredient in a traditional drug. Aspirin is a derivative of a natural product first isolated from willow bark and used as a remedy more than a thousand years ago. The search continues today for therapeutic agents that can be harvested from natural sources and are present in traditional remedies. Biotechnology companies have been established to screen for therapeutic compounds from such interesting sources as: poisonous fish, snakes, and insects; rain forest plants and animals; and marine organisms.

Modern approaches to drug discovery are usually based on a more scientific understanding of a disease or injury process than was possible for our ancestors. Sometimes medical researchers discover that a disease occurs when a protein is missing in the patient (e.g., diabetes and hemophilia). In these cases it is straightforward to imagine treating the disease by supplying patients with the missing protein, either isolated from natural sources or manufactured in genetically modified cells.

Researchers who study the roles of proteins in health and disease have found that a protein often

exerts its effect in a cell, whatever that effect might be, by binding to another molecule. In a normal cell this protein binding promotes the cell's health. In a damaged cell (e.g., a cancer or an infected cell) the binding of one protein to another molecule may trigger a harmful effect. Scientists reason that it might be possible to ameliorate a disease if the binding of involved proteins can be blocked or modified in afflicted cells. Herceptin, discussed in Chapter 1, is an example of a monoclonal antibody drug that was designed to block a disease process by binding to an involved protein.

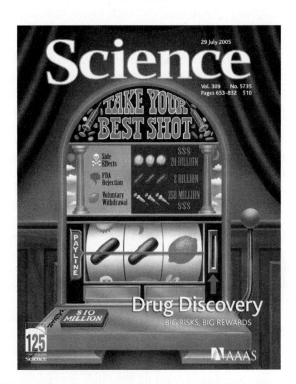

Figure 3.3. **Failure Contributes to the High Cost of Drugs.** Developing new drugs is a huge gamble with high stakes. The vast majority of possible drug candidates fail at some point in the drug development lifecycle, as illustrated in this cartoon from the cover of *Science*. (Illustration by Stephen Wagner. From *Science* 309 no. 721 [2005]. Reprinted with permission from AAAS.)

Statins are a class of small-molecule drugs that prevent disease by targeting, binding, and thereby blocking an enzyme. Medical researchers found that high levels of cholesterol in the blood are correlated with an elevated risk of heart disease. They reasoned that if they could find a drug to lower cholesterol, they could reduce the incidence of heart disease. Statins were discovered by an intensive search of thousands of fungal microorganisms for naturally occurring compounds that bind and inhibit an enzyme necessary for cholesterol production. Chemical modifications of the fungal compounds resulted in statin drugs that are used to lower cholesterol levels in millions of people. Statins are therefore drugs that are based on a scientific understanding of a protein's function combined with the technique of searching natural sources for a useful compound.

Pharmaceutical companies have a sophisticated strategy for searching for new small-molecule drugs that have a desired effect in treating a particular disorder. **High throughput screening (HTS)** *enables scientists to simultaneously test thousands of small chemical compounds for efficacy in a particular disease model.* First researchers discover a specific protein that triggers detrimental effects in a disease; this is called the "target." Then researchers use high throughput screening to search for a compound that binds and modifies the target with a desired effect. Pharmaceutical companies own "libraries" that often contain more than a million distinct chemical compounds. In high throughput screening, scientists test all of these compounds (or a subset) in an automated laboratory test to see if the compounds interact with the target. HTS methods can typically test in excess of 100,000 compounds in a day, enabling a company to screen its entire library against a single target in less than two weeks. In order to do so, however, the company must have an assay (test) that they have proven reproducibly and accurately detect the interaction of the compound with the target. Developing a suitable assay can take months (see Chapter 24 for a discussion of assay development). The outcome of HTS is the identification of one or more "hits"—molecules that have activity against the desired target.

To be a useful drug, an agent must be sufficiently stable in the body to allow it to reach the target site in sufficient quantity and for a long enough time to have a biologically relevant effect. The drug must be safe and not have side effects that would prevent its use for the intended disease. It also must have properties such that it can be manufactured and formulated in a practical manner. Often promising molecules identified in early research studies, such as HTS, have activity but lack these additional requirements. Potential drug molecules therefore go through a process where they are assessed for activity, ability to bind only to the target of interest (selectivity), chemical attributes, solubility, and so on. The original "hit" molecules are then chemically modified and re-assayed. The relationships between various chemical modifications and the activity of the mole-

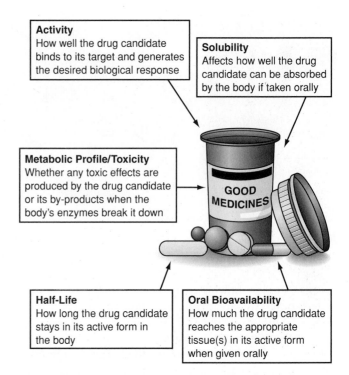

Activity
How well the drug candidate binds to its target and generates the desired biological response

Solubility
Affects how well the drug candidate can be absorbed by the body if taken orally

Metabolic Profile/Toxicity
Whether any toxic effects are produced by the drug candidate or its by-products when the body's enzymes break it down

GOOD MEDICINES

Half-Life
How long the drug candidate stays in its active form in the body

Oral Bioavailability
How much the drug candidate reaches the appropriate tissue(s) in its active form when given orally

Figure 3.4. A Good Drug. This cartoon from a publication by the United States National Institutes of Health shows that a good drug combines a number of attributes, including the best possible activity, solubility, bioavailability, half-life, and metabolic profile. Attempting to improve one of these factors often affects other factors. For example, if scientists structurally alter a lead compound to improve its activity, they may also decrease its solubility or shorten its half-life. The final result must always be the best possible compromise.

cules in the assay are assessed. If several molecules are found that interact with the target, and if they share some common chemical features, then a new chemical(s) with those specific features might be synthesized. This iterative process leads to better understanding of the molecules and the slow optimization of a particular candidate with desirable drug properties that can be further assessed in preclinical testing, see Figure 3.4.

New genetic discoveries play an important role in modern drug discovery. Genetic information can be used to identify subsets of patients who are best served by certain drugs or who are likely to have adverse reactions to certain drugs. A better understanding of cancer biology has led to chemotherapies and other treatments that are tailored to subsets of patients. These advances are expected to lead to more "personalized" medicines with better patient outcomes.

C. Preclinical Development

i. THE OBJECTIVES OF PRECLINICAL DEVELOPMENT

Whatever the means of its discovery, the identification of a candidate drug signals the beginning of the complex development process, see Figure 3.5. The tasks of preclinical development include:

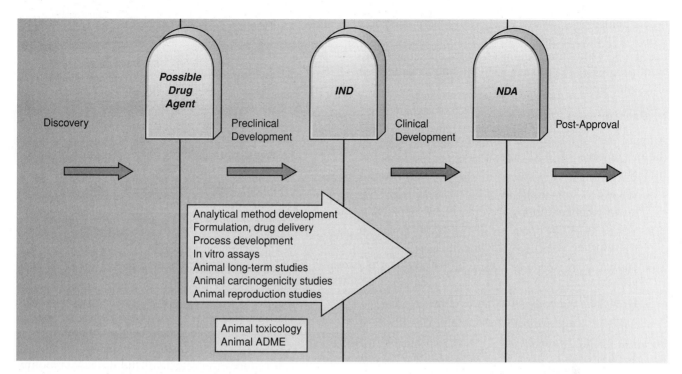

Figure 3.5.   Preclinical Development. A number of activities occur during the preclinical stage to develop the drug and its formulation and test its safety and effects in vitro and in animals.

- Determining the physical and chemical properties of the candidate drug compound
- Testing the candidate drug in vitro
- Determining how to formulate the candidate drug for administration to test subjects and patients
- Developing manufacturing methods for the candidate drug
- Testing the candidate drug in cultured cells
- Testing the candidate drug in animals to study its effects in the body and its safety
- Developing analytical assays to test for the presence of the drug compound, its metabolites, activity, and other features
- Securing intellectual property protection for the potential product, its uses, and its manufacture

ii. PRODUCT DEVELOPMENT

Product development *is the process in which R&D scientists explore and optimize the features of a potential product.* This includes **product characterization,** *the process of studying a therapeutic compound to learn its structure, its interactions with other molecules, and its function.*

Any drug product needs to balance a number of key characteristics. Among these is the requirement that the drug can be formulated it in such a way that it can be administered to a patient. A drug **formulation** *is the form in which a drug is administered to patients including its chemical components, stabilizers, the system of administration (e.g., pill or injection), and the mechanism by which the drug is targeted to its site of action.* Developing the formulation is part of product develop-

ment that begins in the preclinical stage and may continue to be refined until the product is approved for sale. Formulation for large biological molecules means preparing the compound in such a way that it can be injected or otherwise introduced to the body. Small-molecule drugs usually must be soluble or must be chemically modified so that they are soluble.

During product development, studies will be conducted to determine the **stability** of the formulation, *its capacity to remain within its specifications over time.* Laboratory teams test the product's potency, activity, pH, clarity, color, and particulate content over time. Specific analytical instruments may be used to look for breakdown products. It is common with drugs and other products to intentionally subject them to stresses (e.g., high temperature, freezing, exposure to light and air) in the laboratory and check the product for changes. These studies accelerate degradation of the drug compound and are used to investigate degradation pathways and products of degradation. Stability testing is required to determine how expiration dates will be set for a product and to identify methods for its storage, shipping, and handling.

iii. PROCESS DEVELOPMENT

Process development *is the development of methods for producing and purifying the drug substance.* The production process must result in a consistent and pure pharmaceutical substance. The process must be as cost-effective as possible. If the manufacturing process is too complex, then mistakes can occur that might harm patients or might add expense and difficulties to production. Process development is ongoing. It begins in the preclinical stage before animal tests are initiated because there must be a source of drug to administer to the animals. The drug will initially be made in milligram amounts, but as development progresses the process will continue to be refined and scaled-up.

iv. Preclinical Testing

A major part of the drug development process is to test the interactions of the drug with the body. The questions studied are:

- What toxic effects, if any, does the drug have, and under what conditions?
- Where does the drug go in the body; how is it transported; does it accumulate in tissue?
- How does the drug interact with its target in the body?
- How does the body affect and modify the drug?

In vitro *tests are those that are performed using biochemical assays, cultured cells* (cells grown outside the body in dishes), *and other laboratory systems outside animals or humans.* There are many preclinical tests that are performed in vitro. For example, drugs that are ingested must be able to pass through the membranes of the cells lining the digestive tract; otherwise, they will not be taken up by the body. It is possible to assay a compound's ability to do so in vitro by adding the substance to cultured gut cells and testing the ability of the compound to penetrate the cells. Another example is the use of bacteria or cultured cells to test if a compound is **mutagenic,** *that is, whether the compound causes a change in the genetic material of a cell after exposure.*

Animals are likely to have already played a role in the discovery and very early testing of the activity of a potential drug and its interaction with the target in the body. If a candidate drug shows promise, and if it can be consistently manufactured, purified, and formulated in small amounts, then it advances to preclinical tests in animals.

Animal testing plays a key role in assessing the safety of a potential drug and its effects in the body before the agent is ever tested in humans. Animals do not respond exactly as people do to all compounds, but if a drug is not safe for animals, its development will stop—some drug candidates do fail at this stage. In the future, it is likely that computer models and better in vitro assays will reduce our dependence on animal testing.

Animal testing is used largely to look at toxicity, what the drug does to the body. Experimenters administer varying doses of the drug to animals and look for toxic effects. This includes examining all the major organs (heart, lungs, brain, liver, kidney, and digestive system) and other parts of the body that may be relevant (e.g., skin for a drug delivered by a patch). At least two animal species are tested, one of which is almost always a rodent.

Experimenters also use animals to study what the body does to the drug. **Pharmacokinetics** *is the study of how a drug is changed in the body.* Experimenters administer the drug to the animals and test their urine, blood, and feces. They look at:

- **Absorption.** *This is the process by which a drug is transferred from its site of entry in the body to the body fluids, particularly blood.* For drugs that are taken orally, absorption requires that the drug can pass through the membranes of the cells that line the digestive tract. Drugs that are poorly absorbed when taken orally must be administered in another way.
- **Distribution.** *This relates to how the compound spreads through the body to various tissues and organs.*
- **Metabolism.** *This is the process by which compounds are broken down in the body by enzymes.* Small molecule drugs are primarily metabolized in the liver by special enzymes named cytochrome P450 enzymes. Metabolism converts the original molecule to new molecules called *metabolites.* The metabolites may be inert, in which case metabolism reduces the effects of the drug. Metabolites may also be active in the body, with either beneficial or harmful effects.
- **Excretion.** This is the elimination of compounds and their metabolites from the body, usually either in urine or feces.

The term **ADME**, *absorption, distribution, metabolism, excretion* is often used in reference to these animal studies.

The conduct of preclinical animal testing was not scrutinized by the FDA prior to the mid-1970s. In 1975, however, FDA inspections of several pharmaceutical testing laboratories revealed poorly conceived and carelessly executed experiments, inaccurate record-keeping, poorly maintained animal facilities, and a variety of other problems. These deficiencies led the FDA to institute the **Good Laboratory Practice (GLP) regulations,** *which govern preclinical animal studies of pharmaceutical products.* GLPs require that testing laboratories follow written protocols and standard operating procedures (SOPs), have adequate facilities and equipment, provide proper animal care, properly record data, have well-trained and competent personnel, and conduct high-quality, valid toxicity tests. GLPs are also followed when conducting other nonclinical activities in addition to animal studies.

v. The Investigational New Drug Application

If the drug candidate is promising in preclinical testing, then the company *compiles data from preclinical studies and submits a plan for human subject testing to the Food and Drug Administration in the form of an* **Investigational New Drug Application (IND).** According to the FDA website (http://www.fda.gov/cder/regulatory/applications/ind_page_1.htm#Introduction), the IND contains information in three broad areas:

- **Information from animal studies.** Preclinical data is provided for FDA reviewers to assess whether the product is reasonably safe for initial testing in humans. Any previous experience with the drug in humans (e.g., use in other countries) is also included.

- **Information relating to the composition and manufacture of the drug.** This includes the composition of the drug, the manufacturer, the drug's stability, and controls used when manufacturing the drug. Reviewers assess this information to ascertain whether the company can adequately produce and supply consistent batches of the drug.

- **Investigational plan.** Detailed protocols for proposed clinical studies are submitted so that reviewers can assess whether the initial trials will expose human subjects to unnecessary risks. Reviewers also consider whether the trial design is likely to provide necessary information about drug function and safety. The study protocols include the number of subjects to be studied, the criteria by which subjects will be chosen, the tests that will be performed on the subjects, statistical methods by which data will be analyzed, and so on. Information must also be provided on the qualifications of clinical investigators who oversee the administration of the experimental compound. Finally, commitments are required to obtain informed consent from the research subjects, to obtain review of the study by an institutional review board (discussed below), and to adhere to the investigational new drug regulations.

The sponsor must wait 30 calendar days after submitting the IND before initiating any clinical trials. During this time, the FDA has the opportunity to review the IND to help assure that research subjects will not be subjected to unreasonable risk. This IND review serves as one of FDA's "gates." The application becomes effective, and testing in human volunteers can proceed, if the FDA does not disapprove the IND within 30 days.

D. Clinical Development

i. ETHICS

Ultimately, the only way to know if a new compound will be safe and effective in humans is to test it in humans. Clinical trials are the part of development where the safety and efficacy of the product are tested in human volunteers. Clinical trials are subject to stringent regulation to protect the safety and rights of these volunteers. The principles of conduct of clinical trials are set forth in internationally recognized documents, including notably the Declaration of Helsinki, first published by the World Medical Association in 1964. More recently, the International Conference on Harmonisation (http://www

.ich.org) published a document (E6[R1]: Good Clinical Practice: Consolidated Guideline) that "is an international ethical and scientific quality standard for designing, conducting, recording, and reporting trials that involve the participation of human subjects. Compliance with this standard provides public assurance that the rights, safety, and well-being of trial subjects are protected, consistent with the principles that have their origin in the Declaration of Helsinki, and that the clinical trial data are credible." The principles contained in these and similar documents have been translated into regulations enforced by different agencies in different countries. In the United States, **Good Clinical Practices (GCP)** *are regulations that protect the rights and safety of human subjects and ensure the scientific quality of clinical trials of drug safety and efficacy.*

GCPs cover many aspects of clinical trials. They require that clinical trials be designed in a scientifically rigorous fashion, that solid criteria be used to include or exclude potential volunteers, that statistical methods of data analysis are scientifically supported, that valid control groups are used, and so on. GCPs further require that participants are informed about the potential benefits and risks they will experience, that they understand the clinical trial, that their consent is given freely without coercion, and that their privacy is protected.

The company or institution that will perform the studies must also get approval from their institutional review board (IRB). The **IRB** *is a group composed of medical, scientific, and nonscientific community members who are responsible for protecting the rights, safety, and well-being of human subjects in clinical trials.* The IRB reviews the proposed trials, ensures that the method of obtaining informed consent is effective, and that relevant documentation is in order.

Specialized personnel play a key role during clinical trials providing assurance that the trials adhere to written protocols. Later clinical trials typically involve patients and doctors at multiple sites (often international). It is necessary to assure that the selection of patients, administration of the test compounds, follow-up, and measurements of drug activity and safety are all performed in the same way, and that the interpretation of the results is consistent at all sites.

ii. PHASE I, II, AND III CLINICAL TRIALS

Clinical trials are conducted in stages, each of which must be successful before continuing to the next phase, see Figure 3.6 on p. 50. A successful product will take on the order of five years to move through all the phases of clinical trials. **Phase I clinical trials** *are the first introduction of the proposed drug into humans.* Phase I trials normally include 20 to 80 healthy volunteers, but sometimes a small number of patients with the illness to be treated are tested at this point. (Toxic drugs, such as anticancer agents, are first tested in small numbers of patients.) Phase I trials are conducted in dedicated

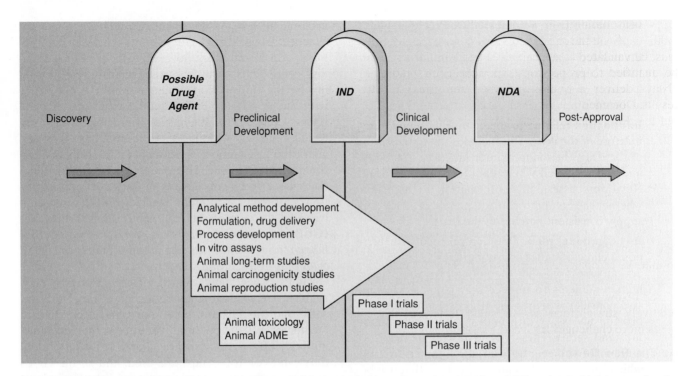

Figure 3.6. Clinical Development. Clinical development is divided into three phases, I, II, and III, each of which is completed before the next begins. Phase I clinical trials are the first introduction of the proposed drug into humans and are generally designed to test the safety of the drug in a small group of healthy volunteers. Phase II trials test the effects of the drug on a small number of patients with the condition the drug is expected to treat. If the drug is successful in phase II, then it progresses to a broader phase III trial in order to establish that a drug has the desired effects.

medical facilities where subjects are monitored during the experiment. The primary purpose of phase I trials is to evaluate the safety of the agent. During phase I trials the drug's pharmacokinetic and pharmacologic properties in healthy volunteers are also examined.

The first tests in humans usually consist of a single dose that is significantly lower (on a weight basis) than the dose at which adverse effects were observed in animals. This is followed by studies in which the dose is escalated to determine the maximum tolerated dose. Unpleasant side effects are common in phase I trials because these studies are intended to evaluate the maximum tolerated dose. If the maximum tolerated dose is below the expected therapeutic dose, then the drug fails and will not be further developed.

If a drug meets the safety requirements of phase I, then it enters phase II clinical trials. **Phase II trials** *are performed on a small number (usually 100 to 300) of patients with the condition the drug is expected to treat.* This phase is used to demonstrate clear drug activity and tolerance of the new compound in patients with mild to moderate disease or conditions. Patients are randomly assigned to receive either the new drug or a placebo (the formulation without the active drug ingredient). The new drug may also be compared to the best existing treatment rather than a placebo. Safety continues to be evaluated in phase II and a dosage regimen for phase III trials is established. If the drug is successful in phase II, then it progresses to a broader **phase III trial** *involving on the order of 1000 to 3000 patients.* Phase III is critical to establish that a drug has the

desired effects. During this phase, the safety of the drug continues to be evaluated, adverse reactions are monitored, an analysis of the drug's risks *versus* its benefits is performed, drug interactions are explored, and other data are collected. Phase III trials require a rigorous statistical demonstration of clinical safety and patient benefit to ensure approval of the therapeutic.

Phase II and III clinical studies are normally **double-blinded**, *meaning that neither the volunteers nor the clinicians know who is receiving the actual drug and who is receiving the placebo or standard treatment.* Double-blinded studies avoid results that are due to the psychological expectations of the participants and limit the potential for the investigators to bias the results, intentionally or unintentionally.

iii. OTHER ACTIVITIES

Even though many drugs fail clinical trials and are abandoned, the sponsor must still invest in a number of other activities during the years in which clinical trials are running. Process developers must work on the manufacturing and purification processes to produce the drug. Manufacturing teams must provide small amounts of the drug for clinical trials. Quality-control analysts must test these clinical-trial materials to ensure their quality. If the product is successful, a team must learn how to scale-up production from levels measured in milligrams to possibly hundreds of kilograms. A manufacturing facility may need to be built or modified, and the production of the drug will need to be transferred from a pilot plant to the manufacturing facility. Laboratory analysts continue to work on developing analytical methods to test the substance. Manufacturing processes must be validated (see Chapter 7) and equipment must be qualified to prove that they work consistently to always deliver a quality product and give reliable results. Documentation systems are optimized throughout the development stages. Longer-term animal studies continue throughout the clinical phases to look at effects on reproduction and offspring, long-term toxicity, and the induction of cancer.

As the product enters phase III clinical trials, its characteristics must be "locked-in" so that the material that is to be sold for patients is demonstrably the same as the material that was tested in the phase III trials. This means that the product must be well-characterized so that the company can prove that it is always making the same substance without variation. As will be discussed below, product characterization, which is usually relatively straightforward for small-molecule drugs, poses more challenges for large biological molecules.

iv. The New Drug Application

If a drug passes all three phases of clinical testing, then the company may submit a New Drug Application (NDA) to the FDA that provides convincing evidence, based on its animal and human testing, that the new product is safe, reliable, and effective. The company must also demonstrate that it can reproducibly manufacture the therapeutic to a scale that supports broad usage by patients. This application is a critical gate controlled by the FDA. Observe that the FDA does not actually test each drug product itself; rather, FDA expert reviewers examine all the test results and information submitted by the company to determine whether a product is acceptable. An application for a new drug can contain hundreds or even thousands of pages of information and requires more than a year, on the average, to review. If the review team decides that the evidence is sufficient, then the new product is approved or licensed by FDA and can be manufactured for commercial sale. This is a major accomplishment and is the beginning of the period where the company can profit from its product.

E. Post-Approval

Once approved, the drug can be manufactured, marketed, and sold, see Figure 3.7. Approved pharmaceutical products must be manufactured in compliance with current Good Manufacturing Practices (cGMP) in the United States and similar regulations in other countries. These regulations are intended to ensure the quality of the products and are described in more detail in Unit II.

Post-market surveillance is used to monitor the long-term safety of the product, product defects, and adverse reactions reported by patients and doctors. Adverse patient reactions must be reported to the FDA and in extreme cases have caused drugs to be withdrawn from the market. There may continue to be monitoring of patients from earlier clinical trials to look for long-term effects and adverse reactions, even after the drug is approved.

A company might want to expand the use of a drug to new applications (e.g., different types of cancer). In this case, the previous safety studies are usually still applicable, but new clinical studies are required to test the efficacy of drug for the new purpose. If successful,

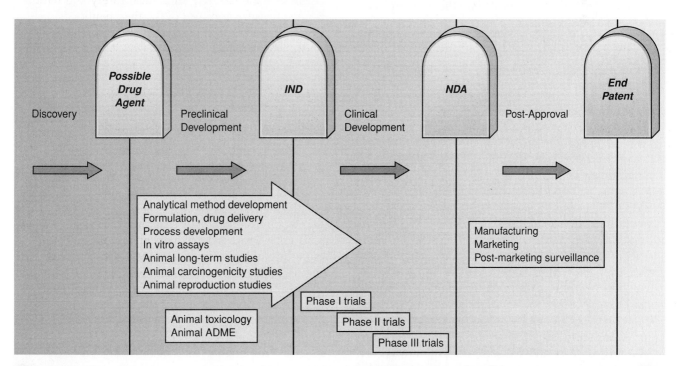

Figure 3.7. Post-Approval. The drug can be manufactured, marketed, and sold after it is approved. It is still subject to FDA oversight and may require additional clinical testing and development. This phase ends when patent protection on the product runs out.

these new studies allow the drug to be labeled for conditions in addition to the one for which the drug was initially approved.

If the manufacturing process for a drug changes substantially, the FDA may require new clinical trials to demonstrate that the product is still safe and effective. This is particularly true for large-molecule biological products. These products are so complex that they are difficult to completely characterize. Furthermore, the properties of a biopharmaceutical are tied to the production process used to make it. Significant changes to the manufacturing process could therefore result in a material not identical to that used in phase III clinical trials and laboratory tests of the product might not detect the difference.

The post-approval phase of the product's lifecycle is time-limited, since patent protection ends after a certain date and other companies may introduce generic versions of the drug. The pathway to introducing small-molecule generic drugs is reasonably straightforward, since these molecules are well characterized. Therefore, a generic manufacturer can demonstrate that their product is the same as that originally developed and tested by the original company. The first biopharmaceutical large-molecule drugs are now coming off-patent and are presenting new challenges for generic manufacturing. This is because the exact nature of each of these drugs depends on the specific cells or organisms in which it was produced, the specific genetic construct used, and the methods by which those cells were grown. These key aspects of the production process are trade secrets protected by the company that originally developed the drug. At the time of writing, there are many questions yet to be resolved relating to the introduction of "generic" biopharmaceuticals into the marketplace.

CASE STUDY

The Discovery of Gleevec

The first commercial success of high throughput screening (HTS) is often considered to be the small molecule drug, Gleevec, which is used to treat chronic myeloid leukemia (CML). The roots of Gleevec go back to basic research performed in the 1970s by scientists who were studying viruses that cause leukemia in animals. Viruses are not known to play a role in causing most common human cancers, but researchers hoped to learn about the fundamental mechanisms of cancer initiation by studying animal models. These animal studies led to the important insight that viruses that cause cancer do so by "taking over" certain normal genes in the animals' cells. The viruses cause the genes to produce too much of their protein product which, in turn, eventually causes the cells to divide malignantly. The gene that is over-expressed in a particular mouse leukemia is called abl.

Meanwhile, in the 1960s and 1970s, other researchers discovered that a type of human cancer, CML, occurs when pieces of two different chromosomes break off and each piece reattaches to the opposite chromosome. This rearrange-

ment fuses part of a specific gene from chromosome 22 (the bcr gene) with part of another gene from chromosome 9 (the abl gene). The resulting chromosome 22 is termed the Philadelphia chromosome, see Figure 3.8. The activity of the protein normally encoded by the abl gene is tightly controlled and limited by the cell. In contrast, the protein produced by the fused bcr-abl gene is over-expressed, eventually causing the cell to divide malignantly.

The work on animal viral cancer and CML merged in 1982 when researchers discovered that abl, the gene that is over-expressed in mouse leukemia, has a counterpart in humans—the same gene that is over-expressed in CML. In mouse cancer the gene is over-expressed because of a viral infection; in human cancer the gene is over-expressed because of the breakage and fusion of the chromosomes. The cause is different, but the malignant result is the same.

The abl gene in both mice and humans encodes an enzyme made in blood cells. This enzyme is a key component of a normal signaling pathway that tells a blood cell it is time to divide. When the abl gene is over-expressed, it makes too much of the encoded enzyme, which turns on the signaling pathway and tells the cell to divide—over and over again—resulting in cancer.

Research in the 1980s showed that the abl enzyme must modify another protein, its substrate, in order to trigger the signaling pathway that tells the cell to divide. This knowledge set the stage for a search for a drug compound that would target the abl enzyme and block it from modifying its substrate.

Pharmaceutical companies initially were unenthused about searching for a CML drug because this is a relatively rare cancer and a drug to treat it seemed unlikely to be profitable. However, a large pharmaceutical company, Novartis, was performing high throughput screening assays looking for small chemical compounds that would block other signaling proteins that are important in common cancers. They found a drug compound, ST1571, that binds to the abl enzyme, see Figure 3.9 on p. 54. Novartis scientists gave some of the compound to

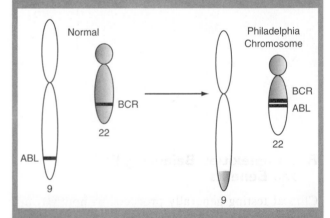

Figure 3.8. The Philadelphia Chromosome. A rearrangement of genetic material between chromosomes 9 and 22 fuses part of a specific gene from chromosome 22 (the *bcr* gene) with part of another gene from chromosome 9 (the *abl* gene). The resulting chromosome 22 is termed the *Philadelphia Chromosome.* The activity of the protein normally encoded by the *abl* gene is tightly controlled by the cell but the protein produced by the fused *bcr-abl* gene is over-expressed.

Dr. Brian J. Druker, a university researcher. Druker tested the compound in cells and animals and found that it was indeed effective in blocking the activity of the CML abl enzyme. Druker's success in early testing led to an exploratory clinical trial with a small number of CML patients, which demonstrated that the compound was effective in humans. Based on this early promising work, Novartis was convinced to continue to develop ST1571. The drug performed well in further testing and it has few side effects compared to traditional chemotherapeutic agents. (Recently, however, it has been found to occasionally have detrimental effects on the heart with long-term use.) ST1571 was approved for patient use in the United States in 2001 where it is marketed as Gleevec.

*Despite its novelty, Gleevec illustrates some common themes of drug discovery. First, the discovery of Gleevec had its roots in years of earlier basic research conducted by scientists in many laboratories, mostly in universities. Second, Gleevec illustrates the value of studying animal models to understand basic principles of health and disease. The work on animal models first showed how genetic changes lead to cancer. Third, the discovery and development of Gleevec required academic researchers, medical researchers, scientists at Novartis, patient advocacy groups (who, at a certain critical juncture encouraged Novartis to continue the development of Gleevec), and businesspeople, all of whom played a role in the discovery and development of the drug. The development of drugs and other medical products is usually so challenging that the efforts of many people and organizations are required. Fourth, the development of most biotechnology products is influenced by the potential of the product to be commercially successful. This has led to efforts by governments and organizations to find strategies to promote the development of drugs that are unlikely to be highly profitable but are necessary to treat life-threatening diseases. Drugs that are used to treat rare disorders, defined as affecting fewer than 200,000 Americans, are called **orphan drugs**, and the markets are correspondingly small. The Orphan Drug Act, which was enacted in 1983 in the United States, provides tax relief and some marketing exclusivity for companies that develop drugs for less-common diseases.*

Sources: Waalen, Jill. "Gleevec's Glory Days," Howard Hughes Medical Institute Bulletin 14, no. 5 (December 2001):10–15. http://hhmi.org/bulletin/dec2001/gleevec/gleevec2.html.

Savage, David G., and Antman, Karen H. "Imatinib Mesylate—A New Oral Targeted Therapy." New England Journal of Medicine 346, no. 9 (2002): 683–93.

F. Complexities: Balancing Risks and Benefits

Clinical testing generally proceeds without harm to volunteers, but it is not without risk, see the Case Study on p. 54. A gene therapy volunteer subject, Jesse Gelsinger, died in 1999 due to an experimental treatment. In March 2006, six healthy volunteers were injected with a new type of monoclonal antibody that was different in its mechanism of action than previous monoclonal antibody therapeutics. This particular antibody triggered a reaction in all six individuals, causing their immune systems to violently overreact and resulting in serious injury to them.

The process for developing a drug also poses financial risks. The drug development process is lengthy and the pharmaceutical industry generally reports that only 1 out of every 10 drugs that enters clinical testing will actually result in a marketed product. The low success rate of drugs coupled with the extensive resources required for their development contribute to the high costs of drugs.

The advantage to the complex, costly, and lengthy drug development process is that it does generally protect patients from overtly harmful new drugs. As we will see in Chapter 5, before animal and human testing of new drug products was required by law, patients often died from dangerous drugs. The system, however, is not perfect. Clinical trials sometimes do not detect adverse effects that only show up after long-term administration of a drug, or adverse effects that harm only a small percent of patients. This is the case, for example, with Gleevec. A study in 2006 indicated that in a small percent of patients the drug appears to cause severe heart disease with long-term use. However, about 90% of CML patients treated with Gleevec survive five years or longer whereas before the drug was approved in 1990, average survival was less than five years. The benefit of Gleevec to CML patients thus outweighs its risks and patients are likely to continue to take the drug.

In a more sensational incident, the drug Vioxx, an anti-inflammatory drug used by roughly 80 million patients to treat pain and arthritis, was found to increase the risk of heart attack and stroke. Studies of Vioxx indicate that it increases the risk of heart disease and stroke, but only in a small percentage of people using it, and only after 18 months of taking the drug daily. This means that the adverse effect was unlikely to turn up in early drug testing. (See, for example, Bresalier, Robert S., Sandler, Robert S., Quan, Hui, et al. "Cardiovascular Events Associated with Rofecoxib in a Colorectal Adenoma Chemoprevention Trial." *New England Journal Medicine* 352, no. 11 (2005): 1092–1102.

The Vioxx case immediately led to hundreds of lawsuits and withdrawal of Vioxx from the market. Both Vioxx and Gleevec elevate heart disease risk with long-term use. CML patients, however, are at very high risk of dying from cancer and so the benefit of the drug is thought to outweigh its risk, whereas Vioxx was used to treat nonfatal, though debilitating, diseases. The Vioxx incident was particularly complex and litigious because there is question as to whether the drug companies knowingly withheld information pointing to the drug's risk. The FDA has also been criticized in the Vioxx case for failing to protect patients from serious risk.

These incidents point to the complexities, risks, and benefits of drug development. It is hoped that as new technologies are created (e.g., methods to use predictive genetic information from patients, computer models of drug interactions) fewer people will suffer unpredicted effects from drugs, and the drug development process will become more reliable.

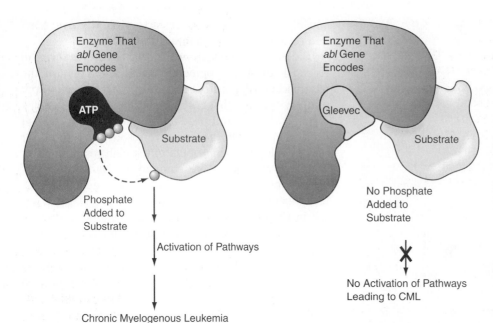

Figure 3.9. How Gleevec Works. On the left, the enzyme encoded by the *abl* gene is shown bound to the molecule adenosine triphosphate (ATP). The enzyme transfers a phosphate from the ATP molecule to the substrate. The addition of the phosphate to the substrate leads to the activation of pathways that cause too many white blood cells to be made in the bone marrow, causing CML. Gleevec takes the place of ATP in the pocket of the enzyme. When Gleevec is present, phosphate is not transferred to the substrate and subsequent events that cause CML do not occur. (Source: Modified from "Targeting the BCR-ABL Tyrosine Kinase in Chronic Myeloid Leukemia," John M. Goldman and Junia V. Melo. *New England Journal of Medicine* 344 no. 14 (2001): 1084–6.)

CASE STUDY

Recent Death in Gene Therapy Study Raises Questions about Informed Consent

Eight years after the death of Jesse Gelsinger, another death occurred during a U.S. gene therapy trial. In February 2007, Jolee Mohr, 36, was enrolled in a Phase I/II study of a gene therapy protocol to potentially treat rheumatoid arthritis, an inflammatory disease of the joints. The gene therapy agent, tgAAC94, consisted of an adeno-associated virus (AAV) designed to deliver a gene that blocks tumor necrosis factor, a substance associated with joint inflammation. AAV is used as a vector in a number of gene therapy protocols, because the virus is only weakly immunogenic and does not cause human disease. The tgAAC94 trial involved the injection of the agent directly into the affected joints; in Mohr's case, the right knee. The study included 120 volunteers, with no previous adverse effects. However, Mohr became ill within 24 hours of her second injection on July 2, 2007, and died 22 days later of massive internal bleeding and multiple organ failure. The gene therapy trial was suspended by Seattle-based Targeted Genetics Corporation, and the FDA was notified.

Mohr's death led to a government investigation by the National Institutes of Health Recombinant DNA Advisory Committee, which determined that Mohr died from complications of a systemic fungal infection. The Committee found no evidence to indicate that her death was a direct result of the gene therapy, although they could not rule out all possibility of an indirect effect. They suggested that a major risk factor for Mohr may have been her long-term drug treatment with Humira (adalimumab), which blocks TNF production systemically by suppressing the immune system. Immunosuppression may have been a key factor in Mohr's development of the opportunistic fungal infection. Since there was no evidence of direct harm from the gene therapy agent, the FDA approved the resumption of the tgAAC94 clinical trial in November 2007.

It is customary in gene therapy trials for patients to continue taking medications that have been effective in treating their disorder. Interestingly, Humira is a recombinant DNA product similar in function to Remicade, which was described in Chapter 1, Table 1.2. Humira, however, is a fully human monoclonal antibody, the first to be approved for human use (in 2002). Abbott Laboratories received the 2007 Galen Prize for Best Biotechnology Product for Humira, which is currently a well-accepted, first-line treatment for severe rheumatoid arthritis and several other autoimmune conditions. Increased risk of infection is a well-known, potential side effect of this medication.

While the gene therapy trial may not have been directly responsible for Mohr's death, it has raised discussion of important bioethical issues that are central to all clinical trials. Many people have questioned whether Mohr, a young woman with a 5-year-old daughter, was a good candidate for a clinical safety study, especially since her arthritis had responded well to Humira treatment. In the cases of Mohr and Gelsinger, there have been much debate about the issue of informed consent. Mohr's husband maintains that he and his wife did not realize that the study had no intended benefit for patients, and that Jolee would not have volunteered had she been aware of this. In turn, Targeted Genetics Corporation and Mohr's doctor point out that they provided the Mohrs with an IRB-approved informed consent document and they had answered the family's questions fully. Mohr clearly signed the consent form (which can be found at http://geneticsandsociety.org/downloads/Mohr%20consent%20form.pdf), but some have questioned whether a patient without a medical background could properly comprehend a 15-page form or ask all of the appropriate questions, even when the form is written in relatively simple language.

These are questions that do not have easy answers. They emphasize the fact that pharmaceutical companies that wish to market new treatments must deal with human issues that extend far beyond drug development and production.

III. BIOPHARMACEUTICALS

A. Overview

As we saw in Chapter 1, the advent of recombinant DNA technology provided a new way to manufacture pharmaceutical agents. The general lifecycle of the resulting biopharmaceuticals is the same as that of other drugs, but there are some details that are of interest to biotechnologists, see Figure 3.10. This section discusses a few of the issues specific to biopharmaceutical development laboratories where genetically engineered cells are used to produce drug products.

This topic is divided into two parts:

- **Product Development.** *This includes tasks relating to the biopharmaceutical molecule itself.*

- **Process Development.** *This includes tasks relating to developing methods for manufacturing and purifying the product.*

B. Product Development

Product characterization begins during the discovery and preclinical phases of drug development. Protein drugs are structurally complex and their structural features are key factors in determining how they will behave in a patient. Every protein has a specific shape, a specific distribution of positive and negative charges on its surface, and a particular glycosylation pattern; these features allow the protein to do its specific work in the cell. The characterization of protein drug compounds therefore requires many laboratory studies to determine their chemical and biological properties. Table 3.2 on p. 56 summarizes some of these studies.

Proteins are very sensitive to their environment. Proteins may degrade over time due to chemical interactions, (e.g., amino acid interactions with oxygen) and physical changes (e.g., unfolding and aggregating). Since the proper function of a protein requires that it retain the correct structure, it is important that protein drugs are formulated in such a way that they are as stable as possible. During pharmaceutical development, laboratory teams will research the most stable formulation for the drug (e.g., pH, buffer constituents) and how it should be handled, transported, and stored, see Figure 3.11 on p. 56. Analysts will also perform stability testing where they will evaluate the drug's potency and general characteristics over time and under various conditions. Analytical instruments, such as high-performance liquid chromatography (described in Chapter 33), are used during stability testing to look for specific protein breakdown products that are associated with product degradation.

C. Process Development

i. CREATION OF MASTER AND WORKING CELL BANKS

Recall from Chapter 1 that biopharmaceutical products are made by creating a genetic construct that codes for the protein of interest. The genetic construct is inserted into host cells (e.g., bacteria, yeast, mammalian cells) that manufacture the protein, which is then isolated from the cells, purified, and formulated into a final product. This production system, which is quite unlike that used when

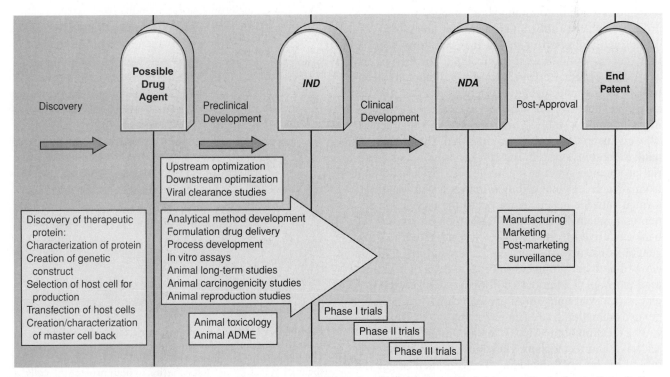

Figure 3.10. Lifecycle of Biopharmaceuticals. The overall lifecycle of biopharmaceuticals is generally similar to that of other products, but there are some additional requirements when using recombinant host cells as a manufacturing system, some of which are shown in this diagram.

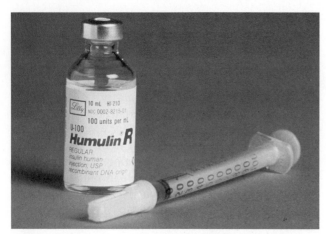

Figure 3.11.   A Final Product Formulation: Humulin, Insulin Manufactured Using Recombinant DNA Technology. (Photo courtesy SIU BioMed/Custom Medical Stock Photo.)

small molecules are synthesized, makes biopharmaceuticals of special interest to us as biotechnologists.

The creation of the genetic construct and the choice of a host cell system must be accomplished early in the discovery and development process. The successful creation of a system to manufacture the protein of interest requires understanding the gene and the protein for which it codes, knowledge that is acquired during the discovery phase of the product lifecycle. This early work is performed by scientists and technicians who are well-versed in molecular biology.

Biotechnology production systems depend on host cells that have been genetically modified to produce a specific desired product. If the host cell mutates or changes in some way, then the product might be altered—which is never desirable. Because of the complexity of biopharmaceutical compounds, a change in the final product might be subtle and difficult for analysts to recognize, yet might harm patients. It is therefore essential that there are methods to protect the stability of the expression system. Early in the history of biopharmaceuticals, scientists and regulatory agencies decided that the creation of a master cell bank (MCB) is a key component in maintaining a stable production system when genetically engineered organisms are involved. The **master cell bank** *is the source of cells used for production.* For bacterial production systems, the master cell bank is derived from a single, original colony of bacteria that was transformed with the genetic construct. If the production system uses mammalian or insect cells, then the master cell bank is derived from a single cell transfected with the genetic construct. In either case, the resulting cells are allowed to divide a number of times and are then distributed into small aliquots (portions), each of which is placed in a vial and stored cryogenically (in cold "suspended animation") to ensure its stability, see Figure 3.12.

The creation of the MCB is an important role of R&D scientists and technicians. They must document the history of the host cells including details of their origin: the species, identity, age, and sex of the donor, if the cells are of animal origin; and the medical history of the cells, if

Table 3.2.   CHARACTERIZING A BIOPHARMACEUTICAL PRODUCT

Identity Testing
- Determining and documenting the amino acid composition of the protein
- Determining and documenting the amino acid sequence (order of the amino acids) for key regions of the protein
- Determining and documenting the size of the protein
- Determining and documenting the conformation (shape) of the protein
- Determining and documenting any protein modifications, such as glycosylation
- Determining and documenting how the protein behaves in electrophoresis, HPLC, and in other test systems

Purity Testing
- Determining impurities from all sources that are associated with the product
- Determining breakdown products that may occur

Potency Testing
- Determining the therapeutic action of the protein
- Finding an assay(s) to measure that activity

Stability Testing
- Determining the stability of the product during storage and setting expiration dates
- Determining the effects of heat, light, time, and other factors on the drug
- Identifying break-down products and how to avoid them
- Determining proper storage conditions

Figure 3.12.   Storage of a Master Cell Bank. Special freezers containing liquid nitrogen provide an extremely low temperature of about −196°C. Note also that the technician is wearing special attire to help protect the cells from contamination. (Photo of BIOMEVA's cell bank storage area courtesy of BIOMEVA.)

they are of human origin. The MCB is prepared using stringent contamination control procedures in specialized facilities by individuals highly skilled in cell culture.

The MCB is the source of all production cells, and hence, is extremely valuable. It is a critically important trade secret. The MCB is protected by the use of security measures, documentation, and alarm systems in the event of freezer malfunction. It is also common to store aliquots in more than one freezer and at more than one site. *Each time a production run begins, an ampoule from the master cell bank is thawed and allowed to multiply to form the* **working cell bank.**

ii. CONTAMINANTS

The cell lines that are used to manufacture biopharmaceuticals are far less likely to carry harmful contaminants than tissues from humans or animals, but there are certain types of contaminants that are of concern when genetically engineered cells are used. Viruses are of particular concern when animal cells are used in production. Endotoxins can be a problem when bacterial cells are used, see Figure 3.13. In addition, the host cells have their own proteins and DNA that must not be allowed to carry over into the product.

The raw materials used to grow the cells are another potential source of contamination. The culture medium used to nourish and sustain the cells may introduce viruses, prions, unwanted proteins, and other contaminants. Raw materials, particularly the culture medium, must be tested for these contaminants. A section in Chapter 30 discusses ongoing efforts to reduce the chance of introducing contaminants via the culture media.

Adventitious contaminants *come from outside the cell culture system, for example, from air, manufacturing operators, or nonsterile equipment.* These contaminants are introduced when Good Manufacturing Practices are not followed. (Good Manufacturing Practices are discussed in Unit II.)

The various types of contaminants that are of importance for biopharmaceuticals are described in Table 3.3 on p. 58.

iii. TESTING THE CELL BANKS

The FDA requires significant testing to "establish all significant properties of the cells and the stability of these properties throughout the manufacturing process." These tests fall largely into three major categories: tests of the identity of the cells and the genetic construct, tests for purity, and tests for stability. Identity tests characterize the cells and the genetic construct. Purity tests look for bacterial, fungal, viral, and sometimes prion contaminants. Stability tests demonstrate that the cells and their constructs do not mutate or change with time. Stability testing requires thoroughly

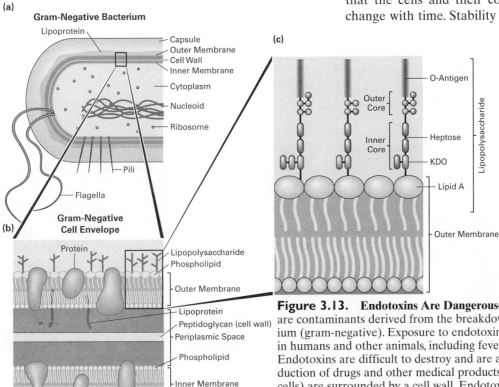

Figure 3.13. Endotoxins Are Dangerous—What Are They? Endotoxins are contaminants derived from the breakdown of a certain type of bacterium (gram-negative). Exposure to endotoxins can cause a severe reaction in humans and other animals, including fever, septic shock, and even death. Endotoxins are difficult to destroy and are a serious concern in the production of drugs and other medical products. Bacterial cells (unlike our cells) are surrounded by a cell wall. Endotoxins are a component of the outer part of the cell wall of gram-negative bacteria. **a.** Half of a bacterium is shown for orientation. **b.** A magnified view of a bacterial cell wall and its surrounding membranes. **c.** On the outside of the outer membrane of the cell wall there is a layer of lipid with associated proteins and sugars. Endotoxins are the lipid and sugar part of the outer membrane, i.e., the lipopolysaccharide. (**a.** and **b.** from http://gsbs.utmb.edu/microbook. **c.** Pizza, M. et al. *Science* 287, no. 5459 (2000): 1816–20, Copyright © The American Association for the Advancement of Science.)

Table 3.3. CONTAMINANTS OF CONCERN IN PRODUCTS MADE USING THE BIOTECHNOLOGY METHODS

- **Adventitious Contaminants.** *Contaminants that come from outside the system that are introduced accidentally and sporadically during processing through a breakdown in good manufacturing practices.*

- **Endogenous Contaminants.** *Contaminants that are naturally found in the host cells.* For example, most mammalian host cell lines are contaminated with viruses that must be removed from any biopharmaceutical products produced by those cells.

- **Bacteria.** *Contaminants that are a concern in any pharmaceutical process.* Bacteria can contaminate a product and harm the patient. They can also infect and destroy the host cells.

- *Mycoplasma.* A type of very small, difficult-to-detect bacterium. *Mycoplasma sp.* can cause disease in humans and animals. They are troublesome contaminants in cell culture because they are difficult to detect, difficult to eradicate, and can subtly alter the properties of infected cells.

- **Viruses.** *Contaminants that can both affect host cells and possibly infect the patient.* Viruses can incorporate themselves into the genome of the host cells and can remain dormant for long periods of time. Mammalian cells used to produce protein products contain endogenous viruses.

- **Pyrogens/Endotoxins.** *Contaminants that are by-products of gram-negative bacteria.* Pyrogens cause fever and elicit a dangerous inflammatory response in mammals; they may remain in a preparation even when all bacteria have been killed.

- **Nucleic Acids (DNA and RNA) from the Host Cells.** There is concern that nucleic acids from the host cells might enter a patient's cells and have a harmful effect.

- **Proteins from the Medium and the Host Cells.** The biopharmaceutical protein of interest is not the only protein in the production system. The culture medium may contain proteins and the host cells also have proteins. These extraneous proteins must be separated from the product. Residual proteins left from the production system may cause allergenic responses in patients. They also might have enzymatic effects, or might affect the regulation of genes in patients.

- **Yeast, Fungi, and Parasites.** These harmful agents might be accidentally introduced into the product from the growth medium, air, workers, and other sources.

- **Prions.** *A type of infectious agent, thought to be composed of misfolded proteins, which cause a variety of neurodegenerative diseases including scrapie, kuru, mad cow disease, and Creutzfeld-Jakob disease.* Prions are a particular concern with any agent that is isolated from an animal source, including fetal bovine serum, which used to be commonly added to cell culture medium to promote healthy cell growth.

- **Residuals from the Cells' Growth Medium.** The host cells require a rich growth medium that contains many substances that should not carry over into the final product; for example, insulin and albumin.

- **Residuals from Processing.** *Substances that contaminate the product during the purification and other processing steps.* All pharmaceuticals are processed in various ways to ensure their purity and to formulate them so they can be administered to patients. Residual materials from processing steps (e.g., chromatography solvents) must be removed.

- **Extractables from Plastic Components and Final Containers.** *Substances that leach out of plastics.* (See the Case Study, below.)

CASE STUDY

Detective Work Reveals the Culprit in Drug Mystery

Pure red cell aplasia (PRCA) is a severe and rare form of anemia. Beginning in 1998, doctors in France, Canada, the United Kingdom, and Spain began to notice an increase in PRCA among patients receiving the drug Eprex, which is used to treat anemia due to renal failure. Eprex is a biopharmaceutical product, erythropoietin, which is manufactured by Johnson & Johnson Company using genetically modified Chinese hamster ovary (CHO) cells. Erythropoietin is a glycoprotein normally produced in the kidney, which stimulates the production and maturation of red blood cells in the bone marrow. Although the incidence of PRCA was always low (about 1 case for every 5000 patient-years of exposure), the increased rate of occurrence of the disease triggered a massive investigation by Johnson & Johnson. More than 100 laboratory scientists, epidemiologists, immunologists, clinicians, and quality-assurance personnel studied the problem over a period of four years.

Recall that immunogenicity is a particular problem with protein therapeutics. According to Fred Bader, a vice president at Johnson & Johnson, nearly all therapeutic proteins elicit antibodies in some fraction of patients receiving the drug. The frequency ranges from 1 case per million doses to over 50% of patients treated. In some cases these antibodies have no clinical effect, but in other cases they inhibit the therapeutic effect of the protein. Patients who were administered Eprex and developed PRCA were found to have mounted an immune response against the recombinant erythropoietin drug. The antibodies produced by these patients not only neutralized the drug, they also attacked their body's natural erythropoietin, thus shutting down their bone marrow's production of red blood cells and causing PRCA. Investigators therefore focused on factors that might trigger this type of immune response in patients receiving Eprex. A scientific report published in 2006 showed that mice exposed only to Eprex did not develop antibodies against it. Investigators therefore looked for adjuvants, agents which might trigger the mice to mount an immune response against the drug. **Adjuvants** are any substances distinct from the protein of interest (in this case, Eprex) that trigger the immune system to respond to the protein of interest. *The administration of various adjuvants combined with Eprex resulted in an immune response in the mice that in some cases mimicked the PRCA seen in humans. Investigators therefore suspected an unknown adjuvant was causing problems in patients.*

There are many agents that can act as adjuvants and were suspects in this case. These include other proteins from the host cells, extraneous chemicals from processing steps, trace minerals, DNA and RNA from the host cells, and other agents in the final formulation. Scientists focused on the most likely adjuvants that might be present in the drug preparation including CHO host cell proteins and polysorbate 80, which was added to the drug formulation to improve its physical properties. They also investigated the possible effects of plasticizers that might have leached from the rubber stoppers of the syringes used to administer Eprex. Their investigations eventually identified leachates from the syringes as the culprits that triggered the serious immune response in human patients. The company switched to coated rubber stoppers to eliminate leaching and this apparently solved the PRCA problem.

In an interesting twist to the case, some evidence implicates polysorbate 80 as a contributor to the leachate problem. Prior to 1998, human serum albumin, a blood product, was used to stabilize the recombinant erythropoietin drug. In 1998, regulatory authorities in the European Union directed Johnson & Johnson to remove human serum albumin from Eprex in order to eliminate the possibility of pathogen contaminants from a human blood product. The company replaced HSA with polysorbate 80 for their European product. Shortly thereafter, the problems with PRCA were reported across Europe, suggesting that polysorbate 80 contributed to the leachate problem. Johnson & Johnson points to this experience as a reason to require new clinical trials whenever any new process to produce biopharmaceuticals is introduced. This issue has serious repercussions for companies that want to create "generic" versions of biopharmaceuticals.

Sources: Ryan, Mary H., et al. "An in vivo model to assess factors that may stimulate the generation of an immune reaction to erythropoietin." *International Immunopharmacology* 6, no. 4 (2006): 647–55.
McCormick, Douglas. "Small Changes, Big Effects in Biological Manufacturing." *Pharmaceutical Technology,* November, 2004, 16.
Bader, Fred. "Immunogenicity of Therapeutic Proteins: A Case Report." *Scientific Considerations Related to Developing Follow-On Protein Products* (FDA Public Workshop. September 14–15, 2004). http://www.fda.gov/cder/meeting/followOn/Bader.ppt.

characterizing the genetic construct and the MCB so that tests can periodically be performed that will be able to detect any alterations in the expression system. The working cell bank used to begin production is tested to see that it has the characteristics it should have. After every production run, the end of production cells are tested to be sure that they have not changed and have not been confused or contaminated with other cells.

Tables 3.4 through 3.6 on pp. 60 and 61 are compiled from the catalogues of several companies that provide cell bank testing services. These tables are included to demonstrate the extensive, varied, and sophisticated laboratory activities required to characterize the cell banks. Observe that the MCB is most extensively tested. The working cell bank is subjected to limited testing, mainly to demonstrate that no adventitious agents have contaminated the system and that the cells are the right ones. End-of-production cells are tested more thoroughly to ensure that they have not become contaminated (e.g., by the culture medium or careless handling), to check the stability of the cells and the genetic construct, and to ensure that viruses that may have been latent and undetected in the MCB have not been induced during culture.

iv. OPTIMIZATION OF PRODUCTION METHODS

The manufacture of a biopharmaceutical product is divided into two parts: "upstream" and "downstream" processing. **Upstream processing** *is the process of growing the cells and their production of the desired product.* **Downstream processing** *is the process of isolating and purifying the product from the cells.* The optimization of upstream and downstream processing begins in the preclinical period of the product lifecycle and continues to be refined throughout development.

Fermentation and cell culture (upstream processing) are complex processes in which cells are grown under controlled conditions. Mammalian cells normally exist in the body where they are bathed by a complex, high-nutrient mixture supplied by the bloodstream. This complex natural environment is absent in a manufacturing environment and cells are supplied with defined (as much as possible) nutrients in the form of growth medium (see also Chapter 30). Cells vary in their nutrient requirements, depending on their type and what product they are expressing. Optimizing the medium for each production system is therefore an important part of process development. Other aspects of the fermentation process must also be optimized, such as how to aerate the medium and how to remove cellular wastes. During this optimization process, scientists must be careful to not only maximize cell growth, but also maximize gene expression—production of product is ultimately what matters. Upstream processing commonly has two phases; a growth phase during which the conditions are optimized to promote cell growth and reproduction, followed by a phase where conditions are optimized for protein production. The details of all these upstream processing conditions are another example of a trade secret.

Downstream processing separates the protein of interest from all other substances, including the cells in which it was produced, the growth medium, and contaminants from any source. Proteins vary greatly from one another, and cells used for production also vary; therefore every downstream process is different and must be optimized. Strategies to isolate a particular biopharmaceutical generally involve a series of steps, each of which employs a different separation method to remove different contaminants. (The fundamental principles of isolating and purifying biological products are covered in more detail in Chapter 33.)

Table 3.4.   TESTING A MASTER CELL BANK

Type of Assay	Assay	Description
PURITY TESTING		
Presence of Microbial Contaminants	General sterility	Tests for the presence of bacterial and fungal contaminants by incubating cell bank material in conditions favorable for contaminant growth.
	Mycoplasma sp.	Tests for the presence of *Mycoplasma sp.*
Presence of Adventitious Viruses	In vitro viral assay	Material from the master cell bank is added to indicator cells in culture dishes. Indicator cells are checked for evidence of viral infection after a specified period of time.
	In vivo viral assay	Animals are injected with material from the master cell bank and observed for clinical signs of viral infection and for serum antibodies against viral agents.
Presence of Endogenous Retroviruses	PERT	Retroviruses produce a characteristic enzyme, reverse transcriptase (RT); the presence of RT is the basis for many retrovirus assays. (RT enables retroviruses to use RNA as a template for the synthesis of a complementary DNA [cDNA] strand.) PERT assays amplify and detect cDNA which is produced only in the presence of RT.
	Transmission electron microscopy	Virus particles are visualized in cells using transmission electron microscopy.
Presence of Specific Viruses	Bovine and porcine viruses	Bovine viruses (e.g., from bovine serum and serum supplements) and porcine viruses (e.g., from the use of the enzyme trypsin) can contaminate the cell line.
	Human virus PCR panel	When human cells are used for production, testing is performed to detect HIV, Epstein-Barr, and other human pathogens. PCR amplification of specific viral nucleic acids is used.
	Rodent virus panel	When rodent cells are used for production, specific tests are done for rodent viruses.
IDENTIFICATION/STABILITY		
Characterization of Cells	Isoenzymes	Isoenzymes are enzymes that differ from one another in a physical property but still catalyze the same reactions. When isoenzymes are analyzed by electrophoresis, different cell lines produce different banding patterns that can be used for identification.
	DNA fingerprinting	DNA fingerprinting produces a banding pattern unique to each cell line.
	Karyology	Karyotyping images the morphology of the cells' chromosomes. Cell lines have characteristic karyotypes that should remain stable over time.
Characterization of Genetic Construct	Copy number	Determines the number of vectors containing the gene of interest that have integrated into the host cells.
	Restriction digest pattern of construct	Characterizes the genetic insert by cleaving the DNA with enzymes. The resulting DNA fragments are separated from one another using gel electrophoresis. The fragments are visualized, resulting in a pattern characteristic of the genetic construct.
	DNA sequencing	The DNA sequence of the protein-coding region of the genetic construct is determined.
OTHER		
Tumorigenicity (ability of cells to cause tumors)	In vitro soft agarose	Many cell lines are able to live and replicate in soft agarose due to a decreased requirement for cell-to-cell and cell-substrate adhesion. Cells that can grow in soft agar or agarose share some qualities with cancer cells.

Table 3.5. TESTING A WORKING CELL BANK

Type of Assay	Assay
PURITY TESTING	
Presence of Microbial Contaminants	General Sterility
	Mycoplasma sp.
Presence of Adventitious Viruses	In vitro viral assay
	In vivo viral assay
IDENTIFICATION/STABILITY	Isoenzymes

Upstream and downstream processing have a somewhat different meaning when transgenic plants and animals are used for production. Upstream processing becomes growing the plants or animals in a controlled environment. Downstream processing might mean isolating a product from milk or harvesting a crop and then purifying a component from part of the plants. Once the product is released from its initial source, the principles of downstream processing are similar, regardless of the method of upstream processing.

v. REMOVING CONTAMINANTS

One of the key concerns during process development is to ensure that contaminants are not present in the final product. Biotechnology products are proteins that cannot be sterilized by heat. Development scientists must therefore find other ways to remove infectious contaminants. In addition, as with all pharmaceutical production, routine steps to prevent contaminating the product during production are required. These routine safeguards include, for example, the use of special clean rooms with very low levels of airborne particulates and the sterilization of processing equipment.

One of the first tasks of the process development team is to identify all of the contaminants that might be present in the host cells and the growth medium and to develop methods of purification to remove those conta-

Table 3.6. TESTING END-OF-PRODUCTION CELLS

Type of Assay	Assay
PURITY TESTING	
Presence of Microbial Contaminants	General Sterility
	Mycoplasma sp.
Presence of Adventitious Viruses	In vitro viral assay
	In vivo viral assay
Presence of Endogenous Retroviruses	PERT
	Transmission Electron Microscopy
IDENTIFICATION/STABILITY	
Characterization of Cells	Isoenzymes
	DNA Fingerprinting
	Karyology
Characterization of Genetic Construct	Copy Number
	Restriction Digest
	Pattern of Construct

minants. As the process is developed, the team must perform **clearance studies** *in which a contaminant is intentionally added to the production system and the ability of the process to remove it is measured.* It is, for example, common to perform viral clearance studies in which viruses are intentionally added to the crude product and their removal by subsequent purification steps is measured. It is recommended that a series of clearance studies are performed using different types and sizes of viruses, DNA, and any additives that might be detrimental and whose removal must be documented.

All these development tasks relating to contamination, and contaminant testing performed by QC analysts, require assays to tell if contaminants are present. There are a number of assays used to test host cells, raw materials, in-process samples, and end-products for contamination. Each assay is generally specific for a certain type of contaminant. For example, one series of assays is used to look for DNA, RNA, and protein contaminants from the host cells. Other series of tests are used to look for viral, bacterial, and *Mycoplasma sp.* contaminants. Some of these assays were described in Table 3.4.

D. Summary: Biopharmaceuticals

We have seen that successful biopharmaceuticals have the same overall lifecycle as other pharmaceutical products, beginning with their birth in research laboratories, proceeding through preclinical development, submission of an IND, clinical development, an NDA (or similar application to the regulatory authorities), followed by approval and manufacturing. There are, however, important differences between small molecule drugs and biopharmaceuticals.

First, biopharmaceuticals are a different sort of molecule than small-molecule drugs. Biopharmaceutical drugs are large, complex molecules whose three-dimensional structure is critical to their activity. As shown in Table 3.2, characterizing biopharmaceuticals is multifaceted and laboratory intensive. Assay methods need to be developed and used to determine the protein's structure, activity, stability, and breakdown products that appear if the protein degrades during storage and handling.

Not only are biopharmaceuticals themselves distinct from small-molecule drugs, their production systems are also different. Small-molecule drugs are often chemically synthesized. Biopharmaceuticals are made by genetically modified living cells that express a protein of interest. Master cell banks are created for biopharmaceuticals to help ensure the stability of the production system and hence the consistency of the drug product. These cell banks are painstakingly created, tested, characterized, and stored.

Using cells as a production system requires careful management of the upstream processing conditions such that the cells survive, thrive, and make product. Optimizing these conditions is another challenging technical task. The methods used to purify biopharmaceuticals are also distinctive because the biopharmaceutical agent is initially mixed in a broth that contains varied contaminants from the host cells. Avoiding adventitious contaminants requires strict adherence to practices that isolate and protect the cells. Removing endogenous contaminants requires developing practical, effective downstream purification methods. These are further challenges for biotechnologists.

The biopharmaceutical industry has grown steadily as the many technical issues associated with these drugs have become better understood. This has resulted not only in increasing numbers of valuable products, but also a deepening of our understanding of cellular processes.

IV. SUMMARY

Chapter 3 continued the discussion of product lifecycles that was introduced in Chapter 2 by exploring the lifecycles of pharmaceutical/biopharmaceutical products. Chapter 3 began with an overview of the lifecycle of pharmaceutical products in general and then turned more specifically to the development of biopharmaceuticals, which are of particular interest to biotechnologists.

The pharmaceutical/biopharmaceutical lifecycle includes a complex development process that costs hundreds of millions of dollars and takes years to complete. Many candidate drug products fail at various points during the process, making drug development a costly enterprise. Drug development begins with preclinical testing, followed by three stages of clinical testing. Regulatory agencies control key "gates" at critical points in this lifecycle and then provide oversight of manufacturing for products that successfully pass through development. The pharmaceutical industry is arguably the most regulated industry. The continuous interaction of government with the industry leads to a distinctive culture and work environment that will be further explored in Unit II.

Looking back at Unit I as a whole, we discussed how the biotechnology industry transforms scientific knowledge and discovery into a wide variety of useful products. Chapter 1 provided a very broad overview of the science and the products of biotechnology. We saw that these products are numerous and include items as diverse as biopharmaceuticals and biofuels. Chapter 2 introduced the business side of biotechnology, including the organization of companies and the role of intellectual property in biotechnology. Chapter 3 focused on products that have therapeutic application in humans.

One of the themes that recurs in this unit is the importance of the laboratory and laboratory personnel in all aspects of biotechnology. Laboratory scientists, technicians, and analysts perform experiments and make discoveries that are the basis for biotechnology. Laboratory personnel transform knowledge from the research laboratory into effective products and perform a wide variety of important laboratory analyses.

Another theme that emerges is that biotechnology fosters interconnections. Research discoveries are almost always the result of the efforts of many people who are connected through publications and personal communications. The staffs of multiple research laboratories and companies often share their expertise as new ideas are brought through development, testing, and into production. The Gleevec case study illustrates how knowledge from various academic research laboratories, together with tools provided by a pharmaceutical laboratory, was used to create an effective drug product. Biotechnology companies are further interconnected with one another in business partnerships. Many biotechnology products take advantage of sophisticated scientific processes that involve patents from multiple parties. Patent licensing agreements therefore connect companies into partnerships. Moreover, the development and production of many products is so complex that companies must partner with one another to be successful.

Unit II continues to look at biotechnology from the perspective of its products and its business side. Unit II discusses product quality and the many systems used to help ensure the quality of products.

PRACTICE PROBLEMS

1. Matching. Match each of the activities below with the lifecycle phase that best fits it.
   i. Research/discovery
   ii. Preclinical development
   iii. Clinical development
   iv. Post-approval
   a. Determining whether a drug candidate is effective by evaluating its effect in several thousand human patients.
   b. Screening 100,000 small chemical compounds to find any that bind to a receptor found on the surface of pancreatic cancer cells.
   c. Performing tests in rats to see how their bodies metabolize a possible drug compound.
   d. Performing tests where a possible drug compound is added to plates containing cultured cells and determining if the compound is toxic to the cells.
   e. Using cultured intestinal cells to determine whether a compound has the ability to cross the cell membrane of the intestine.
   f. Devising and creating the genetic construct used for a monoclonal antibody biopharmaceutical.
   g. Performing surveillance when a drug is administered to millions of patients to see if it causes rare adverse effects.
   h. Developing an assay to detect the drug compound in samples from blood and urine of animal and human subjects.

2. List some similarities and some differences between the drugs Herceptin (see Chapter 1) and Gleevec.

3. The figure on page 63, taken from the FDA's website, is their diagram of the lifecycle of a pharmaceutical product. Begin by relating this diagram to Figure 3.4.

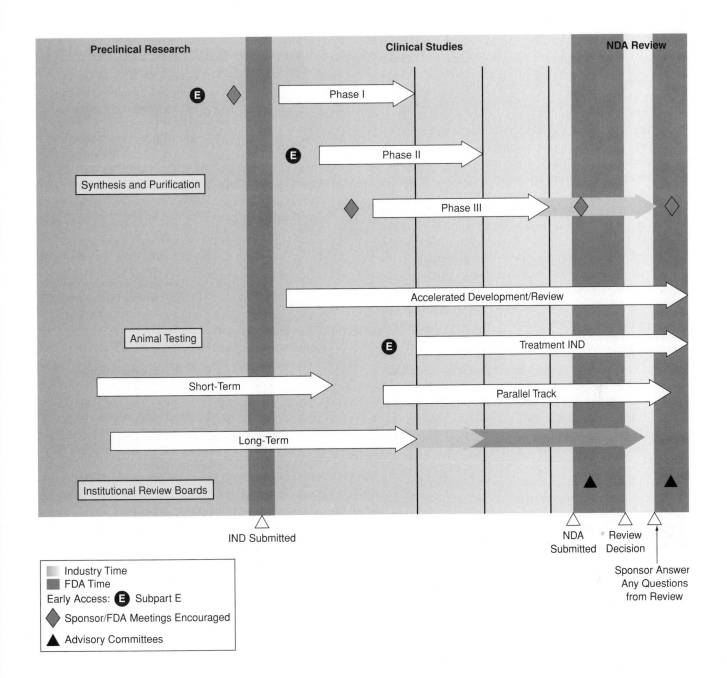

Then, find the FDA's diagram on their website, http://www.fda.gov/cder/handbook/develop.htm. You can click on various terms on the website diagram for more information.

a. Based on information on the website, what is "Accelerated Development Review"?

b. What is a "Treatment IND"?

c. What is meant by "Parallel Track"?

4. Examine Tables 3.4 through 3.6.

a. Based on these tables, list concerns relating to the use of cells for production.

b. Why are end of production cells tested?

5. Which of the following tasks are required when developing and manufacturing any drug, and which are specific to biopharmaceutical products?

a. Genetically engineering a plasmid to carry a gene of interest into host cells.

b. Optimizing the cells' growth medium to maximize protein expression.

c. Avoiding bacterial contamination of the product.

d. Complying with FDA regulations during manufacturing.

e. Complying with FDA regulations during clinical trials.

f. Creating a master cell bank.

g. Removing host cell proteins during purification.

6. Discuss the role of laboratory scientists, analysts, and technicians in the biotechnology industry. Where possible, include examples from Chapters 1–3. (The case studies provide examples that you might include in your answer.)

QUESTIONS FOR DISCUSSION

This question is included to provide subjects for thought, discussion, and further research. It does not have a single answer and no answer key is provided.

During World War II, Nazi physicians conducted brutal "experiments" on human adults, children, and infants in concentration camps. The victims of these "experiments" were frozen to death, mutilated and wounded, sickened, and often suffered greatly before their deaths. The Nuremberg Code is a set of ethical principles for **human experimentation** that was a response to these atrocities. The World Medical Association "Declaration of Helsinki: Recommendations Guiding Medical Doctors in Biomedical Research Involving Human Subjects," which expands on the Nuremberg Code, is a "statement of ethical principles to provide guidance to physicians and other participants in medical research involving human subjects" (http://www.wma.net/e/policy/b3.htm). The principles in the Declaration of Helsinki are encoded in the United States Code of Federal Regulations and therefore are part of the legal landscape of drug testing. (See Title 21-Part 50, "Protection of Human Subjects," and Title 45-Part 46, "Protection of Human Subjects.")

The Basic Principles from the Declaration of Helsinki (as shown on the FDA's website, http://www.fda.gov/oc/health/helsinki83.html) are quoted below. Discuss how these basic principles apply to a company that is conducting clinical trials of a new drug substance. How can a research team ensure that its testing is ethical and conforms to the basic principles?

Section 1.01 I. Basic Principles

1. Biomedical research involving human subjects must conform to generally accepted scientific principles and should be based on adequately performed laboratory and animal experimentation and on a thorough knowledge of the scientific literature.

2. The design and performance of each experimental procedure involving human subjects should be clearly formulated in an experimental protocol, which should be transmitted to a specially appointed independent committee for consideration, comment, and guidance.

3. Biomedical research involving human subjects should be conducted only by scientifically qualified persons and under the supervision of a clinically competent medical person. The responsibility for the human subject must always rest with a medically qualified person and never rest on the subject of the research, even though the subject has given his or her consent.

4. Biomedical research involving human subjects cannot legitimately be carried out unless the importance of the objective is in proportion to the inherent risk to the subject.

5. Every biomedical research project involving human subjects should be preceded by careful assessment of predictable risks in comparison with foreseeable benefits to the subject or to others. Concern for the interests of the subject must always prevail over the interests of science and society.

6. The right of the research subject to safeguard his or her integrity must always be respected. Every precaution should be taken to respect the privacy of the subject and to minimize the impact of the study on the subject's physical and mental integrity and on the personality of the subject.

7. Physicians should abstain from engaging in research projects involving human subjects unless they are satisfied that the hazards involved are believed to be predictable. Physicians should cease any investigation if the hazards are found to outweigh the potential benefits.

8. In publication of the results of his or her research, the physician is obliged to preserve the accuracy of the results. Reports of experimentation not in accordance with the principles laid down in this Declaration should not be accepted for publication.

9. In any research on human beings, each potential subject must be adequately informed of the aims, methods, anticipated benefits, and potential hazards of the study and the discomfort it may entail. He or she should be informed that he or she is at liberty to abstain from participation in the study and that he or she is free to withdraw his or her consent to participation at any time. The physician should then obtain the subject's freely given informed consent, preferably in writing.

10. When obtaining informed consent for the research project the physician should be particularly cautious if the subject is in a dependent relationship to him or her or may consent under duress. In that case the informed consent should be obtained by a physician who is not engaged in the investigation and who is completely independent of this official relationship.

11. In case of legal incompetence, informed consent should be obtained from the legal guardian in accordance with national legislation. Where physical or mental incapacity makes it impossible to obtain informed consent, or when the subject is a minor, permission from the responsible relative replaces that of the subject in accordance with national legislation.

    Whenever the minor child is in fact able to give a consent, the minor's consent must be obtained in addition to the consent of the minor's legal guardian.

12. The research protocol should always contain a statement of the ethical considerations involved and should indicate that the principles enunciated in the present Declaration are complied with.

UNIT II

Product Quality and Biotechnology

This unit is about the means by which people ensure that the products of their work are of good quality. Producing a quality product does not happen by accident—quite the contrary, accidents usually result in an inferior product. The consistent production of quality products requires resources, planning, and commitment. This unit explores how a company, organization, or laboratory translates a commitment to quality into practice.

Chapter 4 is a broad overview of the issues and vocabulary relating to product quality systems.

Chapter 5 begins with a history of medical product regulation in the United States and provides background on the relationship between the biotechnology industry and regulatory agencies.

Chapter 6 discusses documentation, one of the most important aspects of any quality system.

Chapter 7 discusses how the principles of quality systems are applied in biotechnology production facilities.

Chapter 8 discusses how the principles of quality systems are applied in laboratory settings.

BIBLIOGRAPHY FOR PRODUCT QUALITY AND BIOTECHNOLOGY

There is a vast, fluid literature in the area of product quality, particularly as it relates to pharmaceuticals. Since the technology, regulations, and standards that impact biotechnology are changing quickly, the easiest way to enter the literature is through the Internet. Some useful sites are included in these references.

History of Drug Regulation in the United States

Center for Drug Evaluation and Research. "Time Line: Chronology of Drug Regulation in the United States." http://www.fda.gov/cder/about/history/time1.htm.

Chapman, Kenneth G. "A History of Validation in the United States: Part 1." *Pharmaceutical Technology,* October 1991, 82–96. (Historical presentation; includes discussion of some issues and controversies relating to validation.)

Food and Drug Administration. "A Brief History of the Center for Drug Evaluation and Research." http://www.fda.gov/cder/about/history/default.htm

Food and Drug Administration. "FDA History." http://www.fda.gov/oc/history/default.htm.

Food and Drug Administration. "Selected Sources on the History of the FDA." http://www.fda.gov/opacom/morechoices/sources.html.

FDA Office of Women's Health. "FDA Milestones in Women's Health: Looking Back as We Move into the New Millennium." http://www.fda.gov/womens/milesbro.html.

Immel, Barbara. "A Brief History of the GMPs: The Power of Storytelling." *BioPharm International,* August 2000. http://www.immel.com/Whats_New.html#Latest_articles. (A good article about the history of drug regulation in the United States.)

Tetzlaff, Ronald F., Shepherd, Richard E., and LeBlanc, Armand J. "The Validation Story: Perspectives on the Systematic GMP Inspection Approach and Validation Development." *Pharmaceutical Technology,* March 1993, 100–16. (Historical presentation of issues leading to validation.)

The *Barr* Decision

Department of Health and Human Services, U.S. Food and Drug Administration. "Guidance for Industry, Investigating Out of Specification (OOS) Test Results for Pharmaceutical Production." October 2006. http://www.fda.gov/cder/guidance/3634fnl.pdf.

Hoinowski, Alex M., Motola, Sol, Davis, Richard J., and McArdle, James V. "Investigation of Out-of-Specification Results." *Pharmaceutical Technology,* January 2002, 40–50. http://www.pharmtech.com/pharmtech/issue/issueDetail.jsp?id=480.

The GMP Institute. A private organization that conducts GMP training, it has useful resources on the *Barr* decision at http://www.gmp1st.com/barr2.htm.

Documentation and Electronic Signatures

21 CFR Part 11; A Technology Primer, Supplement to Pharmaceutical Technology, Iselin, NJ: Advanstar Publications. 2005. http://www.pharmtech.com/pharmtech/issue/issueDetail.jsp?id=6515.

Department of Health and Human Services, U.S. Food and Drug Administration. "Guidance for Industry Part 11, Electronic Records; Electronic Signatures—Scope and Application." August 2003. (http://www.fda.gov/cder/guidance/5667fnl.htm)

DeSain, Carol, and Sutton, Charmaine Vercimak. *Documentation Practices.* Cleveland, OH: Advantstar Publishers, 1996. (A comprehensive look at documentation with many useful examples.)

Huber, L. "Implementing 21 CFR Part 11 in Analytical Laboratories: Part 1, Overview and Requirements." *BioPharm International,* November 1999, 28–34.

Winter, W., and Huber, L. "Implementing 21 CFR Part 11 in Analytical Laboratories: Part 2, Security Aspects for Systems and Applications." *BioPharm International,* January 2000, 44–50.

Winter, W., and Huber, L. "Implementing 21 CFR Part 11 in Analytical Laboratories: Part 3, Ensuring Data Integrity in Electronic Records." *BioPharm International,* March 2000, 45–9.

Huber, L., and Winter, W. "Implementing 21 CFR Part 11 in Analytical Laboratories: Part 4, Data Migration and Long-Term Archiving for Ready Retrieval." *BioPharm International,* June 2000, 58–64.

Winter, W., and Huber, L. "Implementing 21 CFR Part 11 in Analytical Laboratories: Part 5, The Importance of Instrument Control and Data Acquisition." *BioPharm International,* September 2000, 52–6.

Relevant Portions of the Code of Federal Regulations

21CFR11. Federal government rules that regulate electronic (computer) records, signatures, and other forms of electronic documentation.

21CFR58. Federal government rules that regulate preclinical laboratory studies of pharmaceutical products.

21CFR210. Federal government rules that define current Good Manufacturing Practices for manufacturing, processing, packing, and holding drugs in general.

21CFR211. Federal government rules that define current Good Manufacturing Practices for finished pharmaceuticals.

21CFR820. Federal government rules with which medical device manufacturers must comply, called "Quality Systems Regulation, QSR."

The Code of Federal Regulations is available at: http://www.gpoaccess.gov/cfr/index.html.

The FDA website, http://www.fda.gov/cder/dmpq/ index.htm, has the federal regulations along with notes, proposed changes in the regulations, and other pertinent information concerning cGMP.

The commercial site, http://www.gmp1st.com/gmp.htm, has excellent links to the regulations.

FDA and NIH Documents

Center for Drug Evaluation and Research, Center for Biologics Evaluation and Research, and Center for Devices and Radiological Health, Food and Drug Administration. "Guideline on General Principles of Process Validation." May 1987. http://www.fda .gov/CDER/GUIDANCE/pv.htm.

Department of Health and Human Services, U.S. Food and Drug Administration. "Guidance for Industry: Quality Systems Approach to Pharmaceutical Current Good Manufacturing Practice Regulations." September 2004. http://www.fda.gov/cder/Guidance/. (This draft guidance document is intended to help manufacturers that are implementing modern quality systems and risk management approaches to meet the requirements of the Agency's cGMPs.)

Department of Health and Human Services, U.S. Food and Drug Administration. "Pharmaceutical cGMPs for the 21st Century—A Risk-Based Approach, Final Report." September 2004. http://www.fda.gov/ cder/gmp/gmp2004/GMP_finalreport2004.htm.

Food and Drug Administration, Center for Drug Evaluation and Research, Center for Biologics Evaluation and Research, and ICH. "Guidance for Industry: Q2B Validation of Analytical Procedures: Methodology." November 1996. http://www.fda.gov/ cder/guidance/1320fnl.pdf. (This guideline provides the basis for method validation in the U.S. pharmaceutical and biopharmaceutical industries.)

The National Institutes of Health. "NIH Guidelines: Recombinant DNA and Gene Transfer." http:// www4.od.nih.gov/oba/rac/guidelines/guidelines .html.

ICH and ISO Guidelines

There are a number of ICH documents that are relevant to Biotechnology. ICH Guidelines are divided into four major topic categories and are coded according to those topics (i.e., all "Quality" topics start with "Q").

Q Quality. Topics relating to chemical and pharmaceutical quality assurance. Examples: Q1 Stability Testing, Q3 Impurity Testing.

S Safety. Topics relating to preclinical studies. Examples: S1 Carcinogenicity Testing, S2 Genotoxicity Testing.

E Efficacy. Topics relating to clinical studies in human subjects. Examples: E4 Dose Response Studies, Carcinogenicity Testing; E6 Good Clinical Practices.

M Multidisciplinary. Topics that do not fit into one of the above categories. Examples: M1 Medical Terminology, M2 Electronic Standards for Transmission of Regulatory Information.

ICH. "ICH Harmonised Tripartite Guideline Preclinical Safety Evaluation of Biotechnology-Derived Pharmaceuticals, S6." July 1997. http://www.ICH.org. (This guideline describes preclinical safety testing for most types of biotechnology-derived therapeutics.)

ICH. "ICH Harmonised Tripartite Guideline: Quality Risk Management Q9." November 9, 2005. http:// www.ICH.org.

In the United States, the ISO 9000 standards are prepared by the American Society for Quality Control (ASQC) Standards Committee for the American National Standards (ANSI) Committee on Quality Assurance. The ISO website includes background and ordering information: http://www.iso.org.

Trade Publications

BioPharm International. See references for Unit I.

"GMP Links." This is a comprehensive, annotated portal to dozens of Internet resources, particularly those associated with the FDA. It is maintained by *BioPharm International,* a publication of Advanstar Communications. http://www.biopharminternational.com/biopharm/ static/staticHtml.jsp?id=2455&searchString=links.

Pharmaceutical Technology: Europe. A monthly print magazine with excellent information about the pharmaceutical industry. Back issues are available. http://www.ptemag.com/pharmtecheurope/.

The Pharmaceutical Research and Manufacturers of America (PhRMA) represents the country's leading pharmaceutical research and biotechnology companies. Their publications include the handbook, *Principles on Conduct of Clinical Trials and Communication of Clinical Trial Results.* http://www.phrma.org/clinical_trials/.

Homepages for Agencies and Organizations

The American Association for Laboratory Accreditation: http://www.a2la.org

The Food and Drug Administration: http://www .fda.gov

The Environmental Protection Agency: http://www .epa.gov

The International Conference on Harmonisation: http:// www.ich.org

The United States Department of Agriculture: http:// www.usda.gov

The United States Pharmacopeia: http://www.usp.org

CHAPTER 4

Introduction to Product Quality Systems

I. OVERVIEW

As consumers of products and services we are all familiar with the concept of product quality. At the supermarket we may find certain brands to be superior. Before purchasing a major appliance, we might consult a consumer guide to learn about the performance of various models. In this unit (and in this entire book) we are concerned with product quality, but not from the perspective of being a consumer, rather from the perspective of being the producer of a biotechnology product.

As we saw in Chapter 1, the products of biotechnology are varied. There are tangible products such as drugs, transgenic livestock, modified plants, and enzymes for research. There are laboratories whose products are test results, such as measurements of a product's activity or the performance of an instrument.

The academic research laboratory is a workplace whose product is knowledge. The definition of *quality* varies for these different sorts of products. Government regulations define quality for drugs, medical products, and foods. These regulated products must be safe, effective, reliable, nutritious (for foods), unadulterated, and so on. A quality test result is one that can be relied on when making decisions. A quality research study enhances our understanding of the natural world.

How does a company or a laboratory ensure that its products are of good quality? Consider two events. In 1959 the Pillsbury Company was beginning to manufacture food products for the NASA space program. It was essential that the food be completely pathogen-free to ensure the safety of the astronauts. Scientists considered testing the food for pathogens but analysis convinced them that simply testing the final food products would not ensure their purity. This is because not every food item can be tested; most testing procedures destroy the food or its packaging. Even if nondestructive methods exist, it is too time-consuming to test every item. When final products are tested, therefore, a sample is selected to represent the whole lot. The Pillsbury scientists did an analysis in which they supposed that there was one pathogenic Salmonella organism per 1000 units of food. If 20 of the 1000 units were to be tested, then there would be a 98% chance of missing the single pathogen and accepting the defective lot. The Pillsbury staff concluded that testing samples of the final food products was not an adequate method of ensuring their quality and safety. They decided instead to make sure that the production of the foods was so tightly controlled that no contaminants could enter them.

In the 1960s and 1970s there were a number of septicemia cases in patients caused by intravenous (IV) fluids that were contaminated by bacteria. Many people became ill and some died. FDA inspection teams discovered serious problems in companies manufacturing products for IV use. They found, for example, contaminated cooling water and sterilization equipment that failed to reach sterilizing temperature. These sorts of problems in production led to contaminated products, which in turn, caused illness and deaths in patients.

The pharmaceutical companies responsible for the septicemia outbreaks did have testing programs in place to monitor samples of their final products. If the samples passed inspection, then the quality of the entire batch was assumed to be acceptable for distribution. Unfortunately, as was realized by the Pillsbury scientists, final product testing can miss contaminated items. In the IV fluids case, the acceptance of contaminated lots led to patient deaths.

The Pillsbury analysis and the septicemia deaths illustrate that final product testing, although necessary and important, does not ensure that products are of acceptable quality. As FDA emphatically states:

> Quality, safety and effectiveness are designed and built into a product. The quality of a product does not result from inspecting that product; that is, quality cannot be inspected or tested into the finished product.

How is quality designed and built into a product? There is no single answer to this question, nor is there any simple answer. Producing a quality product requires many coordinated elements (e.g., skilled and knowledgeable personnel, a well-designed and managed facility, and ready access to raw materials). All these coordinated elements together are called a **quality system. A quality system** *is the organizational structure, responsibilities, procedures, processes, and resources that together ensure the quality of a product or service.* All product quality systems have the goal of ensuring a quality product.

Every product is different, companies differ, and the problems that arise during production vary; therefore, there is more than one formal quality system. The general features of the quality systems that are important in biotechnology workplaces are introduced in this chapter and then explored in more detail in Chapters 5–8.

II. QUALITY SYSTEMS IN DIFFERENT WORKPLACES

A. Academic Research Laboratories

Academic scientists performing basic research are unlikely to think of their work using the term *quality system*. Research scientists do, however, have a long tradition of adhering informally to such a system, called "doing good science." "Doing good science" means using consistent, thoughtful, and effective methods; keeping honest and thorough records; verifying all results; and rigorously employing "good laboratory practices." Although the output of an academic research laboratory is not scrutinized by government inspectors, there are methods of monitoring its quality. When research scientists complete a set of work they submit it to a journal to be published. The work is then available to other scientists who consider it critically, evaluating its merits, flaws, and value. In addition, the work of scientists is judged by peers when scientists apply for research funding to support their studies. Scientists whose work is not of high quality have difficulty publishing their work and obtaining grants, and may have difficulty competing for jobs and promotions.

B. Quality Systems in Companies That Are Regulated

The consequences of a poor product are very different in a research laboratory than in a drug or food processing company. Poorly manufactured drugs and contaminated foods have led directly to human injuries and deaths. The consequences of a poorly performed research study tend to be less severe; therefore, in contrast to the informal quality practices of research scientists, the product quality systems used in companies that make drugs, or process foods, are formal and strictly enforced.

The quality system used in companies that make drug and related medical products is called current Good Manufacturing Practices (cGMPs). The cGMPs have evolved within an ongoing relationship between the government (which seeks to protect the consumer) and the pharmaceutical industry. The cGMPs constitute a quality system that has a sweeping effect in every company that produces regulated medical products. The requirements of GMP also radiate throughout the medical products industry to impact those who package, distribute, market, sell, and use these items.

The Good Manufacturing Practices are quality principles formalized into regulations. **Regulations** *are requirements that government-sanctioned agencies, such as the Food and Drug Administration or the Environ-*

mental Protection Agency, impose on an industry and on companies within that industry. Compliance with regulations is required by law.

Regulations are objective and generally focus on safety (or reducing risk), efficacy, and honesty. Regulations tend not to cover subjective areas of product quality, such as the flavor of cookies. Thus, local government agencies are interested in how a chocolate chip cookie baker cleans his bowls—a safety consideration—but do not care about the flavor of his cookies. A pharmaceutical company might similarly add grape flavoring to cough syrup to enhance its appeal. The individual cold sufferer is capable of evaluating whether the flavoring improves the quality of the product, and therefore, this aspect of product quality is not subject to government regulation. In contrast, the safety of that cough syrup is subject to regulation because the consumer cannot perceive whether a medication is safe or not and depends on regulations for protection.

There are other regulation-based quality systems in addition to GMP which are imposed by the government acting on behalf of the consumer; several of these quality systems were mentioned in Chapter 3. For example, The Good Laboratory Practices (GLPs) that govern laboratories using animals to test the safety of drug products is another quality system. The Environmental Protection Agency has a quality system, also called Good Laboratory Practices, that guides the testing of agrochemical products. **The Clinical Laboratory Improvement Amendments of 1988 (CLIA)** *is a system of regulations intended to ensure the quality of results generated by laboratories that perform tests on human specimens.* The CLIA regulations play an important role in ensuring that medical test results can be trusted to make a decision about how a patient should be treated.

C. Quality Systems in Companies That Comply with Voluntary Standards

Not all biotechnology companies make products whose quality is regulated by the government. For example, many biotechnology companies produce enzymes and other molecular biology products that are used in research laboratories. These products are not food or medical products and are not used in humans; therefore, the quality of these materials is not regulated by a government agency. There are, however, **quality standards** that may apply to the production of such products. A **standard,** *broadly defined, is any concept, method, or way of doing things that is established by some authority, by custom, or by general agreement (Merriam-Webster Collegiate Dictionary,* G. & C. Merriam Co., MA, 1977). Quality standards are established by various organizations, agencies, and other entities. Like regulations, quality standards are written down and are followed consistently in different workplaces. Unlike regulations, standards are not imposed by the government and compliance with standards is usually voluntary.

There are many types of standards and many organizations that develop and disseminate standards. Standards may be objective, as, for example, standards that relate to practices in making measurements and performing chemical analyses. Standards may also be subjective, as, for example, those relating to the flavor and other characteristics of foods. Standards can be very useful because they assemble the thinking of many experts into a single, concise, available document. Standards help to ensure consistency in practices among individuals and nations.

Companies and organizations voluntarily comply with standards to improve the quality of their products and to be more competitive in the international marketplace. For example, a chocolate chip cookie baker might voluntarily join an association, such as the (fictitious) "Association of Chocolate Chip Cookie Bakers." The association might have standards with which all members comply, such as a requirement that members use only real chocolate chips (as contrasted with artificially flavored ones).

Some standards are narrow in focus (e.g., there are standards that detail how to correctly place the markings on a laboratory flask). There are also broad systems of standards that are intended to ensure the quality of a final product. These broad quality systems are much like GMP or CLIA in their scope and philosophy. **ISO 9000** *is a series of quality standards published by the* **International Organization for Standardization (ISO).** Companies, including many biotechnology companies, comply with the requirements of ISO 9000 to improve the quality of their products, to make their processes more cost-effective, to demonstrate to potential customers that their products are well-made, and to increase the profitability of their company. ISO 9000 standards are general and so are applicable to any company that makes a product. ISO 9000 standards can also be applied to companies that provide a service, for example, disposing of hazardous waste or repairing equipment.

There is no government agency that enforces and monitors compliance with ISO 9000. If a company or organization decides to voluntarily comply with ISO 9000, then they develop and implement their own quality plan. This plan is formalized and documented in the form of a quality manual. The company may then hire a certified auditor to evaluate whether they are meeting their commitments based on their own plan. If the auditor determines that the company is in compliance with its plan and if the plan includes all the required parts of a quality program, then the company achieves ISO 9000 certification. The company must periodically hire an auditor to conduct inspections to assure that the company remains in compliance with its plan if the company wants to remain certified. Although being ISO 9000 certified does not actually guarantee that a company is producing a high-quality product, it does show that the company has systems in place that support quality.

Table 4.1. *INFORMATION ABOUT THE ISO 9000 SERIES OF STANDARDS*

1. *The ISO 9000 quality standards were first issued in 1987 by the International Organization for Standardization, based in Geneva, Switzerland.* ISO is a worldwide federation of national standards groups from 157 countries.

2. *ISO 9000 standards address quality management and promote international trade and cooperation.* The standards aim to:
    Enhance the quality of goods and services
    Promote standardization of goods and services internationally
    Promote safety practices and environmental protection
    Assure compatibility between goods and services from various nations
    Increase efficiency and decrease costs

3. *ISO 9000 is actually a family of standards with three documents:*
    ISO 9000:2005, Quality management systems—Fundamentals and vocabulary *describes fundamentals of quality management systems, which form the subject of the ISO 9000 family, and defines related terms.*
    ISO 9001, Quality management systems—Requirements *provides a number of requirements that an organization needs to fulfill if it is to achieve consistent products and services that meet customer expectations.*
    ISO 9004, Quality management systems—Guidelines for performance improvements *provides guidelines beyond the requirements in ISO 9001 in order to consider both the effectiveness and efficiency of a quality management system, and consequently the potential for improvement of the performance of an organization.*

4. *ISO 9000 is based on eight quality management principles:*
    Principle 1 Customer focus *states that organizations depend on their customers and therefore should understand current and future customer needs, should meet customer requirements, and strive to exceed customer expectations.*
    Principle 2 Leadership *states that leaders establish unity of purpose and direction of the organization and they should create and maintain the internal environment in which people can become fully involved in achieving the organization's objectives.*
    Principle 3 Involvement of people *states that people at all levels are the essence of an organization and their full involvement enables their abilities to be used for the organization's benefit.*
    Principle 4 Process approach *states that a desired result is achieved more efficiently when activities and related resources are managed as a process.*
    Principle 5 System approach to management *states that identifying, understanding, and managing interrelated processes as a system contributes to the organization's effectiveness and efficiency in achieving its objectives.*
    Principle 6 Continual improvement *states that continual improvement of the organization's overall performance should be a permanent objective of the organization.*
    Principle 7 Factual approach to decision making *states that effective decisions are based on the analysis of data and information.*
    Principle 8 Mutually beneficial supplier relationships *states that an organization and its suppliers are interdependent and a mutually beneficial relationship enhances the ability of both to create value.*

Although ISO 9000 and GMP both have sections that are applicable to obtaining quality results in a laboratory, they focus primarily on manufacturing facilities. **ISO 17025** *is a quality system that is similar to ISO 9000, but focuses on issues specific to laboratories.*

Further information about ISO 9000 is summarized in Table 4.1 and ISO 9000 is contrasted with GMP in Table 4.2 on p. 72.

c. The numbers of peanut chunks in peanut butter

d. The levels of aflatoxins in peanut butter (aflatoxins are potent carcinogens formed by certain molds)

e. The electrical wiring of a food blender

f. The color of a food blender

g. The activity of a DNA cutting enzyme

h. The activity of a substance used to treat asthma

EXAMPLE PROBLEM

Which of the following is likely to be subject to regulatory oversight; which is likely to be the subject of standards; which is likely to be neither?

a. The smoothness of peanut butter

b. The microbial content of peanut butter

ANSWER

a. and c. The smoothness and chunkiness of peanut butter are subjective features and are addressed in a standard (from USDA).

b. and d. Microbial and aflatoxin contamination relate to food safety and therefore are subject to regulatory oversight by FDA.

e. The wiring of devices is covered by electrical codes (regulations).

f. Color is not subject to regulations or standards.

g. Enzymes used in research or teaching laboratories are not subject to regulations or standards. (A company that makes enzymes, however, is likely to voluntarily conform to a quality system like ISO 9000 or to its own internal quality system.)

h. Substances used in the treatment of illness are subject to FDA regulation.

D. Product Quality Systems and the Individual

Depending on where an individual is employed, one or another of these quality systems will impact their everyday work. For example, the work of a production operator in a biopharmaceutical laboratory is stringently monitored and controlled in order to assure the safety and efficacy of the products she helps to produce. This operator will have to follow written procedures meticulously and will have to record each activity in a specific way. An FDA inspector can come in at any time and review her records. A technician working in a testing laboratory may devote considerable attention to recording and tracking incoming samples. He will verify and document that analytical instruments are functioning properly. His activities are directed at ensuring consistent, reliable test results. A research assistant working in an academic research laboratory is not bound by the GMP regulations, nor by ISO 9000, and in fact may not even know they exist. The researcher, however, must diligently record all her work in a laboratory notebook and must be prepared to explain it and justify it to colleagues.

Table 4.2. DIFFERENCES BETWEEN THE *ISO 9000* SERIES OF STANDARDS AND GOOD MANUFACTURING PRACTICES REGULATIONS

ISO 9000	GMP
Compliance is voluntary	Compliance is required by law for companies making regulated products
Compliance is monitored by auditors who are paid by the company. The company complies voluntarily with the auditors' suggestions to improve their product quality.	GMP regulations are enforced by FDA inspectors who have enforcement authority
Standards are generic and can be applied to any manufacturing or service industry	GMP regulations are specific to the pharmaceutical/medical products industry
Requires that companies write a quality manual which provides an overview of their entire quality system	Quality manual is not required
Originated in Europe	Originated in the United States

Thus, whether we are talking about a production operator working to make safe products in a biopharmaceutical company or a research scientist striving to produce "good science," there are quality systems to help achieve the best possible product.

Table 4.3 summarizes quality systems in various workplaces.

Table 4.3. PRODUCT QUALITY SYSTEMS IN DIFFERENT TYPES OF WORK ENVIRONMENTS

Basic Biological Research Laboratories in Academic or Government Settings

In these laboratories scientists investigate fundamental problems in biology. Quality is discussed in terms of "doing good science." The quality of basic research is primarily monitored through review by peers, for example, when a paper summarizing research work is submitted for publication or a grant application is submitted to a granting agency.

Research and Development (R&D) Laboratories Associated with an Industry

In R&D laboratories, researchers investigate questions that are intended to result in commercial products. The regulations and standards that affect individuals in such a laboratory will depend on the nature of their industry. In companies that make drugs and other medical products, the FDA may scrutinize laboratory notebooks and other R&D records. Some companies voluntarily comply with ISO 9000 and/or with ISO 17025, both of which have provisions relating to laboratory work.

Production Facilities

Production facilities must maintain tight control over their processes and systems to ensure that their products are consistent and properly made. In regulated companies and in those certified by ISO or other certification bodies, external government inspectors or auditors will conduct periodic inspections. Individuals who work in production generally follow specific written procedures and are required to record in writing that they did so properly.

Testing Laboratories (Such as Analytical, Quality-Control, Forensic, Microbiology, Metrology, and Clinical Testing Laboratories)

The regulations and standards that apply to testing laboratories vary. Quality-control laboratories in companies that conform to GMP regulations or ISO 9000 standards have strict requirements that they must meet and are accountable to external inspectors or auditors. Clinical laboratory personnel must similarly adhere to quality regulations. Some laboratories voluntarily adhere to quality standards in order to demonstrate to their clients that their test results are trustworthy. Still other testing laboratories may institute their own quality systems to ensure that their work is performed correctly.

E. How Quality Documents Are Written

i. LANGUAGE

Excerpts from the GMP and GLP regulations and from ISO documents are sprinkled through this unit. As you read them, observe that these documents do not provide much detail on implementation. Because every company, laboratory, and organization is different, quality regulations and standards must be written in a general fashion. For example, the GMP regulations contain the statement:

> **21CFR211.63** *Equipment used in the manufacture, processing, packing, or holding of a drug product shall be of appropriate design, adequate size, and suitably located to facilitate operations for its intended use . . .*

This statement does not specify the particular equipment of concern. The terms *appropriate, adequate,* and *suitably* are not clarified. This statement provides no concrete guidance as to how a company might comply with it. In the case of GMP, such generic regulatory statements are interpreted and enforced by the Food and Drug Administration (FDA). FDA publishes their interpretation of the requirements of GMPs in the form of guidance documents called "Guidelines" and "Points to Consider." Guidance documents are not laws, and are intended to help companies apply the general principles of GMP to their own situation. Each company is different and each GMP-regulated company must ultimately develop its own practices and procedures to implement the quality requirements of GMP. The ISO 9000 standards are also written in a generic fashion in order to apply to any product or service. Organizations implementing ISO 9000 standards, therefore, rely on auditors or consultants to provide guidance on how to devise a quality system for their own situation.

ii. COMMON ELEMENTS

Although various quality systems differ from one another (as was shown, for example, in Table 4.2) there are certain elements they tend to have in common. Of these common elements, documentation is probably the most important. **Documentation** *consists of written records that guide activities and substantiate and prove what occurred.* For example, written procedures guide the work of individuals, thus ensuring consistency throughout a facility. Written records show what was done, by whom, and when, thus providing accountability. There is a common saying about the importance of documentation:

> *"If it isn't written down, it wasn't done."*

This saying emphasizes the importance of keeping good records and is applicable in any biotechnology workplace, whether it is a laboratory or production facility. Another common saying is:

> *"Do what you say and say what you do."*

Documentation is part of the job of every bench scientist, technician, and operator in every biotechnology workplace. Because this element of quality is so important, we will devote all of Chapter 6 to it.

Another common element relates to resources. Every company, laboratory, and facility needs resources to produce quality products. If any of these resources is missing or is unsuitable for its purpose, then the final product is likely to be inadequate.

Skilled personnel are one of the most important resources in any company or organization. The employer is responsible for ensuring that all employees have the education and training needed to perform consistently under ideal and unusual circumstances, employees are familiar with all quality requirements pertinent to their work, they are well-supervised, and they receive the information, equipment, and tools to do their jobs properly. In a quality environment, there are usually records for each individual to show their qualifications and to keep track of their ongoing training, education, and acquisition of skills. In almost all companies, employees participate in some sort of safety training so that they know how to deal effectively with hazards.

Employees are responsible for the accuracy and completeness of their work. Employees are required to follow instructions, document their work, observe problems and report them as appropriate, understand the impact and consequences of their actions, and undergo continuous training. The language with which GLP regulations describe personnel is:

> **21CFR58.29** *Each individual engaged in the conduct of or responsible for the supervision of a nonclinical laboratory study shall have education, training, and experience, or combination thereof, to enable that individual to perform the assigned functions . . .*

Facilities, equipment, instruments, and raw materials are other resources that are necessary to make a product. There are various types of biotechnology facilities, including laboratories where research and development occur, laboratories where quality-control testing is performed, greenhouses, animal facilities, fermentation plants, and purification facilities. To produce a quality product, each area must efficiently accommodate the activities that occur there, be suitably heated, cooled, and otherwise have an appropriate environment, and so on. Within the facility there must be properly maintained equipment, instruments, and materials.

III. MANAGING CHANGE, VARIABILITY, AND PROBLEMS

A. Controlling Change

Change is an important topic when discussing product quality systems. In a research laboratory, progress depends on change. Ideas are tested, the results are examined, and the system is changed and tested again. Even though each change should be recorded and its effects evaluated, change is part of the normal, daily routine in a research environment.

In a manufacturing environment, once a product is proven safe, effective, reliable, or otherwise acceptable, the most important feature of its production is consistency. Change, which is the essence of research, can be difficult and complex in a manufacturing environment. Although change is complex in a manufacturing facility, it is inevitable. Raw materials may become unavailable, equipment becomes outdated, computer methods become more powerful, regulatory requirements change, and so on. There is conflict between the need to respond to changing circumstances and the need for control over processes, materials, and documentation. Change management is therefore one of the most important and most difficult aspects of quality systems in a manufacturing environment, particularly where regulated products are made.

The basis of effectively controlling change is that changes are reviewed, evaluated, and approved before and after they are made. Changes require detailed assessment and evaluation. This is because there is the potential that a simple, seemingly unimportant change might alter the way a process runs or the characteristics of a finished product. In regulated companies some changes require approval by the proper regulatory agency.

The need for control of change requires a way of thinking that is alien to most researchers who are continuously looking for new and better ways to accomplish a task. The tension between the need for flexibility during research and the need to control change is very evident when a biotechnology company is making the transition between being involved primarily in research and development to production. During this transition, companies must institute efficient, thorough procedures to manage change, and they must educate personnel in how to use the procedures.

Some points regarding the control of planned changes are shown in Table 4.4.

EXAMPLE PROBLEM

A new quality-control technician is hired in a biopharmaceutical company. She is assigned the task of routinely checking samples of the company's product for the presence of the deadly (fictitious) pathogen, *Badbeastea miserabilis*. The test consists of applying I mL of each sample to a petri dish containing special nutrient medium, allowing the plate to incubate at a specific temperature for two days, and then checking the plate for the characteristic colonies of *B. miserabilis*. The technician (who is eager to make a positive contribution in her new job) reads about a new nutrient medium that supports the growth of *B. miserabilis* better than the old medium, is less costly, and allows the assay to be performed in only one day. Elated, the technician orders the new nutrient medium and begins checking samples of product with the new medium as soon as it arrives. She then tells her supervisor how well this new medium has worked. What do you think about this (imaginary) scenario?

Table 4.4. MANAGING CHANGE

1. *Each laboratory, facility, or organization should have a procedure for making changes.* In a nonresearch environment, this usually involves having an SOP that outlines the process for making a change.
2. *Change should always be justified.*
3. *In a manufacturing facility, proposed changes should be evaluated and preapproved by R&D.*
4. *A technical review of the proposed change should be performed to assess its value, and to address risks associated with the change.*
5. *In a regulated company, the QC/QA unit must review the change to see if it requires approval by regulatory agencies.*
6. *If necessary, the change should be evaluated for financial implications.*
7. *After the change is made its effects should be investigated and documented.*
8. *In a GMP-compliant company, all changes need to be validated (as discussed in Chapter 7) to ensure that the quality of the product or result is unaffected.*

DISCUSSION

Even though the enthusiasm of this new technician is laudable, her new job has not begun auspiciously. The issue in this scenario is change and how change is controlled. The assay for the *B. miserabilis* would have been developed and tested by the R&D unit. During that testing period, R&D scientists would have optimized the assay using the original nutrient medium, checked for potential problems, and written documentation for the assay. Switching to a new nutrient medium could result in an unforeseen difference in the results of the test. For example, certain strains of the pathogen might not grow on the new medium. Before switching to a new nutrient medium, there would need to be testing to see that results with the original and new media are comparable. These tests might require testing samples on both media side by side to see if the results were the same. Changing the nutrient medium would also require that proper forms be completed, that approval from a supervisor(s) and from the Quality Assurance Unit be obtained, and that new directions for performing the test be written. The new technician failed to complete the proper steps.

Note that the new technician made a serious mistake that suggests that her introduction to her job and the way in which she was supervised were inadequate. The company should therefore look seriously at its methods of training and supervising its employees.

Note also that with the appropriate tests and controls, the company may decide to switch to the new medium because of its advantages. The mistake the technician made was to prematurely use the new medium in QC of products, thereby putting a production run at risk.

EXAMPLE PROBLEM

Suppose exactly the same events transpire as in the previous example problem. This time, however, the setting is a research laboratory in a university. What do you think about the scenario now?

DISCUSSION

The technical issues are the same regardless of the type of laboratory. It is important to ascertain that the new nutrient medium provides results that are comparable to the original medium. Testing the new nutrient medium would be well-advised. The difference in a basic research laboratory, as compared with the QC laboratory, is the process for making changes. In a research laboratory there is no Quality Assurance Unit to approve this change. The technician may be able to rely on his own judgment or may need to only consult his direct supervisor. He will need to document the testing he performs of the new nutrient medium in his laboratory notebook. If he changes to the new medium, then that change must be recorded in the notebook and in the written procedure for the assay. He will not, however, need to complete change control forms.

CASE STUDY

Example Relating to Uncontrolled Change

A biotechnology company that makes products for research use was producing an enzyme for use in recombinant DNA procedures. Several customers observed lot-to-lot variation in the performance of the enzyme. The company was concerned about these customer comments and began an investigation of the product. The production records showed that there had been a large increase in demand for the enzyme in recent months and so the company had combined scale-up with production. During this period, production scientists had made substitutions and changes without review. This made it very difficult to identify specific sources of the lot-to-lot variability. Scientists performed a number of analyses on the enzyme. They eventually observed that the enzyme's storage solution, which contained Mg^{++}, was losing the Mg^{++} over time. Mg^{++} is a cofactor required by the enzyme for activity. The loss of magnesium was eventually traced to a change in the plastic tubes in which the enzyme was stored. The tube manufacturer had begun adding an antioxidant to the plastic and the antioxidant reduced the magnesium level in the enzyme solution. In this case, a seemingly trivial change in a raw material caused a problem. Finding the cause of the problem required a costly investigation. Careful control of change aims at preventing this type of predicament, or making it easier to trace the cause of difficulties should they occur.

B. Reducing Variability; Controlling Processes

"Quality and productivity improvement share a common element—reduction in variability through process understanding (e.g., application of knowledge throughout the product life-cycle)."

(U.S. Food and Drug Administration. "Pharmaceutical cGMPs for the 21st Century—A Risk-Based Approach, Final Report." September 2004.)

In the context of product quality, **variability** *is when some characteristic of a product fluctuates.* Product variability is undesirable. Consider, for example, a drug product; it is imperative that its potency is known and does not vary from dose to dose. All quality systems aim to reduce variability in products.

There are multiple sources of variability. Perhaps one person performs a task differently than others do. Perhaps a piece of equipment responds to temperature changes in a facility. Perhaps an incoming raw material is different than in previous shipments. Each source of variability contributes to the overall variability in the final product, see Figure 4.1 on p. 76.

To protect product quality, people attempt to identify and understand all the sources of variability, and to eliminate them whenever possible. The more completely a process is understood, the more the sources of variability can be controlled. A company, for example, might identify the qualities that raw materials must have in order to make a consistent product. The company then would put in place a system to test all incoming raw materials to be sure they have those qualities. In a **controlled process,** *the sources of variability are understood, and, as much as possible, are eliminated.*

Reducing variability is obviously important when tangible products are made, such as drugs, but the principle is also relevant in laboratories. Suppose, for example, a researcher is trying to investigate changes in gene expression at various stages of development. If the researcher uses reagents that are not made consistently, then none of her experimental data will be reliable. If she sees variation in gene expression it might be related to the stage of development of her subjects—but it also might be attributable to variation in the reagents.

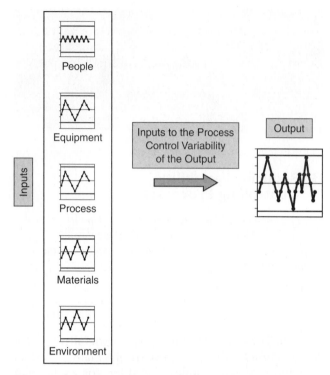

Figure 4.1. Sources of Variability. Variability in how people perform tasks, equipment, materials, processes, and environment (e.g., facility temperature) all contribute to the overall variability of a process. Understanding all of these inputs is necessary to reduce variability. (Illustration concept courtesy of Michael Fino.)

C. When Something Goes Wrong; CAPA

Things go wrong—even in the best-run facilities. Every quality system must therefore include a mechanism to deal with problems. In companies that are regulated or audited, inspectors will look to see that problems are investigated, mistakes are corrected, and a system is in place to prevent recurrences.

The regulations that govern companies making medical devices use the term **CAPA,** *Corrective and Preventive Actions, to refer to the processes by which a company responds to problems and failures.* **Corrective action** *means to fix problems that have already occurred and may happen again.* **Preventive action** *involves looking for problems that have not yet occurred and preventing them.* This term is now commonly used in many companies, not only those that make medical devices, to describe the process by which the company deals with problems.

A general strategy to deal with problems is to:

1. Describe the problem and assess the level of risk it poses.

2. Take short-term action as required to prevent further problems or correct the error.

3. Conduct an investigation to determine the root cause.

4. Develop and implement a long-term solution to prevent recurrence of the problem.

5. Follow up to ensure that the solution was properly implemented and was effective.

Part of the process of evaluating problems is evaluating the risk they pose. In a facility that makes medical products or performs tests of patient samples, a problem might pose the serious risk that patients will be harmed. The more severe the risk posed by the problem, the more rapidly an investigation must occur, and the more resources that should go into fixing the problem. In a research laboratory, problems that are likely to seriously compromise the experimental results of the laboratory are dealt with aggressively.

Root causes *are the "real" or underlying cause(s) of a problem.* If the root cause is not corrected, then the problem might happen again. There may be one or more than one root cause for a problem, and the root cause(s) may be hidden. A root cause might involve, for example, a human error, a malfunctioning instrument, a method or a process that is not effective, a fault in a material, or a problem in the environment (e.g., air, water supply, dirty surfaces). The investigation to find the root cause(s) might include looking at historical data to see if the problem is recurring, reviewing written records from the batch or test affected, and interviewing associated personnel. Once the root cause of a problem is found, it must be fixed. Other potential problems might be uncovered during an investigation that should also be fixed; this is preventive action.

Problem solving in a research laboratory may use an informal approach. Perhaps only the individual affected will investigate the problem, perhaps a few laboratory colleagues will brainstorm and try to identify and fix the problem. In a company an investigation might involve a team including members from QA, QC, production, and R&D. The conduct of the investigation, its results, and actions taken will need to be documented.

The following case studies relate to CAPA systems. The first Case Study, "Analyst Errors," is an example of CAPA analysis in a quality-control laboratory. To understand this case study, recall that quality-control analysts perform tests of final product to see if the products are of suitable quality to release to customers. An **out-of-specification (OOS) result** *is one in which a product fails to meet its requirements.* The cause of an OOS result may be a poorly made product—but it may also be due to an error made in laboratory testing. An OOS result that is invalidated is presumably due to laboratory error. The laboratory in this study is having problems with OOS results that are due to errors made in the laboratory. (This case study is further discussed in Practice Problem 4 at the end of this chapter.)

The second Case Study, "Warning Letter from the FDA Relating to CAPA Violations," relates to a laboratory that was visited by an FDA inspector and was found to have an inadequate CAPA system that was not in compliance with the FDA's requirements. (This case study is further discussed in Practice Problem 5 at the end of this chapter.)

Overincubated ELISA plates	Retrained analyst and updated SOP
Dilution error	Retrained analyst
Expired standard was used	Retrained analyst

This case study is reprinted verbatim by permission from "Navigating CAPA," Paula J. Shadle. BioProcess International. October, 2004, 16.

CASE STUDY

Analyst Errors

As part of efficiency analysis, several QC work units were asked to report metrics on the frequency of out-of-specification (OOS) results and the root causes found. One work unit reported a high rate of OOS results that were not confirmed and resulted in invalidation of test data. Because 70% of the observed OOS results were invalidated, the laboratory performed a great deal of retesting and investigation and its capacity plummeted while cycle time became unpredictable.

The table lists brief summaries of the causes ascribed to several laboratory failures. The pattern was initially missed because it did not correlate with a single analyst, a shift, a test method, or a production sample. When the work unit was compared with others in QC, the invalidation rate stood out, and the cause lists clearly indicated a single problem: failure to follow the procedures exactly. What then were the root causes? Several hypotheses were examined:

- *Analysts not trained*
- *Procedures written poorly*
- *Lack of supervision*
- *Lack of resources causing analysts to rush*

The corrective action taken after each event—retrain the analyst on the specific method—clearly wasn't changing the overall metric. What did affect the OOS rate was additional training and a QA person in plant to support the supervisor, who was spread too thin in overseeing a large number of employees. Many analysts had less than six months on the job and were in various stages of training. Some attitude adjustment was needed to convince the staff that innovation was not acceptable, and group retraining was found to be more effective in changing work habits.

REASONS FOR INVALIDATION OF OOS QC RESULTS

Cause Listed	Corrective Action
Incubation done at 35°C instead of 37°C	Retrained analyst
ELISA plate stored at 4°C for 30 minutes before reading	Retrained analyst
Four instead of five replicates were tested	Retrained analyst
Step four performed before step three	Retrained analyst
Test method run outside of validated range	Retrained analyst

CASE STUDY

Warning Letter from the FDA Relating to CAPA Violations

The Food and Drug Administration has inspectors who periodically inspect companies that make medical products. If the inspectors observe violations, they note them on forms, called "483s," and in official warning letters sent to the company. These letters are posted on the FDA's website. Excerpts from a real warning letter are reprinted below. These excerpts relate to the company's failures to conduct thorough CAPA investigations, initiate corrective actions, and institute a preventive action plan when employees made serious mistakes. As you read this warning letter, consider how the company might improve its CAPA program.

Warning Letter

Dear Mr. T . . . :

We are writing to you because on March 28 through May 20, 2005, the Food and Drug Administration (FDA) conducted an inspection of your . . . facility which revealed serious regulatory problems involving your medical devices, including the implantable . . . . Infusion Ports, . . . drug eluting stents, and . . . balloon dilatation catheters . . . [A list of 6 violations follows. Only those violations relating to CAPA are excerpted here.]

4. Failure to document all activities performed in regards to your corrective and preventive activities, including the investigations of causes of nonconformities, and the actions needed to correct or prevent recurrence of nonconforming product and other quality problems, as required by 21 CFR 820.100(b). During our inspection, we reviewed several . . . CAPAs [investigation reports]. This review indicated that your CAPAs fail to include all the necessary information to describe the incident and/or the nonconforming condition.

For example, [report] CAR-05-004 . . . involved the shipment of . . . units of failed . . . [products] to 5 separate hospitals. This CAPA only states, ". . . product part number H7493897012250, Batch . . . was removed by an operator from a QA quarantined location for shipping. The skid containing these units was labeled 'Pending KDR Test Results' and was also 'S' blocked in SAP. A second operator performed an SAP transaction removing the 'S' block status. This resulted in the units to ship to customers." The CAPA did not include the dates of these serious occurrences, the employees involved, or the number of instances that product was actually either removed from quarantine or overridden in the computer system (SAP). The CAPA also did not list the number of units that were actually shipped, or the number of hospitals that actually received nonconforming product. We learned through interviews with employees, that there were actually 5 separate removal actions of . . . product from the quarantine area. On January 12, 2005 . . . separate batches of . . . product were removed from the quarantine area. These were caught by an employee; however, a CAPA was not generated for the incident. Actual shipment of the . . . units occurred on January 20 and 21, 2005 when it was realized there was a 5th removal of [product] from quarantine. The CAPA for this instance was initiated on January 27, 2005.

A serious event such as the one described above, requires a thorough investigation into the activities that precipitated the actual shipment of adulterated . . . product. Without a thorough investigation into these events, it is difficult to implement an adequate corrective or preventive action that is required by our regulations.

We note in your response letter dated June 20, 2005, that you have supplemented information to the above CAPA and consider the remedial action to be complete and closed. We are concerned that you have not taken adequate action to prevent this type of serious failure from recurring.

5. Failure to establish and maintain an adequate corrective and preventive action procedure which ensures identification of actions needed to correct and prevent the recurrence of nonconforming product and other quality problems, as required by 21 CFR 820.1 00(a)(3). Your CAPA system has failed to identify the necessary actions to correct and prevent the continued distribution of nonconforming product.

For example, we noted that CAPA 04-125 was initiated on November 2, 2004 to address the release of . . . batches of . . . balloon dilatation catheters into finished goods inventory. The only <u>preventive</u> action taken as the result of this CAPA was to "Update procedure S801280-00." This CAPA was closed on February 15, 2005 with a notation that the CAPA plan was effective. This CAPA did not identify any <u>corrective</u> action for the . . . units that were released into finished goods without proper authorization. During our inspection, you had no information or documents to establish that a final release from your . . . manufacturing facility was provided for a final disposition of these . . . batches that had already been distributed to customers . . . .

The specific violations noted in this letter and in the Form FDA-483 issued at the conclusion of the inspection may be symptomatic of serious underlying problems in your establishment's quality system. You are responsible for investigating and determining the causes of the violations identified by the FDA. You also must promptly initiate permanent corrective and preventive action on your Quality System . . . You should know that these serious violations of the law may result in FDA taking regulatory action without further notice to you. These actions include, but are not limited to, seizing your product inventory, obtaining a court injunction against further marketing of the product, or assessing civil money penalties . . .

Sincerely yours,

IV. SUMMARY

This chapter introduced Unit II, Product Quality and Biotechnology, by discussing broad concepts relating to product quality. In every workplace, producing a quality product requires the commitment of everyone who works there, beginning with senior personnel and extending to everyone in the organization.

A quality system is the organizational structure, responsibilities, procedures, processes, and resources that work together to provide a quality product or service. Different biotechnology workplaces produce different types of products, and therefore adhere to somewhat different quality systems. We saw, for example, that academic research scientists strive to "do good science." "Doing good science" is not a formal quality system; it is not spelled out in any single book—although many books discuss it. There is no organization nor any auditors to oversee it. Nonetheless, scientists follow certain practices to help ensure that their research results are of high quality and are trustworthy.

The government steps in to enforce quality practices when the consequences of a poor-quality product are life-threatening, as is the case for food and drugs. For example, the Good Manufacturing Practices, GMPs, are regulations that outline a quality system that pharmaceutical manufacturers are required by law to follow. Other examples of quality systems enforced by the government include the FDA's Good Laboratory Practices, Good Clinical Practices, and quality system regulations, and the EPA's Good Laboratory Practices.

Many companies and organizations are not legally required to comply with quality regulations, but voluntarily adopt quality standards. Companies comply with these voluntary quality systems to improve the quality of their product and therefore, presumably, to be more successful. ISO 9000 is a voluntary quality system that is followed by many organizations around the world.

Although quality systems vary in their details, they share common themes. Methods to reduce variability, control change, and respond to problems are common issues addressed by quality systems. Another commonality is the fundamental role of documentation as a means to guide and record the work of an organization. Documentation is such an important part of a quality system in any biotechnology environment that all of Chapter 6 is devoted to it. Quality systems also address issues relating to resources required to produce a quality product. Chapter 7, which focuses on quality principles and practices in production settings, contains examples of this. Chapter 8 is a short chapter that provides an overview of quality systems in laboratories. This chapter includes some sobering examples of what happens when laboratory workers do not adhere to a quality system. Chapter 8 ends Unit II and provides a transition to the topic of the rest of this textbook—the basic methods used to obtain good quality results in the laboratory.

PRACTICE PROBLEMS

1. Why do companies voluntarily subject themselves to the difficult, time-consuming, and costly process of complying with the ISO 9000 standards?

2. This chapter discussed issues relating to change. A significant change in a biopharmaceutical manufacturing process (for example, a change in culture medium or a change in the genetic construct inside the host cells) can be extremely costly and might have severe consequences. Discuss why change in this environment poses special challenges. Refer to the issues discussed in both Chapters 3 and 4.

3. Personnel who work in research laboratories are not required by law to follow standard procedures for performing most routine tasks (such as preparing laboratory reagents). It is, however, common for research laboratory staff to write and follow standard procedures for various tasks. What advantage is there to having standard procedures in a research environment?

4. Consider the problem and CAPA investigation in the Case Study: "Analyst Errors."

   a. Explain the problem here. As part of your explanation, consider: What is an OOS result? What are the general causes of an OOS result? What do OOS results have to do with QC work units?

   b. Examine the table showing reasons for the erroneous results. What was happening here? What did the laboratory staff do initially to try and fix the problem?

   c. An investigation of the laboratory was performed to look for the root cause. What does "root cause" mean in this situation? What hypotheses were explored in this investigation?

   d. What did the root cause investigation discover? Were any of their hypotheses shown to be true?

   e. The P in CAPA stands for prevention. What did the company do to prevent future problems?

5. Consider the Case Study: "Warning Letter from the FDA Relating to CAPA Violations."

   a. Read carefully the description of events in item number 4, relating to the product number H7493897012250. Explain in your own words what went wrong here and why this is a problem of concern.

   b. The FDA says it is looking for some sort of preventive action. Speculate as to how the company might act to prevent this type of mishap. What would you consider to be an adequate preventive action plan?

   c. What consequences has the company experienced due to these errors? If the company does not adequately respond to the FDA's concerns about its CAPA plan, what might be the consequences in the future?

CHAPTER 5

Biotechnology and the Regulation of Medical and Food Products

I. THE EVOLUTION OF DRUG AND FOOD REGULATION IN THE UNITED STATES

A. Introduction

Previous chapters discussed biotechnology products such as anticancer therapeutics, vaccines, gene therapies, medical diagnostic tools, and genetically modified food crops. Medical products and foods are stringently regulated by the government and these regulations have a profound effect on the biotechnology workplace. This chapter explores issues relating to government regulation of medical and food-related products, the relationship between the government and the biotechnology and pharmaceutical industries, and the history of that relationship. For simplicity, this chapter examines regulation primarily in the United States.

The regulated characteristics of products relate to safety, honesty (as in labeling), and, particularly for medical products, effectiveness and reliability. A drug product, for example, must be correctly labeled, and neither a food or drug product should be contaminated with harmful microorganisms. The use of a mislabeled drug, or consumption of a contaminated food, can cause death. The individual consumer cannot determine whether a drug is properly labeled or whether meat purchased in the supermarket has a suitably low level of microbial contamination. The public therefore looks to government to enact and enforce laws that provide protection from unsafe or ineffective food and drugs.

Although we take for granted that the government regulates the production of drugs and foods, this was not always the case, as is illustrated dramatically in the history of food and drug regulation in the United States. The present system of regulation evolved over the years in response to often tragic events and abuses. Familiarity with these historic events is essential to understanding the regulatory landscape that exists today.

B. The Stories

The quality of drugs and foods in the United States was virtually unregulated until the early 1900s, see Figures 5.1 and 5.2. A major industry emerged at that time to process foods for urban consumers. This bur-geoning food industry was frequently filthy and poorly managed. There were various efforts to regulate food processing; however, no legislation was passed until 1906. At that time, Upton Sinclair published the novel, *The Jungle,* intended to publicize the difficulties facing immigrant workers in the Chicago stockyards. Sinclair vividly described filthy conditions and alarming practices in the food industry. People were so outraged by Sinclair's descriptions that Congress passed a law to help regulate the production of food and, at the same time, drugs. This law was the original **Food, Drug and Cosmetic Act, FDCA,** *which authorized regulations to ensure that manufacturers did not sell adulterated (contaminated) or misleadingly labeled food and drug products. A separate law enforcement agency was formed in 1927 to enforce legislation relating to food and drugs.* It was first known as the Food, Drug, and Insecticide Administration, and then, in 1930, was named the **Food and Drug Administration (FDA).**

The 1906 FDCA failed to deal with the safety or effectiveness of drugs, so in 1933 FDA advisors recommended a complete revision of this inadequate law. A bill was introduced into the Senate, launching a five-year legislative battle between those who wanted drug-law reform and the industry that vigorously fought its passage. During these years, an exhibition of dangerous foods, medicines, medical devices, and cosmetics was prepared to illustrate the shortcomings of the 1906 law. The exhibit showed, for example, an

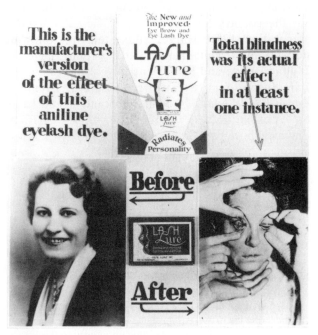

Figure 5.1. Drugs of Dubious Value and Safety Were Widely Sold in the Early 1900s. Many drug products contained alcohol, opium, and morphine, without any labeling to that effect. (Photos courtesy of FDA History Office.)

Figure 5.2. Hazardous Products Persisted Despite the 1906 FDCA. (Photo courtesy of FDA History Office.)

eyelash dye that had blinded women, see Figure 5.2, lotions and creams that caused mercury poisoning, hair dyes that caused lead poisoning, and a weight-loss drug that increased metabolic rate to such an extent that some users died. The FDA was powerless to seize these dangerous products because the 1906 FDCA did not mandate drug safety. First Lady Eleanor Roosevelt borrowed the exhibit and invited congressional wives to the White House to view it. She and the congressional wives were among the most loyal supporters of drug reform throughout the five-year effort.

Eventually a dramatic tragedy facilitated the passage of stronger laws. An antibiotic, sulfanilamide, was introduced in the 1930s. The drug was a major medical advance in the treatment of infectious diseases. However, sulfanilamide is relatively insoluble and therefore was only available in a pill form that was difficult to administer to small children. In 1937, a chemist at the S.E. Massengill Company dissolved a batch of sulfanilamide in the toxic industrial solvent, diethylene glycol, in an effort to obtain a soluble form of the drug. The company marketed the resulting "elixir" resulting in at least 358 poisonings and 107 deaths, mostly of children.

In response to mounting reports of deaths due to the sulfanilamide preparation, the FDA launched the first major recall of a drug product. The agency had the authority to seize the product, not because it caused deaths, but because it was inaccurately labeled an "elixir" when it did not contain alcohol. Most of the FDA's staff of inspectors and chemists, along with many local law enforcement officials, traveled throughout the country looking for the toxic sulfanilamide preparation. James Harvey Young tells the following story about the sulfanilamide recall (Young, James H. "Sulfanilamide and Diethylene Glycol," in *Chemistry and Modern Society: Historical Essays in Honor of Aaron J. Ihde,* edited by John Parascandola and James C. Whorton, 105–25. Washington, D.C.: American Chemical Society, 1983.)

In Atlanta, an elderly FDA inspector, who was also ill, drove through the rain into the north Georgia mountains to a drugstore that could not be reached by phone … A pint bottle of the "elixir" had been sent to the druggist, and the inspector was charged with bringing it back. The druggist had the bottle but four ounces of its red contents were missing, all prescribed by a doctor for one patient, name unknown. When the physician returned from his rounds, he said that he kept no records but believed he had prescribed the medicine for a woman named Lula Rakes. He did not know where she lived, and her surname was not uncommon in the region. The inspector, shouting an inquiry to the deaf druggist, attracted the attention of a bystander who volunteered the guess that Lula lived eight miles over the ridge in Happy Hollow. The inspector drove over the dark mountain road, only to find Lula's home abandoned. A neighbor said that the Rake's family had

moved one valley farther on. Driving onward the inspector at last found the house, and Lula was there. Busy with the moving, she had taken only a few doses of her medicine, but she could not recall what she had done with the bottle. An hour of conversation and exploring finally led to the medicine, with many other articles in a paper sack under the bed. The inspector began his weary way home.

In other, less-fortunate cases, inspectors were not able to find the drug before it was consumed. There were instances of druggists and doctors who falsified their records or lied when asked about the drug. Under the 1906 drug laws, a company had no obligation to prove that their drug was safe, so Massengill was not held responsible for the deaths. The company was, however, fined a small amount for labeling the drug an "elixir."

In response to the sulfanilamide tragedy, drug reform became popular and resistance to regulation was overcome. A revised Food, Drug and Cosmetic Act was passed in 1938 that contained the critical provision requiring that a manufacturer prove the safety of new drugs with animal and clinical studies, thus legislating the overall process of drug testing described in Chapter 3.

In 1941, there were nearly 300 injuries or deaths from sulfathiazole tablets tainted with the sedative phenobarbital. This prompted the FDA to drastically tighten their control over the manufacturing of drugs.

In 1955, a manufacturer did not properly inactivate the virus used to make polio vaccine and 51 people contracted paralytic polio; 10 people died. This incident, and others like it, led to increased FDA factory inspections and increased testing of the safety of products before their release to the public.

One of the great successes of the FDA occurred in the early 1960s. At that time, the drug thalidomide, produced by a German firm, was commonly used in Europe for insomnia and for nausea in pregnant women. The German manufacturer wanted to license and distribute thalidomide in the lucrative U.S. market. The American pharmaceutical firm, William S. Merrell, submitted a New Drug Application (NDA) to the FDA seeking approval of thalidomide in the United States. This application was assigned for review to the agency's newest medical officer, Dr. Frances Oldham Kelsey, who had joined the staff earlier that year. Dr. Kelsey did not think the NDA provided sufficient evidence of the drug's safety and refused to approve the application, thus blocking the legal entry of the drug into the United States, see Figure 5.3. Thalidomide was later found to cause profound birth defects; thousands of malformed children were born in Europe, many of them with no arms or legs. News of the thalidomide tragedy allowed two legislators, Kefauver and Harris, to push stricter drug regulations through Congress that included, for example, provisions controlling testing of drugs in humans.

Figure 5.3. Kelsey Receives Award for Protecting U.S. Consumers from Thalidomide. Kelsey received the President's Distinguished Federal Civilian Service Award in 1962 from President John F. Kennedy, the highest civilian honor available to government employees. (Photo courtesy of FDA History Office.)

In 1963, the first set of Good Manufacturing Practice (GMP) regulations were published to detail how manufacturers should produce safe and effective drugs. A number of injuries and deaths in the 1960s and 1970s caused by contaminated products led to revised GMPs in 1978. These regulations included requirements for standard operating procedures, validated systems (validation is discussed in Chapter 7), and extensive documentation. Other medical products in addition to drugs, such as heart pacemakers, were also included in these regulations.

In the 1970s, inspections of pharmaceutical animal testing laboratories revealed that toxicology studies supporting new drug applications were poorly conceived and improperly conducted. These inadequacies led to the Good Laboratory Practice* regulations of 1976 whose goal is to assure the quality of data submitted to FDA in support of the safety of new products.

In the 1970s, a number of women using the Dalkon Shield, a contraceptive device, were seriously injured. This led to the passage of the Medical Device Amendments in 1976 that strengthened the regulation of medical devices.

*Good Laboratory Practices (GLP), when used in reference to pharmaceuticals, relate to preclinical testing, as was discussed in Chapter 3. The Environmental Protection Agency, EPA, also mandates good laboratory practices. EPA's Good Laboratory Practices guide investigators who are studying the health and environmental effects of agrochemicals. The GLPs from the FDA and the EPA are not identical, but they do serve the same overall purpose of ensuring trustworthy test results. The term good laboratory practices (GLP) is sometimes used much more broadly to refer to good practices and procedures in any type of laboratory.

In 1989, a scandal relating to the manufacture of generic drugs rocked the industry. FDA investigators discovered that some generic drug manufacturers were guilty of illegal practices, fraud, and noncompliance with GMPs leading to defective drug products. These deficiencies led to increased inspections and FDA oversight of the generics industry.

C. The History of Biotechnology Product Regulation

As we saw in Unit I, new biotechnologies have, in the past few decades, dramatically reshaped the modern drug and medical product industries. In 1982, insulin produced by recombinant DNA methods became the first modern biotechnology product to be approved for market and sale. In 1986, the first monoclonal antibody product was approved, and in 1987 the first recombinant DNA product produced in a mammalian cell culture line was approved. Deciding how to regulate these new biotechnology drugs was a challenge to the regulatory system. As we saw in Chapter 1, many people, scientists and the public alike, saw the immense promise of biopharmaceutical products, but were concerned that they might pose unknown risks. This is a common theme with most new technologies; their potential benefits must be balanced against their potential risks.

Biopharmaceuticals like insulin are made using a novel method, but the products themselves are very similar to "traditional" products. For example, prior to 1982 insulin was extracted from animal organs. Genetic engineering allowed companies to produce insulin in genetically modified cells. The product is insulin in both cases, but the manufacturing process is different. The FDA had a regulatory system in place by the 1970s for regulating products isolated from animal tissue. Regulators therefore had to decide whether biotechnology products should be regulated differently than other products (because their method of production is novel) or the same as traditional products (because the products themselves are often similar). This is sometimes phrased as the *product versus process* controversy. At the heart of this debate was uncertainty as to whether the nature of biotechnology processes makes products derived by these methods inherently different from, and potentially more risky than, traditional products.

In the early 1970s, the use of immortal mammalian cell lines in pharmaceutical manufacturing was forbidden. This was because immortal cell lines are similar to cancer cells in that they are not subject to normal controls on growth and they repeatedly divide. It was feared that the use of immortal cells lines might introduce a "factor" into the products that would be carcinogenic in patients. Furthermore, it was feared that mammalian cells might harbor viruses pathogenic to humans that could contaminate products. By the mid-1970s, it became clear that immortal cell lines would be necessary for producing some biotechnology products that could not be made properly by bacteria. It was therefore imperative that the FDA resolve the issue of whether products made by new biotechnology methods were inherently more risky than traditional products, and whether they would require additional or different regulation.

A number of scientific advances in the 1970s led to the acceptance of products made using modern biotechnology methods. Experiments were performed that demonstrated that genetically manipulated organisms could be handled safely. Effective methods of inactivating viruses were developed. The use of continuous cell lines was first tested in other countries without adverse effects, and studies of cancer did not reveal a carcinogenic "factor" in continuous cell lines.

In 1974, the **National Institutes of Health (NIH),** *a federal agency responsible for funding and overseeing research,* assembled an expert group to explore the safety of recombinant DNA technology. This committee, **The Recombinant DNA Advisory Committee (RAC),** reviewed and interpreted what was then known about the safety and risk of genetic manipulation methods. The result of their work was a set of guidelines for conducting experiments involving recombinant DNA. These guidelines, called **Guidelines for Research Involving Recombinant DNA Molecules (NIH Guidelines)** were first published in 1976, and they have since been revised several times. The NIH Guidelines cover such topics as methods of assessing the risk of a recombinant DNA experiment, methods of classifying organisms based on their risk, and methods of containing organisms so that they cannot escape and harm workers or reach the environment. Recent revisions of the Guidelines also discuss methods of working on a large scale with recombinant organisms (e.g., in a production facility), performing human gene therapy, and working with plants and animals. These revised guidelines outline basic practices that are applied to work with recombinant DNA in research and in the pharmaceutical industry in the United States, and sometimes in other countries. The FDA enforces the practices established in the NIH Guidelines when appropriate. (Safety issues related to recombinant DNA research are also discussed in Chapter 12.)

As a result of these developments throughout the 1970s, the FDA decided that its existing regulatory framework for complex molecular products could be extended to include products made by recombinant DNA methods. This means that (1) no additional laws or regulations were created to regulate biotechnology-derived products, (2) biotechnology companies making products for medical use must comply with cGMPs, GLPs, and GCPs, and (3) biotechnology products must undergo the same approval processes as other drug products.

D. International Harmonization Efforts

An important worldwide trend has been the increasing internationalization of commerce. As companies extend their sales to markets outside their native country, they sometimes encounter difficulties meeting the varied regulatory requirements of different nations. There have therefore been important efforts to internationalize both standards and regulations. The development of the ISO 9000 series of standards is an example; ISO 9000 standards are intended to increase consistency internationally in product quality and in production practices.

The medical products industry is stringently regulated in most countries and medical products have therefore been a focus of harmonization efforts. One facet of this effort has been in the field of medical devices, such as pacemakers, medical instruments, and diagnostic kits. In 1996, the FDA announced modifications of the cGMP requirements for medical devices to make them more consistent with the ISO 9000 standards (*Federal Register,* 61 no. 195 (1996): 52601–62.)

An important component of the global harmonization effort is the work of **The International Conference on Harmonisation of Technical Requirements for Registration of Pharmaceuticals for Human Use (ICH). ICH** *is an organization that brings together the regulatory authorities of the European Union, Japan, and the United States and experts from the pharmaceutical industry in the three regions to discuss scientific and technical aspects of pharmaceutical product regulation.* The purpose of the organization, according their website (http://ich.org), is "to make recommendations on ways to achieve greater harmonisation in the interpretation and application of technical guidelines and requirements for product registration in order to reduce or obviate the need to duplicate the testing carried out during the research and development of new medicines." Harmonization facilitates the mutual acceptance of data by regulatory bodies internationally, leading to "a more economical use of human, animal and material resources, and the elimination of unnecessary delay in the global development and availability of new medicines whilst maintaining safeguards on quality, safety and efficacy, and regulatory obligations to protect public health."

ICH has established a number of technical guidelines, many of which are relevant to biotechnology companies that make biomedical products. The 1996 ICH E6 guidance document on Good Clinical Practices is widely used as a guide to performing clinical trials. Other important ICH guidelines cover such areas as: validation of analytical procedures, viral safety evaluation of products, and ensuring genetic stability of biotechnology products. ICH guidelines are steadily becoming more important in guiding industry practice.

E. Risk-Based Quality Systems

We have seen from a historical perspective how and why the development and production of medical products came to be tightly controlled through laws and government oversight. The overall regulatory process that has evolved over the years is now well established and accepted. Even so, the relationship between the government and the pharmaceutical/biopharmaceutical industry is not a static one; it continues to evolve as the world changes. In September, 2002, the FDA rolled out a new initiative called *Pharmaceutical cGMPs for the 21st Century: A Risk-Based Approach*. The motivation behind this initiative was different than others coming from the FDA; it was not a reaction to a tragedy or a scandal. Rather, the new initiative was partly a response to an industry climate in which scientific discoveries and technological advances are being made daily but are slow to be utilized by the pharmaceutical industry. The FDA was therefore looking for a regulatory system that protects patients and yet encourages innovation. The initiative was also a response to the fact that the FDA was finding itself stretched with fewer resources to regulate increasing numbers of companies and products. The FDA was seeking more efficient, cost-effective methods of inspection and oversight. The new initiative was also partly in response to political pressures to lower the cost of drugs. The FDA reasoned that the cost of drugs can be reduced if companies are encouraged to use modern, efficient production methods. The FDA was also trying to align its regulatory practices within its various divisions and with international regulatory bodies.

As the title of the initiative suggests, managing risk is a key to the new initiative. **Risk** *is defined as the combination of the probability of the occurrence of harm and the severity of that harm*. The International Conference on Harmonisation, ICH, says:

> The primary principles of risk management are:
>
> - The evaluation of the risk to quality should be based on scientific knowledge and ultimately linked to the protection of the patient; and
> - The level of effort, formality and documentation of the quality risk management process should be commensurate with the level of risk.

In a risk-based quality system, a company must identify potential risks, analyze the probability of the risk and the severity of harm, then devise a plan to control the risk, focusing on the most critical issues. They must also evaluate whether the plan worked, and if not, modify it. A major goal of risk management is to focus time and attention on the things that matter, and not waste resources on things that do not. For a pharmaceutical company, this means identifying the most critical attributes of their products (e.g., absence of contamination, proper potency),

identifying the aspects of production that affect the critical attributes, and controlling those aspects.

In the risk-based regulatory model, the FDA continues to provide regulatory oversight to the pharmaceutical industry. But, they state:

> FDA resources are used most effectively and efficiently to address the most significant health risks . . . In order to provide the most effective public health protection, FDA must match its level of effort against the magnitude of risk. Resource limitations prevent uniformly intensive coverage of all pharmaceutical products and production. . . . The intensity of FDA oversight needed will be related to several factors including the degrees of a manufacturer's product and process understanding and the robustness of the quality system controlling their process. For example, change to complex products (e.g., proteins . . .) made with complex manufacturing processes may need more oversight.

While the FDA's 2002 initiative appears to be resulting in substantial regulatory changes, the idea of basing decisions on risk is not new. For example, one of the influential events in the biotechnology product-versus-process discussion was a policy statement issued in 1992 by President George H.W. Bush, Four Principles of Regulatory Review for Biotechnology (*Exercise of Federal Oversight Within Scope of Statutory Authority: Planned Introductions of Biotechnology Products Into the Environment*, Office of Science and Technology Policy, Executive Office of the President, 1992). This statement set forth four broad principles of regulation for biotechnology. The first of these principles is:

> Federal government regulatory oversight should focus on the characteristics and risks of the biotechnology product—not the process by which it is created. Products developed through biotechnology do not per se pose risks to human health and the environment; risk depends instead on the characteristics and use of individual products. Biotechnology products that pose little or no risk should not be subject to unnecessary regulatory review during testing and commercialization. This allows agencies to concentrate resources in areas that may pose substantial risks and leaves relatively unfettered the development of biotechnology products posing little or no risk.

Other industries already use a risk-based quality system model. The quality system that is used to regulate food products, called HACCP (discussed in more detail later in this chapter), is based on risk management. In 1997, the FDA announced major revisions to the GMP regulations for medical devices. These requirements, called *Quality System Regulations (QSR)*, direct medical product companies to develop a quality system commensurate with the type of medical device(s) they make, the risk represented by the device(s), the complexity of the device(s), and the complexity of the manufacturing facility. According to this paradigm, the quality system and regulatory oversight required for facilities making tongue depressors is quite different than that required for those making heart pacemakers. The QSRs also focus on building quality into each product from its inception and allow the FDA to look into records from research and development.

II. THE REGULATORY PROCESS

A. Congress Passes Laws

We have seen that the development and production of certain products, including drugs and foods, are regulated by the government. In this section we introduce the mechanism by which regulations are created.

The regulation of food and medical products has been generally driven by a tragedy, a problem, or an advance in science and technology. The public and their elected officials respond by enacting laws that are intended to reduce risks while maximizing benefits. A government agency is empowered to interpret and enforce laws through a system of regulations, see Figure 5.4. For example, the **Food and Drug Administration** *is the regulatory agency responsible for ensuring the safety, effectiveness, and reliability of*

medical products, foods, and cosmetics as set forth in various federal legislative acts, most notably the Food, Drug, and Cosmetic Act of 1938. A brief quote from that act is shown in Figure 5.5. Regulations change as the legislature, regulatory agencies, the public, and the industry respond to events and to one another.

B. The Code of Federal Regulations

The regulations that affect the medical products industry are found in the **Code of Federal Regulations** (CFR), *which is a codification of the rules of the United States federal government.* The CFR contains the complete and official text of the regulations that are enforced by federal agencies. Individuals working in regulated segments of the biotechnology industry must familiarize themselves with the relevant sections and adhere to them in their work.

The CFR is organized as follows:

- The CFR is divided into 50 **titles** that represent broad areas subject to federal regulations.
- Each title is divided into **chapters** that are assigned to various agencies issuing regulations pertaining to that broad subject area.
- Each chapter is divided into **parts** covering specific regulatory areas.
- Large parts may be subdivided into **subparts.**
- Each part or subpart is then divided into **sections**—the basic unit of the CFR.
- Sometimes sections are subdivided further into paragraphs or subsections. Citations pertaining to

REGULATION OF PHARMACEUTICAL PRODUCTS

Congress enacted the Food, Drug, and Cosmetics Act (FDCA) which states that food and drugs should not be "adulterated" and empowers the FDA to enforce food and drug laws.
One of the key clauses in the FDCA is "A drug is adulterated . . . if the methods used in . . . its manufacture . . . do not conform to . . . current good manufacturing practice . . ."

The FDA devises regulations that outline in more detail the meaning of "Good Manufacturing Practices" (GMP). These are published in **The Code of Federal Regulations (CFR)**, a document published annually, which contains all federal regulations.

The FDA periodically publishes **Guidelines** to more fully interpret technical details relevant to GMP. The Guidelines do not have the force of law, but a company should follow them or be prepared to justify any deviation. "**Points to Consider**" are FDA documents that discuss issues relating to innovative technologies.

Figure 5.4. An Example of Regulation. Congress passed an act and authorized a federal agency to interpret and enforce it. The agency devises regulations and enforcement strategies to bring about compliance with the intent of Congress.

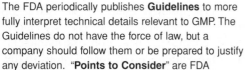

__Federal Food, Drug, and Cosmetic Act__
Adulterated Drugs 501 [351] *A drug or device shall be deemed adulterated—(a) (1) if it consists in whole or in part of any filthy . . . substance; or (2) (A) if it has been prepared, packed, or held under insanitary conditions whereby it may have been contaminated with filth . . . or (B) if it is a drug and the methods used in, or the facilities or controls used for, its manufacture, processing, packing, or holding do not conform to or are not operated . . . in conformity with current good manufacturing practice to assure that such drug meets the requirements of the Act as to safety and has the identity and strength, and meets the quality and purity characteristics, which it purports or is represented to possess . . .*

Figure 5.5. A Brief, but Important Excerpt from the FDCA of 1938. This act forbids adulteration of food and drugs and authorizes the regulation of food and drug quality in the United States.

specific information in the CFR will usually be provided at the section level.

An example of a CFR citation is **21 CFR 58.63(a).** To interpret this:

- The number 21 is the CFR **title.** The broad subject area is Food and Drugs.
- The number 58 is the **part.** Part 58 is entitled "Good Laboratory Practice for Nonclinical Laboratory Studies."
- The number 63 refers to a particular **section.** This section is about "Maintenance and Calibration of Equipment."
- The (a) is the first paragraph or subsection within the section.

The areas of the CFR of primary interest to biotechnologists are Part 58, pertaining to GLP; Parts 210 and 211, pertaining to pharmaceuticals; Part 820, pertaining to medical devices; Part 110 relating to the food industry; and Part 606 relating to the blood products industry. The Code of Federal Regulations is easily accessed from the Internet (see this unit's bibliography, p. 65).

III. THE ORGANIZATION OF THE FDA

The FDA regulates the safety and labeling of all foods except for meat and poultry, which are regulated by the U.S. Department of Agriculture. The FDA also regulates the development, manufacture, marketing, and labeling of prescription and nonprescription drugs, blood products, vaccines, tissues for transplantation, medical devices, radiological products (including cellular telephones), animal drugs and feed, and cosmetics. The FDA is headed by the Commissioner of Food and Drugs, who is appointed by the President of the United States and is confirmed by the U.S. Senate. The FDA is organized into Centers that are described in Table 5.1.

For historical reasons, there are two related, but separate regulatory systems within the FDA for pharmaceuticals. The FDA distinguishes between so-called "drugs" and "biologics." The FDCA defines **drugs** as *"articles intended for use in the diagnosis, cure, mitigation, treatment, or prevention of disease in man or other animals." Drugs are regulated by* **The Center for Drug Evaluation and Research, CDER,** one of the program centers in the FDA. Small, chemically synthesized therapeutic products are likely to be regulated as drugs. Most of the biotechnology products we discussed in previous chapters are regulated by CDER (including some that were handled by The Center for Biologics Evaluation and Research prior to 2003). Biotechnology products regulated by CDER include:

- **Monoclonal antibodies** for therapeutic use
- **Proteins** intended for therapeutic use, including cytokines (e.g., interferons), enzymes (e.g., throm-

Table 5.1. *PROGRAM CENTERS IN THE FDA*

CDER, Center for Drug Evaluation and Research

Regulates prescription and over-the-counter drugs (including small molecules, synthesized compounds, antibiotics, and many biopharmaceuticals).

CBER, Center for Biologics Evaluation and Research

Regulates biologics (e.g., vaccines, blood and blood products, and allergens extracted from natural sources).

CDRH, Center for Devices and Radiological Health

Regulates medical devices (e.g., ultrasonic cleaners for medical instruments, diagnostic kits used in medical laboratories, and ventricular by-pass devices) and many, but not all, in vitro diagnostic kits.

CFSAN, Center for Food Safety and Applied Nutrition

Promotes and protects the public health and economic interest by ensuring that foods are safe, nutritious, and honestly labeled.

NCTR, National Center for Toxicological Research

Conducts research activities that provide the scientific basis for regulatory actions and guide standard-setting.

CVM, Center for Veterinary Medicine

Regulates the manufacture and distribution of food additives and drugs that will be given to animals.

bolytics), and other novel proteins, except for those that are specifically assigned to CBER (e.g., vaccines and blood products). This category includes therapeutic proteins derived from plants, animals, or microorganisms, and recombinant versions of these products

- **Immunomodulators,** which are nonvaccine and nonallergenic products intended to treat disease by inhibiting or modifying a preexisting immune response
- **Growth factors, cytokines, and monoclonal antibodies** intended to mobilize, stimulate, decrease, or otherwise alter the production of blood cells in vivo

According to the FDA, **"Biologics,** *in contrast to drugs that are chemically synthesized, are derived from living sources (such as humans, animals, and microorganisms). Most biologics are complex mixtures that are not easily identified or characterized, and many biologics are manufactured using biotechnology."* Biologics are regulated by the **Center for Biologics Evaluation and Research, CBER.** Substances that are presently regulated as biologics include:

- **Antitoxins, antivenins, and venoms**
- **Blood, blood components, plasma-derived products**
- **Vaccines,** including any future AIDS vaccines
- **Human tissue for transplantation**
- **Allergenic extracts** used for the diagnosis and treatment of allergic diseases and allergen patch tests

- **Cellular products**, including products composed of human, bacterial, or animal cells (such as pancreatic islet cells for transplantation)

- **Gene therapy products**

Marketing new drugs requires FDA approval of a New Drug Application, NDA (as described in Chapter 3), or a New Animal Drug Application (NADA), or, for generic drugs, an Abbreviated New Drug Application (ANDA). These applications are reviewed by CDER. Marketing new biologics requires the company to file a Biologics Licensing Application (BLA), which must be approved by CBER reviewers.

Devices, according to the FDA, *are products, such as tongue depressors and pacemakers, which have a medical purpose, but are not drugs or biologics. Devices are regulated by* **The Center for Devices and Radiological Health (CDRH).** There are a number of biotechnology companies that make diagnostic tools for detecting disease in humans; their products are regulated as devices.

IV. THE REGULATION OF FOOD AND AGRICULTURE IN THE UNITED STATES

A. Introduction to Regulatory Agencies and Relevant Legislation

Food and its production are subject to government regulations pertaining to safety, purity, labeling, and "wholesomeness." The FDA is involved in food safety,

---

Table 5.2. *PRIMARY FEDERAL AGENCIES INVOLVED IN THE REGULATION OF FOOD AND AGRICULTURE IN THE UNITED STATES*

The United States Department of Agriculture (USDA)

The USDA enforces requirements for purity and quality of meat, poultry, and eggs, and it is involved in nutrition research and public education. The USDA regulates genetically engineered food plants through its division of Animal and Plant Health Inspection Service (APHIS).

Food and Drug Administration (FDA)

The FDA's Center for Food Safety and Applied Nutrition (CFSAN) is responsible for ensuring the safety and purity of all foods sold in interstate commerce except for meat, poultry, and eggs. (Foods that are sold only within a state are regulated by that state.) The FDA also regulates the composition, quality, and safety of food and color additives.

Environmental Protection Agency (EPA)

The EPA regulates pesticides, sets tolerance limits for pesticide residues in foods (which the FDA enforces), publishes directives for the safe use of pesticides, and establishes quality standards for drinking water. The EPA is also involved in the regulation of environmental releases of genetically modified organisms.

---

as it is in regulating medical products. The FDA is responsible for the safety of all food except meat, poultry, and frozen and dried eggs. The FDA's regulatory authority for food and cosmetics comes from the Pure Food and Drugs Acts of 1906 and 1938, as well as from other acts that relate to such issues as nutritional labeling and infant formula. Like the centers that regulate medical products, the FDA's **Center for Food Safety and Applied Nutrition (CFSAN)** includes chemists, nutritionists, microbiologists, and toxicologists so that its work has a scientific basis.

In addition to the FDA, **The United States Department of Agriculture (USDA)** and **The Environmental Protection Agency (EPA)** are also empowered to play major roles in the regulation of food and agriculture, see Tables 5.2 and 5.3. The USDA is responsible for the safety of meat, poultry, and eggs. The USDA also publishes voluntary standards for grading foods according to their quality. The EPA regulates pesticides used in agriculture and substances released to the environment.

B. Food Safety

i. CONTAMINANTS

Major issues relating to food safety include avoiding contamination by pathogens, such as bacteria and parasites, minimizing adverse effects of intentional food additives, such as coloring agents and sweeteners, and minimizing adverse effects of unintentional additives, such as pesticide residues. Microbial contamination of foods is probably the most significant food-related problem in the United States. It is not known how many people are affected by food-borne illness, but estimates range as high as 80 million cases of illness and 9000 deaths per year. Many of the problems with foods are traceable to the way they are handled and prepared by the consumer; therefore, the USDA and the FDA have active consumer education programs.

Until recently, prevention of meat and poultry contamination was accomplished primarily through a system of inspectors who inspected all animals slaughtered in an effort to detect diseased animals and the presence of chemical residues. This testing, however, does not prevent contamination from occurring, nor can it detect all contamination that has occurred. The federal government, therefore, recently mandated that meat and poultry producers implement quality systems based on the same principles as ISO 9000 and GMP, that is, that quality cannot be inspected into the final product. Quality must instead be "built into" the product throughout its processing. The quality program for the meat and poultry industry is called **Hazard Analysis Critical Control Points (HACCP). HACCP** *emphasizes reducing hazards throughout the production, slaughter, processing, and distribution of meat and poultry.* For example, hazards include the presence of pathogens, antibiotics, pesticide

Table 5.3. MAJOR FEDERAL LEGISLATIVE ACTS RELATING TO THE REGULATION OF FOOD AND AGRICULTURE

Food Drug and Cosmetics Act (FDCA)

The FDCA gives the FDA the authority to regulate the safety and wholesomeness of foods and food additives.

Federal Meat Inspection Act and the Poultry Products Inspection Act

This Act authorizes the USDA to ensure that meat and poultry products are safe, wholesome, and accurately labeled.

Toxic Substances Control Act (TSCA)

The TSCA is intended to control chemicals that may pose a threat to human health or to the environment. The act requires reporting and record keeping by chemical manufacturers and sets requirements for notifying The EPA when new chemicals are introduced. The provisions of this act primarily affect chemical manufacturers and their R&D laboratories; however, the act also authorizes the EPA to review new chemicals before they are introduced. Under this provision, The EPA has established a program to review microbial products that result from the introduction of genes from one type of bacterium into another type. EPA's rationale for this program is that microorganisms engineered to contain genes from another organism are new "chemicals." This is relevant, for example, if genetically engineered microorganisms are introduced into fields to control pests, or if they are introduced into the environment for purposes of cleaning contaminated soil or water.

The Federal Insecticide, Fungicide, Rodenticide Act (FIFRA)

FIFRA makes EPA responsible for regulating the distribution, sale, use, and testing of pesticides, including those that are the result of genetic modifications to organisms.

The Federal Plant Pest Act (FPPA)

The FPPA regulates the introduction or release to the environment of "plant pests"; that is, any organisms, materials, or infectious substances that can cause damage to any plants or to products of plants. As such, the FPPA relates to genetically engineered organisms if they might be plant pests. This act gives APHIS (part of USDA) the authority to regulate such introductions.

---

and hormone residues, additives, and foreign matter. The implementation of HACCP requires identifying critical points where hazardous substances might be introduced or must be eliminated. There are critical points at the farms where the animals are raised, at the time of slaughter, during the processing and distribution of the product, and during its preparation in homes, restaurants, and institutions. As in pharmaceutical companies, the implementation of HACCP is intended to complement, but not replace, product inspections. It is the food industry's way of building quality into a food product.

ii. FOOD ADDITIVES

People want food to be safe and uncontaminated, and they also want food to be abundant, nutritious, diverse, and economical. These latter objectives are often the result of innovations, such as pesticides that make food more abundant and less expensive, or food additives, such as saccharin, which make it more diverse. Biotechnology methods can also lead to innovations in food products. Innovations sometimes lead to concerns about the safety of food. This has been the case for pesticides, nonnutritive sweeteners, and biotechnology-derived products.

The FDA has the primary responsibility for regulating food additives and new food products, except meats and poultry. As is the case for drugs, new food additives must be tested for safety and then be approved by the FDA before they are marketed; however, there are exceptions for materials that are generally recognized to be safe. The FDA is also responsible for ensuring that food products are produced properly and are not adulterated.

C. Recombinant DNA Methods and the Food Industry

i. SAFETY

The first recombinant DNA food ingredient was approved by the FDA in 1990. It is an enzyme, chymosin (also called rennin), that is produced in *E. coli* containing the bovine chymosin gene. Rennin is the main enzyme used to make milk clot during the production of cheese. The traditional source of rennin is rennet, which is isolated from the fourth stomach of unweaned calves, which are killed for veal. The availability and price of rennet fluctuates depending on the availability of veal.

The FDA's concerns regarding the approval of rennin made by modified bacteria are basically the same as when a drug product made by recombinant DNA methods is introduced to supplement or replace a drug isolated from "natural" sources. The first concern is to establish that the recombinant product is identical to the natural product, or that any differences have no effect on its safety or effectiveness. To demonstrate this for chymosin, the company performed a variety of identity tests on both the gene itself and on the resulting protein. They also performed activity assays that showed that the recombinant chymosin clotted milk. A second aspect of approving the recombinant DNA-derived chymosin was ensuring that the preparation was safe. As with recombinant drug products, this involved demonstrating that the processing steps removed contaminants, such as DNA and protein derived from the host bacterial cells. Investigators also demonstrated that the bacterial cells used in production were themselves nonpathogenic. Once the company producing the chymosin had demonstrated its equivalence to traditional rennin and its safety, the new version was approved.

The FDA's general policy is currently that if chemical, microbiological, and molecular biology studies show that a food or additive is the same as, or substantially equivalent to, an accepted food-use product, then minimal extra testing or regulatory surveillance is required. A special review of a genetically engineered food product is required only when special safety issues exist. For example, special review is required if a biotechnology product can cause allergic reactions, or has different nutritional value than a traditional product.

ii. Environmental Release of Genetically Modified Organisms

A number of transgenic plants have entered the market or have been approved for field testing. There are transgenic plants that can tolerate herbicides, resist insects or viruses, or produce modified fruits and flowers. For example, insect-resistant corn plants have been created by inserting into them a bacterial gene that codes for an insecticidal protein, as discussed in Chapter 1. There are also genetically modified bacteria that have applications in agriculture (e.g., to provide nitrogen to plants). Genetically modified microorganisms are similarly being investigated to clean polluted water and soil.

There are concerns that there will be adverse effects if organisms genetically modified by the methods of biotechnology are released into the environment (as occurs when a modified crop is grown in a field or when modified microorganisms are applied to the soil). For example, it is postulated that transgenic plants could cause the development of new, hardy weed species or contribute to the loss of species diversity. It is feared that insects and other pests will become increasingly resistant to control agents because of the use of genetically modified crops. The regulation of field testing and planting transgenic crops is the responsibility of APHIS and the EPA. Their task has been difficult and particularly controversial. Regulatory agencies will presumably be able to establish increasingly effective policies as they gain more experience with these genetically modified organisms in the environment.

V. Summary

This chapter reviewed many regulations that impact the development and production of pharmaceutical/biopharmaceutical products and foods. These regulations evolved over the years to protect consumers from unsafe and ineffective products. The regulations are enforced by various government agencies, including the FDA, USDA, and EPA. The FDA regulates the safety and labeling of all foods except for meat and poultry. The FDA also regulates the development, manufacture, marketing, and labeling of prescription and nonprescription drugs, blood products, vaccines, tissues for transplantation, medical devices, radiological products, animal drugs and feed, and cosmetics. The USDA enforces requirements for meat, poultry, and eggs, and also plays a key role in regulating genetically modified plants through its APHIS division. The EPA regulates various products including pesticides, and notably is involved in the regulation of environmental releases of genetically modified organisms. Regulations and enforcement agencies have a major impact on the biotechnology industry. It is therefore important for biotechnologists to be at least minimally conversant with regulatory principles and issues.

PRACTICE PROBLEMS

1. Examine Figure 3.7. Discuss how the process outlined in Figure 3.7 relates to the historical incidents described in the beginning of this chapter.

2. **a.** What is the role of Congress in ensuring the safety of food and medical products?

   b. What is the role of the FDA in ensuring the safety of pharmaceutical/biopharmaceutical products?

   c. What is the role of a pharmaceutical/biopharmaceutical company in ensuring drug safety?

   d. What is the role of the consumer in ensuring the safety of food and medical products?

3. Use a web search engine to find the following section of the code of federal regulations: 21 CFR 211.22.

   a. Explain this regulation in your own words.

   b. Is it all right for a company to have the production team take responsibility for testing all incoming materials and all finished product?

CHAPTER 6

Documentation: The Foundation of Quality

I. INTRODUCTION: THE IMPORTANCE OF DOCUMENTATION

Everyone who has taken a science class is familiar with laboratory notebooks. Not every student, however, realizes their tremendous importance in the workplace, nor that laboratory notebooks are one of the basic components of a broader system of documentation. **Documentation,** *which is defined most simply as a system of records,* is essential to any quality system.

FDA's policy regarding documentation is, "if it isn't written down, it wasn't done." If the documentation relating to a particular batch of a regulated product (such as a drug) is lost, accidentally destroyed, or is badly prepared, then that batch of product cannot be sold. Such an error could cost the company millions of dollars. Regulated companies, therefore, have extensive systems in place to ensure that work is recorded, that the documents associated with every product are completed, that all documents are securely stored and can be retrieved from storage, and that documents are protected, just as is the product itself.

Documentation is equally important in a research setting. If a researcher cannot show written evidence of their results, then those results are not credible. In 1986 a scientist in an academic laboratory was accused by a colleague of publishing erroneous data. This accusation led to a series of investigations, beginning at the researcher's institution and eventually winding up before a U.S. Congressional Committee. (The project the researcher was working on was funded by federal grant money.) The Secret Service was called to scrutinize the scientist's laboratory notebooks and raw data. Secret Service analysts found evidence that the researcher did not record her observations at the time they were obtained, that she did not keep all her original observations in her laboratory notebook, but rather used a pad of lined paper, and that dates in her notebooks and on instrument recordings were incorrect. Investigators charged the scientist with intentional fraud. The researcher claimed that even though her notes were not orderly, they were not fraudulent. The investigations culminated in a 1996 hearing that involved 6 weeks of testimony and thousands of pages of written statements. The investigative panel in 1996 concluded that, although she did not follow accepted rules of documentation and had made errors, the researcher was not guilty of fraud. Over 10 years, this incident adversely affected many individuals and caused public concern about the honesty and integrity of scientists. This was an unfortunate situation that likely could have been averted if the scientist had followed accepted documentation practices.

Documents have many important functions. They provide a record of what was done, by whom, when, how, and why. They provide objective evidence that a product was made properly, that all personnel followed proper procedures, and that all equipment was operating correctly. Some of the many functions of documentation are summarized in Table 6.1.

II. TYPES OF DOCUMENTS

A. Overview

There are various types of documents, and a single company or organization is likely to have many documents, each with a particular purpose. It may be helpful to broadly classify documents into three categories to better understand their roles in organizations.

Directive documents *tell personnel how to do something.* Standard operating procedures and protocols, discussed later in this chapter, are examples of directive documents. Information is not added to these documents when work is performed.

Data collection documents *facilitate the recording of data and provide evidence that a directive document has been properly followed.* Information is added to data collection documents during routine operations. Laboratory notebooks, reports, forms, and logbooks, discussed later in this chapter, are examples.

Commitment documents *lay out the organization's goals, standards, and commitments.* A document submitted to the FDA (e.g., a New Drug Application, as described in Chapter 3) is an example.

Although documentation is essential in all biotechnology work environments, the specific types of documents and the systems for documentation vary in different workplaces. In an academic research laboratory, the major documentation requirements are that investigators can reconstruct their work based on their records, solve problems and detect mistakes, prove to the scientific community that their results were properly obtained and were accurately reported, and provide a trustworthy chronological record of their work. The laboratory notebook is the primary document in a research laboratory and will become a matter of public record in patent applications, disputes over who first had an idea, and if there ever should be questions about the correctness or authenticity of reported findings.

Individuals in research and development laboratories likewise rely heavily on laboratory notebooks to document findings, especially when applying for patents. R&D workers also prepare documents that describe how to manufacture the product and the properties of a product.

People in production facilities use documents other than laboratory notebooks. For example, batch records that describe how to make a product are essential in production facilities.

Table 6.2 summarizes various types of documents. The first part of the table focuses on documents that are com-

Table 6.1. THE FUNCTIONS OF DOCUMENTATION

1. *Record what an individual has done and observed.*
2. *Establish ownership for patent purposes.*
3. *Tell workers how to perform particular tasks.*
4. *Establish the specifications by which to evaluate a process or product.*
5. *Demonstrate that a procedure was performed correctly.*
6. *Record operating parameters of a laboratory instrument or a manufacturing vessel.*
7. *Demonstrate by an evidence "trail" that a product meets its requirements.*
8. *Ensure traceability.* Here we define **traceability,** as does ISO, *as the ability to trace the history, applications, and location of a product and to trace the components of a product.* Traceability helps to ensure that if problems arise in a product, then the origin of the problem can be traced to its components, and the product itself can be found and recalled if necessary. Traceability depends on an organized, well-designed system of documentation.
9. *Establish a contract between a company and consumers.* The written specifications, labels, and other documents associated with a product establish that contract.
10. *Establish a contract between a company and regulatory agencies.*

monly found in laboratory environments. These include academic research laboratories, R&D laboratories, testing laboratories, and quality-control laboratories associated with production facilities. With the exception of laboratory notebooks, many of the types of documents in the first part of Table 6.2 also are used in production environments. For example, labels, SOPs, and recordings from instruments are found in both laboratories and production facilities. The second part of the table gives examples of documents that are specific to production facilities and are not used in laboratories.

EXAMPLE

Consider the documentation that might be required when performing a routine laboratory task in a regulated industry, such as mixing a 1 M solution of NaCl:

1. The technician will follow an **approved standard operating procedure** for mixing the solution.
2. The raw materials—clean glassware, NaCl, and purified water—will all have **documents associated with them that show they were tested and found satisfactory prior to being released for use by a technician**.
3. The instruments used to weigh the NaCl and measure the water will have **logbooks** or other documents showing they were properly maintained.
4. The technician will need to record information about the solution on a **form.** The form identifies the person making the solution, the date, the procedure followed, quantities and source of raw materials used,

the storage location and conditions of the solution, the amount made, and the number of the batch.

5. The resulting solution will need an **identifying number** and there will be documentation associated with assigning this number.

6. The solution will need a **distinguishing label**.

7. If the solution is split into more than one container, **documented ID numbers** will need to be assigned to each container referring back to the original solution. Each container will need a **distinguishing label**.

B. Laboratory Notebooks

i. LABORATORY NOTEBOOKS: FUNCTIONS AND REQUIREMENTS

Laboratory notebooks *are assigned to individuals and are a chronological log of everything that individual does and observes in the laboratory.* Of all the documents that are described in this chapter, the laboratory notebook is the most important in research laboratories.

Laboratory notebooks in a biotechnology company are generally distributed to individual investigators by a company representative. The notebook, and the ideas and information recorded in it, are intellectual property that belongs to the company. The ownership rules in academic research institutions vary.

The primary user of a laboratory notebook is the researcher who uses the notebook to track the progress of a project, archive the data generated in experiments, record observations, and record all the details that must be remembered. Researchers use their notebooks as a key tool for trouble-shooting when problems arise.

A laboratory notebook is an important legal document that may be viewed by people in addition to the researcher. A laboratory notebook may provide evidence that is used by the scientific community to assign credit for a research discovery. The notebook documents the honesty and integrity of data that are published in research journals and used in grant applications. Laboratory notebooks can be subpoenaed in litigations and they can be examined by auditors from the FDA, EPA, and other regulatory agencies.

Laboratory notebooks are of particular importance in patent law (see Chapter 2). Laboratory notebooks are the primary evidence by which researchers prove that

---

Table 6.2. TYPES OF DOCUMENTATION

Examples of Documents That Are Common in Laboratories

Directive Documents

1. *Standard Operating Procedures (SOPs)* detail what is to be done to complete a specific task and how to document that the task was done correctly.
2. *Protocols* are similar to SOPs in that they explain how to do a task. The term *protocol,* however, is often reserved for situations where a question or hypothesis is to be investigated (an experiment will be performed) or the procedure is going to be performed only once.
3. *Numbering systems* are used to keep track of materials, equipment, and products.
4. *Labels* are attached to solutions, products, or items to identify them.

Data Collection Documents

5. *Laboratory notebooks* are a chronological log of everything that an individual does in a laboratory.
6. *Forms* contain blanks that are filled out by an analyst to record information. Forms are typically associated with SOPs or other documents.
7. *Reports* are documents generated during the execution of a protocol.
8. *Equipment/Instrument logbooks* keep track of maintenance, calibration, and problems for a given instrument or piece of equipment.
9. *Analytical laboratory documents* record information regarding the testing of a sample.
10. *Recordings from instruments.*
11. *Chain of custody forms* are used to trace the movement of a sample throughout a facility and to keep samples and sample test results from being confused with one another.
12. *Training reports* document that individuals were properly trained to perform particular tasks.
13. *Electronic documents* are explained later in this chapter.

Examples of Documents That Are Specific to Production Facilities

1. *Batch records* are collections of documents associated with a particular batch of a product. (A batch record is both a directive and data collection document.)
2. *Regulatory submissions* are forms filled out and sent to regulatory agencies to inform them of what a company is doing and/or to ask permission to test or sell a product. (These are a type of commitment document.)
3. *Release of final product record* is filled out when a product has been approved for sale. (This is a type of data collection document.)

they were the first to conceive of an invention and in which they document the steps they took to reduce it to practice. Even academic scientists who do not seek commercial gain from their research may find that they want to patent an invention to protect their rights to it, or to ensure that their invention will remain in the public domain. It is therefore essential that all researchers maintain proper laboratory notebooks and that these notebooks unequivocally document the dates at which ideas were conceived and experiments were performed.

There are certain fundamental requirements for keeping a laboratory notebook that allow it to be used for all the purposes described above. These include:

- The laboratory notebook must be complete so that any experiment can be repeated by the researcher or someone else.
- Laboratory notebooks must be chronological to clearly document dates when events occur or ideas are conceived.
- The data in the laboratory notebook must be honestly recorded at the time they are observed.
- Laboratory notebooks should be kept in a manner that makes alterations readily apparent.

Table 6.3 contains standard guidelines that should always be followed to ensure that a laboratory notebook meets these bulleted requirements. Laboratory notebooks are bound, for example, so that pages cannot be added or removed. This helps to ensure that the dates are correct and that the notebook honestly records what happened and when it happened. Every entry in a notebook must be made with a pen that cannot be erased, again to ensure the honesty and the integrity of the data. Mistakes are crossed out with a single line, signed, and

dated. This helps ensure that entries are not obscured, altered, or changed at a later date.

ii. The Content of Laboratory Notebooks

Table 6.4 summarizes the items that are generally recorded in a laboratory notebook. All notebooks include basic information, such as the name of the person to whom the notebook was assigned and a page number and date on every page. A laboratory notebook must be complete enough that the researcher or another individual could exactly repeat the work described based on the information recorded. It is essential that laboratory notebooks include raw data. **Raw data** *are the first records of an original observation.* Depending on the situation, raw data may be written into the notebook with pen by the operator, may be a paper output from an instrument, or, increasingly, may be recorded into a computer medium (discussed later in this chapter). The researcher must save all raw data, ensure that these data are never altered or edited, and ensure that they are retrievable.

Laboratory notebook entries should include ideas as well as experiments. Researchers should explain why each experiment is performed and at the end summarize what the results show, avoiding derogatory comments about their work or ideas (even if something did not go as planned). Documenting ideas is particularly important where patents are potentially involved because the date of conception of an invention is important. Inventors need to have evidence of the date when they first had an idea and they need to show how their experiments and activities were designed to reduce the idea to practice.

One of the challenges in keeping a good laboratory notebook is that it must record clearly what actually

---

Table 6.3. Guidelines for Keeping a Laboratory Notebook

1. *Use only a bound notebook, not a spiral or looseleaf notebook from which pages can be removed or into which pages can be inserted.*
2. *Make sure every page is numbered consecutively before using the notebook.*
3. *Never rip out a page.*
4. *Keep the laboratory notebook in chronological order.* Never skip a page to insert information later.
5. *Blank lines or unused portions of the page should be crossed out with a diagonal line so nothing may be added to the page at a later date.*
6. *Make all entries with indelible ink.*
7. *Be legible, clear, and complete in your entries.* Remember that you, supervisors, colleagues, patent attorneys, and regulatory agency inspectors may review your entries.
8. *Enter all observations and data directly into the notebook—not onto a paper towel or the back of your hand.*
9. *Cross out all errors with a single line so that the underlying text is still clearly legible.* Date when the cross-out was made, explain it briefly, and initial or sign it. In some settings, cross-outs must also be witnessed.
10. *Note all problems; never try to obscure, erase, or ignore a mistake; be honest.* Be objective; avoid derogatory statements about your ideas.
11. *Date and sign each page. In many laboratories a corroborating witness should also sign and date the page.* The scientist and, where relevant, the witness should verify that there are no blank spaces on the pages, that all tables are complete, and that the page is complete. If corrections are later made to an entry, the corrections should be signed and dated by both the scientist and witness.
12. *Be certain that the laboratory notebook is stored in a secure location.*

Table 6.4. TYPICAL COMPONENTS OF A LABORATORY NOTEBOOK

1. *In the front of the notebook: the person to whom the book is assigned, the project, the date of assignment, the company/institution, and any other identifying information.*
2. *A table of contents on the first pages.* The table of contents should include page numbers and descriptions with sufficient detail to allow easy searching of the notebook's contents.
3. *For each project, a listing of the results of any literature search and any experimental information collected from colleagues.*
4. *A page number on every page in consecutive order.*
5. *Dates, titles, and descriptions.* Begin the record of each day's work with the date, title, and description of the objectives for the work. It is common to begin each day on a new page with a diagonal line drawn across the unused part of the previous day's page.
6. *The rationale for each activity performed.* Documenting ideas is important.
7. *Any relevant equations or calculations.*
8. *Complete descriptions of all instrumentation (including models and serial numbers), chemicals used (including manufacturers, catalog and lot numbers, and expiration dates), reagents used (including recipes or references to SOPs), supplies used, samples assayed, standards or reference materials used.*
9. *Procedural details.* If an SOP or protocol from the researcher's institution or company is followed, it should be referenced in a unique fashion (e.g., by title and revision date, and/or by ID number). It is usually not necessary to copy an SOP or protocol into the notebook, but any deviations and their justification should be noted. If a procedure comes from a compendium, journal, book, or manual, the complete reference for the procedure should be cited and it may be necessary to record procedural details in the laboratory notebook.
10. *Sample information.* When samples are tested, information should be provided regarding their source, storage, identifying information, disposal, and so on.
11. *Data.* Data take many forms, for example, values read from instruments, color changes observed, photos, and instrument printouts. Printouts from instruments, photos, and other paper data are generally signed, dated, and securely taped into the laboratory notebook with a permanent adhesive. It is common to sign or write across both the inserted paper and the page on which it is taped in order to authenticate the paper's placement on the page. There should be an explanation of the data provided with it. If data cannot be affixed in the notebook, they may be titled, signed, explained, dated, and filed securely. The data and storage location should be referenced in the laboratory notebook.
12. *Observations.* Observations might include, for example, changes in pH or temperature, humidity readings, and instrument operational parameters.
13. *A brief summary of the work completed.*
14. *A conclusion and brief interpretation of data collected is usually appropriate.* For example, if a particular line of investigation is pursued on the basis of preliminary results, note this in the laboratory notebook. Such decisions may be relevant in patent disputes at a later date.

happened in the laboratory. It is common to outline in a laboratory notebook what one *intends* to do; these plans should be written in the present or future tense. What *actually* occurred must be clearly recorded using the past tense.

Another challenge is that a laboratory notebook must be chronological—it must keep moving forward in time. A researcher might be working on more than one project, in which case it is often simplest to keep a different laboratory notebook for each project. If only one laboratory notebook is used, it is not correct to leave empty pages or spaces on a page in order to fill them in later. Rather, if a researcher returns to a project after recording information on something else, it is correct to state at the top of the page that the work is "continued from page___".

Laboratory notebooks should be witnessed whenever research might lead to a patent. The witness, who is not one of the inventors, reads, signs, and dates each entry. The witness should be sure that he/she understands the entries because the witness is corroborating that the work described really happened. It is prefer-

able that the witness is someone who actually observed the experiments, though this may not be possible in practice. In a company that is regulated, laboratory notebooks are also witnessed. The witness must carefully look for mistakes or omissions, such as an erroneous calculation or a missing date. These mistakes are then corrected, briefly explained, signed by both the researcher and witness, and dated. After a notebook is witnessed, no changes should be made to that page. Ideally notebooks should be witnessed every day.

Figure 6.1 shows an annotated laboratory notebook that illustrates some of these ideas.

C. Other Documents That Are Common in Laboratories

i. STANDARD OPERATING PROCEDURES

Most production facilities and many laboratories use procedures to instruct personnel in how to perform particular tasks. A **procedure** *is a written document that provides a step-by-step outline of how a task is to*

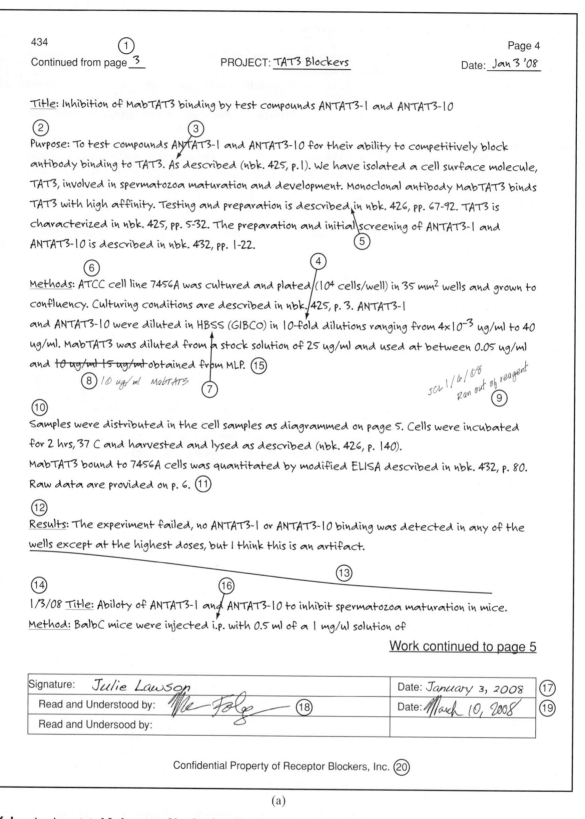

434   Page 4
Continued from page _3_   PROJECT: TAT3 Blockers   Date: _Jan 3 '08_

① ②

Title: Inhibition of MabTAT3 binding by test compounds ANTAT3-1 and ANTAT3-10

Purpose: To test compounds ANTAT3-1 and ANTAT3-10 for their ability to competitively block antibody binding to TAT3. As described (nbk. 425, p.1). We have isolated a cell surface molecule, TAT3, involved in spermatozoa maturation and development. Monoclonal antibody MabTAT3 binds TAT3 with high affinity. Testing and preparation is described in nbk. 426, pp. 67-92. TAT3 is characterized in nbk. 425, pp. 5-32. The preparation and initial screening of ANTAT3-1 and ANTAT3-10 is described in nbk. 432, pp. 1-22.

Methods: ATCC cell line 7456A was cultured and plated (10^4 cells/well) in 35 mm² wells and grown to confluency. Culturing conditions are described in nbk. 425, p. 3. ANTAT3-1 and ANTAT3-10 were diluted in HBSS (GIBCO) in 10-fold dilutions ranging from 4×10^{-3} ug/ml to 40 ug/ml. MabTAT3 was diluted from a stock solution of 25 ug/ml and used at between 0.05 ug/ml and ~~10 ug/ml 15 ug/ml~~ obtained from MLP. ⑮

⑧ 10 ug/ml   MabTAT3

JCL 1/4/08
Ran out of reagent ⑨

Samples were distributed in the cell samples as diagrammed on page 5. Cells were incubated for 2 hrs, 37 C and harvested and lysed as described (nbk. 426, p. 140).
MabTAT3 bound to 7456A cells was quantitated by modified ELISA described in nbk. 432, p. 80.
Raw data are provided on p. 6. ⑪

Results: The experiment failed, no ANTAT3-1 or ANTAT3-10 binding was detected in any of the wells except at the highest doses, but I think this is an artifact.

1/3/08 Title: Abiloty of ANTAT3-1 and ANTAT3-10 to inhibit spermatozoa maturation in mice.
Method: BalbC mice were injected i.p. with 0.5 ml of a 1 mg/ul solution of

Work continued to page 5

Signature: *Julie Lawson*	Date: *January 3, 2008* ⑰
Read and Understood by: ⑱	Date: *March 10, 2008* ⑲
Read and Undersood by:	

Confidential Property of Receptor Blockers, Inc. ⑳

(a)

Figure 6.1.   An Annotated Laboratory Notebook. a. Notebook page. **b.** Legend.

Source: Based on a laboratory manual page and reprinted with permission from Merchant and Gould Legal Firm.

be performed. Such documents are often called **standard operating procedures (SOPs),** although other names are used as well. For example, a research laboratory might simply call these "procedures" (or even "protocols," although this term has another meaning as explained in Table 6.2). Everyone follows the same procedures to assure that tasks are performed consistently and correctly. Standard operating procedures

Legend for Notebook Diagram

1. Top of the page identifies the notebook number, the page the work continued from, the page number, date, and project.

2. The entries are organized and legible.

3. Laboratory abbreviations and designations are defined and referenced.

4. Methods are provided in sufficient detail so that a third party could repeat the experiment using only the references and materials supplied in the notebook.

5. Methods are referenced to earlier notebooks. The term "nbk" should be spelled out in a designated place in each notebook.

6. The cell line is sufficiently identified by its supplier.

7. Well-known abbreviations do not need to be defined further.

8. An initial correction is made with a single line. The corrected text is placed in line, next to the error.

9. A later correction is properly initialed, dated, and explained.

10. The entry is in a single permanent ink.

11. Raw data are identified and entered into the notebook.

12. If the experiment did fail, a simple statement that the experiment will be repeated is sufficient. Here it appears that the results may show that the blockers did work at higher doses. Results should be stated positively and repeated as necessary.

13. Blank regions are blocked out in pen.

14. New entries are re-dated.

15. We do not know who or what MLP is. If it is a supplier, it should be spelled out or referenced to an earlier page. If it is a person, the name should be spelled out or the initials provided in the abbreviations index.

16. "i.p." is a well-known scientific term of art for individuals in this field and need not be further identified. The test is whether an abbreviation could be reasonably interpreted by someone similarly skilled in the art.

17. This notebook page was timely signed and dated.

18. The signature is illegible. Where this is a problem, the name should be printed at least once, beneath the similar signature. Also, if MLP is Mark Peterson, then a question arises as to whether Mark is an inventor. Inventors must not witness notebooks reducing their invention to practice.

19. The witness date is much too late. Preferably, the witnessing signature is provided within the same week or two week period.

20. Each page is labeled as confidential and the property of the particular research organization.

(b)

Figure 6.1.   (*Continued*)

describe what is required to perform a task, who is qualified or responsible for the work, what problems may arise and how to deal with them, and how to document that the task was performed properly. SOPs must be written so they are clear, easy to follow, and can accommodate minor changes in instrumentation.

Table 6.5 on p. 98 outlines the components that are generally expected in an SOP.

Table 6.5. TYPICAL COMPONENTS OF AN SOP

1. *Title.*
2. *ID number, revision number, date of revision.* Most SOPs are revised periodically. It is essential that each worker uses only the most up to date revision.
3. *Statement of purpose* (may restate the title with a little more detail).
4. *Scope* describes when the procedure is relevant. For example, if the procedure is for verifying the performance of a balance, then it might be only for a particular brand, or a particular model.
5. *A statement of responsibility,* who does this task.
6. *Materials required, including manufacturers and identifying information.*
7. *Calculations required, preferably with an example.*
8. *The procedure itself written as a series of steps.* The actions required, how they are performed, and the endpoints of the steps should be included.
9. *References to other documents, as required.*
10. *How to document that the procedure was performed; references to any associated forms.*

EXAMPLE

Discuss the considerations in writing an SOP to make chocolate chip cookies.

A. There is preliminary work that must be completed before an SOP can be written. In this example, a first step might be to list the characteristics in the product that are important, such as: flavor, texture, softness, size, shape, number of chips per cookie.

Each of the preceding properties must then be described (i.e., specified in such a way that it can be measured or evaluated). For example, flavor might be evaluated by individuals trained as tasters.

B. The raw materials and equipment required must be listed, such as: flour, sugar, vanilla, chocolate chips, mixer, bowls, measuring cups, measuring spoons, oven.

There are decisions to make regarding these materials. For example, will a particular brand of flour and chocolate chips be specified? What sorts of measuring devices will be adequate? All of these decisions must be made before the SOP can be written.

C. A standard operating procedure to make cookies with the desired properties can be written. The SOP will include a title, the author, a statement of purpose, ID number and date, statement of scope, who is qualified to make the cookies, raw materials required, and the steps in making the cookies. This is, of course, a formalized "recipe." A form may also be associated with the SOP in which the baker can record information, such as the date, the lot number of the materials used, the time, and the temperature of the oven.

Consider briefly the language used in the SOP. Compare the following alternatives to describe how to mix the batter:

a. Mix the batter for 10 minutes at room temperature.

b. Mix the batter with an electric mixer until the batter is well mixed.

c. Mix the batter with an electric mixer for 10 minutes on setting 1.

d. Mix the batter with an electric mixer until no unmixed flour or large chunks are visible. For example, 10 minutes on an intermediate setting on a brand Z mixer is usually sufficient.

Statement **a** is vague and does not direct the baker in how to accomplish the task or its endpoint. Statement **b** gives an endpoint, but the description is too vague to be useful. Statement **c** is useful only for a particular brand of mixer that has a setting called "1." Statement **d** has the clearest description of an endpoint and is flexible in terms of which mixer is used. The statement "intermediate setting" might be too vague, but it is probably the most applicable to most mixers. In any case, statement **d** explains the outcome that is the important point in this step. Note that if overmixing detracts from the quality of the cookies, then this should be noted in the procedure.

EXAMPLE

Carol DeSain (DeSain, Carol, and Sutton, Charmaine Vercimak. *Documentation Practices,* Duluth, MN: Advantstar Publishers, 1996) cites an example in which the language of a procedure is clear, but the intent is not to a new employee. The statement in the SOP is: "Wash the filter press in mild detergent and rinse with WFI [purified water]. Make sure that the filter support grid is completely dry before placing a new filter on the grid." The writer expects the grid to be air or oven dried. The technician, on the other hand, focuses on the need for the grid to be <u>dry</u>. She therefore retrieves a box of laboratory wipes and wipes the grid dry. This action is not in violation of the SOP, but it results in the contamination of several batches of product with fibers from the laboratory wipes.

This example highlights the importance of knowing the audience. The degree of detail in an SOP and the particular points that are noted will, to some extent, depend on the training and background of the people who will be performing the procedure. It is important when writing an SOP to be conscious of the needs of the reader and when following an SOP to be certain that you understand its intent.

Table 6.6   PROBLEMS TO AVOID RELATING TO SOPs
1. *The SOP says what to do, but not how to do it.*
2. *The procedure was written by someone who does not have experience doing the work.*
3. *The SOP has too much detail or too little detail.*
4. *The procedure is not written in the order in which the tasks are actually performed.*
5. *The SOP is not updated as needed.*
6. *Employees cannot find the right SOP or use an older version.*

Table 6.7.   TYPICAL COMPONENTS OF A PROTOCOL
1. *The hypothesis or question the study is designed to answer.*
2. *A description of the study.*
3. *A plan for how the study is to be conducted.*
4. *Information about how the sample(s) is to be collected, processed, and identified.*
5. *The methods that will be used to test the hypothesis.*
6. *The schedule of testing.*
7. *The way the study results and conclusions will be reported.*
8. *The criteria that will be used to reach conclusions.*

Some of the potential problems with SOPs that must be avoided are summarized in Table 6.6.

In a company, every SOP must be reviewed and accepted before it is used. The person who writes the SOP has responsibility for it and must sign it. A second individual who is knowledgeable about the work also normally shows approval with a signature and, in a company, an individual from the quality-assurance unit also signs it.

SOPs periodically require changes. It is important that the old SOPs are destroyed or made unavailable when changes are made, except for a copy(ies) kept in a historical file. Only the latest revision of the SOP should be available to processing or laboratory technicians. In a regulated setting, depending on the magnitude of the change, revisions may have to be checked and approved by several levels of responsible individuals and by the quality-assurance unit before the change can be implemented. Each revision of an SOP needs a date and revision number so that it can be uniquely identified. An example of an SOP is shown in Figure 6.2.

ii. FORMS

As mentioned in the chocolate chip cookie example, an SOP is often associated with a **form** that is filled in as the procedure is being performed. Filling in the blanks requires the individual performing the task to monitor the process as they go along, thus ensuring that everything is going smoothly. In addition, the form will remind the technician to record information about lot numbers, raw materials, times, temperatures, and other relevant information that is easy to forget to record. In some production laboratories a witness must sign key steps. Figure 6.3 on p. 101 shows an example of a form.

iii. PROTOCOLS

The term **protocol** *is used in some industries to refer to a procedure that tells an operator how to perform a task or* *an experiment that is intended to answer a question or test a hypothesis.* (The term *protocol* may also be used for a procedure that will only be performed one time.) In contrast, SOPs do not lead to the answer to a question. For example, one follows a *procedure* to clean a laminar flow hood, but one follows a *protocol* to investigate the effectiveness of cleaning a laminar flow hood with different cleaning agents.

A protocol must include information on what data are to be collected, how the data are to be gathered, what outcome proves or disproves the hypothesis, and any statistical methods that need to be used. A protocol may reference SOPs. Both research and production facilities use protocols. In research laboratories people obviously investigate questions all the time. In production facilities people investigate whether a product performs as expected, the qualities of the product under certain conditions (such as long storage), the effects of the product in a test population, and so on. Example statements or hypotheses that might be seen in an industrial protocol are:

> This study is designed to demonstrate that the cleaning process for laminar flow hoods does not leave detectable detergent residues on the surface of the hood.

> This study is designed to demonstrate that the antitumor drug XYZ has no adverse medical effects on the livers of test subjects.

The components of a protocol are somewhat different from a procedure and are summarized in Table 6.7. An example of a portion of a protocol is shown in Figure 6.4 on p. 102.

iv. REPORTS

A **report** *is a document that describes the results of an executed protocol.* The report summarizes what was done, by whom, why, the data, and the conclusions. A

Clean Gene, Inc.

STANDARD OPERATING PROCEDURE

SUBJECT: Identification Method for Calcium Sulfate

Prepared By *N. Warren*

Effective Date: 7/1/08

SOP # 2648

Approved By *J. McMillan* ____ Date *6/24/08*

Revision 01

Page 1 of 3

Approved By *J. Lownd* ____ Date *6/26/08*

1. Scope

    Provides the method for assessment of the raw material calcium sulfate.
2. Definitions—NA
3. References:

    3.1 USP, current—General Identification Tests

    3.2 Incoming Raw Material/Component specification: $CaSO_4$, Grade I
4. Reagents

    4.1   3N Hydrochloric Acid, HCl

    4.2   6N Acetic Acid

    4.3   Methyl Red

    4.4   Ammonium Oxalate

    4.5   Ammonia

    4.6   Barium Chloride
5. Responsibility

    5.1 This test is to be performed by an appropriately trained analyst. Analyst training and documentation of training will be conducted per SOP 5688. Approval of analyst data is the responsibility of the Quality Control Manager.

    5.2 The results of the test will be verified by a second qualified analyst.
6. Hazard Communication

    6.1   *3N Hydrochloric Acid, HCl*

    DANGER: Corrosive. Avoid contact with skin and eyes. Avoid inhalation of fumes and mist. Do not mix with caustics or other reactives.

    6.2   *6N Acetic Acid*

    DANGER: Corrosive. Avoid contact with skin and eyes. Avoid inhalation of fumes and mist. Do not mix with caustics or other reactives.

    6.3   *Methyl Red*   **CAUTION: Irritant. Avoid contact with skin and eyes.**

    6.4   *Ammonium Oxalate*

    CAUTION: Irritant. Avoid contact with skin and eyes.

    6.5   *Ammonia*   **CAUTION: Irritant. Avoid contact with skin and eyes.**

    6.6   *Barium Chloride*

    CAUTION: Irritant. Avoid contact with skin and eyes.
7. ATTACHMENTS

    7.1 Attachment I—Form No. 687
8. PROCEDURES

    8.1 Sample Preparation

    8.1.1   In a clean beaker or test tube, add approximately 200 mg of the incoming Calcium Sulfate to be tested to 4 mL of 3N Hydrochloric Acid and 16 mL reagent water.

    8.1.2   Gently warm with low stirring on a hot plate to aid dissolution.

    8.1.3   Portions of the solution should respond to the tests for calcium and sulfate.

    8.2   Calcium Identification

    8.2.1   Place about 10 mL of the solution from 8.1 into a clean test tube.

    8.2.2   Place 2 drops of methyl red into this solution followed by enough ammonia to turn the solution YELLOW.

    [and so on . . . the entire SOP is not shown here]

Figure 6.2. A Portion of an SOP. This SOP is used to test an incoming raw material to ensure it is calcium sulfate.

Clean Gene, Inc. — Page 1 of 1

AGAROSE GEL PREPARATION FORM, REVISION 01

1/31/08 Form # 992

Technician Name _____ Date _____

1. *Add between 1.8 and 2.0 grams of agarose to 100 mL of buffer.*
   Record the ID/lot number for the agarose _____
   Weight of agarose added _____
   Source of buffer used _____
2. *Place the mixture in a microwave oven on its highest setting and heat until the mixture just begins to boil.*
   Time to boiling _____
3. *Let the mixture cool until it is between 58 and 62°C and then pour gel.*
   Temperature when pouring gel _____

Figure 6.3. A Portion of a Form Associated with Preparing an Agarose Solution.

report is written in narrative format. Reports from basic scientific research are published in scientific journals. Reports from investigations performed in a company may or may not be published but must be available for inspection.

v. LOGBOOKS

Logbooks *are used to record information chronologically about the status and maintenance of equipment or instruments.* Logbooks are usually bound and labeled notebooks that are associated with a specific instrument, area, or piece of equipment. When an item is used, calibrated, maintained, or repaired, this is indicated in the logbook.

vi. RECORDINGS FROM INSTRUMENTS

Many analytical instruments generate automatic printouts of results. Printouts are generally considered to be raw data. If the results from the instrument belong in the laboratory notebook, then it is usually acceptable to tape them there. In such cases the laboratory notebook page, signature, and date should be noted on the printout so they can be replaced in the proper place if they become detached. It is common to record on a page how many documents are taped to that page so they can be found if necessary. Alternatively, instrument printouts can also be filed and referred to in the notebook or on a form.

Some instruments monitor themselves and record their operating parameters as they operate. For example, modern autoclaves continuously record the date, time, temperature, and pressure throughout their cycle. This information is important in demonstrating that the instrument performed properly. Files may be kept of

such recordings, or they may be associated with the production records for a particular batch of product (batch records will be discussed later).

Printouts from any instrument must be thoroughly identified (including, for example, the date, product name, batch number, and equipment number), and be signed and dated by the technician.

vii. ANALYTICAL LABORATORY DOCUMENTS

Analytical tests are those that measure a property(ies) of a sample. For example, in an environmental laboratory, analysts might test a sample of lake water to determine the level of cadmium present. In a clinical laboratory, analysts might perform a drug screen of a patient's blood. The product is the test result(s). Documentation is required that provides information both about the method used and the sample being tested. Specific types of information that must be documented are summarized in Table 6.8 on p. 103.

viii. IDENTIFICATION NUMBERS

Identification numbers uniquely identify items. There are many types of items that require identification, including raw materials, documents, equipment, parts, batches of product, chemicals, solutions, and laboratory samples. Identification numbers convey two pieces of information: what the item is, and which one it is. For example, the first part of an ID number might logically tell whether the item is a particular type of instrument or a particular type of solution. The next part of the ID number might tell which particular one of those instruments it is, or which batch of solution it is.

Clean Gene, Inc.

CLEANING PROTOCOL CL 898; REVISION 01

Prepared by *Tania Seid*   Date *3/6/08*

Approved by *J. Mavey*   Date *5/15/08*   Approved by *Jon Reidman*   Date *5/16/08*

PROTOCOL

A TEST OF THE EFFICACY OF ALCOHOL FOR DISINFECTING THE SURFACE OF BIOLOGICAL HOODS

1.0   Study Question

This study is designed to determine the efficacy of alcohol for disinfecting the surface of biological hoods, BH655.

2.0   Objectives

—To determine whether surface cleaning of a contaminated biological hood with ethanol is an effective method of eliminating bacteria.

—If ethanol is effective, to establish the optimal ethanol concentration.

3.0   Plan

—Contaminate the surface of the biological hood with 10^9 cfu of *E. coli* strain 56298.

—Treat the surface with ethanol of different concentrations

—Determine the success of disinfection by using the Standard Wipe Method of detection of bacteria, SOP 87-68.

4.0   Equipment required

—Biological cabinet # 4

—Spectrophotometer # 7

—Overnight culture of *E. coli* strain 56298

—Ethanol at a concentration of: 60, 70, 80, 95%.

—Sterile bacteria wipes (ID No. 932-8)

—Microbial test agar (ID No. 56-84)

—Sterile forceps (ID No. 6-1-97)

—Nutrient agar plates (prepared according to SOP 87-75)

5.0   Procedure

5.1.   Grow overnight culture of *E. coli* (according to SOP 95-17).

5.2   Measure the absorbance of culture at wavelength 520 nm.

5.3.   Calculate the number of cfu/mL.

      1 AU = 10^9 cfu/mL

5.4.   Dilute a portion of the culture to get a concentration of 10^9 cfu/mL

5.5.   Spread 1 mL of diluted culture over a 2 square inch section of the hood.

5.6.   Wait 5 minutes.

5.7.   Spread 10 mL of ethanol across the surface of the contaminated hood.

5.8.   Wait 5 minutes.

5.9.   Using a clean, sterile absorbent wiper, wipe up the ethanol.

5.10.   Allow the area to dry for 5 minutes.

5.11.   As directed in SOP 87-68, using the sterile forceps and wipes, wipe the entire 2 square inch section of the "spill."

5.12.   Place the bacterial test swipe on nutrient medium.

5.13.   Repeat steps 5.5 through 5.12, three times for each concentration of ethanol, for a total of 12 plates.

5.14.   After 36 hours record the results, as directed in SOP 87-68.

[. . . and so on]

Figure 6.4.   A Portion of a Protocol. The protocol directs the activities of the technician in conducting a study.

Table 6.8. **ESSENTIAL INFORMATION TO DOCUMENT IN ANALYTICAL LABORATORIES**

1. *Information regarding each assay method used:*

   The purpose of the test

   The limits of the test (e.g., what is the lowest level of the material of interest that the test can detect)

   The origin of the test method (e.g., whether it came from a compendium of commonly accepted methods)

   Method validation information (method validation is discussed in Chapter 24)

   The suitability of the test for a given purpose

2. *Information regarding each sample:*

   The sample ID number

   How the sample was collected, by whom, and on what date

   Where the sample is stored and the conditions of storage

   How and when the sample is to be discarded

3. *Information regarding each assay:*

   The sample tested

   The date of the test

   Who performed the test

   Reagents and materials used

   The method used to test the sample

   The raw data collected during testing

   Calculations for the sample results

   Reported conclusions based on the test

Table 6.9. *TYPICAL COMPONENTS OF A LABEL*

1. *Date the item was prepared.*
2. *The person responsible for the item.*
3. *The ID number of the item.*
4. *The lot number.*
5. *The identity, name, or composition of the item.*
6. *Safety information.*
7. *The name of the company or institution.*
8. *Storage and stability information.*

not be confused with one another, and that information and results are always associated with the correct sample. **Chain of custody documentation** *provides a chronological history, or "paper trail," for samples.* For example, in a clinical testing laboratory patient samples must be kept in order. An environmental testing laboratory tests samples from many sites and must keep track of them. Forensics laboratories must scrupulously keep evidence from various cases in order; otherwise, their test results will be invalid in court.

Chain of custody documents are a method of organizing information about samples. Each sample must be assigned a unique ID number. Records for each sample must show the source of the sample, who collected the sample, who transported the sample, its condition upon receipt, the date of receipt, how the sample was processed and tested in the laboratory and by whom, how it was stored, and how it was disposed of, if relevant. The sample is logged in and out as it is moved and processed. The format and the exact nature of these records are variable as each organization has its own requirements.

xi. TRAINING REPORTS

Training reports are associated with individuals working in a facility. The training report shows the training the person has completed, the dates, the purpose, and so on. Training reports help to show that individuals are competent to perform their work. In research laboratories,

ix. LABELS

Labels identify equipment, raw materials, products, and other items. Information that may be found on labels is summarized in Table 6.9. An example of a label is shown in Figure 6.5.

x. CHAIN OF CUSTODY DOCUMENTATION

In laboratories that handle many samples that come from diverse subjects or sites, it is critical that samples

---

Clean Gene, Inc.

Solution Name: _____

Concentration: _____

Date Prepared: _____   By: _____   ID No.: _____

Expiration Date: _____

Reagent Source: _____   Lot No: _____

Storage Conditions: _____   Special Precautions: _____

Figure 6.5. **An Example of a Label for a Reagent.**

for example, training records are used to document that employees have been trained in the use of radio-isotopes. An example is shown in Figure 6.6.

D. Documents That Are Specific to Production Facilities

i. BATCH RECORDS

There are many types of documents found in production facilities; a few of these will be explained here. **Batch** records accompany a particular batch of product. A **batch record** *includes step-by-step instructions that detail how to formulate or produce a product, including raw materials required, processing steps, controls, and required testing.* In this sense, it is a directive document, like an SOP. *A batch record also provides blanks in which the operator(s) records information and document activities as they perform them.* In this sense, the batch record is a data collection document. A **master batch record** *is the original signed batch record.* Each time a new production run begins, the master batch record is copied to generate a batch record for routine use.

The batch record is officially issued to the production crew by the quality department. It is essential that the batch record that is issued is complete, readable, and correct.

Table 6.10. ESSENTIAL COMPONENTS OF A BATCH RECORD

1. *Product identification.*
2. *Document identification.*
3. *Company name.*
4. *Dates of manufacturing.*
5. *A step-by-step account of the processing and testing to be done.*
6. *The monitoring specifications—how will the operators know if the process is proceeding properly?*
7. *Raw data that must be collected and blanks to fill in to record it.*
8. *Bill of materials.*
9. *Equipment block.*
10. *Required signatures.*

Table 6.10 describes some of the major components of a batch record. Figure 6.7 shows portions of a batch record.

ii. REGULATORY SUBMISSIONS

Regulatory submissions *are documents completed to meet the requirements of an outside regulatory agency.*

Clean Gene, Inc.

EMPLOYEE SOP TRAINING RECORD

Form # 875
Revision 01, 6/06

EMPLOYEE'S NAME

Rachel Allen

SOP TITLE	REVISION DATE	EMPLOYEE'S INITIALS	TRAINER'S SIGNATURE
Cleaning Hoods	*6/20/07*	*RA*	*JES*
Formulation 10	*1/18/08*	*RA*	*LW*
CaSO₄ Analysis	*2/3/08*	*RA*	*JMM*

Figure 6.6. Portion of an Employee SOP Training Record.

Clean Gene, Inc.

Part No. BR002	Batch Production Record		Doc No. BR002
Lot No.		CO No. C00012	Rev No. 002
Subject: CHO Spinner Daily Sample			Eff Date 28 Feb 07

1.0 REFERENCES:

1.1 BQ005- Nonconforming Product
1.2 BQ001- Line Clearance Procedure
1.3 BP0015- Aseptic Processing Techniques In LFWS
1.4 BQ003- Sampling Inspection Procedure
1.5 BR002- Released Inventory Log Sheet
1.6 BP015- Determination of Cell Viability
1.7 BQ002- Documentation and Verification of GMP Activities

2.0 BILL OF MATERIALS:

Description	Part No.	Lot No.	Planned Quantity	Actual Quantity Used	Expiration Date	Oper.

3.0 EQUIPMENT:

Equipment Description	Equipment ID Numbers (if applicable)	Oper.
Spinner flask		
Micropipettor(s)		
Laminar flow hood		
Microscope		
Hemacytometer		
CO2 Incubator		
Magnetic spinner plate		
BP-400		
Vi-Cell		
Water bath		

4.0 SUPPLIES:

4.1 Pipettes, various sizes
4.2 Gloves, sterile
4.3 Sleeves, sterile
4.4 Pipette tips (various)
4.5 Pipettes (various)
4.6 Wipes

Figure 6.7.   A Portion of a Batch Record. (Courtesy of Michael Fino.)

Clean Gene, Inc.   [not all pages are shown] Page 5 of 5

Part No.		Batch Production Record		Doc No.	
BR002				BR002	
Lot No.			CO No.	Rev No.	
			COOO12	002	
Subject:				Eff Date	
CHO Spinner Daily Sample				28 Feb 07	

Opr/Ver

5.0 PROCEDURE:

Date:_____   Time:_____ am   pm

5.1	Perform line clearance.	_____ / _____
❏	Area and equipment are free of any products from previous operations.	
❏	Area and equipment are free of any packaging materials from previous operations.	
❏	Area and equipment are free from any labels from previous operations.	
❏	Area and equipment are free from any documents from previous operations.	
5.2	Record all appropriate information in the Bill of Materials.	_____ / _____
5.3	Record all equipment being used on the equipment list.	_____ /
5.4	Prepare the biosafety cabinet for aseptic manipulations.	_____ /
5.5	Record the volume of this spinner culture (e.g., 100 mL, 500 mL).	
5.6	Culture volume: _____	_____ /
5.7	Sample the spinner culture.	
	5.7.1 Aseptically remove a small aliquot (3 mL) of the cell culture suspension to perform the following analyses:	
	5.7.1.1 Cell count (automated and manual)	
	5.7.1.2 Percent cell viability (automated and manual)	
	5.7.1.3 Cell culture media chemistries (BP-400)	
	[Three pages of directions are omitted]	
5.14	Clean area.	_____ /

Comments:_____
_____
_____

Document Review

Mfg. Review_____   Date_____   QA Review_____   Date_____

Figure 6.7. (*Continued*)

For example, before testing an experimental drug in humans, a company must submit an application to FDA showing its preliminary research on the drug, its plan for human studies, and other relevant information. (This document is known as an IND, or Investigational New Drug application.)

iii. RELEASE OF FINAL PRODUCT RECORDS

Companies must complete product release documents when a product has been manufactured and tested. The release document certifies the product, shows its specifications, establishes that the product documentation has been reviewed and approved, and states that the product is ready to be sold.

III. ELECTRONIC DOCUMENTATION

Our discussion of documentation so far has assumed that documents are written on paper; indeed, the entire system of documentation in the biotechnology industry was created in a world of paper records. But this assumption does not meet the reality of our "electronic age." Computers are widely used to obtain, analyze, and store information, as discussed in Chapters 34 and 35. Most modern laboratory instruments are connected to computers and/or are controlled by internal microprocessors. Raw materials, products, and samples are routinely bar coded and their movements and disposition tracked by computer.

It is easy to imagine advantages to replacing paper with electronic documentation. For example, manufacturing technicians traditionally record critical information (e.g., material lot numbers, times tasks are performed, temperatures, test results, calculations, equipment identifications) on paper batch records. An electronic batch record system has the potential to use the computer to detect errors such as an omitted lot number, a temperature that is not in the correct range, a test result that is improperly transcribed, and so on. The computer's ability to instantly recognize errors could prevent costly mistakes. Computers similarly have advantages in a research environment. Researchers are beginning to use "electronic notebooks," computers with software designed to perform the roles of a traditional laboratory notebook. Electronic laboratory notebooks can easily store huge amounts of data, search the data, and enable researchers to readily communicate with others via computer.

In March 1997, the FDA issued regulation 21 CFR Part 11 Electronic Records; Electronic Signatures; Final Rule, to address the role of computers in documentation in the pharmaceutical industry. The purpose of the Part 11 regulations is to "provide criteria for acceptance by FDA, under certain circumstances, of electronic [computer] records, electronic signatures, and handwritten signatures executed to electronic records as equivalent to paper records and handwritten signatures executed on paper. These regulations, which apply to all FDA program areas, were intended to permit the widest possible use of electronic technology, compatible with FDA's responsibility to protect the public health" (Food and Drug Administration. "Guidance for Industry Part 11, Electronic Records; Electronic Signatures—Scope and Application." August, 2003. http://www.fda.gov/cder/guidance/5667fnl .pdf). Table 6.11 provides some terminology relating to electronic documentation and 21 CFR 11.

The Part 11 regulations are intended to encourage pharmaceutical companies to adopt modern electronic documentation methods. At the same time, they require that companies validate new electronic documentation methods to prove they are as secure, reliable, and searchable as paper systems. Creating a computer documentation system that is compliant with Part 11 requires a very sophisticated understanding of computer software and hardware—both on the part of computer vendors and companies. The pharmaceutical industry has spent considerable effort and resources developing this understanding and the lessons they have learned are of general interest.

To understand the challenges in using computers for documentation, consider these underlying principles of any good documentation system:

- The proper date must be clearly attached to each event recorded in any document.
- Records should be attributable to a particular individual.
- Documents must be secure, that is, they must be safe from theft, natural disaster, and from access by the wrong people.
- Documents must be readily accessible when needed.
- Documents must provide information for traceability, so that, for example, all the raw materials used in a batch of product and the documents associated with the raw materials can be identified.
- Records should not be capable of being altered either accidentally or intentionally.

As we have seen previously in this chapter, a variety of effective practices have evolved for paper documents in order to meet these bulleted requirements. For example, laboratory notebooks must be signed, dated, and composed in chronological order. These requirements help ensure that the proper dates are attached to events, that the work is attributable to an individual, and that records are kept honestly. Paper documents in a company are organized and stored in rooms with a security system so that they are accessible only to authorized personnel.

Computer systems must have different methods to provide the same controls. Consider, for example, signatures, a key element in a paper documentation system.

Table 6.11. Some Vocabulary Often Used with Reference to 21 CFR Part 11 and Electronic Documentation

Audit Trail. A secure, computer-generated, time- and date-stamped record that allows the reconstruction of a course of events relating to the creation, modification, and deletion of an electronic record.

Biometrics. A method of verifying an individual's identity based on the measurement of physical features or repeatable actions that are unique to that person. For example, fingerprint or retinal scans can be used to identify individuals. A signature can be considered to be a biometric method.

Closed System. A computer system in which access is controlled by the people who are responsible for the content of the system's records. For example, a system of computers that is only accessible to the individuals who work in a company is a closed system.

Electronic Laboratory Notebook, ELN. Any of a wide variety of software program/computer systems designed to fulfill the functions of traditional paper laboratory notebooks. ELNs provide the advantages of computers: They can be electronically searched for specific information, they can archive very large data files, they can hold and display graphics, and they can be used to share information locally and remotely.

Electronic Records. Text, graphics, data, audio, or pictorial information that is created, modified, maintained, archived, retrieved, or distributed by a computer system.

Electronic Signature. A computer equivalent to a handwritten signature. In its simplest form, it can be a combination of a user ID plus password. It may also include identification based on biometric characteristics.

Encryption Software. Software that translates information into a secret code in order to provide security. To read an encrypted file one must have access to a secret key or password.

Hybrid Systems. Systems that use both electronic and paper records. For example, a laboratory instrument might be attached to a computer that retrieves and processes data from the instrument, and then prints out a result that is signed and dated.

LIMS, Laboratory Information Management Systems. Computer-based laboratory systems that automate such activities as tracking work requests, tracking samples, printing analytical worksheets, storing data, analyzing data, performing calculations, providing financial statistics, tracking client requests, and so on.

Metadata. Information that describes the content and context of the data. They help to reconstruct the original raw data. For example, a digital camera produces both a picture and also metadata that includes the camera's shutter speed, f-stop, and other camera settings when the photo was taken.

Open System. A computer system that is not controlled by the persons who are responsible for the content of the system. For example, if a contract laboratory sends data to a company via the Internet, the system is open. Additional security must be in place for open systems as compared to closed systems.

Predicate Rules. The GMP, GCP, GLP, and other regulations (as contrasted with the 21 CFR Part 11 regulations).

Source: Many of these definitions are modified from: Huber, Ludwig. "21 CFR Part 11: Overview of the Final Document and its New Scope." In *21 CFR Part 11; A Technology Primer, Supplement to Pharmaceutical Technology,* Iselin: NJ: Advanstar Publications, 2005, http://www.pharmtech.com/pharmtech/issue/issueDetail.jsp?id=6515.

A signature identifies the signer and generally means that a person consents to something. This is the case, for example, when you sign a credit card authorization. The signature is in ink that permanently binds the signature to the paper so that it's difficult to remove without leaving a trace. Your signature on the form means that you are the proper holder of the credit card and that you agree to the charge. In a paper laboratory notebook, an individual's signature identifies that person and is a method of attesting to the truth of the data recorded. A commonly accepted electronic signature is a log-on procedure in which individuals must enter a unique user ID and secret password. Everything recorded onto the computer while that individual is logged on is attributed to that person. If a person signs on with another's password, it is comparable to forging another person's signature. In situations where more assurance of an individual's identity is required, sophisticated methods of authentication, such as voice recognition and retinal scans, can be used.

There are guidelines (see Table 6.3) to assure the chronology of events recorded in a paper laboratory notebook. Chronology in electronic systems is usually handled with a "time-stamp" that is automatically added by the computer. Time-stamps require software that "knows" the correct time, and that can detect if someone attempts to alter a time-stamp.

It is important that records be private, safe, and that only those with the authority to see them have access. Paper laboratory notebooks can be stored in a locked file cabinet or secure storage facility to meet these requirements. With a computer system, passwords are a primary method of maintaining security and much effort has gone into designing systems to prevent unauthorized access. Data that are sent through an open system (such as the internet) pose a particular security problem that is usually solved with encryption software.

With paper documentation systems, operators record entries in permanent ink to prevent their change. With a computer, a software method of tracking modifications to data must be in place to ensure that original recordings are not erased. If someone legitimately tries to add information to a record, the computer must be programmed to "know" that this act is legitimate and to show both the original record and the revision.

The rules in Part 11 require that pharmaceutical companies prove that the various software controls (such as those mentioned in the above paragraphs) are effective in protecting the integrity of their documents. For example, companies (and software providers) must demonstrate that if someone intentionally tries to change a time-stamp or penetrate a password-protected site that the software is capable of resisting these incursions and recording the attempts. Companies must show that the integrity of data being recorded at the moment a computer crashes can be guaranteed. Computer software and storage devices rapidly become obsolete: This is a major challenge to companies since they must guarantee that their records will be accessible in the future, even if technology changes.

Resolving these technical issues has slowed the adoption of computer-based documentation in regulated industries. In some cases the FDA has decided to be flexible to encourage adoption of electronic documentation. For example, the FDA has decided that companies may archive records in a copied form, such as a PDF file, or a paper record, to avoid problems if an electronic storage medium becomes obsolete. Guidance from the FDA in 2003 indicated that the agency would adopt a risk-based approach to enforcing the requirements of Part 11. This means that documentation that could be expected to impact patient health must be fully validated and compliant with all the requirements of Part 11, but systems of low risk to public safety and health might be less rigorously secured and validated. A critical record would be, for example, the results of quality-control laboratory tests on a final product that are used to decide whether or not the product should be released for sale. These records are critical because the tests help ensure that the product is of high quality. A lower risk record would be, for example, a schedule for employee GMP training sessions.

The general trend in the industry is toward increasing use of electronic documentation in all areas, from laboratory notebooks to sample labeling. Computer technology continues to improve and methods to ensure data integrity are being developed. Remember though that the core principles of documentation are the same, whether paper or computers are used. The principles of keeping a chronological, honest, complete laboratory notebook; writing clear SOPs and forms; ensuring chain of custody for samples; and so on, will remain relevant whether the professional wields a pen or a mouse.

CASE STUDY

Examples Relating to Documentation

The Federal Food and Drug Administration has inspectors who periodically inspect pharmaceutical facilities and biotechnology companies that make regulated products. If the inspectors observe violations of Good Manufacturing Practices, they note the violations on forms, called "483s," and in official Warning Letters sent to the company. If companies fail to correct their deficiencies, then FDA can cause products to be seized and destroyed, fines to be levied, and, in the most extreme cases, individuals in the company may be charged as criminals and may be imprisoned if convicted. The following are excerpts from actual warning letters sent by FDA to various companies. Improper documentation is frequently cited in warning letters. Observe in these warning letters how carefully inspectors checked for proper documentation and the details of their findings.

Warning Letter, Example 1

Dear Sir or Madam:

During an inspection of your drug manufacturing facility ...conducted on August 29 through September 25, 1996, our investigators documented serious deviations from the Current Good Manufacturing Practices Regulations ... Deviations from the GMPs documented during this inspection included:

[A list of 24 violations follows. Only some of those violations relating to documentation are excerpted here.]

(6) Failure to maintain master production records containing: (a) signature and date of the person preparing the record, and date and signature of the person independently checking the record; (b) strength and description of the dosage form; complete manufacturing and control instructions, sampling and testing procedures and specifications for the finished product; ... and (d) a description of the drug product containers, closures and packaging materials, including a specimen or copy of each label or other labeling signed and dated by the person or persons responsible for approval of such labeling ...

(15) Failure to establish written procedures to prevent microbiological contamination of manufactured products....

(21) Failure to document the person performing the analysis, when the analysis was conducted, and/or the name of the active ingredient in ...spectra used as an identity test of active ingredients.

The above identification of violations is not intended to be an all-inclusive list of deficiencies at your facility ...You should take prompt action to correct these deviations.

Failure to promptly correct these deviations may result in regulatory action without further notice . . .

Sincerely,

[Signed by the district director]

Warning Letter, Example 2

Dear . . .

Inspection of your unlicensed hospital blood bank . . . revealed serious violations . . .

Inspection revealed that [prior] blood product disposition records . . . are not available. According to your blood bank supervisor, the missing disposition records were transferred to a computer system and were subsequently "lost" by that system. Your supervisor stated that the computer system has not been validated and is being used only for "practice." Your supervisor also stated the blood bank had written back-up records for the data in the computer system. However, written disposition records could not be produced during the inspection for review . . .

Sincerely . . .

[Signed by the district director]

PRACTICE PROBLEMS

1. Figure 6.8 is a brief portion of a batch record that covers the formulation of a particular product. Note that it consists of a series of steps to be performed by the operator and a series of blanks that the operator fills out as he or she performs each step. The operator initials each step as it is performed and another person verifies that the procedure was properly executed. There are a number of errors in how this batch record was completed. Circle the errors. (Note that it is not necessary to understand the actual procedure in order to detect the errors in how the form is completed.)

2. People commonly explain documentation with the slogan "Do what you say, say what you do." Explain this slogan with reference to the types of documents described in this chapter.

3. Imagine that an entrepreneur opens a new chocolate chip cookie bakery. Discuss the documentation requirements for this new operation.

4. Discuss the warning letters from the FDA in the Case Study Examples. What had these companies failed to document? What are the potential adverse consequences of these problems with documentation?

FORMULATION OF XYZ COMPOUND

Clean Gene, Inc.

3550 Anderson St.
MADISON, WI 54909
Revision 01
Batch Record # 133
Approved by _____*Aaron Reid*_____   _____*Anna Gold*_____   _____*Sam Rothstein*_____
              Date *2/14/08*_____   Date *2/14/08*_____   Date *2/16/08*_____

Issued by: _____*Erin Jane*_____   **Date** *12/3/08*_____   **Lot #** **15.987**

Product Name: Very Good Product

Strength: 10 Units/mL   **Vial Size   10 mL**   **Batch quantity: 350 L**

Reference: Refer to separate Formulation SOP Q75 for quantities of each component.
            Refer to separate instrument/equipment SOP 76 for ID information

---

NOTE: PRODUCT IS TO BE STIRRED CONTINUOUSLY DURING COMPOUNDING AND FILLING

A.   COLLECTION OF WATER FOR INJECTION (WFI), USP.

A1. Collect approximately 370 L of WFI, USP, in a clean, calibrated vessel and cool to 24°C–28°C.
Vessel # ___7___   Amount collected ___*365 L*___   Initial Temperature ___*23°C*___   Time ___*8:00*___ (am)/pm
Final Temperature ___*26°C*___   Time ___*08:15*___ am/pm

                                                    LJ / *JM*
                                                 12/10/08   12/10/08

A2. Close the water for injection valves.
                                                    LJ / *JM*
                                                 12/10/08   12/10/08

A3. Remove about 20 L of the cooled WFI from step 1 and place in a clean, calibrated vessel.
     (This water will be used to bring the final formulation to the proper volume.)

Vessel # _____2_____
                                                    LJ / *JM*
                                                 12/10/08   12/10/08

A4. Remove about 5 L of the cooled WFI from step 1 into a clean, calibrated vessel. (This water will be used to
     prepare the solutions used to adjust the pH.)

B.   PREPARATION OF pH ADJUSTING SOLUTIONS

B1. Collect 750 mL cool WFI from the vessel in step A4 and place into a 1000 mL volumetric flask and add
     100 g of NaOH (Sodium Hydroxide # 875) USP and dissolve.

Amount of WFI collected ___*750 mL*___   Amount NaOH added ___*115 g*___   Lot # _____

Time step completed ___*09:15*___ am/pm
                                                   _____ / *JM*
                                                         12/10/08

B2. Using a water bath containing cold WFI, cool the solution prepared in B1 to 25°C ± 5°C.

Final Temperature ___*31°C*___
                                                    LJ / *JM*
                                                 12/10/08   12/10/08

B3. Bring the solution to 1000 mL with cool WFI from the vessel in step A4.

Approximate volume of WFI added ___*300 mL*___   Time completed ___*09:00*___ (am)/pm
                                                    LJ / *JM*
                                                 12/10/08   12/10/08

     and so on

Figure 6.8.   Practice Problem 1.

CHAPTER 7
Quality Systems in the Production Facility

I. INTRODUCTION

II. ISSUES RELATING TO RESOURCES

    A. Facilities and Equipment

    B. Handling Raw Materials

III. SPECIFICATIONS

IV. PROCESSES

V. VALIDATION OF PROCESSES AND ASSOCIATED EQUIPMENT

    A. Introduction to Validation

    B. When Is Validation Performed?

    C. Process Validation Planning

    D. Activities Involved in Validation

    E. Unplanned Occurrences

I. INTRODUCTION

This chapter provides a brief overview of the issues involved in manufacturing quality biotechnology products. Before discussing how companies make sophisticated products (like recombinant proteins or transgenic plants), however, let us look at a product with which most of us are more familiar.

Suppose that a man who bakes for a hobby decides that his chocolate chip cookies are so well-liked by his friends and family that he wants to open a business producing and selling them. There are many issues to be resolved as he plans his new business. He has completed the initial R&D phase, having experimented with recipes and ingredients, and having tested the quality of his cookies on a small sample of people. As the baker scales up for commercial production, he will need to identify the features that made his cookies suc-

cessful and formalize these features into written specifications. The baker will need to acquire resources, such as a kitchen with large ovens, ample counter space, and refrigerators. He will need to be certain that his processes for making cookies, which were effective in his own kitchen at home, also work in a larger, commercial kitchen. He will need a program to ensure that his facilities and equipment remain clean and operate properly. He will need written recipes and assistant bakers to follow the recipes. He must find ways to ensure that his cookies always bake at exactly the right temperature, for exactly the right length of time. He will need methods to ensure that all his incoming raw materials are of acceptable quality, and will need to have suitable places to store ingredients so they remain fresh. He will need to consider packaging, labeling, and shipping of his completed cookies. Whereas personally sampling the cookies was probably adequate quality control in the early stages of his venture, the baker will need a more formal quality-control program once he begins commercial production. He will have to comply with all regulations relating to food processing, general safety regulations, and environmental regulations, and he might also voluntarily comply with quality standards for cookie makers. Manufacturing quality chocolate chip cookies is a complex goal with many components, all of which need to be planned and coordinated.

The manufacturing issues in a biotechnology company resemble those facing the industrious baker. A biotechnology company must also ensure that adequate resources are available, including facilities, equipment, personnel, and raw materials. Just as the baker must control the temperature of the ovens and the length of baking time, so must a biotechnology company control its processes that turn raw materials into products. Products—whether they are cookies or monoclonal antibodies—must be properly labeled, must be tested for final quality, and must be shipped to their destination. Skilled personnel must be available to carry out all the necessary tasks. Regulations and standards must be met. Although the basic issues are similar in the cookie kitchen and the biotechnology company, there are differences as well. A biotechnology company is likely to produce many different products and at a much larger scale than would the new bakery. Biotechnology products are likely to be far more technically complicated than chocolate chip cookies. The quality requirements in a biotechnology production facility are therefore even more complex and varied than the issues facing the bakery.

II. ISSUES RELATING TO RESOURCES

A. Facilities and Equipment

A production facility must be able to support the production of products. Consider, for example, a facility in which a drug is manufactured. The building layout should be organized so that processing steps flow in an orderly fashion from one place to another. Raw materials that have been tested and accepted for use should be stored in a space separate from unapproved materials to avoid confusing the two. Materials that are nonsterile must be physically separated from sterile ones. The paths of finished products ideally should not cross the paths of raw materials because of the possibility of confusing them or contaminating the finished product. The facility must have controls for environmental factors, such as humidity, temperature, dust, and particulates in the air. The facility must be sufficiently large to accommodate all equipment and personnel safely.

A manufacturing facility must be both well designed and maintained properly. This involves a housekeeping program to keep the facility clean, a pest and rodent con-

trol program, and environmental monitoring and control.

The equipment and instruments in the facility likewise must be suitable for their purposes and must be properly maintained. An important aspect of equipment and facility monitoring is ensuring that all measuring devices operate properly. Measuring instruments are scattered throughout any facility. For example, thermometers are associated with freezers, refrigerators, sterilizing devices, incubators, and rooms. These thermometers must be functioning properly; otherwise, there is no assurance that materials are being held at the proper temperatures.

EXAMPLE

It is determined during the development of a product that the temperature at which the product is produced affects its quality. Temperature, therefore, must be monitored and controlled in the production area of the facility. The program for the control of temperature involves:

1. During the development of the product, laboratory tests are performed to determine the range of temperatures that is acceptable for production. Based on these tests, it is determined that the temperature of the processing area must remain between 65 and 67°F.

2. The temperature in the facility is controlled by the heating and air conditioning system.

3. A thermostat is installed in the production area. The thermostat is designed so that it:

   — monitors the temperature on the production floor

   — automatically prints out a continuous temperature recording

   — connects to the heating and air conditioning system and turns them on or off as needed.

4. A program of scheduled maintenance is implemented to check all parts of the system regularly including the thermostat and the heating and air conditioning equipment. Every week the thermostat is checked and its readings verified to be accurate with a thermometer known to be correct. The heating and air conditioning systems have monthly maintenance routines. The monitoring program is explained in documents. Records are maintained each time equipment is inspected, adjusted, or repaired.

5. If the temperature is about to go out of range in the production area, an alarm is triggered. There is a document that explains to workers what actions to take if the alarm sounds, how to record their activities and observations, and how to follow-up after the problem is resolved. This follow-up should describe what to do with product that was in process when the alarm sounded.

CASE STUDY

Example Relating to Facilities and Equipment Management

Water is one of the most important raw materials in biotechnology production facilities. The following case study describes a situation where a company failed to observe basic quality practices, resulting in contamination of their water. This negligence resulted in patient illness. (The company was subsequently able to correct their deficiencies.)

In 1995 an epidemiologist working at a hospital in Milwaukee noticed that 12 hospital patients, all of whom were on ventilators, developed Pseudomonas cepacia infections. The hospital laboratory was able to find the bacterium responsible for the infection in the patients' sputum and in bottles of "Fresh Moment" mouthwash. Although this bacterium is seldom harmful to healthy people, P. cepacia can be dangerous to people who are ill.

The FDA traced the manufacture of the mouthwash to a particular company, which was subsequently inspected by an FDA investigator. The investigator found six problems thought to have caused the contamination in the mouthwash:

1. *The company's purified water system had not been properly cleaned and tested since earlier in the year. The company's president told FDA officials that the company skipped the scheduled maintenance because it could not afford it.*

2. *The reverse osmosis membranes of the water system had not been changed for more than a year, although this should be done every 4 months, according to the company's written procedure, to prevent microbial buildup. (Water purification systems are discussed in Chapter 28.)*

3. *Employees failed to challenge the system with P. cepacia when they were testing the system. The challenge test would have involved intentionally adding the bacterium to water to see if their purification system could remove it.*

4. *Employees did not use the appropriate hose clamps on equipment. The clamps they used appeared to be the type of hose clamps that are used in the automotive trades, and they were rusty, dirty, and discolored.*

5. *Employees left doors open during production, allowing dust from outside to enter the manufacturing area.*

(This account is excerpted from a report in The FDA Consumer, October 1996.)

B. Handling Raw Materials

Raw materials (e.g., the chemicals, empty vials, and other components that go into making a product) are resources. Unlike facilities and major equipment, raw materials flow into the company, are used and transformed into products, and flow out of the company in another form. Systems must be in place to receive materials, document their arrival, and ensure that the materials received are the ones ordered. Once received, the materials must be quarantined until they have been tested and approved by the Quality Unit. This testing confirms that the material is what it is supposed to be and that it is properly labeled. In a company that makes regulated products, raw materials are chemically tested in the laboratory to be certain of their identity. ISO 9000 has the following requirement relating to raw materials testing:

> **4.10.2 Receiving inspection and testing** *The . . . [company] shall ensure that incoming product is not used . . . until it has been . . . verified as conforming to specified requirements . . .*

Some of the principles that guide handling raw materials are summarized in Table 7.1.

III. SPECIFICATIONS

Specifications *are descriptions that define and characterize properties that a product must possess based on its intended use.* There are specifications for raw materials and for products. Specifications are described in this quote from an FDA document (*Guideline on General Principles of Process Validation,* Food and Drug Administration, 1987).

> During the research and development . . . phase, the desired product should be carefully defined in terms of its characteristics, such as physical, chemical, electrical and performance characteristics . . . It is important to translate the product characteristics into specifications as a basis for description and control of the product . . . For example . . . [c]hemical characteristics would include raw material formulation . . . The product's end use should be a determining factor in the development of product (and component) characteristics and specifications.

We briefly discussed possible specifications for chocolate chip cookies in Chapter 6. Table 7.2, on p. 116 shows another example of specifications, in this case, for salt. The manufacturers' specifications for three salt products are shown: road salt, table salt, and salt for laboratory use. These specifications illustrate several points about specifications:

1. **Manufacturer's specifications describe properties that are important for that product based on its intended use.** Important specifications for all three salt products include their identity (NaCl, FW 58.44) and their purity. The specifications for road salt, however, mention anticaking agents, table salt specifications mention moisture content, and salt for laboratory use specifies maximum allowable levels for a number of trace contaminants. All three products are salt, all are called "sodium chloride," yet because they have different uses, they have specifications for different properties.

   All pertinent aspects of the product that impact its performance, reliability, effectiveness, safety,

Table 7.1. HANDLING RAW MATERIALS

1. *It is essential not to confuse or mislabel items.* Consider, for example, the serious consequences that could occur if a component of a drug product is accidentally mislabeled.

2. *Raw materials intended for production should not be released to production until they have been tested and found to conform with their specifications.* To ensure that materials are not used prematurely, they are quarantined. Materials are not removed from quarantine until they have been approved for use by personnel from the Quality Unit.

3. *Storage conditions should take into consideration hazards associated with the material, its perishability, and its temperature and humidity requirements.* A system should be in place so that materials in storage can be readily located when needed.

4. *Traceability of materials must be assured.* Ensuring traceability requires, for example, that all incoming and outgoing materials are recorded, records of raw material components that go into each product are maintained, and items are properly labeled.

CASE STUDY

Examples Relating to Resources

The Federal Food and Drug Administration has inspectors who periodically inspect pharmaceutical facilities and biotechnology companies that make certain products. If the inspectors observe violations of Good Manufacturing Practices, they note the violations on forms, called "483s," and in official warning letters sent to the company. The following are excerpts from three warning letters that address issues relating to facility design and cross-contamination, equipment maintenance, and raw materials testing. Observe in these warning letters how carefully inspectors checked the facilities and the details of their findings.

Warning Letter, Example 1

Dear . . .
FDA has completed its review of the report on the October 24–31, 1994, inspection of . . . manufacturing sites . . . That . . . inspection revealed a number of significant deviations from FDA's Current Good Manufacturing Practice (cGMP) regulations . . . These observations included . . . the following:
We are concerned that your firm was operating, in an open area, a number of compression machines running different products that provided a potential for product cross-contamination.
Sincerely . . .

Warning Letter, Example 2

Dear . . .
During our August 26, 1996, through September 18, 1996, inspection of your facility . . . our investigator docu-

mented deviations from Current Good Manufacturing Practice . . . Regulations . . . The following deviations pertaining to the manufacture of drug products were found:

1a. Failure to perform and document a specific identity test on a sample of each lot of drug raw material . . .

1b. Failure to test drug raw materials that have been in inventory for extended periods of time to ensure that they will meet specifications for purity, strength, and quality, prior to being used for manufacture of drug products . . .

6. Failure to maintain up-to-date cleaning logs and document cleaning, sanitizing, and maintenance of drug-manufacturing equipment, such as the 300 gallon stainless steel vat #1, 10 gallon stainless steel vat #3, and the bottling machine.

You should take prompt action to correct these deviations . . . Sincerely, . . .

Warning Letter, Example 3

Dear . . .
Inspection of your unlicensed hospital blood bank on April 2–3, 1997 . . . revealed serious violations . . .
Inspection . . . revealed that established written procedures for storage temperature controls, equipment maintenance, and record keeping are not being followed. The investigator documented numerous instances where temperature recorder charts on storage refrigerators exceeded specifications, whereas daily temperature logs show that refrigerator temperatures were within specified limits. Your supervisor stated that one temperature recorder is broken and personnel are allowed to bend the arm/pin back into place in an attempt to maintain the device. No documentation is available to show calibration and maintenance of the recording thermometers.
Sincerely . . .

and stability should be considered when establishing specifications.

2. **The specifications for the same property may vary depending on the intended use.** Although all three salts are NaCl, the salt intended for use in the analytical laboratory is required to be much purer than the other two products. It would be wasteful to spread ultra purified salt on roads and it would be unacceptable to use road salt in the laboratory. None of these products is "better" than the other; they are simply intended for different purposes.

3. **Specifications always are associated with analytical methods.** In order to tell whether a particular batch of salt meets its specifications, it is necessary to have a method to measure its purity, moisture content, and so on. In many cases the evaluation of a product to see if it meets its specifications requires a sophisticated test. For example, it is fairly easy to count the number of chocolate chips in

Table 7.2.   MANUFACTURER'S SPECIFICATIONS FOR ROAD SALT, TABLE SALT, AND SALT INTENDED FOR LABORATORY USE

	Road Salt (NaCl, FW 58.44)	Table Salt (NaCl, FW 58.44)	Analytical Grade Salt (For Laboratory Use) (NaCl, FW 58.44)
Chemical Purity	Minimum 95% NaCl	Minimum 97% NaCl	Minimum 99% NaCl
Color	Clear to white, yellow, red, or black	Clear to white	Clear to white
Maximum Allowed Contaminants	Not specified	As 0.5 ppm Cu 2.0 ppm Pb 2.0 ppm Cd 0.5 ppm Hg 0.10 ppm	Al < 0.0005% As < 0.0001% Ba < 0.0005% Ca < 0.002% Cd < 0.0005% Co < 0.0005% Cr < 0.0005% Cu < 0.0005% Fe < 0.0001% and so on
Physical Requirements	90% of crystals between 2.36 mm and 12.5 mm	90% of crystals between 0.3 mm and 1.4 mm	95% of crystals between 0.18 mm and 0.3 mm
Allowed Additives	Anticaking agents of 5–100 ppm Sodium Ferrocyanide Ferric Ferrocyanide	Coating agents, hydrophobic agents	Not allowed
Moisture	2–3%	≤ 3%	Not specified

each cookie, but it requires an analytical method to analyze the percentage of cocoa content of those chips. Analytical methods used to evaluate specifications must be proven to work properly.

4. **Specifications are written as a range of permissible values because there is inevitably some variation from product to product.** Acceptable ranges or limits should be established for each specified characteristic and should be expressed in readily measurable terms. (Note that the salt specifications express contaminants in terms of limits, that is, the maximum allowable contaminant levels. Crystal size is specified as a range of acceptable values.) The validity of the specification ranges or limits must be verified through testing the product. In addition, as mentioned earlier, the acceptable range may vary for the same compound depending on its intended use.

The preceding examples are of specifications for a final product. Specifications are also required for raw materials, equipment, and processes. For example, the quality of chocolate chips (a raw material) used to make cookies is important in determining the quality of the cookies. The requirements for the chips, therefore, must be specified. Processes also have endpoints that can be described and specified.

Establishing specifications is a key element in a quality program. Specifications document the product quality that must be achieved and are a contract between the producer and the consumer. The term **establishing specifications** *means:*

- *Defining the characteristics of the product, material, or process*
- *Documenting those specifications*
- *Ensuring that the specifications are met*

Establishing specifications occurs during the development phase of a product. During development, the specifications may evolve and change. Once large-scale production begins, specifications should remain fixed.

Deciding what properties should be specified for a product, raw material, or process is not a simple task. Establishing specifications requires knowledge of how the product, material, or process will be used and the properties that will make it suitable for that use. It is also complex to define ranges for specifications. If the range for a specification is set too tightly (i.e., if the acceptable range of values is very narrow), then some adequate product or materials might be rejected. On the other hand, if the range of values for the specifications is too broad, then the quality of the product is not protected.

Because the setting of specifications is both a key component of a quality program and a difficult task, FDA scrutinizes specifications for products it regulates. FDA will not accept specifications if they are not complete, if they are unsuitable for the product, if their range is too broad, if they are unsubstantiated by testing, or if suitable analytical methods to test them are not available.

EXAMPLE

Write specifications for an imaginary enzyme, called enzymase, which is used to destroy all enzymes, other than itself, in a reaction mixture.

The intended purpose of enzymase is to destroy other enzymes. An assay for enzyme destruction, therefore, is developed and the activity of the enzyme preparation is specified using this assay.

The enzyme must be purified before sale. R&D researchers found that if the enzyme is at least 60% pure or better, then it functions well in various applications.

The specifications for enzymase are shown in Figure 7.1.

Enzymase catalyzes the cleavage of any enzyme, destroying its activity. The reaction requires the cofactor Mg^{++}.

MOLECULAR WEIGHT: 68kD

COFACTORS: Mg^{++}

INHIBITION: 50% inhibition by greater than 100 *mM* NaCl and chelators of Mg^{++}

INACTIVATION: Inactivated by heating to 70°C or by the addition of EDTA

SOURCE: Isolated from *B. subtilis* recombinant clone

PURITY: 60–80% pure with 125–130 units/*mg* dry weight

STABILITY: Supplied as a lyophilized powder and stable at least 2 years when stored at 2–8°C. Reconstitute with 10 mM phosphate buffer at pH 5.8.

ASSAY: Enzymase is assayed by the modified enzymase assay at pH 5.8. Enzymase is incubated with purified enzyme for 60 minutes at 37°C. The extent of digestion is determined by the colorimetric ninhydrin method. The amino acids released are expressed as micromoles of leucine per milligram dry weight of enzyme. One unit is equal to one micromole of leucine released in 60 minutes.

Figure 7.1. Specifications for the Imaginary Enzyme, "Enzymase."

IV. PROCESSES

Products are made by processes. ISO 9000 defines a **process** as *"a set of interrelated <u>resources</u> and <u>activities</u> which transform <u>inputs</u> into <u>outputs</u>"* (American National Standard ANSI/ISO/ASQC A8402-1994 *Quality Management and Quality Assurance Vocabulary,* ASQC, Milwaukee, WI, emphasis added). Inputs are raw materials. Outputs are products or intermediates that lead to

products. Resources include personnel, facilities, equipment, and procedures. Activities involve one or more people who use the raw materials, instruments, equipment, and procedures to make a product or intermediate. In order to obtain a quality product, it is essential that processes be carefully designed and developed by the R&D unit. Production operators in a biotechnology company participate in setting up, monitoring, and controlling the processes.

To illustrate a biotechnology process, consider microbial fermentation. Microbial fermentation is a process in which large numbers of microorganisms are grown in nutrient medium contained in large vats, called *fermenters.* During the fermentation process the microorganisms produce a product, such as an enzyme or antibiotic, which can be isolated and purified. The raw materials in a fermentation process include bacteria and the components of the nutrient medium. The activities involved in the fermentation process include preparing the medium, adding the medium and bacteria to the fermenters, monitoring the condition of the microorganisms, and maintaining proper conditions in the fermenters. The resources involved in the fermentation process include raw materials, fermenters, instruments to monitor conditions in the fermenters, personnel to perform the required activities, and procedures for the operators to follow.

The design of a process begins during the research and development phase and is refined as the product enters production. Every process must be designed so that the inputs are efficiently converted into the desired product. An important aspect of designing a process is finding ways to monitor the process as it occurs, and finding ways to control the process so no problems arise. To effectively design, monitor, and control a process, it is necessary to understand it. For example, if there are minor contaminants in the raw materials, will this affect the process? Is the process very sensitive to a particular factor, such as temperature or pH? Do steps of the process need to be completed within a certain time limit? The factors that are found potentially to affect the process adversely are factors that must be monitored and controlled. These factors are sometimes called **critical points** or **control points.**

Monitoring and controlling a process may occur in various ways. Sometimes samples are removed while the process is in progress and are analyzed for a critical characteristic. Such testing is called **in-process testing.** Specifications for the in-process test results are established. If samples do not meet their specifications, then operators may follow a procedure to correct the problem, or they may initiate an investigation.

Sometimes processes are automatically monitored and controlled. For example, temperature and pH probes are routinely inserted into fermenters during production runs. If the temperature or pH exceeds certain limits, then operators may perform corrective actions, or a computer-controlled system may make adjustments. The FDA's 2002 initiative, *Pharmaceutical cGMPs for the 21st Century: A Risk-Based Approach* (see Chapter 5), contains provisions to encourage companies to use modern, automated process control methods. **Process analytical technology, PAT,** *is computer based and relies on sophisticated detection methods that very quickly monitor the nature of a product and its intermediates and adjust manufacturing conditions as needed.* Modern PAT technology is being implemented, for example, to detect microbial contaminants in a drug batch during the manufacturing process and to monitor levels of methionine and glucose (required nutrients) during fermentation. Improvements in real-time process monitoring and control are expected to improve the quality of products and the efficiency of production.

EXAMPLE

a. Discuss the development of a fermentation process to produce the hypothetical enzyme, enzymase.

1. The output of the process is enzymase.

2. An essential early step is to develop a method to analyze the amount of enzymase present. A spectrophotometric assay is developed that is used to measure the amount of enzymase present in samples of the broth.

3. The R&D team performs laboratory experiments to see which bacterial strain makes enzymase most efficiently; the medium formulation that results in the best enzyme yield; the temperature, pH, and aeration conditions under which the bacteria are most productive; and so on. These early experiments provide the information necessary to design the process.

4. The R&D team establishes the fermenter operating parameters, the nutrient medium formulation, and the process endpoint specifications. The endpoint specifications may include, for example, measurements of the number of bacterial cells per milliliter and the amount of enzymase produced per gram of bacteria.

   (Note that the specifications for the endpoint for the fermentation process are not the same as they are for the final product. This is because the enzymase produced by the bacteria must go through a series of isolation and purification steps before it is ready for packaging and sale. These isolation and purification steps are together referred to as **downstream processing.** Thus, the fermentation process produces an intermediate product with its own specifications.)

5. The preparation of standard operating procedures (SOPs) begins in R&D and continues as the process is scaled-up. Once the process is in place and pro-

duction of enzymase begins, changes to the process are tightly controlled.

b. What processing controls are required for the fermentation of enzymase?

1. Preliminary experiments indicate that the hypothetical process is sensitive to pH, temperature, and oxygen levels in the nutrient medium. Early experiments are therefore performed to find the acceptable range for each of these three parameters and to find means to keep them in the correct range. For example, it is found that the optimal pH for enzymase production is 6.6–7.0, and that the pH tends to drop slowly as the fermentation proceeds, eventually reaching pH 5.1. A basic solution must be slowly added to the broth to maintain an acceptable pH during the fermentation.

2. Based on these experiments, the R&D team establishes control limits, expressed as a range, for temperature, pH, and oxygen level. Methods of maintaining each parameter in the desired range are devised. When the process is scaled-up, these control ranges are confirmed to be achievable.

3. Methods of monitoring the three critical parameters are devised. Each parameter is monitored by a separate sterilized probe inserted into the media. The probe output is connected to a computer that displays and stores the values continuously. If any of the three parameters is too high or too low, then a message is displayed and the medium is adjusted: For pH, base is added; for temperature, cooling is initiated or turned off; and for oxygen, aeration is increased or decreased. These activities are documented throughout the process.

Table 7.3 summarizes some of the steps in designing a process.

V. VALIDATION OF PROCESSES AND ASSOCIATED EQUIPMENT

A. Introduction to Validation

In a GMP facility, after processes are developed, they must be **validated.** The goal of **validation** is to ensure that product quality is built into a product as it is produced. **Process validation** *is defined by FDA as "establishing documented evidence which provides a high degree of assurance that a specific process will consistently produce a product meeting its predetermined specifications and quality attributes."*

Validation in the pharmaceutical industry emerged from problems in the 1960s and 1970s. Recall that during this period there were a number of cases of septicemia in patients caused by intravenous (IV) fluids

Table 7.3. DESIGNING A PROCESS

1. *The purpose of the process must be defined, that is, the desired output must be determined.*

2. *An endpoint(s) that demonstrates that the process has been performed satisfactorily must be defined.* A range of accepted values for the endpoint must be established; that is, specifications for the process must be written.

3. *A method to measure the desired endpoint(s) is required.*

4. *Raw materials and their specifications must be established.*

5. *The steps in the process must be determined, usually by experimentation.*

6. *The process must be scaled-up for production.*

7. *An analysis of potential problems must be performed.* This may involve listing steps in the process where things can go wrong. These steps are sometimes called *critical points* or *control points*.

8. *Experiments must be performed to determine how the process must operate at each critical point in order to make a quality product.* For example, it might be established that the temperature during incubation must be between 20 and 25°C.

9. *Methods to monitor the process must be developed.*

10. *Methods to control the process must be developed.*

11. *Effective record-keeping procedures must be developed.*

12. *All SOPs required for the process must be written and approved.*

that were contaminated by bacteria. The contamination was caused by serious problems in companies manufacturing medical products for IV use. The companies had processes in place that made products, and they had programs in place to test samples of their products; however, they still had contamination problems that led to patient deaths.

FDA's present philosophy regarding validation is easily understood in light of the septicemia history. FDA states that quality, safety, and effectiveness are designed and built into the product. Each step in the production process must be controlled to ensure that the finished product meets all quality and design specifications. Process validation is the method by which companies demonstrate that their activities, procedures, and processes consistently produce a quality result. For example, process validation of a sterilization process might involve extensive testing of the effectiveness of the process under varying conditions, when different materials are sterilized, with several operators, and with different contaminants. During

this testing, the effectiveness of contaminant removal would be measured, the temperature and pressure at all locations in the sterilizer would be measured and documented, and any potential difficulties would be identified. Validation thus demonstrates that the process itself is effective. Validation and final product testing are recognized as two separate, complementary, and necessary parts of ensuring quality.

In addition to process validation, there is also validation of major equipment and of tests and assays. Examples of what might be validated are shown in Table 7.4.

B. When Is Validation Performed?

The planning of validation occurs throughout the development of the product. The actual validation process is usually performed before large-scale production and marketing of a product begins. Revalidation is required whenever there are changes in raw materials, equipment, processes, or packaging that could affect the performance of the product.

Validation is important both in "traditional" pharmaceutical manufacturing and in the production of medical products using biotechnology methods. FDA's "Guideline on General Principles of Process Validation" (May 1987) is a general guide that is applicable to most manufacturing situations. There are also specific, detailed guidelines directly applicable to the validation of biotechnology products for medical use. For example, the "Final Guideline on Quality of Biotechnological Products: Analysis of the Expression Construct in Cells Used for Production of r-DNA Derived Protein Products" *International Conference on Harmonisation, in the* Federal Register: February 23, 61(37): 7006–7008, 1996 discusses validation of

Table 7.4. ITEMS THAT ARE VALIDATED

1. *Processes such as:*
   Cleaning and sterilization
   Fermentation
   Purification, packaging, and labeling of a product
   Chromatographic methods used to purify products
   Mixing and blending

2. *Major equipment such as:*
   Heating, ventilation, and air conditioning systems
   Water purification systems
   Autoclaves and steam generators
   Computers that control manufacturing processes
   Freeze dryers

3. *Tests and assays, such as:*
   Microbiological assays
   Contaminant tests
   Raw materials analyses
   Final product tests

processes involving recombinant DNA proteins. Guidelines and resources related to validation are continuously being written and revised. The FDA website is an excellent source of current information.

Validation is a major undertaking that is expensive, time-consuming, and requires extensive planning and knowledge of the system being validated. The advantage to validation is that it helps to assure consistent product quality, greater customer satisfaction, and fewer costly product recalls. Although we are discussing validation primarily from the perspective of the medical products industry, the concept of validation is generally applicable. The cost of a defective batch of product or a product recall is very high in any industry. Ensuring that all production processes function properly to make a quality product makes good business sense.

C. Process Validation Planning

Every process is different. Validation begins, therefore, with an understanding of the process to be validated, how the process works, and what can go wrong. Information required for validation is gathered throughout the development phase. The actual validation of a process does not occur until the process is in place, its specifications have been established, most of the factors that can adversely affect the process have been determined, and methods of monitoring and controlling the process have been identified.

Validation requires a **validation protocol,** *which is a document that details how the validation tests will be conducted.* The components of a validation protocol are summarized in Table 7.5.

Table 7.5.   THE COMPONENTS OF A VALIDATION PROTOCOL

1. *A description of the process to be validated.*
2. *An explanation of the features of the process that are to be evaluated.*
3. *A description of the equipment, raw materials, and intermediate products that are to be evaluated.*
4. *An explanation of what samples are to be collected and how they are to be selected.*
5. *Procedures for the tests and assays to be conducted on those samples.*
6. *An explanation of how the results of the assays are to be analyzed (including statistical methods).*
7. *A statement as to how many times each test is to be repeated.*
8. *SOPs describing how to run the process.*

EXAMPLE

Consider the fermentation process to make the hypothetical enzymase. During R&D the specifications for enzymase were established, an assay for its activity was developed, the process for its production was established, and critical parameters—such as temperature, pH, and aeration—were identified. The requirements for the fermenter equipment used to make enzymase (e.g., the size fermenter required, the presence of ports for insertion of probes, how the fermenter would be cleaned, and so on) were similarly determined during development. If necessary, a suitable fermenter was built. SOPs to perform the fermentation-related activities were written. The process is ready for validation.

D. Activities Involved in Validation

At the beginning of validation, measuring instruments must be calibrated to ensure that their readings are trustworthy. For example, a pH probe used to monitor the pH of a fermentation broth must be calibrated to ensure that it registers pH correctly. Instruments may be rechecked periodically during validation to ensure that their readings remain reliable. When validation is completed, the calibration of each instrument is rechecked to ensure that it is still correct. If any instrument has drifted out of calibration, then validation may need to be repeated.

Processes almost always involve equipment. In order for a process to proceed correctly, the equipment must be of good quality, must be properly installed, regularly maintained, and correctly operated. Equipment must therefore be validated or **qualified,** *that is, checked to ensure that it will function reliably under all the conditions that may occur during production.* Equipment qualification may be considered and performed separately from process validation, but it is also a requirement for process validation.

Equipment qualification is often broken into steps. *First the equipment item is checked to be sure that it meets its design and purchase specifications and is properly installed; this is called* **installation qualification.** Installation qualification includes, for example, checking that instruction manuals, schematic diagrams, and spare parts lists are present; checking that all parts of the device are installed; checking that the materials used in construction were those specified; and making sure that fittings, attachments, cables, plumbing, and wiring are properly connected.

After installation *the equipment can be tested to see that it performs within acceptable limits, which is called* **operational qualification.** For example, an autoclave might be tested to see that it reaches the proper temperature, plus or minus certain limits, in a set period of time; that it reaches the proper pressure, plus or minus certain limits, and so on. The penetration of steam to all parts of the chamber, the temperature of the autoclave in all areas of the chamber, the pressure achieved at various settings, and so on, would all be tested as part of the operational qualification of an autoclave.

Once all measuring instruments are calibrated and all equipment is validated, process validation can be performed. The actual validation of the process will involve assessing the process under all the conditions that can be expected to occur during production. This includes running the process with all the raw materials that will be used and performing all the involved activities according to their SOPs. Testing includes checking the process endpoint(s) under these conditions and establishing that the process consistently meets its specifications.

Process validation also involves challenging the system with unusual circumstances. FDA speaks of the "worst case" situation(s) that might be encountered during production. For example, a sterilization process might be challenged by placing large numbers of an especially heat-resistant, spore-forming bacterium in the corner of the autoclave known to be least accessible to steam. The effectiveness of bacterial killing under these "worst case" conditions must meet the specifications for the process.

After all these validation activities have been performed, the data that were collected must be analyzed, as described in the validation protocol, and a report must be prepared. Successful validation demonstrates that a process is effective and reliable.

EXAMPLE

Discuss briefly how the steps of validation—calibration, equipment qualification, process validation, challenge, and data analysis—apply to the fermentation process used to produce enzymase.

Validation Protocol Purpose: Does the fermentation process for producing crude enzymase reliably and repeatedly produce a crude enzyme that meets all its specifications?

1. **Calibration.** All the measuring instruments must be calibrated to ensure that their readings are correct. This includes, for example, the pH, temperature, and oxygen probes, and the instruments that monitor agitation rates and fluid flows.

2. **Fermenter Equipment Qualification.** The fermenter is a major piece of equipment that requires qualification before the fermentation process is validated. This involves:

   a. **Installation Qualification:** Establishing that the fermenter and its ancillary equipment are properly installed and meet the specifications set by the design team. For example, the following items would be inspected and their presence documented:

   1. The vessel where the bacteria and broth are contained must have features such as smooth walls, proper fittings and seals, and a pressure relief valve.
   2. The various ports from which samples are withdrawn, and into which nutrient medium can be added, must be machined so that contaminants cannot enter the system.
   3. Heating and cooling systems must be in place, their utility requirements must be provided, they must be connected to electricity and water, as required, and so on.
   4. Pumps must be present, have the proper ratings, be connected to power, and so on.
   5. All instruction manuals and schematic diagrams must be present, and spare parts lists must be available.

   b. **Operational Qualification:** Establishing that the fermenter operates as it should. For example, the following items would be tested:

   1. The heating and cooling systems function as specified.
   2. The air and water flow rates are acceptable.
   3. All switches function correctly.
   4. Alarms sound when they should.
   5. A critical feature of a fermenter is that it can be cleaned and sterilized. This requirement is typically tested by intentionally contaminating the fermenter with a test organism, which is a temperature-resistant bacterium. The fermenter is then run through its decontamination cycle and is checked for the number of bacteria remaining.
   6. Any computer associated with the fermentation process must be qualified. It is common to use computers to monitor and control fermenters. For example, a computer might be connected to the pH probe. The computer would continuously record the pH in the fermenter, would display the pH, might control a pump that feeds base into the fermenter as necessary, and might trigger an alarm if the pH exceeds preset limits.

3. **Process Validation.** Bacteria and nutrient medium are introduced to the system and the growth of bacteria and production of enzymase are tested. The fermentation process should be tested under varying conditions as might occur during production. Features to check include:

   1. The ability of the system to maintain the proper temperature, pH, and aeration.
   2. The absence of contamination in the system.
   3. The production of crude enzymase that meets specifications.
   4. Bacterial growth.

4. **Worst-case challenge.** A challenge to the system might include, for example, allowing the fermentation process to run extra days to see whether slow-growing contaminants appear in the broth.

5. **Analysis of Data and Preparation of Report.** The information gathered during the validation is analyzed, summarized, and documented in a report.

CASE STUDY

Examples Relating to Validation

The Federal Food and Drug Administration has inspectors who periodically inspect pharmaceutical facilities and biotechnology companies that make certain products. If the inspectors observe violations of Good Manufacturing Practices, they note the violations on forms, called "483s," and in official warning letters sent to the company. The following are excerpts from two actual warning letters that address issues relating to validation.

Warning Letter, Example 1

Dear Sir or Madam:

During an inspection of your drug manufacturing facility . . . conducted on August 29 through September 25, 1996, our investigators documented serious deviations from the current Good Manufacturing Practices Regulations . . .

Deviations from the GMPs documented during this inspection included:

(8) Failure to validate the cleaning procedure for production and process controls for drug products . . .

(9) Failure to validate the cleaning procedure for the equipment used in manufacturing drug product . . .

The above identification of violations is not intended to be an all-inclusive list of deficiencies at your facility . . . You should take prompt action to correct these deviations.

Failure to promptly correct these deviations may result in regulatory action without further notice . . .

Sincerely,
[Signed by the district director]

Warning Letter, Example 2

Dear . . .

During a comprehensive inspection of your manufacturing facility . . . conducted January 13–March 5, 1997, investigators from this office documented serious deviations from current Good Manufacturing Practice Regulations (cGMPs) . . . as follows:

1) Manufacturing process validation for several product lines were found to be inadequate, for example:

Grifluvin Tablets—four . . . batches made in 1995–1996 were rejected for failure to meet dissolution requirements. As

a result of your investigation, process changes were made and not validated. The original validation, conducted in 1989, did not consider critical process parameters to be evaluated. Other process parameters were not evaluated including the effect of additional drying steps and ranges established for mixing times, granulation, and temperature . . .
Sincerely . . .

E. Unplanned Occurrences

We have so far considered many ways in which problems are avoided by careful planning, design, monitoring, and validation. Even in the best managed facility, however, there are *unexpected occurrences, called* **deviations.** Every company must be prepared for deviations and have a plan to deal with them. This plan includes having forms for documenting the deviation and a review process to decide what action to take. A supervisor and members of the Quality Unit normally review the problem and decide what, if any, corrective action to take. The corrective action and any associated investigation is documented. (See Chapter 4, CAPA.)

When a product or a raw material is out of specification, it is called a **nonconformance.** Nonconformances are also documented. The material is quarantined. Problems with a product often require an investigation to determine the source of the problem and the extent of damage. Problems may have many sources including raw material deviations, human errors, equipment malfunctions, and other unforeseen occurrences. Damage may be minor or may affect an entire batch of product. After investigation, the Quality Unit personnel make a decision as to what to do with the nonconforming material. If it is a final product, then it may need to be destroyed or reprocessed. A raw material may be returned to the vendor. As always, the investigation activities and finding should be documented and follow-up should occur.

PRACTICE PROBLEMS

Problems 1–4 are based on information from FDA publication "Human Drug cGMP Notes" (http://www.fda.gov/cder/dmpq/cgmpnotes.htm). This publication consists of questions asked of FDA and the agency's response. Here, the question asked of the FDA is posed as a problem and the answer is the FDA's comments. See if you can figure out what the FDA's response was.

1. Is it acceptable to place an expiration date on a bottle cap instead of on the bottle label? (September 1996)

2. For drug products formulated with preservatives to inhibit microbial growth, is it necessary to test for preservatives when testing the final product? (June 1996)

3. What is the level of detergent residues that can be left after cleaning on the surface of equipment and glassware? What is the basis for arriving at this level? (June 1995)

4. If the ability of a procedure to clean a piece of equipment made of a particular material, such as 316 stainless steel, is shown to be acceptable and validated, can that material's specific cleaning procedure be used without extensive validation for other pieces of equipment? (June 1995)

QUESTIONS FOR DISCUSSION

1. Consider the warning letters on p. 122 relating to validation.

   a. Why does FDA emphasize validation?

   b. What situations do you think would require a company to revalidate a process?

   c. Procedures for cleaning equipment and work areas are extremely important in a pharmaceutical company. Describe how you think the company that received a warning letter regarding cleaning validation should have validated their procedures for cleaning equipment.

2. Consider the case study on p. 114 relating to the contaminated mouthwash. Discuss how this company might restructure to improve the quality of its operations. For example, how might they prevent mistakes such as open doors and improper hose clamps?

3. The following is an excerpt from an official warning letter sent by FDA to a company:

   Dear . . .
   During a comprehensive inspection of your manufacturing facility . . . conducted January 13–March 5, 1997, investigators from this office documented serious deviations from current Good Manufacturing Practice Regulations (cGMPs) . . . as follows:

   Retest investigations were incomplete in the determination of assignable cause and/or lacked documentation to invalidate initial out-of-range results, for example:

   An initial out-of-specification assay for Grifulvin 500 mg Tablets . . . was invalidated without adequate documentation to support the conclusion that a pipetting error occurred.
   Sincerely . . .

   In this case, laboratory workers apparently tested Grifulvin tablets and the result did not conform to the specifications for the product. Two explanations for this result are: (1) The product is defective. (2) A mistake occurred in the Quality Control laboratory resulting in an erroneous test result. Which of these two possibilities did the company decide was the explanation? To what did the company attribute the out-of-range result? What was the FDA's position? What sort of investigation should the company have performed when the result was found to be out of range?

CHAPTER 8

Quality Systems in the Laboratory

I. GENERAL PRINCIPLES AND LABORATORY MANAGEMENT

   A. Introduction

   B. Laboratory Management and Quality

II. OUT-OF-SPECIFICATION RESULTS AND THE *BARR* DECISION

I. GENERAL PRINCIPLES AND LABORATORY MANAGEMENT

A. Introduction

In 1989 a woman was accused of murdering her baby by feeding him antifreeze in his bottle. The child's blood and his bottle were sent for analysis to a commercial testing laboratory and the hospital laboratory. Workers at both laboratories confirmed that there was ethylene glycol (antifreeze) in the blood and bottle. The woman was convicted of first-degree murder and sentenced to life in prison. Two scientists fortunately became interested in the case after hearing about it on a television broadcast. The scientists proved that the child had not died of poisoning, but rather because he had a rare metabolic disorder. The mother was exonerated (after serving time in prison). When the scientists investigating the case obtained the original laboratory reports, what they saw was, in their words, "scary." One laboratory said that the child's blood contained ethylene glycol even though the sample did not match the profile of a known ethylene glycol standard. The second laboratory found an abnormal component in the child's blood and "just assumed it was ethylene glycol." In fact, samples from the bottle had not showed evidence of anything unusual, yet the laboratory report claimed it contained ethylene glycol. In this case, an innocent person was convicted of murder based on the erroneous statements of laboratory workers (*Science*, Nov. 15, 1991, p. 931). These laboratories were apparently not committed to the principles of quality.

There are various laboratory quality systems, all of which are intended to ensure that data from the laboratory are trustworthy and meaningful. ISO 9000 and GMP include references to quality-control laboratories which test products, raw materials, and in-process samples. An excerpt relating to these laboratory functions in the GMP regulations is as follows:

Laboratory Controls 21CFR211.160
General requirements
Laboratory controls shall include the establishment of scientifically sound and appropriate specifications, standards, sampling plans, and test procedures designed to assure that components . . . in-process materials . . . labeling, and drug products conform to appropriate standards of identity, strength, quality, and purity. Laboratory controls shall include:
(1) Determination of conformance to appropriate written specifications for the acceptance of each lot . . .
(2) Determination of conformance to written specifications and a description of sampling and testing procedures for in-process materials . . .
(4) The calibration of instruments, apparatus, gauges, and recording devices at suitable intervals in accordance with an established written program.

We have previously mentioned other laboratory quality systems including: FDA's GLP (Good Laboratory Practices), which apply to animal studies; the GLP of the Environmental Protection Agency, which apply to studies submitted in support of pesticide approval; and CLIA, which is a quality system for clinical laboratories. ISO 17025 (formerly ISO Guide 25) is a quality system that is similar to ISO 9000, but focuses explicitly on laboratories (as contrasted with production facilities).

Every quality system relies on the commitment of individuals to produce quality results. This commitment must begin with the upper-level managers and supervisors, who have the authority to purchase good equipment, hire skilled staff, and make other investments in quality. The technical tasks in a laboratory are performed by technicians, scientists, media preparation technicians and others, each of whom must be committed to obtaining quality results. The concept of "commitment to good laboratory practice" is formally stated in this brief excerpt from ISO Guide 25:

> <u>Quality System, audit and review</u> **ISO Guide 25-1990**
> *5.1 The laboratory shall establish and maintain a quality system appropriate to the type, range and volume of calibration and testing activities it undertakes. The elements of this system shall be documented . . . The laboratory shall define and document its policies and objectives for, and its commitment to good laboratory practice.*

This chapter discusses quality systems in the laboratory. Chapters 4 through 7 already surveyed many of the basic issues (such as documentation and resources), so just a few points relating to general laboratory management are reviewed here.

B. Laboratory Management and Quality

i. RESOURCES

Technically skilled personnel are one of the most important resources in the laboratory. The laboratory should be managed so as to ensure that there are personnel with the required skills, that each individual's responsibilities are established, and that there is adequate supervision. Each laboratory must establish procedures to train new analysts, and that training should be documented. Experienced analysts require ongoing training to refresh their skills and to learn new technologies as they are developed. An important management issue in testing laboratories is that laboratory technicians and managers should never be placed in a situation where they feel obligated or pressured to report a particular result. Whatever test result is obtained should be exactly recorded and reported (after proper verification).

The laboratory facility must have adequate equipment to perform all required work. In addition, a suitable laboratory environment must be maintained. This includes controlling the temperature, humidity, lighting, dust levels, vibration, and other factors at suitable levels. In some situations, special facilities are required, as when pathogens are studied or when sterility of work areas is required.

Equipment and instruments required to perform tests must be available, be regularly maintained and calibrated, and their maintenance and calibration must be documented. Calibration of instruments needs to be traceable to national or international standards that ensure that the measurements made in one laboratory are consistent with measurements made in other laboratories around the world. (Calibration is addressed in more detail in Unit V.)

ii. MONITORING QUALITY

Laboratories regularly need to review their practices, progress, and results. In a research laboratory, this function might be performed through weekly laboratory meetings and discussions. Outside review occurs when papers are submitted for publication in the scientific literature and when grant proposals are submitted to granting agencies. In a testing laboratory, there are likely to be formal internal and external review processes. Internal review would involve an audit of records, data, and practices by someone within the laboratory or within the company. External review might involve audits by clients or, for laboratories in regulated companies, by inspectors. It is also possible for laboratories to voluntarily become accredited by various accrediting organizations. In this case, external reviewers will audit the laboratory to see that it operates in compliance with quality practices.

In addition to periodic audits of documents and practices, the laboratory should have checks that regu-

---

CASE STUDY

Mix-Up in the Mad Cow Freezer

According to a report from the BBC (British Broadcast Corporation) on November 30, 2001 (http://news.bbc.co.uk/1/low/sci/tech/1684479.stm), a labeling mix-up caused three years of important research on BSE ("mad cow disease") to be scrapped. Researchers at the Institute for Animal Health in Edinburgh were conducting pivotal studies to see if British sheep had become infected with the BSE pathogen. The researchers were studying what they thought was tissue from 3000 sheep brains collected in the early 1990s. Initial findings from the laboratory suggested that sheep brains might be infected with BSE, leading to fears that entire herds of British sheep would need to be destroyed. It was subsequently determined that the experimenters were actually studying tissue from cow brains. Auditors believe that the mix-up occurred due to poor labeling. An audit carried out by the United Kingdom Accreditation Service reported that there was "no formal documented quality system" covering this work and that record-keeping was "inadequate." The director of the research institute disputed the audit report findings but, in any event, everyone agreed that the results of the research were not salvageable.

larly monitor the accuracy of tests and procedures. For example, analysts might periodically verify their assays using reference standards, also called "check samples," which are known to give a particular value in the assay. If the reference standard does not give the proper result, then workers know there is a problem in the method or in its execution. This problem may have many sources, such as an out-of-calibration instrument or an environmental factor (such as temperature or humidity). Issues relating to tests and methods are discussed in Chapter 24.

iii. A NOTE ABOUT QUALITY SYSTEMS AND RESEARCH LABORATORIES

People sometimes think that the quality issues discussed in this unit do not apply to research laboratories. As we have emphasized previously, however, research scientists have always pursued "good science." Adhering to good laboratory practices helps researchers avoid errors, ensures that experimental results can be consistently repeated, and makes the work credible to others.

Suppose a situation arises in a research laboratory where an experimental method that has been used to separate proteins from one another suddenly no longer works. A number of factors may be responsible for the problem. A raw material may have been made improperly by the manufacturer. Reagents may have been mixed improperly or may have degraded. An instrument may be malfunctioning; a technician may have made a mistake. Trouble-shooting this problem will be facilitated if lot numbers for raw materials were recorded, if there are good records of the sources and preparation of samples, if all buffers and reagents were dated and prepared using documented procedures, if the equipment used was consistently maintained and calibrated, and if the analyst kept careful records. Trouble-shooting a problem can be extremely difficult, time-consuming, and expensive if records are not kept and if these good-quality laboratory practices were not followed. If good records are not kept throughout a research program, then there may be gaps that cause problems later on, such as difficulty reproducing experimental results or differences between prototype and production batches. Even a small change in a raw material, a procedure, an instrument, or a process, therefore, should be justified and recorded.

To a great extent, regulations and standards relating to laboratory practice simply formalize and make consistent the traditions of researchers doing "good science." Some of the specific requirements of regulatory agencies, however, are neither mandated nor practical in a research laboratory. For example, it would be unusual to require a supervisor's signature when a routine procedure is completed in a research setting, whereas such a signature may be necessary in a pharmaceutical quality-control environment. The process for changing a procedure or raw material is similarly likely to be informal in a research laboratory and tightly controlled in a testing laboratory.

CASE STUDY

Pathogen Problems

Pathogenic bacteria, viruses, and prions have made the headlines frequently in recent years. In the fall of 2001, for example, anthrax bacterial spores that were intentionally sent through the U.S. postal service sickened 22 people and killed 5. Infectious diseases that in the past would have afflicted only people in a single region can now become global epidemics in a matter of a few months, due to widespread air travel. The rapid spread of Severe Acute Respiratory Syndrome (SARS) was a wake-up call to the international biomedical community. Other pathogens, such as West Nile virus, Ebola virus, and avian influenza (bird flu) are ongoing concerns. Governments and scientists have responded to these natural and terrorist-driven threats by expanding pathogen-research programs.

Pathogen research is conducted by trained personnel in special biocontainment laboratories (see Chapter 12). It is obvious that disregard for quality practices in these biocontainment facilities can have severe consequences. A sobering report in 2007 indicated that personnel in a biocontainment laboratory at Texas A&M University committed a number of safety violations. Inspectors from The Centers for Disease Control and Prevention (CDC) in Atlanta cited violations including the loss of three vials of dangerous Brucella bacteria, an unreported case of an employee diagnosed with brucellosis caused by this type of bacteria, incidents where unauthorized employees worked with pathogenic agents, an instance where a faculty member performed an experiment with recombinant DNA without the necessary CDC approval, concerns about disposal of animals from experiments involving pathogenic agents, and three unreported cases of individuals exposed to the infectious bacterium that causes Q fever. The CDC reports that "There was no evidence that a coordinated response or biosafety assessment was performed as a result" of these exposures. (Recall the discussion of CAPA programs in Chapter 4.)

Although the consequences of poor-quality systems can be dramatic and severe in a facility that deals with deadly pathogens, the underlying cause of the problems is similar in every laboratory. According to Philip Hauck, a biosafety professional in New York City, "People get blasé, I hate to say it. After a while as microbiologists, you're like, 'This thing never bit me.'" In the absence of a robust-quality system, people tend to become lax in their treatment of familiar hazards. Preventing these types of dangerous infractions requires a well-run, supervised system where activities are monitored and documented and problems are investigated and fixed. (See also Practice Problem 3.)

Primary Source: Couzin, Jennifer. "Lapses in Biosafety Spark Concern." Science 317 (Sept. 14, 2007).

II. OUT-OF-SPECIFICATION RESULTS AND THE *BARR* DECISION

Suppose an analyst works in a quality-control (QC) laboratory, runs a test on a final product, and obtains a result that does not meet the specifications for that product. A drug product, for example, is specified to have a potency that is between 2000 and 2500 units/mg, but when potency tested gives a result of 3200 units/mg. This result is called an **out-of-specification (OOS or out-of-spec) result;** *that is, a test result that is outside the range required for the product.* What happens now? There are two possible causes for this out-of-spec result: (1) This batch of drug product really is too potent, or (2) There was an error in the laboratory analysis. These two possibilities have very different implications for the company. The drug cannot be released for sale if it is really too potent, but if there was an error in the laboratory analysis then the drug may actually be fine.

Companies must have procedures to handle out-of-spec results. An analyst's first instinct is often to immediately retest the drug product to see if the same result occurs again. If the result of the retest falls within the specified range, then the analyst might conclude that the original result was due to an unknown mistake, "throw away" the original result, and record the second result. In fact, in some companies analysts have been directed to retest samples over and over again until a passing result is obtained, or to average the results of more than one test to obtain a passing result. The FDA is very concerned with how laboratories handle OOS results in the pharmaceutical industry. If one QC test shows that a product fails to meet its specifications, but a later result shows it passes, how does the company know which result was correct? If laboratory analysts are making mistakes that are just "thrown away," how can the company improve the quality of its testing?

The practice of repeated retesting when OOS results were obtained (along with some other dubious practices) caused one company, Barr Laboratories, to undergo a costly and famous (within the pharmaceutical industry) courtroom battle. FDA investigators inspected Barr's manufacturing plants in 1989 and 1991 and recorded numerous criticisms, including some relating to how Barr handled OOS results. Barr QC analysts would, for example, average OOS values with in-spec values to get a passing value. They would discard raw data and perform multiple retests of samples with no defined endpoint (other than a passing value). They did not perform adequate investigations of failures (see Chapter 4 for a discussion of failure investigations), and they did not adequately validate their testing methods. FDA requested a court injunction requiring Barr to suspend, recall, or revamp numerous products. Barr filed a countersuit alleging that the FDA was applying unfair rules to them. The ensuing courtroom battle generated thousands of pages of testimony, hundreds of exhibits, and numerous lengthy declarations. Judge Alfred Wolin, who presided in the case, was responsible for sorting through the conflicting claims of the various experts called in by both sides. The principles that FDA now enforces relating to OOS results are largely based on Judge Wolin's legal rulings in 1993.

Judge Wolin ruled against Barr Laboratories in a number of instances and demanded recall of certain products, though he stopped short of shutting down their manufacturing operations. Judge Wolin acknowledged in his ruling that the cGMP regulations "create ambiguities" and "the regulations themselves, whose broad and sometimes vague instructions allow conflicting, but plausible, views of the precise requirements, transform what might be a routine evaluation into an arduous task." Judge Wolin ruled that how a company deals with OOS results is often not clear-cut because there are ambiguities and differences from one situation to another.

Judge Wolin did provide a number of specific rulings for the industry, despite the ambiguities relating to OOS results. An important part of his ruling was that a company could not simply retest a sample when an OOS result was obtained. He ruled that if an analyst obtains an OOS result then the analyst must perform the following steps:

1. The analyst must report the result to a supervisor.

2. The analyst and supervisor must conduct an informal lab inspection including: discussing the testing procedure, reviewing all calculations, examining the instruments used, and reviewing the notebooks with the OOS result.

3. The analyst and supervisor must document the investigation results, recording any errors that might have been uncovered and any conclusions that were reached.

This informal investigation may reveal a clear laboratory error, for example, a calculation that was performed incorrectly. If an obvious error is found, then the initial result of the test can be invalidated (not used) and the sample can be retested.

A rigorous formal investigation must be performed if no obvious laboratory error is uncovered in the informal investigation. A formal investigation requires that the quality-assurance (QA) unit be involved and that the investigation team follow a previously written standard operating procedure. During the investigation some limited retesting may occur, not to approve the product, but to determine the cause of the OOS result. If an error is found and the root cause can be documented, then the original test result can be invalidated and the sample can be retested. At this point corrective actions should be taken to avoid further problems of the same nature.

If no laboratory error can be identified, then a more extensive investigation should ensue that will involve looking for errors in the manufacturing process. A report must be written at the end of the investigation that identifies the most probable cause of the OOS result, what actions are to be taken, and by whom. The report must note if a defect in a batch is discovered and if other batches might be affected. If a product has already been released to the public that is found to not conform to its specifications, then the FDA must also be alerted within 72 hours of the discovery. The company must also take preventive actions to prevent recurrences of the problem in addition to taking corrective actions to save the batch, if possible, or to prevent the release of defective product.

A guidance document published by the FDA in 2006 that includes recommendations for laboratory workers is summarized in Table 8.1 (FDA, "Guidance for Industry, Investigating Out of Specification (OOS) Test Results for Pharmaceutical Production." Washington, DC: FDA, 2006).

While the *Barr* case relates to the pharmaceutical industry, there are lessons for any laboratory where samples are tested—which is to say, all laboratories. Most notably, *Barr* highlights the need for well-run quality systems in laboratories. The company spent

huge sums in court costs and recalled products largely as a result of a poorly managed quality system. (See also the costly consequences of poor-quality systems in the Case Study below.) This incident also made clear some of the difficulties in designing a process that works for all situations. In the end, producing quality products relies on sound scientific judgment on the part of analysts and supervisors, coupled with a company's commitment to releasing only high-quality products.

Table 8.1. Avoiding Laboratory Error

- Analysts should verify that all paperwork relating to a product is in order and matches the label before performing a test on it.
- Analysts should verify that the proper test method is being used on a sample.
- Analysts should be certain that all instruments are properly calibrated before beginning the test and that any standards used to test the instruments perform as expected.
- If any standard does not give the expected result, then the test should be suspended and any data already acquired should be investigated.
- If an analyst makes an obvious error, such as spilling some of the sample, that error should be documented and the test begun again.
- An analyst should never finish a test where an error was known to have occurred.
- No samples should be discarded before the results of a test are checked. If the results are out of spec, then the samples must be saved for investigation.

The supervisor has further responsibilities including:

- Ensuring that there is a system to manage samples.
- Ensuring that all analysts have the required training to do their job.
- Notifying QA when necessary.
- Ensuring that all SOPs are available and up to date.
- Ensuring that proper documentation occurs during an investigation.

CASE STUDY

OOS Results

Federal Food and Drug Administration inspectors periodically inspect pharmaceutical facilities. If the inspectors observe violations of Good Manufacturing Practices, they initially report them on Form 483. If the company does not fix the problems, the FDA sends a warning letter to the company. If the company does not correct the violations noted in the warning letter, then serious action follows. This case study reports on a situation where the FDA imposed very serious sanctions on a company that was not in compliance with GMP.

In 2002, Schering-Plough Corporation and two corporate officers signed a consent decree agreeing to take drastic measures to ensure that drug products manufactured at the company's New Jersey and Puerto Rico plants complied with GMP requirements. The company agreed to disgorge profits of $500 million. (Disgorgement of profits is a sanction based on the premise that an individual or corporation is not entitled to profits gained by illegal means.) They also agreed to pay up to $175 million and disgorge additional profits if the company failed to meet certain time frames. The company further reimbursed the government about $500,000 in inspection costs. They agreed to submit comprehensive facility work plans to FDA, have trained FDA-approved personnel at each facility to provide full-time oversight of all operations, and have consultants conduct annual inspections of the facilities for three years. For at least five years the company agreed to conduct regular audits of its operations and make reports to FDA regarding compliance. The company further agreed to periodic FDA inspections. If the company failed to comply with the consent decree they could be held in contempt of court and could face additional penalties, and their plants could be shut down. This consent decree came after several inspections beginning in 1998 found numerous violations of GMP regulations that the company failed to adequately address. Some of these violations relate to OOS results and to investigations following OOS results. Consider the following short excerpts from a Form 483 issued June 5, 2002. (This case is also discussed in Practice Problem 1).

After obtaining an out-of-specification assay result during the validation [of] the K-Dur extended release coating process, you reanalyzed the same sample preparation. When a second OOS result was obtained, you retested the sample using two analysts, this time obtaining two OOS results. These preparations were then further agitated and reanalyzed in duplicate by each analyst obtaining four OOS results for a total of eight OOS results. You then resampled the batch, tested the sample, and obtained passing results, this result was used as the official result to release the batch for further processing.... The investigation states "In a conversation with Technical Services [Personnel], an error in the original sampling process was made. The original results will be voided and resample results will be used as official for this test." You could not provide a description of the sampling error, the specific procedure that was not followed, or any corrective action to prevent a reoccurrence ....

Preliminary variance report 20010219 was initiated to document a low net yield and an interruption during the processing of ... coated pellets.... As part of the investigation, a special request was issued for analytical testing. However, the preliminary report does not list the results but instead states "pending analytical test results." However, the Recommendation for Material Disposition states "Based on the investigation and special request results, approval of the lot is recommended" even though the person writing the report did not have the results of the testing yet.

After obtaining an Out of Specification test result of 121.6% for blend uniformity during the performance qualification of [a product] your firm reanalyzed the same sample preparation. The retest result of 121.9% confirmed the original result. A second sample ... was tested obtaining a result of 122.0%. However this value was invalidated due to chromatography problems. [A] second analysis of the second sample obtained a value of 100%, this value was accepted without question ...

PRACTICE PROBLEMS

1. The fines levied against Schering-Plough in 2002 were among the most severe in the history of FDA enforcement. Explain in your own words what the company did wrong with regard to handling OOS results. Why did the FDA take these errors so seriously?

2. Discuss how the general quality principles discussed in Chapters 4 and 6 relate to laboratories. In your discussion, provide ideas for avoiding problems like the erroneous results reported for the contents of the baby bottle, and the mix-up of tissue samples in the United Kingdom.

3. Consider the case study in this chapter relating to pathogen research at Texas A&M University. Although laboratory facilities for working with pathogens are specially designed and equipped to contain pathogens (as described later in Chapter 12), the same quality principles apply to these laboratories as to any other research facility. Discuss how adherence to standard quality practices might have prevented the problems identified at Texas A&M University. If you like, you can view the CDC's report relating to safety violations at Texas A&M University (http://www.sunshine-project.org/TAMU/CDCTAMUReport.pdf). The CDC report includes corrective action recommendations.

Safety in the Laboratory

Chapters in This Unit

It is essential in a safe workplace to recognize hazards and reduce risks to the workers. **Hazards** *are the equipment, chemicals, and conditions that have a potential to cause harm*, and **risk** *is the probability that a hazard will cause harm*. For example, even though toxic chemicals are hazardous, the risk of working with them is reduced by using smaller working volumes, proper ventilation, shorter working times, and good experimental technique. Laboratory hazards fall into several categories:

- physical hazards
- chemical hazards
- biological hazards

This unit discusses examples of these classes of hazards and specific approaches to risk reduction.

Safety information is critical in the laboratory so that hazard exposure can be reduced or eliminated by Good Laboratory Practices. In addition, in the event of an accident, a quick and appropriate response usually results in less harm to people and property. Knowing about the potential for laboratory injuries can help you anticipate the types of emergencies that are most likely to occur and to plan how you would react.

This unit is intended to provide practical advice that stems from a variety of general information sources, as well as the personal experience of the authors. It is not a substitute for a safety manual, which is specific for an institution or facility.

By OSHA regulation, every laboratory must have either a Chemical Hazard Communication program or a Chemical Hygiene Plan (CHP—see Chapter 9). It is a good practice for every laboratory to have at least one comprehensive reference book covering the specific types of hazards (i.e., biological, radioactive, etc.) found in that setting. There are many excellent books available that can serve as safety references in the laboratory. Each of these tends to focus on specific aspects of safety, so several may be necessary to cover all contingencies.

Unit III is organized as follows:

Chapter 9 discusses general regulatory requirements for laboratory safety management.

Chapter 10 surveys general risk reduction strategies, personal protective equipment, and the most common physical hazards found in laboratories.

Chapter 11 discusses safe handling of chemicals, with special emphasis on those most likely to be found in the biotech laboratory.

Chapter 12 gives an overview of biosafety issues, including universal precautions, containment and sterilization strategies, animal handling, and recombinant DNA guidelines.

BIBLIOGRAPHY FOR UNIT III

There are many excellent books and websites with information on laboratory safety. This list provides basic references, and some books dealing with specific safety topics. Many of the general references also cover the specialized topics.

General Safety References

Cold Spring Harbor Laboratory. *Safety Sense: A Laboratory Guide.* 2nd ed. Cold Spring Harbor, MA: CSH Laboratory Press, 2007. This is a short basic guide for students and individuals new to lab safety.

Furr, A. Keith. *CRC Handbook of Laboratory Safety.* 5th ed. Cleveland, OH: CRC, 2000. This is a comprehensive reference for laboratory safety issues.

Stricoff, R. Scott, and Walters, Douglas B. *Handbook of Laboratory Health and Safety.* 3rd ed. Hoboken, NJ: Wiley-Interscience, 2007.

Industrial Safety References

Goetsch, David L. *Occupational Safety and Health for Technologists, Engineers, and Managers.* 6th ed. Upper Saddle River, NJ: Prentice Hall, 2007. A comprehensive look at industrial safety and health issues and practices.

National Safety Council. *Supervisors' Safety Manual.* 9th ed. Itasca, IL: National Safety Council, 1997. This book discusses the obligations of work supervisors to inform and monitor their employees concerning safety issues.

Plog, Barbara A, and Quinlan, Patricia J. eds. *Fundamentals of Industrial Hygiene.* 5th ed. Itasca, IL: National Safety Council, 2001. This book provides an overview of industrial safety concerns and regulations.

Ergonomic Safety References

Kroemer, Karl H.E., and Grandjean, E. *Fitting the Task to the Human: A Textbook of Occupational Ergonomics.* 5th ed. Cleveland, OH: CRC, 1997.

National Safety Council. *Ergonomics: A Practical Guide.* 2nd ed. Washington, DC: National Safety Council, 1993.

Chemical Safety References

Alaimo, Robert J., ed. *Handbook of Chemical Health and Safety* (ACS Handbooks). Washington, DC: American Chemical Society, 2001.

The Flinn Chemical and Biological Catalog Reference Manual. Batavia, IL: Flinn Scientific. This annual publication contains extremely valuable information on safety issues, particularly chemical safety. They provide accurate information about chemical storage and disposal, proper use of safety equipment, and lab design, among other topics. The free catalog can be obtained from the company at http://www.flinnsci.com/.

National Research Council. *Prudent Practices in the Laboratory. Handling and Disposal of Chemicals.* Washington, DC: National Academy Press, 1998.

Shugar, Gershon J., and Ballinger, Jack T. *Chemical Technicians' Ready Reference Handbook.* 4th ed. Columbus, OH: McGraw-Hill Professional, 1996.

Biological Safety References

CDC/National Institutes of Health, U.S. Department of Health and Human Services, Public Health Service. *Primary Containment for Biohazards: Selection, Installation and Use of Biological Safety Cabinets.* 2nd ed. Washington, DC: CDC, 2000.

Clinical and Laboratory Standards Institute. *Protection of Laboratory Workers from Occupationally Acquired Infections; Approved Guideline.* 3rd ed. Wayne, PA: Clinical and Laboratory Standards Institute, 2005.

Fleming, Diane O., and Hunt, Debra Long, eds. *Biological Safety: Principles and Practices.* 4th ed. Washington, DC: ASM Press, 2006. An excellent general reference on biological safety.

National Institutes of Health. *NIH Guidelines for Research Involving Recombinant DNA Molecules.* Washington, DC: NIH, 2002. http://www4.od.nih .gov/oba/rac/guidelines_02/NIH_Guidelines_Apr_ 02.htm.

National Safety Council. *Bloodborne and Airborne Pathogens.* Itasca, IL: National Safety Council, 2004.

U.S. Department of Labor, Occupational Safety and Health Administration. *Occupational Exposure to Bloodborne Pathogens (29 CFR 1910.1030).* Washington, DC: OSHA, 2006. http://www.osha .gov/comp-links.html.

Cox, C.S., and Wathes, C.M. *Bioaerosols Handbook.* Cleveland, OH: CRC, 1995.

World Health Organization. *Laboratory Biosafety Manual.* 3rd ed. Geneva: World Health Organization, 2004.

U.S. Department of Health and Human Services, Public Health Service, Centers for Disease Control and Prevention, and National Institutes of Health. *Biosafety in Microbiological and Biomedical Laboratories.* 5th ed. Washington, DC: U.S. Government Printing Office, 2007. Contains tables of biosafety level requirements and pathogenicity levels of specific organisms.

Animal Care and Safety References

Institute for Laboratory Animal Research, National Research Council. *Guide for the Care and Use of Laboratory Animals.* Washington, DC: National Academy Press, 1996.

Office of Laboratory Animal Welfare. *Public Health Service Policy on the Humane Care and Use of Laboratory Animals.* Amended 2002. Washington, DC: National Institutes of Health. http://grants .nih.gov/grants/olaw/references/phspol.htm.

Suckow, Mark A., Douglas, Fred A., and Weichbrod, Robert H., eds. *Management of Laboratory Animal Care and Use Programs.* Cleveland, OH: CRC, 2001.

Laboratory Safety Websites

These are current websites that provide helpful information and links to other sites about lab safety issues.

American Biological Safety Association: http://www.absa.org/

Centers for Disease Control: http://www.cdc.gov/

Howard Hughes Medical Institute: http://www.hhmi.org/science/labsafe

MSDS data can be found on the Internet at a variety of sites. For example: http://www.ilpi.com/msds and http://hazard.com/msds

World Health Organization: http://www.who.int/

CHAPTER 9

Introduction to a Safe Workplace

I. INTRODUCTION TO LABORATORY SAFETY

A. Safety in the Workplace

"I think good lab practice, consideration for other people, and safety are three totally related issues."

—David H. Beach, Ph.D.,
Cold Spring Harbor Laboratory

Worker safety is a crucial consideration in the workplace. **Safety** *is defined as the elimination of potential threats to human health and well-being.* While this is an essential goal in every profession, complete safety can never be achieved. All workplaces have the potential for **accidents,** *unexpected and usually sudden events that cause harm.* Even people who work at a desk in an office all day may encounter an **emergency,** *which is a situation requiring immediate action to prevent an accumulation of harm or damage to people or property.* Consider the following Case Study, which occurred in a laboratory where one of the authors worked.

CASE STUDY

Fire in the Workplace

One evening, three laboratory co-workers were finishing experiments when one commented that he smelled smoke. Within one minute, the laboratory filled with black, acrid smoke so thick that the workers could not see their hands 12 inches from their faces. Two workers were together and quickly located the third, who was trying to gather her research notebooks and experimental materials. At this point, the smoke was so thick that conversation was impossible. The workers bent over and immediately moved to the nearest stairwell, evacuating the building. The fire department quickly responded and extinguished the fire, which was traced to electrical wires in an interstitial space. Luckily there were no injuries. Even though no flames reached the laboratory, smoke damage to equipment and materials was extensive. Nothing in this situation was specific to the laboratory. The workers could not have prevented the fire, which started in an inaccessible area; however, these workers responded quickly and appropriately to the circumstances, which could have led to serious injuries from smoke inhalation.

Although laboratories may present special safety challenges, those of us who have worked in laboratories for many years can attest to the fact that most work-related accidents are mundane in nature. They include:

- tripping on unexpected items left on the floor
- falls on slippery floors (especially around sinks and ice machines)
- slamming fingers in cabinet doors
- hallway collisions with co-workers
- minor cuts while picking up pieces of broken glass

It is estimated that 30% of all workplace accidents involve trips, slips, and falls, mostly on flat surfaces. These incidents can usually be prevented by using care and common sense, which are the best approaches to avoiding all accidents. This chapter will address general strategies for workplace safety. Laboratory work, however, does require special precautions for situations that are not found in other workplaces. Every laboratory is different enough to warrant a separate approach to safety. Although it is impossible to address the safety concerns of each individual workplace here, this chapter will address general responsibilities for workplace safety. Detailed information concerning physical, chemical, and biological safety are provided in Chapters 10–12.

B. Hazards and Risk Assessment

All laboratories, by their nature, are filled with hazards. **Hazards** *are the equipment, chemicals, and conditions that have a potential to cause harm.* Heavy equipment, chemicals, electricity, radioisotopes, animals, and infectious agents are examples of hazards that are frequently present in biotechnology laboratories. For this reason, the laboratory setting should always be approached with caution and respect. Hazards can be controlled, however, which is the goal of safety programs.

Because it is impossible to remove all hazards from the biotechnology workplace, the most useful measure of safety in a laboratory is risk. **Risk** *is the probability that a hazard will cause harm.* The risk associated with a hazard can be reduced, thereby reducing the probability of an accident. Although laboratory hazards may seem intimidating, you can significantly reduce the associated risk by being knowledgeable and careful.

Risk assessment *is an attempt to estimate the potential for human injury or property damage from an activity.* Risk assessment is really a measurement of safety. It is difficult to estimate the true risk of specific activities because there are too many variables involved in human actions. Accidents result from both unexpected and predictable errors. Analyzing activities, identifying hazards, and developing strategies for reducing risk lead to the development of safety guidelines. **Safety guidelines** and standards are *procedures that are designed to prevent accidents by reducing the risk of hazards in situations where the hazards cannot be eliminated entirely.*

Given the number of hazards present, it is comforting to know that major accidents are rare in the laboratory. Most injuries are relatively minor because of the natural tendency of workers to be more cautious around major hazards than they are around small hazards or routine tasks, see Figure 9.1. The result is that most accidents happen under mundane circumstances involving routine tasks. A basic premise of risk management is that accidents can be avoided with caution and knowledge, which are the key factors in laboratory safety.

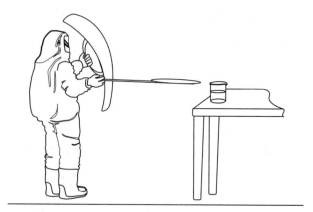

Figure 9.1. An Overzealous Use of Caution for Most Laboratories.

C. Who Is Responsible for Workplace Safety?

Safety is everyone's business. Federal agencies and other organizations are responsible for creating regulations and codes for safe workplaces, see Table 9.1. The employer has the responsibility to provide a safe work environment and a general institutional attitude of "safety first," to train employees to work responsibly, and to develop an emergency response plan. Employees have the right to work in a safe environment, and to be well trained and informed about workplace hazards. It is the responsibility of the employee to apply this training and to implement the safety plan of the institution. This hierarchy of responsibility is illustrated in Figure 9.2.

The key to protecting yourself and others from laboratory hazards is to be continuously aware of the many types of hazards present, and to know how to handle them properly. As an employee, you have both a right and a responsibility to know the hazards in the workplace, and to learn all you can about the work being done at your place of employment.

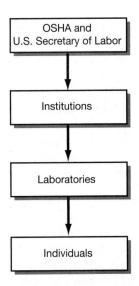

Figure 9.2. Who Is Responsible for a Safe Workplace? (Adapted from Goetsch, David L. *Occupational Safety and Health in the Age of High Technology.* Englewood Cliffs, NJ: Prentice Hall, 1996.)

II. LABORATORY SAFETY MANAGEMENT

A. Regulatory Agencies

There are a vast number of state, federal, and local regulations, as well as industry standards, that affect biotechnology companies. **Standards** *are operating principles or requirements related to many areas in addition to safety.* Many safety standards are voluntary. **Regulations** *are operating principles that are required by law.* The many regulations and standards applied to biotechnology laboratory workers can be arranged into categories:

- **Worker Safety.** For example, there are regulations that require laboratory chemicals to be labeled and that require workers to be informed about hazards (discussed in this chapter).

- **Environmental Protection.** For example, the disposal of hazardous laboratory chemicals is regulated in order to minimize impact to the environment (discussed in this chapter).

- **The Use and Handling of Animals.** For example, there are regulations regarding the cages used to house laboratory animals and regulations aimed at preventing the spread of contagious disease. These

Table 9.1. EXAMPLES OF FEDERAL AGENCIES THAT REGULATE SAFETY AND ENVIRONMENTAL PROTECTION IN BIOTECHNOLOGY ORGANIZATIONS

The Occupational Health and Safety Administration (OSHA)

OSHA is the federal agency charged with ensuring worker safety. OSHA promulgated the Laboratory Standard in 1990 that deals specifically with laboratory safety. The central requirement of the Laboratory Standard is that the employer develop, document, and implement a plan that protects workers from hazards.

The Environmental Protection Agency (EPA)

EPA is responsible for protecting the environment. EPA regulations affect how laboratories and companies handle and dispose of waste, what substances can be emitted into the air and water, the movement, storage, and disposal of hazardous substances, and records relating to chemicals. EPA also regulates certain types of biotechnology field work that involve releasing genetically modified organisms into the environment.

The Department of Transportation (DOT)

DOT regulates the transportation of hazardous materials, such as chemicals, compressed gas cylinders, and hazardous wastes. The regulations cover packaging, labeling, transport, and reporting procedures.

The Nuclear Regulatory Commission (NRC)

NRC is responsible for the safe use of radioactivity. Facilities that use radioactive substances for research purposes, medical applications, or in products must comply with NRC regulations, including those related to worker safety, waste disposal, and record-keeping.

regulations both protect animals from inhumane treatment and prevent faulty experimental results due to sick animals or inconsistent treatment of animal subjects. These regulations will be discussed in Chapter 12.

- **Regulation of Radioisotopes.** For example, these regulations cover such issues as how radioisotopes should be handled and stored, who has access to such compounds, and what documentation is required when radioisotopes are used.

Table 9.1 summarizes the roles of U.S. government agencies whose regulations directly affect biotechnology companies and research laboratories. Even though this book cannot discuss all of the statutes related to laboratory safety, we will discuss some of the most frequently encountered agencies and their regulations. For a more detailed and comprehensive discussion, we recommend starting with the National Research Council's *Prudent Practices in the Laboratory. Handling and Disposal of Chemicals* (Washington, DC: National Academy Press, 1998). Significant amounts of information are also available online from the Occupational Safety and Health Administration and the Environmental Protection Agency (http://www.osha.gov and http://www.epa.gov).

Although occupational safety is regulated primarily by federal agencies, there are many other organizations concerned with safety in the workplace. Many of these organizations establish standards or develop **codes,** *which are sets of standards centered on a specific topic.* A well-known example is the **Underwriters Laboratories (UL)**, *which have developed codes for safe electrical devices.* As with many of these organizations, UL has no enforcement powers. Table 9.2 provides a list of some of the organizations that have developed standards or codes related to worker safety. Even though the standards or codes developed by these organizations are recommendations, they are frequently the basis for federal regulations. For example, OSHA requires that protective eyewear meet American National Standards Institute (ANSI) standards for impact and penetration resistance.

Table 9.2. PROFESSIONAL ORGANIZATIONS CONCERNED WITH WORKPLACE SAFETY

- American Board of Industrial Hygiene (ABIH)
- American College of Occupational and Environmental Medicine (ACOEM)
- American Conference of Governmental Industrial Hygienists (ACGIH)
- American Industrial Hygiene Association (AIHA)
- Institution of Occupational Safety and Health (IOSH)
- The National Association of Safety Professionals (NASP)
- National Safety Council (NSC)

Keep in mind that federal agencies and other organizations cannot provide standards to deal with every possible situation. Comprehensive rules would be prohibitively complex and lengthy. This is one reason why responsibility for safety and the use of good judgment is built into every level of the laboratory, from the organization to the individual.

B. Institutional Responsibility

i. OSHA WORKER SAFETY REGULATIONS

As part of the U.S. Department of Labor, the **Occupational Safety and Health Administration (OSHA)** *is the main federal agency responsible for monitoring workplace safety.*

> OSHA's mission is to assure the safety and health of America's workers by setting and enforcing standards; providing training, outreach, and education; establishing partnerships; and encouraging continual improvement in workplace safety and health.
>
> (From the OSHA website—http://www.osha.gov/oshinfo/mission.html)

Since the passage of the Occupational Safety and Health Act of 1970, OSHA has both developed and enforced safety regulations that encourage employers to reduce hazards in the workplace. In 1983 OSHA created the **Federal Hazard Communication Standard (HCS),** *which regulates the use of hazardous materials in industrial workplaces. It focuses on the availability of information concerning employee hazard exposure and applicable safety measures.* The HCS mandates that employers fulfill specific requirements for worker safety and knowledge. Employers must provide:

- workplace hazard identification
- a written hazard communication plan
- files of Material Safety Data Sheets for all hazardous chemicals
- clear labeling of all chemicals
- worker training for the safe use of all chemicals

Material Safety Data Sheets will be discussed in detail later. The original HCS applied mainly to manufacturing employers until 1987, when it was more broadly applied to all industries where workers are exposed to hazardous chemicals.

After years of development, OSHA provided a set of general safety regulations specifically aimed at laboratories. The 1990 **Occupational Exposure to Hazardous Chemicals in Laboratories Standards (29 CFR Part 1910)** *adapts and expands the HCS to apply to academic, industrial, and clinical laboratories.* The main requirement of these standards is the Chemical Hygiene Plan that each institution must develop for every laboratory. The **Chemical Hygiene Plan (CHP)** *is a written manual that outlines the specific information and procedures necessary to protect workers from hazardous chemicals.*

Although institutions have considerable latitude in developing their CHP, certain issues must be addressed, as outlined in Table 9.3. Another important provision for laboratories is that all work-related injuries and health problems must be reported to OSHA.

These relatively recent regulations reflect an important and positive shift in attitudes about laboratory safety. OSHA regulations demonstrate that laboratory safety is of sufficient concern to warrant the involvement of the federal government. The regulations require that institutions provide resources that help individuals to understand hazards and work more safely. They encourage workers and their supervisors to take extra time if necessary to perform safety-related tasks. They require institutions (which may or may not have actively promoted safety in the past) to invest in safety equipment, training, and protective clothing. Of course, before the existence of these laws, laboratory workers were aware of hazards and wanted to protect themselves and others from injury. Today, however, there are more resources available, people are more knowledgeable about safety issues, and there is more willingness to institute safety practices than there was even 15 years ago.

ii. ENVIRONMENTAL PROTECTION

In addition to regulations governing worker safety, laboratories must also consider environmental issues. Generation and disposal of toxic biological and chemical wastes is a significant factor in laboratory management. The **Environmental Protection Agency (EPA)** *has primary responsibility for enforcement of laws to prevent environmental contamination with hazardous materials.* Some of the legislative regulations enforced by the EPA are the Clean Water Act, the Safe Drinking Water Act, the Clean Air Act, and the Toxic Substance Control Act. The **Toxic Substance Control Act (TSCA)** *was designed to regulate chemicals that pose health or environmental risks.* It has a major impact on the chemical industry and associated laboratories. TSCA establishes chemical inventory and record-keeping requirements, allows the EPA to control or ban hazardous chemicals in commerce, and requires companies to notify the EPA of their intentions to manufacture new chemicals.

Laboratories are also affected by the requirements of the **Resource Conservation and Recovery Act (RCRA)** of 1976, *which provides a system for tracking hazardous waste, including poisonous or reactive chemicals, from creation to disposal.* RCRA provides the EPA with authority for regulating transport, storage, emergency procedures, and waste management plans for toxic materials.

C. Laboratory Responsibility

i. RISK REDUCTION IN THE INDIVIDUAL LABORATORY

Although institutions have policies to implement worker safety, these policies are carried out at the level of the individual laboratory. For example, even though

---

Table 9.3.   REQUIRED ELEMENTS OF A CHEMICAL HYGIENE PLAN

A CHP must provide institutional polices or procedures to address each of the following issues:

- general chemical safety rules and procedures
- purchase, distribution, and storage of chemicals
- environmental monitoring
- availability of medical programs
- maintenance, housekeeping, and inspection procedures
- availability of protective devices and clothing
- record-keeping policies
- training and employee information programs
- chemical labeling requirements
- accident and spill policies
- waste disposal programs

A CHP also generally provides information about emergency response plans, as well as the designation of safety officers.

---

there is usually a CHP for the institution in general, each laboratory is required to have its own additions to the CHP to describe hazards and safety measures unique to that laboratory. Commitment to risk reduction should be a clear and constant goal for all members of the laboratory group. The supervisor or mentor is responsible for setting the tone of the daily operations, as well as for modeling safety and Good Laboratory Practices. Every laboratory with more than three people should have a designated safety officer, who is responsible for monitoring safety practices. In addition, this individual (or a committee, in larger groups) will also:

- serve as safety advisor to the laboratory
- ensure that safety procedures are documented and understood
- act as a liaison with the institution's safety officers
- communicate policy changes to co-workers
- coordinate internal safety inspections
- ensure that equipment is properly maintained
- keep records of hazards and problems within the laboratory

ii. LABELING AND DOCUMENTATION

Labeling of hazardous chemicals is required under the HCS, as well as by common sense. This means that all containers of potentially hazardous chemicals must be labeled to an extent that makes them readily identifiable to new workers or to outsiders in case of a spill or emergency. Lack of proper labeling is one of the most common OSHA citations against laboratories. In addition, the Material Safety Data Sheet (MSDS) for every chemical used in the laboratory must be available to all workers. The **Material Safety Data Sheet**

(MSDS) *is a legally required technical document provided by chemical suppliers that describes the specific properties of a chemical.* The HCS specifies information that must be included in an MSDS. However, OSHA recommends that manufacturers follow the ANSI-recommended format, which includes more information than required by law. Table 9.4 indicates the MSDS elements included in the ANSI standard.

A portion of an MSDS (some are several pages long) for the chemical acrylamide is shown in Figure 9.3. Acrylamide is commonly used in biotechnology laboratories to make polyacrylamide gels. It is a powerful neurotoxin (see p. 167), but it can be used safely with proper precautions, which are indicated in the MSDS.

In addition to labeling chemicals, laboratory rooms and work areas must also be labeled with signs that indicate hazards. These must provide enough information to alert visitors to take appropriate cautions. Any area that is unsafe for visitors without training or specific precautions should be labeled with a "Do Not Enter" sign.

Documentation of laboratory procedures is an essential part of good laboratory practice (see Chapter 6). Every procedure that is repeated should be available in written form to allow consistency among laboratory personnel. One task that is extensively used in industry to provide both safety guidelines for personnel as well as compliance with OSHA regulations is the preparation of a Job Safety Analysis. A **Job Safety Analysis (JSA)** *is a detailed analysis of each step in a procedure, identifying hazards and outlining accident prevention strategies.* An example is provided in Figure 9.4 on p. 140. An effective JSA is usually prepared jointly by safety officers and individuals who perform the procedures, and can be used for both training and documentation of laboratory safety measures.

iii. HOUSEKEEPING

Many hazards can be eliminated or reduced by the simple policy of good housekeeping. The majority of routine maintenance and cleaning in laboratories must be performed by the personnel who are familiar with the hazards present. In most institutions, outside staff do not clean bench tops or equipment. This prevents accidental exposure of staff to hazards that are unknown to them. It protects experimental materials from inadvertent contamination or disposal. It also means, however, that the laboratory workers themselves are responsible for maintaining a clean, orderly work space.

Safety, as well as Good Laboratory Practice, requires that clutter on bench tops and shelves be kept to a minimum. This lessens the risk of reagent mix-ups and potential degradation of old chemicals. Fewer objects in a work area provide fewer opportunities to accidentally contaminate equipment or containers that are not part of the current experiment. Developing a habit of regu-

---

Table 9.4. CONTENTS OF A MATERIAL SAFETY DATA SHEET

An MSDS generally contains the following information, although the format of presentation varies among chemical suppliers.

- Chemical name
- Chemical supplier
- Composition and ingredients information
- Hazard identification
- Exposure levels, with specific concentrations and times
- First aid procedures
- Firefighting procedures
- Accidental release procedures
- Handling and storage procedures
- Recommended personal protection (clothing and equipment)
- Physical and chemical properties
- Stability and reactivity
- Toxicological information
- Environmental impact
- Disposal recommendations
- Transportation information
- Regulatory information

---

larly cleaning your work area will avoid situations where a massive effort must be undertaken to clear working space.

Laboratory bench tops should be routinely cleaned both before and after work sessions. This will guard against any residual chemical or biological contamination that may be present from a previous user, as well as prepare the space for the next user. Cleaning techniques should be based on a worst-case scenario of the hazardous contaminants that may be present. Specific cleaning methods are discussed in Chapters 11 and 12. Cleanup should always include decontaminating and rinsing glassware in preparation for dishwashing procedures. In most institutions, dishwashing facilities will not accept laboratory items that still contain obvious residues of experimental materials. (Figure 9.5, p. 141.)

Even though institutional regulations will require periodic laboratory inspections by external safety officers, a laboratory should not wait for these inspections to identify problems. Regular internal inspections noting housekeeping problems and potential risks can provide timely information about unsuspected hazards as well as the current state of compliance with safety regulations. It is particularly important to perform regular inspections of specific types of hazards, such as gas cylinders and chemical storage and labeling. In this way, small problems can be remedied before accidents occur.

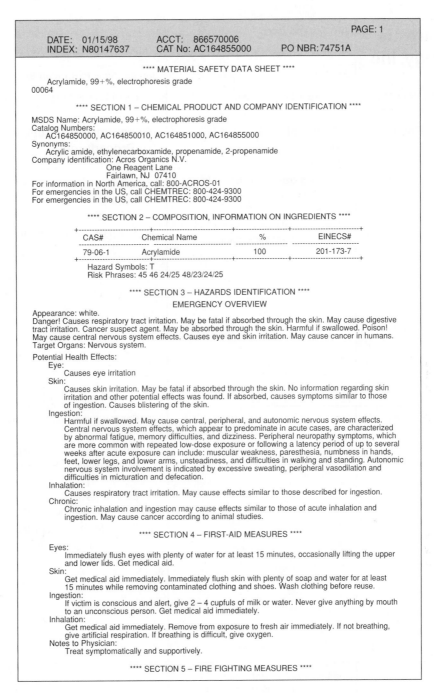

Figure 9.3.   Part of a Material Safety Data Sheet for Acrylamide.

Every laboratory needs to have a system of waste collection and disposal for specific hazards. These may include broken glass and other sharp objects, solid and liquid radioactive waste, and biologically contaminated materials. Institutions must comply with a variety of regulations concerning waste disposal, including detailed labeling of hazardous contents. Proper waste disposal is difficult and extremely expensive for unidentified materials; therefore, every laboratory should have a system for labeling waste at its source. Environmental concerns, as well as common sense, dictate that laboratory waste should be minimized when possible. It is also important that hazardous waste not be mixed with regular trash to avoid serious injuries for housekeeping staff, who may be unknowingly exposed to dangerous chemicals or biological contamination.

iv. Emergency Response

Although all institutions have emergency plans, individual laboratories also need to prepare for potential accidents and emergencies. Some of the basic preparations needed are shown in Table 9.5 on p. 141. Everyone in the laboratory ideally should be trained in basic first aid and CPR.

Genes-Are-Us Technology, Inc JOB SAFETY ANALYSIS	JSA No. and Title: #3—Loading centrifuge		Date: 12/09/08
	Job Title: Lab technician	Job Analysis Performed By: Cynthia Moore	

Sequence of Basic Job Steps	Potential Hazards	Recommended Actions or Procedures
Match adapters to centrifuge tubes	Incorrect size of adapter could create a catastrophic rotor imbalance Damaged adapter could fail and create a catastrophic rotor imbalance	1. Visually match adapter to tube and rotor size. 2. Examine adapters for any signs of wear or chemical contamination. 3. Assemble tubes, adapters, and rotor to determine if the fit is appropriate.
Balance tubes before final loading into rotors	Improperly balanced samples could create a catastrophic rotor failure	1. Confirm that tubes are comparably filled with liquid of the same density. 2. When in doubt, weigh tubes and match pairs with the same weight to be placed on opposite sides of the rotor. 3. Be sure weight does not exceed manufacturer recommendations for the rotor. 4. Distribute tubes symmetrically in the rotor.
Install rotor in the centrifuge	Improperly installed rotor could cause the rotor to come loose during the spin	(Etc.)

Figure 9.4.   Sample Job Safety Analysis Form.

CAUTION
RADIOACTIVITY

Figure 9.5.   I Know I Left My Experiment Here Somewhere.

The laboratory or a nearby area must be equipped with basic safety items such as fire extinguishers and fire alarms, first aid kits, chemical spill kits, a safety shower, and an eye wash station. Each worker needs to know the evacuation plan for the laboratory and the location of emergency telephone numbers and procedures. Remember that most accidents happen very quickly; there is usually no time for carefully planning a response after the accident occurs. This is the reason that emergency procedures need to be understood and practiced in advance.

Laboratory accidents, no matter how minor, must be reported as soon as possible after the occurrence. There is a natural tendency for many people to cover up minor accidents, such as cuts or spills, either from embarrassment, a dread of paperwork, or concerns about being blamed for the incident; however, these are not adequate excuses for secrecy. Accident reporting is a legal requirement and is also essential in order to prevent repeated problems. There are numerous studies that show that people learn to avoid accidents by being informed about other accidents, as well as near misses. If you are injured on the job and require medical treatment, you need to fill out a workers' compensation form. **Workers' compensation** *is a no-fault state insurance system designed to pay for the medical expenses of workers who are injured on the job, or develop work-related medical problems.* It provides a mechanism for your employer to pay for your medical treatments without the necessity of a lawsuit.

D. Personal Responsibility

i. LABORATORY SAFETY AND COMMON COURTESY

Safety is a matter of personal responsibility for every individual who works in a laboratory. These environments are hazardous by nature, but the actual risk is determined to a large extent by the actions of individuals. It is essential to accept responsibility for yourself and those who work around you. We all have a ten-

---

Table 9.5.   BASIC EMERGENCY PRECAUTIONS FOR THE LABORATORY

The following list provides some of the fundamental preparations that every laboratory should have in place for emergency situations.

- Everyone in the laboratory should be aware of basic emergency procedures.
- There should be at least one person trained in first aid and CPR present at all times.
- The first aid kit must be readily accessible and fully stocked.
- All required protective devices, such as fire extinguishers and eyewash stations, must be well marked and easily accessible.
- Emergency telephone numbers and instructions should be prominently posted by every telephone.
- Evacuation routes should be kept clear of boxes or clutter.

---

dency to assume that accidents happen to "other people," who are probably not as smart or careful as ourselves. The fact is, accidents are by definition unexpected and sudden, and they are frequently the result of multiple mistakes or oversights. Be aware at all times of situations that could be modified to provide a safer working environment and report any problems to a supervisor immediately. When in doubt about a situation, always ask someone with experience.

Good laboratory technique and consideration for co-workers are essential parts of laboratory safety. Proper labeling of all materials notifies co-workers of hazardous chemicals. Prompt cleanup of spills, reduction of clutter, and the return to storage of unused chemicals and equipment frees working space and reduces accidental spills and mistakes. Although you may deliberately choose to expose yourself to a known risk under some circumstances, it is inexcusable to expose others who have not made that choice.

It is important that you and each of your co-workers know how to respond to an emergency. Be familiar with the emergency procedures established for the laboratory and know what to do if you or a co-worker needs medical assistance. Acquaint yourself with basic first aid and the location of the first aid materials. Know where to find and how (and when) to use the fire extinguishers, and know the evacuation plan for the building in case of fire or other physical emergency (see the Case Study on p. 134).

One of the primary guidelines for behavior in the laboratory is often referred to as the "rule of reason," where an individual asks themselves how reasonable an action is in the current situation, see Figure 9.6 on p 142. Considering whether an action makes sense under the circumstances frequently suggests an appropriate course of action. When tempted to cut corners, think about what your opinion would be of a co-worker who did the same thing.

Figure 9.6. Inappropriate Laboratory Practices.

ii. PERSONAL HYGIENE

This does not refer to bathing; rather, it refers to personal habits that may increase your risk of hazard exposure. Never eat, drink, smoke, chew gum, or apply cosmetics in a laboratory. Even though you are "not working with anything dangerous," you can never be sure about other hazards in the immediate area. Never use the laboratory or your pockets to store food, beverages, or anything that will be consumed. If you do smoke (outside the laboratory, of course), be aware that cigarettes in open packs can absorb vapors from volatile chemicals. Develop the habit of washing your hands every time you move away from your lab bench, and at regular intervals while you are working.

Drink only from hall fountains. Water from laboratory faucets may not be **potable** *(suitable for human consumption),* due either to water quality or to contamination of the faucet with hazardous materials. For the same reasons, do not consume ice taken from laboratory machines. Even if you are certain that the ice is made from potable water, it is impossible to know exactly what types of containers may have come into contact with the ice. Avoid storing personal items such as coats in the laboratory, where they may become contaminated with hazardous materials.

Long hair must be tied back in the laboratory to prevent a variety of problems. Hair has an unfortunate tendency to fall forward, which can lead to chemical or biological contamination of the hair, or, even worse, ignition due to contact with an open flame. A greater danger that is not often considered is that most people with long hair frequently push the hair back from their faces without conscious thought, leading to potential skin and hair contamination with hazardous substances from a gloved hand. A related issue is the wearing of beards in laboratories where biological agents are used. Many microbiological standards warn against beards because of the possibilities of experimental and personal contamination, and laboratory situations that require the use of fitted respirators may preclude beards entirely. This is an issue that is best judged in the individual situation.

Another somewhat controversial issue is the wearing of contact lenses in the laboratory. Many safety guidelines state that contact lenses should never be worn in the laboratory because of the danger of chemical splashes to the eye and an increased potential for eye damage in accidents. The American Chemical Society Committee on Chemical Safety, however, has concluded that contact lenses are suitable for most laboratory situations, as long as standard eye protection is used, and may actually provide some protection against injuries (*Chemical & Engineering News* 76 no. 22 [1998]: 6). The ACS recommendation has been accepted by OSHA. When wearing contact lenses in the lab, always wear eye protection gear, such as safety glasses or goggles. Never insert or remove contact lenses from your eyes in or near the laboratory. Do not adjust an uncomfortable lens with a possibly contaminated finger. Be sure that your co-workers are aware that you wear contact lenses, in case of an emergency. In case of eye contamination, lenses should be left in place and emergency workers notified of their presence.

iii. WORK HABITS

Many laboratory accidents are the result of simple human carelessness, coupled with the fatigue and distractions that everyone experiences. The obvious prevention measure is to avoid working when tired or distracted, but this of course is not always practical advice. You can address this problem, however, by personally acknowledging your temporarily diminished state of attention, slowing down, and taking extra precautions. Take a few additional moments to plan what you are doing and anticipate any potential problems before they occur.

Do not work alone in the laboratory. Let your co-workers know if you will be in an isolated part of the building, such as a coldroom, for extended periods of time. It is a frequent temptation to finish experiments in the evening or on weekends when the lab is quiet, but this can be a dangerous strategy in the event of a serious accident (see the Case Study below).

CASE STUDY

The Dangers of Working Alone

In 1994, a senior researcher at a major university nearly died while working alone at night. While performing a familiar procedure involving small solvent volumes, a distillation flask exploded, starting a fire. Because he was not wearing a lab coat, his shirt ignited. Even worse, a flying piece of glass severed a major artery in his arm. He collapsed from loss of blood and shock before reaching the emergency phone. His life was saved because a colleague in a nearby office heard the fire alarm, called 911, and then performed first aid until the ambulance arrived. This university officially prohibited working alone in laboratories, but the prohibition was widely ignored. Following this incident, an enforced policy of using a "buddy system" for work at odd hours was developed. Each worker must have at least one co-worker in the immediate vicinity, in case of emergencies.

Another time for concern about hazards is when passing from the laboratory into public areas and back. Do not wear lab apparel from work, especially lab coat and gloves, into public areas. You may know that your hands and coat are uncontaminated by hazardous materials, but the people you encounter will not. If you do feel a need to wear a lab coat in public, keep a clean coat for this purpose (although this will not reassure the people who see you in the lab coat). Never wear safety gloves outside the laboratory and never handle common use items such as telephones, radios, or light switches while wearing gloves. This is an easy way to spread chemical or biological contamination. In labs where radioactive chemicals are used, radiation inspections specifically include checking door knobs, equipment controls, and laboratory desks for radioactive contamination.

If you must transport hazardous materials through public areas, handle them with one gloved hand, leaving the other hand ungloved for opening doors. Samples should be carried in sealed double containers, to prevent spills. Try not to hurry through halls and around corners, to avoid collisions with co-workers. Many experiments have ended up on the floor (or worse, on a person) because of sudden hallway encounters.

Table 9.6 provides a set of general guidelines for safe laboratory practices for the biotechnology worker.

III. SAFETY TRAINING

A. Importance of a Safety Training Program

It is well documented that most injuries occur to workers with less than 2 years experience on the job. This points to the need for safety training programs aimed at all new employees. It is essential that all laboratory personnel participate in these programs because it is not possible to directly supervise lab work at every step.

Training program requirements originate with government agencies and are then applied throughout institutions. The execution of these programs requires the cooperation of both individual laboratories and the workers within them. The key aims of any safety training program are to allow workers to:

- understand the risks inherent in their jobs
- recognize their personal susceptibility to accidents

Table 9.6. PERSONAL LABORATORY SAFETY PRACTICES

- Be sure that you are informed about the hazards that you encounter in the laboratory.
- Be aware of emergency protocols.
- When in doubt about a hazardous material or a procedure, ask.
- Use personal protective wear such as lab coats and safety glasses at all times.
- Do not eat, drink, chew gum, or smoke in the laboratory.
- Avoid practical jokes or horseplay, which can unintentionally create a hazard.
- Use gloves whenever in doubt (see Chapter 10 for guidelines on proper use of gloves).
- Wash your hands regularly, regardless of whether your work requires gloves.
- Always wash your hands thoroughly before leaving the laboratory.
- Read the labels of chemicals and reagents carefully.
- Read procedures before performing them and visualize hazardous steps.
- Minimize use of sharp objects and be sure that you properly dispose of them.
- Clean up spills and pick up any dropped items promptly.
- Label everything clearly.
- Use a fume hood for any chemical or solvent that you can smell, that has known toxic properties, or that is unfamiliar to you.
- Record everything in your lab notebook.
- Always report accidents, however minor, immediately.

- learn about preventive measures that reduce the risk of accidents
- accept personal responsibility for accident prevention

Safety training is not just for new employees. Refresher courses should be available for more-experienced lab workers, allowing them to learn about new policies and resources, and practice safety skills that have not yet been needed.

B. Contents of a Training Program

It is important to remember that even though hazards are always present in the laboratory, the risks to human health can be controlled to some extent. A basic safety training program should answer two questions for each worker:

- What are the specific hazards of this job?
- How can I minimize risks to myself and others?

Minimizing safety risks requires the development and practice of rules that can be applied to the situations that are expected in the laboratory. Many workers complain about "ten pound" institutional safety manuals that attempt to regulate virtually every job process, including breathing rates. These manuals do not tend to be effective in motivating worker compliance. In this situation, individual laboratories may need to develop a more focused and specific safety manual which directly addresses the needs of the lab workers. Table 9.7 provides some guidelines for the development of effective safety manuals.

It is the responsibility of every individual to be certain that they have received adequate training to perform their jobs safely. Ask for a demonstration of equipment use or procedures that are unclear. Be sure you know the location of all exits, safety equipment, and emergency contact numbers. Table 9.8 lists some of the elements that should be covered in a safety training program.

Table 9.7.   *GUIDELINES FOR DEVELOPING AN INDIVIDUAL SET OF LAB RULES*

- Involve lab workers in developing lab rules.

- Write as few rules as possible that will ensure a safe workplace. Too many rules create a *rule overload.*

- Write rules in simple, clear language that can be understood by a wide range of lab workers.

- Write word rules in a positive way, indicating what workers should *do,* rather than what they should *avoid.*

- Provide a mechanism for lab workers to quickly find rules that pertain to their specific situation.

- Do not create rules that will not be monitored and enforced.

Table 9.8.   *ELEMENTS OF A SAFETY TRAINING PROGRAM*

Safety training programs vary widely among institutions. Employees are required by federal law to attend these programs and most institutions require employees to sign a statement indicating that they have received training. Every program should include the following topics:

- Institutional policies—hazard information, inspections, reporting systems, waste disposal

- Safety rules—practices, manual, signs, labels

- Location and use of protective equipment—fire extinguishers, hoods, clothing, safety gear

- Emergency procedures—alarms, injuries, medical assistance

- Chemical hazard awareness, including:

    location of MSDS reference materials

    symptoms of chemical exposure

    detection methods for chemical exposure

    protective mechanisms

    emergency procedures

In addition, there are many laboratory-specific safety issues that may need to be addressed, such as radiation and biological safety.

QUESTIONS FOR DISCUSSION

1. Consider a laboratory situation where large amounts of a toxic and flammable solvent are required for experimental work. How would you approach risk reduction for this hazard?

2. Analyze the Case Study on p. 134 and list the emergency procedures that would apply to this situation.

3. Analyze the Case Study on p. 142. What basic safety precautions were ignored by the researcher involved?

4. Think of an accident that you were involved in or witnessed. What safety standards (if they had been followed) might have prevented this accident?

CHAPTER 10

Working Safely in the Laboratory: General Considerations and Physical Hazards

I. RISK REDUCTION IN THE LABORATORY

There are four general approaches to risk reduction in the laboratory that apply to all categories of hazards:

- Reduce the presence of hazards
- Reduce the risk of inevitable hazards with good laboratory design
- Establish good laboratory practices for handling hazards
- Use personal protective equipment

The best strategy is to start at the top of this list and consider these approaches in sequence. First, the presence of hazards should be reduced as much as possible. For example, amounts of radioactive substances and flammable solvents on the premises should be limited. It may be possible to eliminate hazardous substances or replace them with safer substitutes. Next, the risk of those hazards that cannot be eliminated should be reduced by good laboratory engineering. This means, for example, the installation of properly functioning fume hoods, protective shielding, and fire-resistant chemical storage facilities. Appropriate lab facilities should provide safe separation between personnel and hazards whenever possible. The third approach is to establish good laboratory practices in ways that reduce risk. All personnel must take advantage of the engineered solutions such as fume hoods, be aware of proper procedures for performing hazardous operations, and exercise caution in their work behavior. Adequate personnel training and maintenance of good housekeeping practices are essential. Finally, the provision of personal protective equipment (commonly abbreviated PPE), such as safety goggles, is essential to create a barrier between the worker and hazards, to reduce residual hazards and to guard against unexpected events. Personal protective equipment should be the final consideration after hazards have been minimized as much as possible.

II. PERSONAL PROTECTION IN THE LABORATORY

A. Clothing

i. GENERAL DRESS

Proper clothing is required whenever entering a laboratory. Even though a lab coat will protect you and your clothing from some hazards, what you wear under a lab coat can be just as important, see Figure 10.1. It is important that clothes cover all parts of the body, including legs. For this reason, pants or long skirts are appropriate. Avoid dangling jewelry or ties and long loose hair that can fall into your experiment or get caught in moving equipment. It is also a good idea to refrain from wearing rings, bracelets, or watches in the laboratory. It is easy for chemicals to seep under these items. Any clothing worn in the laboratory should be fire-resistant and easily removable in case of chemical or biological contamination. It should also be appropriate for protection against the types of chemicals that are used in the laboratory (Table 10.1 offers some guidelines). Many experienced laboratory workers keep a spare change of clothing handy in case of spills, or for wearing after work.

ii. LAB COATS

Lab coats should be worn at all times in the laboratory. Even when you are not using hazardous materials yourself, other people's activities in the laboratory might present unexpected hazards. Lab coats provide a barrier against harmful agents and prevent contamination of street clothes. By soaking up spills, they allow more time to recognize contamination problems and protect yourself. They also protect experiments from contaminants outside the laboratory that might be carried in on clothing.

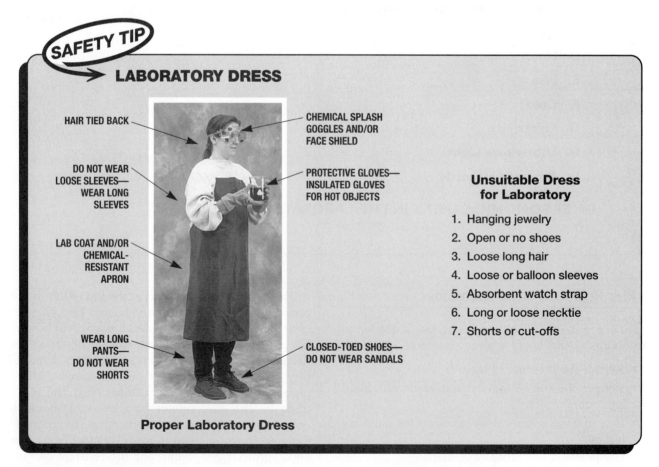

Figure 10.1. Proper Laboratory Dress. (©1998 Flinn Scientific Inc. All Rights Reserved. Reproduced with permission from Flinn Scientific Inc., Batavia, IL, USA.)

Many types of lab coats are available and selection should be based on the hazards that are of most concern. For example, front-buttoning coats are more desirable for protection against chemical spills than for biological hazards because they can be removed quickly. All lab coats should be flame-resistant, with cotton frequently providing the best resistance to both chemicals and heat in a comfortable garment, see Table 10.1. For specialized work, such as pouring large quantities of corrosive chemicals, an impermeable apron may be most appropriate.

To be effective, lab coats must fit properly and remain buttoned at all times in the laboratory. Sleeves must be long enough to provide arm protection and should fit the arm fairly snugly to avoid flapping. Rolling up the sleeves provides a holding area for chemical and biological contaminants and is not recommended. This practice also leaves the wrists and lower arms without protection.

Lab coats should be laundered regularly at your institution, even in the absence of any known contamination. Never take laboratory clothing home for washing. In case of known contamination, the coat can be decontaminated in the laboratory before washing, or be discarded. Do not wear lab coats used in the laboratory into common areas such as lunch rooms or lavatories.

iii. SHOES

Proper footwear for the laboratory includes shoes with covered toes and nonslip soles, which will protect the feet from broken glass and hazardous spills. Sandals, sneakers, or woven shoes provide little protection. Low heels are generally the most comfortable while standing at the lab bench and also protect against falls.

Laboratory workers may want to consider keeping a special pair of shoes to wear only in the lab. Changing to and from street shoes prevents the tracking of hazardous materials from the lab into the outside environment, and also prevents the introduction of potential contaminants into a cleanroom. (A **cleanroom** *is a special laboratory facility where all contaminating materials and any particulate matter in the air must be limited*.) Numerous studies have shown that shoes worn in a bacterially contaminated environment may carry higher concentrations of bacteria on their soles than the floor itself. One method to prevent contamination of shoes is the use of disposable shoe covers, which are routinely used for animal surgery and cleanroom operations, and which are removed before exiting the laboratory. Shoe covers will also prevent your shoes from carrying bacterial contamination into public areas and your home.

B. Gloves

i. CHOICE OF GLOVES

The proper use of gloves in the laboratory provides a significant measure of protection against many types of hazards. One of the most obvious benefits is the creation of a barrier between your skin and chemical or biological contamination.

If you look in a laboratory supply catalog, you will find a bewildering variety of gloves available, in many materials and styles. It is important to remember that although every type of glove provides a barrier, none can protect against all types of hazards. Each glove type will provide at least some protection against one or more of the following:

- corrosive or toxic chemicals
- biological contaminants
- sharps
- extreme temperatures

Table 10.1. PROTECTIVE CLOTHING MATERIALS

Material	Use	Properties
Cotton	Lab coats	Lightweight, degraded by acids
Cotton/polyester blend	Lab coats	Lightweight, neat appearance
Modacrylic	Lab coats	Nonflammable, resistant to most chemicals, easy to clean, low static
Nylon	Lab coats, hair nets	Lightweight, strong, water-resistant, very flammable unless treated
Neoprene	Aprons	Excellent chemical resistance, inflexible
Rubber	Aprons, long gloves	Very good chemical resistance
Vinyl	Aprons, sleeves, shoe covers	Lightweight, high static
Polypropylene	Aprons, full body suits, caps, shoe covers	Chemical resistant, strong, lightweight, water-repellent
Tyvek (high density polyethylene)	Full-body suits, shoe covers, disposable lab coats, aprons, caps, sleeves	Strong, lightweight, excellent barrier protection for user, protects lab materials from human contamination, recyclable material

Table 10.2. PROTECTIVE GLOVE MATERIALS

Type	Advantages	Disadvantages	Recommended for Protection from
Latex	Low cost, good flexibility, comfortable	Poor protection from oils, grease, organic solvents. May trigger allergic reactions.	Bases, alcohols, bloodborne pathogens
Vinyl	Low cost, medium chemical resistance	Protection against some chemicals lower than latex	Strong acids and bases, salts, other aqueous solutions, alcohols
Neoprene	Medium cost, medium chemical resistance, abrasion-resistant	Not as flexible as rubber, can give poor grip	Oxidizing acids, phenol, glycol ethers
Nitrile	Puncture and abrasion-resistant, dexterity, comfortable for longer wear, hypoallergenic, good chemical resistance	Poor protection from benzene, methylene chloride, trichloro-ethylene, many ketones	Oils, aliphatic chemicals, xylene, bloodborne pathogens
Butyl	Specialty glove, resistant to polar organics	Expensive, poor protection from hydrocarbons, chlorinated solvents	Gases, aldehydes, glycol ethers, ketones, esters
Polyvinyl alcohol (PVA)	Specialty glove, resists a very broad range of organic solvents	Very expensive, water-sensitive, poor protection from light alcohols	Aliphatics, aromatics, chlorinated solvents, ketones (except acetone), esters, ethers
Fluoroelastomer (Viton)	Specialty glove, resistant to organic solvents	Extremely expensive, poor physical properties, poor protection from some ketones, esters, amines	Carcinogens, aromatic and chlorinated solvents
Norfoil (silver shield)	Specialty glove, excellent chemical resistance, lightweight, flexible	Poor fit, easily punctured, poor grip	Use as glove liner, good for emergency use in chemical spills

Because no glove can provide all the necessary protective features, most laboratories have several glove types available, including:

- thin-walled gloves for dexterity
- heavy rubber gloves for dishwashing
- insulated gloves for handling hot and cold materials
- puncture-resistant gloves for handling animals

The first step in choosing the right glove for a job is deciding what protection is required. Are you trying to protect yourself or your work materials from contamination? Do you need maximum protection from a highly toxic chemical? Table 10.2 provides an overview of the most common glove materials and their advantages and disadvantages.

When choosing gloves for protection against chemicals, always consult the specific glove manufacturer's chemical resistance chart, which is usually supplied with the gloves or found in the supplier's catalog. This will provide information about the properties of specific glove materials. The information provided generally includes:

- **degradation rate,** *which indicates the tendency of a chemical to physically change the properties of a glove on contact*

- **permeation rate,** *which measures the tendency of a chemical to penetrate the glove material*
- **breakthrough rate,** *which indicates the time required for a chemical that is spilled on the outside of a glove to be detected on the inside of the glove.*

The following example indicates how to use this information.

EXAMPLE: CHOOSING A GLOVE FOR CHEMICAL RESISTANCE

For this example, assume that you are performing an experiment and are concerned about the possibility of acetone spills. After taking precautions to minimize the risk of skin exposure to acetone, you will still want to wear gloves. The two types of gloves you have available are made of either PVC or butyl. You then check the chemical resistance guide from the glove manufacturer and find the following information for acetone:

PVC gloves: Degradation rate > 25% in 30 minutes

Permeation and breakthrough rate < 1 minute

Butyl gloves: Degradation rate—no effect

Permeation and breakthrough rate > 17 hours

This indicates that these PVC gloves are susceptible to chemical breakdown by acetone, and that any acetone spilled on the gloves will contact your skin in less than 1 minute. Butyl, on the other hand, appears to be highly resistant to acetone; the better choice for this situation.

The best option when dealing with highly toxic agents is sometimes to double glove using two different types of gloves (e.g., using chemically resistant gloves under puncture resistant gloves). This provides the benefits of two glove types.

In addition to choosing the proper glove material, you may also have choices in the thickness of glove material. Glove thickness is usually measured in **mils**, *a unit where* 1 mil = 0.001 in. Thinner gloves generally provide more flexibility but less protection. Therefore, use the thickest glove that does not decrease the necessary dexterity for the task.

Do not allow the wearing of gloves to provide a false sense of security when working with highly toxic materials. Because all gloves are permeable to some extent, assume that your gloves may leak and do not rely on them to protect you when a spill occurs. If you are working with potentially hazardous materials that are unfamiliar to you, do not proceed with your experiments until you confirm that you have proper protection, see the Case Study below.

CASE STUDY

Proper Gloves Could Save Your Life

The research world was horrified in June 1997 when Dartmouth Professor Karen Wetterhahn died of mercury poisoning, 10 months after what seemed at the time to be a minor incident. Dr. Wetterhahn, who was considered a careful laboratory worker by her colleagues, was following standard precautions and wearing latex gloves when she spilled one or more drops of highly toxic dimethylmercury on her gloved hand. About 3 months later, she began to develop symptoms of mercury poisoning, starting with nausea and proceeding to neurological problems. Tests revealed that she had been exposed to a single dose of mercury far above toxic levels. Later testing showed that latex disposable gloves offered virtually no protection against dimethylmercury, with a breakthrough rate of 15 seconds or less. The incident has led to improved safety information provided to the users of the chemical, along with a recommendation for double gloving with a silver laminate glove under a heavy-duty neoprene or nitrile glove. This knowledge certainly came at a very high price.

ii. PROPER USE OF GLOVES

Glove choice is only the first step in ensuring safety. Gloves must be used properly in order to provide full protection. In fact, improper use of gloves can actually increase the risk of hazard exposure in the laboratory by spreading contamination or sealing it against the skin of the user.

To provide maximum user protection, before using a pair of gloves, check them for any holes or openings. Disposable gloves are usually mass produced, which means that a certain percent (depending on the manufacturer) will be defective. These should be discarded immediately. Any cuts or abrasions on the hands should be bandaged or covered before donning gloves because these are possible entry sites for contamination. Long or ragged fingernails can easily tear many disposable glove materials. Gloves must be long enough to provide wrist protection; if not, use arm protectors for hazardous work. It is best not to wear wristwatches that can trap contamination against the skin.

When working with hazardous materials, it is important to change gloves regularly. Always have a plentiful supply of disposable gloves nearby for quick changes. Change gloves immediately if you think they might have come into contact with hazardous material, and also change them regularly even if you think that they are clean. Remember that all gloves are permeable to some extent, and the longer you wear them, the more likely they are to develop small holes or tears. When removing fitted disposable gloves, use a removal technique that will not spread contamination from the outside of the glove to the skin. A useful technique is illustrated in Figure 10.2 on p. 150. Never remove thin gloves by pulling on the fingertips; these are likely to tear. Wash your hands thoroughly after glove use and in between glove changes if any contamination is suspected.

In addition to user risk, improper glove use can extend hazards to other members of the laboratory. Once the outside of a glove is contaminated with hazardous material, that contamination will be spread to any surface touched by the glove. Always remove at least one glove when opening refrigerator doors, using laboratory equipment with adjustable controls, touching doorknobs or light switches, answering telephones, or any time you leave the laboratory and pass through common areas. Be careful when writing in lab notebooks while using gloves. The pen used should be kept at the lab bench and considered contaminated. If you need to remove gloves to answer the telephone, for example, do not reuse disposable gloves; get a fresh pair to avoid the risk of a tear or skin contamination.

iii. POTENTIAL HEALTH RISKS FROM DISPOSABLE GLOVES

Many laboratory workers report minor problems from working with disposable gloves. This frequently takes the form of general irritation from the gloves. This can be alleviated by changing gloves often and allowing your hands to dry between changes. Use larger gloves to allow more air circulation around the fingers. It may help to apply a barrier cream under

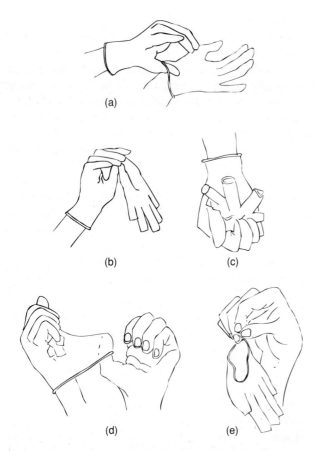

Figure 10.2. Proper Glove Removal Technique. a. Hook a finger on the cuff of one glove, being careful not to contact the skin of the wrist. **b.** Pull the glove from the hand inside out. **c.** Roll up the removed glove in the palm of the remaining gloved hand. **d.** Use an ungloved finger to hook the inside of the glove cuff. **e.** Pull off the glove inside out and discard properly. (Artist: Dana Benedicktus.)

your gloves. These are hand lotions (available in scientific supply catalogs) that are designed to prevent irritation from glove materials and prevent drying of hands. Cotton glove liners that absorb perspiration are also available, although these may be unsuitable for highly hazardous work.

A much more serious problem arises when laboratory workers develop allergies to rubber **latex**, a natural product that is commonly found not only in gloves, but in many pieces of laboratory equipment and household items. **Allergies** *are reactions by the body's immune system to exposure to specific chemicals*; in this case, proteins that are found in latex gloves. Although this problem is less common in the general population, as many as 15 to 17% of health care workers have adverse reactions to latex exposure. The allergy apparently originates in sensitization to proteins found in natural rubber latex, and appears to be more likely to develop in users of powdered gloves. Glove powder, which is generally USP-approved cornstarch, is believed to make skin adherence or inhalation of latex proteins more likely, thereby acting as a sensitizer. A **sensitizer** *is an*

agent that can trigger allergies by itself, or cause an individual to develop an allergic reaction to an accompanying chemical. The resulting reactions can range from minor skin irritation to respiratory shock and even death after exposure.

Current recommendations to avoid the development of latex allergies (assuming that your work includes the use of latex gloves) include:

- choosing gloves marked as **hypoallergenic,** which indicates that the gloves are less likely to trigger allergic reactions than similar gloves
- choosing powder-free gloves
- checking the glove manufacturer's information sheets for data about the levels of latex protein present in the gloves, and choosing gloves with the lowest levels of latex proteins

Individuals who exhibit any symptoms of latex allergies should avoid gloves and all other items containing latex. Other glove materials should be substituted.

C. Eye Protection

According to OSHA, an estimated 2000 eye injuries per day occur in U.S. workplaces. These injuries cost businesses and individuals more than $300 million per year in lost time and money. Of the injured individuals, approximately 60% were not wearing any eye protection, and the remainder were wearing inappropriate devices. The most common type of injuries were the result of small flying particles, and about one in five injuries were caused by chemicals. Laboratory workers and visitors must be provided with adequate means of preventing eye injuries. OSHA regulations require that workplaces provide suitable eye protection gear that:

- protects against the hazards found in that workplace
- fits securely and is reasonably comfortable
- is clean and in good repair

Virtually all protective eyewear sold in scientific catalogs meet strict ANSI standards for impact resistance, but there are many other types of hazards found in biotechnology laboratories, see Table 10.3.

There are three general types of eye and face protection devices that should be available in all laboratories, see Figure 10.3. First, safety glasses are a minimum precaution against small splashes and minor hazards. Safety glasses must have side protection in order to provide splash protection. For this reason, regular eyeglasses are not considered adequate eye protection in the laboratory. The next step up in eye protection is goggles, which are sealed around the eyes, providing good protection against large splashes or caustic agents. As with gloves, the laboratory may

Table 10.3. WHY DO YOU NEED EYE PROTECTION?

Eye protection should be worn at all times in the laboratory. This is also true for visitors, who are not aware of hazards. Chemical goggles may be uncomfortable, but modern safety glasses are often sufficient and are designed to be worn comfortably. Appropriate eye and face protection should always be used because laboratories present many hazards to vision, such as:

- danger of explosion or flying particles
- glassware under vacuum
- corrosive liquids such as acids or bases
- cryogenic materials
- liquids that may splash into the eyes
- compressed gases
- blood and other fluids containing infectious materials, that may form aerosols or splash
- radioactive materials
- ultraviolet light and other radiation

Table 10.4. EMERGENCY EYE WASH PROCEDURES

An important line of defense against eye injuries in the laboratory (or other workplace) is the use of the emergency eye wash station, see Figure 10.4.

- Know the location of the eye wash station and how to operate it.
- If you have an accident involving the eyes or face, yell for help **AND** immediately move to the nearest eye wash station if you can.
- Be prepared to help an accident victim to the eye wash station—seconds count! Never assume that victims can take care of the problem themselves.
- Turn on the water at the station and hold the victim's eyes into the double streams of water. Do not worry about wet clothing.
- Most chemical splashes to the eye and many other injuries involve both eyes, so be sure each eye is properly flushed.
- The eyes should receive a constant stream of warm water for at least 15 minutes. (For this reason personal eye wash bottles are not adequate for an emergency.)
- Help the involved person hold their eyelids open and roll their eyes around to aid in proper flushing.
- Do not attempt to remove any particulate matter from the eye by hand (this includes contact lenses).
- In case of a chemical injury, have someone determine the chemical involved and call for medical help.
- All potentially injured eyes should be examined by a medical professional after the flushing period, even if the person does not feel pain.

stock different types of goggles. Goggle materials may provide protection against chemicals or against ultraviolet radiation, not necessarily both.

Finally, full face shields should be used when working with materials under vacuum or where there is any threat of explosion. It is essential to wear additional eye protection under a face shield, which is not sealed. Shields made of polycarbonate, coupled with appropriate eye wear, can protect against UV radiation, liquid nitrogen, or chemical splashes.

Individuals who require vision correction should wear goggles over their regular glasses, or request prescription safety glasses. Contact lenses are generally considered safe to wear in the lab, but if they are worn, they must not be considered eye protection. If you do wear contact lenses in the lab, be sure that co-workers are aware of this, so that in case of an accident they can inform emergency personnel.

Everyone in the laboratory should be aware of the appropriate procedures to follow in case of an accident involving the eyes or face, Table 10.4. Many studies have shown that a quick emergency response can significantly reduce the possibility of permanent damage to vision. For this reason, ANSI standards recom-

mend that emergency eye wash stations (see Figure 10.4) be installed within "ten second" access from all points in a lab. This may translate to anywhere from 10 to 50 feet from lab benches, depending on possible obstructions. Be certain that these stations are easy to locate, that each worker knows how to use them properly, and that they are tested on a weekly basis, both for function and to wash out any contamination that may have accumulated in the water.

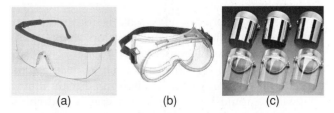

(a)   (b)   (c)

Figure 10.3. Eye And Face Protection. a. Safety glasses with side protection. **b.** Goggles. **c.** Face shields. (Photos courtesy of Fisher Scientific, Pittsburgh, PA.)

Figure 10.4. An Emergency Eye Wash Station. There are two nozzles to direct streams of water at each eye simultaneously. The hand lever should allow the water to stay on with a single push. (Photo courtesy of Fisher Scientific, Pittsburgh, PA.)

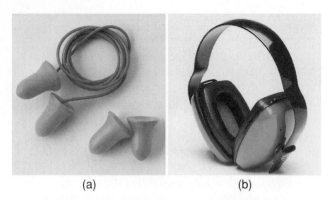

(a)   (b)

Figure 10.5. Two Styles of Ear Protection. a. Ear plugs.
b. Ear muffs. (Photo courtesy of Fisher Scientific,
Pittsburgh, PA.)

D. Ear Protection

Noise-emitting laboratory equipment, such as centrifuges, often produce sound that is uncomfortable or hazardous. Long-term exposure to high noise levels may cause loss of hearing sensitivity. Disposable or personal ear plugs should be available to reduce ambient noise when necessary. **Sonication devices,** *which are used to disrupt cells with high-frequency sound waves,* produce particularly high levels of noise and should only be used with ear protection for everyone in the area. The manufacturer may suggest proper ear protection; otherwise, standard hearing protection devices should worn. Several designs are effective at reducing noise levels, and comfort is a prime consideration when choosing ear protection, see Figure 10.5.

E. Masks and Respirators

Laboratories generally provide access to at least basic respiratory protective equipment. These items may include:

- masks, which filter dirt and large particles from the air, and provide splash protection
- air-purification or filtration respirators
- self-contained breathing systems (in specialized situations)

These items are illustrated in Figure 10.6. The majority of lab personnel never use more respiratory protection than a surgical-type mask, which mainly filters dust and larger aerosols from the air and shields the face from minor splashes. These are ideal for animal work because they can remove allergens from the air. They offer little protection against airborne infectious agents, although they may be helpful in preventing lab personnel from touching their noses or faces while gloved.

Air-purification and self-contained breathing systems are examples of respirators, which are devices that improve air quality for the user. **Respirators,** *breathing devices designed to reduce airborne hazards by manipulating the quality of the air supply,* should not be used by

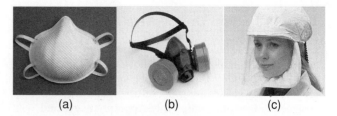

(a)   (b)   (c)

Figure 10.6. Breathing Protection. a. Particle mask.
b. Air-purifying half-mask. **c.** Air purifying hood. (Photos courtesy of Fisher Scientific, Pittsburgh, PA.)

untrained personnel. They are required by OSHA under circumstances where toxic fumes or hazardous air contaminants cannot be removed from the environment by other means. Respirators are ineffective if not used properly, so only trained personnel should be placed in situations where respirators are required.

Air purification respirators *work by filtering the room air through canisters of various adsorbent materials that remove specific contaminants from the air.* When used properly, they can significantly reduce, but not eliminate, airborne contaminants. These respirators require an excellent fit for effectiveness, especially the half-mask models. When fit is a problem, full hood respirators should be used, see Figure 10.6.

A **self-contained breathing apparatus** *contains its own air supply and is required in situations where the user is exposed to highly toxic gases.* These systems are heavy, uncomfortable, and are limited to the short period of protection provided by the gas supply. They are not routinely used and are not appropriate for use by untrained personnel.

III. PHYSICAL HAZARDS IN THE LABORATORY

A. Introduction

There are numerous physical hazards that are encountered in laboratories, which are busy places with many workers sharing the same space and equipment. Being able to work efficiently and safely in crowded spaces is essential. A few of the most common causes of physical injuries in the laboratory will be discussed here.

B. Glassware and Other Sharp Objects

One of the most common injuries in the lab is a minor cut from broken glass or some other sharp item. Even though most laboratory glassware is formulated to resist breakage, they will still develop weak spots, chips, and scratches. Chipped or scratched glassware can break when under pressure, such as during centrifugation, vacuum work, or when filled with liquid. Glassware should be inspected for cracks and chips before being washed and again before laboratory use. Damaged glassware should be discarded or repaired, because it is especially fragile and can easily shatter.

Cut glass plates or glass tubing and rods should be sanded or fire polished to dull the edges.

A significant number of injuries from broken glass occur during the cleaning process. Any sink used to collect dirty glassware should be equipped with a soft mat to prevent breakage. Discard any cracked or chipped glassware after your experiments because these items are especially dangerous for dishwashing personnel. Do not leave broken glass in a sink, where someone else may reach in and be cut. For this and other reasons, never reach into a laboratory sink without hand protection.

Sharps *is a term that describes laboratory items, such as razor blades and needles, that can cause cuts and lacerations.* Razor and scalpel blades should be handled with care, in blade holders whenever possible. Careful covering of the sharp edge with tape when not in use or before discarding can reduce accidental cuts. Never leave these items sitting on lab benches. If using a razorblade to scrape a label from a bottle, be sure to scrape away from yourself, bracing the bottle on a solid surface.

Needles should be handled with caution in the laboratory, especially if they are used with biological materials. It is safest not to reuse needles since many needle punctures are the result of attempts to recap a needle. Gloves will not provide adequate protection against a puncture by a small gauge needle. If it is absolutely necessary to reuse a needle, then recap it by placing the cap on a flat surface and then placing the needle in the cap. Do not hold the needle cap in your hand. All needles, broken glassware, and other sharps should be disposed of in a properly labeled container and not in the general trash, see Figure 10.7.

C. Compressed Gases

Cylinders of compressed gas are commonly found in laboratories because certain laboratory instruments require a supply of specific gases. For example, incubators used to contain cultured mammalian cells require a steady flow of carbon dioxide gas. Gas chromatography instruments require gases such as helium and nitrogen. Gases for laboratory use are stored under high pressure in metal tanks, thus allowing a large amount of gas to be stored in a relatively small volume. Although most gases used in biotechnology laboratories are nontoxic and nonflammable, the gas cylinders themselves can be dangerous because they are under high pressure. Accidents involving the rupture of gas cylinders are rare, but quite dramatic, see Figure 10.8 on p. 154.

Proper storage and handling of compressed gas cylinders is essential to prevent serious accidents. Cylinders should be handled as explosives. Wear eye protection whenever handling tanks, whether they are full or empty. Cylinders should be stored upright, attached to a

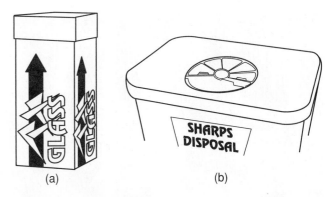

Figure 10.7.   Sharps Disposal. a. Broken glass should be collected in a rigid, clearly marked container. **b.** Proper needle disposal in a marked receptacle. (From *The Medical Laboratory Assistant,* Jacquelyn R. Marshall. Prentice Hall, Inc., Englewood Cliffs, NJ, p. 110, 1990.)

wall or other solid surface with a strong canvas strap or chain. Extra or unneeded gases should not be stored in the laboratory. All cylinders should be delivered with a safety cap over the delivery valve. This cap should be in place at all times when the tank is not in use. It protects the cylinder valve from damage, and it also prevents the accidental opening of the release valve. Fuel and oxygen cylinders must be stored at least 20 feet apart, or separated by a fireproof wall.

When selecting a gas cylinder for use, always read the label. Do not rely on color-coding, which is not uniform among manufacturers. Cylinders should be transported carefully one at a time, using a cart that allows the cylinder to be secured with a strap. Never roll a cylinder on end. These tanks are heavy, and if the cylinder falls, one of the least destructive results would be a broken foot. Cylinders can become high-energy missiles if the gas valve is accidentally knocked off on impact, so be sure that the safety cap is securely fastened when moving tanks.

Compressed gases are dispensed with gas pressure **regulators,** *which decrease and modulate the pressure of the gas leaving the cylinder,* see Figure 10.9 on p. 154. A gas storage tank cannot be directly connected to a laboratory incubator or instrument because of the high pressure of the gas. The gas pressure regulator:

- reduces and controls the pressure of the gas flowing from the storage tank to the instrument
- displays the pressure in the storage tank and the pressure of gas flowing to the instrument

It is the responsibility of the instrument operator to adjust the flow of gas to the instrument, to check that the tank is not depleted, and to replace an empty tank when necessary.

Regulators are threaded to fit the valve outlets of cylinders designed for specific gas types. The gas supplier can help you make sure that you have the correct type of regulator. Never attempt to adapt a regulator to

(a)

(b)

(c)

Figure 10.8.   Compressed Gas Cylinders Are Under Extreme Pressure. This cylinder exploded during storage due to blocking of the pressure release valve and was thrown through the roof of the building, landing 330 feet from the explosion site. **a.** hallway next to laboratory; **b.** laboratory after explosion; **c.** hole in ceiling where cylinder was thrown. (Courtesy Texas State Fire Marshal's Office, Fire Safety Inspection Division.)

an outlet that does not fit. Do not use grease on the valve, washers or O-rings, or regulator fittings. Use a nonadjustable wrench to attach the regulator to the valve. Pliers or other inappropriate devices may damage the cylinder or regulator fittings.

Although there are various types of regulators, the most common type in biology laboratories is the **two-stage gas regulator.** The first stage of a two-stage regulator greatly reduces the pressure of gas leaving the cylinder. The second stage is used to "fine tune" the pressure reaching the instrument. The regulator thus controls the pressure reaching the instrument and protects it from a "blast" of pressurized gas. The parts of a regulator are illustrated in Figure 10.9.

To use a gas cylinder with a two-stage regulator, make sure that the regulator delivery valve is closed, then slowly open the cylinder valve. Never force a valve open. If hand pressure is insufficient to open a valve, it may be damaged and dangerous. Once the cylinder valve is open, which is indicated by a steady reading on the cylinder pressure gauge of the regulator, you can open the regulator valve or handle and adjust the gas flow to the instrument.

Before using any gas, learn about the properties and potential hazards of that gas. Because of the high pressure in gas cylinders, never direct a stream of any gas at another person or yourself.

Cylinders can be checked for leaks with a dilute solution of soap and water applied to the fittings. Leaks will appear as bubbles. Special instruments that detect leaks of certain gases are also available and are recommended by instrument manufacturers under some circumstances.

There is a proper procedure to shut off the flow of gas after use that will decrease the possibility of accidents or damage to regulators. Whenever the gas is not in use, close the cylinder valve completely. Leave the regulator open to empty the line to the gas outlet, then close the regulator. This prevents any residual pressure on the regulator that may cause leakage.

When a tank is almost empty, shut off the gas as described earlier, remove the regulator, and replace the safety cap over the valve. It is best not to drain gas cylinders completely, to avoid contamination of the empty cylinder with air. Mark empty cylinders with the symbol **MT** (empty) to prevent others from trying to guess whether a tank is full or not. Use tape or other large markings. Do not check the fullness of the cylinder by banging on it. Regardless of labels, treat all gas

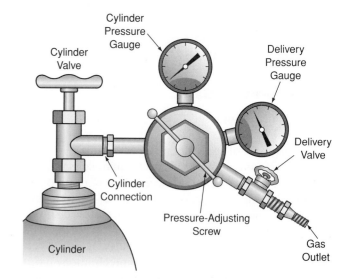

Figure 10.9.   Gas Cylinder Valve and Regulator. The cylinder valve at the top of the high-pressure tank is opened and closed by turning the handwheel. When the valve is open, gas can flow out of the cylinder. The cylinder connection is where the regulator is attached to the tank. The cylinder pressure gauge measures the pressure in the storage tank. The delivery pressure gauge measures the pressure of the gas being delivered to the instrument. The operator controls the flow of gas to the regulator using the pressure-adjusting screw. The delivery valve is then opened to allow flow from the regulator to the instrument; the gas outlet is where the regulator is attached to the instrument.

Table 10.5. Guidelines for Handling Compressed Gas Cylinders

- Be sure all cylinders are clearly labeled.
- Secure cylinders in an upright position at all times, using a strong strap or chain.
- Store only necessary amounts of gases.
- Store gas tanks in a well-ventilated area.
- Never store gas tanks at temperatures above 125°F.
- Do not drop or strike cylinders.
- Use a proper cart with a strap for moving cylinders—do not roll them.
- Do not move a gas tank while the regulator is attached.
- Keep the protective cap on the cylinder head when not in use.
- Use a proper gas pressure regulator.
- Open valves slowly and do not attempt to force a valve open.
- When changing tanks, close the valve between the tank and regulator, and be certain that there is no residual pressure in the gas lines.
- Keep the valves from the tanks closed when the tanks are not in use.
- Never try to repair a cylinder or valve.
- Check cylinders of toxic or flammable gas for leaks on a regular basis.
- Always use safety glasses when handling compressed gases, especially while connecting or disconnecting regulators or supply lines.

cylinders as though they were full. Table 10.5 provides a summary list of guidelines for safely handling compressed gas cylinders.

D. Heat

Laboratory burns due to hot plates, Bunsen burners, and autoclaves are not as unusual as they should be. Any heat-producing equipment can cause burns or a fire and should be treated with respect. Never leave an uncontrolled heat source, such as a Bunsen burner, unattended. Keep the gas regulated on open flames so that they are visible to any observer. Do not place a device with an open flame in a position on the lab bench or hood where you or anyone else might need to reach past it.

Another source of laboratory burns are hot plates that are left on unintentionally, or which have just been turned off. Most people would not touch a hot-plate surface that contained a boiling container, but they might contact a hot surface when it is empty. When you turn off a hot plate, it is courteous to leave a note indicating that the equipment is still hot (and remember to remove the note when the item has cooled).

Boiling or heated liquids are hazards if not handled properly. Use insulated gloves or tongs to handle hot

beakers and flasks. Never heat a sealed container of liquid, because this creates a risk of an explosion and the violent splashing of hot liquid. Be cautious around liquids that may have **superheated,** *which means that they have been heated past their boiling point without the release of the gaseous phase.* Superheated liquids may boil over, sometimes violently, if jarred. This happens regularly with agarose solutions heated in microwave ovens, as well as with liquids that have been autoclaved. It is best to avoid superheating liquids, for example, by heating agarose solutions only enough to melt the gel properly. If you suspect the possibility of superheating, approach the container cautiously, and allow some cooling time before moving the solution.

Most laboratory heat burns are minor. In the case of first-degree burns, which involve reddening of skin, flood the burned area with cold water for 10–15 minutes to reduce the pain and spread of tissue damage. If any blistering occurs or any chemical contamination is involved, the burn should receive prompt medical attention.

E. Fire

Fire is a chemical chain reaction between fuel and oxygen that requires heat or other ignition source, see Figure 10.10. Fuels include any flammable materials, such as paper or solvents. A **flammable** substance *is one that will ignite and burn readily in air*.

The most common source of laboratory fires is the ignition of flammable organic liquids and vapors. This hazard will be discussed more fully in the next chapter on chemical safety. The best overall fire-prevention strategy for labs is to limit sources of flammable materials. This and other general strategies for fire prevention are provided in Table 10.6 on p. 156.

In the event of a fire, it is critical to know what to do. Attempting to extinguish a fire (which may not be the best strategy) requires an understanding of fires and how they spread. The fire triangle is important to remember here, because all fires require fuel, heat, and oxygen. If any of these factors are removed, the fire will be extinguished. The use of appropriate fire extinguishers can control small fires if used properly.

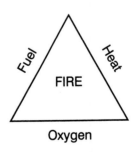

Figure 10.10. The Fire Triangle. Fuel, heat, and oxygen must each be present for fire to occur. (From *Occupational Safety and Health in the Age of High Technology,* David L. Goetsch. Prentice Hall, Englewood Cliffs, NJ, 1996.)

Table 10.6. FIRE PREVENTION IN THE LABORATORY

- Store only minimum amounts of flammable materials.
- Keep open solvent containers away from heat sources.
- Store flammable solvents in appropriate containers.
- Use water baths or hot plates in preference to Bunsen burners.
- Limit other ignition sources, such as sparks or static electricity.
- Never leave an open flame unattended.
- Mark heated hot plates.
- Reduce electrical hazards in the laboratory (see Table 10.7 on p. 158).

There are four types of fires as are summarized in Figure 10.11. Most fires in homes involve Class A combustibles such as paper, cloth, or wood. Class A fires can be extinguished by water or multipurpose dry chemical extinguishers. In the laboratory Class B fires are more likely than Class A. Class B fires usually involve organic solvents and other flammable liquids. Water is more likely to spread than to extinguish this type of fire, which should be smothered with chemical foam or carbon dioxide to remove the oxygen supply from the flames. Class C fires, which involve electrical equipment, also occur in laboratory settings, and water is a poor choice for fighting these fires. Water may increase the possibility of serious electric shock to individuals in the area. Class D fires, which involve combustible metals, are seldom a concern in biotechnology laboratories.

Dry chemical fire extinguishers are common in laboratories. These are highly effective against Class B and C (solvent and electrical) fires. Carbon dioxide extinguishers can be used on the same types of fire, although they tend to have a limited distance of effectiveness. Multipurpose chemical fire extinguishers that can be used for class A + B + C fires are convenient, but most types leave residue behind that requires significant cleanup.

Every laboratory must be equipped with fire extinguishers that are installed close to exits, are easily accessible, and are regularly checked for pressure. Do not try to use these devices, however, unless you have been trained to operate them, and you have practiced recently. Your institutional safety office should be able to provide this training. If you are certain that the fire is contained and minor, and you are qualified to extinguish it, have someone call the fire department before you start. Be sure that all people are out of danger and that they are aware of the fire. Many laboratory flammables will create toxic fumes and heavy smoke as they burn, which makes them difficult to control. Be alert for this possibility. Be sure that you have a clear route of retreat.

When a fire occurs, it is frequently best to evacuate. Everyone working in a laboratory should be familiar with evacuation routes and procedures and it is particularly important that no one stay behind after an evacuation is ordered. Close all fire doors and windows when leaving, if it is safe to do so. Evacuation procedures should include a designated meeting place away from the building. Do not attempt to reenter the building until the appropriate emergency director indicates that it is safe to do so.

Figure 10.11.
Classification of Fires and Rating Portable Fire Extinguishers.
(From *Occupational Safety Management and Engineering*, 4th ed. Willie Hammer, Prentice Hall, Englewood Cliffs, NJ, p. 426, 1989.)

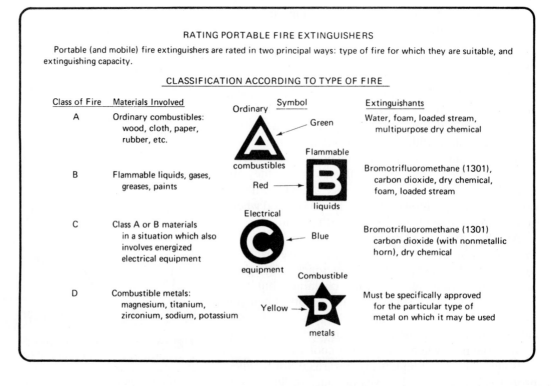

F. Cold

The most common cold hazards encountered in biotechnology laboratories are low temperature freezers, dry ice baths, and liquid nitrogen. Skin contact with **cryogenic,** or *extremely cold (usually defined as temperatures below* $-78°C$*)* substances, or their cooled containers, can result in skin damage similar to heat burns. When handling cryogenic materials, proper hand and eye protection are essential. Gloves should be insulated and loose fitting for quick removal.

Cryogenic liquids are not compatible with ordinary lab glassware or, especially, plasticware. They are usually contained in **Dewar flasks,** *which are heavy multiwalled evacuated metal or glass containers*. While Dewar flasks are similar in appearance to Thermos bottles, they are not the same. Thermos bottles are thin-walled and will shatter if exposed to cryogenic temperatures.

Dry ice, *which is frozen carbon dioxide in solid form,* is frequently used to prepare coolant baths for samples. Most of us have handled dry ice casually at Halloween parties, where it is used to create fuming cauldrons of punch; however, dry ice, which can readily burn wet skin, should always be handled with insulated gloves. Most laboratories that use dry ice keep a supply in a special dry ice chest. If the dry ice is provided as chips, do not use a fragile Dewar flask to scoop up the chips. Use a metal scoop to place the chips in an ice bucket. If the dry ice is furnished as a large block, you will need to make chips of the appropriate size. Do not chip dry ice without goggles for eye protection. A less obvious hazard is the danger of induced **hyperventilation** (*increased respiration rate*) or even **asphyxiation** (*interruption of normal breathing*) from breathing high concentrations of carbon dioxide. Do not lean into the dry ice chest for extended periods of time.

To make a coolant bath, a Dewar flask is usually filled about one third full with an appropriate solvent, and then small pieces of dry ice are added to cool the solvent. Acetone was used historically as the solvent, but this is not recommended for safety reasons, see the Case Study below. Isopropyl alcohol or various solvent mixtures that are relatively nontoxic and nonflammable are recommended (see National Research Council, *Prudent Practices in the Laboratory. Handling and Disposal of Chemicals.* Washington, DC: National Academy Press, 1998). It is essential to wear gloves and use a face shield and eye protection while preparing the coolant bath. The dry ice chips must be added slowly, waiting until the solvent stops bubbling before adding additional dry ice. Once the solvent is suitably cooled, the sample can be introduced to the bath. Always add the item to be cooled slowly. A sudden change in bath temperature can cause the solvent to splatter or overflow the container.

CASE STUDY

Acetone–Dry Ice Bath

A graduate student was making a dry ice and acetone cooling bath on a general use lab bench. Being in a hurry, he added the dry ice chips rapidly and ignored the rapid bubbling of the acetone. He did not notice that the acetone was spilling onto the smooth lab bench and spreading flammable vapors along the surface. When the acetone vapors contacted the motor of a working water bath several feet away, the solvent vapors ignited, causing a fire that completely destroyed the water bath and surrounding materials. Of more immediate concern to the worker himself was the fact that the fire, which started at the water bath, traveled along the stream of solvent vapors back to the dry ice bath. The vapors there ignited as well. The student could have been seriously injured if he had not been standing away from the flask and wearing proper eye protection.

Another cryogenic hazard frequently encountered in laboratories is **liquid nitrogen,** *a liquid form of nitrogen that is supplied in compressed gas cylinders at* $-198°C$. This substance is cold enough to change the physical properties of many materials, and it will cause the equivalent of third degree burns if it contacts skin. It is essential to handle liquid nitrogen with great care. When drawing it from the original cylinder, always wear a face shield and goggles, as well as protective clothing, including heavy leather gloves and leg and foot protection. Because of the high pressure, be certain that you have a firm grip on the dispensing hose, and that the container you are dispensing into is firmly secured in place.

Liquid nitrogen poses both a cryogenic hazard and a hazard from high-pressure gas release. Liquid nitrogen is formed and stored under pressure and can convert to the gaseous form quickly, with a gas expansion ratio of about 700 volumes of nitrogen gas to 1 volume of liquid nitrogen. Special tanks for sample storage have loose fitting caps with vents. Never place liquid nitrogen into a tightly sealed container. Always wear a face shield and eye protection when removing vials from liquid nitrogen. If liquid nitrogen has leaked into a vial during storage, the pressure buildup from nitrogen gas expansion can cause the vial to explode if warmed quickly.

G. Electricity

Laboratories are full of electrical equipment; therefore, they contain electrical hazards. These issues will be discussed in Chapter 18. This is an appropriate place, however, to remind the reader that even small levels of current flowing through the human body can cause harm or even death. An **electrical shock** *is the sudden stimulation of the body by electricity, when the body becomes part of a electrical circuit.*

Table 10.7. REDUCING RISK FROM ELECTRICAL DEVICES

- Use only Underwriters Laboratories (UL)-approved electrical equipment.
- Be sure that all equipment is properly grounded.
- Use GFI circuits whenever appropriate.
- Check electrical cords regularly to be sure they are in good condition.
- Keep your hands dry when handling electrical equipment.
- Never use electrical equipment with puddles of liquid underneath.
- Avoid using extension cords.
- Do not perform repairs inside equipment unless you are qualified to do so.
- Keep equipment unplugged when not in use.
- Plug in fume hood equipment outside the hood.

Table 10.8. GUIDELINES FOR MINIMIZING EXPOSURE TO ULTRAVIOLET RADIATION IN THE LABORATORY

- Wear UV-absorbing eye protection in the presence of UV radiation.
- Protect skin surfaces from UV exposure. Remember your wrists.
- Turn on germicidal lamps in hoods or rooms only when the area is not in use.
- Be careful to avoid reflection from handheld lamps into your eyes.
- Make sure UV light is not directed toward an unsuspecting co-worker.
- Use a UV-absorbing face shield for UV transilluminator work.

Make sure that all electrical equipment is in good working condition and properly grounded. When appropriate, GFI circuits should be installed. **GFI, or ground fault interrupt circuits,** *are safety circuits designed to shut off electrical flow into the circuit if an unintentional grounding is detected.* These are usually installed around sinks and other water sources.

High-voltage power supplies and electrophoresis equipment should be used with caution. Handle power leads one at a time. Table 10.7 provides some general suggestions for avoiding electrical hazards in the laboratory.

H. Ultraviolet Light

Ultraviolet (UV) light *is a form of nonionizing radiation that makes up the light spectrum between visible light and X-rays.* UV radiation is generally divided into three classes:

- UV-A—315–400 nm
- UV-B—280–315 nm
- UV-C—180–280 nm

UV-A, which is also called "black light," increases skin pigmentation and is not usually generated in laboratories. The major sources of UV in labs are transilluminators for visualizing stained DNA bands in electrophoresis gels and germicidal lamps, which operate in the UV-B and UV-C ranges. Hand-held UV lamps for various purposes usually operate around 254 nm. UV-B and UV-C wavelengths damage skin and eyes, and exposure to radiation shorter than 250 nm is considered dangerous.

UV light can cause burns and severe damage to the cornea and conjunctiva of the eyes. These tissues absorb the energies of UV light, with eye irritation developing 3–9 hours after exposure. Skin can also be severely burned by UV radiation, as in a bad sunburn. UV radiation can easily be blocked with certain types of glass and plastic. Eye and skin protection is essential

when UV exposure is a hazard, so check that the materials in the protection devices used will guard against UV. Table 10.8 suggests strategies for minimizing laboratory exposure to UV radiation.

I. Pressure Hazards

Many common laboratory operations, such as filtration, are carried out using a vacuum. All low-pressure work should be carried out inside a shielded enclosure, such as a fume hood, that will provide a partial safety jacket in case of an implosion. An **implosion** *is the collapsing of a vessel under low pressure compared with the outside atmosphere.* To reduce the risk of implosion, only suitably heavy-walled glassware should be used for vacuum work. All vacuum glassware should be rigorously checked for chips or cracks before use. Although the fume hood sash can be used as a partial body shield, you should still use a face shield and goggles when working with systems under strong (high) vacuum.

High vacuums require a vacuum pump, but moderate vacuums suitable for filtration and other purposes can be provided by water aspirators (see Figure 31.1 on p. 591). A **water aspirator** *creates a vacuum through a side arm to a faucet with flowing water.* All vacuum lines should contain a trap or in-line filter for any liquids or vapors that may be pulled into the vacuum source.

A common high-pressure application in biotechnology laboratories is the use of autoclaves for sterilization of glassware and solutions. All modern autoclaves are equipped with safety features to reduce risks but careful operation is still required. Table 10.9 provides some suggestions for the safe use of autoclaves.

IV. ERGONOMIC SAFETY IN THE LABORATORY

While not as dramatic as some safety issues, musculoskeletal disorders—and in particular, repetitive stress injuries (RSIs)—have become the most commonly reported health problems in the laboratory workplace.

Table 10.9.   SAFE USE OF AUTOCLAVES

- Learn correct loading procedures to prevent broken glassware.
- Make sure glassware placed in an autoclave is not cracked or chipped.
- Be sure plastics are autoclavable (Teflon or polypropylene are suitable for this purpose).
- Never autoclave a sealed container.
- Check that the inside is at room pressure before opening the autoclave door.
- Use protection for eyes and hands when opening an autoclave.
- Stand back while opening the autoclave door.
- Release autoclave pressure slowly for liquids to prevent boilover.
- Allow autoclaved liquids to rest for at least 10 minutes before moving the containers to prevent boilover from superheating.

RSIs are caused by **repetitive stress,** *where the same movements and actions are repeated until physical fatigue results.* RSIs are chronic conditions that increase in severity over time. They occur due to muscle and joint stress, inflammation of tendons and associated tissues, and constricted nerves and blood flow to the affected regions.

OSHA data indicate that RSIs account for more than 50% of reported workplace injuries. It is estimated that workers' compensation claims from RSIs cost at least $20 billion each year in lost work time. Employers and employees alike have a strong interest in minimizing these injuries. This is generally achieved through ergonomic safety measures in the workplace. **Ergonomics** *is the study of the effects of environmental factors on worker health and comfort, and the design of environments to increase worker health and productivity.* Ergonomic analysis is used to find the body positions, lab furniture, and equipment that minimize stress and strain on workers. Ergonomic laboratory design can prevent or minimize many health concerns related to the lab.

The two most common musculoskeletal hazards in the laboratory are neck and back injury due to poor posture, and hand and wrist injury due to repetitive hand movements. Neck and back pain can result from the awkward postures dictated by laboratory design. It is impossible for a fixed lab bench to be the ergonomically appropriate height for all workers, given differences in size between individuals. Likewise, biological and fume hoods are generally fixed in height, and require unnatural arm movements to work inside the enclosure. Any task that encourages rounding the shoulders and working with your head and arms in front of the body encourages muscle and joint pain. Considering how many laboratory tasks require leaning forward, such as looking in a microscope, loading samples into electrophoresis gels, performing dissections, and peering at computer screens, it is clear why laboratory workers may experience back and neck pain.

Back and neck problems can frequently be avoided by the use of ergonomically designed chairs, deliberate good posture, and frequent breaks. Any office furniture catalog will offer a variety of well-designed chairs, which should at minimum provide height and tilt adjustment, along with good back support. The best posture to avoid problems is what is called "neutral position," the body alignment that uses the least muscular energy and provides the best possible blood circulation.

The repetitive hand movements of pipetting and computer keyboarding frequently contribute to the development of RSIs. These are two of the most common tasks in a biotechnology laboratory. An experienced technician may dispense as many as sixty or more samples a minute. This task causes repetitive stress and also requires the use of thumb force and awkward bending of the arm and wrist. If a computer keyboard is not located at the optimal height and configuration, the wrists can be held at inappropriate angles. Repetitive pipetting and keyboarding can lead to carpal tunnel syndrome and other problems. **Carpal tunnel syndrome** *is a chronic irritation and swelling of tendons and associated membranes in the wrist.* The irritation and swelling create pressure on the median nerve, resulting in pain and numbness in the thumb and fingers, and loss of manual dexterity. An estimated 4 to 5% of the U.S. population has experienced carpal tunnel syndrome. Another common problem is DeQuervain's tendinitis, which is a painful inflammation of the tendons involved in thumb movement.

Wrist and hand problems may be avoided by purchasing specially designed, ergonomic pipettes and taking frequent breaks. It is good practice to keep your wrists straight and relaxed while using the keyboard and mouse. It may be useful to purchase an inexpensive, soft wrist rest to keep your wrists elevated while typing and to wear a wrist protection device. Most important, do not ignore wrist or arm pain when it starts; take immediate action to prevent the problem from getting worse.

Ergonomic risks are to some extent unavoidable in the laboratory, but you can take steps to protect yourself from health problems. Table 10.10 summarizes general

Table 10.10.   GENERAL ERGONOMIC RECOMMENDATIONS IN THE WORKPLACE

- Maintain good posture at all times.
- Frequently take short breaks from extended activities.
- Change specific tasks every 20 to 30 minutes.
- Perform gentle stretches and exercises for hands, neck, and other stressed body parts on a regular basis throughout the day.
- Request that your employer provide ergonomically designed chairs for working at your desk, the laboratory bench, and in a hood.
- Purchase and use ergonomically designed devices for repetitive tasks.

Table 10.11. EXAMPLES OF PHYSICAL STRESS SOURCES IN THE LABORATORY

Equipment and Activities	Potential Stressors	Possible Solutions
Lab benches	Inappropriate height, hard edges	Use height-adjustable chair with back support; install padding on lab bench edges; use elbow pads
Biological and fume hoods	Inappropriate height, face guard requires awkward movements	Use height-adjustable chair with back support; keep items in hood as close to front as feasible
Centrifuges	Lifting and carrying heavy rotors, bending	Use cart for transporting rotors, use proper lifting techniques
Microscopes	Eyepiece placement too low, awkward placement of hand controls	Purchase ergonomically designed eyepieces that do not require bending; use computer monitor for viewing
Pipetting	Repetitive hand movements, application of force, awkward arm angle	Purchase ergonomically designed pipettors; use automation when possible; experiment with alternate grips; use shorter pipettes
Standing	Back pressure, constriction of blood flow in legs	Avoid standing for more than 20 minutes without walking around; wear comfortable shoes with arch support; install anti-fatigue floor matting
Keyboarding	Repetitive hand movements, application of force, inappropriate height, inadequate wrist support	Install keyboard holder at appropriate height; use wrist support devices to maintain the proper wrist angle; use a good chair with arm rests
Vortex mixing	Vibration	Use minimum speeds; wear elbow pads; utilize closed containers to avoid the need for finger force
Handling vials	Repetitive hand movements, application of force, twisting	Use easy-open vials; use mechanical vial openers

recommendations to avoid musculoskeletal problems. These practices are applicable to activities outside of work as well. Table 10.11 provides additional strategies to deal with individual laboratory tasks, indicating specific stressors and suggesting solutions to minimize the associated physical stress. By developing careful technique, investing in ergonomically designed furniture and equipment, and paying attention to your body's signals, you can minimize your risk of developing a potentially debilitating RSI.

PRACTICE PROBLEMS

1. You are attempting to open the valve on a cylinder of carbon dioxide gas. The valve will not turn despite your best efforts. What should you do?
2. Why are canvas shoes not recommended in the laboratory?
3. Are disposable latex gloves suitable for handling dry ice?
4. How long should an eye be flushed with water after a chemical splash?

QUESTIONS FOR DISCUSSION

1. Think of common situations in the laboratory when gloves are worn. Make a list of possible actions you might take during a procedure and the appropriate times to remove or change your gloves.
2. Consider how you would handle a fire at work. What safety equipment is available? Where are the exits? Are emergency telephone numbers accessible?

CHAPTER 11

Working Safely with Chemicals

I. INTRODUCTION TO CHEMICAL SAFETY

II. CHEMICAL HAZARDS
  - **A. Introduction to Hazardous Chemicals**
  - **B. Flammable Chemicals**
  - **C. Reactive Chemicals**
  - **D. Corrosive Chemicals**
  - **E. Toxic Chemicals**

III. ROUTES OF CHEMICAL EXPOSURE
  - **A. Introduction to Toxicity Measurements**
  - **B. Inhalation**
  - **C. Skin and Eye Contact**
  - **D. Ingestion**
  - **E. Injection**

IV. STRATEGIES FOR MINIMIZING CHEMICAL HAZARDS
  - **A. Preparing a Work Area**
  - **B. Using Chemical Fume Hoods**
  - **C. Limiting Skin Exposure**
  - **D. Storing Chemicals Properly**
  - **E. Handling Waste Materials**

V. RESPONSE TO CHEMICAL HAZARDS
  - **A. Chemical Emergency Response**
  - **B. Chemical Spills**

I. INTRODUCTION TO CHEMICAL SAFETY

Biotechnology laboratories contain a wide variety of chemicals with different health and environmental hazards. It would be an insurmountable task for an individual to memorize all the hazards for each compound. As discussed in Chapter 9, in 1983 OSHA created the Federal Hazard Communication Standard (HCS), which regulates the use of hazardous materials in industrial workplaces. The purpose of this law is to ensure that chemical hazards in the workplace are identified and their risks evaluated, and that this information is communicated to employees. This law requires the employer to:

- identify all chemicals used in the workplace
- have a Material Safety Data Sheet, MSDS, (see Chapter 9) available for each chemical
- label all chemicals
- provide a written program for handling the chemicals
- train employees on the proper use of all chemicals
- provide complete information to health care professionals in emergencies

Chemicals are an integral part of the working environment of the laboratory. All chemicals can be hazardous (e.g., common table salt can be related to high blood

Table 11.1. GOOD PRACTICES FOR WORKING WITH LABORATORY CHEMICALS
• Learn about the physical and toxic properties of a chemical before you start working (e.g., by reading the MSDS).
• Minimize the amount of chemical used.
• Handle, store, and dispose of the chemical according to recommended procedures.
• Work only in well-ventilated areas.
• Label all containers with the chemical name and hazard warnings.
• Wear appropriate personal protective clothing.

pressure in some cases; sugar can be hazardous to an individual with diabetes). When handling any chemicals, even those with no known hazards, the best strategy is to use good general laboratory practices, see Table 11.1.

II. CHEMICAL HAZARDS

A. Introduction to Hazardous Chemicals

Many chemicals found in laboratories can present risks to users who do not practice good laboratory techniques. A chemical is defined as *hazardous* if:

- it has been shown to cause harmful biological effects
- it is flammable, explosive, or highly reactive
- it generates potentially harmful vapors or dust

In addition, many chemicals are considered potential hazards because of their structural similarities to known hazards, even though no data are available about the chemical itself.

The preceding categories are not mutually exclusive, and many chemicals exhibit more than one of these properties. There are a number of labeling systems that chemical suppliers use to indicate hazards on chemical containers. One of the most widespread is the hazard diamond system developed by the National Fire Protection Association (NFPA). The **hazard diamond system** *rates chemicals according to their fire, reactivity, and general health hazards,* see Figure 11.1. This system provides easy-to-read, color-coded information with simple numerical scales. Special hazards, such as radioactivity, are indicated by standard symbols.

It is important to have information about the hazards of laboratory chemicals on hand at all times. One excellent resource for laboratories is *Prudent Practices in the Laboratory. Handling and Disposal of Chemicals* (National Research Council, National Academy Press, Washington, DC, 1998), which provides an overview of the issues related to chemical hazards, provides general information about many laboratory chemicals, and also

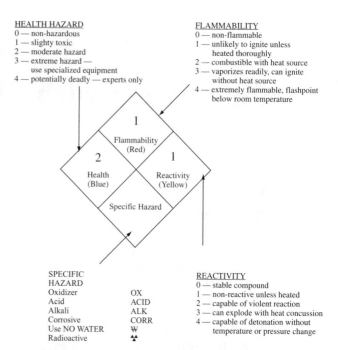

HEALTH HAZARD
0 — non-hazardous
1 — slightly toxic
2 — moderate hazard
3 — extreme hazard — use specialized equipment
4 — potentially deadly — experts only

FLAMMABILITY
0 — non-flammable
1 — unlikely to ignite unless heated thoroughly
2 — combustible with heat source
3 — vaporizes readily, can ignite without heat source
4 — extremely flammable, flashpoint below room temperature

SPECIFIC HAZARD
Oxidizer   OX
Acid   ACID
Alkali   ALK
Corrosive   CORR
Use NO WATER   W
Radioactive

REACTIVITY
0 — stable compound
1 — non-reactive unless heated
2 — capable of violent reaction
3 — can explode with heat concussion
4 — capable of detonation without temperature or pressure change

Figure 11.1. National Fire Protection Association System for Hazard Classification. The hazard diamond has four hazard codes to indicate health hazards, flammability, reactivity, and any special hazards. The numerical rating system uses the numbers 0 to 4, with 0 indicating a non-hazard and 4 denoting the highest hazard level of that type. Special hazards are indicated with symbols as shown in the figure. On the sample label shown, the chemical is moderately toxic, only slightly flammable, and unreactive unless heated.

suggests appropriate reference materials for more detailed information. The Hazard Communication Standard requires that a set of MSDSs be available for all chemicals used in a particular workplace. Many laboratories keep their own set in a notebook. MSDS information is also readily available on the Internet (http://hazard.com/msds/index.html).

B. Flammable Chemicals

As discussed in the previous chapter, laboratory fires frequently involve flammable chemicals (Class B fires). **Flammable** *refers to materials that are relatively easy to ignite and burn.* Note that the term **inflammable** *also refers to flammable materials.* (The term **nonflammable** *is sometimes applied to chemicals that do not readily ignite and burn.*) The most common flammable chemicals found in laboratories are liquid organic **solvents,** *which are chemicals that dissolve other substances.* Many of these liquids are **volatile,** *evaporating quickly at room temperature.* It is the vapor phase of flammable liquids that burns, not the liquid itself.

Flammability is a relative term because all materials can be ignited in the presence of adequate heat. **Flash point** *is the temperature where a chemical produces enough vapor to burn in the presence of an ignition source.* The lower the flash point, the more flammable the compound. Many common laboratory chemicals, such as acetone and hexane, have flash points well below 0°C, see Example 1. This means that flammability is always a concern when working with these materials. Once ignition of a flammable solvent occurs, the increasing temperature increases the rate of vaporization, which in turn provides more fuel to burn. Solvent fires can spread out of control very quickly if there is not prompt corrective action. Another concern with flammable liquids is that solvent vapors can diffuse along lab benches or floors, mixing with air and eventually contacting an ignition source. The flames can then follow the invisible vapor trail back to the original container, igniting the larger source of fuel.

EXAMPLE 1

Acetone

Acetone is an example of a relatively nontoxic but highly flammable solvent that is commonly found in biotechnology laboratories. It has a flash point of −18°C (NFPA Fire Hazard Rating = 3) and it is susceptible to "flash back" because of its volatility (remember the Case Study in Chapter 10, p. 157). The odor of acetone is detectable at concentrations in air well below toxic levels. Acetone can act as an eye and nasal irritant, but these effects are generally short-lived.

There are many systems available for rating the flammability of laboratory chemicals, usually based on their flash points. The NFPA, for example, rates chemicals on a scale of 0–4, with 0 being nonflammable and 4 being extremely flammable. Any chemical with a rating of 3 or 4, which includes those with flash points at room temperature or below, should be treated as a fire hazard. Table 11.2 provides a set of guidelines for handling flammable chemicals to reduce the risk of fire.

Handling chemicals safely depends on knowledge of the properties of individual chemicals. Some powdered metals, for example, are **pyrophoric,** *which means that they will ignite on contact with air.* Other chemicals (e.g., elemental sodium) react violently on contact with water. These types of chemicals must be manipulated only within a controlled environment. Possibly the most dangerous flammable liquid handled regularly in biotechnology laboratories is diethyl ether, with a flash point of −45°C (NFPA Fire Hazard Rating = 4). This chemical should never be placed in proximity to electrical equipment that might produce sparks (e.g., centrifuges or regular refrigerators). It should be handled in fume hoods that will prevent a buildup of flammable vapors.

Fire codes require that 10 gallons or more of flammable liquids be stored in safety cabinets that are designed to minimize the risk of fire or explosion, see Figure 11.2. These cabinets do not require venting in the absence of toxic fumes, and the lack of an outside

Table 11.2. SAFE HANDLING OF FLAMMABLE CHEMICALS

Standard safety practices as described in Table 11.1, plus:

- Keep all flammable substances away from ignition sources.
- Never heat a flammable chemical with an open flame.
- Remember that solvent vapors can mix with air and diffuse to distant ignition sources.
- Know the appropriate fire prevention methods for the chemical.
- Keep containers tightly closed at all times when not in use.
- Never store incompatible chemicals together (keep acids separate from bases, etc.).
- Keep flammable chemicals away from reactive chemicals.
- Work only in fume hoods or other well-ventilated areas.
- Avoid static electricity discharges when working with flammable substances.
- Never pour flammable substances down a drain or into the trash.
- In case of a spill, deal with any skin contamination before beginning the laboratory decontamination process.

air supply will act as a limiting agent for any fire within the cabinet. If venting is introduced, the cabinet system must be constructed with proper ducts and blowers to vent fumes from the building.

Figure 11.2. Storage Cabinet for Flammable Chemicals.
(Photo © 2008, Flinn Scientific, Inc. All Rights Reserved. Reproduced for one-time use with permission from Flinn Scientific, Inc. Batavia, Illinois, U.S.A. No part of this material may be reproduced or transmitted in any form or by any means, electronic or mechanical, including, but not limited to photocopy, recording, or any information storage or retrieval system, without permission in writing from Flinn Scientific, Inc.)

C. Reactive Chemicals

Most laboratory chemicals are reactive to some extent, but those that pose the greatest hazard to laboratory workers are those that undergo violent chemical reactions. Reactive chemicals can be categorized as:

- those that participate in exothermic or gas-generating reactions
- unstable chemicals
- those that form peroxides and other oxidizing agents
- incompatible chemical mixtures

Reactive chemicals may produce an **explosion,** *a sudden release of large amounts of energy and gas within a confined area.* A fire may result, depending on the elements involved. General guidelines for handling reactive chemicals are provided in Table 11.3.

In a laboratory, explosions are likely to occur as a result of combining reactive chemicals in a sealed container. Sealed containers can also explode if used for any **exothermic** (*heat-producing*) or gas-forming chemical reaction, see the Case Study below. This happens with some regularity in mixed chemical waste containers (which are not recommended). Glass bottles should never be used to contain potentially reactive chemical mixtures.

CASE STUDY

Mixed Waste Containers

A graduate student in one of the author's labs was working late and in a hurry. He needed to dispose of some liquid waste and poured it into a glass bottle, designated for organic wastes, inside a fume hood. Noticing that the bottle was full, he replaced the cap, pulled the hood sash 80% closed, and left the room. Within three minutes the bottle, which contained acid as well as organic waste, exploded from a gas-producing chemical reaction. The sash of the fume hood, which was constructed of shatter-proof material, contained much of the flying glass and liquid within the hood, but was sufficiently damaged to require replacement. Several large pieces of chemically contaminated glass flew under the hood sash, through the open lab door, across the hallway, and into the facing lab—a distance of almost 50 feet. It was fortunate that no one was present in the path of the debris. Cleanup and decontamination took several hours. After the incident, only plastic bottles, with the caps removed from the hood, were used as waste containers.

Some chemicals are unstable and are susceptible to chemical breakdown with time, see Example 2. This is one reason why all chemicals should be labeled with the date of receipt. In some cases, the breakdown products are shock-sensitive. The most commonly encountered examples are picric acid, dinitrophenol, and compounds that break down into organic peroxides.

Table 11.3.   SAFE HANDLING OF REACTIVE CHEMICALS

Standard safety practices as described in Table 11.1, plus:

- Know the reactive properties of the chemical (read the MSDS).
- Never mix unknown chemicals together, especially in closed waste containers.
- Label containers of reactive chemicals carefully.
- Store only compatible chemicals in the same area.
- Store oxidizing chemicals away from flammable materials.

A **peroxide former** *is a chemical that produces peroxides or hydroperoxides with age or air contact.* These chemicals, which include a variety of aldehydes, ethers, and ketones, are highly hazardous. They are flammable and may explode on exposure to heat or shock. It is important to refer to the MSDS for information about the hazards of specific chemicals.

EXAMPLE 2

DEPC (Diethyl Pyrocarbonate)

Diethyl pyrocarbonate (DEPC) is a toxic chemical commonly found in biotechnology laboratories. It is used to treat solutions and glassware when isolating RNA because it is an effective agent for inactivating RNA-digesting enzymes. It is also a suspected carcinogen, and should be handled only with gloves. In addition, DEPC breaks down to carbon dioxide gas and ethanol when exposed to moist air. If this decomposition takes place in a sealed bottle, pressure can build to explosive levels. DEPC should be stored under dry conditions in a refrigerator, within a desiccator. If possible, store the bottle in the original metal container to act as an explosion barrier. Always allow refrigerated DEPC to equilibrate to room temperature before opening the bottle. Because DEPC is an explosion hazard, always use goggles and a shield when handling a stock bottle.

Peroxide formation is generally limited to liquids that have evaporated and undergone autooxidation. Diethyl ether and tetrahydrofuran (THF) are the most likely peroxide formers to be found in biotechnology laboratories. These chemicals should be stored away from light and heat in carefully sealed containers. Any containers that show signs of evaporation, especially older containers of ether, should not be handled. Contact your institutional safety office for proper disposal instructions.

In addition to the preceding hazards, certain combinations of chemicals can undergo violent reactions that result in explosions or release of highly toxic gases or other products. The best protection against this phenomenon is to follow standard laboratory procedures

Table 11.4. *Examples of Incompatible Chemicals*

Chemical	Avoid Contact With
Acetic Acid	Aldehydes, bases, carbonates, ethylene glycol, hydroxides, metals, oxidizers, perchloric acid, peroxides, permanganates, phosphates, xylene
Acetone	Acids, amines, oxidizers, plastics
Chlorates	Any finely divided combustible or organic materials
Flammable liquids	Ammonium nitrate, chromic acid, halogens, hydrogen peroxide, nitric acid, oxidizers
Hydrocarbons	Bromine, chlorine, chromic acid, fluorine, sodium peroxide
Hydrogen peroxide	Acetone, acetic acid, alcohols, carboxylic acid, combustible materials, most metals or their salts, nitric acid, phosphorus, sodium, sulfuric acid
Hydrogen sulfide	Metals, nitric acid, oxidizers, sodium
Iodine	Acetylene, ammonia (aqueous or anhydrous), hydrogen gas
Nitric acid	Acetic acid, aniline, chromic acid, flammable gases and liquids, hydrogen sulfide
Perchloric acid	Acetic anhydride, alcohol, paper, sulfuric acid, wood
Potassium permanganate	Benzaldehyde, ethylene glycol, glycerin, sulfuric acid
Sulfuric acid	Potassium chlorate, potassium permanganate (or other compounds with light metals, such as sodium or lithium)

when mixing chemicals. Never combine chemicals without an established set of instructions or without researching the reactive properties of the chemicals involved. Know the hazards associated with the chemicals with which you work. Table 11.4 provides a few examples of incompatible chemical mixtures.

D. Corrosive Chemicals

Corrosive chemicals *are those that can destroy tissue and equipment on contact.* Acids and bases are the most common corrosives found in biotechnology laboratories. Both will cause chemical burns and tissue damage on contact with skin or eyes. An even greater danger is that of inhalation of corrosive vapors, which can irritate or burn mucous membranes and potentially cause serious lung damage. Because of this inhalation hazard, strong solutions of corrosives should always be used in a fume hood. The extent of potential damage caused by a corrosive chemical will depend upon the nature of the chemical and the amount and length of exposure. Table 11.5 offers some guidelines for safe handling of corrosives.

E. Toxic Chemicals

i. Acute versus Chronic Toxicity

"What thing is not a poison? All things are poison and nothing is without poison. It is the dose only that makes a thing not a poison."

—Paracelsus, 16th century

Toxicity *is the term used to describe the capacity of a chemical to act as a poison, creating biological harm to an organism.* A toxic material may alter the function of essential organs of the body; the heart, lungs, liver, nervous system, or kidneys. The level of toxic hazard is

dependent on the nature of the chemical itself, the concentration and length of exposure, the health of the individual, and the speed and success of corrective measures. As recognized by Paracelsus, all chemicals can be toxic at higher dose levels in some individuals, see Table 11.6 on p. 166. Many laboratory chemicals, however, have well-documented toxic effects at low doses. The risk of harmful effects from these substances can be reduced or eliminated by proper laboratory technique and simple precautions.

Toxic materials can act in a variety of ways. An **acute toxic agent** *causes damage in a short period of time, and a single exposure may be adequate to cause harmful effects.* A fast-acting poison like hydrogen cyanide is an example of an acute toxic material. **Chronic toxic agents** *have cumulative effects or may accumulate in the body with multiple small exposures.* These small doses of the toxic material may not produce an immediate effect, but instead

Table 11.5. *Safe Handling of Corrosive Chemicals*

Standard safety practices as described in Table 11.1, plus:

- Always work with corrosive chemicals in a fume hood to avoid respiratory irritation.
- Add acid to water, not the reverse.
- Perform neutralization reactions between acids and bases slowly to minimize gas and heat generation.
- Be sure that protective gear is appropriate to the chemical in use.
- Store acids and bases in separate areas.
- Do not work with hydrofluoric acid without specific training and precautions.

Table 11.6. CLASSES OF RELATIVE TOXICITY

Toxicity Level	Chemical Example	Known Rat Oral LD$_{50}$ (mg/kg)	Approximate Human Oral Lethal Dose (g/70 kg adult)
Almost nontoxic	Sucrose	80,000	5400
Slightly toxic	Ethanol	14,000	918
Toxic	Sodium chloride	3000	258
Highly toxic	Sodium cyanide	6.5	7
Extremely toxic	Strychnine	2.5	0.05

produce injury over time. Lead poisoning in children is an example where cumulative exposure to very low levels of a toxic material can have long-term, serious health effects. Many chemicals can act as both acute and chronic toxic materials, see Example 3. In some cases, exposure to mixtures of chemicals may increase their toxic effects.

EXAMPLE 3

Toxicity of Organic Solvents

Nonchlorinated flammable organic solvents, such as benzene, toluene, and xylene, can exert a wide range of toxic effects. As acutely toxic materials, they can cause headaches and dizziness after inhalation. They are respiratory, skin, and eye irritants. Because of their solvent properties, contact with skin will cause drying, leading to **dermatitis**, *a painful reddening and inflammation of the skin.* Some organic solvents, such as phenol, can act as corrosives as well. Organic solvents also may cause chronic toxicity, with long-term exposure leading to liver, lung, and kidney damage. Chronic exposure to benzene is linked to human leukemia. Virtually all of these chemicals have characteristic odors, and some lab workers mistakenly use these to indicate exposure. The vapor levels necessary for detection by smell, however, have no relationship to toxicity levels in most cases, so this is not an adequate warning system. When working with organic solvents, always use a fume hood and refer to the MSDS for hazard information.

Exposure to toxic chemicals can result in a wide variety of human health problems. The next few sections will discuss some of the types of toxic materials likely to be encountered in the laboratory. Table 11.7 provides some general suggestions for safe handling of toxic chemicals.

ii. IRRITANTS, SENSITIZERS, AND ALLERGENS

Many laboratory chemicals are **irritants,** *which produce unpleasant or painful reactions when they contact the human body.* Skin irritation frequently takes the form of redness, itching, or dermatitis. Chemical irritation can be more serious when the point of contact is the eyes or respiratory tract. It is always a good idea to minimize body contact with all chemicals, even those that seem fairly innocuous. Wear proper protective clothing, includ-

ing gloves and safety glasses, whenever handling chemicals. Some chemicals with acute irritating effects can also exert chronic effects, see Example 4.

EXAMPLE 4

Formaldehyde

Everyone who has taken an anatomy class has probably encountered formaldehyde. Formaldehyde gas is a toxic and highly flammable substance. Diluted formalin, a 37% solution of formaldehyde, has been used historically to preserve tissue specimens for dissection. Formalin contains 7 to 15% methanol to stabilize the formaldehyde. Most suppliers of preserved specimens are now providing propylene glycol–based or other types of tissue preservatives because of the formidable list of the toxic properties documented for formaldehyde, including:

- respiratory and eye irritation
- slow developing burns to eyes and skin
- skin sensitization
- potential carcinogenicity

Table 11.7. SAFE HANDLING OF TOXIC CHEMICALS

Standard safety practices as described in Table 11.1, plus:

- Be aware of both the acute and chronic effects of a known toxic chemical.
- Treat all chemicals as toxic unless otherwise informed.
- Be certain that the gloves you are wearing provide appropriate protection from the chemicals being used.
- Do not rely on odors to warn you of exposure to chemicals.
- Minimize exposures to any toxic substance that can accumulate in the body.
- Maximize precautions when working with any chemical known to be mutagenic or carcinogenic.
- Be alert to symptoms of toxicity or sensitization to chemicals.
- Both men and women should consider reproductive hazards in the workplace. Handle toxic chemicals carefully if you plan on starting or increasing a family, or suspect that you might be pregnant.

Allergies *are reactions by the body's immune system to exposure to a specific chemical,* or **allergen.** Allergies can manifest themselves as skin or respiratory reactions, depending on the route of exposure to the allergen. Just as individuals will have a wide range of sensitivities to environmental allergens, such as dust and pollens, they will also have a wide range of sensitivities and symptoms to chemical allergens and a range of symptoms associated with an allergic response. Reactions can range from a mild rash to nasal congestion to **anaphylactic shock,** *a sudden life-threatening reaction to allergen exposure.*

Before allergies occur, an individual must be sensitized to the allergen. Some chemicals can act as **sensitizers,** *and may trigger allergies themselves, or cause an individual to develop an allergic reaction to an accompanying chemical.* Dimethylsulfoxide (DMSO) is an example of a sensitizing agent that penetrates skin and carries other chemicals with it, as in the Case Study below. Limiting contact with potential allergens and sensitizers, working under well-ventilated conditions, and wearing proper protective clothing are wise precautions to reduce exposure to these agents.

CASE STUDY

Chemical Sensitization

As a beginning graduate student, one of the authors did extensive tissue culture work. This work was performed in an appropriate biological safety cabinet while wearing gloves. After about 4 months, she developed a rash on her left wrist, exactly corresponding to the shape and location of her wristwatch. Because she had worn this same watch almost continuously for more than 6 years, it seemed unlikely that she had suddenly developed an allergy to it. After examining the rash, a dermatologist questioned the author about her work. As soon as tissue culture was mentioned, the doctor asked if she worked with cells treated with DMSO (dimethylsulfoxide). When confirmed, he pointed out that DMSO is a powerful penetrating and sensitizing agent, which probably had triggered a reaction to nickel found in the watchband. Even though the author had never spilled DMSO on her wrist, enough DMSO vapors had reached above the glove line to produce the sensitization reaction. After this, the author wore longer gloves while handling DMSO, and did not wear a watch in the laboratory. The problem disappeared.

---

iii. NEUROTOXINS

Neurotoxins *are compounds that can cause damage to the central nervous system.* In many cases, the neurological effects, which may include loss of coordination and slurred speech, may develop slowly after long-term exposure to small doses of the toxic agent. Organometallic compounds, like methylmercury, act as potent neurotoxins. Acrylamide, which is routinely used to prepare gels for protein separations, is a common neurotoxin in many biotechnology laboratories, as shown in Example 5.

EXAMPLE 5

Acrylamide

Acrylamide is used in the preparation of polyacrylamide gels for protein separation. In its polymerized form, it is considered harmless, and is sold in gardening stores as a water-absorbing agent to be mixed with potting soil for plants. Acrylamide in its unpolymerized form, however, is a potent neurotoxin. It can have both acute and chronic effects. Direct contact can result in eye burns and skin rashes. Symptoms of chronic overexposure include dizziness, slurred speech, and numbness of extremities. Acrylamide is also a suspected human carcinogen. This is a chemical that should always be handled with great respect for its toxic properties. Always wear a lab coat, gloves, and dust mask when handling the solid form. Acrylamide should only be weighed in a designated balance within a fume hood. Given the documented risk from respiration of acrylamide powder, laboratories should consider the extra expense of purchasing premade acrylamide solutions. Although polymerized polyacrylamide gels are nontoxic, do not handle these gels without gloves. There may be residual unpolymerized acrylamide present that can be absorbed through the skin.

---

iv. MUTAGENS AND CARCINOGENS

Mutagens *are compounds that affect the genetic material of a cell.* They cause alterations in DNA that will be inherited by offspring cells. Mutagens are considered chronic toxic agents because the induced damage may not be apparent for years. Mutational damage is cumulative. Because mutagens can affect the genetic material of cells, they may also act as cancer-causing agents. Although the relationship between carcinogenicity and mutagenicity has not been clearly demonstrated for many chemicals, it is prudent to assume that any mutagen is potentially carcinogenic as well. Mutagenicity data are generally derived from animal studies, so the exact human risk is difficult to determine. The MSDS can provide a summary of the available data for individual chemicals. Some common laboratory chemicals that are known to be animal mutagens are shown in Table 11.8 on p. 168.

One of the most commonly used chemicals in biotechnology laboratories is ethidium bromide. **Ethidium bromide (EtBr)** *is a fluorescent dye used to visualize nucleic acids in agarose gels.* It acts by inserting itself into DNA molecules, see Figure 11.3 on p. 168. This provides the mechanism for the strong mutagenicity EtBr demonstrates in animal models, see Example 6.

Table 11.8. EXAMPLES OF KNOWN ANIMAL MUTAGENS

The following are examples of common laboratory chemicals that have been shown to be mutagenic in animal models and are presumed to be human mutagens as well:

- Acridine orange
- Colchicine
- Ethidium bromide
- Formaldehyde
- Hydroquinone
- Osmium tetraoxide
- Potassium permanganate
- Silver nitrate
- Sodium azide
- Sodium nitrate
- Sodium nitrite
- Toluene

EXAMPLE 6

Ethidium Bromide

Ethidium bromide (EtBr) is a known mutagen that is commonly used in biotechnology laboratories to visualize nucleic acids under ultraviolet light. Most laboratories treat solid EtBr and concentrated stock solutions with great respect. However, workers commonly add low concentrations of EtBr to agarose gels and electrophoresis running buffers and these dilute solutions should still be considered potential health hazards. Decontaminating spills or solutions containing EtBr with bleach has been demonstrated to be ineffective and capable of creating breakdown products that may be more harmful than EtBr itself (Lunn, George and Sansone, Eric B. *Destruction of Hazardous Chemicals in the Laboratory.* New York: John Wiley and Sons, 1994, 185.) Several studies indicate that laboratories should follow more effective procedures using special detoxification resins or a deamination procedure with sodium nitrite and hypophosphorous acid. Complete details of these procedures can currently be found at multiple sites on the Internet. The decontamination process can be monitored with a handheld ultraviolet light. All items that come into contact with EtBr, such as gloves, spatulas, or paper towels, should be properly decontaminated or treated as hazardous waste.

Carcinogens, or **cancer-causing agents,** *can initiate and promote the development of malignant growth in tissues.* They act as chronic toxic materials, and exposure to these compounds is often unrecognized for years. Known and suspected laboratory carcinogens are identified and listed by OSHA, and other agencies including the **International Agency for Research on Cancer (IARC),** *an agency that determines the relative cancer hazard of materials.* Carcinogenic substances need to be handled with great care. A list of types of chemicals known to have carcinogenic effects in animals is shown in Table 11.9. Information about carcinogenicity is included in the MSDS for specific chemicals.

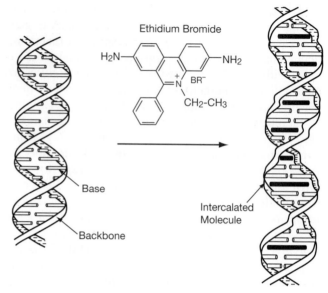

Figure 11.3. Interaction of Ethidium Bromide with DNA.

v. EMBRYOTOXINS AND TERATOGENS

Embryotoxins *are compounds known to be especially toxic to the developing fetus.* For example, organic mercury compounds, some lead compounds, and formamide are known to be embryotoxins. **Teratogens** *are a type of embryotoxin that causes fetal malformation.* They are known to interfere with normal embryonic development, but they do not cause direct harm to the mother. The greatest susceptibility of the fetus to these compounds is generally in the first 12 weeks of pregnancy, sometimes when the woman is unaware of the pregnancy.

Pregnant women and women of child-bearing age should be especially cautious when working with toxic chemicals. Substances that enter the bloodstream of the

Table 11.9. EXAMPLES OF KNOWN TYPES OF ANIMAL CARCINOGENS

Carcinogen Type	Example
Acylating agents	β-Propiolactone
Alkylating agents	Acrolein
	Ethylene oxide
	Ethyl methanesulfonate
Aromatic amines	Benzidine
Aromatic hydrocarbons	Benzene, Benzo[a]pyrene
Hydrazines	Hydrazine
Miscellaneous inorganic compounds	Arsenic and certain arsenic ompounds
Miscellaneous organic compounds	Formaldehyde (gas)
Natural products	Aflatoxins
N-Nitroso compounds	*N*-nitroso-*N*-alkylureas *N*-nitrosodimethylamine
Organohalogen compounds	Carbon tetrachloride Vinyl chloride

Table 11.10.   *SUMMARY OF COMMON CHEMICAL HAZARDS IN BIOTECHNOLOGY LABORATORIES*

Chemical	Flammable	Corrosive	Reactive	Low-Dose Toxicity	Major Route of Exposure
Acetic acid		yes	yes	acute	any
Acetone	yes		yes	chronic	any
Acetonitrile	yes		yes	acute	any
Acrylamide				acute, chronic, neurotoxin	skin, inhalation
Ammonium hydroxide		yes	yes	acute	skin, inhalation
Benzene	yes		yes	chronic, carcinogen	inhalation, skin
Chloroform		yes	yes	possible carcinogen, teratogen	inhalation
Diethyl ether	yes		yes	acute	inhalation
Dimethyl sulfoxide			yes	sensitizer	skin contact with vapors
Ethanol	yes		yes	acute, chronic	ingestion
Ethidium bromide				mutagen	skin
Ethyl acetate	yes		yes	acute, chronic	any
Formaldehyde	yes		yes	acute, sensitizer, carcinogen	any
Hexane	yes		yes	chronic, neurotoxin	inhalation
Hydrochloric acid		yes	yes	acute, corrosive	any
Hydrogen peroxide			yes	irritant	skin
Mercury				acute, chronic, neurotoxin	inhalation, skin
Methanol	yes		yes	acute	ingestion
Nitric acid		yes	yes	acute	skin
Phenol		yes	yes	acute, chronic, neurotoxin, hepatotoxin	skin, inhalation
Pyridine	yes			chronic, hepatotoxin	inhalation, skin
Sodium hydroxide		yes	yes	acute, corrosive	skin, inhalation
Sulfuric acid		yes	yes	acute, corrosive	skin, inhalation
Toluene	yes		yes	chronic, irritant	skin, inhalation

This table was prepared from information provided in *Prudent Practices in the Laboratory. Handling and Disposal of Chemicals,* the National Research Council, National Academy Press, Washington, DC, 1998.

*This list provides a summary of the most significant hazards of some of the chemicals commonly encountered in biotechnology laboratories. Many have additional hazardous properties that are not listed here. Always consult the MSDS and other reference sources when working with an unfamiliar chemical.

mother may be able to pass the placental barrier to the fetus. Always discuss an intended or actual pregnancy with your laboratory supervisor. It may be safest to request alternate duties during at least the first trimester of pregnancy.

Table 11.10 provides a summary of the hazards of some common biotechnology laboratory chemicals.

III. ROUTES OF CHEMICAL EXPOSURE

A. Introduction to Toxicity Measurements

Toxic chemicals can enter the body by four main routes of exposure:

- inhalation
- skin and eye contact
- ingestion
- injection

Types and levels of toxic injuries depend on the exposure route. Some chemicals are especially dangerous when inhaled; others only when ingested. Of the four routes of toxicity, inhalation and skin absorption are most likely in a laboratory workplace.

Every chemical has its own level of toxicity; some are more poisonous than others, see Table 11.6, p. 166. Many chemicals are found in the foods we eat and are considered relatively nontoxic. Some substances, such as hand creams, are good for the skin, but should not be ingested. Many compounds, such as over-the-counter pain relievers, are considered safe at one level, but toxic when the dosage is increased. Other chemicals will cause quick death even in small doses.

One method scientists use to assess the toxic effects of different chemicals is to measure the amount of the compound that will cause a reaction or death in animals. Test animals receive known doses of the chemical and the results are measured. One common measure of chemical toxicity is the LD_{50} level. **LD_{50} (Lethal Dose—50%)** *is the amount of a chemical that will cause death in 50% of test animals.* Animals of different species are used. In order to compare these doses, the amount needed to kill 50% of the animals is recorded in amount (grams or milligrams) per kilogram of the animal's body weight. A larger absolute dose is generally necessary to kill a larger animal, although relative values may be similar (see the Example Problem). LD_{50} studies obviously cannot be performed with humans, but the information from animal studies can be used to indicate the relative toxicity of certain compounds. Animal toxicity is usually, but not always, a good predictor of human toxicity.

EXAMPLE PROBLEM

LD_{50} studies can be performed in a variety of animal species, with the results suggesting relative toxicity levels in humans. For example, consider a chemical that is tested for toxicity in rats and mice. The dose of chemical that kills 50% of the rats tested is 48 mg. The dose of chemical that kills 50% of the mice tested is 5 mg. How would you convert these data to LD_{50} values, and how might you extrapolate the data to humans?

ANSWER

The key additional information needed for calculating LD_{50} values is the mean body weights of the rats and mice tested. In this scenario, assume that the mean body weight of the rats is about 400 g and the mice, about 39 g. The LD_{50} values would be calculated as follows:

For rats—48 mg of chemical divided by 400 g per rat
  = 0.12 mg chemical per gram body weight or
  120 mg/kg body weight = the LD_{50} in rats

For mice—5 mg of chemical divided by 39 g per rat =
  0.13 mg chemical per gram body weight or
  130 mg/kg body weight = the LD_{50} in mice

In order to reach a solid conclusion about human toxicity, you would like to have human data. In its absence, only a hypothetical calculation can be made. Because the preceding LD_{50} values are similar, it would be reasonable to predict that the LD_{50} value for humans could be comparable. Based on this assumption, for a 70 kg (150 lb) human, a lethal dose of the chemical is in the range of 8.4–9.1 grams. This would be considered a mildly toxic chemical.

LD_{50} values depend on the route of chemical exposure. Ingestion, or oral, LD_{50} values may differ significantly from skin exposure LD_{50} values. Although toxicity is a relative term, certain LD_{50} values are generally accepted as indications of high toxicity. These are:

- LD_{50} <500 mg/kg body weight by ingestion or injection
- LD_{50} <1000 mg/kg body weight by skin contact

Toxicity for inhaled chemicals is measured in a similar manner, although the results are expressed differently.

With toxic chemicals that are likely to be inhaled, the **LC_{50} (Lethal Concentration—50%)** is usually provided. The **LC_{50}** *is the chemical concentration in air that will kill 50% of exposed animals.* LC_{50} is usually expressed as **ppm** *(parts per million in air)* or **mg/m³** *(milligrams per cubic meter of air)*. The approximate level indicating high toxicity is an LC_{50} < 2000 ppm inhalation.

B. Inhalation

Toxic vapors, gases, and dusts can all enter the body through inhalation. The respiratory system is wonderfully designed with a large surface area to deliver oxygen and other airborne chemicals to the blood. This unfortunately means that hazardous chemicals in the air can easily penetrate the mucous membranes of the nose, throat, and lungs and be delivered through the bloodstream for distribution to all the tissues of the body. Reactions to inhaled chemicals can range from mild discomfort to burning sensations in the throat to asphyxiation.

For example, exposure to small amounts of the odorless gas carbon monoxide can be lethal. Exposure can cause headache, nausea, and eventually unconsciousness. Only 0.03% in air (300 ppm) can be fatal within 30 minutes. OSHA has set a maximum allowable limit of 50 ppm of carbon monoxide in air for an 8-hour period (time-weighted average, as discussed later in this chapter).

For most volatile chemicals, inhalation exposure levels cannot be estimated by odor or other obvious characteristics. It is important, therefore, to limit exposure to the vapors of chemicals through the use of fume hoods and other methods. OSHA and other agencies have created regulations and guidelines to indicate safe exposure levels to many, although not all, hazardous chemicals. The **American Conference of Governmental Industrial Hygienists (ACGIH)** *is an organization of governmental, academic, and industrial professionals who develop and publish recommended exposure standards for chemical and physical agents.* OSHA has used these standards as the basis for many of their regulations for chemical exposure levels.

ACGIH has established several types of chemical exposure guidelines of importance to laboratory workers. The **threshold limit value (TLV)** *is the airborne concentration for a chemical that most healthy workers can safely be exposed to for 8 hours per day, repeatedly, with no adverse effects.* This is considered the highest safe level for the work environment. Because chemical exposures vary during a day, recommended exposure levels are frequently designated as **TLV-TWA (time-weighted average)** values. *TWA values acknowledge that exposure levels may differ from the recommended TLV during an 8-hour day, but the TLV-TWA will represent the average exposure level.*

For those chemicals that are extremely toxic at certain levels, ACGIH suggests upper limits to short-term

Table 11.11. SUMMARY OF GUIDELINES FOR THE AMOUNTS OF CHEMICALS ALLOWED IN THE AIR

Abbreviation	Name	Definition
TLV	Threshold limit value	Safe airborne concentration of a chemical that healthy workers can be exposed to on a daily basis.
TLV-TWA	Time-weighted average	The average allowable airborne concentration of a chemical within an 8-hour day.
TLV-STEL	Short-term exposure limit	The air concentration allowed for only a short period of time (15 minutes). Only four 15-minute periods of exposure to this concentration are allowed in a day.
TLV-C	Ceiling exposure limit	During the day the exposure to airborne chemicals must never exceed this value.
PEL	Permissible exposure limit (OSHA)	Employers have the legal responsibility to keep employee exposure below this level.

exposures. The **TLV-STEL (short-term exposure limit)** *indicates the air concentration at which only 15 continuous minutes of exposure is allowed, up to four times during an 8-hour day.* Some rapidly acting toxic materials also have **TLV-C (ceiling)** values, indicating the highest concentration allowable. During the day the airborne exposure to the chemical must never exceed this value. Exposures to ceiling levels of chemicals may have biological consequences.

The TLV values established by the ACGIH are merely advisory. OSHA used these values and additional data to set PEL values. **Permissible exposure limit (PEL)** *values are the legal limits set by OSHA for worker exposure to chemicals.* Employers have the legal responsibility not to allow employee exposure to exceed hazardous levels and to provide safety equipment and training to protect workers. PEL values are often (but not always) the same as or similar to TLV values. Table 11.11 summarizes the main types of exposure limit information that are available to laboratory workers.

C. Skin and Eye Contact

Direct skin or eye contact is another frequent route of laboratory injury by toxic chemicals. When working with chemicals, especially in large volumes, it is often difficult to avoid all contact with the chemical. When potentially toxic chemicals contact the skin, there are four possible outcomes:

- the skin will act as a protective barrier—no harm will occur
- the skin surface will react with the chemical—rashes or burns may result
- the chemical will penetrate the skin—allergic sensitization may occur
- the chemical will penetrate the skin and enter the blood stream—systemic toxicity may result

Skin is designed to provide a natural protective barrier for the rest of the body, guarding against both chemical and biological invasion. However, there are natural entry sites for hazards at hair follicles and sweat glands, and any cuts or abrasions of the skin will also provide entry for toxic materials.

Most laboratory workers are naturally cautious when working with known corrosive agents such as acids and bases, which can cause extensive local tissue damage. It is important to remember, however, that many other types of laboratory chemicals can cause skin irritation or burns, and virtually all chemicals will cause problems if splashed into the eyes. Organic solvents can cause direct skin irritation because of their ability to strip natural oils, diminishing the skin's natural barrier properties. This can also lead to increased skin irritation from other chemicals. It is essential to limit chemical contact by wearing proper laboratory attire, including gloves and eye protection.

In many cases, the most serious consequence of skin exposure to chemicals is the absorption of toxic substances into the bloodstream. Potent neurotoxic materials, such as mercury or acrylamide, can easily be absorbed through skin. Phenol, which is frequently found in biotechnology laboratories, can burn the skin directly and also be absorbed systemically. Exposure to phenol is extremely dangerous in large doses, see Example 7.

EXAMPLE 7

Phenol

Phenol is used extensively in biotechnology laboratories to aid in the isolation of nucleic acids. It is corrosive and can cause severe chemical burns. It damages cells by denaturing and precipitating protein. Phenol burns have a characteristic white appearance when they first appear. Because phenol also has anesthetic properties, these injuries may be unnoticed at first. Phenol splashes to the eye are likely to lead to serious burns and possible blindness, so phenol must never be handled without proper eye protection and a readily available eyewash station. Phenol is a respiratory irritant and readily penetrates skin. Whether exposure is by inhalation or skin contact, phenol can enter the bloodstream in significant amounts. From there, it can damage the nervous system, liver, and kidneys. Chronic exposure to phenol vapors can result in chronic symptoms, such as headache, nausea and vomiting, diarrhea, and loss of appetite. Lethal doses of phenol can be inhaled or absorbed through skin relatively easily, so anyone who has significant contact with phenol through a spill or other mishap should seek medical attention. Some symptoms may be delayed in appearance. Phenol fortunately has a characteristic odor that is detectable well below toxic concentrations. This chemical should always be handled in a fume hood to avoid any air contamination of the laboratory.

D. Ingestion

It is common sense that laboratory chemicals taken into the mouth and swallowed can cause harm. Because no one would deliberately consume laboratory chemicals, this is not a common path for biotechnology workplace poisonings. However, chemicals can inadvertently be transferred to the mouth in toxic quantities by contaminated fingers or pencil ends. Develop the habit of keeping your hands and other objects away from your face, mouth, and eyes while in the laboratory. Never pipette by mouth. Separate refrigerators must be designated for food storage, and all food and drinks should be consumed in designated areas where laboratory chemicals are not allowed. Maintaining complete separation of work areas and food areas helps prevent accidental ingestion of chemicals.

Oral toxicity levels for chemicals are determined with animal studies that measure LD_{50} values. Another indicator of chemical toxicity in humans is the **LD_{Lo}** (*Lethal Dose—Low*), *the lowest dose of a compound that has been reported to cause a human death.*

E. Injection

Direct injection of toxic chemicals is a relatively unlikely route of poisoning in the biotechnology laboratory, although it does play a more common role in the spread of biological hazards, Chapter 12. The most likely method of chemical injection is through contact with chemically contaminated broken glass or other sharps. The injection of toxic chemicals that might not otherwise be able to pass the skin or respiratory barrier can be serious, and requires medical attention.

Table 11.12 provides information on the relative toxicities of some of the common chemicals found in biotechnology laboratories.

Table 11.12. *RELATIVE TOXICITIES OF COMMON LABORATORY CHEMICALS*

Chemical	LD_{50}, oral, rat (mg/kg)	TLV (ppm)	TLV (mg/m³)	PEL (ppm)	PEL (mg/m³)
Acetic acid	3310	10	25	10	25
Acetone	5800	750	1780	1000	2400
Acetonitrile	2730	40	70	40	70
Acrylamide	124	—	0.03 skin	—	0.3 skin
Ammonium hydroxide	350	25	17	35	27
Benzene	4894	10	32	1	3
Chloroform	908	10	49	50	240
Diethyl ether	1215	400	—	400	—
Ethanol	7060	1000	1880	1000	1900
Ethyl acetate	6100	400	1440	400	1200
Formaldehyde	500	0.3	0.37	0.75	1.5
Hexane	28,710	50	176	500	1800
Hydrochloric acid	—	5	7.5	5	7
Hydrogen peroxide	75	1	1.4	1	1.4
Mercury	—	—	0.025	—	0.1
Methanol	5628	200	262	200	260
Nitric acid	—	2	5.2	2	5
Phenol	384	5	19	5	19
Sodium hydroxide	140	—	2	—	2
Sulfuric acid	2140	—	1	—	1
Toluene	2650	50	188	200	750

IV. STRATEGIES FOR MINIMIZING CHEMICAL HAZARDS

A. Preparing a Work Area

An important part of good laboratory practice is planning each task, anticipating hazards, and preparing an appropriate work area. For any task, you should have all supplies and small equipment nearby to minimize wasted time and movement. Always wear proper PPE. Read the MSDS and follow the prescribed safety recommendations. Prepare appropriate safety materials before work begins. For example, when working with hazardous liquids, cover work surfaces with absorbent plastic-backed paper that can absorb small spills and be disposed of after your work is finished. Another containment strategy is to perform the work within a tray that can be easily cleaned. The work area should be marked with an appropriate hazard sign while in use.

B. Using Chemical Fume Hoods

i. STRUCTURE AND FUNCTION OF FUME HOODS

Many tasks involving toxic chemicals are best performed in a chemical fume hood. A **chemical fume hood** *is a well-ventilated, enclosed chemical- and fire-resistant work area that provides user access from one side.* The hood isolates and removes toxic or noxious vapors from a working area and protects the user. A fume hood should be used whenever your work involves chemicals with the following characteristics:

- volatility
- unpleasant smells
- a TLV lower than 50 ppm in air

A typical laboratory fume hood is shown in Figure 11.4. The hood usually sits on top of a storage cabinet. The opening is covered by a transparent **sash,** *is a window constructed of impact-resistant material, which can be raised and lowered by the user.* Gas and electrical outlets should be located on the outside of the hood. Inside is a smooth work surface tray, which is designed to contain spills and provide easy cleaning. Most modern fume hoods include interior **baffles,** *which direct the air flow within the hood.* **Airfoils** *located at the bottom and sides of the sash help to reduce air turbulence at the face of the hood.*

There are several types of effective fume hoods available for biotechnology laboratories. The simplest is the **constant air volume (CAV) design,** *where a constant air flow is pulled through the exhaust duct.* In these hoods, raising and lowering the sash changes the face velocity at the sash opening. **Face velocity** *is the rate of air flow into the entrance of the hood, measured in* **linear feet per minute (fpm).** In this conventional hood design, air enters the hood only at the bottom of the sash. When the hood sash is lowered, the face velocity of a CAV

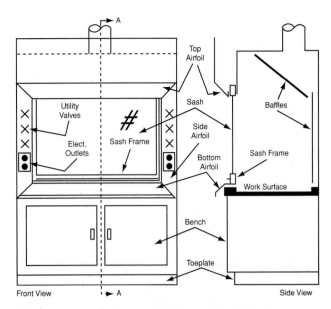

Figure 11.4. A Standard Benchtop Fume Hood. (Reprinted with permission from *Prudent Practices in the Laboratory. Handling and Disposal of Chemicals,* The National Research Council, National Academy Press, Washington, DC, p. 186, 1995.)

hood can increase dramatically, see Figure 11.5. This can have the effect of blowing around paper and small items within the hood when the sash is lowered.

One partial solution to the problem of excessive face velocity is the use of a **bypass hood,** *which has an opening at the top of the hood behind the sash. This allows air to enter the hood and bypass the working face, restricting face velocity when the sash is lowered,* see Figure 11.6 on p. 174. Face velocity will still increase, but not as dramatically.

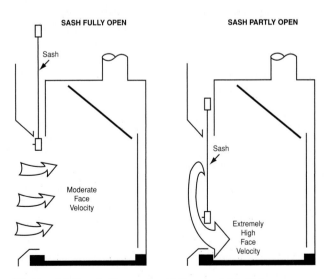

Figure 11.5. A Conventional Constant Air Volume Fume Hood. (Reprinted with permission from *Prudent Practices in the Laboratory. Handling and Disposal of Chemicals,* The National Research Council, National Academy Press, Washington, DC, p. 183, 1995.)

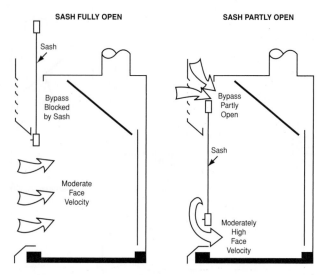

SASH FULLY OPEN   **SASH PARTLY OPEN**

Figure 11.6. A Bypass Fume Hood. (Reprinted with permission from *Prudent Practices in the Laboratory. Handling and Disposal of Chemicals*, The National Research Council, National Academy Press, Washington, DC, p. 184, 1995.)

There are also **variable air volume (VAV) fume hoods** available, *which maintain a relatively constant face velocity by changing the amount of air exhausted from the hood.* VAV hoods are usually designed without an air bypass because air flow into and out of the hood are equalized. This is becoming a preferred design in many laboratories.

ii. PLACEMENT OF FUME HOODS

The location of fume hoods within a laboratory can influence their effectiveness. Modern fume hood design has eliminated much, but not all, of the sensitivity of these hoods to air currents in the room. The movement of materials as well as hands and arms into and from the hood creates air drafts that can pull toxic vapors from the hood into the laboratory. The passage of other workers directly in front of the hood opening can similarly pull vapors through the face opening. Airfoils are designed to minimize this problem, but fume hoods should be installed in locations where they are isolated from traffic and drafts from doors and ventilation fans, see Figure 11.7. Fume hoods should not be located in the corners of rooms, where air currents collide and create turbulence.

iii. TESTING PROCEDURES

Laboratories must test the effectiveness of their fume hoods at least annually, according to OSHA regulations. A complete performance check evaluates three parameters:

- general function and air flow patterns within the hood
- face velocity
- the uniformity of the face velocity

The third measurement is usually performed when a fume hood is first installed, and later estimated using a smoke generator.

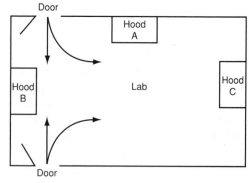

Arrows Indicate Air Currents

Figure 11.7. Placement of Fume Hoods and the Effects of Traffic Air Currents. Hoods B and A are poorly placed to avoid air currents. Hood C is in the better location for this laboratory.

Smoke generators *are small tubes of chemicals, frequently including titanium tetrachloride, that generate highly visible white smoke from a chemical reaction,* see Figure 11.8. They are used to check the general function of the hood. The tube is ignited and slowly moved across the front of the hood and then inside at various locations, to check the direction of air flow. This will indicate whether the hood is drawing air from all parts of the work surface, and whether fumes from inside the hood are entering the laboratory. A smoke generator is also useful for checking the effects of sudden hand motions and room traffic on the effectiveness of the hood.

Smoke generators should be used to determine the most effective locations within the hood for drawing out toxic fumes. For example, in most hoods, volatile chemicals should be handled 5–6 inches back from the sash. Exhaust levels tend to increase toward the back of the hood and decrease somewhat at the edges. During the smoke test, it is helpful to mark the effective parts of the fume hood with tape, indicating to users the optimum placement for experimental materials. This is also a good time to check that large items located in the hood do not obstruct air flow. If they do, they can be relocated to optimize hood function. All equipment within the hood should be elevated at least 2 inches above the work surface to aid in air circulation.

A **velometer (velocity meter)** *is an instrument used to measure the face velocity of the fume hood.* The recommended face velocity for most fume hoods is 100 linear feet per minute (fpm). Face velocities both lower and higher than this value are less effective in containing fumes. Higher velocities create internal turbulence that may be counterproductive, allowing small papers and other items to be drawn into the exhaust vent. This decreases air flow and therefore the effectiveness of the hood. For a standard CAV fume hood, the tester raises or lowers the hood sash to achieve the proper face velocity of 100 fpm. This sash location should be marked with a sign or tape on the hood frame to indicate the optimum sash height.

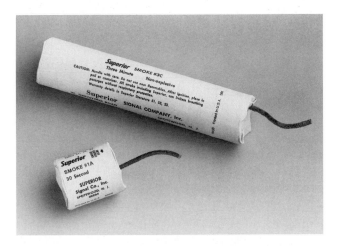

Figure 11.8.   Smoke Generators for Testing Fume Hood Function. (© 1998, Flinn Scientific, Inc. All rights reserved. Reproduced by permission from Flinn Scientific, Inc., Batavia, IL, USA.)

iv. OPTIMAL USE OF A FUME HOOD

Fume hoods cannot protect the user and others in the laboratory unless they are used properly. Before working in the hood, be certain that it is functioning properly. The exhaust fan in the hood should always be running, and in many models the fan cannot be turned off. If yours is adjustable, check that it is turned on and operational. A quick test is to dangle a piece of tissue paper at the hood opening and see that it is drawn toward the hood. All supplies and equipment you will need should be loaded into the hood at the beginning of your work, so that the sash will not have to be raised again while chemicals are in use. Consider chemical compatibility when placing materials within the hood. Table 11.13 summarizes guidelines for proper hood maintenance and use.

It is essential to distinguish between fume hoods and biological safety cabinets. **Biological safety cabinets** *are enclosures designed for the containment of biological hazards.* They are equipped with special filters that remove potentially dangerous particles from the air within the cabinet, and they provide varying degrees of safety and sterility for users and cabinet contents. Fume hoods are not equipped to filter biological hazards from the exhausted air, and do not protect the environment from these agents. Biological safety cabinets can function as fume hoods if they are vented to the outside of the building. However, many of the volatile and toxic chemicals appropriate for fume hoods will destroy the filters in the cabinets, reducing or eliminating the removal of hazardous particles from the air. Biological filters will not detoxify chemically contaminated air and may not provide adequately protect the user from toxic chemicals.

C. Limiting Skin Exposure

As discussed earlier in the chapter, it is essential to avoid skin contact with laboratory chemicals. To protect yourself, always wear a lab coat to protect the body. Wear gloves that are appropriate for the chemicals

Table 11.13.   PROPER USE OF A FUME HOOD

Hood maintenance:

- Regularly test the face velocity and general function of the hood.
- Mark the optimum sash position to attain a face velocity of 100 linear fpm.
- Mark the safe interior working area with tape.
- Keep the sash closed when not in use.
- Do not use the hood work surface for chemical or other long-term storage. Note that most hoods are constructed with storage cabinets underneath.
- Keep the exhaust fan on at all times.
- Do not adjust the interior baffles unless you are trained to do so.

Hood use:

- Be sure that the hood is functioning and drawing air.
- Open the sash and load all necessary items.
- Do not overload the hood, blocking air flow.
- Be sure that all chemicals in the hood are compatible.
- Elevate all equipment at least 2 inches above the work surface.
- Once all items are loaded, move the sash slowly to the optimum position.
- Keep your face outside the hood and behind the sash at all times.
- Use appropriate face and eye protection.
- Move your arms slowly while using the hood.
- Never stack materials on the bottom air foil.
- Work within the marked safe interior working area.
- Avoid allowing paper or other objects to enter the exhaust ducts.
- Decontaminate the working surface properly every day after use.
- Do not use infectious materials within a fume hood.

In case of a spill or fire within the hood:

- Immediately close the sash completely if you can safely.
- Do not turn off the exhaust fan.
- Unplug all equipment within the hood (this assumes that the equipment is plugged into outlets outside the hood).
- Warn other personnel and evacuate the area.

being handled. Remember that gloves differ in materials and in thickness, which can directly affect their ability to protect the hands. Table 11.14 on p. 176 provides data on the chemical compatibility of various types of gloves, measured by breakthrough detection rate (BDT) in minutes. These data were compiled from several different manufacturers, as indicated by the wide range of BDT values for certain glove–chemical combinations. Always check the specific manufacturer's specifications when choosing a glove for toxic chemical work.

Another factor to keep in mind is the length of the gloves (remember the Case Study on p. 167). The skin of the hands is generally thicker and withstands chemical

Table 11.14. CHEMICAL COMPATIBILITY WITH GLOVE MATERIALS

Chemical	Breakthrough Detection Rates (minutes)					
	Rubber	Neoprene	Nitrile	PVC	Butyl	Viton
Acetic acid	31–ND	ND	240–ND	47–300	ND	ND
Acetone	0	12–35	0	0	ND	0
Acetonitrile	0–16	40–65	0	0–24	ND	0
Ammonium hydroxide	58–120	ND	240–ND	60–ND	ND	ND
Benzene	0	15–16	16–27	2–13	30–34	ND
Chloroform	0	14–23	0	0	21	ND
Diethyl ether	0	12–18	33–64	0–14	8–19	12–29
Dimethyl sulfoxide	240	0	0	60	ND	90
Ethanol	ND	ND	225–ND	20–66	ND	ND
Ethyl acetate	0–72	24–34	0–30	0	212–ND	0
Formaldehyde	ND	ND	ND	ND	ND	ND
Hexane	0–21	39–173	234–ND	0–29	0–13	ND
Hydrochloric acid	211–ND	ND	ND	ND	ND	ND
Methanol	60–82	60–226	28–118	3–39	ND	ND
Nitric acid	233–ND	ND	0–72	114–240	ND	ND
Phenol	ND	ND	ND	32	ND	ND
Sodium hydroxide	ND	ND	ND	ND	ND	ND
Sulfuric acid	ND	ND	180–ND	210–ND	ND	ND
Toluene	0	14–25	26–28	3–19	21–22	ND

ND = none detected within 5 hours; 0 = not recommended, less than 10 minutes BDT.
Note: These data are compilations from several glove manufacturers (resulting in ranges of values) and do not constitute specific recommendations. Always check the data for the specific brand and thickness of glove for intended use.

penetration better than the thinner, more sensitive skin of the wrists or forearms. Gloves ideally should be pulled over the cuffs of your lab coat to provide complete protection for your arms, Figure 11.9a. If the available gloves are too short, disposable arm protectors are desirable when working with highly toxic chemicals, Figure 11.9b. Always wash your hands and wrists thoroughly before leaving the laboratory.

D. Storing Chemicals Properly

i. STORAGE FACILITIES

There are numerous state and federal regulations that cover the storage of laboratory chemicals. These vary by location, but certain issues are invariably regulated. Common regulations cover:

- storage of chemicals with fire and explosion risk in approved safety cabinets
- separation of stored chemicals by compatibility
- identification of all chemicals, with appropriate hazard labeling
- limiting access to radioactive materials, explosives, controlled substances, and other special hazards

Laboratory chemicals should be stored in safe locations as close as possible to the point of use. This will minimize the risk of an spill during transport, which is a frequent source of accidents. Flammable and corrosive chemicals should be transported in secondary containers, such as buckets or trays, for protection from spills or breakage. Chemical storage areas must have spill containment materials immediately available that can handle the contents of the largest container of chemical present.

Every laboratory should maintain a current inventory of chemicals, indicating their date of receipt, user, and location. Every chemical should have a specific storage location, where it can easily be found. Chemical containers removed for use should always be returned to their proper location. Only the amounts of chemical in immediate use should be kept at the lab bench or at other work areas. Chemical storage areas should be inspected regularly to be sure no leaking containers or other signs of container damage are present. Laboratories should dispose of old chemicals or those that are no longer needed regularly. This is especially important when personnel change in the laboratory, leaving behind "orphan" chemicals that are unwanted or unidentifiable. Additional guidelines are provided in Table 11.15.

ii. LABELING

All **primary chemical containers** (i.e., those supplied by the manufacturer) should immediately be labeled with

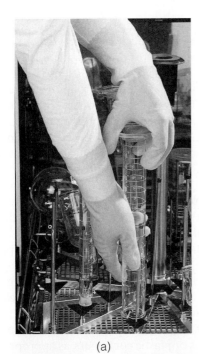

(a)

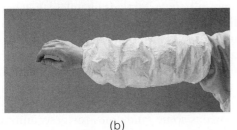

(b)

**Figure 11.9.   Wrist and Forearm Protection Using
a. Gloves or b. Arm Protectors** (Photos courtesy of Fisher
Scientific, Pittsburgh, PA.)

Table 11.15.   GUIDELINES FOR CHEMICAL STORAGE

- Date all chemicals on receipt.
- Be sure all stored chemicals are labeled by contents and general hazards.
- Store only minimum amounts of chemicals.
- Keep only small quantities of chemicals at your work station for immediate use.
- Use proper storage containers and cabinets.
- Do not store flammable chemicals in a standard refrigerator; use a certified spark-free refrigerator or freezer.
- Sort chemicals by hazard and store each type separately. Some of the suggested chemical categories are:
  - acids
  - bases
  - organic oxidizers
  - inorganic oxidizers
  - flammable liquids
  - flammable solids
  - acute poisons
  - water-reactive chemicals

  More information can be obtained by consulting the catalog for Flinn Scientific, Inc., Batavia, IL.
- Maintain a current inventory of chemicals for the laboratory.
- Store every chemical in a specific location.
- Never store chemicals on the floor or above eye level.
- Inspect containers weekly for signs of leakage or deterioration.
- Dispose of old chemicals properly and promptly.
- Be sure that all appropriate spill kits are easily available and fully stocked.

a date and user name when they arrive in the laboratory. The manufacturer will provide information to determine the storage category of each chemical. Chemicals in different categories must always be stored in separate areas. Stock chemicals should not be repackaged in other containers unless the original container is damaged. Containers for immediate use should be clearly labeled with the chemical name, date, user, and appropriate hazard information. Chemical name abbreviations are not considered adequate under OSHA regulations. Handwritten labels should be legible and written in indelible ink; organic solvents are notorious for dissolving writing in nonpermanent ink. Hazard labeling should provide information about the greatest risk from the chemical. For example, strong acids should carry a label for "Corrosives," see Figure 11.10. Laboratories can purchase sets of standard hazard stickers that provide effective warnings for handlers who are unfamiliar with the specific chemical. Experimental materials should also be labeled by hazard.

iii. STORING CHEMICALS BY TYPE

Incompatible chemicals must not be stored next to one another. Chemicals are therefore sorted by type and

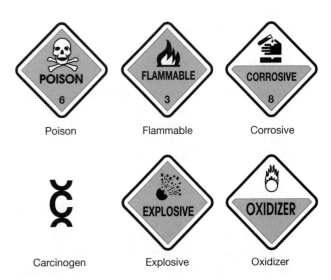

Poison	Flammable	Corrosive
Carcinogen	Explosive	Oxidizer

Figure 11.10.   Common Hazardous Materials Labels.

then compatibility. The most common categories are shown in Table 11.15. While it is easy to identify sulfuric acid as belonging in the acid group, some chemicals are not as easy to classify by name. Chemical manufacturers therefore often provide storage information on labels. Storage compatibility information is always available in the Handling and Storage section of the MSDS provided with each chemical. The MSDS for individual chemicals can also be found on the Internet at many locations. Virtually all chemical supply companies provide an online MSDS for their products in addition to their printed sheets.

Once you have the appropriate information about each chemical's properties and hazards, you can separate them by compatibility. This does not always correspond neatly to the groups. If you look at Table 11.4 on p. 165, which lists incompatible chemical pairings, you can see that there are several combinations of acids that are dangerous. Perchloric acid, for example, should not be stored in proximity to acetic or sulfuric acid. Again, it is important to use the MSDS for each chemical to determine optimal storage locations.

E. Handling Waste Materials

Waste *refers to any laboratory material that has fulfilled its original purpose and is being disposed of permanently.* While OSHA provides the guidelines for labeling chemicals within the laboratory, the U.S. Department of Transportation (DOT) and EPA regulate the handling and disposal of chemicals designated as hazardous waste. Within DOT, the Pipeline and Hazardous Materials Safety Administration (PHMSA) provides oversight for labeling, handling, and packaging hazardous materials for transport within the United States. This includes almost one million shipments each day. The regulations for hazardous waste disposal are complex, and it is essential that laboratories work with the safety office at their institution to ensure that waste is handled and packaged appropriately. Waste disposal has become increasingly expensive, and improper labeling or packaging of various materials can raise the cost significantly.

Biotechnology laboratories must maintain several types of waste containers, to sort materials by hazard. To minimize hazardous waste volume, do not dispose of nonhazardous materials in hazardous waste containers. Receptacles for nonhazardous materials, such as old notes or relatively clean paper towels, should be available away from immediate work areas. No sharps or any hazardous materials should ever be placed in these containers. In most institutions, this nonhazardous waste is handled by the janitorial staff.

Any waste materials that present a health risk should be collected in appropriately labeled containers. To the extent possible, waste should be sorted by type (e.g., separating biological waste from chemical waste). Chemical wastes should similarly be collected in a manner

that keeps hazard groups separate. The EPA classifies chemical wastes by specific characteristics, such as flammability, corrosive properties, and reactive potential. Local waste disposal facilities may also designate chemical compatibility and hazard groups. The safety office at your institution will tell you how to sort waste materials and will place labels on each container for disposal. Always pay attention to these labels. Never put a chemical in a container if you are uncertain that it belongs there (see the Case Study on page 164).

Chemical wastes are generated as the end products of procedures, or as chemical stocks that are no longer needed in the laboratory. Chemical wastes that contain mixtures of compatible chemicals should be labeled with identification of each chemical and its approximate percentage in the mixture. Solid wastes should be divided into those that can be incinerated (e.g., as lightly contaminated paper towels) and those that cannot. Dry wastes should be double bagged for safety and clearly labeled by hazard. Liquid wastes are collected in labeled glass or plastic containers, as appropriate. Never pour chemicals down a sink drain unless they have been neutralized or deactivated in accordance with local regulations for sewer disposal. Never use the drain for flammable solvents or potentially reactive chemicals. Even at very low concentrations, some chemicals, such as sodium azide, can form explosive mixtures within drain pipes, with predictably unpleasant results. Table 11.16 provides some guidelines for determining when stock chemicals should be discarded.

Because of the high expense of toxic waste disposal, it is important to minimize the amounts of waste generated by the laboratory. Table 11.17 contains some sug-

---

Table 11.16. DISPOSAL OF STOCK CHEMICALS

Most laboratories have a tendency to keep chemical stocks "just in case" someone needs them in the future. Although this is a reasonable approach with some chemicals, such as sodium chloride, which is stable and used on a regular basis, it can be a hazardous practice in other cases. Dispose of chemical stocks if any of the following criteria are observed. (Consult with your institutional safety office for proper disposal procedures.)

- The chemical is more than 1 year old and not in current use.
- There is evidence of pressure buildup in a container.
- A formerly clear liquid has turned cloudy.
- Solids have clumped or show other signs of water absorption.
- Chemicals have changed color.
- Containers show signs of damage.
- Chemical identities are unclear.

These stocks should be disposed of in their original containers, with appropriate identity and hazard labeling.

Table 11.17.   MINIMIZATION OF CHEMICAL WASTE

- Maintain a current chemical inventory and avoid duplication of stocks.
- Order as little of a chemical as needed for planned work.
- Date all chemicals on arrival, to eliminate doubts about age.
- Use older chemicals first.
- Be sure all chemical containers are labeled by content.
- Keep all chemicals tightly sealed to prevent deterioration from air exposure.
- Do not contaminate chemical stocks by returning materials to the container.

gestions for reducing the amounts of chemicals that will require disposal. Unknown or unlabeled wastes are by far the most expensive type to dispose of, so all workers should be diligent about chemical labeling.

V. RESPONSE TO CHEMICAL HAZARDS

A. Chemical Emergency Response

Every laboratory is required by OSHA to have an emergency response plan for accidents involving hazardous chemicals. These plans should be explained as part of laboratory safety training, and it is the responsibility of each worker to be familiar with the appropriate actions. Emergency response needs to be an automatic reaction because accidents rarely provide time for detailed analysis and research into procedures (see the Case Study below). It is important to be able to distinguish between a truly minor incident and an emergency. Table 11.18 provides some guidelines that should be considered before they are needed.

In cases where the exposure of a worker to chemicals is obvious, immediately call the Poison Control Center or a designated physician. These telephone numbers should be posted by every telephone in the laboratory area. In the case of a skin or body splash, the victim should move to the emergency shower, remove any contaminated clothing, and drench the affected area for at least 15 minutes. Eye exposure requires flushing of both eyes for at least 15 minutes as well. If chemicals have been inhaled, the person should be moved to an area with fresh air and remain quiet until medical help arrives.

Laboratory workers should be familiar with the symptoms of toxic chemical exposure, which are summarized in Table 11.19. Specific information about individual chemicals can be found in the MSDS, which should always be consulted before chemical use.

It is certainly better to prevent accidents or chemical exposures before they happen. Immediate preventive measures or notification of a supervisor is appropriate

Table 11.18.   WHEN IS A "PROBLEM" INVOLVING CHEMICALS AN EMERGENCY?

Most of us are reluctant to appear to overreact to a potentially dangerous laboratory situation. The following list includes some of the circumstances where the best course of action is to summon help immediately. You probably require assistance from a professional safety officer or emergency response team when a chemical problem:

- causes a serious injury
- involves a public area
- creates a fire hazard
- cannot be isolated or contained by those present
- creates toxic vapors that could spread through the building
- causes property damage
- requires a prolonged cleanup
- involves mercury compounds
- involves hydrofluoric acid
- involves an unknown hazard

Your laboratory should post the telephone numbers of the designated sources of local assistance near all telephones. Some of the numbers that might be useful in an emergency:

- Institutional safety office, in case of spills or questions about safety procedures
- Fire department, in case of fire
- Emergency medical assistance, in case of injury
- Poison control, in case of exposure to toxic chemicals

Table 11.19.   SYMPTOMS OF CHEMICAL EXPOSURE

Any time you become aware of direct skin or eye contact with a laboratory chemical, you should take immediate action to remove any traces of the chemical and seek medical attention as needed. If you or your co-workers notice a chemical odor, especially for low TLV substances, immediate action may be required. In the absence of known contact or inhalation, the following are some of the potential signs of toxic chemical exposure. The MSDS for a chemical will provide specific details.

Acute exposure:
- headache or dizziness
- sudden nausea or vomiting
- coughing spasms
- eye, nose, or throat irritation

especially if:
- the preceding symptoms disappear with fresh air
- the symptoms reappear when work is resumed
- more than one person in the laboratory is affected

Chronic exposure:
- persistent dermatitis
- unusual body or breath odor
- a strange taste in the mouth
- discolored urine or skin
- numbness or tremors

Table 11.20. WHEN TO TAKE SPECIAL PRECAUTIONS

There are many circumstances where a worker should take steps to prevent a problem directly, or to notify an appropriate supervisor. Some examples:

- Chemical leaks or spills are noticed or anticipated.
- Possible symptoms of chemical poisoning are noticed.
- Laboratory workers notice odd smells.
- A fume hood or other safety equipment fails to operate properly.
- A procedure creates chemical exposure that may exceed toxic thresholds.
- A new or altered procedure requires chemicals with unknown properties.

under any circumstances where you have concerns about potential problems. Table 11.20 provides a few examples.

B. Chemical Spills

i. PREVENTING CHEMICAL SPILLS

In laboratory work, chemical spills occur. Most of these will be small and many can be prevented by good laboratory practices. Proper advance preparation for spills ensures that most of these incidents will remain relatively minor in terms of scope and consequences. Many spills are caused by simply knocking over open containers of chemicals. These spills can be avoided by properly arranging experimental materials so that routine arm movements will not endanger chemical containers. Containers should be closed when possible, or anchored in a rack or holder for added stability.

Only small amounts of chemicals, in unbreakable containers when possible, should be kept at lab benches. Any toxic chemicals should have secondary containment, such as spill trays. Benches or fume hood work surfaces should be covered with absorbent, plastic-backed paper for liquid chemical work. This makes clean-up quick and simple.

Many serious spills occur when transporting chemicals through the laboratory. Large reagent containers should be handled one at a time, preferably in a secondary transport container. Never carry glass reagent bottles by the cap or solely by the ring at the top. Support the bottle from underneath with one hand. Always wear gloves when transporting chemicals.

Be aware that the bottom seam of glass bottles is generally the weakest part of the container, see Case Study below. One potential cause of major spills is the transfer of glass bottles from a warm-water bath or freezer directly to a counter top. Thermal shock can weaken the bottom seam and cause it to crack. When thawing reagents in a water bath, always place the bottle in a secondary container that will contain any leaked material.

CASE STUDY

Emergency Response to a Major Chemical Spill

Major chemical spills require a quick response by all members of a laboratory. A biotechnology worker in one of the author's laboratories was thawing a 1 L stock bottle of frozen phenol in a water bath. Following proper procedures, she placed the bottle in a beaker for containment. When thawing was complete, she checked the bottle and beaker for any signs of leakage, and then picked up the phenol bottle at the side and began to transport it across the lab. Halfway across the room, the bottom of the bottle gave way, dumping the entire liter of phenol down her leg and onto the floor. Knowing that phenol is toxic through skin contact and is also corrosive, she immediately called for help and ran for the emergency shower. She was properly attired in a lab coat, gloves, long pants, and solid shoes, which limited her contact area. In the meantime, the lab supervisor arrived on the scene and determined that this was a major spill. She ordered that the room be evacuated, and both doors be closed, locked, and labeled with a "Danger: Phenol Spill" sign. Someone else called the building engineering hotline and ordered an immediate switch of the ventilation system for the floor to total exhaust to prevent the spread of toxic fumes within the building. Another worker was assisting the technician in the shower. The phenol had soaked through her pants fairly quickly and she was convinced to remove them when someone brought a fire blanket to use as a modesty shield. Colleagues contacted the Institution Safety Office and requested an emergency spill team with respirators to clean up the phenol, and then escorted the injured person to the local emergency room. She suffered from nausea and headache for 2 days, but her chemical burns were relatively minor, and there were no apparent long-term effects. Aside from the inconvenience of being evacuated from the contaminated laboratory for several hours, no other personnel were affected. This would have been a very serious accident without the quick response and cooperation of many members of the laboratory team.

ii. HANDLING CHEMICAL SPILLS

When a chemical spill occurs, it is essential to assess the extent of the problem immediately. Even though a small spill is obviously easier to clean up than a larger mess, in many cases the volume of the spill is less significant than the toxicity of the spilled material. Whenever a highly toxic and volatile liquid is spilled, as in the Case Study, it is best to evacuate personnel and inform the safety experts at your institution. Your efforts may be best spent ensuring that others are warned of the hazard and that any injuries are promptly tended.

Small spills of relatively innocuous chemicals can be cleaned up without much concern. For hazardous chemical spills, however, a well-intentioned but inappropriate cleanup effort may cause more harm than good. For example, in the case of a volatile solvent, every liter

Table 11.21.   CHEMICAL SPILL PROCEDURES

It is essential that all laboratory personnel know what to do in advance in the event of a chemical spill. This topic should be discussed at periodic safety meetings, along with other routine safety procedures.

Minor Spills

- Take care of personal contamination first.
- Notify nearby workers and evacuate them as needed.
- Turn off heat or ignition sources if flammable chemicals are involved.
- Prevent the immediate spreading of the spill by layering with towels (do not wipe).
- Avoid breathing any vapors from the spill.
- Dress properly and get the appropriate spill kit.
- First surround the spill with absorbent to contain the spill area.
- Slowly sprinkle absorbent over the spill and follow any directions that accompany the spill kit.
- Collect the contaminated absorbent using a whisk and dust pan, and avoid creating dust or aerosols.
- Place the waste in a disposable bag, seal, and label; the whisk and pan should be thoroughly cleaned or discarded.
- Finish the cleanup by washing the area several times with detergent and water.

Major Spills

- Leave the spill site immediately.
- Warn others about the hazard.
- Tend to any injuries.
- Prevent others from approaching the spill site.
- Call for expert assistance.

spilled can produce up to 600 L of flammable vapors. An attempt to wipe up a spill with paper towels will actually encourage vapor production by increasing the surface area of the spill. Liquid spills should be handled with appropriate chemical absorbents. Absorbent in sufficient quantities to soak up the largest volumes used in the laboratory should be readily available. Dry chemicals should never be swept up, because this will create dust that can be inhaled. Wet mopping is more appropriate in cases where absorbents are unnecessary. Table 11.21 provides a summary of proper procedures for dealing with both minor and major chemical spills.

If hazardous chemicals are spilled on clothing, the clothing should be removed immediately and then rinsed. Any skin under the clothing should be flushed for 15 minutes in an emergency shower. When removing the clothes, be careful not to spread chemical to additional areas of the body. Never pull a contaminated shirt or sweater over the face; if the chemical is toxic, cut the clothing off with scissors. Do not attempt simply to rinse the chemicals out of the clothes while wearing them. A fire blanket or spare lab coat can be used to protect modesty if required.

iii. CHEMICAL SPILL KITS

Properly assembled and conveniently located spill kits can make the cleanup of minor spills significantly easier. **Chemical spill kits** *are preassembled materials for controlling and cleaning up small to medium size laboratory spills.* Every room where chemicals are used or stored should have a kit with sufficient materials to handle a spill of the largest container in the immediate area. A list of suggested contents for a general kit are provided in Table 11.22. Having all necessary materials assembled ahead of time, with all personnel trained to use the kits correctly, can make the difference between a minor inconvenience and a major problem.

Spill control kits may contain both loose absorbents as well as absorbent-filled pads and pillows. The absorbents should be chosen to reduce the vapor pressure from a liquid spill efficiently and to minimize any personal contact with the chemical. The type of absorbent material most applicable in biotechnology laboratories is labeled **universal absorbent.** *These are generally polypropylene or expanded silicates and can absorb virtually any liquid, including some corrosives.* Check the manufacturer's recommendations and be certain that kits are clearly and appropriately labeled. These absorbents control liquid spills but do not reduce the toxic properties of absorbed liquids. Acids and bases, for example, may still require neutralization before disposal.

Loose absorbent works best for small spills. The absorbent should be poured around the edges of the spill to contain it and then sprinkled on the interior. There are special neutralizer absorbents available for acid spills. For larger spills, absorbent pillows can be placed around and on top of the spill. Absorbent should usually be left undisturbed for a short period of time before final cleanup.

Table 11.22.   CONTENTS OF A GENERAL SPILL KIT

You can purchase general chemical spill kits or assemble your own, according to your needs. A general use spill kit should contain:
- chemical-resistant, long-sleeved gloves
- chemical-resistant goggles
- chemical-absorbent materials for any anticipated chemical type
- spill pillows for containment
- small whisk broom and dust pan
- disposable plastic bags for hazardous waste

Most laboratories will require more than one type of spill kit because different absorbents and cleanup procedures are required for acids, bases, organic solvents, and the like. There are also special spill kits that should be purchased if your laboratory contains mercury or hydrofluoric acid hazards.

Any laboratory that uses mercury should keep a special mercury spill kit available. These kits should only be used by trained personnel because of the extreme toxicity of mercury. Mercury is readily absorbed through the skin and by inhalation. Its vapors are odorless and tasteless, so it is difficult to judge mercury exposure levels. Always handle mercury on trays to contain any possible spills. Avoid using mercury thermometers if possible; if unavoidable, purchase thermometer shields to guard against breakage.

PRACTICE PROBLEMS

1. As described in the text, small concentrations of the odorless gas, carbon monoxide, in air can be lethal. Exposure to low levels of carbon monoxide can cause headache, nausea, and eventually unconsciousness. OSHA has set a maximum allowable limit of 50 ppm in air (TWA) for an 8-hour period. What is the significance of the TWA designation?

2. A scientist in your laboratory is carrying a glass bottle of organic solvent across the laboratory when he slips on a wet spot on the floor, loses his grip, and drops the solvent bottle.

   a. What are several precautions that might have prevented this accident?

   b. What immediate steps should this person take to minimize any risk to himself and other laboratory workers?

3. Is the odor of organic solvents a reasonable indicator of safe exposure levels?

4. Material Safety Data Sheets (MSDSs) are a major source of safety information and it is essential to be able to use the information they contain. Excerpts from four Material Safety Data Sheets follow. Based on the information in these MSDSs, answer the questions below. (NOTE: *These are only brief excerpts from the actual MSDSs and are intended only for purposes of study, not for use in the laboratory.*)

Material Safety Data Sheet
General Information
Item Name: ACETONE, REAGENT
Date MSDS Prepared: 01Jan95
Toxicity Information
OSHA PEL: 1000 PPM
ACGIH TLV: 750 PPM/1000 STEL
Physical/Chemical Characteristics
Appearance and Odor: Clear, Colorless, Volatile Liquid with a Characteristic Sweetish Odor
Boiling Point: 133°F, 56°C
Melting Point: −139°F, −95°C
Solubility in Water: Very Soluble
Fire and Explosion Hazard Data
Flash Point: −4°F, −20°C

Extinguishing Media: Water Spray, Dry Chemical, Carbon Dioxide, Alcohol-Resistant Foam

Material Safety Data Sheet
General Information
Item Name: ETHYL ALCOHOL ACS
Date MSDS Prepared: 01Jan95
Toxicity Information
OSHA PEL: 1000 PPM
ACGIH TLV: 1000 PPM
Physical/Chemical Characteristics
Appearance: Clear Colorless Liquid
Fire and Explosion Hazard Data
Flash Point: 55°F

Material Safety Data Sheet
General Information
Item Name: PHENOL, USP
Date MSDS Prepared: 01Jan95
Toxicity Information
OSHA PEL: 5 PPM
ACGIH TLV: 5 PPM
Physical/Chemical Characteristics
Appearance and Odor: Crystal
Fire and Explosion Hazard Data
Flash Point: N/A
Extinguishing Media: Water Spray, CO_2, Dry Chemical or Foam

Material Safety Data Sheet
General Information
Item Name: SODIUM CHLORIDE
Date MSDS Prepared: 01Jan95
Physical/Chemical Characteristics
Appearance and Odor: White Crystal
Boiling Point: 1413°C
Fire and Explosion Hazard Data
Flash Point: N/A
Extinguishing Media: N/A
Special Fire Fighting Proc: N/A

   a. Rank each of the compounds in order from least toxic to most toxic based on their TLV values.

   b. For acetone, the MSDS states "ACGIH TLV 750 ppm/1000 STEL." Explain the difference between the two numbers, 750 ppm and 1000 ppm.

   c. Based on the flash points given in the MSDSs, which of these compounds poses the most significant fire hazard? How should a fire involving that compound be extinguished?

5. Proper chemical storage is an important responsibility of any laboratory. It is critical to determine compatibilities and sort chemicals accordingly. Excerpts from the MSDS of eleven common laboratory substances are shown below. Based on the information given, sort the chemicals into potential storage groups. (NOTE: *These are only brief excerpts from the actual MSDSs and therefore are intended only for purposes of study, not for use in the laboratory. These excerpts are in no way comprehensive.*)

Chemical	Hazard warnings	Incompatibilities
SODIUM CHLORATE	Danger! Strong oxidizer.	Aluminum, strong acids, strong reducing agents, organic matter.
SODIUM HYDROXIDE	Poison! Corrosive. Harmful if inhaled. Reacts with water, acids, and other materials.	Violent reactions with acids and organic halogens.
SODIUM	Danger! Water reactive. Flammable. Corrosive.	Water, oxygen, carbon dioxide, halogens, acetylene, metal halides, oxidizing agents, acids, alcohols, chlorinated organic compounds.
2-PROPANOL	Warning! Flammable liquid and vapor. Harmful if swallowed or inhaled.	Strong oxidizers, acids, acetaldehyde, chlorine, ethylene oxide, hypochlorous acid, aluminum.
D(+)-GALACTOSE	Caution! May cause irritation to skin, eyes, and respiratory tract.	Strong oxidizers.
NITRIC ACID	Poison! Strong oxidizer. Contact with other material may cause fire. Corrosive.	Incompatible with most substances, especially strong bases, metallic powders, carbides, and combustible organics.
NICOTINIC ACID	Avoid unnecessary exposure.	Strong oxidizers.
ETHYL ACETATE	Warning! Flammable liquid and vapor. Harmful if swallowed or inhaled. Affects central nervous system.	Contact with nitrates, strong oxidizers, strong alkalis or acids may cause fire and explosions.
AGAROSE	Avoid unnecessary exposure.	Oxidizers.
ACRYLAMIDE	Warning! Neurotoxin. May cause cancer. Possible teratogen. Thermally unstable.	Acids, oxidizing agents, and bases. Spontaneously reacts with hydroxyl-, amino-, and sulfhydryl-containing compounds.
2-AMINO-2-(HYDROXY-METHYL)-1,3-PROPANEDIOL	Warning! Harmful if swallowed or inhaled. Causes irritation to skin, eyes, and respiratory tract.	Copper, brass, aluminum, and oxidizing agents.

QUESTIONS FOR DISCUSSION

MSDS information for many chemicals is available online. They can currently be found at http://hazard .com/msds/index.html.

Look around your house and identify three to five common household chemicals, such as cleansers or house paint. Look up the MSDS for each of those chemicals and answer the following questions about each one:

a. Is the substance hazardous?

b. If the substance is hazardous, what is the nature of the hazard(s)?

c. What safety precautions should you use when working with each of these household substances?

CHAPTER 12
Working Safely with Biological Materials

I. INTRODUCTION TO BIOLOGICAL SAFETY

A. Biological Hazards

Biological safety is a major issue in biotechnology facilities. These facilities may house a wide range of life forms, ranging from viruses and bacteria to plants, and even to farm animals. For example:

- Bacteria are used for many processes in the food industry, and as tools to explore and produce new protein and enzyme products.

- **Viruses** (*particles containing either DNA or RNA surrounded by a protein coat*) can be used to introduce DNA into cells of bacteria and higher organisms.

- Molds and fungi are capable of producing complex metabolic products (such as penicillin).

- Yeasts are essential for making bread and wine. They can also be used as small eukaryotic factories for synthesis of complex proteins.

- Cells of higher organisms are grown in culture for product testing and producing recombinant proteins.

- Whole plants are studied to develop means of better food production for humans and animals, and as production systems for desirable proteins.

- Whole animals are used in research to test the products of biotechnology, and as factories for the production of antibodies or other complex proteins.

Any of these systems is potentially a **biohazard,** *which is a biological agent with the potential to produce harmful effects in humans.* Examples include microorganisms, recombinant DNA products, cultured human or animal cells, and other agents defined by laws, regulations, or guidelines. **Pathogenicity** *is a term indicating the relative capability of an organism to cause disease in humans or other living organisms.* We commonly think of bacteria and viruses when we think of pathogens, but fungi, single celled protozoa, and even multicellular organisms (such as nematodes and other worms) can be pathogenic.

In order to cause a disease, an organism must be not only pathogenic but also infectious. **Infectious** *refers to the ability of a pathogenic organism to invade a host organism. An organism that causes a specific disease in an infected host is called an* **etiological agent.** For example, the bacterium *Vibrio cholerae* is the etiological agent for cholera. Some organisms can infect a host without causing disease. In this case, the host is called a carrier. A **carrier** *is an infected individual who is capable of spreading the infecting agent to other hosts.*

In order to infect a host, an organism must be able to spread to the host and then penetrate its natural defenses. As with chemical hazards, the primary routes of exposure to biological hazards are inhalation, skin and eye contact, ingestion, and injection. Inhalation of airborne biological agents is the most likely route of laboratory exposure. Many routine laboratory procedures can produce aerosols. **Aerosols** *are very small particles or droplets suspended in the air.* A **bioaerosol** *is an aerosol that includes biologically active materials.* Strategies for the prevention of aerosols and aerosol exposure are discussed in detail later in the chapter.

Many of the same safety precautions that reduce risk from chemical hazards can also be applied to biological agents. The best health protection for a laboratory worker is to learn as much as possible about the biological materials present in their workplace. Table 12.1 provides criteria that can be used to evaluate the risk from a laboratory biohazard.

B. Laboratory-Acquired Infections

Biotechnology laboratories work with different types and amounts of biohazards, each with its own associated risks. There are clear incentives for laboratories to use nonpathogenic organisms because of safety issues related to workers, investment costs, and product safety validation concerns. Therefore, production work, and more than 80% of the research work conducted worldwide, involves the use of nonpathogenic organisms.

There are situations, however, where biotechnologists must work with pathogenic materials. Some laboratory pathogens have caused infections of laboratory workers, resulting in disease, illness, and even death. A major study in 1978 reported on 3921 documented cases of **laboratory-acquired infections (LAIs),** *which are infections that can be traced directly to laboratory organisms handled by or used in the vicinity of the infected individuals.* Of these cases reported between 1924 and 1977, more than 50% were associated with research laboratories, and 168 were ultimately fatal (Pike, R.M. "Laboratory-Associated Infections: Summary and Analysis of 3921 Cases," *Health Lab. Science* 13 (1978): 105–114.). OSHA estimates that as many as 12,000 cases of LAIs with hepatitis B (with 200 associated fatalities) occurred annually among U.S. health care workers prior to the introduction of an effective vaccine.

Table 12.1. Risk Assessment for a Biohazardous Agent

It is essential for every laboratory worker to evaluate the health risks of specific biohazards in their workplace. Your evaluation should address the following issues:

- Is this a known human or primate pathogen?
- What is the history of laboratory use of this organism or agent, and what are the recognized risks?
- Has this agent been associated with laboratory-acquired infections?
- If so, what are the health consequences of these infections?
- Is there an effective treatment or preventive vaccine?
- Does this agent frequently induce sensitivities or allergies in workers?
- What is my potential susceptibility to infection with this agent as a function of age, sex, or medical condition, as documented by previous cases?
- How can I limit my exposure to this agent?
- What are the recommended safety precautions for this agent, and are they being practiced in this laboratory?
- Is the estimated risk acceptable to me?

It can be difficult to predict the health risks to individuals from exposure to low doses of biohazards, so the key to biological safety is prevention of human exposure to potentially harmful agents. Workers who knowingly handle biohazards are not the only individuals who contract LAIs. On average, one out of every four infections associated with a laboratory biohazard occurs among dishwashers, custodians, clerical staff, and maintenance personnel. You should therefore be familiar with the basic safety issues, not only to protect yourself, but also to ensure that others are not exposed to biological hazards. Some precautions are as simple as marking rooms containing these hazards with biohazard warning signs to indicate that only trained personnel should enter the facilities, see Figure 12.1. These signs can also designate the specific biohazards present. Proper waste disposal and decontamination procedures must also be followed to protect other workers.

C. Regulations and Guidelines for Handling Biohazards

There is a wide variety of guidelines, standards, and regulations that govern the use of biological materials in the laboratory. A list of examples is provided in Table 12.2. Much of the regulatory information about biosafety is readily available on the Internet. Many large institutions place copies of their safety manuals on web pages for easy reference. The agencies whose biohazard guidelines are most often cited are the **Centers for Disease Control and Prevention (CDC)** and the **National Institutes of Health (NIH)**. The **CDC** *is an agency of the federal Department of Health and Human Services whose mission is "to promote health and quality of life by preventing and controlling disease, injury, and disability"* (CDC Mission Statement). The **NIH** *is a federal health agency comprising 25 separate*

(a)

(b)

Figure 12.1.   Standard Biohazard Warning Signs. Either the symbol or the background are solid orange or red. **a.** Sign marking entrances to laboratory and storage areas where biohazards are handled and access must be limited. **b.** Symbol for biohazardous waste containers. These containers must be autoclaved or burned before disposal.

centers and institutions. The agency performs and funds major biomedical research initiatives and provides guidelines for laboratory safety. The joint CDC/NIH guidelines form the basis for many of the standard biosafety precautions used in laboratories across the country.

Table 12.2.   EXAMPLES OF GUIDELINES AND REGULATIONS THAT COVER BIOLOGICAL HAZARDS IN THE LABORATORY

Biosafety in Microbiological and Biomedical Laboratories, CDC/NIH: This set of guidelines provides recommendations for facilities, operating procedures, hazard identification, and risk assessment for laboratories working with biological hazards.

OSHA Bloodborne Pathogens Standard, 29CFR1910.1030: This OSHA standard describes the requirements for employers to develop organizational plans for handling human blood and blood-related products. These requirements are discussed later in the text.

Guidelines for Research Involving Recombinant DNA Molecules, National Institutes of Health: These NIH guidelines provide recommendations for facilities, operating procedures, hazard identification, and risk assessment for laboratories involved in recombinant DNA research.

Guide for Care and Use of Laboratory Animals, Institute for Laboratory Animal Research (ILAR): This set of guidelines developed through the National Research Council provides the basic procedures to be used for laboratory animal research, and is considered a primary reference on animal care and use.

Animal Welfare, USDA 9CFR Parts 1, 2, and 3. These regulations of the U.S. Department of Agriculture define the requirements for licensing, registration, identification, records, facilities, health, and husbandry for all animals covered by the Animal Welfare Act.

Most of the above documents are available on the Internet from the appropriate government agency. More references are included at the end of Unit III's introduction. In addition to these federal codes, there are also state, local, and institutional codes, guidelines, and design criteria that apply to biotechnology facilities.

II. STRATEGIES FOR MINIMIZING THE RISKS OF BIOHAZARDS

A. Standard Practices and Containment

Improved sensitivity in many of the assays used in the biotechnology research laboratory have led, in some cases, to a reduction in the quantities of biological agents needed. This is not true in production facilities, however, and it is impossible in many cases to entirely eliminate the use of biohazardous materials. Therefore, the hazards of pathogenic organisms must be reduced by strategies of good laboratory practices and containment. **Standard microbiological practices** *are the basic practices that should be used when working with all microbiological organisms.* These practices have two purposes; to separate the worker from the microorganism, and to maintain the purity of the microbial cultures. These recommended practices, which are similar to those used for chemical safety, are summarized in Table 12.3. Every biotechnologist should be familiar with the practices in this table.

Always wear appropriate PPE when working with biohazards, see Figure 12.2. This minimally includes lab coat, gloves, and eye and face protection. Wash your hands as soon as possible after removal of gloves or other PPE, and immediately after any potential skin contact with biohazardous materials. All PPE must be removed before leaving the laboratory. If lab coats or other nondisposable items become contaminated, they should be disinfected and then laundered on site (not at home).

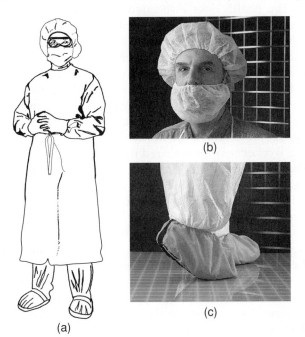

Figure 12.2. Protective Personal Apparel for Routine Work with Body Fluids. a. PPE protection for biohazardous work includes long gown, cap, eye shield, face mask, gloves, and shoe covers. Specialized PPE items are available for specific applications, such as **b.** Beard covers, and **c.** Shoe covers with anti-static strips to prevent static discharges. (Photos courtesy of Fisher Scientific, Pittsburgh, PA.)

Table 12.3.   STANDARD MICROBIOLOGICAL PRACTICES

- Access to the laboratory should be limited to trained individuals, and a standard biohazard sign (see Figure 12.1) must be posted on all entrances.
- Lab coats and eye protection should be worn at all times.
- Workers must wash their hands after any work with microorganisms and whenever they leave the lab.
- Eating, drinking, food storage, and smoking in the laboratory area are prohibited.
- Hand-to-mouth or hand-to-eye contact must be avoided.
- Mouth pipetting of any substance in the laboratory is prohibited.
- Steps must taken to minimize aerosol production.
- Strict policies regarding the use and disposal of sharps must be enforced.
- Work should be performed on a clean hard bench top or other work surface with appropriate disinfectant readily available.
- Work surfaces should be decontaminated after any spill, and at the end of every work session.
- All biological materials must be properly decontaminated before disposal.

Containment *is the control of biohazards by isolation and separation of the organism from the worker.* There are many terms used in the literature to refer to containment practices. In this text, we use the term **primary containment** *to refer to equipment and practices that protect personnel and the immediate laboratory environment from hazardous exposure.* **Secondary containment** *refers to laboratory design features, equipment, and practices that protect the general environment.* **Personal containment** *refers to standard worker practices, such as those outlined in Table 12.3, used to reduce the spread of microorganisms.* The procedures used for work with and disposal of the biohazardous material, along with the practices of proper laboratory hygiene, are examples of personal containment. **Physical containment** *includes laboratory design features and the physical barriers that workers use to isolate biohazards.* Special ventilation systems, PPE, gloves, and biological safety hoods are examples of physical containment.

B. Recommended Biosafety Levels

The NIH classifies organisms into Risk Groups (RGs) according to their health threats to laboratory workers and the community at large. Risk is usually determined based on the pathogenicity of the organism, the infection rate after exposure, and the availability of effective treatment if infection occurs. Table 12.4 on p. 188 provides a summary of the RGs and examples for each group. Based on the RG, specific biosafety levels are required. **Biosafety Levels (BSLs)** *are defined by the*

Table 12.4. CLASSIFICATION OF MICROORGANISMS BY RISK GROUP

Risk Group (RG) and Required Biosafety Level (BSL)	Description	Examples
RG1 (BSL1)	Agents that are not associated with disease in healthy adult humans. No individual or community risk.	*E. coli* strain K12 and other nonpathogenic strains Many non–spore-forming *Bacillus* sp. Adeno-associated virus (AAV) types 1–4
RG2 (BSL2)	Agents that are associated with human disease that is rarely serious and for which preventive or therapeutic interventions are *often* availabale. Moderate individual and low community risk.	*Salmonella* sp. *Streptococcus* sp. *Listeria* sp. Adenoviruses HIV (most protocols) Most poxviruses
RG3 (BSL3)	Agents that are associated with serious or lethal human disease for which preventive or therapeutic interventions *may be* available. High individual risk and low community risk.	*M. tuberculosis* *Brucella* sp. *Yersinia pestis* HIV (some protocols) Retroviruses Some arboviruses
RG4 (BSL4)	Agents that are likely to cause serious or lethal human disease for which preventive or therapeutic interventions are *not usually* available. High individual and community risk.	Ebola virus Lassa virus Marburg virus Other viruses

Adapted from: The National Institutes of Health, Office of Biotechnology Activities. NIH Guidelines for Research Involving Recombinant DNA Molecules. Bethesda, MD: NIH, 2002. World Health Organization. *Laboratory Biosafety Manual.* 3rd ed. Geneva, Switzerland: WHO, 2004.

NIH as the combinations of laboratory facilities, equipment, and practices that protect the laboratory, the public, and the environment from potentially hazardous organisms. While the NIH still designates these risk groups, they are no longer universally used to categorize organisms, because a pathogen's biosafety level can differ according to its specific use. These biosafety levels dictate progressively higher levels of containment for more hazardous microorganisms. **Biosafety level 1 (BSL1)** *is used for well-characterized strains of living microorganisms that are not known to cause disease in healthy adult humans.* BSL1 organisms have been used safely in many laboratories. High school and college students commonly work with BSL1 organisms. Examples of organisms in this category are nonpathogenic strains of *E. coli,* yeasts, and most plants. The fact that BSL1 organisms are generally nonpathogenic does not mean that they can be handled without caution. Standard microbiological practices must be applied to handling these organisms.

Biosafety Level II (BSL2) *is designated when working with agents that may cause human disease and therefore pose a risk to personnel.* Diseases caused by BSL2 organisms are usually treatable or preventive vaccines are available. These organisms are not likely to spread to the external environment and create a general health threat. Organisms in the BSL2 category are pathogenic bacteria such as *Salmonella* and *Clostridium,* molds such as *Penicillium,* and viruses such as polio virus and

rabies virus. As discussed later, human blood and tissue products are handled in a BSL2 environment.

The health risks associated with BSL2 organisms require a higher level of containment than BSL1. Written procedures and special worker training are required. Physical-containment procedures may require that all work be performed in a biological safety cabinet to minimize the release of microorganisms, especially in the form of aerosols. Waste must be decontaminated before disposal. For some infectious agents, the health of the worker may be monitored and the worker may be required to be immunized against the infectious agent. Production of blood antibodies against the microorganism may be checked to determine whether an unvaccinated worker has been exposed to a specific organism.

The majority of laboratories in the United States are designated BSL1 or BSL2. High containment **Biosafety Levels 3 (BSL3)** and **4 (BSL4)** *are generally associated with more dangerous agents that are highly infectious.* These hazardous agents require significantly higher containment barriers in the form of special facilities. The airflow into and out of the room where such organisms are handled is closely monitored. Examples of organisms requiring BSL3 facilities are certain arboviruses (arthropod-borne viruses associated with several human diseases) and large cultures of *Mycobacterium tuberculosis,* the etiological agent for human tuberculosis. Very dangerous BSL4 agents require closed glove boxes, isolated air supplies, and an enclosed breathing apparatus for the operators. BSL4 facilities are found in only a few places in the world. They are required for handling Ebola virus and related organisms that cause rapidly fatal human disease. Table 12.5 provides an overview of the containment practices and equipment required at each of the four biosafety levels. Many of the requirements shown in this table are discussed later in this chapter. There are additional requirements for large-scale (over 10 liters of volume) laboratory operations which use live organisms. For details, see the NIH Guidelines For Research Involving Recombinant DNA Molecules (reference in the Unit III introduction), Appendix K.

Table 12.5. SUMMARY OF BIOSAFETY LEVELS FOR INFECTIOUS AGENTS

Level	Special Safety Practices	Special Equipment	Facilities
BSL1	Standard microbiological practices	No special equipment required	Hand-washing sink required
BSL2	BSL1 practices plus: • Biohazard warning signs • Biosafety manual specific to the laboratory provided • Potentially infectious materials must be kept in a sealed container • Medical surveillance available	Class I or II BSCs to prevent aerosols	BSL1 requirements plus: • Self-closing, locked doors • Autoclave available • Elimination of all porous furniture materials
BSL3	BSL2 practices plus: • Decontamination of all waste • Decontamination of all lab clothing before laundering • Baseline blood serum testing for workers	BSL2 equipment plus: • Physical containment within a BSC for all manipulations • Additional protective clothing and respiratory protection as needed	BSL2 requirements plus: • Physical separation from access corridors • Self-closing, double-door system • Hand-washing sink must operate hands-free • Walls and ceiling must have a smooth, sealed finish for easy decontamination • Exhausted air not recirculated • Negative airflow into laboratory • Facility and procedures must be inspected and documented before use
BSL4	BSL3 practices plus: • Specific and rigorous training for all personnel • A logbook of all entries and exits must be maintained • Special transport protocols for all biological materials • Clothing change before entering • Shower on exit • All material decontaminated on exit from facility	All procedures are conducted in Class III BSCs **or** in Class I or II BSCs in combination with a full-body, air supplied, positive-pressure personnel suit	BSL3 requirements plus: • Separate building or isolated zone • Dedicated air supply and exhaust, vacuum, and decontamination systems • Emergency power supply • All plumbing protected from backflow and all liquids decontaminated before exit from lab

Abbreviations: BSC, Biological Safety Cabinet; **PPE,** Personal Protective Equipment
Adapted from U.S. Department of Health and Human Services, Public Health Service, Centers for Disease Control and Prevention and National Institutes of Health. *Biosafety in Microbiological and Biomedical Laboratories.* 5th ed. Washington, DC: U.S. Government Printing Office. 2007.

C. Laboratory Design

While good laboratory design can provide significant containment of biohazards, absolute containment is not achievable and generally not necessary, except in a BSL4 facility. Figure 12.3 provides examples of primary and secondary containment barriers in the laboratory.

A sampling of some of the recommended physical containment measures that are incorporated into laboratories meeting the CDC/NIH Guidelines for the four BSLs are provided in Table 12.6. The table lists only a few of the laboratory design features addressed in the guidelines.

D. Biological Safety Cabinets

i. WHAT IS A BIOLOGICAL SAFETY CABINET?

Laboratory hoods *are enclosed spaces with separate air supplies or directed air flows.* There are many types of hoods, each designed for specific safety purposes. Chemical fume hoods, for example, are designed to remove volatile gases from the operator. This is accomplished by drawing air from the room past the working space and venting the air outside the building (see Chapter 11).

A cabinet used to handle biohazards superficially resembles a chemical fume hood, but has a different purpose and construction. A **biological safety cabinet (BSC)** *provides containment for aerosols and separates the work material from the operator and the laboratory, while providing clean air within an enclosed area.* They are designed to filter all air that passes through the cabinet using HEPA filters. A **HEPA (High Efficiency Particulate Air) filter** *is a specially constructed filter made of highly pleated glass and paper fibers,* see Figure 12.4. HEPA filters are designed to have a 99.97% efficiency in removing particles of 0.3 µm or larger. This filter size will exclude most bacteria and microorganisms. Viruses found in aerosols are usually associated with particles

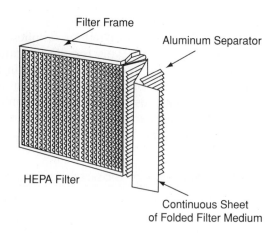

Figure 12.4. Structure of a HEPA Filter.

larger than 0.3 µm and so HEPA filters provide protection against most airborne virus particles. It is important to remember that HEPA filters are designed to remove particulate matter from the air. They are not effective for removing chemical vapors from air.

BSCs contain fans that direct air into a nonturbulent curtain. Air is passed through the HEPA filter and directed over the work surface at a velocity that minimizes disturbances in the interior of the cabinet. This type of airflow is called laminar flow and therefore a **laminar flow cabinet** *is one that has a sterile, directed airflow.* All BSCs have laminar flow designs; however, not all laminar flow cabinets are BSCs. This is because some laminar flow hoods, such as the **clean bench,** or **horizontal laminar flow hood,** *are designed to provide a sterile work surface but not worker protection,* see Figure 12.5. Air passes through the HEPA filter onto

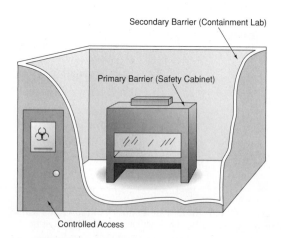

Figure 12.3. The Relationship between Primary and Secondary Barriers for Hazard Containment.

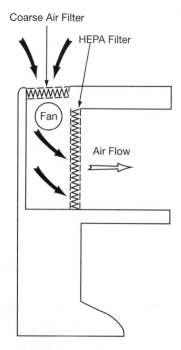

Figure 12.5. Design of a Clean Bench. Nonsterile air is shown as solid arrows, sterile air as open arrows.

Table 12.6. *SAFETY-RELATED LABORATORY DESIGN FEATURES*

Laboratory Design Feature	Biosafety Level			
	1	2	3	4
• Laboratory separated from public areas by a door	yes	yes	yes	yes
• Hand-washing facilities in laboratory	yes	yes	yes	yes
• Dedicated hand-washing facilities with automatic, foot, or knee controls		rec.	yes	yes
• Autoclave in building	rec.	yes	yes	yes
• Autoclave in laboratory			yes	yes
• All laboratory surfaces and furnishings are easily cleaned and disinfected	rec.	rec.	yes	yes
• Laboratory doors labeled with biohazard signs	yes	yes	yes	yes
• Testing of biological safety cabinets meets required specifications after installation		yes	yes	yes
• Biological safety cabinets Class I		rec.	yes	no
• Biological safety cabinets Class II		rec.	yes	yes
• Biological safety cabinets Class III				yes
• Biological safety cabinets equipped with pressure monitoring gauges for all HEPA filters		rec.	yes	yes
• Doors lockable and self-closing, access limited to authorized personnel		yes	yes	yes
• Air exhausted from laboratory at a minimum of ten room volumes per hour		rec.	yes	yes
• Standby electrical generator available for emergency support of essential equipment such as BSCs		rec.	rec.	yes
• Laboratory perimeter sealed to allow gaseous decontamination of whole room			yes	yes
• Clothing change area adjacent to containment area			yes	yes
• Body shower in containment area			rec.	yes
• Independent air supply			rec.	yes
• Ventilated airlock required for the separation of high-containment area				yes

rec. = recommended

the work surface and is vented directly into the room toward the operator. This type of laminar flow cabinet is useful for media preparation or sterile work with nonharmful organisms. It is not a BSC because it does not provide protection for the operator.

ii. CLASSIFICATION OF BIOLOGICAL SAFETY CABINETS

There are three classes of BSCs that provide different levels of protection against biological hazards. A **Class I cabinet** *is designed to protect the operator from airborne material generated at the work surface. The cabinet draws air directly from the room. The air then flows over the working area and is vented back into the room after being filtered.* This cabinet will filter aerosols released from the work materials, and so can be used to reduce aerosol escape to the environment. Because the air flowing over the work surface is room air, this type of hood is not designed to maintain sterility of the work surface. A Class I cabinet is used for routine operations that might generate harmful aerosols, such as vortexing, pouring, or centrifuging solutions containing microorganisms.

A Class II BSC, which is the most common type found in biotechnology laboratories, provides protection for both the operator and the work product. It is designed to draw HEPA-filtered air across the work surface and re-filter the air before venting. Because the air is filtered before it passes over the work surface, this cabinet type will maintain a sterile work area. A Class II cabinet is suitable for microbiology and tissue culture procedures.

Class II cabinets are subdivided into several categories. The most common types are Class IIA and IIB. **Class IIA cabinets** *do not have external ducts, and release filtered air directly into the laboratory,* see Figure 12.6 on p. 192. All airflow across the work surface is also filtered.

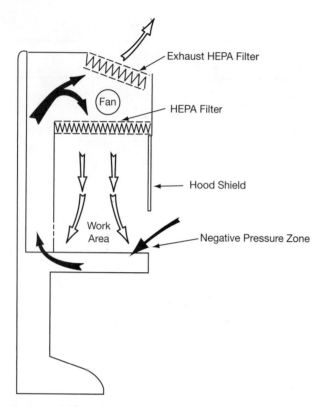

Figure 12.6. **Design of a Class IIA BSC.** Nonsterile air is shown as solid arrows, sterile air as open arrows.

Class IIB cabinets *maintain a faster airflow than Class IIA cabinets, and are ducted to external exhaust systems.* This prevents any cabinet air recirculation in the laboratory. Class IIB cabinets can be used with low levels of volatile chemicals, although it should be noted that this type of work shortens the life of the HEPA filter systems. There are three subtypes of Class IIB cabinets: IIB1, IIB2, and IIB3. These differ in the levels of recirculated filtered air within the cabinet, see Table 12.7.

Class III cabinets, also called **glove boxes,** are gas-tight cabinets designed to provide total containment for extremely hazardous biological agents. They provide the highest level of protection possible for both user and environment. Class III cabinets are designed to isolate the work material completely within a closed system. A Class III cabinet is accessed with air-tight gloves. Materials in these cabinets usually require decontamination before removal from the cabinet. Class III cabinets often have incubators, autoclaves, and air locks directly attached. A Class III cabinet would only be found when working with the most hazardous agents, and special rooms would also be required for proper use.

Table 12.7 summarizes the basic characteristics of the described biological safety cabinets. Proper use of these cabinets is described below.

E. Bioaerosol Prevention

Airborne materials pose a major concern in any laboratory or facility where potentially infectious agents or hazardous materials are used. In the 1978 study of LAIs cited on p. 185, fewer than 20% could be linked to a specific spill or other accident. Researchers believe that about 70% of all LAIs result from inhalation of infectious particles (bioaerosols). Airborne particles (carried by air currents) are highly effective in transmitting certain types of infections. The common cold, for example, can be spread by sneeze-generated aerosols that are inhaled by the next cold sufferer. Airborne organisms can remain suspended for significant periods of time and can be carried through a laboratory by air currents and even throughout buildings by way of the ventilation system. If inhaled, aerosols can carry hazardous materials into the lungs. Aerosols will also eventually settle on laboratory surfaces, where they can deposit hazardous substances.

One of the most important steps that you can take to minimize your risk of LAI is to reduce the presence of aerosols in your working environment. Table 12.8 lists

Table 12.7.   *COMPARISON OF TYPES OF BIOLOGICAL SAFETY CABINETS*

Type	User Protection	Work Surface Protection	Face Velocity (in linear feet per minute)	Airflow Pattern
Clean bench (not a BSC)	No	Yes	Varies	Air in at top, exhausted at front toward operator
Class I	Yes	No	75	Air in at front, exhaust duct within lab
Class IIA	Yes	Yes	100	Air in at front, exhaust duct within lab, 70% air recirculation within cabinet
Class IIB1	Yes	Yes	100	Air in at front, external exhaust duct, 30% air recirculation within cabinet
Class IIB2	Yes	Yes	100	Air in at front, external exhaust duct, no air recirculation within cabinet
Class IIB3	Yes	Yes	100	Air in at front, connected to building exhaust, 70% air recirculation within cabinet
Class III	Yes	Yes	None	Filtered air ducts for supply and exhaust, gas-tight

examples of common laboratory operations that generate airborne particles. Any activity that mechanically disturbs a liquid or a dry powder has the potential to release airborne substances. Care must therefore be used whenever handling any culture containing microorganisms, cells, or infectious agents, and when manipulating solutions containing hazardous substances. Because aerosols are not visible, and because they are created during routine laboratory operations, laboratory workers are often unaware of their presence.

Table 12.8. EXAMPLES OF ROUTINE LABORATORY PROCEDURES THAT GENERATE AEROSOLS

- **Shaking and mixing liquids,** particularly when using vigorous mixing devices like blenders and vortex mixers
- **Pouring liquids** from one container to another
- **Pipetting liquids** generates aerosols when:
  - the last drop of liquid is ejected from a pipette tip
  - a micropipettor tip is ejected into a disposal container
  - liquids splash upon entering a container
  - a used pipette is dropped into a vertical cylinder for cleaning
- **Removing the cap from a tube** can cause aerosol formation if there is liquid under the cap; "snap cap" tubes are particularly likely to cause aerosols.
- **Removing a stopper or cotton plug from a culture bottle or flask**
- **Opening a sealed tube containing a lyophilized (freeze-dried) agent.** When reconstituting lyophilized materials, add liquid to the contents slowly. Mix the contents without bubbling or excessive agitation.
- **Breaking cells open by sonication**
- **Grinding cells or tissues** with a mortar and pestle or homogenizer
- **Bubbling air into a liquid (aeration)**
- **Centrifuging samples** is particularly problematic. Note the following:
  - Centrifugation in uncapped tubes causes aerosols to be released; never centrifuge hazardous materials in uncapped tubes.
  - A centrifuge tube that breaks during spinning will spew its contents into the air.
  - Even if a tube is unbroken and capped, it is possible for the cap to deform under the force of centrifugation, allowing liquid to leak out and form aerosols during spinning.
- **Inserting a hot loop into a bacterial culture.** It is common practice to flame metal loops or needles (to kill any microorganisms) and then to place the loop or needle into a liquid culture. If the loop is hot when inserted into the culture, splattering can occur which leads to aerosol formation. Allow needles and loops to cool before placing them in a culture.

Table 12.9. GOOD PRACTICES THAT REDUCE EXPOSURE TO AEROSOLS

- Whenever possible, use a BSC when performing any operation that may generate hazardous aerosols.
- Before opening containers whose contents have been shaken, blended, or centrifuged, wait a few minutes to allow aerosols to settle.
- Pipetting practices that reduce aerosol formation and protect the operator include:
  - Never pipette by mouth.
  - Plug the tops of pipettes with cotton.
  - Never mix a liquid that contains hazardous material by pipetting up and down.
  - Use pipettes that do not require expulsion of the last drop of liquid.
  - Do not discharge biohazardous material from a pipette in a manner that induces splashing; when possible, allow the discharge to run down the container wall.
  - Place contaminated pipettes horizontally in a pan containing enough decontaminant to allow complete immersion.
- Centrifuge practices that reduce aerosol formation and/or protect the operator include:
  - Use centrifugation rotors and tubes designed to contain hazardous materials.
  - Fill and open centrifuge rotors and tubes in a BSC.
  - Carefully and completely disinfect rotors and tubes used for hazardous materials according to the manufacturer's directions.
  - Discard chipped or cracked centrifuge tubes.
  - Routinely allow the contents of rotors and tubes to settle for a few minutes before opening.
- Avoid pouring liquids or decanting supernatants if possible; it is preferable to use a vacuum system with appropriate in-line safety reservoirs and filters to remove the liquid from a centrifuge tube or other vessel.

Always be careful and deliberate when handling potentially hazardous liquid cultures, and use containment devices as appropriate.

Complete prevention of aerosol formation is not possible. By using careful technique, however, it is possible to greatly reduce their formation rate. For example, it is good practice to wait a few minutes after centrifugation before opening centrifuge tubes and rotors. This allows time for aerosols to settle. Table 12.9 includes examples of other practices that reduce aerosol formation. Even when careful practices are used, some airborne particles are still generated. Most laboratories therefore provide various containment barriers to control unavoidable aerosols. For example, BSCs can be used for operations that are likely to generate aerosols, such as pouring fluids, blending tissue samples, and handling cultures. It is also possible to purchase accessories for centrifugation that safely enclose potentially hazardous materials.

III. INTRODUCTION TO ASEPTIC PROCEDURES

A. Use of Disinfectants and Sterilization

Bioaerosols or work-related splashes can leave contamination on laboratory work surfaces. An effective practice for preventing the spread of biohazards is therefore the use of proper decontamination procedures. Decontamination can be accomplished with the use of germicidal agents. Some of the most common agents are heat (autoclaving), gas exposure (using ethylene oxide, for example), and the use of liquid chemical disinfectants. **Disinfectants** *are chemicals that kill pathogenic microorganisms and other biohazardous particles.* There are many factors that determine the effectiveness of germicidal agents, such as:

- the type of contaminating organism and its susceptibility to disinfecting agents
- the level of contamination
- the chemical composition and concentration of the decontaminating agent
- the length of exposure of the organism to the decontaminant
- the shape and texture of the surface being decontaminated

Biohazardous agents differ in their susceptibility to disinfection procedures, see Table 12.10. For example, most growing bacteria are relatively easy to kill, but bacterial spores are difficult to eradicate. It is essential to know the proper disinfection procedure for the organism you are handling, as discussed in the Case Studies "The Right Disinfectant Choice Can Prevent LAIs," above, and "Isolator Technology Is Only as Effective as the Operator," p. 203.

CASE STUDY

The Right Disinfectant Choice Can Prevent LAIs

While LAIs can result from unforeseen circumstances, many are the result of poor laboratory practices. For example, in 2002 a lab worker in Texas developed cutaneous anthrax after working with cultures of Bacillus anthracis, a spore-forming bacterium. The infection was traced to contaminated freezer vials prepared by the worker, who wiped the outside of the vials with 70% isopropyl alcohol and then handled the containers with bare hands. The lab SOP for decontaminating surfaces required the use of bleach for spore inactivation, but the worker apparently substituted alcohol to avoid removing the vial labels. This LAI, which was treatable and relatively noncontagious, could have been prevented by using the appropriate disinfectant and wearing gloves at all times while handling cultures.

There are distinctions between sanitizing, disinfecting, and sterilizing a surface or object. **Sanitization** *is the general reduction of the number of microorganisms on a surface.* Sanitization is generally the purpose of antiseptics. An **antiseptic** *is a type of mild disinfectant gentle enough to be applied to skin.* Chemical solutions that are suitable for use as antiseptics, such as 3% hydrogen peroxide or 70% isopropyl (rubbing) alcohol, may not provide enough disinfectant strength for work surfaces. **Disinfection** *is the removal of all or almost all pathogenic organisms from a surface.* Disinfectants can be classified as low, intermediate, and high level, according to their capacity to destroy microorganisms. **Sterilization** *is the highest level of disinfection; the killing of all living organisms on a surface.* An item or surface is either sterile or not sterile. High-level disinfectants achieve or approach sterilization, whereas low-level disinfectants are of variable effectiveness, depending on the nature of the contaminating organisms. High-level disinfectants are appropriate for use

---

Table 12.10.   DISINFECTION RESISTANCE OF MICROORGANISMS

Resistance Level to Disinfection	Type of Organism	Examples of Organisms
Least Resistant	Hydrophobic and/or medium-sized viruses	Human immunodeficiency virus Herpes simplex virus Hepatitis B virus
↓	Bacteria	*Escherichia coli* *Staphylococcus aureus*
↓	Fungi	*Candida* sp. *Cryptococcus* sp.
↓	Hydrophilic and/or small viruses	Rhinovirus Poliovirus
↓	Mycobacteria	*Mycobacterium tuberculosis*
Most Resistant	Bacterial spores	*Bacillus subtilis* spores *Clostridium sp.* spores

Table 12.11. STRENGTH OF COMMON GERMICIDAL INGREDIENTS

Ingredient	Concentration for Sterilization	Disinfection Concentration	Disinfection Activity Level	Example of Commercial Product
Alcohols	NA	70%	Intermediate	
Chlorine mixtures	NA	500–5000 ppm chlorine (1–10% bleach solution)	Intermediate	Chlorox
Glutaraldehyde	2% at high pH	2–3%	High	Cidex
Hydrogen peroxide	6–30%	3–6%	High	
Iodophor mixtures	NA	40–50 mg free iodine/liter	Intermediate	Betadine
Phenolic mixtures	NA	0.5–3%	Intermediate	Lysol (some types)
Quaternary ammonium mixtures	NA	0.1–0.2%	Low	Roccal
Formaldehyde*	6–8%	4–8%	High	

NA = not applicable
* Due to its toxic and potentially carcinogenic properties, formaldehyde is used only under special circumstances. The FDA has not formally approved any formaldehyde-based disinfectants.

on medical devices. Low-level disinfectants are frequently called sanitizers to denote their main purpose. Table 12.11 provides a summary of the sterilization and disinfection capabilities of some commonly used agents. Note that many chemical agents cannot be used effectively for sterilization purposes.

The U.S. EPA labels disinfectants as pesticides for regulatory purposes. The Federal Insecticide, Fungicide, and Rodenticide Act (FIFRA) *authorizes EPA to control the distribution, sale, and use of pesticides, and requires all manufacturers to register their pesticides and disinfectants with the EPA before marketing.* Disinfectants that are used on medical devices must be listed with the U.S. Food and Drug Administration. Commonly used disinfectants are also covered under the Toxic Substance Control Act (TSCA), which establishes chemical inventory and record-keeping requirements.

A major consideration when choosing a disinfecting agent is the nature of the biohazardous organism that must be removed from a surface. For human safety reasons, it is best to use the least toxic agent that can get the job done. Table 12.12 shows the relative effectiveness of a selection of chemical disinfecting agents against specific types of biohazardous organisms.

All disinfection and sterilization methods present certain safety hazards to the worker as well as to the targeted microorganisms. Table 12.13 on p. 196 provides examples of some of the human hazards associated with germicidal techniques.

B. Introduction to Aseptic Technique

i. GENERAL PRINCIPLES OF ASEPTIC TECHNIQUE

It is frequently necessary in biotechnology laboratories to culture and grow either prokaryotic or eukaryotic cells. This requires that the cell population of interest remain free of contamination by external microorganisms or foreign cell types. **Aseptic technique** *is a system*

Table 12.12. ANTIMICROBIAL ACTIVITY OF COMMONLY USED CHEMICAL DISINFECTANTS

Antimicrobial Agent	Range of Effectiveness				
	Bacteria	*Mycobacterium Tuberculosis*	Bacterial Spores	Fungal Spores	Viruses
Ethylene oxide gas (500 to 800 mg/L)	4	4	4	4	4
Ethyl alcohol (70%)	2	2	0	2	1
Sodium hypochlorite (5% bleach)	3	2	0	2	2
Glutaraldehyde (2%)	2	2	3	2	2
Iodophors (1%)	2	2	1	1	2
Phenolic derivatives (1% to 3%)	2	2	0	2	1
Quaternary ammonium compounds (3%)	2	0	0	2	1

4: Superior, 3: Very good, 2: Good, 1: Fair, 0: No activity

Table 12.13. HUMAN HAZARDS ASSOCIATED WITH SELECTED ANTIMICROBIAL AGENTS

Antimicrobial Agent	Potential Hazards
UV light	Eye and skin burns; mutagen; carcinogen
Steam (autoclave)	Burns; aerosols and chemical vapors; cuts from imploding glassware
Ethylene oxide	Severe respiratory and eye irritant; sensitizer; mutagen; suspected human carcinogen
Isopropyl alcohol	Flammable; toxic; severe eye irritant
Chlorine (bleach)	Gaseous form highly toxic; oxidizing agent
Formalin (37% formaldehyde)	Strong eye and skin irritant; toxic; sensitizer; mutagen; suspected human carcinogen
Glutaraldehyde	Toxic; skin irritant
Hydrogen peroxide	Oxidizing agent; corrosive; skin irritant
Iodine (iodophors)	Skin irritant; toxic vapors; sensitizer
Phenols	Chemical burns; toxic vapors; strong eye and skin irritant
Quaternary ammonium compounds	Skin irritant

Table 12.14. GENERAL PRINCIPLES OF ASEPTIC TECHNIQUE FOR BACTERIAL AND MAMMALIAN CELL WORK

1. *An object, surface, or solution is sterile only if it contains no living organisms.* If any organisms are present, the item is nonsterile.

2. *Objects or solutions are not sterile unless they have been treated to eliminate microorganisms (e.g., autoclaved, heat-sterilized, irradiated).* All items used in the culture of living cells must be initially sterilized (e.g., growth media, test tubes, culture vessels, pipettes). Some items are sterilized by the user, commonly by autoclaving or filtration. Alternatively, disposable sterile supplies and culture media may be purchased from suppliers.

3. *A sterile object, surface, or solution that comes in contact with a nonsterile item is no longer sterile.* Contaminated items must be replaced with sterile items if the procedure is not complete.

4. *Air is not sterile unless it is sterilized in a closed container.* Whenever a container or object is open to the air, contaminating substances can fall into it. When materials (e.g., growth media, bacteria, mammalian cells) are transferred from one location to another, special practices are used to minimize exposure to air and to avoid contact with nonsterile surfaces and objects.

5. *Human skin and breath are rich sources of contaminating microorganisms.* Proper aseptic technique protects cultures from human contact and aerosols.

of laboratory practices that minimize the risk of biological contamination in cultured cell systems. The term aseptic technique is frequently used interchangeably with *sterile technique,* although sterile technique usually refers to tissue culture practices. **Tissue culture** *is the in vitro propagation of cells derived from tissue of higher organisms.* Aseptic technique has two goals; first, preventing contaminants from the environment from entering a culture, and second, preventing cultured cells from causing adverse effects to humans in the laboratory or escaping to the external environment. This section concentrates on the protection of cultures, although the same practices are essential to protect humans from hazardous culture materials, as discussed later in this chapter. Contaminating organisms that might enter a culture include bacteria, yeast, molds, and viruses, all of which are ubiquitous in the air, on surfaces and objects, on skin, and in a person's breath.

Table 12.14 summarizes key principles that guide aseptic technique. These guiding principles are the same whether one is working with bacterial or mammalian cells. The specific practices for handling mammalian cells are, however, more stringent. A single contaminating bacterial cell can quickly take over a culture of much slower-growing mammalian cells. Bacterial and fungal cultures are less susceptible to major contamination by a few stray organisms because they are typically inoculated with millions of rapidly dividing cells and they are harvested quickly. This does not mean that one should be careless when handling bacteria, but it does mean that nonpathogenic bacteria can frequently be handled on an open lab bench if proper aseptic technique is used. In contrast, it is routine to use protective enclosures, such as laminar flow cabinets, when manipulating mammalian cells and other slow-growing nonpathogenic cultures. The proper use of laminar flow cabinets is discussed in the next section.

One important concern with handling cell cultures is avoiding cross-contamination. **Cross-contamination** *is a type of contamination where cells from one culture accidentally enter another culture.* While cross-contamination of laboratory cultures usually results in loss of the cultures and invalidated research results or products, it can also create unknown hazards for lab workers who are unaware of their exposure to contaminating organisms (see the Case Study "Culture Cross-Contamination Can Be Dangerous"). When working with multiple cell cultures, it is important to only work with one at a time. Also, separate bottles of media and reagents should always be used for different cell types or cultures, to avoid the risk of cross-contamination.

Developing the skills for successful aseptic manipulation of cultured cells is usually a matter of observation and practice. In many laboratories, new personnel learn and practice these techniques using a nonessential (and nonpathogenic) cell sample. The ability to repeatedly manipulate these cells without introducing external contamination usually indicates good aseptic technique. Keep in mind that while some types of contamination are easy to detect, others may not be immediately apparent (for an example, see the discussion of tissue culture contamination with mycoplasma). Because contaminated cultures are almost always useless for laboratory or production purposes, every step of every manipulation must be performed aseptically. It only takes one slip to destroy the purity of a culture. For this reason, one of the "wisdoms" heard in cell culture laboratories is, "When in doubt, throw it out." If you suspect that any object or sample may no longer be sterile, usually due to contact with a nonsterile item or surface, the object should be discarded immediately and replaced with a sterile substitute.

One of the easiest items to contaminate is the tip of sterile pipettes, which can easily contact the bench surface or nonsterile items during manipulations. It is helpful to think of pipettes, inoculating loops, and other sterile items as extensions of your hands and become conscious of their position at all times. Scientists who have excellent aseptic technique cannot always avoid contaminating sterile objects, but they are keenly aware of nonsterile contact and can therefore take steps to minimize any resulting contamination.

ii. HANDLING BACTERIAL CULTURES

Bacteria are frequently grown in glass or plastic **petri dishes** or plates (see Figure 12.7a) filled with a semisolid medium such as nutrient agar to support bacterial growth. These dishes have loose-fitting lids that allow gas circulation but prevent dust from entering the dish.

Most labs purchase plastic disposable dishes, which usually come in sterile bags of stacked plates. Care must be taken when using these plates not to contaminate the rims of the lid or base. Always label the base of the dish with the cell information, medium, and date of culture (lids can sometimes be switched). This should be done before cultures are added to the plates, using an indelible marker.

Additional types of vessels used to culture bacterial cells are shown in Figure 12.7. Bacteria are also frequently grown in liquid media for optimal growth of mass cultures (see Chapter 30 for a more complete discussion of cell media) in vessels of various sizes, from 1 mL (or smaller) test tubes to fermentation vats in production facilities that can hold 25,000 liters or more.

Specific practices for manipulating bacterial cultures on an open lab bench (not surrounded by an enclosure) are summarized in Box 12.1 on p. 198. Examples of such manipulations include: inoculating cells into culture medium, transferring culture medium from a stock bottle into a plate or flask, diluting cells that have outgrown their container, and filter-sterilizing medium.

iii. TISSUE CULTURE PRACTICES

With few exceptions, the same aseptic practices appropriate to bacteria also apply to tissue culture. The term **tissue culture** refers to *the culturing of cells derived from the tissue of multicellular organisms*. The main difference between tissue culture and microbial techniques is that tissue culture is almost always conducted within a relatively sterile enclosure rather than an open bench. As discussed above, cells derived from mammalian tissue (the main animal tissue type used in biotechnology) generally have significantly slower growth rates than bacteria and thus are more susceptible to rapid overgrowth by microbial contaminants. For this reason, the biosafety cabinets, glassware, media, and equipment used for tissue culture should never be used for bacterial cultures.

Mammalian cells are usually **anchorage dependent,** *which means that they require an appropriate solid surface to adhere to*. Anchorage-dependent cells are grown in vessels (usually disposable plastic) that are chemically

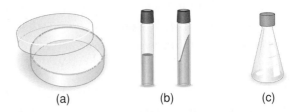

(a)   (b)   (c)

Figure 12.7.   Examples of Bacterial Culture Containers for Laboratories. a. Petri dish. **b.** Capped test tubes used for small liquid cultures (left) or agar slant (right) bacterial cultures. Agar slants are used for culturing small numbers of bacteria on solid medium and are poured on a slant to increase surface area. **c.** Liquid bacterial cultures up to 5 liters are frequently grown in capped Ehrlenmeyer flasks and incubated on shaker platforms for aeration.

BOX 12.1.   SPECIFIC ASEPTIC PRACTICES FOR WORKING ON AN OPEN LAB BENCH

1. *All standard microbiological practices must be applied (see Table 12.3).*

2. *Prepare in advance so that all materials are available and cultures are exposed to the environment for the shortest possible time.*

3. *Avoid contact between any sterile object and any nonsterile item.*

4. *Prepare a clean, clear area in which to work. Swab the surface with 70% alcohol.*

5. *Place all required items in the work area, arranged so that they are accessible and do not require reaching over other items.*

6. *Set up one or more disposal containers with disinfectant for pipettes and any other contaminated items.*

7. *Always have additional disinfectant available in the work area in case of spills.*

8. *Work alongside a lit Bunsen burner, because the flame is thought to create an updraft that helps protect the culture from airborne contaminants.**

9. *Bacteria are generally transferred from one place to another with inoculating loops and needles that are sterilized by holding in a flame until they glow bright red; they are then cooled for 5 or 10 seconds.***

10. *Pipettes are often used for transferring liquids, including medium that contains cells.*

    a. Nondisposable glass pipettes are placed in metal cans tip first and sterilized by autoclaving prior to use.

    b. Disposable sterile pipettes are available either in individual wrappers or in bags with multiple pipettes.

    c. Always open pipette wrappers or bags at the end away from the tip.

    d. When using any pipette, touch only the top of the pipette that never comes in contact with solutions.

    e. Open multi-pipette containers only as needed and avoid touching the unused pipettes that remain in the can or bag.

    f. If the tip of the pipette touches any surface, discard it and its contents.

    g. Briefly pass the tip of a glass (not plastic) pipette through a flame for 2 or 3 seconds. This will not sterilize the tip but is thought to prevent dust from landing on the tip and entering the culture. Be sure the tip is cool before using the pipette.*

    h. Always use a fresh, sterile pipette and never put the same pipette twice into a culture or into medium.

11. *When opening a bottle, flask, or plate:*

    a. Avoid putting the cap or lid down on a surface.

    b. Hold a cap in the crook of your left hand (if right-handed); do not touch the inside of the cap.

    c. When opening petri dishes, hold the lid partway over the bottom of the plate with the open end away from your face, to protect the culture from airborne particles.

    d. Hold open tubes, bottles, and flasks at a slight angle away from your face, to minimize the chance that airborne particles will fall in.

    e. Flame the neck of a bottle or flask briefly after opening it by passing it through a flame for 2 or 3 seconds.*

    f. After use, immediately close the tops of bottles, flasks, plates, tubes, or boxes of sterile pipette tips.

12. *Avoid getting drops of liquid onto the neck of a container underneath a cap that is to be closed.*

13. *Do not pass your arms, hands, or sleeves over any open plate, flask, or bottle, to prevent contamination.*

14. *When finished, clean the work area and swab with 70% alcohol. Be sure to wash your hands thoroughly.*

15. *Perform daily quality control checks.*

    a. Check cultures for contamination.

    b. Check reagents and growth medium for sterility.

16. *Dispose of all contaminated waste according to proper procedure. Handling biological waste is discussed later in this chapter.*

* This practice is usually reserved for open bench work.
** This practice is specific for microbial cultures.

treated to promote cell adherence. Tissue culture dishes look like slightly deeper petri plates, but anchorage-dependent cells will not adhere and grow in untreated plastic plates. Bacteria can be cultured in tissue culture dishes, but because of the special interior coating, these dishes are much more expensive than petri plates. In laboratories where both bacterial and tissue culture are practiced, it is essential to check the labels of culture dish bags. In addition to dishes, there are a variety of other culture containers developed for mammalian cell lines; examples are shown in Figure 12.8. These are available in a variety of coatings to enhance adherence of specific cell types.

Another major difference between the two types of cultures is the typical use of a bicarbonate/CO_2 buffering system for tissue culture (see p. 577 for a detailed discussion). This system requires that mammalian cell cultures be maintained in a CO_2-rich environment, usually an incubator that maintains a 5 to 10% CO_2 atmosphere at 37°C. Because of the need for CO_2 gas exchange, tissue culture containers cannot be tightly sealed during incubation. However, to maintain proper buffering, caps and lids should be tightened when containers are removed from their incubators. Relatively high humidity must be maintained within CO_2 incubators to prevent cell culture media from evaporating, so most incubators have a bottom pan that is partially filled with water. This water can be a fertile medium for the growth of microorganisms, so it is essential that the pan be emptied, cleaned, and disinfected regularly, even if chemicals to inhibit microbial growth are added.

One special concern in mammalian cell culture is contamination with mycoplasma. **Mycoplasma** *are a type of simple bacteria that lack a cell wall and are small enough to pass through regular cell culture sterilization filters* (see Figure 12.12 on p. 204 for a size comparison). They can be either intracellular or extracellular parasites, but in either case, they are not visible using conventional light microscopy and they generally do not outgrow their host cells. This means that the cultures they contaminate do not show typical signs of bacterial contamination, such as pH changes or turbidity in the medium. It is estimated that as many as 10 to 20% of all cell lines have "silent" mycoplasma infections.

Mycoplasma infections are of great concern in a laboratory or production facility because mycoplasma can have wide-ranging effects on gene expression in their host cells. Unusual proteins and metabolic products are often present, invalidating research or production procedures. Mycoplasmal infections can sometimes be treated with specialized antibiotics, but in many cases the infected cells must be discarded, along with any potentially contaminated stocks and reagents.

Many new mycoplasma infections are the result of cross-contamination with previously infected cultures. For this reason, it is essential that different cell cultures be handled separately, with independent media and sterile equipment. Regular mycoplasma testing should be part of any tissue culture operation. Any new cell cultures brought into the laboratory should be quarantined in separate spaces and incubators, and treated as though infected until proven otherwise.

C. Proper Use of Class IIB Biosafety Cabinets

Class II cabinets are appropriate for both microbiology and tissue culture work, although the same cabinet should not be used for both. They provide effective containment and a sterile air supply to the working surface. These cabinets need to be inspected and tested periodically to ensure that the filter is in good condition and the airflow is adequate. A certified technician must decontaminate and replace the filters at regular intervals.

(a)

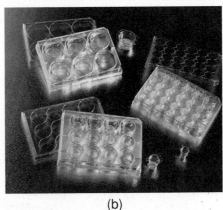
(b)

(c)

Figure 12.8.   Examples of Mammalian Cell Culture Containers for Laboratories. a. Stackable plastic flask. **b.** Microplates with 6–96 wells. **c.** Roller bottles and a roller apparatus. To maximize the number of cells that can be grown in a single vessel, roller bottles of various sizes can be inoculated with cells in a relatively small volume of medium. The bottles are then incubated on their sides in an apparatus that slowly rotates the bottles to bathe the entire interior surface of the bottle with medium. This creates a much larger surface area for growth than using only the bottom of a vessel. (**a.** and **b.** courtesy Fisher Scientific. **c.** CELLROLL roller apparatus by INTEGRA Biosciences.)

Many of these BSCs contain a UV germicidal lamp inside the work area. **Germicidal** *refers to methods that are capable of killing bacteria or other microorganisms.* When using a BSC, turn on the UV lamp for a minimum of 15 minutes before starting work. Because of the radiation, the lamp must be turned off when the cabinet is in use. Serious eye damage and burns to the hands can result if the light is accidentally left on during cabinet use. At least 5 minutes before using the cabinet, start the airflow to purge the hood of unfiltered air. Turn off the germicidal lamp and place all the needed materials in the hood. Only items that are needed in the workspace should be placed in the cabinet. Figure 12.9 shows a recommended layout for materials within a BSC.

All operations in a BSC should be planned to minimize disruptions in the airflow through the cabinet. Move hands and other objects slowly into and out of the cabinet interior. Strong air currents created outside the cabinet by people moving rapidly or laboratory doors opening may also disrupt the airflow pattern and reduce the effectiveness of a BSC. The use of Bunsen burners and open flames should be avoided or kept to a minimum because the heat generated produces air currents that can affect the sterility of the work area, and the added heat may damage the HEPA filter. Box 12.2 provides guidelines for the proper use of Class II BSCs.

D. Cleanrooms in Biotechnology

A **cleanroom** *is an area with a temperature-, pressure-, and humidity-controlled environment in which the levels of contaminants, including dust, aerosols, vapors, and microorganisms, are significantly reduced.* The purpose of most cleanrooms is to protect a product during testing or manufacturing. Cleanrooms vary in size from small enclosures inside a research laboratory to factory floor–size manufacturing areas. They are used in the testing and manufacture of pharmaceutical/biopharmaceutical products in order to limit the presence of microorganisms (i.e., bacteria, viruses, fungi, and spores of any type) that might contaminate the drug product. Cleanrooms are essential in other types of manufacturing as well, for example, electronics, where the slightest particle can destroy the function of miniaturized products such as microprocessors. Cleanrooms are also used for the production and testing of medical devices and foods.

Cleanrooms are subject to standards that specify maximum particulate contamination levels. ISO 14644-1 is the current global standard relating to cleanrooms, and must be adhered to in all ISO 9000-certified organizations. According to the Good Manufacturing Practices (GMP) regulations (21 CFR parts 210 and 211), companies that produce pharmaceuticals and biological products using aseptic processing must provide cleanroom conditions to protect their products from contamination. In **aseptic processing,** *the drug or biological product and its container are sterilized separately, frequently by different methods, and then packaged together under aseptic conditions to create a sterile final product.*

ISO standards designate cleanrooms as ISO Class 1 (cleanest) through 9, depending on the levels of particulates present. Excerpts from these standards are shown in Table 12.15 on p. 202. The ISO Class number represents the logarithm of the number of particles ≥ 0.1 μm in size allowed per cubic meter. This means, for example, that an ISO Class 6 facility will have a maximum of 10^6, or 1,000,000 particles per m^3. To put this in perspective, a Class 9 cleanroom, the least demanding category, has one-third the level of particulate contamination as an average hospital operating room. The cleanliness of a Class 1 facility cannot be found in any natural environment on Earth.

There is a corresponding system of designations for GMP-compliant "Clean Areas," using guidelines from

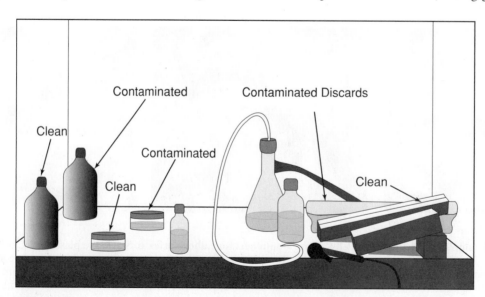

Figure 12.9.   Suggested Layout for Work Materials in a BSC. Note that all materials should be placed 4 to 6 inches back from the front airfoil. Materials should be arranged so that contaminated items do not pass over clean items.

Box 12.2. *SAFE AND EFFECTIVE USE OF CLASS II BIOSAFETY CABINETS*

Class II BSCs are designed to provide protection for workers, the work product, and the environment, when users observe basic guidelines for working in the cabinet.

General Guidelines

- Understand how the cabinet works.
- Do not disrupt the protective airflow pattern of the BSC.
- Plan your work.
- Minimize the storage of materials in and around the BSC.

Cabinet Set-Up Guidelines

- Turn on the germicidal lamp if present; wait at least 15 minutes before proceeding with any activities in the cabinet.
- Turn on the airflow in the BSC at least 5 minutes before starting to place objects in cabinet (if the BSC is turned off when not in use). Newer model BSCs are specifically designed for continuous operation, and turning off the BSC airflow at any time may adversely affect room air balance.
- Be sure that all vents and airfoils are clean and unblocked.
- Before starting to work, wash hands and lower arms thoroughly using a germicidal soap.
- Wipe the work surface in the cabinet with 70% alcohol or other suitable disinfectant.
- Assemble all the materials you will need for your work and wipe each item with 70% alcohol before placing it inside the cabinet.
- Do not overload the cabinet.
- Wait another 5 minutes before beginning to work in the cabinet.

Cabinet Use Guidelines

- Keep all materials at least 4 inches inside the cabinet face, and work with biological materials as deep within the cabinet as feasible.
- Move arms slowly when removing or introducing new items into the BSC.
- Do not place any objects over the front air intake vents or blocking the rear exhaust vents.
- Segregate contaminated and clean items; work from clean areas toward contaminated areas (see Figure 12.9).
- Place a pan with disinfectant and a sharps container (if necessary) inside the BSC for pipette discard; to avoid disrupting the BSC airflow and introducing contaminants to the environment, do not use vertical pipette wash canisters on the floor outside the cabinet.
- Place any equipment that may create air turbulence (such as a microcentrifuge) at the back of the cabinet (but not blocking the rear vents).
- Clean up any spills within the cabinet immediately; wait at least 10 minutes before resuming work.
- When finished, remove all equipment and materials and wipe all interior surfaces of the cabinet with disinfectant.
- Remove PPE and wash hands thoroughly before leaving the laboratory.

---

the FDA. These are Classes 100, 1000, 10,000 and 100,000, defined by the maximum number of $\geq 0.5 \ \mu m$ particles per cubic foot (see Table 12.15). These FDA Classes correspond to ISO Classes 5 through 8. In addition to overall particle levels, the FDA also specifies maximum microbiological contamination levels for each Class to assure aseptic conditions for product handling. The FDA recommends that product handling take place in a Class 100 facility (ISO Class 5 or cleaner) and that areas adjacent to the processing line

be at least Class 10,000 (ISO Class 7). GMP requires a rigorous monitoring program to document that both overall particle and microbiological levels remain below the specified limits. For certification, a cleanroom must pass at least five tests, which may include particle count, temperature, humidity, airflow velocity, air pressure, filter leakage, or structural conductivity. Once certified, a cleanroom is retested at least every two years to confirm the maintenance of appropriate conditions.

Table 12.15. CLEANROOM CLASSES AND STANDARDS

ISO Classification	ISO Maximum Number of Particles per m³ of Air[a]		FDA Clean Area Classification	FDA Recommended Microbiological Levels in Active Air[c]
	Particle size ≥ 0.1 µm	Particle size ≥ 0.5 µm	Number of 0.5 µm particles per ft³ of active air[b]	Colony Forming Units (cfu) per m³
ISO Class 1	10			
ISO Class 2	100	4		
ISO Class 3	1000	35		
ISO Class 4	10,000	352		
ISO Class 5	100,000	3520	100	1[d]
ISO Class 6	1,000,000	35,200	1000	7
ISO Class 7		352,000	10,000	10
ISO Class 8		3,520,000	100,000	100
ISO Class 9		35,200,000		

a. Excerpted from ISO 14644-1.
b. Data measured during activity.
c. Recommended levels.
d. Value normally zero.

Cleanrooms achieve particulate and microbiological control through careful design, operation, and human practices. A sample structure for a cleanroom facility is shown in Figure 12.10, although there are many possible designs for specific applications. The cleanroom is completely sealed off from the outside environment, which allows the operators to control air velocity and pressure, temperature, and humidity. A laminar air flow design is generally used (see p. 190), moving air from filtered ceiling ducts to a recirculating floor exhaust system as shown in Figure 12.10. HEPA and ultra-low penetration air (ULPA) filters are used to remove 99.999% of the airborne particles. Air is exchanged at 10 to 20 times the rate found in the average office. Manufacturing cleanrooms are operated at positive pressure, to prevent particle flow into the clean facility in case of air leaks.

However, for human safety reasons, any cleanrooms used for the handling of biohazards or toxic chemicals are operated under slight negative pressure, to avoid the potential escape of any hazardous materials.

Humans are the greatest source of contamination in the cleanroom environment, contributing approximately 75% of all particles found in a cleanroom. Under normal circumstances, humans generate particles of dead skin, hair, saliva droplets from coughing or talking, microorganisms, even clothing lint. Chemically, humans can introduce skin oils, perspiration, and contaminants from beauty and hygiene products (which are banned from cleanrooms) into the environment. It is estimated that a person sitting at a desk sheds about 100,000 particles per minute, and walking at a moderate pace generates 5,000,000 to 10,000,000 particles per minute! Even

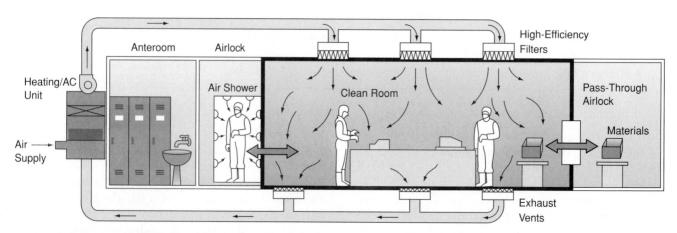

Figure 12.10. A Simplified Cleanroom Design. This is a laminar design facility with an airlock shown on the left side for people to enter and exit the cleanroom. Sterile materials enter through a pass-through airlock on the right side of the figure. The exhaust and pressure-regulating systems are not shown.

slight body movement can shed 4000 bacteria into the air per minute.

Because humans are such a significant source of contamination, cleanroom operators must follow strict practices. For instance, before entering a cleanroom operators must change shoes or put on protective shoe covers to avoid tracking contaminants into the facility. Cleanrooms that maintain high levels of protection have much more complex gowning requirements for operators. Intel, a company that manufactures computer components and other electronic products, has a set of 43 specific instructions that must be followed when entering or leaving their cleanrooms. Cleanrooms that are maintained at very low levels of contamination have a dedicated anteroom outside the main space of the cleanroom where users change clothes and store belongings. The anteroom is connected to the main room through an **airlock** (*a small room with interlocked doors between areas with different cleanliness standards*), which may also contain an air shower (shown in Figure 12.10). Everyone who enters such a cleanroom must wear some type of **bunny suit,** *an outfit of special protective clothing designed to prevent human introduction of particulates into the facility*. The items generally included in these suits are shown in Figure 12.11.

Once inside the cleanroom, humans must follow careful practices to avoid introducing contamination. Proper use of a cleanroom requires training and understanding of the many possible sources of contamination. All activities must be carried out using aseptic techniques in order to maintain product sterility. Air turbulence can cause particles to circulate, so movements within a cleanroom must be uniform, slow, and kept to a minimum. The relatively low humidity conditions in most facilities are conducive to electrostatic discharges, so anti-static mats, clothing, and materials within the facility are essential. All items brought into a cleanroom must be either specially designed to minimize particle shedding or carefully decontaminated before entering. Pencils, ordinary paper, and tissues all generate particles and are therefore not brought into the cleanroom. Watches, jewelry, and keys shed trapped skin particles from their wearers and therefore are removed before donning a bunny suit. Even clicking pens generate particles when used; all writing must be done with one-piece ballpoint pens.

Given that humans are frequently the most disruptive element in a cleanroom, the current trend is toward isolating products and materials that require extreme cleanliness from all human operators. This generally involves the use within a cleanroom of **isolators** *which utilize barrier/isolation technology, where materials occupy an environment that is physically separated from human users.* FDA guidelines recommend that isolators maintain the air quality of at least a Class 100 (ISO 5) cleanroom. Depending on characteristics of airflow within the isolator, these systems can be designed to optimally pre-

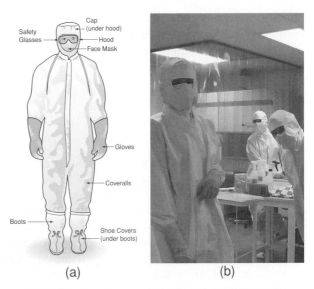

(a)   (b)

Figure 12.11.   Examples of Protective Clothing Used in Cleanrooms. a. The components of a bunny suit. **b.** Bunny suits in use in a small production facility for master cell banks.

vent contamination of materials and products, or to protect humans from hazardous materials. A Class III biological safety cabinet (see p. 192) is an example of the latter type of isolation system. As with much laboratory technology, the equipment is only as effective as the user, as discussed in the Case Study below. Again, knowing the characteristics and hazards of the organisms you handle in the laboratory can help prevent similar occurrences.

CASE STUDY

Isolator Technology Is Only as Effective as the Operator

Another case of lab-acquired SARS occurred in 2003 in a BSL4 facility. In Taiwan. A worker was handling cultures of the SARS virus in a closed glove box appropriate for the facility. As the worker was finishing, he noticed a waste spill in an inaccessible area of the enclosure. Instead of the normal decontamination procedure, he sprayed the spill with alcohol and waited for 10 to 15 minutes before opening the box to transfer materials. As discussed earlier in the chapter, alcohol is not effective for inactivating viruses. When the worker developed SARS, it was discovered that he had just attended a large scientific conference, accidentally exposing dozens of individuals to the virus. Luckily no additional cases of SARS were found, but many people were quarantined as a result.

Isolators have a relatively low air volume compared to full-size cleanrooms, and are particularly well-suited for aseptic operations. While they do not eliminate the need for a surrounding cleanroom, they make it possible for the adjacent larger areas to meet somewhat less stringent contamination standards. The future of barrier/

isolation technology may ultimately lead to isolator systems that can be used effectively without cleanrooms, as well as the development of automated systems to reduce contamination introduced by human operators.

IV. SPECIFIC LABORATORY BIOHAZARDS

A. Microorganisms

Biotechnologists often work with microbiological agents such as bacteria, fungi, protozoan parasites, and viruses. Bacteria, fungi, and viruses are easily spread by aerosols or by hand-to-mouth contamination. Protozoans are more likely to pose a threat when ingested. Skin contact with fungi and parasites may induce allergic reactions, which range from annoying to life-threatening. However, the most common source of LAIs is through broken skin, usually due to injuries from laboratory sharps.

The pathogenic properties of many bacteria have been characterized and bacteria and other microorganisms are classified into risk groups, as shown in Table 12.4. Most biotechnology work uses nonpathogens when possible, and any potential pathogens are usually identified at the onset of work. However, good laboratory practices are necessary when working with both pathogenic and nonpathogenic species. A worker can become infected with a nonpathogenic species and spread the organism to environments outside the laboratory. Good practices will also prevent the spread of unidentified or unknown organisms.

Molds and fungi are ubiquitous airborne organisms, with millions of fungal spores in an average cubic meter of unfiltered air. **Fungi** *are primitive eukaryotic microorganisms that have complex lifecycles and can form spores.* There is an incredible diversity of fungal species. **Molds** *are filamentous forms of fungi, presenting a "fuzzy" appearance when they grow.* The genus *Penicillium* is an example of a mold that produces a useful product; in this case, the antibiotic penicillin. **Yeasts** *are single-celled forms of fungi.* The yeast we are most familiar with is *Saccharomyces cerevisiae*, which is common baker's yeast. *S. cerevisiae* is often used in laboratories as a research organism.

Fungi can sometimes be dangerous to humans through several mechanisms. A **mycosis,** or **mycotic disease,** *is a disease caused by infection with a pathogenic fungus.* Some skin mycoses, such as "athlete's foot," are relatively common and do not pose a serious threat to healthy individuals. Systemic mycoses occur occasionally, especially in individuals with compromised immune systems, although the spread of systemic fungal disease between individuals is rare.

Many fungi produce **mycotoxins,** *which are toxic fungal products that may render the fungus poisonous, or contaminate the environment of the fungus.* The common mold *Aspergillus flavus,* for example, produces aflatoxin, a highly potent carcinogen. Special precau-

tions should be taken to avoid personal contamination when working with mycotoxins or any fungal species known to produce mycotoxins.

One of the most likely problems to arise from exposure to fungi in the laboratory is the development of allergies to the fungal spores. Fungal allergies are common, and are easily triggered in the presence of significant numbers of airborne spores. Because biotechnology facilities are increasingly using fungi to produce enzymes or food products, occupational exposure becomes more likely. Proper containment and reduction of airborne spores in the immediate environment are essential measures to prevent allergies and environmental contamination.

B. Viruses

Viruses *are particles that contain nucleic acid usually surrounded by a protein coat of varying complexity. They require a cellular host for replication.* Viruses can contain either DNA or RNA as their genetic material. Both RNA- and DNA-containing viruses have the ability at times to integrate their nucleic acid sequences into an infected cell's genome. In some instances this viral DNA can direct the cell to make new virus particles, which will then be released as infectious viruses.

Many types of viruses have the ability to remain dormant for long periods of time, and in some cases, to resist disinfection and sterilization procedures (as shown in Table 12.12 on p. 195). Because virus particles are much smaller than other microorganisms, see Figure 12.12, they can aerosolize readily, and are relatively difficult to filter from the air at high concentrations (although HEPA filters are effective at filtering virus particles associated with aerosol particles). It is essential to use standard microbiological practices when handling viruses and virally infected cells. These practices provide the dual benefits of worker protection and protection of the work material from infection with extraneous viruses.

Viruses pose risks of acute and sometimes chronic infection. Acute infection can be transitory and harmless, or it can result in disease. Viruses are classified into

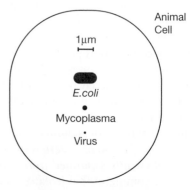

Figure 12.12. Relative Sizes of Biological Organisms Used in Laboratories. These include animal cells, bacteria, and viruses.

the same risk levels as other microorganisms. Always know the properties of and any potential risks posed by viruses that are used in your workplace.

The greatest concern in laboratories that use human tissue or blood products is unknown contamination of materials with HBV or HIV. **HBV, or hepatitis B virus,** *is the etiological agent for human hepatitis B.* **HIV,** or **human immunodeficiency virus,** *is the etiological agent for acquired immunodeficiency syndrome (AIDS).* Most of us are more aware of and concerned about HIV infection, but in reality, the rate of transmission of HIV is much lower than that of HBV. For example, the risk of infection with HIV after skin puncture with an HIV-contaminated needle is approximately 0.3%, based on studies of hospital exposures. The risk of infection with HBV is approximately 30 to 35% after skin puncture with an HBV-contaminated needle (in unvaccinated individuals).

Hepatitis B *is a bloodborne infection that attacks the liver and causes inflammation.* Most people who contract hepatitis B exhibit moderate to severe flu symptoms. Up to 5% of these individuals will develop chronic disease, which can eventually prove fatal. Another 5 to 10% of infected individuals become chronic carriers of HBV. Fortunately there is an effective (85 to 95%) vaccine available that should be administered to all workers who contact human blood or blood products. This vaccine requires a series of three injections. Federal guidelines mandate that this vaccine be offered free of charge to any individual who works with or may be exposed to human tissue or blood products in the workplace.

HIV is an example of a type of RNA virus called a retrovirus. A **retrovirus** *is an RNA-containing virus that also contains* **reverse transcriptase,** *an enzyme that transcribes RNA into DNA.* This DNA copy of the viral genetic material can then integrate into the DNA of a host cell. With some retroviruses, this alteration may be undetectable except by molecular analysis, and does not lead to further problems. In other cases, such as HIV infection, adverse effects become apparent with time.

Without treatment, HIV infection eventually leads to the development of AIDS. **AIDS, or acquired immunodeficiency syndrome,** *is a condition in which the T-cells of the immune system are destroyed, devastating the body's immune defenses and allowing other infectious agents to cause disease and frequently death.* Individuals with AIDS, or other conditions that weaken the immune system, such as chemotherapy, are especially susceptible to infection with microbiological agents that are considered to be relatively nonpathogenic in healthy individuals.

Although there is no vaccine currently available for HIV, there are treatments that help to delay the deterioration of the immune system and therefore the manifestation of immunodeficiency symptoms in HIV-infected individuals. These treatments can slow the development of AIDS and increase both life expectancy and quality of life for infected individuals. Fortunately, HIV is not highly transmissible under laboratory conditions, and the virus is susceptible to inactivation with desiccation or disinfection procedures.

Another potential consequence of exposure to certain types of viruses is the development of cancer. **Cancer** *includes a variety of diseases characterized by the uncontrolled growth of cells and the ability of these cells to spread to other parts of the body* (**metastasis**). **Oncogenic viruses** *are agents that can induce cancer after infecting cells.* This occurs when the viral infection either interferes with normal cellular growth control, or alters the cell's genetic material in a manner that affects growth control.

Potentially oncogenic viruses are classified as of low, moderate, or high-risk to humans. A low-risk virus is one that has the theoretical capability of being oncogenic, but has not yet shown any evidence of a threat. Many viruses that cause cancer in animals fall into this category. A high-risk oncogenic virus would be one that has been shown to cause cancer in humans. A list of viruses that are currently believed to cause human cancer is shown in Table 12.16. It is important to note that many individuals become infected with these viruses without developing cancer as a result. For example, infection with Epstein-Barr virus most commonly results in mononucleosis in the United States. The criteria for designating oncogenic viruses as moderate risk, which are suspected but not proven to be carcinogenic in humans, are shown in Table 12.17 on p. 206.

C. Human Blood Products

Biological materials from humans are a special problem for laboratory workers. There are no test methods available that can guarantee the absence of all infectious agents in biological materials, and infectious agents that are present are likely to be targeted to humans. OSHA therefore requires that all human blood and blood products be handled as though they were contaminated. This

Table 12.16.   VIRUSES CONSIDERED TO BE CAUSATIVE AGENTS IN HUMAN CANCERS

Virus	Associated Human Cancers
Human papilloma virus (HPV)	Cervical cancer (vaccine available)
Epstein-Barr virus (EBV)	Burkitt's lymphoma, nasopharyngeal carcinoma
Hepatitis B and C (HBV and HCV)	Liver cancer (HBV vaccine available)
Human herpes virus 8 (HHV-8)	Kaposi's sarcoma, lymphoma
Human T-cell lymphotropic virus Type 1 (HTLV-1)	Adult T-cell leukemia/lymphoma

Table 12.17. Criteria for Classification as a Moderate Risk Oncogenic Virus
• The virus has been isolated from human cancers.
• The virus causes cancer in nonhuman primates without experimental manipulation of the host.
• The virus can produce tumors in healthy nonprimate mammals.
• The virus can transform human cells in vitro.
• Any virus or genetic material that is a recombinant of an oncogenic animal virus and a microorganism infectious for humans is classified as a moderate oncogenic risk until its human oncogenic potential has been determined.
• Any large-scale or concentrated preparation of infectious virus or viral nucleic acid is classified as a moderate oncogenic risk until its human oncogenic potential has been determined.

involves the routine use of **Universal Precautions,** *which are standard safety practices for handling human blood products,* see Table 12.18. In general, Biosafety Level 2 procedures and a Class II biological safety cabinet should be used. All tissue and waste must be treated as pathogenic and be properly decontaminated.

Table 12.19 provides a list of potentially infectious materials that are considered by OSHA to constitute hazardous human products requiring special precautions. Sometimes commercially purchased human antibodies, cells derived from humans, and blood factors are used in research. The supplier will provide a description of pathogen testing for the product, but the absence of all pathogens cannot be guaranteed. These products should therefore be handled with the same precautions as other human materials.

Institutions and organizations where occupational exposure occurs must develop an Exposure Control Plan to comply with the OSHA Bloodborne Pathogens Standard 29CFR Part 1910.1030. **Occupational exposure** *is defined as reasonably anticipated skin, eye, mucous membrane, or parenteral contact with blood or other potentially infectious materials that may result from the performance of an employee's duties* (U.S. Department of Labor. Occupational Exposure to Bloodborne Pathogens. OSHA 3127, 1996 [revised]). In addition to specifying the use of Universal Precautions, the Exposure Control Plan must also include detailed information about:

• exposure determination
• PPE and first aid materials available for employees
• provision of hepatitis B vaccination
• procedures for any post-exposure follow-up
• documentation requirements
• employee training

Table 12.18. Summary of Universal Precautions
This is a summary of OSHA standards describing safe handling procedures for the materials described in Table 12.19. Notice the similarities between many of these items and those incorporated into standard microbiological practices (Table 12.3).
• Universal precautions must be followed at all times when blood, other body fluids (as described in Table 12.19), and other potentially infectious materials are handled.
• Workers must wash their hands after handling potentially infectious material, whenever changing gloves, and whenever leaving the laboratory.
• Use of needles and other sharps must be avoided whenever possible; if use is necessary, needles must not be recapped, bent, or otherwise manipulated by hand; used needles and other sharps must be disposed of in nearby puncture-resistant containers.
• Gloves and other protective clothing must be worn when there is potential for contact with blood, or other potentially infectious materials; gloves must be replaced when visibly soiled or punctured; fluid-resistant clothing, surgical caps or hoods, face masks, and shoe covers should be worn if there is a potential for splashing or spraying of blood or other potentially infectious materials.
• Blood or other potentially infectious materials must be placed in leak-proof, properly labeled containers during storage or transport.
• Mouth pipetting is not allowed.
• Work surfaces must be clean and decontaminated with an appropriate disinfectant after any known contamination, after completion of planned work, and at the end of the day.
• All contaminated waste material must be decontaminated prior to disposal, following proper institutional procedures.
• Laboratory equipment that has been in contact with blood or other potentially infected material must be decontaminated by laboratory personnel before any servicing or shipping.
• Biohazard warning signs must be prominently placed in facilities where blood or other potentially infected material is present.

Employee training should include information about the OSHA standard, explanations of bloodborne pathogens and Universal Precautions, use of PPE and other equipment, availability of HBV vaccination, and emergency procedures.

It is advisable to be cautious when assisting in a medical emergency that may occur in the laboratory. In an emergency where a loss of blood occurs, after assisting the victim, take time to protect yourself by proper disinfection of your skin and clothing and the contaminated areas. If appropriate, consult a health care professional.

Table 12.19. POTENTIALLY INFECTIOUS HUMAN PRODUCTS

The following materials are potential sources of human bloodborne pathogens:

- Human blood and blood components
- Products made from human blood
- The following human body fluids:
    - semen
    - vaginal secretions
    - cerebrospinal fluid (fluid from the nervous system)
    - synovial fluid (fluid from the joints)
    - pleural fluid (fluid from the lungs)
    - pericardial fluid (fluid from the heart)
    - peritoneal fluid (fluid from the abdomen)
    - amniotic fluid
    - saliva in dental procedures
    - any other body fluid that is visibly contaminated with blood, such as saliva or vomitus
    - all body fluids in situations where it is difficult or impossible to differentiate between body fluids
- Any unfixed tissue or organ (other than intact skin) from a human (living or dead)
- HIV-containing cell or tissue cultures, organ cultures, and HIV- or HBV-containing culture medium or other solutions
- Blood, organs, or other tissues from experimental animals infected with HIV or HBV

Taken from the OSHA Bloodborne Pathogens Standard 29 CFR Part 1910.1030.

D. Tissue Culture

Culturing mammalian cells presents particular areas of biosafety concern. As described previously, these cell cultures generally involve growing a monolayer of cells, covered with a layer of appropriate growth-sustaining liquid medium, in glass or plastic dishes. Cultured cells have a variety of biotechnology applications, some of which are listed in Table 12.20.

Many healthy animal cells are unable to divide and grow in vitro. Cells that do propagate in culture tend to lose some of the normal cell properties exhibited in their natural environments. However, early cell researchers sometimes found that tumor cells are easier to establish in culture. Moreover, many chemical and viral agents can transform normal animal cells so that they have characteristics of tumor cells. In this context, **transform** *means to cause cells to become competent to grow in long-term culture and develop some properties similar to tumor cells.*

Primary cell cultures *are newly isolated cells from tissue or blood, growing outside the body for the first time.* After removal from the body's tissue, the cells are potential carriers for any infectious agents that were

Table 12.20. USES FOR CULTURED CELLS IN BIOTECHNOLOGY LABORATORIES

Cultured cells, especially from mammalian sources, have many uses in biotechnology. These include:

- Tools for biomedical research (e.g., for the study of mechanisms of growth control)
- Sources of nucleic acids for the preparation of DNA libraries and genomic studies
- Sources of species-specific proteins
- Propagation of viruses (e.g., for vaccine production)
- Production of recombinant eukaryotic proteins that cannot be expressed in bacteria (e.g., proteins with extensive post-translational modifications)

present in the organism. Primary cell cultures from humans may therefore contain pathogenic agents that can infect workers. The most likely viral hazards of primary human cell lines are HIV, HBV, and hepatitis C virus. Nonhuman cells, especially from monkeys and other primates, can also carry agents that are serious health hazards.

Cell lines, *which are cells that are established in long-term culture,* are less likely to pose an unknown health threat. Many cell lines used in laboratories have been analyzed and any potential hazards documented. Even these cells, however, may harbor hidden virus particles, which may be released upon unusual circumstances. This is illustrated in the Case Study on p. 197. There are general guidelines for the relative levels of human health risk when handling various types of animal cell lines, shown in Table 12.21. High-risk cell cultures should only be handled by trained personnel using at least a Class II BSC. Note that all human- and primate-derived cell lines are considered high risk. Any procedures that involve these cell lines in manipulations that

Table 12.21. GENERAL RISK LEVELS FOR CULTURED CELL LINES

Relative Health Risk	Types of Cell Lines
Low risk	Well-characterized nonhuman/primate permanent cell lines
Medium risk	Poorly characterized mammalian cell lines
High risk	Cell lines derived from human/primates
	Cell lines with endogenous pathogens (relative risk determined by the nature of the pathogen)
	Cell lines after experimental infection with a pathogen (relative risk determined by the nature of the pathogen)

infect cells with viruses or have the potential to release endogenous viruses should by carried out at BSL3.

Some established cell lines have the ability to induce tumor development when injected into laboratory animals. There have been several reported cases where laboratory workers have developed localized tumors after accidental injuries with sharps contaminated with cancer cell lines. Because one of the most common injuries among tissue culture workers is skin punctures from broken glass or needles, it is essential to handle sharps with care.

When performing tissue culture, a sterile work surface and sterile materials are essential to avoid contaminating the cells with extraneous microorganisms. The same practices that maintain the sterility of cell cultures also serve to protect the worker. Use of a Class II BSC provides a suitable environment for sterile work, although with proper technique, tissue culture can be performed successfully on an open laboratory bench (assuming no pathogenic agents are involved). There are brief discussions of sterile technique and the use of a Class II BSC in earlier sections of this chapter.

All pipettes, culture dishes, dissection instruments, and other tools must be sterilized before and after every use to protect both scientist and culture. When working with biohazardous cell cultures, it is important to keep an autoclavable pan filled with distilled water or disinfectant to one side of the BSC work surface. This can be used to collect all contaminated reusable items, such as pipettes. The pan is covered before removal from the cabinet and autoclaved before reopening. Contaminated liquid culture media for disposal should be treated with a disinfectant such as bleach before discarding. As discussed previously, the disinfectant must be chosen according to the organisms present.

E. Recombinant DNA

Many of the benefits of biotechnology are a result of recombinant DNA methods. **Recombinant DNA (rDNA)** *is a DNA molecule constructed outside of a living cell, joining DNA from more than one source and capable of replication in a host cell.* Recombinant DNA methods are being used around the world in more than 30,000 laboratories. Since 1979, rDNA technology has led to many advances, particularly in the fields of medicine and agriculture.

When the possibilities of rDNA technology were first conceived, many scientists were concerned with the safety aspects of this type of experimentation and the possible introduction of new and unpredictable life forms. In 1975 leading scientists in the field assembled in Asilomar, California, to discuss the safety implications of rDNA methods. This was one of the few times that industry assembled to discuss the safety of a new technology before any major safety incident occurred. The consensus recommendations formulated at this

meeting became the basis for NIH's Guidelines for Research Involving Recombinant DNA Molecules, which were officially established in 1976, and substantially revised several times afterwards. The NIH Guidelines are currently considered the general authority on rDNA safety procedures. While the NIH Guidelines do not have the force of federal regulations, any institution receiving NIH funding must comply with their recommendations. The Guidelines for Research Involving Recombinant DNA Molecules are constantly updated as new information becomes available.

The safety aspects of rDNA have been vigorously debated in both the popular literature and the scientific press. The safety model that is currently applied worldwide assumes that the safety of specific rDNA protocols depends on the DNA fragment used, the properties of the chosen host organism, and the properties of the vector used for delivering the DNA to the host cells. Some of the criteria used for evaluation of rDNA safety are shown in Table 12.22.

The NIH Guidelines for working with rDNA are similar to the regulations for work with infectious agents. The recommended way to protect workers and the environment is with containment barriers. As risks become greater, the procedures and facilities are designed to have a greater level of containment. The NIH Guidelines clas-

---

Table 12.22. SOME FACTORS RELATED TO THE SAFETY OF RECOMBINANT DNA EXPERIMENTS

There are many factors that enter into a risk assessment for planned rDNA work. These factors center on the properties of the host organism and vector, as well as on the nature of the DNA sequence being transferred. Examples of these concerns include:

Host Organism and/or Vector

- Pathogenicity or infectious properties of the host organism or vector

- Potential routes of human infection (aerosols, direct or indirect contact, etc.)

- Ability of an altered host cell or vector to survive outside the biotechnology facility

- Capacity of host cells to transfer genes into other, unintended organisms

Nature of the DNA Sequence

- Level of product expression

- Biological stability of the product

- Coding for:
  - Production of toxins
  - Antibiotic resistance
  - Protein products that trigger allergies
  - Genetic elements that could increase invasive properties of the host

sify rDNA hazards into four risk groups related to their potential pathogenicity for humans. The Guidelines also designate four levels of physical biosafety containment, BL1 through BL4. These levels are very similar in concept to those assigned to microorganisms. All laboratories performing rDNA research must register with their Institutional Biosafety Committee (IBC), which then recommends the appropriate level of containment. In addition to applying NIH guidelines, IBCs implement any local policies regarding rDNA work as well.

The NIH Guidelines provide an "exempt" category, to designate rDNA experiments that do not require containment measures beyond those normally used for the organisms involved. Section III-F-6 of the Guidelines states that exempt experiments are "those that do not present a significant risk to health or the environment" when usual containment procedures are followed. A substantial percentage of all rDNA experiments are in the exempt category. Much of this exempt work is conducted using *E. coli* strain K12, *Saccharomyces cerevisiae*, and non–spore-forming *Bacillus subtilis* host cell systems. In these cases, the host cells themselves contain the rDNA safely.

E. coli strain K12 is a particularly well-characterized bacterium. Despite the sensational news stories about *E. coli*–induced food poisonings, the vast majority of *E. coli* strains have been characterized as nonpathogenic. Strain K12 is a particularly safe variety of *E. coli* that is used in high school and college laboratories, because studies indicate that strain K12 cannot colonize in humans even after deliberate inoculation.

When host cells that require stringent growth conditions and that are generally not infectious (such as *E. coli* K12) are coupled with vectors that only infect this host cell type, biological containment of rDNA is relatively complete. Of course, not every laboratory can conduct their work in these systems, so it is essential to learn about the host cell and vector system used in your laboratory, to better understand any possible biohazards involved.

F. Laboratory Animals

i. HUMANE HANDLING

Animals are often used in biological research and in drug-testing programs (as discussed in Unit I). The decision to use animals is a difficult one, because of ethical concerns relating to their welfare and also because they are expensive, in terms of maintenance and personnel costs. However, there are situations where no other method of obtaining information is adequate. In these cases, there must be a thoughtful evaluation of health and safety issues for both animals and personnel.

There are a variety of regulations and guidelines that address both humane treatment of animals and human safety. The Animal Welfare Act, originally enacted in 1966 and amended several times, is the main federal

Table 12.23.   PRINCIPLES FOR THE CARE AND USE OF VERTEBRATE ANIMALS IN THE UNITED STATES

These are examples of the principles that govern the U.S. government requirements for treatment of animals related to testing, research, or training procedures. Refer to the *Guide for the Care and Use of Laboratory Animals* for further information about these principles and their implementation.

- Transportation, care, and use of animals must conform with the Animal Welfare Act and other applicable federal laws, guidelines, and policies.
- Procedures should be designed and performed with due consideration of their relevance and contribution to scientific knowledge.
- Animals should be an appropriate species and include the minimum number required for valid results. Alternatives to animal use, such as computer modeling and in vitro cell systems, should be considered.
- Animal discomfort, distress, and pain should be avoided or minimized whenever possible.
- Any procedures involving more than slight pain or distress should be performed under anesthesia or other means to minimize these factors.
- Animals in severe or chronic pain or distress that cannot be relieved should be painlessly euthanized when appropriate.
- Animals should be housed under healthy and comfortable conditions appropriate for their species. Veterinary care and consultation must be available.
- All investigators and personnel who handle animals must receive proper training.

Adapted from Public Health Service Policy on the Humane Care and Use of Laboratory Animals (amended 2002).

statute governing the handling and use of animals in the United States. Animal welfare regulations are published each year in *Code of Federal Regulations, Title 9, Chapter 1, Subchapter A—Animal Welfare*, known as 9CFR. The Public Health Service (PHS) Policy on the Humane Care and Use of Laboratory Animals (found at http://grants1.nih.gov/grants/olaw/references/PHSPolicyLabAnimals.pdf), outlines the basic mandatory principles and goals for the care and use of vertebrate animals. This policy is regulated by the Office of Laboratory Animal Welfare (OLAW) and applies to all PHS-supported agencies, research, and animal-related activities. The PHS includes the NIH, FDA, CDC, and many other federal agencies. Some of the basic principles for the care and use of vertebrate animals as outlined by PHS policy are shown in Table 12.23. Current guidelines describing recommended practices for use of animals in the laboratory under these regulations are contained in the *Guide for Care and Use of Laboratory Animals* (NRC, 1996).

All institutions and organizations subject to the Animal Welfare Act and/or receiving PHS support must develop an animal care and use program that includes (among other requirements):

- An Institutional Animal Care and Use Committee (IACUC)
- Appropriate animal housing, maintenance, and support facilities
- A veterinary care program
- Training for all personnel with animal-related duties
- An appropriate program for monitoring activities, record-keeping, and reporting

No animal research can be conducted at PHS-funded institutions without prior approval by the institution's IACUC. An **Institutional Animal Care and Use Committee (IACUC)** *reviews the institution's programs for humane care and use of animals, inspects institutional animal facilities, and reviews and approves all protocols using live vertebrate animals, among other duties.* An IACUC is required to have at least five members, including a scientist familiar with animal research, a veterinarian, a nonscientist such as an ethicist or lawyer, and a member external to the institution.

The PHS recognizes two categories of institutions that perform animal research. Category 2 institutions are evaluated internally by their IACUC, which ensures that the facilities and programs involving animals meet PHS Policy. Category 1 institutions, in addition to meeting all Category 2 criteria, are also voluntarily accredited by the **American Association for Accreditation of Laboratory Animal Care (AAALAC).** AAALAC *is an independent peer-review organization that ensures that companies, universities, hospitals, government agencies, and other research institutions surpass minimal animal-care standards.* Currently, more than 730 institutions in 30 countries are AAALAC-accredited.

One significant requirement of the agencies that oversee animal care is that all personnel who work with animals receive appropriate training. In addition to the considerations of humane treatment, animals must be handled in ways that minimize health threats to both the animals and handlers. Sick or mistreated animals will not provide useful information in any study where they are used. All personnel who perform animal surgery (which must be performed under aseptic conditions) or any other experimental manipulations must receive training in the performance of these techniques.

ii. WORKER SAFETY

Animal work introduces specific safety concerns to biotechnology facilities. Workers who come into contact with laboratory animals can be at risk for allergies, animal bites and scratches, or exposure to infectious dis-

eases. As shown in Figure 12.13, animal research frequently uses large numbers of animals housed in concentrated areas which expose workers to animal-generated bioaerosols. As a consequence, approximately 20% of workers who frequently contact or work in the vicinity of rodents will develop animal-related allergies. These allergies may take months to develop and can be triggered by direct contact with the animals, or by inhalation of the allergens. It is believed that the main causes of animal-related allergies are proteins in the urine and/or saliva of the animals (rather than the fur or dander as previously thought). Worker exposure to these allergens can be reduced (but not entirely eliminated) through the use of cage-top filters that prevent the escape of airborne particles, see Figure 12.13b.

Most individuals with animal-related allergies will experience only symptoms of rhinitis (hay fever), but a small percentage will progress to animal-related asthma, a severe and potentially debilitating condition. For this reason, workers who develop animal-related allergies generally should request a reassignment of duties to minimize animal contact, because allergies tend to worsen with continuing exposure. If reassignment is not an option, the worker must use full body PPEs, including gloves and respiratory protection. Once the symptoms of an allergy occur, it is best to contact a physician with experience in animal-triggered allergies. The physician can advise you about the dangers of continuing to work with animals and any treatments that may be available.

Bites and scratches are another problem that can occur when handling animals. All animal bites must be reported to the safety authorities at your institution, who are required to record all incidents. Bite and scratch areas should be washed immediately and thoroughly. If the injury is deep or extremely painful, further medical attention should be obtained. The biggest health concerns from animal bites and scratches are localized wound infection and tetanus. The probability of infection can be minimized by proper cleansing and bandaging of the wound. Anyone bitten by a laboratory animal should receive a booster injection of tetanus vaccine unless they have a current vaccination. Always keep a record of any immunizations you have received.

Whenever handling animals, it is important to remember that they are a powerful source of aerosols and potential biological contaminants, as illustrated in the Case Study below. **Zoonotic diseases (zoonoses)** *are those that can be passed between different species.* Luckily, zoonotic infections are rare in laboratories not involved in infectious disease research. Most of the laboratory animals you are likely to encounter, such as rats or mice, have been bred and raised in clean environments, and rodents carry few zoonotic disease agents that can be passed to humans. There are some exceptions, such as hantaviruses, which are of concern when handling wild-caught animals. Laboratory primates,

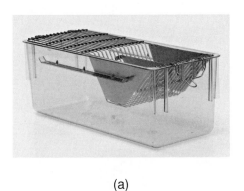
(a)

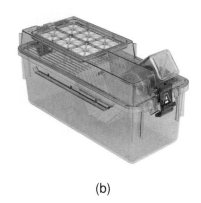
(b)

(c)

Figure 12.13. Rodent Housing in an Animal Facility. a. Cages typically have wire covers to prevent rodents from chewing through the enclosure. The number of animals per cage depends on cage size and experimental design. **b.** Filter tops cover the wire cage top and minimize the spread of bioaerosols and bedding particles. **c.** A rolling rack for multiple animal cages. This arrangement facilitates animal care and can reduce the risk of confusing the cages. (Photos courtesy of Allentown, Inc.)

however, may harbor organisms that can easily infect humans, and any worker who comes into contact with these animals must receive special training that will include information about possible health hazards from zoonotic diseases.

CASE STUDY

Unknown Factors Can Compromise Biological Safety

One common factor in handling cell cultures and animals in the laboratory is the potential for exposure to unknown biological agents. A worker at a cancer research lab in the United States was diagnosed with lymphocytic choriomeningitis, a potentially serious zoonotic infection. When tested for antibodies to the lymphocytic choriomeningitis virus (LCMV), 10% of the approximately 90 facility employees were infected or showed signs of recent infection with the virus. The common factor among the infected employees was handling rodents; in particular, a population of nude mice. Nude mice are an inbred hairless mouse strain that lacks a normal immune system. Because of their inability to fight infection, they must be housed under very clean conditions, making them an unusual source for a zoonotic virus. The problem was soon traced back to a tumor cell line injected into the nude mice to study tumor growth. Unknown to the researchers, the cell line was infected with LCMV. The nude mice were, in essence, acting as viral incubators, shedding LCMV into the bedding and air and exposing any workers who came into proximity to the animals. The infected workers were successfully treated, but the experiment and the infected animals could not be saved.

Special precautions are required whenever laboratory animals are treated with biohazardous materials or infectious agents. In these cases, the worker must consider both animal welfare and biohazard containment.

Only personnel who have specific training in handling pathogenic agents and animals should be allowed in the areas where this work is conducted. In the case of highly pathogenic agents, special animal facilities must be used, with provisions for safe handling of animals and disposal of infectious materials. In these situations, there are defined Animal Biosafety Levels for infected animals, similar to the BSLs described earlier in this chapter. Animal BSLs are described in detail at http://www.cdc.gov/od/ohs/biosfty/bmbl4/bmbl4s4.htm.

All procedures that involve direct handling of biohazardous and infectious materials should be conducted within biological safety cabinets or other physical containment units using appropriate microbiological practices. The generation of bioaerosols and biohazardous waste materials by the animals is a major concern. All bedding, food, excrement, and other materials that contact the infected animals must be treated as biohazardous waste. Animal cages must be disposable or decontaminated before washing. It is absolutely essential that appropriate warning signs be placed on the doors of any containment facilities used for infected animals.

V. HANDLING BIOHAZARDOUS WASTE MATERIALS

Biohazardous waste *consists of biological and biologically contaminated materials that are no longer needed in the laboratory.* These materials include:

- Discarded cultures of bacteria and other cells
- Outdated stocks of organisms
- Human and animal waste, including blood and other body fluids
- Used culture dishes and tubes
- Biologically contaminated sharps

All biohazardous wastes should be placed in labeled, closable, leak-proof containers or bags. These waste

Table 12.24. GUIDELINES FOR SAFE
DECONTAMINATION OF
BIOHAZARDOUS WASTE BY
AUTOCLAVING

- Do not mix chemical and biological wastes for autoclaving.
- All biohazardous waste must be placed in orange/red biohazard bags with a heat-sensitive sterilization indicator.
- Before autoclaving, biohazard waste bags should be kept closed to prevent airborne contamination and odors; while autoclaving, however, the bag must be open to allow the steam to penetrate.
- Add at least 100 mL water to each biohazard bag before autoclaving for effective steam generation.
- Autoclave biohazardous materials for at least 45 minutes at the standard 121°C and 15 PSI for a single bag, and at least 60 minutes for a run with multiple bags.
- After autoclaving, the bag should be closed and disposed of in an opaque regular waste bag.
- All decontamination autoclaves should be tested regularly for sterilization effectiveness (at least annually).

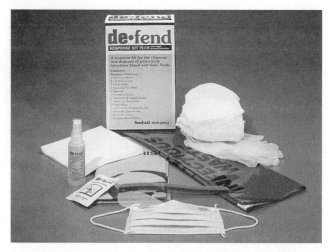

Figure 12.14. Biohazard Emergency Response Kit. These kits typically contain PPE, absorbents with disinfectant, biohazard disposal bags, and instructions for general biohazard cleanup. (Photo courtesy of Fisher Scientific, Pittsburgh, PA.)

containers should usually be placed inside secondary containers in case of accidental punctures. As with chemical wastes, every institution will have specific procedures for sorting and labeling biological wastes.

In most situations, heat sterilization by autoclaving is considered an appropriate decontamination procedure for biohazardous waste. If this method is used, the resulting materials can be discarded as regular laboratory waste (in accordance with local regulations). Table 12.24 provides guidelines for effective sterilization of biological waste using an autoclave.

VI. RESPONSE TO BIOHAZARD SPILLS

Despite precautions, spills of biological materials occur. All laboratories that work with biohazards must have an emergency plan available in case of accidents. This plan should be written and posted at convenient locations for easy access. It is essential that all personnel be familiar with these plans before they are needed. Many of the same precautions that apply to chemical spills are also appropriate for biological spills. Always consider the nature of the hazard before attempting to handle the situation on your own. If a highly pathogenic agent is involved, evacuate the premises and summon help.

In cases where a spill is small, involves a relatively nonpathogenic organism (BSL1 or BSL2), and a biological spill kit is available, the worker can deal with the spill directly. Spill kits are commercially available, or they can be assembled ahead of time in the laboratory, see Figure 12.14. A biological spill kit should include:

- personal protective equipment (e.g., gloves, eye, and face protection)
- absorbent materials (e.g., chemical absorbent or paper towels)
- cleanup tools (e.g., a whisk and dustpan)
- waste container (e.g., biohazard bags)
- disinfectant (e.g., fresh 10% bleach solution)

The bleach solution must be made fresh on at least a weekly basis.

When cleaning up a spill, remember that there are three hazards that must be addressed. The main spill area may be obvious, but there is also a surrounding splash area. This consists of small droplets of liquid that have splashed from the main spill site. It is essential that the cleanup effort does not spread contamination from these splash spots. Perhaps the most significant consideration at the time of the spill is the third element: the generation of aerosols. When the spill is large, or involves a pathogenic organism, it is best to leave the room, close the doors, and wait at least 10 minutes for the aerosols to settle. Keep in mind that every surface where the aerosols settle is also contaminated. Table 12.25 provides a general set of guidelines for cleaning up minor biohazardous spills.

In cases of personal contamination with biohazardous material, it is essential to take immediate steps for decontamination. Remove all contaminated clothing and soak lab coats in a bleach solution to disinfect before washing. Wash any potentially contaminated

Table 12.25.   GENERAL GUIDELINES FOR HANDLING MINOR BIOHAZARD SPILLS

BSL1 Organism

- Be sure you are wearing appropriate PPE, including lab coat, gloves, and eye protection.
- Soak absorbent materials such as towels in an appropriate disinfectant (usually a fresh 10% solution of bleach).
- Use these towels to soak up the spill; do not wipe the spill with dry towels.
- Dispose of the towels as biohazardous waste.
- Clean the spill area with fresh disinfectant and towels.

BSL2 Organism

- Notify all personnel in the area and evacuate for at least 10 minutes to allow aerosols to settle.
- In the meantime, remove any contaminated clothing and decontaminate them.
- Be sure you are wearing appropriate PPE, including lab coat, gloves, and eye protection.
- Soak absorbent material such as towels in an appropriate disinfectant (usually a fresh 10% solution of bleach).
- Cover the main spill area with the towels or other absorbent materials that contain disinfectant.
- Carefully (to avoid creating aerosols) flood the immediately surrounding area with disinfectant.
- Cover all items in the spill area with disinfectant and then remove them from the spill area.
- Remove any broken glassware with forceps and place in a sharps container; never pick up contaminated sharps with your hands.
- Remove the contaminated towels and any other absorbent material and place in a biohazard bag.
- Gently apply additional disinfectant to the spill area and allow at least 20 minutes for decontamination.
- Remove the disinfectant by applying paper towels or other absorbent material, which are then discarded in a biohazard bag.
- Wipe off any residual spilled material and reapply disinfectant for final cleanup.
- Collect all materials in the biohazard bag and seal for autoclaving.
- Reopen area to general use only after spill cleanup and decontamination is complete.

skin areas vigorously for at least 10 minutes. Use an antiseptic such as Betadine if available. In case of eye contamination, use the eye wash station and flush for at least 15 minutes. Your supervisor should be informed immediately of the accident. Depending on the biohazard involved, it may be necessary to seek medical attention and notify proper authorities as designated by your institution's biohazard plan.

PRACTICE PROBLEMS

1. A laboratory scientist drops a glass flask containing pathogenic bacteria. The glass shatters and splatters the scientist with liquid.
   a. What are several precautions the scientist should have taken ahead of time to reduce the risk of this incident?
   b. What should this person do immediately?
2. HEPA filters are designed to remove 99.97% of all particulate matter above the size of 0.3 μm from air. If a cubic meter of air containing 2,880,000 particles of about 0.5 μm passes through the filter, how many particles will remain in the air?
3. What are the similarities and differences between standard microbiological practices and Universal Precautions?

QUESTIONS FOR DISCUSSION

Biological safety issues are of great importance in biotechnology settings. This is reflected in the Skill Standards that have been created for the industry.

1. **Scenario 1 in the Skill Standards for the Bioscience Industry (Education Development Center, Inc., 1995) presents the following topic and question:**

   One part of your laboratory responsibilities is to safely unpack and process biological samples. Demonstrate everything you would do to accomplish this.

   While unpacking samples one morning, you notice that one of the samples is leaking from the container. What should you do?

2. **Scenario 16 in the Skill Standards for the Bioscience Industry (Education Development Center, Inc., 1995) presents the following topic and question:**

   Your job is to assist with cleaning, preparing, sterilizing, and inoculating a bioreactor. Show what tests and procedures you follow to perform these tasks.

   After inoculation, a coworker points out that the bioreactor exit air filter cartridge is not installed. This means that there is no filter between the recombinant cells in the bioreactor and the outside environment. Demonstrate how you would handle this.

UNIT IV

Math in the Biotechnology Laboratory: An Overview

Mathematical descriptions and operations permeate daily life in the laboratory workplace. Math is an important tool for scientists and technicians, just as a hammer is a tool for a carpenter. To be a skillful laboratory biologist you do not need to "like" math any more than a carpenter needs to "like" a hammer. Most carpenters are probably neither happy nor anxious when they encounter a hammer (unless they have recently smashed their finger with one). Students in the sciences, however, sometimes do have to overcome negative feelings about math.* If you are anxious about math, remember that with practice and time people learn to perform the math required in their profession.

The purpose of this unit is to provide a brief review of common math manipulations used in bioscience lab-

oratories and to introduce applications of math relevant to everyday laboratory work. This unit is not intended to replace a math textbook. Rather, these chapters roam through various areas of mathematics, discussing topics that relate to problems commonly encountered in the laboratory. Although these chapters review various basic math concepts, it is assumed that readers can manipulate fractions and decimals, prepare simple graphs, solve an equation with one unknown, and use a scientific calculator.

Readers who are comfortable with the mathematical operations reviewed may still want to work the "Application Problems" in each section. These problems demonstrate the ways in which familiar mathematical tools are applied in the laboratory. Readers who are less comfort-

*Many people believe that they are not good at math and/or have negative feelings about math. Sheila Tobias wrote a book about "math anxiety" (Tobias, Sheila. *Overcoming Math Anxiety*. New York: WW Norton and Co., 1993). Tobias points out that people in Japan and Taiwan believe that hard work leads to good performance in math. In contrast, people in the United States are apt to believe that math ability leads to good performance, that one is either born with this ability or not, and that no amount of hard work can make up for the lack of math ability. Because of these beliefs, Americans may give up on math and try less hard in math classes. In reality, the ability to use math is not a genetic gift, but rather is learned with practice.

able with mathematical calculations will find it beneficial to complete the "Manipulation Problems." Solving the "Manipulation Problems" will increase your comfort and efficiency when performing routine laboratory calculations.

There is often more than one strategy to solve a math problem. In some instances in this unit, we demonstrate two ways to approach a problem. For example, we show how to convert between units using both proportions and a unit canceling strategy. Both strategies, when applied correctly, will provide the correct answer. Individuals may prefer one strategy over another. (Of course, in a classroom situation students are encouraged to use strategies as directed by their instructor.) More experienced scientists and technicians often move fluidly from one problem-solving strategy to another, using whatever approach is most efficient for solving a particular problem. Experienced workers also are likely to check their answers by first using one approach to solve a problem and then testing another approach to see if it gives the same answer.

A few tips for math problem solving include:

1. *Keep track of all information.* Begin with a sheet of paper with plenty of room for each problem. Write down relevant information and record the results of each step.

2. *Use simple sketches, arrows, or other visual aids to help define problems.*

3. *Check that each answer makes sense.*

4. *Keep track of units.* It is tempting to ignore units when performing calculations, but this practice has been the cause of much grief. (We will talk more about units later.)

5. *State the answer clearly; remember to indicate the units.*

Chapter 13 is a brief review of basic mathematical tools used in the biotechnology laboratory, including exponents, scientific notation, logarithms, units of measurement, and equations.

Chapter 14 introduces the use of proportional relationships to solve a variety of practical laboratory calculation problems. Proportions can be used, for example, to convert from one unit of measurement to another or to determine how much solute is required to prepare a solution of a given concentration. The unit canceling method (dimensional analysis), which is commonly taught in chemistry classes, is introduced as another strategy to solve these types of problems.

Chapter 15 introduces graphical methods of data analysis, focusing in particular on linear relationships and also briefly discussing exponential relationships.

Chapter 16 introduces the basic vocabulary, concepts, and mathematical tools of descriptive statistics and how they are used to organize, display, and interpret data in the laboratory. The use of descriptive statistics in a quality-control setting is also introduced.

A BRIEF NOTE ABOUT SIGNIFICANT FIGURES

If I divide 47.0 by 9.0 using my calculator, the answer reads: 5.222222222. I would be unlikely to record my answer in this fashion (with all those 2s). Rather, I would round the value to, perhaps, 5.2. There are conventions that are used to guide decisions about rounding. These conventions are covered under the topic "significant figures."

We will explore significant figures and rounding later when measurements are discussed in detail. This is because the principles of significant figures relate to measurements. For example, suppose a meter stick is used to measure the height of a plant and the value obtained is 64.3 cm. This means that we can be reasonably certain of the height of the plant to the nearest tenth of a centimeter. We do not know the height of the plant to the nearest hundredth place because the meter stick did not provide that much information. When we report the result of this measurement, it is important not to mislead the reader into thinking that the height of the plant is known more "exactly" than it is. For this reason we do not record the plant's height as 64.30, or 64.33 cm. It is also incorrect to "drop" information, for example, by reporting the plant to be 64 cm.

When calculations are performed that involve numbers from measurements, the significant figure conventions provide guidance in how to round the answer. The results of calculations should be rounded in such a way as to indicate the "exactness" with which the original measurements were made. There are many situations in analytical chemistry and in other quantitative applications where it is essential to round values properly after calculations by applying the significant figures conventions. Although significant figures are not the focus of this math unit, some readers may prefer to read the sections on significant figures, pp. 308–311, before completing the problems in this unit in order to better understand rounding conventions.

BIBLIOGRAPHY FOR UNIT IV

General Laboratory Math References

Adams, Dany Spencer. *Lab Math: A Handbook of Measurements, Calculations, and Other Quantitative Skills for Use at the Bench.* New York: Cold Spring Harbor Laboratory Press, 2003.

Campbell, June, and Campbell, Joe. *Laboratory Mathematics: Medical and Biological Applications.* 5th ed. St. Louis: Mosby Co., 1997.

Johnson, Catherine W., Timmons, Daniel L., and Hall, Pamela E. *Essential Laboratory Mathematics.* Clifton Park, NY: Delmar Learning, 2003.

Seidman, Lisa A. *Basic Laboratory Calculations for Biotechnology.* San Francisco: Benjamin Cummings, 2007. (A friendly guide and workbook with multiple solved problems.)

Stephenson, Frank H. *Calculations for Molecular Biology and Biotechnology: A Guide to Mathematics in the Laboratory.* San Diego: Academic Press, 2003. (Includes calculations for some more advanced techniques in molecular biology.)

Zatz, Joel L., and Teixeira, Maria Glaucia. *Pharmaceutical Calculations.* 4th ed. Hoboken, NJ: Wiley-Interscience, 2005.

Basic Statistics References

Ambrose, Harrison W., and Ambrose, Katharine Peckham. *A Handbook of Biological Investigation.* 6th ed. Winston-Salem, NC: Hunter Textbooks, 2002. (This short handbook includes practical statistical information in an easy to understand format.)

Gunter, Bert. "Fundamental Issues in Experimental Design." *Quality Progress* (June 1996): 105–13. (A short, well-written article that addresses some basic principles of statistical thinking.)

Haaland, Perry. *Experimental Design in Biotechnology.* New York and Basel: Marcel Dekker, 1989.

Hampton, Raymond E., and Havel, John E. *Introductory Biological Statistics.* 2nd ed. Long Grove, IL: Waveland Press, 2005.

Kateman G., and Buydens, L. *Quality Control in Analytical Chemistry.* 2nd ed. New York: John Wiley and Sons, 1993.

Kelley, William D., Ratliff, Thomas A. Jr, and Nenadic, Charles. *Basic Statistics for Laboratories: A Primer for Laboratory Workers.* Reinhold, NY: Van Nostrand, 1992.

Miller, J.C., and Miller, J.N. *Statistics for Analytical Chemistry.* 3rd ed. New York: Prentice Hall, 1993. (Contains good explanations of how statistics are applied in the laboratory.)

CHAPTER 13

Basic Math Techniques

I. EXPONENTS AND SCIENTIFIC NOTATION

EXAMPLE APPLICATION: EXPONENTS AND SCIENTIFIC NOTATION

a. The human genome consists of approximately 3×10^9 base pairs of DNA. If the sequence of the human genome base pairs was written down, then it would occupy 200 telephone books of 10^3 pages each. How many base pairs would be on each page?

b. If it costs $0.50 to determine the location (sequence) of a single base pair, how much would it cost to determine the location of all 3×10^9 base pairs?

Answers on p. 222

A. Exponents

i. THE MEANING OF EXPONENTS

We begin this chapter by discussing exponents and scientific notation. Exponents and scientific notation are routinely used in the laboratory and so it is important to be able to manipulate them easily and quickly.

An **exponent** *is used to show that a number is to be multiplied by itself a certain number of times.* For example, 2^4 is read "two raised to the fourth power" and means that 2 is multiplied by itself 4 times:

$$2 \times 2 \times 2 \times 2 = 16$$

Similarly:

$$10^2 \text{ means: } 10 \times 10, \text{ or } 100$$
$$4^5 \text{ means: } 4 \times 4 \times 4 \times 4 \times 4 \text{ or } 1024$$

The number that is multiplied is called the **base** *and the power to which the base is raised is the* **exponent.** In the expression 10^3, the base is 10 and the exponent is 3.

A negative exponent indicates that the reciprocal of the number should be multiplied times itself. For example:

$$10^{-3} = \frac{1}{10} \times \frac{1}{10} \times \frac{1}{10} = \frac{1}{1000} = 0.001$$

Rules that govern the manipulation of exponents in calculations are summarized in Box 13. 1 on p. 219.

EXAMPLE PROBLEM

Perform the operations indicated:

$$\frac{10}{43^2 + 13^3}$$

ANSWER

The denominator involves addition of numbers with exponents. Convert the numbers with exponents to standard notation. Then perform the calculations.

$$\frac{10}{(43)^2 + (13)^3} = \frac{10}{1849 + 2197} = \frac{10}{4046} \approx 0.0025$$

ii. Exponents Where the Base Is 10

In preparation for discussing scientific notation, consider the particular case of exponents where the base is 10. Observe the following rules as illustrated in Figure 13.1:

1. *For numbers greater than 1:*
   - *the exponent represents the number of places after the number (and before the decimal point)*
   - *the exponent is positive*
   - *the larger the positive exponent, the larger the number*

2. *For numbers less than 1:*
   - *the exponent represents the number of places to the right of the decimal point* including *the first nonzero digit*
   - *the exponent is negative*
   - *the larger the negative exponent, the smaller the number*

People commonly use the phrase *orders of magnitude* where **one order of magnitude** *is 10^1*. Using this terminology, 10^2 is said to be two orders of magnitude less than 10^4. Similarly, 10^8 is three orders of magnitude greater than 10^5.

B. Scientific Notation

i. Expressing Numbers in Scientific Notation

Scientific notation *is a tool that uses exponents to simplify handling numbers that are very big or very small.* Consider the number:

0.00000000000000000000000602

$1,000,000 = 10^6$	= one million	(6 places before decimal point)
$100,000 = 10^5$	= one hundred thousand	(5 places before decimal point)
$10,000 = 10^4$	= ten thousand	(4 places before decimal point)
$1,000 = 10^3$	= one thousand	(3 places before decimal point)
$100 = 10^2$	= one hundred	(2 places before decimal point)
$10 = 10^1$	= ten	(1 place before decimal point)
$1 = 10^0$	= one	(no places before decimal point)
$0.1 = 10^{-1}$	= one tenth	(1 place right of the decimal point)
$0.01 = 10^{-2}$	= one hundredth	(2 places right of the decimal point)
$0.001 = 10^{-3}$	= one thousandth	(3 places right of the decimal point)
$0.0001 = 10^{-4}$	= one ten thousandth	(4 places right of the decimal point)
$0.00001 = 10^{-5}$	= one hundred thousandth	(5 places right of the decimal point)
$0.000001 = 10^{-6}$	= one millionth	(6 places right of the decimal point)

Figure 13.1. Using Exponents Where the Base Is 10. (When the decimal point is not written it is assumed to be to the right of the final digit in the number.)

In scientific notation this lengthy number is compactly expressed as:

$$6.02 \times 10^{-23}$$

A value in scientific notation is customarily written as a number between 1 and 10 multiplied by 10 raised to a power. For example:

100 (Standard Notation) $= 10^2 = 1 \times 10^2$ (Scientific Notation)
$$1 \times 10^2 = 1 \times 10 \times 10 = 100$$

300 (Standard Notation) $= 3 \times 10^2$ (Scientific Notation)
$$3 \times 10^2 = 3 \times 10 \times 10 = 300$$

The number of bacterial cells in 1 liter of culture might be 100,000,000,000 (Standard Notation) $=$ 1×10^{11} (Scientific Notation)

A number in scientific notation has two parts. *The first part is sometimes called the* **coefficient.** The second part is 10 raised to some power, the **exponential term.** For example:

	First Part (Coefficient)		Second Part (Exponential Term)
$1000 =$	1	$\times$	10^3
$235 =$	2.35	$\times$	10^2

As shown in Figure 13.1, a negative exponent is used for a number less than 1. Three examples:

$$1 \times 10^{-5} = \frac{1}{10} \times \frac{1}{10} \times \frac{1}{10} \times \frac{1}{10} \times \frac{1}{10}$$
$$= \frac{1}{100,000} = 0.00001$$
$$0.000135 = 1.35 \times 10^{-4}$$

A bacterial cell wall is about 0.00000001 m $= 1 \times 10^{-8}$ m thick

A procedure to convert a number from standard notation to scientific notation is shown in Box 13.2 on p. 220.

EXAMPLE PROBLEM

Convert the number **0.000348** to scientific notation.

ANSWER

Step 1. This number is less than 1. Move the decimal point to the right: **0.000348 → 3.48**

Box 13.1. CALCULATIONS INVOLVING EXPONENTS

1. To multiply two numbers with exponents where the numbers have the same base, add the exponents:

$$a^m \times a^n = a^{m+n}$$

Two examples:

$$5^3 \times 5^6 = 5^9$$
$$10^{-3} \times 10^4 = 10^1$$

To convince yourself that this rule makes sense; consider the following example:

$$2^3 \times 2^2 = 2^5 = (2 \times 2 \times 2) \times (2 \times 2) =$$
$$2 \text{ multiplied 5 times} = 2^5 = 32$$

2. To divide two numbers with exponents where the numbers have the same base, subtract the exponents:

$$\frac{a^m}{a^n} = a^{m-n}$$

Two examples:

$$5^3/5^6 = 5^{3-6} = 5^{-3}$$
$$2^{-3}/2^{-4} = 2^{(-3)-(-4)} = 2^1 = 2$$

You can convince yourself that this rule makes sense by rewriting an example this way:

$$5^3/5^6 = \frac{\cancel{5} \times \cancel{5} \times \cancel{5}}{\cancel{5} \times \cancel{5} \times \cancel{5} \times 5 \times 5 \times 5} = \frac{1}{5 \times 5 \times 5} = \frac{1}{125} = 5^{-3}$$

3. To raise an exponential number to a higher power, multiply the two exponents.

$$(a^m)^n = a^{m \times n}$$

Two examples:

$$(2^3)^2 = 2^6$$
$$(10^3)^{-4} = 10^{-12}$$

To convince yourself that this rule makes sense, examine this example:

$$(2^3)^2 = 2^3 \times 2^3 = (2 \times 2 \times 2) \times (2 \times 2 \times 2) = 2^6$$

4. To multiply or divide numbers with exponents that have different bases, convert the numbers with exponents to their corresponding values without exponents. Then, multiply or divide.

Two examples: Multiply: $3^2 \times 2^4$

$$3^2 = 9 \text{ and } 2^4 = 16,$$
$$\text{so } 9 \times 16 = 144$$

Divide: $4^{-3}/2^3$

$$4^{-3} = \frac{1}{4} \times \frac{1}{4} \times \frac{1}{4} = \frac{1}{64} = 0.015625$$

and $2^3 = 8$

$$\text{so } \frac{0.015625}{8} \approx 0.00195$$

5. To add or subtract numbers with exponents (whether their bases are the same or not), convert the numbers with exponents to their corresponding values without exponents.

For example: $4^3 + 2^3 = 64 + 8 = 72$

6. By definition, any number raised to the 0 power is 1.

For example: $85^0 = 1$

BOX 13.2. *A PROCEDURE TO CONVERT A NUMBER FROM STANDARD NOTATION TO SCIENTIFIC NOTATION*

Step 1. a. If the number in standard notation is greater than 10, then move the decimal point to the left so that there is one nonzero digit to the left of the decimal point. This gives the first part of the notation.
   b. If the number in standard notation is less than 1, then move the decimal point to the right so that there is one nonzero digit to the left of the decimal point. This gives the first part of the notation.
   c. If the number in standard notation is between 1 and 10, then scientific notation is seldom used.

Step 2. Count how many places the decimal was moved in step 1.

Step 3. a. If the decimal was moved to the left, then the number of places it was moved gives the exponent in the second part of the notation.
   b. If the decimal point was moved to the right, then place a − sign in front of the value. This is the exponent for the second part of the notation.

EXAMPLE

Express the number 5467 in scientific notation.

Step 1. This number is greater than 10. Therefore, move the decimal point to the left so that there is only one nonzero digit to the left of the decimal point: = **5467.** *move decimal point 3 places left* → **5.467**

Step 2. The decimal point was moved three places to the left.

Step 3. The exponent for the second part of the notation is therefore 3. This means the number in scientific notation is: 5.467×10^3

Step 2. The decimal point was moved four places to the right.

Step 3. The exponent is −4, so the answer in scientific notation is: 3.48×10^{-4}

So far, we have shown the customary manner of writing numbers in scientific notation, that is, with the coefficient written as a number between 1 and 10. It it is not necessary, however, always to write numbers in scientific notation in this way. For example, the value 205 may be expressed as:

$$205. = 0.205 \times 10^3$$
$$205. = 2.05 \times 10^2$$
$$205. = 20.5 \times 10^1$$
$$205. = 2050 \times 10^{-1}$$
$$205. = 20500 \times 10^{-2}$$

Similarly:

$$1.00 \times 10^4 = 10.0 \times 10^3 = 100. \times 10^2$$
$$3.45 \times 10^{23} = 0.0345 \times 10^{25} = 345 \times 10^{21}$$

There are situations where it is useful to manipulate coefficients and exponents without changing the value of the numbers. This is the case in addition and subtraction of numbers expressed in scientific notation.

EXAMPLE PROBLEM

Fill in the blank so that the numbers on both sides of the = sign are equal. (For example: $2.58 \times 10^{-2} = 25.8 \times 10^{-3}$)

$$0.0055 \times 10^4 = 5500 \times 10^{-}$$

ANSWER

One way to think about this problem:

1. Convert the expression on the left to standard notation; it equals **55.**

2. The number on the right side of the expression, that is, 5500, is larger than the number on the left, that is, 55. The exponent that fills in the blank, therefore, will need to be a negative number (to make 5500 smaller).

3. 5500 times 10^{-2} equals 55. The answer is therefore **−2**.

ii. CALCULATIONS WITH SCIENTIFIC NOTATION

Box 13.3 summarizes methods of performing calculations with scientific notation.

EXAMPLE PROBLEM
$$(3.45 \times 10^{23}) + (4.56 \times 10^{25}) = ?$$

ANSWER

The two numbers in this example would require many zeros if written in standard notation. The strategy of expressing both values in scientific notation with the same exponent, therefore, is preferred over converting both numbers to standard notation.

Step 1. Decide on a common exponent. Suppose we choose 25.

Step 2. Express both numbers in a form with the same exponent.
$$3.45 \times 10^{23} = 0.0345 \times 10^{25}$$

Step 3. Perform the addition.

$$\begin{array}{r} 0.0345 \times 10^{25} \\ + \ 4.56 \ \ \times 10^{25} \\ \hline 4.5945 \times 10^{25} \end{array}$$

(Note: If the coefficient is rounded according to significant figure conventions, then the answer is 4.59×10^{25})

Box 13.3.   *CALCULATIONS INVOLVING NUMBERS IN SCIENTIFIC NOTATION*

1. To multiply numbers in scientific notation use two steps:

   Step 1. Multiply the coefficients together.

   Step 2. Add the exponents to which 10 is raised.

   For example:

   $$(2.34 \times 10^2)(3.50 \times 10^3) = \underset{(multiply\ the\ coefficients)}{(2.34 \times 3.50)} \times \underset{(add\ the\ exponents)}{10^{2+3}} = 8.19 \times 10^5$$

2. To divide numbers in scientific notation, use two steps:

   Step 1. Divide the coefficients.

   Step 2. Subtract the exponents to which 10 is raised.

   For example:

   $$(4.5 \times 10^5)/(2.1 \times 10^3) = \underset{(divide\ the\ coefficients)}{(4.5/2.1)} \times \underset{(subtract\ the\ exponents)}{10^{5-3}} \approx 2.1 \times 10^2$$

3. To add or subtract numbers in scientific notation:

   a. If the numbers being added or subtracted all have 10 raised to the same exponent, then the numbers can be simply added or subtracted as shown in these examples:

   $$(3.0 \times 10^4) + (2.5 \times 10^4) = ? \qquad (7.56 \times 10^{21}) - (6.53 \times 10^{21}) = ?$$

   $$\begin{array}{r} 3.0 \times 10^4 \\ + \underline{2.5 \times 10^4} \\ 5.5 \times 10^4 \end{array} \qquad\qquad \begin{array}{r} 7.56 \times 10^{21} \\ - \underline{6.53 \times 10^{21}} \\ 1.03 \times 10^{21} \end{array}$$

   b. If the numbers being added or subtracted do not all have 10 raised to the same exponent, then there are two strategies for adding and subtracting numbers.

   STRATEGY 1   Convert the numbers to standard notation and then do the addition or subtraction:

   For example: $(2.05 \times 10^2) - (9.05 \times 10^{-1}) = ?$

   Convert both numbers to standard notation:

   $$2.05 \times 10^2 = 205$$
   $$9.05 \times 10^{-1} = 0.905$$

   Perform the calculation:

   $$\begin{array}{r} 205 \\ - \underline{\quad 0.905} \\ 204.095 \end{array}$$

   (Note: If this value is rounded according to significant figure conventions, then the answer is 204.)

   STRATEGY 2   Rewrite the values so they all have 10 raised to the same power:

   For example: $(2.05 \times 10^2) - (9.05 \times 10^{-1}) = ?$

   To convert both numbers to a form such that they both have 10 raised to the same power:

   Step 1. Decide what the common exponent will be. It should be either 2 or −1.
   Suppose we choose 2.

   Step 2. Convert 9.05×10^{-1} to a number in scientific notation with the exponent of 2:

   $$9.05 \times 10^{-1} = 0.00905 \times 10^2.$$

   Step 3. Perform the subtraction:

   $$\begin{array}{r} 2.05 \quad\ \times 10^2 \\ - \underline{0.00905 \times 10^2} \\ 2.04095 \times 10^2 \end{array}$$

   (Note: If the coefficient is rounded according to significant figure conventions, then the answer is 2.04×10^2.)

A calculator will hold a limited number of places and will not accept very large or very small numbers if you try to key in all the zeros. A scientific calculator, however, works easily with large and small numbers expressed in scientific notation. On my calculator, to key in the number 1×10^3, I push the following keys:

1

exp

3

Note that I do not key in the number 10; the base in scientific notation. "exp" tells my calculator that the 10 is present. To key in the number 3×10^{-4} on my calculator, I press:

3

exp

4

+/−

The +/− key tells the calculator I want a negative exponent. "EE" is sometimes used to indicate that 10 is being raised to a certain power. Consult your instruction manual to see how to key in a number in scientific notation.

EXAMPLE APPLICATION ANSWER (from p. 217)

a. The total number of pages needed to record the genome would be

$$200 \times 10^3 = 2 \times 10^5 = 200,000$$

Then, the number of base pairs on each page would be:

$$\frac{3 \times 10^9}{2 \times 10^5} = 1.5 \times 10^4 = 15,000$$

b. $3 \times 10^9 \times \$0.50 = 1.5 \times 10^9 = 1.5$ billion dollars! (There has been much effort in reducing the cost of sequencing DNA.)

MANIPULATION PRACTICE PROBLEMS: EXPONENTS AND SCIENTIFIC NOTATION

1. Give the whole number or fraction that corresponds to these exponential expressions.

a. 2^2 **b.** 3^3 **c.** 2^{-2} **d.** 3^{-3} **e.** 10^2

f. 10^4 **g.** 10^{-2} **h.** 10^{-4} **i.** 5^0

2. Perform the operations indicated:

a. $2^2 \times 3^3$ **b.** $(14^3)(3^6)$ **c.** $5^5 - 2^3$

d. $5^7/8^4$ **e.** $(6^{-2})(3^2)$ **f.** $(-0.4)^3 + 9.6^2$

g. $a^2 \times a^3$ **h.** c^3/c^{-6} **i.** $(3^4)^2$

j. $(c^{-3})^{-5}$ **k.** $\dfrac{13}{43^2 + 13^3}$ **l.** $10^2/10^3$

3. Underline the larger number in each pair. Underline both if their values are equal.

a. 5×10^{-3} cm, 500×10^{-1} cm

b. 300×10^{-3} µL, 3000×10^{-2} µL

c. 3.200×10^{-6} m, 3200×10^{-4} m

d. 0.001×10^1 cm, 1×10^{-3} cm

e. 0.008×10^{-3} L, 0.0008×10^{-4} L

4. Convert the following numbers to scientific notation.

a. 54.0 **b.** 4567 **c.** 0.345000 **d.** 10,000,000

e. 0.009078 **f.** 540 **g.** 0.003040 **h.** 200,567,987

5. Convert the following numbers to standard notation.

a. 12.3×10^3 **b.** 4.56×10^4 **c.** 4.456×10^{-5}

d. 2.300×10^{-3} **e.** 0.56×10^6 **f.** 0.45×10^{-2}

6. Perform the following calculations.

a. $\dfrac{(4.725 \times 10^8)(0.0200)}{(3700)(0.770)}$

b. $\dfrac{(1.93 \times 10^3)(4.22 \times 10^{-2})}{(8.8 \times 10^8)(6.0 \times 10^{-6})}$

c. $(4.5 \times 10^3) + (2.7 \times 10^{-2})$

d. $(35.6 \times 10^4) - (54.6 \times 10^6)$

e. $(5.4 \times 10^{24}) + (3.4 \times 10^{26})$

f. $(5.7 \times 10^{-3}) - (3.4 \times 10^{-6})$

7. Fill in the blanks so that the numbers on both sides of the = sign are equal. For example: $2.58 \times 10^{-2} = 25.8 \times 10^{-3}$

a. $0.0050 \times 10^{-4} = 0.050 \times 10^{\text{---}} = 0.50 \times 10^{\text{---}} = 5.0 \times 10^{\text{---}}$

b. $15.0 \times 10^{-3} = \underline{\quad} \times 10^{-2} = \underline{\quad} \times 10^{-1} = \underline{\quad} \times 10^1$

c. $5.45 \times 10^{-3} = 54.5 \times 10^{\text{---}}$

d. $100.00 \times 10^1 = 1.0000 \times 10^{\text{---}}$

e. $6.78 \times 10^2 = 0.678 \times 10^{\text{---}}$

f. $54.6 \times 10^2 = \underline{\quad} \times 10^6$

g. $45.6 \times 10^8 = \underline{\quad} \times 10^6$

h. $4.5 \times 10^{-3} = \underline{\quad} \times 10^{-5}$

i. $356.98 \times 10^{-3} = \underline{\quad} \times 10^1$

j. $0.0098 \times 10^{-2} = 0.98 \times 10^{\text{---}}$

II. LOGARITHMS

EXAMPLE APPLICATION: LOGARITHMS

The concentration of hydrochloric acid, HCl, secreted by the stomach after a meal is about 1.2×10^{-3} M. What is the pH of stomach acid?

Answer on p. 225

1,000,000	= 10^6	= one million	$\log 10^6 = 6$
100,000	= 10^5	= one hundred thousand	$\log 10^5 = 5$
10,000	= 10^4	= ten thousand	$\log 10^4 = 4$
1,000	= 10^3	= one thousand	$\log 10^3 = 3$
100	= 10^2	= one hundred	$\log 10^2 = 2$
10	= 10^1	= ten	$\log 10^1 = 1$
1	= 10^0	= one	$\log 10^0 = 0$
0.1	= 10^{-1}	= one tenth	$\log 10^{-1} = -1$
0.01	= 10^{-2}	= one hundredth	$\log 10^{-2} = -2$
0.001	= 10^{-3}	= one thousandth	$\log 10^{-3} = -3$
0.0001	= 10^{-4}	= one ten thousandth	$\log 10^{-4} = -4$
0.00001	= 10^{-5}	= one hundred thousandth	$\log 10^{-5} = -5$
0.000001	= 10^{-6}	= one millionth	$\log 10^{-6} = -6$

Figure 13.2. Common Logarithms of Powers of Ten.

A. Common Logarithms

Common logarithms (also called **logs** or **log₁₀**) are closely related to scientific notation. The **common log** *of a number is the power to which 10 must be raised to give that number*, Figure 13.2.

$$100 = 10^2$$
The log of 100 is 2 because 10 raised to the second power is 100
$$\log 10^2 = 2$$

$$1,000,000 = 10^6$$
The log of 1,000,000 is 6 because 10 raised to the sixth power is 1,000,000
$$\log 10^6 = 6$$

$$0.001 = 10^{-3}$$
The log of 0.001 is –3 because 10 raised to the –3 power is 0.001
$$\log 10^{-3} = -3$$

$$1 = 10^0$$
The log of 1 is 0 because 10 raised to the zero power is 1 (by definition)
$$\log 10^0 = 0$$

So, by definition, $\log(10^X) = X$

To what power must 10 be raised to equal the number 5? This is not obvious. We can reason that the number 5 is between 1 and 10 so the log of 5 must be greater than the log of 1 (which is 0) and less than the log of 10 (which is 1). The same is true for the numbers 2, 3, 4, 6, 7, 8, and 9; their logs are decimals between 0 and 1. There is no intuitive way to know the exact log of 5, although we know it is between 0 and 1. Rather, it is necessary to look up the log in a table of logarithms, or to find it using a scientific calculator. The log of 5 is approximately 0.699, which means that $10^{0.699} \approx 5$. Thus:

$$5 = 10^X$$
Take the log of both sides:
$$\log 5 = X$$
From a calculator we find that:

$$X \approx 0.699$$
which means that:
$$5 \approx 10^{0.699}$$

The same logic can be applied to numbers between 10 and 100. For example, what is the log of 48? The log of 10 is 1; the log of 100 is 2. Therefore, the log of the number 48 must fall between 1 and 2. It is, in fact, approximately 1.681.

$$\log 48 = X$$
From a calculator we find that:
$$X \approx 1.681$$
which means that:
$$48 \approx 10^{1.681}$$

All numbers between 10 and 100 have a log between 1 and 2. We can continue and apply the same logic to all positive numbers. For example, the log of 4,987,000 must fall between 6 and 7 because 4,987,000 is between 1 million (10^6) and 10 million (10^7).

$$\log 4,987,000 = X$$
$$X \approx 6.698$$
which means that:
$$10^{6.698} \approx 4,987,000$$

What about numbers between 0 and 1? Observe in Figure 13.2 that the log of 1 is 0 and that the logs of numbers between 0 and 1 are negative numbers. Therefore, all numbers between 0 and 1 have a negative log. For example:

The log of 0.130 is approximately –0.886
which means that:
$$10^{-0.886} \approx 0.130$$

The log of 0.00891 is approximately –2.05
which means that:
$$10^{-2.05} \approx 0.00891$$

Note that 0 and numbers less than 0 do not have logs at all because there is no exponent we can use that will make 10 to that number equal 0 or a negative. Formally, we say that logarithms of 0 or negative numbers are undefined.

B. Antilogarithms

An **antilogarithm (antilog)** *is the number corresponding to a given logarithm*. For example:

The log of 100 is 2; therefore, 100 is the antilog of 2
The log of 5 is approximately 0.699; therefore, 5 is approximately the antilog of 0.699

What is the antilog of 3? Remember that logs are exponents; therefore, to find the antilog of a number, *n*, use *n* as an exponent on a base of 10. If *n* = 3, then antilog 3 = 10^3 = 1000. 1000 is the antilog of 3.

What is the antilog of 2.5?

Antilog $2.5 = 10^{2.5}$

While it is not obvious what $10^{2.5}$ equals, we can reason that the antilog of 2.5 must be a number between 100 and 1000. (If this is not clear, think about the "2" in 2.5.) A scientific calculator is the simplest way to find antilogs. On many calculators the "antilog key" is the second function of the "log key." (Consult your calculator directions to determine how to perform this function.) Using my calculator I can find the antilog of 2.5 in two ways:

2.5

2nd

log

The answer **316.227766** appears.
Alternatively I can press:

10

y^x

2.5

=

The answer **316.227766** appears.
Your calculator might work in a slightly different order; try it out.

EXAMPLE PROBLEM

What is the antilog of 3.58?

ANSWER

We can reason that the answer is between 1000 and 10,000 because of the 3.

Antilog $(3.58) = 10^{3.58} \approx 3801.89$

EXAMPLE PROBLEM

What is the antilog of –0.780?

ANSWER

The antilog of a negative number must be a value between 0 and 1.

Antilog $(-0.780) = 10^{-0.780} \approx 0.166$

C. Natural Logarithms

Common logarithms have a base of 10. *There are also logarithms whose base is ≈ 2.7183, called* **e**. *When e is the base, the log is called a* **natural log**, *abbreviated* **ln.** There are tables of natural logs and there are keys on scientific calculators to find the natural log. Note that the terms *log* and *ln* are not synonyms and should never be used interchangeably. Natural logs are less commonly used in biotechnology laboratories than logs with a base of 10.

D. An Application of Logarithms: pH

The proper function of aqueous biological systems depends on the medium having the correct concentration of hydrogen ions. pH is a convenient means to express the concentration of hydrogen ions in a solution.

The concentration of H^+ ions in aqueous solutions normally varies between 1.0 M and 0.00000000000001 M (where M is a unit of concentration). The number 0.00000000000001 can also be written as 1×10^{-14}. Søren Sørenson devised a scale of pH units in which the wide range of hydrogen ion concentrations can be written as a number between 0 and 14. If the hydrogen ion concentration in a solution is 1×10^{-14} M, then using Sørenson's method we say its pH is 14. If the hydrogen ion concentration of a solution is 0.0000001 M, or 1×10^{-7} M, then we say its pH is 7. If the hydrogen ion concentration of a solution is 1 M, then we say its pH is 0. Thus, to find the pH of a solution:

Step 1. Take the log of the hydrogen ion concentration (expressed as molarity).

Step 2. Take the negative of the log.

For example:

What is the pH of a solution with a H^+ concentration of 0.000234 M?

Step 1. The log of 0.000234 is approximately -3.63.

Step 2. The negative of -3.63 is 3.63. So, 3.63 is the pH of this solution.

We can express the relationship between pH and hydrogen ion concentration more formally using the following expression:

$$pH = -\log [H^+]$$

The symbol [ ] means "concentration." In words, this expression means "pH is equal to the negative log of the hydrogen ion concentration."

If we know the pH and want to find the $[H^+]$ in a solution, then we apply antilogs. To do this:

Step 1. Place a negative sign in front of the pH value.

Step 2. Find the antilog of the resulting number.

For example:

What is the concentration of hydrogen ions in a solution whose pH is 5.60?

Step 1. Place a $-$ sign in front of the pH: -5.60.

Step 2. The antilog of -5.60 is $\approx 2.51 \times 10^{-6}$. The concentration of hydrogen ions in this solution, therefore, is 2.51×10^{-6} M.

The relationship between hydrogen ion concentration and pH is expressed more formally using the following expression:

$$[H^+] = \text{antilog } (-pH)$$

EXAMPLE APPLICATION ANSWER (from p. 222)

The HCl dissociates to release 1.2×10^{-3} M of hydrogen ions. The log of 1.2×10^{-3} is ≈ -2.9. The negative of this log value is 2.9. So the pH of the stomach after a meal is about 2.9.

MANIPULATION PRACTICE PROBLEMS: LOGARITHMS

1. Answer the following without using a calculator:
   a. Is the log of 445 closer to 2 or 4?
   b. Is the log of 1876 closer to 3, 9, or 10?

2. Give the common log of the following numbers.
   a. 100   b. 10,000   c. 1,000,000
   d. 0.0001   e. 0.001

3. The log of 567 must lie between 2.0 and 3.0. For each of the following numbers, state the values the log must lie between.
   a. 7   b. 65.9   c. 89.0   d. 0.45   e. 0.0078

4. Use a scientific calculator to find the log of the following numbers.
   a. 1.50×10^4 b. 345   c. 0.0098
   d. 2.98×10^{-5} e. 1209   f. 0.345

5. Use a scientific calculator to find the antilog of the following numbers.
   a. 4.8990   b. 3.9900   c. −0.5600   d. −0.0089
   e. 9.8999   f. 1.0000   g. 8.9000

6. Use a scientific calculator to convert each of the following [H^+] to a pH value.
   a. [0.45 M]   b. [0.045 M]
   c. [0.0045 M]   d. [0.00000032 M]

7. Use a scientific calculator to determine the hydrogen ion concentration of the solutions with the following pH values.
   a. 4.56   b. 5.67   c. 7.00   d. 1.09   e. 10.1

III. UNITS OF MEASUREMENT

EXAMPLE APPLICATION: UNITS OF MEASUREMENT

The bacterial toxin botulinum is extremely toxic and acts by paralyzing muscles. Exposure to a few micrograms of toxin causes death. The toxin, however, can be used to treat patients with a rare disorder, blepharospasm. These patients are blind because their eyes are squeezed shut. Minute doses of the botulinum toxin allows patients to open their eyes. If a patient is administered 100 pg of toxin, will the patient be able to see or will the patient die?

Answer on p. 226

Scientists and technicians often measure things such as the length of insects, the concentration of pollutants in an air sample, or the amount of air in a patient's lungs. **Measurement** *is basically a process of counting.* For example, consider an insect that is 1.55 cm in length. Length is the property that is measured, 1.55 is the **value** of the measurement, and centimeters (cm) are the **units** of measurement. Another way to write this measurement is: 1.55×1 cm. This latter statement tells us that the length of the insect was determined by comparison with a standard of 1 cm in length and was found to be 1.55 times as long as the standard. Measuring length is therefore comparable to counting how many times the standard must be placed end-to-end to equal the length of the object. Another example is the characteristic of time that can be measured by counting how many times the sun "rises" and "sets." The standard in this case is the apparent motion of the sun and the units are days.

All measurements require the selection of a **standard** and then counting the number of times the standard is contained in the material to be measured. The terms **standard** and **unit** are related, but they can be distinguished from one another. A **unit of measure** *is a precisely defined amount of a property, such as length or mass. A **standard** *is a physical embodiment of a unit.* Centimeters and millimeters are examples of units of measure; a ruler is an example of a standard. Because units are not physical entities they are unaffected by environmental conditions such as temperature or humidity. In contrast, standards are affected by the environment. A strip of metal used to measure meters will differ in length with temperature changes and will therefore only be correct at a particular temperature and in a particular environment.

*A group of units together is a **measurement system.** The measurement system common in the United States, which includes miles, pounds, gallons, inches, and feet, is called* **The United States Customary System (USCS).** In most laboratories and in much of the world, the **metric system,** *with units including meters, grams, and liters, is used,* see Figure 13.3 on p. 226.

The units in the U.S. system are not related to one another in a systematic fashion. For example, for length, 12 in = 1 ft; 3 ft = 1 yd. There is no pattern to these relationships. In contrast, in the metric system, there is only one basic unit to measure length, the meter. The unit of a meter is modified systematically by the addition of prefixes. For example, the prefix **centi** *means 1/100,* so a **centimeter** *is a meter/100.* **Kilo** *means 1000,* so a **kilometer** *is a meter $\times$ 1000.* The same prefixes can be used to modify other basic units, such as grams or liters. The prefixes that modify a basic unit always represent an exponent of 10, see Table 13.1 on p. 226.

The most important basic metric units for biologists are **meters (m),** *for length;* **grams (g),** *for weight;* and **liters (L),** *for volume.* Table 13.2 (p. 227) shows how prefixes are used to modify the basic units. Table 13.3 on p. 227

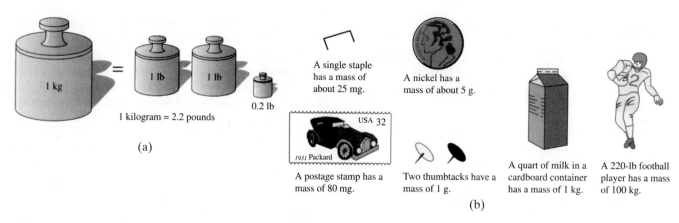

A single staple has a mass of about 25 mg.

A nickel has a mass of about 5 g.

1 kilogram = 2.2 pounds

(a)

A postage stamp has a mass of 80 mg.

Two thumbtacks have a mass of 1 g.

A quart of milk in a cardboard container has a mass of 1 kg.

A 220-lb football player has a mass of 100 kg.

(b)

Figure 13.3.   Units of Measurement. a. A comparison of the metric system kilogram and the USCS pound. **b.** The weight of everyday objects in the metric system.

shows some commonly used conversion factors that relate the USCS to the metric system. In the next chapter, conversions between USCS and the metric system are discussed.

It is always correct to convert between units that measure the same basic property, such as length. Length can be expressed in many units, including miles, meters, centimeters, and nanometers, and these units may be converted from one to the other. In general, it is not correct to convert units from one basic property to another. For example, time is a basic unit that is measured in seconds, minutes, and so on. It is unreasonable to try to convert a number of seconds to centimeters because these units measure different characteristics.

There are a few terms used by biologists that are not part of the "regular" metric system. A microliter is 1/1,000,000 liters and is normally abbreviated μL, but biologists sometimes refer to 1 μL as 1 "lambda" (λ). 10^{-10} meters may be called an "angstrom" (Å). A micrometer is often called a "micron," although micrometer is the preferred term. The terms "cc" and "cm³" (in reference to volume) are both the same as an mL.

The **SI measurement system** *(Système International d'Unités) is an updated version of the metric system that is intended to standardize measurement in all countries.* The SI system was agreed upon in 1960 by a number of international organizations at the Conference Generale des Poids et Measures. The International Organization for Standardization (ISO) has also endorsed the SI system.

There are seven **basic properties** in the SI system: *length, mass, time, electrical current, thermodynamic temperature, luminous intensity,* and *amount of substance.* The units for these basic properties are shown in Table 13.4. To stay strictly within the SI system, the only units that can be used are those in Table 13.4 or those that are a combination of two or more units shown in this table. For example, the SI unit for **volume,** *the amount of space a substance occupies, is: length × length × length in meters, or* **meters³.** Units of liters are not part of the SI system, although liters remain the conventional metric unit for volume measurements in biology laboratories (one liter = 1 dm³).

Table 13.1.   PREFIXES IN THE METRIC SYSTEM

Decimal	Prefix	Symbol	Power of 10
1,000,000,000,000,000,000	exa-	E	10^{18}
1,000,000,000,000,000	peta-	P	10^{15}
1,000,000,000,000	tera-	T	10^{12}
1,000,000,000	giga-	G	10^{9}
1,000,000	mega-	M	10^{6}
1,000	kilo-	K	10^{3}
100	hecto-	h	10^{2}
10	deca-	da	10^{1}
1	Basic unit, no prefix		10^{0}
0.1	deci-	d	10^{-1}
0.01	centi-	c	10^{-2}
0.001	milli-	m	10^{-3}
0.000001	micro-	μ	10^{-6}
0.000000001	nano-	n	10^{-9}
0.000000000001	pico-	p	10^{-12}
0.000000000000001	femto-	f	10^{-15}
0.000000000000000001	atto-	a	10^{-18}

(The prefixes used commonly in biology laboratories are underlined.)

EXAMPLE APPLICATION ANSWER (from p. 225)

A pg = 1×10^{-12} g and
100 pg = 1×10^{-10} g.

A few micrograms would be about 3×10^{-6} g; therefore, 100 picograms is thousands of times less than a few micrograms and this dose of botulinum toxin is likely to be safe.

(100 pg = 0.0001 μg)

Table 13.2.   COMMON METRIC UNITS IN BIOLOGY

Multiple of Ten	Mass	Abbreviation	Volume	Abbreviation	Length	Abbreviation
Basic unit	gram	g	liter	L	meter	m
$\times 10^3$	Kilogram	Kg	Kiloliter	KL	Kilometer	Km
$\times 10^{-1}$	Decigram	dg	Deciliter	dL	Decimeter	dm
$\times 10^{-2}$	Centigram	cg	Centiliter	cL	Centimeter	cm
$\times 10^{-3}$	Milligram	mg	Milliliter	mL	Millimeter	mm
$\times 10^{-6}$	Microgram	μg	Microliter (formerly λ*)	μL	Micrometer (formerly, micron*)	μm
$\times 10^{-9}$	Nanogram	ng	Nanoliter	nL	Nanometer	nm
$\times 10^{-10}$					Angstrom*	Å
$\times 10^{-12}$	Picogram	pg	Picoliter	pL	Picometer	pm

*Older term, not part of SI system
Note: There is inconsistency in capitalization. For example, you may see "Km" or "km" and "mL" or "ml."

Table 13.3.   COMMONLY USED CONVERSION FACTORS FOR UNITS OF MEASUREMENT

Length

1 mm = 0.001 m = 0.039 in
1 cm = 0.01 m = 0.3937 in = 0.0328 ft
1 m = 39.37 in = 3.281 ft = 1.094 yd
1 km = 1000 m = 0.6214 mi

1 in = 2.540 cm
1 ft = 12 in = 0.305 m
1 mi = 5280 ft = 1.609 km

Mass

1 g = 0.0353 oz = 0.0022 lb
1 kg = 1000 g = 35.27 oz = 2.205 lb

1 oz = 28.35 g
1 lb = 16 oz = 453.6 g
1 tn = 2000 lb

Volume

1 mL = 0.001 L = 0.034 fl oz
1 L = 2.113 pt = 1.057 qt = 0.2642 gal

1 fl oz = 0.0313 qt = 0.0296 L
1 pt = 0.4732 L
1 qt = 2 pt = 0.9463 L
1 gal = 8 pt = 3.785 L

Temperature

°C = (°F − 32) 0.556
K = C + 273

°F = (1.8 × °C) + 32
°R = °F + 460   (R is Rankine)

Table 13.4.   THE SI SYSTEM

Basic Property Measured	Unit of Measurement	Abbreviation
Length	meter	m
Mass	gram	g
Time	second	s
Electrical current	ampere	A
Temperature	kelvin	K
Luminous intensity	candela	cd
Amount of substance	mole	mol

MANIPULATION PRACTICE
PROBLEMS: MEASUREMENTS

1. For each pair, underline the larger value. If two values are the same, underline both.

   a. 1 μm, 1 nm
   b. 1 cm, 1 mm
   c. 10 cm, 1000 mm
   d. 10,000 μm, 10 m
   e. 1000 g, 1 kg
   f. 100 nm, 1000 μm
   g. 100 μg, 1 mg
   h. 1 nm, 10 pm
   i. 10 nm, 1 Å
   j. 100 mg, 1 g
   k. 1000 μL, 1 mL
   l. 100,000 μm, 1 m
   m. 1 L, 1000 μL
   n. 1 m, 500 cm
   o. 1 pL, 0.001 nL
   p. 10 nm, 0.1 μm
   q. 100 cm, 0.1 m
   r. 0.0001 m, 10 μm

2. If a sample weighs 100 g, how much does it weigh in kg?

3. What is the approximate volume of a soda can in mL?

IV. INTRODUCTION TO THE USE OF EQUATIONS TO DESCRIBE A RELATIONSHIP

A. Equations

Much of scientific inquiry is about determining the relationship between two or more entities. For example, scientists might study how wolves affect deer populations, how size affects the rate of migration of DNA fragments in electrophoresis, or the relationship between resistance and current in an electrical circuit. Math provides various tools for describing such relationships, including equations, graphs, and statistics.

Equations *are a way to describe relationships using mathematical symbols.* An equation is about a relationship in which two or more things are equal. Letters in a scientific equation represent the items involved in the relationship. Scientific equations may seem intimidating if they use unfamiliar symbols. However, even complicated equations with unfamiliar symbols can tell you about relationships in a system. There are many equations sprinkled throughout this book, all of which convey information about relationships.

For example, consider the equation $V = I\ R$. This equation (to be discussed in Chapter 18) describes the relationship between voltage (V), current (I), and resistance (R) in an electrical circuit. Even if you have little understanding of electricity, you can learn from this equation that:

1. *Voltage, current, and resistance in an electrical circuit are related to each other in a specific way.*

2. *If resistance increases, either the voltage must also increase, or the current must decrease. Similarly, if current increases, then either voltage must increase or resistance must decrease.*

3. *If voltage goes down, then current and/or resistance must also go down.*

For example:

Suppose that initially the voltage = 12, resistance = 4, and current = 3.

(Voltage, current and resistance have units, but we will momentarily ignore them to simplify the example.)

$$V = I\ R$$
$$12 = 3(4)$$

What happens if the resistance increases to 5? Perhaps voltage increases, and becomes 15. Or, perhaps current decreases and becomes 2.4:

$$V = I\ R \qquad\qquad V = I\ R$$
$$15 = 3\ (5) \qquad\qquad 12 = (2.4)(5)$$

In this case, the current did not change, so voltage must increase.

In this case, voltage remained constant.

You will encounter the $V = I\ R$ equation when performing electrophoresis, a technique that uses electricity to separate various proteins and nucleic acids from one another. During electrophoresis the resistance of the system increases as the separation of molecules occurs. The current, therefore, goes down—unless the operator increases the voltage.

Consider another equation:

$$°F = 1.8\ (\text{temperature in }°C) + 32$$

where °F is degrees in the Fahrenheit scale

°C is degrees in the Celsius scale

Celsius and Fahrenheit are scales used to measure temperature. The size of a degree is different in the Celsius and Fahrenheit scales.

1. The numbers 1.8 and 32 in this equation are called **constants** *because they are always present in the equation and always have the same value.* The constant 1.8 is needed in the equation because a Celsius degree is 1.8 times larger than a Fahrenheit degree. The number 32 is necessary because the two scales put the zero point in a different place. Celsius places zero degrees at the temperature of frozen water. Zero degrees Fahrenheit is the lowest temperature that can be obtained by mixing salt and ice.

2. °F and °C are called **variables** *because their value can vary.* (In the previous example, V, I, and R are variables.)

3. Given either of the variables in the equation, it is possible to calculate the other variable. Thus, if we know the temperature in either Celsius or Fahrenheit, the temperature can be converted to the other scale.

B. Units and Mathematical Operations

Units can be manipulated in mathematical operations. Like numbers, units can be multiplied by themselves and can have exponents. For example, the area of a rectangle area is defined as "length times width." In equation form this is:

$$A = (L) \times (W)$$

where A is area L is length and W is width

For example, if the length of a room is 15 feet and its width is 10 feet, then its area is:

$$(L) \times (W) = A$$
$$10 \text{ ft} \times 15 \text{ ft} = 150 \text{ ft}^2$$

The rules for the manipulation of exponents, as shown in Box 13.1, apply to units with exponents. Exponents are added for multiplication and subtracted in division.

For example:

$$\frac{40 \text{ cm}^3}{20 \text{ cm}^2} = \frac{(40) \text{ cm}^{(3-2)}}{(20)} = 2 \text{ cm}^1$$

$$(2.1 \text{ mL})(3.4 \text{ mL}) =$$

$$(2.1 \times 3.4) \times \text{ mL}^{1+1} = 7.1 \text{ mL}^2$$

When working with equations, units can cancel and can be multiplied and divided. For example:

$$2.0 \frac{\text{cm}}{\text{mL}} \times 13 \frac{\text{mg}}{\text{mL}} = 26 \frac{\text{cm(mg)}}{(\text{mL})^2}$$

$$2.00 \frac{\text{mg}}{\text{mL}} \times 13.0 \text{ mL} \times 4.00 = 104 \text{ mg}$$

Another example, which comes from a topic in statistics, is:

$$m = \left[\frac{(4716.20)(\text{mg})(\text{g}) - \left(\dfrac{170 \text{ mg}}{7}\right)(161.03 \text{ g})}{(5750 \text{ mg}^2) - \left(\dfrac{170^2}{7}\right)(\text{mg})^2} \right]$$

At this point, the relationship expressed by this equation is not important, but watch the units as this equation is solved algebraically:

$$m = \frac{(4716.20)(\text{mg})(\text{g}) - 3910.7286(\text{mg})(\text{g})}{5750 \text{ mg}^2 - 4128.57 \text{ mg}^2}$$

$$\approx \frac{805.4714(\text{mg})(\text{g})}{1621.43 \text{ mg}^2} \approx \frac{0.497 \text{ g}}{\text{mg}}$$

EXAMPLE PROBLEM

A box has a length of 3.45 cm, a width of 2.98 cm and a height of 3.00 cm. What is its volume?

ANSWER

$$V = (L)(W)(H)$$

$$V = 3.45 \text{ cm} \times 2.98 \text{ cm} \times 3.00 \text{ cm} =$$
$$= (3.45 \times 2.98 \times 3.00)(\text{cm} \times \text{cm} \times \text{cm}) \approx 30.8 \text{ cm}^3$$

EXAMPLE PROBLEM

Perform the following additions:

a. 23 cm + 56 cm = ?

b. 2 cm + 4 s = ?

ANSWER

a. The correct answer is **79** cm—not just 79

b. The correct answer is: **2** cm + **4** s

This is because s and cm are different units and cannot be added together.

MANIPULATION PRACTICE PROBLEMS: EQUATIONS

---

1. Explain in words what the following relationships mean:

   a. A = 2C   **b.** C = A/D   **c.** Y = 2 (X) + 1

2. An important equation in centrifugation is:

   $$RCF = 11.18 \times r(RPM/1000)^2$$

   It is not necessary to understand the abbreviations or to be familiar with the term "RCF" to answer the following questions:

   a. What happens to RCF if RPM increases?

   b. What happens to RCF if r is decreased?

   c. There are two constants in this equation. What are they?

3. Insert numbers into each equation in Problem 1 that will satisfy the equation. For example, if the equation is $A = 2C$, then $710 = 2(355)$ will satisfy the equation. (So will many other answers.)

4. Solve the following equations; that is, find the value of X. Pay attention to the units, if present.

   a. $3X = 15$

   b. $X = 25(5-4)$

   c. $-X = 3X - 1 \text{ mg}$

   d. $5X = 3X - 5 \text{ mL} + 34 \text{ mL}$

   e. $\dfrac{X}{2} = 25(3)\text{cm}^2$

   f. $X = 3.0 \text{ cm} (2.0 \text{ mg/mL})(2.0)$

   g. $X = \dfrac{25 \text{ mg}}{\text{mL}}(4.0 \text{ mL})\dfrac{3.0 \text{ oz}}{\text{mg}}$

5. Perform the following calculations.

   a. 23.4 pounds × 34.1 pounds = ?

   b. 15.2 g/3.1 g = ?

   c. $\dfrac{25.2 \text{ cm} \times 34.5 \text{ cm}}{3.00}$

   d. $\dfrac{5 \text{ mL} \times 3 \text{ mL} \times 2 \text{ cm}}{2 \text{ cm}}$

CHAPTER 14

Proportional Relationships

I. INTRODUCTION TO RATIOS AND PROPORTIONS

EXAMPLE APPLICATION: PROPORTIONAL RELATIONSHIPS

a. A transgenic animal is one that produces a protein or has a trait from another species as a result of incorporating a foreign gene(s) into its genome. In 1993, transgenic sheep were born that secrete a human protein, AAT, into their milk. AAT is valuable in the treatment of emphysema. A sheep can produce 400 liters of milk each year. If a transgenic sheep secretes 15 g of AAT into each liter of milk she produces, how many grams of AAT can this transgenic sheep produce in a year?

b. If AAT is worth $110/gram, what is the value of a year's production of AAT from this sheep?

Information from: Amato, Ivan (ed.). "A Biotech Bonanza on the Hoof?" *Science* 259 (March 19, 1993): 1698.

Answers on p. 232

*A **ratio** is the relationship between two quantities using division.* For example, we might say a car "gets 30 miles per gallon." This statement describes the relationship between gas consumption and miles traveled. The word "per" means "for every." The preparation of a cake might require 10 ounces of chocolate. The relationship between the cake and the amount of chocolate required is a ratio. Other commonplace examples of ratios are "revolutions per minute" (RPM) and "cost per pound."

A ratio can be expressed with a numerator and denominator, like a fraction. (However, a ratio is not a fraction. Fractions, for example, can be added together and ratios cannot.):

$$\frac{30 \text{ mi}}{1 \text{ gal}} \text{ or, it is also correct to say } \frac{1 \text{ gal}}{30 \text{ mi}}$$

$$\frac{1 \text{ cake}}{10 \text{ oz chocolate}} \text{ or } \frac{10 \text{ oz chocolate}}{1 \text{ cake}}$$

Observe that a ratio is not a type of equation because there is no = sign.

EXAMPLE PROBLEM

A laboratory solution contains 58.5 grams of NaCl per liter. Express this ratio as a fraction.

ANSWER

The relationship can be expressed as:

$$\frac{58.5 \text{ g}}{1 \text{ L}}$$

---

*A **proportion** is a statement that two ratios are equal.* For example, given that a single cake requires 10 ounces of chocolate, 20 ounces of chocolate are needed to bake two cakes. This can be expressed as a proportion equation showing that the two ratios are equal:

$$\frac{1 \text{ cake}}{10 \text{ oz chocolate}} = \frac{2 \text{ cakes}}{20 \text{ oz chocolate}}$$

Read as "1 cake is to 10 ounces of chocolate as
2 cakes is to 20 ounces of chocolate."

A proportion statement is an equation because there is an = sign.

Many situations in the laboratory require the same reasoning as this "chocolate cake" example. For example, if 10 g of glucose are required to make 1 L of a nutrient solution, how many grams of glucose are needed to make 3 L of nutrient solution? Of course, 30 g is the answer.

Even if you are not a great baker, you probably "knew" that if 1 cake requires 10 oz of chocolate, then 2 cakes require 20 oz, but suppose that 16 oz of chocolate are needed for 3 cakes and you want to make 5 cakes. It may not be obvious in this case how much chocolate is

necessary. It is possible to use an equation as a helpful tool to solve this problem:

1. The ratio "there are 16 ounces of chocolate per 3 cakes" can be written as:

$$\frac{16 \text{ oz chocolate}}{3 \text{ cakes}}$$

2. The unknown, ?, is how much chocolate is required for 5 cakes.

3. If 3 cakes require 16 ounces of chocolate then how many ounces of chocolate are required for five cakes? This proportional relationship is written:

$$\frac{16 \text{ oz chocolate}}{3 \text{ cakes}} = \frac{?}{5 \text{ cakes}}$$

(Note that the units must be in the equation.)

4. To solve for the unknown, cross multiply and divide:

Cross Multiply

$$\frac{16 \text{ oz chocolate}}{3 \text{ cakes}} \diagdown\diagup \frac{?}{5 \text{ cakes}}$$

$$(16 \text{ oz chocolate}) (5 \text{ cakes}) = (?) (3 \text{ cakes})$$

Divide

$$\frac{(16 \text{ oz chocolate})(5 \text{ cakes})}{3 \text{ cakes}} = ?$$

The units of cake cancel:

$$\frac{(16 \text{ oz chocolate})(5)}{3} = ?$$

Simplifying further:

$$? \approx 27 \text{ oz chocolate}$$
$$= \text{ how much chocolate 5 cakes requires}$$

5. Thus:

$$\frac{16 \text{ oz chocolate}}{3 \text{ cakes}} \approx \frac{27 \text{ oz chocolate}}{5 \text{ cakes}}$$

(The fractions 16/3 and 27/5 are about equal
although there is a slight difference due to rounding.)

Some important points about proportional relationships include:

1. ***There are many problems in the laboratory that can be easily solved using the same reasoning as the "chocolate cake" example, even though the units may be less familiar and the numbers may be more complex.***

2. ***It is necessary to keep track of units.*** The units in the chocolate cake example are "ounces" (oz) (of chocolate) and "cake" (or "cakes"). In a proportion equation, the units in both denominators

must be the same and the units in both numerators must be the same. For example, if one cake requires 10 oz of chocolate, how much chocolate is needed for two cakes? The **wrong** way to set this up is:

$$\frac{1 \text{ cake}}{10 \text{ oz}} = \frac{?}{2 \text{ cakes}}$$

Cross Multiply and Divide: $? = 0.2 \text{ cakes}^2/\text{oz}$

Clearly, these units are absurd and the answer is wrong.

It is, however, correct to either write:

$$\frac{1 \text{ cake}}{10 \text{ oz}} = \frac{2 \text{ cakes}}{?} \quad \text{or} \quad \frac{10 \text{ oz}}{1 \text{ cake}} = \frac{?}{2 \text{ cakes}}$$

EXAMPLE PROBLEM

If there are about 100 paramecia in a 20 mL water sample, then about how many paramecia would be found in 10^3 mL of this water?

ANSWER

A couple of ways to think about this problem are:

1. Strategy 1 Use a proportion equation as follows:

$$\frac{100 \text{ paramecia}}{20 \text{ mL}} = \frac{?}{1000 \text{ mL}}$$

Cross multiply and divide:

$$(100 \text{ paramecia})(1000 \text{ mL}) = (20 \text{ mL})(?)$$

$$? = \frac{(100 \text{ paramecia})(1000 \text{ mL})}{20 \text{ mL}}$$

$$? = 5000 \text{ paramecia}$$

2. Strategy 2 Use a proportion equation to calculate how many paramecia are in each milliliter:

$$\frac{100 \text{ paramecia}}{20 \text{ mL}} = \frac{?}{1 \text{ mL}}$$

$$? = 5 \text{ paramecia}$$

There are 5 paramecia in each milliliter; therefore, multiply 5 by 1000 to calculate how many paramecia there are altogether. The answer is 5000 paramecia.

EXAMPLE APPLICATION
ANSWER (from p. 230)

a. This is a proportional relationship: if the sheep secretes 15 g of AAT per liter of milk, how much will she secrete into 400 L? Using a proportion equation:

$$\frac{15 \text{ g}}{1.0 \text{ L}} = \frac{?}{400 \text{ L}}$$

$$? = 6000 \text{ g}$$

This is the amount of AAT a single sheep might produce in a year.

b. This is also a proportional relationship:

$$\frac{\$110}{1.0 \text{ g}} = \frac{?}{6000 \text{ g}}$$

$$? = \$660,000$$

This is the potential value of the AAT from one sheep.

MANIPULATION PRACTICE
PROBLEMS: PROPORTIONS

1. Solve for ?.

a. $\dfrac{?}{5} = \dfrac{2}{10}$ **b.** $\dfrac{?}{1 \text{ mL}} = \dfrac{10 \text{ cm}}{5 \text{ mL}}$

c. $\dfrac{0.5 \text{ mg}}{10 \text{ mL}} = \dfrac{30 \text{ mg}}{?}$ **d.** $\dfrac{50}{?} = \dfrac{100}{100}$

e. $\dfrac{?}{30 \text{ in}^2} = \dfrac{15 \text{ lb}}{100 \text{ in}^2}$ **f.** ? is to 15, as 30 is to 90

g. 100 is to 10, as 50 is to ?

2. In the following problems, first state the unknown, set up the proportion equation with the units, then solve for the unknown.

a. If it requires 50 minutes to drive to Denver, then how long does it take to drive to Denver and back?

b. If it requires 1 teaspoon of baking soda to make one loaf of bread, then how many teaspoons of baking soda are required for 33 loaves?

c. If it costs $1.50 to buy one magazine, then how much do 100 magazines cost?

d. If a recipe requires 1/4 cup of margarine for one batch, then how much margarine is required to make 5 batches?

e. If a recipe requires 1/8 teaspoon of baking powder to make one batch, then how much baking powder is required to make a half batch?

f. If one bag of snack chips contains 3 ounces, then how many ounces are contained in 4.5 bags of chips?

APPLICATION PRACTICE
PROBLEMS: PROPORTIONS

The following are all proportion-type problems, like the chocolate cake problem.

1. If there are about 1×10^2 blood cells in a 1.0×10^{-2} mL sample, then about how many blood cells would be in 1.0 mL of this sample?

2. Ten milliliters of buffer are needed to fill a particular size test tube. How many mL are required to fill 37 of these test tubes?

3. If there are about 5×10^1 paramecia in a 20 mL water sample, then about how many paramecia would be in 10^5 mL of this water?

4. If there are about 315 insect larvae in 1×10^{-1} kg of river sediment, then about how many larvae would be in 5.0×10^4 grams of this sediment?

5. A fermenter is a vat in which microorganisms are grown. The microorganisms produce a material, such as an enzyme, that has commercial value. Suppose a certain fermenter holds 1000 L of broth. There are about 1×10^9 bacteria in each milliliter of the broth. About how many bacteria are present in the entire fermenter?

6. Thirty grams NaCl are required to make 1 L of a salt solution. How much NaCl is needed to make 250 mL of this solution?

7. A particular solution requires 100 mL of ethanol per liter final volume. How much ethanol is needed to make 10^5 mL of this solution?

8. A particular solution contains 50 mg/mL of enzyme. How much enzyme is needed to make 10^4 µL of solution?

9. A particular enzyme breaks down proteins by removing one amino acid at a time from the proteins. The enzyme removes 60 amino acids per minute from a protein. If a protein is initially 1000 amino acids long and the enzyme is added to it, how long will the protein chain be after 10 minutes?

10. The components required to make 100 mL of a solution are listed as follows:

Solution Q

Component	Grams
NaCl	20.0
Na azide	0.001
Mg sulfate	1.0
Tris	15.0

(Don't worry at this time about what these components are.)

Prepare a table that shows how to prepare 1.5 L of solution Q.

11. One mole of carbon has a mass of 12 grams. (Although the definition of a "mole" is not important for solving this problem, you can see Chapter 26 for a brief explanation.)

   a. What is the mass of 2 moles of carbon?

   b. What is the mass of 0.5 moles of carbon?

12. One mole of table salt (NaCl) has a mass of 58.5 g.

   a. What is the mass of 1.5 moles of table salt?

   b. What is the mass of 0.75 moles of table salt?

II. PERCENTS

EXAMPLE APPLICATION: PERCENTS

a. There are about 3×10^9 DNA base pairs in the human genome. Human chromosome 21 is the smallest chromosome (besides the Y chromosome) and contains about 2% of the human genome. About how many base pairs comprise chromosome 21?

b. The goal of the Human Genome Project was to find the base pair sequence of the entire human genome. If chromosome 21 is sequenced, and if the sequencing techniques used are correct 99.9% of the time, then how many of the base pairs determined will be incorrect?

c. Approximately one out of every 700 babies born has an extra copy of chromosome 21. Such children have Down syndrome, which is associated with mental retardation and heart disease. About what percent of all children are born with Down syndrome?

Answers on p. 234

A. Basic Manipulations Involving Percents

Percents are a familiar type of ratio. The % sign symbolizes a fraction and the word **percent** means "of every hundred." For example:

10% means 10/100 or "ten out of every hundred"
0.1% means 0.1/100 or "one tenth out of every hundred"

Suppose that 300 students are surveyed to determine their favorite computer game company. Of those, 225 students prefer Brand Z. This information can be expressed as a ratio, that is:

$$\frac{225 \text{ students}}{300 \text{ students}}$$

To convert this information to a percent, it is necessary to remember that percent means "out of every hundred." The ratio 225/300 can be converted to a percent using the logic of proportions:

$$\frac{225 \text{ students}}{300 \text{ students}} = \frac{?}{100}$$
$$? = 75$$

This means "75 out of every hundred" or 75% of students preferred Brand Z

We can generalize from this example and say that the percent of individuals with a particular characteristic is:

$$\frac{\text{number with the characteristic}}{\text{total number}} \times 100\%$$
$$= \% \text{ with characteristic}$$

Thus, the percent of students who prefer Brand Z computer game is:

$$\frac{225 \text{ students}}{300 \text{ students}} \times 100\% = 75\%$$

Observe that the numerator and denominator must have the same units.

Rules for manipulating percents are summarized in Box 14.1.

EXAMPLE PROBLEM

Show two strategies to convert the fraction 34/67 into a percent.

ANSWER

Strategy 1 Use proportions:

$$\frac{34}{67} = \frac{?}{100}$$

Cross multiply and divide

$$? \approx 50.7$$

Therefore, the answer is 50.7%

Strategy 2 Convert the fraction to a decimal and then move the decimal point two places to the right and add a percent sign:

$$\frac{34}{67} \approx 0.507$$

$$\rightarrow 50.7\%$$

B. An Application of Percents: Laboratory Solutions

Percents have many applications in the laboratory, one of which is in expressing the components of a solution. We address "percent solutions" in detail in Chapter 26. In this section, we will introduce some simple example problems relating to laboratory solutions.

EXAMPLE PROBLEM

A laboratory solution is composed of water and ethylene glycol. How could you prepare 500 mL of a 30% ethylene glycol solution?

ANSWER

Strategy 1 We can use the logic of proportions to calculate how much ethylene glycol is required for 500 mL:

$$\frac{30}{100} = \frac{?}{500 \text{ mL}}$$

Cross multiply and divide.

$$? = 150 \text{ mL}$$

This is how much ethylene glycol is needed.

The remainder of the solution will be water. The best way to prepare this solution is to place the 150 mL of ethylene glycol in a piece of glassware that is marked at 500 mL and then to fill it to the 500 mL mark with water.

Strategy 2 Since the total volume of the solution will be 500 mL and 30% of it will be ethylene glycol:

$$500 \text{ mL} \times 0.30 = 150 \text{ mL}$$

This is how much ethylene glycol is needed.

The remainder of the solution will be water. The best way to prepare this solution is to place the 150 mL of ethylene glycol in a piece of glassware that is marked at 500 mL and then to fill it to the 500 mL mark with water.

EXAMPLE APPLICATION ANSWERS (from p. 233)

a. 2 % = 0.02

$3 \times 10^9 \times 0.02 = 6 \times 10^7 = $ the number of base pairs in chromosome 21.

b. If 99.9% of the base pairs are correct, then 0.1% will be incorrect.

0.1% = 0.001

$6 \times 10^7 \times 0.001 = 6 \times 10^4$, which is the number of base pairs that will be incorrect.

c. 1 out of 700 equals $1/700 \approx 0.0014 = 0.14\% = $ percent of children born with Down syndrome.

MANIPULATION PRACTICE PROBLEMS: PERCENTS

1. Without using a calculator, give the approximate percent of each of the following. For example, 35 out of 354 is about 10% because it is close to 35 out of 350 which is close to 10/100 which is 10%.
   a. 98 out of 100
   b. 110 out of 10,004
   c. 3 out of 15
   d. 45 out of 45,002

2. Express the following percents as fractions and as decimals.
   a. 34%
   b. 89.5%
   c. 100%
   d. 250%
   e. 0.45%
   f. 0.001%

3. Calculate the following.
   a. 15% of 450
   b. 25% of 700
   c. 0.01% of 1000
   d. 10% of 100
   e. 12% of 500
   f. 150% of 1000

4. Express the following fractions or decimals as percents.
   a. 15/45
   b. 2/2
   c. 10/100
   d. 1/100
   e. 1/1000
   f. 6/40
   g. 0.1/0.5
   h. 0.003/89
   i. 5/10
   j. 0.05
   k. 0.0034
   l. 0.25
   m. 0.01
   n. 0.10
   o. 0.0001
   p. 0.0078
   q. 0.50

5. Calculate the following without using a calculator.
   a. 20% of 100
   b. 20% of 1000
   c. 20% of 10,000
   d. 10% of 567
   e. 15% of 1000
   f. 50% of 950
   g. 5% of 100
   h. 1% of 876
   i. 30% of 900

6. State the following as percents:
   a. 1 part in 100 total parts
   b. 3 parts in 50 total parts
   c. 15 parts in (15 parts + 45 parts)
   d. 0.05 parts in 1 part total
   e. 1 part in 25 parts total
   f. 2.35 parts in (2.35 parts + 6.50 parts)

Box 14.1. *RULES FOR MANIPULATING PERCENTS*

---

1. *To convert a percent to its decimal equivalent, move the decimal point two places to the left.* For example:

$$10\% = 0.10$$
$$15\% = 0.15$$
$$0.1\% = 0.001$$

2. *To change a decimal to a percent, move the decimal point two places to the right and add a % sign.* For example:

$$0.10 = 10\%$$
$$0.87 = 87\%$$
$$1 = 100\%$$

3. *To change a percent to a fraction, write the percent as a fraction with a denominator of 100.* For example:

$$30\% = 30/100 = 3/10$$
$$1.5\% = 1.5/100$$
$$100\% = 100/100 = 1$$

4. *To change a fraction to a percent:*

    Strategy 1 Use the logic of proportions:

        For example: Change 50/75 to a percent

        a. Set the problem up as a proportion equation

$$\frac{50}{75} = \frac{?}{100}$$

        b. Cross multiply and divide

$$? = 66.66\ldots \quad \text{So } 50/75 \approx 66.67\%$$

    Strategy 2 Convert the fraction to its decimal equivalent and then move the decimal two places to the right and add a % sign:

        Change 50/75 to a percent

        a. Change the fraction to a decimal: $50/75 \approx 0.6667$

        b. Change the decimal to a percent: $0.6667 \approx 66.67\%$

5. *To find a percentage of a particular number convert the percent to its decimal equivalent and multiply the decimal times the number of interest.* For example:

        To find 45% of 900

        a. Convert the percent to a decimal: $45\% = 0.45$

        b. Multiply the decimal times the number of interest: $0.45 \times 900 = 405$.

APPLICATION PRACTICE PROBLEMS: PERCENTS

1. There are 55 students in a class: 25 of them work 20 hours a week or more at jobs outside school; 14 of them work 10–19 hours per week at jobs. The rest work 0–10 hours per week at outside jobs. What percent of the students work 20 hours or more per week at jobs? What percent work 19 hours or less at outside jobs?

2. How much ethanol is present in 100 mL of a 50% solution of ethanol?

3. A laboratory solution is made of water and ethylene glycol. How could you prepare 250 mL of a 30% ethylene glycol solution?

4. A laboratory solution is required that is 10% acetonitrile, 25% methanol, and the rest is water. How could you prepare this solution? (Note that the total volume is unspecified, so you can prepare any volume you want. Also, these components come as liquids.)

5. A particular laboratory solution contains 5 mL of propanol per 100 mL solution. Express this as a percent propanol solution.

6. A solution contains 15 mL of ethanol per 700 mL total volume. Express this as a percent ethanol solution.

7. A solution contains 10 μL of methanol in 1 mL of water. Express this as a percent methanol solution.

8. A solution contains 15 mL of acetone per liter. Express this as a percent solution.

9. Suppose there is a population of insect in which 95% die over the winter, but the rest survive. Each surviving insect produces 100 offspring in the spring, and

then dies. Assuming there is no mortality during the summer and the population has 1000 insects at the beginning of the first winter, how many insects will there be after two winters?

10. Osteoporosis is a disease that leads to brittle bones and is partially caused by a lack of calcium. Each year, 250,000 people suffer osteoporosis-related hip fractures, about 15% of whom die after sustaining such fractures. How many people die each year after fracturing their hip?

11. Double-stranded DNA consists of two long strands. An adenine on one strand is always paired with a thymine on the opposite strand and a guanine on one strand is always paired with a cytosine on the opposite strand. If a purified sample of DNA contains 24% thymine, what are the percentages of the other bases?

12. Olestra is a fat substitute for food products that has been associated with gastrointestinal problems in some preliminary studies. A study was performed in which 1000 subjects were given a free soda and a movie pass. Half the subjects received an unlabeled bag of potato chips made with Olestra; half received chips without Olestra. The incidence of intestinal problems was monitored for 10 days. Of the group fed regular potato chips 17.6% experienced intestinal problems versus 15.8% of the group given Olestra potato chips. How many individuals in each group experienced gastrointestinal problems?

Data from "Olestra at the Movies," *The Scientist* (February 16, 1998) 31.

III. DENSITY

Density, *d,* *is the ratio between the mass and volume of a material.* Thus:

$$density = \frac{mass}{volume}$$

For example, the density of benzene is

$$\frac{0.880\ g}{mL}$$

Various materials have different densities. For example, balsa wood is less dense than lead, see Figure 14.1. A lead brick, therefore, weighs more than a piece of balsa wood that occupies the same volume.

Density can be expressed in various units so it is important to record the units. In addition, the density of a material changes with temperature: Most materials expand when heated and contract when cooled. It is therefore conventional to report the density of a material at a particular temperature. For example:

benzene $d^{20°} = 0.880$ g/mL

This means that the density of benzene is 0.880 g/mL at 20°C.

Balsa Wood (0.160 kg)

Lead Brick (11.3 kg)

Figure 14.1.   Density. A piece of balsa wood is lighter than a lead brick of the same volume.

The densities of solids and liquids are often compared with the density of water. If the density of a material is less than water, then that material will float on water. If the material is more dense than water, then it will sink. Similarly, the densities of gases are often compared with that of air. Materials that are less dense than air (such as balloons filled with helium) rise; materials that are more dense than air do not rise.

Consider different liquids that do not dissolve in one another. If these liquids are mixed in a test tube, then they will separate according to their densities, with the most dense liquid on the bottom, see Figure 14.2.

EXAMPLE PROBLEM

The density for water is:

$$d^{4°} = 1.000\ g/mL$$
$$d^{0°} = 0.917\ g/mL$$

a. What is the volume occupied by 5.6 grams of water at 4°C?

b. Why does ice float?

ANSWER

a. This can easily be solved intuitively. The answer is simply 5.6 mL. This can also be solved using a proportion equation:

$$\frac{1\ g}{1\ mL} = \frac{5.6\ g}{?}$$

Cross multiply and divide:

$$? = 5.6\ mL$$

b. Water is less dense at 0°C than it is at 4°C; therefore, ice (which is 0°C) floats on more dense, unfrozen water.

EXAMPLE PROBLEM

What is the mass of 25.0 mL of benzene at 20°C? (Density = 0.880 g/mL)

ANSWER

Strategy 1 Use a proportion equation:

$$\frac{0.880\ g}{1\ mL} = \frac{?}{25.0\ mL}$$
$$? = 22.0\ g$$

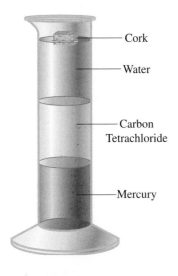

Figure 14.2. Materials of Different Density. The density (at 20°C) of mercury is 13.5 g/mL, of carbon tetrachloride is 1.59 g/mL, and of water is 0.998 g/mL. Because mercury is the most dense, it is found on the bottom of the cylinder. Cork is least dense and so floats on water.

Labels: Cork, Water, Carbon Tetrachloride, Mercury

Strategy 2 Since we know that 1 *mL* has a mass of 0.880 *g*, we can multiply by 25.0 to calculate the mass of 25.0 *mL* of benzene:

$$\frac{0.880\text{ g}}{1\text{ mL}} \times 25.0\text{ mL} = 22.0\text{ g}$$

MANIPULATION PRACTICE PROBLEMS: DENSITY

1. At 25°C, the density of olive oil is 0.92 g/mL and water is 0.997 g/mL. Vinegar has a density similar to that of water. Which is the top layer in vinegar and oil dressing?

2. The mass of a gold bar that is 3.00 cm³ is 57.9 g. What is the density of gold?

3. The density of glycerol at 20°C is 1.26 g/mL. What is the volume of 20.0 g of glycerol?

IV. UNIT CONVERSIONS

EXAMPLE APPLICATION: UNIT CONVERSIONS

a. Phthalates are compounds that make plastics flexible and are used in many products, including food wrap. Studies suggest that phthalates may activate receptors for estrogen, the primary female sex hormone. There is speculation that exposure to such estrogenic compounds may increase breast cancer incidence in women, reduce fertility in men, and adversely affect wildlife. One study suggested that margarine may pick up as much as 45 mg/kg of phthalates from plastic wrap. If you use a half pound of margarine that has absorbed 45 mg/kg of phthalates to bake 3 dozen cookies, how much phthalate will three cookies contain?

b. The pesticide DDT also has estrogenic effects. Although DDT is no longer used in the United States, it is still widely used worldwide. Humans have been shown to accumulate as much as 4 µg/g of body weight of DDT in their tissues. If a woman weighs 148 pounds, how much DDT might her body contain? (Let us simplify the situation by assuming that all tissues in her body accumulate the same maximum amount of DDT. This is probably not correct.)

Information from: "Newest Estrogen Mimics the Commonest?" *Science News.* 148 (July 15, 1995): 47. Raloff, Janet. "Beyond Estrogens: Why Unmasking Hormone-Mimicking Pollutants Proves So Challenging." *Science News* 148 (July 15, 1995): 44–46. Guillette, L.J. Jr. "Endocrine-Disrupting Environmental Contaminant and Reproduction: Lessons from the Study of Wildlife." *Women's Health Today: Perspectives on Current Research and Clinical Practice,* D.R. Popkin and L.J. Peddle (eds.) New York, Parthenon Publishing Group, pp. 201–7. 1994.

Answers on p. 239

A. Overview

A given quantity, amount, or length can be measured in different units. For example, a candy bar that costs $1.00 also costs 100 cents. A dollar is a larger unit of currency than a cent; therefore, it takes many cents (100) to equal the value of a single dollar. A snake that is 1 foot long is also 12 inches or 30.48 cm long. A foot is a larger unit of length than a centimeter. It takes more than 30 cm to equal the length of a single foot.

It is often necessary to convert numbers that have units in the United States Customary System (USCS) to numbers with metric units (e.g., from pounds to kilograms). It is also often necessary to convert from one USCS unit to another USCS unit (e.g., from feet to inches) or from one metric unit to another metric unit (e.g., from centimeters to meters). This section discusses such conversions.

There are two common strategies for performing unit conversions. The first approach is to think of conversion problems in terms of proportions and to use a proportion equation to solve them. The second approach is the use of conversion factors (also called the "unit canceling method" or "dimensional analysis"). Some people prefer one strategy to perform conversions; other people prefer the other. Any strategy is correct as long as it consistently yields the right answers.

B. Proportion Method of Unit Conversion

Let us illustrate the proportion approach with an example of a conversion from the USCS to metric units:

If a student weighs 150 pounds, how much does he weigh in kilograms?

$$(1\text{ kg} \approx 2.2\text{ lb})$$

Using proportions, if 1 kg = 2.2 lb, then how many kilograms is 150 lb?

$$\frac{2.2 \text{ lb}}{1 \text{ kg}} = \frac{150 \text{ lb}}{?}$$

? ≈ 68.2 kg = the student's weight in kilograms

It is a good idea to examine the answer to see if it makes sense. A kilogram is a larger unit than a pound (just as a dollar is a larger unit than a cent). When the student's weight is converted from pounds to kilograms, therefore, the number should be lower—this answer makes sense.

Another example:

A student is 6.00 ft tall. How tall is she in centimeters? (1 in = 2.54 cm, 1 ft = 12 in)

This is a proportion problem that, with the information given, needs to be solved in two steps:

Step 1. Convert the height in feet to inches.

$$\frac{1 \text{ ft}}{12 \text{ in}} = \frac{6.00 \text{ ft}}{?}$$

? = 72.0 in = her height in inches

Step 2. Convert the height in inches to height in centimeters.

$$\frac{1 \text{ in}}{2.54 \text{ cm}} = \frac{72.0 \text{ in}}{?}$$

? ≈ 183 cm = her height in centimeters

Again, it is a good idea to examine the answer to see if it makes sense. A centimeter is a smaller unit than a foot; therefore, when the student's height is converted from feet to centimeters, the number will have to be larger.

Proportions can also be used to convert from one metric unit to another:

How many meters is 345 cm?

There are 100 cm in a meter. So:

$$\frac{1 \text{ m}}{100 \text{ cm}} = \frac{?}{345 \text{ cm}}$$

? = 3.45 m

This answer makes sense. A meter is a larger unit than a centimeter; therefore, when centimeters are converted to meters, the value is a smaller number.

Another example:

Convert 105 cm to nanometers.

(1 meter = 10^2 cm, and 1 m = 10^9 nm)

This problem can be solved using two steps:

Step 1.

$$\frac{1 \text{ m}}{10^2 \text{ cm}} = \frac{?}{105 \text{ cm}}$$

? = 1.05 m

Step 2.

$$\frac{1 \text{ m}}{10^9 \text{ nm}} = \frac{1.05 \text{ m}}{?}$$

? = 1.05×10^9 nm

This answer is reasonable. A nanometer is a very small unit of length; therefore, it will take many nanometers to equal the length of 105 cm. The large answer obtained, 1.05×10^9 nm, makes sense.

C. Conversion Factor Method of Unit Conversion

A second strategy for doing conversion problems is to multiply the number to be converted times the proper conversion factor. For example:

Convert 2.80 kg to pounds.

A pound is 0.454 kg. The conversion factor is expressed as a ratio:

$$\frac{1 \text{ lb}}{0.454 \text{ kg}}$$

Observe that this ratio equals 1. All conversion factors equal 1.

Multiply 2.80 kg by the conversion factor:

$$2.80 \text{ kg} \times \frac{1 \text{ lb}}{0.454 \text{ kg}} \approx 6.17 \text{ lb}$$

The units of kilograms cancel and the answer comes out with the correct units, pounds.

As we demonstrated earlier with proportions, it is good practice to examine the answer to see that it makes sense.

Suppose you had another reference that told you not that 1 lb is 0.454 kg, but rather that 1 kg is 2.205 pounds. If you use this factor directly, you will get the wrong answer:

$$2.80 \text{ kg} \times \frac{1 \text{ kg}}{2.205 \text{ lb}} \approx 1.27 \text{ kg}^2/\text{lb}$$

You can tell that this is the wrong answer because the units are wrong. If you "flip over" the conversion factor, however, the kilogram units cancel, resulting in the correct units at the end:

$$2.80 \text{ kg} \times \frac{2.205 \text{ lb}}{1 \text{ kg}} \approx 6.17 \text{ lb}$$

When using conversion factors, the units guide you in setting up equations. The units must cancel so that the result has the correct units. The term *unit canceling method* is thus a good description for this strategy.

For example:

A student is 6.00 ft tall. How tall is she in centimeters? The conversion factors are:

$$\frac{1 \text{ in}}{2.54 \text{ cm}} \quad \text{and} \quad \frac{1 \text{ ft}}{12 \text{ in}}$$

The inches will have to cancel, the feet will have to cancel, and centimeters must remain. Let us try a couple of ways of setting this up:

$$6.00 \text{ ft} \times \frac{1 \text{ ft}}{12 \text{ in}} \times \frac{1 \text{ in}}{2.54 \text{ cm}} = ?$$

The units of feet do not cancel if the equation is set up this way and cm is in the denominator instead of the numerator; however, if the equation is set up:

$$6.00 \text{ ft} \times \frac{12 \text{ in}}{1 \text{ ft}} \times \frac{2.54 \text{ cm}}{1 \text{ in}} = ?$$

The units of inches and feet cancel leaving the answer with units of centimeters. The answer is 183 cm, which is the same as the answer obtained previously using proportions. Remember to examine the answer to see that it makes sense.

With the unit cancellation method you can string together as many conversion factors as you want into one long equation. If the equation is correct, then the units will cancel, leaving the answer in the desired units.

Consider the conversion of metric units from one another. For example:

How many meters is 345 cm?
The conversion factor is:

$$\frac{1 \text{ m}}{100 \text{ cm}}$$

Multiply $345 \text{ cm} \times \frac{1 \text{ m}}{100 \text{ cm}} = 3.45 \text{ m}$

The units of cm cancel, leaving the answer in the correct units.

The answer makes sense.

Another example:

Convert 105 cm to nanometers.

There are 10^9 nm in one meter and there are 10^2 cm in 1 m. These are the conversion factors. Then:

$$105 \text{ cm} \times \frac{1 \text{ m}}{10^2 \text{ cm}} \times \frac{10^9 \text{ nm}}{1 \text{ m}} = ?$$

The centimeters and meters cancel, leaving the answer in nanometers:

$$105 \text{ cm} = 1.05 \times 10^9 \text{ nm}.$$

This answer makes sense.

Both the proportion and the conversion factor strategies are effective ways to convert numbers from one unit to another. Both strategies require paying attention to the units. Observe that when using proportions to do conversion problems, there is always an equals (=) sign between two ratios. In addition, when using proportions, the units must be the same in both denominators and the same in both numerators. In contrast, with the conversion factor (unit canceling) method, there is not an equals sign between two ratios. There is a multiplication ($\times$) sign between the number to be converted and the conversion factor(s).

EXAMPLE APPLICATION: ANSWERS (from p. 237)

a. Let us solve the first part of the question using the proportion method. First, convert 0.50 lb of margarine to kg:

$$\frac{1 \text{ lb}}{0.454 \text{ kg}} = \frac{0.50 \text{ lb}}{?}$$

? $\approx$ 0.227 kg = weight of margarine in kilogram units.

Margarine can absorb 45 mg of phthalates per kilogram so 0.227 kg of margarine can take up:

$$\frac{45 \text{ mg}}{1 \text{ kg}} = \frac{?}{0.227 \text{ kg}}$$

? $\approx$ 10.22 mg = milligrams of phthalates that 0.227 kg of margarine can absorb.

There are therefore potentially 10.22 mg of phthalates in 36 cookies. To calculate the milligrams of phthalates in three cookies:

$$\frac{10.22 \text{ mg}}{36 \text{ cookies}} = \frac{?}{3 \text{ cookies}}$$

? $\approx$ 0.85 milligrams = mg of phthalates in three cookies.

b. Let us use the conversion factor method to solve part b of the problem. Conversion factors can be strung together into one long equation. First, it is necessary to convert the woman's weight from pounds to grams. A factor must then be included to account for the fact that the woman accumulated 4 μg/g of DDT in all her tissues. Finally, it would be helpful to covert the answer from micrograms to grams. The resulting single equation is:

$$148 \text{ lb} \times \frac{454 \text{ g}}{1 \text{ lb}} \times \frac{4 \text{ } \mu\text{g}}{g} \times \frac{1 \text{ g}}{10^6 \text{ } \mu\text{g}}$$

$\approx$ 0.269 g = accumulated DDT

MANIPULATION PRACTICE PROBLEMS: UNIT CONVERSIONS

(Use either the proportion method or the conversion factor method to solve these problems. Conversion factors are shown in Table 13.3)

1. Convert:
   a. 3.00 feet to cm
   b. 100 mg to g
   c. 12.0 inches to miles (use scientific notation for your answer)
   d. 100 inches to km
   e. 10.0555 pounds to ounces
   f. 18.989 pounds to g
   g. 13 miles to km
   h. 150 mL to liters
   i. 56.7009 cm to nm
   j. 500 nm to μm
   k. 10.0 nm to inches

2. How far is a 10 km race in miles?

3. A marathon is 26.2 miles. Express this in kilometers.

4. How tall is a person who is 5 ft 4 in in meters?

5. In kilometers, how far is a town that is 45 miles away?

6. A car is going 55 mph. How fast is it moving in kilometers per hour?

7. How much does a 3.0 ton elephant weigh in the metric system?

8. Which is the least expensive jar of hot fudge?
   a. $2.50 for 12 oz
   b. $3.67 for 250 g
   c. $4.50 for 0.300 kg
   d. $2.35 for 0.75 lb

   Relationships
   52 weeks = 1 year
   1 week = 7 days
   1 day = 24 hours
   1 hour = 60 minutes
   1 minute = 60 seconds

Fill in the blanks

9. 1 week = ____ days = ____ hours = ____ minutes = ____ seconds

10. ____ year = 1 week = ____ hours = ____ minutes = ____ seconds

11. 1 year = ____ weeks = ____ days = ____ minutes = ____ seconds

   Relationships
   1760 yds = 1 mi
   1 yd = 3 ft
   1 ft = 12 in

12. ____ mi = ____ yds = 1 ft = ____ in

   Relationships
   1 gal = 4 qts
   1 qt = 2 pts
   1 pt = 16 oz

13. ____ gal = ____ pt = 2 oz

   Relationships
   1 km = 1000 m = 10^3 m
   1 m = 100 cm = 10^2 cm
   1 cm = 10 mm
   1 mm = 1000 μm = 10^3 μm
   1 μm = 1000 nm = 10^3 nm
   2.5 cm = 1 in

14. 1 km = ____ m = ____ cm = ____ mm = ____ μm = ____ nm

15. ____ km = ____ m = ____ cm = ____ mm = ____ μm = 1 nm

16. ____ km = ____ m = 2.5 cm = ____ mm = ____ μm = ____ nm

17. a. 6.25 mm = ____ μm
    b. 0.00896 m = ____ mm
    c. 9876000 nm = ____ mm

18. ____ km = 3.0 m = ____ cm = ____ in

   Relationships
   1 g = 10^{-3} kg
   1 mg = 10^{-3} g
   1 μg = 10^{-3} mg

19. 5 kg = ____ g = ____ mg = ____ μg

20. ____ kg = 0.0089 g = ____ mg = ____ μg

21. ____ kg = ____ g = ____ mg = 2×10^{-8} μg

22. a. 0.8657 g = ____ mg
    b. 526 kg = ____ mg
    c. 63 g = ____ μg
    d. 2.63×10^{-6} μg = ____ kg

Relationships

(These units relate to the decay of radioactivity.)
(Ci is a unit called a "Curie")
1 Ci = 3.7×10^{10} dps
(disintegrations per second)
1 Ci = 1000 mCi = 10^3 mCi
1 mCi = 1000 μCi = 10^3 μCi
1 Bq (Becquerel) = 1 dps

23. 1 Ci = ____ dps = ____ dpm (disintegrations per minute)

24. 1 Ci = ____ mCi = ____ μCi = ____ dps

25. ____ Ci = ____ mCi = ____ μCi = 10^5 dps

26. ____ Ci = ____ mCi = 100 μCi = ____ dps = ____ dpm

27. 1 Ci = ____ mCi = ____ μCi = ____ dps = ____ Bq

28. ____ Ci = ____ mCi = 250 μCi = ____ dps = ____ Bq

APPLICATION PRACTICE PROBLEMS: UNIT CONVERSIONS

(Use either the proportion method or the conversion factor method to solve these problems. Conversion factors are shown in Table 13.3)

1. Suppose bacteria are growing in a flask. The growth medium for the bacteria requires 5 g of glucose per liter. A technician has prepared some medium and added 0.24 lb of glucose to 25 L. Did the technician make the broth correctly?

2. A recipe to make 1 L of a laboratory solution is shown as follows:

Solution X

Component	Grams
NaCl	20.00
Na azide	0.001
Mg sulfate	1.000
Tris	15.00

Prepare a table that shows how to prepare 1 mL of the same solution. Express the amounts of each component needed in mg.

3. Suppose a particular enzyme must be added to a nutrient solution used to grow bacteria. The enzyme comes as a freeze-dried powder. The manufacturer of the enzyme states that every gram of enzyme powder actually contains only 680 mg of enzyme, the rest is an inert filler that has no effect. If a recipe calls for 10.0 oz of this enzyme for every 100.0 L of broth, and if you prepare 500.0 L of broth, how much of the enzyme powder will you need to add? Remember to compensate for the inert filler.

V. CONCENTRATION AND DILUTION

A. Concentration

Concentration *is the amount of a particular substance in a stated volume (or sometimes mass) of a solution or mixture.* Concentration is a ratio where the numerator is the amount of the material of interest and the denominator is usually the volume (or sometimes mass) of the entire mixture. For example:

$$\frac{2 \text{ g NaCl}}{1 \text{ L Water}}$$

means that 2 g of NaCl is dissolved in enough water so that the total volume of the solution is 1 L.

The substance that is dissolved is called the **solute.** *The liquid in which the solute is dissolved is called the* **solvent.** In this example, NaCl is the solute and water is the solvent.

Note that the words "concentration" and "amount" are not synonyms. **Amount** *is how much of a substance is present* (e.g., 2 *g*, 4 cups, or one teaspoon). In contrast,

concentration is a ratio with a numerator (amount) and a denominator (usually volume).

Because concentration is a ratio, problems involving concentrations use the same reasoning as other proportion problems. For example, a concentration of 1 mg NaCl in 10 mL of solution is the same as a concentration of 10 mg NaCl in 100 mL of solution. Similarly:

How could you make 300 mL of a solution that has a concentration of 10 *g* of NaCl in 100 mL total solution?

This is a proportion problem:

$$\frac{10 \text{ g}}{100 \text{ mL total}} = \frac{?}{300 \text{ mL total}}$$
$$? = 30 \text{ grams}$$

30 g of NaCl in 300 mL is the same concentration as 10 g of NaCl in 100 mL

EXAMPLE PROBLEM

Which is more pure: a chemical that contains 0.025 g of contaminating material in 10^4 kg or a chemical that contains 10^2 mg of contaminant in 10^4 kg?

ANSWER

There is more than one way to solve this problem. One approach, shown here, is to convert the concentrations of both chemicals into the same units so they can be compared more easily.

$$\text{Chemical 1: } \frac{0.025 \text{ g}}{10^4 \text{ kg}} \text{ of contaminant}$$

Chemical 2: Convert 10^2 mg to grams.
$$10^2 \text{ mg} = 10^{-1} \text{ g}$$

Thus, chemical 2 has a contaminant concentration of
$$\frac{10^{-1} \text{ g}}{10^4 \text{ kg}} = \frac{0.1 \text{ g}}{10^4 \text{ kg}}$$

Chemical 1 is more pure because it has less contaminant per 10^4 kg than chemical 2.

MANIPULATION/APPLICATION PRACTICE PROBLEMS: CONCENTRATION

1. If a solution requires a concentration of 3 *g* of NaCl in 250 mL total volume, how much NaCl is required to make 1000 mL?

2. If the concentration of magnesium sulfate in a solution is 25 g/L, how much magnesium sulfate is present in 100 mL of this solution?

3. If the concentration of magnesium chloride in a

solution needs to be 1 mg/mL, how much magnesium chloride is required to make 15 L of this solution?

4. If a solution requires 0.005 g of Tris base per liter, how much Tris base is required to make 10^{-3} liters of this solution?

5. If there are 300 ng of dioxin in 100 g of baby diapers, how much dioxin is there in 1 kg of diapers?

6. A *mole* is an expression of amount (which is discussed in the Solutions Unit). If the concentration of a solute in a solution is 5.00×10^{-3} moles/L, how many moles are there in 5 mL of this solution?

7. If a solute has a concentration of 0.1 moles/L, how much solute is present in 1 μL of solution?

8. If a solute has a concentration of 10^{-2} g/L, how much solute is present in 78 mL of solution?

9. Enzymes are proteins that catalyze chemical reactions in biological systems. Enzymes are rated on the basis of how active they are or how quickly they can catalyze reactions. Every preparation of enzyme has a certain activity expressed in Units of Activity. Suppose you buy a vial of an enzyme called beta-galactosidase. The vial is labeled: 3 mg solid, 500 Units/mg. How many Units are present in the vial?

10. A vial of the enzyme horseradish peroxidase is labeled: 5 mg solid, 2500 Units/mg. How many Units are present altogether in this vial?

11. If a solution contains 3 g/mL of compound A, how much compound A is present in 1 L of this solution?

12. A solution has 5 μg/L of enzyme Q. How much enzyme Q is present in:
    a. 50 mL of solution
    b. 500 mL of solution
    c. 100 mL of solution
    d. 100 μL of solution

13. A solution has 0.5 mg/mL of the enzyme lysozyme. How much lysozyme is present in:
    a. 5 mL of solution
    b. 0.5 mL of solution
    c. 100 μL of solution
    d. 1000 μL of solution

14. There are analytical instruments in the laboratory that are capable of detecting extremely small amounts of specific chemicals. Suppose a particular instrument can detect as little as a single molecule of benzo(a)pyrene (a carcinogen) out of 10^6 molecules of various compounds. Is the instrument sensitive enough to detect 100 molecules of benzo(a)pyrene out of 10^9 molecules?

15. Which is more pure, a chemical that contains 1 g of contaminating material in 10^6 kg, or a chemical that contains 10^{-2} mg of contaminant in 10^{-3} kg?

B. Introduction to Dilutions: Terminology

EXAMPLE APPLICATION: DILUTIONS

Plasmids are circular DNA molecules that can transport a gene from one bacterium into another bacterium. In nature, plasmids may carry genes that make a bacterium resistant to an antibiotic. As bacteria exchange plasmids carrying resistance genes, resistance to antibiotics spreads among bacterial populations. Plasmids can also be used in the laboratory to transport useful genes into bacteria. Suppose that in a particular experiment, it is necessary to add 0.01 μg of plasmid to a tube. Suppose further that I mL of plasmid in solution at a concentration of I mg plasmid/mL is available. How might the addition of only 0.01 μg of plasmid be accomplished? Assume that it is not possible to accurately measure a volume less than I μL.

Answer on p. 248

There are many situations in the laboratory that require dilutions. A **dilution** *is when one substance (often but not always water) is added to another to reduce the concentration of the first substance. The original substance being diluted may be called the* **stock solution.**

There are various ways to speak about dilutions; unfortunately, this variation in terminology can lead to confusion. Let us illustrate dilution terminology with an example. A baker buys a bottle of food coloring and dilutes it to decorate cookies. The baker takes 1 mL of the concentrated food coloring and adds 9 mL of water so that the total volume of the diluted food coloring solution is 10 mL, Figure 14.3a. Various people might refer to this same dilution using different terminology, Figure 14.3b.

Observe that the word **to** and the symbol **:** are used inconsistently. The word **to** or the symbol **:** is sometimes used before the volume of the *diluting substance,* the **diluent.** (In this example, the diluent is 9 mL of water.) Other times, the word **to** or the symbol **:** is used before the total volume of the final mixture. (In this example, the total volume of the final mixture is 10 mL.) The key

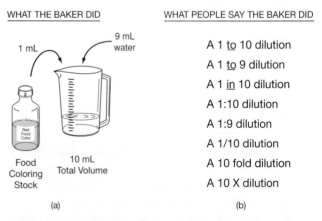

WHAT THE BAKER DID	WHAT PEOPLE SAY THE BAKER DID
	A 1 <u>to</u> 10 dilution
	A 1 <u>to</u> 9 dilution
	A 1 <u>in</u> 10 dilution
	A 1:10 dilution
	A 1:9 dilution
	A 1/10 dilution
	A 10 fold dilution
	A 10 X dilution

(a)   (b)

Figure 14.3.   Dilutions and Terminology. a. What the baker did. **b.** Varied terminology to describe the same dilution.

to avoiding confusion is to keep track of whether you are talking about the *total volume of the final mixture or about the amount of diluting substance.* When reading what other people have written, try to determine what they mean.

In this book, the dilution terminology conforms to that suggested by the American Society for Microbiology (ASM) Style Manual (*ASM Style Manual for Journals and Books, American Society for Microbiology*, Washington, DC, 1991). This terminology is summarized in Box 14.2.

C. Dilutions and Proportional Relationships

The concepts of dilution and proportion are related. For example:

1 mL of food coloring mixed with 9 mL of water is the same dilution as 10 mL of food coloring mixed with 90 mL of water.

$$\frac{1\,mL}{10\,mL} = \frac{10\,mL}{100\,mL} = \frac{1}{10}$$

1 mL in 10 mL total = 10 mL in 100 mL total
= 1/10 dilution

Note that the units in the numerator and the denominator are the same and cancel.

All of the following are the same dilution, **1 in 10 total**:

$$\begin{array}{r} 2\ mL\ food\ coloring \\ +\ \underline{18\ mL\ water} \\ 20\ mL\ total\ volume \end{array}$$

$$\frac{2\,mL}{20\,mL} = \frac{1}{10}$$

$$\begin{array}{r} 100\ \mu L\ enzyme\ solution \\ +\ \underline{900\ \mu L\ buffer\ solution} \\ 1000\ \mu L\ total\ volume \end{array}$$

$$\frac{100\,\mu L}{1000\,\mu L} = \frac{1}{10}$$

$$\begin{array}{r} 17\ fluid\ oz\ juice \\ +\ \underline{153\ fluid\ oz\ water} \\ 170\ fluid\ oz\ total \end{array}$$

$$\frac{17\,oz}{170\,oz} = \frac{1}{10}$$

MANIPULATION PRACTICE PROBLEMS: DILUTIONS (PART A)

(Follow ASM recommendations whenever applicable.)

1. Suppose you dilute 1 oz of orange juice concentrate with 3 oz of water.
   a. Express this dilution using the word *in.*
   b. Express this dilution using the word *to.*
   c. Express this dilution with a : and then with a /.

2. Express each of the following as a dilution (using a /):
   a. 1 mL of original sample + 9 mL of water.
   b. 1 mL sample + 10 mL water.
   c. 3 mL sample in a total volume of 30 mL.
   d. 3 mL sample + 27 mL water.
   e. 0.5 mL sample + 11.0 mL water.

Box 14.2. DILUTION TERMINOLOGY BASED ON ASM RECOMMENDATIONS

1. *1 part food coloring combined with 9 parts water means the food coloring is 1 part in 10 mL total volume or 1/10 food coloring.* The denominator in an expression with a slash (/) is the <u>total volume</u> of the solution, never the amount of the diluting substance.

2. *An undiluted substance, by definition, is called 1/1.*

3. *When talking about dilutions, the symbol : means parts.* If 1 mL of food coloring is combined with 9 mL of water, that is 1 part food coloring plus 9 parts water or 1:9 food coloring **to** water. In this text, a dilution of 1 part plus 9 parts diluent is <u>not</u> referred to as a 1:10 dilution.

   For example:
   A **1:2 dilution** *means there are three parts total volume.*

   A $\frac{1}{2}$ **dilution** *means there are two parts total volume.*

   Therefore:
   1/2 is the same as 1:1
   1:2 is the same as 1/3
   1:3:5 A:B:C means that 1 part A, 3 parts B, and 5 parts C are combined for a total of 9 parts.
   (The parts can be any unit. For example, this might mean 1 mL of A, 3 mL of B, and 5 mL of C. Or, it might mean 1 g of A, 3 g of B, and 5 g of C.)

3. Express each of the following ratios as a dilution (using a /).

  a. 1 part sample: 9 part diluent

  b. 1 part sample: 10 parts diluent

  c. 1:3   **d.** 1:1   **e.** 1:4

4. Express each of the mixtures in problem 2 as a 1:___ ratio.

5. If you take a 0.5 mL sample of blood and add 1.0 mL of water and 3.0 mL of reagents, what is the final dilution of the blood?

D. Calculations for Preparing One Dilution

How could you prepare 10 mL of a 1/10 dilution of food coloring? This is a simple dilution and is easily accomplished as illustrated:

Combine 1 mL of the food coloring with 9 mL of water. The total volume will be 10 mL. Thus, the dilution will be 1/10.

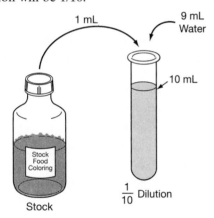

How could you make 10 mL of a 1/5 dilution of food coloring?

Combine 2 mL of food coloring with 8 mL of water. The total volume will be 10 mL. Thus, the dilution will be 2/10 = 1/5.

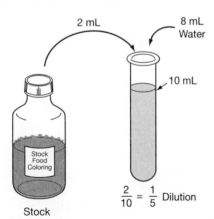

How could you make 100 mL of a 1/10 dilution of food coloring?

Combine 10 mL of food coloring with 90 mL of water. The total volume will be 100 mL. Thus, the dilution will be 10/100.

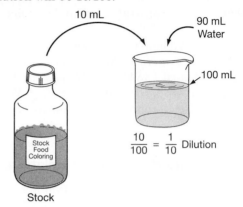

How could you make 250 mL of a 1/4 dilution of food coloring? This may be less obvious than the three previous examples, but it can be solved easily using the logic of proportions:

$$\frac{1}{4} = \frac{?}{250 \text{ mL}}$$

$$? = 62.5 \text{ mL}$$

Take 62.5 mL of food coloring and add enough water to get 250 mL total volume (187.5 mL). The food coloring dilution will be:

$$\frac{62.5 \text{ mL}}{250 \text{ mL}} = \frac{1}{4}$$

Proportions can thus be used to calculate how to make a particular amount of a specific dilution.

MANIPULATION PRACTICE PROBLEMS: DILUTIONS (PART B)

1. How would you prepare 10 mL of a 1/10 dilution of blood?

2. How would you prepare 250 mL of a 1/300 dilution of blood?

3. How would you prepare 1 mL of a 1/50 dilution of blood?

4. How would you prepare 1000 μL of a 1/100 dilution of food coloring?

5. How would you prepare 23 mL of a 3/5 dilution of solution Q?

6. Suppose you have a stock of a buffer solution that is used in experiments. The stock is 10 times more concentrated than it is used (like frozen orange juice that is sold in a concentrated form). How would you prepare 10 mL of the buffer solution at the right concentration?

7. Suppose you have a stock of buffer that is five times more concentrated than it is used. How would you prepare 15 mL at the correct concentration?

8. Suppose you need 10^3 μL of a solution. The solution is stored at a concentration that is 100 times the concentration at which it is normally used. How would you dilute the solution?

9. How would you prepare 50 mL of a 0.01 dilution of buffer?

E. Dilution and Concentration

When dilutions are prepared it is important to keep track of the concentration of the solute in the diluted tubes. Let's look at some examples of the relationship between concentration and dilution.

This example is illustrated in Figure 14.4:

A stock solution contains 10 mg/mL of a particular enzyme. This means that every mL of stock contains 10 mg of enzyme.

The stock solution is diluted by taking 1 mL of the solution (which contains 10 mg of enzyme) and mixing it with 4 mL of water. What is the concentration of enzyme in the diluted solution?

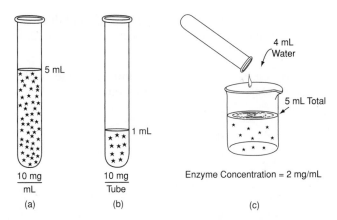

Figure 14.4.   Dilution of an Enzyme Solution. a. The original solution has a concentration of 10 mg enzyme/milliliter. Each ★ represents a mg of enzyme. **b.** 1 milliliter of solution is removed containing 10 mg of enzyme. **c.** The 1 mL of solution is diluted with 4 mL of water. The resulting diluted solution contains 10 mg of enzyme in 5 mL. The concentration of enzyme is therefore 10 mg/5 mL = 2 mg/mL.

Answer:

The resulting diluted solution has a total volume of 5 mL and contains 10 mg of enzyme. The concentration of enzyme is:

$$\frac{10 \text{ mg enzyme}}{5 \text{ mL}} = \frac{2 \text{ mg enzyme}}{1 \text{ mL}}$$

Thus, the stock solution had an enzyme concentration of 10 mg/mL. The diluted solution has an enzyme concentration of 2 mg/mL.

Another example:

A stock solution initially has a concentration of 20 mg of solute per liter.

A diluted solution is prepared by removing 1 mL of stock solution and adding 14 mL of water.

a. What is the concentration of solute in the diluted solution?

b. How much solute is present in 1 mL of the diluted solution?

Answer:

a. The concentration in the stock solution is 20 mg/L. The solution was diluted 1/15, therefore, the concentration in the diluted solution is:

$$\frac{20 \text{ mg}}{1 \text{ L}} \times \frac{1}{15} = \frac{1.3 \text{ mg}}{1 \text{ L}}$$

b. This can be solved using the logic of proportions:

$$\frac{1.3 \text{ mg}}{1000 \text{ mL}} = \frac{?}{1 \text{ mL}}$$

$$? = 0.0013 \text{ mg}$$

This is the amount of solute present in 1 mL of the diluted solution.

Another way to think about concentration and dilution is using this rule:

The concentration of a diluted solution is determined by multiplying the concentration of the original solution times the dilution (expressed as a fraction).

For example:

A solution of 100% ethanol is diluted 1/10. The concentration of solute in the diluted solution is:

$$100\% \text{ ethanol} \times \frac{1}{10} = 10\% \text{ ethanol}$$

The diluted solution has a concentration of 10% ethanol.

(Because a dilution has no units, the result has the same units as the original solution.)

Another example:

A solution has an original concentration of 10 mg/mL of an enzyme.

The solution is diluted 1/5. The concentration of enzyme in the diluted solution is:

$$\frac{10 \text{ mg}}{1 \text{ mL}} \times \frac{1}{5} = \frac{2 \text{ mg}}{1 \text{ mL}}$$

Note that this is the same example as in Figure 14.4, simply described in a different way.

EXAMPLE PROBLEM

A stock solution of enzyme contains 10 mg/mL of enzyme. 100 mL of this stock is diluted with 400 mL of buffer.

a. What is the concentration of enzyme in the resulting diluted solution?

b. How much enzyme will be present in 300 μL of the resulting dilution?

ANSWER

a. When 100 mL of stock solution is mixed with 400 mL of buffer, the dilution can be expressed as:

$$\frac{100 \text{ mL}}{500 \text{ mL}} = \frac{1}{5}$$

The concentration of enzyme in the resulting dilution, therefore, is:

$$\frac{10 \text{ mg}}{1 \text{ mL}} \times \frac{1}{5} = \frac{2 \text{ mg}}{1 \text{ mL}}$$

b. 300 μL = 0.300 mL. Because there are 2 mg/mL enzyme in the diluted solution, in 0.300 mL there are:

$$\frac{2 \text{ mg}}{\text{mL}} \times 0.300 \text{ mL} = 0.6 \text{ mg}$$

It is also possible to use the logic of proportions:

$$\frac{2 \text{ mg}}{1 \text{ mL}} = \frac{?}{0.300 \text{ mL}}$$
$$? = 0.6 \text{ mg}$$

MANIPULATION PRACTICE PROBLEMS: DILUTIONS (PART C)

1. If you prepare a 1/40 dilution of a 50% solution, what is the final concentration of the solution?

2. If you prepare a 1/10 dilution of a 10 mg/mL solution, what is the final concentration of the solution?

3. If you prepare a 1:1 dilution of a 10 mg/mL solution, what is the final concentration of the solution?

4. How much 1/5 diluted solution can be made if you have 1 mL of original solution?

5. To prepare 1000 mL of food coloring at a 1/100 dilution, how much of the original food coloring stock solution is required?

F. Dilution Series

A **dilution series** *is a group of solutions that have the same components but at different concentrations.* One way to prepare a dilution series is to make each diluted solution independently of the others beginning with the initial concentrated stock solution. An example is explained here and is illustrated in Figure 14.5.

An Independent Dilution Series with Three Dilutions

How could you make 1/10, 1/50, and 1/100 dilutions of food coloring from an original bottle of concentrated food coloring?

First, decide how much of each dilution to prepare; for example, 10 mL.

a. To make 10 mL of a 1/10 dilution:

Take 1 mL of the stock and bring it to a volume of 10 mL using water. This is the first diluted solution.

b. To make 10 mL of a 1/50 dilution:

$$\frac{1 \text{ mL}}{50 \text{ mL}} = \frac{?}{10 \text{ mL}}$$
$$? = 0.2 \text{ mL}$$

Thus, take 0.2 mL of the original stock and dilute to a volume of 10 mL. This is the second diluted solution.

c. To make 10 mL of a 1/100 dilution:

$$\frac{1 \text{ mL}}{100 \text{ mL}} = \frac{?}{10 \text{ mL}}$$
$$? = 0.1 \text{ mL}$$

Thus, take 0.1 mL of the original stock and dilute to a volume of 10 mL. This is the third diluted solution.

Thus, this strategy requires removing some of the original stock solution three times, once to make each

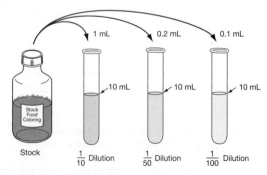

Figure 14.5.   A Dilution Series Where Each Dilution Is Independent of the Others.

dilution. Each of the three dilutions, therefore, is independent of the others. This strategy will work effectively in some situations.

Let us consider a situation where the strategy of preparing independent dilutions is not effective. Suppose you are working with bacteria and there are so many microorganisms in one mL of medium that the broth must be diluted 100,000 times to get a reasonable number of bacteria for counting. Thus, you want a 1/100,000 dilution of the original stock solution of bacteria. Assuming 10 mL final volume is needed, the proportion is:

$$\frac{1 \text{ mL}}{100,000 \text{ mL}} = \frac{?}{10 \text{ mL}}$$
$$? = 0.0001 \text{ mL}$$

To dilute the bacterial broth in one step would require taking 0.0001 mL of bacterial broth and diluting it to 10 mL. It is very difficult, however, to measure 0.0001 mL accurately. The common strategy to prepare such a dilution, therefore, is to use two or more steps. First, a dilution is prepared and then some of this first dilution is removed and used to make a second dilution. Some of the second dilution is used to make a third dilution and so on until the solution is dilute enough. The following example shows how one might prepare 10 mL of a 1/100,000 dilution of bacterial cells in three *steps,* see Figure 14.6.

A Dilution Series Where the Dilutions Are Not Independent of One Another

1. Mix 0.1 mL of the original broth with 9.9 mL of diluent. This will give a dilution of 0.1 mL /10 mL = 1/100.

⇓

2. After thorough mixing, remove 0.1 mL from the diluted bacterial broth prepared in Step 1 and add 9.9 mL of diluent; the total volume in the second dilution is 10 mL. The second dilution has a dilution factor of 1/100, as does the first dilution.

⇓

3. Remove 1 mL from the second dilution tube prepared in Step 2, and add 9 mL of diluent; the total volume in the third dilution tube is 10 mL. The third dilution is 1/10.

The bacteria were therefore diluted 1/100 in the first tube, 1/100 in the second tube, and 1/10 in the third tube. The final, total dilution is:

$$1/100 \times 1/100 \times 1/10 = 1/100,000.$$

The third tube has the desired dilution and volume.

Note in the preceding example that to determine the dilution of solute in the final dilution tube, the dilutions in each intermediate tube were multiplied by one another. Thus, the dilution of solute in the final dilution

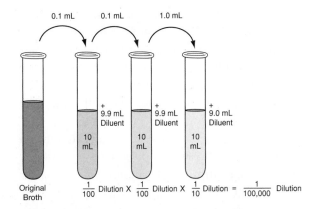

Figure 14.6. A Dilution Series Used to Dilute a Bacterial Stock Solution 1/100,000.

tube was $1/100 \times 1/100 \times 1/10 = 1/100,000$. The procedure to find the concentration of solute in the final tube can be generalized into the following rule:

> **The concentration of a diluted solution in the final tube is determined by multiplying the concentration of the original solution times the dilution in the first tube, times the dilution in the second tube and so on until reaching the last tube.**

Continuing with this bacteria example, suppose that the original broth contained 1×10^9 bacteria per milliliter. What is the concentration of bacteria in the third (final) dilution?

$$\frac{1 \times 10^9 \text{ bacteria}}{1 \text{ mL}} \times \frac{1}{100,000} = \frac{1 \times 10^4 \text{ bacteria}}{1 \text{ mL}}$$

Consider another example of a dilution series where the dilutions are not independent of one another, see Figure 14.7.

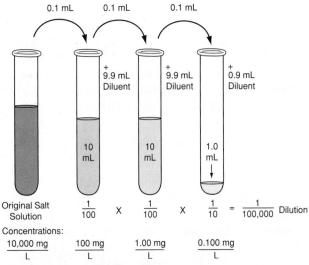

Figure 14.7. Diluting a Salt Solution.

A Dilution Series Where the Dilutions Are Not Independent of One Another

A solution contains 10 g salt/L. How could this solution be diluted to obtain 1 mL of solution with a salt concentration of 0.100 mg/L?

There are various strategies that will work. One example is:

1. Remove 0.1 mL of the original salt solution and add 9.9 mL of water. This is a 1/100 dilution.

   The concentration of salt in this dilution is 100 mg/L because:

original concentration		dilution		concentration after first dilution
$\dfrac{10{,}000 \text{ mg}}{1 \text{ L}}$	$\times$	$\dfrac{1}{100}$	$=$	$\dfrac{100 \text{ mg}}{1 \text{ L}}$

   $$\Downarrow$$

2. Remove 0.1 mL from the first dilution and place it in 9.9 mL of water. This is also a 1/100 dilution.

   The concentration of salt in this dilution tube is 1.00 mg/L because:

concentration after 1st dilution		dilution		concentration after second dilution
$\dfrac{100 \text{ mg}}{1 \text{ L}}$	$\times$	$\dfrac{1}{100}$	$=$	$\dfrac{1.00 \text{ mg}}{1 \text{ L}}$

   $$\Downarrow$$

3. Remove 0.1 mL from the second dilution and place it in 0.9 mL of water. This is a 1/10 dilution. The volume in this tube is 1 mL, as desired.

   The concentration of salt in this tube is 0.100 mg salt/L because:

concentration after 2nd dilution		dilution		concentration after final dilution
$\dfrac{1.00 \text{ mg}}{1 \text{ L}}$	$\times$	$\dfrac{1}{10}$	$=$	$\dfrac{0.100 \text{ mg}}{1 \text{ L}}$

   This is the desired final concentration and volume.

A **serial dilution** *is a series of dilutions that all have the same dilution factor (for example, all are 1/10 dilutions, or all are 1/2 dilutions).* (Note that when people use the phrase *dilution series*, the dilution factor may vary.) Figure 14.8 shows an example of a 1/10 serial dilution.

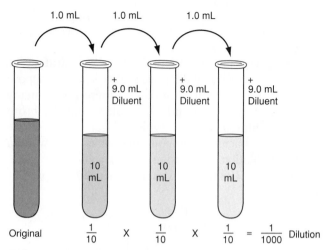

Figure 14.8. A Serial Dilution.

$$\frac{1 \text{ μg}}{1 \text{ μL}} \times \frac{1}{1000} = \frac{0.001 \text{ μg}}{1 \text{ μL}}$$

Then, 10 μL of the diluted plasmid solution will contain 0.01 μg of plasmid.

There are pipettes available that can accurately measure 10 μL volumes.

There are many strategies to dilute the stock plasmid solution 1000 times. One strategy is:

1. Remove 10 μL of the original plasmid stock and add 990 μL of buffer or water resulting in a 1/100 dilution.

2. Remove 10 μL from the dilution in step 1 and add 90 μL of buffer or water resulting in a 1/10 dilution. The total dilution in the second dilution tube is therefore:

$$1/100 \times 1/10 = 1/1000$$

The concentration of plasmid in this tube is 0.001 μg/μL, which is the desired concentration.

This two-step strategy involves volumes that can be measured with reasonable accuracy. This strategy also uses only a small amount of the stock solution, which might be an advantage if the stock is being saved for other experiments.

MANIPULATION PRACTICE PROBLEMS: DILUTIONS (PART D)

1. Dilution series: Explain how to prepare a 1/10 dilution of food coloring. Then, use the 1/10 dilution to prepare a 1/250 dilution of the original stock of food coloring. Use the 1/250 dilution to prepare a 1/1000 dilution of the stock of food coloring.

2. **a.** Explain how a dilution series could be used to prepare a $1/10^6$ dilution of bacterial cells.

   b. Explain how a 1/10 serial dilution could be used to prepare the same dilution of bacterial cells.

3. Explain how 1/5, 1/50, and 1/250 dilutions of a blood sample could be prepared.

EXAMPLE APPLICATION: ANSWERS (from p. 242)

The concentration of plasmid in the stock solution is 1 mg/mL, which is equal to 1 μg/μL. If the plasmid is drawn directly from the stock tube, only 0.01 μL is needed, which is a volume too small to measure accurately. Therefore, the stock must be diluted. If the stock is diluted 1000 ×, its concentration will be:

APPLICATION PRACTICE PROBLEMS: DILUTIONS

1. Suppose you have 20 μL of an expensive enzyme and you cannot afford to purchase more. The enzyme has a concentration of 1000 Units/mL. You are going to do an experiment that requires tubes with a concentration of 1 Units/mL of enzyme and each tube will have 5 mL total volume. How much enzyme does each tube require? _____ How many tubes can you prepare before you run out of enzyme? _____

2. Suppose you have 20 μL of an expensive enzyme that has 1000 Units/mL. You are going to use the enzyme in an experiment that requires tubes with 0.01 Units/mL of enzyme and each tube will have 5 mL total volume. How much enzyme will each tube require? _____ Will you be able to directly measure this amount of enzyme accurately? _____ Show how you can dilute 10 μL of the original 20 μL of enzyme so that you can use it in your experiment. Use a diagram and words to demonstrate your strategy for preparing the enzyme.

3. Antibodies are frequently very concentrated relative to how much is necessary in an experiment. Show how you would use a dilution series to dilute an antibody solution 500,000 ×. Assume you have 1 mL of antibody to begin with, but you want to save at least 0.5 mL of the antibody for future experiments.

4. Counting seems like a simple mathematical process; however, counting microorganisms is not so simple. For one thing, microorganisms, such as bacteria, are not visible to the eye. Another problem is that bacteria may be present in extremely large numbers in a sample. For example, in nutrient broth, there might be 1×10^9 bacterial cells per milliliter. Therefore, microbiologists have devised various methods to count bacterial cells. One such method is called viable cell counting. To perform a viable cell count, a sample of bacterial cells is first diluted in series. Then, 0.1 mL of diluted cells are spread on a petri dish that contains nutrient agar. It is assumed that every living cell in the 0.1 mL placed on the agar divides to form a colony of bacterial cells. A colony contains so many individual cells that it is visible to the eye. It is also assumed that each colony originates from a single cell. It is possible to count the colonies and therefore to estimate the number of bacteria in the 0.1 mL of broth.

   Assume you begin with a culture of bacteria that contains 1×10^9 bacteria/mL. Show how you could dilute the culture so that the final tube has a concentration of 200 bacteria/0.1 mL.

5. You are performing a viable cell count of bacteria. The original broth has an unknown concentration of bacteria. You dilute the culture as shown in the diagram below and plate 0.1 mL of the last three dilutions onto three petri dishes with nutrient agar. The following day you count the number of colonies on the plates with the results shown in the illustration. What was the concentration of bacteria in the original tube?

6. You are performing a viable cell count of bacteria. The original broth has an unknown concentration of bacteria. You dilute the culture as shown in the diagram on p. 250 and plate 0.1 mL of the last three dilutions on three nutrient agar plates. The following day you count the number of colonies on the plates with the results shown in the illustration. What was the concentration of bacteria in the original tube?

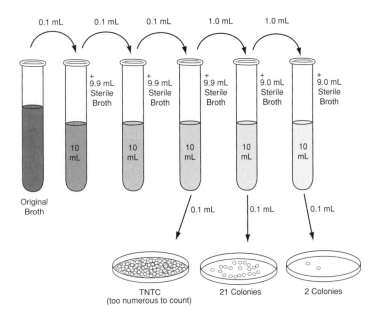

Diagram for Application Practice Problem 5.

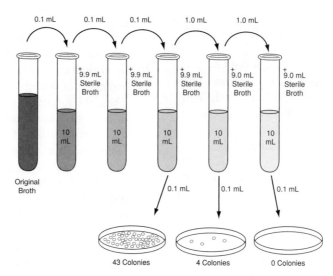

Diagram for Application Practice Problem 6.

7. Suppose you have a bacterial culture with 10^7 cells per milliliter. How would you dilute this culture so that if you plate 0.1 mL of the last dilution, you will get (in theory) 100 colonies?

8. Suppose you do an assay (test) to determine how much protein is present in a sample. To perform the assay, it is necessary to dilute the original sample 1/100. If the assay shows that the concentration of protein in the diluted sample was 50 mg/mL, what was the concentration of protein in the undiluted sample? How much protein was present in 100 mL of the original, undiluted sample?

9. In a protein assay, the amount of protein in 1 mL of diluted sample was 87 mg. If the original sample was diluted 1/50, what was the concentration of protein in the original sample?

10. In a protein assay, the original sample was first diluted 1/5. Then, 5 mL of the diluted sample was mixed with 20 mL of reactants to give 25 mL total. Five milliliters were removed from the 25 mL mixture. The 5 mL contained 3 mg of protein. What was the concentration of protein in the original sample?

11. In a protein assay, the original sample was diluted 1/4. Then, 10 mL of the diluted sample was added to 20 mL of reactants to give 30 mL total. Five milliliters were removed from the 30 mL mixture. The 5 mL contained 10 mg of protein. What was the concentration of protein in the original sample?

CHAPTER 15

Relationships and Graphing

I. GRAPHS AND LINEAR RELATIONSHIPS

A. Brief Review of Basic Techniques of Graphing

Graphs, like equations, are a tool for working with relationships between two (or sometimes more) variables. The general rules regarding graphing will briefly be reviewed in the first part of this chapter. The use of graphing as a tool to perform various tasks in a laboratory will be demonstrated in later sections.

A simple two-dimensional graph, Figure 15.1, consists of a vertical and a horizontal line that intersect at a point called the **origin.** *The horizontal line is the* **X axis,** *the vertical line is the* **Y axis.** The X and Y axes are each divided into evenly spaced subdivisions that are assigned numerical values. To the right of the origin on the X axis the X values are positive numbers; to the left of the origin the X values are negative. On the Y axis values above the origin are positive; below the origin, they are negative.

A basic two-dimensional graph shows the relationship between two variables, one of which is assigned to the X axis, the other to the Y axis. *For any value of a variable on*

the X axis, there is a corresponding value of a variable on the Y axis. The two values are called **coordinates** because they are coordinated or associated with one another. A pair of coordinates can be plotted on the graph as shown in Figure 15.2a on p. 252. *The distance of a point along the X axis is sometimes called the* **abscissa** *and the distance of*

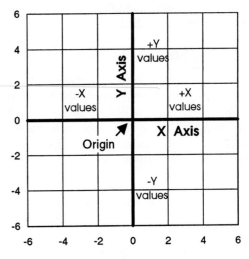

Figure 15.1. A Two-Dimensional Graph.

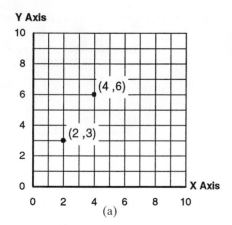

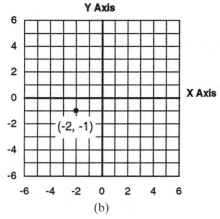

Figure 15.2.   Coordinates and Graphs. a. The points (2,3) and (4,6). **b.** The point $(-2,-1)$.

a point along the Y axis is sometimes called the **ordinate.** The first point shown on the graph has an X coordinate of 2 because it is above the 2 on the X axis, and a Y coordinate of 3, because it is across from the 3 on the Y axis. This point can be written as $X = 2, Y = 3$, or as (2,3). The second point on this graph has the coordinates (4,6). The point in Figure 15.2b has coordinates $X = -2, Y = -1$. To prepare a graph, one marks the locations of a series of points by using their coordinates.

B. Graphing Straight Lines

Two variables may be related to one another in such a way that when plotted on a graph, the points form a straight line. The two variables are then said to have a **linear relationship.** There are many applications in the laboratory that involve linear relationships.

Consider the simple equation:

$$Y = 2X$$

It is possible to find numbers that satisfy this equation. For example:

If $X = 3$, then $Y = 6$
If $X = 4$, then $Y = 8$
and so on

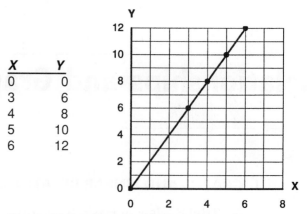

X	Y
0	0
3	6
4	8
5	10
6	12

Figure 15.3.   Graph of the Equation $Y = 2X$.

These numbers are summarized in both tabular and graphical form in Figure 15.3. Each pair of X and Y values are the coordinates of a point on the graph. The points form a straight line when connected.

In the linear equation $Y = 2X$, the value 2 is called the **slope.** In the equation $Y = 3X$ the slope is 3, and in the equation $Y = 4X$ the slope is 4. The equations $Y = 2X$, $Y = 3X$, and $Y = 4X$ are plotted on the same graph in Figure 15.4. Each equation is linear; the difference between the three lines is their steepness. Just as a hill may be more or less steep, so a line has a slope that is more or less steep. The equation $Y = 4X$ defines a steeper line than the other two equations because its slope, 4, is the largest.

$Y = 3X$		$Y = 4X$	
X	Y	X	Y
0	0	0	0
1	3	1	4
2	6	2	8
3	9	3	12
4	12	4	16
5	15	5	20

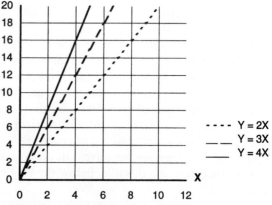

Figure 15.4.   Slope. The greater the value for the slope of a straight line, the steeper the line.

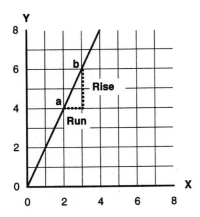

Figure 15.5.   Determining the Slope of a Line. The slope can be calculated based on the coordinates of any two points on the line. Slope = rise/run, which in this case = 2/1 = 2.

Given a straight line plotted on a graph, the slope of the line is calculated by determining how steeply the line rises. For example, consider the line for $Y = 2X$ in Figure 15.5. From point a to point b, X increases by 1 and Y increases by 2. *The amount by which the X coordinate increases is called the* **run;** *the amount by which the Y coordinate increases is called the* **rise.** The rise divided by the run is a numerical measure of the steepness of the slope. To calculate the slope of any straight line, choose any two points on the line. The coordinates for the first point are (X_1, Y_1) and for the second point are (X_2, Y_2). Then:

$$\text{slope} = \frac{\text{rise}}{\text{run}} = \frac{\text{change in } Y}{\text{change in } X} = \frac{Y_2 - Y_1}{X_2 - X_1}$$

For the line $Y = 2X$:

$$\text{slope} = \frac{\text{rise}}{\text{run}} = \frac{2}{1} = 2$$

Any two points on the line $Y = 2X$ will give a slope of 2. Any two points on the line $Y = 3X$ will give a slope of 3 and any two points on the line $Y = 4X$ will yield a slope of 4.

Figure 15.6 shows the graph of a line that goes "downhill" from left to right. For this line, as the X values increase, the Y values decrease. The slope is therefore negative.

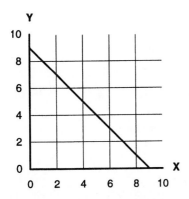

Figure 15.6.   A Line with a Negative Slope. All lines that go "downhill" from left to right have a negative slope.

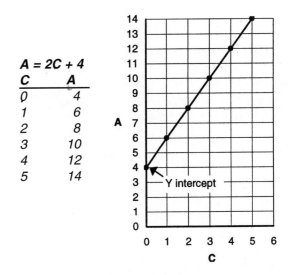

Figure 15.7.   A Graph with a Nonzero Y Intercept.

Figure 15.7 shows the plot for the equation $A = 2C + 4$. (Substitution of other symbols for X and Y does not change the basic meaning of the equation.) The difference between the plot of $Y = 2X$ and $A = 2C + 4$ is that the latter line is higher and does not pass through the origin. The line $A = 2C + 4$ intercepts, or passes through, the Y axis at 4. *The point at which a line passes through the Y axis, that is, where X = 0, is termed the* **Y intercept.** If the line passes through the Y axis at the point where both X and Y are zero, the intercept is often not written. Thus, the Y intercept of the line $Y = 2X$ is 0.

If an equation gives a straight line when it is graphed, then it will have the form:

$$Y = slope(X) + Y\ intercept$$

This general equation for a straight line is sometimes written:

GENERAL EQUATION FOR A STRAIGHT LINE

$$Y = mX + a$$
where
m = the slope and
a = the Y intercept

It is possible to determine the equation for a straight line from its graph. For example, the line drawn in Figure 15.8 on p. 254 has a Y intercept of 3 and a slope of 0.5; therefore the equation for this line is: $Y = 0.5\ (X) + 3$. The general procedure to find the equation for any graphed straight line is shown in Box 15.1 on p. 254.

EXAMPLE PROBLEM

Which of the following equations describe a straight line?

a. $C = 2B$

b. $Q = 25.4T - 5$

c. $Y = X^2 + 3$

d. $C = -V - 34$

ANSWER

All except c are equations for a straight line. The exponent in equation c means that it does not describe a straight line. The rest of the equations have two variables, a slope, and a Y intercept, which may or may not be zero. Equation b fits the linear equation as shown here:

$$Y = mX + a$$
$$Q = 25.4\ T + (-5)$$

Equation d is also the equation for a line:

$$Y = mX + a$$
$$C = (-1)\ V + (-34)$$

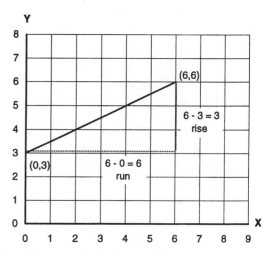

Figure 15.8. Determining the Equation for a Line.
Slope = $(6-3)/(6-0)$ = 0.5. Y intercept = 3. Equation is $Y = 0.5X + 3$.

EXAMPLE PROBLEM

Heating a solution of double-stranded DNA disrupts the hydrogen bonds holding the two strands together and causes the strands to separate from one another. The midpoint of the temperature range over which the strands separate is called the *melting temperature* (T_m). There is a linear relationship between the guanine + cytosine (G + C) content of the DNA and its T_m: The higher the G + C content, the higher the melting temperature. DNA molecules from different organisms have different melting temperatures because their G + C content varies. There are many techniques performed in the laboratory that require the separation of double-stranded DNA. A graph of melting temperature versus G + C content is shown on the next page.

a. What is the melting temperature for DNA that has 62% guanine + cytosine?

b. What is the melting temperature for DNA that is 15% guanine + cytosine?

c. What is the slope of the line?

d. In order to find the equation for this line, it is necessary to know the value for the Y intercept. Observe that, for convenience, the graph is drawn so that the X axis begins at 70°C. It is possible to redraw this graph so that the X axis begins at 0°C. If you were to redraw the graph with the X axis beginning at zero, you would see that the Y intercept is at −172% G + C. In reality, of course, there is no such thing as a DNA molecule with a negative G + C content, nor is there a DNA molecule whose strands separate at 0°C. Nonetheless, the Y intercept for the line on the graph is −172% G + C. What is the equation for the line?

e. Use the equation to determine the melting temperature for DNA that has 50% G + C content.

BOX 15.1. PROCEDURE TO FIND THE EQUATION FOR A STRAIGHT LINE ON A GRAPH

Straight lines can be described by an equation in the form:

$$Y = mX + a$$

where *m* is the slope and
a is the Y intercept

1. Find the Y intercept, that is, the value of Y when $X = 0$. The intercept may be positive or negative.

2. Find the slope by picking any two points and calculating $(Y_2 - Y_1)/(X_2 - X_1)$. The slope may be positive, negative, or zero. (A horizontal line has a slope of zero.)

3. Put the slope and intercept into the proper form by filling in the blanks for slope and Y intercept:

$$Y = \underset{\text{slope}}{\underline{\hspace{1cm}}} (X) + \underset{\text{Y intercept}}{\underline{\hspace{2cm}}}$$

Note: The axes of a graph can have units in which case the slope and the intercept will also have units. Include the units when writing the equation for a line.
Note: This pattern does not fit vertical lines, such as X = 2.

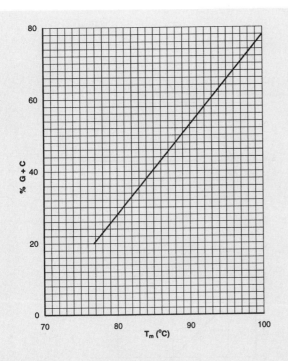

ANSWER

a. From the graph, you can see that the melting temperature for DNA with a G + C content of 62% is about 93.5°C.

b. The graph only extends as low as 20% G + C and therefore it is not possible to use it to determine the melting point for 15% guanine + cytosine. In general, when graphs portray real data collected from samples (as contrasted with theoretical relationships), do not extrapolate past the data provided.

c. The slope of the line can be calculated from any two points that are on the line. For example, 100°C on the X axis corresponds to a G + C content of 78%. 77°C on the X axis corresponds to 20% G + C. The slope is therefore:

$$\frac{(78\% - 20\%)}{(100°C - 77°C)} \approx \frac{2.5\%}{°C}$$

d. The equation for the line is

$$Y = \frac{(2.5\%)}{°C}(X) - 172\%$$

e. Substituting into the equation:

$$50\% = \frac{(2.5\%)}{°C}(X) - 172\%$$

$$\frac{222\%}{\left(\frac{2.5\%}{°C}\right)} = X$$

$$88.8°C = X$$

You can confirm that this is the right answer by looking at the graph.

MANIPULATION PRACTICE PROBLEMS: GRAPHING

1. For each of the following graphs, describe in words the relationship that is displayed.

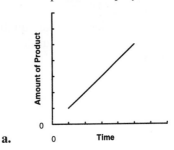

a.

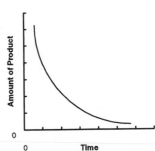

b.

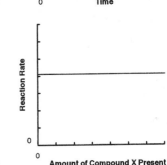

c.

2. The following figure shows a graph with four points, labeled a, b, c, and d. Write the coordinates for each of the four points.

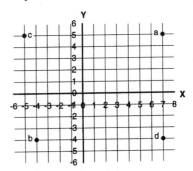

3. In the graph in Problem 2, as you move from point a to point b, by how much does the value of X change? _____ By how much does the value of Y change? _____ As you go from point b to point c, how much does the value of X change? _____ How much does the value of Y change? _____

4. Draw a graph and plot each of these points on the graph:

$$X = 4, Y = 6$$
$$X = 5, Y = -2$$
$$(-4,3)$$
$$(1,1)$$

5. A table showing three values for the equation $Y = 5X + 1$ is given. Fill in the blanks in the table, and then plot the equation.

X	Y
1	6
5	26
10	51
12	__
__	76
__	101

6. Suppose two variables are related by the equation, Variable $A = 3$ (Variable Q) $- 4$. Prepare a table to show this relationship and then graph the relationship.

7. What is the slope and the Y intercept for each of these equations?

    a. $Y = 3X + 2$
    b. $C = 0.2X - 1$
    c. $Y = 0.005X$

8. What is the slope and Y intercept for lines a–d?

c.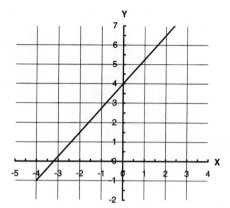

d.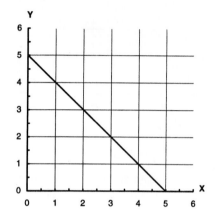

a.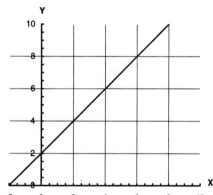

b.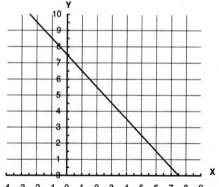

9. Which of these graphs below shows a linear relationship between two variables?

a. 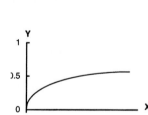   **b.**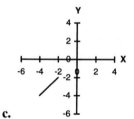

c.   **d.**

10. Which of these equations will form a straight line when plotted on a graph?

    a. $Y = 45X - 1$ **b.** $C = 34D + 17$
    c. $34 + 2 - 4D = E$ **d.** $Q = R$

11. a. For each of the following equations, what is the slope?
  b. For each of the following equations, what is the Y intercept?
  c. Graph the equations.

     i. $Y = (10 \text{ cm/min}) X + 1 \text{ cm}$
    ii. $3 - X = Y$
   iii. $(12 \text{ mg}) + (7 \text{ mg/cm}) X = Y$

12. a. Draw the line that has a Y intercept of 2, that is, $(0, 2)$, and a slope of 0.25.
  b. Draw the line that goes through the point $(-4, 3)$ and has a slope of -2.

13. Find the equations for the lines graphed in each of the following examples.

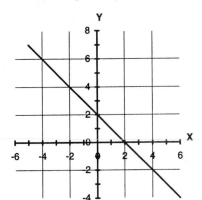

a.

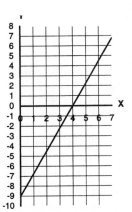

b.

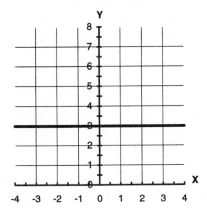

c.

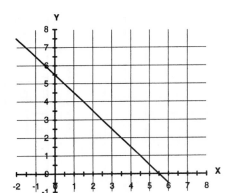

d.

14. The following figure has three lines. Match each line with its equation:

$$Y = \frac{1}{2}X + 2 \qquad Y = \frac{3}{4}X \qquad Y = 2X + 2$$

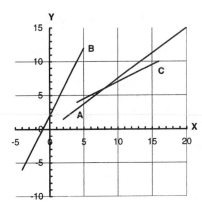

15. Plot the line that contains the following points: $(1,1), (6,6), (9,9)$. What is the equation for this line?

16. The relationship between the temperature in degrees Fahrenheit and degrees Celsius is given by the equation: $°F = 9/5(°C) + 32$.
  a. There are two constants in this equation. What are they?
  b. Which of the two constants is the Y intercept?
  c. Which of the two constants is the slope of the line?
  d. Graph this relationship.

C. An Application of Graphing Linear Relationships: Standard Curves and Quantitative Analysis

This section discusses an important laboratory application of graphs of linear relationships: quantitative analysis. **Quantitative analysis** *is the determination of how much of a particular material is present in a sample.* To make such determinations, a standard curve is used. A **standard curve** *is a graph of the relationship between the concentration of a material of interest and the response of a particular instrument.* Note that the term *standard curve* is used for this sort of graph; however,

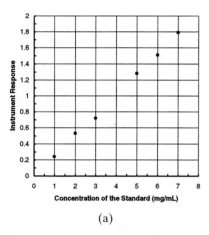
(a)

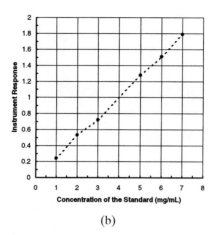
(b)

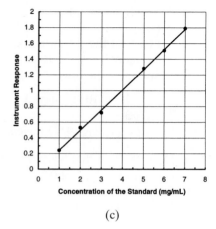
(c)

Figure 15.9. **A Standard Curve. a.** Data from Table 15.1 are graphed. The concentration of compound in the standards is on the X axis, instrument response is on the Y axis. **b.** The points are not connected "dot to dot," as shown here, because this would suggest that slight variations from a straight line are meaningful. **c.** The points are connected into a "best fit line," which best averages all the points.

the desired relationship between concentration and instrument response is usually linear.

A standard curve is constructed as follows: A series of standards containing known concentrations of the material of interest are prepared. The response of an instrument to each standard is measured. A standard curve is plotted with the response of the instrument on the Y axis and the concentration of standard on the X axis. Once graphed, the standard curve is used to determine the concentration of the material of interest in the samples. This process is illustrated in the following example:

1. Standards are prepared containing 1.0 mg/mL, 2.0 mg/mL, 3.0 mg/mL, 5.0 mg/mL, 6.0 mg/mL, and 7.0 mg/mL of a compound of interest.
2. An instrument's response to each standard is measured, see Table 15.1.
3. The resulting data are plotted as points on a graph with the concentration of standard on the X axis and the instrument response on the Y axis, see Figure 15.9a.
4. The points are connected into a line, see Figures 15.9b, c.
5. The standard curve is used to determine the concentration of the material of interest in samples, Figure 15.10.

Observe in Figure 15.9a that the points approximate a line, but they do not all fall exactly on the line. This is

Table 15.1. **Concentration of Standards versus Instrument Response**

Concentration of the Standard (mg/mL)	Instrument Response
1.0	0.24
2.0	0.53
3.0	0.72
5.0	1.28
6.0	1.51
7.0	1.79

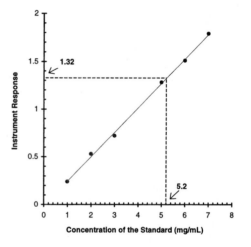

Figure 15.10. **Using a Standard Curve to Determine the Concentration of a Material in a Sample.** In this example a sample gives an instrument reading of 1.32, which corresponds to a concentration of 5.2 mg/mL.

what we would expect due to small errors in measurement. (Measurement errors are discussed in Chapter 17.) In a situation like this, where it is reasonable to assume that the relationship is linear, we connect the points into a straight line. The points are not connected "dot to dot," as illustrated in Figure 15.9b, because this would suggest that the slight variations from the line are meaningful. Rather, the points are connected into a single line that "best fits," or best averages, all the points, as shown in Figure 15.9c.

There are two ways to get a "best fit" line for the points on a graph. One method is to place a ruler over the graph and draw a straight line that appears "by eye" to be closest to all the points. Most people are able to draw a reasonable "best fit" line "by eye," although two people will seldom draw a line with exactly the same slope and Y intercept. *A more accurate method to draw a line of best fit is the statistical technique,* **the Least Squares Method.** (This statistical method is explained in an Appendix to Chapter 25.)

Regardless of whether the points are connected into a line "by eye" or using the Method of Least Squares, the standard curve can be used to determine the level of compound in each unknown sample. In the example illustrated in Figure 15.10, a sample gives an instrument reading of 1.32. From the standard curve, one can see that a reading of 1.32 corresponds to a concentration of 5.2 mg/mL.

EXAMPLE PROBLEM

Biologists commonly use a protein assay to determine the concentration of protein in a solution. Various protein assays are available, many of which measure the color change in a protein solution when it reacts with various dyes. The amount of color appearing is generally proportional to the amount of protein present.

The following graph shows the relationship between the concentration of protein in a series of standards and the amount of color after the standards are reacted with dye. The amount of color is measured in terms of the amount of light absorbed by the dye.

a. In what range of protein concentration does the assay give linear results? What happens at higher concentrations of protein?

b. Suppose you have a sample containing an unknown amount of protein. The sample is reacted with the dye and has an absorbance of 0.70. Based on the standard curve, what is the concentration of protein in the unknown?

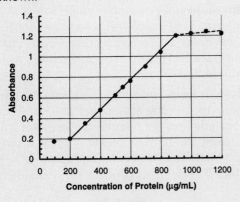

Concentration of Protein (μg/mL)

ANSWER

a. The assay is linear in the middle range; above about 900 μg/mL and below 200 μg/mL the assay is not useful because the absorbance does not change, even if the concentration of protein changes.

b. The concentration of the unknown is about 550 μg/mL.

EXAMPLE PROBLEM

The concentration of compound Z in samples needs to be determined. The response of an instrument to compound Z is related to its concentration. A series of standards are prepared and the instrument's response is measured. The resulting data are shown in the following table.

a. Plot a standard curve for compound Z based on the data in the table. Draw the line that best fits the points.

b. A sample with an unknown concentration of compound Z gives an instrument response of 1.35. Based on the standard curve, what is the concentration of compound Z in the sample?

c. Suppose that the response of the instrument used in this example is known to be inaccurate at readings above 2. If a sample has a reading of 2.67 how can you determine its concentration of compound Z?

d. If a sample is diluted 1/20 and has an instrument reading of 0.45, what was the concentration of compound Z in the original sample?

Concentration of Compound Z in Standard (mg/mL)	Instrument Response
100	0.30
150	0.44
250	0.68
400	1.13
550	1.48
650	1.82
850	2.10

ANSWER

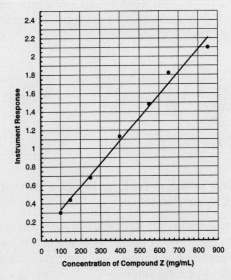

Concentration of Compound Z (mg/mL)

b. Based on the standard curve, the concentration of compound Z in the sample is 500 mg/mL.

c. It is necessary to dilute the sample before reading its absorbance.

d. A value of 0.45 corresponds to a concentration of 150 mg/mL; however, the sample was diluted and this dilution needs to be taken into consideration. We therefore multiply 150 mg/mL times the inverse of the dilution.

$$150 \text{ mg/mL} \times \frac{20}{1} = 3000 \text{ mg/mL}$$

The concentration of compound Z in the original undiluted sample was 3000 mg/mL.

MANIPULATION AND APPLICATION PRACTICE PROBLEMS: QUANTITATIVE ANALYSIS

1. a. Plot a standard curve based on the data in the following table. Draw the line that best fits the values.

   b. Suppose you have a sample that gives an instrument response of 2.35. This value is higher than the reading of the highest standard; therefore, you dilute the sample 1/100. The instrument reading of the diluted sample is 0.78. What amount of the material of interest was present in the original solution?

Amount of Standard (in grams)	Instrument Response
0	0.00
10	0.14
20	0.28
30	0.41
40	0.52
50	0.71
65	0.84
75	1.04
80	1.15

2. A stock solution of copper sulfate has a concentration of 100 mg/mL. How would you dilute the stock solution to prepare each of the following standards?

1 mg/mL
5 mg/mL
15 mg/mL
25 mg/mL
50 mg/mL
75 mg/mL
100 mg/mL

D. Using Graphs to Display the Results of an Experiment

Many experiments involve manipulating one variable and measuring the result of that manipulation on a second variable. For example, an investigator interested in the effect of light intensity on the rate of seedling growth could expose different groups of seedlings to different light intensities and measure their growth rates. The two experimental variables—light intensity and seedling growth rate—can be plotted on a two-dimensional graph.

When plotting data from experiments, one distinguishes between the **dependent variable** and the **independent variable.** In the preceding example, the investigator is looking at whether seedling growth rate is dependent on the light intensity. Growth rate is therefore called the dependent variable. *A variable the investigator controls is an* **independent variable.** *A variable that changes in response to the independent variable is called a* **dependent variable.** It is conventional to plot

the dependent variable on the Y axis and the independent variable on the X axis. In this example, light intensity is plotted on the X axis and seedling growth rate on the Y axis.

Let us consider how the results of a hypothetical experiment might be displayed graphically. Suppose investigators are interested in the effects of a plant hormone on the number of fruits produced by a certain plant. Investigators perform an experiment to determine the relationship, if any, between this hormone and fruit production. The investigators divide plants into 11 groups, each of which is treated identically except for the application of differing levels of the hormone. The investigators count the fruits produced by each plant. The results of this hypothetical experiment are shown in tabular form in Table 15.2 and graphically in Figure 15.11. Fruit production is the dependent variable and is plotted on the Y axis; applied hormone concentration is on the X axis. These data strongly suggest that the hormone boosts fruit production.

There are several important concepts illustrated by this graph:

1. Thresholds. In the central portion of the graph the points appear to form a straight line (i.e., there appears to be a linear relationship between hormone level and fruit production). Below about 5.0 mg/L of hormone and above about 50.0 mg/L of hormone the relationship changes: 5.0 mg/L and 50 mg/L are threshold values. A **threshold** *is a point on a graph where there is a change in the relationship between the variables.* Thresholds at low and high values, as illustrated in Figure 15.11a, are very common when working with biological data.

2. Best Fit Line. The middle points on the graph are close to forming a line, and it is reasonable to conclude that the relationship between fruit production and hormone level is linear between 5.0 and 50.0 mg/L of applied hormone. It is acceptable, therefore, to connect these middle points

Table 15.2. EXPERIMENTAL DATA

Hormone Level (mg/L) (independent variable)	Average Fruit Production per Plant (dependent variable)
0.0	3.2
3.0	3.7
5.0	3.8
10.0	6.5
15.0	12.7
20.0	15.2
25.0	17.0
30.0	24.8
40.0	32.3
50.0	36.0
55.0	36.2
60.0	36.5

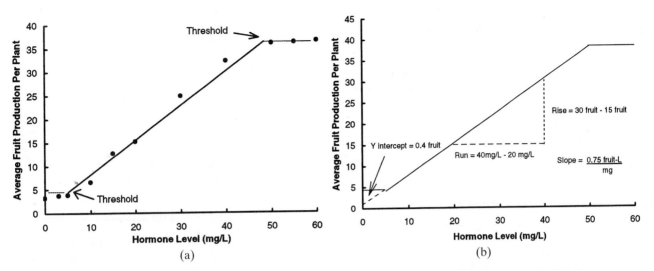

Figure 15.11. Hypothetical Experiment. a. The effect of hormone application on fruit production. **b.** Determining the equation for the line.

into a straight line. A line may be drawn either "by eye" or using the statistical Method of Least Squares.

Once the points are connected it is possible to determine the equation for the line. The slope has units because both the X and Y axes have units. The Y axis has units of "average fruit production per plant," or, more simply, "fruit"; the X axis has units of milligrams per liter. The slope of the line, Figure 15.11b, is:

$$\frac{(0.75 \text{ fruit})L}{mg}$$

To determine the Y intercept of the line, it is necessary to determine where the intercept would be if there were no lower threshold. By using a ruler to extend the line to the Y axis, the Y intercept can be determined to be 0.40 fruit, Figure 15.11b. The equation for this line in Figure 15.11 is therefore:

$$Y = \frac{(0.75 \text{ fruit}) L}{mg}X + 0.40 \text{ fruit}$$

The slope is the amount by which average production of fruit per plant increases with each milligram per liter of increased hormone. The Y intercept is the amount of fruit production expected if there were no hormone added and if the relationship was linear for all values of hormone (which is not the case).

3. **Prediction.** Equations and graphs are tools that can be used to make predictions. For example, the experiment did not involve testing the effect of 13.0 mg/L of hormone on fruit production. We can, however, infer from the graph that if 13.0 mg/L of hormone were to be applied, the average fruit production per plant would be about 10. It is also possible to use the equation to predict the amount of fruit with 13.0 mg/L of hormone:

$$Y = \frac{(0.75 \text{ fruit}) L}{mg} \frac{(13.0 \text{ mg})}{L} + 0.4 \text{ fruit}$$
$$\approx 10.2 \text{ fruit}$$
= predicted average fruit production per plant

The use of equations and graphs for prediction is very powerful; however, when studying natural systems there are often thresholds. We cannot predict how much fruit production there will be with 100 mg/L of hormone because we do not have data at 100 mg/L. It is seldom possible to make predictions about biological phenomena past the range of the data.

4. **Using Graphs to Summarize Data.** The graph of the hypothetical experiment contains the same information as Table 15.2; however, it is usually easier to see a relationship between two variables when the data are displayed on a graph than when they are listed in a table.

So far, we have looked at data where two variables are indeed related to one another. What would a graph look like if the two variables studied are not related to one another? One possibility is shown in Figure 15.12a on p. , where the points appear to be scattered without pattern on the graph. Another possibility is shown in Figure 15.12b, where the graph is "flat." The value of the Y variable is constant regardless of the value for the variable on the X axis. Observe in the hypothetical fruit and hormone experiment, Figure 15.11, the graph of the data is "flat" at high and low hormone levels. We can, therefore, reasonably conclude that at low and high levels of hormone, the production of fruit is controlled by factors other than the level of this hormone.

Figures 15.10 (p. 258) and 15.11 both illustrate situations where the variable on the Y axis clearly appears to be related to the variable on the X axis. The graphs in Figure 15.12 illustrate situations where two variables clearly appear to be unrelated to one another. In practice, it is

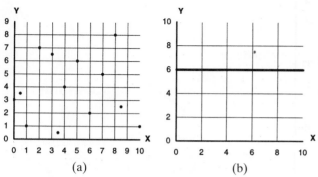

Figure 15.12.   Graphs of Variables That Are Not Related.

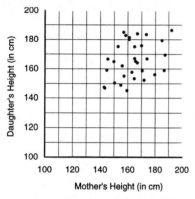

Figure 15.13.   Two Variables Whose Relationship to One Another Is Ambiguous.

sometimes ambiguous whether or not there is a relationship between two variables. For example, in Figure 15.13 there appears to be a weak relationship between the adult height of a daughter and the height of her mother. The relationship is inconsistent, however, because maternal height is not the only factor affecting a woman's adult height. Paternal genes and nutrition also affect height. The plot of daughter versus mother's height, therefore, does not form a very "good" line. (*A graph containing this type of data is called a* **scatter plot** *because the data points are scattered.*)

APPLICATION PRACTICE PROBLEMS: GRAPHING

1. It is important to know whether exposure to agents such as low-level radiation, pesticides, or asbestos is likely to cause cancer. Two different predictions of cancer risk due to exposure to small amounts of cancer-causing materials are shown in the graphs.

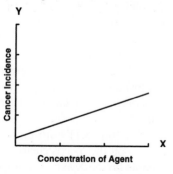

i. No Threshold

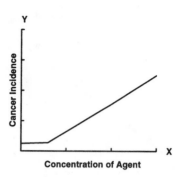

ii. Threshold at Low Exposure Levels

a. If graph *i* is correct (no threshold) and a population of humans is exposed to small amounts of potentially carcinogenic materials, will the incidence of cancer in the population increase? Explain.

b. If graph *ii* is correct, will exposure to small amounts of potentially carcinogenic materials result in an increase in the incidence of cancer?

c. Speculate as to why exposures to small amounts of carcinogenic materials may not increase the likelihood of cancer while exposure to large amounts does increase the probability of cancer.

d. Explain why neither graph *i* nor *ii* intersect the Y axis at zero.

e. Why is it important to know whether graph *i* or *ii* more accurately reflects the truth about a given material? (To learn more about this issue, see, for example, Goldman Marvin, "Cancer Risk of Low-Level Exposure," *Science* 271:1821–22, 1996.)

2. Suppose an investigator is studying the genetics of plant productivity. The investigator determines the mass of 500 parent plants and 1000 of their offspring plants. The investigator plots parent plant mass versus the average of offspring mass.

a. Which of the following graphs (*i* or *ii*) would indicate that there is a relationship between the mass of the parent plant and the mass of the offspring? Explain.

b. Suppose the data suggest that there is no relationship between the mass of the parent plant and the mass of its offspring. Suggest a hypothesis to explain that observation and suggest an experiment to test your hypothesis.

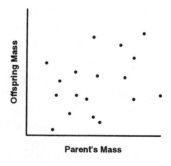

(*i*)

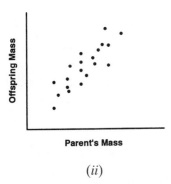

(ii)

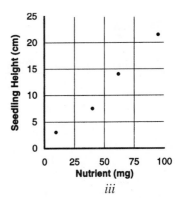

iii

3. a. Draw the best-fit line "by eye" for each of the following graphs. Be careful not to extrapolate (extend the line) past the data.

b. Calculate the slope for each of the lines—do not forget the units.

c. What is the Y intercept for each line?

d. What is the equation for each of these lines? Include the units.

e. Examine graph *i* to determine the mosquito density if there are 5 in of rain.

f. Use the equation for the line in graph *i* to predict mosquito density if there are 5 in of rain. (The answers for 3e and 3f should be the same.)

g. From graph *ii* determine the average shrub's height at 20 months of age. Confirm your determination using the equation for the line in graph *ii*.

h. From graph *iii* determine average seedling height with 50 mg of nutrient. Confirm your determination using the equation for the line in graph *iii*.

4. Ten students took a midterm and final in a course. The scores for each student are plotted in the next graph.

a. What was the approximate average score for the midterm; 20, 40, 60, or 80?

b. What was the approximate average score for the final; 20, 40, 60, or 80?

c. Which exam had lower scores?

d. Was there a clear relationship between a student's midterm grade and their final grade? If the same class and the same exams were given the following year, could the teacher predict the final score of a student based on their midterm? Explain.

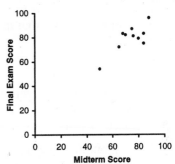

II. GRAPHS AND EXPONENTIAL RELATIONSHIPS

A. Growth of Microorganisms

i. THE NATURE OF EXPONENTIAL RELATIONSHIPS

Previous sections of this chapter focused on relationships that form a straight line when graphed. Although linear relationships are extremely important, not all relationships in the laboratory are linear. This section discusses two important examples of nonlinear relationships: (1) the relationship between the number of bacteria present and time elapsed; (2) the relationship between the amount of radioactivity present and time elapsed.

Suppose there is a single bacterial cell that divides to form two cells. The two cells each divide to form four cells, which divide into eight cells and so on. Suppose that the cells divide every hour and that the number of bacterial cells therefore doubles every hour. We say that these bacteria have a generation time of 1 hour.

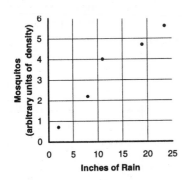

i

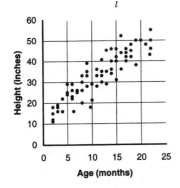

ii

Table 15.3.   BACTERIAL GROWTH

Time Elapsed (hours)	Number of Bacterial Cells Present (N)
0	1
1	2
2	4
3	8
4	16
5	32
6	64
7	128

The relationship between time elapsed and number of bacteria in this example is summarized in Table 15.3 and graphed in Figure 15.14. Note that this relationship does not form a straight line.

How could we write an equation that describes the relationship graphed in Figure 15.14? There are two variables: the time elapsed and the number of bacteria. The equation needs to include both variables to show that the population doubles at a regular interval. The equation that describes this relationship is:

$$Y = 2^x$$

where

x = the number of generations that have elapsed

Y = the number of bacterial cells present

For example:
when $x = 2$, 2 generations have elapsed
$$Y = 2^2 = 4$$
the number of bacteria cells present is 4

when $x = 4$, 4 generations have elapsed
$$Y = 2^4 = 16$$
the number of bacteria cells present is 16

Table 15.4.   BACTERIAL GROWTH BEGINNING WITH 100 CELLS

Time Elapsed (hours)	Number of Bacterial Cells Present (N)
0	100
1	200
2	400
3	800
4	1600
5	3200
6	6400
7	12,800

Now, suppose that there are initially 100 bacterial cells that double as before. The equation that describes these data is:

$$Y = 2^x(100)$$

This example is illustrated in tabular form in Table 15.4 and graphically in Figure 15.15. The graph where there are initially 100 cells present is much like the graph where there is initially only one bacterium, but with a Y intercept of 100.

Being able to calculate bacterial numbers is important in the laboratory in situations where we want to predict approximately how many bacteria will be present in a culture after a certain time, or want to know how long a culture must be incubated to get a certain density of cells. It is possible to write a general growth equation that applies to any bacterial population, regardless of how many cells there are initially and regardless of how long it takes the population to double:

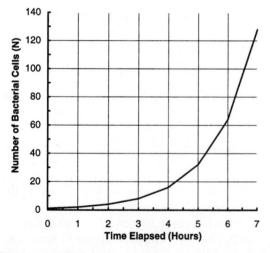

Figure 15.14.   Bacterial Growth. A plot where there is initially only one bacterial cell present and the bacteria population doubles every hour. A graph of this form is called *exponential*.

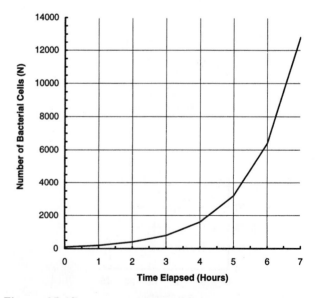

Figure 15.15.   Bacterial Growth. A plot where there are initially 100 bacterial cells present and the bacteria population doubles every hour.

GENERAL EQUATION FOR BACTERIAL POPULATION GROWTH

$$N = 2^x (N_o)$$

where

N_o = the number of bacteria initially
N = the number of cells after x generations
x = the number of generations elapsed
(for example, if the bacteria double every 2 hours, then, after 6 hours three generations have elapsed)

EXAMPLE PROBLEM

A type of bacterium has a doubling time of 20 minutes. There are initially 10,000 bacteria present in a flask. How many bacteria will there be in one hour?

ANSWER

STRATEGY 1: USING A TABLE

Time Elapsed	Number of Generations Elapsed (X)	Number of Bacteria Present (N)
0	0	10,000
20 min	1	20,000
40 min	2	40,000
60 min	3	80,000

Thus, after 60 minutes we predict there will be 80,000 bacteria in the flask.

STRATEGY 2: USING THE EQUATION FOR BACTERIAL POPULATION GROWTH

Three generations have elapsed (the first at 20 minutes, the second at 40 minutes, and the third at 60 minutes). Substituting the number of generations elapsed and the number of cells originally present into the equation gives:

$$N = 2^x (N_o)$$
$$N = 2^3 (10,000)$$
$$N = 8 (10,000)$$
$$N = 80,000$$

The equation for population growth and the "intuitive" approach both give the same answer.

These relationships between the number of bacteria present and the time (or number of generations) elapsed are called **exponential** *because there is an exponent in their equation.* As you can see in Figures 15.14 and 15.15, an exponential relationship is not linear if it is plotted on a regular two-dimensional graph.

ii. SEMILOG PAPER

It is more convenient to work with graphs that are linear in form than with those that are not. It is often

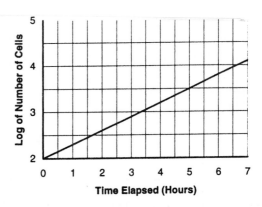

Figure 15.16. A Log Plot of Bacterial Growth. When the log of bacterial cell number is plotted versus time, a straight line is formed.

desirable, therefore, to convert an exponential relationship to a form that is linear when plotted. The way this is accomplished involves logarithms. If, instead of plotting the number of bacterial cells on the Y axis, we plot the log of the number of bacterial cells, then the plot is linear. This approach is illustrated in tabular form in Table 15.5 and graphically in Figure 15.16.

There is an alternative method to graph bacterial growth so that the plot is linear in form. This method involves the use of **semilog paper,** *a type of graph paper that substitutes for calculating logs.* Semilog graph paper has normal, linear subdivisions on the X axis, see Figure 15.17 on p. . The divisions on the Y axis, however, are not even; rather, they are initially widely spaced and then become narrower toward the top of the paper. This spacing takes the place of calculating logs. The graph paper shown in Figure 15.17 is called "semilog" paper because only the Y axis has logarithmic spacing. It is also possible to buy log-log paper where both axes are logarithmic.

Observe in Figure 15.17a that the Y axis has 10 major divisions. These ten divisions together are called a "cycle." The first division in a cycle never begins with 0 because there is no log of 0. Rather, the first division is a power of 10, for example, 0.1, 1, 10, or 100. The second division in each cycle is twice the first, the third division is three times the first, and so on. For example, the Y

Table 15.5.	BACTERIAL GROWTH—LOG OF NUMBER OF CELLS	
Time Elapsed (hours)	**Number of Cells Present (N)**	**Log of Number of Cells**
0	100	2.00
1	200	2.30
2	400	2.60
3	800	2.90
4	1600	3.20
5	3200	3.51
6	6400	3.81
7	12,800	4.11

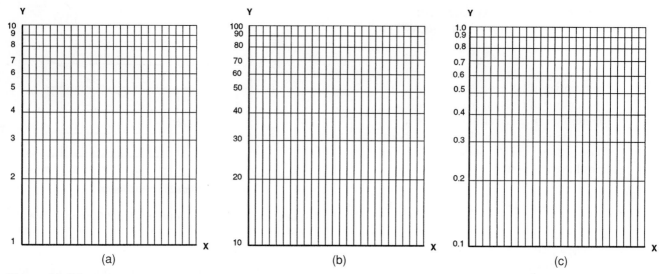

Figure 15.17.   Labeling Semilog Paper. a. Labeling the Y axis when the first division is 1. **b.** Labeling the Y axis when the first division is 10. **c.** Labeling the Y axis when the first division is 0.1.

axis in Figure 15.17a begins at the bottom with 1. The next division, therefore, is 2, the next 3, and so on to 10. The Y axis in Figure 15.17b begins with 10. The next division, therefore, is 20, then 30, and so on to 100. The Y axis in Figure 15.17c begin with 0.1. The next division, therefore, is 0.2, then 0.3 and so on to 1.0.

Figure 15.17 illustrates one cycle semilog paper. The graph paper in Figure 15.18a is called *two cycle semilog paper* because it repeats the same pattern twice. Three cycle semilog paper, Figure 15.18b, repeats the pattern three times. The beginning of each cycle is 10 times greater than the beginning of the previous cycle. For example, if the first cycle goes from 1 to 10, then the

second cycle must range from 10 to 100, the third cycle from 100 to 1000, and so on.

It is possible to purchase semilog paper with various numbers of cycles. To determine how many cycles you need, examine the data to be plotted. For example, in Tables 15.4 on p. 264 and 15.5 on p. 265, the Y values vary between 100 and 12,800. In this case there is no need for a cycle from 1 to 10 or from 10 to 100 so the bottom cycle can begin at 100. The first cycle required to plot these data runs from 100 to 1000. The second cycle ranges from 1000 to 10,000, and a third cycle, from 10,000 to 100,000, is needed to plot the value "12,800." Thus, three cycle semilog paper is required to plot these

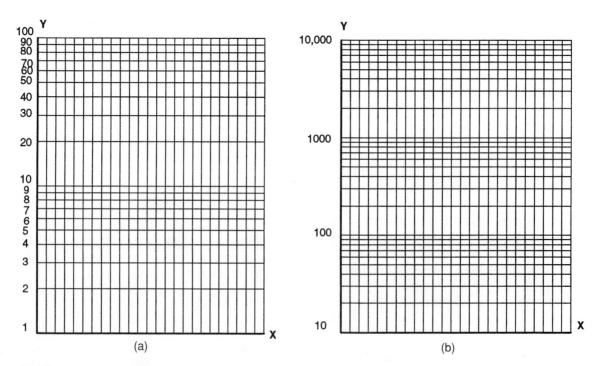

Figure 15.18.   Cycles on Semilog Paper. a. Two cycle semilog paper. **b.** Three cycle semilog paper.

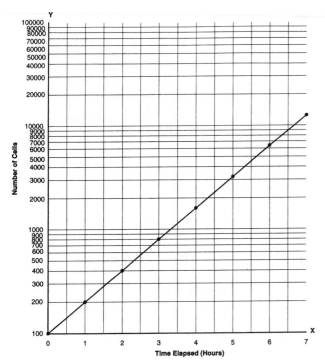

Figure 15.19. Bacterial Growth Graphed on Semilog Paper.

data. Figure 15.19 shows the semilog plot of the bacteria data from Table 15.4. Observe how the Y axis is labeled in this figure. Also, compare the plots in Figures 15.16 on p. 265 and 15.19. Both graphs are linear in form and either is an acceptable way to plot these data.

In principle, all biological populations have the potential to grow exponentially. If any population continued to grow exponentially for a long enough period of time, however, it would cover the earth and eventually the universe. In reality, population growth is limited by space, nutrients, waste buildup, and other factors. Bacterial growth normally looks like the plot in Figure 15.20. When a few bacteria that are not reproducing rapidly are placed in fresh media, there is initially a lag period as they utilize nutrients and prepare to divide. A period of exponential growth follows the lag period.

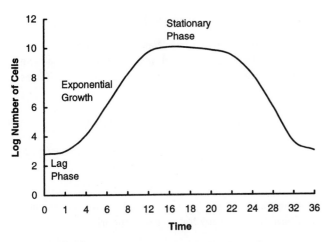

Figure 15.20. The Limits to Bacterial Growth.

The bacteria eventually deplete the nutrients in the broth and generate toxic waste products. Reproduction slows and stops and the population declines.

B. The Decay of Radioisotopes

Radioactive substances decay over time so that there is progressively less and less radioactivity present. The **half-life** *of a radioactive substance is the time it takes for the amount of radioactivity to decay to half its original level.* Radioactivity can be measured in various units including disintegrations/minute, Curies (Ci), and microCuries (μCi). For example, suppose a solution containing a radioactive substance initially undergoes 400 disintegrations/minute and that the half-life of the substance is 1 hour. Then, after 1 hour, there will be 200 disintegrations/minute of radioactivity remaining in the solution. After 2 hours there will be 100 disintegrations per minute, and so on. The relationship between time elapsed and radioactivity for this example is shown in tabular form in Table 15.6 and is graphed in Figure 15.21 on p. .

An exponential equation can be used to calculate the amount of radioactivity remaining after a certain number of half-lives have elapsed:

GENERAL EQUATION FOR RADIOACTIVE DECAY

$$N = (1/2)^t \, N_o$$

where
    N = the amount of radioactivity remaining
    t = the number of half-lives elapsed
    N_o = the amount of radioactivity initially

For example, let us return to the example of the radioactive solution that has a half-life of 1 hour and an activity of 400 disintegrations/minute initially. How much radioactivity will remain after 3 hours? We can see from Table 15.6 that there are 50 disintegrations/minute after 3 hours. It is possible to get the same answer by substituting into the General Equation for Radioactive Decay as follows (3 hours is the same as three half-lives for this substance):

$$N = (\tfrac{1}{2})^3 \, (400 \text{ disintegrations/min})$$
$$= 50 \text{ disintegrations/min}$$

Table 15.6. RADIOACTIVE DECAY

Time Elapsed (hours)	Half-Lives Elapsed	Radioactive Substance Remaining (disintegrations/minute)
0	0	400
1	1	200
2	2	100
3	3	50
4	4	25

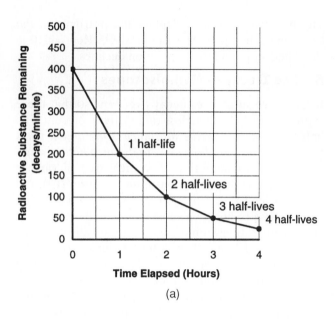

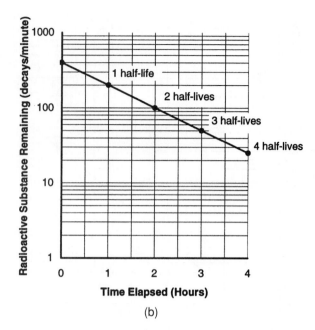

Figure 15.21. **The Relationship between Time Elapsed and Radioactivity Remaining. a.** The relationship on normal graph paper. **b.** The relationship plotted on semilog paper.

EXAMPLE PROBLEM

^{131}Iodine has a half-life of 8 days. If you start with 500 μCi of ^{131}I, how much will remain after 80 days?

ANSWER
STRATEGY 1: USING A TABLE

Time Elapsed (days)	Half-Lives Elapsed	Radioactive Substance Remaining (μCi)
0	0	500
8	1	250
16	2	125
24	3	62.5
32	4	31.25
40	5	15.625
48	6	7.8135
56	7	3.9063
64	8	1.9531
72	9	0.9766
80	10	0.4883

The answer is 0.4883 μCi or about 0.5 μCi remain.

STRATEGY 2: USING THE GENERAL EQUATION FOR RADIOACTIVE DECAY

1. Calculate how many half-lives have passed. The half-life of this isotope is 8 days so the number of half-lives that have passed is: 80/8 = 10.

2. Substitute into the equation:

$$N = \left(\tfrac{1}{2}\right)^t (N_0)$$
$$N = \left(\tfrac{1}{2}\right)^{10} (500 \text{ μCi})$$
$$N \approx 0.4883 \text{ μCi}$$

The answer is that only about 0.5 μCi remain. Eight days is a relatively short half-life. Isotopes with short half-lives rapidly disappear.

EXAMPLE PROBLEM

The half-life for the radioisotope ^{32}P is 14 days. You receive 200 μCi to perform an experiment on March 3. On April 30 you must complete paperwork showing how much ^{32}P activity remains. How much is present?

ANSWER

1. Calculate how many half-lives have passed. There are 58 days between March 3 and April 30. The half-live of this isotope is 14 days so the number of half-lives that have passed is: 58/14 = 4.1.

2. Substitute into the equation: $N = \left(\tfrac{1}{2}\right)^t (N_0)$

$$N = \left(\tfrac{1}{2}\right)^{4.1} (200 \text{ μCi})$$
$$N \approx 11.7 \text{ μCi}$$

The answer to be recorded is 11.7 μCi remain (assuming you have not used any for an experiment).

EXAMPLE

A "Radioactive Material Record" documents the receipt of radioactive material, its use (how much was used, by whom, and on what dates) and its disposal. An example of such a form is shown in Figure 15.22. In this example, the first line shows that on 12/10/07, PBM received 1.0 mCi of material containing the radioactive isotope, ^{32}P. This amount of radioactive material was contained in 1 mL total volume.

On 12/15/07, PBM removed 0.1 mL of the material. This volume contained 0.1 mCi of radioactive material. This left 0.9 mCi in the vial.

On 1/7/08, PBM filled out the disposal information form. She disposed of 0.1 mCi of liquid waste. In addition, at this time, some of the radioactivity had been lost simply due to decay. The half life of ^{32}P is 14 days and so two half-lives (28 days) had elapsed between receipt of the radioisotope on 12/10/07 and 1/7/08 Therefore, of the 0.9 mCi that was in the vial, only 0.225 mCi remained. This means that $0.9 - 0.225 = 0.675$ mCi was "disposed of" due to decay. This is noted on the form. Note that there is still a volume of 0.9 mL on 1/7/08, but this volume only contains 0.225 mCi of radioactivity.

RADIOACTIVE MATERIAL RECORD

USE INFORMATION				DISPOSAL INFORMATION					
Activity (μCi or mCi)		Volume (μL or mL)		Date & Initials	WASTE TYPES AND ACTIVITY (μCi or mCi)					Date & Initials
removed	remaining	removed	remaining		Solid	Liquid	Animal	Decay	Total	
0	1.0	0	1.0	12/10/07 PBM						
0.1	0.9	0.1	0.9	12/15/07 PBM						
						0.1			0.1	1/7/08 PBM
0.675	0.225	0	0.9	1/7/08 PBM				0.675	0.675	1/7/08 PBM

Figure 15.22.   Radioactive Material Record. A form like this is used to document the receipt, use, and disposal of radioactive material.

There is similarity between the equations for the decay of radioactivity and the growth of microorganisms. You can see this similarity by comparing Figures 15.15 (p. 264) and 15.21a (p. 268). Note also the similarity in their equations, which can be expressed as shown here:

A COMPARISON OF THE EQUATIONS FOR BACTERIAL GROWTH AND RADIOACTIVE DECAY

The growth of microorganisms: $N = 2^t (N_o)$

The decline of radioactivity $N = (\frac{1}{2})^t (N_o)$
where
 N = amount (of bacteria or radioactivity) present after a certain period of time
 t = time elapsed (number of generations or number of half-lives)
 N_o = initial amount (of bacteria or radioactivity)

APPLICATION PRACTICE PROBLEMS: EXPONENTIAL EQUATIONS

1. There is an insect population in which each adult leaves 3 offspring and then dies. The generation time is 10 months, that is, the offspring grow up and reproduce every 10 months. In this type of organism, the adults die after they reproduce. The population begins with 20 adults and reproduces for four generations. The resulting population growth is shown in this table:

Generation	Time (months)	Number
0	0	20
1	10	60
2	20	180
3	30	540
4	40	1620

a. Graph this relationship.

b. Determine the equation that describes this relationship.

2. The half-life of the radioisotope ^{32}P is 14 days. If there are initially 300 μCi of this radioisotope:

a. How much will be left after 28 days?

b. How much will be left after 140 days?

c. How much will be left after 200 days?

d. How much will be left after one year?

3. The half-life of the radioisotope ^{22}Na is 2.6 years. If there are initially 600 μCi of this radioisotope, how much will be left after:

a. 1 year

b. 2 years

c. 3 years

4. Fill in the blanks in the form below. The radioisotope involved is ^{32}P with a half-life of 14 days.

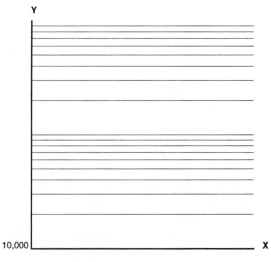

RADIOACTIVE MATERIAL RECORD

USE INFORMATION					DISPOSAL INFORMATION					
Activity (μCi or mCi)		Volume (μL or mL)		Date & Initials	WASTE TYPES AND ACTIVITY (μCi or mCi)					Date & Initials
					Solid	Liquid	Animal	Decay	Total	
removed	remaining	removed	remaining							
0	500	0	1.0	12/3/07 MW						
100	400	0.2	0.8	12/8/07 MW						
100	___	0.2	___	12/12/07 MW						
						200			200	12/17/07 MW
___	___	0	___	12/17/07 MW					___	12/17/07 MW

5. It is difficult to kill all microorganisms in a material. A solution or material to be sterilized is typically heated under pressure. The following graph shows the effect of time of exposure to heat and bacterial death (based on information from Perkins, John J. *Principles and Methods of Sterilization in the Health Sciences.* 2nd ed. Springfield, MA: Charles C. Thomas, 1983.

a. About how many bacteria were there at the beginning of the experiment, before exposure to heat?

b. About how many bacteria were present after 2 minutes of treatment?

c. Why did the investigators show their data on a semilog plot?

6. Label the Y axis of this semilog graph.

CHAPTER 16

Descriptions of Data: Descriptive Statistics

I. INTRODUCTION

A. Populations, Variables, and Samples

The natural world is filled with variability: organisms differ from one another; the weather changes from day to day; the earth is covered by a mosaic of different types of habitats. The natural world is also characterized by uncertainty and chance. The outcome of a card game, the weather, and the personality of a newborn child are all uncertain. Scientists strive to observe and to understand this world, which is characterized by variability and uncertainty.

Statistics is a branch of mathematics that has developed over many years to deal with variability and uncertainty. Statistics provides methods to summarize, analyze, and interpret observations of the natural world. Statistical methods are also used to reach conclusions, based on data, with a certain probability of being right—and a certain probability of being wrong. In this chapter, we will explore a small area of statistics, and its use in summarizing, displaying, and organizing numerical data. We will begin by introducing several fundamental statistical terms.

A **population** *is an entire group of events, objects, results, or individuals, all of whom share some unifying characteristic(s).* Examples of populations include all a person's red blood cells, all the enzyme molecules in a test tube, and all the college students in the United States.

Characteristics of a population often can be observed or measured, such as blood hemoglobin levels, the activity of enzymes, or the test scores of students. *Characteristics of a population that can be measured are called* **variables.** They are called "variables" because there is variation among individuals in the characteristic. A population can have numerous variables that can be studied. For example, the same population of 6-year-old children can be measured for height, shoe size, reading level, or any of a number of other characteristics. *Observations of a variable are called* **data** (singular "datum").

Observations may or may not be numerical. For example, the lengths of insects (in centimeters) are numerical data. In contrast, electron micrographs of mouse kidney cells are nonnumerical observations. The statistical methods discussed in this unit are used to interpret numerical data.

It is seldom possible to determine the value for a given variable for every member of a population. For example, it is virtually impossible to measure the level of hemoglobin in every red blood cell of a patient. Rather, a **sample** of the patient's blood is drawn and the hemoglobin level is measured and evaluated. Investigators draw conclusions about populations based on samples, see Figure 16.1.

Statistical methods are generally based on the assumption that a sample is **representative of its population**—that is, *that the sample truly reflects the variability in the original population*. In order for a sample to be representative of its population, it must meet two requirements: (1) *All members of the population must have an equal chance of being chosen*. If this is the case the sample is called **random.** (2) *The choice of one member of the sample should not influence the choice of another*. If this is the case, the sample is said to have been drawn **independently.**

Note that the statistical meaning of "random" is different from the meaning in common English. In popular usage random means "haphazard," "unplanned," or "without pattern"; this is not its meaning in statistics. Randomness refers to how a sample is taken. For example, a lottery is random if every ticket has an equal chance of being drawn. If a deck of cards is properly shuffled and if a card is picked by chance from that deck, then the choice of the card is random.

To understand independence, consider a coin that is flipped twice. The second toss has a 50% probability of being heads regardless of whether or not the first toss was heads. This is because the outcomes of the two tosses are independent of one another. In contrast, suppose we begin with a full deck of cards and randomly

Population—All 20-year-old men in the United States

Variables—Many variables could be measured for this population such as: height, hair color, college entrance exam score, and birthplace

Sample—A sample of the population of all 20-year-old men is drawn

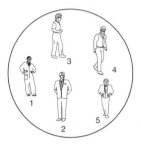

Data—Observations are made of the individuals in the sample

Individual	Height (in cm)	Hair Color	College Exam Score	Birthplace
1	172	brown	1500	WI
2	173	blond	none	IL
3	169	black	1800	NY
4	177	brown	none	WY
5	175	blond	1310	MO

Figure 16.1. Populations, Samples, Variables, and Data. A sample is drawn from the population to represent that population. In this illustration, data for four variables are collected for each individual in the sample.

draw and set aside one card, which turns out to be the five of hearts. A second draw from this deck is not independent because there is now no chance of drawing the five of hearts.

The individuals that comprise a population vary from one another; therefore, even when a sample is drawn randomly and independently, there is uncertainty when a sample is used to represent a whole population. The sample could truly reflect the features of the population from which it was drawn, but perhaps it does not. As you might expect, if a sample is properly drawn, then the larger the sample, the more likely it is to accurately reflect the entire population.

EXAMPLE PROBLEM

a. *In a quality-control setting,* 15 vials of product from a batch were tested. What is the sample? What is the population?

b. *In an experiment,* the effect of a carcinogenic compound was tested on 2000 laboratory rats. What is the sample? What is the population?

c. *A political poll* of 1000 voters was conducted. What is the sample? What is the population?

d. *A clinical study* of the effect of a new drug was tested on 50 patients. What is the sample? What is the population?

ANSWER

a. The sample is the 15 vials tested. The population is all the vials in the batch.

b. The sample is the 2000 rats. The population is presumably all laboratory rats.

c. The sample is the 1000 voters polled. The population is all voters.

d. The sample is the 50 patients tested. The population is all similar patients.

EXAMPLE PROBLEM

What is meant by the statement "two out of three doctors surveyed recommend Brand X ..."? What is the population of interest? What is the sample? Is this sample representative of the population? Does this statement ensure that the product being endorsed is better than any other?

ANSWER

Many abuses of statistics relate to poor sampling. A classic type of abuse is statements such as "two out of three doctors surveyed recommend Brand X ..." The population of interest is all doctors. This statement suggests that two thirds of all doctors prefer Brand X to other brands; however, we have no way to know which doctors were sampled. For example, perhaps the survey only looked at physicians who prescribe Brand X, or only physicians in a certain region. It is possible that the survey sampled doctors who were not representative of all doctors; therefore, the statement does not ensure that most doctors prefer Brand X. In addition, even if two thirds of all doctors really do recommend Brand X, it may or may not be the best brand—it may simply be the most accessible, least expensive, most advertised, and so on.

B. Describing Data Sets: Overview

When a sample is drawn from a population, and the value for a particular variable is measured for each individual in the sample, the resulting values constitute a data set. Because individuals differ from one another, the data set consists of a group of varying values.

A set of data without organization is something like letters that are not arranged into words. Like letters of the alphabet, numerical data can be arranged in ways that are meaningful. (Numerical data, like letters of the alphabet, can also be organized in ways that are confusing, deceptive, or irrelevant. Delving into the many ways that statistics are misused, however, is beyond the scope of this book.) **Descriptive statistics** *is a branch of statistics that provides methods to appropriately organize, summarize, and describe data.**

*There is another branch of statistics, called *inferential statistics,* that provides methods to make predictions about a population based on a sample. This area of statistics is beyond the scope of this book.

Consider, for example, the exam scores for a class of students. The variable that is measured is the exam score for each individual. A list of the scores of all the students constitutes the data set. In order to summarize and describe the performance of the class as a whole, the instructor might calculate the class's average score. The average is an example of a measure that summarizes the data. A measure that describes a sample, such as the average, is sometimes referred to as a "statistic."

The average gives information about the "center" of a data set. There are other measures, such as the median and the mode, which also describe data in terms of its center. *Such measures, which describe the center of a data set, are called* **measures of central tendency.**

Data sets A and B both have the same average:

$$A: 4\ 5\ 5\ 5\ 6\ 6$$
$$B: 1\ 2\ 4\ 7\ 8\ 9$$

Inspection of these values, however, reveals that the two data sets have a different pattern. The data in A are more clumped about the central value than the data in B. We say that the data in B are more **dispersed**, that is, *spread out.*

Measures of central tendency, such as the average, do not describe the dispersion of a set of observations; therefore, there are statistical **measures of dispersion** *which describe how much the values in a data set vary from one another.* Common measures of dispersion are: range, variance, standard deviation, and coefficient of variation.

Measures of central tendency and of dispersion are determined by calculation. The next section discusses the calculation of these measures. It is also possible to organize and effectively display data using graphical techniques. Graphical techniques and their interpretation are discussed in Section III of this chapter. Section IV applies these ideas to the area of controlling product quality in a company.

II. DESCRIBING DATA: MEASURES OF CENTRAL TENDENCY AND DISPERSION

A. Measures of Central Tendency

Consider a hypothetical data set consisting of these values:

2   5   6   7   8   3   9   3   10   4   7   4   6   11   9

The simplest way to organize these values is to order them as follows:

2 3 3 4 4 5 6 6 7 7 8 9 9 10 11

Inspection of the ordered list of numbers reveals that they center somewhere around 6 or 7. The center can be calculated more exactly by determining the average or mean. The **mean** *is the sum of all the values divided by the number of values.* In this example the mean is calculated as:

$$2 + 3 + 3 + 4 + 4 + 5 + 6 + 6 + 7 + 7 + 8 + 9 + 9$$
$$+ 10 + 11 = 94$$

The average or the mean is 94/15 = 6.3

In algebraic notation, the observations are called X_1, X_2, X_3, and so on. In this example there are 15 observations, so the last value is X_{15}. The method to calculate the mean can be written compactly as:

$$\text{mean} = \frac{\sum X}{n}$$

where n = the number of values

$\sum$ (the Greek letter, sigma) is short for "add up." X means that what is to be added is all the values for the variable, X. Thus:

$$\text{mean} = \frac{\sum X}{n} = \frac{X_1 + X_2 + X_3 + \ldots + X_n}{n}$$

Statisticians distinguish between the true mean of an entire population and the mean of a sample from that population, see Figure 16.2. The true mean of a population is represented by μ (the Greek letter, mu). The mean of a sample is represented by $\bar{x}$ (read as "X bar"). In practice, it is rare to know the true mean for an entire population; therefore, the sample mean, or $\bar{x}$ must be used to represent the true population mean.

The median is another measure of central tendency. The **median** *is the middle value of a data set, or the number that is halfway between the highest and lowest value,* when the values are arranged in ascending order. The median for the following values is 6:

2 3 3 4 4 5 6 <u>6</u> 7 7 8 9 9 10 11

When there is an even number of values, the median is the value midway between the two central numbers. The median for these values is 7.5 or the average of the numbers 7 and 8:

3 5 <u>7 8</u> 9 11

Observe that in both of the following data sets the median is 5 even though their means are quite different:

| 2 3 4 <u>5</u> 6 7 8 | $\bar{x} = 5$ |
| 2 3 4 <u>5</u> 6 100 1000 | $\bar{x} = 160$ |

The **mode** *is the value of the variable that occurs most frequently.* For example, the mode in the following data set is "3" since it appears more often than the other numbers:

2 3 3 3 4 5 6 7 7 8

It is possible for a set of data to have more than one mode. *If there are two modes, the data are* **bimodal.** *If there are more than two modes the data are* **polymodal.**

Box 16.1 summarizes calculations of these three measures of central tendency—the mean, median, and mode.

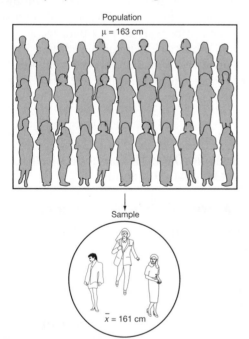

Figure 16.2. The True Mean of a Population Is Distinguished from the Mean of a Sample. The population of interest is all women in the United States; the variable of interest is height. The mean height, μ, for all women is 163 cm. A sample of women is drawn and their heights are measured. The mean height in the sample, $\bar{x}$, is 161 cm.

EXAMPLE PROBLEM

Suppose you are investigating the height of college students. You measure the height of every student in your classes and obtain the data shown in the following table.

a. What is the average height of the men in the class?

b. What is the average height of the women in the class?

c. What is the average height of all students in the class?

Heights of Students in the Class (in cm)					
Men			**Women**		
175.3	176.5	177.8	162.5	166.7	155.6
168.5	165.2	160.2	159.7	163.4	164.2
180.9	188.9	171.6	160.1	168.6	162.6
175.9	174.8	179.2	162.8	158.4	174.9
175.9	175.8	174.8	161.3	166.2	166.7
n_{men} = 15 (number of men)			n_{women} = 15 (number of women)		

ANSWER

a. $\bar{x} = \dfrac{2621.3 \text{ cm}}{15} \approx 174.75 \text{ cm (men)}$

b. $\bar{x} = \dfrac{2453.7 \text{ cm}}{15} \approx 163.58 \text{ cm (women)}$

c. $\bar{x} = \dfrac{5075.0 \text{ cm}}{30} \approx 169.17 \text{ cm (men + women)}$

Box 16.1.   MEASURES OF CENTRAL TENDENCY: CALCULATING THE MEAN, MEDIAN, AND MODE

1. *To calculate the mean, add all the values and divide by the number of values present.*

$$\text{mean} = \frac{\sum X}{n}$$

2. *To find the median:*
   Order the data values.
   If there is an odd number of values, the median is the middle value in the sequence.
   If there is an even number of values, the median is the average of the two middle values.

3. *Mode:* Order the data values. The mode is the most frequent observation. There may be more than one mode.

Example

Given the data set: 12, 15, 13, 12, 13, 14, 16, 19, 21, 13, 15, 14

1. Order the data 12, 12, 13, 13, 13, 14, 14, 15, 15, 16, 19, 21

2. The mean is $\dfrac{12 + 12 + 13 + 13 + 13 + 14 + 14 + 15 + 15 + 16 + 19 + 21}{12} = 14.75$

3. The median is $\dfrac{14 + 14}{2} = 14$

4. The mode is 13.

B. Measures of Dispersion

i. THE RANGE

The **range** is the simplest measure of dispersion. The **range** *is the difference between the lowest value and the highest value in a set of data:*

2 3 3 3 4 4 5 6 7 7 8 9 9 10 11
The range is 2 to 11 = 9

The range is not a particularly informative measure because it is based on only two values from the data set. A single extreme value will have a major effect on the range.

ii. CALCULATING THE VARIANCE AND THE STANDARD DEVIATION

The variance and the standard deviation are more informative measures of dispersion than the range because they summarize information about dispersion based on all the values in the data set. Variance and standard deviation are best explained with a simple example. Assume that the following data are lengths (in millimeters) of eight insects:

4 5 6 7 7 7 9 11

The mean for these data is 7 mm. An intuitive way to calculate how much the data is dispersed is to take each data point, one at a time, and see how far it is from the mean, Table 16.1. *The distance of a data point*

from the mean is its **deviation.** There is a deviation for each data point: The deviation is sometimes positive, sometimes negative, and sometimes zero.

It is useful to summarize all the deviations with a single value. An obvious approach would be to calculate the average of the deviations, but, for any data set, the sum of the deviations from the mean is always zero. Mathematicians therefore devised the approach of squaring each individual deviation value to result in all positive numbers. *The squared deviations can be added together to get the* **total squared deviation,** also called the **sum of squares.** In this example, the sum of squares is 34 mm^2, see Table 16.2, p. 276. The sum of squares is always a positive number that indicates how

Table 16.1.   CALCULATION OF DEVIATION FROM THE MEAN

(Value − Mean)	Deviation
First value − mean =	(4 − 7) = −3 mm
Second value − mean =	(5 − 7) = −2 mm
Third value − mean =	(6 − 7) = −1 mm
Fourth value − mean =	(7 − 7) = 0 mm
Fifth value − mean =	(7 − 7) = 0 mm
Sixth value − mean =	(7 − 7) = 0 mm
Seventh value − mean =	(9 − 7) = +2 mm
Eighth value − mean =	(11 − 7) = +4 mm
	0 = Sum of all the deviations

Table 16.2. CALCULATION OF THE SUM OF SQUARES

(Value − Mean) (mm)	(Deviation) (mm)	(Deviation Squared)
(4 − 7)	−3	9 mm²
(5 − 7)	−2	4 mm²
(6 − 7)	−1	1 mm²
(7 − 7)	0	0 mm²
(7 − 7)	0	0 mm²
(7 − 7)	0	0 mm²
(9 − 7)	+2	4 mm²
(11 − 7)	+4	16 mm²
		34 mm² Total squared deviation = the sum of squares

much the values in a set of data deviate from the mean.

The **variance** *is the total squared deviation divided by the number of measurements:*

$$\text{variance (of a population)}$$

$$= \frac{\text{total squared deviations from the mean}}{n}$$

$$= \frac{\sum (X - \text{mean})^2}{n}$$

where *n* = the number of values

In this example *n* = 8 and the variance is:

$$\text{variance} = \frac{34 \text{ mm}^2}{8} = 4.25 \text{ mm}^2$$

The standard deviation is calculated by taking the square root of the variance:

$$\text{standard deviation} = \sqrt{\text{variance}}$$

In this example:

$$\text{standard deviation} = \sqrt{4.25 \text{ mm}^2} = 2.06 \text{ mm}$$

Note that the standard deviation has the same units as the data.

The standard deviation and the variance are values that summarize the dispersion of a set of data around the mean. The larger the variance and the standard deviation, the more dispersed are the data.

iii. DISTINGUISHING BETWEEN THE VARIANCE AND STANDARD DEVIATION OF A POPULATION AND A SAMPLE

In the preceding discussion we skipped over a significant detail regarding the calculation of the variance and standard deviation. Statisticians distinguish between the variance and standard deviation of a population and the variance and standard deviation of a sample, see Figure 16.3. *The variance of a population is called* σ^2 (read as "sigma squared"). *The*

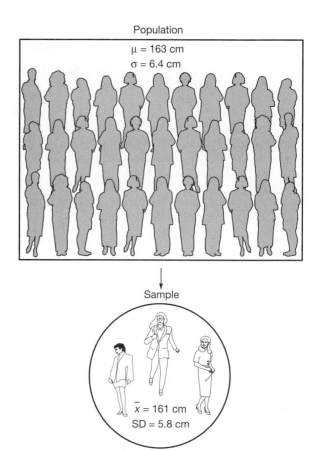

Population

μ = 163 cm
σ = 6.4 cm

Sample

$\bar{x}$ = 161 cm
SD = 5.8 cm

Figure 16.3. The Mean and Standard Deviation of a Population versus a Sample. The population has a true mean and standard deviation. Every sample drawn from a population has its own mean and standard deviation, which are likely to differ from the true mean and standard deviation of the population.

variance of a sample is called S^2. *The standard deviation of a population is called* σ *(sigma). The standard deviation of a sample is sometimes abbreviated* S *and sometimes* SD. Terminology relating to populations and samples is summarized in Table 16.3.

The formulas used to calculate the variance and standard deviation of a sample are slightly differ-

Table 16.3. TERMINOLOGY RELATING TO POPULATIONS AND SAMPLES

Population	Sample
An entire group of events, objects, results, or individuals where each member of the group has some unifying characteristic(s).	A subset of a population that represents the population.
Parameters	**Statistics**
Measures that describe a population.	Measures that describe a sample.
Mean μ	Mean $\bar{x}$
Variance σ^2	Variance S^2
Standard Deviation σ	Standard Deviation S or SD.

ent than those for a population. In the preceding discussion we showed how to calculate the variance and standard deviation of a population. For a sample, the denominator is $n-1$ rather than n. The formulas for variance and standard deviation of a sample are thus:

$$\text{variance (of a sample)} =$$

$$S^2 = \frac{\text{sum of squared deviations from the mean}}{n-1}$$

$$= \frac{\sum(X-\bar{x})^2}{n-1}$$

$$\text{standard deviation} = SD = \sqrt{S^2}$$

$$= \sqrt{\frac{\sum(X-\bar{x})^2}{n-1}}$$

where n = the number of values

In this text we are generally looking at data from a sample and so routinely use $n-1$ in the denominator of the equation.

The standard deviation is a widely used statistic, so being able to calculate it for a given set of data is an important skill. Many scientific calculators are set up to calculate the standard deviation easily. The operator can often enter the data and then press one key to get the standard deviation for a population and a different key to get the standard deviation for a sample. (Refer to the calculator instructions.) There is an alternative equation for standard deviation that is useful if your calculator does not have the standard deviation function built in, see Box 16.2 on p. 278. There are also statistical computer programs that are useful for analyzing large sets of data.

EXAMPLE PROBLEM

The heights for a sample of seven college women are shown in the following table. Calculate the mean ($\bar{x}$), the deviation (d) of each value from the mean, the squared deviation (d^2), the variance (S^2), and the standard deviation (SD).

Measured Value (cm)	d (cm)	d² (cm²)
162.5		
166.7		
155.6		
159.7		
163.4		
164.2		
160.1		

ANSWER

$\bar{x} = 161.74$ cm

Measured Value (cm)	d (cm)	d² (cm²)
162.5	0.76	0.58
166.7	4.96	24.60
155.6	−6.14	37.70
159.7	−2.04	4.16
163.4	1.66	2.76
164.2	2.46	6.05
160.1	−1.64	2.69
		Summation 78.54 cm²

$n - 1 = 6$

$$S^2 = \frac{78.54 \text{ cm}^2}{6} = 13.09 \text{ cm}^2$$

$$SD = \sqrt{13.09 \text{ cm}^2} = 3.62 \text{ cm}$$

iv. THE COEFFICIENT OF VARIATION (RELATIVE STANDARD DEVIATION)

Table 16.4 shows the weights of laboratory mice and rabbits. Observe that the standard deviation for the rabbits' weights has a higher value than that for the mice, but, of course, rabbits weigh more than mice.

The dispersion of a data set can be expressed as the standard deviation divided by the mean. This is called the **coefficient of variation (CV)** or the **relative standard deviation (RSD)**. The formula for the CV is:

$$CV = \frac{\text{standard deviation }(100\%)}{\text{mean}}$$

When the dispersions of the weights of mice and rabbits are expressed in terms of the coefficient of variation, the value is slightly higher for the mouse data than it is for the rabbit data.

Table 16.4. WEIGHTS OF LABORATORY MICE AND RABBITS

Weights of Laboratory Mice (g)	Weights of Laboratory Rabbits (g)
32	3178
34	3500
24	3428
33	2908
36	2757
30	3100
36	2876
36	3369
30	3682
34	2808
$\bar{x} = 32.5$ g	$\bar{x} \approx 3161$ g
SD ≈ 3.75 g	SD ≈ 322.8 g
CV $\approx 11.5\%$	CV $\approx 10.2\%$

Box 16.2. *MEASURES OF DISPERSION: THE VARIANCE, THE STANDARD DEVIATION, AND THE RELATIVE STANDARD DEVIATION*

1. To calculate the variance (S^2) for a sample, use the equation:

$$S^2 = \frac{\sum (X - \bar{x})^2}{n - 1}$$

2. To calculate the standard deviation (SD) for a sample:

$$SD = \sqrt{\frac{\sum (X - \bar{x})^2}{n - 1}}$$

There is an alternative equation that can be used to calculate SD:

$$SD = \sqrt{\frac{\sum X^2 - \frac{(\sum X)^2}{n}}{n - 1}}$$

3. To calculate the coefficient of variation (CV):

$$CV = \frac{\text{Standard deviation } (100\%)}{\text{mean}}$$

Example

Given these data from a sample:

| 1.00 mm | 2.00 mm | 2.00 mm | 3.00 mm | 3.00 mm | 4.00 mm | 4.00 mm | 5.00 mm |

$$\bar{x} = 3.00 \text{ mm}$$

1. $S^2 = \dfrac{((1-3)\text{mm})^2 + ((2-3)\text{mm})^2 + ((2-3)\text{mm})^2 + ((3-3)\text{mm})^2 + ((3-3)\text{mm})^2 + ((4-3)\text{mm})^2 + ((4-3)\text{mm})^2 + ((5-3)\text{mm})^2}{8 - 1}$

$$= \frac{12.00 \text{ mm}^2}{7} \approx 1.71 \text{ mm}^2$$

2. The standard deviation is the square root of the variance = 1.31 mm

Note that the standard deviation and the variance have units if the data have units.

Note that using the alternate formula shown in # 2 gives the same answer:

$$\sum X^2 = 1 + 4 + 4 + 9 + 9 + 16 + 16 + 25 = 84 \text{ mm}^2$$

$$\frac{(\sum X)^2}{n} = 1 + 2 + 2 + 3 + 3 + 4 + 4 + 5 = \frac{(24)^2}{8} = 72 \text{ mm}^2$$

$$SD = \sqrt{\frac{84 \text{ mm}^2 - 72 \text{ mm}^2}{7}} \approx 1.31 \text{ mm}$$

3. The coefficient of variation is:

$$CV = \frac{1.31 \text{ mm} \times 100\%}{3.00 \text{ mm}} \approx 43.67\%$$

Note that the mean and the standard deviation have units, but the units cancel when the CV is calculated. We will come across both the standard deviation and the coefficient of variation in Unit V, where we talk about measurements.

Methods of calculating various measures of dispersion are summarized in Box 16.2.

EXAMPLE PROBLEM

A biotechnology company sells cultures of a particular strain of the bacterium *E. coli*. The bacteria are grown in batches that are freeze dried and packaged into vials. Each vial is expected to contain 200 mg of bacteria. A quality-control technician tests a sample of vials from each batch and reports the mean weight and the standard deviation.

Batch Q-21 has a mean weight of 200 mg and a standard deviation of 12 mg. Batch P-34 has a mean weight of 200 mg and a standard deviation of 4 mg. Which lot appears to have been packaged in a more controlled fashion?

ANSWER

The standard deviation can be interpreted as an indication of consistency. The SD of the weights in Batch P-34 is lower than Batch Q-21. The values for Batch P-34, therefore, are less dispersed than those for Batch Q-21. The packaging of Batch P-34 appears to have been better controlled.

C. Summarizing Data by the Mean and a Measure of Dispersion

It is common in a scientific paper or in other technical literature to see a set of data summarized using the mean followed by a measure of dispersion. The smaller the measure of dispersion, the closer the data points are to one another. Thus, you might encounter a mean reported in any of the following styles:

1. **A mean may be reported as the mean value ± the standard deviation,** see Figure 16.4a. For example, 6.02 ± 0.23 (SD) means the sample mean is 6.02 and the standard deviation of the sample is 0.23.

2. **You may see a mean reported in a scientific paper as the mean value ± standard error of the mean,** see Figure 16.4b. The standard error of the mean (SEM) is not actually a measure of error; rather, it is a measure of dispersion relating to a mean. When encountered in graphs and tables, the simplest way to understand the SEM is to interpret it like the standard deviation: A larger value means there is more variability in the population than a smaller value.

3. **You may see a mean reported in a scientific paper as the mean value and the 95% (or 90% or 99%) confidence interval,** see Figure 16.4c. A

EFFECT OF EXERCISE ON BLOOD PRESSURE: BLOOD PRESSURE BEFORE STARTING AN EXERCISE PROGRAM AND AFTER 16 WEEKS OF SOFT TRAINING

Variable	Baseline mean ± SD n = 43	16 Weeks mean ± SD n = 43
Maximal heart rate	153 ± 15	153 ± 11
Weight (kg)	97 ± 16	97 ± 17
Peak oxygen uptake (mL/kg/min)	21 ± 4	23 ± 4
Submaximal systolic blood pressure (mm Hg)	218 ± 23	187 ± 30
Submaximal diastolic blood pressure (mm Hg)	107 ± 10	94 ± 9

(a)

PHYSICAL TRAITS OF FOUR DIFFERENT SUBSPECIES OF THE WHEAT *TRITICUM MONOCOCCUM*

		Mean Value (± SEM)	
Subspecies	Number Studied	Seed Weight (mg)	Awns per Spikelet
T.m. a	250	21.1 ± 0.2	2.50 ± 0.05
T.m. b	68	30.2 ± 0.7	1.00 ± 0.04
T.m. c	9	22.9 ± 0.7	1.10 ± 0.14
T.m. d	11	21.5 ± 0.9	2.33 ± 0.14

(b)

TOXICITY OF INSECTICIDE PRODUCED BY *E. COLI* RECOMBINANT CELLS

Insect	LC$_{50}$ (mg /mL)*
Cutworm	10.9 (7.0–17.0)
Aphids	17.4 (9.7–29.2)
Beetles	147.8 (79.2–285)

*95% confidence intervals are shown in parentheses

(c)

Figure 16.4. Examples of How the Mean Is Reported in Scientific Literature. a. A study of the effect of exercise on individuals with high blood pressure. Values are reported as the mean ± the standard deviation. **b.** A study of wheat in which physical traits are reported as the mean ± the SEM. **c.** A study of insecticide toxicity on three species of insects. Results are reported as the mean and the 95% confidence interval.

95% confidence interval is a range of values that is constructed in such a manner that if 100 random samples were drawn from the population, we would expect 95 of the 100 confidence intervals to include the true mean of the population. The variability of the data are considered when a confidence interval is calculated: the more variability in the data, the wider the range of the interval.

4. **You may see a mean reported graphically.** The points on the graph in Figure 16.5 represent sample means. The vertical lines are called *error*

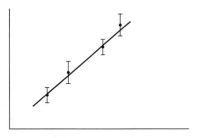

Figure 16.5. The Representation of Sample Means Graphically; Error Bars. Each point on the line in this graph represents the mean of the observations of 20 individuals. The error bar represents ± 1 SD.

Table 16.5.   THE WEIGHTS OF 175 FIELD MICE (IN GRAMS)

21	23	22	19	22	20	24	22	19	24	27	20	21
22	20	22	24	24	21	25	19	21	20	23	25	22
19	17	20	20	21	25	21	22	27	22	19	22	23
22	25	22	24	23	20	21	22	23	21	24	19	21
22	22	25	22	23	20	23	22	22	26	21	24	23
21	25	20	23	20	21	24	23	18	20	23	21	22
22	25	21	23	22	24	20	21	23	21	19	21	24
20	22	23	20	22	19	22	24	20	25	21	22	22
24	21	22	23	25	21	19	19	21	23	22	22	24
21	23	22	23	28	20	23	26	21	22	24	20	21
23	20	22	23	21	19	20	26	22	20	21	22	23
24	20	21	23	22	24	21	23	22	24	21	22	24
20	22	21	23	26	21	22	23	24	21	23	20	20
21	25	22	20	22	21							

III. DESCRIBING DATA: FREQUENCY DISTRIBUTIONS AND GRAPHICAL METHODS

A. Organizing and Displaying Data

Consider a set of data consisting of the times it took 10 students to run a race (in minutes):

9.6   10.3   9.1   7.5   10.9   7.2   6.9   9.5   8.7   8.1

These values are displayed in a **dot diagram,** Figure 16.6.

A dot diagram is a simple graphical technique used to illustrate small data sets. With a dot diagram it is easy to see the general location and center of the data. The dispersion of data is also evident in such diagrams, see Figure 16.7.

bars. The error bars may represent ± one standard deviation. Error bars may also be used to represent ± the standard error of the mean, or they may show the confidence interval. The interpretation of error bars is customarily explained in the figure caption.

A dot diagram is difficult to construct and interpret when a data set is large and there are other graphical methods that are more useful for organizing and displaying large sets of data. For example, suppose a researcher collects data for the weights of a number of field mice, see Table 16.5. An informative way to tabulate these data is to list them by frequency where the **frequency** *is the number of times a particular value is observed.* Table 16.6 shows the values for the mouse weights and the frequency of each value.

Observe in Table 16.6 that the values for mouse weights have a pattern. Most of the mice have weights in the middle of the range, a few are heavier than average, a few are lighter, see Figure 16.8. The term **distribution** *refers to the pattern of variation for a given variable.* It is very important to be aware of the patterns, or distributions, which emerge when data are organized by frequency.

The frequency distribution in Table 16.6 can be illustrated graphically as a **frequency histogram,** see Figure 16.9. A frequency histogram is a graph where the X axis

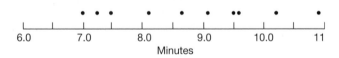

Figure 16.6.   A Data Set Displayed as a Dot Diagram.

Figure 16.7.   Dot Diagrams Illustrate Dispersion. a. Less dispersed data. **b.** More dispersed data.

Table 16.6.   FREQUENCY DISTRIBUTION TABLE OF THE WEIGHTS OF FIELD MICE

Weight (g)	Frequency
17	1
18	1
19	11
20	25
21	34
22	40
23	27
24	19
25	10
26	4
27	2
28	1

Figure 16.8. Distribution of Mouse Weights. Most mice are of about average weight: Some are a bit heavier, some a bit lighter than others. A few mice are substantially heavier, and a few mice are substantially lighter than average.

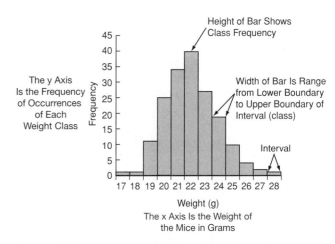

Figure 16.9. Frequency Histogram for Mouse Data.

is the relevant units of measurement. In this example, it is weight in grams. The Y axis is the frequency of occurrence of a particular value. For example, 11 mice are recorded as weighing 19 g. The values for these 11 mice are illustrated as a bar. Note that when the mouse data were collected, a mouse recorded as 19 g might actually have weighed anything between 18.5 and 19.4 g. The bar in the histogram, therefore, spans an interval. In this graph the intervals are 1 g in width. You can think of the bars on the histogram as representing individual mice piled up on the number line with each individual sitting above its score.

It is possible to construct a frequency histogram for any numerical data set consisting of the values for a single variable. The procedure for doing so is given in Box 16.3.

B. The Normal Frequency Distribution

The same data that are shown in Figure 16.9 are graphed in a slightly different form, as a **frequency polygon,** in Figure 16.10a on p. 282. A frequency polygon does not use bars; rather, each class is represented as a single point. The point is placed halfway between the smaller and the larger limit for that class and the points are connected with lines.

Figure 16.10a shows the distribution of weights for a sample of only 175 mice. If the weights of a great many laboratory mice were measured, then it is likely that the frequency distribution would approximate a bell shape, or normal curve, see Figure 16.10b. *The frequency distribution pattern that has a bell shape when graphed is called a* **normal distribution.**

There are many situations in nature and in the laboratory where the frequency distribution for a set of data approximates a normal curve, see Figure 16.11 on p. 282. For example, the height of humans, the weight of animals, and measurements of the same object all tend to be normally distributed.

At this point, it is possible to relate the first part of this chapter, which discussed calculated measures (such as the mean and standard deviation), to the graphical techniques in this section. The center of the peak of a perfect normal curve is the mean, median, and mode. Values are equally spread out on either side of that central high point. Moreover, the width of the normal curve is related to the standard deviation, see Figure 16.12 on p. 283. The more dispersed the data, the higher the value for the standard deviation and the wider the normal curve.

If we know that the pattern of variation for a variable is normally distributed, then we can make certain

---

Box 16.3. CONSTRUCTING A FREQUENCY HISTOGRAM

1. *Divide the range of the data into intervals, or classes.* It is simplest to make each interval the same width. There is no set rule as to how many intervals should be chosen; this will vary depending on the data. (For example, length data could be divided into intervals of: 0–9.9 cm, 10.0–19.9 cm, 20.0–29.9 cm, and so on.)

2. *Count the number of observations that are in each interval.* These counts are the frequencies.

3. *Prepare a frequency table showing each interval and the frequency with which a value fell into that interval.*

4. *Label the axes of a graph with the intervals on the X axis and the frequency on the Y axis.*

5. *Draw in bars where the height of a bar corresponds to the frequency with which a value occurred.* Center the bars above the midpoint of the class interval. For example, if an interval is from 0 to 9 cm, then the bar should be centered at 4.5 cm.

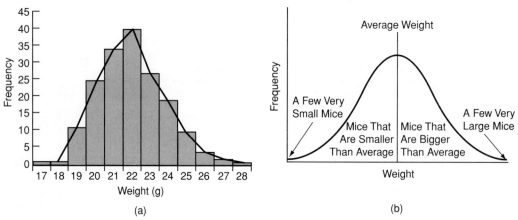

(a)　　　　　　　　　　　　　　　　　(b)

Figure 16.10. The Frequency Distribution for Mouse Weights. a. The weights of 175 mice are graphed both as a frequency histogram and frequency polygon. **b.** If a frequency polygon were prepared for a great many mice, then the shape of the plot would approach a bell shape. The peak in the center is the average weight. Most of the mice have weights somewhere in the middle of the plot, that is, around average weight, while a few mice are substantially heavier or lighter than average. This distribution pattern is called a *normal distribution.*

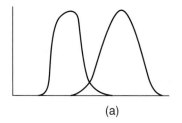

(a)

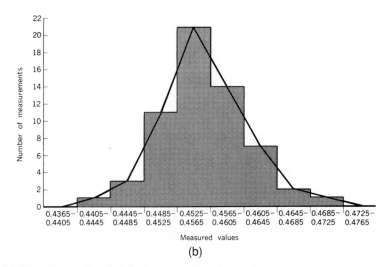

(b)

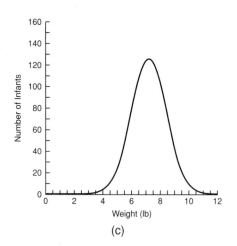

(c)

Figure 16.11. Examples of Relationships That Approximate a Normal Curve. a. An example from chromatography, an analytical technique used to separate and detect various components in a mixture of compounds. Under ideal conditions each compound appears on the chromatographic printout as an approximately normal curve. **b.** The weights of a number of vials from the same batch. The vials vary in weight both because of variability in the manufacturing process and variability that is inherent in repeated measurements. The pattern of variation approximates a normal distribution. If more vials were weighed, then we would expect the pattern to be even closer to a normal curve. **c.** Distribution of birth weight of babies born to teenagers in Portland, Oregon in 1992.

predictions about individuals. For example, consider a 20-year-old man, Robert Smith, who is from a large population of all 20-year-old men. Height in adults is a variable that is normally distributed. This allows us to predict that Robert is most likely about average height. He is equally likely to be either a bit taller or a bit shorter than average. He is unlikely to be substantially taller or shorter than average.

EXAMPLE PROBLEM

A student weighs himself one morning 25 times on his bathroom scale. The resulting weight values (in pounds) are shown.

a. Show these data in a frequency table and graphically.

b. Do these data appear to be approximately normally distributed?

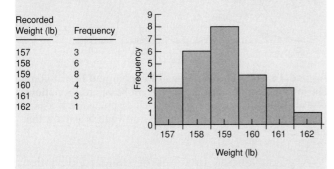

159 159 158 161 160 158 159 157 158 160 159 158 160
159 158 161 160 159 157 159 162 161 159 158 157

ANSWER

These data can be readily summarized using six classes:

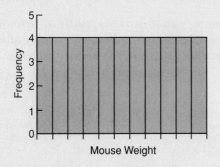

Recorded Weight (lb)	Frequency
157	3
158	6
159	8
160	4
161	3
162	1

b. These data suggest a normal distribution. In fact, it has been found that if a great many measurements of the same item are made, and if those measurements are made properly, then there is variation in the measurements which tends to be normally distributed.

Although data are commonly normally distributed, not all data have a normal distribution. For example, the frequency distribution of a set of data may be **skewed,** *in which case the values tend to be clustered either above or below the mean,* see Figure 16.13. The frequency distribution in Figure 16.14 is **bimodal,** *which means that it has two peaks.* Note that when a distribution is not normal, the mean, the median, and the mode do not coincide.

EXAMPLE PROBLEM

Multiple Choice:

What frequency distribution is illustrated in this histogram?

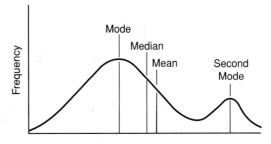

a. All the mice are of the same weight.

b. There are the same number of mice in each weight class.

c. Neither of the above.

ANSWER

b is correct. The histogram illustrates a situation where there are four mice in each weight class.

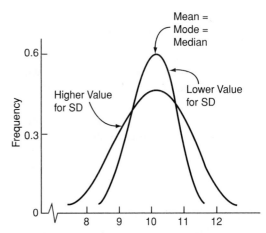

Figure 16.12.   The Normal Curve. When data are perfectly normally distributed, their mean, median, and mode all lie at the peak of the distribution plot and the standard deviation determines the width of the curve.

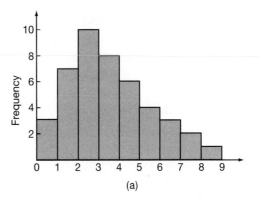

(a)

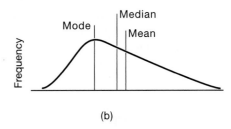

(b)

Figure 16.13.   Distributions That Are Skewed.

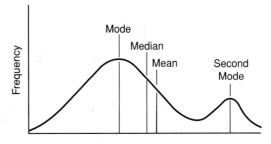

Figure 16.14.   A Bimodal Distribution.

C. The Relationship between the Normal Distribution and the Standard Deviation

The normal distribution is extremely important in interpreting and understanding data; therefore, we will now consider some of the features of this distribution in more detail.

The area under any normal curve, by definition, is equal to 100% or 1.0. The center of the peak of a normal curve is the mean. The normal curve is symmetrical; half of the area under the normal curve lies to the right of the mean, and half lies to the left of the mean, see Figure 16.15.

Figure 16.16 shows four normal probability distributions. The overall pattern of the four curves is the same; each has a peak centered at the mean. Each curve is symmetrical so that half its area lies to the left of the mean and half to the right; however, the four curves are not identical. Three of the four distributions have different means; therefore, the curves lie at different locations along the X axis. The curves also differ in width. Those with smaller standard deviations are narrower than those with larger standard deviations. Every normal curve is thus described by two characteristics:

1. The mean, which locates the center of the curve along the X axis.

2. The standard deviation, which determines how "fat" or "thin" the curve appears.

The normal curve in Figure 16.17 shows the distribution of heights of men in the United States. Observe that this plot is "marked off" in terms of the standard deviation. The mean height for the population is about 175 cm and the standard deviation is about 7.6 cm. The mean +1 SD equals 175 cm +7.6 cm = 182.6 cm. The mean −1 SD is 167.4 cm. Observe the location of these two points on the graph. The mean height +2 SD is 190.2 cm and the mean −2 SD is 159.8 cm. Observe these values on the graph and also the location of the mean ±3 SD.

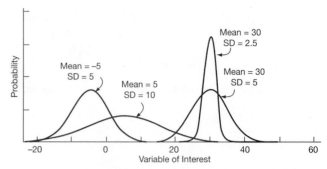

Figure 16.16. The Patterns of Four Normal Distributions Having Different Means and Different Standard Deviations. The mean locates the center of the curve along the X axis. The standard deviation determines how wide or narrow the curve is.

A graph labeled in terms of standard deviations from the mean can be constructed for any set of data that is approximately normally distributed. There is a certain pattern that is always true of such plots, see Figure 16.18. The pattern is such that about 68% of the area under the normal curve lies in the section between the mean +1 SD and the mean −1 SD. About 95.5% of the total area under the curve lies in the section between the mean +2 SD and the mean −2 SD. About 99.7% of the total area under the curve lies in the section between the mean +3 SD and the mean −3 SD. The pattern shown in Figure 16.18 is consistent for any normal curve, regardless of how spread out the curve is, or the value for the mean.

The relationship between the area under a normal curve and the approximate number of standard deviations from the mean is summarized in tabular form in Table 16.7.

So far, the significance of this discussion of standard deviation and the normal curve may seem obscure; however, it is actually of great practical significance.

1. *When a variable is normally distributed, the standard deviation tells us the percent of individuals whose measurements are within a certain*

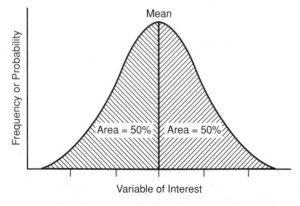

Figure 16.15. The Normal Curve. The center of the peak is the mean. The variable of interest is plotted on the X axis; either probability or frequency is on the Y axis. The normal curve is symmetrical; half its area lies to the left of the mean, half lies to the right.

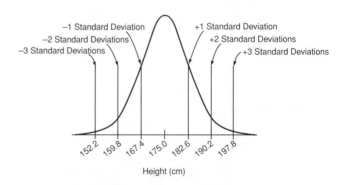

Figure 16.17. The Normal Curve Subdivided in Terms of the Standard Deviation. The frequency distribution for the height of men in the United States in terms of the mean and the standard deviation. (The mean = 175.0 cm; SD = 7.6 cm.)

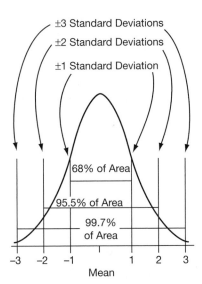

±3 Standard Deviations
±2 Standard Deviations
±1 Standard Deviation

68% of Area
95.5% of Area
99.7% of Area

−3 −2 −1 1 2 3
Mean

Figure 16.18. The Relationship between the Standard Deviation and the Area under the Normal Curve. For any normal distribution, about 68% of the area under the curve lies within ±1 standard deviation from the mean; about 95.5% of the area under the curve lies within ±2 standard deviations from the mean; about 99.7% of the area lies within ±3 standard deviations from the mean.

Table 16.7. THE RELATIONSHIP BETWEEN THE AREA UNDER A NORMAL CURVE AND THE APPROXIMATE NUMBER OF STANDARD DEVIATIONS FROM THE MEAN

Standard Deviations	Approximate Area Under the Curve
±1.0	68.0%
±2.0	95.5%
±2.6	99.0%
±3.0	99.7%

within two standard deviations on either side of the mean. The mean is 175 cm, so there is a 95.5% probability that he is between 159.8 cm and 190.2 cm. If we predict the height of a randomly chosen man to be between 159.8 cm and 190.2 cm, then, about 4.5% of the time, we expect to be wrong, but about 95.5% of the time we expect to be right. Thus, when data are normally distributed, knowing the standard deviation enables us to make predictions and assign them a certain probability of being correct. We can generalize to say that:

Statistics provides tools that enable us to make numerical statements and predictions in the presence of chance and variability. Statistics gives answers that are not expressed as "right" or "wrong"; rather, they are expressed in terms of probability.

range. For example, height in humans is a normally distributed variable; therefore, we know that about 68% of all values measured for height will fall within 1 standard deviation on either side of the mean, see Figure 16.19.

2. *When a variable is normally distributed, the standard deviation tells us the probability of obtaining a particular value if a single member of the population is randomly chosen.* For example, if a man is picked randomly there is a 68% probability that his height will be in the range of the mean ± one standard deviation.

We previously predicted that a 20-year-old man selected at random, Robert Smith, is probably about average height, might be a bit taller or shorter than average but is unlikely to be substantially taller or shorter than average. We can now be much more specific in our predictions. If Robert's name was chosen randomly, then there is about a 95.5% probability that his height is

EXAMPLE PROBLEM

a. What percent of the area under a perfect normal curve lies between the mean and the mean +1 standard deviation, as shown in the shaded area?

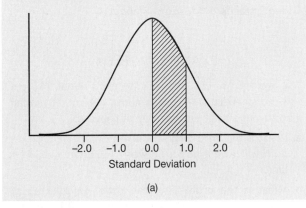

−2.0 −1.0 0.0 1.0 2.0
Standard Deviation

(a)

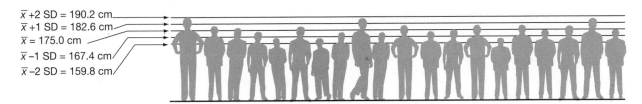

$\bar{x}$ +2 SD = 190.2 cm
$\bar{x}$ +1 SD = 182.6 cm
$\bar{x}$ = 175.0 cm
$\bar{x}$ −1 SD = 167.4 cm
$\bar{x}$ −2 SD = 159.8 cm

Figure 16.19. Men's Height and the Normal Distribution. About 68% of all men are of a height that falls within one standard deviation on either side of the mean. About 95.5% of men are of a height that falls within 2 SD on either side of the mean.

b. What percent of the area under a perfect normal curve lies in the shaded section in this graph?

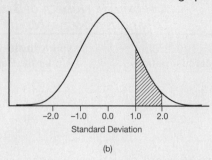

(b)

c. What percent of the area under a perfect normal curve lies in the section shown in this graph?

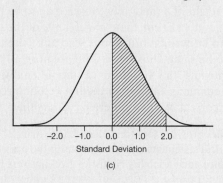

(c)

ANSWER

a. About 34% of the area lies between the mean and +1 standard deviation.

b. About 13.75% of the area lies between +1 and +2 standard deviations.

c. The area under the curve that lies between the mean and 2 standard deviations is about 47.75%.

EXAMPLE PROBLEM

The average for U.S. women's height is about 163 cm and the standard deviation is about 6.4 cm. Assuming that women's height is normally distributed:

a. What is the probability that a woman selected at random will have a height between 156.6 cm and 169.4 cm?

b. What is the probability that a woman selected at random will have a height between 150.2 cm and 175.8 cm?

c. What percent of women have heights between 143.8 cm and 182.2 cm?

d. Show the distribution of women's heights graphically.

ANSWER

a. This is the range of ± one standard deviation (163 cm ± 6.4 cm). We know that about 68% of all values can

be expected to fall within this range, so there is a 68% chance that a woman selected at random will be in this range.

b. This is the range of ± two standard deviations (163 cm ± 12.8 cm). There is therefore a 95.5% chance that a randomly selected woman will fall in this range.

c. This is the range of ± three standard deviations (163 cm ± 19.2 cm). 99.7% of women will have heights in this range.

d.

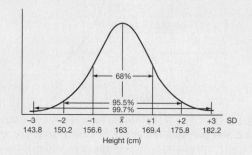

EXAMPLE PROBLEM

The height of adult males is normally distributed with a mean of 175 cm and a standard deviation of 7.6 cm.

a. About what percent of men are taller than 190.2 cm?

b. About what percent of men are shorter than 159.8 cm?

ANSWER

We previously showed that about 95.5% of adult men have heights within two standard deviations of the mean. This means that a total of about 4.5% of men are either taller than 190.2 cm or shorter than 159.8 cm. We expect, therefore, that 2.25% of men will be taller than 190.2 cm and 2.25% of men will be shorter than 159.8 cm.

D. A Brief Summary

There is variability in nature and so there is variability in the data collected when we observe the natural world. Statistics provides methods for organizing, summarizing, and displaying such data. Measures of central tendency and measures of dispersion are calculated values that describe and summarize data. Frequency histograms and frequency polygons display variation graphically.

Patterns emerge when data values are organized by frequency. The normal distribution is a frequency distribution that is often approximated in the natural world. When values are normally distributed, they center around the average. Most individuals have values that are about average, some have values that are a bit higher or lower than average, and relatively few individuals have values that deviate greatly from the average.

Statistics provides tools to organize and help us interpret variability in our observations of the natural world.

IV. AN APPLICATION: CONTROLLING PRODUCT QUALITY

A. Variability

There are many applications of the ideas discussed so far in this chapter. This section focuses on the application of these principles to the area of ensuring product quality.

It might seem that products made by the same process—or samples of product from the same batch, or measurements of the same sample performed in the same way—ought to be identical. In fact, the goal of product quality systems [including ISO 9000 and the Good Manufacturing Practices (GMP), which govern pharmaceutical production] is to reduce variability so that products are consistently of high quality. (See also Chapter 4.) Variability is reduced through such means as the consistent use of standard operating procedures, instrument maintenance programs, and so on. In practice, however, even when product quality systems are implemented there is always some variability in products. This is partly because there is variability inherent in any production process. In addition, measurements vary, even repeated measurements of a single characteristic of a single item. In fact, if there is no variability in a set of measurements relating to a product, then there is likely to be a malfunction or a lack of sensitivity in the system. For example, if the activity of an enzyme product is measured in 10 different vials from a single batch, or if the amount of a drug is measured in 25 tablets, then there will inevitably be some variability in the results. Product variability is not desirable; it is, however, always present.

Statistics provides important tools that can help to maintain the quality of products in the face of this inevitable variability. Statistical tools can be used to:

- determine how much variability is naturally present in a process
- determine whether there is likely a malfunction or problem in a process
- determine how certain we can be of a measurement result in the presence of variability.

Consider this example problem:

EXAMPLE PROBLEM

A technician is responsible for purifying an enzyme product made by fermentation. During the purification process she checks the amount of protein present. The results of six such measurements made during typical runs were (in grams):

23.5 31.0 24.9 26.7 25.7 21.0

One day the technician obtains a value of 15.3 g. Should she be concerned by this value? Another day the technician obtains a value of 32.1 g. Is this value a cause for concern?

ANSWER

Let us begin by looking at these values in an intuitive way. The six typical values appear to hover around about 25 g ± about 5 g. A value of 15.3 g is quite a bit lower than the other values; it is around 10 g below the approximate mean. In contrast, the value 32.1 is not as far from the mean and may therefore be less of a cause for concern.

Let us now apply statistical reasoning to this problem. The six measurements that were made during normal runs are assumed to represent values obtained when the purification process was working properly. The mean for these six values is 25.5 g and their standard deviation is 3.36 g. Based on the normal distribution, we expect 95.5% of the values to fall by chance within 2 SD of the mean. The range that includes two standard deviations is 18.8 g to 32.2 g. Thus, based on the normal distribution, the value of 15.3 g is clearly unexpected on the basis of chance alone and is therefore cause for concern. The value of 32.1 g does not immediately suggest there is a problem (although, if it signals the beginning of an upward trend, there might in fact be a problem in the process).

Note that in this example the technician is applying statistical thinking to her everyday work. She is taking advantage of knowledge about normal distributions and about variability to identify potential problems.

B. Control Charts

There is a common graphical quality-control method that extends the reasoning demonstrated in the preceding example problem. This is the use of **control charts.** We introduce control charts with an example. A biotechnology company uses a fermentation process in which bacteria produce an enzyme that is sold for use in food processing. The pH of the bacterial broth must be controlled over time; otherwise, the pH tends to drop and production of the enzyme diminishes. A preliminary fermentation run that successfully produced enzyme was performed. The pH was monitored throughout the run and the resulting values are shown in Table 16.8 on p. 288.

Let us assume that this fermentation process is tested more times and that the pattern of variability illustrated in Table 16.8 is characteristic of successful runs. The mean hovers around pH 7.05 and the SD is 0.17 pH units. When the company begins to produce this enzyme for sale, the pattern of pH variability should be similar to that observed in the preliminary, successful tests. In a production setting, *a variable that has the same distribution over time is said to be* **in statistical control,** *or, is simply* **in control.** Production processes should be consistent; therefore, having "in control" processes is critical.

Table 16.8. THE pH DURING FERMENTATION IN PRELIMINARY STUDIES

Time	pH
0800	7.22
0830	7.13
0900	6.84
0930	6.93
1000	7.12
1030	7.33
1100	7.04
1130	6.79
1200	6.94
1230	7.03
1300	7.22

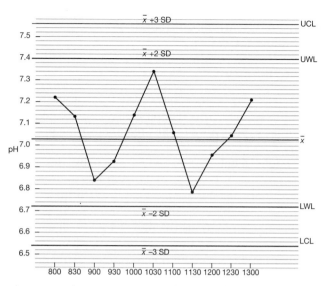

Figure 16.20. A Control Chart for pH Values from a Fermentation Run.

A control chart is a graphical representation of the variability in a process. A control chart for the data in Table 16.8 is shown in Figure 16.20. When a control chart is constructed, a line is drawn across the chart at the mean (in this case, at pH 7.05). A second line is drawn at $\bar{x} +2$ SD (in this example, at 7.39). A third line is drawn on the graph at $\bar{x} -2$ SD (in this example at 6.71). Two more lines are drawn at $\bar{x} +3$ SD (pH 7.56 and 6.54, respectively).

The lines drawn across the control chart have names. The line at $\bar{x} -2$ SD is called the **lower warning limit, LWL.** The line at $\bar{x} +2$ SD is called the **upper warning limit, UWL.** The **lower control limit, LCL,** is $\bar{x} -3$ SD and the **upper control limit, UCL,** is $\bar{x} +3$ SD.

The use of control charts is based on the important assumption that the inherent variability in production processes is approximately normally distributed. Given that this is the case, then 95.5% of all observations are expected to fall within ±2 SD of the mean; therefore, the warning limits define the range in which 95.5% of all values are predicted to lie. One pH observation falling outside the warning limits is not a cause for concern because about 5% of the time values may fall outside these limits simply because of chance variability. Two or more points outside the warning limits, however, may indicate a problem in the process. The control limits define the range in which 99.7% of all points should lie. The probability that an observation, by chance, would fall outside the control lines is only 3 in 1000. It is important, therefore, to stop the process and find the problem if a value outside these action lines is obtained.

When a process is in control, its control chart has features illustrated in Figure 16.20. All the points are between the upper and lower warning limits. About half the points are above the mean and half below. It is also important that there is no trend where the points seem to be increasing in pH over time or decreasing in pH over time.

A control chart helps operators distinguish between variation that is inherent in a process and variation that is due to a problem and should be investigated. Figure 16.21 is a control chart that illustrates several problems. On days 7–10 the process is showing an upward trend. At day 10 there is a high point that is outside the control limits. At day 20 there is a point that exceeds the lower control limit. When these types of problems are observed, the process would probably be stopped and the cause(s) of the deviations would be investigated.

A point that has a value much higher or much lower than the rest of the data is called an **outlier.** In the real world, outliers may or may not signal that a process is out of control. Outliers are sometimes simply errors in how a test was performed. Errors might include, for example, improperly removing and preparing the sample, using a malfunctioning instrument, or inattentiveness on the part of the analyst. For example, an aberrant pH value in the fermentation process might indicate that the bacteria are not growing properly and that enzyme production is likely to be adversely affected. An aberrant pH read-

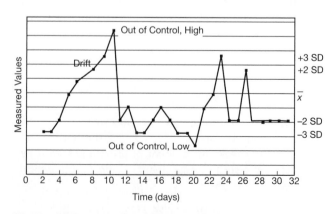

Figure 16.21. A Control Chart Illustrating Possible Problems.

EXAMPLE PROBLEM

The company that produces an enzyme by fermentation performs a production run. The following pH data are obtained. Prepare a control chart for these data. Is the process in control during this run? (Note that the expected pattern of variability in this process was determined in preliminary studies; therefore, use the same mean, standard deviation, UWL, LWL, UCL, and LCL as in Figure 16.20.)

Time	pH
0800	7.34
0830	7.22
0900	7.27
0930	6.88
1000	7.02
1030	7.01
1100	6.78
1130	6.88
1200	6.76
1230	6.57
1300	6.54

ANSWER

The points for this run are plotted on a control chart. Observe that the data shows a downward trend. Moreover, the final point is close to the control limit. These data strongly suggest that the process is not in control.

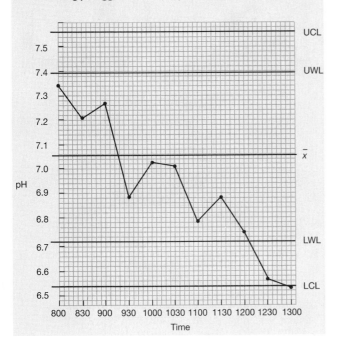

ing might also be due to a malfunction in the pH meter—a testing error. In production settings the initial investigation of an outlier looks at whether it is simply due to testing error.*

Control charts are a statistical tool based on an understanding of variability and the normal distribution. Control charts allow operators to quickly distinguish between inherent, expected variability in measurement values and variation that is caused by a problem in a process. As long as the values remain within the limits of expected variability, the process is considered to be in control and variation is assumed to be due to chance alone. If a value falls outside the range predicted by chance, or if the values begin to be consistently high or low, then there may be a problem in the production process that needs to be examined. There are many types and forms of control charts other than the one illustrated here. The basic purpose of a control chart is the same, however, regardless of the details of its construction.

PRACTICE PROBLEMS

1. a. Ten vials of enzyme were chosen from a batch for quality control testing. What was the sample? What was the population?

b. The glucose level in a blood sample was measured. What was the sample? What was the population?

c. The reading ability of all first graders in Franklin Elementary School was measured. What was the sample? What was the population?

2. a. Ten measurements of a 10 g standard weight were made. Calculate the mean, standard deviation, and coefficient of variation.

10.001 g 10.000 g 10.001 g   9.999 g   9.998 g
10.000 g 10.002 g   9.999 g 10.000 g 10.001 g

b. Calculate the mean, standard deviation, and coefficient of variation for these test scores.

75% 71% 88% 99% 76% 83% 89% 91%

c. The rate of an enzyme reaction was timed in six reaction mixtures. Calculate the mean, standard deviation, and coefficient of variation.

235 sec   287 sec   198 sec
255 sec   234 sec   201 sec

---

*When an aberrant or unexpected test result of a regulated product (such as a pharmaceutical product) is obtained, the company must investigate and respond. How the company must conduct its investigation and the company's response to an unexpected test result has sometimes been a cause of contention between the company and the FDA. This is because an unexpected test result may indicate that a product is defective. An unexpected test value may also indicate that an analyst made a mistake, that a piece of test equipment was malfunctioning, that a value was recorded inaccurately, or other such problems that do not relate to the quality of the product. Moreover, very high or very low measurement results occasionally occur simply by chance and are the result of normal variability in the measurement system. Companies therefore sometimes attribute unexpected test results to laboratory error or to chance while regulators are concerned that the test result might reflect a problem with a product. Disagreements between a company and the FDA in how unexpected test values should be handled have been contested in the courtroom. (See the discussion of the *Barr* decision in Chapter 8.)

d. The analyst who performed the enzyme reactions in problem c tried an alternative assay method and obtained the following results. Calculate the mean and the standard deviation for the alternative method. Compare the two methods.

245 sec   235 sec   256 sec   228 sec
237 sec   244 sec   234 sec   215 sec

3. (Optional: These calculations are best performed using a calculator that automatically performs statistical functions.) Calculate the mean, standard deviation, and relative standard deviation for the following data sets:

a. The weights of 33 vials from a lot were sampled. The results in milligrams are:

115 133 125 124   87   95 136 146 114 133 139
177 193 136 177 193 136 123 171 147 153 149
110 111 121   71 178 173 201 149   94 100 103

b. The weights of 25 samples were measured. The results are:

9.27 g   9.88 g   6.19 g   8.38 g   9.62 g
5.48 g   9.71 g   9.54 g   7.41 g   6.76 g
9.33 g   8.64 g   6.43 g   7.66 g   9.60 g
7.50 g   9.89 g   7.73 g   5.68 g   6.74 g
8.04 g   7.23 g   7.35 g   10.82 g   8.94 g

c. The volumes of 19 samples were measured. The results are:

1022 mL   1045 mL   1103 mL   1200 mL
1189 mL   1234 mL   1043 mL   1089 mL
1103 mL   1020 mL   1058 mL   1197 mL
1201 mL   1189 mL   1098 mL   1155 mL
1056 mL   1023 mL   1109 mL

4. A biotechnology company is developing a new variety of fruit tree and researchers are interested in the number of blossoms produced. The following are the total number of blossoms from 20 trees. Find the mean, median, range, and standard deviation for these data:

34   36   29   27   30   35   32   31   39   30
44   30   33   43   32   21   35   33   40   27

5. A company plans to grow and market Shiitake mushrooms. The yields of mushrooms obtained in preliminary experiments are shown, expressed in kilograms. Calculate the mean, median, range, and standard deviation.

38.2   31.7   28.1   29.1   33.4   25.2
18.2   19.7   41.7   26.2   21.0   52.2

6. The special average test score for a class of thirty students is 75%. One additional student takes the exam and scores 100%. What is the new average for the class?

7. For each of the following three distributions, estimate the mean and mode.

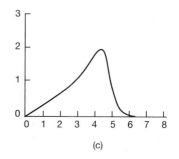

(a)   (b)

(c)

8. Which of the following frequency histograms most closely approximates a normal distribution? Which appears to be bimodal? Which appears skewed?

(a)   (b)

(c)

9. Which of these two distributions is less dispersed?

(a)

(b)

10. Which of these two distributions is less dispersed?

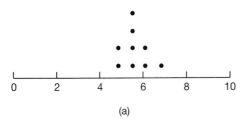

(a)

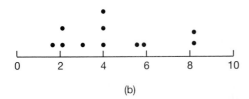

(b)

11. Which of these two distributions is less dispersed?

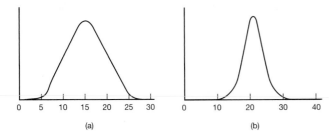

12. Which of these two distributions is less dispersed?

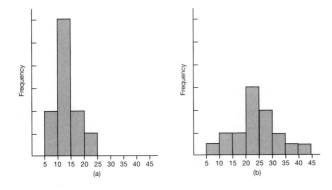

13. Cells in culture are treated in such a way that they are expected to take up a fragment of DNA containing a gene that codes for an enzyme. The activity of the enzyme in cells that take up the gene can be assayed. The more active the enzyme, the better. Suppose a researcher isolates 45 clones of treated cells and measures the enzyme activity in each clone. The results are shown in activity units below.

a. Find the range, median, mean, and standard deviation for the data from these 45 treated clones.

b. Plot these data on a histogram and show on the plot where the mean and median are located.

c. Do you think these cells have taken up the gene fragment containing the enzyme, based on these data? Explain.

10.4	12.2	12.0	9.1	5.8	3.2	9.8	10.1	13.0
2.1	9.8	10.1	1.5	7.8	5.6	2.3	9.8	10.2
9.1	12.3	10.1	12.3	14.2	15.1	13.6	12.1	10.8
9.2	8.9	12.4	11.0	13.1	8.9	9.4	10.2	11.3
12.8	9.0	8.6	12.3	12.0	2.1	0.4	10.6	13.0

14. Suppose that the assay for enzyme activity (discussed in question #13) is performed on cultured cells that have not been treated with the DNA fragment and are not expected to have any enzyme activity at all. In this case, one would expect to see little or no activity when performing the assay. You test this assumption on 20 untreated cell clones that are not expected to have the enzyme and obtain the following results in activity units:

0.0	1.0	0.3	3.0	2.8	1.2	2.3	1.1	0.9
2.5	1.9	0.9	2.2	1.2	0.2	0.1	0.2	0.9
1.0	2.0							

a. Calculate the mean and the standard deviation for these results.

b. Prepare a histogram for these results.

c. Discuss the results shown in question #13 in light of these data in question #14.

15. The following are heights in centimeters of a common prairie aster measured from one field.

a. Determine the range and mean for these data.

b. Plot these data as a frequency histogram. (You will need to divide the data into groups. After finding the range, make 10–12 divisions of equal size and assign each value to the proper class.)

151	182	182	162	177	166	197	144
174	160	131	156	125	170	153	172
146	127	156	140	159	155	158	165
155	141	180	145	145	150	145	135
105	122	180	152	161	170	156	150
122	140	133	145	190	165	176	170
144	160	162	157	155	146	155	143
154	157	141	150	142	139	138	156
130	126	189	138	103	163	135	158
151	129	147	153	150	140	146	138
154	138	179	165	142	140	141	160
125	156	145	159	147	155	189	195
140	141	160	156				

16. Show the following data in the form of a frequency distribution and a frequency histogram and polygon.

23 mg	28 mg	22 mg	23 mg	20 mg	19 mg
22 mg	24 mg	26 mg	23 mg	23 mg	24 mg
21 mg	25 mg	24 mg	25 mg	21 mg	

17. As a trouser manufacturer, you are interested in the average height of adult males in your town. You take a random sample of five male customers and find the mean to be 5 feet 10 inches and the standard deviation to be 3 in. How certain are you of the following statements and why?

   a. The average height of the population is between 0 and 50 ft.

   b. The average height of the population is between 4 ft and 7.5 ft.

   c. The next new customer will be between 5 ft 7 in and 6 ft 1 in.

   d. The average height of the entire male population is 5 ft 10 in.

18. Refer to the Example Problem on p. 286 regarding women's heights.

   a. What is the probability that a woman selected at random will be shorter than 150.2 cm?

   b. What is the probability that a woman selected at random will be taller than 169.4 cm?

19. A pharmaceutical company finds that the average concentration of drug in each vial is 110 μg/vial with a standard deviation of 6.1 μg.

   a. About what percent of all vials can be expected to have between 94.1 μg and 125.9 μg of drug?

   b. What is the likelihood that a vial selected at random will have more than 122.2 μg of drug?

   c. What percent of all the vials are expected to have a values between 103.9 μg and 116.1 μg?

   d. Suppose 100 vials are checked. About how many of them would be predicted to have more than 125.9 μg?

20. A technician customarily performs a certain assay. The results of 8 typical assays are:

   32.0 mg   28.9 mg   23.4 mg   30.7 mg
   23.6 mg   21.5 mg   29.8 mg   27.4 mg

   a. If the technician obtains a value of 18.1 mg, should he be concerned? Base your answer on estimation without performing actual calculations.

   b. Perform statistical calculations to determine whether the value 18.1 mg is out of the range of two standard deviations.

21. A technician customarily counts the number of leaves on cloned plants. The results of nine such counts in successful experiments are:

   75   54   55   61   71   67   51   77   71

   a. If the technician obtains a count of 79, is this a cause for concern? Base your answer on estimation without performing actual calculations.

   b. Perform statistical calculations to determine whether 79 leaves is out of the range of two standard deviations.

22. Examine this control chart. Discuss the points on March 22, April 15, May 31, and June 23.

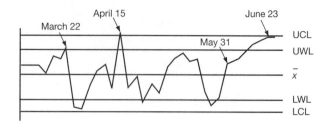

23. Suppose a biotechnology company is using bacteria to produce an antibiotic. During production of the drug, the pH of the bacterial broth must be adjusted so that it remains optimal; otherwise, production of the antibiotic diminishes. Preliminary studies were performed to determine the optimal pH. The results of a successful preliminary study are shown in the following table.

Time	pH
0800	6.12
0830	5.13
0900	5.84
0930	6.53
1000	6.12
1030	6.23
1100	6.04
1130	5.79
1200	5.94
1230	6.03
1300	6.12

   a. Calculate the mean pH and the SD.

   b. Construct a control chart with a central line, UWL, LWL, UCL, and LCL.

The enzyme goes into production and periodic samples of the broth are assayed for pH. The results are shown in the following table.

Time	pH
0800	6.54
0830	6.12
0900	5.87
0930	5.18
1000	4.95
1030	4.89
1100	5.03
1130	5.43
1200	5.34
1230	5.37
1300	5.38

   c. By simply observing these points, what can you say about the process?

   d. Plot these data on the control chart.

   e. Comment on the process; did it ever reach the warning level or action required levels?

UNIT V

Laboratory Measurements

Measurements are *quantitative observations, or numerical descriptions.* Examples of measurements include the weight of an object, the amount of light passing through a solution, and the time required to run a race. Everyone has experience making measurements (e.g., taking the temperature of a child, weighing oneself, or measuring ingredients when cooking).

Measuring properties of samples is an integral part of everyday work in any biology laboratory. For example, solutions are required to support the activity of cells, enzymes, and other biological materials. Preparing solutions involves measuring the weights and volumes of the components. Estimating the quantity of DNA in a test tube may involve measuring how much light passes through the solution. The pH of the media in which bacteria grow during fermentation must be monitored continuously. There are countless measurements

made in most laboratories, each of which must be a "good" measurement.

Although it seems obvious that laboratory measurements should be "good," it is surprisingly difficult to define a "good" measurement. One definition is that a "good" measurement is correct; however, this leads to the question of what is "correct"? Suppose a man weighs himself in the morning on a bathroom scale that reads 165 pounds. Shortly after, he weighs himself at a fitness facility where the scale reads 166 pounds. At this point the man might be somewhat uncertain as to his exact, correct weight; perhaps he weighs 165 pounds, perhaps 166. Perhaps his weight is somewhere between 166 and 165 pounds. He is likely to conclude, however, that he weighs about 165 pounds and leave it at that.

Uncertainty of a pound or so in an adult's weight is seldom of great concern. A bathroom scale that gives a

Non Sequitur

NON SEQUITUR © 1996 Wiley Miller. Dist. By UNIVERSAL PRESS SYNDICATE.
Reprinted with permission. All rights reserved.

weight value within 1 pound of the true weight is a reasonably "correct" instrument. In other situations, however, a measurement must be much more correct. For example, a 1 pound difference in the weight of an infant could mean the difference between a healthy baby and one that is severely dehydrated. In the laboratory, an error of 1 g in a measurement could mean the difference between a successful experiment and a disastrous failure. In a drug product, a 1 mg error in measurement could endanger a patient. In each of these situations a "good" measurement is one that can be trusted to make a decision. A good measurement can be trusted by a physician selecting a treatment for a patient, by a research team drawing conclusions from a study, and by a pharmaceutical company deciding whether to release a drug product to the public.

Laboratory workers play a key role in performing measurements of properties of samples. To make good, trustworthy measurements, laboratory workers must understand the principles of measurement, know how to maintain and operate instruments properly, and be aware of and avoid potential pitfalls in measurement. It takes knowledge and careful technique to produce measurements that can be trusted in a particular situation.

This unit discusses methods of making "good," trustworthy measurements in the laboratory.

Chapter 17 introduces basic principles underlying measurements and terminology relating to metrology (the study of measurements).

Chapter 18 introduces basic principles of electricity and electronics, and instrumental methods of measurement. This chapter serves as a transition to the next five chapters, each of which discusses a specific type of measurement.

Chapter 19 introduces principles and practices relating to weight measurements in the biotechnology laboratory.

Chapter 20 introduces principles and practices relating to volume measurements in the biotechnology laboratory.

Chapter 21 introduces principles and practices relating to temperature measurements in the biotechnology laboratory.

Chapter 22 introduces principles and practices relating to the measurement of pH, selected ions, and conductivity.

Chapter 23 introduces principles and practices relating to the measurement of light transmittance and absorbance.

BIBLIOGRAPHY FOR LABORATORY MEASUREMENTS

- **Manufacturers** are an excellent source of up-to-date information on instrument design, operation, and performance verification.
- **ASTM International and ISO** standards provide specific, technical information. They are available for purchase at http://www.ASTM.org, or http://www.iso.org or at some libraries.
- **NIST publications** provide information on specific measurement topics and are usually available through NIST at http://www.NIST.gov.

Specific examples of these resources are included below.

General Statistics

General statistics books have information about standard deviation, accuracy, precision, and other important concepts in metrology. Examples include:

Freedman, David, Pisani, Robert, and Purves, Roger. *Statistics.* 4th ed. New York: W.W. Norton, 2007.

Miller, J.C., and Miller, J.N. *Statistics for Analytical Chemistry.* 3rd ed. New York: Prentice Hall, 1993.

General Resources

ASTM E456-06. "Standard Terminology Relating to Quality and Statistics." http://www.ASTM.org.

Ewing, Galen Wood., ed. *Analytical Instrumentation Handbook.* 3rd ed. New York: Marcel Dekker, 1997. (Includes chapters on laboratory balances and the use of computers in laboratory equipment.)

Fields, Lawrence D., and Hawkes, Stephen J. "Minimizing Significant Figure Fuzziness." *Journal of College Science Teaching,* Sept/Oct 1986, 30–4.

ISO/IEC 17025, "General Requirements for the Competence of Testing and Calibration Laboratories," 2005. http://www.ISO.org.

ISO/IEC Guide 99:2007, "International vocabulary of metrology—Basic and general concepts and associated terms (VIM)." 2007.

Kenkel, John. *Analytical Chemistry for Technicians.* 3rd ed. Boca Raton, FL: Lewis Publishers, 2003. (Good explanation of instrumental analysis, pH and other measurement topics.)

Pinkerton Richard C., and Gleit, Chester E. "The Significance of Significant Figures." *Journal of Chemical Education* 44(4):232–4, 1967.

Taylor, Barry N., ed. *The International System of Units (SI).* NIST Special Publication 330. Gaithersburg, MD: U.S. Government Printing Office, 2001. (A comprehensive guide to the SI system.)

Taylor, Barry N., and Kuyatt, Chris E. "Guidelines for Evaluating and Expressing the Uncertainty of NIST Measurement Results." NIST Technical Note 1297. Gaithersburg, MD: U.S. Government Printing Office, 1994. This publication gives the NIST policy on expressing measurement uncertainty.

Taylor, John K. *Standard Reference Materials Handbook for SRM Users.* NIST Special Publication 260-100. Gaithersburg, MD: U.S. Government Printing Office, 1993.

Reed, W.P. "Chemical Measurements and the Issues of Quality Comparability and Traceability." *American Laboratory,* December 1994, 18–21.

Weight Measurement

ASTM E617-97, "Standard Specification for Laboratory Weights and Precision Mass Standards." http://www.ASTM.org.

Harris, Georgia L. "Commonly Asked Questions about Mass Standards." NIST. Gaithersburg, MD: U.S. Government Printing Office, 2005. http://ts.nist .gov/WeightsAndMeasures/caqmass.cfm.

Schoonover, Randall M., and Jones, Frank E. "Air Buoyancy Correction in High Accuracy Weighing on an Analytical Balance." *Analytical Chemistry* 53 (1981): 900–2.

Jones, Frank E., and Schoonover, Randall M. *Handbook of Mass Measurement.* Boca Raton, FL: CRC Press, 2002.

Kupper, Walter E. "Honest Weight and True Mass—(They are Not the Same)." *American Laboratory* (December 1990): 8–9.

Mettler-Toledo. *Weighing the Right Way with Mettler: The Proper Way to Work with Electronic Analytical and Microbalances.* Greifensee, Switzerland: Mettler-Toledo, 1989. http://us.mt.com/mt/ed/brochures/Weighing_the_ right_way_0x000246700005761700059b6e.jsp?m= t&key=U1MTg4NjM1NT. (A nicely written booklet on basic principles of weighing.)

Troemner Inc. *Troemner Mass Standards Handbook.* http: //www.troemner.com/literature_lib.php# masshand.

Volume Measurement

ASTM E1044-96, "Standard Specification for Glass Serological Pipets (General Purpose and Kahn)." http://www.ASTM.org.

ASTM E1154-89, "Standard Specification for Piston or Plunger Operated Volumetric Apparatus." http://www.ASTM.org. (Recommended for individuals responsible for checking the calibration of micropipetting devices.)

ASTM E1878-97, "Standard Specification for Laboratory Glass Volumetric Flasks, Special Use." http://www.ASTM.org.

ASTM E288-06, "Standard Specification for Laboratory Glass Volumetric Flasks." http://www.ASTM.org.

ASTM E542-01, "Standard Practice for Calibration of Laboratory Volumetric Apparatus." http://www .ASTM.org.

ASTM E969-02, "Standard Specification for Glass Volumetric (Transfer) Pipets." http://www.ASTM.org.

Gilson, Inc. "Gilson Guide to Pipetting." Madison, WI: Gilson, Inc., 2005. (An excellent primer on micropipettors. Request it from Gilson at sales@Gilson.com.)

Temperature Measurement

ASTM E1-07, "Standard Specification for ASTM Liquid-in-Glass Thermometers." http://www. ASTM.org.

ASTM E344-07, "Terminology Relating to Thermometry and Hydrometry." http://www.ASTM.org.

ASTM E77-98, "Standard Test Method for Inspection and Verification of Thermometers." http://www .ASTM.org.

Nicholas, J.V., and White, D.R. "Traceable Temperatures: An Introduction to Temperature Measurement and Calibration." 2nd ed. Chichester, West Sussex, UK: John Wiley and Sons, 2001. (A good introduction both to temperature measurement and metrology in general.)

pH and Conductivity Measurement

Beckman Corp. *The Beckman Handbook of Applied Electrochemistry.* Beckman Corp. # BR 7739. (Order from Beckman Coulter, http://www .beckmancoulter.com/resourcecenter/literature/ BioLit/BioLitList.asp?ProductCategoryID=EC)

Frant, Martin S. "The Effect of Temperature on pH Measurements." *American Laboratory* (July 1995): 18–23.

Spectrophotometry

Ciurczak Emil W., and Workman, Jr. Jerome. "Getting Started with UV/Vis Spectroscopy." Published as a supplement to *Spectroscopy Magazine,* 1996.

GE Healthcare. "Qualification and Performance Verification Logbook for Biochrom Ltd. UV/Visible Spectrophotometers." GE Healthcare, 2002. http://www4 .gelifesciences.com/aptrix/upp00919.nsf/Content/WD :Qualification+a254283263-B653. (A thorough explanation of performance verification.)

Gore, Michael G., ed. "Spectrophotometry and Spectrofluorimetry: A Practical Approach." Oxford: Oxford University Press, 2000.

Hammond, John, Irish, Doug, and Hartwell, Steve. "Calibration Science for UV/Visible Spectrometry, Part III in a Series on Quality Issues in Spectrophotometry." *Spectroscopy* 13(2) (1998): 64–71.

Levy, Gabor B. "The Editor's Page: A Literature Search." *American Laboratory* (October 1992): 10. (A discussion of who really discovered "Beer's" Law for those with historical interests.)

Mavrodineanu, R., Burke, R.W., Baldwin, J.R., Smith, M.V., Messman, J.D., Travis, J.C., and Colbert, J.C. "Glass Filters as a Standard Reference Material for Spectrophotometry—Selection, Preparation, Certification and Use of SRM 930 and SRM 1930." NIST Special Publication 260-116. Washington, DC: U.S. Department of Commerce/Technology Administration, March 1994. (A good detailed source for those who are verifying the performance of spectrophotometers.)

Owen, Anthony. "Good Laboratory Practice with UV-Visible Spectroscopy System." Publication number 12-5963-5615E. (Application Note.) Germany: Hewlett Packard, 1995.

CHAPTER 17

Introduction to Quality Laboratory Measurements

I. MEASUREMENTS AND EXTERNAL AUTHORITY: STANDARDS, CALIBRATION, AND TRACEABILITY

A. Overview

This chapter discusses general terminology and concepts relating to making "good" measurements. Recall that measurements are numerical descriptions. For example, if an object is said to be "15 grams," then mass is the property that is described, 15 is the value of that property, and grams are the measurement **units. A unit of measure** *is an exactly defined amount of a property.*

We are accustomed to using units of measure, such as grams, pounds, inches, and centimeters. But what, exactly, is the mass of a gram? This may seem like a trivial ques-

tion; obviously, the mass of a gram is a gram. However, exactly defining the meaning of a "gram," or of any other unit of measurement, is anything but trivial. A unit must be defined in some clear way, and everyone who uses the unit must agree on that definition. Establishing the meaning of a unit of measure, therefore, requires international agreement.

Metrologists, *people who work with measurements,* devote much effort toward ensuring international consistency in measurement. One result of this effort is the **SI (System Internationale) measurement system,** *which defines units of measurement* (see pp. 226–227 for more detail on the SI system). The SI definitions of units are an authority to which people in many nations refer when making measurements.

Measurements are always made in conformance with an external authority. As a simple example, we commonly measure length using a ruler. The ruler is our external authority when measuring length. The ruler was marked by the manufacturer so that its lines are correct according to an internationally accepted definition of a "meter."

At this point, it is important to clarify the distinction between a unit of measurement and a standard, such as a ruler. A **standard** *is a physical embodiment of a unit.* Units are not physical entities. Units are unaffected by environmental conditions, but physical standards are affected by the environment. For example, units of centimeters are unaffected by corrosion, but a metal ruler, a physical embodiment of centimeters, may become corroded.

The international efforts to promote "good" measurement practices are intimately associated with the ever-increasing importance of quality systems, such as ISO 9000, ISO Guide 17025, and cGMP (see Chapter 4 for explanations of these terms). Scientists and technicians have always been aware of the importance of "good" measurements. However, as people implement quality systems, they become even more concerned with establishing methods of measurement that are consistent, are widely accepted, and whose accuracy they can document.

The next sections of this chapter explore three key interrelated words in measurement: **standard, calibration,** and **traceability.** Later sections explore four more interrelated terms: **error, precision, accuracy,** and **uncertainty.**

B. Standards

The term *standard* was defined earlier as a physical object that embodies a unit. The most famous physical standard is probably the kilogram standard. The unit of a kilogram has been defined by international treaty to have as much mass as a special platinum-iridium bar* located at the International Bureau of Weights and Measures near Paris. All other mass standards are defined by comparison to this special metal bar. Every country that signed the treaty received a national kilogram prototype whose mass is determined by comparison with the standard in France. The United States' standard is called K_{20} and is housed at The National Institute of Standards and Technology (NIST).

The platinum-iridium bar in France and K_{20} are sometimes called **primary standards,** *standards whose values may be accepted without further verification by the user. A primary standard is used to establish the value for* **secondary standards.** Companies that manufacture

working standards, *standards for use in individual laboratories,* typically use secondary standards as the basis for their products.

For a standard to be useful, its properties must be known with sufficient accuracy to allow it to be used to evaluate another item. The accuracy that is required of a standard depends on the situation. For example, the value for the mass of the platinum-iridium bar in France must be known far more exactly than the value for the mass of a working standard in a teaching laboratory.

The term *standard* does not always refer to a physical embodiment of a unit. Table 17.1 on p. 298 includes several definitions of the word *standard,* some of which are discussed in Chapter 24.

C. Calibration

i. CALIBRATION AS AN ADJUSTMENT TO AN INSTRUMENT OR MEASURING DEVICE

A simple definition of **calibration** *is to adjust a measuring system so that the values it gives are in accordance with an external standard(s).* For example, calibration of an instrument might involve placing a standard in or on the instrument and turning a knob or pressing keys until the instrument displays the value of the standard. After calibration the response of the instrument is in accordance with the standard. The result of calibration (according to this definition) is that the instrument or measuring device is adjusted and it gives more correct values after calibration than before.

Calibration as an adjustment to a measuring system applies to the way a manufacturer makes a measuring device. For example, glassware that is used to measure the volume of liquids is marked with lines that indicate volume. The manufacturer "calibrates" the glassware so that the lines are in the proper place.

Once instruments enter the laboratory they are subject to the effects of aging and their response may be altered by changes in their environment. As a result, the response of instruments in the laboratory drifts. To correct for drift, laboratory equipment must be periodically recalibrated by the user or a service technician. For example, pH meters are very sensitive to the effects of aging and therefore require frequent calibration. Calibration involves adjusting the readings of the pH meter according to the pH of standard solutions of known pH. The calibration of pH meters and other measuring devices will be discussed in more detail in later chapters in this unit.

There is always some error when items are calibrated. The error can be reduced by using more expensive,

---

*There are disadvantages to using a physical object, like a platinum-iridium bar, as the ultimate definition of a unit. The bar could potentially be damaged or destroyed. Its mass changes slightly with dust and with cleaning. It requires a secure and environmentally stable storage site. Many scientists are therefore working toward a definition for the kilogram that relies on a physical constant, not on a physical entity. This has already been done for other units. For example, the meter is defined in terms of the speed of light.

Table 17.1. STANDARDS IN MEASUREMENT

There are 13 definitions for the word *standard* in the *Merriam-Webster Collegiate Dictionary* (G. & C. Merriam Co., MA, 1977). The broadest of these definitions is "something established by authority, custom, or general consent . . ." This meaning encompasses all the others described below.

1. *A standard is a physical object, the properties of which are known with sufficient accuracy to be used to evaluate another item; a physical embodiment of a unit.* For example, a metal object whose mass is accurately known can be used to determine the response of a balance.

    This definition includes the following:

    a. *Primary Standards.* Physical items whose value may be accepted without further verification by the user and which are used to determine the values for secondary standards.

    b. *Secondary Standard.* A standard whose value is determined by direct comparison with a primary standard.

    c. *Working Standard.* A physical standard that is used to make measurements in the laboratory and which is calibrated by comparison with a primary or secondary standard.

2. *In chemical or biological measurements, a standard often describes a substance or a solution that is used to establish the response of an instrument or an assay method to the analyte.* This definition includes the following:

    a. *Standard Reference Material (SRM).* A substance issued from NIST that is accompanied by documentation that shows how its composition was determined and how certain NIST is of given values.

    b. *Certified Reference Material.* A reference material from any source that is issued with documentation.

    c. *Reference Material.* Any substance for which one or more properties are established sufficiently well to allow its use in evaluating a measurement process or an assay.

    d. *Standard.* In common usage, any substance used to determine the response of an instrument or a method to the analyte of interest. The information obtained from a standard solution is often portrayed graphically in a **standard curve** (also called a **calibration curve**).

3. *A standard is a document established by consensus and approved by a recognized body that establishes rules or guidelines to make a procedure consistent among various people.* For example, ASTM specifies standard methods to calibrate volumetric glassware; the U.S. Pharmacopeia specifies methods to perform tests of pharmaceutical products.

---

exacting procedures, but it cannot be eliminated. **Tolerance** *is the amount of error that can be tolerated in the calibration of a particular item.* For example, the tolerance for a "500 g" Class 1 mass standard is ± 1.2 mg, which means the standard must have a true mass between 500.0012 g and 499.9988 g. The tolerances for a standard, a measuring device, or an instrument vary depending on the purpose for which the item is being used. More examples of tolerances for specific devices are shown in later chapters.

If any instrument or piece of apparatus is not properly calibrated, its measurements will deviate unacceptably from their correct values. Improper calibration is a common cause of laboratory error; therefore, maintaining instruments "in calibration" is critical.

ii. CALIBRATION AS A FORMAL ASSESSMENT OF A MEASURING INSTRUMENT

The word *calibration* is sometimes used when an instrument, standard, or measuring device is formally assessed, but is not adjusted. During the assessment, the measuring system being calibrated is compared with a trustworthy standard. Errors in the readings of the item being calibrated are determined and documented. The result of this type of calibration is a document that certifies that the measuring item was functioning in a particular way, what corrections, if any, need to be made to its

readings when it is used, and how certain the user can be of its readings. After this type of calibration the item being calibrated does not perform any better than it did before. However, the calibration document may allow the operator to use the item with better accuracy by applying correction factors to its readings. For example, it may be established by calibration that the readings of a thermometer are consistently 0.1 degree too high. Then, every time the thermometer is used, 0.1 degree must be subtracted from its readings. At present, the word *calibration* is used to sometimes mean an adjustment to a measuring device and to sometimes mean a formal, documented evaluation of an item.

iii. VERIFICATION

It is good laboratory practice to check the performance of instruments regularly to make sure they are functioning properly. **Verification** or **performance verification** *means to check the performance of an instrument or system.* Verification includes checking whether the instrument is properly calibrated. The checking done during verification is a simpler, less rigorous evaluation of the performance of a measuring item than is calibration. Verification is performed in the laboratory of the user and is typically documented in a logbook or on a form, but not with a formal document. For example, in many laboratories balance accuracy is checked regularly by weighing a standard and recording its weight. If the value for the standard's weight does not fall within a particular range, as specified by a quality control procedure, then the balance is repaired. This check of the balance's performance verifies that the balance is properly calibrated, but it should not be mistakenly called "calibrating the balance."

iv. Calibration (Standard) Curves

A calibration curve relates to chemical and biological assays where the response of an instrument to an analyte is determined. For example, it is common to evaluate the amount of protein present in a solution using a spectrophotometer. It is necessary to "calibrate" the response of the spectrophotometer to the amount of protein present. To perform this calibration, the response of the instrument to solutions containing known amounts of protein is plotted on a graph. The solutions with known amounts of protein are *standards*. The resulting graph is a *standard curve* or a *calibration curve* (see pp. 257–259 for more details).

Table 17.2 summarizes terminology relating to calibration.

D. Traceability

i. The Meaning of Traceability

Suppose you use a balance to weigh* an object. How can you be certain, and demonstrate to others, that the balance you used was indeed properly calibrated? This is a quality assurance problem and leads to the concept of **traceability. Traceability** *describes the chain of calibrations that establishes the value of a standard or of a measurement.*

The concept of traceability dates at least back to the ancient Egyptians, who used the pharaoh's arm as the national standard of length. Because the pharaoh could not always be present when measurements were made, the length of his arm was reproduced using a granite

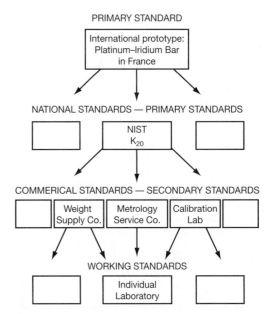

Figure 17.1.   Traceability of Mass Standards Showing Genealogy.

rod. The rod, in turn, was duplicated to make wooden measuring sticks that were used by workers building the pyramids. This system was an early example of using an external standard for measurement and traceability to this standard.

Mass standards provide another example of traceability. A mass standard used in an individual laboratory might be calibrated according to a secondary standard that was in turn calibrated against K_{20}, which was in turn calibrated relative to the international standard in France. Thus, there is a "genealogy" for the standard that is used in a particular laboratory, see Figure 17.1. For traceability purposes, the genealogy for a standard must be documented in a formal certificate.

The purpose of tracing the genealogy of a standard is to ensure that measurements made with that standard,

Table 17.2.   Calibration
1. ***Calibration brings a measuring system into accordance with an external standard(s).*** Calibration may involve adjusting an instrument so that its readings are consistent with an external standard.
2. ***Calibration is sometimes defined as a formal assessment that establishes, under specified conditions, the relationship between values indicated by a measuring instrument or measuring system and known values based on a trustworthy standard.*** The result of such calibration is a document. The instrument is not adjusted according to this definition of calibration.
3. ***Tolerance is the amount of error that can be tolerated in the calibration of a particular item.*** National and international organizations (including ASTM and NIST) specify the tolerances allowed for particular classes of glassware, weight standards, and other measurement items.
4. ***Verification is a check of the performance of an instrument or system.*** Verification is a simpler, less rigorous evaluation of the performance of a measuring item than is calibration.
5. ***Instrumental quantitative analysis frequently involves the construction of a standard (calibration) curve that shows graphically the relationship between the response of the instrument and the amount of reference standard present.***

*Weight and mass are not synonyms, as will be explained in a later chapter. For simplicity, however, we ignore the distinction in this chapter.

ASTM Class 1 calibration masses with Weight Calibration Certificate come with a statement of traceability to NIST standards, including a listing of each weight with its allowable tolerance, the actual mass value of each weight, environmental conditions during calibration, and date of calibration.

(a)

CERTIFICATE OF ACCURACY

This thermometer identified by Serial No. _____
was compared with a standard calibrated at the National Institute of Standards and Technology (NIST), formerly the National Bureau of Standards (NBS), and was found to be accurate within ±1°C. The indications of the thermometer are traced to NIST.
The Standard Serial Number is 55296
The NIST Identification No. is 94415

ERTCO

Ever Ready Thermometer Co., Inc.

(b)

Figure 17.2. **Catalog Specifications Indicating Traceability to NIST. a.** A description of a mass standard. **b.** A description of a thermometer. (b. is reproduced courtesy of Ever Ready Thermometer Co., Inc.)

or calibrations performed using that standard, are trustworthy. A statement of traceability is a quality statement. Manufacturers use the term *traceable to NIST* in their catalogs when they have documented the genealogy of standards used in manufacturing their product, see Figure 17.2.

Table 17.3 summarizes "traceability."

ii. SUMMARY: THE RELATIONSHIP BETWEEN STANDARDS, CALIBRATION, AND TRACEABILITY

To summarize the relationship between standards, calibration, and traceability, consider the example of making a measurement of weight using a balance. The balance was *calibrated* by a technician so that its readings are correct according to an internationally accepted definition of a gram. To calibrate the balance, the technician used secondary mass *standards*. The technician knows that the standards are correct embodiments of the unit called a "gram" because the comparisons of those standards to the K_{20} kilogram standard were properly performed and documented. The secondary standards are therefore *traceable* to NIST.

The ISO 9000 standards refer to this relationship between standards, traceability, and calibration in the following language:

> The supplier [any ISO 9000 compliant company or other entity] shall identify all . . . measuring . . . equipment that can affect product quality, and calibrate and adjust them at prescribed intervals, or prior to use, against certified equipment having a known valid relationship to internationally or nationally recognized standards (Section 4.11.2b).

Table 17.3.   **TRACEABILITY**

1. *Traceability describes the chain of calibrations ("genealogy") that establish the value of a standard or a measurement.*
2. *In the United States, traceability for physical and some chemical standards is often to NIST, since NIST maintains primary standards.*
3. *For the purpose of traceability, measurement values reported must include uncertainties* (discussed later in this chapter).

II. MEASUREMENT ERROR

A. Variability and Error

If you were to weigh the same standard 10 times, then would you get the same value every time? It seems reasonable to expect the 10 values to be identical if you follow the same procedure every time under uniform conditions. In reality, if you were to meticulously weigh the same object repeatedly with a high-quality balance, then the results would vary slightly each time. Twenty such weight measurements are shown in Table 17.4. Note that the first four digits in the weight values for the standard were always the same, but the last three digits varied.

It turns out that there is variability inherent in all observations of nature, including measurements. There is also uncertainty in all measurement values. What does this standard really weigh—exactly 10 g, 9.999591 g, 9.999594 g? Because of the variability in the measurements, we do not know the exact, true value for this standard. In fact, in principle, we can never be certain of the exact "true" value for any measurement, although we can approach that true value.

Error *is responsible for the difference between a measured value for a property and the "true" value for that property.* Statisticians classify measurement errors into types. One such classification scheme is:

1. **Gross errors** *are caused by blunders.* For example, if a technician were to drop a mass standard

---

Table 17.4.   **VALUES OF REPEATED WEIGHT MEASUREMENTS OF A STANDARD OBJECT**

9.999600 g	9.999596 g	9.999600 g
9.999595 g	9.999597 g	9.999600 g
9.999592 g	9.999599 g	9.999601 g
9.999603 g	9.999604 g	9.999598 g
9.999601 g	9.999595 g	9.999593 g
9.999597 g	9.999592 g	9.999590 g
9.999600 g	9.999599 g	

(or drop a balance), these would be gross errors. Gross errors are recognizable and, of course, should be avoided.

2. **Systematic errors** are normally more subtle and harder to detect than gross errors. There are a vast number of causes of systematic errors, such as a contaminated solution, a malfunctioning instrument, an environmental inconsistency (like a change in humidity), and so on.

    An important feature of systematic error is that it results in **bias,** *measurements that are consistently either too high or too low.* For example, if a flask used to measure 500 mL has its 500 mL mark set slightly too low, then it will consistently deliver volumes that are a little less than they should be.

3. **Random errors** *are extremely difficult or impossible to find and eliminate. Random errors cause measurement values that are sometimes too high, and sometimes too low.* Although random error is called "error," in the presence of only random error (and not gross or systematic error) measurements will average the correct "true" value.

What type of error is likely responsible for the variability in weight measurements shown in Table 17.4? If the technician making these measurements blundered or missed subtle factors then the variability would be due to gross or systematic error. If, however, the operator was very skilled and the method of measurement was carefully planned (as we will assume was the case), then the variability in weight measurements was due to random error, which is difficult or impossible to eliminate. Thus, there is error every time a measurement is made. Even if one uses such excellent technique that all gross and systematic errors are eliminated, there will still be random errors.

If we assume that there is no systematic error or bias in the measurements, then we must also conclude that the standard in Table 17.4 weighs slightly less than 10 g. If it was intended to be 10 g, then there was an error made at the time of its manufacture. (Note that if this standard is used to calibrate a balance, in the sense of adjusting the instrument, then all subsequent readings of this balance will be a bit too high and systematic error will be introduced.)

The word *error* has another, somewhat different, usage than we have given so far. Error is sometimes expressed as:

Absolute error = True value − Measured value

Although, in principle, we can never be certain of the exact "true" value for a measurement, in practice, this expression of error has practical application and will be further explored later in this chapter.

B. Accuracy and Precision

At this point, we introduce two important words: **accuracy** and **precision.** Measurement errors lead to a loss of accuracy and precision. These words may be defined as:

> *Precision* **is the consistency of a series of measurements or tests.**

> *Accuracy* **is how close a measurement value is to the true or accepted value.**

It is common to talk about the precision and accuracy of instruments, tests, assays, and methods. For example, we can speak of how consistent (precise) a balance's readings are, or the "correctness" (accuracy) of a test for blood glucose.

Accuracy and precision are not synonyms to scientists, although they are often used interchangeably in nonscientific English. It is possible for a series of measurements to be precise (consistent) but not accurate, see Figure 17.3a. A series of measurements may also average the correct answer yet lack precision, see Figure 17.3b. A series of measurements may be neither accurate nor precise, see Figure 17.3c, or may be both accurate and precise, see Figure 17.3d. One definition of a "good" measurement is that it is both accurate and precise.

Consider precision in more detail. Laboratory workers are aware that measurements repeated in succession on the same day tend to be relatively consistent. In contrast, measurements performed on different days, by different people, and using different materials and equipment, tend to be far more variable. There are different words for precision to make this distinction clear. **Repeatability** *is the precision of measurements made under uniform conditions.* **Reproducibility** *is the precision of measurements made under nonuniform conditions, such as in two different laboratories.* Repeatability and reproducibility are therefore two practical extremes of precision. It is challenging, but important, to work to

Figure 17.3.   Accuracy and Precision.
Precision and accuracy are illustrated using the analogy of a target where the bull's eye is the correct value. **a.** Archer A is precise but not accurate. **b.** Archer B is inconsistent. Archer B's results average the correct value though they are not precise. **c.** Archer C is not very skilled; the results are neither accurate nor precise. **d.** Archer D is expert and is both precise and accurate.

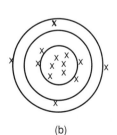
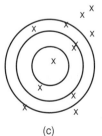

(a)   (b)   (c)   (d)

develop procedures that give results that are as reproducible as possible when performed on different days, by different people, and so on.

EXAMPLE PROBLEM

Four students each completed a laboratory exercise in which they weighed a standard known to be 5.0000 g. The values they obtained (in grams) are shown in the following table. Comment on the accuracy and precision of each student's results.

Juan	Chris	Ilana	Mel
4.9986	5.0021	5.0001	5.1021
5.0020	5.0020	4.9998	4.9987
5.0007	5.0021	4.9999	5.0003
4.9995	5.0022	4.9998	4.9977

ANSWER

Juan's data have a mean value that is reasonably accurate; however, their precision is not as good as the data of other students.

Chris has precise but inaccurate values. It is likely that there is a systematic error in Chris's work which causes all the values to be too high (to be biased).

Ilana's values are both accurate and precise with some random fluctuations affecting only the last figure (the fourth place to the right of the decimal point). These data are considered to be "good"; we expect *some* variability in measurements.

Mel's values are neither accurate nor precise. We may suspect that Mel was careless or had problems with the equipment.

C. The Relationship between Error, Accuracy, and Precision

i. RANDOM ERROR AND LOSS OF PRECISION

Random error leads to a loss of precision because it leads to inconsistency in measurements; due to random error, values are sometimes too high and sometimes too low. We saw the effects of random error on the precision of weight measurements in the example in Table 17.4.

We can also consider a flask that is used to measure a volume of 500.0 mL. Suppose the flask is perfectly marked with a line at exactly 500.0 mL. Although the flask is marked correctly, every time it is used there will

be a tiny variation in how much liquid is delivered due to imperceptible variations in the environment and the person using the flask. This slight variability is random error and will result in volumes being delivered that are sometimes a bit more than 500.0 mL and sometimes a bit less, see Figure 17.4a. In this example, because the flask is perfectly marked at exactly 500.0 mL, if the flask is used many times, then the average volume delivered will equal the true value, 500.0 mL, see Figure 17.4b.

It is good laboratory practice to repeat critical measurements, assays, and tests to determine the impact of random error. As measurements are repeated, you can see their variability. In addition, blunders often show up when repeating measurements. For example, one might accidentally misread a meter or remove the wrong amount of a sample but notice the error when the measurements are repeated.

ii. ERRORS AND LOSS OF ACCURACY

Gross, systematic, and random errors all lead to a loss of accuracy. It is obvious how a gross error could cause a value to be incorrect. A systematic error will also affect the accuracy of a measurement. For example, consider a systematic error where a flask intended to measure out exactly 500.0 mL has its line drawn slightly too low on the flask. Every time this flask is used it will tend to deliver slightly less than 500.0 mL. The volume it delivers is obviously inaccurate. When there is systematic error, the system is **biased,** *and the mean (average) of the measurements will be too high or too low,* see Figure 17.5.

Systematic error can result from many causes. For example, equipment that is improperly calibrated, is not well maintained, or is used improperly; solutions that have degraded; and environmental fluctuations all cause systematic error. Unlike random error, the impact of systematic error is not detected by repeating measurements. If a systematic error is present, then every time the measurement is repeated it will incorporate the same error and tend to be too high or too low. To minimize systematic errors in measurement, it is necessary to be knowledgeable in the methods used, to be attentive to potential problems, and to properly maintain and calibrate equipment.

We know that random error causes a loss of precision. Does random error also cause a loss of accuracy? This question is somewhat more complex. Consider again Figure 17.3b. On the average, the accuracy of

Figure 17.4. Random Error. a. Hypothetical results of repeated measurements using a flask that is correctly marked at 500.0 mL. Each point represents one measurement. **b.** An idealized frequency distribution assuming the same volume measurement was made a great many times with a flask perfectly marked at 500.0 mL. The mean value and the true value are the same.

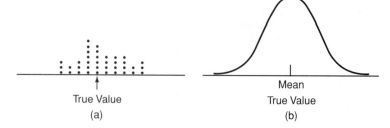

True Value

(a)

Mean
True Value

(b)

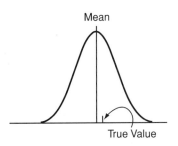

Figure 17.5. Measurements from a Biased System. In the presence of bias, the mean is not the true value for the measurement.

this archer is pretty good; however, if we look only at a single measurement, it may be way off center. Thus, when considering an individual measurement, or only a few measurements, if a system is relatively imprecise, then its accuracy will also be poor. The same idea is illustrated in Figure 17.4a and 17.4b. If a single measurement from Figure 17.4a is selected, that measurement may or may not be close to the correct reading

of 500 mL even though the average, as shown in Figure 17.4b, is accurate. Random error can thus lead to both a loss of precision and a loss of accuracy.

In practice, there are both random and systematic errors present when measurements are made in the laboratory. It is often difficult to tease apart the effects of random errors and systematic errors. Terms relating to accuracy, precision, and error are summarized in Table 17.5.

D. Evaluating the Precision of a Measurement System

Random error leads to variability; it causes repeated measurements to vary from one another. Variability in a series of measurements can be evaluated using statistical tools. For example, consider hypothetical values obtained by a technician analyzing lead levels in paint samples, see Table 17.6. The analyst observes on February 20 that the values from two houses seem to fluctuate more

---

Table 17.5. ERROR, ACCURACY, AND PRECISION

1. *Error is sometimes defined as the difference between a measured value and the true value.*
2. *Error sometimes refers to the cause of variability and inaccuracy. Errors can be classified into three types: gross, systematic, and random.*
   a. *Gross errors. Human blunders.*
   b. *Systematic errors. Errors that cause results to generally be either too high or too low. Repeating measurements is not a useful way to detect systematic errors.*
   c. *Random errors. Errors that cause results to be sometimes too high and sometimes too low and which are difficult or impossible to eliminate completely. Repeating measurements is a useful way to determine the magnitude of random error.*
3. *Errors occur whenever laboratory measurements are performed.*
4. *Errors cause uncertainty in measurements.*
5. *Bias occurs when there is a systematic error(s) that causes measurement values to tend to be too high or too low.* In a biased system, the mean differs from the true value.
6. *Precision is the consistency of a series of measurements.* Precision is affected by random error.
   a. *Repeatability is the precision of measurements repeated in succession.*
   b. *Reproducibility is the precision of measurements performed under varying conditions.*
7. *Accuracy is sometimes defined as how close a measurement is to the true or accepted value.*
8. *Uncertainty is an estimate of the inaccuracy of a measurement that includes both the random and bias components. Uncertainty is also defined as an estimate of the range of values in which the "true" value lies.*

Table 17.6. LEAD LEVELS IN PAINT FROM HOUSES

Feb. 4 House 1 Window (mg/cm^2)	Feb. 4 House 2 Window (mg/cm^2)	Feb. 8 House 3 Window (mg/cm^2)	Feb. 8 House 4 Window (mg/cm^2)	Feb. 8 House 5 Window (mg/cm^2)	Feb. 10 House 6 Window (mg/cm^2)	Feb. 20 House 7 Window (mg/cm^2)	Feb. 20 House 8 Window (mg/cm^2)
1.21	0.43	0.92	0.78	1.51	2.12	0.98	1.23
1.18	0.40	0.93	0.67	1.43	1.99	0.78	0.21
1.31	0.34	0.79	0.71	1.34	2.13	0.21	0.11
1.23	0.41	0.93	0.65	1.47	1.98	1.24	0.89

Some information for this example was taken from Driscoll, J.N., Laliberte, R., and Wood, C. "Field Detection of Lead in Paint and Soil by High-Resolution XRF." *American Laboratory* (March 1995), 34H.

than normal. This observation alerts her to the possibility that there might be a problem in the measurement system. She can analyze the variability in the data using various statistical tools including the **range,** the **standard deviation (SD),** and the **coefficient of variation (CV).**

The **range** *is the difference between the highest and the lowest values of a set of measurements.* Range is a simple indicator of precision: If the range is narrow, then the data are less variable than if the range is wide. Table 17.7 shows the range for the lead measurements. Note that the range of values for February 20 is greater than for the other days.

Standard deviation, SD, is commonly used to evaluate the variability of a group of measurements. (See Chapter 16 for a discussion of SD and CV.) A series of measurements with a lower SD have better precision than a series of measurements from a similar system with a higher SD. The **coefficient of variation (CV)** expresses the standard deviation of a series of measurements in terms of the mean. The standard deviation and the coefficient of variation for the measurement values on February 20 were higher than on the other days, see Table 17.7.

These results suggest that for some reason the data obtained on February 20 were more variable than on the other dates. There are many possible explanations for this variability. The variability might be due to a characteristic of the houses themselves. The windows from which the paint samples were taken may have had many layers of different paints applied, in which case the variability resulted from lack of homogeneity in the samples. An instrument could have been malfunctioning on February 20 and required repair. A reagent involved in the measurement procedure may have degraded. The technician at this point will want to identify the source of variability and correct any problems that may be present.

Another example of the use of standard deviation to express the precision of a measurement system is shown in Figure 17.6. Catalog descriptions of instruments often report the precision of the instrument in terms of the standard deviation of a series of measurements made

Catalog number	pH 20
Type	pH
Range	pH 2.0 to 12.00
Resolution	0.01 pH unit
Repeatability	$\pm$ 0.1 pH unit

Figure 17.6. Catalog Descriptions of a Measurement Instrument. A pH meter description. Repeatability ($\pm$ 0.1 pH unit) is the standard deviation around an undisclosed mean. Note that the term *range* is not an indicator of precision in this case, rather, it indicates the span from the highest to the lowest pH that the meter can read.

with the instrument. The more consistent the instrument, the lower its SD, and, presumably, the better its quality.

EXAMPLE PROBLEM

Which result from a series of measurements is more precise: 15.0 $\pm$ 0.3 g (mean $\pm$ SD) or 15.00 $\pm$ 0.03 g (mean $\pm$ SD).

ANSWER

The SD "0.03" is smaller, or more precise, than 0.3 g; therefore, 15.00 $\pm$ 0.03 g is more precise.

EXAMPLE PROBLEM

A biotechnology company manufactures a particular enzyme that is used to cut DNA strands. The enzyme's activity can be assayed and is reported in terms of "units/mg." Each batch of enzyme is tested before it is sold. The results of repeated tests on four batches of enzyme are shown in the table on the next page.

Table 17.7. STATISTICAL ANALYSIS OF PAINT LEAD LEVELS DATA

Feb. 4 House 1 Window (mg/cm²)	Feb. 4 House 2 Window (mg/cm²)	Feb. 8 House 3 Window (mg/cm²)	Feb. 8 House 4 Window (mg/cm²)	Feb. 8 House 5 Window (mg/cm²)	Feb. 10 House 6 Window (mg/cm²)	Feb. 20 House 7 Window (mg/cm²)	Feb. 20 House 8 Window (mg/cm²)
1.21 1.18	0.43 0.40	0.92 0.93	0.78 0.67	1.51 1.43	2.12 1.99	0.98 0.78	1.23 0.21
1.31 1.23	0.34 0.41	0.79 0.93	0.71 0.65	1.34 1.47	2.13 1.98	0.21 1.24	0.11 0.89
Range = 1.31 − 1.18 = 0.13	Range = 0.09	Range = 0.14	Range = 0.13	Range = 0.17	Range = 0.15	Range = 1.03	Range = 1.12
Mean ≈ 1.23	Mean ≈ 0.40	Mean ≈ 0.89	Mean ≈ 0.70	Mean ≈ 1.44	Mean ≈ 2.06	Mean ≈ 0.80	Mean = 0.61
SD ≈ 0.06	SD ≈ 0.04	SD ≈ 0.07	SD ≈ 0.06	SD ≈ 0.07	SD ≈ 0.08	SD ≈ 0.44	SD ≈ 0.54
CV ≈ 4.5%	CV ≈ 9.8%	CV ≈ 7.7%	CV ≈ 8.2%	CV ≈ 5.1%	CV ≈ 3.9%	CV ≈ 55%	CV ≈ 88%

a. What is the mean activity of the enzyme for each batch?

b. What is the SD for each batch?

c. What is the mean activity for all batches combined?

d. What is the SD for all batches combined?

(See pp. 275–277 for an explanation of SD and p. 276 for the equation for calculating SD.)

ENZYME ACTIVITY (UNITS/MG)

Batch I	Batch 2	Batch 3	Batch 4
100,900	100,800	110,000	123,000
102,000	101,000	108,000	121,000
104,000	100,100	107,000	119,000
104,100	100,800	109,100	121,000

ANSWER

Mean batch I = 102,750 units/mg   SD = 1567 units/mg

Mean batch 2 = 100,675 units/mg   SD = 395 units/mg

Mean batch 3 = 108,525 units/mg   SD = 1305 units/mg

Mean batch 4 = 121,000 units/mg   SD = 1633 units/mg

Mean all batches = 108,238 units/mg

SD all batches = 8254 units/mg

Note that there is, as we might expect, more variability *between* the batches than there is *within* one batch. This is reflected in the fact that the standard deviation for all the batches is higher than the standard deviation of any one batch.

E. Evaluating the Accuracy of a Measurement System

We saw in the last section how the precision of an instrument or measuring system is evaluated by performing a series of measurements and applying statistical tests to the results. Next consider the evaluation of accuracy. Accuracy is the closeness of agreement between a measurement or test result and the true value, or the accepted reference value, for that measurement or test.

We generally do not know the "true" value for a measurement. For example, suppose an analyst is testing the level of glucose in an individual sample of blood. The analyst does not know the *true* blood glucose value for that sample—if the value was known, there would be no point in doing the test. The analyst, therefore, cannot calculate the accuracy of the measurement for that sample. The analyst can, however, evaluate the accuracy of the method itself.

The most obvious way to assess the accuracy of a method or an instrument is to use a standard. For example, the assay method for glucose can be evaluated by testing a blood sample with a known amount of glucose. This sample with a known amount of glucose is a quality-control (or simply control) sample. The analyst assays the level of glucose in the control sample using the same assay method that is used for patient blood samples. The result obtained for the control is compared with the expected result to determine the accuracy of the test procedure.

Two simple ways to express accuracy mathematically are:

Expressing Accuracy as "Absolute Error"

Absolute Error = True Value − Average Measured Value

where "error" is an expression of accuracy and the true value may be the value of an accepted reference material

Expressing Accuracy as "Percent Error"

$$\% \text{ Error} = \frac{\text{True Value} - \text{Average Measured Value}}{\text{True Value}} \times 100\%$$

where "% error" is an expression of accuracy and the true value may be the value of an accepted reference material.

EXAMPLE PROBLEM

Suppose the glucose level in a control sample is stated to be 1000 mg/L. A technician performs a glucose test 10 times on this control sample and gets the following values:

996 mg/L, 1009 mg/L, 1008 mg/L, 998 mg/L, 1001 mg/L, 999 mg/L, 997 mg/L, 1000 mg/L, 1008 mg/L, and 1010 mg/L.

What are the absolute error and percent error based on these data?

ANSWER

The average value from the 10 tests is 1002.6 mg/L. The absolute error, based on this average is:

$$1000 \text{ mg/L} - 1002.6 \text{ mg/L} = -2.6 \text{ mg/L}$$

The percent error, based on this average is:

$$\frac{1000 \text{ mg/L} - 1002.6 \text{ mg/L}}{1000.0 \text{ mg/L}} \times 100\% = -0.26\%$$

It is good practice to check the performance of a test using one or more quality-control samples on a regular basis. This practice of checking methods and instruments with controls is often a documented part of a quality control program. Based on experience, repetition of the test, and knowledge of the system, it is possible to establish a range within which values for the test should fall. If the results of a test lie outside of this range, then it is necessary to look for problems.

Manufacturers establish specifications for their products. Table 17.8 gives the specifications for two models of a volume measuring device, as established by their manufacturer, Gilson, Inc. These specifications guarantee how accurately and how precisely these devices dispense the volumes for which they are set (assuming proper calibration, maintenance, and operation). The specifications are shown in tabular format:

- The first column in the table is the model of the measuring device.
- The second column indicates the volume that the device is set to dispense (i.e., the "true volume").
- The third column is the permissible systematic error, which is a quantitative evaluation of the accuracy of the device. The systematic error is expressed here as the absolute error, the difference between the dispensed volume and the selected volume. It is generally determined based on the average of 10 measurements. Observe that the lower the value specified for systematic error, the better the accuracy and therefore the performance of the device.
- The fourth column is the random error, which is a quantitative evaluation of the precision of the device. The random error is expressed as the standard deviation of a series of measurements, usually 10. Observe that the lower the value specified for standard deviation, the better the precision and performance of the device.
- The fifth column shows the requirements for systematic error for these volume measuring devices as specified in ISO Standard 8655-1:2002 "Piston-operated volumetric apparatus—Part 1: Termi-

nology, general requirements and user recommendations." Observe that the manufacturer specifies a lower (better) value for accuracy than is required by the ISO standard; their devices exceed the requirements of the standard.

- The sixth column shows the minimum requirements for random error as specified in the ISO standard. Observe that the manufacturer guarantees better precision than is required by the ISO standard.

EXAMPLE PROBLEM

Suppose you want to evaluate the precision of a balance. Do you need to use a standard whose mass value is traceable?

Suppose you want to check whether a balance is giving mass readings that are accurate. Do you need to use a standard whose mass value is traceable?

ANSWER

The evaluation of precision does not require a standard whose properties are known. Any object can be used. For example, you could determine the precision of the balance by checking the mass of the same pencil 20 times and calculating the standard deviation for the values.

In contrast, the evaluation of accuracy requires a standard against which the readings of the balance are compared. This is where traceability comes in: Traceability ensures that the mass of the standard used to check the balance is known.

EXAMPLE PROBLEM

An environmental testing laboratory is about to begin testing water for chromium. The company's scientists learn to perform the chromium assay and test their skills by assaying commercially prepared standards with

Table 17.8. CATALOG SPECIFICATIONS FOR VOLUME MEASURING DEVICES DEMONSTRATE THE CONCEPTS OF ACCURACY, PRECISION, SYSTEMATIC ERROR, AND RANDOM ERROR

Model	Volume (μL)	Gilson Maximum Permissible Systematic Error (μL)	Gilson Maximum Permissible Random Error (μL)	ISO 8655 Maximum Permissible Systematic Error (μL)	ISO 8655 Maximum Permissible Random Error (μL)
P-2	0.2	± 0.024	≤ 0.012	± 0.08	≤ 0.04
	0.5	± 0.025	≤ 0.012	± 0.08	≤ 0.04
	2	± 0.030	≤ 0.014	± 0.08	≤ 0.04
P-10	1	± 0.025	≤ 0.012	± 0.12	≤ 0.08
	5	± 0.075	≤ 0.030	± 0.12	≤ 0.08
	10	± 0.100	≤ 0.040	± 0.12	≤ 0.08

Information excerpted from Gilson, Inc. catalog, 2007.

known chromium concentrations. They use four standards and obtain the results below. The scientists want to be able to guarantee their customers that they will be able to analyze chromium in samples with less than ± 2% error. Fill in the table to show the relative percent error for each measurement. Have the scientists met their goal for accuracy based on these results?

Actual Concentration of Chromium in the Standard	Assayed Concentration of Chromium Obtained in the Laboratory	% Error
1.00 μg/L	0.91 μg/L	
5.00 μg/L	4.78 μg/L	
10.00 μg/L	9.89 μg/L	
15.00 μg/L	15.08 μg/L	

ANSWER

The percent errors are, in order: 9.00%, 4.40%, 1.10%, −0.53%

These data suggest that at low concentrations they have not achieved the desired accuracy, although they have at higher concentrations. Refinements in their technique are required.

EXAMPLE PROBLEM

An assay used to measure the amount of protein present in samples is giving erroneous results that are low by about 5 mg. What will be the percent error due to this problem if the actual protein present in a sample is:

i. 30 mg   ii. 50 mg   iii. 100 mg   iv. 550 mg

ANSWER

i. (true − measured value)/true value
= (30 − 25 mg/30 mg) × 100 ≈ 16.7%
ii. 10.0%   **iii.** 5.0%   **iv.** 0.91%

Note that the impact of this error is more pronounced when protein is present in low amounts than when protein is present at higher levels; the 5 mg error is a larger percent of 30 mg than it is of 550 mg.

EXAMPLE PROBLEM

A new balance must be purchased for a teaching laboratory. A teacher compares three competing brands by measuring a 1.0000 g standard five times on each balance.

a. Which balance is most accurate? Show the percent error for each balance.

b. Calculate the SD for the measurements from each balance. Which balance is most precise?

c. What is the CV for each of the balances?

d. Report the mean value for the standard from each balance ± the SD.

e. Which balance would you buy?

Brand A (grams)	Brand B (grams)	Brand C (grams)
1.0004	0.9997	1.0003
1.0005	0.9996	0.9996
1.0004	1.0003	1.0002
1.0003	1.0002	0.9995
1.0005	0.9999	1.0004

ANSWER

Balance A	Balance B	Balance C
Mean ≈ 1.0004 g	Mean ≈ 0.9999 g	Mean = 1.0000 g
% error ≈ −0.04%	% error ≈ 0.01%	% error = 0.00%
SD ≈ 0.0001 g	SD ≈ 0.0003 g	SD ≈ 0.0004 g
CV ≈ 0.01%	CV ≈ 0.03%	CV ≈ 0.04%
Mean ± SD ≈	Mean ± SD ≈	Mean ± SD ≈
1.0004 ± 0.0001 g	0.9999 ± 0.0003 g	1.0000 ± 0.0004 g

The percent error is an indication of the accuracy of the balances. Precision is shown by the SD and CV. Although the mean value of Balance C is correct, its precision is poor compared with the other balances, and any single measurement made on Balance C is likely to be inaccurate. Balance A has the best precision, but it is biased— all the readings are a bit high, which suggests a systematic problem. If Balance A can be recalibrated or otherwise be made accurate, it might be a good choice. Balance B might be an acceptable compromise, particularly if it has other qualities that are desirable, such as ruggedness and ease of use.

III. INTRODUCTION TO UNCERTAINTY ANALYSIS

A. What Is Uncertainty?

The term *uncertainty* is used often in the literature of metrology. To some extent the meaning of *uncertainty* is familiar. For example, consider a statement in the newspaper that "the distance from the earth to the sun is 156,300,000 km." Next, consider a statement that "there are 10 microscopes in our laboratory." You would know from experience that the figure given for the distance to the sun is not likely to be exactly correct; in fact, it could be off by many kilometers. In contrast, the statement that "there are 10 microscopes in the laboratory" is likely to be exactly correct. There is uncertainty in the value reported for the distance to the sun, while there is little, if any, uncertainty in a count of 10 microscopes.

As there is uncertainty in the figure reported for the distance to the sun, so there is uncertainty in the measurements we make in the laboratory. The reason there is uncertainty in laboratory measurements is that error exists. Even when people are very careful, random errors and sometimes systematic errors persist.

Metrologists try to estimate the effect of errors so they can know how much confidence to place in a measurement. Metrologists state that a "good" measurement must include not only the value for the measurement, but also a reasonable estimate of the uncertainty associated with the value. Calibration and traceability documents include statements of uncertainty. Thus, a measurement should properly be of the form:

measured value ± an estimate of uncertainty

Consider, for example, the measurements of the standard weight as shown in Table 17.4 on p. 300. We can be fairly certain that the standard weighs a little less than 10 g although we cannot be sure of its exact true weight. The best estimate of the true weight of the standard is the mean of a large number of measurements. The mean weight of the standard based on the twenty measurements in Table 17.4 is 9.999598 g. Therefore, to the best of our knowledge, the standard's true weight is 9.999598 g —"give or take" something. The "give-or-take" something is the uncertainty in the measurement.* **Uncertainty** *is an estimate of the inaccuracy of a measurement due to all the errors present.*

Estimating uncertainty is complex. It requires identifying to the best of one's knowledge all sources of error, estimating the magnitude of each error, and combining all the effects of error into a single value. Not surprisingly, it is difficult to state, with certainty, how much uncertainty there is in a measurement. How to best evaluate uncertainty is therefore the subject of much discussion.

B. Using Precision as an Estimate of Uncertainty

Although it is difficult to perform an in-depth uncertainty analysis, it is reasonably straightforward to evaluate the uncertainty due to random error. It is common for analysts to repeat a particular measurement and to summarize the uncertainty due to random error using the standard deviation of the repeated measurements.**

Random error is not the only type of error that may be present when a measurement is made. However, it is more difficult to estimate the uncertainty due to other types of errors. In situations where a measurement process is well understood, the analyst may make estimates of the uncertainty caused by a variety of errors. For example, a measuring system may be known to be affected by changes in barometric pressure. The analyst may therefore add in a factor for uncertainty caused by fluctuations in barometric pressure. There are statistical methods that have been developed to deal with these types of uncertainty. (For an explanation of these methods and of current practice in uncertainty measurement, consult, for example, Nicholas and White, or the NIST publication "Guidelines for Evaluating and Expressing the Uncertainty of NIST Measurement Results"; both are referenced at the beginning of this unit.)

C. Using Significant Figures as an Indicator of Uncertainty

i. THE MEANING OF SIGNIFICANT FIGURES

Measurements are properly displayed as a value ± an estimate of uncertainty. We previously discussed methods to estimate the figure that goes after the ± sign. In reality, biologists routinely report measurement values without explicitly adding any ± uncertainty estimate to the value. There are, however, conventional practices used to report measurement values that roughly indicate the certainty of the measurement. These practices are covered under the heading of "significant figures."

Significant figures are the most basic way that scientists show the certainty in a measurement value. A **significant figure** *is a digit within a number that is a reliable indicator of value.* It is easiest to understand significant figures by looking at practical examples from the laboratory and everyday life.

EXAMPLE 1: RULERS

Consider the length of the arrow drawn in Figure 17.7a and 17.7b.

In Figure 17.7a, the ruler's gradations divide each centimeter in half. We know the arrow is somewhat longer than 4 cm and it is reasonable to estimate the tenth place and report the arrow's length as "4.3 cm." Information will be lost if we report the length as simply "4 cm" because it is possible to tell that the arrow is somewhat longer than 4 cm. It would be unreasonable to report that the arrow is "4.35 cm" because the ruler gradations give no way to read the hundredths place. When measurements are recorded, it is customary to record all the digits of which we are certain plus one that is estimated. In this example, the 4 is certain and the .3 is estimated, so the measurement is said to have two significant figures. Thus, by reporting the measurement to be 4.3 cm, we are telling the reader something about how certain we are in the length. We are sure about the 4 and not as sure about the .3.

The subdivisions in the second ruler, see Figure 17.7b, are finer and divide each centimeter into tenths. With the second ruler, we can reliably say the arrow is "4.3 cm," and, in fact, it is reasonable to estimate that the arrow is about "4.35 cm." The 4.3 is certain, the .05 is estimated and so there are three significant figures in the measurement. Thus, the arrow is the same length in both Figure 17.7a and 17.7b, yet, the length recorded is different because the rulers are subdivided differently. The fineness of the measuring instruments, in this case, the rulers, determines our certainty of the length of the arrow. The measurement certainty is reflected by the number of significant figures used in recording the measurement.

---

*In this particular example, if all the uncertainty is due to random error, then it is possible to give a value to the "give or take something". In this case, the standard error of the mean is used to estimate uncertainty. See, for example, *Statistics* by Freedman et al., which is referenced at the beginning of this unit.

**The standard error of the mean, SEM, is used sometimes in a similar fashion, as is a confidence interval for the mean. The calculation of both the SEM and a confidence interval begins with determining the mean and the standard deviation for a series of repeated measurements.

A Note about Terminology

- Some people would say the second ruler is more accurate than the first, because it gives values with more significant figures (i.e., with more certainty).

- Some would say the second ruler is more precise than the first because it allows us to read further past the decimal point. The word *precise* thus has two meanings. Precision may refer to the fineness of increments of a measuring device; the smaller the increments, the better the precision. Precision is also used, as explained previously in this chapter, to refer to the consistency of values: the more consistent a series of measurements, the more precise.

- Some people might say the second ruler has more resolution because it is allows us to discriminate a smaller length change than the first ruler.

EXAMPLE 2: BALANCES

Suppose a particular balance can reliably weigh an object as light as 0.00001 g. On this balance, sample Q is found to weigh "0.12300 g." It would not be correct to report that sample Q weighs "0.123 g" because information about the certainty of the measurement is lost if the last zeros are discarded. In contrast, if sample Q is weighed on a balance that only reads three places past the decimal point, it is correct to report its weight as "0.123 g." It would not be correct to record the weight as "0.1230 g" because the balance could not read the fourth place past the decimal point.

The first balance gives more certainty about the sample weight. The difference in measurement certainty between the two balances is shown by the number of significant figures used: "0.12300 g" has five significant figures, whereas "0.123 g" has only three.

EXAMPLE 3: WEIGHTS OF A STANDARD

Suppose that the technicians who made the weight measurements of the standard in Table 17.4 used a balance that could only weigh objects to the nearest 0.1 g rather than six places to the right of the decimal point. Then, every time they weighed the standard, they would have recorded its weight as "10.0 g," a value with three significant figures. Their results would be consistent, but they would never realize that the standard truly weighs a little less than 10 g. Although consistent, there is therefore less certainty and fewer significant figures in the value "10.0 g" than in the values shown in Table 17.4 on p. 300.

Examples 1–3 show that a basic principle in recording measurements from instruments is to report as much information as is reliable plus one last figure that is estimated and might vary if the measurement were repeated. The number of figures reported by following this principle is the number of significant figures for the measurement. The number of significant figures reported is a rough estimate of the certainty of a measurement.

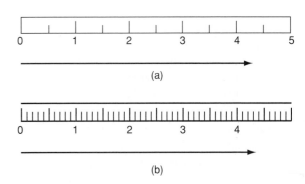

Figure 17.7. Measurements of Length with Two Rulers. a. A ruler where each centimeter is subdivided in half. **b.** A ruler where each centimeter is subdivided into tenths.

Note that most modern electronic instruments show results with a digital display. There is no meter to read. With any instrument having a digital display the last place is assumed to have been estimated by the instrument.

As a final example, let us return to the value for the distance to the sun discussed at the beginning of this section.

EXAMPLE 4: ZEROS AND SIGNIFICANCE

Suppose that one source reports the distance to the sun to be 156,000,000 km, but another reports it as 156,155,300 km. Let us assume that both sources are correct, but the first rounded the number to make it easier to read. The second number, 156,155,300 km, is a more exact figure for the distance, which allows the reader to be more certain about the actual distance than does the first number. We say that the number 156,155,300 has more significant figures (seven) than the number 156,000,000 (which has three significant figures).

This example can be used to illustrate an important point regarding zeros. The zeros in the reported values are essential; without them the distance would be reported as a paltry 156 km or as 1,561,553 km. The zeros are "place holders" that tell us we are talking about a very large distance, but they are not correct indicators of value. Perhaps the exact distance is really 156,155,333 km or 156,155,329 km, or any of a number of other possibilities. In contrast, when I report that there are "10 microscopes in my laboratory" the zero is a reliable indicator of value. The zero shows that there are 10, not 9 or 11, microscopes. Zeros that are place holders are not called significant figures, whereas zeros that indicate value are significant.

Suppose the number given in a report is 45,000. The three zeros in this number each may be place holders or they may be indicators of value. There are various ways to tell the reader whether the zeros at the end of a number are significant. One method is to use scientific (exponential) notation. For example, there are three zeros at the end of the number 45,000. If none of the zeros are significant, then the number could be reported as 4.5×10^4; two significant figures. If there are three significant figures the number could be reported as 4.50 × 10⁴, and so on. If all the zeros in the number 45,000 are significant, this can be shown by placing a decimal point after the number. The number 45,000. and 4.5000 × 10⁴ both have five significant figures.

Table 17.9 summarizes rules regarding how to record measurements with the accepted number of significant figures. As a laboratory professional, it is essential that you record all measured values so that they properly report the number of significant figures provided by the measuring instrument. If you fail to do so, important information may be lost. These rules also show how to decide when a zero is significant and when it is a place holder.

ii. Calculations and Significant Figures

Calculations, such as addition or multiplication, bring together numbers from separate measurements. Each value in the calculation has a particular number of significant figures. The result from the calculation is limited to the certainty (number of significant figures) of the starting number that is least certain.

The result displayed on a calculator is seldom what should be recorded when a calculation is performed because calculators do not give any indication of certainty. The calculator result must be rounded to the proper number of significant figures. When numbers are brought together in calculations it is often not obvious how many significant figures there are. There are various rules to determine the correct number of significant figures (see bibliography pp. 294–295). Table 17.10 summarizes simple rules that are likely to be adequate for situations in biology laboratories. (These rules simplify what is actually a complex procedure.) Remember that these rules are guides to decide where to round the result of a calculation(s).

EXAMPLE PROBLEM

Read the following meter and the digital display. How many significant figures does each measurement have?

ANSWER

The meter reads **"56.6," three significant figures**; the operator estimates the last place. The digital display reads **"56.48," four significant figures**; the instrument estimates the last place.

EXAMPLE PROBLEM

Which of the following are measured values and which are counted (exact) values?

a. The density at 25°C is 1.59 g/mL.

b. The distance from Chicago to Milwaukee is 75 miles.

c. Human body temperature is 37°C.

d. There are four regional campuses in the system.

ANSWER

Only d is counted. The other values are measured and so the rules of significant figures apply to them.

EXAMPLE PROBLEM

A catalogue advertises a particular product to be 99% pure. A competitor advertises their version of the same product to be 98.8% pure. Which one is more pure?

Table 17.9.   Rules to Record Measurements with the Correct Number of Significant Figures

1. ***The number of significant figures is related to the certainty of a measurement.*** (Note that counted values, such as there are "10 microscopes in the laboratory," or "there are 30 students in the class" are not measurements but rather are considered to be "exact." The rules of significant figures do not apply to counted values.)

2. ***When reporting a measurement, record as many digits as are certain plus one digit that is estimated.*** When reading a meter or ruler, estimate the last place. When reading an electronic digital display, assume the instrument estimated the last place.

3. ***All nonzero digits in a number are significant.*** For example, all the digits in the number 98.34 are significant; this number has four significant figures. A reader will assume that the 98.3 is certain and the 4 is estimated.

4. ***All zeros between two nonzero digits are significant.*** For example, in the number 100.4, the zeros are reliable indicators of value and not just place holders.

5. ***Zero digits to the right of a nonzero digit but to the left of an assumed decimal point may or may not be significant.*** Consider the number for the distance to the sun, 156,000,000 km. The decimal point is assumed to be after the last zero. In this case, the zeros are ambiguous and may or may not be reliable indicators of value. Methods of clarifying ambiguous zeros are discussed in the text.

6. ***All zeros to the right of a decimal point and to the right of a nonzero digit before a decimal place are significant.*** The following numbers all have five significant figures: 340.00, 0.34000 (the zero to the left of the decimal point only calls attention to the decimal point), and 3.4000.

7. ***All the zeros to the left of a nonzero digit and to the right of a decimal point are not significant unless there is a significant digit to their left.*** The number 0.0098 has two significant figures because the two zeros before the 98 are place holders. On the other hand, the number 0.4098 has four significant figures.

ANSWER

First, assume that both competitors have faithfully followed the significant figures conventions. Second, assume that the value for purity is based on a calculation(s). Then the first product might reasonably be expected to be anywhere from 98.5 to 99.4% pure because anywhere in that range the value would be rounded to 99%. The second brand would be between 98.75 and 98.84% pure. It is unclear, therefore, which brand is actually most pure. The second manufacturer, however, has presumably been able to ascertain the purity of their product with more certainty.

EXAMPLE PROBLEM

A biotechnology company specifies that the level of RNA impurities in a certain product must be less than or equal to 0.02%. If the level of RNA in a particular lot is 0.024%, does the lot meet the specification?

ANSWER

The specification is set at the hundredth decimal place; therefore, the result is also reported to that place. Rounded to the hundredth place, 0.024% is 0.02%. This lot therefore meets its specification.

PRACTICE PROBLEMS

Summary of Statistical Formulas

The variance for a sample is:

$$\text{Variance for a sample} = \frac{\sum (X - \bar{x})^2}{n - 1}$$

The standard deviation for a sample is the square root of the variance:

$$\text{standard deviation for a sample} = \sqrt{\frac{\sum (X - \bar{x})^2}{n - 1}}$$

The relative standard deviation is:

$$\text{RSD} = \frac{\text{Standard deviation} \times (100\%)}{\text{mean}}$$

1. a. What is the purpose of a mass standard in a laboratory?

   b. Working mass standards need to be periodically recalibrated. What do you think is involved in recalibrating a working mass standard? Why do you think calibration needs to be repeated periodically for a working standard?

Table 17.10. *RULES FOR RECORDING VALUES FROM CALCULATIONS WITH THE "CORRECT" NUMBER OF SIGNIFICANT FIGURES*

---

1. *It is assumed that the last digit of a result from a calculation is rounded.* For example, given the result "45.6," the .6 is assumed to have been rounded; therefore, the calculated value must have been between *45.55* and *45.64.*

2. When rounding:
   a. *If the digit to be dropped is less than 5, then the preceding digit remains the same.*
   b. *If the digit to be dropped is 5 or more, the preceding digit increases by 1.*

   For example: 54.78 is rounded to 54.8
                        54.83 is rounded to 54.8
                        54.65 is rounded to 54.7

   (There are different approaches to rounding when the digit to be dropped is 5. Some people, for example, round a five to the nearest *even* number. Then, 9.65 is rounded to 9.6 and 4.75 is rounded to 4.8.)

3. *Round* **after** *performing a calculation.* If a problem requires a series of calculations, round after the *final* calculation and not after the intermediate calculations.

4. *The rule for expressing the answer after addition or subtraction is different than the rule for multiplication and division. The addition/subtraction rule focuses on the number of places to the right of the decimal point. The answer can retain no more numbers to the right of the decimal point than the number involved in the calculation having the least number of places past the decimal point.* For example, if adding the numbers 98.0008, 7.9878, and 56.2:

   98.0008
   7.9878
   56.2
   ‾‾‾‾‾‾‾
   162.1886 Round to: 162.2

   The answer can be expressed only to the nearest tenth place because the value 56.2 has only that many places past the decimal point.

5. *In multiplication and division, keep as many significant figures as are found in the number with the least significant digits.* For example: $0.54678 \times 0.980 \times 7.899$. A calculator might display the answer as 4.232634916 but the answer should be reported as 4.23, three significant figures. Another example: 7987×12. The answer should have only two significant figures and is therefore not 95,844, but rather 96,000 (the zeros are place holders). (These examples assume that all the values given are measured values.)

6. *Constants are numbers whose value is exactly known.* Constants are assumed to have an infinite number of significant figures. For example, given that "12 in equals a foot," 12 in is a constant.

2. a. If a balance is calibrated with a standard that is supposed to be 100.0000 g but is actually 99.9960 g, how will this affect subsequent results from this balance?

b. If a mercury thermometer is improperly marked so that when the temperature is 0°C the thermometer reads 0.2°C, what effect will this have on subsequent readings?

3. If a balance is improperly calibrated (in the sense of being improperly adjusted or set):

a. Do you expect this problem to affect precision of the instrument? Explain.

b. Will this problem affect the accuracy of the instrument? Explain.

4. If a flask that is used to measure volume is supposed to be marked at 10.00 mL and is actually marked at 9.98 mL, will this affect its accuracy? Will this affect its precision?

5. When a pH probe is placed in a sample, it requires a period of seconds to minutes to stabilize. If a technician using a pH meter sometimes allows the meter to stabilize for a minute or so and other times reads the meter immediately after placing the probe in the sample, will this affect the precision of the technician's results? Will it affect the accuracy of the results? Explain. Is this failure to allow the probe to stabilize an example of a systematic error?

6. Is SD a measure of precision or accuracy? Explain.

7. Suppose you prepare a solution that is intended to have 5.00 mg/mL of protein. You perform a protein assay on samples of this solution three times and obtain the following results:

5.12 mg/mL   4.86 mg/mL   5.13 mg/mL

If you perform a fourth assay of the solution, would you expect the result to be 5.00 mg/mL give or take:

0.03 mg/mL or so   0.15 mg/mL or so
0.06 mg/mL or so

8. The following graphs represent the distributions of measurements from four methods that are being compared with one another. Which method is most accurate? Which method is most precise? (Assume the values on the X axis are the same in all four cases.)

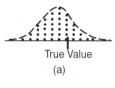
True Value
(a)

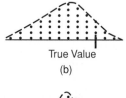
True Value
(b)

True Value
(c)

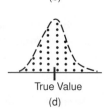
True Value
(d)

9. A standard preparation of human blood serum is prepared that has 38.0 mg/mL of albumin. Technicians from four different laboratories analyze the standard four times in one day and obtain the results (in mg/mL) below. Comment on the various laboratories' accuracy and precision.

Laboratory 1	Laboratory 2	Laboratory 3	Laboratory 4
37.1	38.3	38.6	38.7
37.8	38.0	37.1	39.1
37.7	37.8	39.1	37.5
37.4	38.2	37.0	38.2

10. Technicians from Laboratory 2 repeat the analyses but this time do analyses over a period of 4 months. Their results are:

37.0 mg/mL 37.4 mg/mL 38.1 mg/mL 37.6 mg/mL

Comment on these results in terms of their precision. Are the results more, less, or equally precise when the tests are spread over a period of months?

11. Refer to the Example Problem on p. 302. Calculate the precision of each student's results using the SD (remember the units). Calculate the percent error for the average of each student's results.

12. Antibodies bind to antigens (such as proteins found on viruses and bacteria) and a single antibody molecule can attach to more than one antigen molecule. Monoclonal antibodies are populations of antibodies produced in such a way that they are nearly identical. A population of monoclonal antibodies is prepared that binds to the HIV virus. The number of binding sites per antibody is investigated and the results of five tests are:

3.13   3.11   3.14   3.16   3.12

What is the accuracy and precision of the test used to determine the number of binding sites?

13. The following table shows data from a study of blood calcium levels in several individuals:

Subject	Mean Calcium Level (mg/L)	Number of Observations	Deviation of Results from Mean Values
1	87.5	4	0.13, 0.19, 0.05, 0.11
2	97.6	4	0.18, 0.13, 0.10, 0.02
3	104.8	4	0.09, 0.04, 0.12, 0.06

a. Calculate the SD for each subject's values.

b. Pool the data and calculate the mean value for blood calcium level.

14. A company manufactures buffer solutions for use in calibrating pH meters. A new lot of pH 7.00 buffer was produced. The pH of this new lot was tested on an instrument known to be properly functioning. The results of seven measurements were:

7.12   7.20   7.15   7.17   7.16   7.19   7.15

a. Calculate the mean and SD for these data.

b. Comment on these results if the pH of the buffer is supposed to be 7.00.

15. Samples of air in a particular factory were analyzed for lead. Each individual sample was tested three times and samples were taken on three occasions:

Sample	$\mu g\ Pb/m^3\ Air$
1	1.4, 1.3, 1.2
2	2.3, 2.3, 2.1
3	1.6, 1.5, 1.7

a. Calculate the mean and SD for each sample.

b. Calculate the mean and SD for the pooled set of data.

c. Is the SD higher in a or b? Explain why.

16. A technician at LCJ Associates Environmental Laboratories evaluated a method to determine the method's repeatability and accuracy. The purpose of the method is to test levels of ammonia in water samples. To perform this assessment, the technician carefully prepared two control samples, one with 10 μg/L ammonia in water, the other with 100 μg/L ammonia in water. He tested each sample 7 times using the method being evaluated and documented the results on the form on the next page.

a. Calculate the mean, SD, RSD (CV), and percent error based on these data.

b. Why did the technician prepare two standards at two different concentrations?

c. Explain how these analyses and the completion of this form are part of a quality-control program.

d. How could the technician evaluate the accuracy of the method? What is the accuracy based on these data?

17. For each illustration (below), what measurement should be reported? How many significant figures does the measurement have?

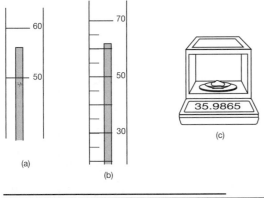

(a)

(b)

35.9865

(c)

(d)

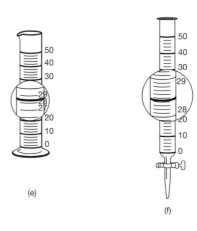

(e)

(f)

18. Put a line through each of the zeros in this problem that are place holders. Place a ? above each zero that is ambiguous (i.e., may be a place holder or may convey value).

a. 2000

b. 1,000,000

c. 0.00677

d. 134,908,098

19. How many significant figures does each of the following numbers have:

a. 45.789

b. 0.00650

c. 10.009

d. 0.000878

20. Round the following numbers to the nearest tenth decimal place.

a. 0.0345

b. 0.98

c. 0.55

d. 0.2245

21. A biotechnology company specifies that a product has at least 10 mg/vial. Do the following lots meet the specification?

lot a.	10.2 mg/vial
lot b.	9.899 mg/vial
lot c.	7.82 mg/vial
lot d.	9.400 mg/vial

22. A biotechnology company specifies that a product has $\leq 0.02\%$ of impurities. Do the following lots meet the specification?

lot a.	0.025%
lot b.	0.015%
lot c.	0.027%
lot d	0.024%

23. A spectrophotometer was used to determine the concentration of protein in a solution. The solution was analyzed six times and the absorbance values were:

0.956 0.948 0.958 0.991 0.963 0.957

According to the spectrophotometer manufacturer, the imprecision of the instrument should not exceed 1% relative standard deviation.

a. Do these results exceed 1% RSD?

b. If so, does this indicate that the spectrophotometer is malfunctioning or does not meet its specifications?

QUESTIONS FOR DISCUSSION

1. If you work in a laboratory, find several examples of instruments or devices that are used to make measurements. Discuss these items in terms of calibration. How are they calibrated (in the sense of being adjusted according to a standard)? Who is responsible for their calibration? Are there standards used to calibrate the items that are traceable to national standards?

2. Examine a catalogue of scientific supplies. What devices are specified to be "NIST traceable"?

3. Obtaining the utmost accuracy and precision in a measurement or assay is generally expensive in terms of money and/or time investment. Discuss the following situations. Is it more important for the test to be as accurate and precise as possible, or is low cost and/or speed a higher priority?

a. Routine testing of the level of pollutant emissions from automobiles in a city where automobiles are required to have functioning antipollution devices.

b. Determination of the level of drug in a blood sample after an overdose to enable physicians to make a rapid decision about treatment.

c. Study of the stability of an enzyme with storage in a freezer to determine how long a product can be stored.

d. Determination of glucose levels in the urine of pregnant women to detect pregnancy-related diabetes.

LCJ Associates Environmental Laboratories

PRECISION AND ACCURACY FORM
QUALITY CONTROL DATA

Parameter _Ammonia NH_3–N_   Date ___5-6-08___

Method ___Am–3___   SOP# ___27931___

Reference ___Water & Wastewater Manual___   Analyst ___N.W.___

	Sample # 72807	Sample # 72808
Concentration of standard	10 μg/L	100 μg/L
n		
1	09.	103.
2	10.	103.
3	08.	104.
4	11.	97.
5	10.	98.
6	13.	102.
7	09.	98.
$\bar{x}$		
SD		
RSD		
% error		

Practice Problem #16.

Introduction to Instrumental Methods and Electricity

I. USING INSTRUMENTS TO MAKE MEASUREMENTS

A. Introduction

The previous chapter discussed fundamental principles of measurement. The five chapters following this one each focus on a specific type of measurement, such as the measurement of volume and of weight. This chapter is a transition that introduces instrumental methods of measurement and basic vocabulary, and also concepts relating to electricity and electronics.

People made measurements before measuring instruments were invented. For example, everyone uses the senses of "feeling" to estimate the air temperature or the weight of an object. Measurements made with senses alone, however, are not very accurate or reproducible; therefore measuring instruments were invented, such as pan balances and mercury thermometers, see Figure 18.1a,c on p. 316. These early measuring devices did not require electricity to make a measurement and are sometimes referred to as "mechanical" instruments.

Mechanical instruments are still used in the laboratory since they are simple to understand, reliable, and often inexpensive. Electronic measuring instruments, however, are now preferred for many applications. Electronic instruments convert a physical or chemical property of a sample into an electrical signal, see Figure 18.1b,d. For example, an electronic balance converts a property of an object, the force of gravity on the object, to an electrical signal. The balance then processes the electrical signal to convert it to a display of weight. Electronic instruments are generally more convenient, faster, and more consistent than mechanical ones. Electronic devices also can measure a smaller change in the property of interest than can mechanical instruments. For example, a mercury thermometer can give a temperature measurement to, at best, the closest tenth of a degree. In contrast, some electronic thermometers can detect a temperature change of a hundredth of a degree or better.

Early electronic instruments were characterized by control knobs, dials, and switches by which the analyst controlled the device. These instruments also had meters in which a needle pointed to the value of the measurement. Modern electronic instruments are usually associated with computers, so they have keypads (in place of knobs, dials, and switches) and display screens and printers (in place of meters with needles). The computer is sometimes an independent unit that is connected to the instrument via a cord or cable; other times, *small computers, called* **microprocessors,** are integrated into the instrument itself. Computers and microprocessors can perform complex manipulations of data, store large amounts of information, control instruments so they operate consistently, and detect instrument malfunctions.

The computers associated with modern laboratory instruments are powerful, so these instruments are often simple to operate and appear "intelligent." Nonetheless, it is important to remember that the human operator, although seemingly removed from the measurement process, is still a key part of making a "good" measurement. It remains the operator's responsibility to understand the measurement system, to be aware of problems, and to be skeptical of the results. Thus, it is essential for laboratory workers to have a basic understanding of how laboratory instruments work.

B. Measurement Systems

A mercury thermometer is a relatively simple measuring device. It contains liquid mercury enclosed in a narrow glass tube or stem. If the thermometer is moved from a cooler to a warmer medium, the mercury expands and so moves up the stem. If the temperature decreases, the liquid contracts down the stem. There is a scale of "degrees" etched onto the glass stem. The oper-

ator reads the temperature of the medium by viewing the height of the mercury column relative to the scale of degrees.

In a mercury thermometer, the mercury "senses" one form of energy, the heat of the medium, and converts it to another form of energy, the mechanical expansion or contraction of mercury. There is an **interface** where the glass surface of the thermometer is in contact with the sample. The mercury acts as a **transducer,** *a device that senses one form of energy and converts it to another form.* The degree scale etched on the stem of the thermometer is the **display** or **readout.** We can generalize from this example and say that a mechanical measurement instrument has these components:

Interface → Transducer (Sensor) → Display (Readout)

Consider a second measuring instrument, an electronic balance. The sample to be weighed is placed on the balance's weighing pan. The weighing pan is the interface. Within the balance an electronic component acts as the transducer and responds to the force of gravity on the sample. The transducer converts that force to an **electrical signal.** A **signal processor** *modifies the electrical signal so that it can be displayed as a weight value.* Thus, in an electronic instrument the components of the measuring system are:

Interface → Transducer (Sensor) →
Signal Processor → Display (Readout)

There are many different transducers. Some respond to physical energies, such as pressure, temperature, and light, while others detect chemical properties such as pH. Examples of electronic measuring instruments are illustrated in Figure 18.2.

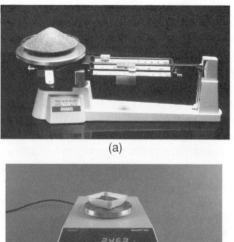

(a)

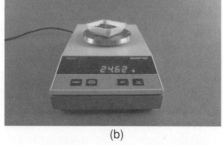

(b)

Fahrenheit (°F)

(c)

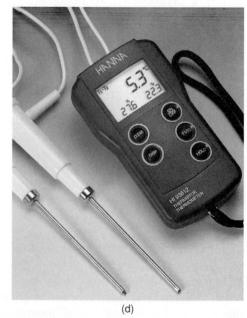

(d)

Figure 18.1. Mechanical versus Electronic Devices. a. A mechanical balance. (Courtesy Pearson Science.) **b.** An electronic balance. (Courtesy Photo Researchers, Inc.) **c.** A mechanical thermometer. **d.** An electronic thermometer.

For a measuring instrument to be useful, certain requirements must be met:

- ***The instrument's response must have a consistent and predictable relationship with the property being measured.*** For example, a certain temperature must predictably cause the mercury to rise to a certain height. A certain weight sample must generate a consistent electrical signal.

- ***The instrument's response must be related to internationally accepted units of measurement.*** For example, there must be a way to relate a particular height of mercury to units of "degrees." A certain electrical signal in a balance must be related to units of "grams."

Calibration *is the process by which the response of an instrument is related to internationally accepted measurement units.* Every measuring instrument needs to be periodically calibrated and to be checked regularly to be sure it is responding predictably and consistently. Calibration and performance verification of specific instruments are discussed in subsequent chapters as these instruments are introduced.

The relationship between the property being measured and the response of an electronic instrument is most often linear. If there is a linear response, adjustment at two points (e.g., zero and full-scale) calibrates the device because two points determine a straight line. If a device is nonlinear, additional points are needed to establish the relationship between instrument response and the property being measured. For example, a pH meter is calibrated in accordance with two standards of known pH. Two standards are sufficient because there is a linear relationship between the response of the instrument and the pH of the sample, see Figure 18.3. **Two-point calibration** *refers to a situation where two standards are used to calibrate an instrument whose response is linear.*

To best understand electronic laboratory instruments, it is necessary to have some knowledge of the principles and vocabulary of electricity and electronics. There are also electrical safety issues with which everyone should be familiar when operating instruments. The rest of this chapter, therefore, introduces electricity, electronics, and safety as they relate to laboratory instruments.

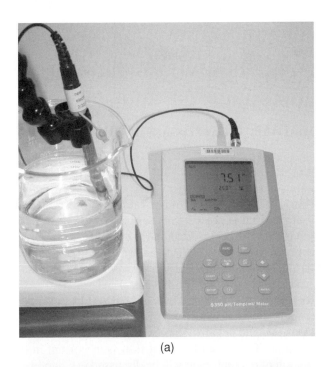

(a)

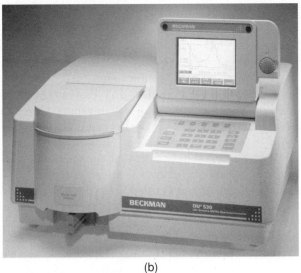

(b)

Figure 18.2. Measurement Instruments. a. A pH Meter System. A probe that generates a signal in response to pH is placed within a liquid sample. The probe is attached via a cable to the signal processing unit and readout device. **b.** Spectrophotometer. Spectrophotometers measure the interaction between light and a sample. A photomultiplier tube located within the spectrophotometer is the transducer that senses light. A liquid sample is placed in a glass container, which is in turn placed inside a light-tight chamber inside the spectrophotometer. (Spectrophotometer image courtesy of Beckman Coulter, Inc.)

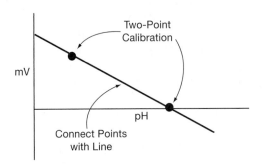

Figure 18.3. Calibration. If there is a linear relationship between instrument response and the property being measured, then two points define the relationship. This is the case for a pH meter.

II. BASIC TERMINOLOGY AND CONCEPTS OF ELECTRICITY

A. Current, Voltage, and Resistance

i. CURRENT

Electricity is understood in terms of the atomic theory of matter. In short, according to this theory, matter consists of atoms that have electrons, protons, and neutrons. Protons have positive charge, electrons have negative charge. A material is said to be **negatively charged** *if electrons are in excess* and **positively charged** *if electrons are depleted.* Objects that have the same charge repel one another, whereas objects that have opposite charges attract one another. Therefore, if two objects are separated, they will be attracted if one is positive and the other is negative. On the other hand, if both are positively or negatively charged, the objects will repel one another.

Metals have a property that is important for electricity. In metals, the outer electrons of atoms are loosely held by the atoms, so these outer electrons move easily and randomly from one atom to another. Suppose there is a metal wire with an excess of electrons at one of its ends and a deficiency of electrons at the other. In this situation, the electrons in the wire will not move randomly. Rather, they will flow away from the end with an excess of electrons and toward the end with a deficiency, see Figure 18.4. It may be helpful to imagine that the flow of electrons in a wire is like water flowing in a river bed. *The flow of either water or electricity is called* **current.**

The flow of electrons, abbreviated I, is measured in units of **amperes,** abbreviated **amps** or **A.** If 6.25×10^{18} electrons pass a certain point every second, the current is said to be 1 amp. An ampere is a rather large unit. The current in many electronic instruments is on the order of **milliamperes (mA)** or even **microamperes (μA).** A **milliampere** *is one thousandth of an ampere;* a **microampere** *is one millionth of an ampere.*

There are two types of electrical current: **alternating current (AC)** and **direct current (DC).** In **direct current** *electrons flow in one direction through a wire.* A battery generates direct current, see Figure 18.5. In **alternating current** *electrons change directions many times per second.* The current that comes from the power company to an outlet in a wall is alternating. In the United States, AC current cycles back and forth with a **frequency** of 60 times per second. In other countries AC current has a frequency of 50 times per second. Frequency is measured

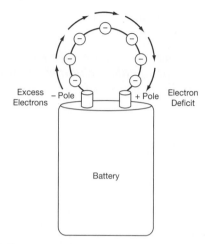

Figure 18.5.   Direct Current. In a battery, current flows directly from a pole with an excess of electrons to a pole with an electron deficit.

in units of **Hertz (Hz),** *where 1 Hz = 1 cycle per second.*

Most laboratory instruments require direct current. Direct current is occasionally supplied to an instrument by batteries. It is more common, however, for a **power supply** *to convert the AC current of the power company to a DC current through a process called* **rectification.**

We are accustomed to electrical current that flows through wires; however, electrical current can also flow through liquids. Electrical current in a liquid is carried by positive and negative ions derived from salts added to the liquid. This type of current flow is important in the function of pH meters, as will be discussed in Chapter 22.

ii. VOLTAGE

Energy *is defined as the ability to do work;* **potential energy** *is stored energy.* For example, water poised at the top of a waterfall has **potential energy.** As the water plummets over the waterfall, its potential energy is converted into the mechanical energy of falling. Gravity is responsible for the potential energy of the water, see Figure 18.6a.

Just as water can have potential energy, there is also **electrical potential energy. Electrical potential** *is the potential energy of charges that are separated from one another and attract or repulse one another because of their unlike or similar charges,* see Figure 18.6b and c. Electrical potential arises because charged objects exert attractive or repulsive forces on one another. For example, in a battery there is a negative terminal with an excess of electrons, and a positive pole with a deficit. The excess electrons at the negative terminal can be thought of as the force that "pushes" electrons toward the positive pole. Thus, as gravity is the force that causes water to flow over a waterfall, so electrical potential is like a force that causes electrons to flow in a wire.

Electrical potential is also called **electromotive force (EMF or ε)** or **voltage.** *The units of voltage are the* **volt (V)** *and the* **millivolt (mV).** The voltage that is supplied to a wall receptacle by the power company is either in the range from 110 to 120 V or is 220 V.

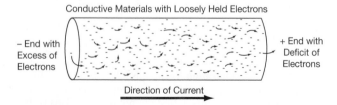

Figure 18.4.   The Flow of Electricity in a Wire. Electrons move from atom to atom.

(a)

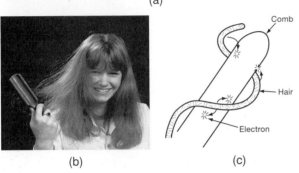

(b)   (c)

Figure 18.6.   Potential Energy. a. The potential energy of water is converted to mechanical energy in a waterfall. (Courtesy: iStockPhoto.) **b.** A familiar example of electrical potential energy. When a comb and hair are rubbed together, electrons are removed from the hair and deposited on the comb. (Courtesy Andrew Lambert Photography/Photo Researchers.) **c.** The comb, therefore, becomes negatively charged; the hair positively charged. The positive and negative charges are thus separated from one another. Opposite charges attract, so the hair and comb attract one another.

The development of voltage requires a method of separating charges from one another so that there is an excess of electrons at one site and a deficiency of electrons at another. There are many ways that voltage can be generated, all of which involve the conversion of some other form of energy into electrical potential energy. Fossil fuels, like coal, gas, and oil, can be burned to create electrical potential. It is also possible to harness the mechanical energy of a waterfall or the wind, the solar power of the sun, the energy of nuclear reactions, and the energy of chemical reactions to make electrical potential.

iii. Resistance

Water in a river cannot flow if its path is blocked by a dam. Similarly, the flow of electricity is impeded if it encounters **resistance** in its path. **Resistance** *is the impedance to electron flow. The units of resistance are* **ohms,** abbreviated, Ω. *One* **ohm** *is the value of resistance through which 1 V maintains a current of 1 A.*

All materials (except so-called superconductors) offer some resistance to current flow. The amount of resistance depends on the material. Electricity flows most readily in **conductors**. **Conductors** *are materials, such as metals,*

whose outer electrons are free to flow from one atom to another. The best conductors are those materials whose outer electrons are most loosely bound. Silver is the best conductor, followed by copper and gold. In contrast, **insulators** *are materials in which the outer electrons of atoms are not free to move and so electricity does not flow readily.* Electricity usually does not flow in the air (lightning is an exception), nor does it flow in glass, plastic, or rubber.

When electricity encounters resistance, some or all of its energy is converted into heat energy. Because even copper wire offers some resistance to current flow, as electricity flows in a wire, some energy is lost as heat. The longer a wire, the more electrical energy is dissipated as heat. This is why the use of extension cords results in a loss of power. A burner on an electrical stove is an application where we take advantage of the heat generated when electricity encounters resistance.

Semiconductors *offer intermediate resistance to electron flow and are used in the manufacture of transistors and other electronic devices.* Semiconductors are composed of crystalline silicon or germanium. Silicon and germanium do not conduct electricity well; however, if impurities are added to their crystals, their structure changes so that electrons can move more easily. Semiconductors play a critical role in electronics because engineers can adjust their resistance and thereby control the flow of current through them.

B. Circuits

i. Simple Circuits

Electricity only flows if there is a complete conductive pathway in which the electrons can move. At one end of the pathway there must be a source providing an excess of electrons; at the other end there must be an electron deficit. *Such a pathway for electricity is called a* **circuit**. Figure 18.5 showed a battery with a negative pole, having an excess of electrons, and a positive pole, having a deficit. The wire connecting the positive and negative poles completed the pathway for electron flow.

In practice, one would not encounter a circuit such as the one in Figure 18.5 because the flow of electrons in the wire would produce no useful work. Electricity is useful when electrical energy flowing in a circuit is transformed into another form, such as heat, light, or mechanical work. Figure 18.7a on p. 320 shows a diagram of a simple circuit that includes a light bulb. As electricity flows through the bulb, a thin wire heats up and emits light. The bulb thus converts electrical energy to light. Items such as burners on a stove, light bulbs, and motors convert electrical energy into useful heat, light, and motion. Motors, bulbs, and other such elements constitute resistance to electron flow. These elements, which provide resistance, are sometimes referred to as the "load" in a circuit. As electricity flows through such devices, work is performed, and heat is generated.

Every practical circuit at a minimum consists of conductors through which electrons can travel, a voltage

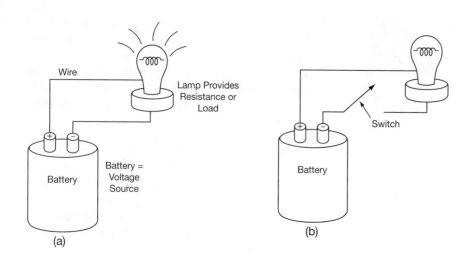

(a)

(b)

**Figure 18.7.   A Simple Circuit.
a.** A simple circuit consisting of a voltage source, a wire, and a resistive element—in this case, a light bulb. **b.** A switch is added to the simple circuit in a.

source, and some type of resistance. In addition, insulation is usually added, serving to confine the current to the desired paths. In Figure 18.7b a **switch** *to control electrical flow* has been added to the circuit. When the switch is closed the circuit is complete and current flows; when the switch is open the circuit is interrupted by air and current ceases to flow.

The circuit shown in Figure 18.7b has been redrawn in Figure 18.8a using schematic symbols instead of pictures. There are a number of symbols used in electrical diagrams; examples are shown in Figure 18.8b.

ii. OHM'S LAW

Ohm's Law (developed by George Simon Ohm) is an equation that relates voltage, current, and resistance in a circuit. Ohm's Law states that the current in a circuit (expressed in amps) is directly proportional to the applied voltage (expressed in volts) and is inversely proportional to the resistance (expressed in ohms). Ohm's Law is written as:

$$\text{Voltage} = (\text{current})(\text{resistance})$$
$$V = IR$$
$$\text{or}$$
$$I = \frac{V}{R}$$

For example, consider a circuit in which there is a motor that provides 5 ohms of resistance. The voltage from the power company that supplies the circuit is 110 V. Assuming there are no other devices in the circuit, it is

possible to calculate the current in the circuit:

$$V = IR$$
$$110 \text{ V} = (?)(5 \text{ ohms})$$
$$? = 22 \text{ amps}$$

Note that the proper units must be used to apply Ohm's Law. For example, if the units in a problem are given as millivolts or microamperes, they must be converted to volts and amps.

EXAMPLE PROBLEM

According to Ohm's Law, if the resistance in a circuit suddenly becomes very low, what will happen to the current in that circuit?

ANSWER

Assuming the voltage remains constant, the current will become very high.

Electrophoresis is a technique commonly used in the biotechnology laboratory that provides a good example of how Ohm's Law is applied in practice. **Electrophoresis** *separates charged biological molecules from one another based on their rate of migration when placed in an electrical field.*

In gel electrophoresis, the sample mixture is placed in a gel matrix. The gel is positioned in a gel box and is covered with a thin layer of buffer. Positive and nega-

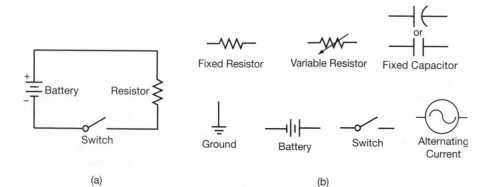

(a)

(b)

Figure 18.8.   An Electrical Diagram. a. The same circuit as shown in Figure 18.7b, but using standard electrical symbols instead of pictures. **b.** Examples of symbols used in electrical schematic diagrams.

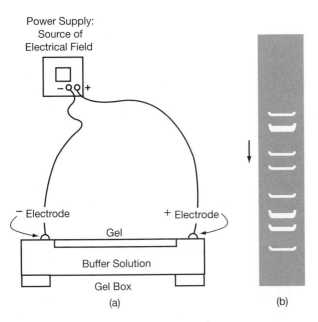

Power Supply:
Source of
Electrical Field

− Electrode
+ Electrode
Gel
Buffer Solution
Gel Box
(a)

(b)

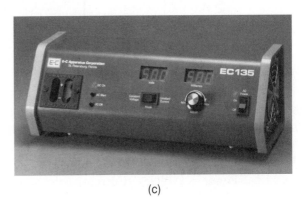

(c)

Figure 18.9. **Electrophoresis. a.** An electrophoresis setup including a gel, a gel box, and a power supply. **b.** A stained gel. The bands represent separated molecules. **c.** An electrophoresis power supply. Constant current or constant voltage can be selected. (Photo courtesy of E-C Apparatus. E-C is a registered trademark of ThermoQuest Corporation.)

tive ions in the gel and the buffer are capable of conducting current. A power supply is used to provide voltage so that there is a positive pole at one end of the gel and a negative pole at the other end, see Figure 18.9a. Negatively charged sample components migrate toward the positive pole; positively charged components move to the negative pole. Their speed of migration is determined by the magnitude of their charge. If two or more components have the same charge, then the larger ones will tend to move more slowly through the gel matrix than the smaller ones. Thus, different biological molecules migrate at different rates depending on their charge and size. Once the components are separated, the gel is removed from the box, a dye is used to stain the biological molecules, and the separated components appear as bands, see Figure 18.9b.

In the early stages of an electrophoresis run, the resistance of the gel increases somewhat as ions are electrophoresed out of the gel. Therefore, if the voltage supplied to the gel remains constant, then the current decreases during the run, in accordance with Ohm's Law. To maintain a constant current during electrophoresis, the voltage must be increased as the run progresses. Most power supplies designed for electrophoresis can be adjusted so that they provide either constant voltage or constant current, depending on the user's preference, see Figure 18.9c. (Electrophoresis power supplies are also made that maintain constant power. Power will be discussed later.)

EXAMPLE PROBLEM

a. At the beginning of an electrophoresis run the voltage is 75 V, and the current is 30 mA. What is the resistance of the gel?

b. The voltage is held constant and the current decreases to 12 mA by the end of the separation. What is the resistance of the gel?

---

c. If an identical gel is run, but this time the current is maintained at 30 mA through the entire run, what will the voltage be at the end of the run?

ANSWER

According to Ohm's Law:

a. $V = IR$
$75\,V = (0.030\,A)(?)$
$? = 2.5\,k\Omega$

b. $V = IR$
$75\,V = (0.012\,A)(?)$
$? = 6.25\,k\Omega$

c. $V = IR$
$? = (0.030\,A)(6250\,\Omega)$
$? = 187.5\,V$

---

iii. SERIES AND PARALLEL CIRCUITS

Most electrical circuits contain more than one resistive element. Circuits with multiple resistant elements (loads) can be arranged in two basic ways: in **series** or in **parallel.**

A circuit with three resistive elements is shown in Figure 18.10a on p. 322. The three resistances are arranged one after the other; hence, they are said to be in series. In a **series circuit,** *current flow has only one possible pathway.* The current passing through resistance 1 is the same as the current passing through resistances 2 and 3. A familiar example of a series circuit is a string of old-fashioned Christmas tree lights. If one bulb burns out, the pathway is no longer complete and all the bulbs go out.

A **parallel circuit** with three resistive elements is shown in Figure 18.10b. The three paths split the total amperage and so the total amperage in the entire circuit is equal to the sum of the currents in the three parallel pathways. In a parallel circuit, if one pathway is broken, current can still flow in the others. Circuits in most modern instruments are complex and consist of both parallel and series pathways.

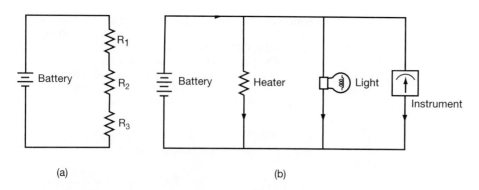

(a) (b)

Figure 18.10. Series versus Parallel Circuits. a. Three resistive elements arranged in series. Current passes through all three elements. **b.** A parallel circuit. The current is split into three paths.

iv. POWER, WORK, AND CIRCUITS

Electricity is used to perform work, such as to generate light or power a pH meter. As an electric device performs work, power is consumed. *The units of power are* **watts,** abbreviated **W. Power** is defined as:

$$\text{Power} = (\text{Voltage})(\text{Current})$$
$$\text{or}$$
$$W = (V)(I)$$

Instrument manufacturers specify how much power their instruments require. The amount of power required by a device is related to how much heat that device will produce during operation. For example, a 100 W light bulb generates more light—and more heat—than a 25 W light bulb. The more power being used in a circuit, the more heat is generated in that circuit. Power ratings are useful to determine whether a circuit is being overloaded (i.e., too much heat is being generated) by devices.

EXAMPLE

A spectrophotometer requiring 300 W, an oven requiring 600 W, and a cell culture hood requiring 200 W are all plugged into the same laboratory circuit and are all turned on. Suppose further that the laboratory is served by a 120 V line with a current of 15 A. Will the load on the circuit exceed its capacity?

The total carrying capacity of the laboratory circuit is:

$$P = IV$$
$$= (15 \text{ A})(120 \text{ V}) = 1800 \text{ W}$$

The total power consumed by the three instruments is:
$$300 \text{ W} + 600 \text{ W} + 200 \text{ W} = 1100 \text{ W}$$

The circuit can therefore handle these three items. If however three more spectrophotometers at 300 W each are plugged into the same circuit and operated at the same time, then the circuit will be overloaded and the circuit breaker or fuse protecting that circuit will blow.

---

v. GROUNDING, SHORT CIRCUITS, FUSES, AND CIRCUIT BREAKERS

Fuses or circuit breakers and grounding wires are the two types of basic electrical safety devices that are incorporated into circuits. Both types of devices are necessary to prevent electrical fires and to protect operators from serious shock.

If an unusually large amount of current flows in a wire, then that wire will become extremely hot—to the point where instruments can be damaged and/or a fire can be ignited. **Fuses** *are simple devices that protect a circuit from excess current.* A fuse contains a thin wire. Heat is generated whenever current flows through the wire; the more current, the more heat. The wire will break if too much current passes through it, thus breaking the circuit and preventing the further flow of current, see Figure 18.11a,b. Fuses are rated according to volts and amps. For example, it is possible to purchase fuses for 10, 20, and 30 amps, 110 V. If a fuse is rated for 20 amps and more than 20 amps passes through it, it will "blow."

A circuit breaker performs the same function as a fuse; however, circuit breakers are more convenient than fuses. A blown fuse must be replaced, whereas a circuit breaker is simply switched back to its original position, once the original problem is solved.

Fuses or circuit breakers are placed in the circuits in houses and buildings. A typical house will receive from 50 to 220 amps from the power company, which is subdivided in the house into individual circuits. There is usually a main fuse or circuit breaker where the electric power line enters the building and another fuse or circuit breaker where each circuit branches from the main circuit, see Figure 18.12.

There are several common causes of excess current leading to blown fuses or tripped circuit breakers:

1. *There may be too many devices plugged into the same circuit.* (Anyone who has lived in an older apartment has probably experienced this problem.)

2. *One device in the circuit may be using too much power.* In this case, an electrician must be called to solve the problem.

3. *There may be a short circuit in a device.* In this case, the device should be unplugged and a service technician called.

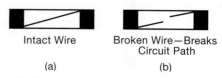

Intact Wire   Broken Wire—Breaks Circuit Path

(a) (b)

Figure 18.11. Fuses and Circuit Breakers. a. Good fuse. **b.** "Blown" fuse.

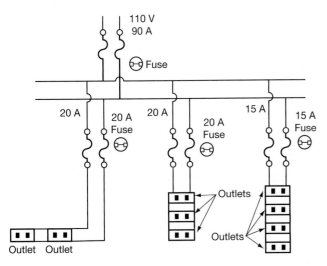

Figure 18.12. Fuses or Circuit Breakers Protect the Circuits in Every Home and Building.

In addition to the fuses and circuit breakers that protect the circuits in a building, most laboratory instruments have their own fuses to protect delicate components. (One of the challenges of laboratory management is keeping track of the spare fuses that go with each instrument so that they are available when needed.) These fuses blow if for any reason too much current flows through the instrument. If an instrument's fuse blows, then it can indicate a problem that requires repair by a qualified person.

Figure 18.13a illustrates a properly functioning instrument which is plugged into a wall outlet. Voltage causes current to flow from the power company,

through the instrument, and then back to the wall outlet, thus completing the circuit. In this example, the electricity powers a motor and so the motor provides resistance to current flow.

Figure 18.13b illustrates the same instrument, but this time the insulation around the power cord is frayed so that the metal wire touches the metal case of the instrument. No current flows through the metal frame because the circuit is not complete. The fuse does not blow, and the motor may be operative. Unfortunately, in this situation, if a person touches the metal casing of the instrument that person may provide a complete path for electrons to flow to the ground, see Figure 18.13c. The person will be shocked; in some cases, such shock can be fatal. This situation is called a **short circuit**, *a situation in which electrical current is able to flow without passing through the resistance, or load.*

Because the situation illustrated in Figure 18.13c is dangerous, modern electrical instruments are provided with a grounding wire, see Figure 18.14 on p. 324. To understand this safety mechanism it is necessary to know that the earth can gain or lose electrical charges and yet remain neutral. The earth, therefore, is considered to be at zero potential and can act as a charge neutralizer. An **electrical ground** *is any conducting material connected to the earth. A* **ground wire** *is attached to the metal frame of the instrument and is connected to the earth via the third prong on the power cords of most modern instruments.* The third prong is a metal conductor that, when plugged into the wall, contacts another wire that is connected to a metal water pipe or other conductor. The water pipe in turn contacts the ground

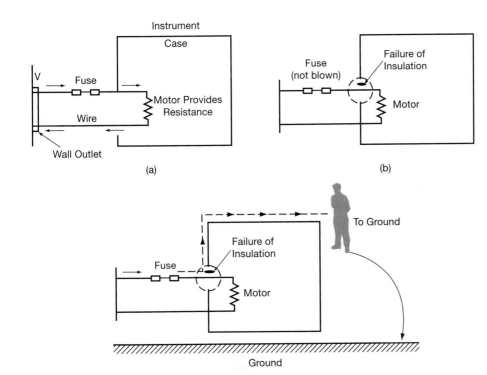

Figure 18.13. Short Circuits and Safety Devices. a. A properly functioning instrument. Voltage causes current to flow through the instrument and then back to the wall outlet. The electricity powers a motor and so the motor provides resistance to current flow. **b.** A short circuit. The insulation is frayed so that the metal wire touches the metal case of the instrument. The fuse does not blow and the instrument may continue to be operative. **c.** If a person touches the instrument in b, current will flow through the person to the ground causing dangerous shock.

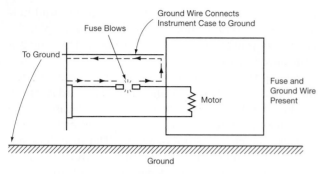

Figure 18.14. A Properly Protected Instrument. The ground wire connects the instrument to the earth, thus harmlessly draining off the current and preventing the user from shock. A fuse is also present to prevent excess current from flowing in the circuit.

under the building. In the event of a short circuit, current flows from the instrument chassis to the third prong and is then discharged harmlessly to the ground. Note also that the fuse will blow in this situation because the current flowing to the ground will be great. The use of a three-prong cord to ground an instrument is an important safety feature and should not be bypassed by using adapters or other methods.

III. BASIC TERMINOLOGY AND CONCEPTS OF ELECTRONICS

A. Electronic Components

i. OVERVIEW

The term **electronic** *refers to electrical devices that have nonmechanical components that generate, amplify, and process electrical signals.* These components are combined and arranged in circuits. Depending on the components used and their arrangement, electronic circuits can perform a vast number of functions ranging from playing music to measuring weight in the laboratory. (The term *electrical* is typically applied to electrically operated equipment like pumps and motors.)

Although the functions that electronic devices perform are numerous, the basic components of their circuits are relatively few. Some of the most common of these components are briefly described in this section, including **resistors, capacitors, diodes,** and **transistors.** The next section discusses how these components are assembled to make functional instruments for measurement.

ii. RESISTORS, CAPACITORS, DIODES, AND TRANSISTORS

Resistors *are electronic components that impede the flow of current.* Resistors are used to control the amount of current in a circuit. There are **fixed resistors** *that have one value of resistance,* and **variable resistors** *that can be adjusted to provide different levels of resis-*

tance. Fixed resistors are manufactured to have a specific resistance and are color coded to indicate what that resistance is.

There are two types of variable resistor: the **rheostat** and the **potentiometer**, abbreviated "pot." There are many situations where it is necessary to be able to vary the amount of current in a circuit. For example, most centrifuges have a speed control adjustment to vary their speed of rotation. Rheostats are frequently part of such speed-adjusting mechanisms. Another example is calibration. Calibration of an electronic instrument usually involves changing the voltage or current in a particular circuit. This can be accomplished by varying the resistance in the circuit with a potentiometer.

A **capacitor** *is a device that stores electrical charge and can be used to provide current in the absence of a battery or current from an outlet.* Capacitors consist of two thin plates made of a conducting material and separated by a layer of insulating material. When current passes through a capacitor one of the plates acquires a positive charge, the other a negative charge. After current flow ceases through the capacitor, the charge can remain on the plates for days. A charged capacitor can function as if it were a DC voltage source.

A defibrillator, which is used medically to shock a patient's heart into a normal rhythm, is an example of how capacitors are used in a device. An electrical charge is stored by a capacitor in the defibrillator. When the defibrillator electrodes are connected across the patient's body, the patient forms a conducting discharge path and a large current flows for an instant.

An **electroporator** is another example of a device based on a capacitor. **Electroporation** *is used to introduce drugs, genetic material, and other molecules into living cells suspended in an aqueous medium.* This method has particular application in biotechnology because it enables scientists to introduce functional DNA into cells. In electroporation a short-duration, high-voltage electrical field is applied to a chamber containing cells. The electric field causes pores to be temporarily created in the cells' membranes through which the molecules enter. The simplest approach to create a high-voltage pulse is to charge a capacitor using a high-voltage power supply and then to discharge the capacitor into the chamber with the cells.

Diodes *are components made of a crystalline material that is capable of conducting electricity in only one direction.* The primary function of **diodes** in instruments is to *convert alternating to direct current, which is called* **rectification.**

There are also diodes, called **light emitting diodes (LEDs),** *that produce light when electrical current flows through them.* LEDs are familiar to us as the displays on digital clocks and other electronic devices.

Transistors *are devices made from semiconductor materials that are used to amplify an electrical signal.* Transistors also function as switches to turn a circuit on or off and can be used to convert direct current to alternating current. Transistors have played a key role in the development of such devices as microprocessors, pocket calculators, home computers, and video games.

iii. INTEGRATED CIRCUITS AND CIRCUIT BOARDS

Two manufacturing processes are used to make electronic circuits, conventional and integrated. Most laboratory instruments combine both circuit types. A conventional circuit contains various electronic components connected to one another by wires and attached to a base. The base is normally a **printed circuit board** *made of a thin piece of plastic or other insulating material.* Copper wires are "printed" onto the board using a special chemical process. Components can be connected to one another on the board in series, parallel, or in a combination of both, see Figure 18.15.

An **integrated circuit (IC)** *is a very small electronic circuit that is assembled on a piece of semiconductor material called a* **chip.** The chip acts both as a base and as a part of the circuit. Integrated circuits contain components such as diodes, transistors, and capacitors linked together by tiny conducting wires. A single chip can contain millions of microscopic parts. The chips are packaged in protective carriers with electrical pin connectors that allow them to be plugged into a circuit board. Integrated circuits are found in almost all modern electronic laboratory instruments. ICs are often combined with one another and with other components onto a printed circuit board.

Electronic circuits are combinations of components connected to one another in various configurations to form **functional units.** Functional units perform the various tasks of electronic instruments. The next section discusses functional units as they relate to instruments that make measurements.

B. Functional Units

i. TRANSFORMERS

Transformers *are used to vary the input voltage entering an instrument.* For example, a transformer can convert the 110 V AC voltage from the power company to

Figure 18.15. Printed Circuit Board.

150 volts for an instrument. It may seem surprising that more voltage can be "created" than originally existed; however, the power supplied to the instrument remains constant. Power equals voltage multiplied by current; therefore, if the voltage supplied to an instrument is increased ("stepped up") by a transformer, then the current to that instrument is reduced. Similarly, if the voltage to the instrument is reduced ("stepped down"), then the current increases.

ii. POWER SUPPLIES

Power supplies were previously mentioned as related to electrophoresis. In electrophoresis the power supply is conspicuous because the operator attaches it to the gel box and selects how much current or voltage is to run through the gel. Most laboratory instruments contain less conspicuous internal power supplies.

The **power supply** plays a variety of roles including:

* *Converting alternating current to direct current.* The power company sends out alternating current, but most instruments require direct current. A power supply is used to convert alternating to direct current.

* *Acting as a transformer.*

* *Regulating the voltage in the event of fluctuations in voltage coming from the supplier.* It is not unusual for the voltage to surge or change unexpectedly. Instruments are protected from these variations by the power supply.

* *Distributing voltage to multiple circuits within an instrument.* Instruments are composed of multiple circuits, each of which requires a source of voltage.

iii. DETECTORS

The term *transducer* tends to be used broadly and so would include simple devices, such as the mercury in a thermometer. The term **detector** *is often used to refer to an electronic transducer that generates an electrical signal in response to a physical or chemical property of a sample.* The electrical output (signal) from a detector can be a change in voltage, current, or resistance. For example, pH meters (Chapter 22) respond to changes in the pH of a sample with a change in voltage. Thermistors (Chapter 21) respond to temperature changes with a change in resistance in a wire. Table 18.1 on p. 326 lists detectors that are described in later chapters in this Unit.

Because the detector generates the signal in a measuring instrument, its capabilities limit the instrument's overall performance. When evaluating detectors it is common to talk about their **detection limit, sensitivity,** and **range.** (These terms are used not only to refer to detectors, but also to describe methods, as is discussed in Chapter 24. Note the context when these terms are used in order to understand the author's intention.)

Table 18.1. EXAMPLES OF DETECTORS

Sample Property That Is Measured	Transducer	Electrical Output
Concentration of H^+ in a Solution	pH Electrode	Voltage
Temperature	Thermistor	Resistance
Temperature	Thermocouple	Voltage
Light intensity	Photomultiplier Tube	Current

The **detection limit** *of a detector is the minimum level of the material or property of interest that causes a detectable signal.* The detection limit is a useful figure when comparing different techniques for measuring an analyte. In practice, the term *detection limit* is sometimes used synonymously with the term **sensitivity,** although these two terms are different. The **sensitivity** *of a detector is its response per amount of sample.* The better the sensitivity, the lower the level of sample the instrument can reliably detect. Figure 18.16 is an advertisement that contrasts a high sensitivity and a low sensitivity detector. Only the high sensitivity detector can generate a detectable signal when trace concentrations of certain materials are present.

Although the sensitivity of a detector is important in determining its lower limit of detection, a second factor is also involved. This factor is **electrical noise.** Electrical noise may have two components. **Short-term electrical noise** *may be defined as random, rapid "spikes" arising in the electronics of an instrument,* see Figure 18.17a. **Long-term electrical noise,** or **drift,** *is a relatively long-term increase or decrease in readings due to*

changes in the instrument and the electronics, see Figure 18.17b. When there is no sample in the instrument, the detector should ideally provide a steady zero response, resulting in a flat baseline on a recorder. In practice, because of short-term noise and drift, the baseline may not be absolutely stable, see Figure 18.17c.

Electrical noise can arise spontaneously in an instrument's electrical circuitry. It can also be caused by other electrical devices, such as nearby power lines. In either event, noise is not related to the sample and can occur even if no sample is present. Electrical noise is a problem because it can interfere with our ability to see a small signal due to the sample, see Figure 18.18.

There are various methods used to express how much electrical noise is present in an instrument. **Signal-to-noise ratio** *expresses the relationship between signal and noise in an instrumental system as the instrument response due to the sample divided by the noise present in the system,* see Figure 18.19. The higher the signal to noise ratio, the better the performance of the instrument. The term **root mean square noise (RMS)** is sometimes used to indicate how much noise is present. The RMS value is based on a statistical calculation that "averages" the noise present over a period of time. The lower the value, the less noise is present in the system.

Thus, the limit of detection of a detector is affected both by its inherent sensitivity and by electrical noise present in the instrument. The **detection limit of a detector** *is therefore often defined in practice as the minimum level of sample that generates a signal at least twice the average noise level.*

The **dynamic range** *of a detector is the range of sample concentrations that can be accurately measured by the detector.* Some detectors have a narrow dynamic range, whereas others have a wide one.

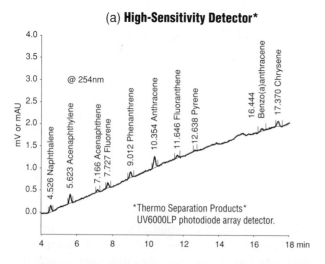

(a) High-Sensitivity Detector*

PAH chromatogram at trace levels clearly shows all compounds in the sample.

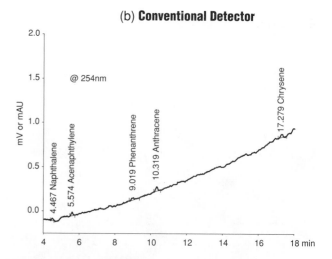

(b) Conventional Detector

PAH chromatogram at trace levels shows only some of the compounds in the sample.

Figure 18.16. Limit of Detection Is an Important Factor in Evaluating a Detector. a. A more sensitive detector generates signal, which appears here as "blips," in response to low levels of the material of interest. **b.** A less sensitive detector does not noticeably respond to low levels of the material of interest. (Data was collected with the Thermo Separation Products UV6000LP Photodiode Array Detector and a conventional photodiode array detector. Reprinted with permission.)

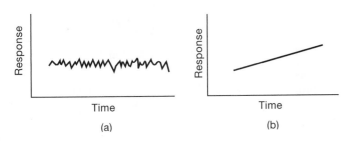

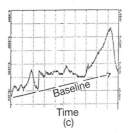

Figure 18.17.   Electrical Noise. a. Electrical noise is brief, short-term spikes that occur even when no sample is present. **b.** Drift causes changes in the baseline over time, such as hours or days. **c.** Both short-term noise and drift may be present in an instrument.

iv. SIGNAL PROCESSING UNITS

a. Overview

The electrical signal that is generated by a detector must be processed before it can be displayed by a readout device.* The processing that is required depends on the detector used and the final form of information required. For example, signal processing units may amplify the signal, count it, or convert it from one form to another, see Table 18.2. A few of these processes will be discussed in more detail in this section.

b. Amplification

An **amplifier** *boosts the voltage or current from a detector in proportion to the size of the original signal.* Amplification is one of the most important types of signal processing because the current or voltage change produced in a detector is often very small. For example, suppose the output of a detector is only 3 μA. It is possible to amplify this current by a thousand times to 3 mA. In order to amplify a current there must be a power supply, so an amplifier is associated with a power supply and with other electronic components.

   Gain *is the degree to which a signal can be increased or decreased.* The amplifier gain is the ratio of the out-

put voltage from the amplifier to the input voltage arriving at the amplifier. For example:

An input voltage of 1 mV is amplified to an output voltage of 100 mV.

$$\text{Gain} = \frac{100 \text{ mV}}{1 \text{ mV}} = 100$$

   In some cases, the operator will need to adjust the gain, or the amplification, of an instrument. If the signal is amplified too little, then the signal will be difficult to see and very small signals may be undetectable on the readout device. If the amplification is set too high, however, the signal may exceed the ability of the readout device to handle it.

   Be aware that once a detector has generated an electrical signal, the signal-to-noise ratio cannot be changed by simple electronic amplification because the amplification of the signal is accompanied by a corresponding boosting of the noise. There are, however, electronic filtering devices and also software programs that extract a

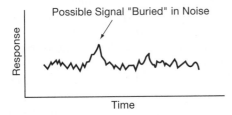

Figure 18.18.   Electrical Noise Can Interfere with Our Ability to See the Signal Arising from the Detector's Response to the Sample.

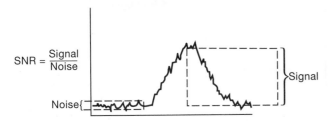

Figure 18.19.   Signal-to-Noise Ratio. The relationship between signal and noise is illustrated by dividing the peak due to the signal by the average height of the spikes due to noise.

Table 18.2.	EXAMPLES OF ELECTRICAL SIGNAL PROCESSING
Type of Transformation	**Purpose**
Amplification	Boosts voltage or current in proportion to the size of the original signal
Analog to Digital Conversion	Converts analog signal to digital signal
Filtering	Separates and removes unwanted noise from the signal generated by a detector
Attenuation	Reduces an amplified signal to make it best fit a readout device
Log to Linear Conversion	Converts a signal which has a logarithmic relationship to the property being measured to a signal with a linear relationship**
Integrator	Calculates the area under a peak (as in chromatography)
Counting	Keeps track of and counts signals from the detector

**Important in spectrophotometry (discussed in later chapters) where transmittance values are converted to absorbance values.

*A short, helpful article on how the signal from a detector is processed is: Hinshaw, John V. "Choosing the Right Detector Settings." *LC/GC* 14, no. 11 (1996): 950–6.

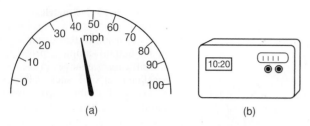

Figure 18.20. Displays of Analog and Digital Signals. **a.** A speedometer is an analog display. **b.** A digital clock radio has a digital display.

signal from noise. These devices are used to process the signal after it leaves the amplifier.

c. Analog to Digital Converter

There are two types of signal: **analog** and **digital.** A signal that can change values continuously is called **analog** *(i.e., "smoothly changing").* In contrast, a **digital signal** *represents information by a variable that can have only a limited number of discrete values.* A meter is often used to display an analog signal, see Figure 18.20a. A meter has a needle that deflects from its zero position when a current is applied. The degree of deflection is related to the electrical signal reaching the meter. Note that the needle can point to any number, and can also point to a value anywhere in between two numbers. The speedometer of a car is a familiar example of such a meter.

The display from a digital clock radio is shown in Figure 18.20b. The clock radio can only display certain values. In this illustration the time reads 10:20. The next value it will display is 10:21. This digital clock radio cannot display any time in between 10:20 and 10:21.

The signal generated by a detector in response to a property of a sample is usually analog. For example, a pH probe develops voltage in response to the H^+ concentration of a solution. This voltage might be 50 mV or 51 mV or any value in between.

Computers are digital instruments that represent all information using only two values: on and off. Each letter in the alphabet and every number processed by a computer is converted into a distinct pattern of on and off signals. A digital system is a collection of compo-

nents that can store, process, and display information in a digital form.

Digital devices, such as computers, cannot directly manipulate an analog signal, such as the signal produced by a detector. When the signal from a detector is to be displayed, stored, or manipulated by a digital device, the analog signal must be converted to a digital signal. An **analog to digital converter (A/D converter)** *is a type of signal processor that very rapidly converts an analog signal into a digital one.* There are also situations where a **digital to analog converter** converts a digitized signal back to an analog signal.

d. Attenuator

An amplified signal must sometimes be reduced in order to be best displayed by a readout device. This is common, for example, with gas chromatography instruments that are attached to strip chart recorders. An operator adjusts an **attenuator** *to reduce a signal.*

v. READOUT DEVICES

Various devices, such as meters, strip chart recorders, and computer screens, are used to display the information from the signal processing unit in a form that is interpretable to a human or to a computer. A display device may be either analog or digital, see Figure 18.21. Table 18.3 lists various types of readout devices, a few of which are discussed in this section.

Analog meters that can display the result of a measurement were mentioned above. Digital devices commonly display individual measurement values with light emitting diodes (LEDs) or with a **liquid crystal display (LCD).** LEDs produce their own light and can therefore be seen in the dark. A digital clock radio is an

Table 18.3. READOUT DEVICES

Analog	Digital
Analog meter	Printer
Strip chart recorder	Digital display
Oscilloscope	Disk (computer)
	Computer screen

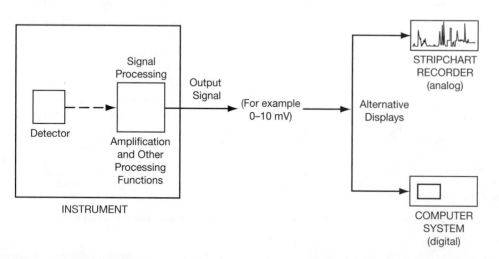

INSTRUMENT

Figure 18.21. A Chromatography System.

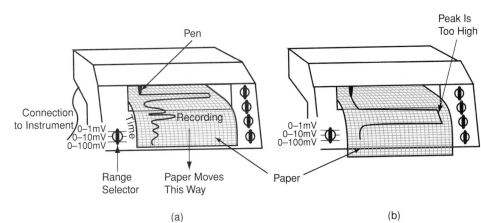

Pen

Connection to Instrument

0–1mV
0–10mV
0–100mV

Time

Recording

Range Selector

Paper Moves This Way

Paper

(a)

Peak Is Too High

0–1mV
0–10mV
0–100mV

(b)

Figure 18.22. Strip Chart Recorders. a. A strip chart recorder has a pen that responds to the signal from the detector, resulting in a continuous recording of instrument response over time. In this recording, the peaks represents biological compounds. The amount of compound present is related to the area under the peak. **b.** The strip chart recorder range should be selected so that the recording fills the paper but does not flatten out, as it does in this illustration.

example of a device with an LED display. LCDs play the same role as LEDs, but they are used where lower power consumption is required. An LCD has a thin layer of a liquid crystalline material sandwiched between two plates of glass. The sandwich normally reflects light; however, if a voltage signal is passed to an area of the LCD, that area darkens. The darkened portions form the letters and numbers of the display. LCDs are used, for example, on digital wrist watches.

Strip chart recorders *are analog display devices that record the output of a detector continuously over time.* A strip chart recorder has a pen that moves up and down over a strip of paper in response to the signal from a detector. The paper slowly moves along a track so the recorder plots the level of the electrical signal versus time. Figure 18.22a shows an example of a strip chart recording generated as a sample flowed through a detector. The detector responded to biological compounds in the sample; the higher the signal, the higher the level of compound in the detector at that moment. In this type of recording, the area under the peak is a measure of the amount of the material being analyzed.

The signal arriving at a strip chart recorder has a range of values. The operator adjusts the recorder according to that range so that the recording fills the paper but does not "run off" the paper. For example, if the lowest value arriving at the recorder is 0 mV and the highest is 80 mV, the operator might turn a switch on the recorder to select a range from 0 to 100 mV. In this case, an input voltage of 100 mV into the strip chart recorder will move the pen all the way from the left-hand edge of the paper to the right; 100 mV is said to be "full scale." If the recorder is set in this way, and a signal of 125 mV is unexpectedly produced, the pen will go "off scale," see Figure 18.22b.

Meters and strip chart recorders are both useful devices and are still sometimes used in the laboratory. Most instruments are now attached to computers or incorporate microprocessors. Computers and microprocessors can take a signal and store it for later retrieval, calculate the concentration of a component in a sample, graph the signal in various ways, make corrections, produce a report, and perform many other functions. Printers are used to display the results after they have been

processed. Because the signal has been processed by a microprocessor or computer, the X axis need not represent time, as it does with a strip chart recorder. It can be, for example, wavelength or concentration. The Y axis similarly is not the direct signal from the detector, but it can be the result of various types of processing.

Printers associated with digital instruments do not usually have range selector knobs like those on strip chart recorders. However, the operator can usually use a keyboard or similar device to "tell" the recorder what will be the lowest and what will be the highest values on the plot.

EXAMPLE

The following figure shows the back panel of a representative instrument illustrating some of the ideas discussed so far.

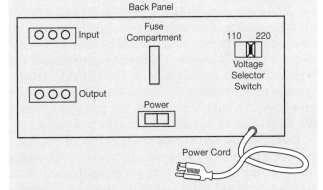

Back Panel

Input

Fuse Compartment

110   220

Voltage Selector Switch

Output

Power

Power Cord

Voltage Selector. The voltage that comes to a particular receptacle from the power company may supply a voltage of either 110 or 220. An instrument can sometimes be used with either input voltage. The voltage selector is a switch that allows the user to select the proper voltage. Note that an instrument may require a fuse to be exchanged if the selected voltage is changed.

Fuse Compartment. Instrument fuses or circuit breakers are often accessible to the user and are often located on the back panel of the instrument.

Output Connectors. Each electronic measuring instrument has some sort of output signal, such as a voltage or a current. This output can be connected via cables or cords to other devices, such as a strip chart recorder, computer, or another instrument.

Input Connector. Some instruments are able to receive electrical information (signal) from other instruments. For example, chromatography instruments may be controlled by computers, in which case connections are required from the computer to the instrument.

Power Connection. This is where the power cord attaches to the instrument.

IV. QUALITY AND SAFETY ISSUES

A. Quality Issues

It is important that laboratory instruments be properly maintained and repaired when malfunctions occur. **Preventive maintenance** *is a program of scheduled inspections of laboratory instruments and equipment that leads to minor adjustments or repairs and ensures that the instruments are functioning properly.* Preventive maintenance programs are intended to:

- ensure correct results
- identify components that require replacement
- ensure that instruments are safe to use
- ensure that instruments will not be shut down by a major problem
- lower the costs of repairs

Recall that **calibration** *is the adjustment of an instrument so that its readings are in accordance with the values of internationally accepted standards.* Every measuring instrument must be calibrated and this calibration may or may not be part of its routine maintenance. The frequency of calibration depends on the type of instrument and its use. The individuals who operate an instrument should ensure that it has been properly calibrated, although the operator may or may not be the one to actually perform the calibration.

Performance verification *is a process of checking that an instrument is performing properly.* One aspect of this process is to check that it is correctly calibrated. Note that we are distinguishing between *calibration* and *performance verification*. We use *calibration* here to mean that adjustments to the instrument are performed so that its readings are correct. *Performance verification* is used to mean that the instrument's performance is checked, but it is not adjusted. Performance verification is usually the responsibility of the individuals who use an instrument.

Instruments can also be **validated**. **Instrument validation** *is a comprehensive set of tests done before an instrument is put in service that demonstrate that it will work and the conditions under which it will function properly.* Instrument validation is a formal process that is required in regulated laboratories and in laboratories that meet certain voluntary standards.

Table 18.4. ENVIRONMENTAL FACTORS THAT COMMONLY AFFECT THE PERFORMANCE OF ELECTRONIC INSTRUMENTS

1. *Temperature.* Changes in ambient temperature often affect instruments; therefore, avoid placing instruments in areas that are subject to direct sunlight or drafts, such as near a window or an open doorway.
2. *Contaminants.* Most instruments are sensitive to smoke, acid fumes, solvent fumes, dust, and dirt. Liquid spills on the top of instruments can also cause damage. Avoid environmental contaminants.
3. *Ventilation.* Most instruments have exhaust vents to dissipate heat generated by the instrument. Do not block the ventilation holes and allow an adequate area around instruments to dissipate heat.
4. *Humidity.* The humidity in the room where instruments are used should ideally be about 40–60%.

An important part of a quality-control program in a laboratory is that all operations performed with an instrument be logged. It is common to have a logbook associated with each instrument in which performance verification checks are recorded, preventive maintenance and repairs are noted, and routine operations are logged.

Table 18.4 lists some environmental factors that may affect the performance of most types of electronic instruments. These factors should be controlled whenever possible in the laboratory.

B. Electrical Safety

The fluids in a person's body can serve as a conducting pathway for current flow. Therefore, if a person touches two objects that differ in electrical potential, current can flow through the person—this is electrical shock. Electrical shock can result in severe internal and external burns, paralysis, muscle contractions leading to falls, and, in extreme cases, to death. The more current that flows through a person's body, the more dangerous. As little as 1 mA is detectable, 15 mA can cause muscles to freeze, and about 75 mA is fatal.

Human skin is an excellent insulator when it is dry. The resistance of dry skin is in the megohm range. Wet skin, unfortunately, offers far less resistance to current flow. The resistance of wet skin is about 300 ohms. If a person with wet skin is exposed to a voltage of 120 V, the current passing through that person can be as high as 0.400 A (400 mA). Take precautions, therefore, to avoid having current pass through your body. Table 18.5 lists common (and common sense) safety rules.

Table 18.5.   ELECTRICAL SAFETY RULES

1. *Use three-pronged power cords and receptacles to ensure proper grounding.* Do not attempt to bypass grounding provided on instruments.

2. *Avoid extension cords if possible.* When required, use only heavy duty, grounded extension cords.

3. *Use properly insulated wires and connections.* Frayed wires or connectors should be repaired by a qualified individual.

4. *Do not handle connections with wet hands or while standing on a wet floor.*

5. *Avoid operating instruments that are placed on metal surfaces, such as metal carts.*

6. *Do not operate wet electrical equipment (unless it is intended to be used in this way) or devices with chemicals spilled on them.*

7. *Be certain that electronic equipment has adequate ventilation to avoid overheating.* Many instruments have fans. Be sure there is enough space around the vents for heat to dissipate.

8. *Never attempt to troubleshoot or repair the electronics of an instrument unless you are trained to do so.*

9. *Never touch the outside of a leaking electrophoresis gel box, or one with a puddle under it, if it is plugged into a power supply.* An electrophoresis gel box normally is closed when the box is connected to a power supply, thus protecting the user from current. If the box is cracked or broken, however, it can leak buffer and therefore also "leak" current.

10. *In case of an electrical fire:*
    a. *If an instrument is smoking or has a burning odor, turn it off and unplug it immediately.*
    b. *Use only a carbon dioxide–type fire extinguisher.* Water and foams conduct electricity and so increase the hazard.
    c. *Some electrical components (specifically selenium rectifiers) emit poisonous fumes when burning.* Avoid the fumes of burning instruments.

Table 18.6.   SIMPLE CHECKS OF AN INSTRUMENT THAT APPEARS TO BE MALFUNCTIONING

1. *Is the instrument plugged in?*

2. *Is the power on?*

3. *If there is a power strip, is it turned on?*

4. *Is the wall outlet functioning? An outlet can be checked by plugging a lamp into it.*

5. *Is the power cord frayed? If so, have it replaced.*

6. *Is there a blown fuse or tripped circuit breaker?*

7. *Is there an accessible bulb that needs replacement?*

8. *Does the instrument have a "reset" button? If so, try pressing it.*

9. *If there is a switch to select either 110 or 220 V, is it properly set?*

10. *It sometimes helps to unplug an instrument and plug it back in.*

11. *Check the instrument manual. There may be a series of troubleshooting steps that the user can easily and safely perform.*

12. *Call the instrument manufacturer and ask their advice.*

Most people who operate laboratory instruments are not specially trained in instrument repair and should never attempt to repair a broken instrument, other than to replace fuses, worn-out bulbs, or other simple tasks as directed by the manufacturer. Capacitors can hold a charge for weeks, so there is the risk of shock from an instrument, even when it is unplugged. Only a trained technician should bleed off the current from a capacitor. An untrained individual seeking to repair an electronic instrument can not only injure her/himself but can also damage an instrument. Table 18.6 lists simple things that can be safely checked by a user when an instrument appears to be malfunctioning.

PRACTICE PROBLEMS

1. In each of the following devices what is the voltage source? What is the resistance or load?

a. Flashlight   **b.** Washing Machine

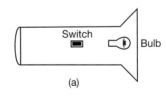

(a)

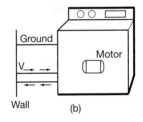

(b)

c. An Electroporator

Switches Control Charging and Discharging of Capacitor

Power Supply — Capacitor — Chamber with Cells Suspended in Buffer

(c)

2. Most household appliances in the United States are designed so that they remain in a "standby" mode in which they continue to draw power, even when they are supposedly turned off. Manufacturers make appliances this way either because it is less expensive or because the device has a feature, such as a memory of previous settings, that requires constant power. This power consumption by devices that are not in use is responsible for billions of dollars of electricity consumption at a cost both to the consumer and to the environment.

Suppose a home has the following devices that are constantly drawing power, 24 hours per day, 7 days a week. (For simplicity, ignore the time that the devices are actually in use and draw more power than is listed.)

Assume that power to this house costs $0.09 per kilowatt-hour. (A kilowatt-hour is equivalent to 1000 W of power used for 1 hour.) How much would the owners of this home pay per year for standby current? How much would a city with 50,000 similar homes pay for standby current? (Information from Janet, Raloff. "Must We Pull the Plug? New Programs Aim to Cut the Juice Drawn by Leaky Appliances." *Science News* 152 [1997]: 226, 227.)

Device	Power Drain While in "Standby"
Telephone Answering Machine	3.2 W
Cordless Phone	2.3 W
Portable Stereo	2.2 W
Color TV/VCR	9.1 W
Compact Audio Unit	9.0 W
Electric Toothbrush	2.2 W
Doorbell	2.2 W
Microwave Oven	3.1 W

3. If the current in a circuit is 12 mA and the voltage is 120 mV, what is the resistance?

4. How many watts of power can be supplied by a 120 V circuit that is able to carry 20 A?

5. A flashlight is supplied with 6 V by batteries. When it is turned on, 400 mA flows through the bulb. What is the resistance of the bulb?

6. Suppose a power supply used in electrophoresis can be set at anywhere from 0 to 500 V and from 0 to 400 mA. What is the maximum amount of power that might be used by this power supply?

7. A professor wants to outfit a teaching laboratory with 15 identical electrophoresis units arranged one after the other on a laboratory bench. Each unit is powered by a power supply like the one described in #6. Suppose that the circuit into which the electrophoresis units will be plugged is served by a 120 V line with a current of 20 A. If students use all 15 devices at the same time, is it possible that the load on the circuit will exceed its power rating?

8. An instrument is rated at 500 W. Assuming that the voltage is 120 V, what is the current in this instrument? What is the resistance of this instrument?

QUESTION FOR DISCUSSION

If you work in a laboratory, list the types of electronic transducers found in the laboratory. List the types of readout devices in the laboratory.

CHAPTER 19

The Measurement of Weight

I. BASIC PRINCIPLES OF WEIGHT MEASUREMENT

Determination of the **weight** of an object is among the most basic measurements made in the laboratory. **Weight** *is the force of gravity on an object.* **Balances** *are instruments used to measure this force.*

The term *weight* is commonly used interchangeably with *mass*, although these words are not synonyms. **Mass** *is the amount of matter in an object expressed in units of grams.* An object's mass does not change when it is moved to a new location, but its weight will change if the force of gravity differs at the new site. For example, an astronaut in space is "weightless" because there is no gravity, but the astronaut's mass does not change with blast-off from earth, see Figure 19.1 on p. 334. The significance of the distinction between mass and weight is explored in Section IV of this chapter.

The process of weighing an object basically involves comparing the pull of gravity on the object with the pull of gravity on standard(s) of established mass. The weighing method illustrated in Figure 19.2a on p. 334 dates back to antiquity, see Figure 19.2b. The object to be weighed and the standard(s) are each placed on a pan hanging from opposite ends of a **lever** or **beam.** When the beam is exactly balanced, gravity is pulling equally on the sample and the standard. They are the same weight. Thus, instruments that weigh objects in the laboratory have come to be called *balances.* *Balances that consist of one beam with two pans are called* **single beam, double pan balances.**

Mechanical balances that employ the principle of balancing objects across a beam are still used in the laboratory. However, modern balances are often electronic and use a somewhat different method of comparison. With electronic balances, the comparison between the sample and the standard is made sequentially. The balance is calibrated to a mass standard at one time and the sample is weighed on the balance afterward.

---

*The word *balance* is usually reserved for laboratory instruments, whereas the common term *scale* is preferred for the less-sensitive weighing instruments used in business, transportation, health care, and the home.

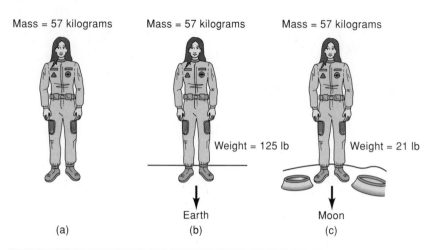

Mass = 57 kilograms   Mass = 57 kilograms   Mass = 57 kilograms

Weight = 125 lb   Weight = 21 lb

Earth   Moon

(a)   (b)   (c)

Figure 19.1. Mass versus Weight. a. An astronaut is composed of matter. Her mass is a measure of that matter. **b.** The weight of the astronaut is a measure of the effect of gravity on her. **c.** On the moon, the astronaut weighs less than on earth because of the moon's weaker gravity, but her mass is unchanged.

II. CHARACTERISTICS AND TYPES OF BALANCES

A. Range, Capacity, and Sensitivity

There are many types and brands of balances with various designs and features. The fundamental characteristics of a balance, however, are its *range, capacity,* and *sensitivity (or readability):*

1. ***Range and Capacity.*** Some laboratory balances are intended to weigh heavier objects; others, lighter ones. The **range** *of a balance is the span from the lightest to the heaviest weight the balance is able to measure.* **Capacity** *is the heaviest sample that the balance can weigh.*

2. ***Sensitivity and Readability.*** **Balance sensitivity** *may be explained as the smallest value of weight that will cause a change in the response of the balance.* Sensitivity determines the number of places to the right of the decimal point that the balance can read accurately and reproducibly, see Figure 19.3. Extremely sensitive laboratory balances can weigh samples accurately to the nearest 0.1 μg (or 0.0000001 g). A less sensitive laboratory balance might read weights to the nearest 0.1 g. In catalog specifications manufacturers express the sensitivity of their balances by their **readability. Readability** *is the value of the smallest unit of weight that can be read; it is the smallest division of the scale dial or the digital readout.*

Range, capacity, and sensitivity are interrelated. A very sensitive balance will not be able to weigh samples in the kilogram range, but a balance intended for heavier samples will not detect a weight change of a microgram.

Analytical balances *are designed to optimize sensitivity and can weigh samples to at least the nearest tenth of a milligram (0.0001 g). The most sensitive analytical balances,* **microbalances** *and* **ultramicrobalances,** *can weigh samples to the nearest microgram (0.000001 g) and the nearest tenth of a microgram (0.0000001 g), respectively.*

Figure 19.4 shows catalog descriptions of various balances. The descriptions show the readability, range, and capacity of each balance. The descriptions also show the repeatability of the instruments and their linearity (which will be discussed later in this chapter).

Most biology laboratories have more than one type of balance. Which balance should be used depends on the weight of the sample, what sensitivity is required, and other characteristics of the sample. For example, it is not reasonable to weigh a person on a balance that has a readability of 0.0001 g, nor is it sensible to use an analytical balance for a laboratory sample that weighs several hundred grams. An analytical balance should be chosen, for example, to weigh milligram amounts of enzyme to use in a reaction.

B. Mechanical Balances

Balances can be broadly classified as either *mechanical* or *electronic.* **Laboratory mechanical balances** *typically have one or more beams; the object to be weighed is placed on a pan attached to a beam and is balanced against standards of known weight.* Mechanical balances do not generate an electronic signal when a

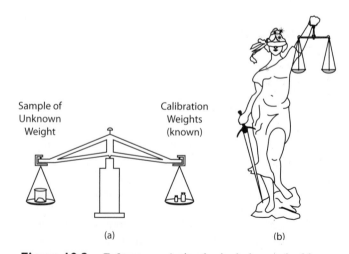

Sample of Unknown Weight   Calibration Weights (known)

(a)   (b)

Figure 19.2. Balances. a. A simple single-beam, double pan balance where a sample on the left pan is balanced by objects of established weight on the right pan. **b.** A double pan balance has been used to symbolize justice since ancient times. In Greek mythology Themis is the goddess of justice. She is usually represented wearing a crown of stars and holding a balance in her hand. (Double pan balance sketch by Sandra Bayna, a biotechnology student.)

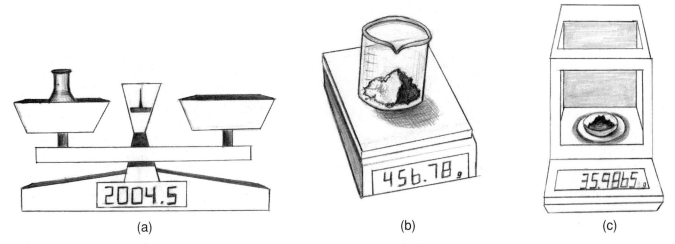

    (a)   (b)   (c)

Figure 19.3. Balance Sensitivity, Capacity, and Range. The balances illustrated in a–c are increasingly sensitive but their range becomes more limited and their capacity is lower as their sensitivity improves. **a.** A balance intended for heavier samples; readability is "0.1 g," range is "0–2200 g." **b.** Readability is "0.01 g" and range is "0–500 g." **c.** Analytical balance. Readability is "0.1 mg," range is "0–60 g." (Sketches by Sandra Bayna, a biotechnology student.)

measurement is made. Although electronic balances have replaced mechanical ones in many settings, manufacturers still produce mechanical balances that are not analytical because they are relatively inexpensive and easy to use. Examples are shown in Figure 19.5 on p. 336.

The analytical laboratory balances manufactured today are electronic; however, older mechanical analytical balances are still found in some laboratories. These mechanical instruments give accurate and reproducible measurements if they are properly maintained and operated, so they will be discussed briefly.

Consider the mechanical analytical balance diagramed in Figure 19.6* on p. 336:

1. A fine knife-shaped edge supports the beam, allowing it to swing freely. There is also a knife edge between the weighing pan and the beam. The beam is balanced on the right side by the **counterweight,** and on the left side by objects of known weight that are associated with the weighing pan. The beam is exactly balanced when the weighing pan is empty.

2. When a sample is added to the weighing pan, the beam is no longer balanced. By turning the appro-

priate knobs, the operator removes weights from the left side to bring the beam back to its resting, balanced position.

3. The amount of weight that must be removed to bring the beam into balance equals the weight of the sample.

4. The beam may remain slightly displaced from its original position by an amount too small to be balanced by the lightest weights. This slight residual displacement is shown on a scale and an image of the scale is projected to the user, hence, the scale is termed an **optical scale.** Thus, the weight of the sample is equal to the total weights removed from the left side plus the amount shown on the optical scale.

Mechanical analytical balances require care to operate. The knife edges and weights are easily dislocated if the balance is jostled. There is usually an "arrested" position and a "half-arrested" position in which the beam is stabilized and cannot swing freely. The balance should be in an "arrested" or "half-arrested" position when the balance is not in use, when removing or returning weights to the beam, and when moving a sample on or off the weighing pan. The balance is fully

CATALOG DESCRIPTIONS OF ANALYTICAL BALANCES				
MODEL NUMBER	CAPACITY (g)	READABILITY	REPEATABILITY (standard deviation)	LINEARITY
BA 50	50	0.1 mg	± 0.1 mg	± 0.2 mg
BA 300	300	0.1 mg	± 0.2 mg	± 0.5 mg
BA 4000	4000	0.1 g	± 0.1 g	± 0.1 g

Figure 19.4. Catalog Specifications for Balances. Catalogs specify the capacity, readability, repeatablity, and linearity for balances.

*The balances illustrated in Figures 19.2a and 19.6 are both mechanical balances but their design is different. Newer mechanical analytical balances usually have just one weighing pan and the beam does not extend equally on both sides of the support. When a sample is added to the weighing pan of the balance illustrated in Figure 19.6, standard weights are *removed* to compensate rather than being added to a second pan. Because the standard weights are removed, this type of mechanical balance is called a *substitution balance.*

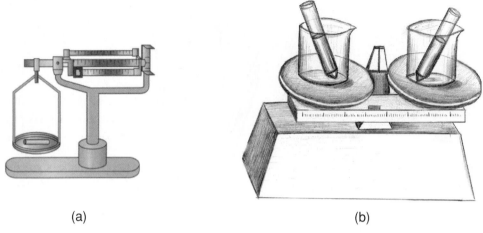

(a)   (b)

Figure 19.5. Nonanalytical Mechanical Balances. a. Triple-beam hanging-pan balance. This balance has three beams. External standard weights are not needed because the weights (sometimes called "riders") slide to the right on the beams to balance the sample. This balance can be read to the nearest 0.01 g, and one additional place may be estimated. **b.** In centrifugation, the tubes and their contents that are opposite one another in the centrifuge must be of equal weight. A single-beam, double-pan balance can be used to compare the weights of centrifuge tubes easily. (Sketch b by Sandra Bayna, a biotechnology student.)

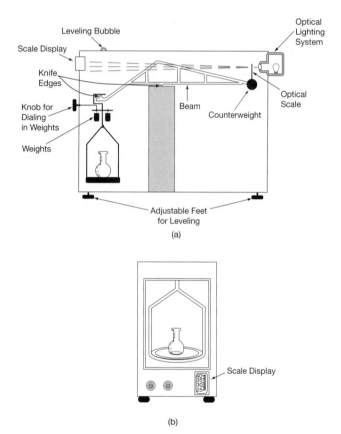

Figure 19.6. A Mechanical Analytical Balance. a. Side view. The pan and the lever both rest on knife edges. A sample placed on the pan displaces the pan from its original balanced position. The operator turns knobs that remove weights until the beam is returned almost to its original position. The residual beam displacement after the weights are removed is shown on a scale, and an image of the scale is projected to the operator. The weight of the sample is equal to the weight removed from the left side plus a small amount shown on the optical scale. **b.** Front view showing knobs and optical scale.

released when a final measurement is being read. A mechanical balance must be perfectly level so that the beam is level. There is often a **leveling bubble** *to show whether the balance is level* and "feet" that can be adjusted up and down if necessary to even the balance, see Figure 19.7.

C. Electronic Balances

Electronic balances do not have beams and use an electromagnetic force rather than internal weights to counterbalance the sample. An electronic balance produces an electrical signal when a sample is placed on the weighing pan, the magnitude of which is related to the sample's weight. The way an electronic balance works is:

1. The weighing pan is depressed by a small amount when an object is placed on it.
2. The balance has a detector that senses the depression of the pan.
3. An electromagnetic force is generated to restore the pan to its original ("null") position.* The amount of electromagnetic force required to counterbalance the object and move the pan back to its original position is proportional to the weight of the object.
4. This electromagnetic force is measured as an electrical signal that is in turn converted to a digital display of weight value. In order to convert an electrical signal to a weight value, the balance ultimately compares the electrical signal of the unknown sample to the signal of standard(s) of known weight.

Electronic balances are easy to use, automatically perform many functions, can be interfaced with

*This type of balance is sometimes called a *null position* balance because the pan is restored to its original position in order to measure the sample's weight.

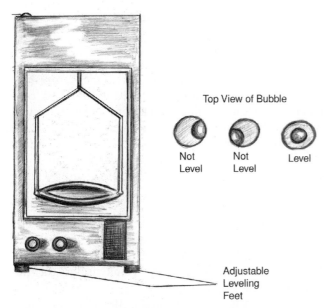

Figure 19.7. Leveling a Balance. The balance is level when the bubble is centered. For example, if the bubble is off-center to the left, then recenter it by raising the right side of the balance (or lowering the left). (Sketch by Sandra Bayna, a biotechnology student.)

computers and other instruments, permit automatic documentation, and are better able to compensate for environmental factors than mechanical balances. For these reasons, electronic balances are in widespread use.

Every model and brand of balance is operated differently; however, there are certain general processes that are required when using most balances. Box 19.1 outlines a generally applicable procedure for operation of an electronic balance.

III. FACTORS THAT AFFECT THE QUALITY OF WEIGHT MEASUREMENTS

A. Environmental Factors and the Operator's Technique

This section discusses factors that affect the accuracy and precision of weight measurements. These consider-

ations are most relevant when working with analytical balances, although they may be important in other situations as well.

Recall that accuracy relates to the correctness of measurements. Precision relates to the reproducibility of a series of measurements. You might assume that the accuracy and precision achievable with a balance are dependent on the quality designed and built into that balance. A balance must be properly manufactured to give accurate, reproducible results. In normal laboratory practice, however, the accuracy and reproducibility of weight measurements are greatly affected by the conditions in that laboratory, by the user's technique, and by the routine maintenance of the balance. For this reason, manufacturers generally do not specify the accuracy of balances in their catalog.

Mechanical balances, and to a lesser extent electronic balances, are affected by environmental conditions, including vibration, drafts, and temperature changes. Individuals who are accustomed to mechanical balances are often aware of the factors that interfere with obtaining accurate, reproducible measurements. In contrast, because electronic balances seem so easy to operate, users of electronic balances may be unaware of potential problems.

Temperature can have a significant effect on weight measurements. If a sample and its surroundings are at different temperatures, then air currents are generated that cause the sample to appear heavier or lighter than it really is. This problem may appear as a weight reading that slowly increases or decreases. Microbalances and ultramicrobalances have thick enclosures to protect them from thermal disturbances outside the balance.

Balances are also affected by the heat from their own electrical components. When a balance is first turned on it goes through a "warming up" period when its temperature is increasing and its readings are unstable. Many balances are therefore left on in a permanent standby mode to avoid delay while the balance warms up.

Temperature not only affects the apparent weight of a sample, it also affects the sample's actual weight. This is because there is always a thin film of moisture around

---

BOX 19.1. *GENERAL PROCEDURE FOR WEIGHING A SAMPLE WITH AN ELECTRONIC ANALYTICAL BALANCE*

1. *Make sure the balance is level.*

2. *Adjust the balance to zero.* Make certain the weighing pan is clean and empty and the chamber doors (if present) are closed to avoid air currents and dust.

3. *Tare the weighing container or weigh the empty container.* Most balances can be **tared,** that is, they *can be set to subtract the container weight from the total weight automatically.* If a balance does not have the tare feature, then it is necessary for the operator to weigh the empty container and subtract its weight from the total weight of the sample plus container.

4. *Place the sample on the weighing pan and read the value for the measurement.*

5. *Remove the sample; clean the balance and area around it.*

objects, which is thicker when the object is colder and thinner when it is warm. Samples are therefore heavier when they are colder. (You can test this phenomenon. Weigh a flask on an analytical balance, then hold it in your hand for one minute and weigh it again. The flask should weigh less the second time. Perspiration and oils from your hands are of minor importance; otherwise, the sample would weigh more the second time.) Even the best electronic balance cannot compensate for this temperature effect because it actually affects the weight of the object.

Plastic and glass weighing containers and samples that are powders or granules are prone to develop a static electrical charge. Static charge occurs when different materials rub against each other. One material acquires an excess of electrons, resulting in a negative charge while the other loses electrons and acquires a positive charge. Nonconducting materials such as glass and plastic retain this static charge. If a sample or weighing vessel and its surroundings have the same charge, they repel each other; if they have opposite charges, they attract one another.

When a sample is charged with static electricity, a field is built up between the sample and the nonmoving parts of the balance. The force of this field can simulate a change in weight that may extend into the gram range. You can recognize this effect by a weight display that rapidly fluctuates. In extreme cases, a charged sample can literally fly out of its container and adhere to the balance chamber walls. Examples of commercial products to reduce static charge are illustrated in Figure 19.8.

Temperature and static charge are only two of the factors that a careful operator must consider when making weight measurements. A number of points relating to good weighing practices are summarized in Table 19.1.

B. Calibration and Maintenance of a Balance

i. OVERVIEW

In order to give accurate readings a balance must be properly calibrated and maintained. Calibration of a balance brings its readings into accordance with the values of internationally accepted standards.

Although balances are calibrated by the manufacturer when they are produced, calibration must be periodically checked in the laboratory of the user. This is because time and use (and abuse) of balances can affect their response. Calibration must also be checked when a balance is moved from one location to another. For measurements in the microgram range, the balance must be recalibrated each time the weather changes. Recalibration of the balance compensates for the effects of location and weather.

Calibration is not performed in the same way for a mechanical and an electronic analytical balance. Inside a mechanical analytical balance there are a number of standard weights, each of which must be of the proper mass and in the proper orientation to ensure accurate weighing results. It is complex to calibrate a mechanical balance with so many internal weights and where all adjustments are made mechanically. A trained technician, therefore, is required to adjust a mechanical balance and to perform routine maintenance. (ASTM Standard E 319–85 describes in detail how to check a mechanical balance.)

In contrast to mechanical balances, it is usually straightforward to calibrate an electronic balance using a mass standard. Most manufacturers include directions for doing so in the instruction manual. Some electronic balances contain internal mass standards; others require that a standard be placed on the weighing pan during calibration. A

(a)

(b)

Figure 19.8. Examples of Commercial Products to Reduce the Effects of Static Charge. a. Ionizing blower "shoots out" ions that neutralize static charge on the item to be weighed. (Photo courtesy of Sartorius Corp.) **b.** Antistatic brush removes electrostatic charge when balance cases and pans are brushed. The brush contains a small amount of radioactive polonium that creates an ionized atmosphere that neutralizes ions when surfaces are brushed. (Photo courtesy of Fisher Scientific, Pittsburgh, PA.).

Table 19.1.   FACTORS THAT AFFECT WEIGHT MEASUREMENTS AND GOOD WEIGHING PRACTICES

1. *Analytical balances, especially mechanical ones, must be level.* Check that analytical balances are level before use.

2. *Vibration will affect balance readings, particularly with mechanical analytical balances.* Mechanical and microanalytical balances are usually placed on special tables designed to prevent vibration. Locate balances away from equipment that vibrates, such as refrigerators and pumps.

3. *Drafts will affect weight measurements.* Do not place balances near fans, doors, or windows. Analytical balances have shielded weighing compartments to eliminate drafts. Keep the weighing compartment doors closed when taking a reading. Whenever possible, situate analytical balances in locations with little activity.

4. *Balances, especially mechanical ones, should not be jostled.* Avoid leaning on the table holding the balance. Avoid moving a balance unless it is stabilized. (Consult the owner's manual for information on how to stabilize the balance.)

5. *Temperature changes will affect the actual and apparent weight of a sample.*
    a. Keep the sample, balance, and surroundings at the same temperature.
    b. Avoid touching samples and containers and placing your hands in the weighing chamber when using an analytical balance. Wearing gloves when handling samples is helpful (and is common practice), but using tongs provides better protection from temperature effects.
    c. Do not locate balances near windows, radiators, air conditioners, or equipment that produces heat.
    d. Allow a balance time to warm up before use or leave it in stand-by mode to avoid warm-up time.

6. *Static charge will affect measurements.* Various sources recommend the following procedures to help reduce static charge.
    a. A variety of products are available commercially to help reduce static, see Figure 19.8. Also new designs of weighing pans help to remove static from samples.
    b. Charging usually occurs when the humidity is low. It is ideal to maintain the humidity between 40 and 60% in a weighing room.
    c. Metal does not tend to become electrostatically charged because it is conductive and charge readily dissipates from it. Metal weighing containers may therefore be the best choice and plastic containers the worst if electrostatic charge is a problem.
    d. Cleaning the walls of the balance chamber with alcohol, or a product intended for cleaning glass, may reduce charging.
    e. A slightly damp clothes softener sheet taped to the inside of the weighing chamber may be helpful.
    f. Balances should not be placed on tables with glass or plastic tops that can build up charge.
    g. Avoid wearing woolen or nylon clothing when weighing samples.

7. *Some samples gain moisture from the air and others lose volatile components to the air.* The weight of such samples will slowly increase or decrease during weighing.
    a. Work quickly but carefully to minimize such changes in the sample's composition.
    b. Maintain the humidity in a weighing room between 40 and 60% if possible because the amount of moisture loss or gain by a sample may be affected by the room's humidity.

8. *Samples that are magnetic may perturb the electromagnetic coil in an electronic balance, leading to incorrect results.* A magnetic sample will appear to have a different weight depending on its position on the weighing pan. It is helpful to distance magnetic samples from the weighing pan by placing a support under the sample.

9. *Electronic balances may give incorrect readings if the load is placed off-center.* Place the sample near the center of the pan.

10. *It is possible to damage electronic balances by placing a load that is too heavy on the weighing pan or by dropping items onto the pan.* Newer balances have features to protect the balance from overload, but it is still good practice not to exceed the capacity of any balance.

11. *Selection of the weighing container may affect the results.* Samples can be weighed in a wide variety of containers including beakers, glassine coated **weighing paper,** and *plastic or metal containers designed for weighing,* called **weigh boats.** If a sample is likely to absorb or lose moisture, or contains volatile components, use a weighing vessel that is capped or has a narrow opening. Weigh samples in the smallest possible container. Be alert to sample particles that may adhere to the weighing vessel when poured out.

12. *Do not touch chemicals being weighed, magnetic stir bars, or the insides of beakers with your fingers.*

13. *Do not return unused chemicals to their storage bottles.*

14. *Balances and weighing rooms should be clean.* Sloppiness can have detrimental effects. Some chemicals are toxic, chemicals left on balances can cause corrosion and damage, and materials from one person's work can contaminate the work of someone else. Remember that balances are expensive instruments and should be well maintained.

BOX 19.2.   *TWO-POINT CALIBRATION OF AN ELECTRONIC BALANCE*

1. ***The balance is set to zero using the appropriate knob, dial, or button when the weighing pan is clean and empty.***
   It is simple to "zero" both electronic and mechanical balances, and they should be "zeroed" every time they are used.

2. ***The second calibration point is usually at the heavier end of the balance's range.*** With an electronic balance, a standard of established mass may be placed on the weighing pan and the balance set to the value of the standard using a lever, button, keypad, or other method of adjustment. Some electronic balances have a button that when pressed causes a calibration standard inside the balance to be weighed and the balance to adjust itself. In fact, there are microprocessor-controlled balances that recalibrate automatically to compensate for environmental changes without any operator intervention.

---

two-point calibration procedure for electronic balances is outlined in Box 19.2 and illustrated in Figure 19.9.

ii. STANDARDS

The accuracy of any weight determination is limited by the accuracy of the standards used for comparison. Weight (mass) standards* are metal objects whose masses are known (within the limits of uncertainty) relative to international standards.

Standards used for calibrating balances can be purchased with a certificate showing their traceability to NIST. This certificate documents that the standard's mass went through a series of comparisons that ultimately links it to the international prototype in France. The checking of standards against one another requires stringently applied methods and specific environmental conditions.

In the United States, NIST houses official mass standards in Gaithersburg, MD. U.S. facilities that manufacture weight standards may send their standards to NIST to be checked against the U.S. standards. A standard that was tested by NIST is sometimes called "directly traceable to NIST" in the catalogs.

Standards for use in weighing can be purchased in different conformations of varying construction and with differing tolerances. The user determines the type of standards required depending on the balance being used and the requirements of the application. In the United States, the specifications for labeling and classification of standards for laboratory balances have been established by ASTM. (These specifications are reported in ASTM Standard E 617, "The Standard Specification for Laboratory Weights and Precision Mass Standards.")

The finest standards, those with the smallest tolerance, are **Class 1**. A "500 g" Class 1 standard, for example, must have a mass within 1.2 mg of 500.0000 g (i.e., it must be between 500.0012 g and 499.9988 g), whereas a **Class 4** standard must be within 10 mg of 500.0000 g. In routine biology laboratory work it is common to use **Class 2, 3, or 4** standard weights. Class 4 is recommended for student use. The calibration standards that manufacturers provide with electronic analytical balances are usually Class 2.

Standard weights should be handled with tongs because they are damaged by skin oils and by cleaners

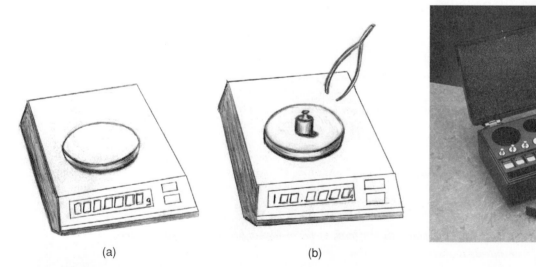

(a)   (b)   (c)

Figure 19.9.   Two-Point Calibration of a Balance. a. The balance is set to zero when the weighing pan is clean and empty. **b.** A calibration standard is placed on the pan. The balance is set to the calibration weight. (Sketches by Sandra Bayna, a biotechnology student.) **c.** Calibration standards. (Photo courtesy of Fisher Scientific, Pittsburgh, PA.)

---

*A note about terminology: Standards used for weighing are often called "weight standards" because they are used during the process of weighing samples. Recall, however, that the weight of any object varies with its location, but its mass is constant. It is the *mass* of the standards that is established by traceability to NIST standards. This is a situation in which the terms *mass* and *weight* tend to be used interchangeably.

that remove such oils. To protect their standards, a laboratory might keep one set for routine use and a second set in storage. Every 6 months or so, the standards for routine use can be checked against the set in storage. Standard weights need to be recertified periodically because even well-maintained weights change over time due to scratches, wear, and corrosion.

Information regarding the classification and features of mass standards is excerpted in the Appendix to this chapter.

C. Quality Programs and Balances

i. Overview

To ensure that balances perform properly, they must be operated correctly. They also must be regularly calibrated, cleaned, and repaired as required. The performance of a balance should be regularly checked to verify and document that it is giving accurate and precise readings. Laboratories that meet the requirements of a quality system, such as ISO 9000 or cGMP, have procedures that detail how to operate each balance and how to maintain, calibrate, and check their performance. Even if a laboratory does not conform to the requirements of a particular quality system, it is still good practice to perform these tasks on a regular basis.

The frequency of balance performance verification and calibration varies among laboratories. In some laboratories the weights of standards are checked each time the balance is used and the values are recorded. In other laboratories, this check is performed less frequently. Some laboratories have the policy that electronic analytical balances are calibrated every time they are used; in other laboratories, calibration is less frequent. Balances might similarly be checked at prescribed intervals for linearity, repeatability, and other qualities of importance for that balance.

Maintenance and quality checks should be documented each time they are performed. Newer balances may have a feature that enables them to automatically record the date, time, and results of calibration for documentation purposes. Consistently checking and recording the weights of standards documents that measurements were made on a properly functioning instrument and alerts the operator to malfunctions that do occur. Whatever process is used in a laboratory to ensure the quality of balance measurements, it is good practice to have a formal, written policy and standard operating procedures that detail the process.

ii. Verifying Balance Accuracy, Precision, and Linearity

Performance verification of a balance involves testing its accuracy, precision, and linearity. *Accuracy* is tested simply by weighing one or more mass standards. The resulting weight must be correct within a tolerance established by a quality-control procedure. This test is simple enough that it can be performed each time a balance is used.

Precision is an important aspect of the quality of a balance's performance. Precision (repeatability) is measured by weighing a sample multiple times and calculating the standard deviation. Manufacturers test the repeatability of their balances under ideal conditions in their testing laboratory and report the results in their balance specifications (Figure 19.4 on p. 335). To determine whether a balance is performing adequately, its repeatability in the laboratory can be compared with the manufacturer's specifications.

Linearity error *occurs when a balance is properly calibrated at zero and full-scale (the top of its range), but the values obtained for weights in the middle of the scale are not exactly correct.* Linearity can be tested by weighing individual subsets of objects and comparing the sum of the subsets to the weight of the objects all together, as shown in Box 19.3 on p. 342. If a balance is discovered to have linearity error, it is necessary to have it repaired professionally. Manufacturers report the linearity error of their balances in the balance specifications, and this figure is an indication of the quality of the instrument.

EXAMPLE PROBLEM

Suppose a technician is verifying the performance of a laboratory balance. The technician weighs a 10 g standard six times.

Weights

10.002 g	10.002 g
10.001 g	9.998 g
9.999 g	10.002 g

a. What is the standard deviation for the balance? (Remember the units.)

b. The balance specifications require that its precision (as measured by the standard deviation) be better than or equal to 0.001 g. Does it meet these specifications? If not, what should be done?

ANSWER

The standard deviation rounds to 0.002 g, which does not meet the specification. There may be a problem with the balance, although it is possible that the operator is not using good weighing practices. For example, the operator might forget to close the doors of the balance chamber or might place the standard in different locations on the weighing pan. The balance should be taken out of use and the cause of the imprecision should be investigated.

---

iii. Operating Balances

Modern electronic balances are easy to operate and so it may seem odd that laboratories have standard operating procedures (SOPs) to detail their operation. SOPs, however, can play a valuable role in ensuring that operators use balances properly, that the performance of each balance is checked regularly, and that these checks are documented. They also remind operators of simple problems that can lead to systematic error. For example,

---

Box 19.3.   *CHECKING THE LINEARITY OF A BALANCE*

Notes

a. Linearity tests do not require standard calibration weights. Any four individual objects that weigh about one fourth the capacity of the balance can be used. For example, if a balance has a capacity of 100 g, then four items with weights of about 25 g each are needed.

b. Each of the following weighings should be repeated, ideally 10 times, and the results of the 10 weighings averaged.

Test 1: Check the Balance Response at Its Midpoint

1. *Select four items whose total weight is about the capacity of the balance and label them A, B, C, D.*
2. *Weigh all four pieces together.* Record this weighing as the "full-scale value."
3. *Weigh pieces A and B together and record their weights.*
4. *Weigh and record the weight of pieces C and D together.*
5. *Add the two values from step 3 and step 4 together.* This is the "weight sum."
6. If the balance is exactly linear at its midpoint, then the weight sum will equal the full scale value. If the two are not equal, divide the difference by two because two measurements were made at this point. The result is the linearity error at the midpoint.

Test 2: Check the Balance Response at 25% of Capacity

1. *Weigh all four pieces individually.* Add their weights together and call this the "weight sum."
2. *If the balance is linear at 25% of its capacity, then the weight sum should equal the full-scale value as determined earlier.* If the two differ, divide the difference by four. This is the linearity error at 25% of capacity.

Test 3: Check the Balance Response at 75% of Capacity

1. *Divide the pieces into the following groups:*
   Group 1: pieces A, B, C
   Group 2: pieces A, B, D
   Group 3: pieces A, C, D
   Group 4: pieces B, C, D
2. *Weigh all four groups one at a time.* Add their weights together and call the result the "weight sum." Multiply the full-scale value by three and call it "75% full scale value."
3. *If the balance is linear at the 75% point, this weight sum will equal the 75% full scale value.* If the two differ, divide the result by four. This is the linearity error at 75% of capacity.

Based on information in Weil, Jerry. "Assuring Balance Accuracy." Product note. Cerritos, CA. Cahn Instruments Inc., November, 1991.

---

the SOP in Figure 19.10 specifies a warm-up period for the balance. Note also in the SOP that the balance is first calibrated and then the weight of a second standard is checked to verify that the balance is reading correctly.

iv. COMPLIANCE WITH THE ELECTRONIC RECORDS REGULATIONS

Recall from Chapter 6 that in 1997 the FDA issued regulation *21 CFR Part 11 Electronic Records; Electronic Signatures; Final Rule* to address the role of computers in documentation in the pharmaceutical industry. Companies that are compliant with the FDA's regulations must consider whether the Part 11 rules apply to their balances. In many cases balances are used as stand-alone instruments that are not interfaced with a computer or with other instruments. The balance generates an electronic signal (as described in Chapter 18) and this signal is converted to a display of weight. The user reads the display and records the reading with a pen into a lab notebook or onto a form. The electronic signal that the balance obtains for the sample is not stored by the instrument—it disappears when the sample is removed from the balance. The documentation with a stand-alone balance is therefore a conventional "paper" (and pen) record and is not subject to the requirements of Part 11.

Balances are, however, sometimes interfaced with an electronic system that receives weight readings, stores the information (e.g., in Excel or in a database), perhaps analyzes the data, and provides an output. In these situations if the company is compliant with FDA regulations and if the weight readings are critical to product safety, efficacy, and quality, then the rules of 21 CFR Part 11 do apply. This means that appropriate controls must be implemented for these weighing records. For example:

- The balance software must stamp each reading with the correct date and time.
- The weighing results must be attributable to a particular person and so a user ID and password is likely to be required.

Clean Gene, Inc.

STANDARD OPERATING PROCEDURE

SUBJECT: Operation of Balance: BRP 88

Balance Location: Media prep lab

Effective Date: 7/22/08

Page 1 of 2

1. Scope: The routine operation and use of Balance model BRP 88
2. Resources: NIST traceable standard mass set (2000 g)
3. References: Manufacturer's bulletin No 2
4. Responsibility: Training level of "General Lab" personnel or greater is required for use of this equipment.
5. Procedure

5.1. <u>Pre-use calibration and verification of balance</u>

Frequency: Daily before use

Form: Balance Instrument record (QF 15.3.6.3)

5.1.1   Before calibration and verification, the balance must be switched on for at least 20 minutes to allow the internal components to come to working temperature.

5.1.2   Check that the leveling bubble is centered.

5.1.2.1   If the bubble is off center, adjust the leveling feet to bring back to center.

5.1.3   Check that the balance is clean.

5.1.4   Remove any load from the pan and press "ON" briefly. When zero "0.00" is displayed, the balance is ready for operation.

5.1.5   Press and hold "CAL" until "CAL" appears in the display, release key. The required standard value flashes in the display.

5.1.6   With tongs, place the traceable standard required from 5.1.5 in the center of the pan. The balance adjusts itself.

5.1.7   When zero "0.00" flashes, remove calibration standard.

5.1.7.1   If "0.00" does not flash, the balance is not verified and must be checked by the metrology department. Refer to SOP QI 15.3.20 for procedure to remove balance from use until metrology can evaluate it.

5.1.8   Choose a NIST traceable standard whose weight is close to that of the sample to be weighed. When "0.00" is displayed the balance is ready.

5.1.9   With tongs, place the traceable standard on the weighing pan.

5.1.10   Wait until stability detector "o" disappears.

5.1.11   Read results.

5.1.12   Record the daily verification on the form QF 15.3.6.3 with date, time, and operator's name.

5.1.13   The balance is now in the weighing mode and is ready for operation. Reverification with standards is not necessary more than once a day.

5.2. <u>Operation</u>

5.2.1   Remove any load from weighing pan and press "ON" briefly. When zero "0.00" is displayed the balance is ready for operation.

5.2.2   Place sample on the weighing pan.

5.2.3   Wait until the stability detector "o" disappears.

5.2.4   Read results.

5.2.5   Switch off by pressing OFF until "OFF" appears in the display. Release key. The balance may be left in this mode for the remainder of the day.

5.3. <u>Taring weighing vessel</u>

5.3.1   Place the empty weighing vessel on the balance, the weight of the container is displayed.

5.3.2   Press "TARE" briefly, the balance will re-zero.

5.3.3   Add sample to the container. The sample weight is displayed.

5.4. <u>Maintenance</u>

5.4.1   Cleaning

5.4.1.1   After working with the balance, the work surface must be checked for residual chemicals or soils. Solid materials must be removed and disposed of properly.

5.4.1.2   Each week, or more often if necessary, the unit should be wiped down with general purpose cleaning agents. Wipe down the weighing pan, baseplate, sides, and lid of the balance.

5.4.1.3   In the event of a chemical spill, assess the hazard. If necessary, call hazard control team (X 3762). If nonhazardous material, remove balance pan and clean thoroughly.

5.4.2   Semi-Annual Calibration

5.4.2.1   Initial and semi-annual calibration is performed by trained service department personnel.

5.4.2.2   If the balance is out of calibration according to 5.1.7.1 it must be re-calibrated immediately by trained personnel in the QC department.

5.4.2.3   Form SOP QI 15.3.20 is to be filled out and the QC department notified for repair or recalibration.

Figure 19.10.   Example of a Standard Operating Procedure for Checking and Operating a Balance.

- The software is likely to prevent unauthorized individuals from using the balance.
- The weighing records that are generated must be secure and must be accessible to people who are authorized to view them.
- The records must provide information for traceability so that weight results can be connected to a particular lot of product or a particular task as well as to the date, time, and individual involved.
- The records must not be capable of being altered either accidentally or intentionally in an uncontrolled manner.

Manufacturers have created newer models of laboratory balance that meet the electronic records requirements. You can find such balances by reading the literature provided by balance manufacturers.

IV. MASS VERSUS WEIGHT

So far we have used the terms *mass* and *weight* without much discussion of the distinction between them. In this section we discuss these two words.

The value we read from a balance is the weight of an object, not its mass. This may seem surprising. After all, the object is directly compared with a standard whose mass is known. For example, consider the objects in Figure 19.11a. The standard is known to have a mass of 1 g and the sample exactly balances the standard. It seems logical, therefore, that the object should also have a mass of 1 g; unfortunately, there is another factor that affects this system—the air.

The major force measured in weighing is the force of gravity pulling down on an object. However, there is also a slight buoyant force from air. The air around a sample or standard gives it a bit of support and makes it appear lighter than it really is. This means that if the same object is weighed in air and in a vacuum, the object will be slightly heavier in the vacuum. This is the **principle of buoyancy:** *Any object will experience a loss in weight equal to the weight of the medium it displaces.* Buoyancy is why ships float, balloons rise, and the weight of an object is different in a vacuum than it is in air, see Figure 19.11b.

Balances are calibrated with metal mass standards. Metal has a relatively high density compared to aqueous solutions and other materials. A 1 g metal mass standard, therefore, displaces less air than does a 1 g mass of water. Because the metal mass standard displaces less air than the water, the metal standard is buoyed less by the air. This difference in buoyancy of a mass standard and a sample explains why the standard and sample in Figure 19.11 can have the same weight yet be of different masses. The sample is buoyed by the air more than the standard. If a balance were calibrated with a mass standard whose density was identical to that of the sample, then the weight of the sample would equal its mass. Because we use metal mass standards and we weigh samples in air, the value we read in the laboratory for a sample is its weight, not its mass.

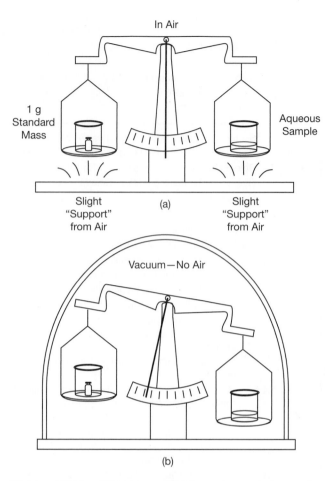

Figure 19.11. **Mass versus Weight. a.** A sample is exactly balanced against a 1g mass standard in air. Although the sample and the standard have the same weight in air, they do not have the same mass. **b.** In a vacuum, the beam will tilt to the side with the sample.

This discrepancy between mass and weight is called the **buoyancy error.** The weight readings for aqueous solutions have a buoyancy error of roughly 1 part in 1000. This buoyancy error is small enough as to be of little concern in most applications. The distinction between mass and weight, therefore, is generally ignored, except when very high-accuracy measurements must be made. (If necessary, there are equations to correct for the buoyancy effect that take into consideration the air density at particular atmospheric conditions and the density of the object being weighed. See, for example, Mettler-Toledo, Inc. *Weighing the Right Way with Mettler: The Proper Way to Work with Electronic Analytical and Microbalances,* Switzerland: Mettler-Toledo, 1989.)

PRACTICE PROBLEMS

1. Suppose you need to prepare a solution with a concentration of 15 mg/mL of a particular enzyme. You try to weigh out 15 mg of the enzyme on the analytical balance, but find that it is extremely difficult to get exactly 15 mg. Suggest a strategy to get the correct concentration even if you cannot weigh exactly 15 mg.

2. How much volume of solution at a concentration of 35 μg/mL can be made in each of these cases?
   a. Sample weighs 0.003560 g
   b. Sample weighs 0.0500 mg
   c. Sample weighs 1.0897 g
   d. Sample weighs 354.8 μg

3. A technician weighs a cell preparation on an analytical balance and observes that the initial weight is 0.0067 g. A while later the weight is 0.0061 g. What might be happening here?

4. Which scale is more sensitive?

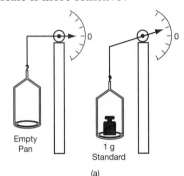

(a)

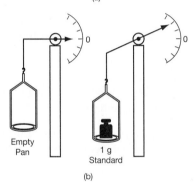

(b)

5. Suppose you are verifying the performance of a laboratory analytical balance. You first calibrate the balance, as directed by an SOP. You then repeatedly weigh a Class 1 standard weight that is 10 g. The results are shown. According to the specifications for the balance, its standard deviation (precision) must be equal to or better than ±0.0001 g and its accuracy must be better than ±0.0001 g. Does the balance meet its performance specifications?

Weight Values Obtained

10.0000 g
10.0001 g
9.9999 g
10.0000 g
9.9998 g
10.0001 g

6. A standard operating procedure to check and operate a balance is shown in Figure 19.10 on p. 343. Read the SOP and answer the following.
   a. How does a user determine whether the balance is operating acceptably?
   b. How does a user document that the balance is operating acceptably?

7. Suppose a 100 g standard mass is accidentally dropped on the floor. As a result, its mass is slightly less than 100 g. If the standard is used to calibrate a balance, what will happen to subsequent readings from that balance? What type of error is this?

8. The linearity of a balance is being checked. The balance has a capacity of 500 g and four weights of about 125 g each are used. The results follow. Calculate the linearity error (if any) at midpoint, 25%, and 75% capacity.

   Midpoint Check
   The four pieces weighed all together weighed 500.001 g.
   A and B together weighed 250.003 g
   C and D together weighed 250.000 g

   Check at 25% of Capacity
   A weighed 125.001 g
   B weighed 125.001 g
   C weighed 125.003 g
   D weighed 124.996 g

   Check at 75% of Capacity
   Group 1: pieces A, B, C weighed 375.001 g
   Group 2: pieces A, B, D weighed 374.998 g
   Group 3: pieces A, C, D weighed 375.001 g
   Group 4: pieces B, C, D weighed 374.996 g

9. (Optional) a. The density of air is about 1.2 mg/cm^3 (depending on the temperature and atmospheric pressure). 1000 mL of water displaces 1000 cm^3 of space (from the definition of a mL). What is the weight of air displaced by 1000 mL of water?

   b. Will the measured weight of a sample change if it is weighed first in air and then in a vacuum? Will its mass change?

QUESTIONS FOR DISCUSSION

1. The ISO 9000 standards state that companies shall "define the process employed for the calibration of inspection, measuring, and test equipment, including details of equipment type, unique identification, location, frequency of checks, check method, acceptance criteria, and the action to be taken when results are unsatisfactory" (section 4.11.2c). Explain how the SOP in Figure 19.10 on p. 343 accomplishes the requirements as stated by ISO 9000.

2. Consider weighing an object in the laboratory.
   a. List as many possible sources of bias as you can.
   b. What would be required to eliminate each source of bias you mentioned in part a?

3. If you work in a laboratory, write an SOP to operate a balance in your laboratory. Be sure to include steps to verify that the balance is responding properly.

Chapter Appendix

Laboratory Mass Standards

Standards are classified by Type and Class. **Type** *refers to the method of construction of the standard.* **Type I standards** *are of one-piece construction and are used when the most accurate and stable standards are required.* **Type II standards** *do not need to be constructed of a single piece of metal and may include additional adjusting material.*

Class indicates the permitted tolerance of the standard. There are two sources of classifications for laboratory mass standards that are commonly cited in the United States:

- ASTM E617-97 "Standard Specification for Laboratory Weights and Precision Mass Standards," with Classes 0–7 where the lower the class number, the smaller the tolerance
- OIML R111 (from The International Organization of Legal Metrology), which identifies Classes E1, E2, F1, F2, M1, M2, M3

Information about these classes is shown below.

Suggested Applications

ASTM Class 0: Used as primary reference standards for calibrating other reference standards and weights.

ASTM Class 1: Can be used as a reference standard in calibrating other weights and is appropriate for calibrating high-precision analytical balances with a readability as low as 0.1 mg to 0.01 mg.

ASTM Class 2: Appropriate for calibrating high-precision, top-loading balances with a readability as low as 0.01 g to 0.001 g.

ASTM Class 3: Appropriate for calibrating balances with moderate precision, with a readability as low as 0.1 g to 0.01 g.

ASTM Class 4: For calibration of semi-analytical balances and for student use.

ASTM Class 5: For student laboratory use.

ASTM Class 6: Student brass weights are typically calibrated to this class. (This class meets the requirements for OIML R 111 Class M2.)

ASTM Class 7: For rough weighing operations in physical and chemical laboratories.

OIML Class E1: Used as primary reference standards for calibrating other reference standards and weights.

OIML Class E2: Can be used as a reference standard in calibrating other weights and is appropriate for calibrating high-precision analytical balances with a readability as low as 0.1 mg to 0.01 mg.

OIML Class F1: Appropriate for calibrating high-precision, top-loading balances with a readability as low as 0.01 g to 0.001 g.

OIML Class F2: For calibration of semi-analytical balances and for student use.

OIML Class M1, M2, M3: Economical weights for general laboratory, industrial, commercial, technical, and educational use. Typically fabricated from cast iron or brass. Class M2 is commonly used for student brass weights.

TOLERANCE FOR MASS STANDARDS* (± MG)

Mass of Standard	Class							
	0	1	2	3	4	5	6	7
1 kg	1.3	2.5	5.0	10	20	50	100	470
500 g	0.60	1.2	2.5	5.0	10	30	50	300
200 g	0.25	0.50	1.0	2.0	4.0	15	20	160
100 g	0.13	0.25	0.50	1.0	2.0	9	10	100
50 g	0.060	0.12	0.25	0.60	1.2	5.6	7	62
10 g	0.025	0.050	0.074	0.25	0.50	2.0	2	21
1 g	0.017	0.034	0.054	0.10	0.20	0.50	2.0	4.5

* Excerpted from ASTM E617-97 "Standard Specification for Laboratory Weights and Precision Mass Standards."

EXAMPLES

When a balance is being calibrated and the utmost accuracy possible is required, it is necessary to choose the correct mass standard. It is recommended that the tolerance of the standard be four times more accurate than the readability of the balance. For example:

a. A balance has a readability of 1 mg and is to be calibrated with a 100 g standard weight. What type of mass standard should be chosen? To make this decision, divide 1 mg by 4; the result is 0.25 mg. This is the tolerance that is needed in the 100 g standard. Refer to the tolerance chart for weight standards. For a 100 g standard, the tolerance of a class 1 standard is 0.25 mg. A Class 1 standard, therefore, should be chosen.

b. A balance has a capacity of 60 g and a readability of 1 mg. It is to be calibrated with a 50 g standard. What class standard mass is required? The readability is 1 mg; therefore, the tolerance of the standard should be ± 0.25 mg. From the previous table, observe that a 50 g standard of class 2 has a tolerance of 0.25 mg.

c. A balance has a capacity of 500 g, a readability of 0.1 g, and is to be calibrated with a 500 g standard mass. What class standard is required? The readability is 0.1 g; therefore, the tolerance of the standard should be 0.025 g or 25 mg. A class 4 standard is adequate.

CHAPTER 20

The Measurement of Volume

I. PRINCIPLES OF MEASURING THE VOLUME OF LIQUIDS

A. Overview

Volume *is the amount of space a substance occupies.* The **liter** *is the basic unit of volume.** Various devices are used to measure the volume of liquids, depending on the volume being measured and the accuracy required. For larger volumes, biologists use glass and plastic vessels, such as **graduated cylinders** and **volumetric flasks.** **Pipettes** are usually preferred for volumes in the

1–25 mL range. Various **micropipetting devices** are commonly used to measure volumes in the microliter range. These various methods of measuring volume are discussed in this chapter.

B. Basic Principles of Glassware Calibration

Manufacturers of glassware and plasticware for volume measurements (such as graduated cylinders, volumetric flasks, and pipettes) are responsible for the calibration of those items. **Capacity marks** and **graduations** *are*

*The proper unit for volume in the SI system is the cubic meter, m³. Due to its large size, it is rarely used in the biology laboratory. The cubic centimeter, cm³, is used in the SI system. The milliliter, mL, is equivalent to the cm³.

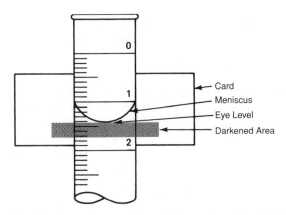

Figure 20.1.   The Meniscus. A dark card may be placed behind the meniscus to make it more visible. When reading volume, the eye should be level with the bottom of the meniscus.

lines marked on volume measuring devices that indicate volume. Calibration of glass and plasticware involves placing capacity marks and graduations in such a way that they correctly indicate volume. The American Society for Testing and Materials (ASTM) distributes standards that specify exacting procedures for how certain volume measuring devices are to be calibrated, checked for accuracy, and labeled. When manufacturers calibrate devices in accordance with ASTM standards, they generally state this in their catalog.

The meniscus must be considered when glassware is calibrated. The **meniscus** (Greek for "crescent moon") is a curve formed by the surface of liquids confined in narrow spaces, such as in measuring devices, see Figure 20.1. Conventional practice and ASTM standards specify that the lowest point of the meniscus is used as the point of reference in calibrating a volume measuring device; therefore, the lowest part of the meniscus is also the reference when reading the volume of a measuring device. It may be helpful to place a dark background, such as a piece of black paper, behind the device to facilitate reading the meniscus accurately. It is important to hold your eyes level with the liquid surface when reading a meniscus.

Glassware for measuring liquid volume is calibrated either **to contain (TC)** or **to deliver (TD).** A device that is **TC** will contain the specified amount when filled to the capacity mark. It will not deliver that amount if the liquid is poured out because some of the liquid will adhere to the sides of the container. A **TD** device is marked slightly differently so that it does deliver the specified amount, assuming the liquid is water at 20°C and it is poured using specific techniques. Consider the logic behind the two methods of calibration. A solution made to a specific concentration of solute can be prepared in a volumetric flask. The volume of water in the flask is critical in obtaining the right solute concentration, so a flask calibrated "to contain" is properly used. In contrast, if the volumetric flask is to be used to pour an exact volume of liquid to another container, then a flask calibrated "to deliver" should be used. (Plastics are con-

sidered to be nonwetting—that is, water does not stick to them—so there is no difference between TC and TD for plasticware.)

There is inevitably a small amount of error in the volume calibration of glassware, plasticware, and pipettes, even when they are manufactured properly. The **tolerance,** or how much error is allowed in the calibration of a volume measuring item, depends on the volume of the item and the purpose of the measurements to be made with it. The most accurately calibrated glassware is termed **volumetric;** volumetric glassware therefore has the narrowest tolerance allowed in its calibration. For example, Class A volumetric flasks are used where high accuracy is required, such as in the preparation of standard solutions. The tolerance of a 100 mL Class A volumetric flask that meets ASTM standards is ± 0.08 mL (i.e., its true marked volume must be between 99.92 mL and 100.08 mL). A Class B 100 mL volumetric flask, which is used where slightly less accuracy is required, must be marked within 0.16 mL of 100 mL. The manufacturer must therefore calibrate a Class A flask more exactly than a Class B flask, and one might expect to pay more for it.

For high-accuracy volume measurements, it is important to consider two effects of temperature. First, the glass or plastic forming a volume measuring device will expand and contract as the temperature changes, thus affecting the accuracy of its markings. Second, the volume of materials, including aqueous solutions, changes as the temperature changes. ASTM standards specify that manufacturers calibrate devices at 20°C using water at the same temperature. In laboratory situations where the utmost accuracy is required, volume measurements should be made at 20°C or correction factors to account for temperature should be applied (see ASTM Standard E 542).

To ensure accuracy when using volume measuring devices, it is necessary to use the device in the same manner as it was calibrated. Recalibration is required, or some accuracy may be lost, when devices calibrated with water are used to deliver other liquids. If labware is distorted by heating or is contaminated, then the calibration marks will not be accurate. Systematic error will occur if for any reason a piece of glassware does not read the correct volume. Repetition of the measurement will not reveal this error.

II. GLASS AND PLASTIC LABWARE USED TO MEASURE VOLUME

A. Beakers, Erlenmeyer Flasks, Graduated Cylinders, and Burettes

Beakers and Erlenmeyer flasks are containers intended primarily to hold liquids, not to measure volumes. These vessels are marked with graduation lines, but they are at best calibrated with a tolerance of ± 5%. This means, for example, that when the manufacturer marks a beaker at

"100 mL," it may actually indicate anywhere between 95 and 105 mL.

Graduated cylinders *are cylindrical vessels calibrated with sufficient accuracy for most volume measurements in biology laboratories.* A moderately priced 100 mL graduated cylinder is calibrated with a tolerance of ± 0.6 mL.

Graduated cylinders are graduated with a number of equal subdivisions. For example, a 100 mL graduated cylinder might be etched with 100 subdivisions, one for each mL, whereas a 500 mL cylinder might have a mark every 5 mL. This means that a single graduated cylinder can be used to measure various volumes; however, the correct cylinder should be chosen depending on the volume desired. For example, it is not possible to measure 83 mL accurately with a 500 mL graduated cylinder; a 100 mL graduated cylinder is more appropriate, see Figure 20.2a–c. Note also that graduated cylinders are not designed as containers for mixing and for storing solutions because they are unstable and easily knocked over, they are expensive compared with storage bottles, and they often do not come with caps.

Burettes (also spelled *buret*) *are long graduated tubes with a stopcock at one end that are used to dispense known volumes accurately,* see Figure 20.2d. Burettes have a long tradition of use in chemistry laboratories for making accurate volume measurements.

B. Volumetric Flasks

Volumetric flasks *are vessels used to measure specific volumes where more accuracy is required than is attainable from a graduated cylinder.* Each **volumetric flask** *is calibrated either "to contain" or "to deliver" a single volume,* see Figure 20.3. Although volumetric flasks are calibrated with high accuracy, they have the disadvantages that they are relatively expensive and that each is calibrated for only one volume (e.g., 10, 50, or 1000 mL). A volumetric flask cannot be used, for example, to measure 33 mL. Volumetric flasks, therefore, are typically used in biology laboratories only in situations where high-accuracy volume measurements are required.

The ASTM tolerances for volumetric flasks of different sizes are shown in Table 20.1. Observe that there is less error tolerated in the calibration of Class A flasks than in the calibration of Class B flasks.

Manufacturers sell special, serialized (numbered) volumetric flasks, see Figure 20.4. These flasks are individually calibrated using equipment whose calibration is traceable to NIST and are accompanied by a certificate of traceability. The certificate documents that a given flask is accurate within the tolerance specified for its type and class. Certified, serialized glassware may be required in facilities meeting ISO 9000 or GMP requirements.

Volumetric flasks must be used correctly to ensure accurate volume measurements. Box 20.1 contains points relating to the proper use of a volumetric flask.

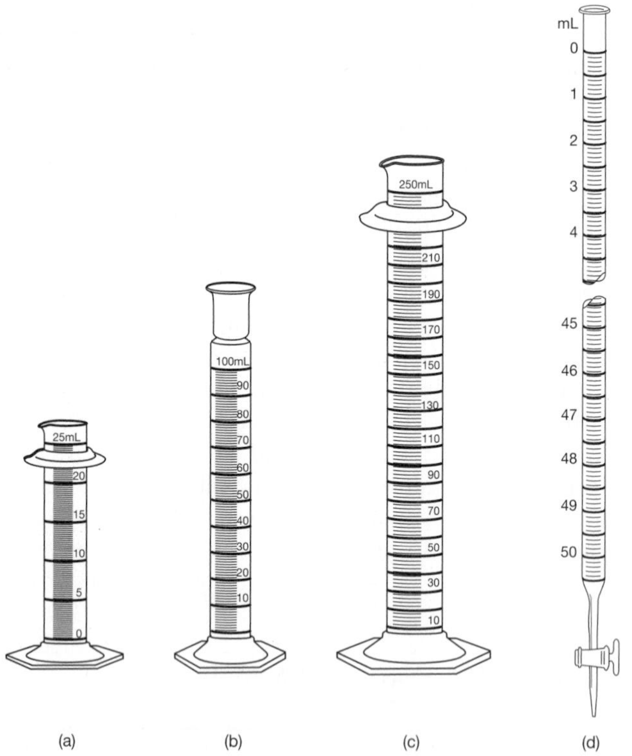

Figure 20.2.   Graduated Cylinders and Burettes. a. A 25 mL graduated cylinder has a graduation mark every 0.5 mL and is useful for volumes from 1 to 25 mL. **b.** A 100 mL cylinder has a graduation mark every 1 mL and is most useful for volumes from 10 to 100 mL. **c.** A 250 mL cylinder is marked every 2 mL and is most useful for volumes from 100 to 250 mL. For volumes less than or equal to 100 mL, a 100 mL graduated cylinder can be used with better accuracy. **d.** A burette that is calibrated every 0.1 mL.

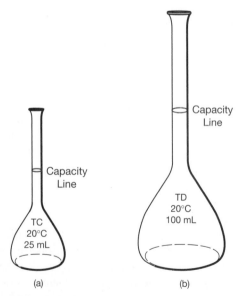

Figure 20.3.   Volumetric Flasks. a. A 25 mL volumetric flask that is calibrated to contain 25 mL when filled exactly to its capacity line. **b.** A volumetric flask that is calibrated to deliver 100 mL when it is filled exactly to its capacity line and the liquid is then poured out.

Table 20.1.   *ASTM STANDARDS FOR THE TOLERANCE OF VOLUMETRIC FLASKS*

Capacity (mL)	Class A Tolerances (± mL)	Class B Tolerances (± mL)
5	0.02	0.04
10	0.02	0.04
25	0.03	0.06
50	0.05	0.10
100	0.08	0.16
200	0.10	0.20
250	0.12	0.24
500	0.20	0.40
1000	0.30	0.60
2000	0.50	1.00

From ASTM Standard E 288-06, "Standard Specification for Laboratory Glass Volumetric Flasks."

Pyrex Brand Serialized and Certified with $ Barrelhead Stopper

Individually serialized and supplied with a Certificate of Identification and Capacity. Individually calibrated and certified against equipment whose calibration is traceable to NIST. Calibrated "To Contain" within Class A tolerances

Figure 20.4. **Catalog Description for a Serialized and Certified Volumetric Flask that is Individually Calibrated According to ASTM Standards.** (Courtesy of Fisher Scientific, Pittsburgh, PA.)

III. PIPETTES

A. Pipettes and Pipette-Aids

Pipettes (also spelled "pipet") *are hollow tubes that allow liquids to be drawn into one end and are generally used to measure volumes in the 0.1–25 mL range.* Pipettes may be made of glass or plastic and can be disposable or intended for multiple use. Presterilized disposable pipettes with cotton plugged tops are available for microbiology and tissue culture work.

Pipette-aids *are devices used to draw liquid into and expel it from pipettes,* see Figure 20.5 on p. 352. In the past pipette-aids were less common and people would suck liquid into pipettes, like sucking a drink into a straw. This procedure, called *mouth-pipetting,* is dangerous because it is easy to accidentally inhale or swallow radioactive substances, pathogens, hazardous chemicals, or the like. Most older laboratory workers have stories of aspirating a mouthful of some nauseating material. Although mouth-pipetting is now forbidden by safety regulations, its terminology persists and pipettes are often calibrated as "blow out." A pipette calibrated as "blow out" was previously literally blown out with one's mouth. Now, the last drop is ejected from a "blow out" pipette using a pipette-aid.

Pipettes come in different styles and sizes. Manufacturers place colored bands at the top of pipettes to indicate their capacity and graduation interval. These colored bands facilitate choosing the proper pipette and sorting pipettes after washing.

B. Measuring Pipettes

A measuring **pipette** *is calibrated with a series of graduation lines to allow the measurement of more than one volume.* Figure 20.6 on p. 352 illustrates how to read the meniscus on a measuring pipette.

Box 20.1.   *PROPER USE OF VOLUMETRIC FLASKS*

1. *Choose the proper type of flask for the application: either "to contain" or "to deliver," either Class A or Class B, either serialized or not serialized.*

2. *Be sure the flask is completely clean before use.*

3. *Read the meniscus with your eyes even with the liquid surface.* The bottom of the meniscus should exactly touch the capacity line.

4. *If the flask is calibrated "to deliver," then pour as follows:*

   a. Incline the flask to pour the liquid; avoid splashing on the walls as much as possible.

   b. When the main drainage stream has ceased, the flask should be nearly vertical.

   c. Hold the flask in this vertical position for 30 seconds and touch off the drop of water adhering to the top of the flask by touching it to the receiving vessel.

5. *Never expose volumetric glassware to high temperatures because heat causes expansion and contraction that can alter its calibration.*

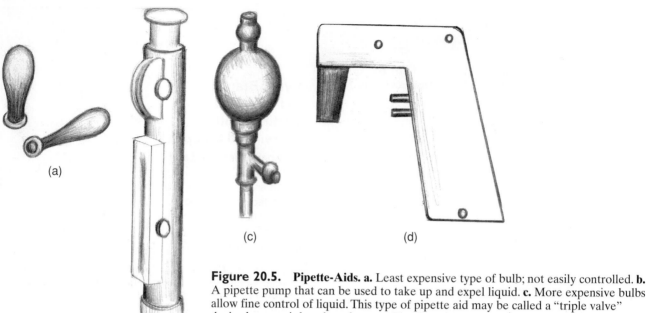

Figure 20.5.   Pipette-Aids. a. Least expensive type of bulb; not easily controlled. **b.** A pipette pump that can be used to take up and expel liquid. **c.** More expensive bulbs allow fine control of liquid. This type of pipette aid may be called a "triple valve" device because it has three buttons: The first displaces air from the bulb, the second is used to draw liquid into the pipette, and the third is used to expel the liquid. **d.** Electronic pipette-aid that allows fine control and ease of use. (Sketches by Sandra Bayna, a biotechnology student.)

Measuring pipettes are sometimes classified as **serological** or **Mohr.** Both these types are calibrated "to deliver." Pipettes termed **serological** *are usually calibrated so that the last drop in the tip needs to be "blown out" to deliver the full volume of the pipette,* see Figure 20.7a,b. Manufacturers place one wide or two narrow bands at the top of a pipette to indicate that it is calibrated to be "blown out." This "blow out" band(s) appears above the color coding band that indicates the capacity and graduation interval for the pipette. The proper way to use a serological pipette is summarized in Box 20.2.

Mohr pipettes *are calibrated "to deliver" but, unlike the serological pipettes described earlier, the liquid in the tip is not part of the measurement and the pipette is not blown out,* see Figure 20.8a,b. Serialized Mohr pipettes that have a certificate of traceability to NIST can be purchased.

C. Volumetric (Transfer) Pipettes

Volumetric (transfer) pipettes *are made of borosilicate glass and are calibrated "to deliver" a single volume when filled to their capacity line at 20°C,* see Figure 20.9. Volumetric pipettes are the most accurately calibrated

Figure 20.6.   Reading the Meniscus on a Measuring Pipette. Liquid was drawn up to exactly the zero mark and was then dispensed. Reading the value at the bottom of the meniscus shows that 3.19 mL of liquid was delivered.

Box 20.2.   USING A SEROLOGICAL PIPETTE

1. *Check that the pipette is calibrated to be "blown out" by looking for the band(s) at the top.*
2. *Examine the pipette to be sure the tip is not cracked or chipped.*
3. *Fill the pipette about 10 mm above the capacity line desired and remove any water on the outside of the tip by a downward wipe with lint-free tissue.*
4. *Place the tip in contact with a waste beaker and slowly lower the meniscus to the capacity line.* Do not remove any water remaining on the tip at this time.
5. *Deliver the contents into the receiving vessel by placing the tip in contact with the wall of the vessel.*
6. *When the liquid ceases to flow, "blow out" the remaining liquid in the tip with one firm "puff" with the tip in contact with the vessel wall, if possible.* Because we no longer actually "puff" on pipettes, the last drop is ejected with a pipette-aid.

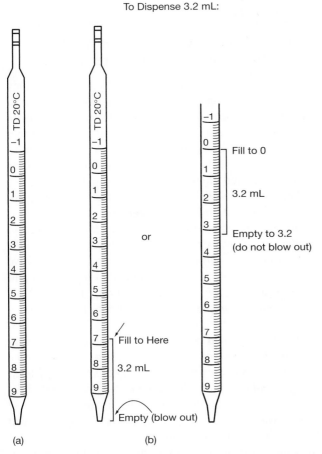

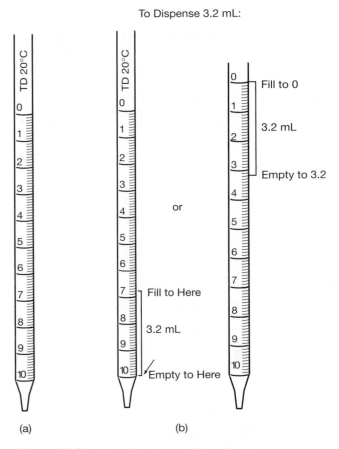

Figure 20.7. **Using Serological Pipettes. a.** A serological pipette calibrated so that the tip includes the last milliliter. The bands at the top indicate that this pipette is to be "blown-out." Note that this pipette has a scale that extends above zero to expand the calibrated capacity of the pipette. **b.** Using a serological pipette to dispense 3.2 mL.

Figure 20.8. **Mohr Pipettes. a.** Mohr pipettes are calibrated so that the tip does not include the last milliliter. **b.** Using a Mohr pipette to dispense 3.2 mL.

pipette type, see Table 20.2 on p. 354. Serialized volumetric pipettes with a certificate of traceability to NIST can be purchased.

Volumetric pipettes are calibrated so that "delivery of the contents into the receiving vessel is made with the tip in contact with the wall of the vessel and no afterdrainage period is allowed" (ASTM Standard 969). These pipettes are not "blown out." Remember to check that the tip is not cracked or chipped before using a volumetric pipette.

D. Other Types of Pipettes and Related Devices

There are various types of pipettes in addition to those described above. For example, **Pasteur pipettes** *are used to transfer liquids from one place to another.* Pasteur pipettes are not actually volume-measuring devices because they have no capacity lines, but they are convenient for transferring liquids, see Figure 20.10a on p. 354. There are glass pipettes for measuring volumes less than 1 mL. These small volume, reusable

"micropipettes," however, are not widely used in biotechnology laboratories because their use has been supplanted by micropipetting instruments, as described in Section IV of this chapter. Disposable pipettes for measuring less than 1 mL volumes are so convenient that they are sometimes used when high accuracy is unnecessary, see Figure 20.10b. Inexpensive capillary glass pipettes that deliver in the microliter range are sometimes used (e.g., in teaching laboratories), see Figure 20.10c.

Manual dispensers for reagent bottles *are devices placed in a reagent bottle with a tube that extends to the bottom of the bottle. The dispenser has a plunger that is depressed to deliver a set volume of liquid,* see Figure 20.10d. These devices effectively take the place of using a pipette or other device to remove liquids from reagent bottles.

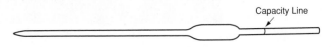

Figure 20.9. **A Volumetric Pipette.**

Table 20.2.   TOLERANCES FOR VOLUMETRIC CLASS A AND CLASS B PIPETTES COMPARED WITH GENERAL PURPOSE SEROLOGICAL PIPETTES

Capacity (mL)	Volumetric Class A Tolerance	Volumetric Class B Tolerance	Glass Serological Tolerance
0.1	—	—	0.005
0.2	—	—	0.008
0.5	0.006	0.012	0.01
1.0	0.006	0.012	0.02
2.0	0.006	0.012	0.02
3.0	0.01	0.02	—
4.0	0.01	0.02	—
5.0	0.01	0.02	0.04
6.0	0.01	0.03	—
7.0	0.01	0.03	—
8.0	0.02	0.04	—
9.0	0.02	0.04	—
10.0	0.02	0.04	0.06
15.0	0.03	0.06	—
20.0	0.03	0.06	—
25.0	0.03	0.06	0.10
50.0	0.05	0.10	—
100.0	0.08	0.16	—

Tolerances are expressed as ± mL.
Information from ASTM Standard E 969-02 "Standard Specification for Volumetric (Transfer) Pipets" and ASTM Standard E 1044-96 (reapproved 1990) "Standard Specification for Glass Serological Pipets (General Purpose and Kahn)."

IV. MICROPIPETTING DEVICES

A. Positive Displacement and Air Displacement Micropipettors

Glass and plastic pipettes are typically used in biology laboratories to measure volumes as small as about 1 mL. **Micropipettors** *are devices commonly used to measure smaller volumes, in the 1 to 1000 μL range.* These volume-delivering devices go by many names in addition to micropipettor, such as **microliter pipette, piston or plunger operated pipette,** and, simply, **pipettor.**

There are two distinct designs of micropipettor: **positive displacement** and **air displacement,** see Figure 20.11. **Positive displacement micropipettors** *include syringes and similar devices where the sample comes in contact with the plunger and the walls of the pipetting instrument.* Positive displacement devices are recommended for viscous and volatile samples. Syringes are commonly used to inject small volume samples into instruments for chromatographic analysis.

Air displacement micropipettors *are designed so that there is an air cushion between the pipette and the sample such that the sample only comes in contact with a disposable tip and does not touch the micropipettor itself.* Disposable tips reduce the chance that material from one sample will contaminate another or that the operator will be exposed to a hazardous material. Air displacement micropipettors accurately measure the volume of aqueous samples and are among the most common instruments currently used in biotechnology laboratories.

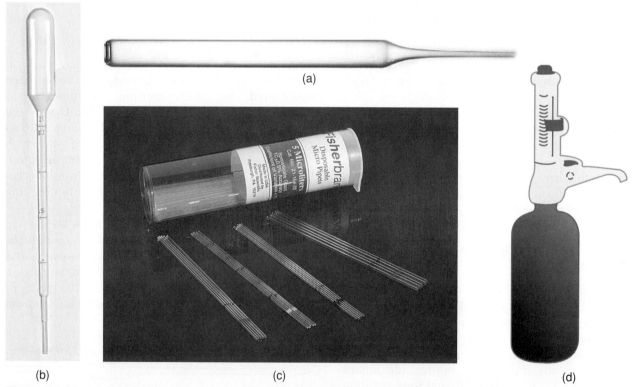

(a)

(b)   (c)   (d)

Figure 20.10.   **Examples of Various Types of Pipette and Liquid Dispensing Devices. a.** A Pasteur pipette. **b.** Disposable plastic pipette to measure microliter volumes with an accuracy of about ± 10%. **c.** Glass capillary pipettes for measuring microliter volumes. **d.** A manual dispenser for a reagent bottle. (a–c courtesy of Fisher Scientific, Pittsburgh, PA.)

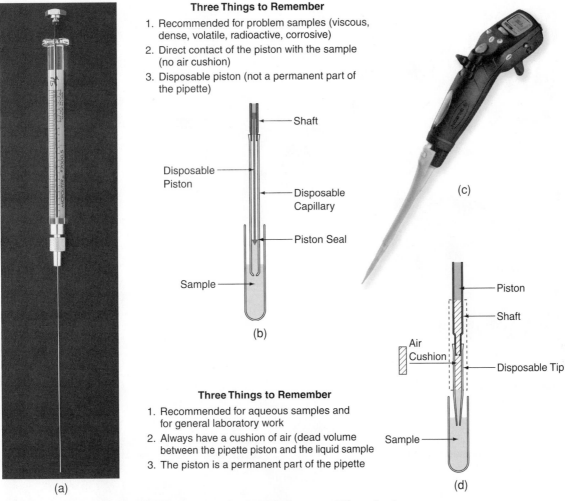

Three Things to Remember

1. Recommended for problem samples (viscous, dense, volatile, radioactive, corrosive)
2. Direct contact of the piston with the sample (no air cushion)
3. Disposable piston (not a permanent part of the pipette)

Shaft

Disposable Piston

Disposable Capillary

Piston Seal

Sample

(b)

(c)

Piston

Shaft

Air Cushion

Disposable Tip

Sample

Three Things to Remember

1. Recommended for aqueous samples and for general laboratory work
2. Always have a cushion of air (dead volume between the pipette piston and the liquid sample)
3. The piston is a permanent part of the pipette

(a)

(d)

Figure 20.11. Positive Displacement versus Air Displacement Micropipettors.
a. Syringe used to inject small volume samples into chromatography instrument; positive displacement. (Courtesy of Fisher Scientific, Pittsburgh, PA) **b.** Diagram of a positive displacement device. **c.** Air displacement micropipettor. **d.** Diagram of an air displacement device. (Illustrations b, c, and d are provided courtesy of Gilson, Inc.)

Air displacement micropipettors come in an assortment of styles, see Figure 20.12 on p. 356. Some deliver only one volume, such as 100 μL; others, called **digital microliter pipettors,** can be adjusted to deliver different volumes over a range, such as 10–100 μL. Micropipettors can be manually or electronically driven, and some are microprocessor controlled. Repetitive pipetting can lead to injury, so, in laboratories where repetitious pipetting is performed, it is worthwhile to investigate micropipettors designed to reduce operator fatigue.

B. Obtaining Accurate Measurements from Air Displacement Manual Micropipettors

i. Procedure for Operation

It is important to operate micropipettors properly or they will not deliver the correct volumes. Manual micropipettors have **plungers** *by which the operator controls the uptake and expulsion of liquids.* As the operator depresses the plunger, different "stop" levels can be felt. Although the detailed operating directions vary depending on the type and brand of micropipet-

tor, the general operation of most manual micropipettors is similar and is summarized in Figure 20.13 and in Box 20.3, both on p. 357.

ii. Factors That Affect the Accuracy of Manual Micropipettors

A variety of factors affect the accuracy of volume measurements by micropipettors:

1. **The operator's technique is the most important factor affecting the performance of a manual micropipetting device.** If the operator does not operate the micropipettor consistently, smoothly, and correctly, it will not accurately deliver the specified volume. (See Box 20.3 on p. 357.)

2. **The physical and chemical properties of the liquids being measured will affect the volumes delivered.** (See Table 20.3 on p. 359.)

3. **Measurements are affected by the environment in which they are made.** (See Table 20.3 on p. 359.)

4. **The condition of a micropipettor will affect its performance.** Basic micropipettor maintenance will be discussed shortly.

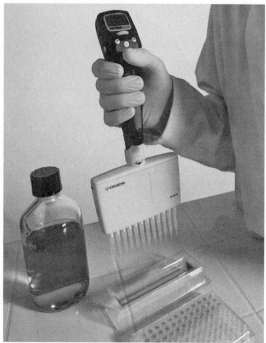

(a)

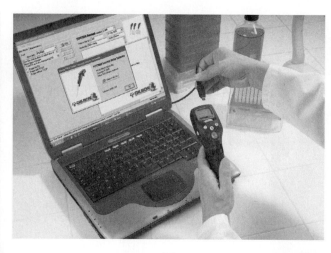

(b)

(c)

Figure 20.12. Various Types of Air Displacement Micropipettors and Tips. a. A multichannel micropipettor used to simultaneously deliver the same volume to 12 wells of a 96-well plate. **b.** An electronic, motorized, microprocessor-controlled micropipettor. This device reduces operator fatigue and allows the user to program the device, for example, to repeatedly dispense particular volumes. The pipette's internal software also allows the user to set a maintenance interval with an alarm that sounds when it is time for service. A computer can be used with this micropipettor to store and organize maintenance reports and data, thus assisting in adhering to GLP requirements. **c.** Disposable tips are stored in racks inside autoclavable plastic boxes; racked tips do not need to be touched with one's fingers to attach them to a micropipettor. (Photos provided courtesy of Gilson, Inc.)

iii. TIPS

Air displacement micropipettes require a disposable tip; the choice of tip can have a major impact on pipetting results. Some items to consider when choosing tips:

- There are different sizes of tips to match different models of micropipettor; be sure to choose the right one. Some manufacturers color code their tips and micropipettors to avoid confusion.

- Micropipettor manufacturers sell tips to match their devices. It is possible to purchase less-expensive, generic tips in bulk, but these may not seal properly on every model of micropipettor, may leak, and may not dispense liquids as accurately as tips recommended by the manufacturer.

- Special wide-bore tips are useful to reduce shearing when pipetting DNA and other large mole-

cules and intact cells. These tips may, however, be susceptible to error due to changes in barometric pressure. Smaller bore tips may be better for reaching inside small tubes or vials.

- Tips can be purchased loose in bags, or mounted in racks that sit inside plastic boxes, see Figure 20.12c. Racked tips can be mounted on the end of a micropipettor without touching them, which is desirable to avoid contamination.

- It is common practice to autoclave boxes of tips to sterilize them. When removing an autoclaved tip from its box, quickly remove the tip and shut the box to reduce the exposure of the remaining tips to the environment.

- Longer tips are available with small flat ends that are especially designed for loading samples into electrophoresis gels.

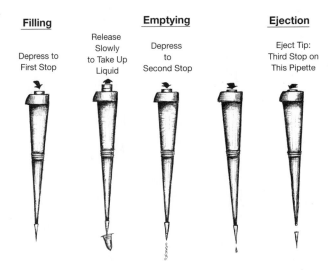

Filling

Depress to
First Stop

Emptying

Release
Slowly
to Take Up
Liquid

Depress
to
Second Stop

Ejection

Eject Tip:
Third Stop on
This Pipette

Figure 20.13. Using a Manual Micropipettor. Note that not all micropipettors have a third stop to eject the tip. (Sketches by Sandra Bayna, a biotechnology student.)

Box 20.3. PROCEDURE TO OPERATE A MANUAL MICROPIPETTOR (IN THE USUAL, FORWARD MODE)

1. *Set the micropipettor (if it is adjustable to different volumes) to the desired volume.*

2. *Attach a disposable tip to the micropipettor shaft.* Press firmly to ensure an airtight seal.

3. **Optional: Prewet the tip by aspirating and dispensing the solution to be measured into a waste container or back into the original solution.**

   Note: Prewetting leaves a thin film of the liquid to be measured on the inside of the tip. Some operators prewet the tip, others do not. There is evidence that better accuracy is obtained with a prewetted tip. Since prewetting the tip will affect its performance, it is important that operators are consistent in whether they prewet the tip or not.

4. *Hold the micropipettor vertically and, while observing the tip and the sample, depress the plunger to the first stop, and place the tip in the liquid.*

   Note: For the utmost accuracy, it is recommended that the tip be immersed into the sample a specified distance. This distance may be specified by the manufacturer. Otherwise, general guidelines are:

   for volumes of 1–100 μL, immerse the tip 2–3 mm
   for volumes of 101–1000 μL, immerse the tip 2–4 mm
   for volumes of 1001–10,000 μL, immerse the tip 3–6 mm

5. *Allow the plunger to slowly return to its undepressed position as the sample is drawn into the tip.* Wait 1 second before removing the tip from the liquid.

   Note 1: *Never allow the plunger to snap up!* (If the plunger "snaps," fluid can be aspirated into the interior of the micropipettor. Also, the volume measured will be incorrect.)

   Note 2: If any liquid remains on the outside of the tip remove it carefully with a lint-free tissue, taking care not to wick liquid from the tip orifice.

   Note 3: Pull the micropipettor straight out of the container after aspirating sample. Do not allow the tip to touch the side of the container.

6. *Place the tip so that it touches the side of the container into which the sample will be expelled; depress the plunger to the second stop.* Watch as the sample is expelled to the container. Wait about 2 seconds and be certain all the liquid is expelled. Remove the tip from the vessel carefully, *with the plunger still fully depressed.*

   Note: It is also correct to dispense the sample directly into a solution already in the tube and to rinse the tip with the solution. Discard the tip after such use.

7. *Eject the tip using the third stop, tip ejector button, or other mechanism.*

The procedure outlined in this box is the most common method of pipetting, called *forward mode* pipetting. Forward mode pipetting is recommended for aqueous samples. There is also *reverse mode* pipetting, which is less commonly used but is recommended for volatile and viscous samples. Reverse mode pipetting is described in Table 20.3. The following diagram shows the steps in forward mode pipetting.

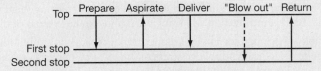

Top   Prepare   Aspirate   Deliver   "Blow out"   Return

First stop
Second stop

(Figure modified from one in ASTM E 1154-89 (reapproved 1993) "Standard Specification for Piston or Plunger Operated Volumetric Apparatus.")

- Some tips have calibration lines to indicate volume.
- For molecular biology applications, it may be desirable to buy tips that are certified to be free of endotoxins, DNase (an enzyme that degrades DNA), and RNase (an enzyme that degrades RNA). This is particularly important in pharmaceutical and biomedical research and testing facilities.
- Tips with filters are available to avoid contamination, as described in more detail below.

C. Contamination and Micropipettors

Improper use of micropipettors can result in liquids being accidentally drawn into the micropipettor assembly. Such liquids may contaminate later experiments, damage the micropipettor, or put the operator at risk. **Aerosols** *are fine liquid droplets that remain suspended in the air.* Aerosols are readily produced during pipetting and even proper use of a micropipettor can result in aerosol contaminants being drawn into the instrument. Aerosols can be a serious problem if they contain pathogens, toxic materials, or radioactive substances.

The polymerase chain reaction (PCR) is an extremely sensitive method of amplifying and detecting DNA present in samples in tiny amounts. If DNA from one PCR sample accidentally gets into the mixture for another sample, it can have serious consequences. *DNA contamination from one sample to another sample is called* **carryover** and micropipettors are a major source of such contamination.

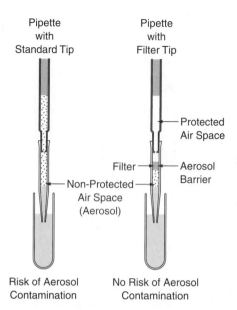

Figure 20.14. Avoiding Carryover and Contamination of a Micropipettor. A normal tip, shown of the left, is not designed to prevent aerosol contamination. The tip on the right has a barrier filter.

To avoid micropipettor contamination and carryover consider the following:

1. **Manufacturers produce special tips that have a barrier that keeps aerosols and liquids from entering the micropipettor assembly, see Figure 20.14.** These tips are more expensive than normal tips, but they may be a good investment in certain circumstances.

Box 20.4. PIPETTING METHODICALLY AND ERGONOMICALLY

1. *Plan exactly what will go into each tube.* A table is a convenient way to display the ingredients of each tube.

2. *For GLP purposes, record the identity of the micropipettor to be used, its history (e.g., dates of service, repairs, calibration checks), and environmental conditions.*

3. *Obtain all materials required and arrange them for easy access.* Remove unnecessary clutter. The opening for the recipient of used tips should be at the same height as the end of the pipette. Frequently used materials should be right in front of you to avoid reaching.

4. *Sit at a height such that you can rest your elbow on the work surface.*

5. *Label all tubes carefully before the addition of liquids.*

6. *Place tubes to be filled in the front of the rack; move each one back after material is added to it.*

7. *Check off components that have been added.*

8. *Change tips each time a new material is pipetted and whenever a tip touches components previously added to a tube.* Discard a tip that does not drain completely.

9. *After adding all the components to a tube, be sure the liquids are well-mixed, are not adhering to the tube walls, and are in the bottom of the tubes.* Mixing may be accomplished by inverting the tubes, tapping them with your finger, or using a vortex mixer. Some materials, like long strands of DNA, are damaged by vortexing and rough shaking. Such samples can be mixed by gently inverting the tubes or gently passing them up and down through a wide-bore micropipettor tip. Brief centrifugation will bring the liquids to the bottom of the tube.

10. *Take time to relax.* If possible, switch periodically to different types of work to avoid fatigue from repetitive pipetting. Let go of the pipette periodically to relieve stress on fingers and hands. Take short breaks when possible to change sitting position and to relax arms and shoulders.

11. *If using micropipettors frequently, consider investing in a motorized, ergonomically designed device.*

Table 20.3.   AVOIDING ERROR IN THE OPERATION OF AIR DISPLACEMENT DIGITAL MICROPIPETTORS

1. *Avoid damaging the micropipettor:*
   Never drop a micropipettor.
   Never rotate the volume adjuster of an adjustable micropipettor beyond the upper or lower range of the instrument.
   Never pass a micropipettor through a flame.
   Never use a micropipettor without a tip, thus allowing liquid to contaminate the shaft assembly.
   Never lay a filled micropipettor on its side, thus allowing liquid to contaminate the shaft assembly.
   Never immerse the barrel of a micropipettor in liquid; only immerse the tip.
   Never allow the plunger to snap up when liquid is being aspirated.
   Store micropipettors vertically in a proper stand.

2. *Use the proper disposable tip:*
   Make sure the tip matches the micropipettor. Tips come in various sizes for different size and brand pipettes. A poorly fitting tip will leak and deliver inaccurate volumes.
   Be sure the tip is firmly on the micropipettor. If the tip is not firmly attached the sample may leak and the volume dispensed will be too low.
   Use an autoclaved (sterile) tip where appropriate.
   Specialized tips are available for specific purposes, such as loading electrophoresis gels.
   Replace the tip if liquid adheres inside it and does not drain easily.

3. *Be sure the plunger moves slowly and evenly.*

4. *Wait a second or two after taking up sample to allow complete filling.*

5. *Hold the pipette as straight up and down as possible when taking up sample. If the micropipettor is tilted, too much liquid may be drawn in.*

6. *Watch the sample as it comes into and leaves the tip.* The liquid sometimes does not successfully enter the tip or is not expelled completely. If an air bubble is observed, return the sample to its original vessel and try again, immersing the tip to the proper depth and pipetting more slowly. If an air bubble returns, try a new tip. Check for drops remaining in the tip after the liquid is expelled. It may be possible to rinse the tip with the liquid into which the sample is being expelled.

7. *Avoid cross-contamination (i.e., contaminating one sample with material from another).* Change tips every time a different material is pipetted. If the same material is being added to a number of tubes, it is possible to use the same tip repeatedly. To avoid cross-contamination, expel the drop onto the side of each tube and watch to be sure the tip does not touch the liquid already in the tube.

8. *Micropipettors are calibrated with water and therefore are not accurate when used to measure liquids with densities that differ greatly from that of water.* It is possible to determine the actual volume of a liquid of any density delivered by a micropipettor at a particular setting. This determination is done empirically by setting the pipette to a particular volume, dispensing the liquid, weighing it, and determining the actual volume of that liquid dispensed at that pipettor setting. The weight of the liquid is converted to its volume based on its density. (See Lide, David R. [ed. in chief] *The Handbook of Chemistry and Physics.* Boca Raton, FL: CRC Press, Inc. to determine the density of a liquid.) A table can then be prepared for the liquid of interest that shows the volume dispensed at each setting of that micropipettor.

9. *Maintain the room humidity between 40 and 60%, if possible.* If the room humidity is too low, evaporation can occur quickly and may affect small volume measurements.

10. *Maintain the micropipettor and sample at room temperature, if possible.* If the micropipettor and the sample are not at room temperature (e.g., if the measurements are made in a cold box) then the volumes delivered by the micropipettor will be inaccurate. For the most accurate measurements, the micropipettor, tips, and sample must all be at the same temperature.

11. *Air displacement micropipettors are not very accurate when dispensing volatile liquids (like acetone) or viscous materials (like protein solutions). Accuracy may be improved by the following:*
    If possible, use a positive displacement micropipettor.
    Prewet the tip when pipetting volatile liquids and viscous samples. Fill the tip and empty it once or twice before the desired volume is taken up and dispensed.
    If an air displacement pipette must be used for a volatile or viscous liquid, a technique called "reverse mode" pipetting may be effective. This is accomplished by depressing the plunger to the *second stop* to take up the sample. Release the plunger after taking up the sample. The tip is then touched to the side of the receiving container and the sample is expelled by depressing the plunger to the *first stop*. In this method, the liquid remaining in the tip is not "blown-out" and is discarded. Reverse mode pipetting is shown in the following diagram.

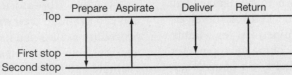

(Figure modified from one in ASTM Standard E-1154.)

2. It is helpful to have micropipettors dedicated to a particular technique and used by only one operator.

3. Proper operation of the micropipettor, as described in Table 20.3 on p. 359, is essential to avoid contamination problems.

Safety note: Be certain to safely remove any radioisotopes or pathogens from an instrument before calibrating or repairing it.

D. Pipetting Methodically

It is frequently necessary to pipette more than one liquid into a series of tubes. Each tube may receive the same ingredients or one or more components may vary from tube to tube. Although this task seems simple, it is easy to make a mistake when doing repetitious work. Moreover, it is necessary to work quickly if unstable ingredients are being added. There are electronic, computer-controlled pipettes for repeated pipetting that help ensure consistency in volume and reduce operator fatigue. The operator, however, must still be methodical to avoid errors. Box 20.4 on p. 358 includes suggestions on how to pipette multiple ingredients into many tubes methodically.

E. Verifying That Micropipettors Are Performing According to Their Specifications

Micropipettors are calibrated by the manufacturer when they are made, but they can become less accurate as they are used. The performance of micropipettors, therefore, should be verified periodically. In some laboratories users check their own micropipettors, some organizations regularly send their instruments back to the manufacturer to be inspected and repaired, and some facilities have in-house specialists and laboratory facilities for this purpose. Laboratories that are GMP or ISO 9000 compliant have written policies for micropipettor performance evaluation and routine maintenance.

There are no set rules for how often micropipettor calibration should be checked. Some laboratories routinely check and possibly calibrate each micropipettor every month, other laboratories every three or six months, and some laboratories once a year. The ideal frequency depends on how much the device is used, the applications for which it is used, the number of operators who use it, the nature of the liquids dispensed, and the recommendations of the manufacturer. (If a micropipette is dropped or known to be damaged it should be taken out of use and checked.) A survey of companies that calibrate micropipettors for clients revealed that, on the average, 36% of the micropipettes sent in for routine calibration failed a test of accuracy and precision when they arrived (Pavlis, Robert. "Sur-

Pipetman Specifications

Model	Volume µL	Increment µL	Accuracy (mean error)		Precision (repeatability)	
			Relative %	Absolute µL	Relative %	Absolute µL
P-2	0.2		12	0.024	6	0.012
	1.0	0.002	2.7	0.027	1.3	0.013
	2.0		1.5	0.030	0.7	0.014
P-10	1		2.5	0.025	1.2	0.012
	5	0.02	1.5	0.075	0.6	0.03
	10		1	0.1	0.4	0.04
P-20	2		7.5	0.15	2	0.04
	5	0.02	3	0.15	0.9	0.045
	10		1.5	0.15	0.5	0.05
	20		1	0.2	0.3	0.06
P-100	10		3.5	0.35	1	0.1
	50	0.2	0.8	0.4	0.24	0.12
	100		0.8	0.8	0.15	0.15
P-200	50		1.0	0.5	0.4	0.2
	100	0.2	0.8	0.8	0.25	0.25
	200		0.8	1.6	0.415	0.3
P-1000	100		3	3	0.6	0.6
	500	2	0.8	4	0.2	1
	1000		0.8	8	0.16	1.5
P-5000	500		2.4	12	0.6	3
	2500	2	0.6	15	0.2	5
	5000		0.6	30	0.16	8
P-10ML	1mL		5	50	0.6	6
	5mL	20	1	50	0.2	10
	10mL		0.6	60	0.16	16

Specifications subject to change without notice.

Figure 20.15. **Specifications for Micropipettes.** These specifications include values for accuracy and precision for eight models of micropipette, each at more than one volume. Accuracy is reported both as percent (relative) error and as absolute error; precision is reported both as the coefficient of variability and as standard deviation (see Box 20.5 for more information on these calculations). (Reproduced courtesy of Gilson, Inc.)

prising Statistics on Pipet Performance." *American Laboratory,* March 2004, 8–10.). This means that these poorly performing micropipettors were unknowingly used in the clients' laboratories for an indefinite period of time.

Micropipettor performance is most often evaluated using a **gravimetric** *(which means "relating to weight")* procedure. Gravimetric methods take advantage of the high accuracy and precision attainable with modern balances. During gravimetric testing a desired volume of purified water is placed into or is dispensed from the device to be checked. *This desired volume is termed the* **nominal volume.** The volume measurement is then checked by weighing the liquid on a high-quality, well-maintained balance. A calculation is performed to convert the weight of the water to a volume value. If the volume measuring device was properly operating, then the nominal volume will be identical to the volume as calculated from the weight of the water. Most laboratories have high-quality balances and can evaluate their micropipettors by a gravimetric procedure as outlined in Box 20.5 on pp. 361–363. (It is also possible to use spectrophotometric methods to evaluate micropipettors, but those methods are not covered here.)

Performance evaluation procedures evaluate the accuracy and precision of the micropipettor. These accuracy and precision values can be compared with the manufacturer's specifications, see Figure 20.15, or with the requirements of the laboratory's quality control procedure. If a micropipettor is performing outside of its specifications for accuracy, then it may be possible to recalibrate it using directions from the manufacturer. In other cases, the instrument will need to be repaired by the manufacturer or an in-house specialist.

F. Cleaning and Maintaining Micropipettors

Undamaged micropipettors require periodic performance evaluation but minimal maintenance. Routine maintenance includes cleaning the exterior of the micropipettor and some manufacturers suggest replacing internal seals and o-rings periodically, see Figure 20.16 on p. 365.

Micropipettors that are heavily used or are roughly handled may become damaged. Common symptoms of damage include:

1. **The micropipettor leaks or drips.**
2. **Visual inspection shows that the tip does not take up the right volume of liquid.**
3. **The plunger jams, sticks, or moves erratically.**
4. **The micropipettor is blocked and will not aspirate liquid.**

Micropipettors should be frequently tested for leaks. A simple method to do so is to place a disposable tip on the device and adjust it to its highest volume setting. Aspirate purified water that is at the same temperature as the micropipettor. Place the device vertically in a vibration-free stand. No droplets should appear on the tip before a period of 20 seconds has elapsed; if they do, there is a leak problem.

Leaking is frequently caused simply by poorly fitting pipette tips. Replacing the tip will sometimes help; other times switching brands of tip will resolve this problem. A poorly fitting tip is also a common reason a micropipettor aspirates the wrong volume.

A "sticky" plunger may be bent and needs replacement. All four of these symptoms may also be caused by a contaminated or damaged piston. Damaged seals or a loose shaft will also cause the micropipettor to leak or take up the wrong volume of liquid.

Many users return damaged micropipettors to the manufacturer for repair. If you decide to service a micropipettor yourself, determine whether the instrument is under warranty, and, if so, whether servicing it will invalidate the warranty. Consult the manual for your micropipettor if you decide to repair the device yourself. Table 20.4, on p. 365 has tips regarding cleaning and simple repairs that are applicable to various styles of micropipettor. Table 20.5 on p. 366 has troubleshooting tips.

After a micropipettor is serviced, it must be checked to see that it is performing according to its specifications. The gravimetric method described in Box 20.5 is suitable for this purpose.

Box 20.5.   *A Gravimetric Procedure to Determine the Accuracy and Precision of a Micropipettor*

(Based primarily on ASTM Method E 1154 "Standard Specification for Piston or Plunger Operated Volumetric Apparatus")

SUMMARY OF METHOD: This is a general procedure to verify that a micropipettor is functioning properly by checking the accuracy and precision of the volumes it delivers. The procedure is based upon the determination of the weights of water samples delivered by the instrument. The weight of the water is converted to a volume based on the density of water.

NOTES

a. Clean and check the micropipettor according to the manufacturer's instructions before checking its performance.
b. The analytical balance used to check the micropipettor must be well-maintained and properly calibrated. It must also be in a draft-free, vibration-free environment. It is recommended that the balance meet the minimum requirements shown in the table at the end of this box.
c. The water used should be purified and degassed. Discard water after one use.
d. Document all relevant information including date, micropipettor serial number, temperature when the check was performed, name of person performing evaluation, and so on.
e. Allow at least 2 hours for the micropipettor, tips, vials, and water to equilibrate together to room temperature.
f. This gravimetric procedure depends on converting a weight measurement to a volume value. To make this conversion with the utmost accuracy, it is necessary to correct for the following: the evaporation of water during the test procedure, the exact temperature of the water, the barometric pressure at the time of measurement, and the buoyancy effect. These corrections are found in the notes at the end of this box. The calculation for converting weight to volume that is shown in the body of this box is a commonly used approximation that does not correct for all these factors. If the performance verification is being performed to meet external quality requirements, it may be necessary to apply the corrections.

Box 20.5. *(continued)*

g. Replicate number: For a more thorough check of the micropipettor performance, 10 replicate measurements at each volume are recommended. For a quick check of the micropipettor, 4 or 6 replicate measurements are sufficient.

h. Perform the following procedure as quickly as possible, but do not compromise the consistency of volume delivery and technique. The time required to make each measurement should be as consistent as possible because evaporation will affect the results.

i. To reduce the effects of evaporation it is recommended to:
   i. Keep the humidity of the room between 50 and 60%.
   ii. Surround the weighing vessel with a reservoir of water.

j. Computer programs are available that automatically perform the required calculations. Computer-based calibration systems that interface directly with the balance used for calibration are also commercially available.

PROCEDURE

1. *Set the micropipettor to deliver a particular volume, referred to as the "nominal volume."*
2. *Optional: If corrections are to be applied, (see Note f), then measure and record the barometric pressure and the temperature of the water.*
3. *Place a small amount of water in a weighing vessel, such as a vial. The exact amount of water does not matter, but should be about 0.5 mL.*
4. *Tare the balance to the vial.* (Handle the vial with tongs.)
5. *Optional: If you customarily prewet the tip, then prewet the tip by aspirating one volume of purified water and dispensing it into a waste container.*
6. *Pipette the selected nominal volume of water into the preweighed vial using proper technique.*
7. *Weigh the vial and record the weight of the delivered water.*
8. *Repeat the measurement 4 or 10 times by repeating Steps 3–7.*
9. *Optional: If a correction for evaporation is to be applied, then perform a control blank by repeating steps 3–7 exactly as in a normal weighing, but without actually delivering any liquid to the weighing vessel.* It is suggested that this evaporation control check be performed at the beginning and end of each series of measurements and between each group of 10 samples in a larger series. Some sources recommend repeating four simulated weighing cycles and calculating the mean weight loss per cycle (in mg). Under normal conditions, this value is reported to be between 0.01 mg and 0.03 mg.
10. *For adjustable micropipettors that can be set to different volumes, it is good practice to check the micropipettor at three volumes: the maximum capacity of the micropipettor, 50% of maximum capacity, and about 10% of maximum capacity.* Perform the entire procedure at each of the three volumes.

CALCULATIONS

For each nominal volume checked:
1. *Calculate the mean weight of the water.*
2. *Convert the mean water weight to the mean volume measured.*
   Assume that the density of water is 0.9982 g/mL (its density at 20°C)
   1 µL of water weighs 0.9982 mg.
   Then:

$$\text{Mean Volume} = \frac{\text{Mean Weight}}{\text{Density}} = \frac{\text{Mean Weight (in mg)}}{0.9982 \text{ mg}/\mu L}$$

   Alternatively, for the highest accuracy, apply the corrections in Step 5.

3. *Determine the accuracy of the micropipettor as follows:*

$$\text{Absolute Error} = \text{Mean Volume Measured} - \text{Nominal Volume}$$

or

$$\text{Percent Error} = \frac{\text{Mean Volume Measured} - \text{Nominal Volume}}{\text{Nominal Volume}} \times (100\%)$$

Box 20.5. *(continued)*

4. *Determine the precision of the micropipettor by calculating the standard deviation (SD) or the coefficient of variation (CV) as follows:*

$$SD = \pm \sqrt{\sum (X_i - \bar{x})^2 / n - 1}$$

where n = the number of measurements, X_i = each volume measurement, $\bar{x}$ = mean volume measured

$$CV = \pm \frac{SD \times 100\%}{\bar{x}}$$

5. **ALTERNATIVE TO STEP 2:** *Calculations to convert the mean water weight to the mean volume taking into consideration corrections for evaporation, barometric pressure, temperature, and buoyancy:*

   i. *Calculate the evaporation, e, based on the loss in weight of the blank.*

   ii. *Calculate the mean volume of the water as follows:*

   $$\text{Mean volume} = (\text{mean weight} + e)\,(z)$$

   where: mean weight = mean weight of the 4 or 10 repeats at a given nominal volume; e = the evaporation; z = a conversion factor in microliters per milligram that incorporates the buoyancy correction for air at the test temperature and barometric pressure. The values for z are found in the following table.

   iii. *Continue with the calculations for accuracy and precision as shown earlier.*

Z VALUES: CONVERSION FACTOR VALUES (µL/MG), AS A FUNCTION OF TEMPERATURE AND PRESSURE, FOR DISTILLED WATER

Temperature (°C)	Air pressure (kPa)						Temperature (°C)	Air pressure (kPa)					
	800	853	907	960	1013	1067		800	853	907	960	1013	1067
15	1.0016	1.0018	1.0019	1.0019	1.0020	1.0020	22.5	1.0032	1.0032	1.0033	1.0033	1.0034	1.0035
15.5	1.0018	1.0019	1.0019	1.0020	1.0020	1.0021	23	1.0033	1.0033	1.0034	1.0035	1.0035	1.0036
16	1.0019	1.0020	1.0020	1.0021	1.0021	1.0022	23.5	1.0034	1.0035	1.0035	1.0036	1.0036	1.0037
16.5	1.0020	1.0020	1.0021	1.0022	1.0022	1.0023	24	1.0035	1.0036	1.0036	1.0037	1.0038	1.0038
17	1.0021	1.0021	1.0022	1.0022	1.0023	1.0023	24.5	1.0037	1.0037	1.0038	1.0038	1.0039	1.0039
17.5	1.0022	1.0022	1.0023	1.0023	1.0024	1.0024	25	1.0038	1.0038	1.0039	1.0039	1.0040	1.0041
18	1.0022	1.0023	1.0024	1.0024	1.0025	1.0025	25.5	1.0039	1.0040	1.0040	1.0041	1.0041	1.0042
18.5	1.0023	1.0024	1.0025	1.0025	1.0026	1.0026	26	1.0040	1.0041	1.0042	1.0042	1.0043	1.0043
19	1.0024	1.0025	1.0025	1.0026	1.0027	1.0027	26.5	1.0042	1.0042	1.0043	1.0043	1.0044	1.0045
19.5	1.0025	1.0026	1.0026	1.0027	1.0028	1.0028	27	1.0043	1.0044	1.0044	1.0045	1.0045	1.0046
20	1.0026	1.0027	1.0027	1.0028	1.0029	1.0029	27.5	1.0044	1.0045	1.0046	1.0046	1.0047	1.0047
20.5	1.0027	1.0028	1.0028	1.0029	1.0030	1.0030	28	1.0046	1.0046	1.0047	1.0048	1.0048	1.0049
21	1.0028	1.0029	1.0030	1.0030	1.0031	1.0031	28.5	1.0047	1.0048	1.0048	1.0049	1.0050	1.0050
21.5	1.0030	1.0030	1.0031	1.0031	1.0032	1.0032	29	1.0048	1.0049	1.0050	1.0050	1.0051	1.0052
22	1.0031	1.0031	1.0032	1.0032	1.0033	1.0033	29.5	1.0050	1.0051	1.0051	1.0052	1.0052	1.0053
							30	1.0052	1.0052	1.0053	1.0053	1.0054	1.0055

MINIMUM BALANCE REQUIREMENTS FOR GRAVIMETRIC TEST OF A MICROPIPETTOR

Test Volume (µL)	Balance Readability (mg)	Standard Deviation (mg)
≥1	≤0.001	≤0.002
≥11	≤0.01	≤0.02
≥101	≤0.1	≤0.1
≥1000	≤0.1	≤0.2

EXAMPLE

A digital micropipettor with a range of 100–1000 μL is evaluated for performance.

The pipettor is set for 1000 μL and high purity water at 20°C is dispensed four times. The dispensed water is weighed on an analytical balance. The four aliquots of water weigh:

$$0.9931 \text{ g} \quad 0.9948 \text{ g} \quad 0.9928 \text{ g} \quad 0.9953 \text{ g}$$

The micropipettor is then set for 500 μL and four aliquots of water are dispensed that weigh:

$$0.4960 \text{ g} \quad 0.4964 \text{ g} \quad 0.4970 \text{ g} \quad 0.4968 \text{ g}$$

The micropipettor is then set at 100 μL and four aliquots of water are dispensed that weigh:

$$0.1010 \text{ g} \quad 0.0901 \text{ g} \quad 0.0959 \text{ g} \quad 0.0947 \text{ g}$$

Determine the accuracy and the precision of this pipette (as defined in Box 20.5) at all three volumes using the approximation shown in Box 20.5 on pp. 361–363. Compare them with the specifications in Figure 20.15 on p. 360.

For a Nominal Volume of 1000 μL:

Step 1. Calculate the mean weight of the water from the four aliquots.

$$\frac{0.9931 \text{ g} + 0.9948 \text{ g} + 0.9928 \text{ g} + 0.9953 \text{ g}}{4} = 0.9940 \text{ g} = 994.0 \text{ mg}$$

Step 2. Calculate the mean volume of the water: $\text{Volume} \approx \dfrac{994.0 \text{ mg}}{0.9982 \text{ mg/}\mu\text{L}} = 995.8 \text{ μL}$

Step 3. Calculate the percent error: $\dfrac{995.8 \text{ μL} - 1000 \text{ μL} (100\%)}{1000 \text{ μL}} = -0.42\%$

Step 4. Calculate the standard deviation. The four weight measurements converted to volumes are:

$$\frac{993.1 \text{ mg}}{0.9982 \text{ mg/}\mu\text{L}} \quad \frac{994.8 \text{ mg}}{0.9982 \text{ mg/}\mu\text{L}} \quad \frac{992.8 \text{ mg}}{0.9982 \text{ mg/}\mu\text{L}} \quad \frac{995.3 \text{ mg}}{0.9982 \text{ mg/}\mu\text{L}}$$

$$\approx 994.9 \text{ μL} \quad \approx 996.6 \text{ μL} \quad \approx 994.6 \text{ μL} \quad \approx 997.1 \text{ μL}$$

$$\text{SD} = \sqrt{\frac{(994.9 - 995.8)^2 + (996.6 - 995.8)^2 + (994.6 - 995.8)^2 + (997.1 - 995.8)^2}{3}}$$

$$= \sqrt{\frac{0.81 + 0.64 + 1.44 + 1.69}{3}} \approx 1.24 \text{ μL}$$

$$\text{CV} = 1.24 \text{ μL}/995.8 \text{ μL} \times 100\% = 0.12\%$$

For a Nominal Volume of 500 μL:

Mean Volume ≈ 497.4 μL   % Error ≈ −0.51%   SD ≈ 0.44 μL   CV ≈ 0.09%

For a Nominal Volume of 100 μL:

Mean Volume ≈ 95.6 μL   % Error ≈ −4.4%   SD ≈ 4.49 μL   CV ≈ 4.69%

Summary of Results for All Three Volumes

	% Mean Error (Relative)	Mean Error (Absolute*)	Precision (CV)	Precision (SD)
1000 μL	−0.42%	−4.2 μL	0.12%	1.24 μL
500 μL	−0.51%	−2.6 μL	0.09%	0.44 μL
100 μL	−4.4%	−4.4 μL	4.69%	4.49 μL

*Mean Error Absolute = (mean volume − nominal volume)

Comparing these values with Figure 20.15 shows that the results at 1000 μL and 500 μL are within the manufacturer's specifications but the results at 100 μL are not. How this problem is handled depends on the policies in the laboratory.

EXAMPLE PROBLEM

Suppose you are going to prepare 100 mL of physiological saline solution, which is 0.9% NaCl in water.

a. How much solute is necessary? What is a good way to measure the solute?

b. How much solvent is necessary? What is a good way to measure it?

c. List some common measurement errors that must be avoided to ensure that the solution is made properly.

ANSWER

a. The solute is 0.9 g of NaCl. This would be measured on a balance that reads at least to the nearest 0.01 g. (Recall that the last digit on an instrument may be estimated.)

b. Close to 100 mL of water will be required to bring the solution to a volume of 100 mL. A graduated cylinder or volumetric flask would commonly be used to make this solution.

c. There are many potential sources of error in preparing this solution, such as improper reading of the meniscus, use of dirty glassware, use of an improperly maintained balance, failure to pour all the salt from the weighing container into the mixing container, and so on. Care is required to avoid these major sources of error. There is also likely to be a small amount of error due to the fact that some error is tolerated in the calibration of a graduated cylinder or volumetric flask. This error, however, is very small compared with the potential error of the mistakes listed above. If a balance is properly calibrated, the amount of error due to the balance is insignificant compared to these other factors.

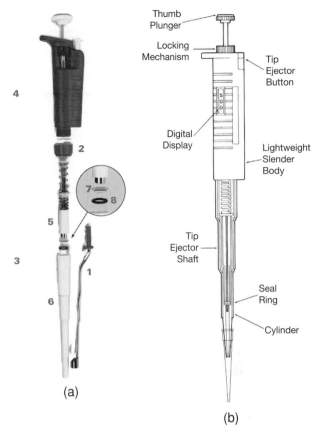

Figure 20.16. The Parts of a Manual Micropipettor. a. Dissembled parts: (1) tip ejector, (2) connecting nuts, (3) bottom part of device, (4) handle, (5) piston, (6) tip holder (nose cone), (7) seal, (8) o-ring. The user does not dissemble the upper part. Consult the manufacturer's guide for other models. (courtesy of Gilson, Inc.) **b.** A schematic illustration of an Oxford Brand Micropipettor. (Courtesy of Fisher Scientific, Pittsburgh, PA.)

G. Automated Liquid Handling Devices

Biological systems (e.g., cells, proteins, DNA, RNA) are active in an aqueous environment. Liquids are therefore involved in almost every biotechnology procedure. A typical biotechnology procedure might involve adding microliter volumes of multiple aqueous solutions to each sample—which is why biotechnologists must be able to properly operate a micropipettor. However, when there are hundreds or thousands of samples, as is often the case, a human wielding a micropipettor is not ideal. It is not easy for a person to keep track of and accurately dispense liquids to thousands of samples, nor is it healthy for a per-

---

Table 20.4. NOTES ON CLEANING AND SIMPLE REPAIR OF MICROPIPETTORS

1. *Consult the manufacturer's manual before attempting to clean or repair a micropipettor.* Follow the manufacturer's recommendations, as applicable.
2. *Use safety precautions when handling instruments that have been used to pipette hazardous materials.*
3. *Manufacturers recommend cleaning both the exterior of micropipettors and internal parts with distilled water alone, with soap solution followed by a rinse in distilled water, and/or with isopropanol (depending on the type of contamination).* Dry with a lint-free tissue.
4. *If you disassemble a micropipettor, keep track of the order and orientation of the parts as you remove each one.* There are springs inside the micropipettor assembly, so when the micropipettor is unscrewed, small parts may spring loose.
5. *If liquids are aspirated into the micropipettor accidentally, it is most effective to clean the device immediately and not let liquids dry inside it.* The micropipettor can usually be unscrewed and interior parts can be cleaned with distilled water. Dry the instrument after cleaning and then reassemble it.
6. *Many types of micropipettors have internal seals and o-rings that can become worn and can be replaced by the user.*
7. *Volatile organic compounds can cause the seals to swell, which in turn makes the piston move erratically.* Replace the seals.
8. *Manufacturers may recommend that the piston, seals, and o-rings be greased after cleaning.* Check the manual.

Table 20.5. MANUFACTURERS' TROUBLESHOOTING TIPS

Problem	Possible Cause	Solution
Aspirated volume too low	Damaged seal and/or piston	Clean or replace piston, replace seal
Leaking	Unsuitable tip	Use manufacturer's tip
	Tip not firmly seated	Press tightly
	Nose cone scratched	Replace nose cone
	Worn o-ring and seal	Replace o-ring and seal
	Piston contaminated	Clean piston, replace o-ring and seal
	Organic solvent is being dispensed	Use a positive displacement pipette.
		Alternatively, with an air-displacement device, saturate the air cushion of the pipette with solvent vapor by aspirating and distributing solvent repeatedly. The leak will stop when the air space is saturated with vapor.
Plunger jams, sticks, moves erratically	Seal is swollen by reagent vapors	Open micropipette and ventilate, clean piston if necessary, replace seal and o-ring
	Piston contaminated or damaged	Clean or replace piston and seal and o-ring

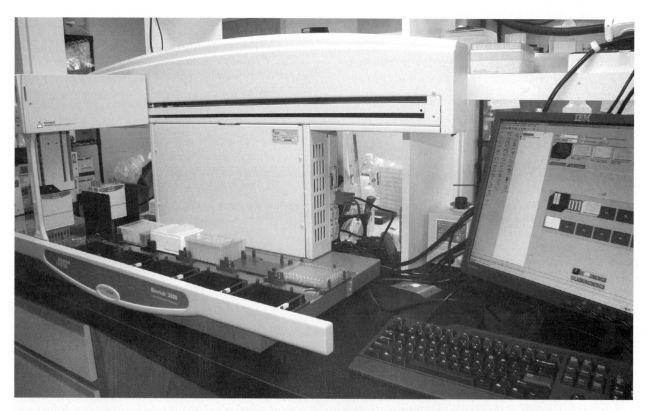

Figure 20.17.   An Automated Liquid Handling Device in a Research and Development Laboratory. The 96-well plates sit on a track in the front of the device. In the back are boxes of fresh tips. The moving head with eight pipette tips is positioned on the left end of the track. The computer on the right is monitoring the progress of the robot.

son to perform repetitive micropipetting over a long period of time. Engineers therefore invented **liquid handling robots** *that can be programmed to accurately dispense multiple liquids into numerous samples.* Liquid handling robots are becoming increasingly common in testing laboratories, production settings, and research laboratories. Robots are routinely used in drug discovery laboratories, where thousands of drugs are tested on cells or other samples, in genomics laboratories where thousands of DNA samples are sequenced, in microarray production, where thousands of nanogram spots of specific DNA fragments are applied to small chips, and in many other biotechnology applications.

There are various types of liquid handling robots designed for different applications. There are, for example, robots with varying numbers of channels (where each channel has a pipette tip). Some robots allow each channel to be individually programmed to dispense different liquids or different volumes, while other robots do not have that flexibility. Some robots have a self-correcting feature that enables them to sense whether the correct volume has been aspirated. The robot shown in Figure 20.17 is analogous to an 8 channel micropipettor. This robot is designed to add liquids to samples in 96-well plates (where each well is like a tiny test tube). The robot is programmed with a computer to aspirate liquids from a reservoir and then move along a track dispensing a set volume of the liquid to specific wells on every plate.

CASE STUDY

Analyzing Cancer at the Molecular Level

Cancer treatments tailored to an individual patient are an emerging application of modern genomic research. Genomic Health is a company that produces a diagnostic test kit to quantify the likelihood of breast cancer recurrence in women with newly diagnosed, early stage breast cancer. This information is used by women and their physicians to help chart an individualized treatment strategy. Genomic Health's diagnostic test involves analyzing tissue from the patients for the expression of 21 genes that are associated with breast cancer (e.g., Her2, a gene discussed in Chapter 1). Recall from Chapter 1 that when a gene is expressed, mRNA is transcribed by the cell using that gene as a template; the mRNA directs the synthesis of the protein encoded by that gene. The test method developed by Genomic Health analyzes the levels of mRNA for the 21 genes of interest in the patients' biopsy tissue. As is the case with many molecular biology procedures, their process involves a series of reactions; each reaction requires combining small volumes of liquids. The volumes of each liquid must be accurately dispensed in order to arrive at an accurate measurement of the expression levels of the genes. An inaccurate measurement of expression level can result in an erroneous diagnosis and treatment for a patient. The company is aware of the potentially

severe consequences of error. They therefore use automated liquid handlers to help ensure consistency and they regularly verify the precision and accuracy of the volumes dispensed by the robots. The company further verifies the performance of the system for each type and brand of tip that is used during dispensing. Attention to technical detail—such as the accuracy of dispensing—is critical when translating state-of-the-art research discoveries into practical clinical tools that help real patients.

Primary sources: "When Analyzing Cancer at the Molecular Level, Microliters Matter," American Biotechnology Laboratory, November/December 2007.

Harris, L., Fritsche, H., Mennel, R., Norton, L, Ravdin, P., Taube, S., Somerfield, M.R., Hayes, D. F., Bast Jr., R.C. "American Society of Clinical Oncology 2007 Update of Recommendations for the Use of Tumor Markers in Breast Cancer." Journal of Clinical Oncology, October 2007.

PRACTICE PROBLEMS

1. What is the volume of each of the following liquids?

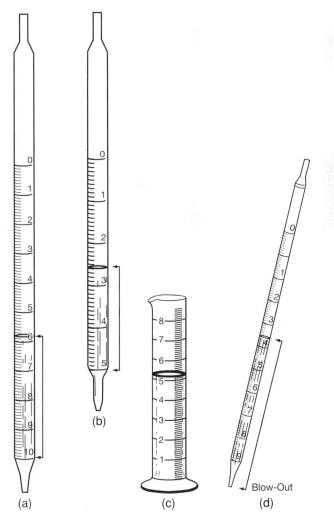

(a)   (b)   (c)   (d)   Blow-Out

2. What type of device(s) is used commonly in biology laboratories to measure a volume of:

    a. 100 mL   **b.** 95 mL   **c.** 2 mL

    d. 100 μL   **e.** 5 μL   **f.** 500 μL

3. List common sources of error that might affect each of the measurements in question #2.

4. Each of the devices given as an answer in question #2 has a certain allowed error, or tolerance associated with it. Because of this tolerance there is some uncertainty in its measurements.

   Based on information given in this chapter, what is the tolerance of each device you gave as an answer to question #2?

5. Suppose you are planning to purchase a new micropipettor to pipette volumes in the 100–200 μL range. You consult a catalog and find the following information for 3 brands of pipettor: Brand A, Brand B, and Brand C. Based on the catalog specifications, which micropipettor is most accurate? Which micropipettor is most precise? Which would you purchase?

	Volume Range	Accuracy (Expressed as % Error)	Precision (CV)
Brand A	40 to 200 μL	± 1%	0.5%
Brand B	100 to 200 μL	± 0.5%	0.3%
Brand C	100 to 200 μL	± 0.3%	0.4%

6. Suppose you are checking the accuracy of a micropipettor. Working at 20°C you weigh 10 empty tubes. You set the micropipettor to deliver 100 μL and carefully dispense that volume to each preweighed tube. You weigh the filled tubes on a calibrated balance and determine the weight of the water in each tube. The mean value for the weight of the water is 0.0994 g.

   a. The nominal volume is 100 μL. What do you expect the mean value for the weight of the water to be?

   b. What is the accuracy of your micropipettor (expressed as percent error) based on this result?

   c. The manufacturer specifies that the micropipettor should have an accuracy (expressed as % error) at least as good as ±2%. Does this pipettor meet the specification for accuracy?

7. The accuracy and precision of a micropipettor was checked using the procedure in Box 20.5. The micropipettor was set to deliver 500 μL. A standard form was used to record the results of the evaluation. A portion of the form and the results are shown below.

   a. Fill in the rest of the form.

   b. Based on the information in Figure 20.15, does this micropipettor meet its specifications? Place the answer in the pass/fail blank on the form.

Clean Gene, Inc.

VERIFICATION OF PERFORMANCE OF MICROPIPETTOR REPORT
FORM 232

CALIBRATION TECHNICIAN ____J.E.S.____

CALIBRATION DATE ____9/22/07____ NEW DUE DATE FOR NEXT CALIBRATION ____3/22/2008____

MICROPIPETTOR ID NUMBER __2127__ MANUFACTURER ____Finestt Pipettes Inc.____

MODEL NUMBER __micro B__ LOCATION ____Lab #27____ TEMPERATURE ____20°C____

RANGE __100–1000μL__ PRIMARY USER ____R.E.S.____

SUMMARY:

PREVERIFICATION CLEANING AND ADJUSTMENTS ____Changed Seals & O-rings____

_____ PASS/FAIL _____

STATUS _____

Volume 1

NOMINAL VOLUME (μL)	TUBE NUMBER	INITIAL WEIGHT OF TUBE	WEIGHT AFTER H₂O DISPENSED	NET WEIGHT OF DISPENSED H₂O	H₂O VOLUME
500	1	1.9355 g	2.4352 g		
500	2	1.9877 g	2.4876 g		
500	3	1.9787 g	2.4789 g		
500	4	1.9850 g	2.4873 g		
500	5	1.9755 g	2.4763 g		
500	6	1.9387 g	2.4387 g		

Mean Water Volume ($\bar{x}$) _____ % INACCURACY _____ SD _____ CV _____

CHAPTER 21

The Measurement of Temperature

I. INTRODUCTION TO TEMPERATURE MEASUREMENT

A. The Importance and Definition of Temperature

Biological systems in the laboratory are sensitive to temperature. For example, enzymes require a specific temperature for optimal activity. The polymerase chain reaction (PCR, an enzymatic process that is used to greatly amplify specific sequences of DNA) requires that a reaction mixture cycle between certain specific temperatures. Microorganisms and cells growing in culture must be incubated at a specific temperature. The calibration and operation of laboratory instruments requires temperature control. Temperature is one of the key process control variables in the fermentation industry and is closely monitored and controlled in the pharmaceutical industry. The laboratory contains many devices to control temperature, including ovens, heaters, incubators, water baths, refrigerators, freezers, freeze dryers, and PCR thermocyclers. It is thus essential that there be convenient and accurate methods to measure temperature in the laboratory.

Although temperature is familiar to everyone and has been studied for hundreds of years, its physical basis is not inherently obvious. Physicists now explain temperature as being related to the continuous, random motion of the molecules that make up substances. **Heat** *is a form of energy that is associated with this disordered molecular motion.* **Heating** *is the transfer of energy from an object with more random internal energy to an object with less.* For example, a burn results when a large amount of energy from a hot object is transferred to a person's skin.

Temperature *is a measure of the average energy of the randomly moving molecules in a substance. Temperature tells us which way heat will flow:* Objects with a higher temperature lose heat to objects with a lower temperature. *When two objects are at the same temperature, they are said to be in* **thermal equilibrium;** there is no net transfer of energy from one to the other. Temperature is measured by allowing a thermometer to come to thermal equilibrium with the material whose temperature is being determined. For example, taking a person's temperature involves placing a thermometer in contact with that person and waiting until the thermometer and person reach thermal equilibrium. The thermometer reading then indicates the person's temperature.

B. Temperature Scales

i. A BIT OF HISTORY

There are three temperature scales most commonly used in biology laboratories: Fahrenheit, Celsius (also previously called centigrade), and Kelvin. It is informative to discuss briefly how there came to be three scales in widespread use.

The Fahrenheit scale was invented in the early 1700s by Daniel Gabriel Fahrenheit. Fahrenheit made thermometers similar to the mercury thermometers used today. It is not entirely clear how Fahrenheit calibrated his thermometers, but it appears to have been more or less as follows. Fahrenheit chose two reference points. The first point was the coldest system he could produce, that is, a mixture of salt, water, and ice. He called the temperature of this mixture "zero." The second reference point was the body temperature of a person, which he called "96." (His methods were slightly in error because the temperature of a healthy person is 98.6 degrees on the Fahrenheit scale.) Fahrenheit placed mercury inside thin glass tubes. He positioned this assembly in the salt, water, and ice mixture and placed a mark on the glass tube to indicate the height of the mercury. He then placed the assembly in contact with a person and similarly marked the glass tube to indicate the height to which the mercury rose. Next, he subdivided the length of the glass tube between the "zero" and "96" marks into 96 equal divisions, each of which he called a **degree.** The result was a thermometer marked in degrees according to the **Fahrenheit temperature scale.** Note that this method is based on the assumption that the relationship between mercury expansion and temperature is linear.

Anders Celsius also made mercury thermometers in the 1700s. Celsius chose as his fixed reference points the boiling point of water and the freezing point of water. Celsius divided the interval between his two reference points into 100 divisions, each called a degree. The result is a thermometer marked according to the **Celsius temperature scale.** It is interesting to note that Celsius originally assigned the boiling point of water "0"

and the freezing point "100." This was, of course, later reversed so we now think of temperature as going down when it gets colder, rather than up.

Over the years, some people preferred and used Celsius thermometers, some used Fahrenheit's, and some people invented other temperature scales based on other reference temperatures. For example, there were temperature scales based on the temperature at the first frost and the temperature of underground caverns.

The Kelvin scale was devised in the early 1800s by various scientists including William Thomson (later, Lord Kelvin). *The Kelvin scale uses a unit, called a* **kelvin,** *which is the same size as a degree in the Celsius scale.* However, the Kelvin scale sets the zero point at **absolute zero,** *the temperature at which, theoretically, molecules stop their internal random motion.* Zero kelvin is $-273.15°C$; the zero point on the Celsius scale corresponds to 273.15 K. (The degree sign, "°," is not used when temperature is expressed in kelvins.) There are 100 K between the freezing and boiling points of water, just as there are 100°C between these two points.

ii. CONVERSIONS FROM ONE TEMPERATURE SCALE TO ANOTHER

As a result of various people's work in temperature measurement, there are now three temperature scales which you are likely to encounter, each of which assigns different temperature values to materials that are the same temperature. For example, the temperature at which water freezes is variously called 32°F, 0°C, and 273.15 K, see Table 21.1 and Figure 21.1.

Although Celsius is the most common scale used in biology, laboratory procedures may be written using any of the three temperature scales. It is therefore necessary to be able to convert temperatures from one scale to another. Simple equations can be used for such conversions, as shown in Box 21.1.

Table 21.1.   COMPARISON OF THE FAHRENHEIT, CELSIUS, AND KELVIN TEMPERATURE SCALES

Absolute Zero	$-460°F$	$-273°C$	0 K
Freezing Point of Water	32°F	0°C	273 K
Average Room Temperature	68°F	20°C	293 K
Normal Human Temperature	98.6°F	37°C	310 K
Boiling Point of Water	212°F	100°C	373 K

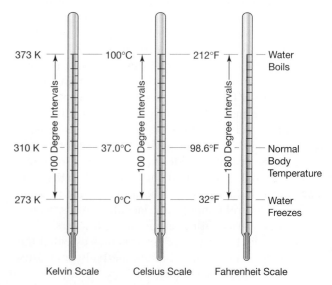

Figure 21.1.   Comparison of Kelvin, Celsius, and Fahrenheit Temperature Scales.

Box 21.1.   CONVERSION FROM ONE TEMPERATURE SCALE TO ANOTHER

1. *To convert from degrees Fahrenheit to degrees Celsius:*

$$°C = (°F − 32°)\, 0.556$$

Subtract 32° from the Fahrenheit reading
Multiply the result by 0.556 or 5/9

Example: Convert human body temperature, 98.6°F, to degrees Celsius

$$°C = (98.6° − 32°)\, 0.556 ≈ 37.0°C$$

2. *To convert from degrees Celsius to degrees Fahrenheit:*

$$°F = (°C × 1.8) + 32°$$

Multiply the Celsius reading by 1.8 or 9/5
Add 32 degrees

Example: Convert human body temperature, 37.0°C, to degrees Fahrenheit

$$°F = (37.0° × 1.8) + 32° = 98.6°F$$

3. *To convert from degrees Celsius to kelvin and from kelvin to degrees Celsius:*

$$°C = K − 273$$
$$kelvin = °C + 273$$

Example: Convert human body temperature, 37.0°C, to kelvin.

$$kelvin = 37.0°C + 273 = 310\ K$$

4. *To convert degrees Fahrenheit to kelvin:*
Convert Fahrenheit to Celsius
Convert Celsius to kelvin

Example: Convert human body temperature, 98.6°F to kelvin.

$$°C = (98.6° − 32°)\, 0.556 ≈ 37.0°C$$
$$kelvin = 37.0°C + 273 = 310K$$

---

iii. FIXED REFERENCE POINTS AND THERMOMETER CALIBRATION

We have seen that temperature scales are based on two or more fixed reference points. **Fixed reference points** *are systems whose temperatures are determined by some physical process and hence are universal and repeatable.*

Like Celsius, metrologists today use **phase transitions** as the references for calibration of laboratory thermometers and verification of their performance. A **phase transition** *is where a liquid turns to a solid or a vapor, or a vapor turns to liquid.* **Freezing point** *is the temperature at which a substance goes from the liquid phase to the solid phase.* **Boiling point** *is the temperature at which a substance in the liquid phase transforms to the gaseous phase* (under specified conditions of

pressure). People also use the **triple point** of various substances to define temperatures. The **triple point of water** *is a single temperature and pressure at which ice, water, and water-vapor coexist with one another in a closed container.* The triple point temperature for water (at atmospheric pressure) is 0.01°C. At all other temperatures and pressures, only two phases can coexist (i.e., either water and ice or water and vapor).

The boiling and freezing points of water are adequate references for thermometers used for most biological applications. Scientists who study nonliving systems use the freezing points, boiling points, and triple points not only of water but also of oxygen, hydrogen, and various metals to extend the range of temperatures at which there are fixed references.

The reason that phase transitions are used for thermometer calibration is that given the proper apparatus and conditions, phase transitions always occur at a predictable temperature and pressure. It is possible, therefore, to consistently calibrate thermometers anywhere in the world.

II. THE PRINCIPLES AND METHODS OF TEMPERATURE MEASUREMENT

A. Overview

There are various transducers (sensors) used to measure temperature, all of which respond to the effects of temperature on a physical system. For example, mercury thermometers, like those described earlier, measure the expansion or contraction of mercury in a glass tube in response to temperature changes. Types of temperature transducers include:

1. *Liquid expansion devices*
2. *Bimetallic expansion devices*
3. *Change-of-state indicators*
4. *Metallic resistance devices*
5. *Thermistors*
6. *Thermocouples*

The first two types of sensors—liquid expansion and bimetallic devices—are similar in that they are both based on the fact that most materials expand as the temperature increases and contract when it decreases. Mercury thermometers are liquid expansion devices. Liquid expansion thermometers are simple to use and relatively inexpensive.

Liquid expansion thermometers, bimetallic devices, and change-of-state indicators are mechanical measuring instruments (i.e., they do not require electricity to make a measurement nor do they generate an electrical signal). These devices are usually read by human eyes, so they cannot be directly interfaced with computers or electronic instrumentation. In contrast, resistance thermometers, thermistors, and thermocouples are electronic devices that generate an electrical signal and are readily interfaced with other equipment. Electronic thermometers are accurate, precise, versatile, compact, can be used at a wide range of temperatures, and can be used for remote sensing. As these electronic thermometers decrease in cost and become increasingly available, they are taking the place of liquid expansion thermometers in many laboratories.

Because liquid expansion thermometers are the type of thermometer traditionally most often handled in the biology laboratory, they will be discussed first. Bimetallic expansion thermometers and change-of-state indicators will then be described. The latter part of this section will outline the basic features of electronic temperature sensing devices.

B. Liquid Expansion (Liquid-in-Glass) Thermometers

i. DESCRIPTION

Liquid expansion thermometers *contain liquid that expands or contracts with temperature changes and moves up or down within a narrow glass capillary tube, or* **stem.** When the temperature of the surrounding medium increases, the liquid expands more than the glass, so the liquid moves up the stem. When the liquid and the substance whose temperature is being measured are in thermal equilibrium, the liquid stops rising. The temperature of the bulb is read by noting where the top of the liquid column coincides with a scale marked on the stem. If the temperature decreases, the liquid contracts down the stem until thermal equilibrium is reached. As was previously discussed, the scale of a liquid-in-glass thermometer is determined by marking the stem at two (or more) reference points and subdividing the interval between the marks. Liquid-in-glass thermometers are typically used in the range from −38°C to 250°C.

Mercury and alcohol are the liquids commonly used in liquid expansion thermometers. Mercury is used because it does not adhere to glass and its silvery appearance makes it easy to read. Alcohol (also called "spirit"), which is dyed red to make it easier to read the meniscus, is a safer material and poses less threat to the environment than mercury. Alcohol can be used at temperatures below −38.6°C, where mercury freezes. Mercury thermometers have a more linear response with temperature than alcohol and for this reason are more often used in the laboratory.

> *Safety note:* **A mercury spill kit should always be available in laboratories where mercury thermometers are used. See pp. 180–182 for information on spill cleanup.**

Liquid-in-glass thermometers have certain components, see Figure 21.2. These components are:

1. **Liquid,** *usually mercury or alcohol*
2. **Stem,** *a glass capillary tube through which the mercury or organic liquid moves as temperature changes*

3. **Bulb,** *a thin glass container at the bottom of the thermometer that is a reservoir for the liquid*
4. **Scale,** *graduations that indicate degrees, fractions of degrees, or multiples of degrees*
5. **Contraction chamber** *(not present on all thermometers), an enlargement of the capillary bore that takes up some of the volume of the liquid thereby allowing the overall length of the thermometer to be reduced*
6. **Expansion chamber,** *an enlargement of the capillary bore at the top of the thermometer that prevents buildup of excessive pressure*
7. **Stem enlargement** *(not present in all thermometers), a thickening of the stem that assists in the proper placement of the thermometer in a device (e.g., in an oven)*
8. **Immersion line** *(not present on all thermometers), a line etched onto the stem to show how far the thermometer should be immersed into the material whose temperature is to be measured*
9. **Upper and lower auxiliary scales** *(not present on all thermometers), extra scale markings at the zero degree and 100 degree areas to assist in calibration and verification of the performance of the thermometer*

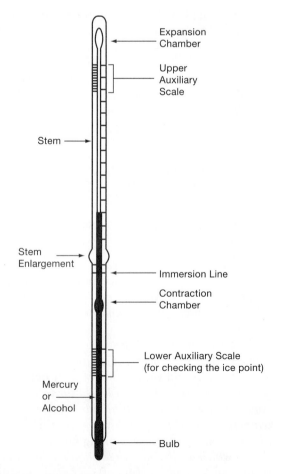

Figure 21.2.   The Parts of a Mercury Thermometer.

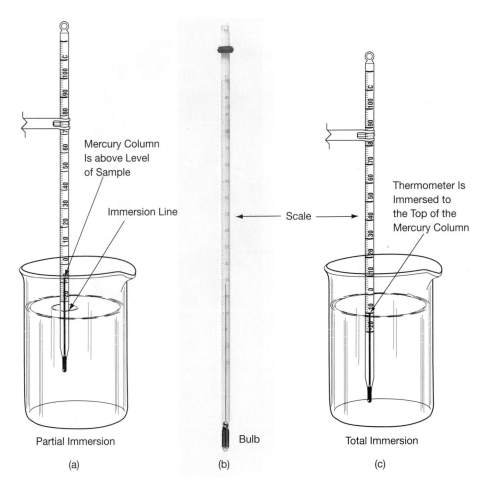

Mercury Column
Is above Level
of Sample

Immersion Line

Scale

Thermometer Is
Immersed to
the Top of the
Mercury Column

Partial Immersion

Bulb

Total Immersion

(a)

(b)

(c)

Figure 21.3.   Partial and Total Immersion Thermometers. a. Partial immersion thermometer. A line on the stem indicates the required immersion depth. The thermometer is immersed only to the immersion line. **b.** Total immersion thermometer. **c.** A total immersion thermometer is positioned so that all the mercury (or alcohol) within the thermometer is immersed in the material whose temperature is being measured. (Photo courtesy of Fisher Scientific, Pittsburgh, PA.)

Liquid-in-glass thermometers come in many styles that differ in their scale divisions, range, length, accuracy, intended applications, and other qualities. For example, a thermometer for use in a refrigerator or freezer might have a range from −40°C to 25°C. A thermometer for monitoring the temperature in an autoclave might measure temperatures from 85–135°C and may be specially designed to hold and display the maximum temperature reached during sterilization, even after the autoclave has cooled.

ii. Immersion

The depth of immersion of a liquid thermometer into the material whose temperature is to be measured affects the accuracy of the measurement. This is because the liquid in the part of the stem that is not immersed is not at the same temperature as the liquid in the bulb. The manufacturer takes this difference into account when the thermometer is calibrated and indicates on the thermometer how far it is to be immersed. There are two basic immersion types of the thermometer: partial immersion and total immersion. A third type, complete immersion, is less common.

A **partial immersion thermometer** *is manufactured to indicate temperature correctly when the bulb and a specified part of the stem are exposed to the temperature being measured,* see Figure 21.3a. Partial immer-

sion thermometers are easiest to read because their liquid column is not fully immersed in the medium. They are ideal for low-volume solutions. A partial immersion thermometer is marked with a line to indicate how far it should be immersed.

A **total immersion thermometer** *is designed to indicate temperature correctly when that portion of the thermometer containing the liquid mercury or alcohol is exposed to the medium whose temperature is being measured,* see Figure 21.3b,c. Total immersion thermometers are more accurate than partial immersion thermometers and are suggested for applications where accuracy is critical or when a thermometer is used for calibrating other thermometers or equipment.

If a total immersion thermometer is not immersed to the proper depth there will be inaccuracies in its readings. In general, at the temperatures used in a biology laboratory (e.g., between 0 and 100°C), the correction is less than 1 degree. Nonetheless, it is good practice to always immerse a thermometer properly.

There are several ways to distinguish a partial from total immersion thermometer. A partial immersion thermometer has an immersion line marked on the stem; a total immersion thermometer does not. The immersion depth, in millimeters, is sometimes written on the stem of a partial immersion thermometer. Man-

Temp Range (°C)	Div (°C)	Total Length (mm)	Immersion Depth (mm)
24 to 38	0.05	305	76
−1 to 51	0.1	460	Total
−1 to 101	0.1	610	Total
−20 to 110	1	305	76
−100 to 50	1	305	76
−1 to 201	0.2	610	Total

Figure 21.4. Catalog Descriptions Specify How a Thermometer Should Be Immersed.

ufacturers always specify the immersion type for each thermometer in their catalog, see Figure 21.4.

A total immersion thermometer properly indicates temperature when that portion of the thermometer containing the liquid mercury or alcohol is exposed to the medium whose temperature is being measured. A total immersion thermometer is not intended to actually be placed inside an oven or to be immersed *completely* in a hot liquid. A total immersion thermometer will be inaccurate if it is completely immersed, and there is the possibility that the thermometer will break at high temperatures if placed inside an oven or incubator. Most laboratory ovens have a slot in which the thermometer is inserted so that the proper length of the thermometer is inside the oven and the top part protrudes visibly outside the oven. There are **complete immersion thermometers** *that are designed so that the entire thermometer, from top to bottom, is exposed to the medium whose temperature is being measured.* Figure 21.5 shows a complete immersion thermometer specifically designed to be placed inside an oven, incubator, or freezer. Observe that this thermometer is protected in case of breakage.

iii. LIMITATIONS OF LIQUID EXPANSION THERMOMETERS

Although liquid expansion thermometers are routinely used with acceptable results, they have inherent limitations. Liquid expansion thermometers have fairly wide tolerances (i.e., they may be in error by a fair amount). It is not unusual for manufacturers to specify that the tolerance for a general purpose mercury thermometer is ± 1°C or ± 5°C (± 2°F or ± 9°F). A thermometer may be in error by as much as its maximum tolerance. This means, for example, that if you are using a thermometer whose tolerance is ± 5°C, and you measure the temperature of a solution as 37°C, the actual temperature of that solution may be anywhere from 32 to 42°C.

ASTM classifies thermometers and specifies how much error is tolerable for a particular class. Tolerances for partial immersion thermometers are greater

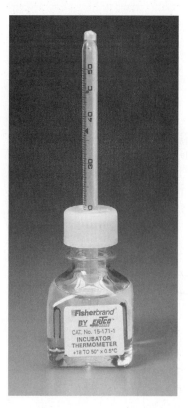

Figure 21.5. Complete Immersion Thermometer Intended for Use in a Freezer, Oven, or Incubator. (Courtesy of Fisher Scientific, Pittsburgh PA.)

than for total immersion thermometers. More expensive "precision thermometers" have narrower tolerances than other thermometers. To meet ASTM requirements, precision thermometers must have an error of less than 0.1–0.5°C, depending on their range. A precision thermometer whose range does not include 0°C will have a second, **auxiliary scale** *that shows the zero point so that the performance of the thermometer can be verified with an ice point bath* (discussed later).

Other limitations of liquid expansion thermometers are described in Table 21.2.

iv. PROPER USE OF LIQUID EXPANSION THERMOMETERS

Many of the problems that arise with liquid expansion thermometers are easy to diagnose. The most common problem is breakage. Another common problem with liquid thermometers is that the liquid column develops separations due to thermal or mechanical shock during shipping or handling. If the liquid column is discontinuous there will be substantial errors in the thermometer's readings. Thermometers should be inspected before use to be sure the liquid column is continuous. Box 21.2 shows several procedures that are used to reunite a liquid column.

Thermometers must be used properly to give correct measurements. Table 21.3 on p. 376 summarizes points

Table 21.2. LIMITATIONS OF LIQUID EXPANSION THERMOMETERS

1. *The capillary bore within the thermometer in which the liquid moves must be very smooth so that the liquid is not impeded at any spot.* This is difficult to achieve, and less expensive thermometers may have slight imperfections leading to inaccuracy in their measurement.

2. *The scale must be very carefully etched onto the stem to ensure accuracy.* Not all thermometers are accurately subdivided into degrees.

3. *The working range of liquid thermometers is limited.* Mercury freezes at −38.6°C and alcohol at −200°C. In practice, liquid thermometers are seldom used above about 250°C. The limited range of liquid expansion thermometers is not usually a problem for biologists because life exists only in a fairly narrow range of temperatures.

4. *Glass thermometers are fragile.* The glass comprising the bulb provides a thin interface between the liquid and the substance whose temperature is to be measured. This means the glass is thin and delicate and may easily be deformed or broken.

5. *Mercury, commonly used in liquid expansion thermometers, is hazardous to human health and the environment.* Mercury fumes and solids are toxic, and mercury from a broken thermometer can contaminate both the air and a water bath, a counter, or other site.

regarding the proper use of liquid expansion thermometers.

C. Bimetallic Expansion Thermometers

Materials generally expand as the temperature increases and contract when it decreases. Each material has a characteristic *thermal expansion coefficient that measures how much the material expands for a specific increase in temperature.* If two metals with different coefficients of expansion are bonded together and are then subjected to heat, the strips of metal will distort and bend. The amount of bending is related to the temperature, see Figure 21.6a on p. 376.

A bimetallic dial thermometer is constructed by fusing together two metal strips, typically one of brass and one of iron, which have different coefficients of expansion. When the temperature changes, the two metals respond unequally, resulting in bending. One end of the strips is fastened and cannot move, the other moves along a scale to display the temperature, see Figure 21.6b on p 376.

Bimetallic expansion thermometers are simple and convenient, but they are generally not intended to be as accurate as liquid expansion thermometers. They do have many applications, however, such as for checking the temperature of chemicals in a photography darkroom and as oven thermometers in the home. Note that bimetallic laboratory thermometers should be discarded if the stem is bent or deformed.

Box 21.2. REUNITING THE LIQUID COLUMN OF LIQUID EXPANSION THERMOMETERS

1. *Lightly tap the thermometer by making a fist around the bulb with one hand and gently hitting your fist into the palm of the other hand.*

2. *Clamp the thermometer vertically. Slowly immerse the bulb of the thermometer in a freezing bath of salt, ice, and water.* The liquid should all slowly contract into the bulb, causing it to reunite. Be careful to cool the bulb only, not the stem. It may be helpful to remove the bulb periodically to slow the motion of the liquid. Remove and allow the thermometer to warm while it is supported in an upright position. Repeat several times if necessary.

3. *If 1 and 2 do not work, and if the separation is in the lower part of the column, then try cooling the thermometer, as described in 2, but use a mixture of dry ice and either acetone or alcohol.* Wear cold-resistant gloves. Cool only the bulb, not the stem. Remove the bulb from the cold. Do not touch the bulb until it has warmed to room temperature; it might shatter.

4. *As a last resort, particularly if the separation is in the upper part of the mercury column, warm the thermometer.*
   a. Use this method only for thermometers with expansion chambers above the scale.
   b. Place the thermometer in a beaker of water or other nonflammable liquid and heat it slowly.
   c. As the thermometer heats, the liquid should expand into the upper expansion chamber.
   d. When the liquid has moved into the expansion chamber, gently tap the thermometer with a gloved hand to reunite the column.
   e. Watch to be sure the expansion chamber does not overfill since if it does, the thermometer will break.
   f. Allow the thermometer to cool slowly.
   g. *Never use an open flame to heat the bulb.*

5. *If the column appears to be successfully reunited, check the ice point. If the ice point is shifted from its previous position, discard the thermometer.*

6. *If the liquid column cannot be reunited, discard the thermometer.*

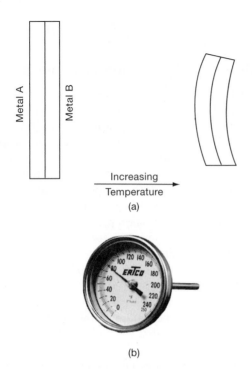

Figure 21.6. Bimetallic Expansion Thermometers. a. The unequal bending of two metals. **b.** A bimetallic thermometer. (Photo courtesy of Ever Ready Thermometer Co.)

D. Change-of-State Indicators

Change-of-state indicators *are varied products, all of which change color or form when exposed to heat.* For example, items can be painted with lacquers that change appearance when heated or marked with crayons that melt when a particular temperature is reached. Labels are available that can be attached to items and change colors at specific temperatures, see Figure 21.7a. There are encapsulated liquid crystals that change to various colors when exposed to particular temperatures. Crystals change colors reversibly and so can be reused. A color guide is supplied with the crystals to indicate the temperature reached, see Figure 21.7b. Crystals have a fairly narrow temperature range, but are useful at temperatures applicable to biological systems.

E. Resistance Thermometry: Metallic Resistance Thermometers and Thermistors

Metallic resistance thermometers and **thermistors** are two types of thermometers based on the principle that the electrical resistance of materials changes as their temperature changes. **Metallic resistance thermometers** or **resistance temperature detectors (RTDs)** *use metallic wires in which resistance increases as the temperature increases.* **Thermistors** *use semiconductor materials in which the resistance decreases as the temperature rises.*

Platinum wire resistance thermometers are considered to be the most accurate type of thermometer available. They are frequently used as references against which other thermometers are calibrated.

Table 21.3. PROPER USE OF LIQUID EXPANSION THERMOMETERS

1. *Visually inspect a thermometer before use for:*
   breaks in the liquid column or bubbles
   foreign material in the stem
   distortions in the scale

   If the liquid is separated, reunite it as described in Box 21.2. If the thermometer contains foreign material or has a distorted scale, it should be discarded.

2. *Avoid tapping a glass thermometer against a hard surface to prevent small fractures.*

3. *Support the thermometer by its stem; do not allow the bulb to rest on a surface.* The bulb is fragile.

4. *Do not stir solutions with thermometers.*

5. *Never subject a glass thermometer to rapid temperature changes.* For example, do not move a thermometer from an ice bath and place in a boiling water bath. This may cause the thermometer to break or the liquid filling solution to separate.

6. *Never place a thermometer in an environment above its maximum indicated temperature.*

7. *Immerse the thermometer to the depth indicated by the manufacturer.*

8. *Read the meniscus with your eyes at an even level with the top of the liquid column.* For mercury, the top of the meniscus is used as the indicator. For organic liquids, the bottom of the meniscus is the part that is used for a reading.

9. *Store thermometers as instructed by the manufacturer.* Mercury thermometers are most often stored horizontally in a protective case to avoid damage. It is sometimes recommended that alcohol thermometers and thermometers used at temperatures below 0°C be stored vertically in a rack.

10. *Place thermometers properly in equipment whose temperature is being measured.*
    a. Do not allow thermometers to touch the sides of equipment whose temperature is being measured.
    b. Place the thermometer in a manner to avoid breakage.
    c. Safely mounted thermometers for use in refrigerators, incubators, and other equipment are available commercially. Use these when possible.
    d. For freezers and refrigerators place the thermometer so that the stem does not touch any metal surface or ice. Air must be able to flow unobstructed around the stem of the thermometer.
    e. For large freezers or refrigerators it is good practice to check the temperature at more than one location.

Thermistors are constructed of a hard, ceramic-like material made up of a compressed mixture of metal oxides. The material can be molded into many shapes. Thermistors respond rapidly to temperature changes, can be miniaturized, function in a wide range of temperatures, and are very accurate. Thermistors are often used in a production setting as part of a temperature control device.

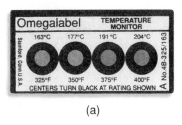

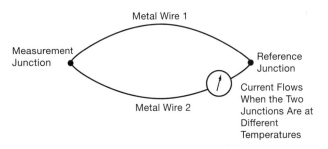

Figure 21.8.   A Simplified Thermocouple.

Figure 21.7.   Change-of-State Indicators.
a. A label that changes colors at specific temperatures. **b.** Liquid crystal display. (Photos © copyright. Thermographic Measurements, reprint permission courtesy of Omega Engineering, Inc., www.omega.com.)
(See color insert C-1.)

F. Thermocouples

A **thermocouple** *consists of two wires made of different metals that are joined together,* see Figure 21.8. Two dissimilar wires joined as shown in Figure 21.8 generate a voltage. The magnitude of the voltage depends on the temperature at the junction between the wires. The voltage that develops in the wires, therefore, can be converted to a temperature measurement.

Thermocouples are rugged and versatile. They come in various types that are assigned a letter designation depending on the types of metals used in their construction. For example, a Type T thermocouple has a copper wire and a wire made from a copper-nickel alloy. This type of thermocouple is used from 0 to 350°C. Each type of thermocouple is useful in different applications and has a different temperature range.

Figure 21.9 shows a thermocouple probe that is connected to a continuous recording device, thus providing continuous temperature documentation.

III. VERIFYING THE PERFORMANCE OF LABORATORY THERMOMETERS

Thermometers have many uses in biology laboratories, including checking the temperatures of solutions and monitoring the temperatures of water baths, incubators, ovens, refrigerators, and freezers. It is good laboratory practice to verify that the thermometers used to perform these tasks are performing acceptably. Performance verification of a thermometer demonstrates that its readings are correct within its specified tolerance range.

For most thermometers, about 95% of all possible malfunctions will affect the ice-point reading (Nicholas, J.V., and White, D.R. *Traceable Temperatures: An Introduction to Temperature Measurement and Calibration,* Chichester, England: John Wiley and Sons, 1994). At the ice point, a Celsius thermometer should read at the "0" mark and a Fahrenheit thermometer should read at the "32" mark; otherwise, the thermometer is not performing

adequately. Part A of Box 21.3 on p. 378 describes a procedure for checking whether a thermometer is reading correctly at the ice point.

An ice point check is useful, but it only verifies the performance of a thermometer at one temperature. It is possible to check a thermometer more thoroughly by comparing its readings with those of a standard thermometer that is certified to be "NIST-traceable." A NIST-traceable thermometer is manufacturer-calibrated against a standard thermometer that, in turn, was calibrated at NIST. The readings of the thermometer that is to be checked are compared with the readings of the NIST-traceable thermometer at several temperatures spanning the thermometer's range. A procedure to check the performance of a thermometer by comparing it with a NIST-traceable thermometer is shown in part B of Box 21.3.

A thermometer should give the correct readings at the ice point and/or when compared with a standard thermometer. (Recall, however, that a given thermometer may have a fairly wide tolerance range.) Liquid expansion thermometers cannot be repaired or adjusted (except to reunite the liquid column). If the readings of a general purpose liquid expansion

Figure 21.9.   A Thermocouple Connected to a Continuous Recording Device. (© Copyright by Cole-Parmer Instrument Co.; used by permission.)

Box 21.3. VERIFYING THE PERFORMANCE OF A THERMOMETER

A. Checking the Ice Point* of the Thermometer

1. *Visually inspect the thermometer whose performance is to be tested.* Check the column for separation or bubbles. Reunite the column and remove bubbles present.

2. *Make certain the outside of the thermometer is clean.*

3. *Prepare an ice point apparatus:*
   a. *A Dewar flask (a vacuum-insulated flask for holding cold materials) is preferable to contain the ice point apparatus.*
   b. *Fill the flask one third full with distilled water. Then add ice shaved or smashed into tiny chips.* The ice should be made from distilled water.
   c. *Compress the ice water mixture into a tightly packed slush.* It is necessary to maintain tightly packed ice so that the thermometer is in close thermal contact with the ice-water mixture. Lack of good thermal contact is the major potential source of error in this procedure. Use only enough water to maintain good contact with the thermometer. The ice should not float. Remove excess water. It may be helpful to place a siphon tube that touches the bottom of the flask. The purpose of the siphon is to drain off excess water that collects at the bottom of the Dewar. The ice should appear clear and not white. White ice is a sign that the ice is at a temperature slightly below zero degrees. Avoid handling the ice to avoid contaminating it.
   d. *Wait 15–20 minutes for the mixture to reach equilibrium.*
   e. Periodically remove water from the bottom of the flask and add more ice.

4. *Immerse the thermometer in the apparatus to a depth approximately one scale division below the 0°C gradation.* Allow the thermometer to equilibrate to the temperature of the ice bath. The thermometer should read 0°C. It is recommended to use a 10X magnifying glass to read the scale display. Be certain your eyes are level with the display. The correct temperature is read from the top of the meniscus for a mercury thermometer.

B. Checking a Thermometer against a Certified Reference Thermometer at Temperatures other Than the Ice Point

If a certified reference thermometer is available, then the thermometer can be checked against the reference at various temperatures. The certified standard thermometer should have calibrated intervals ten times more precise than the working thermometer. (For example, to verify a thermometer with 0.1°C intervals, use a reference with intervals of 0.01°C.)

1. *Adjust a stable water heating bath to the temperature required for the analysis of interest.*

2. *Place the reference thermometer and the thermometer to be checked in the water.* The thermometers should be placed close to one another, but with sufficient space between them to ensure adequate circulation in the bath. Immerse the thermometers to their proper depth and allow time for them to equilibrate to the temperature of the bath.

3. *After thermal equilibrium is reached (several minutes for liquid-in-glass thermometers) determine the temperature reading for both thermometers.* Note that reference thermometers come with a report of calibration that may include a correction factor. If there is a correction, then add or subtract that factor to or from the reading of the reference thermometer to obtain the actual temperature. It is good practice to take several readings of each thermometer and average them.

4. *Using a heated liquid bath, sequentially adjust the temperature of the bath to read each temperature for which the reference thermometer is calibrated.* For example, if the reference thermometer is calibrated at 25, 30, and 37°C, then sequentially adjust the water bath to these temperatures, as shown on the reference thermometer. Place the thermometer(s) whose performance is being checked into the water. The reading of the thermometer being checked should be the temperature of the water bath within the tolerance for that thermometer.

5. *Document the results of performance verification in a logbook or on appropriate forms.*

Based on ASTM Standard E 77-07 "Standard Test Method for Inspection and Verification of Thermometers." Consult this standard for more details.

*The freezing (ice) point of water is essentially independent of the environment, but the boiling point of water depends on the atmospheric pressure; therefore, the freezing point is more convenient for verification than the boiling point.

thermometer fall outside its tolerance range, it is recommended that the thermometer be discarded.

Note that standard mercury thermometers used to verify the performance of other thermometers may have correction factors. This means that the reading of the thermometer was tested at a specific temperature and was found to be slightly "off" by the amount of the correction factor. The correction factor is added to or subtracted from the reading of the standard thermometer to get the correct temperature. Thus, standard thermometers that have known corrections are not discarded even if they are marked slightly imperfectly.

PRACTICE PROBLEMS

1. A thermocycler (an instrument used when DNA is amplified by the polymerase chain reaction method) is adjusted to cycle a reaction mixture between the temperatures shown. Convert these temperatures to °F.

   94°C   (for 1 minute)

   37°C   (for 1 minute)

   72°C   (for 1 minute)

2. A thermophilic bacterium is found in nature at temperatures near 81°C. Convert this to °F.

3. A child is running a fever of 103.2°F. Convert this to °C.

4. **a.** The relationship between resistance and temperature in a platinum (Pt) wire is shown in the following graph. If a platinum resistance thermometer is calibrated using two reference points (e.g., the freezing and boiling points of water), would you expect the calibration to result in accurate results?

   b. The relationship between resistance and temperature in a nickel (Ni) wire is shown in the graph. If a resistance thermometer were made using a nickel wire and if it was calibrated using two reference points (e.g., the freezing and boiling points of water), would you expect the calibration to result in accurate results?

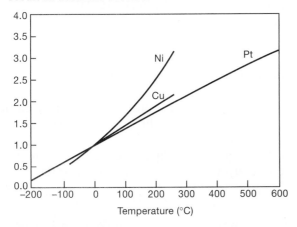

(From "Traceable Temperatures: An Introduction to Temperature Measurement and Calibration," J.V. Nicholas and P.R. White. Copyright John Wiley and Sons Ltd. Reproduced by permission)

5. A 24-hour temperature recording from a fermenter is shown in the following figure. What was the temperature at 1:00 A.M.? What was the lowest temperature in the fermenter? What was the highest temperature? Is this an analog or a digital form of a display?

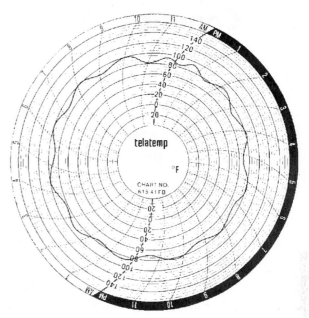

(Courtesy of Telatemp Corp., Fullerton, CA.)

QUESTIONS FOR DISCUSSION

If you work in a laboratory, examine the laboratory thermometers and answer these questions. For liquid expansion thermometers, what is the filling liquid? What is the thermometer's range? Identify each thermometer as either partial or total immersion. If a thermometer is a partial immersion type, what is its specified immersion depth? Do any of your thermometers have an auxiliary scale? Determine whether the laboratory thermometers have an expansion bulb.

Are there electronic thermometers in your laboratory? If so, what is their purpose?

CHAPTER 22

The Measurement of pH, Selected Ions, and Conductivity

I. THE IMPORTANCE AND DEFINITION OF pH

The chemistry of life is based on water. Life evolved in the presence of water and cells contain 80 to 90% water. **pH,** *which is a measure of the acidity of an aqueous solution,* is intrinsically related to water chemistry. Maintaining the proper pH is essential for living systems. For example, plants do not grow if the soil pH is wrong, animals die if their blood pH deviates from normal, and microorganisms require their growth medium to be a particular pH. Many industrial processes, such as food processing, sewage treatment, water purification, and pharmaceutical production, are likewise sensitive to pH. Biological systems and many industrial processes, therefore, must be monitored and maintained at the proper pH. This chapter briefly defines pH and discusses how it is measured.

Pure water naturally dissociates to a limited extent to form **hydrogen (H^+)** and **hydroxide (OH^-)** ions:

$$H_2O \leftrightarrows H^+ + OH^-$$

In pure water at $25°C$, an equilibrium is established such that the concentration of H^+ ions is the same as the concentration of OH^- ions; it is 1×10^{-7} mole/L.

(See pp. 496–497 for an explanation of molarity.) Pure water, and other solutions with a H^+ ion concentration of 1×10^{-7} mole/L, are called **neutral.**

When **acids** *are added to water, they release hydrogen ions into solution.* On the other hand, when **bases** *are added to water, they cause hydrogen ions to be removed from solution.* Thus, the addition of acids to water causes the concentration of hydrogen ions to be greater than 1×10^{-7} mole/L. The addition of bases to water causes the concentration of hydrogen ions to be less than 1×10^{-7} mole/L.

Strong acids and **strong bases** completely dissociate when dissolved in water and therefore have a strong effect on the hydrogen ion concentration. For example, hydrochloric acid (HCl) dissociates completely to release H^+ ions and Cl^- ions, thus increasing the H^+ ion concentration of the solution:

$$HCl \text{ in water} \rightarrow H^+ + Cl^-$$
(all is in this form)

NaOH is an example of a strong base that dissociates completely in water as follows:

$$NaOH \text{ in water} \rightarrow Na^+ + OH^-$$
(all is in this form)

The OH^- ions react with H^+ ions to form water, thereby lowering the concentration of hydrogen ions.

Weak acids and **weak bases** do not completely dissociate in water and therefore have a smaller effect on the concentration of hydrogen ions in the solution. For example, acetic acid forms about one hydrogen ion for every 100 molecules:

$$CH_3COOH \text{ in water} \leftrightarrows H^+ + CH_3COO^-$$
(most of the acid is in this form)

The weak base, ammonia (NH_3) removes hydrogen ions from water with the formation of OH^- ions:

$$NH_3 + H_2O \leftrightarrows NH_4^+ + OH^-$$
(most of the base is in this form)

pH is a convenient way to express the hydrogen ion concentration* in a solution, or the solution's acidity. *pH is the negative log of the H^+ concentration when concentration is expressed in moles per liter.*

To illustrate the definition of pH, consider the pH of pure water:

The H^+ concentration of pure water is 1×10^{-7} mole/L.
The log of 1×10^{-7} is -7.
The negative log of 10^{-7} is $-(-7) = 7$.
The pH of pure water is 7.

Thus, the H^+ concentration of pure water is 10^{-7} mole/L and its pH is 7. By definition, the pH of any neutral solution is 7. The practical range of pH is between 0 and 14. (Very concentrated solutions of acid or base can fall outside this pH range and require special care when handling to protect both the technician and the equipment.) Solutions with a pH less than 7 are acidic while those with a pH greater than 7 are basic.

The ultimate importance of H^+ concentration is that it profoundly affects the properties of aqueous solutions. For example, a strongly acidic solution, such as concentrated sulfuric acid, can dissolve iron nails. (Strong acids and bases thus require safety precautions in the laboratory.) A weaker acid, citric acid, contributes to the taste of foods like lemon and orange juice, see Table 22.1.

The pH scale is logarithmic. For example, a solution with a pH of 5.0 has 10 times more hydrogen ions than a solution with a pH of 6.0 and a solution of pH 7.0 is 100 times less acidic than a solution at pH 5.0. As hydrogen ion concentration, acidity, increases, the pH value decreases, see Table 22.2 on p. 382.

Because maintaining the proper pH is critical to biological systems, methods are required to measure the concentration of hydrogen ions in solutions. Two common approaches to pH measurement in the laboratory are the use of **indicator dyes** and the use of **pH meter/electrode** measuring systems. We discuss first indicator dyes and then pH meter measuring systems.

Table 22.1.   THE pH OF COMMON COMPOUNDS

pH Value	Compounds
0.3	Sulfuric acid (battery acid)
2	Lemon juice
3	Wines
3–4	Oranges
4–5	Beers
5–6	Measured pH of purified laboratory water
7	Pure water
7.3–7.5	Blood (human)
7–8	Egg whites
8	Sea water
8.4	Sodium bicarbonate
10	Milk of magnesia
11.6	Household ammonia
12.6	Bleach
14	1 M Sodium hydroxide

*The term *pH* properly should be defined as the negative log of the H^+ activity, not its concentration. The activity of hydrogen ions is primarily related to their concentration, but it is also affected by other substances in the solution, by the solvent, and by the temperature of the solution. Defining pH in terms of the apparent concentration of hydrogen ions is an approximation that is very commonly used and is adequate for most purposes in a biology laboratory.

EXAMPLE PROBLEM

a. What is the pH of a solution with an H^+ ion concentration of 10^{-4} mole/L?

b. What is the pH of a solution with an H^+ ion concentration of 5.0×10^{-6} mole/L?

ANSWER

a. $pH = -\log [H^+] = -\log 10^{-4} = -(-4) = 4$

b. $pH = -\log [H^+] = -\log 5.0 \times 10^{-6} \approx -(-5.3) = 5.3$

EXAMPLE PROBLEM

What is the concentration of H^+ ions in a solution with a pH of 9.0?

ANSWER

$$pH = -\log [H^+]$$
$$9.0 = -\log [H^+]$$
$$-9.0 = \log [H^+]$$
$$\text{antilog } (-9.0) = 1 \times 10^{-9} \text{ mole/L}$$

Table 22.2. THE RELATIONSHIP BETWEEN $[H^+]$ AND pH

Hydrogen Ion Concentration (mole/L)		pH
Acidic		
10^0	1.0	0
10^{-1}	0.1	1
10^{-2}	0.01	2
10^{-3}	0.001	3
10^{-4}	0.0001	4
10^{-5}	0.00001	5
10^{-6}	0.000001	6
Neutral		
10^{-7}	0.0000001	7
Basic		
10^{-8}	0.00000001	8
10^{-9}	0.000000001	9
10^{-10}	0.0000000001	10
10^{-11}	0.00000000001	11
10^{-12}	0.000000000001	12
10^{-13}	0.0000000000001	13
10^{-14}	0.00000000000001	14

II. pH INDICATORS

pH indicators *are dyes whose color is pH dependent and that change colors at certain pH values.* Indicators can be directly dissolved in a solution or can be impregnated into strips of paper that are then dipped into the solution to be tested. **Phenolphthalein** is a well-known indicator dye (and formerly the main ingredient in some laxatives). Phenolphthalein changes from colorless to red between pH 8 and 10. **Litmus,** which is extracted from lichens, is probably the oldest pH indicator. Litmus paper is pink in acidic solutions and blue in basic ones.

Indicators do not change colors sharply at a single pH; rather, they change over a range of one or two pH units. Thus, indicators are often used to broadly categorize a solution as acidic, basic, or neutral, or to determine roughly the pH of a solution. It is also possible to purchase pH indicators or pH paper that contain mixtures of dyes. These mixtures exhibit a range of colors and can be used to determine the pH of a sample to ± 0.3 pH units.

Indicators are a useful means of monitoring pH for some applications. For example, human cells can be grown in the laboratory in nutrient medium. Proper pH of the medium is critical for cell survival but is easily altered by metabolites from the cells, contamination from microorganisms, and improper culture conditions. Therefore a nontoxic, nonreactive pH indicator dye, phenol red, is added to cell culture media. Changes in the indicator dye color provide immediate, visible information about the condition of the culture without opening the culture plates and exposing them to potential contaminants. When the medium is a certain shade of pink, the culture is healthy. A yellowish tinge indicates that the medium is becoming acidic and requires attention. Another example is the use of pH paper to measure the pH of toxic or radioactive solutions. The paper can be dipped in the hazardous solution and discarded, thus avoiding contamination of a nondisposable pH measuring device.

pH indicators are inexpensive and simple to use; however, pH indicators show, at best, the pH of a solution within a range of about 0.3 pH units. In addition, indicators cannot be used in certain types of solutions. pH meter measuring systems should be used in situations where pH indicators provide inadequate information. Table 22.3 summarizes potential sources of error when using pH indicators. Table 22.4 lists common pH indicators.

III. THE DESIGN OF pH METER/ELECTRODE MEASURING SYSTEMS

A. Overview

pH meter/electrode *systems are the most common method of measuring pH in the biology laboratory.* Although more expensive and more complicated to use than chemical indicators, pH meters provide greater accuracy, sensitivity, and flexibility. When used properly, pH meters can measure the pH of a solution to the nearest 0.1 pH unit or better, and they can be used with a variety of samples.

A pH meter/electrode measuring system consists of (1) a **voltmeter** that measures voltage, (2) two **electrodes** connected to one another through the meter,

Table 22.3. POTENTIAL SOURCES OF ERROR WHEN USING pH INDICATORS

1. *Acid–base error.* Indicators are themselves acids or bases. When indicators are added to unbuffered or weakly buffered solutions, such as distilled water or very dilute solutions of strong acids or bases, the indicator causes a change in pH of the sample solution.

2. *Salt error.* Salts at concentrations above about 0.2 M can affect the color of pH indicators, leading to inaccuracy.

3. *Protein error.* Proteins react with indicators and can have a significant effect on their color. Indicators, therefore, should not be used to measure the pH of protein solutions.

4. *Alcohol error.* The solvent in which a sample is dissolved can affect the color of an indicator; therefore, a sample dissolved in alcohol may be a different color than if it is dissolved in buffer.

Table 22.4. EXAMPLES OF COMMON INDICATORS

Indicator	Color Change
Blue Litmus	Changes from blue to red denoting a change from alkaline to acid
Brilliant Yellow	Changes from yellow at pH 6.7 to red at pH 7.9
Bromocresol Green	Changes from yellow at pH 4.0 to blue at pH 5.4
Bromocresol Purple	Changes from yellow at pH 5.2 to purple at pH 6.8
Congo Red	Changes from red at pH 3.0 to blue at pH 5.0
Cresol Red	Changes from yellow at pH 7.2 to red at pH 8.8
Methyl Red	Changes from red at pH 4.2 to yellow at pH 6.2
Neutral Litmus	Red in acid conditions and blue in alkaline conditions
Neutral Red	Changes from red at pH 6.8 to orange at pH 8.0
Phenol Red	Changes from yellow at pH 6.8 to red at pH 8–8.2
Phenolphthalein	Changes from colorless at pH 8.0 to red at pH 10.0
Red Litmus	Changes from red to blue denoting a change from acid to alkaline

The U.S. Pharmacopeia ("U.S. Pharmacopeia/National Formulary," The United States Pharmacopeial Convention, Rand McNally Publishers, 2006) is a good source of information about pH indicators, their chemical nature, solubility, and preparation, because indicators are used in testing drugs.

and (3) the **sample** whose pH is being measured. The term **pH meter** *is commonly used to refer both to the voltage meter and its accompanying electrodes.*

When the two electrodes are immersed in a sample, they develop an electrical potential (voltage) that is measured by the voltmeter. (The Appendix to this chapter has a brief explanation of how electrodes develop potential.) The magnitude of the measured voltage depends on the hydrogen ion concentration in the solution; therefore the voltage reading can be converted by the instrument to a pH value, see Figure 22.1.

B. The Basic Design and Function of Electrodes

i. ELECTRODES AND THE MEASUREMENT OF pH

The electrodes are the heart of the pH measuring system and require some discussion. A pH meter is connected to two electrodes. One is a pH measuring electrode, the other is a reference electrode. The **pH measuring electrode** has a thin, fragile glass bulb at its tip and is therefore often called a **glass electrode,** see Figure 22.2 on p. 384. The glass that composes the bulb is of a special type that is sensitive to the concentration of H^+ ions in the surrounding medium. An electrical potential develops between the inner and outer surfaces of the glass bulb when the measuring electrode is immersed in a sample solution. It is this potential whose magnitude varies depending on the concentration of hydrogen ions in the solution. The glass electrode contains a buffered chloride solution. A metal wire inside the electrode acts as a lead connecting the glass bulb to the voltmeter.

You might wonder why two electrodes are required because the glass measuring electrode by itself is sensitive to pH. In practice, it has been found that it is not possible to measure the potential of a single electrode. The pH measuring electrode must be paired with a second, **reference electrode,** *which has a stable, constant voltage.*

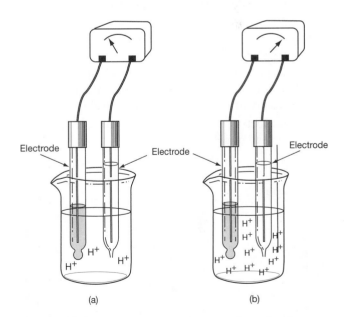

Figure 22.1. Overview of a pH Meter Measuring System. The system consists of two electrodes that are immersed in the sample and which are connected to one another through a meter. **a.** The concentration of H^+ ions in the solution is relatively low. **b.** A higher concentration of H^+ ions leads to a change in the voltage difference between the two electrodes.

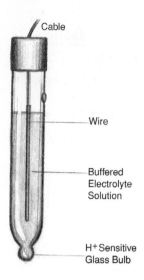

Figure 22.2.   pH Measuring Electrode (glass electrode). The bulb of this electrode is composed of glass that develops an electrical potential when placed in solution. The magnitude of the potential depends on the concentration of hydrogen ions. (Sketch by Sandra Bayna, a biotechnology student.)

The reference electrode has a strip of metal and an **electrolyte solution** (filling solution) enclosed in a glass or plastic tube. **Electrolytes** *are substances, such as acids, bases, and salts, that release ions when dissolved in water.* In the reference electrode, the metal and electrolyte react with one another to develop a constant voltage.

The two electrodes are placed in the sample whose pH is to be measured. The following events occur when electrodes are in the sample solution:

1. The electrical potential (voltage) of the glass electrode varies depending on the sample's hydrogen ion concentration.

2. The electrical potential of the reference electrode is constant.

3. The two electrodes are connected to one another through the voltage meter.

4. The meter measures the voltage difference between the reference electrode and the measuring electrode. This voltage is an electrical signal.

5. The electrical signal is amplified and converted to a display of pH.

ii. MORE ABOUT REFERENCE ELECTRODES

The reference electrode is the component of the pH measuring system that is most likely to cause difficulties; therefore, it is important to be familiar with reference electrode design. Observe in Figure 22.3a that the end of the reference electrode that is submerged in the sample has a *porous plug, called a* **junction,** or **salt bridge.** The filling solution, containing salts, flows slowly out of the reference electrode, through the junction, into the sample. This slow flow of ions through the junction is necessary to maintain electrical contact between the reference electrode and the sample solution whose pH is being measured. This flow of ions constitutes electrical current. (Recall that electrical current is a flow of charge. When current flows in a solution, it consists of ions. This is in contrast to the flow of electricity in a wire, where electrical current is due to the movement of electrons.) Because electrolyte flows out of the electrode, the sample solution cannot enter the junction and contaminate the electrode. The flow of filling solution is normally slow enough that it does not significantly contaminate the sample whose pH is being measured. There are different classes of junctions that vary in their flow rates and design. Classes of junction are described in Table 22.5 on p. 386.

Reference electrode filling solution is lost through the junction, so many electrodes have a **filling hole** that is used to replenish the electrolyte, see Figure 22.3a. Other reference electrodes are manufactured with a gel-like filling solution that cannot be replenished. With any reference electrode, pH cannot be properly measured if the electrolyte is depleted or if the junction is blocked so that electrolyte does not contact the sample.

There are two different classes of reference electrodes for pH meters: silver/silver chloride (abbreviated Ag/AgCl) and calomel. An **Ag/AgCl electrode** *contains a strip of silver coated with silver chloride.* The strip is immersed in an electrolyte solution that is normally saturated in KCl and silver chloride, see Figure 22.3b. A

Figure 22.3.   Reference Electrodes. a. The basic features of a reference electrode. **b.** A silver/silver chloride reference electrode. **c.** A calomel reference electrode. (Sketches by Sandra Bayna, a biotechnology student.)

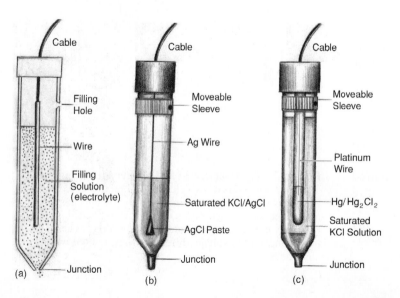

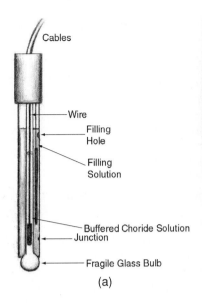

Cables

Wire

Filling Hole

Filling Solution

Buffered Choride Solution

Junction

Fragile Glass Bulb

(a)

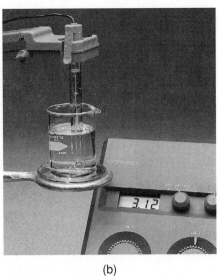

(b)

Figure 22.4.   Combination Electrodes. a. Diagram of a combination electrode. **b.** A combination electrode and pH meter. (Sketch by Sandra Bayna, a biotechnology student.)

calomel electrode *contains a platinum wire coated with calomel, which is a paste consisting of mercury metal and mercurous chloride.* The calomel is in contact with a filling solution that is normally saturated KCl, see Figure 22.3c.

Reference electrodes must maintain a constant electrical potential. Both Ag/AgCl and calomel reference electrodes are designed so that their potential is sensitive to the concentration of Cl^- in their electrolyte filling solution. The concentration of Cl^- in the solution is usually held constant by using a filling solution that contains saturated KCl. Because the filling solution is saturated it is not unusual to see KCl crystals on the end of reference electrodes.

The pH meters in biology laboratories usually appear to have only one electrode, not two. This is because they have convenient, compact, **combination electrodes. Combination electrodes** *are made by combining the reference and pH measuring electrodes into one housing,* see Figure 22.4 a,b. Combination electrodes work efficiently for most purposes but are not suitable for measuring the pH of certain "difficult" samples. (Difficult samples will be discussed later in this chapter.)

pH meters can be made still more compact by combining the electrodes and the meter into a single housing, see Figure 22.5. These compact pH meter systems are useful for field work and for making rapid measurements.

C. Selecting and Maintaining Electrodes

i. SELECTING ELECTRODES

There are many variations of the basic electrode designs discussed above. Electrodes come in different sizes, shapes, and brands and have features that vary from one electrode to another. Some electrodes are "general purpose"; others are intended for specific applications. It is best to consult the manufacturers and purchase electrodes that are recommended for the types of samples whose pH is measured in your laboratory. General considerations regarding selecting electrodes are summarized in Table 22.5.

ii. MAINTAINING ELECTRODES

Electrodes require proper care. Manufacturers provide specific information regarding care of their electrodes. General considerations regarding maintenance of electrodes are summarized in Table 22.6.

IV. OPERATION OF A pH METER SYSTEM

A. Calibration

Calibration, also called **standardization,** is a critical step in operating a pH meter because **calibration** *tells the meter how to translate the voltage difference between the*

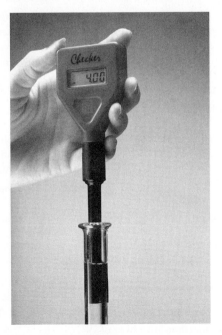

Figure 22.5.   A Compact, Portable pH Meter. (Courtesy Hanna Instruments, Inc.)

Table 22.5. CONSIDERATIONS IN SELECTING ELECTRODES

1. *Size and shape.* It is possible to purchase electrodes in various lengths and shapes to accommodate different samples.

2. *Connections.* Electrodes are connected to pH meters with plug-in cables that must match the meter, see Figure 22.6.

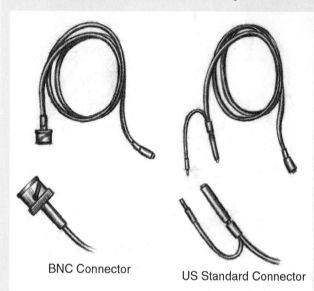

BNC Connector

US Standard Connector

Figure 22.6. Examples of pH Meter Cables and Connectors.

3. *Measuring electrodes.* There are several types of measuring electrode that vary in the type of glass used to make the bulb. A "general purpose" measuring electrode is acceptable unless one is working with difficult samples, such as very pure water, solutions that are poorly conducting, or nonaqueous liquids.

   At pH values above about 9 or 10, measuring electrodes begin to respond to Na^+ and K^+, as well as H^+. This response is termed **alkaline error.** There are specially designed measuring electrodes available that minimize alkaline error at high pH.

4. *Reference electrodes.* "General purpose" reference electrodes are usually Ag/AgCl.

   Calomel electrodes are more resistant than Ag/AgCl to certain solutions, such as Tris buffers, which are frequently used in molecular biology. Calomel electrodes will therefore last longer and are recommended in laboratories where Tris is frequently used.

   Calomel electrodes cannot be used in solutions hotter than about 80°C and they contain mercury, which is hazardous if the electrode is broken.

5. *Junction.* The junction is a part of the reference electrode. Junctions come in various styles, sizes, and shapes to allow faster or slower flow of filling solution. (Flow rates vary from about 0.5 to 100 μL/hr.) The choice of junction depends on the sample. Combination electrodes have a general purpose junction with a moderate flow rate. The classes of junctions include:

A fibrous junction is composed of a fibrous material such as quartz or asbestos. The fibers may be packed loosely or tightly; the looser the packing, the more rapid the flow of filling solution. Quartz fiber junctions are commonly used as a general purpose junction.

A fritted junction is a white ceramic material consisting of many small particles pressed closely together. Filling solution can leak through open spaces between the particles. The flow rate across a fritted junction is relatively slow. This type of junction is recommended when the sample should not be contaminated by filling solution or has a pH greater than 13.

A sleeve junction is made by placing a hole in the side of the glass or plastic housing of the electrode. The hole is covered with a glass or plastic sleeve. The sleeve junction is the fastest flowing type, is the least likely to become clogged, and is easiest to clean, since the sleeve is removable. The sleeve junction is recommended for samples that are viscous or have a high concentration of solids; and for brines, colloidal suspensions, nonaqueous samples, solutions of low ionic strength, strong acids and bases, and strong oxidizing and reducing agents. Faster flowing junctions have the disadvantage that the sample is more contaminated by filling solution than with slower flowing junctions.

A double junction is a less common and more complex reference electrode configuration, see Figure 22.7. It is used when the sample is incompatible with the reference electrode filling solution. The double junction allows for a second filling solution that contacts the sample to be placed outside the normal reference electrode junction. This configuration separates the reference electrode filling solution from the sample. This type of electrode is useful for applications such as wastewater and testing at chemical processing plants.

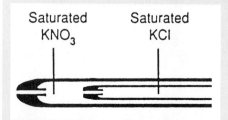

Saturated KNO₃

Saturated KCl

Figure 22.7. Reference Electrode with a Double Junction. (Courtesy of Radiometer Analytical A/S.)

6. *Combination versus separate electrodes.* Combination electrodes are more convenient than separate electrodes; however, there are samples whose pH is difficult to determine the pH and require a type of junction or a type of glass electrode that is not available in the combination configuration. Separate electrodes are necessary to determine the pH of such samples.

7. *Solid state electrodes.* Electrode/meter systems exist that incorporate a small silicon chip sensitive to H^+ ions into the head of a solid state electrode. These electrodes are different than conventional glass bulb electrodes. They do not require filling solution, are easily cleaned, and can measure very small volume samples.

Table 22.6.   CONSIDERATIONS IN MAINTAINING ELECTRODES

1. ***Filling solution.*** Reference electrodes contain filling solution, generally saturated KCl or AgCl/KCl depending on the type of electrode and the sample.

   Gel-filled electrodes contain gelled filling solution that is never refilled, so the electrode must be discarded when the electrolyte is depleted.

   Refillable electrodes are periodically refilled through the filling hole. Use the filling solution recommended by the manufacturer for the electrode. Fill the electrode nearly to the top of the filling solution chamber to ensure proper flow out the junction.

   The filling hole must be open when a sample's pH is measured so that filling solution can flow from the junction properly. The filling hole must be closed during storage so that filling solution is not lost.

2. ***Storage.*** There is no single rule for how to store all types of electrodes; therefore, consult the manufacturer's instructions. Some considerations include:

   Silver/silver chloride reference electrodes can be stored in a saturated KCl solution.

   Calomel electrodes may be stored in a 1/10 dilution of their filling solution.

   Glass electrodes should not be stored in solutions that are very high in sodium or potassium chloride because the glass bulb can exchange H^+ ions for potassium or sodium, thus reducing its sensitivity to hydrogen ions.

   A combination electrode must be stored in such a way as to protect both the reference electrode and the glass electrode. A combination electrode with a Ag/AgCl reference should not be stored in a solution with low chloride. The glass bulb should not be stored in solutions with high levels of potassium or sodium chloride. Manufacturers therefore sometimes recommend that combination electrodes stored for long periods of time be stored dry.

   Do not store electrodes in distilled water.

   New electrodes, an electrode stored dry, and electrodes that have been cleaned should be conditioned before use by soaking for at least 8 hours in pH 7 buffer.

3. ***Cleaning.*** The measuring electrode bulb, the reference junction, and filling solution must be kept clean. Cleaning protocols are in the section "Trouble-Shooting pH Measurements."

4. ***Measuring the pH of strongly acidic or basic solutions.*** Take a reading of such solutions quickly and then rinse the electrodes thoroughly.

5. ***Nonaqueous samples.*** Frequent immersion in nonaqueous samples will cause the glass measuring bulb to dry and no longer function. If measuring the pH of nonaqueous samples, periodically rehydrate the bulb by soaking it in distilled water. Consult the manufacturer for details.

6. ***Fragility.*** The measuring bulb of an electrode is fragile and cannot be repaired if it is broken or cracked. Keep the electrode bulb away from moving stir bars.

measuring and reference electrodes into units of pH. pH meter systems need to be calibrated every day or every time the system is used because the response of electrodes declines over time.

There is a linear relationship between the voltage measured by the system and the pH of the sample. Because the relationship is linear, two buffers of known pH are used to calibrate a pH meter, see Figure 22.8 on p. 388. The first buffer is normally pH 7.00. The second is typically pH 4.00, 10.00, or 12.00, although other standardization buffers are available. If the sample to be measured is acidic, an acidic calibration buffer is used and if the sample is basic, a basic calibration buffer is chosen. If very high accuracy is required, the second standard buffer pH should be within 1.5 pH units of the sample pH.

The general method of calibration is to first immerse the electrodes in pH 7.00 buffer. Using an appropriate knob, dial, keys, or button, the meter is set to display pH 7.00. The meter internally adjusts itself so that a pH of 7.00 corresponds to 0 millivolts (mV). The adjustment for pH 7.00 buffer may be called "set," "zero offset," or

"standardize" because the meter "sets to" 0 mV. The electrodes are then placed in the second buffer and by using the appropriate knob, dial, keys, or button the meter is adjusted to display the pH of the buffer. The adjustment for the second buffer may be called "slope," "calibrate," or "gain" because internally the meter uses the second reading to establish a calibration line with a particular slope. The result of this process is that the meter constructs a calibration line that can be used to convert any voltage to its corresponding pH.

The slope of the calibration line is a measure of the response of the electrodes to pH. The steeper the slope, the more sensitive the electrodes are to hydrogen ions. The slope ideally should be close to -59.2 mV/pH unit when the temperature is 25°C, see Figure 22.9a on p. 388. As electrodes age or become dirty, they are less able to generate a potential and the slope of the line declines, see Figure 22.9b on p. 388. As the electrode response deteriorates, the system is eventually no longer able to adjust to the value of both buffers (as discussed in the section "Trouble-Shooting pH Measurements" starting on p. 392).

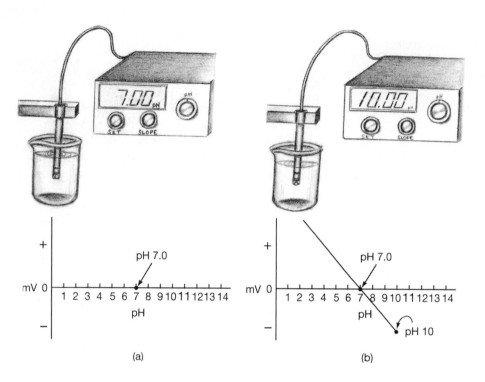

Figure 22.8. pH Meter Calibration. The process of calibration establishes the relationship between pH and mV so that the meter can convert any mV reading to a pH value. **a.** The meter is adjusted to display pH 7 when the electrodes are immersed in pH 7 calibration buffer. **b.** The meter is adjusted to display the value of a second calibration buffer. The slope of the resulting line ideally is −59.2 mV/pH unit at 25°C. (Sketches by Sandra Bayna, a biotechnology student.)

Some pH meters display the slope of the calibration line as an indication of the condition of the electrodes. The slope is often expressed as a percent of the theoretical value. For example, a 97% slope is equivalent to a slope of −57.4 mV/pH unit.

When operating an older pH meter, the first calibration buffer should be pH 7.00. This means, for example, that it is not recommended to calibrate an older meter using standard buffers of pH 4.00 and pH 1.68, as might be desirable when measuring the pH of acidic samples. Newer, microprocessor-controlled meters often have a feature so that the first buffer does not need to be pH 7.00 and any two standard buffers that bracket the pH of the samples may be used for calibration. Moreover, some microprocessor-controlled meters allow the use of more than two standard buffers to determine the calibration line. These features of newer instruments can improve the accuracy of pH measurements. Consult the manufacturer's instructions to determine the features of a specific meter.

Good-quality calibration buffers are essential for calibration of the pH measuring system. Calibration buffers are commonly purchased as ready to use solutions, often in various colors where each color is a different pH. Tips regarding calibration buffers are in Table 22.7.

B. Effect of Temperature on pH

Temperature is very important in the measurement of pH. Temperature has two effects:

1. The **measuring electrode's response to pH** (the potential that develops) is affected by the temperature, see Figure 22.10.

2. The **pH of the solution** that is being measured may increase or decrease as its temperature changes.

It is important to consider both temperature effects when operating a pH meter.

Most pH meters can compensate for the first factor (i.e., temperature-dependent changes in the electrode response). The meter needs to "know" the temperature of the solution to make this adjustment. It is possible to measure the temperature of a solution with a thermometer and, using the proper knob or dial, "tell" the pH meter the solution's temperature. There are also *electronic temperature probes,* **automatic temperature compensating (ATC) probes,** that can be placed in the sample alongside the pH electrodes. The probes are connected to the pH meter and automatically measure the sample temperature and report it to the meter, which compensates accordingly. Compact devices may even have a temperature probe built into the electrode housing.

The second temperature effect is that the pH of some solutions changes as their temperature changes. For

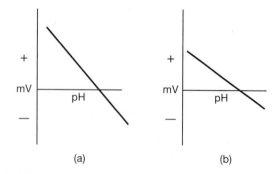

Figure 22.9. Electrode Response and Slope. a. Normal electrode response. **b.** Effect of aging or dirt on electrode response.

Table 22.7.   CALIBRATION BUFFERS

1. ***Reaction with CO_2.*** Some buffers react with CO_2 from the air, resulting in a change in their pH. Buffer containers, therefore, should be kept closed. Remove a small amount of buffer each time an instrument is calibrated and throw it away after use. pH 10.0 buffer is particularly sensitive to CO_2.

2. ***Expiration date.*** Premixed buffers come with an expiration date. For best accuracy, do not use buffers after their expiration date.

3. ***Temperature.*** The pH of a buffer will change as its temperature changes. For example, a common standardization buffer has a pH of 7.00 at 25°C, but at 0°C, its pH is 7.12. When calibrating with this buffer at 25°C, set the meter to exactly 7.00, but if the buffer is at 0°, set the meter to 7.12. The pH listed on the container of a commercially purchased buffer is generally its pH at 25°C. Manufacturers can provide a table that lists the pH of a standardization buffer at various temperatures. Allow sufficient time for the electrode to equilibrate to the temperature of the buffer. If the change in temperature is extreme, or the pH of the buffer is very high or low, equilibration may take as long as 3 or 4 minutes.

4. ***Contamination.*** Buffers that are visibly contaminated by mold growth or contain precipitates must be discarded. Never pour buffer back into its original bottle because this will contaminate the entire stock.

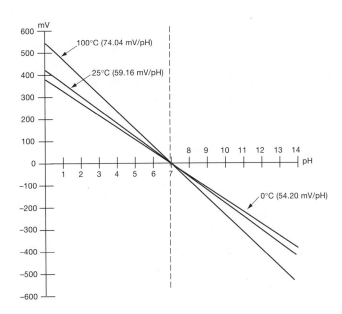

Figure 22.10.   The Effect of Temperature on Electrode Response. Observe that the slope of the calibration line changes at different temperatures. (© Copyright Omega Engineering, Inc. All rights reserved. Reproduced with permission Omega Engineering, Inc., Stamford, CT 06907, www.omega.com.)

example, if Tris buffer is prepared to pH 8.0 at room temperature, its pH will be close to 7.6 when used at body temperature and 8.8 when placed on ice. It is best, therefore, to measure the pH of a solution when the solution is at the temperature at which it will be used.

When measuring the pH of samples that are not at room temperature, it is common practice to calibrate the meter at room temperature using standardization buffers at room temperature. Then, when it is time to measure the pH of the sample solution, the meter is adjusted to the temperature of the sample and its pH is determined. This practice generally gives sufficiently accurate results.*

C. Summary of the Steps in Operating a pH Meter

Although each brand and style of pH meter differs, there are certain steps that are usually performed when measuring the pH of a sample. These steps are summarized in Box 22.1 on pp. 390–391.

D. Measuring the pH of "Difficult Samples"

The sample is a key part of the measurement system. The sample must be homogeneous to avoid variation

and drift as pH is measured. The sample must also be equilibrated to a particular temperature. Some samples undergo chemical reactions that cause their pH to change over time, either among substances in the sample itself, or with CO_2 from the air.

Some samples are inherently difficult to pH and may require special techniques or specific types of electrodes. Difficult sample types include those that:

- **Are nonaqueous**
- **Are high purity water, containing little salt**
- **Contain high salt concentrations**
- **Contain high protein concentrations**
- **Contain S^{-2}, Br^-, or I^-**
- **Contain Tris buffer**
- **Are viscous**
- **Are turbid**

Table 22.8 on p. 392 includes tips for handling difficult samples.

E. Quality Control and Performance Verification of a pH Meter

As with any instrument, pH meters require a quality-control plan. At the least this includes keeping a log

---

*It is also possible to calibrate a pH meter with two calibration buffers that are equilibrated to be the same temperature as the sample. The pH meter temperature setting is adjusted to this temperature before calibration (or an ATC probe is used). Because the pH of calibration buffers varies with temperature, the meter must be adjusted to the pH of each calibration buffer at the temperature of interest. (For example, the pH of a typical pH 7.00 calibration buffer is 7.00 at 25°C and it is 6.98 at 37°C.)

Box 22.1.   CONVENTIONAL METHOD FOR MEASURING THE pH OF A SOLUTION

1. *Allow the meter to warm up as directed by the manufacturer.*
2. *Open the filling hole; be certain that the filling solution is nearly to the top.* (Refillable electrodes only.)
3. *If the meter has a "standby" mode, use it when the electrodes are not immersed in the sample. Use the "pH" mode to read the pH of a sample or standard.*
4. *Calibrate the system each day or before use:*
   a. *Adjust the meter temperature setting to room temperature with the appropriate control, or use an ATC probe.*
   b. *Obtain two standard buffers; one that has a pH of 7.00 at room temperature, the other that is acidic if the sample is acidic and basic if the sample is basic.*
   c. *Rinse the electrodes with distilled water and blot dry.* Do not wipe the electrodes as this may create a static charge leading to an erroneous reading.
   d. *Immerse the electrodes in room temperature, pH 7.00 calibration buffer.* Be certain that the junction is immersed and that the level of sample is below the level of the filling solution, see Figure 22.11. Disengage the standby or follow the manufacturer's instructions. Allow the reading to stabilize.

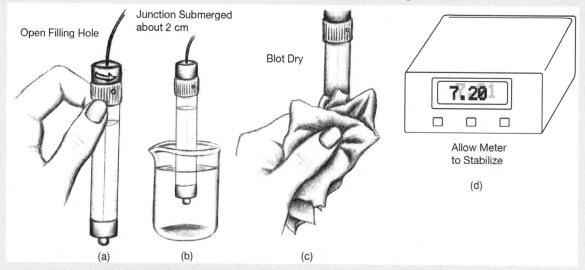

Open Filling Hole

Junction Submerged about 2 cm

Blot Dry

7.20

Allow Meter to Stabilize

(d)

(a)   (b)   (c)

Figure 22.11   Points to Remember in Using a pH Meter. a. Open the filling hole before measuring pH; close it afterward. **b.** Make sure the junction is submerged in the solution whose pH is being measured. **c.** Rinse and blot the electrode. **d.** Allow time for the reading to stabilize. (Sketches a–c by Sandra Bayna, a biotechnology student.)

   e. *Adjust the meter to read 7.00 using the correct knob, dial, button, or other method of adjustment.* Reengage the standby mode, if present.
   f. *Remove the electrodes, rinse with distilled water, and blot dry.* You can also rinse the electrodes with the next solution and do not dry.
   g. *Place the electrodes in the second standardization buffer, set the meter to read pH, and allow the reading to stabilize.* Adjust the meter to the pH of the second buffer with the proper method of adjustment. Remove, rinse, and blot the electrodes.
   h. *(This step is not necessary with newer, microprocessor-controlled pH meters.) Recheck the pH 7.00 buffer as in step d and readjust as necessary.* Then, recheck the second buffer and readjust the meter as necessary. Continue to rinse the electrodes after each solution.
   i. *(This step is not necessary with newer, microprocessor-controlled pH meters.) Read both calibration buffers and readjust as needed up to three times.* If the readings are not within 0.05 pH units of what they should be after three adjustments, the electrode probably needs cleaning, as described later.
5. *Optional: It is possible to perform quality-control checks at this point.*
   a. *Linearity Check. To check the linearity of the measuring system, take the reading of a third calibration buffer.* For example, if the meter was calibrated with pH 7.00 and pH 10.00 buffers, check the pH of a pH 4.00 calibration buffer. Immerse the electrodes in the third buffer, place the meter in pH mode, allow the reading to stabilize, and record the value. *Do not adjust the meter to this third calibration buffer; the purpose of this third*

Box 22.1. *(continued)*

buffer is to check the system's linearity. If the reading is not within the proper range, as defined by your laboratory's quality-control procedures, the electrodes should be serviced as described in the section "Trouble-Shooting pH Measurements" starting on p. 392.

   b. *A second quality-control check is to test the pH of a control buffer whose pH is known and that has a pH close to the pH of the sample.* It is common to set the maximum allowable error of the control buffer to be ± 0.10 pH units. Do not adjust the meter to the pH of the control buffer. The purpose of this buffer is to check the accuracy of the system. If the pH reading of the control buffer is not within the required tolerance, the electrodes require maintenance.

6. ***Set the meter to the temperature of the sample or place an automatic temperature probe in the sample.***

7. ***Place the electrodes in the sample.***

8. ***Allow the pH reading to stabilize.*** In general, the reading should stabilize within a minute. The pH reading initially changes rapidly, then it slowly stabilizes, see Figure 22.12. One of the causes of error in using pH meters is not allowing the reading to stabilize.

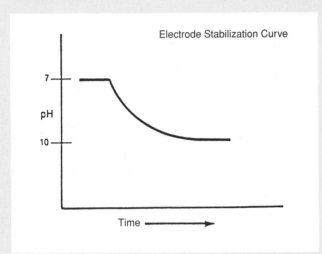

Figure 22.12   Stabilization of a pH Reading.
(Courtesy of Beckman Coulter, Inc.)

Note the following points:

   a. If you wait *too* long, the pH of some samples will change due to exposure to air, chemical reactions, or other factors.

   b. Solutions of high or low ionic strength and nonaqueous solutions may require longer to stabilize.

   c. Many new pH meters have an "autoread" feature that automatically determines when the electrode has stabilized and locks on this reading. For example, the meter may lock on the reading if it changes less than 0.004 pH units over 10 seconds. The autoread feature may help standardize readings made in a laboratory because it is consistent from reading to reading and between operators. It also ensures that impatient operators will wait long enough when taking a reading.

9. ***Record all relevant information.*** Be sure to note the temperature of the solution when its pH was measured.

10. ***Remove the electrodes from the sample, rinse them, and store them properly.*** Close the filling hole (refillable electrodes only).

book, performing regular calibration, and cleaning and replacing electrodes as needed. Although each laboratory must develop its own plan, Box 22.2 on p. 393 summarizes common, routine quality-control procedures for pH meter measuring systems.

   Like balances and other laboratory instruments, pH meters are often operated in a stand-alone mode where the analyst reads and manually records a displayed pH value. Sometimes, however, pH meters are interfaced with computers that are able to store, analyze, and print data. Vendors who manufacture pH meters now provide software that is compatible with the requirements of the FDA's regulations in 21 CFR Part 11. Figure 22.13 on p. 392 shows an example of a printout from a pH meter that has the capability of storing and printing data. This system has these features:

   • Date and time stamp are provided.

   • The system documents when the meter was last calibrated.

   • The calibration information is linked to the samples' data that were recorded.

   • The system documents calibration data, including standards used, temperature, mV, and slope.

   • The software can be programmed to remind the user to recalibrate the instrument after a specified time has elapsed.

Table 22.8. DIFFICULT SAMPLES

1. *Nonaqueous solvents.* pH relates to aqueous solutions and it is difficult to obtain a meaningful pH reading for a nonaqueous sample. Nonaqueous solvents, like acetone, can dry the glass electrode membrane, which must be hydrated to function properly. After measuring the pH of solvents, it may be necessary to soak the measuring electrode in an aqueous solution. Consult the manufacturers for further advice.

2. *High-purity water.* High-purity water (e.g., acid rain, well water, distilled, or boiler feedwater) is very difficult to pH because it does not readily conduct current and it absorbs CO_2 from the atmosphere, which affects its pH. Tips for measuring the pH of such samples include:

   Avoid the absorption of CO_2 by taking a reading soon after immersing the electrodes deep within the sample.

   Cover the sample container with a stopper that has holes for the electrodes.

   Blanket the surface of the sample with nitrogen gas.

   Use special, low ionic strength buffers for calibration. (Such buffers are available commercially.)

   Use additives to increase the ionic strength of the sample without affecting its pH. (Advertised, for example, by Orion, Boston, MA).

   Use a refillable electrode with a low resistance glass measuring bulb.

   Add 0.30 mL of high-purity, saturated potassium chloride to 100 mL of test solution and measure the pH immediately without agitation.

3. *High-salt samples.* In samples with large amounts of salt, sample ions compete with the reference filling solution ions. This competition affects the voltage read by the meter. It may be possible to compensate for this effect by using calibration buffers that also have a high salt concentration. An electrode with a double junction may also be helpful.

4. *Sample–electrode compatibility.* The sample must be compatible with the reference electrode. Tris buffer, sulfides, proteins, Br^-, and I^- can precipitate or complex with the silver in Ag/AgCl electrodes, leading to a clogged junction. A calomel electrode is therefore recommended when working with these materials.

5. *Slurries, sludges, viscous, and colloidal samples.* These types of sample may require a fast-flowing junction, usually the sleeve type. These samples also may coat the electrode bulb. Cleaning procedures are shown in the section "Trouble-Shooting pH Measurements" below.

   Consult manufacturers for more information on specific electrodes to match particular types of samples.

V. TROUBLE-SHOOTING pH MEASUREMENTS

A. General Considerations

i. OVERVIEW

A pH meter system must be able to respond to a range of hydrogen ion concentrations from 10^0 to 10^{-14} moles/L, must be responsive to very low concentrations

```
----- CALIBRATION -----
03  Dec  2007   16:30
Standard Buffers
1:   pH       4.00     25.6 C
     mV      175.9
     Slope 1–2          99.2 %
2:   pH       7.00     25.7 C
     mV        0.0
     Slope 2–3          97.8 %
3:   pH      10.01     25.6 C
     mV     −176.9
--------------------
Sample 1
     03  Dec  2007   16:52
     pH      7.432    25.6 C
Sample 2
     03  Dec  2007   16:54
     pH      5.567    25.6 C
Sample 3
     03  Dec  2007   16:55
     pH      9.701    25.6 C
```

Figure 22.13. A Printout from a pH Meter That Is Interfaced with a Computer.

of hydrogen ions and to very small changes in hydrogen ion concentration, and must be able to detect hydrogen ions selectively in the presence of other ionic species. Compared with many other measuring instruments, therefore, a pH meter must have a wide range of response and be very sensitive and specific. Considering these stringent requirements, it is not surprising that problems sometimes arise when using a pH meter. This section discusses sources of common problems and general strategies for identifying and solving them.

The first step in trouble-shooting is recognizing that there is a problem. Although this seems obvious, problems with pH meter systems may be subtle and are sometimes overlooked. Symptoms of pH system problems include:

1. **The reading drifts in one direction and either will not stabilize or takes an unusually long time to stabilize.**

2. **The reading fluctuates.**

3. **The meter cannot be adjusted to both calibration buffers.**

4. **The apparent pH value for a buffer or sample seems to be wrong.**

5. **There is no reading at all.**

ii. THE COMPONENTS OF A pH MEASURING SYSTEM AND THEIR POTENTIAL PROBLEMS

Problems with a pH meter system can arise in any of its components: (1) the reference electrode, (2) the measuring electrode, (3) the meter, (4) the calibration buffers, or (5) the sample.

The reference electrode junction is probably the most common source of problems. If the junction is dirty and partially occluded, then sample ions can

---

Box 22.2.   COMMON ROUTINE QUALITY CONTROL PROCEDURES FOR pH METER SYSTEMS

Quality Control Activity	Frequency		
	Each Use	Each Day	As Needed
Calibrate with two buffers, pH 7.00 and one other	✔		
Perform linearity check (as described in Box 22.1 on p. 390)	✔		
Read and record the pH of a control buffer whose pH is close to that of the sample	✔		
Check level of filling solution		✔	
Clean and recondition electrodes			✔

---

migrate into the junction and set up new electrical potentials that interfere with the proper reading of the electrode. The first indication that the junction is occluded is a long stabilization time where the reading drifts slowly toward the correct pH. Slow equilibration, however, can also be caused by changes in the temperature of the sample (such as is caused by a warm stir plate), by reactions occurring within the sample, or by incompatibility between the reference electrode and the sample. Methods of checking for junction occlusion are given in Table 22.9.

If the junction is completely plugged, the reading may never stabilize. This is also caused by broken electrodes and by some problems within the meter. The best way to check which component is at fault is to substitute in a new reference or combination electrode. If the reading stabilizes with a new electrode, then the old one was faulty.

The measuring electrode bulb must be clean so it can be in close contact with the solution. Certain samples (e.g., those with high levels of protein) can leave deposits on the bulb that must be periodically cleaned, as described in Box 22.5 on p. 394.

Table 22.9.   TROUBLE-SHOOTING pH METERS: SYMPTOM I—THE READING DRIFTS, WILL NOT STABILIZE, OR TAKES AN UNUSUALLY LONG TIME TO STABILIZE

Drift is often due to (1) changes in the sample, (2) problems with the reference electrode, or (3) problems with the measuring electrode.

1. *The temperature of the sample may be unstable, resulting in a fluctuating or drifting pH reading.* Wait until the temperature is stable and adjust the meter to the sample temperature.

2. *The sample may not be homogeneous, leading to a fluctuating or drifting reading.* Stir the sample.

3. *The sample may be undergoing chemical reactions or may be reacting with CO_2 from the atmosphere.* Consider the chemistry of the sample. Read the pH as soon as the reading appears to stabilize.

4. *The sample may be inherently difficult to pH and may require special techniques or specific types of electrodes.* See Table 22.8 on p. 392.

5. *The filling solution may be contaminated.* (Refillable electrodes only.) If so, replace the filling solution with fresh electrolyte (Box 22.3 on p. 394).

   Excessive crystallization can occur if the electrode was stored for a long time with the filling hole open or was stored in the cold.

   If the electrode was used to determine the pH of a sample when the filling solution was low, then the filling solution might have been contaminated by the entrance of sample.

6. *The reference electrode junction may be occluded by crystals or may be dirty.* This is a common cause of sluggish or drifting response. With a liquid filled reference electrode, a clogged junction can be detected by applying air pressure to the filling hole. A bead of electrolyte should appear readily on the junction. If not, the junction is likely fully or partially occluded. If the junction is occluded, clean it as described in Box 22.4 on p. 394.

7. *Long immersion in high-purity water can damage the junction, resulting in reduced or no ion flow and drifting or fluctuating readings.* The reference electrode may need replacement.

8. *The silver or calomel element is damaged.* This is a less-likely problem, but it can occur due to long exposure to sulfides, proteins, heavy metals, and pure water. If this is the problem, it is necessary to replace the electrode.

9. *The measuring electrode bulb may be dirty.* To clean the bulb, see Box 22.5 on p. 394.

10. *The response of the glass that constitutes the measuring bulb can decline due to prolonged use, excessive alkaline immersion, or high-temperature operation.* This can be tested by performing a "linearity check" as described in Box 22.1 on pp. 390–391. If the reading is out of range, then try rejuvenating the glass electrode as shown in Box 22.6 on p. 395.

BOX 22.3.   REPLACING ELECTROLYTE IN A REFERENCE ELECTRODE

1. *Wash the electrode in warm water.*
2. *Shake until the crystals dissolve.*
3. *Remove the old filling solution.* It may be helpful to use a narrow plastic pipette or a syringe attached to plastic tubing to remove the solution.
4. *Insert fresh solution of the proper type.*
5. *It may be helpful to soak the electrode overnight in 1 M KCl before using.*

BOX 22.4.   CLEANING THE JUNCTION OF A REFERENCE ELECTRODE

These methods are suitable for combination electrodes or individual reference electrodes. The first methods are least drastic and should be tried first. Continue to the next step only if the previous one has failed.

1. *Inspect the reference cavity for excessive crystallization.* If there are excessive crystals, replace the old filling solution as discussed in Box 22.3. (Refillable electrodes only.)
2. *For gel-filled and calomel references, soak the electrode tip in warm water (about 60°C) for 5–10 minutes.* Be careful not to heat the water much above 60°C.
3. *Dissolve crystals from the end of the electrode by soaking it in a solution of 10% saturated KCl and 90% distilled water for between 20 minutes and 3 hours.* Warm the solution to about 50°C. Immerse the electrode about 2 in into the solution. Refillable electrodes can be soaked overnight in this solution.
4. *Use a commercially available junction cleaner.*
5. *Remove any proteins that coat the outside of electrodes or penetrate into the junction.* Recommendations for removing protein deposits include rinsing in enzyme detergent like Terg-A-Zyme (from Alconox), using pepsin/HCl cleaners (5% pepsin in 0.1 M HCl), or soaking the electrode in 8 M urea for about 2 hours. After removing the protein deposits, rinse the electrode and empty and replace the filling solution.
6. *Immersion in ammonium hydroxide may be helpful for refillable Ag/AgCl electrodes that are clogged with silver chloride.* Empty the filling solution and immerse the electrode in 3 to 4 M NH_4OH for only 10 minutes. Ammonia can damage the internal element of the electrode if the electrode is soaked in it too long. (NH_4OH is very caustic and the fumes are irritating. Use it in a fume hood.) Rinse the inside and outside of the electrode thoroughly with distilled water and add fresh filling solution. If the electrode is a combination electrode, soak it 10–15 minutes in pH 4 buffer.
7. *Immersion in concentrated hydrochloric acid for 5–10 minutes may be helpful for calomel electrodes.* Use adequate ventilation and safety precautions to avoid injury by the acid.
8. *It may be possible to clear the junction by forcing filling solution through the junction by applying pressure to the filling hole or vacuum to the junction tip.*
9. *As a last resort, for ceramic junctions only, sand the junction with #600 emery paper.* For combination electrodes, be careful not to contact the glass bulb.

(A "sleeve junction" has a collar or sleeve that can be loosened to unclog the junction. When the sleeve is loosened, the filling solution empties. With this type of junction, simply loosen the collar, clean the junction, and refill the electrode.)

BOX 22.5.   CLEANING THE MEASURING ELECTRODE BULB

Use a soft toothbrush, Q-tip, or tissue to clean the bulb. Gently wipe or pat the bulb; do not rub.

1. *Remove dirt with warm soapy water.* A mild dishwashing detergent without hand lotion can be used.
2. *Remove protein contaminants by washing in warm water with an enzyme detergent like Terg-A-Zyme (from Alconox) or 5% pepsin in 0.1 N HCl followed by rinsing in water.* If your application routinely involves many pH measurements of protein solutions, it is good practice to clean the bulb on a regular basis.
3. *Remove inorganic deposits by washing with EDTA, ammonia, or 0.1 N HCl.* Rinse the bulb with water and soak it in dilute KCl.
4. *Remove grease and oil with methanol or acetone.* Rinse the bulb with water and soak it in dilute KCl.
5. *To remove fingerprints, wipe the bulb gently with a 50/50 mixture of acetone and isopropyl alcohol.* Soak the electrode in a commercial soaking solution or pH 4 buffer for 1 hour.

Box 22.6.   "REJUVENATING" A MEASURING ELECTRODE BULB

Note that solutions used for soaking and conditioning a pH measuring electrode are usually slightly acidic. This is because H^+ ions from the soaking solution replace contaminants in the glass of the bulb.

Some manufacturers recommend the following steps when the linearity of the system is poor or the system response is sluggish. The first methods are least drastic and should be tried first. Continue to the next step only if the previous one has failed.

1. *Soak the bulb in pH 4.00 calibration buffer overnight.*

2. *Soak the bulb in 1M HCl for 30 minutes.*

3. *Immerse the bulb in 0.1 M HCl for about 15 seconds, rinse with water, then immerse in 0.1 M KOH (or 0.1 M NaOH) for 15 seconds.* Repeat several times and rinse. Soak the electrode in pH 4 buffer overnight.

SAFETY NOTE: Manufacturers sometimes recommend rejuvenating glass bulbs with ammonium bifluoride or with hydrofluoric acid. Ammonium bifluoride is hazardous and should be used with caution. Hydrofluoric acid is extremely hazardous! It will degrade glass containers. It can cause serious or fatal injuries even at low concentrations. We do not recommend its use except by those who are specially trained and have access to appropriate safety equipment.

---

pH measuring electrodes age, even if they are properly maintained, so they have a finite lifespan. Aqueous solutions slowly attack the glass membrane and form a gel layer on the bulb surface. As the gel layer thickens, the response of the electrode becomes sluggish and the slope of the calibration line decreases.

There is a great deal of resistance to current flow in a pH meter measuring system and the potential developed by the pH electrode bulb is very small. The voltmeters that are used to measure pH, therefore, are specially designed for these conditions. Complete lack of response (or sometimes, dramatically unexpected readings) is likely caused by problems with the meter itself (in contrast to the electrodes). The meter, however, is the least likely component of the system to cause problems.

If the buffers used to calibrate the instrument are not correct, then all readings made will be incorrect. This problem may be detected if a solution of known pH is available and its pH is checked.

If the sample is not homogenous, or if its temperature is not stable, then the pH readings will fluctuate or drift. If the sample is "difficult," the readings may be unusually slow to stabilize and/or may be incorrect.

Table 22.10.   TROUBLE-SHOOTING pH METERS: SYMPTOM 2—DIFFICULTY ADJUSTING THE METER TO BOTH CALIBRATION BUFFERS

1. *The measuring electrode may be dirty.* Clean the bulb as described in Box 22.5.

2. *The response of the glass that constitutes the measuring bulb can decline.* Rejuvenate the electrode as shown in Box 22.6.

Table 22.11.   TROUBLE-SHOOTING pH METERS: SYMPTOM 3—THE VALUE READ FOR A BUFFER OR SAMPLE SEEMS TO BE WRONG

1. *The calibration buffers have deteriorated.* This is the most likely problem; try fresh calibration buffers.

2. *The sample may not be homogenous.* Stir the sample.

3. *The measuring electrode may be dirty.* Clean the bulb as described in Box 22.5.

4. *The response of the glass that constitutes the measuring bulb has declined.* Rejuvenate the electrode as described in Box 22.6.

B. Trouble-Shooting Tips

i. SIMPLE MISTAKES

Once a problem is noted, the first step in trouble-shooting a measurement instrument is always to look for and correct simple (embarrassing) mistakes. For pH meter systems these include:

- **The electrode measuring bulb and the junction are not immersed in the sample.**
- **The meter is not turned on or plugged in, or the electrode cables are not connected to the meter.**
- **The reference electrode is not filled with electrolyte.**
- **The reference electrode filling hole is closed.**
- **The sample is not well-stirred.**

Table 22.12.   TROUBLE-SHOOTING pH METERS: SYMPTOM 4—NO RESPONSE AT ALL

1. *Check that the meter is plugged in, turned on, and the electrode cables are properly connected.*

2. *Consult the pH meter manufacturer.* The operating manual may have a trouble-shooting guide.

- **The calibration buffers are not fresh.**
- **An electrode is cracked or broken and must be replaced.**

ii. Trouble-Shooting Problems Based on Symptoms

If no simple mistake is discovered, then the symptoms observed provide clues for trouble-shooting. Tables 22.9–22.12 provide tips for trouble-shooting pH meter problems based on symptoms noted.

C. Procedures for Cleaning and Maintaining Electrodes

Follow the manufacturer's instructions if available. The procedures outlined in Boxes 22.3 through 22.6 on pp. 394–395 are general, commonly recommended suggestions.

VI. Other Types of Selective Electrodes

An **indicator electrode** *responds selectively to a specific molecule in solution.* For example, there are indicator electrodes that respond to dissolved CO_2 and that respond to dissolved O_2. Biologists use dissolved O_2 electrodes to monitor the oxygen level in the broth during fermentation because the growth of most microorganisms is sensitive to oxygen level.

Ion-selective electrodes (ISE) *are a class of indicator electrode, each of which responds to a specific ion.* A pH measuring electrode is an example of an ion-selective electrode. Other ISEs include those that respond to Ca^{++}, Na^+, K^+, Li^+, F^-, and Cl^-. ISEs have many applications. For example, sodium levels in human blood are severely disturbed by cardiac failure and liver disease. Sodium levels in a patient's blood serum can be tested using a sodium ion selective electrode. ISEs can be used to measure fluoride levels in drinking water, nitrates in soils, plants, or foods, calcium in milk, and other applications. Figure 22.14 shows a catalog listing of different types of ISEs. There are many methods

published in official compendia that utilize ISEs. For example, ASTM, EPA, and AOAC all have ISE methods for measuring levels of fluoride in water.

An ion-selective electrode measuring system is diagramed in Figure 22.15. An ISE has a membrane that, like the glass bulb of a pH electrode, develops an electrical potential in response to a particular ion. The ISE is connected to a reference electrode through a voltage meter. Although the electrodes are different, a standard pH meter can be used for these measurements, as long as it is able to display mV as well as pH.

When working with indicator electrodes it is necessary to determine the relationship between electrical potential and the concentration of the ion of interest. A calibration curve must be constructed, as shown in Figure 22.16.

Each type of indicator electrode is different; therefore, it is important to read the manual that comes with the electrode. Table 22.13 lists some general points of concern regarding indicator electrodes.

VII. The Measurement of Conductivity

A. Background

There are situations where it is useful to measure the combined concentration of all the ions in an aqueous solution, see Table 22.14 on page 398. The total ion concentration is determined by measuring the **conductivity** of the solution.

Conductivity *is the ability of a material to conduct electrical current.* Metals are extremely conductive; electrons carry current through metals at almost the speed of light. Electrical current can also flow through solutions by the movement of ions, although this movement is not as rapid as current flow in metals. Thus, aqueous solutions containing ions are conductive. The higher the concentration of mobile ions in a solution, the more conductive the solution; therefore, conductivity is an indicator of total ion concentration.

Specifications and Ordering Information: Orion ISE's

Type	Concentration Range	Interferences
Gas-Sensing Models		
Ammonia	1 to 5×10^{-7}M; 17,000 to 0.01 ppm	Volatile amines
Carbon Dioxide	10^{-2} to 10^{-4}M; 440 to 4.4 ppm	Volatile weak acids
Nitrogen Oxide	5×10^{-3} to 4×10^{-6}M; 230 to 0.18 ppm	CO_2, volatile weak acids
Solid-State Models		
Bromide	1 to 5×10^{-6}M; 79,900 to 0.40 ppm	S^-, I^-, CN^-, high levels of Cl^-, NH_3
Cadmium	10^{-1} to 10^{-7}M; 11,200 to 0.01 ppm	Ag^+, Hg^{++}, Cu^{++}, high levels of Pb^{++}, Fe^{++}
Cadmium Comb.	10^{-1} to 10^{-7}M; 11,200 to 0.01 ppm	Ag^+, Hg^{++}, Cu^{++}, high levels of Pb^{++}, Fe^{++}
Chloride	1 to 5×10^{-5}M; 35,500 to 1.8 ppm	OH^-, S^- Br^-, I^-, CN^-
Chloride Comb.	1 to 5×10^{-5}M; 35,500 to 1.8 ppm	OH^-, S^- Br^-, I^-, CN^-
Chlorine Comb.	3×10^{-4} to 10^{-7}M; 20.0 to 0.01 ppm	Strong oxidizing agents
Cupric	10^{-1} to 10^{-8}M; 6350 to 6.4×10^{-4} ppm	Ag^+, Hg^{++}, high levels of Cl^-, Br^-, Fe^{++}
Cyanide	10^{-2} to 8×10^{-6}M; 260 to 0.2 ppm	S^-, I^-, Br^-, Cl^-

Figure 22.14.   A Catalog Listing of Various Types of Ion-Selective Electrodes. (Courtesy of Fisher Scientific, Pittsburgh, PA.)

Table 22.13. PRACTICAL CONSIDERATIONS FOR INDICATOR ELECTRODES

1. *The sensing tips of the electrodes are fragile and should be protected from breakage and drying.*

2. *The response of the electrodes may be affected by the combined concentration of all the ions in the solution.* The standards used to prepare the calibration curve and the samples should be roughly comparable in their total ionic concentration. It is possible to purchase ionic strength adjusting buffers that are used to prepare standards and to dilute samples in order to achieve comparable ionic strength.

3. *Temperature affects the response of ISEs so the calibration curve must be prepared at the same temperature as the samples are measured.*

4. *The calibration standards should have the same pH as the samples.*

5. *Although indicator electrodes are usually specific for one substance, under certain conditions other substances may interfere with the measurement of interest.* Be aware of contaminants that might adversely affect the measurement.

6. *The sample must be well stirred.*

7. *Ion-selective electrodes are generally slower to stabilize than pH electrodes.* At least 15 minutes is recommended for stabilization.

8. *Consult the manufacturer's manual for storage directions.*

We are accustomed to thinking of water as a good conductor of electricity (and so would avoid being on a lake in a metal canoe during a thunderstorm). However, because it is ions that actually conduct electricity in a solution, ultrapure water has a very low conductivity.

Some substances ionize more completely than others when dissolved in water, so their solutions are more conductive. **Electrolytes** *are substances that release ions when dissolved in water.* **Strong electrolytes** *ionize completely in solution, whereas* **weak electrolytes** *partially ionize.* **Nonelectrolytes** *(e.g., sugar, alcohols, and lipids) do not ionize at all when dissolved.* The magnitude of current that will flow through an aqueous solution depends on the com-

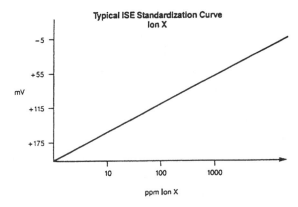

Figure 22.16. **A Calibration Curve for an Ion-Selective Electrode.**

bined concentration of each electrolyte present and the degree to which each ionizes.

B. The Measurement of Conductivity

i. THE DESIGN OF CONDUCTIVITY INSTRUMENTS

The measurement of conductivity, like the measurement of pH, involves the use of an electrochemical system. A simple device to measure conductivity consists of two flat electrodes that are dipped into the sample solution and are connected to a battery and to an **ammeter,** *a current measuring instrument*, see Figure 22.17. The battery generates electric current that can flow only if the sample solution is conductive and completes the circuit path. The electrode connected to the positive pole of the battery is positive, whereas the electrode connected to the negative pole is negative. Positive ions in the solution are attracted to the negative electrode; similarly, negative ions are attracted to the positive electrode. At the positive electrode negative ions give up electrons that flow in the wire to the positive pole of the battery. At the negative

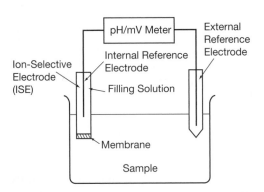

Figure 22.15. **An Ion-Selective Electrode Measuring System.**

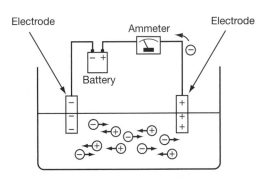

Figure 22.17. **The Measurement of Conductivity.** The meter consists of two flat electrodes, a battery, and a meter that measures current. Positive ions flow toward the negative electrode and negative ions toward the positive electrode. Electrons flow in the wires between the electrodes and the battery and meter. This makes a complete circuit for current flow. The more ions present in the solution to carry current, the more current registered by the ammeter.

Table 22.14.   SOME USES OF CONDUCTIVITY MEASUREMENTS

1. *If there is only one type of salt in a solution, it is possible to determine its concentration by comparing its conductivity with standards of known salt concentration.*

2. *Pure water does not readily conduct electricity, so a measurement of conductivity indicates the amount of impurities in the water.* Very low conductivity indicates pure water. There are a vast number of applications where the purity of water is important, such as in the production of pharmaceuticals, maintaining cells in culture, and ensuring the proper operation of industrial heating and cooling systems.

3. *Detectors that measure the conductivity of a solution can be attached to chromatography instruments and used to monitor the elution of sample components, as shown in Figure 22.18.*

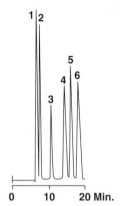

Organic Acid Standards

1. Oxalic Acid
2. Maleic Acid
3. Malic Acid
4. Succinic Acid
5. Formic Acid
6. Acetic Acid

Column:	Wescan Anion Exclusion, 100 x 7.5 mm
Mobile Phase:	1mM Sulfuric Acid
Flow Rate:	0.6mL/min
Detector:	Conductivity

Figure 22.18.   Display Formed by a Conductivity Detector Used to Monitor the Elution of Sample Components from a Chromatography Column. (Chromatogram courtesy of Alltech Associates, Inc.)

4. *Conductivity can be used clinically to indicate electrolyte levels in body fluids.*

5. *Conductivity can be used to approximate "total dissolved solids" in a solution.* This measurement is frequently used, for example, in the pharmaceutical industry as an indicator of purity.

6. *Some enzyme-catalyzed reactions release ions, so changes in conductivity can be used to monitor some enzyme reactions.*

7. *Electrophoresis, a method commonly used to separate charged biological macromolecules, is dependent on using buffers with the proper conductivity.*

8. *In industrial processes, conductivity may be monitored as an indication of the concentration of an important chemical.*

9. *Conductivity may be monitored in fish tanks as an indicator of water quality.*

10. *Conductivity can be used as a routine quality-control check of laboratory reagents to see that they have been prepared properly. The conductivity of a given reagent should be close to the same each time it is prepared. If it is not, then there was likely a mistake in the reagent's preparation or a problem with an ingredient.*

---

electrode positive ions take electrons. These processes result in continuing electron flow toward the negative electrode and continuing movement of ions through the solution. The more ions present, the more current is registered by the ammeter.

ii. THE CELL CONSTANT, K

The amount of current flow that is measured by a conductivity instrument will vary depending on *the distance between the two electrodes,* **d,** *and the area of the electrodes,* **A,** see Figure 22.19. The farther apart the electrodes, the less current is registered. Conversely, the larger the electrode area, the more current that is measured. For samples with low conductivity it is best to configure the measuring device so that it is as sensitive as possible by having a small distance between the electrodes and/or a large electrode surface. If the sample is very conductive, however, then a wider spacing between electrodes and/or smaller electrodes are more effective.

Manufacturers make conductivity instruments with sample holders, or **sample cells,** having different configurations of distances and electrode surface areas. The

configuration is described by the **cell constant** for the device. The **cell constant,** *abbreviated* K, *is defined as:*

$$K = d/A$$

where d = the distance between the electrodes
A = their area

The units of K are therefore $cm/cm^2 = 1/cm$.

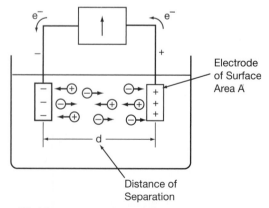

Electrode of Surface Area A

Distance of Separation

Figure 22.19.   The Configuration of a Conductivity Measuring System. The surface area of the electrodes and the distance between them will affect the measurement.

Box 22.7.   UNITS OF MEASUREMENT OF CONDUCTIVITY AND RESISTIVITY

1. **Resistance.** The more readily a solution conducts current, the less resistance there is to current flow.
   Resistance is measured in ohms (abbreviated Ω)
   1 megohm (MΩ) = 1,000,000 Ω

2. **Conductance.** The conductance of a solution is the inverse of its resistance.
   Conductance = 1/resistance
   1/ohm = 1 mho = unit of conductance
   1 mho = 1 Siemen (abbreviated S) = another unit of conductance
   1 micromho (μmho) = 1 microSiemen (μS) = 1 mho /1,000,000

3. **Conductivity.** Conductance is the property that is measured, but to compare the results from various instruments, or from one laboratory, with another, it is necessary to convert conductance to conductivity. This is done using the following equation:
   Conductivity = conductance (K)
   The units of conductivity therefore are: (Siemens)(1/cm) = Siemens/cm

   In practice, a S/cm is too large, so most instruments display values as **μS/cm** or **mS/cm**. In catalogs, **μS/cm** and **mS/cm** may be abbreviated as **μS** and **mS**, respectively.
   1 μS/cm = 0.001 mS/cm = 0.000001 S/cm = 1 μmho/cm

4. **Resistivity.** This term is:
   1/conductivity
   The units of resistivity, therefore, are ohm-cm

   Resistivity is often used in the measurement of ultrapure water due to its extremely low conductivity.

5. It is possible to convert between conductance and resistance
   For example:
   Convert a conductance of 40 μS to resistance
   Conductance = 40 μS = 40 $\times$ 10^{-6} S = 4 $\times$ 10^{-5} S = 4 $\times$ 10^{-5} mho = 4 $\times$ 10^{-5}/1 ohm
   Resistance = 1/conductance
   Therefore, resistance = (4 $\times$ 10^{-5}/1 ohm)$^{-1}$ = 1 ohm/4 $\times$ 10^{-5} = 25,000 ohm

---

The smaller the value for *K,* the shorter the distance between electrodes and/or the larger the electrodes. A device with a lower *K* value, therefore, is more sensitive than a device with a higher *K* value. For measuring the conductivity of pure water, a sensitive device with a *K* value of about 0.1/cm is preferred. For samples of moderate conductivity, a cell with a *K* value of around 0.4/cm is useful, and for samples with very high conductivities, a cell with a *K* value of around 10/cm is chosen. Figure 22.20 shows a conductivity instrument that has three probes available to match different type of samples.

iii. "CONDUCTIVITY" AND "CONDUCTANCE"; UNITS USED FOR "CONDUCTIVITY" MEASUREMENTS

There is a distinction between the terms *conductivity* and *conductance.* **Conductivity** is an inherent property of a material. **Conductance** *is a measured value that depends on the configuration of the conductivity measuring device used.* Conductance is sometimes also called "measured conductivity." For example, the *conductance* of a sample of pure water will vary depending on the value for *K* in the measuring instrument used. The *conductivity* of the water sample is the same, regardless of the instrument used to measure it. (The terms *resistivity* and *resistance* can similarly be distinguished from one another.)

Thus, there are four terms and four slightly different units used to describe a solution's ability to conduct current. These various units, all of which are used to describe the ability of a solution to carry current, are summarized in Box 22.7. Table 22.15 on p. 400 shows the resistivity and the conductivity of some common types of solutions.

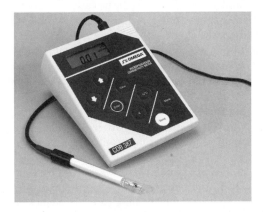

Figure 22.20.   A Conductivity Meter. This microprocessor-based benchtop conductivity meter is able to measure a wide range of conductivities, from a few μs/cm all the way to 200 mS/cm. It comes with a probe with a *K* value of 1.0. Other probes are available for this meter with *K* values of 0.1 and 10.0. (copyright© EDT Instruments, LTD. Reprinted with permission; courtesy of Omega Engineering, Inc., www.omega.com.)

Table 22.15.   THE CONDUCTIVITIES AND RESISTIVITIES OF COMMON TYPES OF SOLUTIONS

Solution	Conductivity	Resistivity
Ultrapure Water	0.055 µS/cm	18,000,000 ohm-cm = 18M Ω-cm
Distilled Water	1 µS/cm	1,000,000 ohm-cm = 1 M Ω-cm
Deionized Water	80 µS/cm	12,500 ohm-cm
0.05% NaCl	1000 µS/cm	1,000 ohm-cm
Seawater	50,000 µS/cm	20 ohm-cm
30% H_2SO_4	1,000,000 µS/cm	1 ohm-cm

Values shown are approximate.

iv. CONDUCTIVITY MEASUREMENTS

As with any instrument, conductivity meters require calibration. Calibration standards for conductivity measurements are commercially available.

Temperature has a large effect on conductivity. Conductivity standards will have a table to show their conductivity at various temperatures. Further notes about calibration and other practical comments about conductivity measurements are summarized in Table 22.16.

PRACTICE PROBLEMS

1. Calculate the hydrogen ion concentration of a solution whose pH is 9.0.
2. Calculate the hydrogen ion concentration of blood whose pH is 7.3.
3. The concentration of hydrochloric acid, HCl, secreted by the stomach after a meal is about 1.2×10^{-3} moles/liter. What is the pH of stomach acid?
4. Calculate the pH of 0.01 M HNO_3, nitric acid. Nitric acid is a strong acid that dissociates completely in dilute solutions. (Clue: because the nitric acid dissociates completely, the concentration of H^+ ions will be 0.01 M.)
5. (Optional) What is the pH of 0.0001 M NaOH? (Sodium hydroxide is a strong base.) Clue: in aqueous solutions, $[H^+][OH^-] = 10^{-14}$.
6. Consider the response of an electrode to a change in pH as shown in the graph on the next page.
   a. What is the slope of the line at 25°C?
   b. When the pH changes by one pH unit, for example, goes from 5.0 to 6.0, how much does the voltage change at 25°C?

Table 22.16.   PRACTICAL CONSIDERATIONS FOR CONDUCTIVITY MEASUREMENTS

1. *Standards.* Conductivity meters should be calibrated with a standard before use. Choose a standard whose conductivity is similar to that of the solution to be measured. Standards readily pick up contaminants from the air. Keep the standards' containers closed to avoid contamination and evaporation. A standard solution exposed to air will degrade within a day. Remember, too, that evaporation will increase the conductivity of a standard.
2. *Temperature.* Raising the temperature increases the conductivity of a solution. The effect of temperature is different for every ion, but the conductivity typically increases by about 1–5% per °C. Many conductivity meters have temperature-sensing elements and are able to compensate automatically for temperature.
3. *Depth of immersion.* The depth to which the electrodes are immersed can affect the current readings. Do not immerse to the bottom of the solution if there are particulates settled there.
4. *Air bubbles.* Make sure that air bubbles are not trapped in the probe when making measurements.
5. *Electrode maintenance.* Probes should be rinsed with distilled water after each series of measurements. Probes may be allowed to dry during storage, but they may require several minutes to stabilize when placed in a solution. Do not allow dried salts or particulates to build up on the probe.
6. *Meter configuration.* Conductivity meters come with various designs that affect the size of the electrodes and the distance between the electrodes, see Figure 22.19 on p. 398. It is necessary to choose a conductivity meter that has the proper configuration to match the sample.
7. *Acids and bases.* Most acids and bases are considerably more conductive than their salts because H^+ and OH^- ions are extremely mobile. For example, NaOH is more conductive than NaCl.
8. *Particulates.* For greatest accuracy, make sure that there are no particulates suspended in the sample. Filter or settle out particulates.
9. *Linearity.* The relationship between conductivity and ion concentration is generally, but not always, linear. In highly concentrated solutions, the relationship between conductivity and concentration may be nonlinear due to interactions between ions.
10. *Total dissolved solids (TDS) measurement.* Conductivity is related to the concentration of all dissolved electrolytes in a solution. Conductivity, therefore, is sometimes used to indicate the concentration of "total dissolved solids" (often minerals like $CaCO_3$) in a solution. In these cases, the meter should be calibrated with a dissolved solids standard that contains the same type of solids as the sample to be tested. Otherwise, there will be serious errors in the measurement because each type of dissolved substance has a different effect on conductivity. Commercially available TDS standards include calcium carbonate (often used to test boiler water), sodium chloride (often used to test brines), potassium chloride, and a mixture of 40% sodium sulfate, 40% sodium bicarbonate, and 20% sodium chloride (often used to test water from lakes, streams, wells, and boilers).
11. *Cleaning the electrodes.* Clean cells with mild liquid detergent. Dilute HCl may also be used. Rinse the cells several times with distilled water and recalibrate before use.

c. What is the slope of the line at 0°C?

d. When the pH changes by one pH unit, for example, goes from 5.0 to 6.0, how much does the voltage change at 0°C?

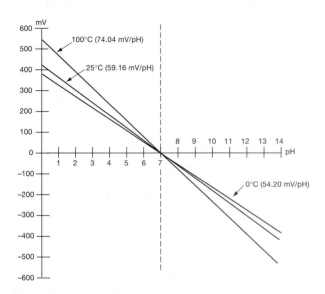

Note: We consider the slope to be negative (see Chapter 15). However, in this question, we are concerned with the magnitude of the slope and not its direction.

7. A recording from an industrial fermentation run is illustrated above, right. During this run, bacteria reproduced many times over and were harvested to be sold for use in commercial food production. A pH probe was present in the fermentation broth and a continuous recording of the pH was made. When the microorganisms (the seed culture) were first added to the broth, the pH was about 6.8. The pH of the culture changed during the active growth phase. Answer the following questions about this recording.

a. What happened to the pH as the bacterial population grew?

b. Speculate as to what might be the difficulties in designing pH probes for use in fermentation.

c. Why do you think the company generates a continuous recording of pH during each fermentation run?

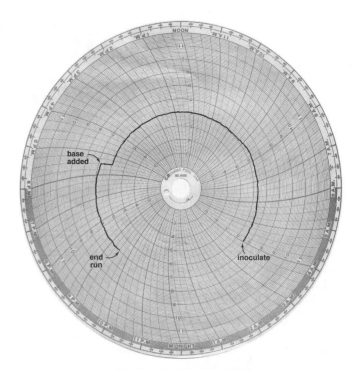

For Practice Problem #7.

2. How does temperature affect the measurement of volume, weight, and pH?

3. Suggest some likely causes for the following pH measurement problems.

a. The pH reading slowly drifts.

b. The pH reading fluctuates.

c. There is no reading at all.

4. Suggest a way to check each possibility you suggested in question 3.

5. If you work in a laboratory, write a standard operating procedure (SOP) to calibrate and operate a particular pH meter that is available in your laboratory.

6. a. Discuss how you would go about demonstrating that the pH meter is giving consistent, accurate results when measuring the pH of samples from a sewage treatment plant.

b. Write an SOP to measure the pH of samples from a sewage treatment plant. Include safety precautions to protect the operator. Include precautions to avoid contamination of the meter and to avoid cross contamination (where one sample contaminates another sample).

QUESTIONS FOR DISCUSSION

1. Compare and contrast the calibration of electrodes, pipettes, and balances. Based on this comparison, define the term *calibration* in your own words.

Chapter Appendix

A Brief Introduction to Electrochemistry

Understanding how electrodes generate electrical potential requires some knowledge of electrochemical reactions. Although the theory of electrochemistry is complex, it may be helpful to realize that this theory was devised by scientists to explain concrete, empirical observations of nature. This section briefly describes some of the early experimental observations in electrochemistry and the theory that arose to explain these observations.

Voltage (electrical potential) *is the potential energy of charges that are separated from one another.* Electrical potential arises because charged objects exert attractive or repulsive forces on one another. The development of voltage therefore requires a method of separating charges. Electrochemical reactions are one of the ways by which positive and negative charges become separated.

Some of the most important observations relating to electrochemistry were made by Alessandro Volta in the early 1800s (e.g., see Bordeau, Sanford P. *Volts to Hertz: The Rise of Electricity.* Minneapolis, MN: Burgess Publishing, 1982). Volta placed disks of two different metals adjacent to one another, but separated by a piece of cloth soaked in an electrolyte solution. Volta observed that one of the metal disks would acquire a positive charge and the other a negative charge. Volta assembled large stacks of disks separated by pieces of felt soaked in salt water. He found that if he touched the top and bottom disks in the stack at the same time, he received an electrical shock.

Scientists explain the shock Volta received as being due to **electrochemical reactions** *in which metals either release electrons or take up electrons, becoming either positively or negatively charged.* Electrochemical reactions caused separation of positive and negative charges from one another, leading to the development of voltage. When Volta touched the top and bottom disks, the electrical potential caused current to flow between the disks through his body so that he was shocked.

Although Volta did not understand electricity as it is understood today, he was able experimentally to learn a great deal about electrochemical reactions. Volta made stacks with different metals and discovered that metals could be arranged in order of how likely they are to become positively charged. For example, he found that when zinc and copper were used in the stacks, the zinc became positively charged and the copper negatively charged. Iron and copper also left the copper negatively charged, but the shock Volta experienced was milder from stacks made with iron than with zinc. If copper and silver were used, the copper became positively charged. Scientists now interpret Volta's results to mean that metals differ in their tendency to lose electrons; some metals lose electrons more readily than others. The electrochemical reactions in Volta's stacks resulted in movement of electrons from the metal that had more tendency to lose electrons to the other metal. The ordering of metals according to their tendency to lose electrons is now taught as the "activity series of metals." Scientists have investigated many metals and given numerical values to their relative tendency to lose electrons.

The idea of combining a piece of metal and an electrolyte solution in order to develop voltage has useful applications. The most familiar example is a **battery,** *which harnesses the voltage due to electrochemical reactions to do useful work.* A battery consists of two **electrodes,** *metal strips in contact with an electrolyte solution, where electrochemical reactions occur.* As in Volta's stacks, the reactions lead to the development of voltage. When the two electrodes are connected via a metal wire, current flows.

Ion selective electrodes are another application that takes advantage of electrochemical reactions. Volta observed that the magnitude of the potential developed by his stacks was controlled by both the nature of the metals involved and the concentration of certain ions in the electrolyte solution. The higher the concentration of the ions, the higher the potential. This relationship between ion concentration and potential is the basis for ion-selective electrodes, such as pH measuring electrodes, which develop a greater or lesser potential depending on the concentration of a particular ion in a solution. Thus, the design of pH meter measuring systems is based on the observations of Volta and other scientists.

CHAPTER 23

Measurements Involving Light A:
Basic Principles and Instrumentation

I. LIGHT

A. Introduction

Previous chapters discussed the measurement of certain physical properties of a sample, such as its weight, volume, and temperature. This chapter discuss measurement of another physical characteristic of a sample: its ability to absorb light.

Light is essential for life. Almost every living creature on earth is reliant on the energy from sunlight; for example, light enables humans to see, and light sustains plant growth. Light is able to play these roles because it interacts with molecules.

In the laboratory, we take advantage of the interaction of light with materials in samples to obtain information about those samples. **Spectrophotometers** *are instruments used to measure the effect of a sample on a beam of light.* Spectrophotometric assays are used, for example, to determine how much DNA is present in a cellular extract, the purity of protein in an enzyme preparation, the activity of an enzyme, and to confirm the identity of an ingredient in a drug formulation.

This chapter begins by exploring the nature of light. It will then discuss the design, operation, and performance of instruments that can detect and measure light. Chapter 25 will explain how measurements involving light are used to obtain meaningful information about biological systems.

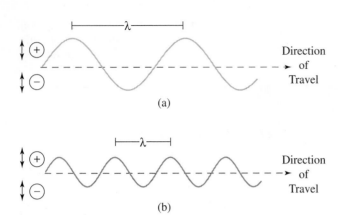

(a)

(b)

Figure 23.1.   Electromagnetic Radiation Traveling as Waves. Electromagnetic radiation is often visualized with peaks and troughs, similar to waves of water. Unlike water waves, however, electromagnetic radiation is not associated with the motion of matter. Rather, electromagnetic radiation is associated with regular changes in electric and magnetic fields. The electromagnetic radiation in **a.** has a longer wavelength and carries less energy than the radiation in **b.**

B. Electromagnetic Radiation

i. THE ELECTROMAGNETIC SPECTRUM

Light is a type of **electromagnetic radiation,** that is, *energy that travels through space at high speeds*. There are different categories of electromagnetic radiation including gamma rays, x-rays, ultraviolet (UV) light, visible (Vis) light, infrared (IR) light, microwaves, and radio waves. *All the types of electromagnetic radiation together are termed the* **electromagnetic spectrum.** Visible, ultraviolet, and infrared radiation are all called "light" even though our eyes are only sensitive to visible light.

Each type of radiation within the electromagnetic spectrum has a characteristic way of interacting with matter. For example X-rays, which have great energy, can easily penetrate matter. We take advantage of this characteristic when producing X-ray images for medical use. Cells in our eyes are able to interact with light in the visible region of the electromagnetic spectrum, leading to vision.

Scientists describe electromagnetic radiation as having a "dual nature." This is because sometimes electromagnetic radiation is best explained as consisting of waves moving through space with crests and troughs, much like waves in a pond. Other times electromagnetic radiation is portrayed as "packages of energy" moving through space. *Each "package of energy" is called a* **photon.**

Figure 23.1 illustrates the wavelike nature of light. *The distance from the crest of one wave to the crest of the next is the* **wavelength** *(abbreviated λ) of the radiation*. The wavelength of electromagnetic radiation varies greatly depending on its type. Some types of electromagnetic radiation have wavelengths so short they are measured in nanometers, but other types of electromagnetic radiation have wavelengths thousands of meters in length. For example, the wavelengths of X-rays are measured in nanometers, whereas radio waves can approach 10,000 meters, see Figure 23.2.

Different types of electromagnetic radiation not only have different wavelengths, they also carry different amounts of energy. The shorter the wavelength of electromagnetic radiation, the more energy is carried by its photons. X-rays have very short wavelengths and consist of photons which carry a great deal of energy. In contrast, radio waves, which have long wavelengths, have photons with much less energy.

ii. THE VISIBLE PORTION OF THE ELECTROMAGNETIC SPECTRUM

Consider the portion of the electromagnetic spectrum to which the human eye is sensitive. Humans perceive different wavelengths of visible light as being different colors, see Table 23.1. For example, if a light source emits light with wavelengths between about 505 and 555 nm, that light is perceived as green. "White" light emitted by the sun or an electric light bulb is **polychromatic,** *it is a mixture of many wavelengths.*

Figure 23.2.   Types of Electromagnetic Radiation and Their Wavelengths. Visible light occupies a small portion of the electromagnetic spectrum. Within the visible portion of the spectrum, different wavelengths are associated with different colors. *(See color insert C-1.)*

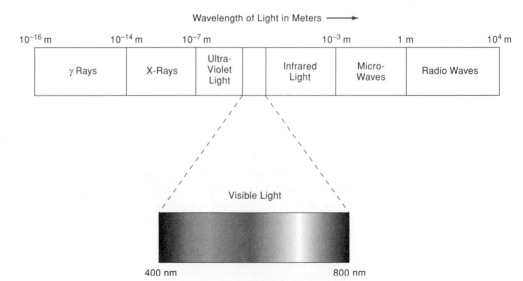

Table 23.1. *THE RELATIONSHIP BETWEEN THE WAVELENGTH OF VISIBLE LIGHT AND THE COLOR PERCEIVED*

λ of the Light (nm)	Color Our Eyes Perceive
380–430	Violet
430–475	Blue
475–495	Greenish blue
495–505	Bluish green
505–555	Green
555–575	Yellowish green
575–600	Yellow
600–650	Orange
650–780	Red

Because color is subjective, these wavelengths are approximations. The particular ranges here are based on various sources.

iii. THE UV PORTION OF THE ELECTROMAGNETIC SPECTRUM

Ultraviolet light is categorized into three types: UV-A (315–400 nm), UV-B (280–315 nm), and UV-C (180–280 nm), see Table 23.2. UV-A is the type used in tanning booths and "black lights." The illumination from a "black light" causes certain pigments to emit visible light; these materials are called **fluorescent.**

UV-B light has sufficient energy to damage biological tissues and is known to cause skin cancer. The atmosphere blocks most UV-B light; however, as the ozone layer thins, this form of radiation has become more of a health concern. UV-B light is routinely used in the biotechnology laboratory when visualizing DNA. The dye, ethidium bromide, intercalates into the grooves of DNA, forming a complex that fluoresces when exposed to UV-B radiation. The fluorescence makes the DNA visible to the eye and allows it to be photographed. There are other dyes that similarly fluoresce when exposed to UV light and can be attached to biological molecules. Fluorescence emitted by molecules allows investigators to visualize their location.

UV-C, with high energy and short wavelengths, is rarely observed in nature because it is absorbed by

Table 23.2. *ULTRAVIOLET LIGHT*

Type	Wavelength	Examples of Applications
UV-A ("long-wave")	315–400 nm	Used in "black lights" and tanning booths.
UV-B ("medium-wave")	280–315 nm	Used in electrophoresis to visualize DNA tagged with ethidium bromide.
UV-C ("short-wave")	180–280 nm	Used in bacteriocidal lamps and for inducing mutations in cells.

oxygen in the air. UV-C is used in germicidal lamps to kill bacteria. Most hoods manufactured for culturing cells are equipped with UV-C lamps to help prevent bacterial contamination of the cultures. UV-C light damages DNA in cells and therefore can be used in research to intentionally induce mutations in bacteria and cultured cells.

> **Safety Note: There are sources of UV light in the laboratory and it is important to avoid exposing your skin or eyes to this type of radiation. Safety glasses and goggles that specifically block UV radiation are available from scientific supply companies.**

C. Interactions of Light with Matter

Various events can occur when light strikes matter, see Figure 23.3. (1) The light can be **transmitted** *(i.e., pass without interaction through the material, as when light passes through glass).* (2) The light can be **reflected** *(i.e., change directions, as is light reflected by a mirror).* (3) The light can be **scattered,** *in which case the light is deflected into many different directions.* Scattering occurs when light strikes a substance composed of many individual, tiny particles. (4) The light can be **absorbed,** *in which case it gives up some or all of its energy to the material.* When light is absorbed, energy carried by photons is transferred to the electrons within a material. As this transfer occurs, light energy is converted to heat energy. Light absorption provides the basis for spectrophotometric measurements.

Because absorption is the basis of spectrophotometry, we consider it in more detail. Every material has a particular arrangement of electrons and of bonds involving electrons. For this reason, different materials vary from one another in terms of which wavelengths of

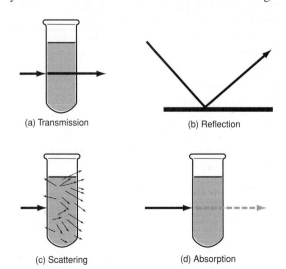

(a) Transmission (b) Reflection (c) Scattering (d) Absorption

Figure 23.3. Events That Can Occur When Light Strikes Matter. a. The light may be transmitted through the object without interaction. **b.** The light may be reflected by the surface of the object. **c.** The light may be scattered by small individual particles. **d.** The light may lose energy to the matter in which case it is absorbed.

Figure 23.4.
The Colors of Solid Objects.
a. An object appears black if it absorbs all colors of light. **b.** An object appears white if it reflects all colors. **c.** An object appears orange if it reflects only this color and absorbs all others. **d.** An object also appears orange if it reflects all colors except blue, which is complementary to orange. *(See color insert C-1.)*

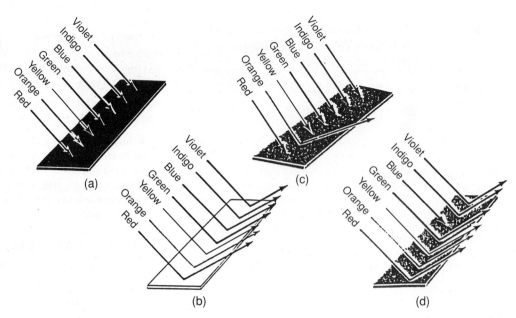

light they absorb. The color of an object is determined by which wavelengths of light it absorbs. If an object absorbs light of a particular color, then that color does not reach our eyes when we look at the object. If an object absorbs all colors of light, then that object appears to be black, as in Figure 23.4a. If an object reflects visible light of a particular color, we see that color when we look at the object. An object appears white if it reflects all colors of light, see Figure 23.4b. The object in Figure 23.4c appears to be orange because it reflects orange light and absorbs all other colors.*

The object in Figure 23.4d reflects all colors except blue. When blue light is removed from the mixture of wavelengths, our eyes perceive the object to be orange. Orange and blue are called *complementary colors.* Figure 23.5 shows a color wheel in which complementary colors are across from one another. Looking at the color wheel, we can determine, for example, that if green is removed from a beam of light, then the light will appear to be red.

In this discussion of color, we have so far considered only solid objects that either absorb or *reflect* light. When we use a spectrophotometer, we are almost always working with samples that are in liquid form. These liquids may either absorb or *transmit* light of a particular wavelength. The spectrophotometer enables us to quantify how much light is absorbed and how much is transmitted by a solution at a specific wavelength.

Figure 23.5. A Color Wheel Showing Complementary Colors. *(See color insert C-2.)*

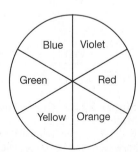

Table 23.3 is a guide that shows the general relationship between the color of light absorbed by a liquid solution and the color the solution appears to the eye. For example, if a solution absorbs blue light (light in the range from about 430 to 475 nm), then the solution will appear to be its complementary color, orange.

EXAMPLE PROBLEM

What color will a solution appear to be if it absorbs primarily light that is 590 nm?

ANSWER

We know that 590 nm is yellow light. From Table 23.3 you can see that a solution that absorbs primarily yellow light will appear to be violet. Yellow and violet are complementary colors.

Table 23.3. *THE ABSORPTION OF LIGHT OF PARTICULAR WAVELENGTHS AND THE COLOR OF SOLUTIONS*

Light Absorbed by the Solution

Wavelength (nm)	Color	Color the Solution Appears to Be
380–430	Violet	yellow
430–475	Blue	orange
475–495	Greenish blue	red-orange
495–505	Bluish green	orange-red
505–555	Green	red
555–575	Yellowish green	violet-red
575–600	Yellow	violet
600–650	Orange	blue
650–780	Red	green

Note: Colors that appear alongside one another on this table are said to be "complementary" colors.

Color is not actually a property of light, nor of objects that reflect light. It is actually a sensation that arises in the brain. Specialized cone cells in the eye transmit signals to the brain when stimulated by light of specific wavelengths.

Consider the example illustrated in Figure 23.6. In Figure 23.6a, white light is shining on a test tube that contains a solution of red food coloring. The food coloring absorbs incoming light that is green. It also absorbs some light that is bluish-green and greenish-blue. Light of other colors is transmitted through the food coloring. Because blue and greenish light is removed from the white light shining on the red food coloring, our eyes perceive the solution to be orange-red (see Table 23.3). A spectrophotometer enables us to graphically display this pattern of light absorption by the red food coloring

solution, see Figure 23.6b. In this display, wavelength is on the X axis. A measure of the amount of light absorbed by the food coloring is on the Y axis. Observe on the graph that there is a peak between about 490 nm and 530 nm. This peak represents the absorption of light at those wavelengths. The graph in Figure 23.6b, *that shows the extent to which a particular material absorbs different wavelengths of light, is called an* **absorbance spectrum.**

Note that red food coloring also absorbs light that has wavelengths shorter than 350 nm. This light, however, is in the ultraviolet range and therefore has no effect on the solution's apparent color.

Figure 23.7a on p. 408 shows the absorbance spectrum, as determined using a spectrophotometer, for the compound riboflavin. The absorbance spectrum of riboflavin has a pattern of "peaks" and "valleys" that is different than that of red food coloring. Every compound has its own characteristic absorbance spectrum. The absorbance spectrum of a particular compound always has the same pattern of peaks and valleys, regardless of how much of that compound is present, Figure 23.7b.

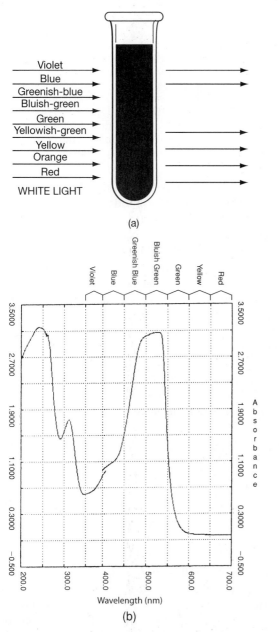

(a)

(b)

Figure 23.6.   The Absorption of Light by a Solution of Red Food Coloring. a. When white light is shined on a tube containing red food coloring, blue to green colored light is absorbed. **b.** The absorbance spectrum for red food coloring, generated by a spectrophotometer. The X axis of an absorbance spectrum is wavelength, the Y axis is the amount of light absorbance. A peak of visible light absorption occurs between about 490 and 530 nm. *(See color insert C-2.)*

EXAMPLE PROBLEM

Solutions of proteins and of nucleic acids are colorless. Does this mean they do not absorb any light?

ANSWER

It means they do not absorb any light in the *visible* range of the electromagnetic spectrum. Proteins and nucleic acids do, in fact, absorb light in the *UV* range of the spectrum.

EXAMPLE PROBLEM

What color is the solution whose absorbance spectrum is shown in the following graph?

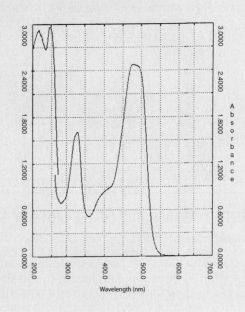

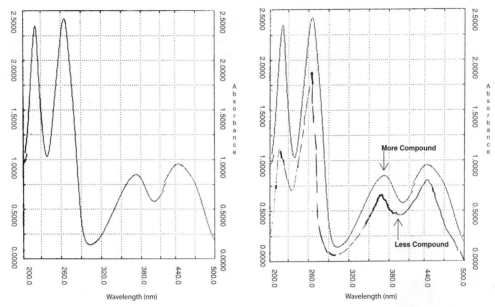

Figure 23.7. The Absorbance Spectrum for Riboflavin. a. The peaks and valleys of this absorbance spectrum are characteristic of riboflavin. **b.** Increasing amounts of riboflavin have the same spectral pattern; however, the height of the peaks is greater if more compound is present.

ANSWER

This compound has an absorbance peak in the greenish-blue region of the spectrum. From Table 23.3 on p. 406 we would expect it to be orange and it is, in fact, the dye Orange G.

EXAMPLE PROBLEM

Chlorophyll is one of the pigments responsible for the green color of leaves. Chlorophyll enables the leaves to absorb light from the sun and convert the energy of the light into food. Does chlorophyll capture all wavelengths of light from the sun?

ANSWER

No. Leaves are green because chlorophyll does not absorb much green light and those wavelengths are reflected.

EXAMPLE PROBLEM

What happens if red light is shined on a red solution?

ANSWER

The solution will not absorb the light; rather, it will transmit the light.

II. THE BASIC DESIGN OF A SPECTROPHOTOMETER

A. Transmittance and Absorbance of Light

i. TRANSMITTANCE

Figure 23.8 illustrates the basic principles underlying the operation of a spectrophotometer. In this illustration, two solutions are placed inside a spectropho-

tometer. The first is red food coloring dissolved in water. The analyte is red food coloring and water is the solvent. The second solution is pure water (i.e., the solvent used to dissolve the red food coloring). The water is called the **blank.** (We consider the meaning of a blank in more detail later.) A beam of green light is incident on (shines on) both solutions. As we know from Figure 23.6, the red food coloring will transmit little or no green light. In contrast, the water will readily transmit green light. A spectrophotometer electronically compares the amount of light transmitted through the sample with that transmitted through the blank. *The ratio of the amount of light transmitted through a sample to that transmitted through the blank is called the* **transmittance,** *abbreviated* ***t:***

$$\frac{\text{light transmitted through the sample}}{\text{light transmitted through the blank}} =$$
$$\text{transmittance} = t$$

It is also common to speak of **percent transmittance, %T,** *which is the transmittance times 100%:*

$$\%T = t \times 100\%$$

In the example illustrated in Figure 23.8, the transmittance is less than 1 because the amount of light transmitted through the sample is less than the amount transmitted through the blank. (The red food coloring in the sample absorbs some of the light.) In another situation, where both the sample and the blank transmit the same amount of light, the transmittance is 1, or %T is 100%. If the sample transmits no light at all, then the transmittance is 0. Thus, transmittance can range from 0 to 1 and percent transmittance from 0%

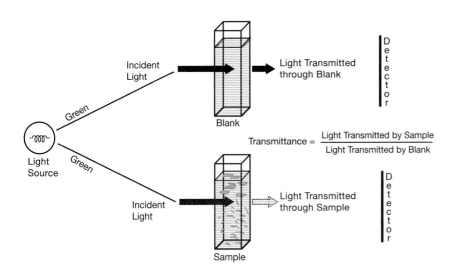

Transmittance = $\dfrac{\text{Light Transmitted by Sample}}{\text{Light Transmitted by Blank}}$

Figure 23.8.   Transmittance. Two solutions are placed inside a spectrophotometer: red food coloring and a blank consisting of pure water. Green light is shined on the two solutions and the amount of light transmitted through each is measured by a detector. The ratio of the amount of light transmitted through the sample to that transmitted through the blank is called the *transmittance.*

(no light passed through the sample relative to the blank) to 100% (all light was transmitted).

ii. ABSORBANCE

All spectrophotometers measure transmittance (i.e., the ratio between the amount of light transmitted through a sample and the amount of light transmitted through a blank). Analysts, however, are generally interested in the amount of light absorbed by a sample, for example as illustrated in Figure 23.6 on p. 407 and Figure 23.7 on p. 408 . It is customary, therefore, to convert transmittance to a measure of light absorption, called **absorbance. Absorbance** *is calculated based on the transmittance* as:*

$$A = -\log_{10}(t)$$

Absorbance is also called **optical density, OD.**

Absorbance does not have units, but absorbance values are sometimes followed by the abbreviation A or AU. For example, an absorbance value might be written as "1.6," or "1.6 A," or "1.6 AU," or "OD 1.6." Transmittance and absorbance are measured at a particular wavelength, so it is customary to record both the absorbance and the wavelength. For example,

"$A_{260} = 1.6$," means "an absorbance of 1.6 was measured at an analytical wavelength of 260 nm."

Figure 23.9a shows the relationship between how much light is transmitted through a solution and the concentration of analyte in that solution. Observe that the relationship between analyte concentration and transmittance is not linear, see Figure 23.9a. In contrast, there is a linear relationship between the absorbance of light by a sample and the concentration of analyte, see Figure 23.9b. For this reason, absorbance values are more convenient than transmittance values.

As one would expect, when the absorbance of light increases, the transmittance decreases, and vice versa, Table 23.4. For example, if the percent transmittance is 100%, then there is no light absorbance by the sample relative to the blank. Suppose in another case that a sample absorbs so much light that the percent transmittance is only 0.1%. In this example, the transmittance (*t*) is 0.001 and the absorbance is:

$$A = -\log(0.001) = -\log(10^{-3}) = -(-3) = 3$$

Most spectrophotometers can convert transmittance values to absorbances, so they can display either absorbance or transmittance. It is also possible to use a calculator to convert transmittance to absorbance, and vice versa.

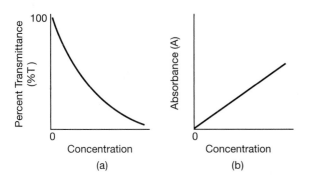

Figure 23.9.   The Relationship between % Transmittance and Absorbance of Light and the Concentration of Analyte.
a. Concentration of analyte in a sample versus percent transmittance is not linear. **b.** In contrast, the relationship between absorbance and analyte concentration is linear.

Table 23.4.   THE RELATIONSHIP BETWEEN ABSORBANCE AND TRANSMITTANCE, $A = -\text{LOG}_{10}(t)$

t	%T	A
0.000	0.00	∞
0.001	0.10%	3
0.010	1.00%	2
0.100	10.0%	1
1.000	100%	0

*An alternative equation to calculate the absorbance is $A = 2 - \log_{10}(\%T)$. Both this equation and the one in the body of the text will give the same answer.

EXAMPLE PROBLEM

Convert a transmittance value of 0.63 to absorbance.

ANSWER

$$A = -\log(t)$$
$$A = -\log(0.63)$$
$$A = -(-0.20) = 0.20$$

EXAMPLE PROBLEM

Convert an absorbance value of 0.380 to transmittance and % transmittance.

ANSWER

$$0.380 = -\log(t)$$
$$\text{antilog}(-0.380) = t$$
$$0.417 = t \text{ or } 41.7\% = \%T$$

EXAMPLE PROBLEM

A solution contains 2 mg/mL of a light absorbing substance. The solution transmits 50% of incident light relative to a blank. a. What is the absorbance in this case? b. What is the expected absorbance of a solution containing 8 mg/mL of the same substance? c. What is the expected transmittance of a solution containing 8 mg/mL of the same substance?

ANSWER

a. The absorbance of 2 mg/mL of this substance is equal to $-\log(0.50) = 0.30$.

b. The relationship between absorbance and concentration is linear (within a certain range, as discussed in the next chapter). Assuming linearity (and assuming that if there is no analyte in the sample, the absorbance is zero), if 2 mg/mL has an absorbance of 0.30, then 8 mg/mL should have an absorbance of 1.20.

c. If the absorbance is 1.20, the transmittance = antilog $(-1.20) = 0.063 = 6.3\%T$. Observe that the relationship between concentration and transmittance is not linear.

iii. THE BLANK

A spectrophotometer is used to determine the absorbance of light by the analyte in a sample. In order to make this determination, a spectrophotometer always compares the interaction of light by a sample with its interaction with a blank, as was illustrated in Figure 23.8. Let us consider in more detail the composition of the blank.

An analyte is seldom alone in a sample solution; rather, it is dissolved in a solvent and sometimes various reagents are added to the sample as well. When the sample is placed in the spectrophotometer, not only the analyte may absorb light, but the solvent and added reagents may absorb light as well. Moreover, the sample is contained in a *holder, called a* **cuvette** (or **cell**), which may absorb a small amount of light. The **blank** *should contain no analyte, but it should contain the solvent and any reagents that are intentionally added to the sample.* The blank is held in a cuvette that is identical to that used for the sample, or the blank is alternately placed in the same cuvette as the sample. In the example in Figure 23.8, the blank is pure water because the analyte, red food coloring, was dissolved in water. In another situation, the blank may have a different composition.

B. Basic Components of a Visible/UV Spectrophotometer

i. OVERVIEW

We can now consider in more detail the various components that make up a spectrophotometer. A basic spectrophotometer consists of a source of light, a wavelength selector, a sample holder, a detector, and a display, see Figure 23.10.

ii. LIGHT SOURCE

The **light source,** also called a **bulb** or **lamp,** *produces the light used to illuminate the sample.* There are various types of lamps available for spectrophotometry. All lamps produce polychromatic light, but each type emits a different range of wavelengths. *The light used for visible spectrophotometry is usually produced by bulbs with a* **tungsten filament.** *A* **deuterium arc lamp** *is commonly used to produce light in the UV range.* Spectrophotometers that can be used in both the visible and UV ranges have two light sources. Alternatively, a **xenon lamp** can be used *that produces light in both the UV and visible ranges.*

Every type of lamp emits more light of some wavelengths than others. For example, a tungsten lamp has maximum emission at wavelengths above about 650 nm. The UV emission maximum for deuterium arc lamps is at about 240 nm.

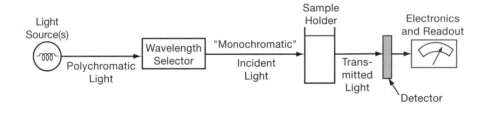

Figure 23.10. Simplified Design of a Spectrophotometer. Polychromatic light is emitted by the light source. A narrow band of light is selected in the wavelength (radiation) selector. The light passes through the sample (or the blank), where some or all of it may be absorbed. The light that is not absorbed falls on the detector.

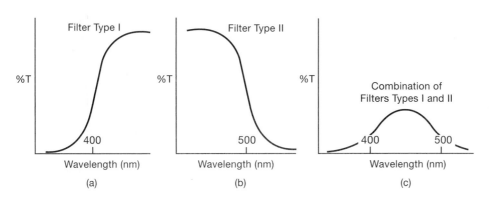

Figure 23.11. Filters. a. A filter that blocks the transmission of light below 400 nm. **b.** A filter that blocks the transmission of light above 500 nm. **c.** Two filters can be combined so as to allow only light in a certain wavelength range to pass. In this example, light between 400 and 500 nm is transmitted through the filter.

iii. WAVELENGTH SELECTOR

Light emitted by the spectrophotometer light source is polychromatic, but the absorption of light by a sample is wavelength-specific. Spectrophotometry measures how much light of a particular wavelength is absorbed by a sample. Spectrophotometers, therefore, require a wavelength-selecting device. The wavelength-selecting component of a spectrophotometer isolates light of a particular wavelength (in practice, it isolates a narrow range of wavelengths) from the polychromatic light emitted by the source lamp.

Some instruments, called **photometers,** use filters as an inexpensive wavelength selector. A **filter** *blocks all light above or below a certain cutoff,* see Figure 23.11a,b. Two filters can be combined, as shown in Figure 23.11c, to make a filter that transmits light in a certain range.

Filters cannot produce **monochromatic light** *(i.e, light of a single wavelength).* It is more effective to use a **monochromator** to produce light of a very narrow range of wavelengths. **Monochromators** *are devices that can disperse (separate) polychromatic light from the light source into its component wavelengths, and that can be used to select light of desired wavelengths.* A monochromator disperses light, much as light is dispersed into a rainbow by a prism, see Figure 23.12.

The three main components of a monochromator, see Figure 23.13, are:

1. **Entrance slit.** Light from the lamp is emitted in all directions. *The entrance slit shapes the light entering the monochromator into a parallel beam that strikes the dispersing element.*

2. **Dispersing element.** In earlier spectrophotometers, prisms were used to disperse light into its components. In modern instruments, **diffraction gratings** are the most common type of dispersing element. **Diffraction** *is the bending of light that occurs when light passes an obstacle.* **Diffraction gratings** *consist of a series of microscopic grooves etched onto a mirrored surface.* The grooves diffract light that strikes them in such a way that polychromatic light is dispersed into its component wavelengths.

3. **Exit slit.** *The exit slit permits selected light to exit the monochromator.*

After light is dispersed by the diffraction grating, it is necessary to select light of the desired wavelength. This is accomplished by rotating the grating so that the light that reaches the exit slit is of the desired wavelength, see Figure 23.14 on p. 412. The accuracy of wavelength selection in a spectrophotometer depends on the correct mechanical operation of the diffraction grating.

If a perfect monochromator could be produced, and if the spectrophotometer were set to select a particular wavelength, then only light of that single wavelength

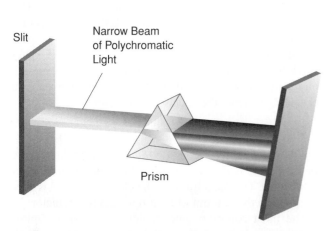

Figure 23.12. The Dispersion of Light into its Component Wavelengths by a Prism. *(See color insert C-2.)*

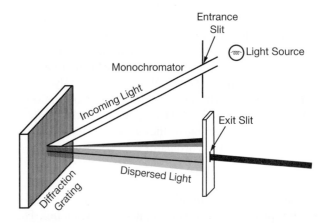

Figure 23.13. The Components of a Monochromator. The monochromator consists of an entrance slit, a dispersing element, and an exit slit. The dispersing element is a diffraction grating which consists of a series of microscopic grooves etched on a mirrored surface.

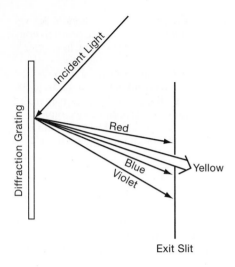

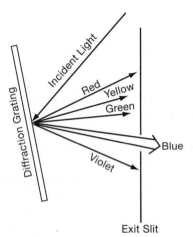

Figure 23.14. Rotating the Grating so That Light of a Selected Wavelength Falls on the Exit Slit.

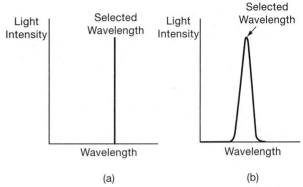

Figure 23.15. The Light Emerging from a Monochromator. **a.** If the light from a monochromator were truly monochromatic, then when the instrument was set for a certain wavelength, only that wavelength would be incident on the sample. **b.** In reality, the light incident on a sample consists of a narrow band of wavelengths, centered on the selected wavelength. (Note that this figure is not an illustration of a sample spectrum, as are Figure 23.6b on p. 407 and Figure 23.7 on p. 408. Rather, it illustrates the wavelengths of light that are incident on the sample.)

would illuminate the sample, see Figure 23.15a. In reality, the light that emerges from a monochromator consists of light spanning a narrow range of wavelengths, centered at the chosen wavelength, see Figure 23.15b.

Modern spectrophotometers are well-designed so that, in most situations, the light emerging from the monochromator has a sufficiently narrow span that it can be treated as if it were truly monochromatic. There are situations, however, where the "monochromaticity" of the light incident on the sample affects an analysis and more expensive instruments with higher-quality monochromators are preferred. The Appendix to this chapter discusses issues relating to the "monochromaticity" of light exiting the monochromator.

iv. SAMPLE CHAMBER

The sample (or the blank) is contained in a cuvette, which is placed inside a light-tight **sample chamber.** The **path length** *is the distance that the light travels through the sample or blank; it is determined by the interior dimension of the cuvette.* It is customary to use a 1 cm

path length when working with sample volumes of 1–5 mL. Manufacturers make systems, including cuvettes and cuvette holders, that permit the analysis of small volumes, down to the microliter range. Being able to work with small volumes is useful in molecular biology applications where reaction mixes and samples are often less than 50 μL. Note that the path length in small volume systems may not be 1 cm.

The use of a fiber-optic probe to replace a cuvette held in a sample chamber is a recent development in spectrophotometry. Fiber-optic cables are widely used in the telecommunications industry to transmit light over long distances without intensity losses or signal dispersion. A similar technology can be used in spectrophotometers to guide light from the source lamp through a fiber-optic cable to a probe. The end of the probe is dipped into the sample, which may be contained in almost any type of vessel; hence, these are called *dip probes.* Light interacts with the sample at the end of the dip probe tip. The path length of the dip probe may be 1 cm in length, or it may be narrower. Depending on the nature of the molecules in the sample, some of the light will be absorbed. The remaining light is then reflected back through a second fiber optic cable that leads to a detector inside the spectrophotometer. The probe is connected to the spectrophotometer via a cable that may be a couple of meters long; this has the advantage that the sample can be located apart from the spectrophotometer.

v. DETECTOR

The **detector** *senses the light coming through the sample (or the blank) and converts this information into an electrical signal.* The magnitude of the electrical signal is proportional to the amount of light reaching the detector.

The most common type of detector in spectrophotometers is called a **photomultiplier tube (PMT).** A PMT contains a series of metal plates coated with a thin layer of a **photoemissive material** *that emits electrons when it is struck by photons.* Electrons are held loosely by the photoemissive material so the energy of light is sufficient to "knock" electrons from the plates. Many electrons are released by each incident photon, and each electron can knock out even more electrons, resulting in a cascade that amplifies the original light signal. The stream of emitted electrons constitutes an electrical signal that is processed by the detector assembly and is converted to a reading of transmittance and/or absorbance.

All detectors have limitations. For one thing, detectors are not accurate at very high or very low levels of transmittance. It is best, therefore, to work with samples whose absorbance is in the midrange of the instrument. Second, detectors are more responsive to some signals than others. In spectrophotometry, this means that the detector will respond more strongly to some wavelengths of light than others.

vi. DISPLAY

There are many ways that the signal from the detector can be manipulated and displayed. The simplest method is to use a meter to display the resulting value. A meter has a needle that moves to indicate the value either on a scale of transmittance or absorbance. The signal can also be converted into a digital display. Modern instruments send the signal to a printer and/or to a computer for display, storage, and analysis.

III. MAKING MEASUREMENTS WITH SPECTROPHOTOMETERS

A. More about the Blank

Recall that the transmission of light through a sample is always compared with the transmission of light through a blank. This is accomplished in a simple spectrophotometer by first inserting the blank into the light beam and, using the appropriate knob, dial, or key strokes, setting the instrument to read 100% transmittance (or the absorbance is set to read zero). After the instrument has been adjusted with the blank, the sample is placed in the light beam and transmittance and/or absorbance is measured and displayed. In the normal operation of a spectrophotometer, it is never correct to attempt to measure the absorbance of a sample without first "blanking" the instrument.

A spectrophotometer needs to be set to 100% T (or zero absorbance) using the blank every time the wavelength is changed. This is because, as noted previously, the lamp emission intensity varies at different wavelengths. The monochromator efficiency and the detector's sensitivity to light also change as the wavelength changes. It is necessary to compensate for these instrumental variations by resetting the instrument to 100% T (or zero absorbance) using a blank every time the

wavelength is changed. Moreover, even if the wavelength of an instrument were never changed, the spectrophotometer's responsiveness to a sample would change over time due to power line fluctuations, aging of the bulb, and other factors. Setting the instrument with the blank compensates for all these factors. Thus, "blanking" the instrument compensates for the effects of:

- absorption of light by materials in the sample other than the analyte, such as the solvent and reagents
- absorption of light by the cuvette
- variations in light emission as the light source ages
- fluctuations in power levels
- variations in light emission intensity with wavelength
- variations in the sensitivity of the detector to different wavelengths of light

B. The Cuvette

Glass and certain types of plastic are transparent to light in the visible range, but they block UV light. These materials are therefore used to make cuvettes for work in the visible range, see Figure 23.16a on p. 414. Quartz glass is transparent to visible and UV light, so it is used to make cuvettes for both visible and ultraviolet work, see Figure 23.16b. There are also disposable plastic cuvettes that may be used down to about 285 nm, see Figure 23.16c. Note that the optical quality of plastic cuvettes may not be suitable for some applications.

Even the highest-quality cuvettes absorb a small amount of light. It is possible, therefore, to purchase paired cuvettes that are manufactured to be identical in their light absorbing properties. One of the pair is used for the blank and the other for the sample.

Cuvettes are carefully manufactured, precision optical devices that must be properly maintained. Small scratches and dirt will absorb or scatter light and therefore damage the cuvette. Table 23.5 summarizes considerations relating to cuvettes. Box 23.1 on p. 414 outlines methods for determining whether two cuvettes match, for checking the cleanliness of cuvettes, and for cuvette cleaning.

C. The Sample

Samples should be well mixed and homogenous without air bubbles. Particulates should be avoided (except in certain applications where turbid samples, such as bacterial suspensions, are assayed).

The solvent is an important part of the sample. Although solvents, such as water, buffers, or alcohol, appear transparent, they absorb light at certain wavelengths in the UV range. At wavelengths below a particular cutoff value, solvents absorb so much light that they interfere with the analysis of a sample. Table 23.6 on p. 415 is a general guide to typical solvents for UV/Vis

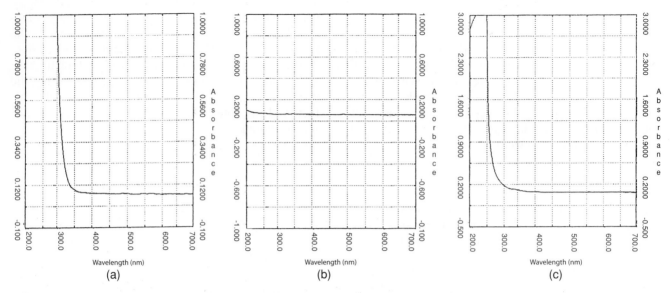

Figure 23.16. Absorbance of Light by Cuvettes Made of Various Materials. a. The absorbance spectrum of a glass cuvette intended for work in the visible range. This cuvette is unsuitable for work below about 330 nm. **b.** The absorbance spectrum of a quartz cuvette that is suitable for work down to least 200 nm. **c.** The absorbance spectrum of a disposable polyacrylate plastic cuvette sold for work in the UV range down to about 290 nm. (Air was used as the blank for all three spectra.)

Box 23.1. CLEANING AND CHECKING CUVETTES

1. *Be certain to only use clean cuvettes.*

2. *Wash cuvettes immediately after use. Recommended washing methods include:*

    Rinse with distilled water immediately after use.

    Wipe the outside of glass and quartz cuvettes with lens paper.

    After an aqueous sample, wash in warm water, rinse with dilute detergent, thoroughly rinse with distilled water. Avoid detergents with lotions. Be certain all detergent is removed after washing.

    For organic samples, rinse thoroughly with a spectroscopy grade solvent.

    If simple rinses are not sufficient to remove all traces of samples, an acid cuvette washing solution may be used. For example:

    > To make 1 liter of washing solution
    > 425 mL of distilled water
    > 525 mL of ethanol
    > 50 mL of concentrated acetic acid

    Do not allow acid to remain in the cuvette for more than 1 hour. Rinse thoroughly with distilled water.

    An enzyme-containing detergent, such as Tergazyme, is sometimes recommended to remove dried proteins from cuvettes.

    Avoid blowing air into cuvettes to dry them.

3. *For high-accuracy work, the cleanliness of cuvettes and matching response of two cuvettes can be checked as follows (from ASTM E-275 "Standard Practice for Describing and Measuring Performance of Ultraviolet, Visible, and Near-Infrared Spectrophotometers"):*

    1. Set the instrument to 100% T (or 0 A) with only air in the sample holder.

    2. Fill the cuvette with distilled water and measure its absorbance at 240 nm (quartz cuvette) or 650 nm (glass). For scanning instruments scan across the spectral region of interest. The absorbance should not be greater than 0.093 for 1 cm quartz cuvettes or 0.035 for glass.

    3. Rotate the cuvette 180 degrees in its holder and measure the absorbance again. Rotating the cuvette should give an absorbance difference less than 0.005.

    4. If an instrument is to be used with two cuvettes, one for the sample and one for the blank, check how well the two cuvettes are matched. Fill each with solvent and measure the absorbance of the sample cuvette. The absorbance difference between the two cuvettes should be less than 0.01.

work and the approximate cutoffs at which they begin to absorb light. The actual absorbance cutoffs for solvents vary depending on their grade and purity, the pH, the calibration of the spectrophotometer, and other factors. When working in the UV range, therefore, it may be necessary to prepare an absorbance spectrum of the solvent to check whether it absorbs appreciable amounts of light at the analytical wavelength(s). Solvents that are manufactured specifically for use in spectrophotometry are labeled "Spectroscopy" or "Spectro-Grade."

Figure 23.17a on p. 416 shows the absorbance spectrum for a commonly used biological buffer, Tris. Tris absorbs light in the UV region of the spectrum. Figure 23.17b shows the absorbance for denatured alcohol (ethanol). Note the strong absorbance peak at around 270 nm, a region of importance in the analysis of DNA, RNA, and proteins. Figure 23.17c is the spectrum for absolute (pure) ethanol, which has little absorbance above about 240 nm. Figure 23.17d is the spectrum for acetone.

EXAMPLE PROBLEM

You place a blank in a spectrophotometer but the blank has so much absorbance that the instrument cannot be set to zero absorbance. How should you proceed?

ANSWER

Think of factors that might account for this absorbance, such as:

1. The blank might be cloudy or turbid. Check its appearance. Cloudy or turbid blanks (and samples) generally should be avoided.

2. If the wavelength the instrument is set to read is in the ultraviolet range, make sure the cuvette is quartz or plastic suitable for UV use. If not, use a different cuvette.

3. Check that the cuvette is properly placed in the sample holder and is in the correct orientation. Many cuvettes have two sides that are transparent to light and two sides that are not transparent. The cuvette must be placed properly in the sample holder so that light can penetrate the cuvette.

4. Check the cuvette to make certain it is clean, without fingerprints or liquid droplets on the outside.

5. Some solvents appear transparent but absorb light in the UV range. Be certain that the solvent does not absorb light at the wavelength being used, see Table 23.6.

IV. DIFFERENT SPECTROPHOTOMETER DESIGNS

A. Scanning Spectrophotometers

The spectrophotometer illustrated in Figure 23.10 on p. 410 is a simple, **single-beam** instrument. Instruments

---

Table 23.5.   PROPER SELECTION AND CARE OF CUVETTES

1. *Use quartz cuvettes for UV work; glass, plastic, or quartz are acceptable for work in the visible range.*

2. *Matched cuvettes are manufactured to absorb light identically so that one of the pair can be used for the sample and the other for the blank.*

3. *Make sure the cuvette is properly aligned in the spectrophotometer.*

4. *Do not touch the base of a cuvette or the sides through which light is directed.*

5. *Do not scratch cuvettes; do not store them in wire racks or clean with brushes or abrasives.*

6. *Do not allow samples to sit in a cuvette for a long period of time.*

7. *Disposable cuvettes are often recommended for colorimetric protein assays because dyes used for proteins tend to stain cuvettes and are difficult to remove.*

---

with this basic design have been used successfully for years. Basic single-beam spectrophotometers, however, have the disadvantage that the operator must reset the instrument to 100% transmittance (or zero absorbance) with the blank each time the wavelength is changed. This process of manually selecting each wavelength and repeatedly "blanking" the instrument is tedious when an absorbance spectrum is being prepared. Manufacturers, therefore, have developed instruments called **scanning spectrophotometers** *that are capable of rapidly scanning through a range of wavelengths and constructing an absorbance spectrum.*

A **double-beam scanning spectrophotometer** is illustrated in Figure 23.18 on p. 417. Double-beam instruments have two sample holders. Both the sample and the blank are placed in the instrument at the same time so that the absorbance of the sample can be continuously

---

Table 23.6.   APPROXIMATE UV CUTOFF WAVELENGTHS FOR COMMONLY USED SOLVENTS

Solvent	UV Cutoff (nm)
Acetone	320
Acetonitrile	190
Benzene	280
Carbon tetrachloride	260
Chloroform	240
Cyclohexane	195
95% Ethanol	205
Hexane	200
Methanol	205
Water	190
Xylene	280

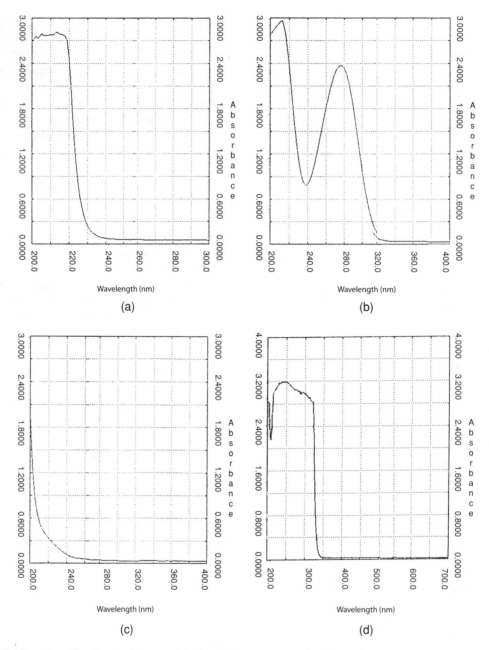

Figure 23.17. The Absorbance Spectra of Various Solvents. a. 1 *M* Tris (Sigma) buffer absorbs strongly below about 220 nm. **b.** Denatured ethanol absorbs strongly in the UV region. (Fisher Scientific ethyl alcohol containing 1% ethyl acetate, 1% methyl isobutyl ketone, 1% hydrocarbon solvent, and methanol.) **c.** The absorbance spectrum of pure ethanol (Fisher Scientific). **d.** The absorbance spectrum for acetone (general use grade, Caledon Laboratories Limited).

and automatically compared with that of the blank. The instrument illustrated alternates the light beam between the blank and the sample many times per second. (Note that Figure 23.8 on p. 409 also illustrates the principle of a double-beam instrument.)

Another type of scanning spectrophotometer takes advantage of microprocessor technology. In this type of instrument, the operator first places the blank in the sample compartment. The spectrophotometer scans the absorbance of the blank at all the wavelengths of interest as a microprocessor inside the instrument "memorizes" the values. The blank is then

removed from the sample compartment and the sample is similarly scanned across the wavelength range. The microprocessor compares the absorbance of the blank to the absorbance of the sample at each wavelength and generates a spectrum automatically.

Scanning spectrophotometers often allow the operator to control the *rate at which the sample is scanned, the* **scan speed.** The scan speed is restricted to the rate at which the instrument can respond. A scan speed that is too fast creates an effect called *tracking error*. When **tracking error** *occurs, the absorbance peaks recorded for a sample are slightly shifted from their true locations.*

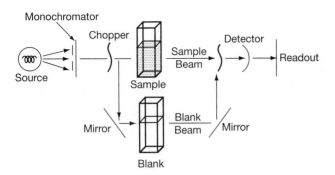

Figure 23.18. A Double-Beam Scanning Spectrophotometer. Light emitted by the source passes through a monochromator. After the monochromator, the "chopper" rapidly alternates the beam between the sample and the blank. The output of the detector is proportional to the ratio between the blank and the sample beams and so is a measure of transmittance.

B. Instruments with Photodiode Array Detectors

All the single- and double-beam instruments described so far have a photomultiplier tube (PMT) detector. Instruments with PMTs measure the transmittance (and absorbance) of a sample one wavelength at a time and therefore acquire a spectrum over a period of seconds or minutes. Manufacturers also make "higher end" spectrophotometers that are equipped with a different type of detector called a **photodiode array detector (PDA).** A **PDA** *can determine the absorbance of a sample over the entire UV/Vis range virtually instantaneously.*

Photodiode array detectors are based on the electronic components, diodes. Each diode in a PDA is calibrated to respond to a specific wavelength. When struck with light of that wavelength, the voltage of the diode changes. This voltage change can be converted to an electrical signal. A PDA contains a two-dimensional array of hundreds of diodes spaced closely together.

In an instrument with a PDA, all the light from the source is directed at the sample rather than being dispersed in a monochromator. There is a monochromator located *after* the sample chamber that disperses the light that has been transmitted through the sample or the blank, see Figure 23.19. The dispersed, transmitted light strikes the photodiode array. Each diode responds

to light at its calibrated wavelength and sends a signal to a computer. For every wavelength in the spectrum, the computer compares the amount of light transmitted by the sample and the blank. The computer converts this information to a complete absorbance spectrum for the sample.

C. Microprocessors and Spectrophotometers

Microprocessors play a wide range of roles in modern spectrophotometers. Microprocessors control wavelength scanning, generate absorbance spectra, store and retrieve information, perform calculations and statistics on data, and plot data on graphs. A variety of software is available that simplifies routine operations. Many of the applications described in Chapter 25 are now automated. As has been discussed in previous chapters on balances and pH meters, spectrophotometer manufacturers have modified their software so that their spectrophotometers can be operated in compliance with 21 CFR part II requirements.

V. QUALITY CONTROL AND PERFORMANCE VERIFICATION FOR A SPECTROPHOTOMETER

A. Performance Verification

i. OVERVIEW

The performance of a spectrophotometer needs to be periodically checked and the results documented. If necessary, the instrument must be repaired and/or calibrated to bring its readings into accordance with the values of accepted standards. To decide whether an instrument is performing satisfactorily, its performance is compared with specifications established by the manufacturer for each model. (In some cases, spectrophotometers will also need to meet specifications established by regulatory authorities). The methods that are used for verifying the performance of an instrument come from various sources including manufacturers, the U.S. Pharmacopeia, ASTM, and other standards and regulatory authorities.

The frequency with which the performance of a spectrophotometer is verified depends on the laboratory, the uses of the instrument, and whatever regulations or

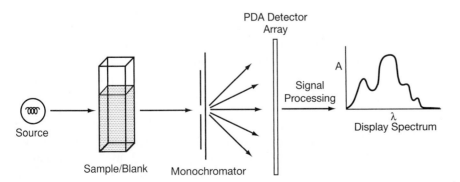

Figure 23.19. The Optical Design of an Instrument with a PDA. The monochromator is located after the sample compartment. Each diode in the PDA array responds to light at its calibrated wavelength and sends a signal to a computer. The computer compares the amount of light transmitted by the sample and the blank and converts this information to a complete absorbance spectrum for the sample.

standards apply. Instruments that are handled roughly, or are used in environments that are dusty, have chemical vapors, or vibration will require more frequent calibration and maintenance than other instruments.

Modern spectrophotometers are sufficiently complex that the individuals who operate the instruments are usually not the ones who check their performance. Performance verification is typically performed by trained service personnel, as is most repair and adjustment. Regardless of who performs the performance checks and repairs on a spectrophotometer, however, the operator should be aware of the characteristics of the instrument that affect its capabilities, range, and its ability to give accurate, reproducible results. Understanding these characteristics is important, for example, in the following situations:

- *When purchasing a spectrophotometer it is important to choose a model whose capabilities match the applications for which it will be used.*

- *When developing a spectrophotometry method for others to follow, the originator must describe the instrument characteristics required to duplicate the method.*

- *When following standard methods written by others, an analyst must ascertain that her/his spectrophotometer meets the requirements of the method.*

- *In situations where "difficult" samples are analyzed or high accuracy is required, it is important to have an instrument with the necessary features.* For example, if the absorbance of samples with very high levels of analyte is to be measured, some models of spectrophotometer will perform better than others.

The performance characteristics that determine the accuracy, precision, and range of operation of a spectrophotometer include (but are not limited to):

Calibration, which has two components:
    Wavelength Accuracy
    Photometric Accuracy
Linearity of Response
Stray Light
Noise
Baseline Stability
Resolution (*discussed briefly here and in more detail in the Appendix to this chapter*)

Note: For microprocessor-controlled instruments, the performance of the software should also be verified as described by the manufacturer.

This section explains performance characteristics of a spectrophotometer. More information on calibration and performance verification is provided in the articles listed in this unit's bibliography.

ii. CALIBRATION

Calibration of a spectrophotometer brings the readings of an individual instrument into accordance with nationally accepted values. Calibration is therefore part of routine quality control/maintenance for a spectrophotometer. There are two parts to calibrating a spectrophotometer: **wavelength accuracy** and **photometric accuracy. Wavelength accuracy** *is the agreement between the wavelength the operator selects and the actual wavelength that exits the monochromator and shines on the sample.* **Photometric accuracy,** or **absorbance scale accuracy,** *is the extent to which a measured absorbance or transmittance value agrees with the value of an accepted reference standard.*

Consider wavelength accuracy. Recall that a monochromator always generates a small range of wavelengths, not a single wavelength, see Figure 23.15 on p. 412. If an instrument is well calibrated with respect to wavelength accuracy, then when it is set to 550 nm, the center of the wavelength peak incident on the sample will be 550 nm plus or minus a certain tolerance.

Wavelength accuracy is determined using certified standard reference materials (SRMs) available through NIST, or standards that are traceable to NIST. An absorbance spectrum for the reference standard is prepared in the instrument whose performance is being checked. The absorbance peaks for reference standards are known, so the wavelengths of the peaks generated by the instrument can be checked for accuracy. The manufacturer specifies the wavelength accuracy of a given instrument. For example, a high-performance instrument may be specified to a wavelength accuracy with a tolerance of ± 0.5 nm. A less-expensive instrument may be specified to have a wavelength accuracy with a tolerance of ± 3 nm. Figure 23.20 shows the spectrum for a commonly used reference standard, a **holmium oxide filter.**

Wavelength accuracy needs to be periodically checked. Wavelength accuracy can deteriorate if the source lamp, associated mirrors, and other parts are not properly aligned. This might occur, for example, after replacing a bulb. Problems leading to wavelength inaccuracy can also arise in the monochromator and in the display device.

Photometric accuracy ensures that if the absorbance of a given sample is measured in two different spectrophotometers at the identical wavelength and under the same conditions, the readings will be the same, and will correspond to accepted values. This is a difficult objective to achieve, even with properly calibrated instruments, because spectrophotometers differ in their optics and designs.

Absorbance scale accuracy is determined using SRMs available through NIST, or standards that are traceable to NIST. NIST establishes the transmittance of its standards using a special high performance spectrophotometer. Thus, in the United States, the

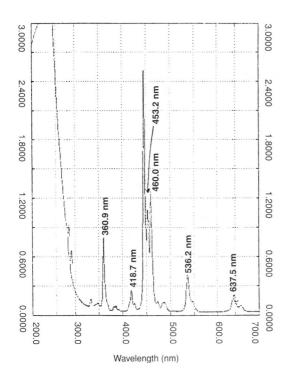

Figure 23.20. The Spectrum of a Holmium Oxide Filter Used to Test the Wavelength Accuracy of a Spectrophotometer. The filter is placed in the spectrophotometer and its spectrum is generated. If the instrument is properly calibrated with respect to wavelength, then the peaks in the spectrum will correspond to the specified wavelengths. (Based on wavelengths in ASTM E275 "Standard Practice for Describing and Measuring Performance of Ultraviolet, Visible, and Near-Infrared Spectrophotometers.")

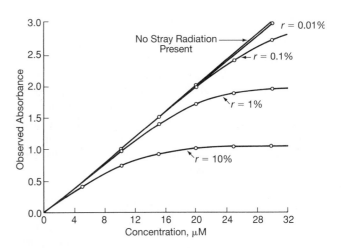

Figure 23.21. The Effect of Stray Light. In the absence of stray light, the relationship between concentration and absorbance is perfectly linear. When stray light is present, it causes the relationship to deviate from linearity at higher absorbance levels. The more stray light (r) present (expressed here as a percent), the lower the absorbance level at which this deviation occurs.

meaning of a transmittance value (and therefore of an absorbance value) is defined by NIST.

For many biological applications of spectrophotometry, photometric accuracy is not critical. As long as an individual instrument is internally consistent and its readings are linear and reproducible, it will work effectively to measure the concentration of analytes, as is discussed in detail in Chapter 25. There are, however, a few applications where a properly calibrated instrument is essential. This is the case, for example, where spectrophotometric measurements from various laboratories are being compared.

iii. STRAY LIGHT (STRAY RADIANT ENERGY)

Stray light *is radiation that reaches the detector without interacting with the sample.* This may occur if light is scattered by various optical components of the monochromator or sample chamber. This scattered light does not pass through or interact with the sample, but some of it may fall on the detector, resulting in an erroneous transmittance value. Scratches, dust, and fingerprints on the surface of cuvettes can also deflect light from its proper path. Stray radiation is reduced by keeping surfaces and cuvettes clean and by avoiding fingerprints and scratches. Another source of unwanted light is leaks of room light into the sample chamber.

When the absorbance of a sample is very high, the amount of transmitted light reaching the detector becomes vanishingly small. If there is stray light present, that stray light will be erroneously considered as transmitted light. The presence of stray light, therefore, is a factor that limits the accuracy of a spectrophotometer at high absorbance values.

Figure 23.21 shows the effect of stray radiation on the relationship between concentration and absorbance. That relationship should ideally be linear at all absorbance values. In Figure 23.21, r is a value that represents the level of stray light: The higher the value for r, the more stray light present. Given an instrument with high levels of stray light (i.e., where $r = 10\%$), the relationship between absorbance and concentration begins to be nonlinear at absorbances of less than 1. A moderately priced spectrophotometer might be expected to have a stray light value on the order of 1%; therefore, it can be expected to be linear in response until the absorbance approaches 2. More expensive spectrophotometers may have negligible levels of stray light and may respond in a linear fashion at absorbances of 3 or more.

EXAMPLE PROBLEM

A sample is expected to have an absorbance of 2 absorbance units. If a significant level of stray light is present (e.g., 1%), will the absorbance of the sample be greater than, less, than or equal to 2?

ANSWER

The absorbance will appear to be lower than 2 because stray light reaching the detector is interpreted as transmitted light.

iv. PHOTOMETRIC LINEARITY

Linearity of detector response *is the ability of a spectrophotometer to yield a linear relationship between the intensity of light hitting the detector and the detector's response.* It is critical that the detector's response is proportional to the amount of light incident on it. A spectrophotometer may fail to respond in a linear fashion due to stray light, problems in the detector, the amplifier, the readout device, or a monochromator exit slit that is too wide.

v. NOISE AND SPECTROPHOTOMETRY

Recall from Chapter 18 that electrical noise includes background electrical signal arising from random, short-term "spikes" in the electronics of an instrument. These electrical "spikes" occur in the absence of a sample.

Signal in a spectrophotometer is electrical current arising in the photodetector that is related to the interaction of light with a sample (or with a blank). Signal is what we are interested in when using an instrument. At low concentrations of analyte, the electrical noise in a spectrophotometer can interfere with the measurement of absorbance, as illustrated in Figure 23.22. Figure 23.22a shows the absorbance spectrum for an undiluted sample of methylene blue. Methylene blue has three main peaks at about 240 nm, 280 nm, and 665 nm; the latter peak has a "shoulder" at about 620 nm. This spectrum, with its peaks and valleys, is the "signal." Note that in Figure 23.22a the absorbance scale on the Y axis runs from 0.0000 to 3.0000. In Figure 23.22a electronic noise is not evident because there is ample signal from the sample. Figure 23.22b is the spectrum of a diluted solution of methylene blue that was prepared by taking 1 part methylene blue and adding 24 parts of water. The pattern of peaks and valleys characteristic of methylene blue is difficult to distinguish. In order to better visualize the spectrum of this diluted sample, it is replotted in

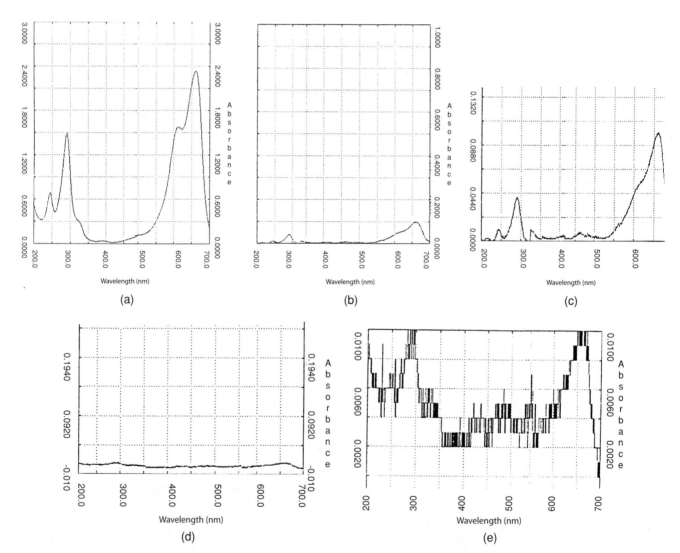

(a)   (b)   (c)

(d)   (e)

Figure 23.22.   The Relationship between Signal and Electrical Noise Illustrated with Methylene Blue. a. The absorbance spectrum for methylene blue. Electrical noise is not evident. **b.** The spectrum of methylene blue diluted 1/25. **c.** The same plot as shown in (b), but with the Y axis expanded to run from zero to 0.1320. Noise becomes visible. **d.** The spectrum of the methylene blue diluted 1/150. **e.** The same plot as in (d), but the Y axis is expanded. The signal of the methylene blue peaks can be barely discerned "buried" in the electrical noise. (All spectra were plotted on a Beckman DU 64 spectrophotometer.)

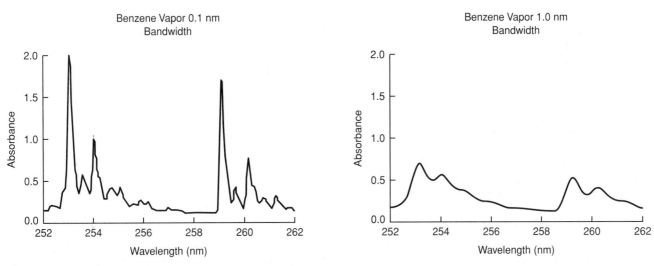

Figure 23.23. Resolution and Spectral Bandwidth. The better the resolution of an instrument, the better the peaks of an absorbance spectrum can be detected. These are absorbance spectra of benzene vapor in a sealed cuvette. In a higher resolution spectrophotometer (with a relatively narrow spectral band width), left, benzene's peaks are clearly resolved. The peaks at 253.49 nm and 259.56 nm are visible if the spectral bandwidth is 0.2 nm or less. In an instrument with less resolution, right, the peaks are not resolved. (Reproduced courtesy of the Starna group of companies, www.starna.com.)

Figure 23.22c, with the Y axis expanded to run from zero to 0.1320 absorbance. In the expanded plot in Figure 23.22c the peaks and valleys characteristic of methylene blue are reasonably clear, but the "spikes" due to electrical noise also become barely visible. Figure 23.22d shows the spectrum of a very dilute sample of methylene blue that was prepared by taking 1 part of the original solution of methylene blue and adding 149 parts of water. The characteristic pattern of methylene blue is not detectable. If the Y axis is expanded so that it runs from zero to 0.0100, then the signal of the methylene blue peaks can be barely discerned "buried" in the electrical noise. Thus, at low levels of sample, electrical noise limits our ability to distinguish the pattern due to the sample.

vi. RESOLUTION

Figure 23.23 shows two absorbance spectra for benzene; however, the two spectra have a different appearance. In the first spectrum the peaks can be clearly distinguished from one another. In the second spectrum they largely have "run together." We say that the first spectrum has better **resolution,** *the individual peaks can be better distinguished from one another.*

Resolution is a property of an instrument. A more expensive, high-performance spectrophotometer is able to provide better resolution for a given sample than is a less-expensive instrument. Having an instrument with "good" resolution is important in qualitative analysis where distinctive peaks are used to identify a substance. "Good" resolution is also important in the analysis of a sample containing more than one compound whose absorbance peaks are close together. The better the resolution of the instrument, the better substances with close absorbance peaks can be distinguished.

The resolution of an instrument is primarily determined by how "monochromatic" the light exiting the

monochromator is. The narrower the range of wavelengths incident on the sample, the better the resolution of the instrument. High-performance spectrophotometers are designed to minimize the wavelength range incident on the sample. In an instrument's specifications, the "monochromaticity" of the instrument and therefore its resolution may be expressed in terms of the "spectral band width" or "spectral slit width." The smaller the value given, the more monochromatic the light and the better the resolution of the instrument. Resolution and spectral band width are discussed in more detail in the Appendix to this chapter.

A practical way to check the resolution of an instrument is to prepare an absorbance spectrum of a reference material that has two or more peaks whose wavelengths are close together. The instrument's ability to distinguish these peaks is an indication of its resolution. For example, a holmium oxide filter has three peaks between 440 and 470 nm. If a holmium oxide filter is placed in the sample compartment and scanned, the three peaks should be distinct, given a properly functioning instrument with "good" resolution (Figure 23.20 on p. 419).

B. Performance Specifications

The choice of a spectrophotometer depends on the applications for which it will be used. For research and for qualitative work, a more expensive instrument with better resolution is desirable. A spectrophotometer that can scan an absorbance spectrum and that has both UV and visible capabilities is often required. In laboratories where a spectrophotometer is used only for routine colorimetric assays, less-expensive models may be sufficient. The samples analyzed will also determine any special features required. Molecular biology experiments frequently involve very low volume samples, so microvolume cuvette systems become valuable. Another important issue in choosing a new spectrophotometer is its software capabilities. Ease of operation and availability of software to simplify biological applications are factors to consider. Table 23.7 on p. 423 explains the performance specifications for a high-quality, high-resolution spectrophotometer.

PRACTICE PROBLEMS: SPECTROPHOTOMETRY, PART A

1. Microwaves involve wavelengths in the range from 100 μm to 30 cm. Express the wavelengths of microwave radiation in terms of nm.

2. Which has the most energy: X-rays, green light, or ultraviolet light? Which has with the least energy?

3. Are spectra A, B, and C probably from the same compound, or are they probably the spectra of different compounds? Explain.

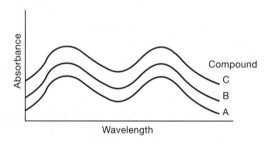

4. If blue light is shined on a solution of orange food coloring, will the light be absorbed?

5. What color are the dyes whose absorbance spectra are shown here? Explain.

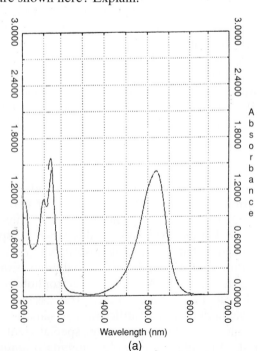

(a)

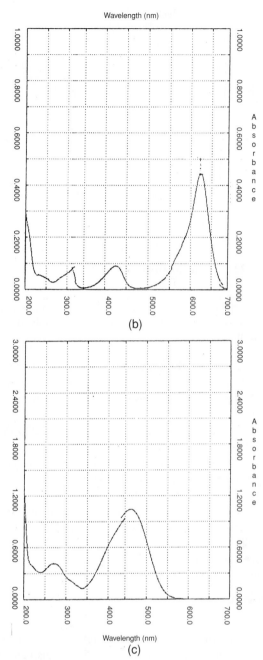

(b)

(c)

6. A spectrum for the plant pigment α-carotene, is shown in the following graph. Like chlorophyll, this pigment absorbs light, allowing the plants to make carbohydrates.

   a. What wavelengths of light does this pigment allow the plants to use?

   b. What color would a plant be if it contained only α-carotene?

   (The peaks for α-carotene are at 420, 440, and 470.)

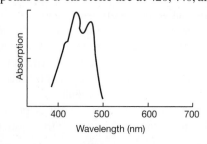

Table 23.7.   PERFORMANCE SPECIFICATIONS FOR A SPECTROPHOTOMETER

Specification	Example
Capability (Minimally, states whether it has both UV and visible capability and scanning capability.)	**Scanning UV/Vis with microcomputer electronics**
Optics (Describes whether single- or double-beam. The monochromator type is also specified.)	**Double-beam with concave holographic grating with 1053 lines/nm**
Wavelength range (Indicates the wavelengths at which the spectrophotometer can be used.)	**190-325 (UV) 325–900 (Vis)**
Sources (A UV/Vis spectrophotometer has a source to produce light in both spectral ranges.)	**Pre-aligned deuterium and tungsten-halogen lamps**
Wavelength accuracy (Agreement between displayed and actual wavelength.)	**± 0.1 nm**
Wavelength repeatability (A measure of the ability of a spectrophotometer to return to the same spectral position.)	**± 0.1 nm**
Spectral slit width (An indication of the resolution attainable. The lower the value for spectral slit width, the better the resolution.)	**± 1 nm**
Photometric accuracy (The ability of the detector to respond correctly to transmitted light.)	**± 0.003A at 1.00 A measured with NIST 930 Filters**
Photometric stability (An indication of drift.)	**± 0.002 A/hr at 0.00 A at 500 nm**
Noise (Random electrical signals. RMS is a statistical method of measuring noise. The lower the value, the less noise.)	**0.00015 A RMS**
Stray light (The amount of light reaching the detector that was not transmitted through the sample.)	**0.05% at 220 nm and 340 nm**
Scanning speed (The rapidity at which the instrument changes from one wavelength to another during scanning.)	**750 nm/min**
Microprocessor capabilities: **Data handling programs** (Specific programs to simplify operation or expand the data analysis capabilities of the instrument.)	**Standard curves, linear and nonlinear; kinetics; spectral scanning**
Accessories (Special features.)	**Special sample holder for small volumes available** **Gel scanning accessories available** **Temperature-controlled sample chamber accessories available**

7. The absorbance spectrum for the plant pigment phycocyanin is shown in the following graph.

   a. What color light does this pigment absorb?

   b. What color would a type of algae be if it contained primarily phycocyanin?

   c. Do you think this pigment would be more common in red algae or in blue-green algae?

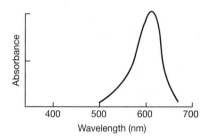

8. a. Plot an absorbance spectrum based on the following data.

   b. What color is this substance based on the absorbance spectrum?

   c. Optional: What is the natural band width for this substance? (See the Chapter Appendix on pp. 426–428.)

Wavelength (nm)	Absorbance
380	0.01
382	0.00
384	0.01
386	0.02
388	0.07
390	0.10
392	0.15
394	0.38
396	0.46
398	0.54
400	0.55
402	0.52
404	0.42
406	0.37
408	0.26
410	0.11
412	0.08
414	0.03

9. The values for the transmittance of a compound at different wavelengths are given.

   a. Graph these data.

   b. What color is this compound?

Wavelength (nm)	t
550	0.01
552	1.06
554	1.02
556	1.01
558	1.08
560	0.98
562	0.91
564	0.45
566	0.36
568	0.20
570	0.09
572	0.06
574	0.06
576	0.21
578	0.35
580	0.46
582	0.98
584	1.04

10. Convert the following transmittance values to %T.

   0.876   0.776   0.45   1.00

11. Convert the following values to absorbance values.

   $t = 0.876$ %T $= 25\%$

   $t = 0.776$ %T $= 15\%$

   $t = 0.45$ %T $= 95\%$

   $t = 1.00$ %T $= 45\%$

12. Fill in the following table:

Absorbance	Transmittance (t)	% Transmittance (T)
0.01	___	___
___	0.56	___
___	___	1.0%

13. Convert the following absorbance values to transmittance.

   1.24   0.95   1.10   2.25

14. The specifications for two spectrophotometers are given on the following page. Answer the following questions about these spectrophotometers.

   a. Which spectrophotometer should be purchased to do qualitative analysis in the ultraviolet range?

   b. Which spectrophotometer would be expected to be best able to measure a sample with very high absorbances?

   c. Which spectrophotometer should be used to measure the density of bands in electrophoresis gels?

Instrument A

Specification

Capability	Scanning Vis range with microcomputer electronics
Optics	Double beam with concave holographic grating with 1053 lines/nm
Wavelength range	325–850 (Vis)
Sources	Pre-aligned tungsten-halogen lamps
Wavelength accuracy	± 0.5 nm
Wavelength repeatability	± 0.5 nm
Spectral slit width	± 12 nm
Photometric accuracy	± 0.0013 A at 1.00 A measured with NIST 930 filters
Photometric stability	± 0.006 A/hr at 0.00 A at 500 nm
Noise	0.00030 A, RMS
Stray light	0.15% at 220 nm and 340 nm
Scanning speed	750 nm /min
Microprocessor capabilities: Data handling programs	Standard curves, linear and nonlinear; kinetics; spectral scanning
Accessories	Small volume holder available

Instrument B

Specification

Capability	Scanning UV/Vis with microcomputer electronics
Optics	Double beam with concave holographic grating with 1053 lines/nm
Wavelength range	170–325 (UV) 325–850 (vis)
Sources	Pre-aligned deuterium and tungsten-halogen lamps
Wavelength accuracy	± 0.1 nm
Wavelength repeatability	± 0.1 nm
Spectral slit width	± 1 nm
Photometric accuracy	± 0.0005 A at 1.00 A measured with NIST 930 filters
Photometric stability	± 0.002 A/hr at 0.00 A at 500 nm
Noise	0.00014 A, RMS
Stray light	0.05% at 220 nm and 340 nm
Scanning speed	550 nm/min
Microprocessor capabilities: Data handling programs	Standard curves, linear and nonlinear; kinetics; spectral scanning
Accessories	Small volume holder available
	Gel scanning
	Temperature control

d. Which would probably be better as a low-cost instrument for routine assays or for a teaching instrument.

e. Which spectrophotometer would be expected to have the best resolution?

f. Which spectrophotometer would be best able to distinguish the absorbance from compound A, which has a peak absorbance at 550 nm and compound B, whose peak absorbance is at 557 nm?

15. Two laboratories are comparing the performance of their spectrophotometers. They take the identical sample and measure its absorbance. (Assume the temperature is the same in both laboratories.) One laboratory gets an average absorbance reading of 0.72 AU; the other laboratory gets 0.87 AU. What factors might account for the difference between the two laboratories?

16. 50 μg/mL of a certain compound is expected to have an absorbance of 1.95 on a certain spectro- photometer. If significant stray light is present, will the apparent absorbance be greater than, less than, or equal to 1.95?

QUESTIONS FOR DISCUSSION

If you work in a laboratory with a spectrophotometer, find the answers to the following questions:

a. Does the spectrophotometer have the capability to measure absorbance in both the UV and visible range?

b. What is the smallest volume sample that can be measured using the cuvettes and sample holders available in your laboratory?

c. What is the instrument's specification for stray light?

d. Optional: What is the instrument's spectral slit width?

Chapter Appendix
Spectral Band Width and Resolution

i. SPECTRAL BAND WIDTH AND THE ACCURACY OF A SPECTROPHOTOMETER

Spectral band width *is a measure of the range of wavelengths emerging from the monochromator when a particular wavelength is selected.* Recall that although a spectrophotometer may be set to a single, specific wavelength, in practice a narrow range of wavelengths emerges from the monochromator. Spectral band width is a value that describes the ability of the spectrophotometer to isolate a portion of the electromagnetic spectrum.*

The spectral band width is an instrument characteristic that affects the accuracy of the measurements and the resolution obtainable from a particular spectrophotometer. In general, the narrower the range of wavelengths emerging from the monochromator, the better the instrument.

Spectral band width is determined by replacing the instrument's normal light source with a special line source (i.e., a lamp that emits light only at specific, individual wavelengths). (A method for measuring spectral band width using a mercury lamp is detailed in ASTM Standard E-958 *"Standard Practice for Measuring Spectral Bandwidth of Ultraviolet-Visible Spectrophotometers."*) The intensity of light reaching the detector from the line source is plotted versus wavelength. The width of the emission peak at half the peak's height is measured. In the example illustrated in Figure 23.24,

the wavelength at which the peak light intensity occurs is 550 nm. The peak height at this position is 4.0 cm above the baseline. Half the peak height is 2.0 cm. The width of the peak at half its height is 2 nm, so the spectral band width is 2 nm.

Spectral band width should not be confused with natural band width. **Natural band width** *describes the range of wavelengths absorbed by a particular substance.* For example, the natural band width of DNA is around 45 nm, as shown in Figure 23.25. The natural band width of a compound is an intrinsic property of that compound and is independent of the instrument's characteristics.

The fact that the light incident on a sample is not truly monochromatic leads to some inaccuracy in measurements of absorbance or transmittance. The amount of inaccuracy depends on both the spectral band width and the natural band width. The narrower the spectral band width (i.e., the more monochromatic the light) the more accurate the instrument. If the natural band width of a compound is relatively wide (as is, for example, the natural band width of DNA), then the inaccuracy caused even by a wide spectral band width is relatively insignificant. In contrast, if the natural band width of a compound is comparatively narrow, then a narrow spectral band width is required to get accurate measurements. A general rule is that for the most accurate measurements at an absorbance peak, the spectropho-

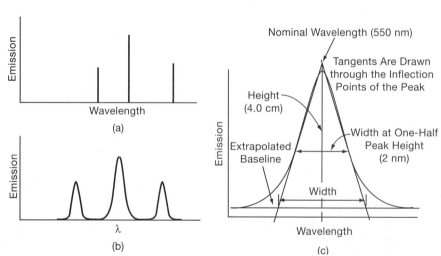

Figure 23.24. Spectral Band Width.
a. The actual emission spectrum of a line source as it would appear in a "perfect" spectrophotometer. **b.** The spectrum of a line source as it appears in practice. **c.** The wavelength span is described by the spectral band width that is measured at half the height of the peak.

*There are several terms used to describe the "monochromaticity" of light. These include *spectral band width, bandwidth or band width, effective band width,* and *bandpass or band pass.* These terms are not always used in the same way by different authors. The definitions in this text are consistent with those of ASTM (ASTM Standard E-131 "Standard Definition of Terms and Symbols Relating to Molecular Spectroscopy").

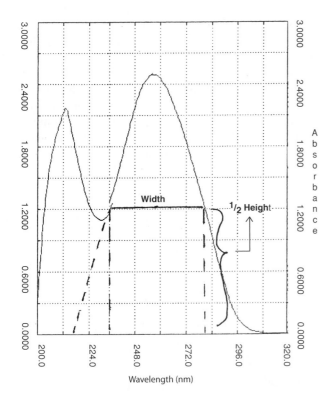

Figure 23.25. The Natural Band Width of DNA.

slit of the monochromator. The narrower the slit, the narrower the range of wavelengths selected.

The second feature of the monochromator that controls the range of wavelengths selected is the **linear dispersion of the monochromator. Linear dispersion** *is a measure of the dispersion of light by the diffraction grating.* The linear dispersion of the monochromator is not controlled by the user; rather, it depends on the manufacturing process used to produce the grating. Values for linear dispersion are obtained from the manufacturer of a spectrophotometer. Linear dispersion is generally expressed in terms of how many wavelengths (in nanometers) are included per millimeter at the exit slit.

Spectral slit width combines the value for the slit width of an instrument and its linear dispersion into a single value. **Spectral slit width** *is defined as the physical width of the monochromator exit slit, divided by the linear dispersion of the diffraction grating.* For example, consider the spectrophotometer in Figure 23.26. The width of the exit slit is 0.2 mm. The monochromator disperses light in such a way that at the plane of the exit slit there are 0.25 mm/nm. Thus, the spectral slit width is:

$$\frac{0.2\ \text{mm}}{0.25\ \text{mm/nm}} = 0.8\ \text{nm}$$

The term *spectral slit width* is not synonymous with *spectral band width.* Both terms, however, relate to the "monochromaticity" of light emerging from the mono-

tometer must have a spectral band width equal to or less than one tenth of the natural band width of the compound to be measured. (For DNA, this means the spectral band width of the instrument should be less than or equal to 4.5 nm.)

Suppose the absorbance of the same sample is measured in two instruments. The first instrument has a very narrow spectral band width; the second has a broader spectral band width. It seems reasonable to suppose that the sample would have the same absorbance in both instruments. In fact, if all other factors are equal, the first instrument will give a higher absorbance reading for the sample than would the second. This is because the apparent absorptivity (absorptivity and absorptivity constants are discussed in Chapter 25) of an analyte approaches its maximum, "true," value as the spectral band width decreases. The "true" absorptivity constant for a particular compound at a particular wavelength is a theoretical concept measurable only in the presence of truly monochromatic light.

ii. THE FACTORS THAT DETERMINE THE SPECTRAL BAND WIDTH OF AN INSTRUMENT

The spectral band width is controlled by the design of the monochromator. Some monochromators can provide more monochromatic light than others. There are two features of the monochromator that affect the "monochromaticity" of selected light. The first is the **slit width,** *which is the physical width of the entrance or exit*

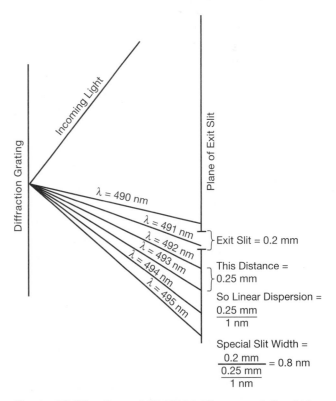

Figure 23.26. Spectral Slit Width. The spectral slit width in this example is 0.8 nm.

chromator. Spectral band width is difficult to measure so, in practice, the spectral slit width is used as an indicator of light "monochromaticity."

We have seen that the more monochromatic the light from the monochromator, the more accurate the readings of the spectrophotometer and the better its resolution. It seems reasonable, therefore, that manufacturers would always design instruments with the narrowest possible spectral slit width. In fact, it is true that more sophisticated spectrophotometers have narrower spectral slit widths than do less sophisticated instruments; however, there are practical limitations to how narrow the spectral slit width can be.

One way to make an instrument with a narrow spectral slit width is to make the exit slit of the monochromator very narrow. Although it is possible to manufacture an extremely narrow exit slit there are two problems with this approach. First, as the slit gets narrower and narrower, less and less light passes through it. Eventually, so little light is incident on the sample that electronic noise becomes significant relative to the amount of transmitted light. Noise then limits the ability of the spectrophotometer to make accurate measurements. Second, if the exit slit is too narrow, then its edges diffract light and the slit acts as a second, unwanted diffraction grating.

iii. RESOLUTION AND SPECTRAL BAND WIDTH

Recall that resolution affects our ability to distinguish peaks from one another if their wavelengths are close together. Spectral band width affects resolution. A

high-performance instrument has a narrow spectral band width; therefore, it also has optimal resolution. Less-expensive instruments tend to have wider spectral band widths and less ideal resolution.

When reading instrument specifications, the spectral slit width of the monochromator is commonly used as an indication of the resolution of a spectrophotometer. The lower the value for spectral slit width, the more monochromatic the light and the better the resolution, see Figure 23.27.

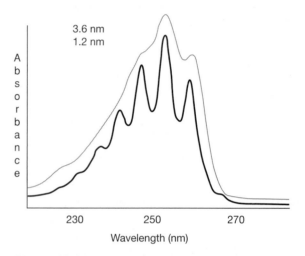

Figure 23.27. Spectral Slit Width and Resolution. A narrower spectral slit width results in much improved resolution. (Courtesy of Waters Corp.)

UNIT VI

Quality Assays and Tests

Chapters in This Unit

✦ Chapter 24: Introduction to Quality Laboratory Tests
and Assays

✦ Chapter 25: Measurements Involving Light B: Applications
and Methods

Producing data is an essential function of laboratory scientists and technicians. **Data** *are observations of the natural world that may take the form of numerical values, photographs, words, graphs, diagrams, or other expressions.* This unit discusses acquiring data through **assays** and **tests,** *which are measurements of a property of a sample.* A measurement of the amount of cholesterol in a blood sample, the lead levels in paint chips, or the activity of a DNA-cutting enzyme are all examples of assays.

Chapter 24 discusses general principles, terminology, and quality considerations relating to assays and tests.

Chapter 25 focuses on spectrophotometric tests and assays that are common in biotechnology laboratories.

BIBLIOGRAPHY FOR UNIT VI

Method Validation

Various organizations provide information relating to the validation of assays, analytical procedures, methods, and tests. The International Conference on Harmonisation of Technical Requirements for the Registration of

Pharmaceuticals for Human Use (ICH) has been helpful in harmonizing definitions and in determining the basic requirements for validation in the pharmaceutical area (although some differences still exist between organizations).

A few key documents and interesting articles are listed here:

Ermer, Joachim, and Miller, John H. McB. (eds.)
Method Validation in Pharmaceutical Analysis.
Weinheim, Germany: Wiley-VCH, 2005.

"Chapter <1225> Validation of Compendial Methods."
United States Phamacopeia 30. Rockville, MD: The
United States Pharmacopeial Convention, Inc.,
2006.

"ICH Q2A: Validation of Analytical Methods
(Definitions and Terminology)." International
Conference on the Harmonisation of Technical
Requirements for the Registration of
Pharmaceuticals for Human Use, 1994.

"ICH Q2B: Analytical Validation—Methodology."
International Conference on the Harmonisation of
Technical Requirements for the Registration of
Pharmaceuticals for Human Use, 1996.

Kanarek, Alex D. "Method Validation Guidelines." *BioPharm International,* September 15, 2005. http://www.biopharminternational.com/biopharm/. A good summary of the status of method validation in the pharmaceutical industry and related terminology.

Lundblad, Roger L., and Price, Nicholas C. "Protein Concentration Determination: The Achilles' Heel of cGMP?" *BioProcess International,* January 2004, 38–47.

Mehta, Shivani, and Keer, Jacquie T. "Performance Characteristics of Host-Cell DNA Quantification Methods." *BioProcess International,* October 2007, 44–58.

Scientific Working Group on DNA Analysis Methods (SWGDAM). "Revised Validation Guidelines." *Forensic Science Communications* 6, no. 3 (2004). http://www.fbi.gov/hq/lab/fsc/backissu/july2004/standards/2004_03_standards02.htm.

Spectrophotometric Assays

Copeland, Robert A. *Methods for Protein Analysis.* New York: Chapman and Hall, 1994.

Glasel, Jay A. "Validity of Nucleic Acid Purities Monitored by 260 nm/280 nm Absorbance Ratios." *BioTechniques* 18, no. 1 (1995): 62–3.

Müller, Hanswilly, and Schweizer, Bettina. "Biochemical Applications for UV/Vis Spectroscopy: DNA, Protein and Kinetic Analysis." Waltham, MA: Perkin-Elmer Corp., 1996.

Manchester, Keith L. "Use of UV Methods for Measurement of Protein and Nucleic Acid Concentrations." *BioTechniques* 20, no. 6 (1996): 968–70.

Manchester, Keith L. "Value of A_{260}/A_{280} Ratios for Measurement of Purity of Nucleic Acids." *BioTechniques* 19, no. 2 (1995): 208–10.

Weber, Klaus, Bernard, J.L., and Jamutowski, R.J., eds. "Spectrophotometric Multicomponent Analysis (MCA): An Ideal Tool for the Pharmaceutical and Life Science Laboratory." Beckman Applications Data Sheet ADS 7794, 1989. Good summary of applications of this methodology; also has analysis of math involved in MCA.

Willfinger, William W., Mackey, Karol, and Chomczynski, Piotr. "Effect of pH and Ionic Strength on the Spectrophotometric Assessment of Nucleic Acid Purity." *BioTechniques* 22, no. 3 (1997): 474–81.

PCR Assays

Maurer, John., ed. *PCR Methods in Foods.* New York: Springer, 2006.

CHAPTER 24

Introduction to Quality Laboratory Tests and Assays

I. INTRODUCTION

The products of a laboratory are knowledge, data, and information. Assays are one of the primary means by which laboratory analysts obtain data or information. We broadly define an **assay** *as any test used to analyze a characteristic of a sample, such as its composition, purity, or activity.* As is true for measurements, a "good" assay might be defined as one that provides data that can be trusted when making decisions or reaching conclusions. This chapter introduces consider-

ations that are important in obtaining "good," reliable results from assays.

There are a vast number of assays and tests that are performed in biotechnology laboratories. Research scientists use assays to look for effects of treatments on their experimental subjects. Pharmaceutical analysts test samples from animal and human subjects to determine what happens to a drug in the body. The discovery of Gleevec (see Chapter 3) is an example where researchers used a sophisticated assay method to look for the binding of a disease-related protein to various potential drug compounds. Development scientists in a biotechnology company might use an assay to determine whether bacteria transformed with a gene of interest are making the protein for which it codes. Forensic scientists use DNA "fingerprinting," an assay method that gives information about the identity of a person. Quality-control analysts might use a series of assays to evaluate a product to see whether it meets its specifications. The series might include, for example, separate tests for bacterial, viral, DNA, and protein contaminants. QC analysts might use yet another assay to measure the concentration or potency of an active substance in a product. Observe that some assays are **quantitative** *and provide numerical data* (e.g., a measure of the concentration or potency of a substance), while others provide *non-numerical* **qualitative** *information* (e.g., whether or not binding of two molecules occurs, whether a substance is present, whether a suspect might have been at a crime scene, the chemical identities of the components of a mixture).

Performing assays is such an integral part of all laboratory work that it is not surprising to find that there are many relevant and interrelated terms; Table 24.1 on pp. 433–434 discusses this terminology. Table 24.2 on pp. 434–436 provides examples to illustrate the wide diversity of assay methods used in biotechnology laboratories.

Many of the concepts and vocabulary that were discussed in previous chapters also apply to assays. Chapters 15 and 16 include math tools that can be used for analyzing the results of assays. Chapter 17 discusses related topics including accuracy, precision, and error. Chapter 25 continues the discussion that begins in this chapter by delving into spectrophotometric assays, which are a class of assays that are important to biotechnologists.

II. THE COMPONENTS OF AN ASSAY

A. Overview

Figure 24.1 provides a simple overview of an assay. Assays are performed on samples where a **sample** *is a part of the whole that represents the whole.* When a clinical laboratory technician analyzes the level of glucose in a blood sample, for example, the sample represents all the blood in the patient. It is not possible to test all the blood in a person, so a small sample is used instead. A forensic scientist might analyze hair samples from a crime scene. A food microbiologist might test a small meat sample to look for contaminants. A researcher might test a sample of cells subjected to an experimental treatment.

The samples in an assay undergo some procedure (method) in which the property of interest is analyzed. Often a laboratory instrument is involved in testing the samples. Information from the samples may be compared to information from a reference standard or reference material that is well-characterized with respect to the property of interest. The data collected in the assay

are analyzed, displayed, and/or stored. Statistical methods are often used in the analysis of data from assays.

B. The Sample

i. OBTAINING A REPRESENTATIVE SAMPLE

The samples that are analyzed in a laboratory assay or test represent the "whole." The "whole" is the object, batch, material, area, or population of individuals under investigation. The "whole" might be, for example, a batch of a biopharmaceutical product, all patients with a disease, or a field of genetically modified corn.

It may or may not be straightforward to obtain a sample that truly represents the whole. A blood sample taken from the arm of a patient is generally acknowledged to be representative of the whole of that person's blood. Devising a sampling strategy can be challenging, however, if the "whole" is heterogeneous. A classic problem in biological experimentation is how to sample an experimental agricultural plot. A field is likely to be heterogeneous with respect to its topography, soil moisture, soil type, sunlight, insects, and a host of other factors. Suppose that scientists who are performing field tests of a new strain of corn go to the field and collect a sample consisting of plants located near their parked car. These corn plants probably experienced different conditions than plants elsewhere in the field, and so are not a representative sample of the entire corn crop.

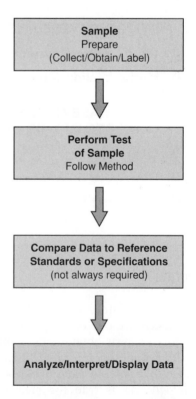

Figure 24.1. Overview of an Assay. Assays or tests are performed on samples that undergo some process in which a property of the sample is measured. Information from the sample may be compared to a reference standard. The data collected from the sample(s) are analyzed, displayed, and stored.

Table 24.1.   TERMINOLOGY RELATING TO ASSAYS

Accuracy. *The closeness of a test result to the true or accepted value.*

Analyte. *A substance whose presence and/or level is evaluated using an assay.*

Analytical Method. *A laboratory method used to determine what is present in a sample mixture, compound, or other substance.*

Artifact. *A distortion or error in the data.* For example, in electron microscopy an artifact might be a substance that appears to be a component of a cell but, in fact, was accidentally created by the analyst during the preparation of the tissue sample.

Assay. *A test used to analyze a characteristic of a sample, such as its composition, purity, or activity.* The terms *assay, test,* and *method* are often used interchangeably, though the term *assay* is not generally applied to a test of an instrument's performance. *Assay* can also be used as a verb, meaning "determine"; for example, "The technician *assayed* the protein concentration in a sample."

Bioanalytical Method. *A term that sometimes loosely refers to any test of a biological material.* The FDA defines the term more narrowly (in Guidance for Industry, Bioanalytical Method Validation) as a method used for the "quantitative determination of drugs and/or metabolites in biological matrices such as blood, serum, plasma, or urine . . . tissue and skin samples" taken from animal and human subjects.

Bioassay or **Biological Assay.** *1. Analysis, as of a drug, to quantify the biological activity(ies) of one or more of its components by determining its capacity for producing an expected biological activity. 2. Any assay that involves cells, tissues, or organisms as test subjects.* Bioassays are used for many purposes, for example, for testing the potency of a drug.

Cell-Based Assay. *Commonly describes any assay that involves living cells as the test subjects; a type of bioassay.* Cell-based assays are used, for example, to test a material for viral contaminants, to look for a response to a drug product, and to monitor the toxic effects of a compound.

Characterization (in the context of drugs). *Process in which a molecular entity's physical, chemical, and functional properties are determined using specific assays.* The qualities of the entity are then defined in terms of the results of those assays.

Limit of Detection (LOD). *The lowest concentration of the material of interest that can be detected by the method.*

Limit of Quantitation (LOQ). *The lowest concentration of the material of interest that the method can quantitate with acceptable accuracy and precision.*

Linearity. *The ability of a method to give test results that are directly proportional to the concentration of the material of interest (within a given concentration range).*

Matrix. *The physical material in which a sample or analyte is located or from which it must be isolated, for example, blood, soil, or fiber.*

Method. *Detailed description of the means of performing an assay.* A method describes the steps necessary to perform an analysis and related details, such as how the sample should be obtained and prepared, the reagents that are required, the set-up and use of instruments, comparisons with reference materials, calculations, and so on. This term is often used as a synonym for *procedure, analytical procedure, assay,* or *test.*

Method Validation. *Process used to demonstrate the ability of an assay method to obtain a desired result reliably.* Method validation is a formal process required by regulatory authorities in some situations.

Precision. *The degree of agreement between individual test results when a procedure is applied over and over again to portions of the same sample.*

  Repeatability. *The precision of measurements made under uniform conditions.*

  Reproducibility (sometimes called Ruggedness). *The precision of measurements made under nonuniform conditions, such as in two different laboratories.*

Preparative Method (in contrast to an analytical method). *Method that produces a material or product for further use (perhaps for commercial sale, or perhaps for further experimentation).* Chromatography, for example, is used to separate the components of a sample from one another. If the components are separated in order to purify an enzyme that will later be used in experiments, then chromatography is being used as a preparative method. Chromatography is also frequently used to help identify the components present in a mixture, in which case, chromatography is used as an analytical method.

Protocol. *1. Synonym for* procedure *or* method. *2. The formal design or action plan of a research study.*

Qualitative Analysis. *A test to determine the nature of the component(s) of a sample.*

Quantitative Analysis. *A test of how much of a particular analyte(s) is in a sample.*

Range. *The range of concentrations, from the lowest to the highest, that a method can measure with acceptable results.*

Resolution. *The smallest difference between two entities that can be separated and detected.*

Robustness. *The capacity of a method to remain unaffected by small, but deliberate variations in method parameters (based on the definition in ICH Q2A [1a]).*

(Continued)

Table 24.1. **Terminology Relating to Assays** *(Continued)*

Sample. *A subset of the whole that represents the whole (e.g., a patient's blood sample represents all the blood in the patient's body).*

Specificity (sometimes called ***selectivity***). *A measure of the extent to which a method can unequivocally determine the presence of a particular compound in a sample in the presence of other materials that may be expected to be present.*

System Suitability (according to FDA). *"Determination of instrument performance (e.g., sensitivity and chromatographic retention) by analysis of a reference standard prior to running the analytical batch."*

Test (according to ISO Guide 17025). *A technical operation that consists of the determination of one or more characteristics of performance of a given product, material, equipment, organism, physical phenomenon, process, or service according to a specified procedure.* The result of a test is normally recorded in a document sometimes called a test report or a test certificate. Note that a "test" may or may not involve a "sample."

Table 24.2.   **Examples of Assay Methods and Associated Instrumentation Common in Biotechnology Laboratories**

Microbial Testing, Bioburden. *Testing the microbial population associated with an unsterilized product or component; relates to the cleanliness of the production environment and the handling of the product.* Bioburden testing is required in GMP compliant facilities.

    Pour Plate Method: A specific quantity of test sample (e.g., 10 g or 10 mL) is transferred to 90 mL of buffer, a portion of which is transferred to a pour plate that contains nutrient medium. The medium is then incubated, typically for 48 hours, and checked for microbial growth. Another portion is transferred to a medium that is appropriate for growing fungi and mold; this plate is incubated for up to 7 days and checked for growth.

    Membrane Filtration: Contaminants in the test material are first concentrated by running the sample through filters that trap microbes. The filters are then placed onto growth media that are incubated and monitored for microbial growth.

Microbiological Testing, Sterility. *Tests for the presence of bacterial and fungal contaminants in sterilized preparations.* The U.S. Pharmacopeia sterility test is a common method in which the test sample is incubated in nutrient rich growth medium for a specified number of days and the medium is monitored daily for turbidity, an indicator of microbial growth.

Viral Contaminants Testing. *Tests for the presence of viral contaminants.*

    Infectivity Assays: The test sample is inoculated onto susceptible cultured cells. The susceptible cells are incubated and observed over a period of time (e.g., 28 days). If virus is present, the cells become infected and damaged, with visible morphological changes.

    Transmission Electron Microscopy: Viral particles are very small, but they can be visualized using the high magnifications achievable with a transmission electron microscope. This method is used, for example, to look for endogenous virus in host cells used for biopharmaceutical production.

    Polymerase Chain Reaction (PCR): PCR is used for the detection of DNA that is specific to a particular virus or class of virus.

Pyrogen/Endotoxin Assays. *Tests for pyrogens/endotoxins, which are bacterial byproducts that elicit a dangerous immune response in mammals when present at very low levels.*

    ***Limulus* Amoebocyte Lysate Test (LAL):** A sensitive test for the presence of endotoxin contaminants that is based on the ability of endotoxin to cause a coagulation reaction in a reagent purified from the blood of the horseshoe crab.

    USP Rabbit Pyrogen Test: Sample is injected into rabbits that are monitored to see if the test mixture elicits a fever response.

Amino Acid Analysis. *Used to help identify or characterize specific proteins.* The protein or peptide is broken down into its individual free amino acids.

Peptide Mapping. *Used to profile a protein's structure by breaking apart the protein with highly specific enzymes and separating the resultant peptides by HPLC or electrophoresis to produce a "map" or "profile."* The peptide map for a sample can be compared to a map done on a reference standard to confirm the identity of a product. Together with amino acid analysis, peptide mapping is used to help identify or characterize specific proteins, for example, to confirm the batch to batch consistency of biopharmaceutical product and the genetic stability of cells that produce the product.

Polymerase Chain Reaction (PCR). *A technique for greatly amplifying the amount of DNA with a specific, desired sequence.* PCR has a vast number of assay applications, such as testing drug products for specific contaminants, testing food to see if it has been genetically modified, and human identity testing.

    Quantitative Real Time PCR (qRT-PCR): A version of PCR that quantifies the amount of DNA present at the same time as it is amplified.

(Continued)

Table 24.2. EXAMPLES OF ASSAY METHODS AND ASSOCIATED INSTRUMENTATION COMMON IN BIOTECHNOLOGY LABORATORIES (Continued)

Reverse Transcriptase PCR (RT-PCR): A version of PCR that is used for the detection and quantitation of specific messenger RNA sequences.

A note about terminology: There is confusion in the acronym for real-time PCR; it may be called *RT-PCR, qRT-PCR, or q-PCR.* The term *RT-PCR* is problematic because this acronym was originally coined to refer to reverse transcriptase PCR and is still used in that way. *Q-PCR* is not ideal because there are many different ways that PCR is being used quantitatively. Some sources suggest avoiding confusion by using *qRT-PCR* to mean quantitative real-time PCR.

Enzyme Assays. *Assays that test the activity of an enzyme.*

Immunological Methods. *Methods that depend on the specific interaction of an antigen and its antibody.* These methods are used for many purposes, for example, testing the identity of a biopharmaceutical product, testing a material for the presence of specific contaminants, studying the localization of a cellular protein, and, in a clinical setting, determining whether a patient has been exposed to a particular pathogen (e.g., the HIV virus).

Enzyme-Linked Immunosorbent Assay (ELISA): A technique that utilizes antibodies specific to the substance of interest (the antigen); these antibodies are linked to an enzyme that causes a detectable colored signal when the antigen is present.

Western Blotting: Proteins are first separated from one another by gel electrophoresis and then are transferred to a membrane. A specific protein on the membrane is identified through its reaction with a labeled antibody that is detectable, usually by fluorescence, color, or radioactivity.

Ultraviolet and Visible Light Spectrophotometric Methods. *Methods that rely on the ability of samples to absorb visible or ultraviolet light of specific wavelengths.* A wide variety of spectrophotometric methods are used to measure analyte quantity, activity, identity, and purity in samples. (These assays are discussed in Chapter 25.)

Chromatographic Methods. *A large class of methods used to separate the components of a mixture from one another and identify them.* Chromatography is used, for example, to determine the identity, quantity, and purity of a material; and during stability testing to look for degradation products. (Chromatography is explained in more detail in Chapter 33.)

Electrophoresis. *Methods that separate molecules from one another based on their relative ability to migrate when placed in an electrical field.* Electrophoretic methods are widely used to separate mixtures of proteins and mixtures of DNA fragments, for example, to confirm the identity of a DNA fragment, to look for protein contaminants in a product, and for roughly quantifying a protein or DNA product. (Electrophoresis is explained in more detail in Chapter 33.)

Agarose Gel Electrophoresis. *Method that uses a gel matrix composed of a highly purified form of agar; commonly used to separate DNA fragments on the basis of base pair length.* DNA is applied to wells in the agarose gel. An electrical field is applied to the gel and the negatively charged DNA fragments migrate toward the positively charged end of the gel. Because the agarose acts as a molecular sieve, smaller DNA fragments move faster than larger ones, thus fragments separate from one another. The DNA fragments are invisible and therefore must be stained for visualization.

Polyacrylamide Gel Electrophoresis, PAGE. *Electrophoresis method that uses polymerized polyacrylamide as the matrix.* A common adaptation of PAGE, sodium dodecyl sulfate-PAGE (SDS-PAGE) is used to separate proteins on the basis of molecular weight. Protein samples are treated with the negatively charged detergent, SDS. SDS denatures the proteins and confers a negative charge on them so that they migrate in the electrical field toward the positively charged end of the gel. Smaller proteins migrate through the matrix more quickly than larger ones. The proteins are invisible and therefore, after separation, must be stained for visualization or be detected by Western Blotting, as described above in this table.

Isoelectric Focusing, IEF. *Method capable of detecting a single-charge difference between protein molecules; molecules are separated as they run through a pH gradient in an electrical field.* IEF detects post-translational modifications that are made to proteins and that result in net charge differences. It is used, for example, to evaluate protein identity and purity and to look for protein changes during stability testing. (See, for example, the Case Study below, "Testing for Illicit Use of EPO by Elite Athletes.")

Capillary Electrophoresis. *A miniaturized instrumental version of electrophoresis that is amenable to automation, requires very little sample, and provides quantitative information.* Capillary electrophoresis is widely used for analyzing food samples, forensic samples, oligonucleotides, proteins, peptides, and DNA. It is used in genome sequencing. It is also used in the pharmaceutical industry to help determine the identity, purity, and structure of small molecule and protein-based drugs.

Two-Dimensional Gel Electrophoresis. *A type of high-resolution gel electrophoresis in which proteins are first separated by isoelectric focusing and then are separated based on differences in their molecular weight.* This method is important in proteomics. It can resolve hundreds of proteins on a single gel and can also be used to detect post-translational modifications, which cannot be predicted from genome sequences.

Mass Spectrometry (Mass Spec). *A powerful instrumental analytical technique used to identify and measure a wide variety of biological and chemical compounds based on differences in both their mass and charge.* Mass spec can be used, for example, to identify exactly what compounds are present in a sample and to determine the structure of a protein.

Electron Microscopy (EM). *Microscopic methods that use electrons instead of light to illuminate a sample and enable biologists to view extremely small structures, including cellular organelles, bacteria, and viruses.* EM has many applications, such as researching the cellular effects of a compound or disease state and checking cells for viral or microbial contamination.

Transmission Electron Microscopy (TEM). *Utilizes a type of electron microscope in which a narrow beam of electrons is passed through a very thin section of sample, (usually 1 to 200 nm thick).* The electrons that penetrate the sample form an image on a fluorescent screen or on photographic film. This type of microscope can attain very high magnifications, of several hundred thousand times or more.

Scanning Electron Microscopy (SEM). *Utilizes a type of electron microscope that provides a three-dimensional image of the surface of an object magnified from about 200X to 35,000X.* A finely focused beam of electrons is scanned back and forth across the object's surface; the object interacts with the beam and this interaction is used to construct an image.

Similarly, a fermenter is heterogeneous over time. As a fermentation process proceeds, changes occur in the number of cells present, the chemical composition of the broth, oxygen availability, pH, and so on. Obtaining a representative sample requires understanding the heterogeneity of the system of interest and then devising a sampling strategy.

Another issue in sampling is that humans usually have preferences when sampling. For example, a person casually sampling an experimental field might take only plants that are easily accessible. Even when people try not to be biased, they unknowingly have preferences. For example, when people try to list a series of numbers with no inclination for particular numbers, there still are patterns in their choices because of subconscious preferences.

Statisticians may be consulted to develop a sampling strategy when large and complex studies are performed or when a sampling process is being developed for a new product. In a smaller or simpler study certain basic sampling guidelines are often employed. Box 24.1 illustrates some common methods used to help obtain a representative sample in several situations.

ii. SAMPLE PREPARATION

Biological samples frequently require extensive preparation before they can be analyzed. These preparations may greatly affect the results of a measurement or an assay. Prior to electron microscopy, for example, samples are chemically preserved, dehydrated with alcohol, embedded in plastic, thinly sliced into sections, and stained with heavy metals. The sections are then viewed with the electron microscope. The structures that are visualized depend on the way that the sections were prepared. In molecular biology, cellular extracts might be treated with harsh chemicals, centrifuged, stored, and otherwise manipulated. There is an example on pp. 558–559 of this text of a study in which investigators were comparing cells that contained an oncogene (a cancer-causing gene) with cells that did not have the oncogene. The scientists found that their results varied depending on how the cells were prepared. The effects of sample preparation thus had the potential to obscure the effects due to the oncogene.

It is not uncommon during sample preparation that certain components of a sample will be selectively lost while others are retained. Similarly, biological structures may be altered and biological activities, such as enzymatic activity, may be lost. All such alterations in the samples are causes of systematic error. Sample preparation is thus a major issue in the analysis of biological samples. It is often necessary to perform preliminary investigations of the effects of sample preparation before conducting assays or experiments. A few basic strategies to deal with sample preparation problems are summarized in Box 24.2.

C. The Method

Assay and test methods are often invented by research scientists who need to assess some feature of their research subjects. Scientists publish their methods in scientific research literature and then other people adapt the methods for various purposes. Sometimes research and development scientists in a biotechnology company invent new methods or modify existing methods as commercial products. In these cases, the companies sell kits with the materials needed to perform the method along with instructions to do it. There are, for example, biotechnology companies that sell kits for detecting DNA and for quantifying proteins in samples.

The methods by which tests and assays are performed sometimes must be standardized, among individuals, laboratories, and nations. In certain disciplines there are organizations that make recommendations for standard methods. Table 24.3 on p. 438 lists a few of the organizations involved in establishing standards and accepted methods that are of interest to biotechnologists. A **standard method** *is a document established by consensus and approved by a recognized body that establishes rules or guidelines to make a procedure consistent among various people.* Often standard methods for particular applications are collected and published in a compendium. The **U.S. Pharmacopeia (USP)** *is a compendium that contains hundreds of accepted methods for tests commonly performed in the pharmaceutical industry.* The

Box 24.1. OBTAINING A REPRESENTATIVE SAMPLE

1. **When selecting a sample of individuals from a population:**
   a. *Devise a numbering system to assign an identification number to all the potential members of the sample.* For example, if a sample is to be 40 students chosen at random from all the students in a college, then the students' college ID numbers can be used.
   b. *Throw a die or pick numbers from a bowl to choose the individuals who will actually be in the sample.*
   c. *As an alternative to picking numbers from a bowl or throwing a die, refer to a random number table.* A random number table is a list of numbers with no preferences for certain numbers or patterns. Close your eyes and point anywhere in the table to choose the starting point. Then, read off numbers in the table.
   d. *Computer programs that generate random numbers can be used to pick members of a sample.* Computers are particularly helpful when a sample is large.

2. **When selecting a sample of an area:**
   a. *Draw or map the area on a piece of paper.*
   b. *Place a grid with horizontal and vertical lines over the map.*
   c. *Number the boxes formed by the grid.*
   d. *Use a random number table to choose grid boxes.* Sample the areas under the chosen grid boxes.

3. **When sampling biological solutions in the laboratory:**
   a. *Be certain liquids are well-mixed before withdrawing a sample.*
   b. *When a liquid is removed from a freezer, be sure all the ice is melted before removing a sample.* Mix the liquid. Different components of the liquid may thaw at different rates.
   c. *When taking a sample from a suspension, such as a bacterial culture, make sure the suspension is evenly distributed.*
   d. *When sampling a powder, either mix it or take a random part of the solid.*
   e. *If a material is known to be heterogeneous, it may be advisable to take a series of samples from different parts of the material and to combine them into one sample. This is called a* **composite sample.**

Box 24.2. AVOIDING SYSTEMATIC ERRORS IN SAMPLE PREPARATION

1. **Prepare all the samples, controls, and standards in a study identically.**
   a. If there are samples from experimental and control groups, be certain that they are treated in the same way.
   b. Create written procedures for sample handling that are followed by all analysts.
   c. Devise methods to ensure that all reagents, materials, and equipment used in sample preparation are consistent.

2. **Test for the loss of specific components.** This can often be accomplished by adding a known amount of the material of interest to a sample. (This is called *spiking* the sample.) The sample is then taken through the sample preparation steps. The amount of the material of recovered after sample preparation is compared to the amount initially added.

3. **Test different sample preparation procedures.** Experiment with different sample preparation methods to determine their effects on the results.

USP methods have been validated (a term defined later in this chapter) and are accepted for use by the Food and Drug Administration (FDA). Figure 24.2 shows an example of a U.S. Pharmacopeia standard assay method to confirm the identity of sodium chloride tablets. Another compendium is the **Official**

Assay for sodium citrate—

*Cation-exchange column—*Mix 10 g of styrene-divinylbenzene cation-exchange resin with 50 mL of water in a suitable beaker. Allow the resin to settle, and decant the supernatant liquid until a slurry of resin remains. Pour the slurry into a 15-mm × 30-cm glass chromatagraphic tube (having a sealed-in, coarse-porosity fritted disk and fitted with a stopcock), and allow to settle as a homogeneous bed. Wash the resin bed with about 100 mL of water, closing the stopcock when the water level is about 2 mm above the resin bed.

*Procedure—*Transfer an accurately measured volume of Oral Solution, equivalent to about 1 g of sodium citrate dihydrate, to a 100-mL volumetric flask, dilute with water to volume, and mix. Pipet 5 mL of this solution carefully onto the top of the resin bed in the Cation-exchange column. Place a 250-mL conical flask below the column, open the stopcock, and allow to flow until the solution has entered the resin bed. Elute the column with 60 mL of water at a flow rate of about 5 mL per minute, collecting about 65 mL of the eluate. Add 5 drops of phenolphthalein TS to the eluate, swirl the flask, and titrate with 0.02 N sodium hydroxide VS. Record the buret reading, and calculate the volume (B) of 0.02 N sodium hydroxide consumed. Calculate the quantity, in mg, of $C_6H_5Na_3O_7 \cdot 2H_2O$ in each mL of the Oral Solution taken by the formula:

$$[1.961B(20/V)] - [(294.10/210.14)C]$$

in which 1.961 is the equivalent, in mg, of $C_6H_5Na_3O_7 \cdot 2H_2O$, of each mL of 0.02 N sodium hydroxide. V is the volume, in mL, of Oral Solution taken, 294.10 and 210.14 are the molecular weights of sodium citrate dihydrate and citric acid monohydrate, respectively, and C is the concentration, in mg per mL, of citric acid monohydrate in the Oral Suspension, as obtained in the Assay for citric acid.

Figure 24.2. An Example of a Standard Method from the U.S. Pharmacopeia. (Reprinted with permission from USP 23-NF16. All rights reserved. Copyright © 1995 The United States Pharmacopeial Convention, Inc.)

Table 24.3. *A Selected List of Agencies and Organizations That Are Involved in the Standardization of Measurements and Assays*

ANSI (American National Standards Institute)

Administrator and coordinator of the U.S. private sector voluntary standardization system. ANSI does not itself develop American National Standards; rather it facilitates their development by establishing consensus among qualified groups. ANSI is the sole U.S. representative to ISO. Examples of services provided by ANSI are:

1. Promoting the use of U.S. standards internationally.

2. Playing an active role in the governance of ISO.

3. Promoting the use of standards, including ISO 9000 in the United States.

AOAC International (Formerly the Association of Official Analytical Chemists)

An independent association of scientists devoted to promoting methods validation and quality measurements in the analytical sciences. Examples of services provided by AOAC are:

1. Developing validated standard methods of analysis in microbiology and chemistry.

2. Publishing validated standard methods in the compendium, *Official Methods of Analysis of AOAC International.*

ASTM (American Society for Testing and Materials)

Coordinates efforts by manufacturers, consumers, and representatives of government and academia to develop by consensus standards for materials, products, systems, and services. Examples of services provided by ASTM are:

1. Developing and publishing technical standards (e.g., those covering the requirements for volume-measuring glassware). More than 10,000 ASTM standards are used worldwide.

2. Providing technical publications and training courses.

CLSI (Clinical and Laboratory Standards Institute) (Formerly NCCLS)

An organization that promotes voluntary consensus standards relating to laboratory procedures, methods, and protocols applicable to clinical laboratories.

ISO (International Organization for Standardization)

A worldwide federation of national standards bodies from some 157 countries that promotes the development of standards and related activities in the world with a view to facilitating the international exchange of goods and services. Examples of services provided by ISO are:

1. Publishing and updating the SI (metric) system of units. The SI system is covered by a series of 14 international standards

2. Developing by consensus international standards including the ISO 9000 quality series.

NIST (National Institute for Standards and Technology) (Formerly the National Bureau of Standards)

A U.S. federal agency that works with industry and government to advance measurement science and develop standards. Examples of services provided by NIST are:

1. Providing calibration services and primary standards for mass, temperature, humidity, fluid flow, and other physical properties.

2. Providing more than 1300 standard reference materials to be used in biological, chemical, and physical assays

3. Performing research and development of new standards, including standards for biotechnology.

USP (United States Pharmacopeia)

An organization that promotes public health by establishing and disseminating officially recognized standards relating to medicines and other health care technologies. The methods in the USP are mandatory in the United States Pharmaceutical industry and in other situations where a company is cGMP compliant. (See Unit II for a description of cGMP requirements.) Examples of services provided by USP are:

1. Developing validated standard methods of analysis for pharmaceutical and health-related products.

2. Publishing validated standard methods in the compendia *The United States Pharmacopeia.*

3. Distributing reference materials to be used with the methods outlined in the *Pharmacopeia.*

Methods of Analysis of AOAC International, *which contains microbiological and chemical analysis methods.* Table 24.4 lists a few major compendia used in bioscience laboratories.

Biotechnology companies that make products must have assays to test their final products, incoming raw materials, and in-process samples. Sometimes research and development scientists must create new assays for testing new products. This might be the case in a biopharmaceutical company that is developing a novel therapeutic. More commonly, development scientists modify and optimize existing methods, including those from compendia. The creation and optimization of test methods is an important part of product development. As the product and the processes by which a product is made are refined, so too are the methods to support it.

D. The Reference Materials

Many assays and tests require *chemical or biological substances whose compositions are reasonably well established and that are used for comparison;* these substances are called **reference materials, reference standards, controls**, or simply, **standards.** For example, when a sample is tested using a chromatography or electrophoresis method the result is a profile or "fingerprint" that varies depending on the composition of the sample. Different materials result in different pro-

files. The profiles of samples can be matched to the profiles of standards of known composition. If the standard and the sample profiles match, then they may be the same material. (Note, however, that two or more different materials sometimes have very similar profiles using electrophoresis or chromatography.)

Obtaining and characterizing reference materials is critical in the development of many assay methods. Standards sometimes have to be prepared in the laboratory that is developing the assay, in which case considerable effort will be required to establish and document their composition, stability, and other qualities. In many cases, reference standards of acceptable quality can be purchased. The National Institute of Standards and Technology (NIST) maintains more than 1300 different materials, called **standard reference materials (SRMs),** whose compositions are determined as exactly as possible with modern methods. A wide variety of SRMs for chemistry, physics, and manufacturing are available. A biotechnology subdivision at NIST sells SRMs for biotechnology applications, such as DNA standards for quality control in forensic DNA testing. The U.S. Pharmacopeia organization also sells well-characterized, certified reference standards that the FDA accepts for use by pharmaceutical manufacturers.

Chemical and biological reference materials are available from commercial suppliers, as well as agen-

---

Table 24.4. *EXAMPLES OF METHODS COMPENDIA*

Bacteriological Analytical Manual

Food and Drug Administration and AOAC International
Contains methods for the microbiological analysis of foods. Available online at http://www.cfsan.fda.gov/~ebam/bam-toc.html.

British Pharmacopeia, BP

British Pharmacopoeia Commission Secretariat
Contains validated standard methods of analysis for pharmaceutical and health-related products; accepted in the United Kingdom.

European Pharmacopeia, EP

European Directorate for the Quality of Medicines—Council of Europe
Contains validated standard methods of analysis for pharmaceutical and health-related products; accepted in Europe outside the United Kingdom.

Official Methods of Analysis of AOAC International

AOAC International
Contains a variety of microbiological and chemical analysis methods.

Standard Methods for the Examination of Water and Wastewater

A joint publication of the American Public Health Association, the American Water Works Association, and the Water Environment Federation
Contains current practices for the analysis of water.

The United States Pharmacopeia/National Formulary, USP-NF

United States Pharmacopeia
Contains validated standard methods of analysis for pharmaceutical and health-related products; accepted by the U.S. Food and Drug Administration.

cies like NIST and USP. Commercially manufactured reference materials often have associated documentation that shows how their properties were determined and the certainty that the manufacturer has in their values. Table 24.5 shows the documentation for a reference standard containing a mix of minerals. Note the care that was used to determine the levels of each mineral and the description of the variability associated with the measurements. A well-characterized standard like this mineral mixture might be used, for example, in a situation where regulatory decisions will be made based on test results.

Biologists commonly use products of biological derivation that are difficult and expensive to characterize as fully as the mineral mixture just described. Biological materials that are reasonably well-characterized are used to determine the response of a method or an instrument to the substance of interest. Figure 24.3 shows examples of catalog specifications for standards used in biological and clinical assays.

CASE STUDY

Purity Testing in Food

It is illegal to add commercial sweeteners (such as corn syrup) to foods without proper labeling; such sweetened foods are considered to be adulterated. Since commercial sweeteners are far less expensive than fruit juice, maple syrup, or honey, producers sometimes illegally substitute these inexpensive sugars for more expensive ones. Two brothers who owned a decades-old Mississippi honey and syrup business were sent to prison after years of adulterating their products, selling them cheaply, and thereby undercutting legitimate honey and syrup producers (FDA Consumer, April 1997). Standard chemical analyses do

not reveal adulteration because the chemical composition of corn syrup is very similar to that of honey, maple syrup, and fruit juice. Investigators therefore developed a gas chromatography method that is capable of distinguishing sweetened from pure products. In this example, adulterated apple juice has a distinctive "fingerprint" when compared with a pure apple juice reference standard.

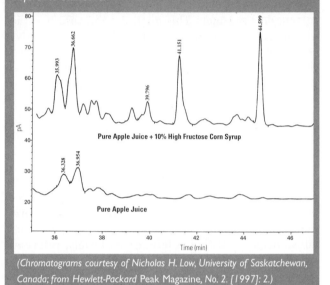

(Chromatograms courtesy of Nicholas H. Low, University of Saskatchewan, Canada; from Hewlett-Packard Peak Magazine, No. 2. *[1997]: 2.)*

Regardless of the source of reference materials, they must be carefully and securely stored to avoid loss, degradation, or damage. Well-characterized reference materials are a valuable resource in a laboratory.

E. Documentation

Issues relating to documentation are shown in Box 24.3. This box is derived from the Food and Drug

---

Table 24.5. SPECIFICATIONS FOR A MINERAL MIXTURE

Property	Units	Consensus Value*	Std. Deviation**
Alkalinity	mg/L	93.56	4.35
Calcium	mg/L	27.21	2.20
Chloride	mg/L	91.74	4.09
Conductivity	µmho/cm	661.00	15.60
Fluoride	mg/L	2.39	0.18
Magnesium	mg/L	12.18	0.66
Potassium	mg/L	9.08	0.68
Sodium	mg/L	83.28	4.14
Sulfate	mg/L	81.10	6.10
Hardness	mg/L	116.37	5.22

*The amount of each mineral in the mixture was rigorously analyzed in 230 laboratories and the mean of the results is reported as the "Consensus Value."
**The standard deviation column indicates the variability in the data from the various laboratories.

A 2153 **Fraction V Powder**
2–8°C **Minimum 96%**
(electrophoresis)
Initial fractionation by cold
alcohol precipitation
Remainder mostly globulins.
Prepared by a modification of
Cohn, using cold ethanol, pH, and
low temperature precipitation, followed by additional
pH adjustment step prior to final drying.
The pH of a 1% (w/v) aqueous solution is
approx. 7.
Ref.: Cohn, E.J., et al., J. Am Chem. Soc., **68**, 459
(1946).

(a)

ACCUSET™ LIQUID
A 2539 **CALIBRATOR**
–0°C **Level 1**
◆ Human serum preparation
stabilized with ethylene glycol for use in calibrating
automated analyzers. Values assigned for the
following constituents: albumin, bilirubin, calcium,
carbon dioxide, chloride, cholesterol, creatinine,
glucose, iron, magnesium, phosphorus, protein,
triglycerides, urea nitrogen and uric acid. Calibrator
values for Technicon RA-1000, Electro-Nucleonics,
Gemeni and Gemstar, BMD Hitachi 705, Gilford and
other instruments.
R: 20/21/22-36/37/38 S: 45-26-36/37/39

(b)

Figure 24.3. Catalog Descriptions of Materials Used as Standards in Biological and Clinical Assays. a. Description of bovine serum albumin, a protein that is often used as a standard in assays that determine the amount of protein in a preparation. **b.** A human serum preparation with known values for a number of clinically relevant substances; used to calibrate clinical instruments. (Descriptions reproduced by permission of Sigma-Aldrich Co.)

Administration's Good Manufacturing Practices regulations and describes the items that the FDA says should be recorded when conducting tests on samples in quality-control laboratories in pharmaceutical companies. Although the information in this box is intended for pharmaceutical analysts, it is relevant in many other settings as well.

III. AVOIDING ERRORS IN ASSAYS AND TESTS

A. Methods Are Based on Assumptions

Assay methods are based on certain assumptions and have potential inaccuracies and uncertainties. Consider, for example, a conventional test used to deter-

---

Box 24.3. RULES FOR LABORATORY RECORDS FROM GMP REGULATIONS 21 CFR 211.194

a. Laboratory records shall include complete data derived from all tests necessary to assure compliance with established specifications and standards, including examinations and assays, as follows:
  1. A description of the sample received for testing with identification of source (that is, location from which the sample was taken), quantity, lot number or other distinctive code, date sample was taken, and date sample was received for testing.
  2. A statement of each method used in the testing of the sample. The statement shall indicate the location of data that establish that the methods used in the testing of the sample meet proper standards of accuracy and reliability . . .
  3. A statement of the weight or measure of sample used for each test, where appropriate.
  4. A complete record of all data secured in the course of each test, including all graphs, charts, and spectra from laboratory instrumentation, properly identified to show the particular component . . . tested.
  5. A record of all calculations performed in connection with the test, including units of measure, [and] conversion factors . . .
  6. A statement of the results of tests and how the results compare with established standards . . .
  7. The initials or signature of the person who performs the test and the date(s) the tests were performed.
  8. The initials or signature of a second person showing that the original records have been reviewed for accuracy, completeness, and compliance with established standards.
b. Complete records shall be maintained of any modification of an established method employed in testing. Such records shall include the reason for the modification and data to verify that the modification produced results that are at least as accurate and reliable for the material being tested as the established method.
c. Complete records shall be maintained of any testing and standardization of laboratory reference standards, reagents, and standard solutions.
d. Complete records shall be maintained of the periodic calibration of laboratory instruments, apparatus, gauges, and recording devices . . .

mine if a substance is contaminated by bacteria. Bacteria are too small to see with our eyes and are usually too numerous to count individually. It is necessary, therefore, to use an indirect assay to visualize and count bacteria in a substance. One such method is the spread plate method:

1. Obtain a sample from the substance to be tested.

2. Prepare an appropriate serial dilution of this sample.

3. Apply 0.1 mL of the diluted sample onto petri plates that contain sterile bacterial nutrient medium and distribute the sample evenly over the plate.

4. Incubate the plates overnight to allow bacterial growth.

5. If bacteria were present in the sample, bacterial colonies will grow on the plates; count these colonies. Each colony is called a *colony forming unit (CFU)*.

6. The number of bacteria in the original substance is calculated based on the assumptions that:

   a. Each visible colony is derived from a single bacterium that was present in the sample.

   b. Each living bacterium in the diluted sample forms a colony on the plate.

This counting method is indirect; the number of colonies on the plate is used as an indicator of the number of bacteria in the original substance. Although the spread plate counting method is simple and widely accepted, there are cautions in its use including:

1. The bacteria (if present) in the original substance must be uniformly distributed when the sample is taken so that the sample represents the whole substance.

2. The method assumes that dead or nonreproductive bacteria are unimportant, since these will not be counted.

3. The method is inaccurate if adjacent bacteria merge to form a single colony.

4. The method is inaccurate if too few or too many bacteria are plated.

5. The method is inaccurate if the plates are incubated too briefly or too long.

The spread plate method is widely used and its limitations are well understood in the scientific community. When using less-common assay methods and new methods, it is important for the analyst to evaluate the method's limitations and understand its applications.

B. The Use of Positive and Negative Controls

Positive and negative controls are critical components of most assays used in biotechnology laboratories. For example, if one were using the plate count method to indicate the number of bacteria in a sample, it would be good practice to remove 0.1 mL of the diluting solution (step 2) and apply it to another petri plate that contains nutrient medium. The diluting solution should be sterile, as should be the petri plate and the nutrient medium. Therefore, no colonies should grow on this plate. This is a *negative control*. If any colonies appear on the negative control plate, it means that the diluting material or the plates were contaminated with bacteria to begin with, or that contaminants were introduced accidentally sometime during the procedure. In the event of contamination, none of the results of the test are valid and it must be repeated with fresh, uncontaminated materials. A positive control would be to plate 0.1 mL of a substance known to contain bacteria; bacteria should grow. If they do not, the growth medium or conditions are not correct and none of the results of the test are valid.

Let's further consider positive and negative controls by discussing an important class of assay methods, those that involve antibodies. *Methods involving antibodies are called* **immunochemical methods**. Antibodies have evolved over millennia to find, recognize, and bind specific invading substances (e.g., viruses, bacteria) from amidst a myriad of molecules. (See Chapter 1, p. 16 for an explanation of antigens and antibodies). Different antibodies bind to different antigen targets. Analysts take advantage of the remarkable specificity of antibodies to target and visualize a substance of interest. For example, a method called *Western Blotting* uses antibodies to identify a particular protein that has been separated from other proteins by electrophoresis. Antibodies can be applied to microscope slides of tissue to find and identify a particular cellular component. Antibodies can also be applied to cellular material that has been prepared for electron microscopy to find and identify a subcellular structure. An example of the latter would be the use of antibodies to find a protein that resides in the membrane of the cell and recognizes hormones.

The complex formed when an antibody binds its target is invisible and therefore by itself is not useful. Antibodies that are used for immunochemical assays, however, are labeled with chemical tags that make them visible to the eye or to an instrument. Common tags used to label antibodies include: radioactive compounds that expose photographic film, thereby leaving a detectable spot; enzymes that create visible deposits when exposed to their substrate; and fluorescent compounds that are visible when viewed with the proper light source. To see, for example, whether a particular hormone receptor is present on the surface of cells, it is possible to obtain antibodies that selectively recognize and bind to this receptor. These specific antibodies are labeled with a tag that forms a visible deposit in the presence of the proper substrate. The antibody thus finds and binds the hormone receptor; the complex is treated with the substrate; a visible deposit forms. The deposit is viewed under a light or electron microscope and the hormone receptor is visible. Figure 24.4a shows tissue that has been treated in this way.

Immunochemical methods are a powerful means to visualize biological structures that would otherwise be invisible. They are important tools in many research laboratories. Immunochemical methods are integral in the development and testing of biopharmaceutical products. Antibody methods can be used in conjunction with microscopic techniques, as shown in Figure 24.4. They can be used with cells in culture and with proteins immobilized on a plate or on a plastic surface. They can be used in conjunction with electrophoresis.

In all immunological applications, however, there are problems that can arise. For example, antibodies may bind to molecules that are similar to, but not the same as, their target. This is called *nonspecific binding* and it may occur if incubation conditions are not optimized. If too little antibody is used, the target may not be visible. If the assay conditions are wrong, then a deposit may form where it should not, or may fail to form where it should. The sample preparation steps may affect the antibody's ability to bind to its target. All these problems can cause systematic errors that seriously affect the results of the assay. Developing a reliable immunoassay requires understanding the potential problems, performing preliminary tests to identify as many problems as possible, optimizing the sample preparation steps, and optimizing the conditions used during the assay. An important strategy to avoid errors when performing immunochemical assays is to use positive and negative controls. These types of controls are helpful in determining whether an unknown variable is affecting the results. Such controls are best explained by example.

Suppose we are using an immunochemical method to see whether a particular tissue contains a hormone receptor of interest. There are four possible outcomes when the tissue is treated:

1. The tissue might contain the receptor and might form a visible deposit when treated by immunological methods. This is called a **true positive result** *because the result is correct and it is positive for the receptor.*

2. The tissue might contain the receptor, but because of some problem in the procedure, might fail to form a deposit when treated. This is called a **false negative result.** *The result is negative and it is incorrect.*

3. The tissue might not contain the receptor and not form a visible deposit when treated. This is called a **true negative result.** *The result is negative for the receptor and it is correct.*

4. The tissue might not contain the receptor, but it might form a visible deposit because of nonselective binding or another problem in the procedure. This is called a **false positive result**. *The result is positive and it is incorrect.*

In order to help avoid false results, as in situation 2 and 4 above, positive and negative controls can be used. A **negative control** *is tissue that is like the test sample but is known not to contain the target of interest and therefore should not form a visible deposit,* see Figure 24.4b. The negative control tissue is treated exactly the same

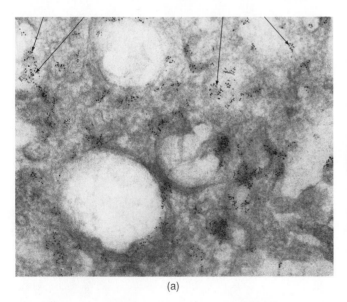
(a)

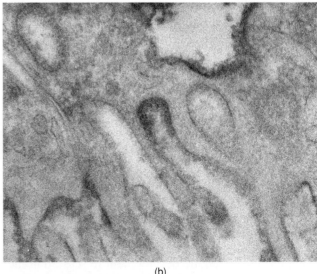
(b)

Figure 24.4. Immunocytochemistry of Kidney Cells. a. The small, round, dark deposits in this electron micrograph (arrows) represent antibodies bound to a specific receptor protein. **b.** A negative control. This is tissue known not to have the receptor of interest. The negative control tissue is processed alongside the experimental samples and is treated exactly the same as experimental tissue. No antibody deposits appear in this tissue.

way as the other samples. If the negative control tissue does show a positive result, then we know there was too much antibody or there was another problem in the procedure and so we would not trust the results on the samples. The routine use of negative controls helps avoid false positive results in assays.

A **positive control** *is tissue known to contain the target of interest and which therefore should form a visible deposit if the reagents and procedures used were good.* Thus, if the positive control fails to form a visible deposit then we know there was a problem in the procedure. The routine use of positive controls helps avoid false negative results in assays.

In addition to the preceding possibilities, the results of an assay might be ambiguous. For example, the deposit might be difficult to see, the tissue may appear to be labeled in some areas, but not in other similar areas, various tissue samples might unexpectedly differ, and so on. Ambiguous results are frustrating and typically mean that the experimental test system has not been optimized adequately. More preliminary testing may reveal conditions that can be slightly altered to give better results.

The use of positive and negative controls is applicable in many situations in addition to immunochemical assays. For example, positive and negative controls are used in clinical testing of patient samples. Consider a blood test used to see if a patient is infected with a certain parasite. A *positive control* would be blood that is known to test positive for the parasite. The positive control is run alongside the patient's sample and helps detect false negative results. (A *false negative* occurs when a person who is infected shows up as uninfected.) A negative control (i.e., blood that is known not to react with the test) is used to avoid a false positive result. (A false positive occurs when a person who is uninfected shows up as infected.)

A note about terminology: When an assay or a test is performed, there are often more materials tested than just the "experimental" samples. There are also the positive and negative controls and the reference materials. All of these are often referred to as *samples.* A negative control, for example, may be called a *negative control sample.* The samples that are the actual subject of the analysis may be called the *unknowns,* the *unknown samples,* the *experimental samples,* or the *test samples.* It is essential that all the *samples* are processed and analyzed in the same way.

In summary, when performing tests and assays on biological samples, it is important to be aware of the assumptions, limitations, and potential errors inherent in those tests. Most biological tests and assays are complex and so require development (discussed in the next section) and practice to get the best results. It is also important to assiduously use positive and negative controls each time an assay is run to detect problems in the procedure.

CASE STUDY

Testing for Illicit Use of EPO by Elite Athletes

Erythropoietin (EPO) is a glycoprotein hormone naturally produced by the kidney that boosts the production of red blood cells. A recombinant form of EPO, rhEPO, is used medically as a drug to treat patients with anemia. Athletes sometimes take rhEPO to enhance their performance, although this practice is unfair, dangerous to the athlete, and illegal. Athletes use rhEPO illicitly because the oxygen-carrying capacity of their blood is the main factor limiting their performance; rhEPO boosts their production of red blood cells. An assay exists to test the urine of athletes to see if they have taken rhEPO. There is controversy, however, as to whether this is a "good" assay that can be trusted.

Lance Armstrong, see Figure 24.5, is an athlete who overcame life-threatening cancer to become the seven-time winner of the prestigious Tour de France bicycle race. An article published in 2005 in a French newspaper alleged that urine samples taken from Armstrong and stored since 1999 tested positive for rhEPO. Armstrong denies that he has ever used any banned substances to enhance his performance. The news story predictably resulted in an international controversy with some people indignantly claiming that Armstrong's astonishing athletic career was a sham and others equally adamant that the allegations were concocted to discredit him.

Armstrong is not the only athlete who disputes a positive rhEPO test result. Various athletes claim that the assay yields false positive results at an unknown frequency. The stakes of drug testing are high for individual athletes, fans, officials, and competitors. Although rhEPO testing of athletes has been used since the early 2000s, the question of its trustworthiness continues to be a topic of investigation into 2008.

Figure 24.5. *Lance Armstrong, Seven-Time Winner of the Tour de France.* (Courtesy AFP/Getty Images.)

Urine testing for rhEPO is based on the fact that there are subtle differences in the glycosylation patterns of natural and recombinant EPO (glycosylation is illustrated in Figure 1.8); these differences affect the charge on the molecule. The assay for rhEPO has two major steps. The proteins in a urine sample from the athlete are first separated from one another using isoelectric focusing. In this method, the proteins migrate under the influence of an electrical current through a gel that has a pH gradient. Proteins move through the gel until they arrive at the point in the pH gradient at which they have no net charge (the isoelectric point). Natural EPO and rhEPO separate from one another because of the differences in their charges. At this point the EPO molecules are separated, but invisible. The next step in the process is to transfer the proteins to a polymer sheet where they are exposed to a monoclonal antibody that recognizes human EPO. The antibody finds and binds to the natural and rhEPO molecules. The antibody is attached to a substance that makes it visible. The result is a visible pattern of bands; each band consists of EPO molecules with identical charge that have bound to the antibody. The pattern of bands from individuals who have only natural EPO is slightly different than the pattern when rhEPO is also present.

How could this test result in false positives? One possibility is cross-reactivity between the antibody and a protein that is not EPO. Such cross-reactivity would result in additional bands that might mimic the appearance of rhEPO. Indeed, according to a study published in 2007, some athletes produce protein-rich urine that cross-reacts with the antibody used for EPO testing. According to this study: "We show here that this widely used test can occasionally lead to the false-positive detection of rhEPO . . . in postexercise, protein-rich urine, probably because the adopted monoclonal anti-EPO antibodies are not monospecific." Beullens, Monique, Delanghe, Joris R., and Bollen, Mathieu. ("False-Positive Detection of Recombinant Human Erythropoietin in Urine Following Strenuous Physical Exercise." Blood 107 no. 12 [2006]: 4711–13.)

In the case of Armstrong's samples that appeared to be positive for rhEPO, it has been suggested that bacterial growth might have contaminated the sample during its years of storage, producing proteins that cross-react with the antibody. There is also the possibility that given the long storage period and the absence of a clear chain of custody process, a labeling or handling error caused the wrong samples to be attributed to Armstrong.

The finding that the EPO test can, in at least a few athletes, result in false positives leads to many questions. How common are these false positive results? Is there some way to identify athletes who are likely to produce a false positive result? How common are false negative results? How many samples were used to validate the test before it was accepted for use? Was the original development and validation of the assay sufficiently rigorous? These are all questions that are being discussed by athletes, officials, and sports enthusiasts around the world.

IV. METHOD DEVELOPMENT AND VALIDATION

A. Method Development

A test method is not a tangible product, like a therapeutic drug or a cloned sheep, but it is still a product and requires development, as does any other product. Method development is often complex and requires experimentation and optimization of many factors.

Finding the basic concept for an assay occurs during the discovery phase of new product development. The concept must involve some sort of interaction or effect that can be detected and that is relevant to the property being analyzed. For example, a scientist interested in studying the functions of a hormone receptor might devise an immunochemical method to detect the receptor. A scientist interested in studying chemotherapeutic agents might devise a cell-based assay that detects biochemical changes in cultured cells exposed to the drugs.

Scientists experiment with a variety of factors during assay development to optimize the assay's performance and to eliminate interferences and ambiguous results. Development scientists must decide, for example, what chemical tag to use to label antibodies when optimizing an immunochemical assay. Scientists will likely experiment with more than one type of label to see which is best. Immunological assays also require optimization of incubation times, temperatures, and antibody concentrations. Assays that monitor the activity of enzymes require optimizing the temperature, the buffer, and the concentrations of substrates and cofactors. The polymerase chain reaction, which is discussed in more detail in a later section of this chapter, requires an instrument that cycles through a series of temperatures in a controlled fashion. The temperatures and their duration must be optimized. Cell-based assays require optimal conditions for maintaining the cells. Many assays require an instrument for detection and/or to separate the components of a mixture. Instrumental methods always require optimization of the equipment parameters. Assays require positive and negative controls that must be optimized. During method development the conditions of the method, the materials required, the controls, and the steps to perform the assay are all established.

Data analysis is another aspect of assays that needs to be developed. Data analysis may involve calculations, comparisons with standards, and comparisons with published data. Computers are now often used to help eliminate artifacts, annotate data to make it easier for a human viewer to interpret, store data, and report data. Computers make it easier for an operator to work with data, but computer software must be extensively tested to ensure its validity.

The case study below provides an example of method development. (For another example, see the Case Study in Chapter 2, "Crime Scene Investigation: A Behind-the-Scenes Story." on pp. 31–32.)

CASE STUDY

Assay Development at an Early Stage Biotechnology Company

BellBrook Labs is an early stage biotechnology company located in Madison, Wisconsin that is working to create assay systems that can be sold to pharmaceutical companies. Pharmaceutical companies will use the assays for high throughput screening, HTS. As we saw in Chapter 3, HTS assays are used to identify potential drugs by looking for compounds that inhibit the pathological effects of an aberrant protein associated with a specific disease. Pharmaceutical companies have "libraries" with hundreds of thousands of compounds that are possible drug candidates. The limiting factor for the pharmaceutical companies is finding reliable, rapid methods of screening all their compounds to find the few that might have a therapeutic effect—which is where BellBrook Labs finds its business niche.

One of BellBrook's patented discoveries is an assay system that targets phosphodiesterase proteins. These proteins are normal enzymes that in certain situations behave aberrantly and become associated with many diseases including rheumatoid arthritis, diabetes, and erectile dysfunction. The phosphodiesterases convert the molecule cyclic AMP (cAMP) into AMP as part of their normal mode of action. BellBrook scientists are therefore developing an assay system that detects the conversion of cAMP to AMP as a measure of the activity of the phosphodiesterase enzymes. The assay depends on an antibody that BellBrook manufactures. The antibody is exquisitely sensitive and specific and can distinguish cAMP from AMP. It binds AMP, which is the product of the reaction of interest. BellBrook also has an instrument, called a fluorescence polarization instrument, that detects the AMP-antibody complex, see Figure 24.6.

Pharmaceutical companies use this assay system to test their thousands of potential drug compounds by creating thousands of reaction mixtures containing the phosphodiesterase enzymes and cAMP, along with various ions and other molecules that are required by the reaction. After a set length of time, they place the reaction mixtures in a fluorescence polarization instrument to detect whether cAMP has been converted to AMP. If cAMP is converted to AMP then the potential drug candidate had no effect. If the conversion of cAMP to AMP is inhibited by a particular compound, then that compound can be further explored as a potential drug to treat various diseases.

Before the BellBrook assay system is ready for sale, months of experimentation and optimization are required. This task has fallen to Dr. Rebecca Josvai, an R&D scientist at BellBrook Labs, and her colleagues. Dr. Josvai's tasks include:

- testing the various components of the reaction mixture at many concentrations
- testing different preparations of antibody to see which is most sensitive and specific
- testing for potential causes of erroneous results, for example, spontaneous conversion of cAMP to AMP in the absence of enzyme
- optimizing the conditions for the reaction, such as temperature and time
- optimizing settings of the fluorescence polarization instrument
- testing the assay using molecules known to be inhibitory (positive controls) and molecules known not to be inhibitory (negative controls) to be sure it always performs as expected
- scaling up the assay so that thousands of compounds can be rapidly tested robotically

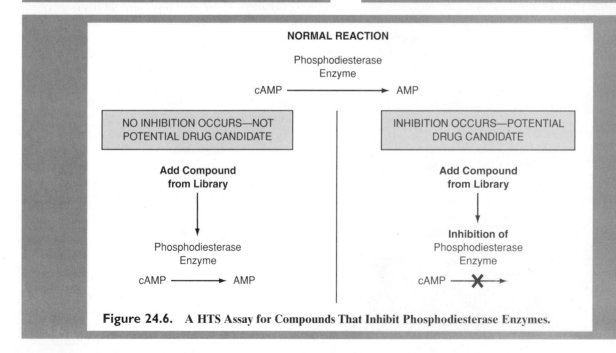

Figure 24.6. **A HTS Assay for Compounds That Inhibit Phosphodiesterase Enzymes.**

- *determining what controls should be used every time the assay is run in order to detect potential errors*

Dr. Josvai is shown in Figure 24.7 setting up a series of reactions to determine the effect of enzyme concentration. She uses a 384-well plate; each well is like a tiny test tube. To avoid errors, Dr. Josvai follows a spreadsheet that she previously prepared that tells her how much of each reaction component to place in each well.

After the reactions have occurred, Dr. Josavi places the 384-well plate in the fluorescence polarization instrument, which collects data from all 384 wells and sends it to a computer. Once the assay has been successfully optimized, it will be tested using a small robotic apparatus that automatically dispenses set volumes of liquids into the wells of plates. In a pharmaceutical company, robotic apparatuses are used to screen many thousands of compounds far faster than a human could do so (see Figure 20.17 on p. 366).

The information from each of Dr. Josvai's experiments is saved on a computer spreadsheet and documented in a laboratory notebook, which is signed and dated daily, and is countersigned by a witness at regular intervals.

The optimization of this assay illustrates some of the many efforts required for assay development. It is important to predict (and prevent) potential pitfalls of the assay. All aspects of assay development must be documented including the conditions of each assay test, the results of various experiments, the exact materials used, suppliers of reagents, and instrument parameters. Observe that each of Dr. Josvai's experiments is actually 384 experiments, each conducted in its own tiny well. With so much data coming from a single experiment, a computer is required to store, process, and display the information. For BellBrook Labs, all this experimentation and optimization is necessary to create a reliable assay that will be useful to pharmaceutical customers—and will make the company profitable.

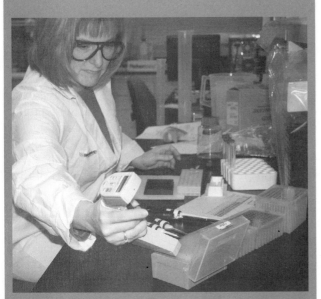

Figure 24.7. Dr. Josvai Preparing Reactions.

B. Method Validation

i. OVERVIEW

According to the International Conference on Harmonisation of Technical Requirements for Registration of Pharmaceuticals for Human Use (ICH), "the objective of validation of an analytical procedure is to demonstrate that it is suitable for its intended purpose." Method validation is a formal process used to verify the ability of method to obtain a desired result reliably. Validation is an essential part of the overall quality assurance program in a laboratory. As stated by John Butler, from the National Institute of Standards and Technology,

> *Validation* refers to the process of demonstrating that a laboratory procedure is robust, reliable, and reproducible in the hands of the personnel performing the test in that laboratory. A *robust method* is one in which successful results are obtained a high percentage of the time and few, if any, samples need to be repeated. A *reliable method* refers to one in which the obtained results are accurate and correctly reflect the sample being tested. A *reproducible method* means that the same or very similar results are obtained each time a sample is tested.

Validation involves documenting, through the use of specific laboratory investigations, that the performance characteristics of the method are suitable and reliable for the intended analytical applications. Method validation occurs after method development.

Formal validation of test methods is required in the pharmaceutical industry. Forensics laboratories also validate DNA testing methods. Although the process of method validation discussed in this chapter is based primarily on requirements in the pharmaceutical industry, the more general principles of validating a method are relevant in any laboratory. Validation helps ensure that an assay method provides trustworthy information.

A variety of parameters are tested during validation; the particular parameters depend on the method and the organization involved. For example, the International Conference on Harmonisation (ICH) and the United States Pharmacopeia (USP) both provide guidelines for method validation for the pharmaceutical industry. Their guidelines are similar, but not identical. Generally the tests performed during validation fall into one of the following categories that are described in more detail below: (1) **accuracy,** (2) **precision,** (3) **limit of detection,** (4) **limit of quantitation,** (5) **specificity,** (6) **linearity,** (7) **range,** (8) **robustness,** (9) **ruggedness,** and (10) **system suitability.**

ii. ACCURACY

It seems obvious that an assay or test method should give the "right" answer. **Accuracy** *is the closeness of a test result to the true or accepted value.* Accuracy is sometimes evaluated by testing a reference standard whose true or expected value for the test is known. Accuracy is also sometimes evaluated by comparing the results of the method being validated with another method that was previously validated and is considered to be correct. Accuracy can be expressed as the difference between the result obtained for the test and the known or accepted value expected.

In the pharmaceutical industry it is common to test accuracy by adding known amounts of the analyte to samples—called *spiking* the samples—followed by a determination of the percent recovery of the added material. If the accuracy of the assay is "perfect," the assay will detect 100% of the spiked material. The ICH recommends collecting data from a minimum of nine determinations over a minimum of three concentration levels covering the specified range (for example, three concentrations, three replicates each).

iii. PRECISION AND REPRODUCIBILITY

Precision *is the degree of agreement between individual test results when the procedure is applied over and over again to portions of the same sample.* There are different types of precision. Measurements repeated in succession on the same day tend to be relatively consistent. In contrast, measurements performed on different days, by different people, and using different materials and equipment, tend to be far more variable. **Repeatability** *is the precision of measurements made under uniform conditions.* **Reproducibility** *(sometimes called ruggedness) is the precision of measurements made under nonuniform conditions, such as in two different laboratories.* Repeatability and reproducibility are therefore two practical extremes of precision.

iv. LIMIT OF DETECTION

Limit of detection (LOD) *is the lowest concentration of the material of interest that can be detected* by the method. (The term **sensitivity** sometimes is used synonymously with limit of detection.) Every method has a limit below which it cannot detect the analyte of interest. Analytical methods, therefore, can never prove that a particular substance is not present in a sample; rather they show that it is not present at a concentration above a certain detection limit. For some situations, the limit of detection of a test is very important. For example, if one develops a test for purpose of detecting viral contamination in a drug product, one wants to be sure of the minimum number of viral particles that must be present to give a positive result. In this case the limit of detection of the test must be validated. In other situations, this parameter may not require validation. For example, an assay that is designed solely to confirm the identity of a drug substance assumes that the substance being analyzed constitutes the bulk of the material present. In this case the limit of detection may not require validation.

v. LIMIT OF QUANTITATION

Limit of quantitation (LOQ) *is the lowest concentration of the material of interest that the method can quantify with acceptable accuracy and precision.* The limit of quantitation is usually a higher concentration than the limit of detection. As for the limit of detection, the limit of quantitation may or may not require validation, depending on the nature of the test.

vi. SPECIFICITY

Specificity (some sources call this parameter **selectivity**) *is a measure of the extent to which a method can unequivocally determine the presence of a particular compound in a sample in the presence of other materials that may be expected to be present.* Other materials might include impurities, degradation products, matrix, and so on. A very specific test will only give a positive result to the compound of interest. For example, the polymerase chain reaction can be used to detect a particular sequence of DNA with high specificity, regardless of whether other DNA is present in the sample. A less-specific test might erroneously give a positive result when a substance similar to the compound of interest is present. This issue is discussed, for example, earlier in this chapter in the Case Study "Testing for the Illicit Use of EPO by Elite Athletes," pp. 444–445.

Testing specificity might involve adding the substance of interest to different relevant matrices (e.g., blood, soil, and buffer) and adding possible interfering agents (e.g., metabolites, detergents, and decomposition products) and seeing if the same results are obtained. Validating specificity is very important in a method such as DNA fingerprinting that must identify a single individual in a sample that may include many impurities.

vii. LINEARITY AND RANGE

Linearity *is the ability of a method to give test results that are directly proportional to the concentration of the material of interest (within a given concentration range).* **Range** *is the range of concentrations, from the lowest to the highest, that a method can measure with acceptable results.* Tests and assays have a particular, limited range in which they are linear. (This idea is discussed in more detail in Chapter 25.) When an assay or a test is developed, it is customary to determine the linear range. Validating linearity is important for many quantitative methods.

EXAMPLE

Based on the graph on the next page, this assay is approximately linear between 0 mg/mL of protein and about 25 mg/mL of protein. At higher protein concentrations, the assay does not give a linear response.

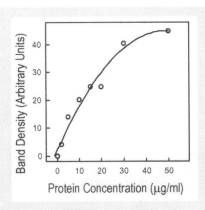

(Reprinted with permission from Funderburgh, James L., and Prakash, Sujatha. "SDS-Polyacrylamide Gel Electrophoretic Analysis of Proteins in the Presence of Guanidinium Hydrochloride." *BioTechniques* 20 (March 1996): 376–8.)

viii. ROBUSTNESS

Robustness *is the capacity of a method to remain unaffected by small, but deliberate variations in method parameters (based on the definition in ICH Q2A [1a]).* Robustness provides an indication of the method's reliability during normal use. For example, chromatographic analytical methods usually require a particular buffer pH. A chromatographic method that has good specificity in a range from pH 6.5 to 6.9 is more robust (relative to pH) than a method that requires a pH between 7.2 and 7.3. Robustness is important because there are often subtle and not so subtle differences between laboratories and individuals performing a particular method.

EXAMPLE

A chromatographic method of analyzing tricyclic antidepressants was evaluated for its robustness relative to pH. The amount of time it took a particular drug to run through the column was checked at pH 6.9, 7.0, and 7.1. The results were:

pH	Running Time
6.9	19.28
7.0	21.34
7.1	23.48

The percent difference between the running time for this drug at pH 6.9 and 7.0 is 9.65% and between pH 7.0 and 7.1 is 10.03%. This test would likely be considered to have poor robustness relative to pH. This is because it is difficult in the laboratory to absolutely ensure that a solution has a pH within 0.1 of a pH unit; it is not unusual for a solution to vary 0.1 pH units under normal laboratory conditions. In chromatography, 10% differences in running time could adversely affect the results of the test by causing sample components to shift positions and interfere with one another.

(This example is derived from "Contemporary Issues in Regulatory Compliance" by Michael E. Swartz, presented in a workshop by Waters Corporation, October 10, 1996.)

ix. RUGGEDNESS

The U.S. Pharmacopoeia defines robustness as given above. The USP further describes a related parameter, called "ruggedness," which is the degree of reproducibility of the results obtained by the analysis of the same samples under a variety of conditions (e.g., in different laboratories; performed by different analysts; using different instruments and reagents). Ruggedness provides a measure of the reproducibility of the test results given the variation in conditions normally expected from laboratory to laboratory and analyst to analyst. Ruggedness and robustness are similar parameters (and are sometimes used as synonyms) but measuring ruggedness according the USP requires that aliquots of the same batch of sample are tested in different laboratories.

x. SYSTEM SUITABILITY

A method can enter into routine use after it has been developed and validated. System suitability tests may be thought of as a way to periodically verify that the method is still effective for routine use. The FDA defines **system suitability** as: "Determination of instrument performance (e.g., sensitivity and chromatographic retention) by analysis of a reference standard prior to running the analytical batch." System suitability requires that all the components of a system be tested together. If a method involves chromatography, for example, then the instrument, analyst, reagents, and samples must all be tested together to be sure the system as a whole functions properly. As defined by the FDA, system suitability testing requires a characterized reference standard that contains the substances present in the actual samples along with expected impurities. The reference standard is run before the actual samples. Parameters, such as resolution between peaks and reproducibility, are determined for the reference standard and compared against specifications previously established for the method. Standards of known absorbance can similarly be used to periodically test a spectrophotometric method.

xi. VALIDATION PARAMETERS REQUIRED

It is not always necessary to evaluate every analytical performance parameter. The type of assay method and its intended use determine which parameters need to be investigated. For example, the USP classifies analytical methods into three categories:

- Category I, assays that quantitate the major components or active ingredients of a drug product.
- Category II, assays that detect impurities and degradation products.
- Category III, a variety of other assays that test performance.

For assays in category 1, it is not necessary to validate the limit of detection (LOD) or the limit of quantitation

(LOQ) of the assay because the major component or active ingredient to be measured should be present at high levels. The remaining analytical performance parameters are relevant. Assays in category 2 are divided into two subcategories: *Quantitative* and *limit tests*. *Quantitative tests* ask the question: How much of a particular substance or contaminant is present? *Limit tests* ask the question: Is the substance or contaminant present in an amount above a certain threshold? For quantitative tests the limit of detection need not be evaluated, but the remaining parameters are required. Limit tests, in contrast, require determination of the limit of detection but since quantitation is not required, it is sufficient to measure the LOD and demonstrate specificity and ruggedness. For category III assays, the requirements depend on the nature and purpose of the assay, see Table 24.6.

It is also common to distinguish between full validation and partial validation. **Full validation** is necessary when "inventing" a new assay and when developing methods to test a new drug entity. **Partial validation** is used for methods that have already been validated. Partial validation would be appropriate, for example, when analysts in a laboratory begin using an assay that was successfully used elsewhere, or begin using an assay that is from the USP. Partial validation may also be appropriate when a new instrument is purchased for an existing assay, or a new type of sample is to be tested with an existing method.

Forensic scientists distinguish between developmental and internal validation where **developmental validation** *involves new technologies.* Many of the tasks of developmental validation may be performed by a biotechnology company that sells kits, equipment, and reagents to perform an assay. **Internal validation**, *on the other hand, involves verifying that established procedures examined previously during developmental validation (often by another laboratory) will work effectively in one's own laboratory.* Developmental validation is typically performed by commercial kit manufacturers and large labs, such as the FBI Laboratory, while internal validation is the primary form of validation performed in smaller local and state forensic DNA laboratories.

CASE STUDY

Crime Labs and DNA Fingerprinting

Crime laboratories use DNA fingerprinting to identify or exclude suspects. An organization called the Technical Working Group on DNA Analysis Methods (TWGDAM) developed the original "Guidelines for a Quality Assurance Program for DNA Analysis" for use in forensic laboratories (Crime Laboratory Digest 22 no. 2 [1995]: 21–38.).* The guidelines outline the validation of DNA fingerprinting methods. Some of the studies required by these guidelines are:

a. The DNA fingerprinting procedure must be evaluated using fresh tissues and fluids. *DNA isolated from different tissues of the same individual must yield the same fingerprint. This evaluates the robustness of the method.*

b. "Using specimens obtained from donors of known type, evaluate the reproducibility of the technique both within the laboratory and among different laboratories." *This evaluates the reproducibility of the method.*

c. "Prepare dried stains using body fluids from donors of known types and analyze to ensure that the stain specimens exhibit accurate, interpretable, and reproducible DNA types or profiles that match those obtained on liquid specimens." *This evaluates the accuracy, reproducibility, and robustness of the technique.*

d. Mixed Specimen Studies—"Investigate the ability of the system to detect the components of mixed specimens ..." *Crime scene samples often include DNA from more than one individual, and sometimes from animals as well as humans. The method needs to be reliable in the presence of mixtures. This is a test of specificity.*

e. Matrix studies—"Examine prepared body fluids mixed with a variety of commonly encountered substances (e.g., dyes, soil) and deposited on commonly encountered substrates (e.g., leather, denim)." *Matrix studies are sometimes considered tests of the accuracy of the method. They also test its robustness and specificity.*

f. Minimum sample—"Where appropriate, establish quantity of DNA needed to obtain a reliable typing result." *This is an evaluation of the limit of detection of the method.*

*The guidelines for validation of DNA testing methods have been revised, but the ideas in this example are still relevant. The citation for the revised guidelines is found in the introduction to this unit.

Table 24.6. USP Data Elements Required for Assay Validation

Analytical Parameter	Category I	Category II Quantitative Test	Category II Limit Test	Category III
Accuracy	Yes	Yes	*	*
Precision	Yes	Yes	No	Yes
Specificity	Yes	Yes	Yes	*
LOD	No	No	Yes	*
LOQ	No	Yes	No	*
Linearity	Yes	Yes	No	*
Range	Yes	Yes	*	*
Ruggedness	Yes	Yes	Yes	Yes

*May or may not be required depending on the nature of the test.

CASE STUDY

Testing Home Pregnancy Test Kits

The hormone hCG appears around the 4th day after conception and so provides an early indication of pregnancy. Pregnancy test kits are based on an immunological assay for the hormone, hCG, in urine or serum.

Companies that manufacture pregnancy test kits must submit an application to the Food and Drug Administration (FDA) that contains evidence demonstrating that their kit is effective and reliable in detecting pregnancy. The FDA provides advice, in the form of a guidance document (Review Criteria for Assessment of Human Chorionic Gonadotropin (hCG) in Vitro Diagnostic Devices [IVDs]), as to how companies should acquire such evidence. This document outlines studies that should be performed, the types of data that should be collected and the statistical analyses to be performed on those data. The procedures outlined by the FDA emphasize proper sampling and data analysis and so provide a good example of many of the principles discussed in this chapter. A few sections from this FDA guidance document are quoted and discussed in this box.

1. Pregnancy test kits detect hCG in either urine or blood serum and so investigators must collect samples of serum and urine on which to test their kits. The FDA guidance document states:

Whole blood specimens should be collected into suitable tubes with/without anticoagulants, allowed to clot at room temperature and centrifuged. All specimens not tested within 48 hours of collection should be stored at −20ºC. Repeated freezing and thawing should be avoided. Serum specimens showing gross hemolysis, gross lipemia, or turbidity may give false results.

A first morning urine is recommended because hCG concentration is highest at this time. If specimens cannot be assayed immediately, they should be stored at 2–8ºC for up to 48 hours.

Note: If claims are being made to "use any time of day" then samples collected any time of day should be used in the study.

Proper sample preparation is essential since the preparation and handling of a sample affects the results of assays. If a company claims that their test does not require first morning urine, this claim must be backed by collecting samples at all times of the day.

2. An important aspect of demonstrating the effectiveness of a test kit is showing that the intended users are able to get proper results. Thus, if the test kit is to be used in physician's offices, then the FDA specifies that it should be tested in physician's offices. Similarly, if the kit is intended for use at home, then it should be tested in homes. The following are excerpts from the directions for evaluating a kit intended for home use:

At least 100 urine specimens should be used ... FDA recommends that the home users collect the samples and perform the tests ... For devices utilizing more than one testing procedure, i.e., urine stream or dip procedure, data must be provided to validate the equivalency of both procedures.

Home users should be selected on a random basis as they present themselves at clinics and/or physicians offices or via advertisements. They should represent diversity of age, background, and education.

The kit must be tested by a number of people (at least 100). It would not be acceptable for a company to pick test subjects who are well educated, have more background or training, or in some other way are not representative of the market as a whole. An improper selection of subjects might bias the results to make it appear that the kit works better (or worse) than it actually does. These studies validate the robustness of the kit under conditions that are as realistic as possible.

3. The company must test the specificity of the kit when low levels of hCG are present.

LH (luteinizing hormone) is a hormone whose structure is somewhat similar to hCG but that does not indicate pregnancy. If the kit responds to LH, then false positives will result. Assay optimization is necessary so that the kit is sensitive to low levels of hCG (such as at the very beginning of pregnancy), yet does not detect LH.

4. They must test the sensitivity of the kit. Spiking means that known amounts of hCG are added to specimens. This is done to test the kit's ability to give true positive results, even in the presence of very low levels of hCG.

Sensitivity should be evaluated by spiking at least 20 clinical samples from normal, nonpregnant females or males with five different concentrations of hCG below, at and above stated sensitivity. For example, use 0, 20, 25, 50, 100 mIU/mL for a kit with a detection limit of 25 mIU/mL .... Assay sensitivity should be such that small quantities of hCG will be detected while false-positive results due to the presence of LH will be minimized. Additionally, to support the detection limit 100 percent (%) of the samples tested must be positive and must have reacted within the specified time frame.

The guidance document specifies more studies that we will not discuss here. These other studies are intended to ensure that the kit works consistently, that it is properly labeled, that the instructions to the users are clear, and so on.

(From the Center for Devices and Radiological Health, FDA. January 14, 1998. http://www.fda.gov/cdrh/ode/odecl592.html)

VI. POLYMERASE CHAIN REACTION ASSAYS ILLUSTRATE THE CONCEPTS IN THIS CHAPTER

A. Overview

This section discusses PCR-based assays in order to illustrate the ideas introduced in this chapter. **PCR**, which stands for the **polymerase chain reaction**, *allows analysts to select a specific, target sequence of DNA that is present in low amounts and create millions of copies of that sequence.* The PCR method was introduced by Kary Mullis in 1983. In the ensuing years, PCR has been widely adapted to become one of the most important techniques in molecular biology. PCR can be used as a preparative tool to generate significant amounts of a specific fragment of DNA to use for another purpose, for example, for cloning or for transfecting cells. In this chapter, however, we are interested in PCR as it is used in assays, that is, to obtain information about a sample.

PCR is used in assays to see if a target DNA sequence, usually a particular gene, is present in a sample. Suppose, for example, a patient is suspected to be suffering from infection by a particular pathogenic bacterium. A PCR diagnostic test may be used to confirm the diagnosis. For a PCR assay to be useful for this purpose, there must be a particular gene or DNA sequence, called the *target*, which is found only in that pathogenic bacterium. A PCR diagnostic test can be then be used to determine whether this target sequence is present in a sample (e.g., blood or sputum) taken from the patient. DNA is extracted from the sample and is subjected to PCR. If the target gene is present it is amplified to create many copies; this amplified DNA can then be detected. If the target gene is present, the test is positive. (Without PCR, not enough of the target gene from the bacterium would be present to be detectable, even if the patient is infected by this organism.) PCR methods are usually much faster than other methods of pathogen diagnosis, such as growing up the infectious agent in culture and then identifying it using biochemical or immunological methods. PCR assays in the clinic can therefore help a patient receive proper treatment sooner than other diagnostic methods.

Another example is the use of PCR in food safety testing to look for pathogen contaminants in meat, produce, and dairy products. Suppose, for example, there is an outbreak of food poisoning thought to be associated with bagged spinach. Spinach samples from bags suspected to be involved can be rapidly tested for various possible pathogens using PCR. PCR assays can therefore allow many spinach samples to be rapidly screened before the food is distributed to consumers. As in medical diagnosis, PCR's speed and sensitivity are great advantages in this situation where waiting a few days for results could result in deaths.

PCR is a vital part of modern methods of DNA fingerprinting where it allows investigators to prepare DNA fingerprints from evidence containing only trace amounts of DNA. Residual DNA from a hair root or small drops of blood at a crime scene can yield enough DNA for a fingerprint when PCR is used for amplification. If the fingerprint matches a suspect, then it provides evidence that the suspect was at the crime scene.

B. How Does PCR Work?

i. AMPLIFICATION

PCR creates duplicate copies of a sequence of DNA by mimicking the cell's natural ability to replicate (copy) DNA. DNA **replication** *is an enzymatic process in which a cell that is getting ready to divide duplicates its*

DNA to make two identical copies, one of which will be passed along to each daughter cell. To understand replication (and thus PCR), recall that DNA is composed of strands consisting of a sequence of four nucleotide building blocks: *A, T, G,* and *C.* (Refer to Chapter 1 and particularly Figure 1.1 for more information about DNA structure). Double-stranded DNA consists of two strands of DNA. An *A* on one strand always pairs with a *T* on the opposite strand, and a *G* always pairs with a *C.* We say that *A* is "complementary" to *T,* and *G* is "complementary" to *C.* Because an *A* must always be across from a *T,* and a *G* across from a *C,* the sequence of one strand specifies the sequence of the opposite strand. During replication, the two original strands of DNA separate from one another. Each strand then serves as the **template,** or *guide,* for the manufacture of a new, complementary strand, see Figure 24.8. Replication requires an enzyme, DNA polymerase, which performs the work of linking nucleotide subunits one after another to form the new DNA strands.

In PCR, the analyst simulates DNA replication by combining the required ingredients in a test tube and providing suitable conditions for a polymerase enzyme

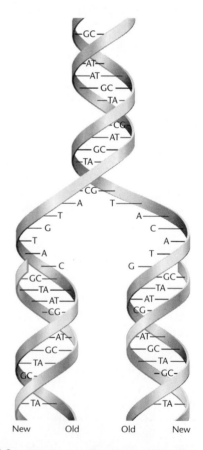

Figure 24.8. Replication Duplicates DNA When a Cell Divides. At the beginning of replication there is one double-stranded DNA molecule. The strands separate from one another and each strand serves as the template to make a new, complementary strand. After replication there are two identical double-stranded DNA molecules.

to synthesize new strands of DNA. The ingredients of the reaction mixture for PCR are:

- The four nucleotide subunits, *A*, *T*, *G*, and *C*.
- Template DNA. The template DNA may or may not contain the target sequence.
- Two DNA primers, to be discussed below.
- A DNA polymerase enzyme. In PCR, the enzyme must be heat stable for reasons that will become clear shortly. *Taq* DNA polymerase, an enzyme isolated from bacteria that live in hot springs, is commonly used for this purpose.
- Buffer and cofactor(s) to provide suitable conditions for the polymerase enzyme.

Although the PCR method simulates DNA replication, there are a couple of significant differences between PCR and replication as it occurs naturally in a cell. First, in PCR, only a short section of the DNA template is copied (typically up to around 10,000 base pairs). The section of DNA that is replicated is the target sequence. How is it that in PCR the DNA polymerase enzyme only duplicates this target rather than all of the DNA present? The answer lies in an important characteristic of DNA polymerases. These polymerases require a very short section of double-stranded DNA to initiate, or *prime* synthesis. A **primer** *is a short piece of DNA (or RNA) complementary to one end of a section of the DNA that is to be replicated.* Primer binds to single-stranded DNA creating a short double-stranded section. The DNA polymerase enzyme finds this double-stranded section and begins there to synthesize a new DNA strand. Observe in Figure 24.9 how the primers function in a PCR reaction. The target region of DNA that is to be amplified is a portion of the template DNA. There is a short primer that is complementary to a sequence at the beginning of the target region on one strand of the DNA. There is a second short primer that is complementary to a sequence at the end of the target region on the opposite strand of DNA. Observe how the primers thus recognize and bind to the ends of the target sequence so that they bracket the target portion of the DNA template that will be amplified. The polymerase enzyme always extends the new strand in only one direction, as shown by arrows in the figure. The analyst adds the primers to the PCR reaction mixture. The analyst thus controls the starting and ending points for DNA replication in PCR by the choice of specific primers. In order to design these primers, the analyst must know the DNA sequence of the target DNA, or at least must know the DNA sequence of the ends of the target region. PCR assays therefore rely on our knowledge of the DNA sequences of specific organisms and viruses—knowledge that is increasing at an exponential rate.

A second important difference between PCR and natural DNA replication is that, in nature, replication occurs once when a cell divides. In the laboratory during

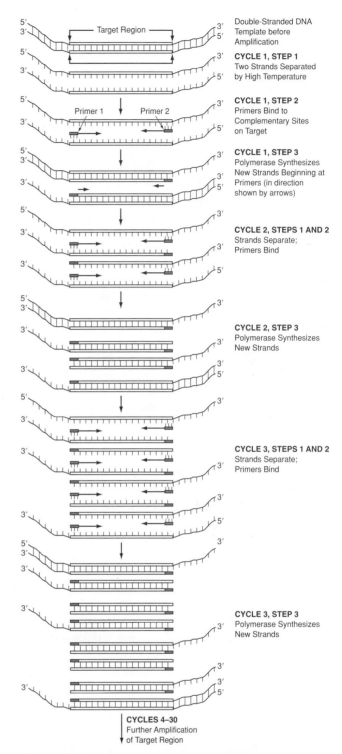

Figure 24.9.   The Polymerase Chain Reaction. In the first step of PCR, the two strands of the DNA template are separated by high temperature. Next, the temperature is lowered so that the primers complementary to the ends of the target sequence can bind to the template. The primers bracket the ends of the target region of DNA. In the final step of PCR, a DNA polymerase enzyme builds new double-stranded DNA molecules by adding nucleotides complementary to the target sequence, beginning with each primer and moving only in one direction, as shown by the arrows. This process is repeated for 20 to 40 cycles; assuming 100% efficiency, the number of copies of the target sequence doubles after every cycle.

PCR, the target DNA is duplicated over and over again to generate millions of copies. This is accomplished by providing temperature conditions that promote repeated cycles of replication. After combining the components of the reaction mixture together in a tube, the analyst puts the tube in a device, called a **thermocycler,** *which subjects the mixture to a series of repeated temperature cycles.* As you can see in Figure 24.9 on p. 453, at the beginning of each cycle, step 1, the template DNA is separated into its two strands. This is accomplished by subjecting it to elevated temperatures, usually between 92 and 94°C. Next, step 2, the temperature is brought down to a temperature in the range of 37–72°C. At this lower temperature, the primers associate with (anneal to) complementary regions of the target DNA (if the target is present). In the final step, step 3, usually at 72°C, *Taq* DNA polymerase copies each strand starting from the primer on that strand. To do this, it takes the needed nucleotide from the reaction mixture and attaches it to the growing strand. Because the *Taq* polymerase comes from a bacterium that lives in hot springs, the enzyme is heat stable at the temperatures used in PCR.

These three steps constitute one cycle of PCR, at the end of which there are two double-stranded DNA target molecules where originally there was only one. The process is then repeated, resulting in four target molecules, and again to give eight, and so on. It is this repetition that results in amplification of the target DNA sequence. Usually the process is continued for 25 to 40 cycles until millions of copies of the target sequence have been created. Note that if the target sequence is not present in the DNA template, then the primers do not bind the template and no amplification occurs.

ii. DETECTION OF THE PCR PRODUCT

The tube containing the reaction mixture is removed from the thermocycler after it has completed the proper number of cycles. If the amplification was successful, then there is a product, called the **amplicon,** *consisting of millions of copies of the target sequence.* The amplicon must then be detected. The conventional method of detection is to analyze a portion of the tube's contents using agarose gel electrophoresis followed by staining with the dye, ethidium bromide (see p. 321). Agarose gel electrophoresis separates DNA fragments based on differences in their sizes, with smaller fragments moving farther through an agarose gel support than larger fragments. Ethidium bromide allows DNA to be visualized and photographed because it fluoresces when excited with UV light. PCR-amplified DNA is thus detectable as a band on an agarose gel, as in Figure 24.10.

A molecular weight marker that contains DNA fragments of known sizes is run alongside the PCR samples during electrophoresis. The different fragments in the molecular weight marker move different distances through the gel. The size of the expected amplicon is known based on the length of the sequence bracketed

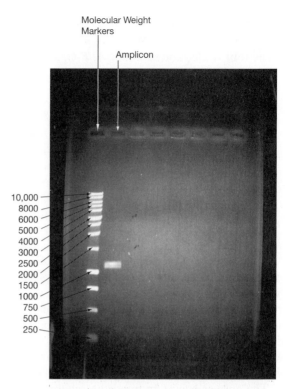

Figure 24.10. Detecting a PCR Product. The bright bands on the gel represent DNA fragments. Lane one contains a molecular weight marker consisting of a number of DNA fragments whose sizes are shown on the left. The higher the band is on the gel, the larger the fragment. The band observable in lane 2 is a PCR product, the amplicon. It is expected to be 1100 base pairs (bp) in size. The amplicon migrated a little less distance in the gel than the 1000 bp fragment in the molecular weight marker lane, which is consistent with the amplicon being 1100 bp in size.

by the primers, as was shown in Figure 24.9. If there is a band of DNA visible on the gel in the sample lane, then its mobility through the gel is compared to the mobility of the molecular weight markers. This allows the analyst to estimate the size of the DNA in the sample. If the sample produces a DNA band of the expected amplicon size, it is likely to be the DNA of interest, as is shown in Figure 24.10. The sample must produce an amplicon of the expected size to be considered a positive result. If a band appears on a gel, but it is not the size expected for the amplicon, then the result is negative. (Proper controls, discussed below, must also be included in the gel.)

It is also possible to use nonelectrophoresis methods to detect the amplicon. For simplicity, however, our discussion assumes that an electrophoretic detection system is used.

C. Obtaining Trustworthy Results from a PCR Assay

i. PCR HAS POTENTIAL PITFALLS

PCR can seem deceptively simple to perform. An analyst isolates DNA from a sample following a standard method, purchases quality-assured reagents from commercial suppliers, pipettes each reagent into a tube, and places the tube in the thermocycler. The analyst turns on the thermocycler and returns a few hours later to remove a tube laden with amplified DNA. In reality, like all assays, PCR is not so simple, and there are many pitfalls that can lead to deceptive results, even in a routine, optimized PCR-based assay. It is therefore important to understand this method in order to get trustworthy results.

A conventional PCR assay is used to answer a question, such as, was a particular suspect at a crime scene? Is a patient infected with a particular pathogen? Is a particular microorganism present in a batch of cheese? These are yes/no questions. If a band is visible on a gel in a PCR assay, one might consider this to be a positive result. If no band is visible, this might be called a negative result. But, how trustworthy is the result? Is it possible that a visible band is a false positive? Is it possible that no band appears, but this is a false-negative result? In fact, both false-positive and false-negative results can occur in PCR assays. In order to assure that a result is trustworthy it is necessary to have controls. In this context, the word *control* takes on two slightly different meanings. First, there are positive and negative controls that are used with PCR, in the sense that was discussed in section III B of this chapter. Second, the assay must be "controlled" in the sense that the analyst uses proper techniques to avoid introducing contaminants and errors into the system. We will consider both types of *control* in more detail.

ii. SAMPLE PREPARATION

As with all assays, PCR assays require that the sample is chosen correctly and prepared properly. Sample selection and preparation vary greatly depending on the purpose of the assay. At a crime scene, for example, the selection of sample may involve searching the scene for evidence from which DNA can be isolated. In food-safety testing, a subset of food items must be chosen in some reasonable way, since obviously all the food from a given farm, store, or batch cannot be tested.

Once the sample is chosen, template DNA must be prepared from it. The preparation method is likely to involve separating the DNA from the matrix in which it is found and removing contaminants that might interfere with the analysis. There are many methods of isolating template DNA for PCR. In general, sample preparation begins by lysing the cells containing the template DNA. Sometimes this involves mechanical grinding or shearing. Detergents are often added to open the cell and nuclear membranes. Enzymes may be used to break open cells and also to destroy proteins from the sample. Proteinase K, which breaks the bonds holding amino acids together in proteins, is commonly used for this purpose. Sometimes the template DNA is extracted with organic solvents followed by precipitation with ethanol. Alternatively, molecular biology companies sell proprietary products that conveniently extract and cleanup DNA for PCR.

Some DNA preparation methods extensively purify the template away from all contaminants in the sample, but the trend is toward rapid methods that often provide minimal DNA cleanup. This is because the PCR reaction is usually sensitive to the target DNA, even if it is present in the midst of other cellular constituents and nontarget DNA sequences. Note, however, that there can be inhibitory substances present that may bind required cofactors or otherwise reduce the activity of the polymerase enzyme. Inhibitors can come from many sources including the matrix from which the DNA was isolated. Foods, for example, have many potential inhibitory agents including proteases, nucleases, collagen, and fatty acids. If inhibitors are found to be a problem in a particular PCR method, then it is important to add steps to purify the template away from the inhibitory substances. It may require a period of experimentation to find a method that effectively prepares template DNA from a specific source.

iii. AVOIDING CONTAMINATION

PCR is extremely sensitive—this is one of its major advantages. Even minuscule amounts of DNA can be amplified so that they are detectable on an electrophoresis gel after PCR. This sensitivity, however, also makes PCR extremely vulnerable to contamination. Unsuspecting investigators have thought that they were investigating DNA from experimental samples, when they were, in fact, investigating their own DNA that fell into the test tube in minute amounts, or other contaminating DNA from a pipette or other object. PCR requires carefully avoiding contamination to avoid false positives. Sometimes contaminating DNA does not cause problems because it does not contain the sequence recognized by the primers and so is not amplified. Other contaminating DNA can totally invalidate an assay. For example, DNA typing in forensic laboratories involves primers that recognize human DNA. Therefore, if a few skin cells from the analyst fall into the reaction mixture, then the analyst will be fingerprinting him- or herself—not a person present at the crime scene.

The amplicon from previous assays is an important potential source of DNA contamination; this is called cross-contamination. If even the smallest amount of previously amplified DNA finds its way into a new reaction mixture, it will serve as template in the subsequent reaction, thus leading to a false positive result. The tubes containing amplified product must therefore be treated carefully. Simply opening these tubes is problematic because it creates aerosols that may contain millions of copies of the target DNA. These copies may float in the air, fall on the counter, and be picked up by an analyst's sleeve, contaminate an analyst's glove, or fall on a piece of equipment. It is essential to use proper practices to avoid cross-contamination between samples.

Ideally there should be at least four distinct, dedicated rooms or areas in a PCR laboratory, see Figure 24.11 on p. 456. One is a space for preparing the sample, including extracting the DNA. This space may also be used for preparing buffers and other reagents. A second space should be used for combining the ingredients of the reaction mixture. A third space is set aside for the thermocycler. A fourth space is used to open the tubes after amplification and for performing detection by electrophoresis or another method. In an ideal situation, people and materials would always move in one direction, from the sample preparation area to the detection area. Before moving to the next area, personnel should change lab coats and gloves. Personnel would ideally begin in the morning in the preparation area and end the day in the detection area. The purpose of this one-way traffic is to prevent the product of a previous PCR assay from entering the area where new reactions are set up.

Dedicated PCR workstations provide a relatively economical means to isolate a small area for a specific PCR-related task. PCR workstations are enclosed cabinets that protect a work area from drafts. They usually contain germicidal UV lights that can be turned on to denature DNA contaminants, see Figure 24.12.

Equipment can be a source of contaminants. It is essential to have dedicated micropipettors that are only used for preparing PCR mixtures. These pipettors should be used with tips that have aerosol barriers (as shown in Figure 20.14 on p. 358). The surfaces of lab benches and equipment, such as micropipettors and PCR tube racks, should be routinely cleaned with a bleach solution to destroy DNA contaminants.

iv. POSITIVE AND NEGATIVE CONTROLS

There are various controls that are used to help recognize false positive and negative PCR results. One way to check for contamination is to routinely run negative controls alongside the test samples. The simplest PCR negative control is a tube that has all the components of the reaction mixture except one (usually template DNA). This tube should not have an amplicon after the PCR reaction, see Figure 24.13. If, however, contaminants enter the PCR reaction (e.g., from the analyst's skin, aerosols in the air, contaminated pipettes, contaminated reagents), then a band(s) may appear in this control lane, alerting the analyst to the presence of contamination.

DNA from a different, related organism can also be used as a negative control. In food testing, for example, a negative control might be template DNA from a non-pathogenic organism that is related to the pathogen of interest. If the primers are properly designed, the DNA from the related nonpathogenic organism will not be amplified. This negative control therefore tests for the specificity of the primers. It also tests for nonoptimized replication conditions (e.g., temperatures that are too low) that allow the primers to bind nonspecifically to the "wrong" DNA sequence.

A false negative result can occur in PCR for a variety of reasons. Inhibitors, for example, can cause false negative results. False negative results can also occur

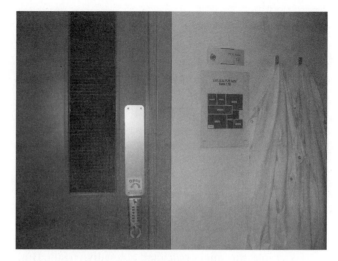

Figure 24.11. Use of Facility Design to Help Avoid Contamination in PCR. It is not always possible to have a separate PCR facility, but a facility such as this can reduce contamination. This is a picture of the entrance to a separate PCR suite located inside of a contract research laboratory that performs preclinical drug testing. (See Chapter 3 for more information on this type of testing.) There is a lock on the door to restrict access to only individuals trained in PCR. The locking system will not allow a person to enter the suite more than once a day to prevent an analyst from introducing DNA contamination from previous assays on their hands or body. People change their lab coat before entering the facility and then move in one direction through the PCR suite to avoid contaminating samples with already-amplified DNA. Once inside the suite, there are separate rooms for setting up the PCR reaction, running the reaction, and performing electrophoresis. Doors only open in one direction so a person cannot take materials backward through the facility. All personnel change their lab coats every time they move to a different room.

Figure 24.12. An Analyst Preparing a PCR Reaction Mixture inside a PCR Workstation.

Table 24.7. COMMON CONTROLS AND STANDARDS IN PCR

Control or Standard	Type	Purpose
No template in reaction mixture	Negative control	Detect contaminating DNA if present; avoid false positives
Template DNA from related organism	Negative control	Detect nonspecific primer binding; avoid false positives
Purified template known to contain the target sequence	External positive control	Detect inhibition of the reaction, failure to add all reaction components, problems with reaction mixture components, problems with cycling conditions; avoid false negatives
Purified template containing a shortened target sequence	Internal positive control	Detect inhibition of the reaction, failure to add all reaction components, problems with reaction mixture components, problems with cycling conditions; avoid false negatives
Molecular weight markers	Standards	Analyze size of amplicon

because one or more components of the reaction mixture were improperly prepared or because the reaction conditions were suboptimal. It is therefore common to use an external positive control. An external positive control is a sample that is known to contain the target DNA. The positive control is amplified in a separate tube alongside the test samples. The positive control DNA should produce an amplicon. If it does not,

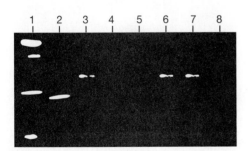

Figure 24.13. Positive and Negative Controls in PCR.
The gel above is from a clinical test where PCR is used diagnostically to detect a genetic disease. **Lane 1 contains a molecular weight marker** where each of the bright bands is a DNA fragment of a different, known length. **Lanes 2 and 3 are positive controls.** Lane 2 is amplified DNA from a patient known to have the disease of interest. Lane 3 is amplified DNA from a healthy person. Observe that the DNA band in Lane 2 has moved farther than the one in Lane 3. Clinicians know where the bands in healthy people and affected people should occur relative to the reference markers in Lane 1. If the expected results for healthy and affected people had not appeared, then the analysts would know there was a problem in the procedure. **Lanes 4, 5, and 8 are negative controls.** Each was lacking some necessary component of the reaction mixture and therefore should have no bright bands at all after PCR. The presence of any bands in any of these negative control lanes would indicate that contamination had occurred. As expected, the negative control lanes are dark. **Lanes 6 and 7 are patient samples.** Both patients are considered to be healthy since their DNA band is at the same level in the gel as that of the healthy control.

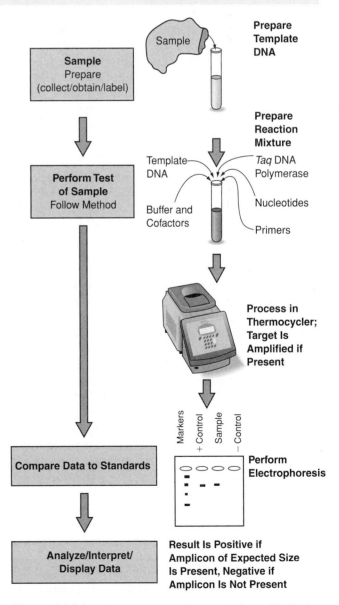

Figure 24.14. Overview of a PCR-Based Assay Method.

inhibitors might be present, one or more reagents might be improperly prepared, the cycling conditions might be suboptimal, or another problem may have occurred. Figure 24.13 on p. 457 shows the use of positive controls.

Some analysts use internal amplification controls as an alternative to an external positive control. An internal amplification control is template DNA that is like the target of interest, but a section of its DNA has been deleted. This type of control is added into the reaction mixture in the same tube as the test sample. In a properly functioning assay, the internal control target DNA will bind to the primers, just as the target of interest does. The internal control DNA will also be amplified, just as the target is amplified. But, the internal control template will produce a smaller-sized amplicon because it is missing part of the DNA sequence of interest. The smaller-sized amplicon should always appear, even if the sample is negative for the target DNA. Its presence proves that the reagents worked properly and that inhibitors did not prevent amplification. If the assay is positive for the target DNA, then two amplicons will be present that are easily distinguished. If neither amplicon is present, then the assay result cannot be trusted.

As a further control, one must always run a molecular weight marker if identification of the amplicon is based on its electrophoretic mobility. If a band is present, but is the wrong size, then the assay is negative and the band is attributed to nonspecific primer binding. The consistent use of a molecular weight marker when using detection by electrophoresis is one of the guards against false positives. The molecular weight markers may also be considered to be standards, in the sense discussed in section II D of this chapter.

Table 24.7 on p. 457 summarizes these controls and standards.

V. DESIGN AND OPTIMIZATION OF A PCR ASSAY

The method development phase for a PCR assay involves a number of activities. First, it is necessary to find a sequence or partial sequence for the target DNA in order to design the primers. The specificity of the assay depends on the specificity of the primers; if they bind to DNA other than the sequence of interest, then the assay will not be specific. It is also necessary to optimize the sample selection and preparation of template DNA. This step will vary greatly depending on the nature of the sample and the purpose of the assay. It is also necessary to optimize the reaction components and conditions. Magnesium, for example, is a required cofactor for the polymerase enzyme. If too little magnesium is present, then less amplicon may form and the assay will lack sensitivity. If, however, too much magnesium is present, it may promote nonspecific binding, thus reducing the specificity of the assay. Similarly, the cycling conditions must be optimized. If the temperature of the primer extension phase is too low, nonspecific binding may occur, again reducing assay specificity. If

the temperature of the primer extension phase is too high, the sensitivity of the assay may be impaired.

The controls for the assay must be selected. A no-template negative control is easy to prepare; positive controls are sometimes more difficult to obtain. An internal amplification control may be created. A detection method must be devised. If agarose gel electrophoresis is chosen, then the electrophoresis conditions and the proper molecular weight markers must be selected. Like all assays, sophisticated PCR-based assays will not produce trustworthy results unless the analyst understands their pitfalls and carefully provides the proper controls.

Figure 24.14 on p. 457 summarizes the steps in a PCR-based assay. The case study below provides a further summary and illustration of PCR by discussing validation of a method to detect a pathogenic bacterium in cheese.

CASE STUDY

A PCR Assay for Listeria Monocytogenes, a Cause of Potentially Deadly Food Poisoning

Listeria is a group of bacteria that are widely found in the environment on decaying vegetation, soils, sewage, silage, and water. Most Listeria are not pathogenic to humans, but one type, Listeria monocytogenes can cause a potentially fatal disease called listeriosis. Listeriosis may cause sepsis (infection that spreads through the blood) and meningitis (infection of the lining of the brain). Maternal infection during pregnancy sometimes results in abortion of the fetus or severe septic illness of the newborn. Humans are occasionally sickened by this bacterium after ingesting contaminated food, particularly dairy products.

Conventional tests to screen foods for Listeria sp. (and for other bacterial pathogens as well) involve culturing samples in specific culture media over a period of time (often days) to allow the bacterium to grow and reproduce. Once many bacteria are present, it is possible to use biochemical or immunological methods to identify their type. Conventional culture methods are time-consuming and contaminated foods can be widely disseminated before a contamination problem is detected. In the case of Listeria, the traditional culture method is not well suited to distinguish pathogenic from nonpathogenic strains.

Because of the limitations of conventional screening methods, researchers have been developing rapid genetic methods to screen food for pathogens, including Listeria monocytogenes. In this case study, investigators reported on a new PCR screening method. Although the authors did not use the term validation, they did examine two validation parameters that are of critical importance in PCR assays: specificity and sensitivity. They also looked at factors relating to the robustness of the assay. The authors performed each test three times, which provides an indication of the reproducibility of the method; they did not, however, report their results for this parameter. It is standard prac-

tice in validation to compare a new method, in this case, the PCR method, to the traditional "gold standard" method. The authors therefore compared the results of their PCR assay to results obtained using a conventional culture method. They note briefly that the conventional culture method confirms their PCR results.

Specificity

An assay that is not specific produces false positive results. In PCR assays, the primers, which bracket the area that is amplified, are the critical component that determines the method's specificity. By 2006, the gene sequences of more than 90 bacterial pathogens had been determined, and more are being sequenced all the time. This means that it is possible to use a computer and the tools of bioinformatics to look for gene sequences that are unique to a particular pathogen. These unique genes are often those that cause a pathogen to harm a cell. This is the case for L. monocytogenes. According to the authors, this pathogen causes invasive disease by crossing the cell barrier that lines the intestine. If the bacterium could not cross the intestinal cell lining, then it could not enter the body to cause infection of the blood or brain lining. It has been discovered that L. monocytogenes is able to move within the cytosol of infected cells by interacting with the cells' cytoskeleton. A gene present in L. monocytogenes, called actA, has been implicated in the interaction of the bacterium with the host cell cytoskeleton. actA is characteristic of L. monocytogenes. It turns out that this gene is not found in other closely related nonpathogenic bacteria. The authors therefore selected primers that recognize the actA gene in order to make their assay specific to L. monocytogenes.

The authors demonstrated the specificity of their assay by comparing the results for 17 different strains of L. monocytogenes to 5 strains of related, but nonpathogenic Listeria. The results are shown in Figure 24.15. In Figure 24.15a you can see a strong band for all strains of L. monocytogenes. Observe that sometimes the bands are 300 base pairs (bp) in size and sometimes they are 400 bp in size. This is because there are two forms (alleles) of the actA gene. Some pathogenic L. monocytogenes strains have one form of the gene and a 300 bp segment is amplified by the primers; others have the other form and a 400 bp segment is amplified. Figure 24.15b shows that none of the nonpathogenic strains of Listeria that were tested had a band at 300 or 400 bp. This is a good indication of specificity. You can see a fainter band at the bottom of all the lanes in Figure 24.15b. These bands are not the right size to be the target gene. They are most likely "primer-dimers," a common artifact that occurs when primers act as their own template to make a small PCR product. Note also that Figure 24.15b includes a positive control in Lane 8, which helps assure us that the negative results in the other lanes are trustworthy.

Sensitivity in this assay is the lowest number of bacteria that can be detected. The more sensitive the assay, the fewer false negative results it will generate. It is important that tests for food pathogens be very sensitive because ingesting food contaminated by only a few microorganisms can lead to ill-

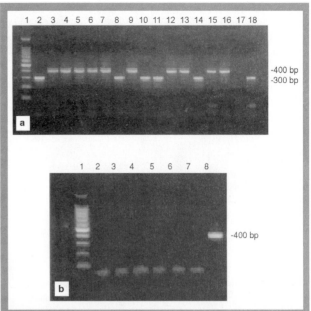

Figure 24.15. *Electrophoresis Gels Demonstrating the Specificity of the* **L.** monocytogenes *PCR Assay. a. The first lane contains a molecular weight marker. Subsequent lanes are assay results from foods contaminated by different pathogenic strains of* L. monocytogenes. *All gave a positive result, as expected, and produced a band that corresponds in size to the actA gene target.* **b.** *The first lane contains a molecular weight marker and lane 8 is a positive control. All the other lanes are assay results from foods contaminated by various nonpathogenic strains of* Listeria. *All gave a negative result, as expected, and did not produce a band that corresponds in size to the actA gene target. Image from: C. Longhi, A. Maffeo, M. Penta, G. Petrone, L. Seganti, M.P. Conte. Detection of Listeria Monocytogenes in Italian-Style Soft Cheeses.* Journal of Applied Microbiology *94, no. 5 (2003), Fig. 1.*

ness. In order to test the sensitivity of their assay, the researchers obtained soft cheeses from the grocery store. Twenty-five grams of each sample was homogenized and then centrifuged for 10 minutes at a low speed to remove large particulate matter. The cheese samples were then artificially contaminated with six different levels of bacteria: 1, 10^1, 10^2, 10^3, 10^4, and 10^5 CFU per mL of cheese. The cheeses were then subjected to the PCR assay method to see if the contaminants could be detected. The authors report that their PCR assay could detect as few as 1 to 10 CFU/mL in mozzarella, 10^2 CFU/mL in crescenza, and 10^3 CFU/mL in ricotta.

Robustness

The authors compared the effectiveness of their assay on three different types of soft cheese: mozzarella, crescenza, and ricotta. They found that the detection limit was as much as 1000 times higher in ricotta than in mozzarella, meaning that the assay worked better for mozzarella. They suggested that inhibitory factors present in the ricotta might have interfered with the assay. If this is the case, then further work in purifying the template DNA might eliminate this problem, although the authors did not discuss how this might be done.

The authors looked at the effect of refrigerator storage on the sensitivity of the assay. They inoculated cheeses with known amounts of L. monocytogenes and then incubated the samples for 7 days in the refrigerator. They found that after storage they were much less likely to detect bacteria. The authors attribute this result either to a loss of viability of the bacteria or to an increase in inhibitory factors in the cheese. If the bacteria indeed become nonviable with storage, then it appears that refrigeration causes the cheese to become safer to eat. In contrast, if inhibitors cause the assay to be negative, then these are false negatives and do not indicate that the cheese is safer. This issue seems worth exploring in more detail, although the authors did not indicate any planned follow-up. (The authors may have been less concerned with this result because they reside in Italy where they note that most soft cheeses are consumed within 48 hours of manufacture. In the United States, however, it is common to consume cheese that is not nearly this fresh.)

This study illustrates the value of a rapid PCR assay for detection of Listeria monocytogenes and the methods by which such an assay can be validated. It also indicates that an analyst using this assay must be knowledgeable to obtain trustworthy results. Refrigeration, for example, affects the assay and the sensitivity of the assay was not the same for all types of cheese. Analysts must consider these factors (and probably others as well) when adopting this assay for use in their own laboratories.

This discussion is based on an article, Longhi, C., Maffeo, A., Penta, M., Petrone, G., Seganti, L., and Conte, M.P. "Detection of Listeria Monocytogenes in Italian-Style Soft Cheeses." Journal of Applied Microbiology, 94 (2003): 879–85.

PRACTICE PROBLEMS

1. A company produces a product that is packaged into vials that are then loaded onto trays. A technician tests the first vial in each tray as part of a quality-control program. Is this a good method of obtaining a random sample?

2. A research laboratory is developing a chromatographic method to check blood samples for the presence of a particular hormone. The scientists take blood and add to it some of the hormone of interest and a second hormone that is structurally similar to see if the method can distinguish between the two. Which of the following characteristics of the assay are they checking? (1) accuracy, (2) precision, (3) limit of detection, (4) limit of quantitation, (5) specificity (selectivity), (6) linearity, (7) range, (8) robustness.

3. A research team is developing a method of DNA fingerprinting. The method involves cleaving DNA samples with restriction enzymes that recognize and cut at specific base sequences. The resulting fragments of DNA are separated from one another using electrophoresis and are made visible with dye. DNA from different sources forms different patterns when treated with this method.

   a. DNA restriction enzymes are sensitive to the level of salt in their buffer. The researchers therefore test the method using three buffers with different salt concentrations. Which of the characteristics of an assay listed in problem 2 are they checking?

   b. If too little DNA is used in this method, then the pattern is not visible. If too much DNA is used, then the pattern is diffuse and is not useful. The researchers therefore evaluated the method to discover the minimum and maximum amount of DNA that can be used. Which of the characteristics of an assay listed in problem 2 are they checking?

4. The term *sensitivity* with reference to a biological assay is related to how low a level of target the assay can detect. The more sensitive the assay, the lower the level that can be detected. Which of the following statements is true?

   a. There is a risk associated with optimizing an assay to increase its sensitivity. This risk is that the number of false negative results will increase.

   b. There is a risk associated with optimizing an assay to increase its sensitivity. This risk is that the number of false positive results will increase.

5. Multiple choice: In an immunological assay, nonspecific binding tends to cause which of the following:

   a. False negative results
   b. False positive results
   c. The positive control comes up negative
   d. The negative control comes up positive
   e. b and d
   f. a and c

6. The directions for a home pregnancy test kit include the following instructions: "A pinkish-purplish line will form in the control (upper) window to tell you the test is working correctly. A positive result is two lines, one in each of the two windows (the upper and lower). A negative pregnancy test is a single line in the upper (control) window and no line in the lower window." Explain the upper window; why does a line form there? What is the significance of this line?

Practice problems 7-11 relate to a situation where the interpretation of the results of an assay can be difficult. In this case the data relate to an imaginary assay for an imaginary disease, called philolaxis disease.

The serum level of the imaginary protein, protein PHX, is elevated in patients suffering from philolaxis disease. The level of this protein varies from individual to individual and is approximately normally distributed. (You may want to review the section on the normal distribution in Chapter 16.) Figure **a** shows the distribution of the level of this protein in healthy patients, Figure **b** shows the distribution in ill patients.

The distributions for healthy and ill patients are super-imposed in Figure **c.** The level of protein PHX is used diagnostically to test whether a person is sick with philolaxis disease. You can see that there is variability in the serum levels of PHX for both affected and unaffected individuals and there is overlap in the two distributions. For this reason, there can be difficulty in classifying some patients as having the disease or not, based on this assay.

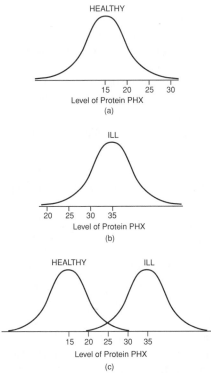

7. **a.** Does a person with a serum level of protein PHX of 10 have philolaxis disease?
   b. Does a person with a level of 20 have philolaxis disease?
   c. Does a person with a level of 25 have philolaxis disease?
   d. Does a person with a level of 40 have philolaxis disease?

8. When clinicians test for philolaxis disease based on serum levels of protein PHX, they select a cutoff score. If a patient's level of protein PHX is above the cutoff score the result is considered to be positive for philolaxis disease. If the patient's level is below the cutoff score the result is considered to be negative for the disease.
   a. Multiple choice: If the clinicians decide that the cutoff score should be 20, about what percent of the time can they expect to get a false positive result?
   (i) 0%   **(ii)** 15%   **(iii)** 50%   **(iv)** 100%
   b. If the clinicians decide that the cutoff score should be 20, about what percent of the time can they expect to get a false negative result?
   (i) 0%   **(ii)** 15%   **(iii)** 50%   **(iv)** 100%

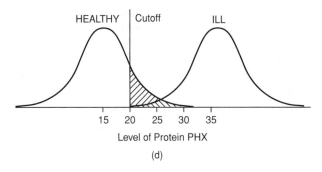

9. Multiple choice:
   a. If the clinicians decide that the cutoff score should be 25, about what percent of the time can they expect a false positive result?
   (i) 0%   **(ii)** 2.5%   **(iii)** 50%   **(iv)** 100%
   b. If the clinicians decide that the cutoff score should be 25, about what percent of the time can they expect a false negative result?
   (i) 0%   **(ii)** 2.5%   **(iii)** 50%   **(iv)** 100%

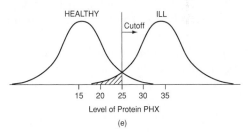

10. Multiple choice:
   a. If the clinicians decide that the cutoff score should be 30, about what percent of the time can they expect to get a false positive result?
   (i) 0%   **(ii)** 15%   **(iii)** 50%   **(iv)** 100%
   b. If the clinicians decide that the cutoff score should be 30, about what percent of the time can they expect to get a false negative result?
   (i) 0%   **(ii)** 15%   **(iii)** 50%   **(iv)** 100%

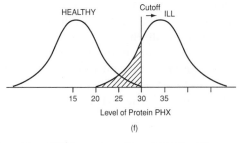

11. Consider HIV testing (for AIDS). If a person is told they do not have the virus, when in fact they do, they might unknowingly infect other people and also would not receive treatment. It is therefore common practice to use two tests for AIDS. The first test has a high rate of false positives and a low rate of false negatives. This test is relatively inexpensive and is used to screen for all samples that

might be positive. All blood that tests positive with the first method is retested with a second assay that is more expensive but is better able to distinguish a true positive result.

In the case of philolaxis disease, if clinicians want the lowest possible level of false negatives, what cutoff score should be chosen?

Note that often in the laboratory there is more than one method available to test for a certain material or phenomenon. Each test is likely to have its own strengths and weaknesses, but together, the tests provide more reliable information than any test by itself. For this reason, when possible, it is good practice to use more than one test to confirm critical results.

12. Benedict's test is used to determine if a sample contains reducing sugars. (Reducing sugars are those that contain –CHO or –C=O groups. Glucose is an example of a reducing sugar; sucrose is a nonreducing sugar.) Benedict's test is performed by adding Benedict's reagent to the sample and heating the mixture. The Benedict's reagent turns from blue to green in the presence of small amounts of reducing sugars and turns reddish-orange in the presence of an abundance of reducing sugars. Suppose this test is to be used to test apple, orange, and pineapple juice samples for the presence of reducing sugars. List the likely samples for this experiment, including experimental and control samples.

13. Starch is a coiled chain of glucose molecules. Iodine can be used to test for the presence of starch in a sample. Iodine turns from its normal color of yellowish-brown to a bluish-black color when starch is present. Suppose iodine is to be used to test samples extracted from onion, potato, and green peppers for the presence of starch. List the likely samples for this experiment, including experimental and control samples.

14. You may want to review Figure 1.3 on p. 7 to understand the process described in this question. Researchers isolate the human gene that codes for a protein that is important in triggering cell division. The researchers want to insert the gene into bacteria and then use the bacteria to produce large quantities of the protein. These are the tasks they perform:

- Isolate the DNA of interest.
- Insert this DNA into a plasmid vector. The vector also contains a gene for resistance to the antibiotic ampicillin.
- Cause bacteria to take up the plasmid vector. (The process in which the bacteria take up the plasmid is called *transformation*.)
- Prepare the following plates on which to culture bacteria:

Plate A: Contains nutrient medium and ampicillin
Plate B: Contains nutrient medium and no ampicillin
Plate C: Contains nutrient medium and ampicillin
Plate D: Contains nutrient medium and no ampicillin

- Add bacteria that are presumed to have taken up the plasmid vector to Plate A and Plate B.
- Add bacteria that were not exposed to the plasmid vector to Plate C and Plate D.
- Incubate the plates overnight to allow the growth of bacteria.
- Examine the plates, looking for visible colonies of bacteria.

Answer the following questions about this procedure:

a. If the procedure works as the experimenters plan, then on which plate(s) do they expect colonies to grow?

b. If the procedure works as the experimenters plan, then on which plate(s) do they expect colonies will not grow?

c. Which of the plates are controls? Why are they used?

CHAPTER 25

Measurements Involving Light B: Applications and Methods

I.   INTRODUCTION

Chapter 23 described the nature of light and the design, operation, and performance verification of spectrophotometers that measure light. This chapter explains how measurements involving light are used in assays to obtain information about biological samples.

Spectrophotometers can be used to help answer essential questions in the laboratory, including: (1) What is the identity or nature of the component(s) of a sample? (2) How much of an analyte is present in a sample? *Assays that identify the components of a sample are* **qualitative,** *whereas those that determine how much analyte is present are* **quantitative,** see Figure 25.1.

There are many spectrophotometric assay methods. For example, there are spectrophotometric methods to confirm the identity of drug products (a qualitative application), to measure levels of water pollutants (quantitative), to determine the levels of proteins in a sample (quantitative), and to study the activity of enzymes (quantitative). These spectrophotometric analytical methods must be developed, optimized, and validated. Note that ensuring that an assay method works effectively is a separate task from verifying that the instrument involved is properly functioning.

This chapter begins by explaining the general principles of qualitative and quantitative analysis by spectrophotometry. The chapter then looks at the development, optimization, and validation of spectrophotometric methods. In addition, there are three sections that briefly introduce instrumental methods related to spectrophotometry.

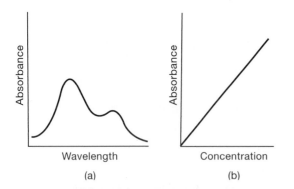

Figure 25.1.   Applications of Spectrophotometry.
a. Qualitative applications use absorbance spectra to give information about the component(s) of a sample. (Recall that an absorbance spectrum is the absorbance measured for a sample [Y axis] over a range of different wavelengths [X axis]). **b.** Quantitative applications determine the concentration of a substance in a sample. Concentration (or amount) is on the X axis, and, as for qualitative applications, absorbance is on the Y axis. (See pp. 257–259 for an introduction to standard curves.)

II.   QUALITATIVE APPLICATIONS OF SPECTROPHOTOMETRY

Qualitative applications of spectrophotometry use the spectral features of a sample to obtain information about the nature of the sample's component(s). It is sometimes possible to use an absorbance spectrum like a "fingerprint" to identify an unknown substance. The infrared (IR) spectra of organic compounds are complex and form particularly distinctive "fingerprints." Organic chemists therefore frequently use IR spectra to identify compounds.

UV/visible spectra are less complex and distinctive than IR spectra, so they are used less commonly for identification of unknown substances. There are, however, situations where information about a sample can be obtained from its UV/Vis spectrum. For example, the U.S. Pharmacopeia includes a number of tests of drug identity in which an absorbance spectrum from 200 to 400 nm is used to confirm the identity of a drug. In these identity tests, the spectrum of the test sample is compared with the spectrum of a pure reference material. If the sample and reference have identical spectra, then it is likely that they are the same compound. UV/Vis identity tests are frequently used because they are simple, rapid, and the equipment to prepare an absorbance spectrum is widely available. A problem with identifying substances using UV/Vis spectrophotometry is that some compounds whose structures are similar have spectra that cannot be distinguished. UV spectra, therefore, are frequently used in conjunction with other identification methods.

UV/Vis spectra can also be used to obtain information about structural characteristics of a substance. For example, a spectrophotometer can be used to monitor hemoglobin as it binds to oxygen and other compounds. Figure 25.2a shows the difference between the spectra of hemoglobin bound to oxygen and hemoglobin bound to carbon monoxide. Similarly, the spectrum of the protein bovine serum albumin shifts when the protein becomes denatured (loses its normal structure). This shift can be monitored with spectrophotometry, see Figure 25.2b.

III.   INTRODUCTION TO QUANTITATION WITH SPECTROPHOTOMETRY: STANDARD CURVES AND BEER'S LAW

A.   Constructing a Standard Curve

Quantitative applications of spectrophotometry have the purpose of determining the concentration (or amount) of an analyte in a sample. Quantitative analysis with a spectrophotometer typically involves a "standard curve." A **standard curve (calibration curve)** *for spectrophotometric analysis is a graph of analyte concentration (X axis) versus absorbance (Y axis).*

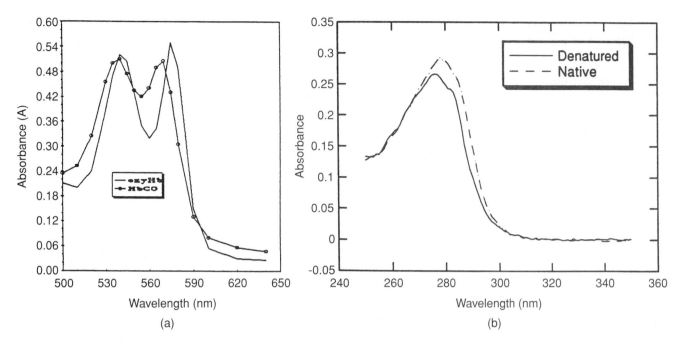

Figure 25.2. Absorbance Spectra and Structure. a. The difference in the absorbance spectra of hemoglobin bound to oxygen and to carbon monoxide. (Used by permission from "Spectrophotometric Properties of Hemoglobin: Classroom Applications," Roger Frary. *The American Biology Teacher* 59 (2):104–7, 1997.) **b.** The absorbance spectrum of bovine serum albumin changes when the protein loses its normal structure. (Used by permission from Copeland, Robert A. *Methods for Protein Analysis.* New York: Chapman and Hall, 1994:175.)

To construct a calibration curve, standards are prepared with known concentrations of analyte. The absorbances of the standards are determined at a specified wavelength and the results are graphed. Given a standard curve, it is possible to determine the concentration of an analyte in a sample based on the sample's absorbance.

EXAMPLE

a. Construct a standard curve for red food coloring.

b. Determine the concentration of red food coloring in a sample with an absorbance of 0.50.

Step 1. Prepare a series of standards of known concentration by diluting a stock solution with distilled water.

Step 2: Prepare one tube that has no dye. This is the blank. (In this case, the blank is simply distilled water.)

Step 3. Place the blank in the spectrophotometer at the specified wavelength and adjust the spectrophotometer to 0 absorbance or 100% transmittance.

Step 4. Read the absorbance of each standard at the specified wavelength. Example results are shown in the following table.

Step 5. Plot the data on a graph with standard concentration on the X axis and absorbance on the Y axis, as shown in the following graph.

Step 6. Draw a best fit line to connect the points. (See Chapter 15 for a discussion of best fit lines and the Appendix to this chapter for a statistical method to draw a best fit line.)

Step 7. Read the absorbance of the unknown at the specified wavelength.

Step 8. Determine the concentration of the unknown based on the standard curve.

EXAMPLE DATA

Concentration of Standard	Absorbance
0.0 ppm	0.00
2.0 ppm	0.14
4.0 ppm	0.26
6.0 ppm	0.45
8.0 ppm	0.59
10.0 ppm	0.75
12.0 ppm	0.91
unknown	0.50

a. The standard curve is:

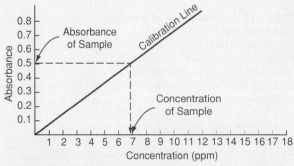

b. Based on the standard curve, the unknown has a concentration of 6.9 ppm.

B. The Equation for the Calibration Line; Beer's Law

Information about how an analyte interacts with light can be obtained from the line on a calibration curve. This section explores the features of that line in more detail.

The line on a calibration curve has an equation, as does every line. (See pp. 251–257 for a review of lines and linear equations.) Recall that the general equation for a line is:

$$Y = mX + a$$

where
m = the slope
a = the Y intercept

For the line on a calibration plot, the Y axis is absorbance and the X axis is concentration (or amount). Therefore, in the equation for the line on a spectrophotometry calibration curve, we can substitute A (for absorbance) and C (for concentration) as follows:

$$A = mC + a$$

For a calibration line, the Y intercept is normally zero. (A Y intercept of zero means that both the absorbance and analyte concentration are zero. A blank, containing no analyte, is used routinely to set the instrument to zero absorbance.) Thus, the equation for the line on a calibration curve is:

$$A = mC + 0$$
or simply
$$A = mC$$

in words:

Absorbance =
(slope of calibration line)(Concentration of analyte)

Consider the slope of a calibration line. If the slope is relatively steep, it means that there is a dramatic change in absorbance as the concentration increases. On the other hand, if the slope is not very steep, then as the concentration of analyte increases, the absorbance does not increase dramatically.

What determines the steepness of the slope of a standard curve? One factor is the nature of analyte. Recall from the previous chapter that different compounds have differing patterns of absorbance of light at different wavelengths. (For compounds that absorb light in the visible range, this pattern relates to the color of the compound.) For example, observe in Figure 25.3a that Compound A absorbs more light at 550 nm than does Compound B. (Assume that the concentrations of both compounds were equal and that the conditions were the same when the spectra were constructed.) If we plot a standard curve for Compounds A and B at 550 nm, as is shown in Figure 25.3b, the line for Compound A will have a steeper slope than for Compound B. This is because Compound A has a greater inherent tendency to absorb light at 550 nm than does Compound B. Thus, at a given wavelength, the slope of a calibration line will vary from one compound to another.

A second factor that affects the slope of the calibration line is the path length. If the light has to pass through a longer path, then there is more absorbing material present and more of the light will be absorbed. Thus, the two main factors that affect the slope are:

1. The tendency of the compound of interest to absorb light at the wavelength used

2. The path length

The inherent tendency of a material to absorb light at a particular wavelength is called its **absorptivity.** The greater the absorptivity of a material, the more it absorbs light at that wavelength. The absorptivity of a specific compound at a particular wavelength is constant.

The value of the absorptivity constant of a particular compound at a specific wavelength has many names, including the **absorptivity constant** (abbreviated with a Greek alpha, α) and the **absorption coefficient.** It is sometimes termed the **extinction coefficient** because it is an indication of the tendency of a compound to "extinguish" or absorb light. The absorptivity constant has units that vary depending on the units in which the concentration of the analyte is expressed. Units for expressing concentration include moles per liter, milligrams per milliliter, and parts per million. If concentration is expressed in moles per liter, then the absorptivity constant is variously termed: **the molar absorptivity constant, ε (the Greek letter epsilon), the molar extinction coefficient,** or **the molar absorption**

Figure 25.3. The Slope of the Calibration Line. a. Compound A absorbs more light at 550 nm than does compound B. **b.** The calibration line for Compound A is steeper than for Compound B at 550 nm.

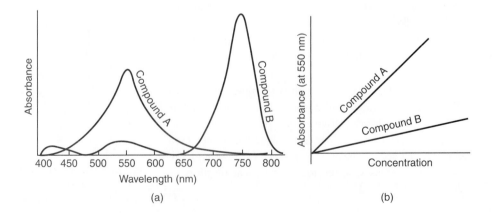

(a)

(b)

coefficient. Because the terminology varies, it is necessary to note the units in your own calculations and to be aware of the units in the calculations of others.

Returning to the equation for the calibration line, we can rewrite the equation to include the path length and the analyte's absorptivity, both of which contribute to the slope. The equation thus becomes:

$$A = mC$$
$$A = (\alpha b)C$$

where

    A = the absorbance
    α = the absorptivity for that compound at that wavelength
    b = the path length
    C = the concentration
  (αb) = m = the slope

This equation, which shows the relationship between absorbance, concentration, absorptivity, and path length, is famous. It forms the basis for quantitative analysis by absorption spectrophotometry. Its discovery is variously attributed to Beer, Lambert, and Bouguer, but it is generally referred to simply as Beer's Law. **Beer's Law** *states that the amount of light emerging from a sample is reduced by three things:*

1. The concentration of absorbing substance in the sample (C in the equation)

2. The distance the light travels through the sample (path length or b)

3. The probability that a photon of a particular wavelength will be absorbed by the specific material in the sample (the absorptivity or α)

EXAMPLE PROBLEM

a. Using Beer's Law, what is the concentration of analyte in a sample whose molar absorptivity constant is 15,000 L/mole-cm and whose absorbance is 1.30 AU in a 1-cm cuvette?

b. Suppose that the absorbance of the preceding analyte is measured in a 1.25-cm cuvette. Will its absorbance be less than, more than, or equal to 1.30?

c. Will changing from a 1-cm cuvette to a 1.25-cm cuvette affect the molar absorptivity constant?

ANSWER

a. Substituting into the equation for Beer's Law:

$$A \quad = \quad \alpha \qquad b \quad C$$
$$1.30 = \frac{(15{,}000\ \text{L})\,(1\ \text{cm})\,C}{\text{mole–cm}}$$

The cm cancel:

$$1.30 = \frac{(15{,}000\ \text{L})\,C}{\text{mole}}$$

Solving for the concentration:

$$C = \frac{1.30}{\dfrac{(15{,}000\ \text{L})}{\text{mole}}}$$

$$\approx 8.67 \times 10^{-5}\ \frac{\text{mole}}{\text{L}}$$

b. This change in cuvette will affect the path length. Because the path length is longer, the absorbance will be greater than 1.30.

c. The absorptivity constant is an intrinsic property of the analyte and is unaffected by the cuvette.

EXAMPLE PROBLEM

DNA polymerase is an enzyme that participates in the assembly of DNA strands from building blocks of nucleotides. The absorptivity constant of this enzyme at 280 nm (under specified pH conditions) is 0.85 mL/(mg)cm. (Worthington, Von, ed. *Worthington Enzyme Manual: Enzymes and Related Biochemicals.* Freehold, NJ: Worthington Biochemical Corp., 1993.) If a solution of DNA polymerase has an absorbance of 0.60 at 280 nm (under the proper conditions), what is its concentration?

ANSWER

Substituting into the equation for Beer's Law:

$$A \quad = \quad \alpha \qquad\qquad b \quad C$$
$$0.60 = \frac{0.85\ \text{mL}}{\text{cm(mg)}} \times 1\ \text{cm} \times C$$

$$C = \frac{0.60}{\dfrac{0.85\ \text{mL}}{\text{mg}}}$$

$$\approx 0.71\frac{\text{mg}}{\text{mL}}$$

C. Calculating the Absorptivity Constant from a Standard Curve

The absorptivity constant is an important property of an analyte. It is a useful skill to be able to calculate and report the absorptivity constant for a particular compound of interest based on data from your own spectrophotometer and in your own laboratory. This section shows two strategies to determine the absorptivity constant for a particular compound at a specific wavelength. The first strategy requires measuring the absorbance of a single standard and then calculating the absorptivity constant based on that single measurement. The second strategy is more reliable and involves constructing a standard curve and calculating the absorptivity constant based on its slope.

STRATEGY 1: CALCULATING AN ABSORPTIVITY CONSTANT BASED ON A SINGLE STANDARD

Beer's Equation can be rearranged as follows:

$$\text{absorptivity constant} = \frac{\text{Absorbance}}{(\text{path length})(\text{concentration})}$$

If the absorbance and concentration of a single standard are known, these values can be entered into the preceding equation to determine the absorptivity constant. This method is illustrated in the following example:

EXAMPLE

A standard containing 75 ppm of Compound Y is placed in a 1 cm cuvette. The absorbance of the standard is 1.20 at 450 nm. Assuming that there is a linear relationship between the concentration of Compound Y and the absorbance, and assuming that the standard was diluted properly, what is the absorptivity constant for this compound at 450 nm?

Substituting into the equation for Beer's Law:

$$\text{absorptivity constant} = \frac{\text{Absorbance}}{(\text{concentration})(\text{path length})}$$

$$\alpha = \frac{1.20}{75 \text{ ppm } (1 \text{ cm})}$$

$$= 0.016/\text{ppm-cm}$$

The absorptivity constant at 450 nm, based on this single standard, therefore, is 0.016/ppm-cm.

This first strategy will give a value for the absorptivity constant. However, it is not the best method to use because it is based on only one standard. If that single standard is diluted incorrectly or if for some reason the absorbance value is slightly off what it should be, then the absorptivity constant calculated will also be inaccurate.

STRATEGY 2: CALCULATING THE ABSORPTIVITY CONSTANT FROM A STANDARD CURVE

A better way to calculate the absorptivity constant for a particular compound at a specified wavelength is to base it on the absorbance of a series of standards. This is readily accomplished using a standard curve. Recall that the slope of the calibration line is *the absorptivity constant* (α) multiplied by *the path length*. Further, recall that the equation for the slope of a line is:

$$m = \frac{Y_2 - Y_1}{X_2 - X_1}$$

where X_1 and X_2 are the X coordinates for any two points on the line and Y_1 and Y_2 are the corresponding Y coordinates for the same two points.

Determining the absorptivity constant based on the slope of a standard curve is illustrated in the following example and is summarized in Box 25.1.

EXAMPLE

The calibration curve at 550 nm for Compound Q is shown. What is the absorptivity constant for this compound at 550 nm? (Assume the path length is 1 cm.)

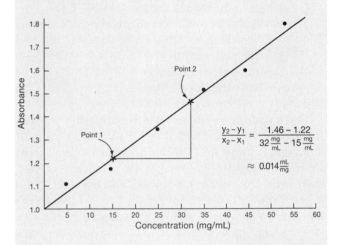

Step 1. Calculate the slope of the calibration line. Note the units. In this case, the slope is about 0.014 mL/mg.

Step 2. The path length is 1 cm.

Step 3. From Beer's Law, the slope of the calibration line is:

$$\text{slope} = (\text{path length})(\text{absorptivity constant})$$

so the absorptivity constant $= \dfrac{\text{slope}}{\text{path length}}$

In this example:

$$\text{absorptivity constant} = \frac{0.014 \text{ mL}}{(1 \text{ cm}) (\text{mg})}$$

D. Variations on a Theme

i. Overview

Now that we have explored the Beer's Law equation, we will consider more examples of how it is used to determine the concentration (or amount) of analyte in a sample. In general, the best way to determine the concentration of an analyte in a sample is to construct a calibration curve based on a series of standards. There are, however, alternatives to constructing a calibration curve each time a sample is analyzed. These variations can be used successfully, but it is imperative to be aware of (and avoid) their potential pitfalls.

Box 25.1. DETERMINATION OF THE ABSORPTIVITY CONSTANT FROM A CALIBRATION CURVE

1. *Prepare a calibration line with concentration on the X axis and absorbance on the Y axis.*
2. *Calculate the slope of the calibration line using the equation:*

$$m = \frac{Y_2 - Y_1}{X_2 - X_1}$$

3. *Determine the path length for the instrument.* (The path length is generally 1 cm, assuming a standard type of cuvette and holder is used.)
4. *Solve the equation:*

$$\text{Absorptivity constant} = \frac{\text{slope}}{\text{path length}}$$

---

ii. VARIATION 1: USING A CALIBRATION CURVE FROM A PREVIOUS ANALYSIS

In laboratories where a particular quantitative analysis is done routinely, it may be determined that a calibration curve needs to be produced and checked only once in a while. The same calibration curve is then used regularly for all samples. This method is accurate only if the following assumptions are true:

- **The instrument must remain calibrated and in good working order over time.** For example, if the spectrophotometer becomes unaligned so the wavelength is not the same as it was when the curve was first constructed, then using the standard curve will lead to inaccurate results.

- **The solvents and reagents must be consistent.** Reagents may change with storage. If a batch of reagents used when making the standard curve varies from a batch used for samples at a later time, then the results may be inaccurate.

- **Assay conditions, such as incubation times and temperatures, must be consistent.**

- **A proper blank must be used when constructing the standard curve and when measuring the absorbances of the samples.**

Note that it is good practice to regularly check the absorbance of a control sample with a known analyte concentration. If the control has an unexpected reading, then the analyst is alerted that there may be a problem.

iii. VARIATION 2: DETERMINING CONCENTRATION BASED ON THE ABSORPTIVITY CONSTANT

A second alternative is to determine the absorptivity constant as shown in Box 25.1. Then, for subsequent

samples the concentration of analyte in a sample can be determined based on Beer's Law:

$$A = \alpha bC$$
therefore
$$C = A/\alpha b$$

This method is illustrated in the example problems on p. 467. This method is accurate only if the following assumptions are true:

- **The relationship between absorbance and concentration must be linear.**
- **The instrument must remain calibrated and in good working order over time.**
- **The solvents and reagents must be consistent.**
- **Assay conditions, such as incubation times and temperatures, must be consistent.**
- **A proper blank must be used.**
- **The sample's concentration must be in the linear range of the assay (discussed in more detail later).**

iv. VARIATION 3: DETERMINING CONCENTRATION BASED ON AN ABSORPTIVITY CONSTANT FROM THE LITERATURE

A still less-accurate method to determine the concentration of analyte in a sample is to use a value for the absorptivity constant published in the literature. Absorptivity constants for various compounds at various wavelengths are frequently reported in articles, catalogs, and other technical literature. In principle, spectrophotometers can be calibrated so that if the absorbance of the same sample is measured in various instruments, the same absorbance values will be obtained. In practice, two instruments often do not read identical absorbance values for a given sample. Moreover, the conditions when spectrophotometric measurements are made (such as solutions used, temperature, pH, etc.) are likely to vary from one laboratory to another. An absorptivity constant derived in one laboratory is therefore unlikely to be exactly reproducible in other circumstances. Nonetheless, there are situations where constants from the literature can be used to give estimates of concentrations. For example, if an analyst is comparing two methods of purifying a protein to see which is more efficient, it is reasonable to estimate and compare the amount of protein in the two preparations based on extinction coefficients from the literature. In contrast, if the analyst wants to know how much protein is actually in each of the preparations, then a standard curve for that protein must be prepared by the analyst.

EXAMPLE PROBLEM

Suppose a manufacturer provides an extinction coefficient for an enzyme.

a. What features of your spectrophotometer must match those of the manufacturer in order for you to get the same extinction coefficient for this enzyme?

b. If your instrument does not match the company's instrument, is the extinction coefficient listed in the catalog useful information?

ANSWER

a. To reproduce an extinction coefficient obtained on one instrument using another instrument:

Both instruments must be calibrated for wavelength and photometric accuracy. In addition, the degree of "monochromaticity" of light exiting the monochromator must be the same. (The more monochromatic the light, the higher the absorptivity constant will be for a given compound.)

b. An extinction coefficient from the literature can be useful as an approximation. In addition, if the enzyme is measured consistently on the same instrument, then it can be compared in a relative way from batch to batch or from assay to assay.

v. VARIATION 4: USING A SINGLE STANDARD RATHER THAN A SERIES OF STANDARDS

It is possible to use a single standard each time samples are analyzed. Then, the amount of analyte in the sample is proportional to the amount in the standard, as illustrated in the following example.

EXAMPLE

A standard is prepared with 10 mg/mL of analyte.

The absorbance of the standard = 1.6.

The sample's absorbance is 0.8.

What is the concentration of analyte in the sample?

$$\frac{10 \text{ mg/mL}}{1.6} = \frac{?}{0.8}$$
$$? = 5 \text{ mg/mL}$$

This method is accurate only if the following assumptions are true:

- **The single standard must be prepared accurately.** Manufacturers sometimes provide carefully tested standards for a particular method.

- **The relationship between analyte concentration and absorbance must be linear.**

- **The instrument must be set to zero absorbance (or 100% T) using a properly made blank containing no analyte.**

- **The sample and the standard must have absorbances in the linear range of the assay.**

EXAMPLE PROBLEM

A standard containing 20 mg/mL of Compound Z is placed in a 1 cm cuvette. The absorbance of the standard is 1.20 at 600 nm. A sample containing Compound Z has an absorbance of 0.50. Assuming that there is a linear relationship between the concentration of Compound Z and the absorbance, and assuming that the standard was diluted properly, what is the concentration of Compound Z in the sample?

There are two ways to solve this.

STRATEGY 1: USING PROPORTIONS

This is a proportional relationship; therefore,

$$\frac{20 \text{ mg/mL}}{1.2} = \frac{?}{0.5}$$
$$? \approx 8.33 \text{ mg/mL}$$

STRATEGY 2: BASED ON THE ABSORPTIVITY CONSTANT

Calculate the absorptivity constant based on this one standard. Substituting into the equation for Beer's Law:

$$A = \alpha \, b \, C$$
$$1.20 = \alpha \, (1 \text{ cm}) \, (20 \text{ mg/mL})$$
$$\alpha = \frac{1.20}{20 \, \frac{\text{mg(cm)}}{\text{mL}}}$$

The absorptivity constant, based on this single standard, is:

$$\alpha = 0.06 \text{ mL/mg(cm)}$$

From Beer's Law, therefore, if the absorbance of the sample is 0.50, its concentration is:

$$A = \alpha \, b \, C$$
$$0.50 = \frac{0.06 \text{ mL}}{\text{cm(mg)}} \, (1 \text{ cm}) \, C$$
$$C = \frac{0.50}{0.06 \, \frac{\text{mL}}{\text{mg}}}$$
$$C \approx 8.33 \text{ mg/mL}$$

E. Deviations from Beer's Law

We have now seen how Beer's Law, which states that the absorbance of a compound is directly proportional to its concentration, is the basis for quantitation using UV/Vis methods. In many systems, Beer's Law holds true and it is therefore a very useful equation. There are situations, however, where real systems deviate from Beer's Law. The causes of deviation from Beer's Law include:

1. **At high absorbance levels, stray light causes spectrophotometers to have a nonlinear response.**

2. **At very low levels of absorbance a spectrophotometer may be inaccurate.**

3. **There may be one or more components in the sample that interfere with the measurement of absorbance of a particular analyte.** We will discuss sample components that interfere with analysis in a later section of this chapter.

Consider what occurs when absorbance is very high or low. Observe that the calibration line drawn in Figure 25.4 has two thresholds. At very low absorbance levels, the calibration curve deviates from linearity. At high absorbance levels the instrument is not able to measure the small amount of light passing through the sample accurately. Thus, at high and low concentrations of analyte, the relationship between absorbance and concentration is not linear; therefore, Beer's Law does not apply.

All quantitative spectrophotometric assays have a range of concentration within which the values obtained are reliable. Above this concentration range the absorbance readings are too high to be useful and below this range the absorbance values are too low. It is essential that the standards and samples all have concentrations such that their absorbance falls in the range where the values obtained are accurate. Thus, a very important cause of deviation from Beer's Law is failure to stay within the linear range of the assay.

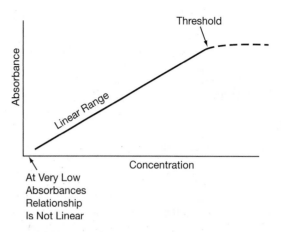

Figure 25.4.   The Relationship between Absorbance and Concentration Is Linear at Intermediate Concentrations of Analyte, But Not at High or Low Concentrations.

EXAMPLE PROBLEM

Suppose you repeat the assay as discussed in the preceding Example Problem, this time taking 1 mL of the sample and diluting it with 9 mL of distilled water. Now, the absorbance of the sample is 1.2. What is the concentration of analyte in this sample?

ANSWER

Reading from the standard curve, the concentration of the sample is about 40 mg/L. However, the sample was diluted 10X. Therefore, multiply by ten, to give an answer of 400 mg/L.

Although there is a lower limit of concentration below which an analyte cannot be detected, spectrophotometry is generally a method used to evaluate analytes that are present at very low levels in a sample. You can easily demonstrate that this is so by taking food coloring and trying to measure its absorbance. You will find that the absorbance of undiluted food coloring is far too intense to measure in a spectrophotometer. In fact, food coloring must be diluted a great deal before its absorbance falls into the linear range of a spectrophotometer. Samples often must be diluted before they can be analyzed spectrophotometrically.

EXAMPLE PROBLEM

Suppose you have prepared a standard curve, as shown in the following graph. What is the concentration of analyte in a sample whose absorbance is 1.95?

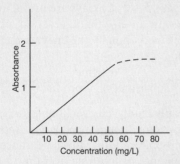

ANSWER

The absorbance of the sample is too high; it does not fall on the linear range of the standard curve. All we can tell is that the concentration of analyte in the sample is above about 50 mg/L. To determine the concentration of analyte, it is necessary to dilute some of the sample and repeat the assay.

EXAMPLE PROBLEM

Assume that a spectrophotometer is able to read accurately in the range from 0.1 to 1.8 AU. The molar absorptivity constant for NADH is 15,000 L/mole-cm at 260 nm. Using Beer's Law, calculate the concentration range of NADH that can be quantitated accurately at this wavelength based on the limits of the spectrophotometer.

ANSWER

This involves calculation of the molar concentrations that will produce absorbances of 0.1 and 1.8. From Beer's Law:

$$C = \frac{A}{\alpha\, b}$$

Substituting 0.1 and 1.8 into the equation (and assuming a 1.0 cm path length):

$$C = \frac{0.1}{\dfrac{(15{,}000\ \text{L})}{(\text{cm})\,\text{mole}}\,(1.0\ \text{cm})} \approx 6.7 \times 10^{-6}\ \text{mole/L}$$

$$C = \frac{1.8}{\dfrac{(15{,}000\ \text{L})}{(\text{cm})\,\text{mole}}\,(1.0\ \text{cm})} = 120 \times 10^{-6}\ \text{mole/L}$$

The range of NADH concentrations that can be detected at this wavelength with this spectrophotometer is from 6.7×10^{-6} mole/L to 120×10^{-6} mole/L. These are dilute solutions of NADH.

IV. QUANTITATION WITH SPECTROPHOTOMETRY: COLORIMETRIC ASSAYS, TURBIDOMETRY, AND KINETIC ASSAYS

A. Colorimetric Assays

A visible spectrophotometer can only be used to analyze a material that absorbs visible light. Biological materials by themselves are usually colorless (i.e., they do not absorb visible light). **Colorimetric assays** are used to analyze materials that are naturally colorless. A **colorimetric assay** *is one in which a colorless substance of interest is exposed to another compound and/or to conditions that cause it to become colored:*

$$\text{substance without color} + \text{reagent(s)} \xrightarrow{\substack{\text{proper} \\ \text{conditions}}} \text{product with color}$$

For example, there are several commonly used colorimetric assays to measure the concentration of proteins in a sample. One such assay, the Biuret method, involves dissolving copper sulfate in an alkaline solution and adding it to the protein sample. Complexes form between the copper ions and nitrogen atoms in the proteins. These complexes produce a blue color that is measured in a spectrophotometer at 550 nm. The more protein present, the more intense the blue color.

When performing a colorimetric assay, the standards, the samples, and the blank should be handled identically. For example, if the samples are subjected to heating, cooling, the addition of various reagents, or other treatments, then the standards and the blank should also be treated in these ways. Note also that when performing colorimetric assays, if a sample is too concen-

trated to be in the linear range of the instrument, then it must be diluted before the reagent(s) are added. Diluting the final reaction mixture will not produce accurate results.

B. Turbidometry and the Analysis of Bacterial Suspensions

Turbid solutions *contain small suspended particles that both absorb and scatter light.* Scattered light is generally deflected away from the detector, so it appears to have been absorbed when, in reality, it was not. Turbidity can also cause an apparent shift in absorbance peaks because shorter wavelengths are scattered more readily than longer ones. How much a beam is attenuated by scattering depends on the optics of the instrument, the orientation of the cuvette, and the uniformity of the suspension. Any inconsistencies in a turbid system, therefore, will give inconsistent absorbance readings. Turbid samples should normally be avoided in spectrophotometry. Suspended materials can be removed by centrifugation or filtration of the sample.

A useful exception to the rule of avoiding turbid samples is in evaluating the concentration of microorganisms in a sample. The greater the concentration of bacteria in a sample, the more they scatter light and the higher the apparent absorbance of the solution. Although absorbance is not actually being measured, within limits, the relationship between the apparent absorbance reading and the concentration of microorganisms is linear. Apparent absorbance, therefore, can be used as a measure of cell density in a suspension.

C. Kinetic Assays

Kinetic spectrophotometric assays *measure the changes over time in concentration of reactants or products in a chemical reaction.* Kinetic assays are useful in the analysis of enzymes. In an enzymatic reaction, one or more substrates are acted upon by the enzyme and converted to product(s):

$$\text{substrate(s)} \xrightarrow{\text{enzyme}} \text{product(s)}$$

As enzymatically catalyzed reactions proceed, the substrate(s) is consumed and product(s) appear. In an enzyme assay, therefore, either the appearance of product or the disappearance of substrate over time is monitored.

Although enzyme assays may be considered as a class of quantitative assay, the quantitation of enzymes is different from that of other proteins. Proteins other than enzymes are typically measured in terms of their concentration, for example, in terms of milligrams per milliliter. Enzymes can be measured not only in terms of their concentration but also in units of activity. Activity is a measure of the amount of substrate that is converted to product by the enzyme in a specified amount of time under certain conditions. Thus, time is an important factor in enzyme assays. Figure 25.5 shows a plot of a kinetic assay.

V. SPECTROPHOTOMETRIC METHODS

A. Developing Effective Methods for Spectrophotometry

i. OVERVIEW

Spectrophotometric methods delineate the steps for analyzing a sample using a spectrophotometer. For example, a colorimetric method might involve (1) adding a reagent to samples, standards, and the blank, (2) heating the mixtures at a certain temperature for a certain time, (3) cooling them for a set period, (4) reading their absorbances, (5) constructing a standard curve, and (6) determining the concentration of analyte in the samples based on the standard curve. A spectrophotometric identification method might involve scanning a sample over a range of wavelengths and comparing peaks in the spectrum to those of a standard. There are thousands of published spectrophotometric analytical methods available. Sources of methods include laboratory manuals, research articles, the U.S. Pharmacopeia, and the manuals of AOAC and ASTM.

Spectrophotometric methods are widely used partly because spectrophotometers are present in most laboratories and are relatively easy to operate. Another important advantage to spectrophotometric methods is that an individual substance in a mixture can be analyzed without separating the components of the sample from one another. For quantitative analysis, the presence of multiple substances in a sample is not a problem as long as a wavelength exists where only the material of interest absorbs light. Spectral analysis of samples with multiple components may be possible if features of their spectra can be distinguished.

Developing and optimizing a spectrophotometric method requires finding a basis for the method (e.g., a colorimetric reaction) that is specific for the material of

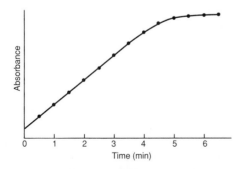

Figure 25.5.   An Example of a Kinetic Assay. Ingested alcohol is rapidly distributed throughout the bloodstream. The removal of alcohol from the body is a time-dependent, enzymatically catalyzed process that occurs in the liver. Alcohol dehydrogenase is the enzyme responsible for the initial metabolism of alcohol. The enzymatic reaction catalyzed by alcohol dehydrogenase results in the formation of NADH, which absorbs light at 340 nm. This graph shows the results of an assay for alcohol dehydrogenase in which the activity of alcohol dehydrogenase is monitored by following the change in absorbance at 340 nm over time.

interest, finding the optimum wavelength for analysis (for quantitative assays), determining the type of sample for which the method is appropriate, determining the concentrations of sample for which the assay is accurate, discovering characteristics of the sample that might interfere with the accuracy of the method, and determining what instrument performance features are necessary. Some of these factors are discussed in more detail in the upcoming sections. Examples of methods that are commonly used in biology laboratories are explored to illustrate some of the issues that arise when using spectrophotometric assay methods.

ii. THE SAMPLE AND INTERFERENCES

The sample is a key part of a spectrophotometric method. As was already discussed, each method is accurate only within a certain range of analyte concentrations. The analyte concentration in the sample, therefore, must match the requirements of the assay. Another important requirement is that the sample not contain substances that interfere with the assay. An **interfering substance,** *in its broadest sense, may be considered as any material in the sample that leads to an inaccurate result in the analysis.*

Interferences may cause absorbance readings to be incorrectly high or low. In absorption spectrophotometry, interferences are often compounds that absorb light at the same wavelength as the analyte and therefore result in an absorbance value that is too high. Impurities that reduce (quench) light readings are often encountered in fluorescence assays (discussed later).

Figure 25.6 on p. 474 illustrates the effect of an interfering substance in a sample. The interfering substance in this example is a detergent, Triton X-100, which is sometimes used in biological solutions to solubilize membrane proteins and to prevent protein aggregation. Detergents are also commonly used as cleaning agents. Both proteins and Triton X-100 absorb light at 280 nm. Residual detergent, therefore, must be removed from protein preparations before they are analyzed by spectrophotometry at 280 nm.

Colorimetric reactions are frequently used to visualize an analyte in the presence of other substances. For example, consider a sample containing a mixture of proteins, nucleic acids, and other cellular components. To analyze only the proteins, reagents can be added to the sample that selectively react with the proteins to form a colored product. There are situations, however, where a color-forming reagent(s) reacts with a substance in the sample in addition to the analyte. A substance that also reacts in an undesired fashion with colorimetric reagents is an interference. Table 25.1 on p. 474 summarizes practical points relating to interfering substances in colorimetric analyses.

It is not possible to compensate for the effect of an interfering substance by using a blank. A blank contains the solvent and reagents that are intentionally added to the sample. Interfering materials, by their nature, are variable and unpredictable substances in the sample that cannot be intentionally included in the blank.

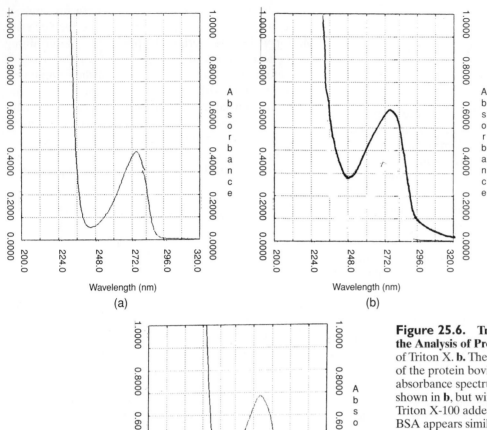

Figure 25.6. Triton X-100 as an Interference in the Analysis of Protein. a. The absorbance spectrum of Triton X. **b.** The absorbance spectrum of 1 mg/mL of the protein bovine serum albumin (BSA). **c.** The absorbance spectrum of the same sample of BSA as shown in **b**, but with a small amount (0.03%) of Triton X-100 added. Although the spectrum for BSA appears similar to that in **b**, in fact, the absorbance at 280 nm is raised due to the contamination of Triton X.

The methods used to handle interferences vary depending on the situation. In some cases, it is possible to find an analytical wavelength where the analyte absorbs light and the interfering substance(s) does not. In other situations, where the nature of an interfering substance is known, it is possible to analyze more than one compound: the analyte, and the interference(s). Because each component individually obeys Beer's Law, and because both compounds are known, the concentration of each can be determined. A simple example of this approach will be discussed later in the section on UV methods. An alternative spectrophotometric method sometimes exists that is insensitive to the interference(s) present in a sample. In other situations, where the nature of an interfering substance is unknown and a suitable analytical wavelength cannot be found, the sample must be purified to eliminate the interference.

Table 25.1. SYMPTOMS THAT AN INTERFERING SUBSTANCE IS PRESENT IN A COLORIMETRIC ASSAY

1. *Interference is sometimes detectable if the color formed by a sample mixture is not of the same hue as that of the standards.*

2. *Interference is indicated if a precipitate forms or turbidity occurs when the sample and reagent are mixed together, but not when the standards and reagent are mixed together.* (This behavior is occasionally due to a concentration of analyte in the sample that is too high. In this case, simply diluting the sample will eliminate the problem.)

3. *If color does not form at the same rate or with the same stability in a sample as in the standards, an interfering substance(s) may be present in the sample.*

iii. CHOOSING THE PROPER WAVELENGTH FOR QUANTITATIVE ANALYSIS

Quantitative analyses are performed at a specific wavelength. Method development therefore requires finding the optimal wavelength based on the absorbance spectra for standards containing the analyte and based on representative samples.

The most important requirement when choosing the analytical wavelength is that the substance of interest absorb light at that wavelength. The more strongly the analyte absorbs light at the chosen wavelength, the better the assay method will be able to detect low levels of compound. A second factor to consider is the nature of the absorbance peak. It is desirable that the peak of absorbance not be too narrow (i.e., the natural band width of the peak should be fairly wide). If the peak is narrow, then any small error in the wavelength setting of the spectrophotometer will result in a large change in absorbance. For example, observe in Figure 25.7 that at λ_2 this compound strongly absorbs light, but the peak is very sharp, so a small error in wavelength setting will result in a large change in the absorbance measured. At λ_1 there is a peak with a broader natural band width that is probably the best choice of wavelength for analysis of this compound. If very low levels of analyte must be measured, then the maximum sensitivity is required and λ_2 might be used for analysis.

The presence of interfering substances in the sample is a complicating factor in choosing an analytical wavelength. If interfering substances are present, it may be possible to choose a wavelength for analysis at which the analyte absorbs less light but the interfering substances are not absorptive.

Thus, the factors that are important in choosing the optimal wavelength are:

1. **The analyte should have a peak of absorbance at the wavelength chosen.**

2. **The absorbance peak ideally should be broad.**

3. **Interfering substances should not be present that absorb light at the chosen wavelength.**

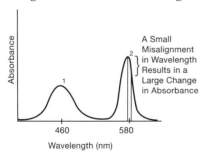

Figure 25.7.   Determining the Optimal Wavelength for Quantitative Analysis. The wavelength chosen must be one at which the analyte absorbs light. There is a strong absorbance peak at 580 nm; however, because the peak is sharp, a small misalignment in the wavelength adjustment will cause a major change in absorbance. Unless very low levels of analyte must be measured, 460 nm therefore appears to be a better choice.

EXAMPLE PROBLEM

Based on the following absorbance spectrum of Compound A, what is the optimal wavelength for performing quantitative analysis of Compound A?

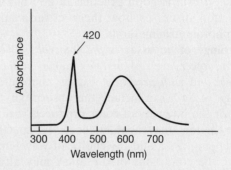

ANSWER

600 nm appears to be a good choice because the compound absorbs light at that wavelength and the peak is not as "sharp" as it is at 420 nm.

EXAMPLE PROBLEM

Based on the following absorbance spectra, what is the optimal wavelength for performing quantitative analysis of Compound A in samples?

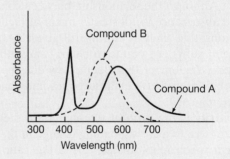

ANSWER

There is an interfering substance, Compound B, that absorbs light at 600 nm. 420 nm appears to be a suitable wavelength where Compound A absorbs light and there is minimal interference.

B. Method Validation

Validation of an analytical method is intended to provide documented evidence that a method will perform as is required. The criteria used to evaluate methods include linearity and range, limit of detection and limit of quantitation, selectivity, ruggedness, accuracy, and precision. These criteria address the limitations of an assay, its ability to give accurate results, the types of samples for which it is suitable, and the conditions under which it is useful. Although formal validation of analytical methods is required only in laboratories that are regulated or meet certain standards, evaluating an analytical method based on relevant criteria is good practice in any laboratory. This section discusses how these criteria apply to spectrophotometric methods.

The **range of an assay** *is the interval between the upper and lower levels of analyte that can be measured accurately and with precision.* Factors that determine the range of an assay include the absorptivity of the analyte or the intensity of color produced in a colorimetric reaction, the capabilities of the spectrophotometer being used, and the choice of wavelength. In addition, the conditions of the assay may affect the range. For example, the intensity of color of some solutions is pH-dependent.

The range of a spectrophotometric assay is usually considered to be the range in which the relationship between absorbance and concentration is linear. Figure 25.8 shows a standard curve for a spectrophotometric assay. At concentrations of analyte above about 130 μg/mL the relationship between concentration and absorbance is not linear. The linear range for this assay is from about 5 to 130 μg/mL.

There are situations where a calibration curve can be used that is not linear. Figure 25.9 illustrates a nonlinear calibration curve from an assay used to quantate proteins. Although the relationship between concentration and absorbance is not strictly linear at higher concentrations, this assay is still useful given a standard curve with sufficient points to clearly show the relationship.* (Remember that Beer's Law does not apply if the relationship between absorbance and concentration is nonlinear.)

Sensitivity is the ratio of the change in the instrument response to a corresponding change in the stimulus. For a spectrophotometer, this is the ratio of the change in absorbance to a corresponding change in analyte concentration:

$$\text{Sensitivity} = \frac{\text{Change in absorbance}}{\text{Change in analyte concentration}}$$

This concept was illustrated in Figure 25.3 on p. 466. Compound A had more inherent tendency to absorb light at 550 nm than Compound B. At 550 nm, therefore, a spectrophotometer is more sensitive to Compound A than Compound B. Thus, in spectrophotometry, the sensitivity of a method is affected largely by the absorptivity of the analyte or by the intensity of color developed in a colorimetric assay.

*Many spectrophotometers are equipped with software that automatically draws the best fit line connecting the points of a standard curve. If an assay is known to be nonlinear, these automatic line fitting programs should not be used.

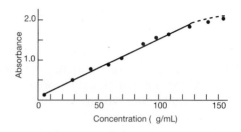

Figure 25.8.   Linear Range of an Assay. This assay is useful in the linear range from about 5 to 130 μg/mL.

The **limit of detection** of an assay *is the lowest concentration of analyte in a sample that can be detected, but not necessarily quantified.* **The limit of quantitation** *is the lowest concentration of analyte that can be reliably quantified.* The limits of detection and quantitation of an assay are affected by the sensitivity of a method. The more sensitive, the lower the level of analyte that can be detected. In addition, the limits are affected by noise. As the concentration of the analyte approaches zero, the signal becomes buried in the noise, and the level of analyte falls below the detection limit of the instrument. Thus, if a method needs to be able to detect trace levels of analyte, the method should be very sensitive and the noise in the spectrophotometer should be as low as possible.

Selectivity *is the ability of an assay to distinguish the compound of interest in the presence of other materials.* Selectivity in spectrophotometry, therefore, relates to the ability of a method to measure an analyte accurately in the presence of potentially interfering substances in the sample.

Ruggedness *is the degree of reproducibility of test results obtained by the analysis of the same samples under a variety of normal test conditions.* A rugged method gives accurate results over a range of conditions. Many spectrophotometric assays are affected by solution conditions, such as pH, temperature, and salt concentration. This is because the properties and

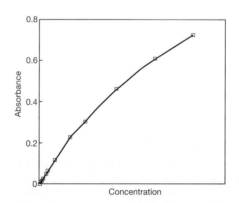

Figure 25.9.   The Standard Curve from a Lowry's Assay for Protein. The assay gave linear results at the lower concentrations, but it was not linear at higher concentrations.

structures of biological molecules change as their solution conditions change. As the molecules change form, so too, their absorptivity may change. Colorimetric reactions similarly may be affected by solution conditions; therefore, these conditions need to be controlled in spectrophotometric methods.

Accuracy is the closeness of a test result to the "true" value. One of the most important and basic ways to ensure accuracy in a quantitative spectrophotometric assay is to prepare a standard curve each time an assay is performed, using carefully prepared standards.

Precision and **long-term reproducibility** are often the most important aspects of an assay. For example, suppose spectrophotometry is being used to monitor the amount of protein present in a preparation as the protein is purified. Protein purification has multiple steps that may take place over a number of days. An assay method must give consistent results over a period of time; otherwise, it is useless to monitor the purification.

C. Colorimetric Protein Assays; An Example

Colorimetric assays of proteins are an important application of spectrophotometry in biology. Most proteins do not naturally absorb light in the visible range. Colorimetric methods, therefore, have been devised in which proteins are reacted with certain dyes with the development of an intense color. Protein assays illustrate some of the issues that arise in using colorimetric methods.

Three common colorimetric protein assay methods are summarized in Table 25.2. The methods vary in features such as their range and their tendency to be affected by interfering substances. It is necessary, there-

Table 25.2. PROTEIN QUANTITATION METHODS

Method	Principle	Approximate Concentration Range	Linearity	Interfering Substances	Comments
Biuret	Peptides react with Cu^{2+} in alkaline solution to yield a purple complex that has an absorption maximum at 540 nm.	500–8000 µg/mL	Not linear at all concentrations	Some materials interfere, including Tris buffer.	The Biuret method is least susceptible to protein-to-protein variation because it measures the peptide bonds in a protein. The major disadvantage of the Biuret method is its relative insensitivity to low levels of proteins.
Lowry	Color results from the reaction of Folin phenol reagent with amino acids in proteins.	1–300 µg/mL	May be nonlinear above about 40 µg/mL	Many compounds interfere including phenols, glycine, ammonium sulfate, and Tris buffer.	This method is sensitive. The Lowry method is not totally specific for proteins; substances beside proteins also react with the reagent and thus interfere. The amount of color development depends on the percentage of tyrosine and tryptophan amino acids in the protein; therefore, some proteins react more intensely than others.
Bradford	Reaction under acidic conditions with Coomassie Brilliant Blue G-250 reagent. Reagent has two color forms: red and blue. The red form turns to the blue form when the dye binds to protein.	25–1400 µg/mL	May become nonlinear at higher concentrations.	Relatively few interfering substances. Some detergents interfere.	A specific and sensitive assay.

Some of the information in this table is from Copeland, Robert A. *Methods for Protein Analysis*. New York: Chapman and Hall, 1994.

fore, to choose the assay method that best suits your samples. It is important that the range of analyte concentration that can be measured by the assay matches the range of concentrations expected in the samples. Thus, if low levels of protein are to be measured, the Lowry or the Bradford methods are preferred over the Biuret method.

Protein assays are often not linear. As shown in Figure 25.9, these assays can still be used, but a standard curve is always required to establish the relationship between absorbance and concentration. Colorimetric protein assays are all affected to some extent by interfering substances. As can be seen in Table 25.2, interferences are particularly problematic for the Lowry assay.

The Lowry and the Bradford methods both develop more intense color with some proteins than with others. To get accurate results with these methods, therefore, it is necessary to construct a standard curve for each protein of interest. For example, if the protein to be measured is DNA polymerase, then the standard curve should be constructed with DNA polymerase to get the best accuracy. In practice, it is common to construct a standard curve for a protein assay using bovine serum albumin (BSA) because the protein of interest is often expensive or difficult to obtain in large quantities, whereas BSA is readily available and is relatively inexpensive. Using BSA as the standard will give results that are not accurate; however, using BSA to make the standard curve will give a reasonable estimate of protein concentration in many situations. For example, proteins are purified in a series of steps. After each step, the amount of protein present is measured. If this measurement is consistently made with reference to BSA, then the relative effectiveness of each purification step can be determined. In contrast, if an analyst wants to know the absolute amount of a specific protein in a preparation, then a standard curve for that protein must be prepared.

EXAMPLE PROBLEM

A Lowry assay was performed. A standard curve was constructed as shown using the protein bovine serum albumin. Three samples of another protein, Protein X, were then analyzed. The absorbances of the samples were:

    Sample A,   0.10
    Sample B,   0.58
    Sample C,   1.3

a. What is the concentration of protein in the three samples based in the standard curve?

b. Is it reasonable to use BSA to prepare the standard curve when the protein of interest is Protein X?

c. Would it affect the reproducibility of the assay if a new batch of Folin reagent were used? (Folin phenol reagent is the color-producing reagent in the Lowry assay.)

d. Would it affect the reproducibility of the assay if the bulb on the spectrophotometer became less bright between assays?

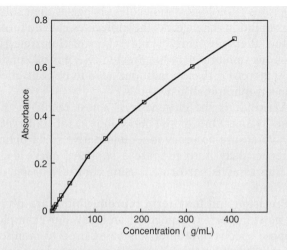

ANSWER

a. Sample A has a concentration of about 40 μg/mL. Sample B has a concentration of 300 μg/mL.
The absorbance of Sample C exceeds the range of the standard curve. In this case, it is necessary to repeat the assay, diluting the sample before the color-developing reagents are added.

b. Because the Lowry method primarily detects tyrosine and tryptophan amino acids, and because the percent tyrosine and tryptophan varies from protein to protein, it is not desirable to prepare a standard curve with a protein different from the one being evaluated. This practice is necessary, however, in situations where sufficient amounts of purified protein of interest to make a standard curve are unavailable or excessively costly.

c. As long as a new standard curve is prepared each time the assay is performed, and both batches of reagent are properly made, there should be no problem. (The results may not be accurate if a standard curve is not prepared each time the assay is run.)

d. Changes in the bulb will not affect the reproducibility of the assay as long as a new standard curve is prepared each time the assay is performed.

D. Analysis of DNA, RNA, and Proteins; Examples of UV Spectrophotometry and Multicomponent Analysis

i. THE UV ABSORBANCE SPECTRA OF NUCLEIC ACIDS AND PROTEINS

Although most biological molecules do not intrinsically absorb light in the visible range, they do absorb ultraviolet light. Biologists take advantage of UV absorbance to quickly estimate the concentration and relative purity of DNA, RNA, and proteins in a sample. Points of general importance regarding spectrophotometric methods are illustrated by these applications.

Proteins have two UV absorbance peaks: one between 215 and 230 nm, where peptide bonds absorb,

Table 25.3. **WAVELENGTHS THAT ARE RELEVANT TO THE MEASUREMENT OF NUCLEIC ACIDS AND PROTEINS**

Wavelength	Significance	Comments
215–230 nm	Minimum absorbance for nucleic acids Peptide bonds in proteins absorb light	Measurements are generally not performed at this wavelength because commonly used buffers and solvents, such as Tris, also absorb at these wavelengths.
260 nm	Nucleic acids have maximum absorbance	Purines absorbance maximum is slightly below 260; pyrimidines maximum is slightly above 260. Purines have a higher molar absorptivity than pyrimidines. The absorbance maximum and absorptivity of a segment of DNA, therefore, depends on its base composition. Proteins have lower absorbance at this wavelength.
270 nm	Phenol absorbs strongly	Phenol may be a contaminant in nucleic acid preparations.
280 nm	Aromatic amino acids absorb light	Nucleic acids also have some absorbance at this wavelength.
320 nm	Neither proteins nor nucleic acids absorb	Used for background correction because neither nucleic acids nor proteins absorb at this wavelength.

and another at about 280 nm due to light absorption by aromatic amino acids (tyrosine, tryptophan, and phenylalanine). DNA and RNA have an absorbance maximum at approximately 260 nm and an absorbance minimum at about 230 nm. Certain subunits of nucleic acids (purines) have an absorbance maximum slightly below 260 nm, whereas others (pyrimidines) have a maximum slightly above 260 nm. Therefore, although it is common to say that the absorbance peak of nucleic acids is 260 nm, in reality the absorbance maxima of different fragments of DNA vary somewhat depending on their subunit composition. Table 25.3 summarizes the UV wavelengths relevant to the measurement of nucleic acids and proteins.

Figure 25.10 on p. 480 shows the absorbance spectra for DNA and proteins. Note that although proteins have lower absorbance at the absorbance peak of nucleic acids, 260 nm, both proteins and nucleic acids absorb light at 280 nm. Therefore, if nucleic acids and proteins are mixed in the same sample, their spectra interfere with one another.

ii. CONCENTRATION MEASUREMENTS OF NUCLEIC ACIDS AND PROTEINS

Consider two UV methods of determining concentration in a "pure" sample containing only proteins or nucleic acids. The first method involves constructing a standard curve; the second is a "short-cut" method based on absorptivity constants from the literature.

It is possible to determine the concentration of nucleic acids or proteins based on their absorbance at a wavelength of 260 nm or 280 nm, respectively. A calibration curve using standards of known concentration can be constructed. For accurate results, the standard curve should be prepared using the protein of interest or DNA that is similar to that in the sample being measured. The linear range for DNA values is reported to be from about 5 to 100 μg/mL. Depending on the protein, UV analysis of proteins at 280 nm has a linear range from about 0.1 to 5 mg/mL.

Biologists commonly use a "short-cut" to provide a rough estimate of the concentration of nucleic acid or protein in a sample based on the sample's absorbance at 260 or 280 nm. This short-cut method uses absorptivity constants. Recall that given an absorptivity constant, it is possible to bypass the preparation of a standard curve by applying Beer's Law (see p. 469).

The absorptivity constant for a particular protein at 280 nm depends on its composition. Proteins that contain a higher percentage of aromatic amino acids have higher absorptivities at 280 nm than do those with fewer. The absorptivity constant for a nucleic acid depends on its base composition and on whether it is single-stranded or double-stranded. Despite the fact that different proteins and nucleic acid fragments vary in their absorptivity, analysts commonly use "average" absorptivity constants to estimate the concentration of nucleic acid or protein in a sample. The average absorptivity constants for proteins and nucleic acids lead to the following relationships:

- **If a sample containing pure double-stranded DNA has an absorbance of 1 at 260 nm, then it contains approximately 50 μg/mL of double-stranded DNA.**

- **If a sample containing pure single-stranded DNA has an absorbance of 1 at 260 nm, then it contains approximately 33 μg/mL of DNA.**

- **If a sample containing pure RNA has an absorbance of 1 at 260 nm, then it contains approximately 40 μg/mL of RNA.**

- **Values for proteins vary. A very rough rule is that if a sample containing pure protein has an absorbance of 1 at 280 nm, then it contains approximately 1 mg/mL of protein.** For example, 1 mg/mL of bovine serum albumin is reported to have an A_{280} value of 0.7. Antibodies (which are a type of protein) at a concentration of 1 mg/mL are reported to have an A_{280} between 1.35 and 1.2. (Values from Harlow, E., and Lane, D. *Antibodies: A Laboratory Manual.* New York: Academic Press, 1988, 673.)

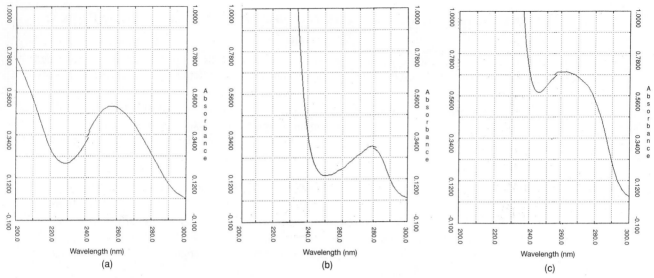

Figure 25.10. Absorbance Spectra for DNA and Protein. a. The absorbance spectrum for DNA. **b.** The absorbance spectrum for protein (BSA). **c.** The absorbance spectrum for a mixture of DNA and protein. Distinct peaks for DNA and protein cannot be resolved.

EXAMPLE PROBLEM

A laboratory scientist isolated DNA from bacterial cells and was interested in estimating the amount of DNA present in the preparation. The preparation had a volume of 2 mL. The scientist removed 50 μL from the preparation, added 450 μL of buffer, and then read the absorbance of the dilution at 260 nm. The absorbance was 0.65. Assuming the sample was pure, about how much double-stranded DNA was present in the original 2000 μL preparation?

ANSWER

A proportion equation can be set up based on the relationship that a sample containing 50 μg/mL of pure double-stranded DNA has an absorbance of 1 at 260 nm:

$$\frac{1}{50\ \mu g/mL} = \frac{0.65}{?}$$

$$? = 32.5\ \mu g/mL$$

Taking into account that the preparation was diluted, the concentration was:

$$32.5\ \mu g/mL \times 10 = 325\ \mu g/mL$$

To estimate the amount of DNA in the original preparation, note that there were 2 mL of isolated product. The concentration of DNA in that preparation was about 325 μg/mL. The 2 mL of the original preparation, therefore, had about:

$$2\ mL \times 325\ \mu g/mL = 650\ \mu g\ \text{double-stranded DNA}$$

iii. ESTIMATION OF THE PURITY OF A NUCLEIC ACID PREPARATION

It is possible to use UV spectrophotometry to estimate the purity of a solution of nucleic acids. This method involves measuring the absorbance of the solution at two wavelengths, usually 260 nm and 280 nm, and calculating the ratio of the two absorbances:

- **An A_{260}/A_{280} ratio of 2.0 is characteristic of pure RNA.**

- **An A_{260}/A_{280} of 1.8 is characteristic of pure DNA.**

- **An A_{260}/A_{280} ratio of about 0.6 is characteristic of pure protein.**

A ratio of 1.8–2.0 is therefore desired when purifying nucleic acids. (Note that this method does not actually distinguish DNA and RNA from one another.) A ratio less than 1.7 means there is probably a contaminant in the solution, typically either protein or phenol.

EXAMPLE PROBLEM

Returning to the previous Example Problem, the laboratory scientist wants to determine whether the DNA preparation is contaminated by proteins. The absorbance at 260 nm was 0.65 and at 280 nm was 0.36. Is this solution likely to contain pure nucleic acids?

ANSWER

The A_{260}/A_{280} ratio is $0.65/0.36 \approx 1.81$. This ratio is consistent with a pure preparation of nucleic acids.

iv. MULTICOMPONENT ANALYSIS; THE WARBURG-CHRISTIAN ASSAY METHOD

(Multicomponent analysis is a more complex application of spectrophotometry and some readers may prefer to skip this section.)

It is sometimes desirable to estimate the concentrations of protein and of nucleic acids in a sample that contains both. The basic premise of an analysis such as this, a **multicomponent analysis,** is that each substance in the mixture individually obeys Beer's Law; therefore, the absorbances of two or more components in a mixture add together. For example, if nucleic acids and proteins are mixed in a sample, then the total absorbances at 280 nm and 260 nm are:

$$A_{280} = A_{280} \text{ proteins} + A_{280} \text{ nucleic acids}$$
$$A_{260} = A_{260} \text{ proteins} + A_{260} \text{ nucleic acids}$$

To relate the absorbances at 260 nm and 280 nm to the concentrations of proteins and nucleic acids, these equations can be rewritten in terms of absorptivity constants for proteins and nucleic acids at each wavelength. Warburg and Christian performed these calculations using absorptivity constants for a yeast protein and for yeast RNA and published the results in 1941 (Warburg, Otto, and Christian, Walter. "Isolierung und Kristallisation des Gärungsferments Enolase." *Biochem. Z.* 310 [1941/1942]: 384–5). They derived these two equations:

$$[\text{nucleic acid}] \approx 62.9 \, A_{260} - 36.0 \, A_{280} \text{ (in units of } \mu g/mL)$$
$$[\text{protein}] \approx 1.55 \, A_{280} - 0.757 \, A_{260} \text{ (in units of } mg/mL)$$

where the brackets indicate "concentration."

To determine the concentrations of proteins and nucleic acids in a mixture, one measures the absorbances at 260 and 280 nm and solves the preceding equations.

EXAMPLE PROBLEM

A technician wants to get a rough estimate of the concentration of proteins and nucleic acids in a cell preparation. The absorbance of the preparation at 260 nm is 0.940 and at 280 nm is 0.820. Based on the Warburg-Christian equations, what is the approximate concentration of nucleic acids and proteins in the preparation?

ANSWER

Substituting into the Warburg-Christian equations:

$$[\text{nucleic acid}] \approx 62.9 \, (0.940) - 36.0 \, (0.820) \approx 29.6 \, \mu g/mL$$
$$[\text{protein}] \approx 1.55(0.820) - 0.757(0.940) \approx 0.559 \, mg/mL$$

Thus, based on the Warburg-Christian equations, the concentration of nucleic acids in the preparation is estimated to be 29.6 μg/mL and of proteins to be 0.559 mg/mL.

The Warburg-Christian equations are approximations based on absorptivity constants derived for the yeast protein, enolase, and yeast RNA. These equations, therefore, cannot be expected to give accurate results when applied to other proteins and nucleic acids. It is possible, however, to generalize the logic used by Warburg and Christian by substituting into the equations absorptivity constants determined for your analytes at specific wavelengths. If the correct absorptivity constants and the optimal wavelengths are used, it is possible to quantitate multiple components in a sample, even if their spectra overlap to some degree. (For more information about multicomponent analysis, see, for example, C.T. Kenner, *Instrumental and Separation Analysis.* Columbus, OH: Charles E. Merrill, 1973.) In addition, spectrophotometer manufacturers have developed software and methods to facilitate multicomponent analyses.

v. CONSIDERATIONS AND CAUTIONS

It is common for analysts to calculate and report purities based on A_{260}/A_{280} ratios and concentrations based on Warburg-Christian calculations. In fact, these calculations are performed so routinely that many spectrophotometers can perform them automatically and display the resulting values. It is important, however, to be aware that values obtained by these methods are only approximations and sometimes may be deceptively inaccurate. The reasons for inaccuracy include:

1. The A_{260}/A_{280} method is based on the spectral characteristics of "average" proteins and nucleic acids. In reality, proteins and nucleic acids vary from one another, so their spectra may vary from one another.

2. The Warburg-Christian method is based on absorptivity constants derived from one protein and nucleic acid from one organism. The absorptivity constants for other proteins and nucleic acids vary.

3. These UV methods assume that the spectrophotometer is accurately calibrated. If the spectrophotometer is displaced by as little as 1 nm, the values may be significantly affected. (See, for example, Manchester, Keith L. "Value of A260/A280 Ratios for Measurement of Purity of Nucleic Acids." *BioTechniques* 19 no. 2 [1995]: 208–10.)

4. These UV methods are affected by the pH and ionic strength (salt concentration) of the buffer, resulting in variability. (See, for example, Wilfinger, William W. "Effect of pH and Ionic Strength on the Spectrophotometric Assessment of Nucleic Acid Purity." *BioTechniques* 22 no. 3 [1997]: 474–80.)

These cautions apply not only to UV analysis of nucleic acids and proteins, but to other spectrophotometric methods. It is always important to be aware of potential inaccuracy when using absorptivity constants reported in the literature. The type of spectrophotometer used and its calibration will affect results in almost any method. It is also important to consider matrix effects (i.e., effects due to the solvent and other components in the sample).

EXAMPLE PROBLEM (and a true-life tale)

An analyst in our laboratory was attempting to check the absorbance spectrum of a solvent to see whether it was suitable for use in the UV range. She filled a cuvette with the solvent and prepared an absorbance spectrum, which is shown in the graph. There is a problem here—can you figure out what it is?

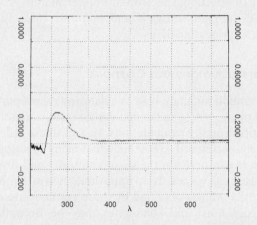

ANSWER

Based on this spectrum, she could not reach any conclusion about the solvent—the cuvette was dirty. The peak at 280 nm was due to protein that was left in the cuvette by another user and had dried onto the cuvette walls. The analyst soaked the cuvette for a couple of hours in distilled water and rinsed it thoroughly. She then obtained a spectrum with no peak at 280 nm, indicating that the solvent has little absorption in the UV range (and that soaking and rinsing the cuvette with distilled water cleaned it effectively in this case).

Box 25.2 on p. 484 summarizes the various UV methods described in this section.

E. Avoiding Errors in Spectrophotometric Analyses

The accuracy of spectrophotometric assays is affected by systematic errors. For example, in colorimetric assays, how long color is allowed to develop may affect the results. Color may be slow to develop in samples of low concentration and fast where concentration is high. Inconsistency in timing or a poor choice of development time may therefore lead to inaccurate results. Errors may occur if too little color developing reagent is added to samples so

that the assay underestimates the amount of analyte present in concentrated samples. Using a blank that contains a different solvent than the samples will lead to error. Many (but not all) errors can be avoided by preparing a standard curve each time an assay is performed rather than relying on an extinction coefficient from previous assays, a single standard, or a standard curve prepared at a different time.

Table 25.4 summarizes practical considerations relating to obtaining good results with spectrophotometric methods.

EXAMPLE PROBLEM

Suppose you are performing a colorimetric assay that includes a boiling step. Five minutes of boiling are necessary for complete color development. You become distracted and accidentally stop the boiling at only 3 minutes. How will this affect your results?

ANSWER

Assuming you prepare a standard curve each time you do the assay, and assuming the reagent still acts in such a way that color is proportional to the amount of analyte, then the standards should be affected by the shortened boiling time in the same way as the samples. The results, therefore, will still be accurate. You may observe, however, that the range for the assay is different than usual. Lower concentrations of analyte that are normally in the range of the assay may in this case not cause a detectable color change. (When you are familiar with an assay, any change in its performance should alert you to the fact that there is a problem.) In addition, note that if you do not prepare a standard curve each time the assay is performed, this type of mistake will lead to undetected errors in the results.

VI. ASSOCIATED SPECTROPHOTOMETRIC AND FLUORESCENCE METHODS

A. Introduction

There are instruments in addition to spectrophotometers that measure interactions between light and a sample to obtain information about that sample. Some of these instruments allow samples to be studied that are not liquids in a cuvette. Other instruments are based on principles that are related to, but are not the same as, absorption spectrophotometry. There are also modified spectrophotometers that find a variety of uses in the laboratory. These instruments and associated technologies expand the applications that can be performed using light measurements. This section discusses three examples of light-measuring instrumentation that are of particular importance in the biotechnology laboratory.

Table 25.4.   CONSIDERATIONS IN OBTAINING ACCURATE SPECTROPHOTOMETRIC MEASUREMENTS

FACTORS RELATING TO THE INSTRUMENT AND ITS OPERATION

1. *Always use a well-maintained and calibrated instrument.*

2. *Use the correct wavelength.* Assays are optimized at a particular wavelength and irreproducibility or inaccuracy may occur at the improper wavelength.

3. *Allow the spectrophotometer light source to warm up, as directed by the manufacturer.* In most laboratories the lamps are turned off when the instrument is not in use because lamps have a finite life span and are relatively expensive to replace. Note that the UV source may take longer to warm up than the visible light source.

4. *Use the proper cuvettes.*

   a. For high-accuracy work, use high-quality cuvettes that are clean, unscratched, and whose sides are exactly parallel.

   b. Use quartz cuvettes for work in the UV range.

   c. Use matched cuvettes for the sample and the blank, or use the same cuvette sequentially for both the sample and blank.

5. *Make certain cuvettes and glassware are clean.* In addition, drops or liquid smeared on the outside of the cuvette can cause erratic readings.

6. *Avoid fogging.* When the air is humid and the sample or blank is cold, moisture can condense on the cuvette leading to fogging.

FACTORS RELATING TO THE ASSAY METHOD AND THE SAMPLE

7. *Avoid samples and standards whose absorbance is very high or very low.* Depending on the instrument, the recommended range of absorbance will probably be about 0.5–1.8.

8. *Be aware of sample characteristics that affect absorbance.*

   a. The color and the absorbance of some samples varies with the pH of the solution; in such cases, make sure the pH of standards and samples is the same.

   b. Sample absorbance may vary with temperature; in such cases, water-jacketed accessories are available for some spectrophotometers to hold the temperature constant during measurement. Otherwise, use a water bath or other device to control the temperature of samples and standards when they are not in the spectrophotometer.

   c. In principle, the absorptivity of an analyte should be constant regardless of its concentration. In reality, some analytes change absorptivity as their concentration increases. In these cases, it is difficult to get a linear response.

   d. The ultraviolet absorption of a compound may vary in different solvents. Be sure, therefore, to record the solvent used and, for aqueous solvents, the pH. When following established methods, use the specified solvent.

9. *Be aware of time as a factor in analysis.* The color intensity of samples reacted with color-developing reagents may increase or decrease over time, or the color may shift in absorbance peak. Time is a particularly important factor in assays involving enzymatic reactions.

10. *Avoid deteriorated reagents and standards.* Many color-developing reagents deteriorate over time. It may be helpful to check reagents regularly by measuring their absorbance versus a blank of distilled water. Initial deterioration of reagents can often be recognized by changes in the reagent's absorbance. Discard reagents that become turbid or cloudy, that change colors, or that contain precipitate.

11. *Be aware that some samples are naturally fluorescent.* There are materials that fluoresce when irradiated with certain wavelengths of light. This results in erroneous transmittance values if the fluorescence falls on the detector. Fluorescent samples can be evaluated if a filter is available that blocks light of the fluorescent wavelength while allowing transmitted light to pass. The filter is placed between the sample and the detector.

12. *Avoid turbidity (except in some special cases where measurements of turbidity provide useful information).* Turbidity is a very common source of error (e.g., environmental water samples are often turbid). Particles in suspension affect the transmittance of light and may cause there to appear to be more absorbance than there really is. If the sample is not homogeneous, turbidity will result in inconsistent and incorrect results. Suspended material should generally be removed from samples prior to spectrophotometry by centrifugation or filtration. Buffers and water used for spectrophotometry may also require filtering.

13. *Use spectroscopy-grade solvents.*

14. *Avoid substances that interfere with the assay.*

BOX 25.2. *APPROXIMATING THE CONCENTRATION AND PURITY OF DNA, RNA, OR PROTEIN IN A SAMPLE*

1. **Concentration of double-stranded DNA $\approx$ 50 μg/mL $\times$ the absorbance at 260 nm**
2. **Concentration of single-stranded DNA $\approx$ 33 μg/mL $\times$ the absorbance at 260 nm**
3. **Concentration of RNA $\approx$ 40 μg/mL $\times$ the absorbance at 260 nm**

Turbidity causes an apparent increase in the absorbance of a sample, leading to incorrect readings. To compensate for slight turbidity, a background correction can be used. Proteins and nucleic acids do not absorb at 320 nm. If a sample absorbs at 320 nm, the absorbance is due to turbidity. The absorbance at 320 nm can be subtracted from the readings at 260 nm and 280 nm:

4. **Concentration of double-stranded DNA $\approx$ 50 μg/mL $(A_{260} - A_{320})$**
5. **Concentration of single-stranded DNA $\approx$ 33 μg/mL $(A_{260} - A_{320})$**
6. **Concentration of RNA $\approx$ 40 μg/mL $(A_{260} - A_{320})$**

---

7. **Concentration of protein $\approx$ 1 mg/mL $\times$ the absorbance at 280 nm**
8. **Concentration of protein $\approx$ 15 mg/mL $\times$ the absorbance at 215 nm**

Notes: Values for proteins vary. The rule that 1 mg/mL of protein has an absorbance of 1 is approximate.

Tris and other common solvents also absorb light at 215 nm. For this reason, 280 nm is far more commonly used for protein measurements.

---

9. *Purity*

An A_{260}/A_{280} ratio of 2.0 is characteristic of pure RNA

An A_{260}/A_{280} of 1.8 is characteristic of pure DNA

An A_{260}/A_{280} of 0.6 is characteristic of pure protein

Notes: A ratio of 1.8–2.0 is desired when purifying nucleic acids.

This method does not distinguish DNA and RNA from one another.

A ratio of less than 1.7 means there is probably a contaminant in the solution, usually either protein or phenol.

Warburg-Christian Equations

For Solutions Containing a Mixture of Nucleic Acids and Proteins

10. **[nucleic acid] $\approx$ 62.9 A_{260} − 36.0 A_{280} (in units of μg/mL)**
11. **[protein] $\approx$ 1.55 A_{280} − 0.757 A_{260} (in units of mg/mL)**

B. Gel Scanning/Densitometry

Electrophoresis is a widely used method in which mixtures of proteins or nucleic acids are separated from one another by allowing them to migrate through a gel matrix in the presence of an electrical field. The result of electrophoresis is that the macromolecules of different sizes are separated into individual bands in the gel. The separated bands are invisible until they are stained with some type of dye. After staining, the band pattern on the gel, or a photograph of the gel, can be examined and used to obtain information about the sample.

Sometimes only the pattern of the bands in a gel is analyzed; other times, analysts are interested not only in the pattern, but also in quantifying the amount of nucleic acid or protein in each band. **Densitometry** *allows investigators to determine the amount of material in a band based on the intensity of its stain.* Densitometry uses light to measure how much material is present in a specific place on a gel or other matrix. When light is shined on a band in a gel, the amount of light absorbed is related to how much stain is present, which in turn depends on how much material is in the band, see Figure 25.11. The absorbance of light by a band is compared with the absorbance of light by the support (gel) alone. Thus, an area of the support without stain is the blank.

The source of light in dedicated densitometers is sometimes a laser that produces a very high-intensity red light. Laser densitometers can pick up very fine details in gels, assuming the bands absorb red light, as most do.

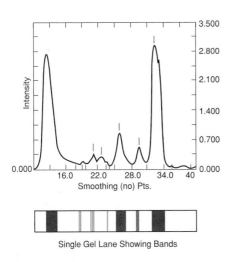

Single Gel Lane Showing Bands

Figure 25.11. Densitometry of the Bands on an Electrophoresis Gel. The peaks represent the amount of material in each band. (The Y axis represents absorbance.) (Graph courtesy of Beckman Coulter, Inc.)

Densitometry can be performed in various ways. The gel itself can be analyzed or a photograph or slide of the gel can be scanned. Densitometry can be performed either using conventional spectrophotometers with special adaptations or by using instruments dedicated to densitometry. When densitometry is performed with a conventional spectrophotometer, a special holder is required to place the gel into the sample chamber. The holder must be moveable so that each band can be exposed to the light beam, one at a time. Optical arrangements are needed to limit the light beam to a specific band, and a mechanism must be provided to move the gel in the sample chamber.

C. Spectrophotometers or Photometers as Detectors for Chromatography Instruments

There are many instances in biology where the components of a sample mixture are separated from one another before analysis. Chromatography is a commonly used separation method. During chromatography samples plus solvents flow through a matrix held in cylindrical columns or on flat plates. Depending on their chemical and physical nature, some components of a mixture move more quickly through the column or across the plate than others, so the components are separated from one another by their rate of motion.

As drops of sample and solvent leave a chromatography column, they can be routed to flow through a detector. There are many types of detectors available, but spectrophotometers and photometers are most common. A spectrophotometer or photometer used for chromatography does not involve cuvettes. Rather, there is a **flowcell** *through which the sample continuously flows*. As the sample moves through the

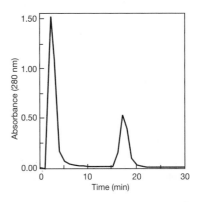

Figure 25.12. A Spectrophotometer as a Detector in Chromatography. The absorbance of a sample eluted from a chromatography column. Time is on the X axis and absorbance on the Y axis. The peaks represent the components of the sample that absorb light at 280 nm.

flowcell, a continuous absorbance reading is produced, see Figure 25.12. Because most biological materials absorb light in the UV range, these spectrophotometric detectors are frequently set to a UV wavelength (e.g., 280 nm).

Chromatography instruments may use a photometer that can measure light absorbance at only a single wavelength, or a few discrete wavelengths. Variable wavelength detectors have a monochromator and can be adjusted to any wavelength in a certain range. In some cases, spectrophotometers with photodiode array detectors (PDAs) are used in conjunction with a chromatography instrument. The advantage of a PDA is that the absorbance spectrum of each material can be instantaneously generated as it elutes from the column. The resulting spectra are useful in identifying the separated components of the sample.

D. Fluorescence Spectroscopy

So far, we have discussed primarily the absorption of light by a sample. **Fluorescence** is another type of interaction that can give useful information about a sample. In **fluorescence,** *light interacts with matter in such a way that photons excite electrons to a higher energy state. The electrons almost immediately drop back to their original energy level with the reemission of light.* The fluorescence of clothing or posters when exposed to "black lights" are familiar examples of this phenomenon.

Fluorophores *are molecules that fluoresce when exposed to light.* Most biological molecules are not naturally fluorescent. Fluorescent compounds, therefore, are used to label or tag desired biological structures. For example, ethidium bromide is a fluorophore that is hydrophobic, planar, and positively charged. Because of these properties, ethidium bromide spontaneously inserts into DNA molecules. One of the most common molecular biology techniques is to soak an electrophoresis gel containing fragments of DNA in a solu-

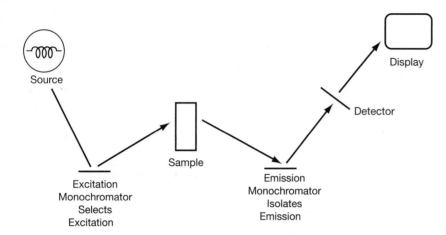

Figure 25.13. A Fluorescence Spectrometer. The excitation beam is directed at the sample. Fluorophores in the sample are excited by the excitation beam and emit fluorescent light, which passes through a second monochromator. The second monochromator isolates light of a wavelength matching that emitted by the fluorophore. A photomultiplier tube detects the light emitted by the sample.

tion containing ethidium bromide. The ethidium bromide intercalates into the DNA. Ethidium bromide complexed with DNA fluoresces when exposed to light in the UV-B range, so it lights up the bands of DNA when the gel is placed on a UV light source.

Fluorophores can be used to tag antibodies that bind to their target biological structures with high affinity. Fluorescent tags have also been used to label cancer cells, genes, proteins coded by specific genes, intracellular proteins, pathogens, and a host of other materials of biological importance.

The basic design of a fluorimeter (fluorescence spectrometer) is shown in Figure 25.13. Like a spectrophotometer, the fluorimeter has a light source that emits polychromatic light. The light enters a monochromator that functions to isolate light of a narrow range of wavelengths. In fluorescence spectroscopy, *the light that shines on the sample is called the* **excitation beam.** The wavelength of the excitation beam must be selected by the operator because different fluorophores are excited by different wavelengths of light. The excitation beam

shines on the sample. Fluorophores in the sample are excited by the light and emit fluorescence. At this point, a fluorescent spectrophotometer differs from a UV/Vis spectrophotometer. The emitted light passes through a second monochromator that is adjusted by the operator to isolate the light that was emitted by the fluorophore. A photomultiplier tube detects the light emitted from the sample.

Table 25.5 summarizes issues relating to fluorescence measurements.

PRACTICE PROBLEMS

1. a. Title the X and Y axis for each of the two graphs on the next page.

b. Which one is called an absorbance spectrum? Which one is called a standard curve?

c. Which is associated with quantitative analysis? Which is associated with qualitative analysis?

Table 25.5. ISSUES RELATING TO FLUORESCENCE MEASUREMENTS

1. **Quenching** *occurs when an interference in the sample prevents the analyte from fluorescing.* This is a difficult problem to detect and often shows up as less sensitivity in an assay than is expected. Some fluorescent systems are quenched by oxygen from the air. Low concentrations of metal ions, as are present in some cleaning agents, may cause quenching.

2. *Fluorescence is much more sensitive to changes in the environment than is absorption.* Fluorescence increases as temperature decreases. On warm days, therefore, there may be noticeably less fluorescence than on cool days. Standards and samples should be at the same temperature. Some instruments control temperature to eliminate this variable.

3. *pH may affect the fluorescence of a sample.*

4. *There are fluorescent additives in plastics.* These additives may leach into samples that are stored in plastic containers.

5. *Most detergents used to wash glassware are strongly fluorescent.* Check that the detergent used in the laboratory does not contribute to fluorescence at the wavelength being measured.

6. *Stopcock grease and some filter papers contain fluorescent contaminants.*

7. *Microorganisms contaminating a solution may contribute fluorescence or may scatter emitted light.*

8. *The strong light sources used to excite the sample may bleach the fluorescent compounds.* Bleaching may be reduced by the choice of a different excitation wavelength, or by reducing the intensity of the excitation beam.

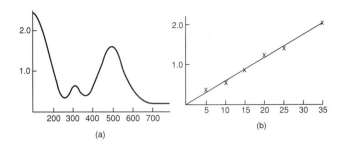

(a)

(b)

2. Label each of the following as a qualitative assay or a quantitative assay.

   a. Using an assay to determine the concentration of DNA in a sample.

   b. Using an infrared spectrum to identify an organic compound.

   c. Testing the amount of alcohol in a driver's blood.

   d. Determining the levels of lead in a child's blood.

   e. Determining whether a blood sample contains amphetamines.

   f. Determining whether or not a compound is oxygenated.

   g. Measuring the amount of product formed in an enzymatic reaction.

3. The following are absorption spectra for three compounds, A, B, and C, respectively. Assume all three compounds were present at the same concentration when the spectra were plotted and that the analysis conditions were the same.

   a. What color is each compound?

   b. Which of these compounds will have the highest absorptivity constant at a wavelength of 540 nm? Explain.

   c. Which of these will be detectable in the lowest concentration at 540 nm? Explain.

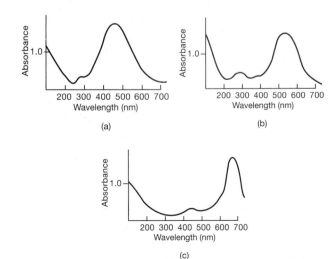

4. What is the linear range for the following assays?

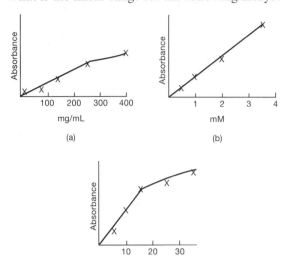

mg/mL

(a)

mM

(b)

ppm

(c)

5. a. Prepare a standard curve from the data below.

   b. Is this graph linear? If all or part of the graph is linear, what is the equation for the line?

   c. If all or part of the graph is linear, determine the value of the absorptivity constant based on the line. Remember to include the units. (Assume the path length is 1 cm.)

   d. Does this graph show a threshold? If so, where? If so, why is there a threshold?

Standard #	Concentration in mg/liter	Absorbance
1	2	0.22
2	4	0.45
3	10	1.11
4	12	1.34
5	20	2.23
6	30	2.21
7	40	2.22
8	50	2.21

6. A certain compound is detected at 520 nm. A series of dilutions were prepared and the data below were collected.

   a. Plot the data.

   b. Calculate the absorptivity constant based on the slope of the line. (Assume a path length of 1 cm.)

   c. Determine the amount of compound in the samples.

Standard #	Concentration in mM	Absorbance
1	0	0.0
2	80	0.2
3	160	0.4
4	240	0.8
5	320	1.0

Sample	Absorbance
A	0.18
B	0.31
C	0.96
D	1.0

7. a. Prepare a standard curve from the data below.

b. Determine the absorptivity constant. (Assume the path length is 1 cm.)

c. What is the concentration of an unknown which is diluted 5-fold and has an absorbance of 0.30?

Standard #	Concentration in mM	Absorbance
1	1	0.22
2	2	0.46
3	5	1.08
4	6	1.34
5	10	2.18

For problems 8–18, assume Beer's Law applies.

8. A certain sample has an absorbance of 1.6 when measured in a 1-cm cuvette.

a. If the same sample's absorbance is measured in a 0.5-cm cuvette, will its absorbance be greater than, less than, or equal to 1.6?

b. What is the absorbance of this sample in the 0.5-cm cuvette?

9. A 25 ppm solution of a certain compound has an absorbance of 0.87. What do we expect will be the absorbance of a 40 ppm solution? Explain.

10. A 10 mg/mL solution of a compound has a transmittance of 0.680.

a. What is the absorbance of this solution?

b. What will be the absorbance of a 16 mg/mL solution of this compound?

c. What will be the transmittance of a 16 mg/mL solution of this compound?

11. A solution contains 30 mg/mL of a light-absorbing substance. In a 1-cm cuvette its transmittance is 75% at a certain wavelength.

a. What will be the transmittance of 60 mg/mL of the same substance?

b. What will be the absorbance of 60 mg/mL of the same substance?

c. What will be the transmittance of 90 mg/mL of the same substance?

d. What will be the absorbance of 90 mg/mL of the same substance?

12. A standard solution of bovine serum albumin containing 1.0 mg/mL has an absorbance of 0.65 at 280 nm.

a. Based on the BSA result, what is the protein concentration in a partially purified protein preparation if the absorbance of the preparation is 0.15 at 280 nm?

b. Suggest two possible sources of error in this analysis.

13. The molar absorptivity constant of ATP is 15,400 L/mole-cm at 260 nm. (Value from Segel, I.H. *Biochemical Calculations.* 2nd ed. New York: John Wiley and Sons, 1976.) A solution of purified ATP has an absorbance of 1.6 in a 1-cm cuvette. What is its concentration?

14. The molar absorptivity constant for NADH is 15,000 L/mole-cm at 260 nm. (Value from Segel, I.H. *Biochemical Calculations.* 2nd ed. New York: John Wiley and Sons, 1976.) A sample of purified NADH has an absorbance of 0.98 in a 1.2-cm cuvette. What is its concentration?

15. The molar absorptivity constant for NADH is 15,000 L/mole-cm at 260 nm and is 6220 L/mole-cm at 340 nm. The molar extinction coefficient of ATP is 15,400 L/mole-cm at 260 nm and zero at 340 nm. (Values from Segel, I.H. *Biochemical Calculations.* 2nd ed. New York: John Wiley and Sons, 1976.) How could you determine the concentration of NADH in a sample that contains ATP as well as NADH?

16. Calculate the range of molar concentrations that can be used to produce absorbance readings between 0.1 and 1.8 if a 1-cm cuvette is used and the compound has a molar absorptivity of 10,000 L/mole-cm.

17. a. A sample of purified double-stranded DNA has an absorbance of 4 at 260 nm. The technician therefore dilutes it, mixing 1 mL of the DNA solution with 9.0 mL of buffer. This time the A_{260} reading is 1.25. What is the approximate concentration of the DNA in the original sample based on the equations in Box 25.2?

b. A sample of purified RNA has an absorbance at 260 nm of 2.4. The technician therefore dilutes it by mixing 1 mL of the RNA solution with 4.0 mL of buffer. This time the A_{260} reading is 0.63. Approximately how much RNA was in the original sample?

18. a. A sample of reasonably pure double-stranded DNA has an A_{260} of 0.85. What is the approximate concentration of DNA in the solution?

b. A sample of reasonably pure RNA has an A_{260} of 3.8. Because a spectrophotometer is not accurate reading absorbances that are so high, the analyst diluted the sample by adding 1 part sample to 9 parts buffer. The absorbance was then 0.69. What is the concentration of RNA in the solution?

19. Suppose a spectrophotometer has not been recently calibrated and its photometric accuracy is incorrect.

a. Will the results of a quantitative assay be inaccurate if a standard curve is used?

b. Will the results of a quantitative assay be inaccurate if a single standard is used?

c. Will the results of a quantitative assay be inaccurate if the concentration of sample is calculated based on an absorptivity constant from the literature?

20. Suppose in a colorimetric assay, depending on the concentration of analyte in the sample, full color development requires anywhere from 3 to 5 minutes. You do not realize that there is a timing effect and allow the reaction to proceed for only 2 minutes. Will this cause an error in your results?

21. Suppose you are using a colorimetric assay that involves a color-developing reagent that deteriorates over time. You accidentally use an old bottle of the reagent that is less effective than it should be in causing a color change. What effect will this have on your results?

22. You are performing a spectrophotometer assay in which you use a buffer to prepare the standards and for the blank; however, the samples are centrifuged blood. Is this a potential source of error?

23. (Note that this problem is based on data obtained in our laboratory during a study of UV estimation methods for nucleic acids and proteins.)

An analyst was evaluating the methods summarized in Box 25.2 on p. 484 to establish whether they would be useful in her investigations. She prepared three standards: (1) a standard containing pure double-stranded DNA, (2) a standard containing pure bovine serum albumin, and (3) a standard containing both double-stranded DNA and BSA mixed together. She measured the absorbance of the three standards at 260 and 280 nm and prepared an absorbance spectrum for all three standards. She then applied the equations shown in Box 25.2 to her data. Because she had prepared the standards from pure DNA and protein, she knew what the results of the calculations should have been. The steps she performed and the data she obtained are shown:

Step 1. The technician dissolved 0.06 mg of DNA in 1 mL of a low salt buffer. (Low salt buffer was used as the blank also.) The DNA spectrum she obtained is shown in Figure 25.10a on p. 480.
  The A_{280} for the DNA was 0.260.
  The A_{260} for the DNA was 0.497.

Step 2. She dissolved 0.5 mg of BSA in 1 mL of a low salt buffer. The protein spectrum is shown in Figure 25.10b on p. 480.
  The A_{280} for the protein was 0.276.
  The A_{260} for the protein was 0.176.

Step 3. She prepared a solution containing 0.06 mg/mL DNA and 0.5 mg/mL BSA in a low salt buffer. The spectrum for the mixture is shown in Figure 25.10c.
  The A_{280} for the mixture was 0.547.
  The A_{260} for the mixture was 0.685.

a. Estimate the concentration of DNA in the DNA standard based on its UV absorbance and on Equation 1 in Box 25.2 on p. 484. The concentration of DNA in the standard was 0.06 mg/mL (based on its weight). Comment on the accuracy of the UV estimation method.

b. Estimate the concentration of protein based on its UV absorbance. Use the reported value that 1 mg/mL of BSA has an absorbance of 0.7 at 280 nm. The concentration of protein in the standard was 0.5 mg/mL (based on its weight). Comment on the accuracy of the UV estimation method.

c. Check whether the absorbances at 260 nm and at 280 nm were additive as predicted by the equations:

$$A_{260} = A_{260} \text{ proteins} + A_{260} \text{ nucleic acids}$$
$$A_{280} = A_{280} \text{ proteins} + A_{280} \text{ nucleic acids}$$

Substitute into the first equation the A_{280} reading from the pure BSA sample and the A_{280} reading from the DNA sample. Substitute into the second equation the A_{260} readings for both pure samples. Comment as to whether the absorbances were additive.

d. Calculate the A_{260}/A_{280} ratios for all three standards. Comment as to whether the ratios are as expected.

e. (Optional) Substitute the A_{260} and A_{280} values obtained for the mixture into the Warburg-Christian equations, 9 and 10 in Box 25.2 on p. 484. Comment as to the accuracy of the Warburg-Christian equations.

f. Comment on all the preceding results.

QUESTIONS FOR DISCUSSION

A mixture of phenol, chloroform, and isoamyl alcohol (25:24:1) is used in the separation of nucleic acids from cellular extracts (see page 556 for a brief discussion of phenol extractions). Phenol is normally removed from the DNA preparations by precipitating the DNA from solution and then resuspending it in buffer. The absorbance spectrum for a dilute solution of the phenol mixture (0.1%) is shown in the graph. Explain the significance of this spectrum to analysts working with nucleic acids and proteins.

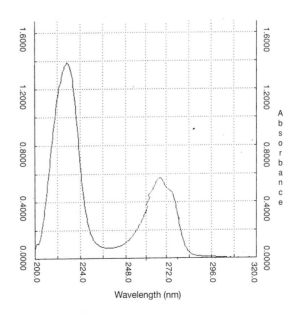

Chapter Appendix
Calculating the Line of Best Fit

i. OVERVIEW

A **standard curve** *displays the relationship between the concentration (or amount) of an analyte present and an instrument's response to that analyte.* Figure 25.14 shows a standard curve involving a spectrophotometer. Recall that the variable that is controlled by the investigator (in this example the concentration of compound) is termed the *independent variable* and is plotted on the X axis. The variable that varies in response to the independent variable (in this example the absorbance) is the *dependent variable* and is plotted on the Y axis. When a standard curve is prepared, we expect that the response of the instrument will be entirely determined by the concentration of analyte in the standard. If this expectation is met (and if the relationship between concentration and instrument response is linear), then all the points will lie exactly on a line; however, there are sometimes slight errors or inconsistencies when a standard curve is constructed. These small errors cause the points to not lie exactly on a line, as shown in Figure 25.14.

As is shown in Figure 25.14, the points can easily be represented with a line, although the points do not all fall exactly on the line. In a situation like this, where a series of points approximate a line, we can use a ruler to draw "by eye" a line that seems to best describe the points. Drawing a line "by eye," however, is not the most accurate method, nor does it give consistent results, see Figure 25.15. It is more accurate to use a statistical method to calculate the equation for the line that best describes the relationship between the two variables. *The statistical method used to determine the equation for a line is called* **the method of least squares.** The least

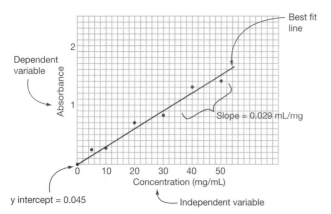

Figure 25.14.   The Relationship between the Concentration of Analyte and the Response of an Instrument; a Standard Curve. Due to slight inconsistencies, the points do not all lie exactly on a line.

y intercept = 0.045

Slope = 0.029 mL/mg

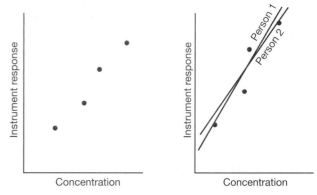

Figure 25.15.   Drawing a Best Fit Line "By Eye." Two people connect the concentration points into a best fit line by eye with slightly different results.

squares method calculates the equation for a line such that the total deviation of the points from the line is minimized. *The equation for the line calculated using the method of least squares is called the* **regression equation.** *The line itself is called the* **best fit line.**

Recall that the equation for a line has the form:

$$Y = \text{slope}(X) + Y \text{ intercept}$$
$$\text{or}$$
$$Y = mX + a$$

where
m = the slope
a = the Y intercept
X = the independent variable
Y = the dependent variable

The least squares method is used to calculate: (1) the slope and (2) the Y intercept of the line that best fits the points. (The equations to perform these calculations are shown later.)

When the least squares method is applied to the data graphed in Figure 25.14, the slope for the linear regression line is calculated to be 0.0285 mL/mg. The Y intercept is calculated to be 0.045. Therefore, the equation for the line best fitting the points in Figure 25.14 is:

$$Y = \left(0.0285 \, \frac{\text{mL}}{\text{mg}}\right)X + 0.045$$

Note that the slope and the Y intercept can have units. (In this example, absorbance has no units; therefore, the Y intercept has no units.)

Most modern spectrophotometers can automatically calculate the equation for the line of best fit for a series of points. Computers and some calculators can also be used to facilitate the determination of the equation for a line of best fit.

ii. CALCULATING THE LINE OF BEST FIT

This section describes the method of least squares that is used to calculate the linear regression equation. It is based on the following assumptions:

- *The values of the independent variable,* X, *are controlled by the investigator.*

- *The dependent variable,* Y, *must be approximately normally distributed.* (For example, if the same sample of compound is measured many times with a spectrophotometer, the absorbance values will vary slightly from reading to reading due to random error. These various measurements are expected to be normally distributed.)

- *If there is a relationship between the two variables, it is linear.*

(1) To calculate the slope of the line of best fit, the equation is:

$$m = \frac{\sum XY - \dfrac{\sum X \sum Y}{N}}{\sum X^2 - \dfrac{(\sum X)^2}{N}}$$

(2) To calculate the Y intercept, the equation is:

$$a = \overline{Y} - m\overline{X}$$

where $\overline{X}$ and $\overline{Y}$ are the means of the X and Y values, respectively

$$\overline{X} = \frac{\sum X}{N} \text{ and } \overline{Y} = \frac{\sum Y}{N}$$

N = the sample size

These calculations are summarized in Box 25.3.

As an example we show how the linear regression equation is calculated for the data graphed in Figure 25.14. Table 25.6 shows the data for this example along with preliminary calculations.

The values from Table 25.6 can be "plugged into" the equations for the best fit line. The slope of the line is:

$$m = \frac{164.45 \text{ mg/mL} - \dfrac{733.15 \text{ mg}}{7 \text{ mL}}}{5525 \dfrac{\text{mg}^2}{\text{mL}^2} - \dfrac{24025 \text{ mg}^2}{7 \text{ mL}^2}} \approx 0.0285 \text{ mL/mg}$$

The Y intercept is:

$$a = 0.676 - (0.0285 \text{ mL/mg})(22.14 \text{ mg/mL}) \approx 0.045$$

The regression equation for this line, therefore, is:

$$Y \approx \left(0.0285 \frac{\text{mL}}{\text{mg}}\right)X + 0.045$$

Box 25.3.   CALCULATION OF THE LINEAR REGRESSION EQUATION

Note: Keep track of the units when calculating and expressing the equation of a line.

1. **Arrange the data into a table of X and Y pairs.**
2. **Calculate the sum of the X values, $\sum$X.**
3. **Calculate $\overline{X}$.**
4. **Calculate $\sum$X² by squaring each X value and adding the results together.**
5. **Calculate $(\sum$X$)^2$ by squaring the sum of the X values.**
6. **Calculate $\sum$Y.**
7. **Calculate $\overline{Y}$.**
8. **Calculate $\sum$X$\sum$Y.**
9. **Multiply each X by its corresponding Y value and add the products to obtain $\sum$XY.**
10. **Calculate the slope, m, using the preceding equation.**
11. **Calculate the Y intercept using the preceding equation.**
12. **Place the values calculated for the slope and the Y intercept into the equation:**
    Y = mX + a

Table 25.6.   CALCULATING THE LINE OF BEST FIT

Concentration of Analyte X (mg/mL)	Absorbance Y (Absorbance has no units)	XY (mg/mL)	X² (mg²/mL²)
0	0	0	0
5	0.25	1.25	25
10	0.26	2.60	100
20	0.70	14.00	400
30	0.82	24.60	900
40	1.30	52.00	1600
50	1.40	70.00	2500

$\sum X = 155$ mg/mL $\qquad$ $\sum Y = 4.73$ $\qquad$ $\sum XY = 164.45$ mg/mL $\qquad$ $\sum X^2 = 5525$ mg²/mL²

$\overline{x} \approx 22.14$ mg/mL $\qquad$ $\overline{Y} \approx 0.676$ $\qquad$ $N = 7$ $\qquad$ $\sum X \sum Y = 733.15$ mg/mL $\qquad$ $(\sum X)^2 = 24025$ mg²/mL²

UNIT VII

Laboratory Solutions

Chapters in This Unit

- ✦ Chapter 26: Preparation of Laboratory Solutions A: Concentration Expressions and Calculations

- ✦ Chapter 27: Preparation of Laboratory Solutions B: Basic Procedures and Practical Information

- ✦ Chapter 28: Solutions: Associated Procedures and Information

- ✦ Chapter 29: Laboratory Solutions to Support the Activity of Biological Macromolecules

- ✦ Chapter 30: Culture Media for Intact Cells

Biological systems evolved in an aqueous environment. The recombinant DNA techniques that allow investigators to manipulate genes are based on enzymatic reactions that occur only in the proper aqueous solutions. Mammalian cells grown in culture require a carefully controlled aqueous medium. The first step in nearly every biological procedure, whether for research or production purposes, is the preparation of aqueous solutions containing the proper components to support the biological system of interest.

A **solution** *can be defined as a homogeneous mixture in which one or more substances is (are) dissolved in another.* **Solutes** *are the substances that are dissolved in a solution. The substance in which the solutes are dissolved is called the* **solvent.** Although solutes and solvents may be gases, liquids, or solids, in biological applications, the solutes are solids or liquids and the solvent is a liquid—most often, water. In this unit, the term **biological solution** *refers broadly to biological materials and the liquid environments in which they are contained in the laboratory.*

This unit discusses a number of considerations relevant to the preparation and use of solutions in biology laboratories.

Chapter 26 discusses how solute concentrations are expressed in "recipes" and introduces the calculations associated with solution preparation.

Chapter 27 introduces practical considerations relating to the mixing of solutions; buffers; and calculations for solutions that contain more than one solute.

Chapter 28 discusses the purification of water for laboratory use, cleaning of glassware and plasticware, sterilization of solutions, and storage of solutions containing biological materials.

Chapter 29 explores the reasons why biological solutions for proteins and nucleic acids contain certain components.

Chapter 30 introduces general considerations relating to the preparation of media that are used to support cultured bacterial and mammalian cells.

BIBLIOGRAPHY FOR UNIT VII

For general information about molarity, molality, normality, and percent solutions, consult any basic chemistry textbook.

Manufacturers' catalogs are a valuable source of information about chemicals. The Sigma-Aldrich Company catalog, for example, contains formula weights, compound names, and formulas for many chemicals (www.Sigmaaldrich.com).

General References

Fasman, Gerald D., ed. *The Practical Handbook of Biochemistry and Molecular Biology*. Boca Raton, FL: CRC Press, 1989. (Contains extensive information about biological materials including buffers for biological systems.)

The Merck Index: An Encyclopedia of Chemicals, Drugs, and Biologicals. 14th ed. Whitehouse Station, NJ: Merck, 2006. (An encyclopedia that contains information about thousands of biologically relevant chemicals including their structures, densities, molecular weights, and solubilities.)

Shugar, Gershon, and Ballinger, Jack T. *Chemical Technicians' Ready Reference Handbook*. 4th ed. New York: McGraw-Hill, 1996.

Water Purification and Ancillary Methods

McLaughlin, Malcolm. "Stopping Residue Interference on Labware and Equipment." *American Laboratory News Edition*, June 1992.

Perkins, John J. *Principles and Methods of Sterilization in the Health Sciences*. 2nd ed. Springfield, IL: Charles C. Thomas, 1983. (A classic in-depth reference on sterilization methods.)

Pure Water Handbook. 2nd ed. Minnetonka, MN: Osmonics. osmolabstore.com/documents/pwh-s.pdf.

The Water Book. Dubuque, IA: Barnstead/Thermolyne Corp.

The Water Purification Primer. Bedford, MA: Millipore Corporation.

Components of Solutions

Ausubel, Fred M., Brent, Roger, Kingston, Robert E., et al., eds. *Current Protocols in Molecular Biology*. New York: John Wiley and Sons, 2007.

Deutscher, Murray P., ed. "Guide to Protein Purification," in *Methods in Enzymology,* vol. 182. San Diego, CA: Academic Press, 1990.

Sambrook, Joseph, and Russell, David W. *Molecular Cloning: A Laboratory Manual*. 3rd ed. Cold Spring Harbor, MD: Cold Spring Harbor Laboratory Press, 2001.

Scopes, Robert K. *Protein Purification: Principles and Practice*. 3rd ed. New York: Springer-Verlag, 1994.

Culture Media

BACTERIAL MEDIA

Atlas, Ronald M. *Handbook of Microbiological Media*. 3rd ed. Boca Raton, FL: CRC Press, 2004. (Contains the formulations, methods of preparation, sources, and uses for several thousand different media.)

Difco and BBL Manual of Microbiological Culture Media. BD Diagnostic Systems. http://www.bd.com/ds/techincalcenter/difcoBblManual. asp.

MAMMALIAN MEDIA

ATCC supplies cells and provides a number of resources relating to the culture of all types of cells, including published references, media formulations, and web-based technical information, http://www.atcc.org.

Bertani, G. "Lysogeny at Mid-Twentieth Century: P1, P2, and Other Experimental Systems." *Journal of Bacteriology* 186 (2004): 595–600. (Contains, among other topics, a historical note relating to the naming of LB broth.)

Butler, M., ed. *Mammalian Cell Biotechnology: A Practical Approach*. Oxford, UK: IRL Press, 1991.

Butler, M. and Dawson, M., eds. *Cell Culture LabFax*. Oxford, UK: Bios Scientific, 1992. (Contains extensive recipes and technical information.)

"Culture Media: A Growing Concern in Biotechnology," Supplement to *BioProcess International Magazine*, June 2005. (Contains ample information about culture media from a production standpoint. Also contains basic cell biology background.)

Darling, D.C., and Morgan, S.J. *Animal Cells Culture and Media: Essential Data*. Chichester, UK: Wiley, 1994. (Small handbook with extensive technical information.)

Eagle, Harry. "The Specific Amino Acid Requirements of a Mammalian Cell (Strain L) in Tissue Culture." *Journal of Biological Chemistry* 214, no. 2 (1955): 839. (A seminal paper in mammalian tissue culture.)

Freshny, R. Ian. *Culture of Animal Cells: A Manual of Basic Technique*. 5th ed. Hoboken, NJ: Wiley-Liss, 2005. (A commonly referenced manual of mammalian cell culture techniques with extensive information for both beginners and more experienced culturists.)

Martin, Bernice M. *Tissue Culture Techniques*. Boston: Birkhäuser Press, 1994. (A handy introduction to the basic principles of cell culture.)

COMPANIES THAT DISTRIBUTE CELL CULTURE PRODUCTS PROVIDE INFORMATION ON THEIR WEBSITES

BD Diagnostic Systems, http://www.bd.com/ds/productCenter/DCM-Ingredients.asp

Neogen Corporation, http://neogen.com/acumedia.htm

Promocell, http://www.promocell.com/us/homepage.htm

Sigma-Aldrich, http://www.sigmaaldrich.com/Area_of_Interest/Life_Science/Cell_Culture.html

Preparation of Laboratory Solutions A: Concentration Expressions and Calculations

I. OVERVIEW

Solution preparation involves various basic laboratory procedures, such as weighing compounds and measuring the volumes of liquids. Solution preparation also requires interpreting the "recipe" (procedure) for the solution. The following two chapters discuss the basic procedures and the calculations that are the cornerstones of solution making.

Preparing laboratory solutions can be compared with baking. Suppose you decide to bake a batch of apple crisp. The first step is to find a recipe, Figure 26.1, that states:

1. The *components* of apple crisp
2. *How much* of each component is needed
3. *Advice for preparing* (mixing and baking) the components properly

To make apple crisp successfully you need to understand the terminology in the recipe and to measure, combine, and prepare the ingredients in the proper fashion.

Preparing a laboratory solution has much in common with baking. You need a recipe (procedure) for

preparing the solution. You follow the procedure and combine the right *amount(s)* of each *component* (solute) in the right volume of solvent to get the correct *concentration* of each solute at the end. You may also need to adjust the solution to the proper pH, sterilize it, or perform other manipulations. There are differences, however, between baking and preparing laboratory solutions. One important difference is that it is appropriate to creatively modify a recipe when baking, but laboratory procedures should be strictly followed.

The recipe for apple crisp lists the *amount* of each component needed. A procedure to make a laboratory solution may similarly list the amount of each solute to use. For example, laboratory recipe I, Figure 26.1, lists the amount of each of the four solutes required: Na_2HPO_4, KH_2PO_4, $NaCl$, and NH_4Cl. Water is the solvent. The total solution has a volume of 1 L. In overview, the procedure for preparing laboratory recipe I is:

1. Weigh out the needed amount of each solute.
2. Dissolve the solutes in less than 1 L of water.
3. Add enough water so that the final volume is 1 L.

KITCHEN RECIPE I
APPLE CRISP

Preheat oven to 375°F
    12 oz can frozen
        lemonade
    3 lb sliced apples
Reconstitute juice with water.
Cover apples with juice and
refrigerate.

Blend together: 3 c rolled oats
              1/2 c honey
              1 c flour
              3/4 c butter
              1/2 tsp cinnamon

Place apples in bottom of 9 x 13 inch
baking pan. Crumble oat mixture
over apples. Pour juice over apples.
Bake 1 hour.

LABORATORY RECIPE I

Na_2HPO_4	6 g
KH_2PO_4	3 g
NaCl	0.5 g
NH_4Cl	1 g

Dissolve in water
Bring to a volume of 1 L

Figure 26.1. **Comparison of a Kitchen Recipe and a Laboratory Recipe.**

Laboratory recipe I is easy to interpret. Many recipes for laboratory solutions, however, do not show the *amount* of each solute required; rather, they show the *concentration* of each solute needed. **Amount** and **concentration** are not synonyms. **Amount** *refers to how much of a component is present.* For example, 10 g, 2 cups, and 30 mL are amounts. **Concentration** *is an amount per volume; it is a fraction.* The numerator is the amount of the solute of interest. The denominator is usually the volume of the entire solution, the solvent and the solute(s) together. For example, "2 g of NaCl/L" means there are 2 g of NaCl dissolved in enough liquid so that the total volume of the solution is 1 L.

Concentration is always a fraction, although concentrations are sometimes expressed in ways that do not look like fractions. For example, 2% milk is an expression of concentration that does not at first glance appear to be a fraction. But, in fact, the % symbol represents a numerator and denominator separated by a line. "2%" milk has 2 g fat/100 mL of liquid. ("Percent" comes from Latin and means *per hundred*.) Molarity, molality, and normality, which are covered later in this chapter, are also expressions of concentration where the numerator and denominator are not explicitly shown.

There are various ways to express concentration, each of which is associated with certain mathematical calculations. The next sections discuss these concentration expressions and calculations and give examples that biologists are likely to encounter.

II. Types of Concentration Expressions and the Calculations Associated with Each Type

A. Weight per Volume

The simplest way to express the concentration of a solution is as a fraction with the amount of solute, expressed as a weight, in the numerator and the volume in the denominator. This type of concentration expression is often used for small amounts of chemicals and specialized biological reagents. For example, a solution of 2 mg/mL proteinase K (an enzyme) contains 2 mg of proteinase K for each milliliter of solution.

When concentrations are expressed as fractions, proportions can easily be used to determine how much solute is required to make the solution. For example:

How much proteinase K is needed to make 50 mL of proteinase K solution at a concentration of 2 mg/mL?

$$\frac{?}{50\ \text{mL}} = \frac{2\ \text{mg proteinase K}}{1\ \text{mL}}$$

? = 100 mg = amount proteinase K needed.

EXAMPLE PROBLEM

How many micrograms of DNA are needed to make 100 µL of a 100 µg/mL solution?

ANSWER

The amount of DNA required is found by using a proportion, but first 100 µL must be converted to milliliters, or milliliters must be converted to microliters so the units are consistent.

$$100\ \mu L = 0.1\ \text{mL}$$

$$\frac{?}{0.1\ \text{mL of solution}} = \frac{100\ \mu g\ \text{DNA}}{1\ \text{mL solution}}$$

? = 10 µg = amount of DNA needed.

Table 26.1 summarizes some practical considerations regarding the preparation of solutions.

Table 26.1. Practical Information

1. **SOLVENT.** When the solvent is not stated, assume it is distilled or otherwise purified water. (Water purification is discussed in Chapter 28.)

2. **"BRINGING A SOLUTION TO VOLUME."** In the proteinase K example, the enzyme will take up some room in the solution. If you add the proteinase K to 50 mL of water, the volume will be slightly more than 50 mL. The most accurate way to prepare the solution, therefore, is to dissolve the proteinase K in *less* than 50 mL of water and then bring the solution to a final volume of exactly 50 mL using a volumetric flask or graduated cylinder.

   This procedure is called **bringing the solution to the desired final volume (BTV or "bring to volume").** Figure 26.2 on p. 496 illustrates a method of bringing a solution to volume.

   Note that in practice 100 mg of proteinase K takes up so little volume that it is sensible simply to add the enzyme to exactly 50 mL of water. In general, however, the correct way to make a solution is to "bring it to volume."

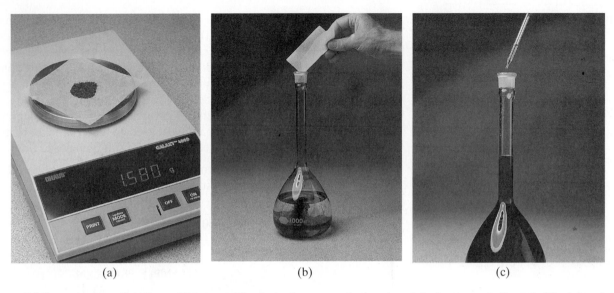

Figure 26.2. Bringing a Solution to Volume. a. The desired amount of solute is weighed out on a balance. **b.** The solute is dissolved with mixing in less than the total solution volume. (A beaker is often used when stirring the solution at this step.) **c.** The solution is brought to its final volume by slowly filling the flask to its mark.

PRACTICE PROBLEMS: WEIGHT PER VOLUME

1. How would you prepare 100 mL of an aqueous solution of $AgNO_3$ of strength 0.1 g $AgNO_3$/mL?

2. How many milligrams of NaCl is present in 50 mL of a solution that is 2 mg/mL NaCl?

3. How would you prepare 5 mL of proteinase K at a concentration of 100 μg/mL?

B. Molarity

Molarity *is a concentration expression that is equal to the number of moles of a solute that are dissolved per liter of solution.* Molarity is used to express concentration when the number of molecules in a solution is important. For example, in an enzyme-catalyzed reaction the numbers of molecules of each reactant and the enzyme are important. If there is too little enzyme relative to the number of molecules of reactants present, the reaction may be incomplete, whereas adding too much enzyme is costly and inefficient.

A brief review of terminology: A **mole** *of any element always contains* 6.02×10^{23} *(Avogadro's number) atoms.* Because some atoms are heavier than others, a mole of one element weighs a different amount than a mole of another element. *The weight of a mole of a given element is equal to its atomic weight in grams, or its* **gram atomic weight.*** Consult a periodic table of elements to find the atomic weight of an element. For example, one mole of the element carbon weighs 12.0 g.

Compounds *are composed of atoms of two or more elements that are bonded together.* A mole of a compound contains 6.02×10^{23} molecules of that compound. The

gram formula weight (**FW**) or **gram molecular weight** (**MW**) *of a compound is the weight in grams of 1 mole of the compound,* see Figure 26.3. The *FW* is calculated by adding the atomic weights of the atoms that make up the compound. For example, the gram molecular weight of sodium sulfate (Na_2SO_4) is 142.04 g:

2 sodium atoms 2×22.99 g = 45.98 g
1 sulfur atom 1×32.06 g   = 32.06 g
4 oxygen atoms 4×16.00 g = 64.00 g
                                   142.04 g

A I molar solution of a compound contains I *mole*
of that compound dissolved in I L of total solution.

For example, a 1 molar solution of sodium sulfate contains 142.04 g of sodium sulfate in 1 liter of total solution, see Figure 26.4. We also say that a "1 molar"

Figure 26.3. 1 Mole of Various Substances.

---

*Weight and mass are not synonyms and it is correct to speak of gram molecular *mass* in the context of molarity. It is common practice, however, to speak of "atomic weight," "molecular weight," and "formula weight," as we do in this book.

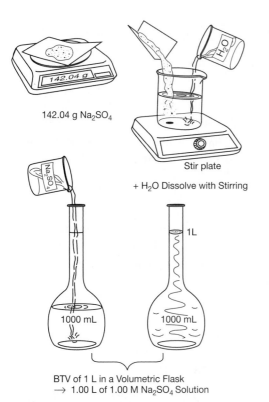

Figure 26.4. A 1 M Solution of Sodium Sulfate (Na$_2$SO$_4$). The FW of Na$_2$SO$_4$ is 142.04. This amount is dissolved in water so that the total volume is 1.00 L.

solution has a "molarity of 1." Note that a "mole" is an expression of *amount* and "molarity" and "molar" are words referring to the *concentration* of a solution.

The word *molar* is abbreviated with an upper case *M*. It is also common in biology to speak of "millimolar" (mM) and "micromolar" (μM) solutions. A **millimole** *is 1/1000 of a mole;* a **micromole** *is 1/1,000,000 of a mole.* For example:

1 M NaCl   = 1 *mole* or 58.44 g of NaCl in 1 L of solution

1 mM NaCl = 1 *mmole* or 0.05844 g of NaCl in 1 L of solution

1 μM NaCl = 1 *μmole* or 0.00005844 g of NaCl in 1 L of solution

---

EXAMPLE PROBLEM

How much solute is required to prepare 1 L of a 1 M solution of copper sulfate (CuSO$_4$) solution?

ANSWER

Calculate the *FW* or find it on the chemical's bottle. By calculation:

$$1 \text{ copper } = 1 \times 63.55 \text{ g} = 63.55 \text{ g}$$
$$1 \text{ sulfur } = 1 \times 32.06 \text{ g} = 32.06 \text{ g}$$
$$4 \text{ oxygens} = 4 \times 16.00 \text{ g} = 64.00 \text{ g}$$

$$FW = 159.61 \text{ g}$$

159.61 g of CuSO$_4$ is required to make 1 L of 1 M CuSO$_4$.

EXAMPLE PROBLEM

How much solute is required to prepare 1 L of a 1 mM solution of CuSO$_4$?

ANSWER

The formula weight of copper sulfate is 159.61; therefore,

$$1 \text{ mmole} = \frac{159.61 \text{ g}}{1000} = 0.15961 \text{ g}$$

1 L of 1 mM CuSO$_4$ requires 0.15961 g.

EXAMPLE PROBLEM

How many micromoles are there in 17.4 mg of NAD, FW 663.4? (NAD, nicotinamide adenine dinucleotide, is an organic compound that is important in metabolism.)

ANSWER

First, convert 17.4 mg to grams. 17.4 mg = 0.0174 g. Using a proportion:

$$\frac{1 \text{ mole}}{663.4 \text{ g}} = \frac{?}{0.0174 \text{ g}}$$

$$? \approx 2.62 \times 10^{-5} \text{ moles} = 26.2 \text{ micromoles}$$

Molarity is a concentration expression; concentration expressions are ratios. We can determine how to prepare solutions of different molarities or different volumes by using straightforward proportional relationships. For example:

How much solute is required to prepare 1 L of 0.25 M sodium chloride solution? Using the reasoning of proportions, if a 1 M solution of NaCl requires 58.44 g of solute, then:

$$\frac{?}{0.25 \text{ M}} = \frac{58.44 \text{ g}}{1 \text{ M}}$$

? = 14.61 g = amount of solute to make 1 L of 0.25 M NaCl.

EXAMPLE PROBLEM

How much solute is required to prepare a 1 L solution of NaCl of molarity 2.5?

ANSWER

If a 1 M solution of sodium chloride requires 58.44 g of solute, then:

$$\frac{58.44 \text{ g}}{1 \text{ M}} = \frac{?}{2.5 \text{ M}}$$

? = 146.1 g = amount of solute required to make 1 L of 2.5 M NaCl.

It is sometimes necessary to make more or less than 1 L of a given solution. In these cases, proportions can again be used to determine how much solute is required. For example:

How much solute is required to make 500 mL of a 0.25 M solution of NaCl?

We know that 14.61 g of solute is required to make 1 L of 0.25 M NaCl.

$$\frac{?}{500 \text{ mL}} = \frac{14.61 \text{ g}}{1000 \text{ mL}}$$

? ≈ 7.31 g NaCl = amount of solute to make 500 mL of 0.25 M NaCl.

Thus, it is possible to use proportions to calculate how much solute is needed to make a solution with a molarity other than 1 M and a volume other than 1 L. In practice, however, most often people use a formula to calculate the amount of solute required to make a solution of a particular volume and a particular molarity. This can be written as shown in Formula 26.1.

The following example problem illustrates both the proportion strategy and the use of Formula 26.1 to calculate how much solute is required to make a molar solution.

Formula 26.1. CALCULATION OF HOW MUCH SOLUTE IS REQUIRED FOR A SOLUTION OF A PARTICULAR MOLARITY AND VOLUME

Solute Required = (Grams/Mole)(Molarity)(Volume)

where:
Solute Required is the amount of solute needed, in grams
Grams/Mole (grams per mole) is the number of grams in 1 mole of solute
Molarity is the desired molarity of the solution expressed in moles per liter
Volume is the final volume of the solution, in liters

Note: To use this formula, volume must be expressed in units of liters. To convert a volume expressed in milliliters to liters, simply divide the volume in milliliters by 1000. For example:

To convert 100 mL to liters:

Divide by 1000

$$\frac{100}{1000} = 0.1$$

Therefore, 100 mL = 0.1 L

Similarly, 3420 mL = $\frac{3420}{1000}$ = 3.420 L

EXAMPLE PROBLEM

How much solute is required to prepare 300 mL of a 0.800 M solution of calcium chloride? (FW = 111.0)

ANSWER

Strategy 1—Proportions

First, determine how much solute is needed to make 1 L of 0.800 M. We know that 111.0 g is required to make 1 L of a 1 M solution. So:

$$\frac{?}{0.800 \text{ M}} = \frac{111.0 \text{ g}}{1 \text{ M}}$$

? = 88.8 g = amount of solute to make 1 L of 0.800 M solution.

Second, determine how much solute is needed to make 300 mL.

$$\frac{?}{300 \text{ mL}} = \frac{88.8 \text{ g}}{1000 \text{ mL}}$$

? = 26.64 g = grams of solute required to make 300 mL of 0.800 M solution.

Strategy 2—Formula

First, convert 300 mL to liters by dividing by 1000:

$$\frac{300}{1000} = 0.300 \text{ Therefore, 300 mL} = 0.300 \text{ L}$$

Plug into Formula 1:

Solute Required =

$$\frac{(111.0 \text{ g})}{\text{mole}} \quad \frac{(0.800 \text{ mole})}{\text{L}} \quad (0.300 \text{ L})$$

= 26.64 g = grams of solute required.

Observe that the units cancel, leaving the answer expressed in grams.

We have now shown how to calculate the amount of solute required to prepare a solution of a given molarity and volume using two different strategies. Either strategy when performed correctly will give the right answer. Box 26.1 outlines the procedure for making a solution of a particular volume and molarity.

EXAMPLE PROBLEM

How would you prepare 150 mL of a 10 mM solution of $Na_2SO_4 \cdot 10H_2O$ using the procedure outlined in Box 26.1?

ANSWER

Step 1. Find the FW of the solute. The FW is 322.04.

Step 2. Determine the molarity required. The molarity required is 10 mM, which is equal to 0.010 M.

Step 3. Determine the volume required. The volume required is 150 mL, which is equal to 0.150 L.

Step 4. Determine how much solute is necessary.

Box 26.1. PROCEDURE TO MAKE A SOLUTION OF A PARTICULAR VOLUME AND MOLARITY

1. ***Find the FW of the solute.*** *It is possible to calculate the FW of a compound by adding the atomic weights of its constituents. For compounds with complicated formulas or that come in more than one form, it is better to find the FW on the label on the chemical's container or by looking in the manufacturer's catalog.*
2. ***Determine the molarity required.***
3. ***Determine the volume required.***
4. ***Determine how much solute is necessary either by using proportions or Formula 26.1.***
5. ***Weigh out the amount of solute required as calculated in step 4.***
6. ***Dissolve the weighed out compound in less than the desired final volume of solvent.***
7. ***Place the solution in a volumetric flask or graduated cylinder. Add solvent until exactly the right volume is reached (BTV).***

Notes

i **Hydrates** *are compounds that contain chemically bound water.* The bound water does not make the compounds liquid; rather, they remain powders or granules. The weight of the bound water is included in the FW of hydrates. For example, calcium chloride can be purchased either as an anhydrous form with no bound water, or as a dihydrate. Anhydrous calcium chloride, $CaCl_2$, has a formula weight of 111.0. The dihydrate form, $CaCl_2 \cdot 2H_2O$, has a formula weight of 147.0 (111.0 plus the weight of two waters, 18.0 each). When hydrated compounds are dissolved, the water is released from the compound and becomes indistinguishable from the water that is added as solvent.

ii Compounds that come as liquids can be weighed out, but it may be easier to measure their volume with a pipette or graduated cylinder. Convert the required weight to a volume based on the density of the compound. (See pp. 236–237 for a discussion of density.)

Strategy 1—Using Proportions to Calculate How Much Solute is Necessary

If it requires 322.04 g to make 1 L of a 1 M solution, then:

$$\frac{322.04 \text{ g}}{1.000 \text{ M}} = \frac{?}{0.010 \text{ M}}$$

? = 3.2204 g = solute needed to make 1 L of 10 mM (i.e., 0.010 M) solution.

To make 150 mL:

$$\frac{?}{150 \text{ mL}} = \frac{3.2204 \text{ g}}{1000 \text{ mL}}$$

? ≈ 0.4831 g = amount of solute required to make 150 mL of 10 mM solution.

Strategy 2—Using Formula 26.1 to Calculate How Much Solute is Necessary

Convert 150 mL to liters, 150/1000 = 0.150 L.

Substitute values into Formula 26.1:

Solute Required = (Grams/Mole) (Molarity) (Volume)

$$\frac{322.04 \text{ g}}{1 \text{ mole}} \times \frac{0.010 \text{ mole}}{\text{L}} \times 0.150 \text{ L}$$

$$\approx 0.4831 \text{ g}$$

Step 5. Weigh out the amount of solute required as calculated in step 4, that is, 0.4831 g of $Na_2SO_4 \cdot 10H_2O$.

Step 6. Dissolve the weighed out compound in less than the desired final volume (150 mL) of solvent.

Step 7. Place the solution in a volumetric flask or graduated cylinder. Add solvent until exactly 150 mL is reached (BTV).

PRACTICE PROBLEMS: MOLARITY

Assume water is the solvent for all questions.

1. If you have 3 L of a solution of potassium chloride at a concentration of 2 M, what is the solute? _____ What is the solvent? _____ What is the volume of the solution? _____ Express 2 M as a fraction. _____

2. How much solute is required to prepare 250 mL of a 1 molar solution of KCl?

3. How would you prepare 10 L of 0.3 M KH_2PO_4?

4. How would you prepare 450 mL of a 100 mM solution of K_2HPO_4?

5. How much solute is required to make 600 mL of a 0.4 M solution of Tris buffer (FW of Tris base = 121.10)?

6. Suppose you are preparing a solution that calls for 25 g of $FeCl_3 \cdot 6H_2O$. You look on the shelf in your laboratory and find anhydrous ferric chloride, $FeCl_3$. What should you do?

7. How many micromoles are there in 150 mg of D-ribose 5-phosphate, FW = 230.11?

C. Percents

When concentration is expressed in terms of percent, the numerator is the amount of solute and the denominator is 100 units of total solution. There are three types of percent expressions that vary in their units.

TYPE I: *WEIGHT PER VOLUME PERCENT.*
GRAMS OF SOLUTE PER 100 mL OF SOLUTION.

A **weight per volume** *expression is the weight of the solute (in grams) per 100 mL of total solution,* see Figure 26.5. This is the most common way to express a percent concentration in biology manuals. If a recipe uses the term % and does not specify type, assume it is a weight per volume percent. This type of expression is abbreviated as **w/v**. For example:

20 g of NaCl in 100 mL of total solution
is a 20%, w/v, solution.

Box 26.2 shows the procedure for preparing a w/v solution.

Box 26.2. PROCEDURE FOR PREPARING A
WEIGHT PER VOLUME PERCENT SOLUTION
_____

1. *Determine the percent strength and volume of solution required.*
2. *Express the percent strength desired as a fraction (g/100 mL).*
3. *Multiply the total volume desired (Step 1) by the fraction in Step 2.*
4. *Dissolve the amount of material needed as calculated in Step 3. (Assume water is the solvent if solvent is not specified.)*
5. *Bring the solution to the desired final volume.*

EXAMPLE PROBLEM

How would you prepare 500 mL of a 5% (w/v) solution of NaCl?

ANSWER

1. Percent strength is 5% (w/v). Total volume required is 500 mL.
2. Expressed as a fraction, $5\% = \dfrac{5\ g}{100\ mL}$
3. $\dfrac{(5\ g)}{100\ mL}(500\ mL) = 25\ g$

   = amount of NaCl needed
4. Weigh out 25 g of NaCl. Dissolve it in less than 500 mL of water.
5. In a graduated cylinder or volumetric flask, bring the solution to 500 mL.

TYPE II: *VOLUME PERCENT.*
MILLILITERS OF SOLUTE PER 100 mL OF SOLUTION.

In a **percent by volume expression**, abbreviated **v/v**, *both the amount of solute and the total solution are expressed in volume units.* This type of percent expression may be used when two compounds that are liquid at room temperature are being combined. For example:

100 mL of methanol in 1000 mL of total
solution is a 10% by volume solution.

Box 26.3 shows the procedure for making a v/v solution.

20.0 g Na$_2$SO$_4$

+ H$_2$O Dissolve with Stirring

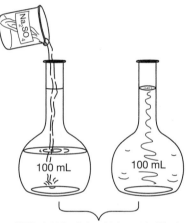

100 mL   100 mL

BTV of 100 mL in a Volumetric Flask
→ 100 mL of a 20% w/v Na$_2$SO$_4$ Solution

Figure 26.5. **A 20% Weight per Volume Percent Solution of Sodium Sulfate (Na$_2$SO$_4$).** Twenty grams of Na$_2$SO$_4$ is dissolved in water and brought to a volume of 100 mL.

Box 26.3. PROCEDURE FOR PREPARING A
PERCENT BY VOLUME SOLUTION
_____

1. *Determine the percent strength and volume required.*
2. *Express the percent desired as a fraction (mL/100 mL).*
3. *Multiply the fraction from Step 2 by the total volume desired (Step 1) to get the volume of solute needed.*
4. *Place the volume of the material desired in a graduated cylinder or volumetric flask. Add solvent to BTV.*

EXAMPLE PROBLEM

How would you make 100 mL of a 10% by volume solution of ethanol in water (v/v)?

ANSWER

1. Percent strength is 10% (v/v). Total volume wanted = 100 mL.

2. $10\% = \dfrac{10\ mL}{100\ mL}$

3. $\dfrac{10\ mL}{100\ mL} \times 100\ mL$

= 10 mL of ethanol needed.

4. Place 10 mL of ethanol in a 100 mL volumetric flask. Add water to 100 mL.

Note: You cannot assume that 10 mL of ethanol + 90 mL of water will give 100 mL total volume; their combined volume may be slightly less than 100 mL. The most accurate way to prepare this solution, therefore, is to bring it to volume.

TYPE III: *WEIGHT PERCENT.*
GRAMS OF SOLUTE PER 100 GRAMS OF SOLUTION

Weight (mass) percent, **w/w**, *is an expression of concentration in which the weight of solute is in the numerator and the weight of the total solution is in the denominator.* This type of expression is uncommon in biology manuals, but you may encounter it if you work with thick, viscous solvents whose volumes are difficult to measure. For example:

5 g of NaCl plus 20 g of water is a 20% by mass solution.

Because:

The weight of the NaCl is 5 g.

The total weight of the solution = 20 g + 5 g = 25 g

$$\text{Weight percent} = \frac{\text{weight of solute}}{\text{total weight of solution}} \times 100$$

$$\frac{(5\ g\ NaCl)}{25\ g} \times 100 = 20\%$$

Box 26.4 shows the procedure for preparing a w/w solution.

EXAMPLE PROBLEM

How would you make 500 g of a 5% NaCl solution by weight (w/w)?

ANSWER

1. Percent strength is 5% (w/w). Total weight of solution desired is 500 g.

2. $5\% = \dfrac{5\ g}{100\ g}$

3. $\dfrac{5\ g}{100\ g} \times 500\ g = 25\ g = $ NaCl needed.

Box 26.4.   PROCEDURE FOR PREPARING A PERCENT BY WEIGHT SOLUTION

1. *Determine the percent and weight of solution desired.*
2. *Change the percent to a fraction (g/100 g).*
3. *Multiply the total weight of the solution from Step 1 by the fraction in Step 2 to get the weight of the solute needed to make the solution.*
4. *Subtract the weight of the material obtained in Step 3 from the total weight of the solution (Step 1) to get the weight of the solvent needed to make the desired solution.*
5. *Dissolve the amount of the solute found in Step 3 in the amount of solvent from Step 4.*

---

4. 500 g − 25 g = 475 g = the amount of water needed.
5. Dissolve 25 g of NaCl in 475 g of water to get a 5% NaCl solution by weight.

PRACTICE PROBLEMS: PERCENTS

1. How would you prepare 35 mL of a 95% (v/v) solution of ethanol?
2. How would you prepare 200 g of a 75% (w/w) solution of resin in acetone?
3. How would you prepare 600 mL of a 15% (w/v) solution of NaCl?
4. Suppose you have 50 g of solute in 500 mL of solution.
   a. Express this solution as a %.
   b. Is this a w/w, w/v, or v/v solution?
5. Suppose you have 100 mg of solute in 1 L of solution.
   a. Express this as a %.
   b. Is this a w/w, w/v, or v/v solution?
6. What molarity is a 25% w/v solution of NaCl?

D. Parts

i. THE MEANING OF "PARTS"

Parts solutions tell you how many parts of each component to mix together. The parts may have any units but must be the same for all components of the mixture. For example:

A solution that is 3:2:1
ethylene:chloroform:isoamyl alcohol is:
3 parts ethylene, 2 parts chloroform,
1 part isoamyl alcohol.

There are many ways to prepare this solution. Two examples are:

Combine:		**Combine:**
3 L ethylene	or	3 mL ethylene
2 L chloroform		2 mL chloroform
1 L isoamyl alcohol		1 mL isoamyl alcohol

EXAMPLE PROBLEM

How could you prepare 50 mL of a solution that is 3:2:1 ethylene:chloroform:isoamyl alcohol? (Clue: The amount required of each component is calculated using a proportion equation.)

ANSWER

1. Add up all the parts required. In this case, there are $3 + 2 + 1$ parts = 6 parts.

2. Use a proportion to figure out each component. If you need 3 parts of ethylene out of 6 parts total, then:

Ethylene: $\dfrac{3}{6} = \dfrac{?}{50 \text{ mL}}$ $? = 25 \text{ mL}$

You need 25 mL of ethylene.

If you need 2 parts of chloroform out of 6 parts total, then:

Chloroform: $\dfrac{2}{6} = \dfrac{?}{50 \text{ mL}}$ $? \approx 16.7 \text{ mL}$

You need 16.7 mL of chloroform.

If you need 1 part of isoamyl alcohol out of 6 parts total, then:

Isoamyl alcohol: $\dfrac{1}{6} = \dfrac{?}{50 \text{ mL}}$ $? \approx 8.3 \text{ mL}$

You need 8.3 mL of isoamyl alcohol.

Thus, this solution requires 25.0 mL of ethylene
16.7 mL of chloroform
<u>8.3</u> mL of isoamyl alcohol
For a total of 50.0 mL

ii. PARTS PER MILLION AND PARTS PER BILLION

A variation on the theme of parts is the concentration expression "parts per million (ppm)." **Parts per million** *is the number of parts of solute per 1 million parts of total solution.* Any units may be used but must be the same for the solute and total solution. **Parts per billion (ppb)** *is the number of parts of solute per billion parts of solution.* (Percents are the same class of expression as ppm and ppb. Percents are "parts per hundred.") For example:

5 ppm chlorine might be:
5 g of chlorine in 1 million g of solution
5 mg chlorine in 1 million mg of solution
5 lb of chlorine in 1 million lb of solution
and so on

iii. CONVERSIONS

Concentration is most often expressed in terms of ppm (or ppb) in environmental applications. This expression of concentration is useful when a very small amount of something (such as a pollutant) is dissolved in a large volume of solvent. For example, an environmental scientist might speak of a pollutant in a lake as being present at a concentration of 5 ppm.

To prepare a 5 ppm solution in the laboratory, you must convert the term 5 ppm to a simple fraction expression such as grams per liter or milligrams per milliliter to determine how much of the solute to weigh out. Grams per liter, however, has units of weight in the numerator and volume in the denominator, but ppm and ppb expressions have the same units in the numerator and denominator.* To get around this problem, convert the weight of the water into milliliters based on the conversion factor that 1 mL of pure water at 20°C weighs 1 g. For example:

$$5 \text{ ppm chlorine} = \frac{5 \text{ g chlorine}}{1 \text{ million g water}}$$

$$= \frac{5 \text{ g chlorine}}{1 \text{ million mL water}} = \frac{5 \text{ g}}{1000 \text{ L}}$$

To make the expression simpler, it is possible to divide the numerator and denominator both by 1 million:

$$5 \text{ ppm chlorine} = \frac{5 \times 10^{-6} \text{ g chlorine}}{1 \text{ mL water}}$$

5×10^{-6} g is the same thing as 5 μg.

So, 5 ppm chlorine in water is the same as $\dfrac{5 \text{ μg}}{\text{mL}}$ of chlorine in water.

For any solute:

$$1 \text{ ppm in water} = \frac{1 \text{ μg}}{\text{mL}} = \frac{1 \text{ mg}}{\text{L}}$$

$$\text{Also, } 1 \text{ ppb in water} = \frac{1 \text{ ng}}{\text{mL}} = \frac{1 \text{ μg}}{\text{L}}$$

EXAMPLE

How would you prepare a 500 ppm solution of compound A?

Recall that:

$$1 \text{ ppm in water} = \frac{1 \text{ mg}}{\text{L}}$$

So 500 ppm must be equal to $\dfrac{500 \text{ mg}}{\text{L}}$

$$500 \text{ mg} = 0.500 \text{ g}$$

Thus, an acceptable method to prepare 500 ppm of compound A is to weigh out 0.500 g, dissolve it in water, and BTV 1 L.

PRACTICE PROBLEMS: PARTS

1. Suppose you have a recipe that calls for 1 part salt solution to 3 parts water. How would you mix 10 mL of this solution?

2. Suppose you have a recipe that calls for 1 : 3.5 : 0.6 chloroform:phenol:isoamyl alcohol. How would you prepare 200 mL of this solution?

---

*Percent expressions are actually "parts per hundred." Thus, technically the units should be the same in the numerator and denominator. When we use % expressions with units of weight in the numerator and volume in the denominator, we are using the approximation that 1 mL of solvent weighs 1 g, which is reasonable if the solvent is water.

3. How would you mix 45 μL of a solution which is 5 : 3 : 0.1 Solution A:Solution B:Solution C?

Assume water is the solvent in problems 4–6.

4. Convert 3 ppm to:

a. milligrams per milliliter

b. milligrams per liter

5. Convert 10 ppb to milligrams per liter.

6. How would you prepare a solution that has 100 ppm cadmium?

E. Molality and Normality

i. MOLALITY

Molality and **normality** are two ways to express concentration that are used much more commonly in chemistry than biology manuals. We will define them briefly here, but refer to a chemistry text for more information.

Molality, abbreviated with a lowercase *m, means the number of molecular weights of solute (in grams) per kilogram of solvent*, see Figure 26.6. Molality has units of weight in both the numerator and denominator.

> **A 1 molal solution of a compound contains 1 mole of the compound dissolved in 1 kg of water.**

For example:

58.44 g of sodium chloride (FW = 58.44)
in 1 kg of water, is a 1 m solution.

In practice, the difference between a 1 molar solution and a 1 molal solution is:

• When preparing a 1 molal (1 m) solution, add the solute(s) to 1 kg of water.

• When preparing a 1 molar (1 M) solution, add water to the solute(s) to bring to the final volume of 1 liter.

ii. NORMALITY

Normality, abbreviated *N*, is another way to express concentration. **Normality** *is the number of "equivalent weights" of solute per liter of solution*, see Figure 26.7.

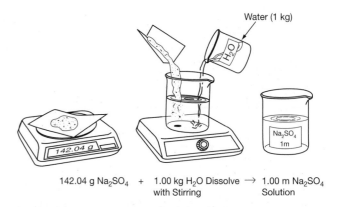

142.04 g Na₂SO₄   +   1.00 kg H₂O Dissolve   →   1.00 m Na₂SO₄
with Stirring   Solution

Figure 26.6.   A 1 m (1 Molal) Solution of Sodium Sulfate (Na₂SO₄). 142.04 g of Na_2SO_4 is dissolved in 1 kg of water.

Normality is used when it is important to know how many "reactive groups" are present in a solution rather than how many molecules.

The key to understanding normality expressions is knowing the terms *reactive groups* and *equivalent weight*. Because biologists are most likely to encounter normality in reference to acids and bases, consider these terms as they relate to acids and bases. Recall that acids dissociate in solution to release H^+ ions, and bases, like NaOH, dissociate to release OH^- ions. The H^+ and OH^- ions can participate in various reactions; hence, they are "reactive groups." *For an acid*, **1 equivalent weight** *is equal to the number of grams of that acid that reacts to yield 1 mole of H^+ ions, that is, 1 mole of reactive groups. For a base*, **1 equivalent weight** *is equal to the number of grams of that base that supplies 1 mole of OH^-, that is, 1 mole of reactive groups.* For example:

$$NaOH \rightarrow Na^+ + OH^-$$

$$H_2SO_4 \rightarrow SO_4^{-2} + 2H^+$$

In the first reaction, 1 mole of NaOH dissociates to produce 1 mole of Na^+ and 1 mole of OH^-. Because 1 equivalent weight of NaOH is the number of grams that will produce 1 mole of OH^- ions, the equivalent weight of NaOH is the same as its molecular weight. The number of grams of sodium hydroxide that will

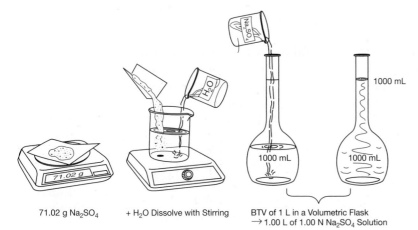

71.02 g Na₂SO₄   + H₂O Dissolve with Stirring   BTV of 1 L in a Volumetric Flask
                                                         → 1.00 L of 1.00 N Na₂SO₄ Solution

Figure 26.7.   A 1 N Solution of Sodium Sulfate (Na₂SO₄). When 1 mole of Na_2SO_4 dissolves, 2 moles of Na^+ ions are formed. The FW of Na_2SO_4 is 142.04; the equivalent weight is 71.02. (In this case, an equivalent is the number of grams of solute that will produce 1 mole of ionic charge in solution.)

produce 1 mole of OH^- is 40.0 g. In contrast, 1 mole of sulfuric acid dissociates to form 2 moles of H^+ ions. It only requires 0.5 mole of H_2SO_4 to produce 1 mole of H^+ ions. Therefore, the FW of H_2SO_4 is 98.1, but its equivalent weight is $98.1 \times 0.5 = 49.1$.

> A I normal solution of a compound contains I equivalent weight of compound dissolved in enough water to get I L.

Thus:

To make a 1 N solution of NaOH dissolve 40.0 g of solute in enough water to make 1 L— the same as a 1 M solution.

To make a 1 N solution of H_2SO_4, however, dissolve 49.1 g of solute in water to make 1 L of solution—not the same as a 1 M solution.

Biology laboratory manuals seldom express concentration in terms of normality with the exceptions of HCl and NaOH. Because HCl dissociates to produce one H^+ ion and NaOH dissociates to produce one OH^- ion, their equivalent weight and FW are the same. For NaOH and HCl, 1 M = 1 N.

III. Summary

Recipes in biology manuals sometimes list the *amounts* of each component required, other times, the *concentration* of each component required. There are various methods of expressing concentration. These methods are summarized in Tables 26.2 and 26.3.

A couple of helpful rules are:

1. **When do you need to know the FW of a compound?**

   When you have a recipe expressed in molarity, molality, or normality; or when you are converting to or from molarity, molality, or normality. With % solutions you do not need to know the FW of the compound (unless you are converting a % to molarity).

2. **How do you convert molarity expressions to percents (w/v), and vice versa?**

   There are several strategies to do these conversions. Some people find the following equation useful:

   $$\text{Molarity} = \frac{\text{w/v \% } \times 10}{\text{FW}}$$

   (This equation works because % is g/100 mL. Multiplying the percentage by 10, therefore, gives g/1000 mL or grams per liter.)

For example,

Convert 20% (w/v) NaCl to molarity
In this case, *w/v* % is 20
The FW is 58.44
Substituting into the equation gives:

$$\text{Molarity} = \frac{20 \times 10}{58.44} \approx 3.42 \text{ M}$$

Another example:

Convert 3 M NaCl to a percent
The unknown, ?, is the percent (w/v)

$$3 \text{ M} = \frac{? \times 10}{58.44}$$

$? \approx 17.53\%$

Table 26.2. Summary of Methods Used to Express Concentration

A. Weight per Volume

A fraction with the weight of the solute in the numerator and the total volume of the solution in the denominator.

B. Molarity

The number of moles of solute per liter of solution.

C. Percents

i. MASS PER VOLUME PERCENT (w/v %)
   Grams of solute per 100 mL of solution.

ii. VOLUME PERCENT (v/v %)
   Milliliters of solute per 100 mL of solution.

iii. MASS PERCENT (w/w %)
   Grams of solute per 100 g of solution.

D. Parts

i. Amounts of solutes are listed as "parts." The parts may have any units, but the units must be the same for all components of the mixture.

ii. PARTS PER MILLION AND PARTS PER BILLION Parts per million: the number of parts of solute per 1 million parts of total solution. Parts per billion: the number of parts of solute per 1 billion parts of solution.

E. Expressions of Concentration Not Common in Biology Manuals

i. MOLALITY Moles of solute per kilogram of solvent.

ii. NORMALITY Number of gram-equivalents of solute per liter of solution. (Normality is reaction-dependent.)

Table 26.3. A Comparison of Methods of Expressing the Concentration of a Solute

Concentration of Solute (Na_2SO_4)	Amount of Solute	Amount of Water
1 M	142.04 g Na_2SO_4	BTV 1 L with water
1 m	142.04 g Na_2SO_4	Add 1.00 kg of water
1 N	71.02 g Na_2SO_4	BTV 1 L with water
1%	1 g Na_2SO_4	BTV 100 mL with water

Chapter Appendix
Molarity Calculations Relating to DNA and RNA

A. DALTONS

The mass of one atom of an element is equal to the combined mass of its protons and neutrons (the electrons' masses are so slight that they are usually ignored). A periodic table provides the atomic mass for every element reported in units of atomic mass units or daltons, abbreviated D or Da. A **dalton** is *defined to be 1/12th the mass of a ^{12}carbon nucleus, which is about the same as the mass of a hydrogen atom.* The mass in grams of one hydrogen atom is about 1.6605×10^{-24} g. So there are $\approx 6.0222 \times 10^{23}$ D per gram.

The number 6.0222×10^{23} is **Avogadro's number,** *the number of molecules or atoms in a mole of any substance.* The mass of a *single* atom or molecule is its atomic mass or its formula mass in units of *daltons.* The mass of a *mole* of any atom or compound has the same numerical value, but in units of *grams.*

B. THE MOLECULAR WEIGHTS OF NUCLEIC ACIDS

It is necessary to know the molecular or formula weight of a substance of interest in order to perform molarity calculations. The formula weight of a specific chemical compound is always the same so it is usually straightforward to find the formula weights of chemicals by looking at the label on their containers. DNA is different, however, because its sequence and length vary depending on the source. A DNA molecule may be single-stranded (ss) or double-stranded (ds) and it may consist of anywhere from a few nucleotides to billions of base pairs. There is therefore no single MW for all DNA molecules. RNA molecules are similarly variable.

Let us consider how to find the MW of DNA, first for oligonucleotides of known sequence, and second for other types of DNA. Oligonucleotides are short segments of DNA that are often synthesized in a laboratory. To find the molecular weight of synthesized oligonucleotides, add the weights of all the nucleotides present. Table 26.4 shows the MW of individual nucleotides that have been incorporated into a nucleic acid. (When nucleotides form phosphodiester bonds, water is lost so the MW of the incorporated bases is about 18 D less than that of solitary nucleotides.) When oligonucleotides are synthesized, the last nucleotide in the chain is generally lacking a phosphate group; rather it has a single H in that position. Synthesized oligonucleotides also usually have an OH group at the other end of the chain. The MW of the missing phosphate must be subtracted from the weight of the oligonucleotide, and the weight of the OH must be added. Therefore, 61.96 is subtracted from the total weight of the nucleotides. (Some sources subtract 61.) Formula 26.2 on p. 506 is used to determine the MW of a DNA oligonucleotide, based on the values in Table 26.4.

Sometimes the exact sequence of a DNA or RNA molecule is unknown but its length in bases or base pairs is given. In these cases conversion factors can be used to convert between length and MW. (These conversion factors provide estimates since the exact MW of a DNA molecule depends on its exact sequence.) A single nucleotide, on the average, has a molecular weight of 330 D and a base pair on the average has a weight of 660 D. (These values are from Sambrook, J., Fritsch, E.F., and Maniatis, T. *Molecular Cloning: A Laboratory Manual.* Cold Springs Harbor, NY: Cold Spring Harbor Press, 1989, C1. Other references use slightly different average molecular weights, for example, 649 D for double-stranded and 325 D for single-stranded DNA.)

Table 26.4.	MOLECULAR WEIGHTS OF NUCLEOTIDES INCORPORATED INTO NUCLEIC ACIDS
A in DNA	313.22 D
C in DNA	289.18 D
T in DNA	304.21 D
G in DNA	329.22 D
A + T	$313.22 + 304.21 = 617.43$
G + C	$289.18 + 329.22 = 618.40$
A in RNA	329.22 D
C in RNA	305.18 D
U in RNA	306.20 D
G in RNA	345.22 D

Values from Roskams, Jane, and Rodgers, Linda., eds. *Lab Ref: A Handbook of Recipes, Reagents, and Other Reference Tools for Use at the Bench.* Cold Spring Harbor, NY: Cold Spring Harbor Laboratory Press, 2002.

Formula 26.2. To Determine the MW of a DNA Oligonucleotide

$$MW = (N_c \times 289.18) + (N_a \times 313.22) + (N_t \times 304.21) + (N_g \times 329.22) - 61.96$$

where
N_c = number of cytosines
N_a = number of adenines
N_t = number of thymines
N_g = number of guanines

EXAMPLE PROBLEM

What is the MW of a double-stranded DNA molecule that is 100 bp long?

ANSWER

100 bp $\times$ 660 D/bp = 66,000 D

Thus, one mole of this oligonucleotide would weigh about 66,000 g.

C. CONVERTING BETWEEN MICROGRAMS AND PICOMOLES

There are helpful formulas, 26.3 and 26.4, to use in situations where it is necessary to convert an amount of DNA from micrograms into picomoles or vice versa. Remember that for DNA:

dsDNA	**ssDNA**
1 mol $\approx$ (660 g) $\times$ (# base pairs)	1 mol $\approx$ (330 g) $\times$ (# bases)
1 pmol $\approx$ (660 pg) $\times$ (# base pairs)	1 pmol $\approx$ (330 pg) $\times$ (# bases)

EXAMPLE PROBLEM

A vial contains 10 μg of pBR322 vector, which is 4361 bp in length. How many picomoles are present?

ANSWER

$$10 \text{ μg} \times \frac{1 \text{ pmol}}{660 \text{ pg}} \times \frac{1}{4361} \times \frac{10^6 \text{ pg}}{1 \text{ μg}} \approx 3.47 \text{ pmol}$$

EXAMPLE PROBLEM

A vial contains 10 μg of a 1000 base DNA fragment. How many picomoles are present?

ANSWER

$$10 \text{ μg} \times \frac{1 \text{ pmol}}{330 \text{ pg}} \times \frac{1}{1000} \times \frac{10^6 \text{ pg}}{1 \text{ μg}} \approx 30.30 \text{ pmol}$$

EXAMPLE PROBLEM

A vial contains 1 pmol of a 1000 bp fragment. How many μg are present?

ANSWER

$$1 \text{ pmol} \times \frac{660 \text{ pg}}{\text{pmol}} \times 1000 \times \frac{1 \text{ μg}}{10^6 \text{ pg}} = 0.66 \text{ μg}$$

These formulas relating to calculations for DNA and RNA solutions are summarized in Table 26.5.

Formula 26.3. TO CONVERT FROM MICROGRAMS TO PICOMOLES DNA

$$\text{pmol of dsDNA} \approx \text{μg (of dsDNA)} \times \frac{1 \text{ pmol}}{660 \text{ pg}} \times \frac{1}{\# \text{ bp}} \times \frac{10^6 \text{ pg}}{1 \text{ μg}}$$

$$\text{pmol of ssDNA} \approx \text{μg (of ssDNA)} \times \frac{1 \text{ pmol}}{330 \text{ pg}} \times \frac{1}{\# \text{ bases}} \times \frac{10^6 \text{ pg}}{1 \text{ μg}}$$

Formula 26.4. To Convert from Picomoles to Micrograms DNA

$$\mu g \text{ of dsDNA} \approx \rho mol \text{ (of dsDNA)} \times \frac{660 \text{ pg}}{\rho mol} \times \# \text{ bp} \times \frac{1 \,\mu g}{10^6 \text{ pg}}$$

$$\mu g \text{ of ssDNA} \approx \rho mol \text{ (of ssDNA)} \times \frac{330 \text{ pg}}{\rho mol} \times \# \text{bases} \times \frac{1 \,\mu g}{10^6 \text{ pg}}$$

Table 26.5. *COMMONLY USED FORMULAS RELATING TO DNA AND RNA SOLUTIONS*

Average Molecular Weights:

1. *Average MW of a deoxynucleotide ≈ 330 D*
2. *Average MW of a DNA base pair ≈ 660 D*
3. *Average MW of a ribonucleotide ≈ 340 D**
4. *Average MW of ds DNA ≈ (number of bp)(660 D)*
5. *Average MW of ss DNA ≈ (number of bases)(330 D)*
6. *Average MW of ss RNA ≈ (number of bases)(340 D)**

To Calculate the Molecular Weight of a Synthesized Oligonucleotide of Known Sequence:

7. $MW = (N_c \times 289.18) + (N_a \times 313.22) + (N_t \times 304.21) + (N_g \times 329.22) - 61.96$

To convert μg to ρmol:

8. $\rho mol \text{ of dsDNA} \approx \mu g \text{ (of dsDNA)} \times \frac{1 \,\rho mol}{660 \text{ pg}} \times \frac{1}{\# \text{ bp}} \times \frac{10^6 \text{pg}}{1 \mu g}$

9. $\rho mol \text{ of ssDNA} \approx \mu g \text{ (of ssDNA)} \times \frac{1 \,\rho mol}{330 \text{ pg}} \times \frac{1}{\# \text{ bases}} \times \frac{10^6 \text{pg}}{1 \mu g}$

10. $\rho mol \text{ of ssRNA} \approx (\mu g \text{ of ssRNA}) \times \frac{1 \,\rho mol}{340 \text{ pg}} \times \frac{1}{\# \text{ bases}} \times \frac{10^6 \text{pg}}{1 \mu g}$

To convert pmol to μg:

11. $\mu g \text{ of dsDNA} \approx \rho mol \text{ (of dsDNA)} \times \frac{660 \text{ pg}}{\rho mol} \times \# \text{ bp} \times \frac{1 \mu g}{10^6 \text{ pg}}$

12. $\mu g \text{ of ssDNA} \approx \rho mol \text{ (of ssDNA)} \times \frac{330 \text{ pg}}{\rho mol} \times \# \text{ bases} \times \frac{1 \mu g}{10^6 \text{ pg}}$

13. $\mu g \text{ of ssRNA} \approx \rho mol \text{ (of ssRNA)} \times \frac{340 \text{ pg}}{\rho mol} \times \# \text{ bases} \times \frac{1 \mu g}{10^6 \text{ pg}}$

---

*Value from Roche Technical Resources. "Roche Applied Science Lab FAQs." 3rd ed. http://www.roche-applied-science.com.

CHAPTER 27

Preparation of Laboratory Solutions B: Basic Procedures and Practical Information

I. PREPARING DILUTE SOLUTIONS FROM CONCENTRATED SOLUTIONS

A. The $C_1V_1 = C_2V_2$ Equation

In the laboratory we frequently use concentrated solutions that are diluted before use. *The concentrated solutions are sometimes called* **stock solutions.** A familiar example is frozen orange juice that is purchased as a concentrate in a can. The orange juice is prepared by diluting it with a specific amount of water. In the laboratory, a 2 M stock solution of Tris buffer might be prepared that will be used at concentrations of 1 M of less. The Tris stock would then be diluted to the proper concentration whenever needed.

There is an extremely helpful equation that is used to determine how much stock solution and how much diluent (usually water) to combine to get a final solution of a desired concentration.

> **The $C_1V_1 = C_2V_2$ Equation:**
> **How to Make a Less-Concentrated Solution from a More-Concentrated Solution**
> $Concentration_{stock} \times Volume_{stock} = Concentration_{final} \times Volume_{final}$
> **This equation can be abbreviated: $C_1V_1 = C_2V_2$**

Box 27.1 outlines the procedure to use the $C_1V_1 = C_2V_2$ equation and the example problems illustrate its application.

Box 27.1. USING THE $C_1V_1 = C_2V_2$ EQUATION

---

1. *Determine the initial concentration of stock solution. This is C_1.*
2. *Determine the volume of stock solution required. This is usually the unknown, ?.*
3. *Determine the final concentration required. This is C_2.*
4. *Determine the final volume required. This is V_2.*
5. *Insert the values determined in Steps 1–4 into the formula and solve for the unknown, ?.*

Note: The $C_1V_1 = C_2V_2$ equation is only used when calculating how to prepare a **less-concentrated solution from a more-concentrated solution.** Do not try to use this equation to calculate amounts of solute needed to prepare a solution of a given molarity or percent.

EXAMPLE PROBLEM

How would you prepare 100 mL of a 1 M solution of Tris buffer from a 2 M stock of Tris buffer?

ANSWER

Consider first, is this a situation where a less-concentrated solution is being made from a more-concentrated solution? Yes. It is appropriate, therefore, to apply the $C_1V_1 = C_2V_2$ equation:

1. Concentrated solution = 2 M = C_1.
2. The volume of concentrated stock necessary is? (what you want to calculate) = V_1.
3. The concentration you want to prepare is 1 M = C_2.
4. The volume you want to prepare is 100 mL = V_2.

$$C_1\,V_1 = C_2\,V_2$$
$$2\,M(?) = 1\,M\,(100\,mL)$$
$$2\,M(?) = 100\,M\,(mL)$$
$$? = \frac{100\,M(mL)}{2\,M}$$
$$? = 50\,mL$$

5. Take 50 mL of the concentrated stock solution and bring to volume (BTV) of 100 mL.

A stock solution is sometimes written as an "X" solution where X means "times"; *how many times more concentrated the stock is than normal.* A 10 X solution is 10 times more concentrated than the solution is normally prepared. To work with "X" solutions, use the $C_1V_1 = C_2V_2$ equation.

EXAMPLE PROBLEM

A recipe says to mix:

10 X buffer Q	1 μL
Solution A	2 μL
Water	7 μL

What is the concentration of buffer Q in the final solution?

ANSWER

First consider whether this is a situation where a less-concentrated solution is being made from a more-

concentrated solution. Yes. It is therefore appropriate to apply the $C_1V_1 = C_2V_2$ equation:

$$C_1 = 10\,X \qquad C_2 = \text{unknown}$$
$$V_1 = 1\,\mu L \qquad V_2 = 10\,\mu L = \text{total solution volume}$$
$$C_1\,V_1 = C_2\,V_2$$
$$10\,X\,(1\,\mu L) = ?\,(10\,\mu L)$$
$$? = 1\,X$$

This means that 1 μL of a 10 X stock in a final volume of 10 μL will give a final concentration of 1 X. (Similarly, 1 mL of the 10 X buffer in a solution with a final volume of 10 mL would give a final concentration of 1 X.)

In the following example, Recipe Version A begins with buffer that is 10 times more concentrated than it is usually used. In Recipe Version B, the buffer is 5 X more concentrated than its usual concentration. Either 5 X or 10 X buffer stock can be used, but how much is needed varies. If 10 X buffer is used, then only 1 μL is required; however, if 5 X buffer is used, then 2 μL are required. The volume of the total solution must not vary, so 7 μL of water is used in Recipe Version A. In Recipe Version B, only 6 μL of water is required.

Recipe Version A		*Recipe Version B*	
10 X buffer Q	1 μL	5 X buffer Q	2 μL
Solution A	2 μL	Solution A	2 μL
Water	7 μL	Water	6 μL
	10 μL		10 μL

B. Dilutions Expressed as Fractions

Dilutions of concentrated stocks are sometimes expressed in terms of *fractions.* (Refer to Chapter 14 for a detailed discussion of such dilutions.) When a dilution is expressed as a fraction, most people do not use the $C_1V_1 = C_2V_2$ expression.

For example:

A 95% solution is diluted 1/10. What is its final concentration? $95\% \times 1/10 = 9.5\%$ so the final concentration is 9.5%.

A 1 M solution is diluted 1/5. What is its final concentration?

$1\,M \times 1/5 = 0.2\,M$ so the final concentration is 0.2 M.

Suppose you want to make 10 mL of a 1/50 dilution of food coloring from an original bottle of concentrated food coloring. Then,

Set up a proportion:
$$\frac{1}{50} = \frac{?}{10 \text{ mL}}$$
$$? = 0.2 \text{ mL}$$

So, take 0.2 mL of the original stock and dilute to a volume of 10 mL.

II. MAINTAINING A SOLUTION AT THE PROPER pH: BIOLOGICAL BUFFERS

A. Practical Considerations Regarding Buffers for Biological Systems

i. THE USEFUL pH RANGE

Biological systems require a particular pH, and so the interior of cells, blood, and other fluids are buffered *in vivo* (in the living organism). In the laboratory, comparable buffering systems must be artificially reproduced. **Laboratory buffers** *are laboratory solutions that help maintain a biological system at the proper pH*. When a solution is buffered, it resists changes in pH, even if H^+ ions are added to or lost from the system.

There are many combinations of chemicals and some individual chemicals that have the ability to act as buffers. Each of these buffers has chemical characteristics that make it more or less useful for a particular purpose. Some of these chemical properties that distinguish buffers will briefly be described.

The behavior of a buffer is displayed graphically in Figure 27.1. To prepare this graph, NaOH was slowly added to a buffer solution and the resulting pH was measured. Observe that the buffer tends to resist a change in pH at around pH 4.8. In other words, there is a pH at which OH^- or H^+ can be added to this solution with little change in its pH. The **pK_a** *of a buffer is the pH at which the buffer experiences little change in pH upon addition of acids or bases*. The buffer illustrated in Figure 27.1 is an acetate buffer that has a pK_a of 4.76. As shown in Table 27.1, different buffers have different pK_as. Buffers are effective in resisting a change in pH within a range of about 1 pH unit above and below their pK_a. For example, the pK_a for acetate buffer is 4.8 and acetate buffer is effective in the range from about pH 3.8 to 5.8. Thus, different buffers are effective in different pH ranges. Note also that a buffer may have more than one pK_a. For example, phosphate buffer has a pK_a at 2.15, 7.20, and 12.33.

Every biological system has an optimal pH. Stomach enzymes work at around pH 2, whereas DNA is synthesized at pH 6.9. In the laboratory, it is necessary to choose a buffer whose pK_a matches that of the system of interest. If one wants to work with a DNA solution at pH 6.9, acetate buffer should not be chosen.

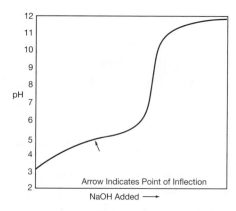

Figure 27.1.   The Effect of Adding OH^- to Acetate Buffer. At around pH 4.8, the pK_a for acetate buffer, adding base has relatively little effect on the pH.

ii. CHEMICAL INTERACTIONS WITH THE SYSTEM BEING STUDIED

A laboratory buffer should be inert in the system being studied. For example, Tris buffer is unsuitable for some protein assays because it reacts with the assay components, giving erroneous results. Phosphate buffers contribute phosphate ions to a solution, which inhibit some types of enzyme reactions. An important biotechnology example is the use of Tris-Borate-EDTA (TBE) buffer versus Tris-Acetate-EDTA (TAE) buffer when DNA fragments are separated from one another by gel electrophoresis. TBE is most commonly used for electrophoresis of DNA because it gives good resolution of DNA fragments and has good buffering capacity. It is difficult, however, to recover DNA from gels that were run with TBE buffer because the borate interacts with oligosaccharides in the gel. TAE, therefore, is the buffer of choice when DNA is to be extracted from gels after electrophoresis.

Some buffers bind ions, such as Ca^{++} or Mg^{++}, making these ions inaccessible. If the biological system of interest requires such ions, then a buffer that binds them is inappropriate; in other cases, metal ion chelation is desirable. (For more information about specific buffers, see, for example, Chambers J.A.A., and Rickwood, D., eds. *Biochemistry LabFax*. Oxford, UK: Bios Scientific, 1993.)

iii. CHANGES IN pH WITH CHANGES IN CONCENTRATION AND TEMPERATURE

It is convenient to prepare buffers as concentrated stock solutions that are diluted before use. Some buffers, however, change pH when diluted. Tris buffer is fairly stable when diluted. Its pH decreases about 0.1 pH units for every 10-fold dilution—which is acceptable in most applications. If the change with dilution of a buffer is not acceptable in an application, then the use of stock solutions should be avoided.

Temperature changes may markedly affect the pH of a buffer. For example, the pH of a Tris buffer will decrease approximately 0.028 pH units with every degree Celsius increase in temperature. Thus, Tris buffer prepared to pH 8.00 at room temperature (25°C)

Table 27.1. THE pK_A VALUES FOR SOME COMMON BUFFERS AND THE EFFECT OF TEMPERATURE CHANGE ON THE pH OF THESE BUFFERS*

Trivial Name	Buffer Name	pK_a	dpk_a/dt
Phosphate (pK_1)	—	2.15	0.0044
Malate (pK_1)	—	3.40	—
Formate	—	3.75	0.0
Succinate (pK_1)	—	4.21	−0.0018
Citrate (pK_2)	—	4.76	−0.0016
Acetate	—	4.76	0.0002
Malate	—	5.13	—
Pyridine	—	5.23	−0.014
Succinate (pK_2)	—	5.64	0.0
MES	2-(N-Morpholino)ethanesulfonic acid	6.10	−0.011
Cacodylate	Dimethylarsinic acid	6.27	—
Dimethylglutarate	3,3-Dimethylglutarate (pK_2)	6.34	0.0060
Carbonate (pK_1)	—	6.35	−0.0055
Citrate (pK_3)	—	6.40	0.0
Bis-Tris	[Bis(2-hydroxyethyl)imino]tris(hydroxymethyl)methane	6.46	0.0
ADA	N-2-Acetamidoiminodiacetic acid	6.59	−0.011
Pyrophosphate		6.60	—
EDPS (pK_1)	N,N'-Bis(3-sulfopropyl)ethylenediamine	6.65	
Bis-Tris propane	1,3-Bis[tris(hydroxymethyl)methylamino] propane	6.80	—
PIPES	Piperazine-N,N'-bis(2-ethanesulfonic acid)	6.76	−0.0085
ACES	N-2-Acetamido-2-hydroxyethanesulfonic acid	6.78	−0.020
MOPSO	3-(N-Morpholino)-2-hydroxypropane-sulfonic acid	6.95	−0.015
Imidazole	—	6.95	−0.020
BES	N,N-Bis-(2-hydroxyethyl)-2-aminoethanesulfonic acid	7.09	−0.016
MOPS	3-(N-Morpholino)propanesulfonic acid	7.20	0.015
Phosphate (pK_2)	—	7.20	−0.0028
EMTA	3,6-Endomethylene-1,2,3,6-tetrahydrophthalic acid	7.23	—
TES	2-[Tris(hydroxymethyl)methylamino]ethanesulfonic acid	7.40	−0.020
HEPES	N-2-Hydroxyethylpiperazine-N'-2-ethanesulfonic acid	7.48	−0.014
DIPSO	3-[N-Bis(hydroxyethyl)amino]-2-hydroxypropanesulfonic acid	7.60	−0.015
TEA	Triethanolamine	7.76	−0.020
POPSO	Piperazine-N,N'-bis(2-hydroxypropanesulfonic acid)	7.85	−0.013
EPPS, HEPPS	N-2-Hydroxyethylpiperazine-N'-3-propanesulfonic acid	8.00	—
Tris	Tris(hydroxymethyl)aminomethane	8.06	−0.028
Tricine	N-[Tris(hydroxymethyl)methyl]glycine	8.05	−0.021
Glycinamide	—	8.06	−0.029
PIPPS	1,4-Bis(3-sulfopropyl)piperazine	8.10	—
Glycylglycine	—	8.25	−0.025
Bicine	N,N-Bis(2-hydroxyethyl)glycine	8.26	−0.018
TAPS	3-{[Tris(hydroxymethyl)methyl]amino}propanesulfonic acid	8.40	0.018
Morpholine	—	8.49	—
PIBS	1,4-Bis(4-sulfobutyl)piperazine	8.60	—
AES	2-Aminoethylsulfonic acid, taurine	9.06	−0.022
Borate	—	9.23	−0.008
Ammonia	—	9.25	−0.031
Ethanolamine	—	9.50	−0.029
CHES	Cyclohexylaminoethanesulfonic acid	9.55	0.029
Glycine (pK_2)	—	9.78	−0.025
EDPS	N,N'-Bis(3-sulfopropyl)ethylenediamine	9.80	—
APS	3-Aminopropanesulfonic acid	9.89	—
Carbonate (pK_2)	—	10.33	−0.009
CAPS	3-(Cyclohexylamino)propanesulfonic acid	10.40	0.032
Piperidine	—	11.12	—
Phosphate (pK_3)	—	12.33	−0.026

Source: Reprinted by permission from *Guide to Protein Purification,* Murray P. Deutscher, ed. Academic Press, 1990. *Methods in Enzymology* 182.
*Selected buffers and their pK_a values at 25°C.

will be close to pH 7.66 when used at body temperature (37°C) and pH 8.70 when placed on ice. Buffers that are temperature-sensitive should be prepared at the temperature at which they will be used.

Some buffers are more sensitive than others to temperature. Table 27.1 gives values for various buffers that are used to calculate the change in pH with each degree change in temperature. The larger the numeric value for "dpK_a/dt," the more the buffer will change pH with each degree of temperature change. Negative values for dpK_a/dt mean that there is a decrease in pH with an increase in temperature, and vice versa.

Note that there are factors that affect the temperature of a solution while its pH is being measured. Magnetic stirring devices generate heat. Some chemical reactions are endothermic or exothermic (remove or generate heat); so as chemicals are combined, their temperature may change.

iv. OTHER CONSIDERATIONS

Buffer solutions should always be prepared with high-quality water. It is often good practice to sterilize buffers by autoclaving or ultrafiltration to prevent bacterial and fungal growth that can degrade organic buffers. Note, however, that some solution components are sensitive to heat and cannot be autoclaved.

Note also that in biology manuals, solutions that are called "buffers" may have functions in addition to maintaining the proper pH. For example, "buffers" may also contain salts, EDTA, β-mercaptoethanol, and so on. Thus, biologists use the term *buffer solution* in a general way, as compared with a strict chemical definition.

B. Adjusting the pH of a Buffer

i. OVERVIEW

When a buffer is prepared, the correct amount of the buffer compound(s) is weighed out and dissolved in water to get the desired concentration. For example, Tris buffer comes as a crystalline solid. (Tris that is not conjugated to any other chemical is called *Tris base.*) Tris base has a FW of 121.1 and a pK_a of 8.06. A 1 M solution of Tris can be prepared by dissolving 121.1 g of Tris base in 1 L of water. The pH of such a solution, however, will be greater than 10. This pH is far from the pK_a for Tris, so the solution will have little buffering capacity. Moreover, few biological systems function at such a high pH. Before the Tris solution is useful as a biological buffer, its pH therefore must be lowered. Most buffering compounds are not at their pK_a when they are dissolved and must be adjusted to the desired pH. There are several strategies to obtain a buffer that has both the correct pH and the correct concentration. This section describes these strategies and illustrates them using Tris and phosphate buffers as examples.

Strategy 1.

The buffer is titrated with a strong acid or base. The buffer is dissolved in water, but it is not brought to its final volume. If the pH of the solution is lower than the

desired pH, then a strong base (often NaOH) is added to raise the pH. Conversely, if the pH of the solution is above the desired pH, then a strong acid (often HCl) is added to lower the pH. Once the desired pH is reached, the solution is brought to its final volume. This procedure is summarized in Box 27.2 and an example of this method (Tris buffer) is given.

There are situations where the sodium from NaOH, or the chloride from HCl, interferes with the system of interest. In such cases, other acids or bases are substituted.

Strategy 2.

The buffer is prepared from two reagents, one of which is more acidic, the other of which is more basic. By combining the two forms in the correct proportions, a solution with the desired pH can be obtained. This method will be illustrated for Tris buffer.

Strategy 3.

The buffer is prepared as two stock solutions, one of which is more acidic and the other more basic. The two stocks are combined in the proper proportion to get the desired pH. An example of this method, phosphate buffer, will be discussed.

ii. TRIS BUFFER

Tris, (Tris(hydroxymethyl)aminomethane), is one of the most common buffers in biotechnology laboratories. Tris is used because it buffers over the normal biological range (pH 7 to pH 9), is nontoxic to cells, and is relatively inexpensive. Table 27.2 summarizes some considerations regarding Tris buffers.

As mentioned earlier, the pK_a of Tris is 8.06 but when dissolved in water its pH is greater than 10. Two strategies to bring Tris to the correct pH are as follows:

Strategy 1.

The buffer is titrated with a strong acid until it is the desired pH. Because Tris base has a basic pH when dissolved, it is necessary to add acid to bring the pH down to a value suitable for biological work. If the acid required to bring Tris to the proper pH is unspecified, it

---

Table 27.2. **CONSIDERATIONS WHEN USING TRIS BUFFERS**

1. *Tris buffers change pH significantly with changes in temperature.* Final pH adjustments to Tris buffers, therefore, should be made when the solution is at the temperature at which it will be used.
2. *Tris buffers require a compatible electrode for accurate pH determination.* A calomel (mercury) reference electrode rather than a silver/silver chloride electrode is recommended.
3. *Tris is subject to fungal contamination.* Contamination can be avoided by autoclaving the Tris solution or filter sterilizing it before long-term storage.
4. *Tris is minimally sensitive to concentration changes.* Tris can therefore be stored as a concentrated stock solution that is diluted before use.

is most common to use HCl for titration, but other acids may be used as well. For example, boric acid, citric acid, and acetic acid can be used to prepare Tris-borate, Tris-citrate, and Tris-acetate buffers, respectively.

Strategy 2.

Combine Tris base with a conjugated form of Tris called Tris-HCl. Tris-HCl is a solid form of Tris that has been conjugated to HCl. If Tris-HCL is dissolved in water, its pH is less than 5. It is possible to combine Tris base with Tris-HCl to get a pH between 7 and 9. Depending on the pH desired, different amounts of the two forms of Tris are combined, as shown in Table 27.3 on p. 514. An upcoming example problem illustrates this strategy. Note also that Figure 27.4b on p. 522 contains a procedure to prepare Tris buffer based on this strategy.

EXAMPLE PROBLEM

Outline a procedure to make 300 mL of a 0.8 M solution of Tris buffer, pH 7.6 at 4°C. (FW of Tris base = 121.1) Use the strategy in Box 27.2.

ANSWER

1. First, it is necessary to calculate the amount of solute (Tris base) required. This may be accomplished using Formula 26.1 on p. 498.

Solute Required =
(Grams/Mole) (Molarity) (Volume)
(121.1 g/mole) (0.80 mole/L)(0.300 L) = 29.064 g

2. Dissolve 29.064 g of Tris base in about 250 mL of 4°C, purified water. The pH will be above 10.

3. Add concentrated HCl (remember to use safety precautions) while stirring and reading the pH. (Use a pH meter equipped with a calomel electrode.)

4. Check the temperature of the solution. The temperature may change both during the initial dissolution of the Tris and during the addition of acid to the solution. Ensure that the temperature of the solution is 4°C before performing the final adjustment of its pH.

5. Continue adding HCl until the solution is exactly pH 7.6.

6. Bring the solution to 300 mL with water. (Because the pH of Tris is relatively insensitive to dilution, the pH should still be close to 7.6 after the solution is brought to volume.)

Note: There is an alternate strategy that allows this buffer to be prepared at room temperature. From Table 27.1, we know that the change in pH with temperature for Tris is −0.028 pH units per degree Celsius. Suppose room temperature is 22°C. Then the difference between room temperature and 4°C is −18°C. Multiplying −18°C × −0.028 gives a change of +0.504 pH units. The solution, therefore, can be prepared at room temperature but be brought to pH 7.096 (round to 7.1). When the buffer is used at 4°C, its pH should rise to be very close to pH 7.6. The advantage to this method is that it avoids the requirement of holding the temperature at 4°C as the solution is prepared.

BOX 27.2. *GENERAL PROCEDURE TO BRING A SOLUTION TO THE CORRECT pH USING A STRONG ACID OR A STRONG BASE*

1. *Determine the amount of solute(s) required to make a buffer of the correct concentration.*

2. *Mix the solute(s) with most, but not all, the solvent required. Do not bring the solution to volume.* (To ensure consistency, a standard operating procedure will usually specify how much water to dissolve the solute(s) initially.)

3. *Place the solution on a magnetic stir plate, add a clean stir bar, and stir.*

4. *Check the pH. (Be careful not to crash the stir bar into the electrode.)*

5. *Add a small amount of acid or base, whichever is needed to bring the solution toward the desired pH.* The recipe will often specify which acid or base to use; if the recipe does not specify, it is usually safe to add HCl or NaOH.

6. *Stir again and then check the pH.*

7. *Repeat Steps 5 and 6 until the pH is correct.*

8. *Bring the solution to the proper volume once the pH is exactly correct, then recheck the pH.* (With many buffers, when you bring the solution to its final volume, the pH does not change significantly.)

NOTES ABOUT ADJUSTING THE pH OF A SOLUTION

i *Temperature.* pH the solution when it is at the temperature at which you plan to use it. Note also that some solutions change temperature during mixing. For example, the addition of acid to Tris buffer may cause its temperature to change. Restore the buffer to the correct temperature before making the final adjustment of its pH.

ii *Overshooting the pH.* If you accidentally overshoot the pH the correct procedure is to remake the entire solution. It is not recommended to compensate by adding some extra acid or base. This is because the additional acid or base adds extra ions to the solution; these ions will change the composition of your solution and may adversely affect it. In addition, if you add more acid or base sometimes (because of a mistake), but not other times, there will be an inconsistency that will be very difficult to diagnose if problems arise later.

iii *Overshooting the Volume.* If you accidentally overshoot the volume, begin again to avoid changing the concentration of solute(s) and to avoid introducing inconsistency. (Using more concentrated acid or base may help avoid this problem, but be careful not to overshoot the pH.)

Table 27.3. COMBINING TRIS BASE AND TRIS-HCL TO GET A BUFFER OF THE DESIRED pH*

pH at			0.05 M Grams/liter	
5°C	**25°C**	**37°C**	**Tris HCl**	**Tris Base**
7.55	7.00	6.70	7.28	0.47
7.66	7.10	6.80	7.13	0.57
7.76	7.20	6.91	7.02	0.67
7.89	7.30	7.02	6.85	0.80
7.97	7.40	7.12	6.61	0.97
8.07	7.50	7.22	6.35	1.18
8.18	7.60	7.30	6.06	1.39
8.26	7.70	7.40	5.72	1.66
8.37	7.80	7.52	5.32	1.97
8.48	7.90	7.62	4.88	2.30
8.58	8.00	7.71	4.44	2.65
8.68	8.10	7.80	4.02	2.97
8.78	8.20	7.91	3.54	3.34
8.86	8.30	8.01	3.07	3.70
8.98	8.40	8.10	2.64	4.03
9.09	8.50	8.22	2.21	4.36

To make 0.05 M Tris buffer: Dissolve the indicated amounts of Tris-HCl and Tris base in water to a final volume of 1 L. Tris-HCl and Tris base are somewhat hygroscopic at high humidity. For precise work, desiccation before weighing is recommended. Tris solution can be autoclaved.
Source: "Sigma Catalogue," Sigma Company, P.O. Box 14508, St. Louis, MO 63178. Reprinted by permission from Sigma-Aldrich Co.
*Note that Sigma company calls their Tris "Trizma."

EXAMPLE PROBLEM

Outline a procedure to make 300 mL of a 0.05 M solution of Tris buffer, pH 7.62 at 37°C. Use the information in Table 27.3.

ANSWER

1. Table 27.3 tells us that 1 L of 0.05 M buffer at pH 7.62 and at 37°C requires 4.88 g of Tris-HCl and 2.30 g of Tris base.
2. Because we do not want to make a liter of buffer, it is necessary to calculate how much of each chemical is required to make 300 mL. This can be easily accomplished using the logic of proportions:

Tris-HCl

$$\frac{4.88 \text{ g}}{1000 \text{ mL}} = \frac{?}{300 \text{ mL}}$$

$$? = 1.464 \text{ g}$$

Tris base

$$\frac{2.30 \text{ g}}{1000 \text{ mL}} = \frac{?}{300 \text{ mL}}$$

$$? = 0.690 \text{ g}$$

3. Combine 1.464 g of Tris-HCl and 0.690 g of Tris base in 250 mL of purified water that is at 37°C.
4. Ensure that the temperature of the solution is 37°C.
5. Check the pH of the solution with a pH meter having a calomel electrode.
6. If the pH is within 0.05 units of 7.62, then bring it to volume with water. (If the pH is not sufficiently close to 7.62, it may be acceptable to add small amounts of either acid or base to bring the pH to the exact pH.)

iii. PHOSPHATE BUFFER

Phosphate is another buffer commonly used for biological systems. Phosphate buffer is typically prepared by mixing solutions of sodium phosphate dibasic (Na_2HPO_4) and sodium phosphate monobasic (NaH_2PO_4). At 25°C, a 5% solution of dibasic sodium phosphate has a pH of 9.1, whereas a 5% solution of the monobasic form has a pH of 4.2. Mixtures of these two solutions will make a buffer with a pH between the two extremes.

The conventional way to prepare phosphate buffer is to prepare stock solutions of the monobasic and dibasic compounds that have the same molar concentration, for example, 0.2 M. The two solutions are then combined in the proportions given in Table 27.4 to achieve the proper pH. The concentration of the resulting buffer is determined by the molarity of the phosphate ions. This means that if 0.2 M sodium phosphate monobasic and 0.2 M sodium phosphate dibasic are combined, the resulting concentration will still be 0.2 M. Phosphate buffer prepared with a mixture of the monobasic and dibasic forms has a pK_a of 7.2 and so its buffering range is about pH 6 to pH 8.

Phosphate buffers can also be prepared with the potassium salt of phosphate and occasionally with a mixture of the sodium and potassium salts. Phosphate buffers are less sensitive than are Tris buffers to the effect of temperature, but are more sensitive to dilution. Phosphate buffers, therefore, should not be prepared as concentrated stock solutions. They should instead be prepared to the concentration at which they will be used.

EXAMPLE PROBLEM

Using Table 27.4, outline a procedure to make 0.1 M sodium phosphate buffer, pH 7.0.

ANSWER

1. Prepare the two stock solutions, Solution A and Solution B, as directed in Table 27.4. These stocks have a molarity of 0.2 M.
2. Mix the amounts indicated on the chart, that is, 39.0 mL of Solution A and 61.0 mL of Solution B. This solution has a concentration of 0.2 M.
3. Bring the solution to a volume of 200 mL as directed in the chart. Dilution of the buffer with water to 200 mL means that the final concentration of phosphate in the solution is 0.1 M. Check the pH.

III. PREPARING SOLUTIONS WITH MORE THAN ONE SOLUTE

A. Introduction

Chapter 26 discussed how to prepare a solution containing a single solute in such a way that the solute is present at the correct concentration. This chapter has so far added another issue to the preparation of a labora-

Table 27.4. PREPARING PHOSPHATE BUFFER

To Prepare 0.1 M Sodium Phosphate Buffer of a chosen pH

Step 1. **Prepare Solution A,** 0.2 M $NaH_2PO_4 \cdot H_2O$ (monobasic sodium phosphate monohydrate, FW = 138.0): Dissolve **27.6 g $NaH_2PO_4 \cdot H_2O$** in purified water. Bring to volume of 1000 mL.

Step 2. **Prepare Solution B,** 0.2 M Na_2HPO_4 (Dibasic Sodium Phosphate):
Use either: **28.4 g Na_2HPO_4** (anhydrous dibasic sodium phosphate, FW = 142.0)
or **53.6 g $Na_2HPO_4 \cdot 7H_2O$** (dibasic sodium phosphate heptahydrate, FW = 268.1)
Dissolve in purified water and bring to volume of 1000 mL.

Step 3. Combine Solution A + Solution B in the amounts listed in the following table.

Step 4. Bring the mixture of Solution A + Solution B to a final volume of 200 mL.

Volume A (mL)	Volume B (mL)	pH
93.5	6.5	5.7
92.0	8.0	5.8
90.0	10.0	5.9
87.7	12.3	6.0
85.0	15.0	6.1
81.5	18.5	6.2
77.5	22.5	6.3
73.5	26.5	6.4
68.5	31.5	6.5
62.5	37.5	6.6
56.5	43.5	6.7
51.0	49.0	6.8
45.0	55.0	6.9
39.0	61.0	7.0
33.0	67.0	7.1
28.0	72.0	7.2
23.0	77.0	7.3
19.0	81.0	7.4
16.0	84.0	7.5
13.0	87.0	7.6
10.5	89.5	7.7
8.5	91.5	7.8
7.0	93.0	7.9
5.3	94.7	8.0

Tabular information from *Biochemistry LabFax,* J.A.A. Chambers and D. Rickwood, eds. Bios Scientific Publishers, Oxford, UK, 1993; and *Buffers: A Guide for the Preparation and Use of Buffers in Biological Systems,* Calbiochem Corporation, San Diego, CA.

Recipe I		Recipe II		Recipe III
Na_2HPO_4	6 g	1 M $MgCl_2$	5 mL	0.1 M Tris
KH_2PO_4	3 g	0.4% thymidine	10 mL	0.01 M EDTA
NaCl	0.5 g	20% glucose	25 mL	1% SDS
NH_4Cl	1 g			

Dissolve in water. Bring to a volume of 1 liter.

Figure 27.2. Examples of Three Recipes for Laboratory Solutions That Contain More Than One Solute.

Recipe II also states how much of each solute is required; however, the solutes are prepared as individual solutions before being combined. To prepare Recipe II, first make a 1 M solution of $MgCl_2$, a 0.4% solution of thymidine, and a 20% solution of glucose. Then, mix 5 mL of the $MgCl_2$ solution, 10 mL of the thymidine solution, and 25 mL of the glucose solution. Additional water is not added to the mixture of the three solutes.

Recipe III is different in that it does not state the *amount* of each solute required. Rather, the recipe states the *final concentration* of each solute. If the recipe or procedure gives only the final concentration(s) needed for the solute(s), you will have to *calculate the amount(s)* you need of each. The reason authors often record concentrations instead of amounts is they do not know what volume you will need. Recipes like Recipe III are discussed in detail in this section.

B. A Recipe with Multiple Solutes: SM Buffer

Consider the following recipe:

SM Buffer

0.2 M Tris, pH 7.5
1 mM $MgSO_4$
0.1 M NaCl
0.01% gelatin

This is the entire recipe as it might appear in a manual. SM buffer contains four solutes that are combined in water. The recipe lists the *final concentration* of each solute; Tris is present in the final solution at a concentration of 0.2 M, $MgSO_4$ at a final concentration of 1 mM, and so on. The first instinct people sometimes have when looking at a recipe like this is to prepare four separate solutions of 0.2 M Tris, 1 mM magnesium sulfate, 0.1 M NaCl, and 0.01% gelatin, and then to mix them together; however, this instinct is wrong. If you combine any volume of 0.2 M Tris with any volume of 1 mM $MgSO_4$ or any of the other solutes, they will dilute one another, giving the wrong final concentrations.

Consider two correct strategies to prepare this solution.

Strategy 1: Preparing SM Buffer without Stock Solutions

In overview, first prepare 0.2 M Tris, pH 7.5 but do not BTV. Then, calculate how many grams of each of the other solutes is required. These solutes are weighed out

tory solution; ensuring that the solution is at the correct pH as well as the correct concentration. We will now discuss yet another issue relating to solution preparation, the preparation of solutions with more than one solute. Figure 27.2 shows three examples of recipes for solutions with more than one solute.

Recipe I is relatively simple to interpret. It states the *amounts* of four solutes to weigh out and combine in water.

and dissolved directly in the Tris buffer. Note that when the temperature is not specified, assume a solution is to be at room temperature. The steps are:

1. Decide how much volume to prepare, for example, 1 L.

2. One liter of 0.2 M Tris requires 24.2 g of Tris base (FW = 121.1). Dissolve the Tris in about 700 mL of water and bring the pH to 7.5. Do not bring the Tris to volume yet.

   (The Tris buffer is brought to the proper pH before the other solutes are added because the recipe shows Tris to be a particular pH.)

3. One liter of 0.1 M NaCl requires 5.84 g of NaCl. Add this amount to the Tris buffer.

4. One liter of 1 mM $MgSO_4$ requires 1/1000 of its FW. (Magnesium sulfate comes in more than one hydrated form. Read the container to determine the correct molecular weight.) Weigh out the correct amount, and add it to the Tris buffer. (For example, $MgSO_4 \cdot 7\ H_2O$ has a FW of 246.5. Add 0.246 g of this form of $MgSO_4$.)

5. 0.01% gelatin is 0.1 g in 1 L. Weigh out 0.1 g of gelatin and add to the Tris buffer.

6. Stir the mixture until all the components are dissolved, then BTV of 1 L with water.

7. It is good practice to check the pH at the end: It should be close to 7.5 because the Tris is fairly stable with dilution. Record the final pH, but do not readjust it (unless directed to do so by a procedure you are following).

Strategy 2: Preparing SM Buffer with Stock Solutions

In this strategy, the four solutes are each prepared separately as *concentrated stock solutions.* Then, when the four stocks are combined, they dilute one another to the proper final concentrations.

1. Prepare a stock solution of Tris buffer at pH 7.5. There is no set rule as to what concentration a stock solution should be. In this case, a useful concentrated stock solution might be 1 M (i.e., 5 X more concentrated than needed). To make 1 L of 1 M stock, dissolve 121.1 g of Tris base in about 900 mL of water. Bring the solution to pH 7.5 and then to a volume of 1 L.

2. Prepare a stock solution of magnesium sulfate (e.g., 1 M). To make 100 mL of this stock, dissolve 0.1 FW of $MgSO_4$ in water and bring to a volume of 100 mL. (For example, $MgSO_4 \cdot 7\ H_2O$ has a FW of 246.5. Use 24.6 g of this form of $MgSO_4$.)

3. Prepare a stock solution of NaCl, for example, 1 M. To make 100 mL of stock, dissolve 5.84 g in water and bring to a volume of 100 mL.

4. Prepare a stock solution of gelatin, for example, 1%, by dissolving 1 g in a final volume of 100 mL of water.

5. To make the final solution it is necessary to combine the right amounts of each stock. Because this is a situation where concentrated solutions are being diluted, use the $C_1 V_1 = C_2 V_2$ equation four times, once for each solute. For example, to figure out how much Tris stock is required:

$$C_1 \qquad V_1 \quad = \qquad C_2 \qquad V_2$$
$$(1\ M)\ (?) \quad = \quad (0.2\ M)\ (1000\ mL)$$
$$? = 200\ mL$$

To make 1000 mL of SM buffer, therefore, combine:

Components	Final Concentration
200 mL of the 1 M Tris stock, pH 7.5	200 mM Tris (0.2 M)
1 mL of the 1 M magnesium sulfate stock	1 mM $MgSO_4$
100 mL of the 1 M NaCl stock	0.1 M NaCl
10 mL of the 1% gelatin stock	0.01% gelatin

6. BTV 1000 mL and check the pH.

Both Strategy 1 and Strategy 2 are correct. It is generally efficient to make stock solutions of solutes that are used often because weighing out chemicals is time-consuming. For example, many solutions require NaCl; therefore, it might be useful to keep a 1 M stock solution of NaCl on hand. Chemicals that are used infrequently should not be kept as stock solutions because microorganisms can grow in them and they may slowly degrade. (Sterilizing solutions can sometimes extend their shelf life.)

C. A Recipe with Two Solutes: TE Buffer

Consider another example:

TE Buffer

Component	Final Concentration
Tris, pH 7.6	10 mM
Na_2-EDTA	1 mM

This buffer contains Tris at a pH of 7.6 and a final concentration of 10 mM and Na_2-EDTA at a final concentration of 1 mM. As for SM buffer, it is incorrect to prepare 10 mM Tris buffer and 1 mM Na_2-EDTA and then to mix them together because they will dilute one another. It is conventional to prepare TE buffer by combining stock solutions of Tris, at the desired final pH, and EDTA. (For example, this strategy is outlined in Sambrook, J., Fritsch, E.F., and Maniatis, T., *Molecular Cloning: A Laboratory Manual.* 2nd ed. Cold Spring Harbor, NY: Cold Spring Harbor Laboratory Press, 1989.)

To make TE buffer:

1. Decide how much volume to prepare, for example, 100 mL.

2. Na_2-EDTA does not go into solution easily and is relatively insoluble in water if the pH is less than 8.0. Na_2-EDTA, therefore, is commonly prepared

as a concentrated stock solution with a pH of 8.0. To make 100 mL of a 0.5 M stock of Na_2-EDTA, add 18.6 g of EDTA, disodium salt, dihydrate (FW = 372.2) to 70 mL of water. Adjust the pH to 8.0 slowly with stirring by adding pellets of NaOH or concentrated NaOH solution. When the Na_2-EDTA is dissolved, bring the solution to volume.

3. The Tris buffer needs to be dissolved and brought to the proper pH before it is combined with the Na_2-EDTA. Like the Na_2-EDTA the Tris should also be prepared as a concentrated stock solution (e.g., 0.1 M). To make 100 mL of Tris stock, dissolve 1.21 g of Tris in about 70 mL of water, adjust it to pH to 7.6 with HCl, and then bring it to 100 mL final volume.

4. Use the $C_1V_1 = C_2V_2$ equation twice, once for each solute, to calculate how much of each stock will be required to make a solution of TE with the proper concentration of both solutes.

To calculate how much 0.1 M stock of Tris is required:

$$C_1 \quad V_1 \quad = \quad C_2 \quad V_2$$
$$(0.1\,M)(?) \quad = \quad (0.010\,M)(100\,mL)$$
$$? = 10\,mL$$

(Note that the units must be consistent on both sides of the equation so 10 mM was converted to 0.010 M.)

To calculate how much 0.5 M Na_2-EDTA stock is required:

$$C_1 \quad V_1 \quad = \quad C_2 \quad V_2$$
$$(0.5\,M)(?) \quad = \quad (0.001\,M)(100\,mL)$$
$$? = 0.2\,mL = 200\,\mu L$$

5. Combine 10 mL of Tris stock and 200 μL of Na_2-EDTA stock and bring the solution to the final volume, 100 mL, with water.

6. Check the pH.

IV. PREPARATION OF SOLUTIONS IN THE LABORATORY

A. General Considerations

Box 27.3 outlines a general procedure for making a solution in the laboratory and summarizes many of the issues discussed so far in these two chapters on solutions.

Box 27.3.   A GENERAL PROCEDURE TO MAKE A SOLUTION IN THE LABORATORY

1. *Read and understand the procedure; identify and collect materials and equipment needed.*
2. *Determine the amount of each solute or stock solution that is required.*
3. *Weigh the amount of each solute required, or measure the volume of each stock solution.*
4. *Dissolve the weighed compounds, or mix the stock solutions, in less than the desired final volume of solvent.*
5. *Bring the solution to the correct pH, if necessary.*
6. *Bring the solution to volume. Check the pH, if necessary.*
7. *It may be necessary to sterilize the solution.*

Notes:

i **Water.** Unless instructed otherwise, use the highest quality (most purified) water available to make solutions.

ii **Grades of Chemicals.** Be sure that the purity and form of chemicals match your requirements. See Table 27.5 on p. 518.

iii **Contamination.** Do not return chemicals to stock containers so as not to contaminate the stocks of chemicals. In addition, avoid putting spatulas into stock containers, although it may be acceptable to use a clean spatula in some situations. To protect yourself and to avoid contaminating solutions, never touch any chemicals or the interior of glassware containers.

iv **Labeling.** ALWAYS LABEL THE CONTAINER OF EVERY SOLUTION. At a minimum include the following on the label: the name and/or formula of the solution, the concentration of the solution, the solvent used, the date prepared, your name or initials, and safety information such as whether the solution is toxic, radioactive, acidic, basic, and so on. Additional information, such as the lot number of the solution, may be required.

v **Concentrated and Dilute Acids and Bases.** The term *concentrated* refers to acids and bases in water at the concentration that is their maximum solubility at room temperature, see Table 27.6 on p. 518. For example, concentrated hydrochloric acid has a molarity of 11.6 and is 36% acid. In contrast, concentrated nitric acid is 15.7 M and 72% acid. When working with concentrated acids, put the acid into water and not vice versa to avoid hazardous splashing and overheating.

vi **Hygroscopic Compounds. Hygroscopic compounds** *absorb moisture from the air.* A compound that has absorbed an unknown amount of water from the air may not be acceptable for use. Hygroscopic compounds, therefore, should be kept in tightly closed containers and should be exposed to air as little as possible. It may be advisable to purchase such compounds in small amounts so that once their container is opened they are rapidly used and are not stored. These compounds may be stored in desiccators to help protect them from moisture. In some laboratories, it is common practice to record the date a bottle of a chemical is first opened. A chemical that has been opened and stored too long may need to be discarded.

Table 27.5. GRADES OF CHEMICAL PURITY

Chemicals are manufactured with varying degrees of purity. More purified preparations are usually more expensive to purchase.

1. *Reagent (ACS)* or *Analytical Reagent (ACS)* grade chemicals are certified to contain impurities below the specifications of The American Chemical Society Committee on Analytical Reagents. If the bottle was not contaminated by previous use, these chemicals are usually acceptable for making solutions.

2. *Reagent* grade. If the American Chemical Society has not established specifications for a particular chemical, the manufacturer may do so and label the chemical Reagent grade. The manufacturer must then list the maximum allowable impurities for the chemical.

3. *U.S.P.* grade chemicals meet the standards of the U.S. Pharmacopeia, the official publication for drug product standards. Such chemicals are acceptable for drug use and are often also acceptable for general laboratory use.

4. *CP* means chemically pure and is usually similar to, or slightly less pure than, reagent grade. **Purified** means that the manufacturer has removed many contaminants. Read the manufacturer's specifications to determine whether these chemicals are acceptable for your purposes.

5. *Commercial* or *Technical* grade chemicals are intended for industrial use. **Practical** grade is intended for use in organic syntheses. In general, commercial, technical, and practical grade chemicals are not used for laboratory solutions.

6. *Spectroscopic* grade chemicals are intended for use in spectrophotometry because they have very little absorbance in the UV or IR range.

7. *HPLC* grade reagents are organic solvents that are very pure and are intended for use as the mobile phase in HPLC.

8. *Primary Standard* grade chemicals are usually the most highly purified chemicals and are intended to be used as standards.

9. *Other.* Chemical manufacturers often have their own grades, such as "tissue culture grade." Read their catalog to determine their specifications.

Table 27.6. ACIDS AND BASES

Acid or Base	Formula	FW	Moles per Liter	Density (g/mL)	% Solution (w/w)
Acetic acid, glacial	CH_3COOH	60.1	17.4	1.05	99.5
Hydrochloric acid, concentrated	HCl	36.5	11.6	1.18	36
Nitric acid, concentrated	HNO_3	63.0	16.0	1.42	71
Sulfuric acid, concentrated	H_2SO_4	98.1	18.0	1.84	96
Sodium hydroxide, pellets	NaOH	40.0	Solid	—	—
Sodium hydroxide, diluted	—	—	6.1	1.22	20

For more information on acids and bases, see also O'Neil, Maryadele J., ed. *The Merck Index: An Encyclopedia of Chemicals, Drugs, and Biologicals.* 14th ed. Whitehouse Station, NJ: Merck, 2006.

B. Assuring the Quality of a Solution

Quality practices relevant to laboratory solutions include the items discussed in the next sections.

i. DOCUMENTATION

All steps of solution preparation require documentation, including procedure(s) followed, the details of the preparation (weights, volumes, and measurements), problems or deviations from the procedure, components of the solution including the manufacturer's lot numbers, who prepared the solution, date of preparation, handling after preparation, maintenance and calibration of equipment used, and safety considerations. It is always good practice to check and record the pH of a biological solution. Such documentation is important, even in laboratories that do not adhere to a formal quality system.

Figure 27.3 is an example of a form used in a biotechnology company that has a centralized solution-making facility. The side of the form in Figure 27.3a is completed when a solution is requested by a laboratory scientist and the side in Figure 27.3b is filled in by the person who prepares the solution. In many companies, such information is entered directly into a computer system, but paper can also be used.

ii. TRACEABILITY

In production facilities, it is essential that every component used in producing a product be traceable. If a solution is used in production, documentation that identifies each component of that solution must be connected to the product. The solution itself is assigned an identification number that travels with the solution as it is used for various purposes. This system ensures that if a problem arises, every component of every solution that went into production of that product can be identified. Labeling the solution itself, is, of course, essential.

iii. STANDARD OPERATING PROCEDURES

SOPs tell laboratory workers in detail how to maintain and use instruments and how to prepare a particular solution. An SOP might describe, for example, exactly how a particular brand of balance is to be maintained and calibrated, and how the preparer of a solution is to document that maintenance and calibration were performed. A solution SOP will usually show any calculations required (how much weight or volume of each component is needed) or example calculations, what type of purified water is to be used, whether the solution should be brought to a particular pH and, if so, how, whether the pH should be checked, whether the solution should be sterilized and, if so, how, and any other specific details. Each time a solution is made, the procedure must be followed and the particular procedure that was followed must be documented. Examples of industry procedures to make solutions are shown in Figure 27.4, on pp. 521–522.

Note that there is often some ambiguity in published solution recipes. For example, the temperature of the solution may be unstated, or it may be unclear as to whether a specified pH refers to a particular solute or whether it is the pH of the final solution. SOPs, therefore, play a key role in ensuring consistency and clarity in

---

MEDIA PREPARATION FORM (*SECTION A*)

ITEM NUMBER	LOT NUMBER

A-1.

1. NAME OF MEDIA/SOLUTION (*and pH @ Temp.*): _____

2. Requested By _____ Room No. _____ Dept. No. _____

3. Project # (*if applicable*)_____ 4. PIP # (*if applicable*)_____

5. Date and Time Ordered_____ Date and time Required _____

NOTE: Please allow a minimum of 24 hours for buffers and solutions, and 48 hours for agar-based plates and New Formulations.

A-2. TOTAL QUANTITY REQUIRED: _____

A-3. Check Product Type:

1. MEDIA PREP PRODUCT ☐   2. PRODUCT ☐   3. SHOP ORDER COMPONENT ☐

4. NEW FORMULATION ☐ *Check with Media Prep if unsure if request is a New Formulation*

    1) If a MEDIA PREP PRODUCT, write BPCS ITEM NUMBER in the box above. Media Prep will fill in LOT NUMBER.

    2) If a PRODUCT, write Production BPCS ITEM NUMBER and SHOP ORDER (LOT) NUMBER in the boxes above. Attach (staple) a copy of the CURRENT Procedure to the Media Preparation Form.

    3) If PRODUCTION SHOP ORDER COMPONENT, list PRODUCT ITEM # and SHOP ORDER # below.

        ITEM #: _____   SHOP ORDER #: _____

    4) If a NEW FORMULATION, identify the source of the new formulation. e.g., Notebook # and Page #, Journal article (Journal name, vol., article, and page), Reference Book (Title and page), etc. Attach a copy to this form.

    SOURCE: _____

    Complete the following sections of the Bill Of Materials on the reverse side (*SECTION B*) for NEW FORMULATIONS: Component, Component Item No., and Final Conc. for Media/Solution Components. If BPCS Item No.'s are unavailable, write vendor name (after the component) in Component Column, and vendor Item No. in the Item Coiumn.

A-4. Check items to be included in prep   *A-5.* DISTRIBUTION

☐ Autoclave	# of Plates:	Vol. Liquid/Per Vol. Container:
☐ Filter Sterilize	# of Flasks:	Vol. Liquid/Per Vol. Container:
☐ Filter (to remove extraneous debris)	# of Bottles:	Vol. Liquid/Per Vol. Container:
WATER SOURCE:	# of Tubes:	Vol. Liquid/Per Vol. Container:
☐ Deionized (DI.)	# of Slants:	Vol. Liquid/Per Vol. Container:
☐ NANOpure	# of Carboys:	Vol. Liquid/Per Vol. Container:
☐ Other (Describe):	Other:	

A-6. SPECIAL INSTRUCTIONS/COMMENTS: _____

_____

A-7. DELIVERY INSTRUCTIONS: _____

☐ Check this box if you wish to receive a copy of the completed Media Preparation Form.

(a)

Figure 27.3. Media Preparation Form. a. This part of the form is completed by the person requesting a solution. **b.** This information is entered into the computer system as the solution is made.

MEDIA PREPARATION FORM (*SECTION B*)   page ___ of ___
(*Complete all sections. Write a line or dash in sections that do not apply.*)

ITEM NUMBER	LOT NUMBER

NAME OF MEDIA/SOLUTION: _____

TARGET pH @ Temp. _____ (1 X pH @ Temp.) _____

CONDUCTIVITY RANGE/(Dilution): _____

SOP NUMBER FOLLOWED: _____

BILL OF MATERIALS   QUANTITY PREPARED: _____

Component (vendor)	Item Number	Lot Number	Final Conc.	Amount per Liter	Total Amount	Equipment	Amount Added

WATER (*circle source and write location*) Deionized/NANOpure/Ultrafiltered_____

EQUIPMENT AND MEASUREMENTS:

Balance(s) *ID #:* (1) _____ (2) _____ (3) _____

pH Meter *ID #:*_____ Conductivity Meter *ID #:*_____

Mixing Vessel: _____ Dispensing Equipment: _____

Filtered with: _____

Filter Sterilized with: _____

Autoclave(s) *ID #, cycle time, and run #:* _____

CALIBRATE pH ELECTRODE

Standard Buffers:_____ Slope/Efficiency @ Temp: _____

Initial pH: _____ pH adjusted with: _____

Final pH @ Temp: _____ 1 X pH @ Temp:_____

CONDUCTIVITY/Dilution: _____

DEVIATIONS FROM PROCEDURE (*list components, critical steps, etc., that differ from the procedure*)_____
_____

COMMENTS:_____
_____
_____

Location _____ Hours_____ Prepared by_____ Date _____

(b)

Figure 27.3. *continued.*

solution preparation. Even in laboratories that do not adhere to a particular quality system, it is good practice to develop standard procedures for preparation of solutions. For solutions that are not made routinely, be certain to record in a laboratory notebook all calculations and all details of preparation.

iv. INSTRUMENTATION

Balances, pH meters, and other instruments used in the preparation of a solution must be properly maintained and calibrated, and maintenance and calibration records must be kept for each instrument. When a solution is made, the instruments used are recorded. Note that the form in Figure 27.3b has blanks that relate to instrumentation.

v. STABILITY AND EXPIRATION DATE

Every solution should be labeled with an expiration date based on its stability during storage.

vi. USING CONDUCTIVITY FOR QUALITY CONTROL

A solution that is made consistently should always have about the same conductivity. Therefore, based on experience, the range within which the conductivity should lie can be specified. The conductivity can then be mea-

1.0 FORMULATION   DNA BUFFER   DAB-00S

Component	Item Number	FW or Conc.	Amount/L	Final Concentration
2 MTris-HCl, pH 7.3	TRIS-73S	2 M	5 mL	10 mM
5 MNaCl	NACL-05S	5 M	2 mL	10 mM
0.25 MEDTA, pH 8.0	EDTA-25S	0.25M	4 mL	1 mM
D.I. H$_2$O		—	Q.S.	—

Target pH: 7.50 ± 0.10 @ 25°C
Conductivity Range: 55–75 µS/cm @ 25°C, 1:40 dilution (2.5 mL/97.5 mL D.I. H$_2$O)

2.0 SAFETY PRECAUTIONS

   2.1 Wear safety glasses, labcoat, and disposable laboratory gloves at all times during the preparation.

3.0 PROCEDURE

   3.1 Read referenced SOPs.

   3.2 Use only preapproved components as specified by the item number listed in the Formulation.

   3.3 Add components to the mixing vessel in the order listed above.

   3.4 Q.S. to final volume with D.I. H$_2$O.

   3.5 Allow to mix for minimum of 10 minutes on a stir plate or with a motorized lab stirrer (e.g., Lightning Mixer).

   3.6 Check final pH with a *calibrated* pH electrode.

   3.7 Filter sterilize (if requested) or autoclave (if requested).

   3.8 Store at room temperature (20–25°C).

(a)

Figure 27.4. Procedures to Make Solutions. a. Procedure to make DNA buffer. This procedure uses the strategy of preparing concentrated stocks and diluting them to make the desired solution. Q.S. is the same as "bring to volume." **b.** Procedure to make 2 *M* Tris-HCl, pH 7.3. This is one of the components of the DNA buffer. *continued*

sured each time the solution is prepared for quality-control purposes. An unexpected conductivity reading is a warning that there is a problem.

PRACTICE PROBLEMS

1. Suppose you start with 5 X solution A and you want a final volume of 10 µL of 1 X solution A. How much solution A do you need?

2. How would you mix 75 mL of 95% ethanol if you have 250 mL of 100% ethanol?

3. How would you mix 25 mL of 35% acetic acid from 25% acetic acid?

4. If you have only 65 mL of 0.3 M Na$_2$SO$_4$, how much 0.1 mM solution can you make?

5. **a.** How would you prepare a 0.1 M stock solution of Tris buffer, pH 7.9? The FW of Tris is 121.1.

   b. How would you use this Tris stock to prepare a 10 mM solution of Tris, pH 7.9?

6. Using the information in Table 27.1, suggest a buffer to maintain a system at pH 8.5.

7. Calculate the pH of a Tris buffer solution that is pH 7.5 at 25°C and then is brought to a temperature of 65°C.

8. Write a procedure to first prepare stock solutions and then to use these stock solutions to prepare Breaking Buffer.

Stock Solutions

1 M Tris, pH 7.6 at 4°C
1 M Mg acetate
1 M NaCl

Breaking Buffer

0.2 M Tris, pH 7.6 at 4°C
0.2 M NaCl
0.01 M Mg acetate
0.01 M β-mercaptoethanol
5%(v/v) Glycerol

NaCl FW = 58.44

β-Mercaptoethanol FW = 78.13, comes as a liquid, density = 1.1 g/mL

Tris FW = 121.1

Magnesium acetate•4H$_2$O FW = 214.40

Notes:

  i There are a couple of points to note regarding the Tris. The recipe specifies that the Tris buffer should be pH 7.6 at 4°C; therefore, bring it to the proper pH at this specified temperature.

  ii β-Mercaptoethanol comes as a liquid. It can be weighed out in grams, but it is much easier to measure β-mercaptoethanol with a pipette. The density

1.0 FORMULATION

2 M TRIS-HCL, pH 7.3

Component	Item Number	FW or Conc.	Amount/L	Final Concentration
Trizma HCl	014426	157.6	274.0 g	1.74 M
Trizma Base	014393	121.1	32.0 g	0.26 M
D.I. H₂O		—	Q.S.	—

Target pH: 7.30–7.35 @ 25°C

Conductivity Range: 150–230 μS/cm @ 25°C, 1:1000 dilution (100 μL/100 mL D.I. H_2O)

2.0 SAFETY PRECAUTIONS

    2.1 Wear Safety Glasses, Labcoat, and disposable laboratory gloves at all times during the preparation.

3.0 PROCEDURE

Note: Quantities of Tris HCl and Tris Base are obtained from the Trizma Mixing Table (see Sigma Trizma Technical Bulletin No. 106B). Values given in the table are specific for a 0.05 M solution. Multiply values by 40 to determine quantities of Tris HCl and Tris base needed for a 2 M solution. Increasing the total Tris concentration from 0.05 M to 0.5 M will increase the pH by about 0.05 pH units, and increasing the total Tris concentration to 2 M generally causes an increase of about 0.05 to 0.10 pH units.

Note: For Tris buffers, *as temperature increases pH decreases.* As a general rule for Tris buffers: *pH changes an average of 0.03 pH units per °C below 25°, and pH changes an average of 0.025 pH units per °C above 25°C.*

    3.1 Read referenced SOPs.

    3.2 Use only preapproved components as specified by the item number listed in the Formulation.

    3.3 Start with D.I. H₂O at approximately 60% of the total volume to be prepared. Add components in the order listed above, and allow to mix until dissolved. *Note:* Mixing of Tris components in water results in an endothermic reaction, and the temperature of the solution cools significantly.

    3.4 Bring volume up to approximately 90% of the final volume, allow to mix and to warm up to near room temperature (20–25°C).

    3.5 Check pH with a *calibrated* pH electrode near room temperature (20–25°C).

    3.6 If the pH is within ± 0.05 units @ 25°C, Q.S. to final volume with D.I. H₂O and allow to mix for a minimum of 5 minutes.

        3.6.1 If pH is not within acceptable limits, adjust carefully with concentrated HCl until pH is within acceptable limits. Q.S. to final volume with D.I. H₂O and allow to mix for a minimum of 5 minutes.

    3.7 Check final pH with a *calibrated* pH electrode.

    3.8 Check pH of a 0.05 M sample (2.5 mL of Tris buffer diluted to 100 mL).

    3.9 Check conductivity of the 2 M solution @ 1:1000 dilution.

    3.10 Filter to remove extraneous particles and dirt.

    3.11 Store at room temperature (20–25°C).

    3.12 *Note:* For large volume preparations (100 liters or more), heat approximately 30 liters of D.I. H₂O per 100 liters to approximately 80–90°C. Weigh out Tris powders and add to mix tank. Add approximately 10 liters of D.I. H₂O to the Tris powders. Add the 30 liters of heated D.I. H₂O. Bring to 90% of the total volume to be prepared. Perform steps 3.3 through 3.9 above.

(b)

Figure 27.4. *continued.*

of β-mercaptoethanol is 1.1 g/mL. This means that 1 mL of β-mercaptoethanol weighs 1.1 g.

    iii β-Mercaptoethanol is extremely volatile, so there are safety precautions involved in its use. Refer to its MSDS (Material Safety Data Sheet) to find out how to use this chemical safely. It is a good policy to check safety precautions whenever working with a new chemical (see Chapter 9).

    iv Magnesium acetate comes as a tetrahydrate with four bound waters. The molecular weight given in the problem is for the tetrahydrate form.

9. Deoxynucleotide triphosphates (dNTPs) are the building blocks of DNA. There are four deoxynucleotide triphosphates: dATP, dTTP, dGTP, and dCTP. Suppose that you are going to repeatedly perform a procedure that calls for 10 μL of a solution that contains a mixture of:

    1.25 mM each of dATP, dTTP, dGTP, and dCTP

You do not have any deoxynucleotide triphosphates in your laboratory, so you are going to buy some. You look in a catalog and find the following entry:

DESCRIPTION: Deoxynucleotide triphosphates are provided at a concentration of 100 mM in water.

Product	Amount in Vial	Catalog #
dATP	40 μmoles	XXXA
dCTP	40 μmoles	XXXC
dGTP	40 μmoles	XXXG
dTTP	40 μmoles	XXXT

a. Observe that the four nucleotides are purchased separately and will then need to be combined into one solution. Decide how many vials of each nucleotide to purchase. Assume you will make enough of the solution to perform the procedure 100 times.

b. Outline how you would prepare the nucleotide solution.

QUESTIONS FOR DISCUSSION

1. Explain the concept of traceability as illustrated by the form in Figure 27.3 on pp. 519–520.

2. Explain how the buffer in Figure 27.4b on p. 521 is brought to the proper pH.

3. The following question is modified from Scenario 4 in the "Skill Standard for the Biosciences Industries" document (Educational Development Corporation, 1995).

   a. You have been given a procedure to prepare a solution. Discuss the considerations important in preparing a solution.

   b. One of the materials that you require to make the solution must be purchased from a specific vendor. You find out, however, that the vendor is out of this particular chemical. What should you do?

CHAPTER 28

Solutions: Associated Procedures and Information

I. INTRODUCTION

This chapter will discuss a variety of practical topics that are related to laboratory solutions. These topics include water purification, properly choosing and cleaning glass and plasticware, and sterilization and storage of solutions and biological samples.

Water is critical for life and is the most important reagent used in biology. For example, it has been estimated that a biotechnology company might require 30,000 L of purified water to produce a single kilogram of recombinant product (Hodgson, John. "Still Waters Run Deep." *Bio/Technology*, 12 October 1994 983–7). Obtaining acceptably purified water is an essential task in any laboratory or biotechnology facility.

Glass and plastic labware must be appropriate for its application and must be cleaned properly. The individuals in a laboratory who are responsible for clean glassware play a key role. Dirty glassware may lead to invalid chemical analyses, alter experimental results, inhibit the growth of cells and microorganisms, or add contaminants to a product. Determining how to properly clean glassware and equipment and validation of cleaning methods are critical concerns in the biotechnology/pharmaceutical industry. FDA frequently cites companies for improper cleaning.

Once prepared, solutions often must be sterilized—without affecting the activity of their components. Various methods are available for this purpose. Finally, solutions and biological samples often must be stored in a way that prevents deterioration.

There are many methods used to purify water, clean glassware, store solutions, and perform other related tasks. This chapter will introduce some of the general concerns relating to these topics.

II. WATER PURIFICATION

A. Water and Its Contaminants

The quality of water used to prepare solutions is of utmost importance. For example, in vitro fertilization—the joining of egg and sperm in a petri dish—will not occur if the water used to make solutions has minuscule amounts of contaminants. In electron microscopy, small particulate contaminants can be magnified hundreds of thousands of times to become major nuisances. Cells in culture will die if extremely small amounts of contaminants are present. Minor levels of impurities in water can introduce major errors into analytical procedures.

Water is an excellent solvent and it readily dissolves contaminants from a wide variety of sources, see Table 28.1. In the environment, water receives contaminants from industry, municipalities, and agriculture. Water dissolves minerals from rocks and soil that make it "hard."

A variety of organic materials from dead plants, animals, and microorganisms are present in water. Gases from the air dissolve in water. Suspended particles enter water from silt, pipe scale, dust, and other sources. In the laboratory, contaminants may leach into water from glass, plastic, and metal containers.

If water is not sterilized, unwanted microorganisms will readily grow in it and may release toxic bacterial byproducts. These byproducts are termed *endotoxins* and *pyrogens*. **Pyrogens** *are derived from lipopolysaccharides that are found in the cell walls of gram-negative bacteria. Pyrogen levels are measured in* **endotoxin units (EU/mL).** Pyrogens were thought to be toxic, so they are often called *endotoxins.* In fact, pyrogens are not directly toxic; rather, they induce a dangerous immune response in mammals. This immune response leads to fever (hence the name *pyrogen,* "heat producing"), shock, and cardiovascular failure. Pyrogens induce adverse effects when present at minute levels; therefore, their elimination is of critical importance in the manufacture of drugs for human and animal use. Even sterile water can contain pyrogens. The term *endotoxin* may be used synonymously with *pyrogen,* or it is sometimes used more broadly to refer to any contaminating byproduct of bacteria.

Removing these varied contaminants from water is an important task in almost any laboratory and is a major operation and expense in bioprocessing industries. To meet these requirements for purified water, laboratories and industrial facilities use a variety of carefully engineered processes.

Table 28.1. CONTAMINANTS OF WATER

Types of Contaminants

1. **Dissolved Inorganics.** This term refers to matter not derived from plants, animals, or microorganisms that usually dissociates in water to form ions. Calcium and magnesium, the minerals that make water "hard," are examples of dissolved inorganics, as are the heavy metals cadmium and mercury.
2. **Dissolved Organics.** Organic compounds are broadly defined to be any materials that contain carbon and hydrogen. Organic materials in water may be the result of natural vegetative decay processes, and they may also be human-made substances such as pesticides.
3. **Suspended Particles.** This category includes diverse materials, such as silt, dust, undissolved rock particles, and metal scale from pipes.
4. **Dissolved Gases.** Water can dissolve naturally occurring gases, such as carbon dioxide, as well as gases from industrial processes, such as nitric and sulfuric oxides.
5. **Microorganisms.** This category includes bacteria, viruses, and other small agents.
6. **Pyrogens/Endotoxins.** Pyrogens and endotoxins are substances, derived from bacteria, that are of special concern in industries that manufacture products for use in humans and animals. See text for an explanation of these terms.

Related Terminology

1. **Total Organic Carbon (TOC).** A measure of the concentration of organic contaminants in water.
2. **Total Solids (TS).** A measure of the concentration of both the dissolved (TDS) and suspended solids (TSS) in a solution. Total dissolved solids include both organics (such as pyrogens and tannin) and inorganics (such as ions and silica). Total suspended solids include organics (such as algae and bacteria) and inorganics (such as silt). It is common to approximate TDS by measuring the conductivity, although conductivity is actually a measure only of conductive species.

B. What Is Pure Water?

There actually is no such thing as "pure water." Although a variety of effective methods are used to remove contaminants from water, regardless of how stringently it is purified it is never totally free of contaminants. Professional societies therefore publish standards that contain specifications for the relative purity of water to be used in certain applications. The American Society for Testing and Materials (ASTM) establishes specifications for reagent (laboratory) water. The College of American Pathologists (CAP) and the Clinical and Laboratory Standards Institute, (CLSI, formerly NCCLS) establish specifications for water used in clinical laboratories. The Japanese, European Union, and United States Pharmacopeias contain specifications for water used in manufacturing and testing pharmaceutical products.

Water purity standards historically were often linked to the method by which the water was purified; distillation in particular was often required to produce high-quality grades of water. The requirement to use a particular purification method limited the use of modern technologies that are often more efficient and cost effective than older methods. There has therefore been a recent shift in the water purity standards so that they no longer mandate particular methods of purification. Current standards do require that certain key contaminants are measured and that their levels not exceed specified limits.

ASTM identifies four types of laboratory water. Type I is the highest class of purity. Type I water is used for most analytical procedures (e.g., trace metal analysis) because it has very low levels of contaminants. Type I water is also used for tissue culture, high performance liquid chromatography (HPLC), electrophoresis buffers, immunology assays, preparation of standard solutions, and reconstitution of lyophilized materials. Type II water may have somewhat higher levels of impurities than Type I, but it is still suitable for most routine laboratory work, such as microbiology procedures. Type III and Type IV waters are acceptable for some applications, such as initial glassware rinses (final rinses should be in whatever type of water is required for the procedure to be performed) and preparing microscope slides to examine tissue.

ASTM further classifies water according to grade. There are three grades, A, B, and C, that can be applied to the four types of water. These grades vary in their permissible levels of bacterial contaminants and endotoxin levels. ASTM specifications are shown in Tables 28.2 and 28.3.

There are applications where water that exceeds Type I standards is required. Analytical tests using high performance liquid chromatography, for example, are extremely sensitive and can detect organic compounds in samples at levels as low as a few parts per billion. Type I water may have more impurities than this and therefore is not acceptable for extremely sensitive HPLC work. Cell culture techniques may require water with lower pyrogen levels and fewer contaminants than are specified by Type I standards. There are water purification systems available that meet these stringent requirements, but they must be properly operated and maintained to deliver water of the highest quality.

CLSI identifies clinical laboratory reagent water (CLRW) that is analogous to ASTM's Type I and Type II water. They also identify other types of water for

Table 28.2. ASTM REQUIREMENTS FOR PURITY OF PURIFIED WATER

Type	Grade	Maximum Conductivity at 25°C μS/cm	Minimum Resistivity at 25°C MΩ-cm	Maximum Total Organic Carbon (μg/L)	Maximum Sodium (μg/L)	Maximum Chloride (μg/L)	Maximum Total Silica (μg/L)	Maximum Bacteria (colony forming units/mL)	Maximum Endotoxin (endotoxin units/mL)
I		0.0555	18	50	1	1	3		
I	A	0.0555	18	50	1	1	3	10/1000	0.03
I	B	0.0555	18	50	1	1	3	10/100	0.25
I	C	0.0555	18	50	1	1	3	100/10	
II		1.0	1.0	50	5	5	3		
II	A	1.0	1.0	50	5	5	3	10/1000	0.03
II	B	1.0	1.0	50	5	5	3	10/100	0.25
II	C	1.0	1.0	50	5	5	3	100/10	
III		0.25	4.0	200	10	10	500		
III	A	0.25	4.0	200	10	10	500	10/1000	0.03
III	B	0.25	4.0	200	10	10	500	10/100	0.25
III	C	0.25	4.0	200	10	10	500	1000/100	
IV		5.0	0.2		50	50			
IV	A	5.0	0.2		50	50		10/1000	0.03
IV	B	5.0	0.2		50	50		10/100	0.25
IV	C	5.0	0.2		50	50		100/10	

pH is specified only for Type IV water and must be between 5.0 and 8.0.

Table 28.3.   ASTM WATER: METHODS OF PRODUCTION

Type	Production Process
I	Purify to a conductivity of 20 µS/cm or less at 25°C by distillation followed by polishing with a mixed-bed of ion-exchange materials and a 0.2 µm membrane filter. Type I reagent water may be produced with alternate technologies as long as the specifications are met and the water is shown to be appropriate for its application.
II	Purify by distillation to a conductivity of less than 1.0 µS/cm at 25°C. Ion exchange, distillation, or reverse osmosis and organic absorption may be required prior to distillation if the purity required cannot be obtained in a single distillation. Type II reagent water may be produced with alternate technologies as long as the specifications are met and the water is shown to be appropriate for its application.
III	Purify by distillation, ion exchange, continuous electrodeionization, reverse osmosis, or a combination of these, followed by polishing with a 0.45 µm membrane filter. Type III reagent water may be produced with alternate technologies as long as the specifications are met and the water is shown to be appropriate for its application.
IV	Purify by distillation, ion exchange, continuous electrodeionization, reverse osmosis, or a combination of these. Type IV reagent water may be produced with alternate technologies as long as the specifications are met and the water is shown to be appropriate for its application.

other purposes, such as autoclaving and washing laboratory materials.

The water used for producing pharmaceutical products must meet different standards than laboratory water. These standards are found in the pharmacopeias for various countries. The European Union, Japan, and the United States distinguish between purified water (PW), water for injection (WFI), and sterile purified water. **Purified water** *is produced using a suitable process from source water that complies with EPA (Environmental Protection Agency) drinking water regulations or comparable regulations in the EU and Japan.* PW cannot be used for injection, but can be used for nonsterile drugs. PW is also used for laboratory tests and assays, unless otherwise stated, and can be sterilized and used for sterile, noninjected drugs. PW can be made by any method but must meet certain purity standards. WFI *is defined as water purified by distillation or a purification process that is equivalent or superior to distillation in the removal of chemicals and microorganisms.* It must meet the same requirements as purified water or sterilized purified water and, in addition, it must not contain more than 0.25 USP endotoxin units/mL and bacterial levels lower than 10 colony forming units (CFU)/100 mL. Table 28.4 summarizes European Union and United States Pharmacopoeia standards for water.

C. Methods of Water Purification

i. OVERVIEW

Source or **feed water** *is the water that is to be purified.* Most laboratories and biotechnology facilities begin with municipal (tap) water that is partially purified to make it safe for drinking. Such water has reduced levels of microorganisms and other contaminants, but contains high levels of chlorine and varying levels of many other contaminants. The quality of municipal water varies greatly depending on geographic location, the season, treatment methods used, and other factors. One of the oldest biotechnology industries, the brewing industry, has a tradition of choosing manufacturing sites based on the quality of local water available.

Some water sources contain contaminants that are difficult to remove, even with the best treatment methods. Because source water is so variable, and because each application has different purity requirements, water-purification systems are tailored to meet the needs of a particular laboratory or industrial facility. There are a variety of methods of purifying water that vary in their costliness and ability to remove different kinds of contaminants. These are summarized in Table 28.5 on p. 528. Because each purification method removes some, but not all, types of contaminants, it is common for water-purification systems to use more than one purification method.

Table 28.4.   PHARMACOPOEIA REQUIREMENTS FOR PURITY OF PURIFIED WATER

Properties	European Union Pharmacopeia	United States Pharmacopeia
Nitrates	< 0.2 ppm	—
Heavy metals	< 0.1 ppm	—
TOC (a measure of organics)	< 500 µg/L	< 500 µg/L
Conductivity	< 4.3 µS/cm at 20°C	< 1.3 µS/cm at 25°C
Bacteria (guideline)	< 100 CFU/mL	< 100 CFU/mL
Additional requirements for WFI	Conductivity < 1.1 µS/cm at 20°C Bacteria: < 10 CFU/100 mL Endotoxins: < 0.25 USP International units/mL	Bacteria: < 10 CFU/100 mL Endotoxins: < 0.25 Endotoxin units/mL

Table 28.5. A COMPARISON OF WATER-PURIFICATION METHODS

	Distillation	Deionization	Filtration	Reverse Osmosis	Adsorption	Ultrafiltration	UV Oxidation
Dissolved ionized solids	E/G	E	P	G	P	P	P
Dissolved organics	G	P	P	G	E	G	G
Dissolved ionized gases	P	E	P	P	P	P	P
Particulates	E	P	E	E	P	E	P
Bacteria	E	P	E	E	P	E	G
Pyrogens	E	P	P	E	P	E	P

E = Excellent; G = Good; P = Poor (Courtesy of Barnstead/Thermolyne.)

ii. DISTILLATION

Distillation was invented hundreds of years ago and is still sometimes used for preparing laboratory water. **Distilled water** *is prepared by heating water until it vaporizes,* see Figure 28.1. The water vapor rises until it reaches a **condenser** *where cooling water lowers the temperature of the vapor and it condenses back to the liquid form.* The condensed liquid flows into a collection container. Most contaminants remain behind in the original vessel when the water is vaporized. Ionized solids, organic contaminants that boil at a temperature higher than water, pyrogens, and microorganisms are all removed from distilled water. A few contaminants, however, cannot be removed from water by distillation. These include dissolved gases such as carbon dioxide, chlorine, ammonia, and organic contaminants that volatilize at a low temperature.

Distillation is a well-established method of water purification that is the traditional "gold standard." Distillation, however, is relatively expensive compared

with other methods because of the substantial energy input required to heat the water. Distillation is slow and therefore distilled water is generally prepared in advance and stored for later use. This storage of the distillate can be problematic since plasticizers and other substances will leach out of plastic storage containers and re-contaminate the water. In addition, bacteria grow very well in water that has been standing. Stills require frequent maintenance to remove scale build-up when the source water is hard. For all these reasons, distillation is being replaced in many facilities by newer technologies, as has been acknowledged in recent changes to water purity standards.

iii. ION EXCHANGE

Ion-exchange *removes ionic contaminants from water.* During ion exchange water flows through cartridges packed with bead-shaped resins, called *ion exchange resins.* As the water flows past the beads, ions in the water are exchanged for ions that are bound to the beads. There are two types of resins. **Cationic resins** *bind and exchange positive ions.* **Anionic resins** *bind and exchange negative ions.*

Water softeners (such as many people have in their homes) are a type of ion exchanger. **Water softeners** *exchange positive ions that make water hard, particularly calcium and magnesium, with sodium ions.* Water softeners thus remove calcium and magnesium from water and add sodium to it. Water softeners use cationic beads that initially have Na^+ ions loosely bound to them. As water flows past these beads, positive ionic contaminants in the water exchange places with the Na^+ ions. The contaminants remain linked to the beads while the sodium ions enter the water stream. Water softening may be used as a preliminary step in water purification, but it is not adequate for laboratory water purification because it leaves sodium ions and all negative ions in the water.

Deionization *is an ion exchange process used in the laboratory to remove "all" ionic contaminants from water,* see Figure 28.2. Deionization involves both cation and anion exchange beads. The anionic resin beads ini-

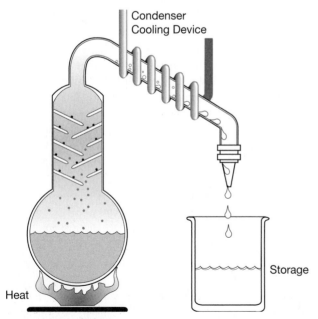

Condenser
Cooling Device

Heat

Storage

Figure 28.1. Distillation. (Illustration courtesy of Millipore Corporation.)

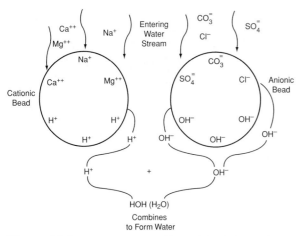

Figure 28.2. Deionization. (Courtesy of Osmonics, Inc., Minnetonka, MN.)

tially have loosely bound hydroxyl ions, whereas the cationic beads have numerous hydrogen ions loosely bound to their surfaces. As water passes by the anion exchange beads, negatively charged ions in the solution take the place of the OH^- ions on the beads. The OH^- ions enter the water stream, but the contaminants remain bound to the resin beads. Positive ionic contaminants in the water similarly exchange places with H^+ ions that are attached to the cationic beads. The hydrogen ions enter the water; the positive ionic contaminants remain bound to the beads. The H^+ ions and the OH^- ions that enter the water stream combine with one another to form more water molecules.

After a period of use, cation and anion contaminants from the water will have replaced most of the active hydrogen and hydroxyl sites in the resins and the cartridges will need to be replaced or regenerated. Alternatively, **electrodeionization, EDI,** *is a newer variant of deionization that uses electrical power to continuously split water molecules, eliminating the need for regenerating the resin.* EDI has the advantages that it is continuous and therefore provides more consistent water purity than conventional methods.

Deionized water is more pure than tap or softened water because the ionic contaminants have been removed, but it is generally not acceptable for making laboratory solutions because it may contain organic contaminants, pyrogens, and microorganisms.

iv. CARBON ADSORPTION

Carbon adsorption *is an effective method for removing dissolved organic compounds from water.* Carbon adsorption uses **activated carbon,** *which is traditionally made by charring wood by-products at high temperatures, then "activating" it by exposure to high temperature steam.* There is also a synthetic form of activated carbon that is produced from styrene beads. Natural activated carbon can release some ionic contaminants into the water being treated; the synthetic version is less likely to release ionic contaminants.

Activated carbon has an extremely porous, honeycomb-like, structure. As water passes through the activated carbon, chlorine and organic impurities adsorb (stick) to the carbon and are thus removed, see Figure 28.3. Activated carbon is sometimes used before deionization to protect the resins from contamination by organic materials and chlorine.

v. FILTRATION METHODS

Filtration *removes particles as the water passes through the pores or spaces of a filter.* There are five types of filters used in water treatment:

1. **Depth filters** *are made of sand or matted fibers and retain particles through their entire depth by entrapping them,* see Figure 28.4a on p. 530. Depth filters are useful for removing relatively large particles, greater than about 10 μm, from the water stream. (Pollen and some yeast cells are examples of substances in this size range.) Depth filters may be used at the beginning of a purification system as an inexpensive method to remove large particles and debris that might otherwise clog and foul the downstream elements of the system.

2. **Microfiltration membrane filters (also called screen filters or microporous membranes)** *are made by carefully polymerizing cellulose esters or other materials so that a membrane with a specific pore size is formed,* see Figure 28.4b on p. 530. Water treatment filters often have pore sizes on the order of 0.20 μm because bacteria are too large to penetrate such pores. Microfiltration membrane filters are useful for removing bacteria and particles above a specified pore size, but they are not useful for removing smaller, dissolved molecules.

3. **Ultrafiltration membranes** *do not have measurable pores and will prevent the passage of dissolved molecules, including most organics,* see Figure 28.4c on p. 530. Materials down to 1000 D in molecular weight (smaller than most proteins) can be separated from water using ultrafilters. Two important applications of ultrafilters are the removal of

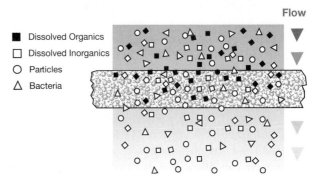

■ Dissolved Organics
□ Dissolved Inorganics
○ Particles
△ Bacteria

Figure 28.3. Activated Carbon. As water flows through the porous activated carbon, chlorine and dissolved organics are selectively adsorbed by the carbon. (Courtesy of Millipore Corporation.)

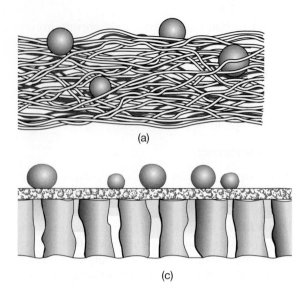

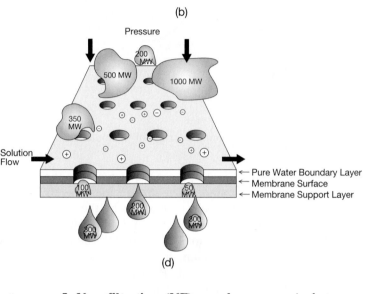

Figure 28.4. Filtration Methods. a. Depth filter. **b.** Membrane filter with pores of a particular size. **c.** Ultrafilter used to retain molecules that are above a particular molecular weight. **d.** Reverse osmosis. (**a–c.** Illlustrations courtesy of Millipore Corporation. **d.** Illustration courtesy of Osmonics, Inc., Minnetonka, MN.)

viruses and pyrogens from water. Small, dissolved inorganic molecules can pass through ultrafilters.

4. **Reverse osmosis (RO) membranes** *are still more restrictive than ultrafiltration membranes,* see Figure 28.4d. Materials down to 300 D in molecular weight can be removed from water using RO and so this method is effective in removing viruses, bacteria, and pyrogens. Moreover, RO membranes also reject ions and very small dissolved particles, such as sugars.

In reverse osmosis, water under pressure flows over a special, thin RO membrane. The membrane allows water to pass through, but rejects a large proportion (on the order of 95–99%) of all types of impurities (particles, pyrogens, microorganisms, colloids, and dissolved organic and inorganic materials). The permeate (water that has passed through the membrane) will contain very low levels of contaminants that are able to pass through even an RO membrane, but most types of contaminants are greatly reduced. Because RO membranes are very restrictive, the flow rate through them is slow.

Reverse osmosis is similar to ultrafiltration in that both involve membranes that reject small materials. The major differences between ultrafiltration and RO include: (1) RO can retain very small solutes, including salt ions. Ultrafiltration membranes allow such very small solutes to pass through. (2) Higher pressures are used in RO to force water through the membrane. (3) An ultrafiltration membrane retains particles based almost solely on their size. An RO membrane retains materials based both on their size and on ionic charge.

5. **Nanofiltration (NF) membranes** *are in between ultrafilters and reverse osmosis membranes.* RO can remove the smallest solute molecules, in the range of 0.0001 micrometer in diameter and smaller; nanofiltration removes molecules in the 0.001 micrometer range. NF is essentially a lower-pressure version of reverse osmosis that is used for applications where the utmost purity of product water is not required. Like RO, NF is also capable of removing bacteria, viruses, and other organics. Since NF requires lower pressure than does RO, energy costs are lower than for a comparable RO treatment system.

vi. OTHER METHODS: ULTRAVIOLET OXIDATION AND OZONE STERILIZATION

Ultraviolet (UV) oxidation *is a method for the removal of organic contaminants from water.* Water is passed continuously (for example for 30 minutes) across the light path of an ultraviolet lamp that emits light at a wavelength of 185 nm. The organic compounds in the water oxidize to simple compounds such as CO_2. UV light at a wavelength of 254 nm will kill bacteria and is therefore sometimes used to sterilize water.

In industrial settings ozone is also sometimes used to sterilize water. Ozone kills bacteria by rupturing their membranes and acts much more quickly than chlorine. Both ozone and chlorine must be removed from the water by later purification steps.

vii. WATER PURIFICATION SYSTEMS

Many laboratories and most biotechnology and pharmaceutical facilities use water that is purified by a combina-

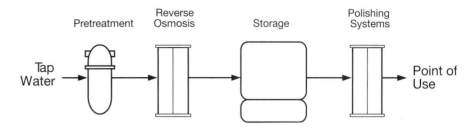

Figure 28.5.   A Multistep Water Purification System. (Courtesy of Millipore Corporation.)

tion of methods, see Figure 28.5. The choice of methods and system design will depend on the applications for which the water is to be used and the quality of source water available. For example, a water-purification system to prepare Type I water might include the following steps:

1. *Tap (source) water is softened or distilled.*
2. *The water then passes through activated carbon to remove organics and chlorine.*
3. *Reverse osmosis is used next and the treated water is moved to a storage tank until needed.*
4. *When a user wants purified water, water from the RO tank is run through a series of ion exchange cartridges in which all ionized contaminants are removed.*
5. *The water might be filtered through an ultra-filtration membrane.* This final filtration step removes almost all microorganisms including those that the water picked up while traveling through the purification system.

viii. HANDLING OF REAGENT WATER

Highly purified water is an extremely aggressive solvent. It will readily leach contaminants from any vessel in which it is stored and will also dissolve CO_2 from the air. In addition, bacteria can multiply in purified water during storage. Type I water, therefore, cannot be stored and must be used immediately after it is produced. Type II water can be stored for short periods of time.

ix. A NOTE ABOUT TERMINOLOGY

There are a variety of methods used to purify water. Distillation, however, is the traditional method, and it is common for people to speak of highly purified water as being "distilled" (DI) whether it is purified by RO, ultrafiltration, or actually by distillation. Deionized water is less pure and should never be referred to as "distilled" water. In this book, the term *purified water* is used to mean water that has been distilled or that has been run through a system as shown in Figure 28.5. Because the terminology of water purification is used inconsistently, standard operating procedures should explicitly state the source of water required.

D. Operation and Maintenance of Water Purification Systems

i. MONITORING THE SYSTEM FOR QUALITY

Purified water must be monitored periodically to be certain that it meets quality requirements. Because there are a variety of contaminants in water, no single test is adequate to measure water purity. The following are some quality parameters that can be monitored:

1. **Resistivity (resistance).** Ionic contaminants in water are easy to detect because they make water more conductive (i.e., less resistant to current flow). Resistance, therefore, is a measure of water purity: The more water resists electrical current flow, the fewer ionic contaminants it contains.

   Water purification systems typically have an attached meter to read the water resistance. A user who wants to draw water flushes the system for a minute or so, discarding the rinse water. As water is flushed through the system the resistivity increases until it reaches an acceptable level. The theoretical maximum ionic purity for water is 18.3 megohm-cm and 17.0 megohm-cm is considered an acceptable value for Type I water in many laboratories. Nonionic contaminants do not affect resistivity, so even if the resistance of water is 17.0 megohm-cm, it may contain contaminants.

2. **Bacterial counts.** Bacterial counts are used to monitor levels of microorganisms in water. A particular volume of the water is filtered through a membrane filter that traps bacteria on its surface. The filter is incubated on nutrient medium. *The numbers of colonies growing on the medium are counted to give the number of* **colony forming units (CFU)** *per volume of water.* The various standards for water quality specify a maximum number of colony forming units that are acceptable.

3. **Pyrogens.** Most clinical, analytical, and research laboratories do not check for pyrogen contamination but this assay is essential in the pharmaceutical industry. The ***Limulus* Amebocyte Lysate Test (LAL)** *is used to monitor pyrogens in a water sample.* The LAL test requires an extract of blood from the horseshoe crab, *Limulus polyphemus.* Different dilutions of the water to be tested are mixed with the *Limulus* blood extract and any pyrogens present cause the extract to clot. The clotting results can be converted to **endotoxin units per milliliter (EU/mL).**

4. **Total organic carbon.** Organic contaminants can interfere with analyses and can indicate that bacteria have contaminated the water treatment system. Organic carbon can be monitored using a chemical method (involving potassium permanganate) or instrumental methods.

5. pH. Ultrapure water that is exposed to air reacts with carbon dioxide, resulting in a pH of about 5.7. If it is in a covered container, pure water will have a pH of about 6.0. In some situations the pH of the water is important and is monitored.

Other water quality parameters, such as particulate and silicate levels, can be tested where necessary. Whether various tests are performed, and if so, how often, depends on the needs of the laboratory or industrial facility, see Figure 28.6.

ii. MAINTENANCE

Water purification systems need to be operated and maintained properly to ensure high-quality water. The maintenance required and its frequency will vary depending on the type of laboratory, the contaminants present in the source water, and any regulations that are applicable. Some general considerations include:

1. **Distilled water apparatus.** A distilled water apparatus requires frequent, sometimes daily, cleaning; otherwise, it can become contaminated by microorganisms and their by-products. Distilled water should be prepared freshly each day because it can absorb contaminants from storage vessels and microorganisms can grow in it.

2. **Ion exchange resins.** Conventional ion exchange resins must be regenerated once they have exchanged all their Na^+, H^+, or OH^- ions for contaminants. Regeneration is sometimes performed on-site, and sometimes cartridges are sent off-site for maintenance. Microorganisms can attach to deionization beads, thereby adding contaminants to the water. Deionization systems are therefore periodically sanitized.

3. **Filters.** Filters must be tested when installed to be certain they have no small holes and are effective. They must be periodically flushed and sanitized to remove contaminants from their surfaces. Filters must be replaced occasionally because they can develop defects over time and bacteria can actually grow through their pores.

4. **Activated Carbon.** Activated carbon may become colonized by bacteria and may become clogged by particulates. Activated carbon beds must therefore be washed to remove particu-

---

Water

H$_2$O   M$_r$ 18.02   [7732-18-5]   EEC No 2317912

95284	*BioChemika* for molecular biology; filtered through a 0.2 μm membrane filter; DEPC-treated and autoclaved; DNases, RNases, proteases and phosphatases; none detected; d$_4^{20}$ 1.00; specific resistance at 25°C: 1.8·10^7 Ωcm, measured at production
95285	*BioChemika* for endotoxin trace analysis; ultrafiltered and autoclaved; Endotoxins: none detected (LAL-test); d$_4^{20}$ 1.00; specific resistance at 25°C: 1.8·10^7 Ωcm, measured at production
95289	*BioChemika* for cell biology; ultrafiltered and autoclaved; mycoplasma, chlamydiae and viruses: none detected (DAPI-test): d$_4^{20}$ 1.00; specific resistance at 25°C: 1.8·10^7 Ωcm, measured at production
95301	for ^{2}H-NMR-spectroscopy; deuterium-depleted; Isotopic purity: ^{2}H$_2$O < 0.0001%; d$_4^{20}$ 1.00
95311	Selectophore®; d$_4^{20}$ 1.00
95303	for UV-spectroscopy; d$_4^{20}$ 1.00

Fluka-Guarantee:

A:		λ(nm):	200
d: 1 cm, Ref. cell: air		A$_{max}$:	0.01

95306	B&J Brand (product line of Burdick & Jackson)
95305	for inorganic trace analysis; filtered through a 0.2 μm membrane filter; d$_4^{20}$ 1.00; specific resistance at 25°C: 1.8·10^7 Ωcm, measured during production

Fluka-Guarantee:

Residue on evaporation .... < 0.001%   Residue on ignition .... < 0.0005%   KMnO$_4$-reducing matter (as O) .... < 0.000003%

Bromide (Br)......... < 0.000001%	Ca............... < 0.000001%	Mn .... < 0.0000005%
Chloride (Cl).......... < 0.000001%	Cd............... < 0.0000005%	Mo .... < 0.0000005%
Iodide (I)........... < 0.000001%	Co............... < 0.0000005%	Na .... < 0.000001%
Nitrate (NO$_3$) .... < 0.000001%	Cr .... < 0.0000005%	NH$_4$ .... < 0.0000005%
Phosphate (PO$_4$) ...... < 0.000001%	Cu............... < 0.0000005%	Ni .... < 0.0000005%
Sulfate (SO$_4$) .... < 0.000001%	Fe .... < 0.0000005%	Pb .... < 0.0000005%
Al .... < 0.0000005%	K................ < 0.000001%	Sr .... < 0.0000005%
Ba................. < 0.0000005%	Li............... < 0.0000005%	Zn .... < 0.0000005%
Bi .... < 0.0000005%	Mg............. < 0.0000005%	

95286	for organic trace analysis; trace organics < 0.000003% (GC); d$_4^{20}$ 1.00

Figure 28.6.   Catalog Specifications for Purified Water. This manufacturer produces purified water for sale. Observe that depending on the application, different characteristics of the water are featured and different purification methods are specified. (Reprinted with permission of Fluka Chemie AG.)

lates. Activated carbon must also be periodically recharged or replaced as active sites are filled by contaminants.

5. **UV lamps.** As UV lamps age, their light output decreases. Bulbs must therefore be replaced about once a year.

iii. DOCUMENTATION

As with any other aspect of a quality-control program, the operation, maintenance, and monitoring of water purification systems requires planning and documentation. This includes developing standard operating procedures, training operators, planning routine maintenance and monitoring, establishing logbooks, developing written procedures to deal with problems, and documenting corrective actions taken.

EXAMPLE PROBLEM

a. Ultrapure, sterile water is stored in a very clean, sterile glass bottle for 4 days. Speculate as to the quality of this water after storage.

b. Ultrapure, sterile water is stored in a very clean, sterile plastic bottle for 4 days. Speculate as to the quality of this water after storage.

ANSWER

a, b. Although the bottles are clean and the conditions are sterile, ultrapure water is an extremely aggressive solvent and will leach heavy metals (which are minor contaminants of glass) from the glass container into the water. Ultrapure water will similarly leach organic carbon from the plastic and will become contaminated.

III. GLASS AND PLASTIC LABWARE

A. The Characteristics of Glass and Plastic Labware

Glass and plastic labware is used in laboratories for preparing and storing solutions, for containing cells and microorganisms, for holding experimental reaction mixtures, and so on. There is a wide variety of glass and plastic items available for laboratory use, so these items can be matched to their application.

Various grades of glass differing in strength, heat resistance, and other factors are used to make beakers, flasks, graduated cylinders, and pipettes. Glass is relatively inert when in contact with chemicals and is therefore used to make containers for chemicals. Glass, however, may adsorb metal ions and proteins from solutions and is not recommended for the storage of metal standards. Also, trace amounts of contaminants, such as sodium, barium, and aluminum can leach out of glass containers. In normal laboratory work these trace contaminants are not significant. However, in some analytical and pharmaceutical applications, they are problematic.

Borosilicate glass *is used to make general purpose laboratory glassware that is strong, resistant to heat and cold shock, and does not contain discernable contamination by heavy metals.* Borosilicate glass is sold, for example, under the trade names of Pyrex (Corning Glass Works, Corning, NY) and Kimax (Kimble Glass Company, Vineland, NJ). Corex (Corning) is a type of glass that is resistant to physical stress and is therefore used to make centrifuge tubes. Type I glass refers to highly resistant borosilicate glass used in the manufacture of ampules, vials, and other containers for pharmaceutical applications. Additional information about types of glass is shown in Table 28.6.

Table 28.6. *COMMON TYPES OF GLASS*

Type	Qualities	Purpose
Borosilicate (Pyrex, Kimax)	High resistance to heat and cold shock. Low in metal contaminants.	General purpose
Corex	Aluminosilicate glass, high resistance to pressure and scratching.	Centrifugation
High silica (quartz)	Greater than 96% silicate. Excellent optical properties.	Optical devices, mirrors, cuvettes
Low actinic*	Red-tinted to reduce light exposure of contents.	To contain light-sensitive compounds
Flint	Soda-lime glass containing oxides of sodium, silicon, and calcium. Poor resistance to temperature changes and high temperature, poor chemical resistance.	Disposable glassware, such as pipettes
Optical	Soda-lime, lead, and borosilicate.	Used in prisms, lenses, and mirrors

*Actinic light is defined as light capable of producing a photochemical reaction. This is usually near ultraviolet or blue visible light with wavelengths between 290 and 450 nm.
(Information in this table is primarily adapted from *Clinical Chemistry: Theory, Analysis, Correlation*, Lawrence A. Kaplan and Amadeo J. Pesce. 3rd ed. St. Louis: Mosby, 1996.)

Table 28.7. COMMON PLASTICS AND THEIR QUALITIES

Type	Qualities	Purpose
Polyethylene	Biologically and chemically inert, resistant to solvents. Strong oxidizing agents will eventually make polyethylene brittle. Not resistant to autoclaving.	Used for disposable plasticware
Polypropylene	Translucent, can be autoclaved, resistant to solvents. More susceptible than polyethylene to strong oxidizers. Biologically inert. Thin-wall products permeable to CO_2 and O_2.	General purpose
Polystyrene	Rigid, clear, brittle. Biologically inert. Sensitive to organic solvents. Used where transparency is an advantage. Melts in autoclave. Permeable to CO_2.	Used for disposable plasticware and tissue culture products
Polyvinyl chloride (PVC)	Can be made soft and pliable.	Often used for laboratory tubing.
Polycarbonate	Transparent and strong. Sensitive to bases, concentrated acids, and some solvents.	Used for centrifuge-ware
Teflon (DuPont) (PTFE)	Highly resistant to chemicals and temperature extremes. Inert, comparatively costly.	Wide variety of uses including fittings in instruments, stir bars, stopcocks, tubing, bottle cap liners, water distribution systems
Nylon	Strong, resistant to abrasion, resistant to organic solvents. Poor resistance to acids, oxidizing agents, and certain salts.	Widely used for membranes and filters in biology

Based primarily on information from Nalgene Catalog and Falcon Labware Catalog.

Plastics are less breakable than glass, and they come in a variety of types that vary in their properties and applications. For example, some types of plastics are more resistant than are others to heat, acids, organic solvents, and the forces in a centrifuge. Various plastics are incompatible with certain chemicals and may darken, become brittle, crack, or dissolve when exposed to them. Some plastics are porous and stored liquids may evaporate from them. Some plastics are transparent, others are not. Table 28.7 briefly summarizes the qualities of common types of plastics and their uses. Figure 28.7 summarizes information regarding the resistance of various types of plastics to various classes of chemicals. Consult a chemical compatibility chart when using plastics to hold specific chemicals, as shown, for example, in the Nalgene Labware Company catalog.

B. Cleaning Glass and Plastic Labware

i. OVERVIEW

There are many procedures used to clean labware. Which procedure is best depends on the items to be cleaned, the soils contaminating the labware, and the application for which they will be used. Proteinaceous deposits, organic residues, metals, greases, and microorganisms are examples of contaminants that may need to be removed. Note that safety precautions are required when washing labware that was used for hazardous materials, or if the cleaning agents used are dangerous.

In general, labware washing has five steps:

1. **Prerinse.** A prerinse or soak immediately after use prevents contaminants from drying onto labware.
2. **Contaminant Removal.** Contaminants are typically removed using detergents or solvents in conjunction with a physical method such as scrubbing or exposure to jets of water.
3. **Rinse.** All traces of detergent and cleaning solvents are rinsed away.
4. **Final Rinse.** Any residues from the rinse in Step 3 are removed. Purified water is used for final rinses.
5. **Drying.**

The effectiveness of cleaning depends on four key factors: The proper choice of detergent, the effectiveness of the mechanical action, how long the items are subjected to mechanical washing, and the temperature of the wash water. Hotter water is more effective than is cooler water.

ii. DETERGENTS

There are various types of detergents used for cleaning labware that vary in their pH, additives, and other qualities. It is necessary to choose the correct detergent and to dilute and use it properly. The choice of detergent is based on what is to be cleaned, what contaminants must be removed, and the physical method of cleaning that will be used. Table 28.8 lists some general guidelines to help determine which detergent to use.

Chromic acid washes were often routinely used in the past to remove proteinaceous and lipid contaminants from glassware. The use of chromic acid, how-

CHEMICAL RESISTANCE SUMMARY*

E = Excellent Resistance
G = Good Resistance
F = Fair Resistance
N = Not Resistant

Classes of Substances at 20°C	ECTFE/ETFE	FEP/TFE/PFA	FLPE	FLPP	HDPE	LDPE	PC	PETG	PMMA	PMP	PP/PPCO	PS	PSF	PUR	PVC (Bottle)	Flexible PVC Tubing	PVDF	ResMer	TPE***
Acids, Dilute or Weak	E	E	E	E	E	E	E	G	G	E	E	E	E	F	E	G	E	E	G
Acids,**Strong and Concentrated	E	E	G	G	G	G	N	N	N	E	G	F	G	N	G	F	E	G	F
Alcohols, Aliphatic	E	E	E	E	E	E	G	G	N	E	E	G	G	N	G	F	E	E	E
Aldehydes	E	E	G	G	G	G	F	G	F	G	G	F	F	N	G	N	G	G	G
Bases/Alkali	E	E	F	E	E	E	N	N	F	E	E	E	E	F	E	F	G	E	F
Esters	G	E	G	G	G	G	N	F	N	E	G	N	N	N	N	N	G	F	N
Hydrocarbons, Aliphatic	E	E	E	G	G	F	G	G	G	G	G	F	G	G	G	F	E	G	E
Hydrocarbons, Aromatic	G	E	E	N	N	N	N	N	N	N	N	N	N	N	N	N	E	F	N
Hydrocarbons, Halogenated	G	E	G	F	N	N	N	N	N	N	N	N	N	N	N	N	F	F	F
Ketones, Aromatic	G	E	G	G	N	N	N	N	N	F	N	N	N	N	F	N	F	F	N
Oxidizing Agents, Strong	E	E	F	F	F	F	F	F	N	G	F	G	G	N	G	F	G	G	N

*For tubing chemical resistance, other than PVC, see tubing section.
**Except for oxidizing acids: for oxidizing acids, see "Oxidizing Agents, Strong."
***TPE gaskets.

Figure 28.7.   Summary of Resistances of Various Plastics to Various Classes of Chemical Agent. (Reproduced courtesy of Nalge Nune International.)

ever, has serious disadvantages. It is extremely corrosive and hazardous, so extreme caution is necessary to avoid injury when using such washes. Chromic acid contains chromium, a heavy metal, which requires costly disposal in a special hazardous waste site. Because of the hazards and costs associated with acid baths, they are now less common. Proper use of detergents, milder acids, such as 1 M HCl and 1 M HNO$_3$, and organic solvents usually eliminate the need for strong acid washes.

iii. PHYSICAL CLEANING METHODS

Detergents are used in conjunction with physical methods of cleaning. Depending on the situation, a variety of physical methods are used to remove contaminants from labware. These include:

Table 28.8.   GENERAL GUIDELINES FOR CHOOSING A DETERGENT

1. **Alkaline Cleaners.** A broad range of organic and inorganic contaminants are readily removed from glassware by mildly alkaline cleaners.
2. **Acid Cleaners.** Metallic and inorganic residues are often solubilized and removed by acid cleaners.
3. **Protease Enzyme Cleaners.** Proteinaceous residues are effectively digested by protease enzyme cleaners.
4. **Cleaners for Radioactive Residues.** Radioactive residues are often removed by detergents with high chelating (binding) capacity. Detergents are available that are specifically designed to remove radioactive contaminants.
5. **Plasticware.** Plasticware is generally cleaned with mild, neutral pH detergents.
6. **Dishwashers.** Laboratory dishwasher detergents should be low foaming.
7. **Manual Cleaning.** For manual, ultrasonic, and soak cleaning, a mildly alkaline (pH 8–10), foaming detergent will often work well.
8. **Corrosive Cleaners.** Some detergents are corrosive and some require expensive means of disposal. Consult the manufacturer about corrosiveness and disposal.
9. **Instructions.** Prepare and use a detergent according to the manufacturer's instructions.
10. **Phosphorus and Nitrogen Analysis.** Glassware to be used for analyzing these elements should not be washed with detergent. NaHCO$_3$ solution is recommended in the *Water and Wastewater Examination Manual* (V. Dean Adams. Boca Raton, FL: Lewis Publ., 1990), followed by a soak in 6 N HCl.
11. **Tissue Culture.** Use detergents specifically formulated for cleaning items used in tissue culture.

1. **Soaking.** This simply involves submerging the items to be cleaned in a cleaning solution. The items are sometimes rinsed and used immediately, but further cleaning is usually required.

2. **Manual cleaning.** Items are washed by hand using a cloth, sponge, or brush followed by thorough rinsing. This method is not suitable for washing large numbers of items. Gloves and eye protection should be used as described by the detergent manufacturer, or if labware was used for hazardous substances. Centrifuge glassware and tubes, metal equipment, and most plasticware should not be cleaned with metal brushes or other abrasives. Scratches on centrifuge tubes can lead to material fatigue and breakage when the tube experiences the force of centrifugation.

3. **Machine Washing.** Most laboratories and production facilities wash labware in commercial dishwashers. Glassware washers have a single chamber that is sequentially filled and emptied through its cycle. The cycle typically consists of a prewash with room temperature water, a wash cycle of 2 to 15 minutes in a detergent solution at 60 to 90°C, and multiple rinses. The items may be dried in the washer or in a separate drying oven. General directions include:

   - Load items so that their open ends face the spray nozzles.
   - Place difficult-to-clean articles in the center of the rack, preferably with the spray nozzles pointing directly into them.
   - Place small items in baskets.
   - Use low-foaming detergents as directed by the manufacturer.
   - Use hot water, above 60°C (140°F).

4. **Automatic Siphon Pipette Washing.** Reusable glass pipettes are usually washed in long cylinders designed for this purpose. The cylinders are connected to a water supply so that the pipettes can alternately be filled with cleaning solution and drained. General directions include:

   - Presoak the pipettes by complete immersion in washing solution immediately after use.
   - Place detergent in the bottom of the washer.
   - Place the pipettes in a holder in the washer.
   - Attach the washer to the water supply and run the water so that the pipettes are filled and drained completely.
   - Continue to wash until all the detergent has washed and drained through the pipettes.
   - Use a final rinse as appropriate.

5. **Ultrasonic washing.** Ultrasonic washers expose items to high-pitched sound waves that effectively penetrate and clean crevices, narrow spaces, and other difficult-to-reach surfaces. Metal and glass are usually safely washed in these cleaners, but some types of plastic are weakened by ultrasonic cleaning. General directions include:

   - Dilute the cleaning solution and add to cleaning tank, run the machine several minutes to degas the solution, and allow the heater to come to temperature.
   - Place items in racks or baskets in the machine.
   - Align irregularly shaped items so that the long axis faces the transducer that generates sound waves.
   - Immerse the articles for 2 to 10 minutes.

6. **Clean in Place (CIP).** Large, nonmovable items, such as fermentation vessels or pipes, must be cleaned in place. These items are often designed so that cleaning agents can be pumped into and circulated through the device.

iv. Rinsing

Rinsing is a key step in washing labware. Standard washing procedures often specify that items should be rinsed three times with water. For machine washing, this means three rinse cycles are used. For manual washing, the item is filled and emptied at least three times. Tap water may be used in certain applications for initial rinses, but it should be followed by several rinses in purified water. Deionized water heated to 60 to 70°C is an effective rinsing agent. The type of water for the final rinse depends on the application for which the item will be used.

v. Drying

Do not dry labware by wiping with any type of towel or tissue. Towels and tissues can contaminate the item with fibrous and chemical residues. Items can be allowed to air dry on racks or can be more rapidly dried in heated drying ovens at temperatures below 140°C. Note, however, that volumetric glassware should not be heated because heat causes expansion and contraction of the glass that may alter its calibration. Some types of plastics are also sensitive to heat.

vi. Standard Procedures for Washing Glass and Plastic Labware

There are many appropriate procedures used to wash glassware. A common, general purpose glassware washing procedure is shown in Box 28.1. (This procedure is the U.S. Environmental Protection Agency Procedure for Containers, Protocol B.) Table 28.9 lists some tips regarding glassware washing.

vii. Quality Control of Labware Washing

It is very important that labware is properly cleaned. Each production facility and laboratory must have procedures for washing items according to the applications for which those items will be used. Laboratories that meet GMP and ISO 9000 requirements must demonstrate that their cleaning procedures are, in fact, effective. Determining the effectiveness of a cleaning

---

Box 28.1. A GENERAL METHOD FOR CLEANING LABWARE

1. *Wash containers in hot water with laboratory-grade, nonphosphate detergent.*
2. *Rinse three times with copious amounts of tap water to remove detergent.*
3. *Rinse three times with Type I water.*
4. *Oven dry.*
5. *Cool in enclosed, contaminant-free environment.*
6. *Cover to avoid exposure to dust.*

---

procedure can be difficult and there is no standard method for doing so. Some general considerations for assuring the quality of washing operations include:

- Machine washers can be monitored to see that they maintain the proper temperature throughout their cycle, have no blocked spindles, and that the quality of the incoming water is adequate.
- The effectiveness of a cleaning method can be evaluated by intentionally soiling items with substances that are visible. After cleaning, residues can be visually detected.
- In pharmaceutical facilities and other sites where consistently and stringently clean labware is required, testing methods may be very sophisticated. Items may be periodically swabbed and the swabs tested by HPLC, spectrophotometry, or other methods to detect contaminants.

As with any other aspect of a quality-control program, the operation, maintenance, and monitoring of washing systems requires planning and documentation.

Standard operating procedures, operator training, plans for routine maintenance and monitoring, logbooks, written procedures to deal with problems, and documentation of corrective actions taken are components of a quality program.

IV. STERILIZATION OF SOLUTIONS

There are two methods commonly used to sterilize solutions in the laboratory: **autoclaving** and **filtration.** An **autoclave** *is a pressure cooker; materials to be sterilized are heated by steam under pressure.* The pressure in an autoclave is usually held at 15–20 lb/in^2 (PSI) above normal pressure. This moist heat is very effective at destroying microorganisms; when water is present, bacteria are killed at lower temperatures than when heat alone is used. Sterilization with an autoclave is relatively fast and efficient and is the method of choice in many applications. Figure 28.8 on p. 538 shows the basic features of an autoclave and Table 28.10 on p. 539 summarizes general autoclaving guidelines.

After items have been subjected to autoclaving, it is important to know whether all microorganisms were indeed killed. The best way to test whether sterilization has occurred during autoclaving is to use spore strips. **Spores** *are dormant microorganisms.* Autoclaving should involve sufficient heat and sufficient time to kill spores. **Spore strips** *are dried pieces of paper to which large numbers of nonpathogenic, heat-resistant* Bacillus thermopholus *spores are adhered.* Spore strips are placed in the autoclave in the area that is most inaccessible to steam, such as in a flask or bottle. After autoclaving, the strips are transferred to growth medium, being careful not to contaminate them with organisms from the air or laboratory. If sterilization in the autoclave was successful, the

---

Table 28.9. GUIDELINES REGARDING GLASS AND PLASTICWARE WASHING

1. *After washing, glassware should be visually inspected to be sure it is clean and not cracked.* Cracked glassware should be repaired or discarded.
2. *A simple, common test of the cleanliness of glassware is that it wets uniformly with distilled water.* Grease, oils, and other contaminants cause water to bead, or flow unevenly across the glass surface. Uneven wetting of the surface of volumetric glassware will alter the volumes measured and may distort the meniscus.
3. *Plastic containers may be preferable to glass for trace-metal analysis.* Soaking plastic for a few hours (not more than 8 hours) in 1 M HCl or 1 M HNO_3 followed by a rinse with high-purity water will remove trace metals from plastics. It may also be helpful to clean plastics with alcohol or chloroform followed by a rinse in 1 M HCl prior to trace metal analysis. This treatment removes trace levels of organic compounds from the plastic surface. Low levels of organic compounds can contribute to the adsorption of trace elements. (Be certain the plastic is compatible with the cleaning process.)
4. *It is common to bake glassware after washing when working with RNA, in order to destroy RNA-degrading enzymes.*
5. *Labware is frequently autoclaved after washing when sterility is required.* Note that some types of plastic cannot withstand autoclaving. In addition, plastics may require longer autoclave cycle times than glass because the heat transfer properties of plastic and glass are different. Cycle lengths for plastics need to be determined empirically for each liquid and container.
6. *Solvents such as acetone, alcohols, and methylene chloride can be used with glassware to remove greases and oils.* Plasticware, however, may be dissolved by acetone and other organic solvents, so these should be used cautiously on plastic.
7. *Newly purchased glass and plastic items contain contaminants and should be washed before use.*
8. *Some plastics, including polycarbonate and polystyrene, are weakened by the heat in machine washers.* These plastics should be washed by hand in water that is less than 57°C.

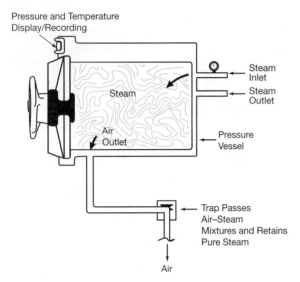

Figure 28.8. The Basic Features of an Autoclave. Items to be sterilized are placed in the chamber and the chamber door is tightly sealed. Air is purged from the chamber at the beginning of the sterilization cycle. After the air is removed, steam under pressure is introduced into the chamber and the temperature rises to the desired level. The autoclave is maintained under pressure at a high temperature for a predetermined time. The steam is then exhausted from the chamber and the sterile materials are removed.

spores will not grow. It is important when performing spore tests to use a positive control and a negative control. The positive control is a spore strip that was not autoclaved and should grow if the spores are viable and the medium is properly prepared. The negative control is to place strips without spores in medium. No growth should occur in the negative control unless there is contamination.

Although spore strips are recommended for testing autoclave function, they require time to incubate, so they do not provide instantaneous information regarding an autoclave cycle. There are a number of convenient products used to identify materials that have been through a cycle in an autoclave. For example, autoclave tape is a material that can be wrapped on objects and changes colors when subjected to pressurized steam. Such devices are typically used in research laboratories to show that an item went through a sterilization cycle. Note that autoclave tape and other such materials may change colors before all microorganisms are killed.

Some solution components are heat sensitive and must be sterilized by methods that do not involve heat. Heat-sensitive substances include proteins, vitamins, antibiotics, animal serum, and volatile chemicals. These materials are usually prepared as concentrated stock solutions and are then sterilized by filtration through a sterile filter with pores that are less than 0.2 μm in diameter. The filter-sterilized components are then added to the rest of the solution using care to maintain the sterility of the mixture. Filtration through membrane filters is effective for removing microorganisms, and is standard practice in many laboratories. Note that

viruses will penetrate microfilters. In situations where viruses and pyrogens must be removed, as in the preparation of pharmaceuticals, ultrafilters can be used.

As with any other aspect of a quality program, the operation, maintenance, and monitoring of sterilization systems requires planning and documentation. Most modern autoclaves automatically record information for every cycle such as date, temperature reached, pressure, time, and so on. Other aspects of quality, such as logbooks and operator training, are also essential.

V. STORAGE OF BIOLOGICAL SOLUTIONS AND MATERIALS

A. General

Biological materials and solutions must be properly stored to avoid degradation and contamination. Materials that may be stored in the laboratory include: DNA, RNA, and protein samples; inorganic solutions (e.g., buffers and other reagents); intact cells and tissues; and products (e.g., drug substances). There is no single way to store all materials and the best method of storage of a particular substance is often determined by experimentation.

Buffers, salt solutions, and other chemical reagents are often stored indefinitely at room temperature. Be aware, however, that microorganisms grow vigorously in buffers and salt solutions. The solution components also may precipitate over time. Examine solutions before use for visible particulates that may indicate contamination or precipitation. Solutions with some particulates may be acceptable in some applications; in other applications they are not. Buffer and salt solutions that have been autoclaved are usually stable at room temperature. Alternatively, small amounts of bactericidal agents are sometimes added to solutions to prevent microbial contamination.

B. Different Degrees of Cold

Cold protects biological materials by inhibiting the growth of contaminating microorganisms, slowing the activity of degradative enzymes, and slowing spontaneous degradation. Temperatures that are commonly used for storage are:

- 4°C (refrigerator)
- −20°C (freezer)
- −70 to −80°C (low temperature freezer)
- −196°C (liquid nitrogen)

Laboratory refrigerators are set to 4°C. Water and biological materials do not freeze at this temperature and some biological activities continue, but at a slower rate. This condition is adequate for the storage of some materials, particularly for short periods.

Freezing is commonly used for long-term storage. Although water freezes at 0°C, this temperature does not prevent biological and chemical activity. Laboratory freezers therefore go down to −20°C for routine storage.

Table 28.10. AUTOCLAVING GUIDELINES

1. *Steam sterilization requires that the steam directly contact every surface and article being sterilized.* For this reason:

   a. Do not overcrowd an autoclave. Allow spaces between items so that steam can contact all surfaces. When sterilizing instruments or fabrics, the steam must penetrate to and contact each and every surface and fiber. Anhydrous materials like oils, greases, and powders resist contact with water and therefore cannot be sterilized in an autoclave.

   b. Air can impede the contact of steam with a material. Autoclaves therefore evacuate air before introducing steam into the sterilization chamber. It is possible for air to be trapped in the material being autoclaved. For example, if a tube or needle is plugged, then air cannot escape, steam cannot enter, and that tube or needle will not be sterilized.

   c. Aqueous solutions do not require direct steam contact but they do require adequate heating.

2. *Containers should always be loosely capped.* This ensures that steam can penetrate to all surfaces. Loose caps also allow pressure to equilibrate inside and outside the container. Containers can shatter if the pressure is not equilibrated. Flasks are often covered by cotton plugs or loosely twisted caps.

3. *It is important that the autoclave be held at the proper temperature for the proper amount of time.* The minimum temperature at which sterilization occurs is 121°C (250°F). Too little time at this temperature leads to failure of sterilization; too much time can degrade media containing carbohydrates, agar, and other sensitive materials. It is standard practice to hold the autoclave at high temperature for 20 minutes. Depending on what is being autoclaved, however, 20 minutes may not be adequate. For example, plastics are poor conductors of heat. It may take much longer than 20 minutes to sterilize a 1 L solution in a plastic container. The minimum time required for sterilization of materials in Pyrex vessels similarly varies depending on the container. Containers larger than 1 L are likely to require more than 20 minutes sterilization time. It is best to autoclave materials that are similar together, so that the optimal sterilization conditions will be the same for all items in the autoclave.

4. *The quality of the water used to make steam is important.* Volatile impurities in the water will be carried into all autoclaved items and solutions. It is common to use water that was deionized or treated by reverse osmosis.

5. *Use a "slow exhaust" setting when autoclaving solutions.* Solutions do not boil during autoclaving even though the temperature in the autoclave is above the boiling point of water. This is because the pressure of the steam prevents boiling. After sterilization, however, if the steam is allowed to exit the chamber quickly, then solutions will boil violently out of their containers. When solutions are autoclaved, the autoclave should therefore be set to exhaust the steam slowly.

6. *Note that even when steam is released slowly, solutions will lose about 3 to 5% of their liquid due to evaporation.* People sometimes add an extra 5% of distilled water to each flask being autoclaved to compensate for this loss.

7. *Containers with liquids should not be filled more than 75% full to allow for fluid expansion and to prevent overflow.*

8. *Some materials should be autoclaved in separate containers because they will alter one another in some way.* For example, phosphates and glucose can interact with one another when subjected to heat so that the solution is degraded.

9. *Autoclaves attain very high temperatures and pressures and must be safely operated.* Safe operation requires that the instrument be properly maintained and the operators understand how to use it safely. Note that older autoclaves sometimes lack the safety features of newer models.

Even −20°C, however, is not sufficiently cold to freeze concentrated solutions of biological molecules. Moreover, degradative enzymes that may be present in a biological sample can slowly destroy proteins and other cellular materials stored at −20°C. Freezers that attain temperatures in the range of −70 to −80°C are therefore preferred for storing substances extracted from cells.

Although freezing is a good way to store laboratory materials, potentially destructive physical events occur during the freezing and thawing processes. Different components in a solution can freeze and thaw at different rates allowing materials to be exposed to extremes of pH and salt concentration. Another problem is that ice crystals, which can damage biological substances, form during freezing and melt during thawing. It is important to avoid multiple freeze-thaw cycles in order to limit damage related to ice crystals. To avoid multiple freeze-thaw cycles, *solutions are divided into small volumes that are frozen in individual containers, called* **aliquots.** Only as many aliquots as are needed at a given time are thawed; any unused, thawed material is discarded. This technique ensures that a substance is

thawed only one time. (It is critical that all aliquots are labeled carefully to avoid later confusion.) Biological materials should be stored in freezers that are not "frost-free" to avoid repeated freeze-thaw cycles. It is possible to purchase insulated containers to help prevent thawing if a frost-free freezer must be used.

C. Storing Proteins

Proteins vary in their storage requirements and experimentation at different temperatures and in different storage solutions may be required to determine the best conditions, see Table 28.11 on p. 540. A few comments about protein storage:

- Proteins are seldom stored at room temperature because they rapidly become degraded and inactive, often as a result of microbial activity.

- Proteins are sometimes stored refrigerated in the form of stable salt precipitates. Some proteins are stably stored in buffer for a few days to a few weeks at refrigerator temperature (e.g., some concentrated stocks of unconjugated antibodies).

Table 28.11. PROTEIN STORAGE CONDITIONS

Temperature	Maximum Recommended Time	Comments	Number of Times Sample May Be Removed from Storage
4°C	Hours to days	Requires sterile conditions or addition of antimicrobial agent	Many
−20°C (with 25 to 50% glycerol or ethylene glycol)	1 year	Usually requires sterile conditions or addition of antimicrobial agent	Many
−70 to −80°C or in liquid nitrogen	Years	Does not require sterile conditions or addition of antimicrobial agent	Once; repeated freeze-thaw cycles generally degrade proteins
Lyophilized (usually also frozen)	Years	Does not require sterile conditions or addition of antimicrobial agent	Once; it is impractical to lyophilize a sample multiple times

Primary Source: Pierce Technical Resource, "Protein stability and storage." http://www.piercenet.com.

- Some sources recommend that proteins be rapidly frozen and later thawed. Rapid freezing and thawing is thought to reduce exposure of the proteins to extremes of pH or salt concentration. Vials of materials can be rapidly frozen by immersion in a dry ice bath and thawed rapidly in lukewarm water.

- Whenever possible, proteins should be prepared at a relatively high concentration (e.g., 1 mg/mL) in buffer. High concentration stabilizes the protein's structure and helps minimize loss due to the protein adsorbing to the surface of its storage container.

- Various additives are used to extend the storage life of proteins. Care must be taken, however, since these additives contaminate the sample. Check that no deleterious interaction with the protein of interest will occur when using additives.

Carrier proteins (e.g., 0.1 to 1% bovine serum albumin, BSA), are sometimes added to stabilize dilute protein solutions and avoid the adsorption of scarce protein to the surface of a storage container.

Cryoprotectants, used at concentrations from 10 to 50%, stabilize protein solutions during freezing and thawing. Glycerol and ethylene glycol are common protectants that keep solutions from actually freezing, prevent ice crystal formation, and improve protein stability. It is common to receive enzymes that have been stabilized with glycerol from suppliers. Note that glycerol may affect subsequent steps (e.g., chromatography or restriction enzyme digests) when the protein is removed from storage.

Protease inhibitors (e.g. pepstatin A, EDTA, EGTA) may be added to inhibit the activity of enzymes that digest proteins.

Antimicrobial agents (e.g., sodium azide, NaN_3, at a final concentration of 0.02 to 0.05% (w/v) or thimerosal at a final concentration of 0.01% (w/v), inhibit microbial growth.

Reducing agents (e.g., dithiothreitol, DTT, or 2-mercaptoethanol, 2-ME, at final concentrations of 1 to 5 mM) help to maintain the protein in the reduced state by preventing oxidation of cysteines.

Nonionic detergents at low concentrations (e.g., 0.15%) may be added to prevent protein aggregation and adsorption to the surface of the storage container.

Proprietary products are sold by manufacturers (e.g., Pierce) to help stabilize proteins during storage.

D. Storing DNA and RNA

DNA is relatively stable as long as nucleases are absent. Some sources recommend storing genomic DNA, plasmids, and other small DNA molecules at 4°C for short periods. For long-term storage, DNA may be more stable if stored frozen. EDTA is often added to DNA storage solutions because it chelates magnesium ions that are required for the activity of endonucleases, enzymes that break down nucleic acids.

Ambion, a biotechnology company that specializes in RNA products, reports that RNA can be stored for short periods at −20°C. For long-term storage they suggest a −80°C freezer. They state that RNA can be stored in water or buffer but is most stable in an NH_4OAc/ethanol mixture at −80°C. They also recommend aliquoting RNA to avoid multiple freeze-thaw cycles and to reduce the risk of introducing RNases into the stock.

E. Storing Intact Cells

Virtually all biological activity ceases at −130°C. Living eukaryotic cells (e.g., cultured mammalian cells, sperm, eggs, embryonic stem cells, and embryos) are routinely stored below this temperature in a state of "suspended animation." This very low temperature is usually achieved by placing the cells in liquid nitrogen or, less commonly, in specialized mechanical freezers.

Bacterial cultures on agar (see Chapter 30) may be stable for several months at 4°C. Agar plates should be sealed to prevent drying. For longer-term storage, bacteria may be rapidly frozen in liquid nitrogen and stored in a −70°C freezer. Stabilizers are sometimes added to sensitive bacterial strains. Bacterial cultures may also be freeze-dried (discussed in the next section) for long term storage.

F. Freeze Drying

Freeze drying, also known as **lyophilization,** *is a process in which a material is first frozen and then dried under a vacuum.* The purpose of freeze drying is to halt biological activity and degradation by removing all water. Vacuum is used during the freeze-drying process to cause ice to change directly from solid to vapor without passing through a liquid phase. Removing water from a substance in this way leaves the basic structure and composition of the substance intact while inhibiting enzymatic and microbial degradation. Freeze-dried substances in tightly sealed vials can be stably stored for long periods of time.

Freeze drying is familiar outside the laboratory as a method of preserving foods for astronauts and hikers. Freeze drying is used in the laboratory to preserve proteins and other materials (e.g., antibodies, enzymes, and oligonucleotide primers for PCR). It is common to receive materials from manufacturers that have been lyophilized for storage and shipment.

When it is time to use a freeze-dried material, water or buffer is added and the original properties of the material are restored quickly. (If the material does not go into solution easily, it may have denatured during the freeze-drying process.) Manufacturers who provide freeze-dried products usually provide guidance for their rehydration. If not used immediately, rehydrated materials are usually aliquoted and frozen. Note that some freeze-dried products are lightweight and can easily be lost unless care is exercised when opening their vial and handling the product.

PRACTICE PROBLEMS

1. Pyrogens are negatively charged molecules that range in molecular weight from about 1000 D to more than 10,000 D. What type of filtration method can remove pyrogens?
2. Viruses are about the same size as the protein albumin and are slightly bigger than pyrogens. What type of filtration method will remove them from a solution?
3. Is table salt, NaCl, removed from a solution that is passed through an ultrafilter? Is it removed by deionization?
4. Many water-purification systems recirculate already purified water back through the purification steps. Why?
5. The term *log reduction value* (*LRV*) describes mathematically the ability of an ultrafiltration membrane to remove pyrogens from water. The higher the LRV, the greater the rejection of pyrogens by the membrane. The equation for LRV is:

$$\text{log reduction value} = \log\left(\frac{\text{feed water pyrogen concentration}}{\text{product water concentration}}\right)$$

For example, a filter was challenged by running water through it containing pyrogens at a concentration of 10,715 EU/mL. The endotoxin concentration in the product water was 0.0023 EU/mL. What was the LRV?

$$\text{LRV} = \log\left(\frac{10,715 \text{ EU/mL}}{0.0023 \text{ EU/mL}}\right) \approx 6.7$$

In another challenge, the same membrane was used to filter water containing pyrogens at a level of 4.5 EU/mL. After filtration, the water had less than 0.001 EU/mL. What was the LRV?

(The information in this example is from "Reducing Pyrogen Levels in Ultrapure Water, Part 2: Producing Pyrogen-Free Water Using Hollow Fiber Ultrafiltration." Millipore Technical Brief. Bedford, MA: Millipore Corp.)

QUESTIONS FOR DISCUSSION

1. You are placed in charge of purchasing a water-purification system for a laboratory. For each of the following types of laboratories, what would be important considerations to discuss with manufacturers? What standards would your purified water need to meet? What purification methods do you think would be most appropriate for your water?

   a. You work in an analytical testing laboratory where you test air and water samples for pollutants.

   b. You work in a biotechnology research laboratory on a variety of genetic engineering projects.

2. A biotechnology company has been successfully purifying its water for several years. Workers suddenly discover that one of their products is no longer effective and they trace the problem to the water used in production. Discuss factors that might have led to their sudden difficulties with water quality. Suggest changes they could make in their operating procedures to help avoid similar crises in the future.

3. If you have a water-purification system in your laboratory, find out how it is designed and what routine maintenance it requires.

4. Write an SOP to clean glassware in your laboratory manually. If you do not work in a laboratory, write an SOP to clean glassware in a research laboratory where proteins are purified and analyzed.

5. If there is a machine washer in your laboratory, read its manual. Describe potential problems that can arise and how these problems can be recognized.

6. If you have a dishwasher at home, examine the quality of its performance. Devise a method to evaluate how effectively your dishes are cleaned.

CHAPTER 29

Laboratory Solutions to Support the Activity of Biological Macromolecules

I. INTRODUCTION

Chapters 26, 27, and 28 discussed many principles, calculations, and other details regarding how to prepare laboratory solutions. For example, pp. 515–516 explained how to calculate the required amount of each component of SM buffer. We have not, however, addressed the question of *why* SM buffer has the various components that it does. This chapter explores the purposes of the components used in solutions in biology laboratories.

Biologists work in the laboratory with materials derived from organisms. These materials include various proteins, DNA, RNA, and cellular organelles. The activity of all these biological materials is intimately associated with their structure; in fact, the relationship between structure and function is one of the most important themes in biology. For example, the function of enzymes is to catalyze reactions. For catalysis to occur, the enzyme must recognize and bind the reactants involved. Recognition and binding occur because the enzyme's structure complements that of the reactants, see Figure 29.1. Enzymes do not function if their normal structure is degraded. Thus, it is critical that laboratory solutions provide conditions that preserve the structural integrity of biological molecules.

542

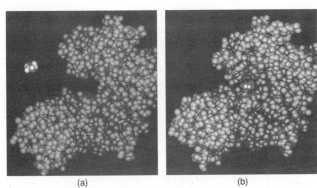

(a)   **(b)**

Figure 29.1. Example of the Relationship between Protein Structure and Function. An enzyme and its substrate. The smaller substrate molecule **a.** binds within a groove in the larger enzyme that is complementary in shape and ionic properties **b.** Many proteins bind other molecules at sites whose shapes are complementary. *(See color insert C-2.)*

Table 29.1. PROTEIN FUNCTIONS

Function	Examples
Metabolism	*Enzymes* catalyze reactions
Cellular Control	*Hormones* affect metabolism of target cells
Structure	*Keratin* forms the structure of hair and nails
Storage	*Ovalbumin* (egg white) stores nutrients
Transport	*Hemoglobin* transports O_2 and CO_2
Mobility	*Actin* and *myosin* cause muscle contraction
Defense	*Antibodies* destroy invading microorganisms
Gene Regulation	*Repressor proteins* in bacteria "turn off" genes
Recognition	*Receptors* recognize specific hormones

There are many different solutions used in the biology laboratory. These solutions differ from one another because the requirements of various biological systems vary. For example, a solution to support the structure and activity of an enzyme is likely to differ from a solution for DNA. A solution that is used to store a material is likely to differ from a solution for isolating that material.

This chapter begins by briefly describing the structure and function of proteins. Then, the types of laboratory solutions that support proteins are explored. The second part of this chapter briefly discusses the structure and function of nucleic acids (DNA and RNA) followed by an exploration of how nucleic acids are supported in biological solutions.

II. MAINTAINING THE STRUCTURE AND FUNCTION OF PROTEINS IN LABORATORY SOLUTIONS

A. An Overview of Protein Function and Structure

i. PROTEIN FUNCTION AND AMINO ACIDS: THE SUBUNITS OF PROTEINS

The first part of this chapter discusses the solutions used when working with proteins in the laboratory. In order to understand why these solutions have the components they do, it is necessary to know something about protein function and structure.

Proteins perform a myriad of essential functions in cells, see Table 29.1. This functional diversity is possible because proteins are structurally diverse. As with the English language, where an immense number of words with varied meaning are formed using only 26 letters, so diverse proteins are formed using primarily 20 different amino acid subunits. Each of the 20 amino acids has the same core structure consisting of an **amino group,** NH_2, and a **carboxyl group,** $COOH$, attached to a central carbon atom, Figure 29.2a. What makes each amino acid distinct is the presence of a characteristic **side chain,** or **R**

group, which has a particular structure and chemical properties. The diversity and versatility of proteins is due to the wide variety of characteristics conferred on proteins by the various amino acid R groups.

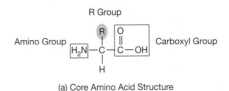

(a) Core Amino Acid Structure

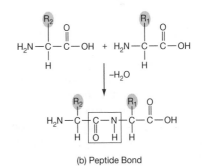

(b) Peptide Bond

(c) Polymer of Amino Acids

Figure 29.2. Protein Structure. a. The core structure of an amino acid consists of an amino and a carboxyl group attached to a central carbon atom. Each amino acid has a different R group attached. **b.** Proteins are polymers of amino acids joined by peptide bonds. A peptide bond is a covalent bond formed when the carboxyl group of one amino acid joins with the amino group of a second amino acid. **c.** Every protein has a different sequence of amino acids that determines the structure and function of the protein.

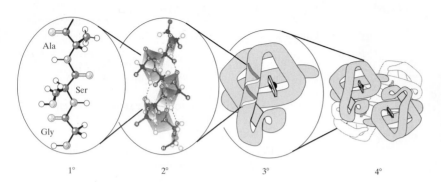

1° 2° 3° 4°

Figure 29.3. **The Four Levels of Protein Structural Organization:** Primary (1°) structure is the sequence of amino acids, represented here by three-letter codes. Certain proteins or sections of proteins form regularly repeating patterns, called secondary (2°) structure. Many proteins require complex tertiary (3°) folding to provide the right structure and conformation for recognition, binding, and solubility. Quaternary (4°) structure occurs when folded chains associate with one another to form complexes.

A protein is formed by amino acid subunits linked together in a chain. *The bonds that connect the amino acids to one another are* **covalent peptide bonds,** see Figure 29.2b. Every protein has a different sequence of amino acids that determines the final structure and function of the protein, see Figure 29.2c.

ii. The Four Levels of Organization of Protein Structure

Every protein has a distinct, complex three-dimensional structure. To describe these varied protein structures, people speak of four levels of organization: primary, secondary, tertiary, and quaternary, see Figure 29.3. Secondary, tertiary, and quaternary structure are sometimes collectively referred to as *higher order* structure.

1. **Primary Structure** *refers to the linear sequence of amino acids that comprise the protein.*

2. **Secondary Structure** *refers to regularly repeating patterns of twists or kinks of the amino acid chain.* Two common types of secondary structure are called the **α-helix** and **β-pleated sheet.** *Regions of proteins that do not have regularly repeating structures are often said to have a* **random coil** secondary structure, although their folding patterns are not truly random. Secondary structure is held together by weak, noncovalent, molecular interactions, called *hydrogen bonds*, between nonadjacent amino acids. **Hydrogen bonds** *form when a hydrogen atom that is bonded to an electronegative atom (like F, O, or N) is also partially bonded to another atom (usually also F, O, or N).* Thus, a hydrogen bond forms because a hydrogen atom is shared by two other atoms.

3. **Tertiary Structure** *refers to the three-dimensional globular structure formed by bending and twisting of the protein.* Globular structures include regions of α-helices, β-sheets, and other secondary structures that are folded together in a way that is characteristic of that protein. Tertiary structure is generally stabilized by multiple weak, noncovalent interactions. These include:

   • **Hydrogen bonds** that form when a hydrogen atom is shared by two other atoms.

   • **Electrostatic interactions** that occur between charged amino acid side chains. **Electrostatic interactions** *are attractions between positive and negative sites on macromolecules.* These weak interactions bring together amino acids and stabilize protein folding, see Figure 29.4.

   • **Hydrophobic interactions** *that reflect the tendency of the aqueous environment of the cell to exclude hydrophobic amino acids.* Proteins spontaneously fold in such a way that hydrophobic amino acids are inside, protected from the aqueous environment. In contrast, hydrophilic amino acids tend to be found on the protein's surface.

In addition to these weak, noncovalent interactions, some proteins have covalent **disulfide bonds** which stabilize their three-dimensional structure. **Disulfide bonds** *form when sulfurs from two cysteine molecules (cysteine is an amino acid) bind to one another with the loss of two hydrogens*, see Figure 29.5. Disulfide bond formation is an example of oxidation. **Oxidation** *is defined as the removal of electrons from a substance.* In biological systems, oxidation often involves the removal of an electron accompanied by a proton (i.e., a hydrogen atom).

4. **Quaternary Structure** *refers to the complex formed when two or more protein chains associate with one another.* Some proteins have separately synthesized subunit chains that join together to form a functional protein complex.

The covalent bonds that link amino acids into their primary chain structure are much stronger than the noncovalent interactions that stabilize higher order structure. Relatively little energy (e.g., in the form of heat) is needed to disrupt the weaker noncovalent bonds. Higher-order structures persist because a large number of weak interactions stabilize them.

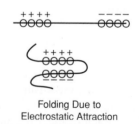

Folding Due to
Electrostatic Attraction

Figure 29.4. Electrostatic Interactions and Protein Folding. Interactions between positively and negatively charged amino acids are called electrostatic interactions. Proteins have various arrangements of positive and negative charges due to different amino acid side groups. Electrostatic interactions between charged side groups lead to folding of the protein.

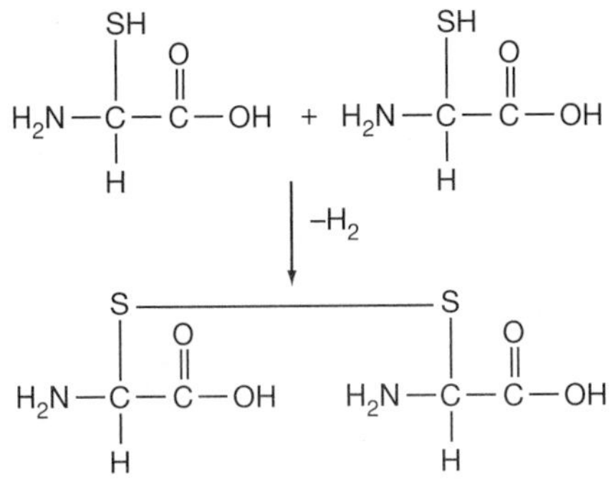

Figure 29.5. Disulfide Bonds. Disulfide bonds contribute to tertiary folding. A disulfide bond is formed by two cysteine molecules.

The "weakness" of the noncovalent interactions confers two important characteristics on proteins. First, proteins are flexible, and this flexibility is critical to their biological function. For example, enzymes change shape in order to bind to their substrates. Second, the normal folding and shape of proteins is readily disrupted by changes in temperature, pH, and other conditions. There are many circumstances in the laboratory, therefore, where the primary structure of the amino acid chain remains intact but the complex shape of the protein is lost.

B. How Proteins Lose Their Structure and Function

Laboratory solutions for proteins usually must provide suitable conditions so that the proteins' normal structure, and therefore their normal activity, is maintained. To understand how the structure of proteins is protected in laboratory solutions, it is necessary to understand how that structure can be destroyed. There are various ways by which proteins lose their normal structure and therefore their ability to function normally:

1. **Proteins can *denature* or unfold so that their three-dimensional structure is altered but their primary structure remains intact.** Because the amino acid side chain interactions that stabilize protein folding are relatively weak, environmental factors including high temperature, low or high pH, and high ionic strength can cause denaturation.

    There are many laboratory situations where denaturation occurs reversibly under one set of circumstances and proteins regain their normal conformation when suitable conditions are restored. Denaturation can also be irreversible, as when eggs are hard boiled. When boiled, the translucent egg protein, albumin, is denatured and becomes hard egg white.

2. **The primary structure of proteins can be broken apart by enzymes, called *proteases,* that digest the covalent peptide bonds between amino acids.** *This digestive process is called* **proteolysis.** Cells contain proteases sequestered in membrane-bound sacs called *lysosomes.* When cells are disrupted, the lysosomes break and release their proteases. In the laboratory, it is therefore necessary to minimize the activities of cellular proteases to protect proteins from proteolysis. Methods used to minimize proteolysis include low temperature, short working times, and addition of chemicals that inhibit proteases. Microbes also release proteases and, therefore, antimicrobial agents are often added to solutions to protect proteins.

3. **Sulfur groups on cysteines may undergo oxidation to form disulfide bonds that are not normally present.** Extra disulfide bonds can form when proteins are removed from their normal environment and are exposed to the oxidizing conditions of the air. Because disulfide bonds affect protein folding, additional disulfide bonds may produce undesirable changes in protein tertiary structure. Disulfide bond formation is an oxidation reaction; therefore, antioxidizing chemicals, called **reducing agents,** are often added to protein solutions to prevent unwanted disulfide bonding.

4. **Proteins can aggregate with one another, leading to precipitation from solution.** Each protein requires specific conditions in order to remain in solution.

5. **Proteins readily adsorb (stick to) surfaces and their activity is then lost.**

C. The Components of Laboratory Solutions for Proteins

Laboratory solutions for proteins must provide them with suitable conditions. What constitutes "suitable conditions," however, varies greatly. Because proteins differ from one another the conditions for manipulating one protein may not work for another. Even when considering a single protein, the optimal conditions differ depending on the situation. For example, extracting proteins from intact tissue requires disrupting the cells and may require extracting the proteins from membranes and/or tight associations with other molecules. Special *lysis buffers* are used for this type of cellular disruption (see the Case Study on pp. 558–559). The solutions that are used when one wants a protein to function properly in a test tube are different than the lysis buffers used to disrupt cells. Yet a different set of conditions is usually required for storing a protein. Thus, the solution requirements for each protein and each task must be determined individually, often by experimentation.

Chemical agents that are used in protein solutions may perform more than one role depending on the situation. Moreover, the same agent may be desirable in one situation and problematic in another. For example, detergents play a familiar, useful role as cleaning agents. Detergents can also act to solubilize or denature proteins. Detergents are sometimes intentionally added to solutions because their effects are useful. At other times, detergents are scrupulously removed from solutions because their effects are detrimental.

Although laboratory solutions for proteins vary significantly depending on the circumstances, there are classes of agents commonly found in protein solutions. These classes of agents are summarized in Table 29.2 and general principles regarding these agents will be discussed.

i. Proteins and pH: The Importance of Buffers

Proteins that reside in the cell's cytoplasm normally have many charged amino acid side groups that interact with one another and affect the structure of the protein. The charges on these side groups are affected by the pH of their solution. The solution pH, therefore, affects the three-dimensional structure of proteins, their activity, and their solubility. There is no universal rule as to what pH is optimum, although pH extremes generally denature proteins. As a rule, the pH of a laboratory solution should be maintained as closely as possible to the pH found in the protein's native environment, which is typically somewhere between pH 6 and 8. A **buffer** is used in solutions to maintain the proper pH.

ii. Salts

Salts *are compounds made up of positive and negative ions.* Biological systems frequently require salts, so many laboratory solutions contain salts at a specific concentration.* **Ionic strength** *is a measure of the charges from ions in an aqueous solution.* The general importance of ionic strength to biological molecules in laboratory solutions can be described as follows. Due to electrostatic interactions, two macromolecules in solution tend to attract one another if they have opposite charges, tend to repulse one another if they have the same charges, and have neither interaction if the molecules are neutral. Ions in a solution also interact with macromolecules based on charge. High concentrations of ions in a solution will tend to counteract or shield opposite macromolecular charges.

In protein solutions, therefore, salts can affect the electrostatic interactions between the side chains of amino acids. At low ionic strengths, the ions in solution have little effect on these interactions, see Figure 29.6a. Increased ionic strength tends to stabilize protein structures by shielding unpaired charged groups on protein

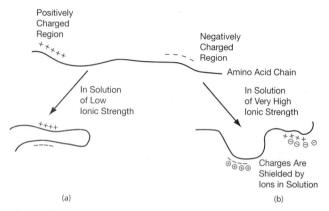

Figure 29.6. The Effect of Ionic Strength on Macromolecular Surface Charges. a. A solution of low ionic strength. **b.** A solution of high ionic strength.

surfaces. At very high ionic strengths, many proteins are denatured, presumably because normal sites of electrostatic interaction between charged groups on amino acids are neutralized by the salt ions, see Figure 29.6b.

The ionic strength of a solution affects both the three-dimensional structure of proteins and their solubility. Most cytoplasmic proteins are soluble at salt concentrations of about 0.15–0.20 M but often precipitate at higher or lower salt concentrations. This phenomenon is frequently exploited in the laboratory to selectively isolate proteins from other solution components. Salt is added to the solution until the proteins of interest precipitate. The precipitated proteins can then be retrieved. Salt-precipitated proteins usually regain their normal structure and solubility when the salt is removed. High salt concentrations retard microbial action; therefore, proteins are sometimes stored as salt precipitates.

The salt concentration in a solution can also be manipulated to control the binding of proteins to a solid matrix. Certain solid materials are designed so that they bind proteins when particular conditions of salt concentration and pH are met. Under these conditions, a desired protein may selectively bind to the solid matrix, leaving contaminants behind in the solution. The isolated protein is then recovered by changing the salt concentration so that the protein leaves the matrix and rinses into fresh buffer solution.

The exact effects of ionic strength on a protein depend on the protein, the type of salts involved, and the pH of the solution. Recipes for protein solutions therefore specify the pH, the type of salts, and their concentrations.

A summary of the roles that salts play in protein solutions includes:

1. **Salts affect the three-dimensional structure of proteins and may be manipulated to stabilize structure or to intentionally denature a protein.**
2. **Salts affect the solubility of proteins and may be manipulated to keep proteins in solution or to cause them to precipitate from solution.**

Table 29.2. Classes of Agents Used in Laboratory Solutions for Proteins

Buffers	Precipitants
Salts	Reducing agents
Cofactors	Metal chelators
Detergents	Antimicrobial agents
Organic solvents	Protease inhibitors
Denaturants and solubilizing agents	Storage stabilizers

*It is not surprising that salts play many roles in biological systems; after all, life evolved in the sea. This ancestry is reflected in the fact that salts perform a number of essential roles in organisms and salt levels are rigorously controlled in living systems.

3. Salts may be used to control the binding of a protein to a solid matrix.

4. Salts may be used to control the binding of proteins to other macromolecules (discussed on pp. 555–556).

5. Salts may be added as cofactors for enzymatic reactions, as discussed in the next section.

iii. COFACTORS

Cofactors are *chemical substances that many enzymes require for activity.* Cofactors include metal ions, such as Fe^{++}, Mg^{++}, and Zn^{++}. Cofactors may also be complex organic molecules called **coenzymes,** such as **nicotinamide adenine dinucleotide (NAD)** and **coenzyme A.** Cofactors are sometimes added to or removed from protein solutions. For example, Mg^{++} is a cofactor for nucleases, enzymes that digest DNA and RNA. In some situations Mg^{++} is added to solutions while in other applications it is carefully avoided. Magnesium is also a necessary cofactor for the polymerase enzyme that amplifies DNA in the polymerase chain reaction (PCR). If there is too little magnesium in the solution mixture for a PCR reaction, then the yield of product is reduced. If there is too much magnesium in the reaction mixture, then there can be undesired, nonspecific background amplification.

iv. DETERGENTS

Many types of detergents are used in the laboratory. **Detergents** *are a class of chemicals that have both a hydrophobic "tail" and a hydrophilic "head."* Detergents therefore have two natures: They can be both water-soluble and lipid-soluble. *Some detergents have hydrophilic portions that are ionized in solution* (**ionic detergents**) *whereas others have hydrophilic sections that are not ionized in solution* (**nonionic detergents**). **SDS (sodium dodecyl sulfate)** is an example of an ionic detergent; **Triton X-100** is a nonionic detergent.

Detergents interact with and modify the structure of proteins. Detergents are sometimes intentionally added to protein solutions for several reasons:

1. Detergents may be used to solubilize proteins that are associated with lipid membranes. Proteins that are associated with lipid membranes tend to be hydrophobic and difficult to bring into solution. Because detergents are both soluble in water and have hydrophobic properties, they can interact with and increase the water solubility of hydrophobic proteins. Nonionic detergents are generally preferred for solubilizing membrane-bound proteins because they tend not to be denaturing.

2. Detergents are sometimes added to solutions to intentionally denature proteins. Ionic detergents tend to effectively denature proteins. Denaturants are discussed below.

3. Detergents may be added in low levels to solutions to prevent protein aggregation and prevent proteins from adsorbing to surfaces, such as glass.

Some cautions regarding detergents are:

- *Some detergents, such as Triton X-100, absorb UV light and can interfere with spectrophotometric assays.*
- *Commercial detergents intended for cleaning may contain contaminating substances that affect proteins.* To avoid such contaminants, manufacturers purify detergents that are intended for protein work.
- *Detergents denature proteins and therefore detergent residues from cleaning glassware and equipment must be removed.*

v. ORGANIC SOLVENTS

Organic solvents, such as **ethanol** and **acetone,** are sometimes added to biological solutions. Organic solvents may be used to solubilize membrane proteins and occasionally to cause them to precipitate from a solution as part of a purification strategy. Organic solvents, however, cause most proteins to denature irreversibly and should generally be avoided when protein activity must be retained.

vi. DENATURANTS, SOLUBILIZING AGENTS, AND PRECIPITANTS

There are situations where it is desirable to denature the higher order structure of proteins. For example, polyacrylamide gel electrophoresis (PAGE) is a method used to separate proteins from one another by forcing them to migrate through a gel-like matrix in the presence of an electrical current. Before electrophoresis, the proteins are intentionally denatured with the detergent SDS so that their three-dimensional structures do not affect their movement in the gel. Ionic detergent is preferred for this task because it is a more effective denaturing agent than nonionic detergent. (In addition to denaturing proteins, SDS also confers a uniform negative charge on all the proteins so that differences in their charges do not affect their migration in the gel. When SDS is used with PAGE, proteins separate from one another based on differences in their size.) Proteins do not need to maintain their normal activity during electrophoresis, so denaturation is not a problem.

Urea at high concentrations (4–8 M) and **guanidinium salts** are other effective denaturants. Urea is frequently used when proteins are separated from one another by a method called *isoelectric focusing.* Like PAGE, isoelectric focusing requires that the proteins be completely unfolded. SDS, however, is not used in isoelectric focusing because SDS confers a charge on the proteins that interferes with the focusing method.

Urea, SDS, and guanidinium salts not only denature proteins, they also cause them to be soluble in aqueous solutions. Some proteins are not normally soluble in laboratory solutions and are therefore very difficult to manipulate in the laboratory. For example, bacteria may produce valuable recombinant proteins but sequester them in insoluble inclusion bodies. Guanidinium hydrochloride can be used in recovering the desired recombinant proteins from the bacterial inclusion bodies (see, for example, Funderburgh, James L., and Prakash, Sujatha. "SDS-Polyacrylamide Gel Electrophoretic Analysis of Proteins in the Presence of Guanidinium Hydrochloride." *BioTechniques* 20 (March 1996): 376–8.).

We have just seen that some agents that denature proteins cause the proteins to be soluble in aqueous solutions. There are also agents that simultaneously denature and cause proteins to precipitate out of solution. **TCA (trichloroacetic acid)** and the organic solvents ethanol, acetone, methanol, and chloroform are examples of agents that both denature and precipitate proteins. Such precipitation can be exploited in the laboratory to separate large macromolecules (including proteins, DNA, and RNA) from smaller molecules such as salts, amino acids, and nucleotides. Simultaneous denaturation and precipitation is useful, for example, when a protein is to be analyzed for its amino acid content. In this case, maintaining the protein's activity and three-dimensional structure is unnecessary.

There are situations where it is desirable to precipitate a protein, but the protein will later be required to exhibit normal activity. In such cases, precipitants such as **ammonium sulfate** and **PEG (polyethylene glycol)** are used. These agents cause proteins to precipitate without irreversibly denaturing them. Ammonium sulfate is inexpensive and is commonly used to separate proteins from one another during protein purification procedures. As noted earlier, salt-precipitated proteins generally recover their normal activity when the salt is removed.

vii. REDUCING AGENTS

Recall that when proteins are exposed to the oxidizing conditions of the air, sulfur groups on cysteines may undergo oxidation to form unwanted disulfide bonds. Laboratory workers sometimes add **reducing agents** to solutions to simulate the reduced intracellular environment and prevent unwanted disulfide bond formation. Common reducing agents are β-**mercaptoethanol** and **DTT (dithiothreitol).**

viii. METALS AND CHELATING AGENTS

Metal ions, such as Ca^{++} and Mg^{++}, are frequently present when cells are disrupted and are often introduced as contaminants in reagents and water. Metals can accelerate the formation of undesired disulfide bonds and can act as cofactors for proteases. **Chelating agents,** therefore, are often added to laboratory solutions *to bind and remove metal ions from solution.* The most common chelator is **EDTA (ethylenediaminetetraacetic acid).**

ix. ANTIMICROBIAL AGENTS

Low concentrations of antimicrobial agents, such as **sodium azide,** are used to kill microbes and to prevent proteolysis caused by microbes. Sodium azide is toxic and may affect the activity of some proteins.

x. PROTEASE INHIBITORS

Proteases are enzymes that are released into the cell extract when cells are disrupted. Until the proteins of interest are purified from the extract, protease inhibitors may be necessary to prevent protein degradation. EDTA is considered a protease inhibitor because it removes metal ion cofactors from solution. **PMSF (phenymethanesulfonyl fluoride)** and **pepstatin** are examples of agents used to inhibit specific classes of proteases. Protease activity is also reduced by keeping protein solutions cold and working quickly to isolate proteins of interest from the cell extract.

D. Physical Factors Relating to Proteins in Solution

i. TEMPERATURE

Temperature has profound effects on proteins. Most proteins exhibit their maximum biological activity at physiological temperatures (30–37°C). Extremely high temperatures denature proteins, as when eggs are hard-boiled. The optimal temperature for a protein depends on the situation. For example, the optimal storage temperature is seldom the same temperature at which a protein has maximal activity. Cold temperatures can be used to reduce microbial growth and slow the degradation of proteins by proteolytic enzymes. Ice water baths are used in the laboratory to maintain a biological system at 4°C. Refrigerator and freezer temperatures are used to hold biological solutions when not in use. A cold room or large refrigerator is frequently used to perform operations like enzyme purification and column separations.

Table 29.3 shows some temperatures that are significant in the laboratory.

ii. ADSORPTION

Surface charges on proteins make them "sticky," and many proteins will **adsorb,** *stick to,* almost any surface.

Table 29.3. *SIGNIFICANT BIOLOGY LABORATORY TEMPERATURES*

121	Autoclave temperature used to sterilize equipment and solutions
100	Water boils
90	DNA separates into two strands
65–70	Activity range of certain thermophilic bacteria
37	Temperature of warm-blooded mammals
20–25	Room temperature
4–8	Refrigerator temperature
0	Ice water bath
0	Water freezes/melts
−20	Freezer temperature
−70	Ultra-low temperature freezer
−196	Temperature of liquid nitrogen

Temperatures in °C.

Glassware and plasticware provide large surfaces that can adsorb proteins from solution, thus reducing protein activity. In addition, when protein solutions are filtered (as is sometimes done to sterilize the solution), proteins may adsorb to the filters and be lost. Some ways to reduce the problem of adsorption are shown in Table 29.4.

There are situations where protein adsorption can be exploited in the laboratory. For example, Western blotting is a technique used to identify proteins based on their interactions with specific antibodies. Western blotting requires immobilizing proteins on a solid surface so that they are not washed away during rinses. Protein immobilization is accomplished by adsorbing the proteins to a

---

Table 29.4. METHODS TO REDUCE PROTEIN ADSORPTION

1. *Use labware coated with silanes, agents that make the labware "slippery."* (See, for example, Sambrook, J., Fritsch, E.F., and Maniatis, T. *Molecular Cloning: A Laboratory Manual.* 2nd ed. Cold Spring Harbor, NY: Cold Spring Harbor Laboratory Press, 1989.)

2. *Use plastic containers and filters that are made from materials that do not adsorb proteins.* Consult the manufacturers.

3. *If possible, maintain a high concentration of protein in a solution (for example, 1 mg/mL).* In such cases a small proportion of the protein may stick to the glassware, but most is left in solution.

4. *Add inert carrier proteins, such as BSA (bovine serum albumin), to otherwise dilute protein solutions.* The inert protein sticks to and covers up the sticky sites on the glass or plasticware so that the protein of interest remains in solution.

Table 29.5. GENERAL RULES FOR PROTEIN HANDLING TO REDUCE DEGRADATION

1. *Wear gloves when handling protein solutions to avoid introducing contaminating proteases from your hands.*

2. *Mix protein solutions gently since some proteins are denatured by vigorous shaking.*

3. *Use only clean glassware.* All glassware and plasticware should be well washed and rinsed. Residual detergents or metal ions may have deleterious effects on proteins. A final rinse with EDTA may be used to eliminate metal ions.

4. *Use sterile buffers, solutions, and glassware to reduce bacterial contamination when working with or storing proteins for long times.* Consider the use of antimicrobial agents such as sodium azide.

5. *In general, keep protein-containing solutions cold in the laboratory.*

6. *Use only high-purity water to prepare solutions.*

nylon or plastic membrane. Another example is the selective binding of a desired protein to a solid matrix.

E. Methods of Handling Protein-Containing Solutions

Some considerations in handling solutions containing proteins are summarized in Table 29.5.

F. Summary of Protein Solution Components

Table 29.6 summarizes the components of protein solutions in tabular form.

Table 29.6. SUMMARY OF COMMON COMPONENTS OF SOLUTIONS USED TO MAINTAIN PROTEIN STRUCTURE AND FUNCTION

Type of Agent	Function	Examples
Buffering agents	Maintain pH	Tris, phosphate, acetate buffers
Salts	Control ionic strength	NaCl, $MgCl_2$
	Cofactors	
Antimicrobial agents	Prevent microbial contamination	Na-azide
Antioxidants	Prevent unwanted disulfide bond formation	β-mercaptoethanol, DTT
Protease inhibitors	Prevent proteolysis	EDTA, PMSF, pepstatin
Chelators	Remove metals from solution	EDTA, EGTA*
Detergents	Solubilize membrane proteins	Triton X-100, SDS
	Prevent aggregation and adsorption	
Denaturants	Denature proteins	
solubilizing		SDS, urea, guanidinium salts
precipitating		acetone, phenol/chloroform
Precipitants	Precipitate proteins	Ammonium sulfate, PEG, TCA
Cofactors	Required for enzyme activity	Mg^{++}, Fe^{++}, CoA
Storage stabilizers	Protect proteins during freezing	Glycerol

*EGTA is similar to EDTA, but it has a higher binding capacity for calcium than for magnesium. This chelator is commonly used when the regulation of calcium is desired.

PRACTICE PROBLEMS: PROTEINS

1. *EcoR1* is an enzyme that is used in the laboratory to cleave DNA at specific sites. The first recipe below is for the storage buffer used by the manufacturer to store and ship *EcoR1* enzyme. The second recipe is for the buffer used when the enzyme cleaves DNA—when maximal activity is required. When the enzyme is used in the laboratory, a very small amount of the enzyme (in storage buffer) is removed and diluted in fresh activity buffer.

   a. Suggest what the purpose of each solution component might be.

   b. Explain the differences between the solution for storage and the solution for activity. Why are the two solutions not identical?

Storage Buffer (store at −20°C)	Activity Buffer (incubate enzyme with DNA at 37°C for 30 min)
10 mM Tris-HCL, pH 7.4	90 mM Tris-HCl, pH 7.5
400 mM NaCl	50 mM NaCl
0.1 mM EDTA	10 mM $MgCl_2$
1 mM DTT	
0.15% Triton X-100	
0.5 mg/mL BSA	
50% glycerol	

2. Examine the following recipes that come from various protein manuals. Suggest a function for the various components.

   a. Sample Buffer Used When Loading Proteins into Electrophoresis Gel

       60 mM Tris-HCl (pH 6.8)

       2% SDS

       14.4 mM β-mercaptoethanol

       0.1% bromophenol blue (dye that moves ahead of proteins, used to decide when to halt electrophoresis)

       25% glycerol (increases the solution density causing the proteins to sink into the gel)

   b. Restriction Enzyme Buffer (Low Salt)

       10 mM Tris-HCl (pH 7.5)

       10 mM $MgCl_2$

       1 mM DTT

3. Based on the answers to questions 1 and 2, what are common components of protein solutions? What is the purpose of these components?

III. MAINTAINING THE STRUCTURE AND FUNCTION OF NUCLEIC ACIDS IN LABORATORY SOLUTIONS

A. An Overview of Nucleic Acid Structure and Function

i. DNA STRUCTURE

DNA (deoxyribonucleic acid) is a biomolecule that comprises genes. **Genes** *encode the amino acid sequences for proteins.* Progress in understanding genes is leading to numerous biotechnology products and major research advances. At the same time, a flood of techniques have been and are being developed for working with nucleic acids in the laboratory. As with proteins, it is essential to understand the physical and chemical nature of nucleic acids when manipulating them in laboratory solutions.

DNA is a linear polymer of **deoxyribonucleotide** ("nucleotide") subunits. **Nucleotides** *are composed of a phosphate group, a five-carbon sugar (deoxyribose), and one of four nitrogenous bases (adenine, guanine, cytosine, or thymine),* see Figure 29.7. **DNA** *consists of nucleotides connected into strands by covalent* **phosphodiester bonds.** The bonds link the phosphate group from one nucleotide to the sugar of another. The result is a "backbone" of alternating phosphates and sugars with the nitrogenous bases sticking out to the side.

The phosphate groups on each nucleotide readily give up H^+ ions; in fact, DNA is called an *organic acid.* Thus, in solution at neutral pH, negatively charged oxygens are exposed along the backbone of DNA strands. Nucleic acids dissolve readily in aqueous solutions because of the negatively charged backbone. The bases themselves are hydrophobic, but they face inward and are protected from the aqueous environment in double-stranded DNA.

Chromosomal DNA exists in the cell as a *double helix* of two strands of DNA, see Figure 29.7. The bases of the two strands pair with one another so that a cytosine on one strand is always across from a guanine on the opposite strand and an adenine is always across from a thymine. Cytosine and guanine are said to be *complementary*; adenine and thymine are also *complementary*. Double-stranded DNA is like a ladder with the bases forming the rungs on the inside and the phosphate–sugar backbone forming the outside rails. Imagine that the ladder is taken and twisted—this twist causes DNA to be called a double *helix*.

Complementary pairs of bases are held together by hydrogen bonds. Recall that hydrogen bonds are relatively weak compared to covalent bonds. Therefore, the phosphodiester bonds that link the nucleotide backbone are much stronger than the bonds holding the two strands together. This means that there are many conditions in the laboratory where the two DNA strands separate from one another, yet the strands themselves remain intact.

ii. RNA STRUCTURE AND FUNCTION

RNA (ribonucleic acid) *is a polymer of nucleotides similar to DNA except that the sugar is ribose and the nitrogenous base uracil is present in place of thymine.* There are various types of RNA with different structures and functions, including: messenger RNA (mRNA), transfer RNA (tRNA), and ribosomal RNA (rRNA). *Messenger RNA* carries information about the amino acid sequence of a protein from the nucleus to the cytoplasm. *Transfer RNA* shuttles amino acids to

Molecular model   **Stylized diagram**   **Chemical structure**

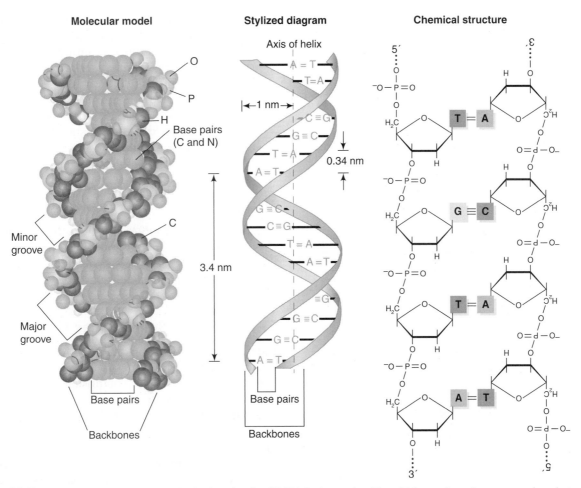

Figure 29.7.   The Structure of DNA. The basic subunit of DNA is the nucleotide, which consists of a sugar, a phosphate group, and a nitrogenous base. The subunits of DNA are connected by covalent phosphodiester bonds between the phosphate group on one nucleotide and the sugar on the next. In solution, the phosphate groups lose a H^+ and so have a negative charge. Chromosomal DNA is a double helix consisting of two strands of DNA held together by hydrogen bonds between complementary bases. Observe that three hydrogen bonds stabilize each G–C linkage, but only two hydrogen bonds stabilize each A–T linkage.

the ribosomes during protein synthesis. *Ribosomal RNA is part of the structure of the ribosomes.* Other types of RNA play various roles in controlling gene expression.

RNA is single stranded and is shorter than chromosomal DNA. Although RNA is normally single stranded, complementary bases *within* an RNA strand sometimes pair with one another. These bonds and other weak interactions cause RNA to fold into various complex conformations, see Figure 29.8 on p. 552. As with proteins, the varied conformations of RNA molecules are important in controlling their functions. RNA can also pair with complementary strands of DNA.

B. Loss of Nucleic Acid Structure

Like proteins, nucleic acids can lose their normal structure in several ways:

1. **DNA can be denatured.** When speaking of DNA, **denaturation** *means that the two complementary strands separate from one another.* The relatively weak hydrogen bonds that hold together the two strands can be disrupted by changes in pH (pH

greater than 10.5 will denature DNA), high temperature, or addition of organic compounds such as **urea** and **formamide.** Low salt concentrations also promote denaturation. Denaturation of DNA is reversible under suitable conditions; **renaturation** *occurs when complementary base sequences reestablish hydrogen bonds.*

The base composition of double-stranded DNA determines *the temperature at which it denatures, its* **melting temperature** *or* T_m. Three hydrogen bonds form between guanine and cytosine base pairs, while only two form between adenine and thymine base pairs. Therefore, the total bond strength between G-C pairs is stronger than between A-Ts and a DNA molecule that is comparatively rich in G-Cs denatures at a higher temperature than one that is rich in A-Ts.

At pH 7.0 and room temperature, DNA solutions are highly viscous. When DNA is denatured, it becomes noticeably less viscous and it absorbs more UV light at a wavelength of 260 nm.

Local denaturation of DNA is an essential step in transcription (when RNA is copied from tem-

RNA

Cytidine

Uridine

Adenosine

Guanosine

(a)

(b)

Figure 29.8. The Structure of RNA. a. Linear polymer of RNA. **b.** Example of RNA with secondary structure.

plate DNA) and replication (when a cell divides and makes a copy of its DNA for the daughter cell). There are situations in the laboratory where denaturation is similarly desirable and other situations where denaturation should be avoided.

2. **The covalent bonds that connect nucleotides into DNA and RNA strands can be broken by enzymes called** *nucleases.* Nucleases are found on human skin, in cells, and in cellular extracts.

DNA degrading nucleases are relatively fragile, require Mg^{++} as a cofactor, and are commonly destroyed during routine DNA isolation procedures. In contrast, RNA nucleases are ubiquitous, difficult to destroy, and generally do not require metal ion cofactors to be active. The RNA nuclease, **RNase A,** can even survive periods of boiling or autoclaving. RNA nucleases frequently contaminate glassware and other laboratory items, and make RNA difficult to manipulate in the laboratory. Rules typically cited to help reduce the loss of RNA are shown in Table 29.7.

3. **Chromosomal DNA is a long, fragile molecule that is easily sheared into shorter lengths.** Vigorous vortexing, stirring, or passing a DNA solution through a small orifice, such as a hypodermic needle, will shear DNA.

C. Laboratory Solutions for Nucleic Acids

The structure and function of nucleic acids is not as variable as is that of proteins. For example, there are only four different nucleotides comprising DNA, but there are 20 different amino acids commonly found in proteins. There are, however, a multitude of different tasks performed in bioscience laboratories that involve DNA, and there are numerous techniques used to manipulate nucleic acids; therefore, there are many solutions used for DNA work. A summary of the basic classes of agents found in DNA solutions are listed in Table 29.8 and then are discussed in more detail. Table 29.9 on p. 554 summarizes a number of solution components in terms of their functions in laboratory procedures.

When working with proteins in laboratory solutions, it is common to remove proteases carefully and to avoid conditions that will damage the proteins. In contrast, when working with nucleic acids, the first agents used are those that denature, precipitate, and destroy proteins. This is because nucleases (which are proteins) are released when cells are disrupted. These endogenous nucleases must be destroyed or the nucleic acids will be degraded. Later, certain proteins may be added to nucleic acid solutions to perform a particular task. For example, *DNA may be intentionally cleaved during cloning procedures using nucleases called* **restriction endonucleases.**

i. NUCLEIC ACIDS AND pH: THE IMPORTANCE OF BUFFERS

The pH of the immediate environment can have a profound effect on the structure of nucleic acids; therefore, buffers are added to nucleic acid solutions. For example:

1. **Manipulation of pH is a technique used for controlling DNA denaturation and renaturation in certain procedures.**

2. **Treatment with mild acids, such as 5% trichloroacetic acid (TCA) at or below room temperature, will precipitate nucleic acids.**

Table 29.7. COMMONLY CITED GUIDELINES FOR PREVENTING THE LOSS OF RNA

1. *Ribonucleases are released into solution when cells are disrupted.* Strong protein denaturing agents are used to destroy these endogenous RNases when cells are disrupted for RNA isolation. Protein denaturing agents used in RNA solutions include **6 M urea, SDS,** and **guanidinium salts (guanidine thiocyanate** and **guanidine hydrochloride),** sometimes in conjunction with a reducing agent such as β-mercaptoethanol. Tissue may also be rapidly frozen to inactivate endogenous ribonucleases. However, freezing temperatures disrupt cells and can cause the release of ribonucleases that become active when the material is thawed.
2. *Peoples' hands are a major source of RNase contamination and gloves should be worn when working with RNA.* Once gloves have come in contact with a surface that was touched by skin (e.g., a pen, notebook, laboratory bench, pipette, etc.) the gloves should be changed.
3. *Sterile technique should always be used when working with RNA because microorganisms on airborne dust particles are a major source of ribonuclease contamination.*
4. *Disposable sterile plasticware is seldom contaminated with RNases.* Once packages of plasticware are opened, the contents must be protected from dust and contaminants.
5. *Nondisposable glassware can be baked at 200°C overnight to inactivate RNases.*
6. *The active site of RNase A contains a histidine amino acid; this active site can be destroyed by modifying the histidine.* The chemical **DEPC (diethyl pyrocarbonate)** in low concentrations (around 0.05–1%) can inactivate the histidine binding site. DEPC is therefore frequently used to treat equipment and solutions that will be used for RNA work. Some points regarding the use of DEPC are:
   a. DEPC is carcinogenic and should be used in a hood and handled while wearing gloves.
   b. DEPC can inhibit enzymatic reactions and can interact with nucleic acids. After DEPC inactivates RNases, it therefore needs to be completely degraded. DEPC is usually degraded by autoclaving for 30 minutes to 1 hour, which breaks the DEPC down to carbon dioxide and ethanol.
   c. Glassware can be soaked in DEPC overnight. The glassware is then autoclaved for 30 minutes to inactivate the DEPC.
   d. DEPC can be used to treat water by mixing 0.1% DEPC with the water and allowing the mixture to sit at least 6 hours at room temperature. The solution is then autoclaved to destroy the DEPC.
   e. DEPC can be used to treat various solutions *(but not Tris buffer)*. 0.1% DEPC is added to the solution and the mixture is incubated overnight. The solution is then autoclaved 1 hour.
7. *It is useful to have a separate laboratory space and separate equipment for working with RNA.* This is particularly important if DNA work is also performed in the laboratory because it is common to add RNase to DNA solutions intentionally to destroy unwanted, contaminating RNA. Once RNase is present in the laboratory, it is difficult to eliminate from pipettes, centrifuges, pH electrodes, and other devices.
8. *Products (such as water, tips, tubes, buffers, enzymes, and bovine serum albumin) are all potential sources of RNase contamination.* Manufacturers produce many products that have been treated to remove ribonucleases and certify them to be RNase-free.
9. *Enzymes that are ribonuclease inhibitors, RNasins, can be purchased from molecular biology suppliers.* These inhibitors bind to and inactivate some classes of RNase.
10. *Clean Surfaces.* Commercial products are available to remove RNases from benchtops and other exposed surfaces.
11. *Storage.* One source recommends performing a salt/alcohol precipitation of RNA and freezing it as a precipitate. The low temperature and alcohol inhibit the activity of trace RNases.

3. **Prolonged exposure to dilute acid at or above room temperature, or briefer exposure to more concentrated acid (e.g., 15 minutes in 1 M HCl at 100°C) will cause depurination of DNA and RNA (removal of most of the purines, adenine and guanine).** The pyrimidines (cytosine, uracil, and thymine) in these nucleic acids can also be removed by exposure to more concentrated acids or by longer periods of exposure.

4. **Very strong acid treatments will break the phosphodiester bonds that join nucleotides.**

5. **Scientists can exploit the fact that DNA and RNA react differently under mildly alkaline conditions.** For example, 0.3 M KOH at 37°C for 1 hour will cleave the phosphodiester bonds in RNA, but the phosphodiester bonds of DNA will remain intact under these conditions.

ii. SALTS

Like proteins, nucleic acids are sensitive to the ionic strength of their solution. Consider the following three important cases:

1. **Denaturation of double-stranded DNA is controlled in the laboratory by the ionic strength of the solution in conjunction with its temperature and pH.** Consider DNA dissolved in a solution of

Table 29.8. CLASSES OF AGENTS USED IN LABORATORY SOLUTIONS FOR NUCLEIC ACIDS

Buffers	Nuclease inhibitors
Salts	Metal chelators
Organic solvents	Coprecipitants
Nucleases	Antimicrobial agents

Table 29.9.   SUMMARY OF SOLUTION COMPONENTS AND PROCEDURES FOR MANIPULATING DNA

Procedure	Components of Solutions
Isolation and Purification of Nucleic Acids from Cells	
Before DNA can be studied, sequenced, or recombined, it must be isolated and purified.	
Degrade the cell wall	Enzymes: Lysozyme for bacteria
Remove membranes	Detergents: SDS, Sarkosyl
Remove proteins and RNA	Enzymes: Proteinase K, RNase A
Denature and remove proteins	Phenol/chloroform extraction
Preferentially precipitate DNA	Ethanol (65–70%) in high salt
	Isopropanol with salt
Concentrate DNA in aqueous solution	*sec*-butanol
Purify specific types of nucleic acid	CsCl density gradient
	Commercial spin columns, glass beads
	Ion exchange resins
Preferentially degrade chromosomal DNA to isolate plasmids	NaOH + SDS
Nucleic Acid Storage Buffers	
Protection during storage of DNA.	
Maintain proper pH	Buffers: Tris
Chelate metals (inhibit nucleases)	EDTA
Maintain ionic strength	NaCl
Enzymatic Digestion and Modification	
Enzymes are used to digest, recombine, and modify DNA. These enzymes require proper solution conditions.	
Maintain proper pH	Buffers: Tris
Prevent the loss of enzyme due to adsorption to container	Carriers: BSA
Enzyme cofactors	Divalent cations: Mg^{++}, Mn^{++}, Fe^{++}
Stabilize protein and lower freezing temperature	Glycerol
Prevent aggregation and sticking	Low levels of detergents such as Triton X
Inhibit proteases	EDTA
Electrophoresis	
DNA molecules can be separated by size during migration in an electrical field.	
Gel matrix	Agarose, acrylamide
Running buffers	Tris borate EDTA, Tris acetate EDTA
Denaturing gels (lower the melting temperature of the double-stranded DNA)	Formamide, urea
Tracking dyes to visualize the migration of DNA	Xylene cyanol, bromophenol blue
Dyes to detect the presence of DNA	Ethidium bromide, methylene blue
Transformation/Transfection of Purified DNA into Cells	
Introduction of foreign DNA into a cell. In bacteria it is necessary to prepare the cell; in eukaryotic cells it is necessary to prepare the DNA.	
Preparation of bacterial cells to be "competent"	$CaCl_2$, RbCl, DMSO
Uptake of precipitated DNA by eukaryotic cells	$CaPO_4$ precipitation of DNA
Fusion of eukaryotic cells with other membranes	Liposomes, lipid-encased DNA
Permeabilization of eukaryote membranes	DEAE dextran, DMSO, glycerol
Hybridization	
Process in which denatured DNA binds (hybridizes) with complementary strands of DNA or RNA.	
Separation of DNA strands	NaOH, heat
Hybridization solution components:	
"Blocking agents" to fill nonspecific binding sites	BSA, nonfat dry milk, casein, gelatin
"Crowding agents"—inert compounds that effectively reduce the volume of the solution	Denhardt's Reagent: Ficoll, PVP, PEG
Nonspecific nucleic acid to reduce background by binding nonspecific sites	Salmon sperm DNA, tRNA

low ionic strength, see Figure 29.9a. The negatively charged phosphate groups cause the two DNA strands to repel one another. In contrast, in a solution of moderate ionic strength, such as 0.4 M NaCl, Na$^+$ ions are available that are attracted to negatively charged phosphate groups. The negative charges of the phosphate groups are thus neutralized, so the two DNA strands do not repel one another, see Figure 29.9b. Double-stranded DNA, therefore, is less stable and is more easily denatured in a solution with 0.01 M salt than it is in a solution of 0.4 M salt. As a result DNA denatures at a lower temperature in 0.01 M NaCl than it does in 0.4 M NaCl. In solutions with salt concentrations between 0.01 M NaCl and 0.4 M NaCl, the likelihood of denaturation is intermediate.

2. **Hybridization *(binding)* of single-stranded DNA with short strands of complementary DNA or RNA is affected by the ionic strength of the solution. Stringency** *refers to the reaction conditions used when single-stranded, complementary nucleic acids are allowed to hybridize.* At high stringency, binding occurs only between strands with perfect complementarity. (Perfect complementarity means that every guanine is base-paired with a cytosine and every adenine is base-paired with a thymine.) At lower stringency, there can be some mismatch of bases across the strands and hybridization still

occurs. There are situations in the laboratory where high stringency is required and other situations where lower stringency is desirable.

Strands come together more easily when the salt concentration is high and the temperature is relatively low. Under these conditions, hybridization can occur even if there are some base pair mismatches; the conditions have lower stringency, see Figure 29.10 on p. 556. In contrast, when the temperature is higher and the salt concentration is lowered, the conditions are more stringent.

3. **The ionic strength of the solution affects interactions between proteins and DNA.** Protein–DNA interactions are of critical importance in the cell. For example, such interactions are involved in transcription and replication. Electrostatic interactions between negatively charged phosphates on DNA and positively charged amino acids promote protein–DNA binding. However, electrostatic interactions are not specific. This means that electrostatic interactions do not cause an enzyme to find and bind to a *specific sequence* of DNA. Electrostatic interactions are affected by the concentration of salt in the solution. As salt increases, the strength of electrostatic interactions decreases. For example:

  • **Chromatin** *is the native form of DNA, which is composed of DNA bound to many different proteins.* In 0.01 M NaCl chromatin is stable, but in 1 M NaCl it dissociates almost com-

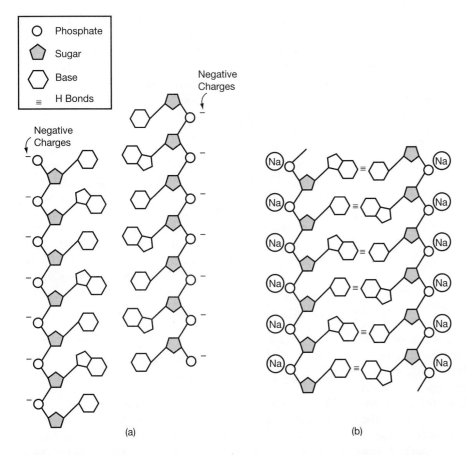

Figure 29.9.   The Effect of Ionic Strength on Nucleic Acid Denaturation and Renaturation. a. In a solution of low ionic strength the two DNA strands are negatively charged and repel one another, making denaturation more likely. **b.** In a solution of moderate ionic strength, positive ions shield negatively charged phosphates.

(a)   (b)

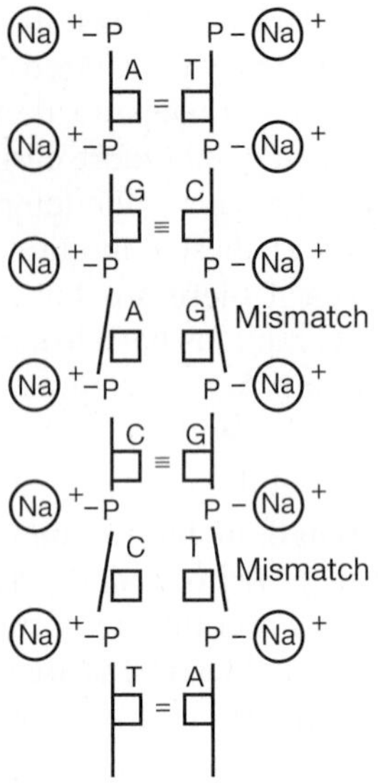

Figure 29.10. The Effect of Ionic Strength on Stringency. At low temperatures and high ionic strength, some mismatches can be tolerated; stringency is lower than at higher temperatures and lower ionic strength.

pletely to yield free DNA molecules and proteins. This effect is thought to be due to the loss of electrostatic interactions that normally stabilize chromatin. When NaCl is added, the salt ions shield the various charged sites and disrupt the electrostatic bonding.

- *Restriction enzymes that cut DNA at specific sites require the correct concentration of salts to cut DNA selectively.* At lower salt concentrations, electrostatic interactions increase, allowing the restriction enzyme to nonspecifically bind to and cut the DNA. A restriction enzyme is said to exhibit **star activity** *when it "mistakenly" cleaves DNA sequences that are not its proper specific target.* Star activity is an example of nonspecific protein–DNA interaction. Reduced salt concentration is one of several factors that promote star activity.

Specific interactions between proteins and particular base sequences of DNA are not electrostatic. Unlike electrostatic interactions, specific protein–DNA interactions are frequently stable in higher salt concentrations.

iii. Organic Solvents

Organic solvents are commonly used to isolate DNA and RNA from other components of cellular extracts. A standard extraction procedure involves addition of the organic solvents **phenol** and **chloroform,** which simultaneously disrupt cell membranes and denature proteins (including endogenous nucleases). Phenol and chloroform are not water-soluble, so the mixture is centrifuged to separate it into an organic phenol/chloroform phase and an aqueous phase. Denatured proteins move into the organic phase or the interface between the two phases. The nucleic acids remain in the aqueous phase. A low

concentration of **isoamyl alcohol** is usually added to aid in clean separation of the organic and aqueous phases. Phenol is relatively unstable in storage, so **8-hydroxyquinoline** is usually added to phenol as an antioxidant.

iv. Precipitants

Ethanol plays an important role in working with nucleic acids because it precipitates DNA and RNA, physically isolating them from a solution containing other components. Nucleic acids do not lose their structural or functional integrity when isolated with phenol/chloroform and/or ethanol.

Isopropanol is sometimes substituted for ethanol when the final solution volume must be minimized because less isopropanol is needed than ethanol to precipitate DNA.

Salt is added when nucleic acids are ethanol-precipitated because relatively high concentrations of monovalent cations (0.1 to 0.5 M) promote aggregation and precipitation of nucleic acid molecules. Various salts are used in different situations; however, sodium acetate is most common.

v. Coprecipitants

Alcohol precipitation is ineffective if only small amounts of nucleic acids are present in a solution. Inert coprecipitants, such as **yeast tRNA** or **glycogen,** are therefore sometimes added to solutions to help precipitate nucleic acids.

vi. Nucleases

There are many uses for nucleases in the laboratory. For example, when isolating DNA it is common to use **RNase A** to intentionally degrade contaminating RNA. Restriction endonucleases are frequently added to DNA solutions to recognize and cleave DNA phosphodiester bonds at specific sites.

vii. Nuclease Inhibitors

Unwanted nucleases are released when cells are disrupted. As mentioned earlier, organic solvents used to isolate DNA and RNA denature nucleases and other proteins. **Proteinase K** is an enzyme that is also added to solutions to help degrade nucleases and histones (proteins associated with DNA in the chromosome). Proteinase K is active over a wide range of pH, salt, detergent, and temperature conditions. Chelators are also used as nuclease inhibitors because many nucleases require a metal ion cofactor for activity.

viii. Antimicrobial agents

Antimicrobial agents are often added to DNA solutions to inhibit the growth of microorganisms during storage.

ix. Chelators and Metal Ions

Unwanted nucleases can degrade nucleic acids, particularly during long periods of storage. There are many situations, however, when nucleases are used to intentionally break chains of nucleotides in a controlled fashion. Most nucleases require Mg^{++} as a cofactor. The levels of chelators and metal ions, therefore, are controlled in nucleic acid solutions. For example, EDTA is typically added to nucleic acid solutions during storage to reduce unwanted nuclease activity. Excess Mg^{++} is added when nuclease activity is desired.

D. Physical Factors

i. TEMPERATURE

Temperature is frequently used in the laboratory to control denaturation and renaturation of DNA. For example, DNA must be denatured whenever a procedure requires the DNA to hybridize to complementary nucleic acid fragments. Temperature is critical in the polymerase chain reaction (PCR) technique that is used to amplify the amount of a specific sequence of DNA. High temperature (around 90°C) is used to denature the DNA to be amplified. The temperature is then lowered to allow replication of the single-stranded DNA (see also Practice Problem 4, p. 560).

Proteins are frequently used to manipulate nucleic acids in the laboratory. The temperature of a nucleic acid solution, therefore, may be chosen to meet the requirements of the proteins involved. For example, restriction endonucleases are most active at 37°C, so this temperature is used to digest DNA.

ii. SHEARING

Because of their long strands, nucleic acids are more prone to shearing than proteins. Mixing and pipetting, therefore, must be particularly gentle to isolate high-molecular-weight nucleic acids.

CASE STUDY

Isolating Plasmids: The Roles of Solution Components

Let us consider as an example the solutions used in a common procedure: the isolation of plasmids from bacteria. **Plasmids** *are small circular molecules of DNA found in bacteria that exist separate from the bacterial chromosome. In nature, plasmids sometimes carry genes that are responsible for the spread of antibiotic resistance among bacteria.*

In the laboratory, scientists can modify plasmids and use them to carry genes of interest into host bacteria. Once inside the bacteria the plasmids multiply as the bacteria divide. The plasmids can be isolated from the host bacteria using various techniques, such as the "plasmid minipreparation procedure." The conventional minipreparation procedure takes advantage of the small, coiled nature of the plasmids to separate them from the larger, linear chromosomal DNA. The minipreparation procedure will be discussed. As we go through each step, we will explain the roles various solution components are thought to play in the separation.

Minipreparation Procedure to Isolate Plasmids from Host Bacteria

The Solutions

LB (Luria Broth) (1 L)
10 g tryptone
5 g yeast extract
5 g NaCl
Sterilize by autoclaving

GTE Buffer
25 mM Tris-HCl, pH 8.0
10 mM EDTA
50 mM glucose

Potassium Acetate/Acetic Acid
60 mL of 5 M potassium acetate
11.5 mL of glacial acetic acid
28.5 mL of H_2O

SDS/NaOH
1% SDS
0.2 N NaOH

TE Buffer
10 mM Tris-HCl, pH 7.5–8.0
1 mM EDTA

The Procedure

1. **Grow the plasmid-containing bacterial culture overnight in LB medium.** *LB medium contains nutrients for bacterial growth.*

2. **Remove 1.5 mL of the culture and spin down the cells in a microfuge.**

3. **Discard the supernatant and completely resuspend the cells in 100 μL of ice-cold GTE buffer.**
   The buffer is kept cold to inhibit nucleases released during the procedure. GTE Buffer contains: Tris to maintain the proper pH (pH 8.0), EDTA to bind divalent cations (Mg^{++}, Ca^{++}) in the lipid bilayer (thus weakening the cell membrane), and glucose, which may act as an osmotic support or may help to prevent cells from clumping.

4. **Add 200 μL of SDS/NaOH and gently mix.**
   The bacterial cells are lysed by the SDS, which solubilizes the membrane lipids and cellular proteins. The base denatures both the chromosomal and plasmid DNA. The strands of the circular plasmid DNA remain intertwined. The suspension is mixed gently to avoid shearing the chromosomal DNA.

5. **Set on ice for 10 minutes.**
   This "resting step" allows time for the SDS/NaOH to contact all the cells and the processes in step 4 to go to completion.

6. **Add 150 μL of cold potassium acetate/glacial acetic acid and mix gently. A white precipitate will appear.**
   The acetic acid neutralizes the sodium hydroxide added in step 4, allowing the DNA to renature. The potassium acetate causes the SDS to precipitate, along with associated proteins and lipids. The large chromosomal DNA strands renature partially into a tangled web that precipitates along with the SDS complex. In contrast, the small, coiled renatured plasmids remain suspended. In this step, therefore, they are separated from chromosomal DNA, many proteins, and the detergent.

7. **Centrifuge the mixture; save the supernatant.**
   The discarded precipitate contains cellular material, linear DNA, protein, and detergent, whereas the supernatant contains the suspended plasmids.

8. **Remove 400 μL of the supernatant, add an equal volume of isopropanol, and mix.**

   Isopropanol, like ethanol, precipitates DNA. Isopropanol is used here because a smaller volume of isopropanol is required than ethanol. The alcohol quickly precipitates DNA and more slowly precipitates proteins. This step is therefore done quickly.

9. **Centrifuge and save the pellet.**

   The alcohol-precipitated plasmid DNA is in the pellet.

10. **Wash the pellet in 200 μL of 100% ethanol.**

    Although isopropanol precipitated the DNA in the preceding step, an ethanol wash helps remove extra salts and other unwanted contamination.

11. **Centrifuge and save the pellet.**

12. **Resuspend the pellet in 15 μL of Tris/EDTA buffer (TE).**

    TE is often used for short-term storage of DNA. Tris maintains the pH, whereas EDTA chelates divalent cations to inhibit nuclease activity.

Based primarily on information in Micklos, David A., and Freyer, Greg A. DNA Science. Cold Spring Harbor, NY: Cold Spring Harbor Press, 1990.

EXAMPLE PROBLEM

Southern blotting and probe hybridization is a method in which DNA fragments separated from one another by electrophoresis are adhered to a plastic or nylon membrane and are then exposed to a single-stranded nucleic acid probe. The probe will bind to any single-stranded fragments adhered to the membrane that have a nucleotide sequence complementary to the probe. Before the DNA is adhered to the membrane, it is soaked in a solution containing NaOH. What is the function of the NaOH in this soaking solution?

ANSWER

In order for the probe to bind to DNA fragments, the DNA fragments (and the probe) must be single stranded. The complementary bases will then be available for hybridization. The NaOH is added to the solution to denature the double-stranded DNA.

CASE STUDIES

Optimizing Solutions and Avoiding Problems with Solutions

Solutions often have multiple components, each of which must be present at the proper concentration. The early phases of research often include finding optimal solution components, sometimes by trying various combinations of agents until the best results are obtained. It is essential to understand the roles of the various components when opti- mizing solution recipes, as the following examples from the literature illustrate.

Example 1

An illustration of a study that looked at the effects of laboratory solutions on the results of an experiment was reported by Favre and Rudin (Favre, Nicolas, and Rudin, Werner. "Salt-Dependent Performance Variation of DNA Polymerases in Co-Amplification PCR." BioTechniques 21 [July 1996]: 28–30). These authors were using the technique of reverse transcriptase PCR to amplify and detect specific mRNAs. The scientists were attempting to amplify two different mRNAs simultaneously in the same reaction tube. They tested reverse transcriptase enzymes from four manufacturers, in each case using the buffer solution provided by the manufacturers. They found that the enzymes/buffers from all four manufacturers were effective when the two different mRNAs were amplified individually in separate tubes. When they attempted to amplify both types of mRNA together in the same tube, however, only one manufacturer's kit was effective. Further study showed that the effective kit contained buffer with 200 mM Tris-HCl, whereas the other manufacturers provided buffer with 100 mM Tris-HCl. They concluded that the buffer concentration was critical in the amplification and detection of two mRNAs simultaneously.

Example 2

Another illustration of a study that investigated the effects of buffer components was reported by Ignatoski and Verderame (Ignatoski, Kathleen M. Woods, and Verderame, Michael F. "Lysis Buffer Composition Dramatically Affects Extraction of Phosphotyrosine-Containing Proteins." BioTechniques 20 [May 1996]: 794–6.). In order to understand this example, it is necessary to understand the role of lysis buffers:

All cells are surrounded by a cell membrane that maintains the structural integrity of the cell and also controls the interactions of the cell with its external environment. Cell membranes are composed of **phospholipids** *with embedded proteins and glycoproteins. Phospholipids have a negatively charged, hydrophilic phosphate group and "tails" that are hydrophobic hydrocarbon chains. The hydrophilic "heads" face out to the cytoplasm or the exterior of the cell, whereas the hydrophobic tails face one another.*

Cell membranes and cell walls must be broken (lysed) to get access to the molecules inside. Solutions whose primary function is to lyse these structures are called **lysis buffers.** *Membranes are usually lysed by detergents that, like the membranes, have both hydrophilic and hydrophobic portions. Mild detergents are preferred, such as* **Triton X-100** *or* **sarkosyl (N-lauroylsarcosine).**

Plant and bacterial cells also have cell walls that surround their cell membranes. Cell walls have varied structures, but they are usually more difficult to break than membranes. A variety of methods are used to disrupt cell walls including physical ones, such as grinding or crushing, or the addition of enzymes, such as **lysozyme** (for lysing bacteria).

Ignatoski and Verderame were comparing proteins in cultured cells that were transformed with an oncogene (a cancer-causing gene) with proteins in cultured cells that were not transformed. The cells were broken open with lysis buffer, releasing their proteins into solution. The released solubilized proteins were analyzed. The scientists discovered that the components of the lysis buffer had a significant effect on what proteins they detected and therefore, their conclusions regarding differences between transformed and nontransformed cells.

Ignatoski and Verderame compared the effects of three different lysis buffers. All three buffers included protease inhibitors and also the following ingredients:

Lysis Buffer A	Lysis Buffer B
50 mM Tris-HCl, pH 7.5	10 mM Tris-HCl, pH 7.4
150 mM NaCl	50 mM NaCl
1% Nonidet P-40 (NP-40)	50 mM NaF
(detergent)	1% Triton X-100
0.25% Na$^+$ deoxycholate	(detergent)
(detergent)	5 mM EDTA
10 mg/mL BSA	150 mM Na$_3$VO$_4$
	30 mM Na$_4$P$_2$O$_7$

Lysis Buffer C
30 mM Tris-HCl pH 6.8
150 mM NaCl
1% NP40 (detergent)
0.5% Na$^+$ deoxycholate (detergent)
0.1% SDS (detergent)

The authors observed that the three lysis buffers appear to "be solubilizing different subsets of cellular proteins" and they postulated that:

1. **The type and concentration of detergent could have an effect on which proteins are detected.** Ionic detergents, like Na$^+$ deoxycholate, disrupt the cell membrane more completely than do nonionic detergents, such as NP40 or Triton X-100. Thus, more proteins might have been released into solution when ionic detergents were used.

2. **The ionic strength of the buffer plays a critical role in protein solubilization.** Proteins need to be soluble in aqueous solutions to be detected. High salt concentrations might disrupt interactions of proteins with insoluble cellular components, thereby promoting protein solubility. The three buffers contained different salts, which may have caused the proteins solubilized in each buffer to have been different.

3. **pH can affect detergent solubility, especially of Na$^+$ deoxycholate, thereby changing which pro-**

teins are solubilized.** The pH of Buffer C was different from the other two buffers; therefore, the proteins that were detected may have been affected.

4. **The buffer effectiveness may have been affected by the presence of phosphate.** Buffer B was the only buffer containing phosphate, and this buffer resulted in the solubilization of a different subset of proteins than the other buffers.

The authors of the preceding study concluded that the lysis buffer components can play a role in affecting experimental results and therefore "several different lysis buffers . . . should be tested to obtain optimal experimental conditions."

As the preceding examples illustrate, the composition of solutions has a critical effect on results. Once solution recipes have been developed for a particular task, it is therefore essential to follow these recipes exactly as written (or to control and document alterations in solution components carefully).

In a production setting, optimizing a solution requires consideration not only of the effectiveness of the solution, but also the cost and availability of the components. Large quantities of raw materials will be needed when the product goes into mass production. Therefore, raw materials should be chosen that are as inexpensive as possible and yet meet the requirements of the process. Also, it is necessary to find suppliers of the solution components who can provide sufficient raw materials of consistent high quality. In a production setting, once the solution recipes have been established and suppliers of raw materials have been found, it is particularly important not to modify any aspect of the solution without careful consideration and control of the changes.

Example 3

An example of the type of problem that occurs when solution recipes are not followed was reported by McCarty (McCarty, Maclyn. The Transforming Principle: Discovering That Genes Are Made of DNA. New York: W.W. Norton, 1985.) in a book describing research that led to the discovery that genes are made of DNA. He reported:

I don't mean to imply that things always went smoothly. . . . There were a number of hitches along the way. . . . We even had some problems with simple fundamental operations, like growing the type III pneumococci for extraction [of DNA]. Early in the fall the central media department supplied us with a few 75-L lots of broth that either sustained the growth of the organisms very poorly or not at all. I spent some time trying to find the source of the trouble and ended up participating in the preparation of our next batches of media. The difficulty was never pinpointed, but fortunately the difficulties disappeared with careful attention to the details of the cookbook-type recipe that had to be followed.

PRACTICE PROBLEMS: DNA

1. Suggest the possible purpose of the components of the following commonly used solutions.

   a. **TBE buffer—electrophoresis buffer**

      0.089 M Tris base (pH 8.3)
      0.089 M boric acid
      0.002 M EDTA

   b. **TE buffer**

      10 mM Tris-HCl (pH 8.0)
      1 mM EDTA

   c. **Hybridization buffer**

      (From Sambrook, et al., see bibliography, p. 493)
      40 mM PIPES Buffer (pH 6.4)
      1 mM EDTA
      0.4 M NaCl
      80% formamide

2. Examine the recipe for hybridization buffer, shown above. Comment on the salt concentration in terms of stringency.

3. (Modified from Freifelder, David. *Principles of Physical Chemistry With Applications to the Biological Sciences.* 2nd ed. Boston, MA: Jones and Bartlett, 1985.) The enzyme RNA polymerase binds to DNA and controls the synthesis of a complementary strand of RNA from the DNA template. The binding of RNA polymerase to DNA in cells is specific in that only a particular required sequence of DNA is copied into RNA. It has been observed that in a test tube in **0.2 M NaCl,** four RNA polymerase molecules bind to a particular DNA molecule and synthesize the same, specific RNA molecule that is found in living organisms. In contrast, in a test tube in **0.01 M NaCl,** about 50 RNA polymerase molecules bind to the DNA and a great many different RNA molecules are synthesized.

   a. What does this tell us about the interactions between DNA and RNA polymerase?

   b. What are the implications of these observations for working with RNA polymerase in the laboratory?

4. **PCR, the polymerase chain reaction,** is a method used to copy a specific region of DNA many thousands of times. PCR requires **primers** that are short, single strands of DNA. The primers are complementary to regions flanking the DNA region of interest. There are three phases to the PCR cycle:

   Phase 1: The two strands of the parent DNA molecule that includes the sequence of interest are separated by heating to about 92°C.

   Phase 2: The solution is quickly cooled to around 72°C and the single-stranded DNA hybridizes with the primers.

   Phase 3: Once the primers have annealed, the thermostable enzyme, *Taq 1* DNA polymerase, is able to copy the DNA region of interest, beginning from the location where the primers have bound.

The PCR cycle is repeated many times, greatly amplifying the number of copies of the DNA of interest.

   a. Examine the following recipe for PCR reaction buffer. Discuss the salt concentration in terms of DNA denaturation and hybridization of the primer to the DNA.

   b. Explain the role of the $MgCl_2$ in the reaction mix.

   c. *Taq 1* polymerase is functional at high temperatures. Why is this type of DNA polymerase essential for the PCR method to work?

      PCR Reaction Buffer

      500 mM KCl
      100 mM Tris-HCl (pH 8.0)
      20 mM $MgCl_2$

5. What do you think would happen to a bacterial culture with the addition of dish-washing detergent? Consider the membranes, proteins, and nucleic acids.

6. You have a sample of DNA (10 μg in 50 μL) that is stored in TE buffer (10 mM Tris pH 7.6 and 1 mM EDTA). Upon addition of 100 μL of ethanol, no precipitate is seen. Why not? What do you do to recover the DNA?

QUESTIONS FOR DISCUSSION

1. Suppose you are working in a university research laboratory and are taking over a project that was initiated by another investigator. The project involves isolating DNA from a particular type of fungus. One of the problems the previous investigator had begun to explore, but had not resolved, was that the DNA isolated was inconsistent in its qualities. The DNA sometimes seemed to be fragmented, sometimes more DNA was isolated than other times, and sometimes, when the DNA was digested with restriction endonucleases, the digestion did not appear to go to completion, even when the enzymes were known to be effective. Discuss components of the various solutions involved that might be affecting the results. How could you investigate the various components of the solutions in a systematic, well-documented way?

2. Suppose you are now working in a production facility producing the DNA described in Question 1. Discuss the particular concerns you would have in this situation.

Chapter Appendix

Alphabetical Listing of Common Components of Biological Solutions

Acrylamide. Monomers that polymerize to form a polyacrylamide matrix used in gel electrophoresis (PAGE).

bis-acrylamide. Cross-linker added to acrylamide monomers before polymerization. Controls the porosity of the resulting gel.

Agarose. Matrix composed of complex carbohydrates derived from seaweed, used for gel electrophoresis.

Ammonium acetate. Salt added to DNA in solution to aid in ethanol precipitation. At a concentration of 2 M it will preferentially facilitate ethanol precipitation of larger DNA and allow nucleotides to remain in solution.

Ammonium persulfate. A free radical generator added to unpolymerized acrylamide to stimulate polymerization.

Ampicillin. Antibiotic that inhibits bacterial cell wall synthesis.

Avidin. Egg white protein that has a high affinity for biotin; may be conjugated with an enzyme or other compound as part of a detection system.

BCIP (5-Bromo-4-chloro-3-indolyl phosphate *p*-toluidine salt). A substrate for the enzyme alkaline phosphatase. In conjunction with NBT produces a blue color indicating the presence of this enzyme.

Biotin. Vitamin that has a high affinity for avidin. Biotin can be conjugated to various molecules, then bound to and detected by an indicator-linked avidin molecule.

Bromphenol blue. Dye used to visualize sample movement in gel electrophoresis.

BSA (bovine serum albumin). A protein often used as a carrier to help protect proteins; also used to block nonspecific binding sites.

sec-Butanol. Water-insoluble alcohol used to reduce the water in aqueous DNA solutions, thus concentrating DNA into a smaller volume.

CAA (casamino acids). Digested protein added to bacterial growth media as a source of amino acids.

CaCl$_2$ (calcium chloride). Salt used to increase permeability of cellular membranes. Bacteria incubated in CaCl$_2$ at cold temperatures become competent to take up DNA.

Calcium phosphate. Salt that forms a precipitate with DNA. The complex is introduced into cells in culture.

Chloramphenicol. Antibiotic used to amplify plasmids in bacteria by restricting protein synthesis but not nucleic acid synthesis. Also the substrate for CAT (chloramphenicol acetyl transferase) enzyme assay.

Coomassie (brilliant) blue stain. Blue stain that binds to proteins. Used to visualize protein bands in gels after electrophoresis and in protein assays.

CsCl (cesium chloride). Salt used to form density gradients during ultracentrifugation to separate DNA by buoyant density.

CTAB (cetyltrimethylammonium bromide). Nonionic detergent. Precipitates DNA at salt concentrations below 0.5 M; at high salt concentrations, complexes with polysaccharides and protein.

DEAE dextran (diethylaminoethyl-dextran). Used to permeabilize mammalian cells to facilitate DNA uptake.

Denhardt's reagent. Solution used in hybridization reactions as a buffer and as a blocking agent. Contains BSA, Ficoll, and PVP.

DEPC (diethyl pyrocarbonate). A RNase inhibitor. Added to water, solutions, and glassware to inhibit the action of nonspecific RNases.

Detergents. Water-soluble compounds that have a hydrophobic tail. Detergents have multiple uses, including solubilization of membranes and membrane proteins, cell lysis, denaturing proteins, wetting surfaces, and emulsification of chemicals.

Dextran sulfate. Used as a "crowding agent" to effectively increase the concentration of nucleic acids in a hybridization mixture.

DMSO (dimethyl sulfoxide). Compound used to solubilize chemicals and to permeabilize cells.

DNase. Any member of a class of naturally occurring enzymes that digest DNA.

dNTP (deoxyribonucleoside triphosphate). Designates any of the four deoxyribonucleotides or a mixture of the four deoxyribonucleotides that comprise DNA.

DTT (dithiothreitol). A reducing agent added to protein solutions to inhibit the formation of unwanted disulfide bonds.

EDTA (ethylenediaminetetraacetic acid). Chelator of divalent ions. Added to DNA solutions to reduce the activity of nucleases and modifying enzymes by binding to and removing cofactors.

EGTA (ethylene glycol-bis[β-aminoethyl ether]-N, N, N', N'-tetraacetic acid). Chelator of divalent ions; binds preferentially to Ca^{++}.

Ethanol. An alcohol that dehydrates and precipitates DNA when in high salt buffer.

Ether. Volatile organic solvent; used to remove residual phenol from aqueous solutions.

Ethidium bromide. Dye that binds to nucleic acids and fluoresces when exposed to UV light. Used to visualize and quantify nucleic acids.

Ficoll. Inert polymer that acts as a "blocking agent" and/or "crowding agent" when included in a hybridization solution. A component of Denhardt's Reagent.

Formamide. Chemical added to destabilize nucleic acid duplexes.

Glycerol. Viscous, syrupy liquid added to solutions to reduce their freezing temperature, allowing solutions to be very cold without freezing. Increases the permeability of membranes and helps to stabilize proteins.

Glycogen. Inert carrier added to help ethanol precipitate low concentrations of DNA.

Guanidine isothiocyanate (GITC) also **Guanidine thiocyanate (GTC).** Salt that denatures proteins; added to preparations when isolating RNA to denature and remove contaminating RNases and to free RNA from proteins.

HEPES. Commercially available nontoxic buffer often used in cell culture.

8-hydroxyquinoline. Antioxidant, added to redistilled phenol to prevent unwanted oxidation products.

Isoamyl alcohol. Added to phenol/chloroform during extraction of protein from DNA to reduce interface foaming.

Isopropanol. Alcohol that will precipitate DNA when in high salt solution.

KCl (potassium chloride). Commonly used salt.

LiCl (lithium chloride). Salt which at 4 M will preferentially precipitate high molecular weight RNA.

β-mercaptoethanol (Also, **2-mercaptoethanol**). Reducing agent; added to protein solutions to restrict the formation of unwanted disulfide bonds. Can be used to help inactivate RNases.

MgCl$_2$ (magnesium chloride). Source of Mg^{++}, essential cofactor of many enzymes that act on nucleic acids.

NaCl (sodium chloride). Table salt, most common salt used to increase the ionic strength of a solution.

NaOH (sodium hydroxide). Strong base; at 0.2 M will denature DNA.

NBT (nitro blue tetrazolium). Used with BCIP as a color indicator for detection of alkaline phosphatase activity.

Nitrocellulose. Membrane; used to immobilize nucleic acids during a Southern blot.

PBS (phosphate buffered saline). Isotonic buffer used for intact cells.

PCA (perchloric acid). Acid used to denature and precipitate nucleic acids and proteins.

PEG (polyethylene glycol). Waxy polymer available in varying sizes used to precipitate virus particles and protein. Added to hybridization solutions to reduce the liquid concentration in the solution.

Phenol. Water-insoluble organic solvent used as a protein denaturant. When added to an aqueous solution of DNA and protein, the protein is denatured and separates into the phenol phase and interface, and the nucleic acids are left in the aqueous phase.

Phenol/chloroform. Addition of chloroform to phenol increases the phenol solubility of denatured proteins.

Phenol/chloroform/isoamyl alcohol. Mixture of solvents used in DNA isolation to denature and remove proteins from the solution. Isoamyl alcohol is added to improve phase separation.

PMSF (phenylmethylsulfonyl fluoride). A protease inhibitor; added to crude cell extracts to reduce degradation of protein by proteases.

Potassium acetate. Salt commonly added with ethanol to precipitate DNA.

Protease inhibitors. Compounds that inhibit activity of unwanted proteases.

PVP (polyvinylpyrrolidone). Inert polymer added to Denhardt's solution as a "crowding agent."

RNase. Any one of a class of enzymes that degrade RNA.

SDS (sodium dodecyl sulfate; also known as **lauryl sulfate, sodium salt).** Ionic detergent that is effective as a protein denaturant. Commonly used to denature proteins before separation by electrophoresis.

Sodium azide (NaN$_3$). Antimicrobial agent added to some solutions to reduce contamination and degradation by microorganisms.

Sorbitol. Inert sugar sometimes used as an osmotic support to protect cells after their cell wall is removed.

Spermine. Added to stimulate T4 kinase activity.

SSC (standard saline citrate). Buffer used in Southern blotting.

Streptomycin sulfate. Antibiotic that inhibits protein synthesis in bacteria.

TAE (tris acetate EDTA buffer). Buffer for gel electrophoresis of nucleic acids.

TBE (tris borate EDTA buffer). Buffer for gel electrophoresis of nucleic acids.

TCA (Trichloroacetic acid). Protein and nucleic acid denaturant and precipitant.

TEMED (N,N,N′,N′-Tetramethylethylenediamine). Stimulates polymerization of polyacrylamide.

Tetracycline. Antibiotic that inhibits protein synthesis in bacteria.

Tris (tris[hydroxymethyl]aminomethane). Commercial buffer used commonly in molecular biology.

Triton X-100. Nonionic detergent that solubilizes cells and cell membranes.

Urea. RNA denaturant; destabilizes DNA by reducing the temperature at which the DNA double helix denatures; protein denaturant.

X-gal (5-bromo-4-chloro-3-indolyl β-D-galactopyranoside). Color indicator for the presence of the enzyme β-galactosidase.

Xylene cyanol. Tracking dye used in nucleic acid electrophoresis.

CHAPTER 30

Culture Media for Intact Cells

I. INTRODUCTION

Many applications in biotechnology require cultured, living cells, that is, cells maintained inside plates, flasks, vials, dishes, bioreactors, or fermenters. Molecular biologists, for example, routinely grow bacteria for genetic transformation procedures. Researchers use cultured mammalian cells to study questions, for example: How do cells communicate? How do disease agents damage cells? What are the mechanisms by which drugs act on cells? What causes cells to become malignant? Human and animal tissues are grown outside the body for therapeutic purposes, such as to provide skin grafts for burn victims. Cultured animal cells are used to grow viruses for vaccine production. Cultured cells are used in pharmaceutical testing, for example, to study the effects and possible toxicity of new drugs, cosmetics, and chemicals. At a much larger scale, genetically modified cells are used to manufacture biopharmaceuticals and industrial enzymes. In all these applications, the cells need to survive, grow, and reproduce while in an artificial, aqueous, "culture" environment. This chapter therefore extends our discussion of biological solutions to include those *that support living cells;* these solutions are called **nutrient media, culture media, growth media, cell culture media,** or simply **media.**

Like the biological macromolecules DNA, RNA, and proteins, discussed in Chapter 29, cultured cells require a buffered aqueous environment that contains specific dissolved solutes in the proper concentrations. Working with living cells, however, introduces some solution-related issues that we have not yet addressed. Perhaps the most obvious is that cells require nutrients if they are maintained in culture for more than a few hours. Another feature of intact cells is that they are surrounded by a cell membrane that is permeable to the flow of water and some, but not all, solutes. Intact cells therefore require an osmotically (defined below) balanced environment. Living cells also require a culture environment that is sterile, except for the cells of interest (see Chapter 12 for a discussion of aseptic technique).

This chapter discusses culture media for maintaining intact bacterial and mammalian cells. We first discuss osmotic considerations and isotonic solutions that are used for short-term (minutes to hours) maintenance of cells. The next section discusses media that support the nutritional and physical requirements for growth and reproduction of microbial cells, particularly bacteria. The final section in this chapter discusses the medium requirements for cells derived from animals, particularly those of mammalian origin. (Plant tissue culture is outside the scope of this discussion.)

II. ISOTONIC SOLUTIONS

Intact cells contain water, salts, and various biological molecules enclosed by a cell membrane. The presence of a cell membrane adds an important dimension to the preparation of laboratory solutions. The cell membrane is **semipermeable,** *which means that it allows water and some small molecules to flow through unimpeded, whereas other molecules are restricted.* Cells exist in an aqueous environment so water is constantly moving back and forth across the cell membrane. The net amount of water that flows into a cell must equal the net amount of water that leaves because if more water enters a cell than exits, then the cell swells and may burst. Conversely, if more water exits than enters, then the cell shrinks. A cell is said to be in **osmotic equilibrium** *when the net rates of water flow into and out of the cell are equal.*

The flow of water across the cell membrane is controlled by the concentrations of solutes inside and outside the cell. The net flow of water across the cell membrane is from the side that has the lower solute concentration to the side that has the higher solute concentration. If there is a higher solute concentration outside the cell than inside, then more water flows out of the cell than enters it. Conversely, when there is a lower solute concentration outside the cell, then more water flows in. **Osmosis** *is the net movement of water through a semipermeable membrane from a region of lesser solute concentration to a region of greater solute concentration.* **Osmotic pressure** *is the amount of pressure that needs to be exerted to halt the water's movement.*

In Figure 30.1a there is initially a lower solute concentration on the left side of the U-tube than on the right. The two sides are separated by a semipermeable membrane that allows water to flow freely, but is impermeable to the solute particles. On the average, more

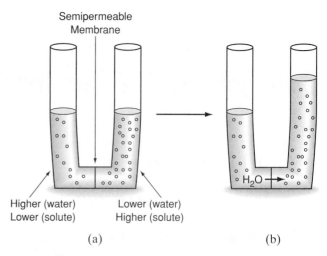

Figure 30.1. **Osmotic Pressure.** The net flow of water is from the left side, where the solute concentration is lower, to the right side where the solute concentration is higher. The water level rises on the right side and eventually equilibrates when the osmotic pressure equals the force of gravity.

water will flow from the left side into the right side than vice versa. The level of water will therefore rise on the right side, 30.1b. Eventually equilibrium will be achieved such that the gravitational pressure on the water equals the osmotic pressure.

Osmotic pressure is usually expressed in terms of the concentration of particles in a solution in units either of **milliosmolality** *or* **milliosmolarity** *where:*

milliosmolality *is milliosmoles/kg of solvent*
milliosmolarity *is milliosmoles/L of solution*

Consider these two concentration expressions:

Numerators: The units in both numerators are milliosmoles (mOsm). **Osmoles** *are determined by the number of moles of particles in a solution.* Accordingly, **milliosmoles** *are determined by the number of millimoles of particles in a solution.* Glucose, for example, does not dissociate when dissolved in water. Therefore, 1 millimole of glucose provides 1 millimole of solute particles in a solution.

For glucose: *1 millimole = 1 milliosmole*

In contrast, salts like NaCl and KCl dissociate in water. One mole of NaCl dissociates to form one mole of Na^+ ions and one mole of Cl^- ions to make two moles of osmotically active solute particles.

Therefore, for these salts: *1 millimole = 2 milliosmoles.*

Denominators: Observe that the units in the denominator are either *kg,* for osmolality, or *L* for osmolarity. The volume of a solution may vary if the temperature changes, but the weight of a solution does not vary with temperature. The measurement of osmotic pressure often involves changing the temperature of the solution. Therefore, osmolality, which has units of kg, is more commonly used than osmolarity, which has units of L.

The most common method of measuring the osmotic pressure of a solution is with a **freezing point depression osmometer.** This device is based on the fact that particles dissolved in water depress the temperature at which the water freezes; the more particles that are dissolved, the more the freezing point is depressed. A sample of the solution is frozen in the osmometer and the temperature at which it freezes is measured. This is compared to standards of known osmolality to determine the osmolality of the sample.

Cultured cells need to be surrounded by solutions that have about the same solute concentration as the interior of the cells to prevent them from swelling or shrinking. An **isotonic solution** *is one that has the same solute concentration as the interior of the cell,* see Figure 30.2. **Physiological solutions** *are isotonic solutions that support intact cells in the laboratory.* Salts are commonly used in laboratory solutions to maintain the proper osmotic equilibrium for cells. According to literature from ATCC (American Type Culture Collection, an organization that maintains and distributes cultured cells), the useful range of osmolality of cell culture media for mammalian cells is 260 to 320 mOsm/kg. Human blood plasma is about 290 mOsm/kg. Culture media used for growing human stem cells have a relatively tight tolerance for osmolality and should be 290 to 300 mOsmol/kg. Expressed in units of molarity, isotonic solutions for mammalian cells should be about 150 mM salt (assuming a salt that dissociates to form two ions in solution). Nonmammalian animal cell types, like those from insects, vary in their osmotic requirements from about 155 mOsm/kg, to as much as 375 mOsm/kg.

The optimal osmolality for bacterial culture is variable since bacteria live in many different environments. Bacteria from the ocean, for example, have evolved in a different osmotic environment that those in fresh water. Bacteria often live in changeable environments (e.g., soil) and therefore are often more tolerant of variation in osmolality than mammalian cells. Nonetheless, extremes of osmolality do affect bacteria. Bacterial cells (unlike mammalian cells) are surrounded by a rigid cell wall. If there is a higher solute concentration outside a bacterial cell than inside, then water flows out of the cell and the cell membrane and contents shrink away from the cell wall. This condition inhibits bacterial growth and reproduction. For this reason, since ancient times, foods have been preserved by the addition of high levels of salts and sugars.

The simplest solution commonly used to maintain osmotic equilibrium is **standard (physiological) saline,** *which is 0.9% NaCl.* Most physiological solutions also have a buffering component in addition to salt. Table 30.1 shows examples of **balanced salt solutions** and **buffered salt solutions** *that are intended to maintain cells for short periods* (minutes to hours).

For longer culture, cells require a medium that is not only osmotically balanced, but also contains nutrients and various other substances to support growth and reproduction. Cell culture media are formulated to provide the proper osmotic environment for cells while also providing these nutrients and growth promoting constituents.

Table 30.1. EXAMPLES OF BALANCED SALT SOLUTIONS

Physiological Saline (mammalian cells)	
NaCl	0.9% w/v
Phosphate Buffered Saline	
NaCl	7.20 g/L
Na_2HPO_4 (anhydrous)	1.48 g/L
KH_2PO_4	0.43 g/L
Tris Buffered Saline	
NaCl	0.9%
Tris pH 7.2	10–50 mM
Ringer's Solution (mammalian cells)	
$CaCl_2$	0.25 g/L
KCl	0.42 g/L
NaCl	9.00 g/L
Dulbecco's Phosphate Buffered Saline (DPBS)	
$CaCl_2$	0.10 g/L
$MgCl_2 \bullet 6H_2O$	0.10 g/L
KCl	0.20 g/L
KH_2PO_4 (anhydrous)	0.20 g/L
NaCl	8.00 g/L
Na_2HPO_4 (anhydrous)	1.150 g/L

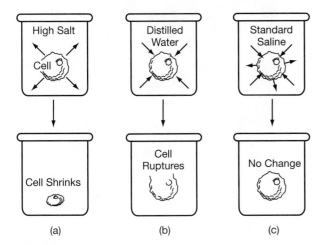

Figure 30.2. The Flow of Water across the Cell Membrane. a. If there is a higher solute concentration outside the cell than inside, then water flows out of the cell and the cell shrinks. **b.** If there is a lower solute concentration outside the cell, water flows into the cell and it may burst. **c.** An isotonic solution is in equilibrium with the interior of the cell.

III. BACTERIAL CULTURE MEDIA

A. Background

Bacterial growth media are aqueous mixtures of nutrients used to support the growth and reproduction of bacteria. The use of bacterial growth media is traced back to the 1800s when scientists, including Louis Pasteur and Robert Koch, conducted pioneering studies of the role of bacteria in causing disease. Early microbiologists grew bacteria on gelatin substrates in crude broths with nutrients derived from extracts of plant and animal tissues (e.g., beef brain and heart). Gelatin was not an ideal substrate because many bacteria digested the gelatin and it liquefied above room temperature. One of the advances that emerged from Koch's laboratory was the discovery that agar, a substance derived from seaweed, could be used to make a solid substrate on which bacteria readily grow. The use of agar was suggested by Angelina Hesse, the wife of a physician who was working in Robert Koch's laboratory. Hesse was assisting her husband in the laboratory as an illustrator and laboratory technician. Hesse was aware of the use of agar in the East Indies as a cooking ingredient to help harden foods. She suggested using agar to harden the culture media in the laboratory—an idea that Koch and his colleagues tried with great success. The new agar medium proved to be far superior to extracts mixed with gelatin and enabled Koch and other scientists to isolate colonies of specific disease causing organisms.

In the early 1900s, microbiologists discovered that the components of bacterial culture media can be prepared in large batches which are then dehydrated, a form in which they are conveniently stored until needed. Commercial vendors began to manufacture dehydrated ingredients and combinations of ingredients for microbial media. Microbiologists learned to combine these readily available ingredients in different ways to best promote the growth of certain types of microorganisms, eventually creating hundreds of different media formulations optimized for different applications. Preparing culture media for bacteria today is often a straightforward procedure in which specified amounts of purchased, dehydrated components are dissolved in water, according to a recipe tailored for a specific bacterium and application. The resulting mixture is sterilized, usually by autoclaving. Compendia and handbooks are available with standard formulations and methods for food microbiology, medical microbiology, and environmental microbiology (examples of compendia are provided in the Unit VII bibliography). Although the culture media used today are easier to prepare and more versatile than they were in the nineteenth century, extracts from plant and animal tissue and agar from seaweed are still staples of the microbiologist's pantry.

B. Types of Bacterial Growth Media

i. LIQUID VERSUS SOLID

Bacterial culture media can be prepared to be liquid, solid, or semisolid. **Liquid culture media,** called **bacterial broths,** *are aqueous based mixtures of nutrients that do not contain a hardening agent.* Bacteria grow suspended throughout a liquid broth. Broths are generally used for propagating large numbers of bacteria, either in the laboratory or for industrial-scale fermentation.

Solid and **semisolid media** *are aqueous based mixtures of nutrients that contain agar as a hardening agent to provide a solid substrate for bacterial growth.* Bacteria can be grown as distinct colonies on top of solid media, see Figure 30.3, or sometimes within the medium. Solid media are used to observe bacterial colony appearance, for isolating colonies derived from a single organism, and for storage. Semisolid media may be used for determining bacterial motility and for promoting growth of bacteria in the absence of oxygen.

ii. CHEMICALLY DEFINED VERSUS CHEMICALLY UNDEFINED

Both liquid and solid culture media may be chemically defined or undefined. A **chemically defined medium** *is one in which all the ingredients and their quantities are known.* These media typically contain purified organic salts and simple organic compounds, such as glucose and amino acids. Defined media may also include such substances as trace elements, nucleotides, and vitamins.

Minimal media *are a type of defined medium that contain only the minimum nutrients required for a par-*

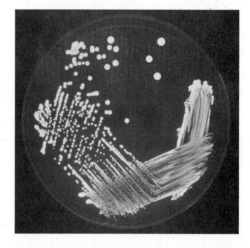

Figure 30.3. *Legionella sp.* **Colonies Growing on the Surface of an Agar Plate.** Nutrient agar provides nutrients and a surface for the propagation and isolation of individual colonies. Each round white object at the top of this agar plate is a colony consisting of millions of bacterial cells all probably derived from a single parent cell. *Legionella* is a type of bacterium that caused an outbreak of pneumonia during the 1976 convention of the American Legion, sickening at least 180 people and killing 29. (Photo Courtesy of James Gathany, Centers for Disease Control and Prevention Public Health Image Library.)

ticular microbe to survive and reproduce. Table 30.2 is an example of a formulation for a defined, minimal medium that is used to culture naturally occurring *Escherichia coli. E. coli* is a common bacterium normally found in the intestine of humans and other animals that has the ability to synthesize all its nutrients from relatively simple precursors.

A **complex or undefined medium** *is one in which the exact chemical constitution of the medium is not known.* Complex media are usually derived from materials of biological origin such as blood, milk, yeast, beef, or vegetables. These sources supply a mixture of amino acids and low-molecular-weight peptides, carbohydrates, vitamins, minerals, and trace metals.

Defined media have the advantage that they are consistent from batch to batch and do not contribute unknown impurities to experiments or production systems. Defined media, however, are often expensive because their components are purified. Complex media have the advantages that they usually provide all the growth factors that are required by bacteria, they tend to be less expensive than defined media, and they are broadly useful for culturing a variety of bacteria. Most medically important bacteria are cultured in complex media containing blood, serum, and tissue extracts. The major disadvantages to complex media are that they introduce lot to lot variability and unknown components into the system.

Table 30.2. MINIMAL AGAR, DAVIS

Ingredient	Amount (per liter)	Function of Component
Agar	15.0 g	Harden medium
K_2HPO_4	7.0 g	Buffer pH; provide P and K
KH_2PO_4	2.0 g	Buffer pH; provide P and K
$(NH_4)_2SO_4$	1.0 g	Buffer pH; provide N and P
Glucose	1.0 g	C and energy source
Sodium citrate	0.5 g	Buffer pH
$MgSO_4 \cdot 7H_2O$	0.1 g	Provide S and Mg^{++}

pH will be 7.0 ± 0.2 at 25°C

Preparation:

1. Add components to cold purified water.
2. Bring to volume of 1 L.
3. Mix thoroughly.
4. Gently heat and bring to boiling.
5. Distribute into tubes or flasks.
6. Autoclave 15 min at 15 PSI pressure at 121°C.
7. Pour into sterile Petri dishes or leave in tubes.

Source: Atlas, Ronald M. Handbook of Microbiological Media 3rd ed. Boca Raton, FL: CRC Press, 2004.

iii. SELECTIVE AND DIFFERENTIAL MEDIA

Selective media *inhibit the reproduction of unwanted organisms and/or encourage the reproduction of specific organisms.* For example, if a bacterium is resistant to a certain antibiotic, then that antibiotic can be added to the medium in order to prevent other cells, which are not resistant, from growing. Selective media are useful for the isolation of specific microorganisms from mixed populations.

Differential media *are designed to reveal differences among microorganisms or groups of microorganisms that are growing on the same medium.* Differential media usually contain a chemical that is utilized or altered by some microorganisms and not by others. When particular bacteria grow on differential medium, they cause a visible change, such as the production of a colored product. While selective media are used to allow the growth of only selected microorganisms, differential media allow the growth of multiple types, but cause them to have distinguishing characteristics. A growth medium can be both selective and differential. Selective and differential media may be either undefined or defined.

C. Ingredients

i. AGAR

Agar, *which is derived from seaweed, is the most common hardening agent for bacterial media.* Bacteria do not digest agar and it is relatively inexpensive. Agar does not melt until its temperature is about 84°C, but once melted it does not solidify again until its temperature drops to about 38°C. This is important because agar remains solid at most practical temperatures and, after melting, it can be poured into dishes or tubes at a convenient temperature. The typical concentration of agar is about 15.0 g/L (1–2%). Lower concentrations of 7.5 to 10.0 g/L are used in soft or semisolid media.

ii. FORMULATIONS REFLECT THE COMPOSITION OF BACTERIA

Media formulations must supply the nutrients that bacteria need to grow and reproduce. Bacteria are about 70% water, so purified water (see Chapter 28) is the primary ingredient in culture media. On a dry weight basis, bacteria are about half carbon, with lesser amounts of oxygen, nitrogen, hydrogen, and phosphorus, and 1% or less of sulfur, potassium, sodium, calcium, magnesium, chloride, and iron, see Table 30.3 on p. 568. Growth media for bacteria provide nutrients in roughly the same percentages. Growth media additionally provide an energy source to the cells (unless they are able to use photosynthesis).

Once nutrients are inside a cell, they are chemically modified by metabolic processes to meet the nutritional needs of the cell. Some, but not all, bacteria are able to survive on a simple growth medium because they are able to synthesize everything they need from one or two carbon sources, a few salts, and a nitrogen source.

Table 30.3. NUTRIENT REQUIREMENTS OF BACTERIA

Element	% of Dry Weight	Function	Chemical Form Commonly Added to Culture Media
Carbon	≈ 50	Main building material of cells	**Organic;** simple sugars (e.g., glucose, acetate, or pyruvate; extracts such as peptone, tryptone, yeast extract, etc.) **Inorganic;** carbon dioxide (CO_2) or hydrogen carbonate salts (HCO_3^-)*
Oxygen	≈ 20	Constituent of water; electron acceptor in aerobic respiration	
Nitrogen	≈ 14	Constituent of amino acids, nucleic acids, and coenzymes	**Organic;** amino acids, nitrogenous bases, peptones **Inorganic;** NH_4Cl, $(NH_4)_2SO_4$, KNO_3, and for dinitrogen fixers, N_2
Hydrogen	≈ 8	Constituent of organic compounds and water	Present in most chemicals added to medium
Phosphorus	≈ 3	Constituent of nucleic acids and phospholipids	KH_2PO_4, Na_2HPO_4*
Sulfur	≈ 1	Constituent of certain amino acids and several coenzymes	Na_2SO_4, H_2S
Potassium	≈ 1	Main cellular inorganic cation and cofactor for certain enzymes	KCl, K_2HPO_4*
Magnesium	≈ 0.5	Inorganic cellular cation, cofactor for certain enzymes	$MgCl_2$, $MgSO_4$
Calcium	≈ 0.5	Inorganic cellular cation, cofactor for certain enzymes	$CaCl_2$, $Ca(HCO_3)_2$*
Iron	≈ 0.2	Component of cytochromes and certain nonheme iron-proteins and a cofactor for certain enzymes	$FeCl_3$, $Fe(NH_4)(SO_4)_2$, Fe-chelates
Trace elements		Required in trace amounts for a variety of functions, for example, as cofactors	$CoCl_2$, $ZnCl_2$, Na_2MoO_4, $CuCl_2$, $MnSO_4$, $NiCl_2$, Na_2SeO_4, Na_2WO_4, Na_2VO_4
Organic growth factors			Vitamins, amino acids, purines, pyrimidines

*also act as buffers

Sources: "Nutrition and Growth of Bacteria" in *Todar's On-Line Textbook of Bacteriology.* http://www.textbookofbacteriology.net/. Sigma-Aldrich "Microbiology Technical Reference." http://www.sigmaaldrich.com/Area_of_Interest/Analytical Chromatography/Microbiology.html.

iii. SOURCES OF NUTRIENTS

Meat extracts have long been used in complex media to supply nutrients including carbon, nitrogen, vitamins, and trace minerals. Yeast extracts are another common constituent of complex media. **Yeast extract** *is the water-soluble portion of yeast cells that have been allowed to die so that the yeasts' digestive enzymes break down their proteins into simpler compounds while preserving the vitamins from the yeast.* Yeast extract is a rich source of amino acids and vitamins. It is typically used at a concentration of 0.3 to 0.5% and has a pH of about 6.6 in media.

Peptones *are hydrolyzed (cleaved) proteins formed by enzymatic digestion or acid hydrolysis of natural*

substances including milk, meats, and vegetables. Many complex media contain peptones as the main source of nitrogen-containing compounds (amino acids, peptides, and proteins). Peptones also provide vitamins, minerals, and carbohydrates. Casein, the principal protein in milk, is a common protein substrate for forming peptones, and the term *tryptone* has traditionally meant a digest of casein. Many microbiologists are now trying to avoid animal-derived substances in growth media, so vegetable sources, such as soybean meal, may be substituted to make "tryptones." Soytone, Bacto Peptone, and Bacto Tryptone are examples of the more than 50 commercial peptones available. The enzymes used in the past for digestion of the substrate were derived from animal sources, particularly pork pancreas. Digestive enzymes for making peptones now are often manufactured by genetically modified organisms.

Defined media do not contain such substances as meat extracts or peptones. Nitrogen is commonly added to defined media in chemical form as ammonium, nitrate, or as amino acids. Glucose is a common carbon and energy source found in both undefined and defined media. Other carbon/energy sources in both types of medium include: acetate, glycerol, certain lipids, and proteins. Sodium and potassium phosphates are used in defined and undefined media to provide phosphorus and to serve as buffers. Defined media contain sulfur in the form of inorganic salts of sulfate, hydrogen sulfide, sulfur granules, or thiosulfate, or in the form of organic compounds like cysteine and methionine (both of which are amino acids).

iv. TRACE ELEMENTS

Trace elements are those substances required by cells in small amounts. Magnesium, calcium, and iron may be placed in this category, although these elements are required in higher amounts than substances such as cobalt. Trace elements are required for the activity of a number of enzymes. Iron is an essential component of cytochromes, which are molecules required for producing energy in cells. Defined media include trace elements supplied as mineral salts (e.g., $CaCl_2$, $CuSO_4$). Complex media may rely on extracts and peptones to supply trace elements, but complex media may also be supplemented with mineral salts. Some trace elements are required in such small amounts that they are present naturally in sufficient amounts in agar and other media components.

v. GROWTH FACTORS

Naturally occurring *E. coli* can grow on simple, minimal media because they have biosynthetic pathways to convert glucose, nitrogen, sulfur, and other simple building blocks into all their required cellular molecules (e.g., proteins, vitamins, and nucleic acids). Many bacteria, however, cannot synthesize all their own molecular components. Those components that cannot be synthesized by bacterial cells and must be obtained from the environment are broadly termed *growth factors*. Vitamins, amino acids, purines, and pyrimidines (required for synthesis of DNA and RNA) frequently fall into this category. All these components are added to defined media in purified forms (hence adding expense). Growth factors in undefined media may be supplied by peptones and various extracts and/or may be added in purified form directly to the medium.

vi. SUPPLEMENTS

Bacterial culture media may contain other substances in addition to those listed above that are loosely clas-

sified as *supplements*. Examples include: selective dyes, pH indicators, antibiotics, and chelating agents that bind to metals.

Molecular biologists frequently supplement their bacterial culture media with antibiotics. This is because antibiotic resistance is often used to select for those bacteria that have been transformed with a gene of interest. Molecular biologists use plasmids to carry a gene of interest into bacteria. The plasmids are engineered to carry both the gene of interest and also a gene that makes bacteria resistant to a particular antibiotic. Bacteria that are transformed by taking up the plasmid therefore have an antibiotic resistance gene and will grow on a culture medium that contains the antibiotic. Nontransformed bacteria that did not take up the plasmid will not grow when exposed to the antibiotic. Bacterial strains that carry plasmids with antibiotic selection markers should be cultured in the presence of the selective agent.

Note that antibiotics are often sensitive to light and should be protected from it. They are also sensitive to heat and therefore are sterilized by filtration (see Chapter 31). Agar-containing media should be cooled to 50°C or lower before adding antibiotics. ATCC recommends that sterile stock solutions of antibiotics be stored in a –20°C freezer in the dark. ATCC further recommends that bacterial plates containing antibiotics should be stored no longer than three months at 4°C due to degradation of antibiotics.

D. Preparing Bacterial Culture Media

Like other biological solutions, bacterial growth media are made by combining the proper amounts of ingredients in the proper final volume of water so that each ingredient is present at the correct final concentration. The ingredients for bacterial media are readily available from commercial suppliers. Box 30.1 describes some common practices for making culture media. Box 30.2 and Figure 30.4 show common practices for pouring agar plates. Refer also to Chapter 12 for information about aseptic technique.

E. LB Agar/Broth

LB broth/agar is a complex bacterial medium commonly used by molecular biologists for culturing *Escherichia coli* in genetic transformation procedures. Lysogeny broth (LB) was created in the 1950s by Giuseppe Bertani, who was studying viruses that infect bacteria. (Lysogeny is a condition in which the viruses survive within a host bacterium but do not reproduce and burst the bacterial cell.)

Naturally occurring *E. coli* can grow on minimal medium, but the strains used for recombinant DNA work have been mutated so that they require a complex growth medium and will not grow readily outside

---

Box 30.1.   COMMON PROCEDURES AND TIPS FOR BACTERIAL CULTURE MEDIA

Storage of Dehydrated Media

Dehydrated media are hygroscopic (absorb water) and are sensitive to heat, light, and extreme fluctuations in temperature.

1. *Store dehydrated media according to the manufacturer's directions, usually below 25°C in a dry area, away from direct sunlight and heat sources.* Some media are stored refrigerated.

2. *Record the date of receipt of each container of media on its label and the date that the container is first opened.*

3. *Check the expiration date; some media have longer shelf lives than others. Discard after expiration date.*

4. *Use media in the order of receipt; finish a bottle before opening a fresh one.*

5. *After use, tightly close the bottle. Whenever possible avoid repeated opening and closing of media bottles since this will cause the media to degrade more quickly.*

6. *Discard medium if it is not free-flowing, if its color has changed, or if it appears abnormal.*

Preparation of Media

1. *Water for reconstituting dehydrated microbial media should be freshly prepared by distillation, deionization, or reverse osmosis.*

2. *Weigh out the required amount of dehydrated medium, or the various components of the medium; place in a clean, dry flask that is 2 to 3 times larger than the final volume desired.* Avoid inhaling the powder or prolonged skin contact.

3. *Add half the appropriate amount of water to the flask, swirl to mix. Pour the rest of the water down the sides of the flask to wash any powder adhering to the sides of the flask.* (Dried powder that adheres above the liquid level may not be sterilized during autoclaving and may cause contamination.)

    a. *Most broths are clear at this point and do not require boiling to dissolve the components.*

    b. *If required by a procedure or manufacturer's directions, bring the mixture to the proper pH. Some procedures call for adjusting the pH before bringing the mixture to final volume, while others do so afterward.*

4. **Agar: *When agar is used, most manufacturers recommend melting the agar before autoclaving using a microwave, hot plate, boiling water, or steam bath. Heat the medium to boiling with frequent stirring or swirling to prevent overheating. Caution is required as agar media may boil unexpectedly.*** The purpose of this step is to ensure that the powdered agar completely dissolves and is uniformly distributed throughout the medium. This reduces the risk of contamination that may occur if any dry powder remains above the level of the liquid.

    a. Some procedures do not call for melting the agar before autoclaving since agar will dissolve in the auto-clave. Be sure the agar is completely dissolved and uniformly mixed after autoclaving if following this practice.

    b. Culture media with a pH below 6.0 will hydrolyze agar and reduce its gelling properties. If using an acidic medium, avoid re-melting the agar once it has hardened.

    c. Procedures may call for agar to be autoclaved separately from the other components, especially when making minimal media, to avoid formation of a precipitate. In these procedures, the other ingredients, such as salts, are prepared in a concentrated form and autoclaved. Glucose is prepared separately as a concentrate and is sterilized. Magnesium sulfate also is often prepared separately from other components. After sterilization, the agar, salts, and glucose (and other heat labile ingredients, such as antibiotics) are combined aseptically.

5. *When glucose (also called dextrose) is required, add it to water slowly with mixing, otherwise the glucose will clump and resist dissolving.*

    a. Glucose will turn brown if autoclaved with other ingredients so some procedures call for glucose to be filter sterilized.

Box 30.1. *(continued)*

   b. Some procedures call for glucose to be autoclaved separately as a concentrated stock solution and added to the rest of the ingredients after autoclaving.

   c. Some sources recommend that glucose not be autoclaved at a pH above 7.0 and that it not be autoclaved in the presence of amino acids.

6. *Autoclave the solution(s). (Refer also to Chapter 28 for more information on autoclaving.)*

   a. *Loosely cap flasks or tubes with cotton plugs, plastic foam plugs, plastic, or metal caps.*

   b. *Place tubes in racks.*

   c. *Be sure flasks are less than two-thirds full.*

   d. *Sterilize for 15 to 20 minutes at 15 PSI, 121°C for quantities of liquid media up to one liter. If larger volumes are sterilized in one container, especially if the medium was not hot when placed in the autoclave, then a longer period should be employed.* (Recommendations for sterilization times, temperatures, and pressures are typically included in media recipes and/or in information from the manufacturers of dehydrated media.)

   e. *Avoid prolonged heating or oversterilization, which, may adversely affect some medium components and may cause agar to begin to precipitate.* Repeated melting of solidified agar or long holding of melted agar at high temperature may also cause a precipitate to form.

   f. *Use a slow exhaust setting on the autoclave to release the pressure slowly at the end of sterilization, otherwise, the medium will boil over, blowing the plugs from the tubes or flasks.* The pressure, however, should drop rapidly enough to avoid excessive exposure to heat after the sterilization period. Around 10 minutes is recommended as the time to reach atmospheric pressure.

   g. *Remove media from autoclave as soon as sterilization is complete.*

   h. *Tighten caps once the medium is cool (< 40°C).*

7. *Add supplemental, heat-sensitive ingredients just before use.*

   a. *If agar is present, cool the medium to 50 to 55°C before adding other components.* (A flask containing liquid at 50°C feels hot but can be held continuously with bare hands.) Broths can be cooled to room temperature before adding supplements.

   b. *Sterilize antibiotics by filtration through a filter unit with a pore size of 0.1 or 0.2 μm. If antibiotics are not used right after preparation, distribute them into aliquots and store in the dark at –20°C.* (See also Practice Problem 4.)

   c. *Aseptically add supplements to the medium.* Swirl to mix; avoid bubbles.

8. *Aseptically dispense sterile medium into sterile tubes, flasks, or petri dishes.*

9. *Store prepared medium at the temperature indicated in product description.* The length of time that prepared media can be stored varies depending on the type. Some manufacturers recommend storing simple broths and agars no longer than 6 months. Selective media may degrade more quickly.

10. *For quality control, test the medium.*

   a. *Test a portion of prepared tubes or plates by incubating them at an appropriate temperature (e.g., 30 to 37°C, or 20 to 25°C) for 2 to 5 days to check for sterility; do not reuse these tubes or plates.*

   b. *Test the prepared medium for growth performance by inoculating it with a stock culture.*

   c. *Periodically test stored medium to see that it continues to support growth as it did when freshly prepared and that it is not contaminated.*

   d. *Visually inspect prepared media before use.* Determine that color, clarity, pH, and other characteristics of the culture medium are typical as listed in the product description.

Main Sources: Difco Laboratories. *The Difco Manual.* 10th ed. 1984.
Invitrogen Corporation. "Media Preparation and Bacteriological Tools."
https://catalog.invitrogen.com/index.cfm?fuseaction=iProtocol.home.
Oxoid Ltd. a division of Thermo Fisher Scientific. "FAQ's." 2006.
http://www.oxoid.com/uk/blue/techsupport/its.asp?itsp=faq+c=UK+lang=EN.

Box 30.2.   Pouring Nutrient Agar Plates

---

1. *Prepare and autoclave nutrient agar; bring to a temperature of 50 to 55°C.*
2. *Label the bottom of sterile 100 mm plastic culture plates.*
3. *Spread the plates out on lab bench; turn on Bunsen burner.*
4. *Open the flask containing the nutrient agar and pass the top through the Bunsen burner flame.*
5. *Pour plates:*
   a. Carefully lift the lid of each culture plate; do not set the lid on the lab bench.
   b. Quickly pour in agar until the bottom plate is about 1/3 full (1/4 inch of agar). Immediately replace the lid after pouring the medium.
   c. Repeat for each plate, occasionally flaming the top of the flask.
   d. Hold the flask slightly tilted to prevent airborne contaminants from falling into the flask.
   e. Work quickly, but carefully, so that the agar does not harden before all the plates are poured.
6. *Let plates cool with the lids on to solidify agar.* Freshly poured plates are wet, which can allow bacteria to spread where they should not and can interfere with other applications. Many sources advocate allowing the plates to dry inverted at room temperature for 12 to 48 hours. They can also be dried in an inverted position in a 37°C incubator for 0.5 to 3 hours.
7. *For storage, return dried plates to their plastic sleeves and turn them upside down to prevent condensate from pooling on the surface. Store at 4°C.*
   a. Plates that were stored in the cold should be warmed to room temperature for a few hours before use to avoid condensate.
   b. Store plates that contain antibiotics (which are light sensitive) in a dark room or wrapped in aluminum foil at 4°C for not more than three months. Plates should be inverted.

---

the laboratory. A complex medium is also useful since cells grow more slowly and/or produce less DNA and protein in minimal medium than they do in complex medium. LB is ideal for these strains.

There are several different versions of LB broth that vary in the amount of sodium chloride they contain, thus providing different osmotic conditions. The concentration of salt can affect the yield of plasmid obtained when the culture is being used for the purpose

of isolating plasmids from transformed bacteria. The low-salt formulations, Lennox and Luria, are useful for cultures requiring salt-sensitive antibiotics.

- LB-Miller (10 g/L NaCl)
- LB-Lennox (5 g/L NaCl)
- LB-Luria (0.5 g/L NaCl)

Box 30.3 provides an example of an LB formulation that is used for making agar plates.

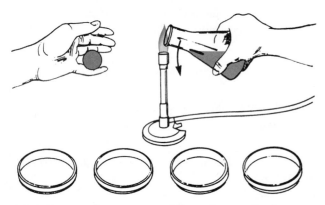

(a)   Remove the stopper, and flame the mouth of the flask.

(b)   Remove the cover from one dish, and pour nutrient agar into the dish bottom.

Figure 30.4.   Pouring Agar Plates. a. The cap is removed from the nutrient agar flask and the mouth is passed briefly across the flame. If all of the nutrient agar will be poured into plates, the cap can be set aside. If the flask will need to be closed again, then the cap is held in the left hand. **b.** The cover is removed from each dish, one at a time, and agar is poured into the plate. Observe how the principles of aseptic technique are applied. The flask is flamed to create an updraft away from the lip. The plates are kept closed until needed, and the lid of the plates is held above the open plate to protect it from the air. The cap is not placed down on the lab bench, because the bench is assumed to be contaminated.

Box 30.3. *LB AGAR*

Ingredient	Amount	Function of Component
Agar	15.0 g	Hardens medium
Tryptone	10.0 g	Supplies varied nutrients, including nitrogen, sulfur, carbon, vitamins
Yeast extract	5.0 g	Supplies varied nutrients, including nitrogen, sulfur, and carbon, also vitamins and trace elements
Sodium chloride	10.0 g	Supplies sodium ions for membrane transport and osmotic balance
Water	Bring to 1 L	

Directions:

1. Mix ingredients in purified water.

2. Some procedures call for adjusting the pH of LB to 7.5 or 8 with sodium hydroxide. A disadvantage to using sodium hydroxide for this purpose is that the pH will not be buffered in the medium, which means that metabolic byproducts from the bacteria will change the medium's pH during growth. A less-commonly used option is to adjust the pH with 5 to 10 mM Tris buffer that is at the desired pH.

3. Sterilize mixture by autoclaving at 121°C for 15 minutes.

4. Cool medium to about 55°C before adding antibiotics and before pouring plates. A water bath can be used to bring the medium to the correct temperature. (The bottom of the flask will be warm to the touch but will not burn the hand at this temperature.)

5. Add antibiotics, if required, from sterile stock solutions. (See also Practice Problem 4.)

6. Pour plates as directed in Box 30.2 on p. 572.

IV. MEDIA FOR CULTURED MAMMALIAN CELLS

A. Introduction

Next we consider culture media for cells from multicellular organisms, focusing in particular on mammalian cells. Mammalian cells—derived from such sources as monkeys, mice, hamsters, and humans—are very important in biotechnology, both in research and in production.

We briefly begin with some terminology relating to mammalian cell culture. **Primary cultures** *are derived directly from tissue removed from an animal.* The tissue is cut into small fragments and placed in sterile culture medium in a culture plate. The tissue may also be treated with enzymes to break it apart into individual cells.

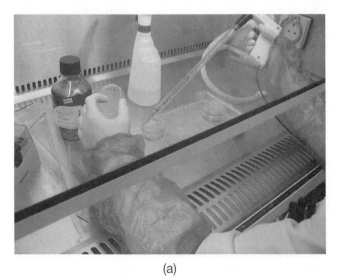

(a)

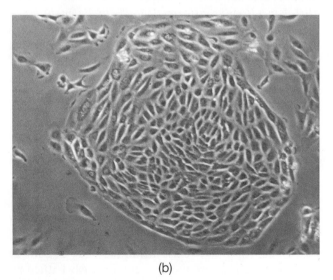

(b)

Figure 30.5. Subculturing Cells. a. A technologist is shown pipetting fresh culture medium from the bottle on the left into small culture plates containing mammalian cells. The same guiding principles of aseptic technique apply as were previously discussed (see Chapter 12). The technologist is using aseptic procedures inside a laminar flow hood that helps protect the cultures from airborne contaminants. The technologist is wearing disposable sleeve protectors and gloves to further protect the cells from contaminants on his skin and lab coat. **b.** Cultured cells as seen under a microscope. This field of view shows cells that are adhered to the bottom surface of a culture plate. The nuclei are visible as round structures located in the middle region of each cell. Photo **b** from L. D. Dunfield, T. G. Shepherd, and M. W. Nachtigal. Primary culture and mRNA analysis of human ovarian cells. *Biol. Proced. Online* 2002;4(1): 55–61; Fig. 1a. Image provided by Mark W. Nachtigal.

Certain types of cultured cells grow **in suspension** *as single cells or as small clumps of cells floating in their culture medium.* Cells derived from blood, for example, tend to grow in suspension. Alternatively, cultured mammalian cells may grow as a **monolayer of adherent cells** *that are attached to the surface of their vessel.* (Monolayer means cells do not grow on top of one another.) Primary cultures derived from solid tissue typically grow as a monolayer.

As cultured cells grow and divide, the nutrients in their medium are depleted and cellular metabolic waste products accumulate. The spent medium is therefore periodically removed and the cells are fed with fresh medium. If treated properly, cells will divide and multiply in culture and will eventually run out of space in their culture vessel. They must then be **subcultured** or **passaged,** *that is, divided by dilution and transferred into more plates with fresh culture medium*, see Figure 30.5 on p. 573. Suspended cells can simply be diluted with culture medium and placed into more culture vessels. Adherent cells must be gently removed from the surface of their culture plate and then diluted and placed in new plates or other larger culture vessels (e.g., flasks or bottles). Cells that are derived from tissue can be subcultured a number of times, but eventually they age and stop dividing.

Some cells can be **transformed;** *these cells undergo genetic changes that change their growth properties.* An important characteristic of transformed cells is that they acquire the ability to grow and divide indefinitely in culture. Transformation occasionally occurs spontaneously but is more commonly induced by exposing cells to certain chemicals, viruses, or radiation. A **continuous cell line** *consists of cells from a single source that have been transformed and can be maintained indefinitely in culture.* Continuous cell lines are commonly used in research and production, even though transformed cells have lost many of the growth controls and properties of normal cells. Chinese hamster ovary cells (CHO)—the most common type of cells used for producing biopharmaceutical products (see Chapter 3)—are a transformed continuous cell line derived from the ovary cells of a hamster.

The cell culture medium used throughout the cell culture process must be carefully formulated and prepared. Some of the same ideas discussed for bacterial media also apply to mammalian culture media, but mammalian cells generally have more complex and stringent growth requirements than bacteria. Mammalian cells normally exist inside the body in contact with other cells where they are continuously bathed by blood with a complex, high-protein nutrient mixture. It is challenging to simulate this natural environment in culture vessels or bioreactors. The water used for mammalian cell culture must be of very high purity; the cells' gaseous environment must be closely controlled; pH and osmolality must be maintained in narrow ranges; the culture medium must provide a complex mix of nutrients. Mammalian cell culture is particularly threatened by contamination because common contaminating organisms, such as bacteria and molds, multiply much more rapidly than mammalian cells and can quickly take over a cell culture. For all these reasons, performing mammalian cell culture requires training and practice.

Different cell types have different requirements for their growth media. The early cell culture technologists had to develop formulations to support the growth of their cells, but now there are existing recipes for media for routine culture of many cell types. Organizations that supply cells (e.g., ATCC) provide information on the cell culture medium recommended for culturing those cells. It is also possible to find information on cell culture media for specific cell lines in the scientific literature describing the origin of those cell lines.

Cell culture technologists in some facilities mix culture media themselves from the individual components, but the complexity of the formulations makes this practice relatively unusual in smaller-scale operations. High-quality, premixed mammalian cell culture media are conveniently (if expensively) available from commercial suppliers.

Biotechnology companies may perform cell culture at a very large scale using large volumes of cells to manufacture a product of interest. The use of cultured cells for production introduces some special concerns. The culture medium must not only provide for healthy cell growth and reproduction, but must also provide the building blocks (e.g., amino acids and sugars) for synthesis of the product. The particular composition of the medium will therefore have a major impact on the yield of product made by those cells. Moreover, the constituents of the cell culture medium must be removed from the final product; this affects the downstream purification process. The product may be used therapeutically in humans; this means it must be free of harmful contaminants that might be introduced from the medium. For all these reasons, finding a suitable culture medium for production is an important part of the development of a product that is manufactured by cells. Research and development scientists in industrial settings commonly perform extensive experimentation to optimize the culture medium formulation so that it provides the best yield of purified product most cost effectively.

This chapter focuses on the basic requirements of mammalian cell culture medium. We discuss common cell culture media formulations, their general characteristics, and usual preparation from commercially available products. Preparation of medium from its individual constituents, and optimization of media for particular applications are outside the scope of this discussion. For more information on cell culture and cell culture media, consult this unit's bibliography.

B. Basal Cell Culture Media

i. A FAMILY OF BASAL MEDIA

Mammalian cell culture is a more recent technology than bacterial culture. Many of the cell culture media and methods used today were developed in the 1950s and 1960s by pioneers, such as Harry Eagle, who systematically investigated the nutritional requirements of cultured cells. Eagle's studies led to the formulation of a defined **basal liquid medium** *that contains a mixture of nutrients dissolved in a buffered, isotonic salt solution.*

Eagle's original basal medium formulation (BME) was designed to grow mouse fibroblasts and HeLa cells (a cell line from a human cervical carcinoma). This formulation was later modified in various ways to suit a wide range of cells, thus creating a "family" of basal media.

Some common basal media, such as minimal essential medium (MEM) and Dulbecco's modification of Eagle's medium (DME), are relatively simple. Others are enriched with more constituents, such as additional vitamins and amino acids, trace elements, lipids,

and nucleic acids. Ham's F12 is an example of a richer medium. These richer media were developed to support specific cell types and also to provide the base for serum-free formulations (discussed in more detail below). The richer basal media are often termed *complex* because they have additional ingredients. Note that the word *complex* has a different meaning in bacterial culture, where it means that the exact components of the medium are not known. Some examples of common basal media and their characteristics are shown in Table 30.4.

Eagle's basal medium and others like it are incomplete; they are designed to be supplemented with animal blood serum. **Blood serum** *is the liquid component of blood from which blood cells and most clotting factors have been removed.* Serum provides a rich and complex source of proteins, polypeptides, amino acids, growth factors, lipids, carbohydrates, salts, hormones, and vitamins. This means that, when complete, these traditional (classic) mammalian cell culture media are undefined. Serum is discussed in more detail in a later section of this chapter.

Table 30.4. *EXAMPLES OF BASAL CELL CULTURE MEDIA*

Name	Features
Eagle's Minimal Essential Medium (MEM)	A commonly used modification of Eagle's original formulation. Contains a higher concentration of amino acids than BME; supports cell growth for several days.
Dulbecco's Modified Eagle's Medium (DME)	A commonly used modification of Eagle's original formulation. Contains four times the concentration of vitamins and twice the concentration of amino acids as BME. Contains twice as much HCO_3^- and CO_2 as BME to improve buffering capacity. Contains iron. Energy sources are optimized for protein production and nucleic acid metabolism. Originally contained 1 g/L of glucose. Available now with high glucose, (4.5 mg/L) and low glucose (1 g/L).
RPMI-1640 (Roswell Park Memorial Institute)	An enriched formulation originally derived for human leukemia cells; now used for many mammalian cell lines.
McCoy's 5A	Developed in 1959 by McCoy and coworkers for cultivation of cells from a liver tumor. Subsequently modified to create a medium that supports the growth of cells derived from a wide variety of tissues.
Ham's F-10 and F-12	Ham's nutrient mixtures were originally developed to support growth of specific cell types including Chinese hamster ovary and HeLa cells. Both mixtures were formulated for use with or without serum supplementation or with low serum concentration, depending on the cell type being cultured.
Dulbecco's Modified Eagle's Medium/Ham's F-12 Nutrient	During the past decade, researchers have learned to culture a variety of cell lines in medium that contains reduced serum or no serum. These media are supplemented with nutrients, growth factors, and hormones (e.g., insulin, transferrin, and epidermal growth factor). A 1:1 mixture of DME and Ham's F-12 nutrient mixture often provides the base for these low-serum and serum-free media. HEPES, an organic buffer, is included at a final concentration of 15 mM to compensate for the loss of buffering capacity that results from eliminating serum.
Iscove's Modified Dulbecco's Medium (IMD)	A modification of DME containing additional amino acids and vitamins, selenium, sodium pyruvate, HEPES, and potassium nitrate instead of ferric nitrate. Has been useful in growing hybridomas and has been used as the base for serum-free formulations (discussed later in this chapter).

Sources: ATCC. *Media Brochure.* http://www.ATCC.org.
Sigma-Aldrich technical information, http://www.sigmaaldrich.com.

ii. COMPONENTS OF BASAL CELL CULTURE MEDIA

Table 30.5 shows the ingredients of Dulbecco's Modified Eagle's Medium (DME), an example of a classic basal medium formulation. (Formulations for other common media can be found in the references cited in this unit's bibliography.) DME contains a defined mixture of many nutrients including:

- inorganic salts
- amino acids
- vitamins
- glucose (as a source of energy and carbon)

A **balanced salt solution (BSS)** *is a mixture of inorganic salts in specific concentrations.* As discussed earlier in this chapter, these solutions are isotonic. Balanced salt solutions may be used by themselves to support the survival of mammalian cells for a few hours or so, but lack the nutrients required for longer periods of survival and growth. Cell culture media therefore contain a balanced salt solution plus nutrients.

Salts play multiple roles in mammalian cell culture media. They help maintain osmotic equilibrium. They help regulate the transport of materials across the cell membrane; sodium, potassium, and calcium are particularly important in membrane transport. Salts function as cofactors for enzymes. Salts also play a role in the attachment of adherent cells to the surface of their plate or flask. In situations where culturists want to grow cells in suspension, rather than adhered to the surface of their vessel, the calcium and magnesium levels are reduced in the medium.

There are a variety of salt solution recipes used in different media. Earle's and Hank's are two specific BSS formulations that are sometimes used to supply salts in basal cell culture media. Both Earle's and Hank's contain phosphate and bicarbonate for buffering purposes (discussed below) and also contain calcium, magnesium, potassium, sodium, and phosphate. They differ in their concentrations of these salts. Hank's buffered salt solution contains a substantially lower concentration of sodium bicarbonate than does Earle's BSS. Manufacturers will sometimes provide basal media with a choice of BSS, for example, Eagle's MEM with either Earle's salts or Hank's salts.

Amino acids are supplied to cultured cells for the synthesis of proteins and to supply energy. There are two types of amino acids, essential and nonessential. **Essential amino acids** *are those that cells cannot synthesize, plus cysteine and tyrosine.* Culture media must supply essential amino acids. The essential amino acids are: L-arginine, L-cysteine, L-glutamine, L-histidine, L-isoleucine, L-leucine, L-lysine, L-methionine, L-phenylalanine, L-threonine, L-tryptophan, L-tyrosine, and L-valine. L-Proline is required by Chinese hamster ovary cells, which are important in biopharmaceutical manufacturing. Nonessential amino acids may be added to culture media, either because a particular cell type cannot make them, or to enrich the medium to improve the growth of the cells.

Minimal basal media contain the B vitamins (folic acid, biotin, and pantothenate) plus choline, folic acid, inositol, and nicotinamide. Other vitamins are supplied

Table 30.5. COMPONENTS OF DME

Inorganic Salts (mg/L)

NaCl	6400.00
KCl	400.00
$CaCl_2 \cdot 2H_2O$	264.90
$MgSO_4 \cdot 7H_2O$	200.00
$NaH_2PO_4 \cdot 2H_2O$	140.00
$NaHCO_3$	3700.00

Amino Acids (mg/L)

Arginine•HCl	84.00
Cystine•diNa	56.78
Glutamine	584.60
Glycine	30.00
Histidine•HCl•H_2O	42.00
Isoleucine	104.80
Leucine	104.80
Lysine•HCl	146.20
Methionine	30.00
Phenylalanine	66.00
Threonine	95.20
Tryptophan	16.00
Tyrosine•diNa	89.46
Valine	93.60

Trace Element (mg/L)

$Fe(NO_3)_3 \cdot 9H_2O$	0.10

Vitamins (mg/L)

Choline•Cl	4.00
Folic acid	4.00
Inositol	7.00
Nicotinamide	4.00
Pantothenate•Ca	4.00
Pyridoxal•HCl	4.00
Riboflavin	0.40
Thiamine•HCl	4.00

Other Components (mg/L)

Phenol Red	10.00
Glucose	4500.00
Pyruvate•Na	110.00
CO_2 (gas phase %)	10

by serum. Richer medium formulations supply other vitamins in addition to these, such as vitamin C (ascorbic acid) and vitamin E (alpha-tocopherol).

Most media formulations use glucose to supply energy and as a source of carbon. A few media substitute galactose or another sugar for the glucose. Glucose was originally thought to be the only energy source in basal media, but studies have shown that cultured cells also use amino acids for energy. The amino acid L-glutamine is especially important for this purpose and is included at a high concentration relative to other amino acids.

Some basal culture media, including DME, contain sodium pyruvate. Pyruvate is part of one of the chemical pathways that cells use to create energy. It is added to media as an energy source and to provide a carbon skeleton for the synthesis of other molecules. Its addition helps the growth of some cell types and it is particularly important when serum concentration is reduced in the medium.

There are richer basal medium formulations that provide additional nutrients, such as additional amino acids, proteins, vitamins, fatty acids, and lipids. These formulations are used for specific cell types and applications, and as a base for serum-free media.

iii. Sodium Bicarbonate/CO_2 Buffering System

Mammalian cells require a pH-controlled environment. Many cell types are maintained at a pH of 7.2 to 7.4, but some cells prefer a pH closer to 7.0 (e.g., Chinese hamster ovary cells) and others a somewhat higher pH of 7.4 to 7.7 (e.g., fibroblasts). Cells produce lactic acid during metabolism and so a buffer is required to maintain the pH of their medium at a constant value.

Bicarbonate/CO_2 is the most common buffering system for mammalian cell culture because it is relatively inexpensive and nontoxic, and bicarbonate has nutritional value. The bicarbonate/CO_2 buffer system functions both to bring the pH of the medium into a specified range and also to resist a change in pH if acids or bases are added to the medium.

The bicarbonate/CO_2 buffer system requires addition of both carbon dioxide and bicarbonate to the culture medium. Sodium bicarbonate, $NaHCO_3$, is directly dissolved in the culture medium. Carbon dioxide, a gas, is most commonly provided by a CO_2 tank that is attached to the cell culture incubator and is adjusted with a regulator to deliver a steady flow of 5 to 10% CO_2 in air. The caps for culture vessels are designed in such a way that contaminants are excluded but gases can exchange with the culture medium, thus providing both carbon dioxide and oxygen. Sometimes, depending on the cell type, the culture medium, and the cell density, sufficient CO_2 is generated by the cells during metabolism and an external CO_2 tank is not required.

To understand this buffer system, consider first what happens when sodium bicarbonate is added to the culture medium. The sodium bicarbonate dissociates to

form HCO_3^-, which undergoes these reactions:

$$HCO_3^- + H^+ \rightarrow H_2CO_3 \rightarrow CO_2 + H_2O$$

If low concentrations of bicarbonate alone are added to the culture medium, the HCO_3^- will tend to remove hydrogen ions, leaving the solution somewhat basic. Next consider what happens when carbon dioxide is added to the medium. Carbon dioxide from the air reacts with the water in the culture medium to form H_2CO_3, carbonic acid. The carbonic acid in turn ionizes:

$$H_2O + CO_2 \rightarrow H_2CO_3 \rightarrow H^+ + HCO_3^-$$

The net result of increasing the level of carbon dioxide is to add hydrogen ions, which makes the solution somewhat acidic.

In cell culture, both CO_2 and bicarbonate are added to the medium together. In this case equilibrium is reached between the reactions; the exact pH at which equilibrium occurs depends on the percent CO_2 and the concentration of bicarbonate provided. For example, for DME, 44 mM bicarbonate and 10% carbon dioxide are typically used. This concentration of bicarbonate by itself would bring the medium to a pH of about 7.8. The CO_2 alone would bring the medium to a pH of about 4.4. When both are present, the medium equilibrates at about pH 7.2.

This system is buffered in that if moderate amounts of either an acid or a base are added to the medium, the system shifts back to the desired equilibrium pH. If H^+ ions are added to the medium, then:

$$H^+ + HCO_3^- \rightarrow H_2CO_3 \rightarrow CO_2 + H_2O$$

If a base is added to the medium then H^+ ions are removed from the system. In this case, more carbon dioxide reacts with water to produce more hydrogen ions:

$$H_2O + CO_2 \rightarrow H_2CO_3 \rightarrow H^+ + HCO_3^-$$

The system can thus compensate for small additions of acid or base and return the pH to the proper value.

There are limitations to the CO_2/bicarbonate buffering system. The medium may become basic very quickly if removed from the CO_2 incubator. An organic buffer, usually HEPES, is often added to the medium at concentrations of 10 to 20 mM to compensate for this problem or to control the pH for cells that are very sensitive to pH change. HEPES provides good control of pH, but it is more expensive than a bicarbonate buffer system and may be toxic to cells when present at higher concentrations. Note also that if HEPES is added to the medium, the equivalent concentration of NaCl is usually omitted to control the osmolality of the solution.

Most cell culture media include the pH indicator, phenol red, which allows the culturist to quickly judge the condition of a culture by the color of the medium. Below pH 7.0 the indicator is orange, becoming yellow at about pH 6.5. At pH 7.2 to 7.4 it is red, and above 7.5 it is reddish blue. Cells produce lactic acid during metabolism. When the amount of lactic acid rises above the buffering capacity of the medium, the HCO_3^- is depleted and the pH drops. The pH indicator turns orange indicating that it is time to feed the cells with fresh medium.

iv. WATER

Water is the primary ingredient in cell culture media and it must be highly purified and endotoxin free. Culturists working in laboratories that require only small volumes of medium often purchase water purified for cell culture. Laboratories and production facilities where large volumes of culture media are required usually have onsite water purification systems. The water should minimally meet Type I water purity standards, as described in Chapter 28. A multistep purification process is used to prepare highly purified cell culture water. A typical system might include a first stage of reverse osmosis or distillation followed by a second stage of carbon filtration to remove organic and inorganic contaminants. Mixed-bed deionization to remove all ionized substances provides a third stage of purification. Just before use, purified water is typically filtered through a microfilter that removes contaminating microorganisms and contaminants released by the deionizing apparatus. Ultrapurification systems usually continuously recycle the water through the system because bacteria and fungi readily contaminate standing water; substances from the storage container also leach into highly purified water.

C. Supplements to Basal Culture Media

i. SERUM

Mammalian cell culture supplements *can be defined as materials that are sometimes added to basal media after the medium is prepared and/or that are added in varying concentrations.* Serum has traditionally been the most important supplement used for mammalian cell culture. Early cell culture technologists found that serum greatly enhances the growth of cultured cells. Serum contains a rich and complex array of many micronutrients and growth factors that promote cell reproduction. Serum also increases the buffering capacity of the medium and helps protect cells against mechanical damage, which can occur when cells are stirred or scraped from the surface of a plate.

The blood of fetal calves (fetal bovine serum, FBS), is a common source of serum because it is rich in embryonic growth factors. Other less-expensive sources of serum include calves and horses. Serum is processed from its animal source, sterilized by filtration, dispensed into aliquots, frozen, and stored at $-20°C$ until use. When needed, serum is thawed and added aseptically to the culture medium, usually at a concentration of 5 to 20% (v/v).

Procedures from different sources vary in their approach to adding serum (and other supplements). When 10% serum is specified, for example, some procedures call for adding 100 mL of serum to 1000 mL of prepared medium. The result is that the concentration of serum is actually about 9.1% and the concentration of other components in the medium are also slightly altered from their reported values. The reason to prepare the medium this way is to avoid extra manipulation, for example, by removing 900 mL of a previously prepared medium and adding 100 mL of serum. Each manipulation provides an opportunity for contaminants to enter the medium. Many commonly used continuous cell lines are tolerant of variation in the exact concentration of the constituents of their medium, so a serum level that ranges between 9 and 10% will support their growth. This practice is problematic, however, when concentrations are communicated between technologists. Also, some cells are exacting in their growth requirements (e.g., primary cells) and concentrations of media constituents must be maintained in a narrow range.

Some procedures call for heat inactivation of serum by incubating it at 56°C for 30 minutes. It is then dispensed into aliquots and frozen. Heat inactivation of serum is thought to aid cell culture by inactivating complement (a group of blood proteins associated with the immune system). Heat inactivation is useful for assays or procedures where complement is not desired (e.g., when cells will be used to prepare or assay viruses). Heat has also been used in the past to destroy *Mycoplasma,* but this is no longer necessary because most serum suppliers use 0.1 μm filters to remove this small bacterial contaminant. It is advisable to avoid heat inactivation unless it is required because heat can reduce or destroy growth factors, thus reducing the potency of the serum. Boxes 30.4 and 30.5 outline procedures for thawing and inactivating serum.

ii. L-GLUTAMINE AND BICARBONATE

L-glutamine and glucose are common constituents of cell culture media that are often added as supplements right before use. L-glutamine is an essential amino acid that is unstable in liquid solutions at temperatures of 4°C or higher. It dissociates to form toxic ammonium. Therefore glutamine is prepared as a sterile, concentrated 100X stock, aliquoted, and stored frozen so that it does not degrade. L-glutamine is typically used at a final concentration of 0.1 to 0.6 g/L. Cell culture media should be changed frequently to ensure availability of this amino acid and to avoid toxicity.

L-alanyl-L-glutamine is sometimes used as a substitute for L-glutamine because it is stable over time. It is a dipeptide, that is, two amino acids chemically bonded together. Cells can retrieve L-glutamine by cleaving the peptide bond.

Sodium bicarbonate is usually included in liquid basal media preparations, but is often left out of dry preparations because it can liberate CO_2 during storage. Sodium bicarbonate at a concentration of 0.4 to 4.0 g/L is usually added to powders at the time of reconstitution and preparation.

Box 30.4. PROCEDURE TO THAW SERUM

Serum can be stored for at least 2 years at −20°C or −70°C with little deterioration in growth-promoting activity.

1. **Remove serum from freezer and refrigerate overnight at 2°C to 6°C.**
2. **Preheat a water bath to 37°C.**
3. **Transfer the serum bottles to the 37°C water bath.**
4. **Agitate the bottles from time to time in order to mix the solutes that tend to concentrate at the bottom of the bottle.**
   a. Do not keep the serum at 37°C any longer than necessary to completely thaw it. It is not recommended to thaw serum at a higher temperature.
   b. Thawing serum in a bath above 40°C without mixing may lead to the formation of a precipitate inside the bottle.
   c. Do not subject serum to repeated freezing and thawing.

Note: According to ATCC, serum may be cloudy after thawing. This is because serum may contain small amounts of fibrinogen, a blood protein that may be converted to insoluble fibrin. ATCC has tested serum after this has happened and their studies indicate that the serum is still suitable as a supplement for cell culture media. If the presence of this flocculent material is a concern, it can be removed by filtration through a 0.45 μm filter. A precipitate can also form in serum that is incubated at 37°C for prolonged periods of time. Electron microscopy and X-ray microanalysis indicate that the precipitate may include crystals of calcium phosphate. The formation of a calcium phosphate precipitate does not alter the performance of the serum as a supplement for cell culture.

Source: ATCC. http://www.atcc.org/common/technicalInfo/faqCellBiology.cfm#Q39.

Box 30.5. PROCEDURE TO HEAT INACTIVATE SERUM

1. **Preheat water bath to 56°C.** Ensure that there is sufficient water to immerse the bottle above the level of serum.
2. **Mix thawed serum by gentle inversion and place serum bottle in the 56°C water bath.**
3. **After the temperature of the water bath reaches 56°C again, continue to heat for an additional 30 min.** Mix gently every 5 min to insure uniform heating.
4. **Remove serum from water bath and cool.** ATCC recommends storing serum at −70°C if possible, otherwise at −20°C.

Source: ATCC. http://www.atcc.org/common/technicalInfo/faqCellBiology.cfm#Q39.

---

and gram negative bacteria. Combinations of more specific antibiotics, such as penicillin and streptomycin, are also used.

Mycoplasma is a type of very small, difficult to detect bacterium. In humans and animals, *Mycoplasma* can cause disease. These bacteria are troublesome contaminants in cell culture because they are difficult to detect, difficult to eradicate, and can subtly alter the properties of infected cells. Antimycoplasma agents include gentamicin and kanamycin. These are more commonly used with human-derived cells where *Mycoplasma* contamination can be a significant problem. Antifungal agents are also sometimes added but should be used sparingly.

The use of antibiotics is not recommended for "curing" contaminated cultures. Unless contaminated cultures are irreplaceable, they should be discarded and an intensive decontamination program should be initiated. Note that Chapter 3 discussed the use of master and working cell banks in industry. If a working cell bank becomes contaminated, a new vial from the protected master cell bank can be retrieved, thus avoiding the disastrous loss of a valuable cell line.

iv. OTHER SUPPLEMENTS

There are a number of substances with specialized biological functions that are added to culture media for special purposes and for specific cell types. For example, MEM is a simple medium that is often supplemented with a solution of nonessential amino acids. This supplement can be prepared or purchased separately as a sterile stock (often 10 mM; 100X) that is aseptically added to the medium for a final concentration of 0.1 mM each. MEM can also be purchased with nonessential amino acids already added.

iii. ANTIBIOTICS

Some cell culture technologists add antibiotics to cell culture media in low concentrations to help avoid contamination, although most cell culture technologists avoid antibiotics whenever possible. The use of antibiotics can cover up poor sterile technique and can lead to low-level bacterial contamination that goes undetected. Antibiotic use can also encourage the growth of antibiotic-resistant bacteria. The use of antibiotics can therefore result in spurious results in cell culture experiments and is avoided in industry.

It is most common to add antibiotics for short periods when performing primary culture or when establishing cultures of valuable cells. If antibiotics must be used, broad spectrum antibiotics (e.g., ampicillin, gentamicin, kanamycin, and neomycin) are sometimes preferred because they are toxic to both gram positive

Selection agents are supplements used when cells are genetically transfected to cause the death of untransfected cells, as was described above in the section on antibiotics in bacterial media. When cells are genetically transformed with a gene of interest they are also transformed with a gene that enables them to detoxify a toxic agent. The toxic agent is added to the culture medium so that only genetically transformed cells survive.

D. Preparing Standard Media

i. PURCHASING MEDIA

Premixed, quality-controlled mammalian cell culture media are available from commercial suppliers. This section discusses the use of standard commercially prepared cell culture media.

Media are supplied from manufacturers in three forms:

- 1X sterile liquids
- 10X sterile concentrates
- powdered, dehydrated media

The most convenient and expensive media are the 1X liquid media; these do not require the addition of water. The 10X concentrates must be diluted by the user with sterile, highly purified water. Powdered media are less expensive than liquid, but require dissolution and sterilization by the user.

Most manufacturers provide standard media in several forms to provide flexibility for the user. It is possible, for example, to purchase DME with either 1.0 or 4.5 g/L of glucose. Some basal media can be purchased with either Hank's BSS or Earle's BSS. Sodium pyruvate and L-glutamine might be included in a medium mixture or might be left out.

The details of preparing media vary somewhat depending on which type is chosen. Some supplements are added before the medium is sterilized, while others are sterilized separately and are added aseptically to the rest of the medium.

ii. STERILIZING MAMMALIAN CELL CULTURE MEDIA

Liquid 1X and 10X media are sterile when received from the manufacturer, but powdered media must be dissolved and sterilized by the user. In contrast to most bacterial media, mammalian cell culture media usually contain components that are heat sensitive and cannot be autoclaved. Mammalian culture media are therefore sterilized by filtration. Filtration removes microorganisms because they are too large to pass through the pores of the filter. A pore size of 0.2 μm is traditionally used because fungi and most bacteria are removed by this size pore. *Mycoplasma,* however, are very small and require 0.1 μm filters for removal.

For filtering volumes in the milliliter range, a small filter can be attached to a syringe; pressure is applied to the syringe plunger to force the liquid through the filter (see Figure 31.5 on p. 597). Syringe filtration is often used to sterilize concentrated supplements that are added to media in small volumes.

Laboratory-scale filtration of cell culture medium is commonly performed by placing a filtration unit on top of a sterile bottle and *applying vacuum to the underside of the filter unit; this is called* **negative pressure filtration** (see Figure 31.4 on p. 596). Negative pressure filtration is commonly used for volumes up to a few liters. Alternatively, *liquid culture medium can be forced through a filter with a pump that applies pressure to the top of the unit; this is called* **positive pressure filtration.** Positive pressure filtration is usually preferred when larger volumes of medium are filtered.

Table 30.6 summarizes general considerations relating to the preparation of mammalian cell culture media. Box 30.6 on p. 582 is an example of a procedure to prepare cell culture medium from a 1X liquid; Box 30.7 on p. 582 is an example of a procedure to prepare medium from a 10X liquid concentrate; and Box 30.8 on p. 583 is an example of a procedure to prepare medium from a commercially available powder mix.

Table 30.6. *GENERAL CONSIDERATIONS RELATING TO MAMMALIAN CELL CULTURE MEDIA*

1. *Water for cell culture can be sterilized by autoclaving.* Use glass or plastic bottles designed for autoclaving. Loosen the caps and include 10% extra volume to allow for evaporation. Caps should not be tightened until bottles have cooled, otherwise, the vacuum resulting from cooling can cause breakage.

2. *Filtration (0.1 to 0.2 μm pore size) is used to sterilize heat-labile substances; most mammalian cell culture media contain such ingredients.* To filter protein supplements (e.g., hormones, growth factors), use only filters that are specified to be low-protein binding.

3. *Media containing HEPES, riboflavin, and tryptophan can be photoactivated by normal fluorescent lighting producing toxic hydrogen peroxide and free radicals.* Short-term exposure is not normally a problem, but media should not be stored in lighted walk-in cold rooms or refrigerators with glass doors that allow room light to reach the media. The addition of sodium pyruvate to media is said to reduce or eliminate this problem. (Ryan, John. "Understanding Cell Culture Contaminants." *Cell Culture Manual.* 2nd ed. St. Louis, MO: Sigma-Aldrich, 2006–2007).

4. *Common additions to media:*
   a. 1X liquid media are typically provided by the manufacturer without serum or L-glutamine.
   b. 10X concentrated liquid media are typically provided by the manufacturer without serum, L-glutamine, or sodium bicarbonate.

Table 30.6.   *(continued)*

c. Powdered media are typically provided by the manufacturer without serum or sodium bicarbonate.

d. These substances are added to the medium just before use. (See Practice Problems 6 through 9.)

5. ***Sodium bicarbonate may be added to media either in solid form prior to sterilization (e.g., for powdered media) or as a 7.5% sterile solution after sterilization.*** Follow the manufacturer's directions for the recommended amounts.

6. ***L-glutamine is an essential amino acid, required by virtually all mammalian and invertebrate cell lines, that is somewhat unstable in liquid media.***

a. L-glutamine is usually omitted from commercial liquid media and must be aseptically added from a sterile, concentrated stock solution prior to use.

b. Additional L-glutamine can be added to media to extend its shelf life.

c. Liquid L-glutamine stock can be purchased from most commercial vendors of cell culture supplies.

d. L-glutamine concentrations for mammalian cells vary from 0.68 mM for Medium 199 to 4 mM for DME.

e. Dipeptides containing L-glutamine are a stable alternative to L-glutamine itself. Dipeptides are two amino acids linked by a chemical bond. Cells can retrieve L-glutamine by cleaving the peptide bond.

7. ***Pyruvate is sometimes added to culture media. It is an intermediate in a metabolic pathway that provides energy to cells.***

a. Pyruvate can pass readily into or out of cells and its addition to culture media provides both an energy source and a carbon skeleton for the synthesis of other molecules.

b. Pyruvate addition may be advantageous when maintaining certain specialized cells, when maintaining cells at low density, and when the serum concentration is reduced in the medium. It may also help reduce fluorescent light-induced phototoxicity.

c. Pyruvate is usually added to a final concentration of 0.1 mM and is commercially available as a 10 mM (100X) stock solution.

8. ***Antibiotics.***

a. Antibiotics should be avoided for routine culture work because they can mask subtle bacterial or fungal contamination and may affect the growth of sensitive cells.

b. Antibiotics are sometimes added for short periods to primary cultures or as a safeguard while expanding valuable cultures to produce cell banks.

c. Typical concentrations in medium are 50 to 100 units penicillin G, 50 to 100 µg/mL of gentamicin sulfate, or 2.5 µg/mL of amphotericin B.

9. ***HEPES is an effective organic buffer that is commonly used for cell culture.*** It can be toxic to sensitive cells and has also been shown to increase the sensitivity of media to fluorescent light.

10. ***Concentrated media are manufactured at a low pH because their components are less likely to precipitate when the pH is acidic.*** After dilution the medium might be too acidic or possibly too basic (depending on the formulation) and must be brought to the correct pH. Powdered media are usually brought to a pH 0.1 or 0.2 pH units below the desired final pH because their pH rises slightly during filtration.

11. ***Adjusting the pH of media.*** Sterile NaOH and HCl are used to bring media to the correct pH. They can be sterilized by filtration through a 0.2 µm filter; check that the filter membranes are compatible with these substances before use.

12. ***Sterilized, complete media are commonly stored at 4°C in the dark if not used immediately.***

a. Serum is stored separately and added to otherwise complete medium just before use.

b. Consult the manufacturer to determine how long a medium can be stored.

13. ***Commercially prepared media are isotonic.*** It is advisable to check the osmolality of the medium if it is supplemented with extra salt solutions or large volumes of buffering substances. Some drugs and hormones used to supplement media are initially dissolved in an acidic or basic solution and the medium is brought to the correct pH after their addition. These acids and bases can significantly alter the osmolality of the final medium. It is therefore important to be conscious of osmotic effects when adding supplements to a medium (e.g., when adding a drug substance to test its effect or adding a supplement to enhance growth).

14. ***Since the osmolality of culture media can rise as a result of evaporation, CO_2 incubators must be well humidified in order to prevent evaporation.***

15. ***Media that contain Earle's salts are intended to be used with 5% carbon dioxide.***

a. This is typically accomplished in an incubator provided with CO_2 gas and culture vessels that allow gas exchange.

b. Alternatively, the vessels may be gassed with 5% CO_2 after filling.

c. Bicarbonate levels of 2.2 g/L are typical with Earle's salts.

16. ***Hank's salts have a lower bicarbonate level (0.35 g/L) and less buffering capacity than Earle's salts and are designed to be used without addition of carbon dioxide to the gas phase.*** A culture vessel that does not allow gas exchange should be used.

Primary source: ATCC. http://www.atcc.org/common/technicalInfo/faqCellBiology.cfm#Q39.

Box 30.6. AN EXAMPLE OF A PROCEDURE FOR PREPARING A CELL CULTURE MEDIUM FROM 1X LIQUID

Use disposable, sterile cell culture plasticware or well-cleaned, sterilized glassware dedicated to cell culture.

1. *Check that serum, L-glutamine stock solution, and any other required supplements are ready.*
2. *Perform calculations (see Practice Problems for examples) to determine amounts required for serum, glutamine, and any other additions required.*
3. *Thaw serum, glutamine, and any other frozen supplements by placing in a 37°C water bath.*
4. *Prepare laminar flow cabinet.*
5. *Swab the outside of the bottle of medium and any supplement tubes with 70% alcohol; place in laminar flow cabinet.*
6. *Using aseptic technique, add the correct amount of supplements to medium.*
7. *Check that the pH equilibrates at the proper temperature.* If it does not, alter the CO_2 concentration or adjust the pH of the medium.
8. *Label the bottles with name of medium, supplements, your name, and date.*
9. *The complete medium is ready to use and has a limited shelf life.* If stored, place in the dark at the temperature recommended by the manufacturer.
10. *Quality control:*
    a. Incubate one or more bottles under cell culture conditions for three days to check for visible contamination (e.g., cloudiness, precipitate, flocculent matter, change in pH). It is advisable to check both the basal medium and the complete medium with serum. Alternatively, remove 1 mL aliquots from each bottle of medium and transfer to 12-well culture plates. Incubate.
    b. Record how and when all glassware and equipment were sterilized, the lot numbers of filters used, all media component lot numbers and vendors, and how and when media were sterilized. Every bottle of culture medium should be individually labeled.
    c. If a separate microbiology area is available, it is possible to check the medium for contamination by applying some to a bacterial growth plate and incubating the plate; nothing should grow.

Box 30.7. A PROCEDURE FOR PREPARING 1 L OF MEDIUM FROM A 10X LIQUID CONCENTRATE

- Use only highly purified, sterile water.
- Use disposable, sterile cell culture plasticware or well-cleaned, sterilized glassware dedicated to cell culture.

1. *Check for precipitate.* Precipitates will normally redissolve with dilution. If however, a precipitate has formed due to degradation of a component of the medium, then the quality of the medium may be reduced and it should be tested for its ability to support cell growth.
2. *Check that sterile, highly purified water, serum, sterile 7.5% bicarbonate stock solution, glutamine stock solution, and any other required supplements are ready.*
3. *Perform calculations (see Practice Problems on p. 585 for examples) to determine amounts required for water, serum, glutamine, and any other additions required.*
4. *Prepare laminar flow cabinet.*
5. *Thaw serum, glutamine, bicarbonate, and any other frozen supplements in a 37°C waterbath.*
6. *Swab the outside of the bottle of medium concentrate and any tubes containing supplements with 70% alcohol; place in laminar flow cabinet.*
7. *To make 1 L, using aseptic technique, measure 700 to 750 mL of sterile, highly purified water into a suitable sterile container.*
8. *Using aseptic technique, gently stir and add 100 mL of 10X concentrate.*
9. *Using aseptic technique, add the correct amount of 7.5% sodium bicarbonate based on manufacturer's instructions.*
10. *Using aseptic technique, add sterile L-glutamine and other supplements.*
11. *Adjust the pH with sterile NaOH or HCl.*
12. *Add sterile, highly purified water to 1 L.*

Box 30.7. *(continued)*

---

13. ***The complete medium is ready for use and has a limited shelf life.*** If stored, place in the dark at the temperature recommended by the manufacturer.

14. ***If medium is not used immediately, add serum to the desired final concentration at the time of use.***

    a. Store the basal nutrient medium and serum individually. Prepare the complete medium at the time of use and only in the volume necessary.

    b. Add serum at 1 to 20% v/v to otherwise complete medium.

15. ***Quality control, as in Box 30.6.***

Box 30.8. *A PROCEDURE FOR PREPARING 1 L OF MEDIUM FROM POWDER*

- Use only highly purified water.
- Use disposable, sterile cell culture plasticware or well-cleaned, sterilized glassware dedicated to cell culture.
- Powdered media are extremely hygroscopic and must be protected from atmospheric moisture.
- The preparation of medium in concentrated form is not recommended because some of the amino acids have limited solubility and may precipitate in concentrated solutions to form insoluble salt complexes.
- Filter sterilize the medium immediately after mixing to avoid microbial growth.

1. ***Select a container as close in size to the final volume as possible. Measure out 90% of the final volume of water required.*** Some sources recommend using sterile water and autoclaved containers at this point, although the medium will be filter sterilized later.

2. ***Slowly add the powdered medium to the water with gentle stirring.***
    a. Rinse the original package with a small amount of water to remove all traces of powder.
    b. Cover and stir with a sterile stir bar until dissolved.
    c. Do not use heat.

3. **Optional:** ***Add the amount of HEPES that yields a concentration of 15 mM in the final volume of medium.*** Omit this step if the powdered medium is formulated with HEPES or if HEPES is not desired.

4. ***Add the amount of sodium bicarbonate recommended by the supplier for use in a CO_2-controlled atmosphere.***

5. ***Add NaOH or HCl with gentle stirring to adjust the pH.*** Adjust the medium to 0.1 to 0.3 pH units below the desired final pH because the pH may rise this much with filter sterilization.

6. ***Bring to the final volume with highly purified water*** (some sources use sterile water).

7. ***Sterilize the medium in a laminar flow hood by filtration through a 0.1 or 0.2 μm filter into sterile media bottles.***

8. ***Cap tightly with sterile closures and store in the dark at the temperature recommended for the product.***

9. ***At the time of use, remove appropriate amount of medium using aseptic technique.***
    a. Add glutamine from a sterile stock solution to give a final concentration of 2 mM.
    b. Add serum to the desired final concentration at the time of use.

10. ***Quality control, as in Box 30.6, see p. 582.***

Primary source: Bonifacino, Juan S., Dasso, Mary, Harford, Joe B., Lippincott-Schwartz, Jennifer, and Yamada, Kenneth M. (Eds.), "Media for Culture of Mammalian Cells." *Current Protocols in Cell Biology.* New York: John Wiley and Sons, 1998.

E. Serum-Free Media, Animal Product Free–Media, Protein-Free Media, and Defined Media

The use of serum in cell culture is problematic, despite its many benefits. Serum is expensive. Serum introduces lot to lot variability into the cell culture process because the individual animals from which it is taken vary. It is common for technologists in cell culture facilities to test several lots of sera for compatibility with a particular cell line and buy entire lots of the best sera. Even when this practice is used, the lot is eventually gone and a different lot of serum must be introduced. This variability is a problem both in production settings and in research studies where it introduces unknown variables into the experiments.

Two additional serum-related problems arise when mammalian cells are used for biopharmaceutical production. First, serum introduces impurities that need to be removed from products during downstream processing, thus adding complexity and cost. Second, serum may be contaminated with pathogens (e.g., viruses and prions) that are difficult to remove even with the best modern methods. These pathogens may or may not be a risk to the cells themselves, but can contaminate a product made by the cells. In the 1980s and 1990s some people in Europe died of a new variant of the neurological disease, Creutzfeldt-Jacob disease, thought to be transmitted from beef infected with a bovine pathogen. In response, in 1993 the FDA recommended that pharmaceutical manufacturers not use bovine-derived materials from cattle that resided in, or originated from, countries where this bovine disease had been diagnosed. This concern about bovine pathogens has now been extended to other animal-derived materials; if cows harbor pathogens that threaten humans, so can other animals.

Serum therefore raises a variety of concerns:

- Expense
- Lot to lot variability
- Introduction of unknown substances into the culture
- Product purification issues
- Pathogen contamination

The first three of these concerns affect those who use cells either for research or production purposes. The latter two concerns are of particular concern when cells are used for biopharmaceutical manufacturing.

In response to these concerns about serum, researchers have been trying to determine the identities and functions of the many components of serum in order to develop serum-free media (SFM). It has been a challenging task to develop SFM that nourish cultured cells as effectively as serum-supplemented media. The main ingredients of serum are well known (e.g., the proteins albumin and transferrin), but serum contains numerous, diverse substances, many of them active at very low concentrations. Serum-free media therefore require a great many additions, including:

- **Lipids**
- **Vitamins**
- **Essential trace metals** (including iron, zinc, copper, selenium, manganese, molybdenum, aluminum, silver, and nickel)
- **Attachment factors** (aid cells in attaching to substrates)
- **Cytokines** (small messenger proteins released by blood cells that are used for communication between cells)
- **Hormones** (e.g., insulin, growth hormone, and steroids)
- **Growth factors** (chemicals that play numerous roles in the promotion of cell growth and cell maintenance)

The first approach to creating SFM was to supplement enriched basal media with animal-derived protein hydrolysates (peptones) and/or with purified proteins from animal or human sources. Although some of the concerns relating to serum are alleviated by the use of these first generation serum-free formulations, their high protein content can cause problems with downstream purification. Also, these formulations contain substances derived from animal sources and so raise the same concerns about animal pathogens as does serum. Furthermore, the use of hydrolysates introduces variability and undefined substances into the medium. Bovine serum albumin, BSA, for example, is an important serum protein that is often included in SFM. It is isolated from blood and usually contains a variable mixture of chemicals including fatty acids and other proteins.

As a next step in medium development, scientists substituted yeast and plant-based hydrolysates (e.g., from soy, wheat, and cottonseed) for animal-based hydrolysates. These substitutes have been suitable for many cell lines, including CHO and other production cell lines. Biopharmaceutical products made with these plant-based media have reached the market. Although these media do not contain materials directly isolated from animals, they nonetheless are undefined and may vary from lot to lot.

A third step in medium development is the creation of completely defined SFM in which all the medium requirements of the cells are provided in purified form. The result of these developments is that there is now an array of media optimized for various applications, see Figure 30.6, that have one or more of the following characteristics:

- **Low serum.** Enriched formulations designed primarily to save money by reducing, but not eliminating, serum (see Practice Problem 11 on p. 587).
- **Serum-free.** Enriched formulations that eliminate the need for serum supplementation but may con-

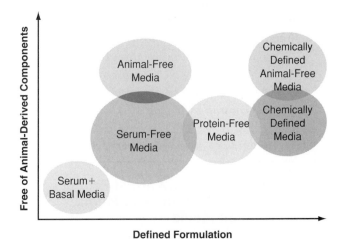

Figure 30.6. Trends in Cell Culture Media. The trend in cell culture media is toward more defined media and media whose components are not isolated from animals. (Reprinted with permission from Hodge, Geoffrey. "Media Development for Mammalian Cell Culture." *Biopharm International*, May 1, 2005.)

tain animal-derived constituents and may be undefined.

- **Animal-derived component-free media (ADCF).** Enriched formulations that do not require serum and whose constituents are not derived from animal sources. These media address safety concerns associated with animal-derived raw materials but they may provide an uncharacterized mixture of peptides and trace elements derived from plant sources and may introduce lot to lot variability.
- **Protein-free media.** Formulations that do not contain proteins (e.g., BSA, insulin, attachment factors). These media are a step toward completely defined media, but may contain both animal-derived components (depending on the manufacturer) and protein hydrolysates.
- **Defined media.** Contain only known, purified constituents. May or may not be ADCF.

These newer cell culture media have many applications. In industry, new, optimized media formulations have contributed to increased productivity of biopharmaceutical production systems and have alleviated many of the concerns about animal pathogens. Scientists have benefited from the extensive research into serum and its properties because it has led to a better understanding of the many and often subtle factors that affect cell growth and division. There are, however, costs associated with the use of newer cell culture media. Unlike classic formulations (e.g., DME) which are readily adaptable to a variety of cell types and applications, the new formulations are tailored for specific cell types and specific applications. They can therefore be expensive and difficult to develop. Moreover, animal-free, defined media require the addition of a number of highly purified constituents, each of which may be costly. Entirely avoiding animal sources in the production of these purified constituents may require producing dozens of substances in genetically modified organisms. Much remains to be learned before it becomes routine to use entirely defined, animal product–free, serum-free media for cell culture.

PRACTICE PROBLEMS

1. What is the milliosmolarity of 0.9% NaCl w/v? (FW = 58.44)

2. A kilogram of a particular solution contains 5% glucose monohydrate (w/w) (MW = 198) and 20 millimoles of NaCl.
   a. How many milliosmoles are in the solution?
   b. What is the milliosmolality of the solution?

3. This is the recipe for a defined bacterial medium, as given in a manual of bacteriology.

M63 medium, 5X

10 g $(NH_4)_2SO_4$
68 g KH_2PO_4
2.5 mg $FeSO_4 \cdot 7H_2O$
Adjust to pH 7 with KOH

Directions: Dilute concentrated medium to 1X with sterile water. The following sterile solutions should be included, per liter:
1 mL of 1 M $MgSO_4 \cdot 7H_2O$
10 mL of 20% carbon source (glucose or glycerol)
0.1 ml of 0.5% vitamin B1 (thiamine)

   a. How much concentrate should be used to make a 1X medium if 1 liter is required?
   b. What is the concentration of glucose in the final solution?
   c. What is the concentration of $MgSO_4 \cdot 7H_2O$ in the final solution?
   d. What is the concentration of thiamine in the final solution?

4. Many molecular biology procedures require the addition of antibiotics to nutrient agar plates to select for bacteria that have taken up an antibiotic resistance gene. Antibiotics are not sterilized by autoclaving. They are prepared as concentrated stock solutions, filter sterilized, and added to the nutrient agar medium after it has cooled to about 55°C. Table 30.7 on p. 586 shows typical values for antibiotic concentrations in stock solutions and for final concentrations of antibiotics in the nutrient agar. Fill in the final column of the table with the volume of stock solution needed to make 1 L of nutrient agar. The first answer is filled in as an example.

5. The recipe for acetate differential agar from the website of BD Diagnostic Systems (http://www.bd.com/ds/) is shown below. This is a bacterial medium that is used to differentiate between *Shigella* and *E. coli* bacteria. *E. coli* can grow on this medium but *Shigella* cannot. (Many species of *Shigella* cause a severe intestinal illness in humans called shigellosis, which causes high fever and acute diarrhea. Most species of *E. coli* are normal inhabitants of the healthy intestine in humans and other animals.)

Acetate Differential Agar (per liter)

Sodium acetate	2.0 g
Magnesium sulfate	0.1 g
Sodium chloride	5.0 g
Monoammonium phosphate	1.0 g
Dipotassium phosphate	1.0 g
Bromthymol blue	0.08 g
Agar	20.0 g

   a. What is the source of carbon in this medium?
   b. What is the source of nitrogen in this medium?
   c. What is the likely purpose of the NaCl?
   d. What is the likely purpose of the phosphate compounds?

Table 30.7. ANTIBIOTICS

Antibiotic	Stock Solution Concentration	Recommended Working Concentration	Amount Stock Required (per L)
Ampicillin* (sodium salt)	50 mg/mL in water	100 µg/mL	2 mL
Chloramphenicol	34 mg/mL in ethanol	170 µg/mL	___
Kanamycin	10 mg/mL in water	50 µg/mL	___
Streptomycin	10 mg/mL in water	50 µg/mL	___
Tetracycline HCl	5 mg/mL in ethanol	50 µg/mL	___

*Ampicillin is commonly used as a selective agent but note that beta-lactam-antibiotics (including ampicillin) are not very stable and will slowly degrade when dissolved, even when frozen at $-20°C$. It is therefore best to reconstitute ampicillin shortly before use. The sodium salt is more soluble than ampicillin alone and is therefore used for culture media.

Source: "The QIAGEN Guide to Good Microbiological Practice Part II — Storage of *E. Coli* Strains." *Qiagen News* 5, 1998.

e. Given that this is a differential medium, what is the likely purpose of the bromthymol blue?

f. What is the purpose of the agar? What is the percent concentration of the agar?

6. A total volume of 500 mL of culture medium for mammalian cell culture is desired, including 20% serum. The culture medium is provided as a 1X liquid and has all required components, except serum. What volumes of medium and serum are required?

7. Suppose that the medium from question 6 does not contain glutamine and so it must be added. The glutamine is stored as a concentrated stock solution that is 100X. What volumes of medium, serum, and glutamine are required?

8. One liter of mammalian cell culture medium is to be prepared from 10X concentrate. Glutamine is available in a 200 mM stock solution. Undiluted serum is also available. Sodium bicarbonate $(NaHCO_3)$ is available as a sterile 7.5% solution (which is the same as 0.89 M). You want the final concentration of the medium to be 1X, the glutamine to be 2 mM, the serum to be 10%, and the sodium bicarbonate to be 26 mM. How much is required of each of the following?

a. 10X concentrate

b. Glutamine stock

c. Serum stock

d. Sodium bicarbonate stock

e. Sterile, highly purified water

9. Different mammalian cell culture media require different concentrations of sodium bicarbonate. Table 30.8 shows the levels of sodium bicarbonate recommended by a manufacturer for different media. Fill in the blanks in the table; the first row has been completed as an example.

10. Different mammalian cell culture media require different concentrations of glutamine. Table 30.9 shows the levels of sodium bicarbonate recommended by a manufacturer for different media. Fill in the

Table 30.8. RECOMMENDED ADDITIONS OF SODIUM BICARBONATE

Medium	mL of NaHCO₃ Required, 7.5% Stock (per L)	g of Solid NaHCO₃ Required (per L)	Final Concentration NaHCO3 (mg/L)
DME	49.3	3.70	3700
DME/Ham's F12	32.5	___	___
Ham's F12	___	___	1176
MEM Earle's salts	___	2.20	___
MEM Hank's salts	4.7	___	___
RPMI 1640	___	___	2000
McCoy's 5A	29.3	___	___
MEM Alpha	___	___	2200

Source: Promocell. http://www.promocell.com.

Table 30.9.   RECOMMENDED ADDITIONS OF L-GLUTAMINE

Medium	mg/L in Final Medium	mM in Final Medium	mL/L of Stock Required
AMEM	292.3	___	10
BME	___	2.0	___
DME	___	___	20
F12K	___	2.0	___
Ham's F10	___	___	5
Ham's F12	146.2	___	___
Iscove's DME	584.6	___	___
EMEM	___	___	10
RPMI 1640	___	2.05	10.25

Source: Mediatech, Inc., http://www.cellgro.com.

blanks in the table; the first row has been completed as an example. The L-glutamine is assumed to be in a 200 mM stock solution. Its FW is 146.15.

11. Assume that serum costs $250 for a 500 mL bottle and that medium costs $15 for a 500 mL bottle.

   a. How much would it cost to make 500 mL of cell culture medium that includes 10% serum?

   b. How much would it cost to make 500 mL of cell culture medium that includes 2% serum?

For simplicity, assume in both cases that the entire bottle of medium is used, but only as much serum as is needed is removed.

(Data are from Ryan, John A. "Reducing Cell Culture Costs by Reducing Serum Levels." *American Biotechnology Laboratory* (Nov./Dec. 2006)18.

12. List similarities and differences between mammalian cell culture media and bacterial media.

UNIT VIII

Basic Separation Methods

Chapters in This Unit

- ✦ Chapter 31: Introduction to Filtration
- ✦ Chapter 32: Introduction to Centrifugation
- ✦ Chapter 33: Introduction to Bioseparations

This unit provides an introduction to the basic principles of the most common molecular separation methods used in biotechnology laboratories. The ability to separate biological materials from one another is essential in any biotechnology setting. Consider, for example, the production of insulin to treat diabetic individuals. Insulin can be obtained either from animal pancreatic tissue or from *E. coli* bacteria that have been genetically modified to produce insulin. If pancreatic insulin is used, it must first be extracted from the tissue. Then, a series of steps must be performed which sequentially remove contaminants (such as other proteins, lipids, and cellular debris) from the insulin before it is suitable to inject into a human. Similarly, insulin isolated from bacteria must be purified of all bacterial substances. The techniques used to extract, isolate, and purify the insulin are called *bioseparation methods*.

There are many different types of bioseparation methods, each of which separates materials based on differences in their molecular properties. For example, filtration separates materials based on differences in their sizes—as agarose gel electrophoresis does for DNA fragments. Centrifugation separates materials based on differences in their sizes or densities. Various forms of chromatography separate materials based on properties such as size, charge, or solvent solubility.

Applications of these separation techniques are common in the biotechnology laboratory. The separation of DNA fragments of different sizes from one another using agarose gel electrophoresis is key to all recombinant DNA technology. Another common example is the use of centrifugation to separate cellular organelles (such as mitochondria and nuclei) from one another. Common examples of applications of filtration are found in the removal of contaminants from biological preparations, water purification, and sterilization of solutions.

As you can see from the examples above, separation techniques are used whenever a biological material is purified for further study and analysis, as when gene segments are purified for later sequencing. Separation techniques are also essential whenever specific biological products are isolated and purified for commercial sale in large-scale, production settings. Many separation techniques can be applied to both small- and large-scale preparations.

In order to completely isolate and purify a particular biomolecule (e.g., insulin), it is virtually always necessary to perform a series of separation techniques in succession. Each separation technique removes certain contaminants from the product of interest. One of the challenges to researchers is to develop a bioseparation strategy, consisting of a series of separation techniques, which most effectively isolates and purifies a biological product of interest. This unit addresses in more detail two of the most routinely used separation methods: filtration and centrifugation.

Chapter 31 introduces the basic principles of filtration and their applications in both small- and large-scale operations.

Chapter 32 introduces the basic principles of centrifugation and their applications. It includes information about instrumentation and safety issues related to centrifugation.

Chapter 33 is an overview of the principles and strategies of product purification applied in both research and production laboratories, along with discussion of additional techniques.

BIBLIOGRAPHY FOR UNIT VIII

There are many references available that provide detailed information about general principles of separation and about specific bioseparation techniques, such as centrifugation. The list below provides starting materials for a more in-depth study of purification methods.

General Bioseparations

Although each of these references mentions proteins in their titles, many of the techniques and approaches described could be applied to other types of biomolecules as well.

Rosenberg, Ian. *Protein Analysis and Purification: Benchtop Techniques*. 2nd ed. Boston: Birkhäuser, 2004.

Cutler, Paul. *Protein Purification Protocols (Methods in Molecular Biology)*. 2nd ed. Totowa, NJ: Humana Press, 2003.

Roe, Simon. *Protein Purification Techniques: A Practical Approach*. 2nd ed. New York: Oxford University Press, 2001.

Centrifugation

Ford, T.C., and Graham, J.M. *An Introduction to Centrifugation*. Oxford, UK: Bios Publishers, 1991. Hard to find, but a readable basic primer on centrifuges.

Graham, John. *Biological Centrifugation (The Basics)*. New York: Garland Science, 2001. Overview of basic principles.

Lindley, John. *User Guide for the Safe Operation of Centrifuges*. 2nd ed. Rugby, UK: Institution of Chemical Engineers, 1987.

Rickwood David, and Graham, John. *Biological Centrifugation*. Berlin: Springer Verlag, 2001.

A number of useful technical brochures and documents can be obtained from centrifuge manufacturers and supply companies using the Internet. Two examples are:

Beckman Coulter, Inc., Fullerton, CA. http://www .beckmancoulter.com. They have a number of publications about applications and rotor safety.

Cole-Parmer Instrument Company, Vernon Hills, IL. http://www.coleparmer.com/.

Specialized Bioseparation Techniques

Many techniques such as electrophoresis and chromatography are sufficiently specialized to be described in separate volumes. Below are some suggested references for detailed information beyond that provided by the general references above. In addition, the manufacturers and suppliers of materials for these techniques are usually excellent sources of information.

Dong, Michael W. *Modern HPLC for Practicing Scientists*. Hoboken, NJ: Wiley-Interscience, 2006.

Miller, James M. *Chromatography: Concepts and Contrasts*. Hoboken, NJ: Wiley-Interscience, 2004.

Rathor, Anurag S., and Velayudhan, Ajoy, eds. *Scale-Up and Optimization in Preparative Chromatography: Principles and Biopharmaceutical Applications*. Boca Raton, FL: CRC, 2002.

Westermeier, Reiner. *Electrophoresis in Practice: A Guide to Methods and Applications of DNA and Protein Separations*. 4th ed. Hoboken, NJ: Wiley-VCH, 2005.

CHAPTER 31

Introduction to Filtration

I. INTRODUCTION

A. The Basic Principles of Filtration

Filtration is a common separation method based on a simple principle: Particles smaller than a certain size pass through a porous filter material; particles larger than a certain size are trapped by the filter. Anyone who has strained spaghetti through a colander, made coffee with coffee filters, or played with a sieve at the beach has used filters. Gases, such as air, can be filtered as well as liquids. The air filters in cars and furnaces are commonplace examples.

The fluid and particles that pass through a filter are called the **filtrate** *or* **permeate.** *The materials trapped by the filter are sometimes called the* **retentate.** We are sometimes interested in the filtrate, as is the case with a coffee filter. We are interested other times in the material retained on the top of the filter, as is the case with spaghetti. We occasionally want to collect both the retained particles and the filtrate.

Filtration is also commonly observed in nature. Water is cleared of particulates as it passes through sandy soil to the groundwater. The kidneys are effective filtration devices that allow small, unwanted metabolites to pass from blood into the urine while retaining the relatively large blood cells.

Just as filtration plays various roles in the home and in nature, so it has a long history in the laboratory. Filtration was traditionally accomplished by folding a filter paper and placing it in a funnel for support. Liquids poured into the funnel would drip through the filter and solids would remain on the surface. To speed the process, analysts could add a vacuum pump to the flask. These simple filter paper systems are still used in laboratories to remove coarse particles from a solution, see Figure 31.1.

Although the principle of filtration is simple, the study of filtration is complicated by its wide range of applications and the many types of filtration devices available. Filtration is used from the smallest scale in the laboratory, where samples of only a few microliters

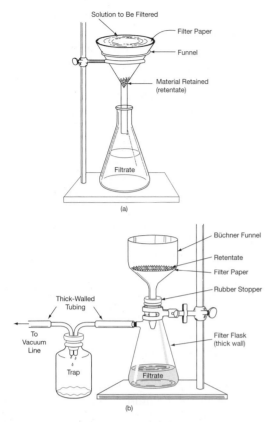

Figure 31.1. Laboratory Filtration Systems. a. The simplest laboratory filtration system consists of a folded piece of filter paper supported in a conical funnel. Liquid moves through the filter due to the force of gravity. **b.** Vacuum can be used to facilitate the movement of fluid through the filter. In this example, a Büchner style funnel is used to support the filter paper. Note that it is good practice to place a trap bottle between the vacuum flask and the source of vacuum to trap materials that may be pulled into the tubing. An alternative approach is to use an in-line disk filter to protect the vacuum source.

may be processed, to major industrial processes involving thousands of liters.

Depending on the application, the sample type, and the scale, filtration systems can vary greatly. Regardless of the simplicity or complexity of a filtration system, however, the following four components are present:

1. **The filter itself**
2. **A support for the filter (like the funnels in Figure 31.1)**
3. **A vessel to receive the filtrate**
4. **A driving force (such as gravity or vacuum) that drives the movement of fluids and particles through the filter**

These four components are present both in a simple coffeemaker and in a complex process filtration system in a pharmaceutical company.

B. Overview of Issues in Filtration

There are several issues that affect filtration and which must be considered in designing or selecting a filtration system. The most obvious is clogging. Filters clog if they are covered by large particles, or by aggregations of smaller particles. Oils, lipids, and fats can similarly form a film on the surface of a filter that prevents filtration. In some situations clogging is "cured" simply by replacing the clogged filter with a new one. In more sophisticated systems, particularly in industrial settings, clogging is reduced by moving the liquid to be filtered across the filter surface. Such systems are discussed later in this chapter.

Some substances, like proteins, bind to certain filter materials. *When a component binds to the surface of a medium, such as a filter, it is called* **adsorption.** Adsorption tends to block the pores of a membrane, resulting in a lower rate of filtration. Adsorption also leads to loss of the adsorbing component in the sample. Some materials used in the manufacture of filters bind macromolecules readily, whereas others minimize binding.

Adsorption is different than absorption. **Absorption** *is when liquids are taken up into the entire depth of a material, as when water is absorbed by a sponge.* There is no chemical selectivity associated with absorption as there is with adsorption. Both adsorption and absorption can occur in filtration.

Extractability is another issue in filtration. **Extractables** *are compounds from the filter that leach out and enter the sample being filtered.* Fibers from the filter may similarly enter the sample being filtered. The contamination of a sample by materials from the same filter can be a serious problem in some applications. For example, in cell culture, extractable substances from some filters have been shown to inhibit the growth of cells. In the pharmaceutical industry, filters have many applications, such as removing bacteria and viruses from products, and purifying water. It is essential that these filtration processes not introduce impurities into the drugs. Another application of filtration is to remove particulates from samples before they are injected into instruments for analysis. Many types of instruments used in analyzing samples detect minute amounts of compounds, so impurities from filters must be meticulously avoided.

II. TYPES OF FILTRATION AND FILTERS

A. Overview

Filters are generally classified into three types according to the size of particles they retain, see Figure 31.2 on p. 592.

1. **Macrofilters or general filters (depth filters)** *are used for the separation of particles on the order of 10 μm or larger.* A coffee filter and a common laboratory paper filter are examples of macrofilters.

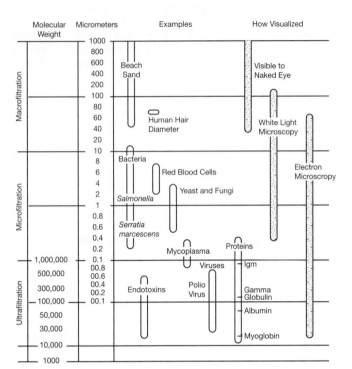

Figure 31.2. **Filtration and Size.**

2. **Microfilters** *are typically used to separate particles whose sizes range from about 0.1 µm to 10 µm.* Bacteria and whole cells fall in this size range.

3. **Ultrafilters** *are used to separate macromolecules on the basis of their molecular weight.* For example, large proteins can be separated from smaller ones using ultrafiltration.

Each of these classes of filtration is discussed in more detail in the next sections. In addition, we will briefly consider two related techniques: dialysis and reverse osmosis.

B. Macrofiltration

General laboratory filters are inexpensive devices used to remove relatively large particles, above about 10 µm from a liquid. These filters are made of materials such as paper, woven polymer fibers, stainless steel mesh, or sand. General filters do not have pores or holes of a particular size; rather, they consist of a convoluted granular or fibrous matrix. Particles are trapped both on the filter surface and within the sinuous matrix. Because these filters trap particles throughout their depth, they are also called **depth filters.**

Macrofilters for the laboratory are usually made of cellulose (paper) or of glass or woven polymer fibers. In practice, both types may be referred to as *filter paper.* Glass-fiber filters consist of long strands of borosilicate glass that are formed under high heat conditions. In contrast to filters made of paper, glass- and polymer-fiber filters allow faster flow rates, are stable under a wide range of temperatures, and are compatible with corrosive chemicals.

Manufacturers make a variety of paper filters and polymer- and glass-fiber filters. These filters vary in the density of their mesh and therefore in the sizes of particles they tend to trap. Filters that retain smaller particles tend to have slower flow rates than those that trap only larger particles. Different types of paper filters are also graded by the amount of ash residue they leave when burned. Low ash level is important in some applications in analytical chemistry where a sample solution is filtered and then the filter paper is burned to leave behind only the sample. The ash level is also a measure of the purity of the fibers used to make the filter: the lower the ash level, the higher the purity. The term **quantitative grade** *refers to filter papers that leave little ash when burned and are intended for use in certain chemical analyses.* **Qualitative grade** *papers leave significant amounts of ash.* **Hardened grade** *papers are treated to be stronger and better able to withstand vacuum filtration than unhardened papers.*

General filters are often used as "prefilters" to remove large particulates from a solution before it is filtered with a microfilter or ultrafilter. The prefilter prevents clogging of the more expensive, finer filter.

C. Microfiltration

i. MICROFILTERS

Microfiltration typically separates particles in the range of about 0.1–10 µm from a liquid or gas medium. Microfiltration filters are called **membranes** *and are manufactured to have a particular pore size.* Particles larger than the rated size are retained on the surface of the membrane and smaller particles pass through. When a manufacturer specifies the **pore size (absolute)** *they mean that 100% of particles above that size will be retained by the membrane under specified conditions.* **Pore size (nominal)** *means that particles of that size will be retained with an efficiency below 100% (typically 90–98%).* The methods of rating the nominal pore size can vary between manufacturers. Note that, in practice, pore size ratings refer to the size of particles retained, not to the actual physical dimensions of the pores.

Microfiltration membranes can be manufactured from a variety of plastic, cellulose derivatives, metals, and ceramic materials, as shown in Table 31.1. While the first factor to consider in choosing a microfilter is pore size, different types of filter membranes differ in important properties. These properties include:

1. **Resistance to organic solvents.** Some membranes are dissolved by organic solvents, others are not. When organic solvents are being filtered, it is necessary to choose a membrane made of a solvent-resistant material.

Table 31.1. *EXAMPLES OF MATERIALS COMMONLY USED TO MAKE MICROFILTER MEMBRANES*

Material	Features	Examples of Applications
Cellulose acetate	Hydrophilic Very low aqueous extractability Very low binding Not resistant to organic solvents except low molecular weight alcohols Resistant to heat	General filtration Sterilizing cell culture media General sterilization
Nitrocellulose	Hydrophilic Fast flow rates Readily binds nucleic acids and proteins	General filtration Sterility testing
Polyethersulfone, PES, and polysulfone	Hydrophilic Low extractability Low protein binding Wide chemical compatibility High throughput, autoclavable	Sterilization of culture media and other solutions
Nylon	Hydrophilic Strong Fast aqueous flow rates Readily binds proteins Compatible with alcohols and many solvents used in HPLC	Filtering organic solvents Filtering HPLC samples
PTFE	Naturally hydrophobic Inert to most chemically aggressive solvents, strong acids, and bases Expensive	Ideal for filtering gases, air Filtering organic solvents
PVDF	Must be rendered hydrophilic Highly resistant to solvents Very low binding properties	Filtering samples before HPLC
Polypropylene	Hydrophilic Low fiber release Resistant to many solvents	Filtering samples before HPLC

2. **Binding properties.** Some membrane materials readily bind (adsorb) single-stranded DNA, RNA, and proteins. There are applications in molecular biology where this binding is desirable because it immobilizes the macromolecules. For filtration, this binding is often undesirable because the macromolecules are effectively "lost" from the sample.

3. **Surface smoothness.** Some membranes, called *screen membranes,* have a smooth surface with regularly spaced and evenly sized pores. Other membranes are more irregular in their pore structure. Smooth membranes are useful, for example, when the particles trapped on a membrane are to be viewed with a microscope.

4. **Extractables.** Some membranes have extremely low levels of extractables, whereas with others there is extraction from the membrane.

5. **Wetting properties.** Most membranes are hydrophilic, but some are hydrophobic. Almost all biological samples are aqueous and are filtered through hydrophilic membranes (which are readily wetted by water). Gases and organic solvents are almost always filtered with hydrophobic membranes (which are not readily wetted by water).

6. **Other characteristics.** These include the size of the filtration area with respect to sample size, strength of the filter, its resistance to heat, the rate at which fluids flow through it, whether it can be autoclaved, and so on.

Examples of key properties of some of the most common microfiltration membrane materials are shown in Table 31.1. While there are a variety of factors to consider when choosing a filter, the choice is simplified for routine applications because manufacturers provide filtration units tailored specifically for particular purposes.

ii. APPLICATIONS OF MICROFILTRATION

One of the most important applications of microfiltration, both in the laboratory and in industry, is to remove contaminating bacteria, yeast, and fungi from solutions. Other methods of sterilization require heat or chemicals that are destructive to certain types of materials. For example, filtration is often used to sterilize cell culture media because these media contain heat-sensitive solutes, such as vitamins and antibiotics.

Various pore sizes of microfiltration membrane will retain various sized microorganisms:

- **0.10 μm,** recommended by some manufacturers to remove Mycoplasma, which is a very small type of bacterium that can contaminate cell cultures.
- **0.22 μm,** standard pore size for removing *E. coli* bacteria
- **0.65 μm,** used to remove fungi and yeast
- **0.45–0.80 μm,** used for general particle removal
- **1.0, or 2.5 or 5.0 μm,** for "coarse" particles

Microfiltration is used both in the laboratory and in industry to sterilize both liquids and also gases, such as air and carbon dioxide. In many bacterial fermentations, air is supplied continuously to the fermentation vessel to provide agitation and oxygen. Hydrophobic microfilters are placed in the air stream to remove contaminating particles and microorganisms from the air before it enters the fermentation vessel. Filters may also be attached to supply lines for carbon dioxide and air running to animal cell culture vessels. Filters are used not only to protect the contents of fermentation vessels from contamination by outside air, but also to protect the facility from the vessel contents.

Microbiologists use microfiltration membranes when they monitor the levels of coliform bacteria in drinking water. A sample of water is filtered through a smooth membrane that retains bacteria on its surface. The membrane is then incubated on a petri dish with nutrient medium. Nutrients pass through the filter allowing the bacteria to grow into colonies. The colonies are counted to provide an estimate of the quantity of bacteria in the water supply. In an analogous way, particulates in air can be assessed by filtering large volumes of air through a membrane filter. The captured particulates can be analyzed, for example, by microscopic examination.

iii. HEPA Filters

High Efficiency Particulate Air (HEPA) *filters are used to remove particulates, including microorganisms, from air.* HEPA filters are manufactured to retain particles as small as 0.3 μm. Unlike microfilter membranes, however, HEPA filters are depth filters made of glass microfibers that are formed into a flat sheet. The sheets are then pleated to increase the overall surface area. The pleated sheets are separated and supported by aluminum baffles, see p. 190.

HEPA filters have many applications both in the laboratory and in industry. HEPA filters are used in laboratory biological safety hoods to protect products from contamination and/or personnel from exposure to hazardous substances. In animal care facilities, HEPA fil-

ters are used on the tops of animal cages to protect valuable laboratory animals from infection with microorganisms. In industry, HEPA filters may be used to filter the air in entire rooms to protect products from contaminants.

D. Ultrafiltration

i. Ultrafilters

Ultrafilters are membranes that separate sample components on the basis of their molecular weight. Membranes used for ultrafiltration have pore diameters from 0.1 to 0.001 μm and can separate particles with molecular weights ranging from about 1000 to 1,000,000.

The term *molecular weight cutoff* (MWCO) is used to describe the sizes of particles separated by an ultrafilter. The **molecular weight cutoff** *is the lowest molecular weight solute that is 90 to 95% retained by the membrane.* MWCO values are not absolute because the degree to which a particular solute is retained by an ultrafiltration membrane is not entirely dependent on its molecular weight. The shape of the solute, its association with water, and its charge also affect its permeability through an ultrafiltration membrane. For example, a membrane is less likely to retain a linear molecule than a coiled, spherical molecule of the same molecular weight. In addition, the nature of the solvent, its pH, ionic strength, and temperature, all affect the movement of solutes through membranes. By convention, if a membrane is rated to have a MWCO of 10,000, this means that the membrane will retain at least 90% of globular-shaped molecules whose molecular weight is 10,000 or greater.

When selecting an ultrafiltration membrane, many of the same considerations previously discussed come into play. The most common materials for these membranes are PES (polyethersulfone), polysulfone, and cellulose acetate. The MWCO you choose should generally be 3 to 5 times smaller than the molecular weight of the molecules to be retained on the membrane. If flow rate is a main consideration, a membrane with a higher MWCO will maximize flow rate in the filtration system, while a lower MWCO will maximize retention. Relative retention rates for specific molecules will also depend on factors such as molecular shape, pH, polarity, and other molecules present.

ii. Applications of Ultrafiltration

The applications of ultrafiltration can be classified as either fractionation, concentration, or desalting. **Fractionation** *is the separation of larger particles from smaller ones.* For example, proteins that are significantly different in size can be separated from one another by ultrafiltration. A protein product that is made by cells in culture can be separated from other components of the cell culture medium in this way. Intact strands of DNA can similarly be separated from nucleotides.

In **concentration,** *solvent is forced through a filter while solute is retained. The initial volume of the sample is thus reduced and the high molecular weight species are concentrated above the filter.* For example, gel electrophoresis is used to separate and visualize proteins. Before electrophoresis, the proteins must be concentrated because only a very small volume can be applied to the gel. Ultrafiltration can be used for this purpose.

In **desalting,** *low molecular weight salt ions are removed from a sample solution.* Ultrafiltration is a simple method to remove salts because they readily penetrate the membranes, leaving the solutes of interest on the membrane surface.

E. Dialysis and Reverse Osmosis

Dialysis and **reverse osmosis** are separation processes that use membranes similar to those used for ultrafiltration. ***Dialysis*** *is based on differences in the concentrations of solutes between one side of the membrane and the other.* Solute molecules that are small enough to pass through the pores of the membrane will diffuse from the side with a higher concentration to the side with a lower concentration. The distinctive feature of dialysis is that differences in solute concentration provide the "driving force"; dialysis does not require pumps or a vacuum to force materials through the pores of the membrane.

Dialysis membranes are typically made of a thin mesh of cross-linked cellulose esters or PVDF (polyvinylidene fluoride). They can be purchased as a roll of tubing, which is cut to length and washed thoroughly before use. As in ultrafiltration, dialysis membranes are chosen on the basis of their MWCO, which in most cases will be two to three times smaller than the molecules to be retained by the membrane.

Desalting is an example of the use of dialysis in the laboratory. During the process of purifying a protein, it is common to cause the protein to precipitate from solution by adding high concentrations of salt. Subsequent steps in the protein purification process require that the salt be removed; this is called *desalting.* The salts can be removed from the protein by placing the sample in a bag made of dialysis membrane, see Figure 31.3a. The dialysis bag containing the sample is sealed at both ends and is suspended in a large volume of water or buffer solution. Thus, the concentration of salts is much higher inside the bag than outside. The relatively large protein molecules cannot penetrate the pores of the dialysis membrane and so remain inside the bag, but small molecules, including salt, readily move through the membrane. Over time, low molecular weight salt molecules inside the bag diffuse out to the water or buffer solution. The concentration of salt inside the bag and outside the bag eventually equalizes and the system reaches equilib-

rium. The salt molecules that were originally inside the dialysis bag are now distributed throughout both the large volume of buffer (or water) and the dialysis bag. The salts are not completely removed from the sample, but their concentration is much reduced. In order to further reduce the concentration of salt in the sample, the dialysis bag can be moved into fresh water or buffer solution and the process can be repeated.

Dialysis is relatively inexpensive (compared with ultrafiltration), simple, and gentle. Because dialysis relies on passive diffusion, however, it is a relatively slow process. A number of special devices are available from manufacturers to make dialysis more efficient and convenient. For example, dialysis cassettes are sealed containers that eliminate potential sample leakage, as shown in Figure 31.3b. The sample is introduced into the cassette through a gasket using a needle and syringe. The cassette is then treated in the same manner as described for the tubing. Cassettes provide a much greater surface area than plain tubing and can therefore shorten dialysis time significantly. They can also provide better recovery rates for small volume samples.

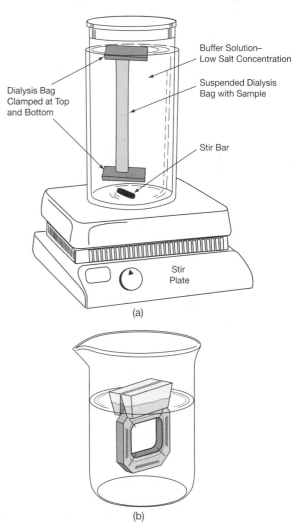

Figure 31.3.   Dialysis. a. Using standard tubing. **b.** Using a dialysis cassette.

Reverse osmosis (RO) *is used to remove very low molecular weight materials, including salts, from a liquid (usually water).* Reverse osmosis is important in water purification systems (see Chapter 28). In reverse osmosis, water under pressure flows over a special, thin RO membrane. The membrane allows water to pass through, but rejects a large proportion (on the order of 95–99%) of all types of impurities (viruses, particles, pyrogens, microorganisms, colloids, and dissolved organic and inorganic materials). The permeate will contain very low levels of contaminants that are able to get by even an RO membrane, but most types of contaminants are greatly reduced. An RO membrane retains materials based both on their size and on ionic charge and it can retain smaller solutes than an ultrafiltration membrane. RO processes can be scaled from preparation of home drinking water to massive water-desalination projects.

III. FILTRATION SYSTEMS

A. Small-Scale Laboratory Filtration Systems

The principles of filtration are the same, whether the sample is 10 μL in the laboratory or 10,000 L in industry, but the design of filtration systems depends on the scale involved. The filter's size and shape, the support of the filter, the type of force used to move fluids through the filter, and the vessels involved can vary greatly depending on the scale.

Filtration requires a force to cause materials to flow through the filter. As was illustrated in Figure 31.1 on p. 591, it is conventional to use vacuum filtration in the laboratory. Figure 31.4 shows laboratory vacuum filtration systems that take advantage of membrane filters.

Another laboratory method to gently force samples through a filter is to use a syringe. The sample is loaded into the syringe and the disk-shaped filter unit is mounted on the end. The sample fluid is forced through the filter by depressing the plunger, see Figure 31.5. Syringe filtration units contain microfilters of varying pore sizes. Sterile syringe filter units are available and are popular for sterilizing small volumes of sample, such as milliliter solutions of an antibiotic for cell culture. Very small syringe filter units are used for removing particulates from microliter volume samples prior to HPLC.

Spin filtration is a type of laboratory filtration system that uses a centrifuge to speed filtration. Spin filtration units contain ultrafilters, or sometimes microfilters, that are housed within a centrifuge tube. The sample is placed in the tube on top of the filter and the unit is spun in a centrifuge. During centrifugation, the liquid and smaller particles are forced through the filter and are captured in the bottom of the centrifuge tube. Larger molecules remain behind on the surface of

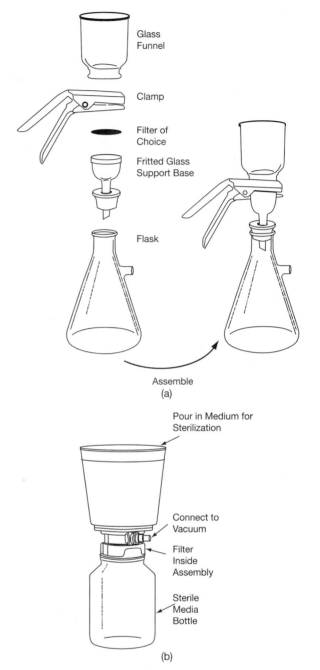

Figure 31.4. Laboratory Filtration Systems. a. A membrane filter is mounted in a holder that clamps to the top of a flask. A vacuum is used to pull the material through the filter. The clamping structure allows the membrane to be firmly sealed in the reusable glass and steel supports. **b.** A sterile filtration system used to conveniently sterilize cell culture media.

the filter, see Figure 31.6. Spin filters can be used to fractionate, concentrate, and desalt samples. They are often used in the molecular biology laboratory for small volume samples of proteins, nucleic acids, antibodies, and viruses.

Ultrafiltration plates, shown in Figure 31.7, are ideal when high sample through-put is required. These are typically found in 96- and 384-well configurations. In most

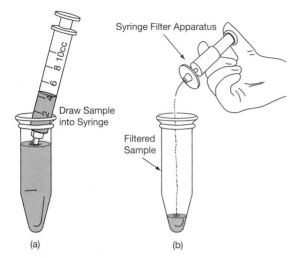

Figure 31.5. Syringe Filter Device. a. The sample is drawn into the syringe. **b.** The filter is attached to the syringe. Sample is forced through the filter by depressing the plunger.

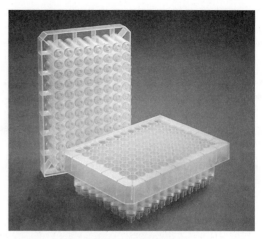

Figure 31.7. Multi-Well Filtration Plates. Note the sample collection tubes under the filters. (Courtesy of Millipore Corporation.)

cases, individual membranes are sealed into each well to prevent cross-contamination. Samples are added to the wells, and filtration can be accomplished by either plate centrifugation or application of a vacuum. These plates are well-suited for laboratory automation systems.

B. Large-Scale Filtration Systems

Membrane filtration is widely used in the biotechnology industry, in both upstream and downstream applications, for collecting and purifying products. **Process filtration** *involves large volumes of liquids, as are found in biotechnology production facilities, pharmaceutical companies, and food production facilities.* Ultrafiltration is frequently used for sterilization of large-scale biological samples, and to separate cell products from media and cell contaminants.

Upstream filters sterilize all components entering the bioreactor, including the air supply. The effective life-span of the sterilization filters is increased with the use of prefilters that remove larger contaminants from the incoming air and liquids. The products of the

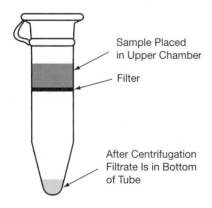

Figure 31.6. A Simple Spin Filtration System. Sample is loaded into the top of the tube. Sample is forced through the filter by the force of centrifugation.

bioreactor then proceed to downstream processing (see Chapter 3).

Many biopharmaceuticals, such as erythropoietin and tissue plasminogen activator, are made by large-scale cell culture and the desired product is often secreted by the cells into the surrounding liquid medium. Therefore, the first step in downstream processing is to separate the cells and cell debris from the liquid containing the product. This separation step is called *clarification*. Clarification is often a costly and challenging process involving thousands of liters of material and a fragile protein product. Filtration is usually preferred over centrifugation for clarification because filtration tends to be less expensive, gentler, and more convenient. There is a trend in industrial biotechnology toward using depth filtration to separate cells from their products. Clarification can be accomplished relatively quickly using cellulose-derived depth filters, which have a high capacity. Disposable filter units are increasingly common, eliminating the need for costly cleaning, sterilization, and validation procedures.

Filtration systems for industry must be designed to handle large volumes in a reasonably short time. Large-scale systems have sophisticated designs to maximize the surface area for filtration. The simplest way to do this is by using large sheets of filter membrane. More complex systems form the membranes into tubes, spirals, or pleats to maximize surface area. For example, **hollow-fiber ultrafilters** *are cylindrical cartridges packed with ultrafiltration membranes formed into hollow fibers.* The liquid to be filtered flows through the lumen of the fibers. As molecules pass through the fiber core, substances smaller than the MWCO penetrate the membrane, whereas those that are larger are concentrated in the center, see Figure 31.8 on p. 598.

Membrane clogging is a major problem in industrial-volume applications. One method to reduce clog-

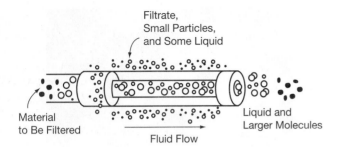

Filtrate,
Small Particles,
and Some Liquid

Material
to Be Filtered

Fluid Flow

Liquid and
Larger Molecules

Figure 31.8. Hollow Fiber Ultrafilter System. Sample flows through the lumen of the tubular filter. Small molecules pass through the filter.

ging is to use **tangential flow,** or **cross-flow filtration,** *where the fluid to be filtered flows over the surface of the filter as well as through the filter.* The sweeping motion of the fluid clears the surface of the membrane, thus reducing clogging. Tangential flow filtration is mainly used for downstream ultrafiltration applications.

Current research in new filtration technologies is focused on high-resolution product purification and removal of viral contaminants from process streams, which is increasingly accomplished with virus filters. Downstream virus filtration is essential in the manufacture of many biopharmaceuticals and other products which are derived from cells or blood plasma (such as fetal calf serum), where the materials may harbor endogenous or cross-contaminating viruses. Filter manufacturers are developing more efficient membranes to remove virus particles from media without damage to or loss of protein products.

PRACTICE PROBLEMS

1. Order the following items in terms of size, from the smallest to the largest.

   pollen grains (about 30 μm)   sand grains

   polio virus   red blood cells

   albumin (a protein)   NaCl ions

   Serratia marcescens (a type of bacteria)

2. For each of the following separations, state whether it will involve macrofiltration, microfiltration, or ultrafiltration.

   a. Purifying antibodies from a liquid medium.

   b. Removing viruses from a vaccine.

   c. Removing salts from a solution containing DNA.

   d. Sieving large particulates from water before it is treated in a sewage treatment plant.

   e. Sterilizing cell culture medium by removing bacteria.

   f. Removing pyrogens (fever causing agents) from a drug product.

   g. Harvesting mammalian cells from a fermenter.

   h. Removing *Mycoplasma* from bovine serum (serum is sometimes added to cell culture media).

3. The specifications for three filters are shown. Match the filters with the applications that follow:

   Ready Separation Hollow Fiber Filtration System

   The "Ready Separation" System is a compact hollow fiber filtration system capable of processing from 5 to 100 L. The system is sterilizable and can be purchased with microfiltration membranes from 0.1 to 0.80 μm or with ultrafiltration membranes from 300 to 500,000 MWCO.

   Filter Type XYZ

   The XYZ unit contains a hydrophobic, solvent-resistant, 0.2 μm membrane designed for use as a sterilizing filter for gases and liquids. The unit is a pyrogen-free, sterile, single use device.

   Filter Type ABC

   The ABC filtration unit is a low protein-binding, sterile filter for aqueous, proteinaceous substances. It is intended for applications where minimal sample loss is desired. It is a 0.22 μm filter, single use product for use with syringes.

   a. A fermentation process is being designed in which microorganisms produce an antibiotic that will be isolated from the broth. The microorganisms generate carbon dioxide, which is vented from the fermenter via a plastic tube. On the end of the tube is a filtration unit that keeps organisms from the outside air from contaminating the fermenter. What type of filtration unit is used?

   b. In the laboratory, an antibiotic solution needs to be added to cell culture plates. It is heat sensitive and is therefore sterilized by filtration. Which filter would be used?

   c. An organic solvent is to be used with HPLC. It needs to be filtered to remove particulates. Which filter would be used?

   d. A biotechnology company uses a large-scale cell culture process to produce a valuable protein that is being tested as a drug. The protein is secreted by the cells into the culture medium. Which type of filter might be used in the process of isolating the protein product from the cell culture medium?

CHAPTER 32

Introduction to Centrifugation

A brief note about safety: Centrifuges pose a variety of hazards to laboratory personnel. Moreover, although centrifuges and their components appear solid and sturdy, they are in fact expensive devices that can easily be damaged by mishandling. Safe and proper handling of centrifuges is discussed throughout this chapter.

I. INTRODUCTION TO CENTRIFUGATION: PRINCIPLES AND INSTRUMENTATION

A. Basic Principles

i. SEDIMENTATION

To begin thinking about centrifugation, consider a cylinder filled with a slurry of gravel, sand, and water, see Figure 32.1. We would expect that, over time, gravity would

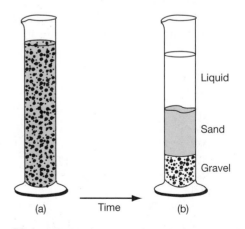

Figure 32.1. Sedimentation Due to Gravity. a. A slurry consisting of gravel, sand, and water is placed in a cylinder. **b.** The particles sediment according to their size and density under the influence of gravity.

cause the particles to **sediment** *(i.e., to move through the liquid and settle to the bottom of the cylinder)*. The larger gravel particles will sediment more quickly than the smaller sand particles so that there will eventually be a bottom layer of mostly gravel, a layer of mostly sand, and relatively clear liquid in the upper part of the cylinder. Thus, a separation, or fractionation process will occur in which the particles separate from the liquid medium and from one another due to the force of gravity.

Biologists seldom need to separate gravel and sand from one another; rather, they are interested in such materials as cells, organelles, bacteria, and viruses. These biological materials, called *particles* in the context of centrifugation, are much smaller than sand and would take a long time to sediment due to the force of gravity alone. A **centrifuge** *is a piece of equipment that accelerates the rate of sedimentation by rapidly spinning the samples, thus creating a force many times that of gravity.*

A simple centrifuge is illustrated in Figure 32.2. There is a central drive shaft that rotates during centrifugation. A **rotor** *sits on top of the drive shaft and holds tubes, bottles, or other sample containers.* As the drive shaft rotates, the sample containers spin rapidly, creating the force that facilitates the sedimentation of particles.

ii. FORCE IN A CENTRIFUGE

To appreciate the nature of the force in a centrifuge, imagine that you tie a stone to the end of a string and whirl it rapidly in a circle above your head. You will feel a pull on your hand as the stone rotates. Whenever an object is forced to move in a circular path (as in a centrifuge) force is generated. Now, consider what would happen if you whirled the stone faster—you would feel more pull. Similarly, the faster a sample is spun in a centrifuge, the more force is experienced by that sample. The speed of rotation in a centrifuge is expressed as **revolutions per minute (RPM).**

In addition to the speed of rotation, there is another (somewhat less obvious) factor that affects how much force is experienced by a sample in a centrifuge. If you whirl a stone tied to a long string, there will be more force on the stone than if you use a

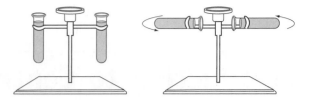

Figure 32.2. **A Simple Centrifuge.** Tubes containing a sample are rapidly spun about a central shaft, creating a force many times that of gravity.

shorter string. In a centrifuge, the further a particle is from the center of rotation, the more force the particle experiences. *The distance from the center of rotation to the material of interest is called* **the radius of rotation (r),** see Figure 32.3. In a centrifuge, the radius of rotation varies depending on the particular equipment being used.

Thus, there are two factors that determine the force experienced by a particle in a centrifuge:

1. **The speed of rotation, expressed as RPM**

2. **The distance of the particle from the center of rotation, *r***

The force acting on samples in a centrifuge is expressed as the **relative centrifugal field (RCF).** The relative centrifugal field has units that are multiples of the earth's gravitational field (g). You may see the force in a centrifuge expressed as, for example, 10,000 × g or 10,000 RCF. Both these expressions mean that the force in the centrifuge is 10,000 times greater than the normal force of gravity.

It is possible to calculate the relative centrifugal field in a centrifuge if one knows the speed of rotation and the radius of rotation. The equation for the relative centrifugal field is:

$$RCF = 11.2 \times r \left(\frac{RPM}{1000} \right)^2$$

where

RCF = the relative centrifugal field in units of × g

RPM = the speed of rotation in revolutions per minute

r = the radius of rotation in centimeters

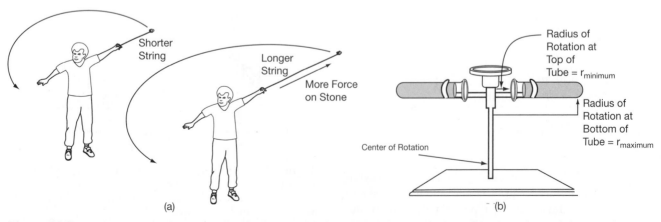

Figure 32.3. **The Radius of Rotation. a.** The longer a string being whirled, the greater the radius of rotation and the more force that is experienced by the stone. **b.** In a centrifuge, the further a particle is from the center of rotation, the more force it experiences.

You may also see this equation rearranged as:

$$\text{RCF} = 1.12 \times 10^{-5} \, (\text{RPM})^2 \, (r)$$

Observe that two centrifuges may be spinning at the same rate, yet they might each be subjecting the samples to a different force. This is because the radius of rotation varies. For this reason it is more useful to report to others the relative centrifugal field when a sample is centrifuged, rather than simply the speed of rotation.

EXAMPLE PROBLEM

Your centrifuge equipment has a maximum radius of rotation of 9.2 cm. If you spin a sample at a speed of 15,000 RPMs, what is the relative centrifugal field?

ANSWER

Substituting into the RCF equation gives:

$$\text{RCF} = 11.2 \times 9.2 \left(\frac{15,000}{1000} \right)^2$$

$$\text{RCF} = 11.2 \times 9.2 \, (225) = 23,184 \times g$$

The relative centrifugal field is **23,184 × g**.

Substituting into the rearranged form of the equation gives the same answer:

$$\text{RCF} = 1.12 \times 10^{-5} \, (\text{RPM})^2 \, (r)$$

$$\text{RCF} = 1.12 \times 10^{-5} \, (15,000)^2 \, (9.2)$$

$$= 23,184 \times g$$

EXAMPLE PROBLEM

A colleague tells you that a sample was centrifuged for a certain time at a speed of 15,000 RPM and that the force of centrifugation was 16,000 × g. Your centrifuge equipment has a radius of rotation of 9.2 cm. What speed will you use to achieve a force of 16,000 × g?

ANSWER

The unknown in this case is the speed of rotation. It is possible to algebraically rearrange the RCF equation to solve for RPMs as follows:

$$\text{RCF} = 11.2 \times r \left(\frac{\text{RPM}}{1000} \right)^2$$

$$\frac{\text{RCF}}{11.2 \, r} = \left(\frac{\text{RPM}}{1000} \right)^2$$

$$\sqrt{\frac{\text{RCF}}{11.2 \, r}} = \left(\frac{\text{RPM}}{1000} \right)$$

$$1000 \sqrt{\frac{\text{RCF}}{11.2 \, r}} = \text{RPM}$$

Then, substituting into this equation gives:

$$1000 \sqrt{\frac{16,000 \times g}{11.2 \, (9.2 \text{ cm})}} = \text{RPM}$$

$$12,461 = \text{RPM}$$

Thus, in order to achieve a force of 16,000 RCF, you will need to use a speed of 12,461 RPM. Observe that to obtain a force of 16,000 × g, the two centrifuges must be run at different speeds. **It is the force in the centrifuge that ultimately matters, not the speed of rotation.**

One tool for approximating these values without directly performing the calculations is to use a nomogram. A centrifugation **nomogram (or nomograph)** *is a chart that can be used to determine either RCF, RPM, or rotor radius values (rarely), if the other two values are known.* An example is shown in Figure 32.4 on p. 602. A nomogram has three parallel lines, each with a different log scale representing rotor radius in cm, RCF in g units, or RPM values. Taking a straight edge and connecting the two known values on the appropriate scales, the analyst can determine the third value by noting where the straight edge line crosses the additional line.

EXAMPLE

A colleague tells you that a sample was centrifuged at 30,000 × g for 15 minutes. Your rotor can achieve a maximum RCF of only 20,000 × g. What can you do?

It may be possible to compensate for the reduced force in your rotor by using a longer centrifugation run (but see the cautions at the end of this example). The time required in your rotor can be calculated using the following equation:

$$t_a = \frac{t_s \times \text{RCF}_s}{\text{RCF}_a}$$

where

t_a = time required in alternate rotor

t_s = time specified in the procedure

RCF_a = RCF of alternate rotor

RCF_s = RCF specified in the procedure

Thus:

$$t_a = \frac{15 \text{ min} \times 30,000 \times g}{20,000 \times g} = 22.5 \text{ min}$$

Spinning the sample for 22.5 minutes at 20,000 × g will duplicate the required conditions.

Note, however, that changing the time or the force of centrifugation may affect the results. Harder pelleting (more force) may damage sensitive biological materials. Longer spin times may lead to deterioration of some samples, particularly if the centrifuge is not refrigerated. Density gradient separations (discussed in a later section) are more complex than simple sedimentation and may not work if the time or RCFs are changed. For all these reasons, before changing centrifugation conditions, check that the changes will have no adverse effects.

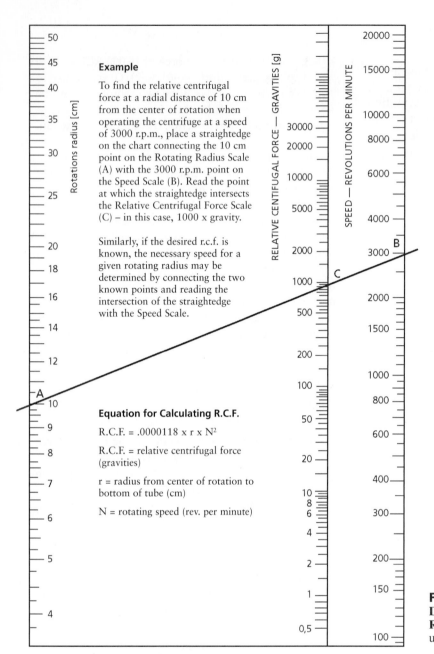

Example

To find the relative centrifugal force at a radial distance of 10 cm from the center of rotation when operating the centrifuge at a speed of 3000 r.p.m., place a straightedge on the chart connecting the 10 cm point on the Rotating Radius Scale (A) with the 3000 r.p.m. point on the Speed Scale (B). Read the point at which the straightedge intersects the Relative Centrifugal Force Scale (C) – in this case, 1000 x gravity.

Similarly, if the desired r.c.f. is known, the necessary speed for a given rotating radius may be determined by connecting the two known points and reading the intersection of the straightedge with the Speed Scale.

Equation for Calculating R.C.F.

R.C.F. = .0000118 x r x N^2

R.C.F. = relative centrifugal force (gravities)

r = radius from center of rotation to bottom of tube (cm)

N = rotating speed (rev. per minute)

Figure 32.4. Example of a Nomogram for Determining RCF or RPM Values for Specific Rotors. See text for an explanation of how to use a nomogram.

III. FACTORS THAT DETERMINE THE RATE OF SEDIMENTATION OF A PARTICLE

It is reasonable that the greater the force during centrifugation, the more quickly particles sediment. Another factor that affects the sedimentation rate is the viscosity of the liquid medium through which the particles are moving. The less viscous the medium, the faster the rate of sedimentation.

The rate of sedimentation of a particle also depends on its own characteristics. Larger particles sediment faster than smaller ones (as was also shown earlier in the example of sand and gravel sedimenting in a cylinder). It is also the case that the sedimentation rate of a particle depends on its density relative to the density of the surrounding liquid medium. Recall that density is the mass per unit volume. A stone, for example, is more

dense than a beach ball. If a stone and a beach ball are thrown into a pond, the stone will sink because it is more dense than the water, but the ball will float. The density of particles similarly affects their movement in a liquid medium during centrifugation.

If a particle is more dense than the liquid medium, then the particle sediments (sinks) during centrifugation. If the densities of the particle and the liquid medium are equal, the particle does not move relative to the medium—no matter how strong the relative centrifugal force is. Rather, the particle remains stationary in the medium. There are also situations where the medium is more dense than a particle (as water is more dense than a beach ball). In these situations, a particle can actually "float" in a centrifuge tube, or move in the opposite direction of the force of centrifugation. Thus:

1. **If the density of a particle and the liquid medium are equal, the particle does not move.**

2. **If the density of a particle is greater than the density of the medium, the particle sediments (moves away from the center of rotation).**

3. **If the density of a particle is less than the density of the medium, the particle moves toward the center of rotation (floats "upward" in the tube).**

If the liquid medium in centrifugation is water, or a water-based buffer, then the density of most biological materials, such as cells, organelles, and macromolecules, is greater than the density of the medium. These particles move toward the bottom of the tube. Lipids and fats, which are less dense than water, are found at the top of the tube after centrifugation.

In summary* the factors that influence the rate of sedimentation of a particular particle include:

1. **The relative centrifugal field.** The higher the relative centrifugal field, the faster the rate of sedimentation.

2. **The viscosity of the medium.** The less viscous the medium through which the particles must move, the faster the rate of sedimentation.

3. **The size of the particle.** The larger a particle, the faster its rate of sedimentation.

4. **The difference in density between the particle and the medium.**

iv. SEDIMENTATION COEFFICIENTS

You may see the term **sedimentation coefficient, or S value. The sedimentation coefficient** *is a value for a particular type of particle that describes its rate of sedimentation in a centrifuge.* The larger the S value, the faster the particle will sediment. The sedimentation coefficient is a physical constant that refers to a particular biological particle. Various proteins, different organelles, and nucleic acids each have their own sedimentation coefficient. Sedimentation coefficients have been measured for a variety of biological particles; some examples are shown in Table 32.1. Observe in this table that larger particles generally have larger sedimentation coefficients.

Table 32.1.   EXAMPLES OF SEDIMENTATION COEFFICIENTS

Biological Particle	Sedimentation Coefficient, S
Soluble Proteins	
Ribonuclease (MW = 13,683)	1.6
Ovalbumin (MW = 45,000)	3.6
Hemoglobin (MW = 68,000)	4.3
Urease (MW = 480,000)	18.6
Other Materials	
E. coli rRNA	20
Calf liver DNA	20
Ribosomal subunits	40
Ribosomes	80
Tobacco mosaic virus	200
Bacteriophage T2	1000
Mitochondria	15,000–70,000

Notes: 1. The sedimentation coefficient depends on the liquid medium and the temperature. Standard conditions for these coefficients are in water at 20°C.
2. Sedimentation coefficients are expressed in Svedberg units, S, where one Svedberg unit = 10^{-13} seconds.

v. APPLICATIONS OF CENTRIFUGATION

Centrifugation has a long history in the biology research laboratory, where it is used in the separation, isolation, purification, and analysis of biological materials. In the biotechnology industry, centrifugation often plays an essential role in the separation of cells from broth after fermentation.

Centrifugation applications may be broadly classified as either preparative or analytical. **Preparative centrifugation** *provides separated materials for later use.* Examples of applications of preparative centrifugation are in Table 32.2 on p. 604.

Analytical centrifugation *is used in specialized research settings to determine the molecular weight, purity, shape, and other physical characteristics of proteins and other macromolecules.* Analytical centrifugation involves specially designed, computer-controlled centrifuges that allow the operator to monitor the movement of particles. Analytical centrifugation is not nearly as common as preparative centrifugation and is not considered further in this text.

*You may come across the following equation, which combines these various factors as follows:

$$v = \frac{d^2(\rho_p - \rho_m)\, g}{18\,\mu}$$

where
v = velocity of sedimentation
d = the diameter of the particle
ρ_p = density of the particle
ρ_m = density of the medium through which the particle is sedimenting
g = the relative centrifugal force
μ = the viscosity of the medium

Although this equation may seem formidable because it has several unfamiliar symbols, it is possible to obtain information from it fairly easily. Observe that the viscosity of the medium is in the denominator—the more viscous the medium, the slower the velocity of sedimentation. The g force is in the numerator—the higher the g force, the faster the sedimentation velocity. The difference in density between the particle and the medium is also in the equation. When a particle and the medium are of the same density, $(\rho_p - \rho_m) = 0$, which means the sedimentation velocity is zero—the particle does not move. If the particle is more dense than the medium the particle moves with a certain velocity toward the bottom of the tube. It is also possible for the particle to be less dense than the medium. In this case, the velocity is a negative number—that means the particle moves upward in the tube.

Table 32.2.   EXAMPLES OF APPLICATIONS OF PREPARATIVE CENTRIFUGATION

Separation of intact, single-cell suspensions such as bacteria, viruses, and blood cells from a liquid medium

Separation of cellular organelles, such as mitochondria and ribosomes from disrupted cells

Separation of biological macromolecules, including DNA, RNA, and protein from a solution

Separation of plasma (the fluid portion of blood) from blood cells

Separation of immiscible liquids from one another

Separation of cells from broth after fermentation

B. Design of a Centrifuge: Overview

The typical components of a centrifuge are shown in Figure 32.5. The motor, which turns the drive shaft, is located at the base of the unit. A rotor sits on top of the drive shaft and holds the sample containers. With the exception of older, low-speed instruments, centrifuges have a protective enclosure so that the rotor sits in a chamber. In modern centrifuges the chamber has a cover that is usually locked when the centrifuge is spinning. In higher speed centrifuges, the enclosure and cover are made of reinforced steel which, in the event of a serious mishap, can withstand the tremendous force of a rotor that flies off the drive shaft. Note that older high-speed centrifuges may lack the protective safety features of newer models. Most centrifuges also have a braking device to slow the rotor, a microprocessor control device to set the speed, and a timer.

There are many styles and models of centrifuges, so it is convenient to classify them into types. The two main physical types are floor models and benchtop

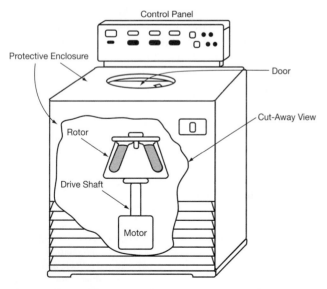

Figure 32.5.   Basic Components of a Centrifuge. A floor model is shown.

instruments. While benchtop centrifuges are particularly convenient, floor models frequently offer higher sample capacities and sometimes higher speeds as well. A simple classification by speed is summarized in Table 32.3 and in the following list:

- **Low-speed centrifuges** typically attain rotation speeds less than 10,000 RPM and forces less than $8000 \times g$.
- **High-speed centrifuges** typically attain rotation speeds up to about 30,000 RPM and generate forces up to about $100,000 \times g$.
- **Ultraspeed centrifuges**, or **ultracentrifuges**, typically attain rotation speeds up to about 120,000 RPM and generate forces up to $800,000 \times g$.

Heat is generated as a rotor turns. High-speed centrifuges and ultracentrifuges, therefore, are usually refrigerated. Refrigeration is preferred for most biological samples to protect them from adverse effects due to heat, although intact cultured cells are usually centrifuged gently at room temperature. Some benchtop centrifuges can be installed in a cold room to protect samples during longer runs.

Ultracentrifuges have vacuum pumps that evacuate air from the chamber. This is necessary because the friction of air rubbing against the spinning rotor creates drag and heat. Vacuum is commonly achieved using a system with two pumps. The first pump is a "roughing pump" that draws air out of the system beginning at atmospheric pressure. A diffusion pump takes over when some vacuum has been established. Diffusion pumps generate heat and so require a cooling system.

Centrifuges can be classified both on the basis of speed and by their capacity or by the sizes and types of sample containers they can spin. For example, **microfuges** or **microcentrifuges** *are small centrifuges that are intended for small volume samples in the microliter to 1 or 2 mL range.* Microfuges are useful in molecular biology laboratories for conveniently pelleting small volume samples. At the other end of the spectrum, there are industrial, large-capacity centrifuges that are used in production facilities (e.g., to separate large volumes of cells from nutrient broth).

II. MODES OF CENTRIFUGE OPERATION

A. Differential Centrifugation

Preparative centrifugation can be performed in several different modes. Of these modes, *differential centrifugation,* also called *differential pelleting,* is the simplest and probably the most common in the biology laboratory. **Differential centrifugation** *separates a sample into two phases: a* **pellet** *consisting of sedimented materials,* and a **supernatant** *consisting of liquid and unsedimented particles.* A given type of particle may move into the pellet,

Table 32.3.   CENTRIFUGES

Characteristics	Type			
	Low Speed	**High Speed**	**Ultracentrifuge**	**Microfuge**
Maximum Speed (RPM)	10,000	30,000	120,000	15,000
Maximum Force (× g)	8000	100,000	800,000	21,000
Refrigeration	Sometimes	Yes	Yes	Sometimes
Vacuum	No	Sometimes	Yes	No
Examples of applications	Harvest intact plant and animal cells. Separate blood components. Harvest larger organelles (such as nuclei). Separate immiscible liquids.	Purify viruses. Isolate mitochondria, lysosomes, and other moderate size organelles.	Purify small organelles such as ribosomal subunits. Purify dissolved DNA, RNA, and proteins.	Purify precipitated nucleic acids. Perform small volume separations.

remain in the supernatant, or be found in both phases. Which sample components move into the pellet and which remain in the supernatant depends on the nature of the components, on the force, and on the duration of centrifugation.

In differential centrifugation, a uniform sample is poured into a centrifuge tube or bottle. The sample is centrifuged for a certain time with a particular relative centrifugal force. As the sample spins, larger particles sediment faster than smaller ones. By using a suitable combination of g force and time, a pellet enriched in a particular component can be obtained, see Figure 32.6.

After centrifugation, the supernatant can readily be decanted from the tube. It is possible to continue by centrifuging the supernatant again at a higher force, thus obtaining a second pellet. This process can be repeated, resulting in a series of pellets containing progressively

smaller particles. For example, a common application of differential centrifugation is the isolation of cellular organelles from a cellular homogenate. (A cellular homogenate is produced by disrupting cells and releasing their organelles, membranes, and soluble macromolecules into a liquid medium.) Nuclei are the largest organelle. It is possible to obtain a pellet enriched in nuclei by centrifuging a cellular homogenate for 10 minutes at 600 × g. Other organelles tend not to move into the pellet given this relatively low centrifugal field and short run time. The supernatant is then removed and spun at a higher g force. This process is continued until the smallest subcellular particles have been pelleted, see Figure 32.7 on p. 606.

Although differential centrifugation is simple and widely used, it has disadvantages. When a sample mixture is initially poured into a centrifuge tube, the various sample components are dispersed throughout the tube. Thus, each pellet consists of sedimented components contaminated with whatever unsedimented particles were at the bottom of the tube initially. Moreover, the yield of smaller particles is lowered by their loss in early pellets. Differential centrifugation may also require starting and stopping the centrifuge repeatedly, which is time-consuming for the operator. Density gradient centrifugation is another mode of centrifugation that reduces or eliminates these problems.

B. Density Centrifugation

i. THE FORMATION OF DENSITY GRADIENTS

A **density gradient** *is a column of fluid that increases in density from the top of the centrifuge tube to the bottom.* **Density gradient centrifugation** *includes several techniques for separating molecules based on their rate*

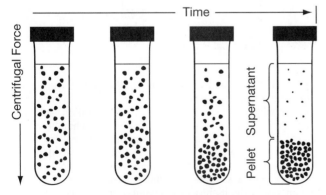

Figure 32.6.   Differential Centrifugation. A suspension of particles in a liquid medium is initially uniformly distributed in the centrifuge tube. During centrifugation, particles sediment into a pellet at differing rates depending mainly on their size.

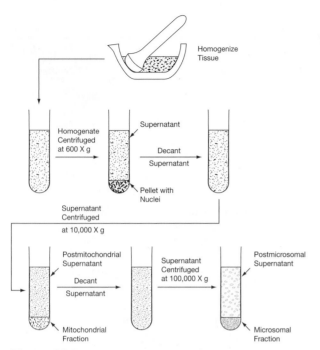

Figure 32.7.   Example of a Scheme to Isolate Cellular Organelles Using Differential Centrifugation. Various organelles are separated from one another as the sample is centrifuged at progressively higher forces. The final microsomal fraction includes fragments of the endoplasmic reticulum and attached ribosomes.

of sedimentation or their buoyancy in a density gradient. As different particles move through the density gradient, they separate into discrete bands, see Figure 32.8.

Density gradients can be formed in three ways. Some high molecular weight solutes, such as cesium chloride (CsCl), form gradients spontaneously during centrifugation. With these media, the sample and the gradient material are mixed together and centrifuged. As centrifugation proceeds, a gradient forms.

Many density gradient media do not spontaneously form gradients, so the operator prepares the gradient before the separation is begun. A preformed gradient may either be a step gradient or a continuous gradient. A **step gradient** *is formed in layers with the densest layer on the bottom. Each layer has a different density and is clearly demarcated from the layers above and below.* A step gradient is prepared from separate solutions of the gradient medium, each of which has a different concentration and therefore a different density. The different solutions are carefully layered one on top of the other using a pipette or syringe, see Figure 32.9a. The sample is loaded on top of the preformed gradient.

In **continuous gradients** *there is a smooth decrease in density from bottom to top with no sharp boundaries between layers.* Preformed continuous gradients are often produced by continuously mixing together two solutions of different density while simultaneously delivering the mixture to the centrifuge tube. This

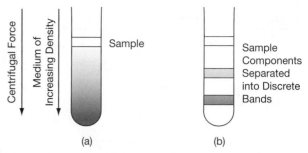

Figure 32.8.   Density Gradient Centrifugation. a. A sample is layered on top of a preformed density gradient. **b.** During centrifugation sample components separate into discrete bands as they move through the gradient.

requires a special apparatus, called a **gradient maker,** see Figure 32.9b. Alternatively, if a gradient maker is unavailable, a preformed step gradient will diffuse into a continuous gradient over time.

There is a wide variety of materials used to make density gradients. Sucrose and Ficoll are examples of media often used for preformed gradients. Percoll, a colloidal silica, is commonly used for separation of cells and organelles. Cesium chloride and other cesium salts form a gradient spontaneously when centrifuged. The medium chosen must be compatible with the sample, must have a sufficient range of densities to separate all the components of the sample, must not interfere with later steps in the procedure, must not be corrosive to the particular rotor being used, and so on. Because the

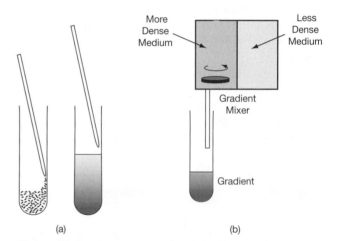

Figure 32.9.   Preparing Density Gradients. a. Layering a step gradient. **b.** Preparing a continuous gradient using a gradient maker. A high-density (high-concentration) solution of the gradient medium is placed in one of the chambers of the gradient maker; this chamber contains a stir bar for mixing the solution. A lower-density solution is placed in the other chamber. A narrow tube connecting the two chambers is opened. At the same time, liquid is allowed to begin flowing out of the high-density chamber into the centrifuge tube. Only dense solution initially reaches the centrifuge tube. As time goes on, more and more of the less dense solution flows into the mixing chamber, progressively diluting the medium. As a result, the medium flowing into the centrifuge tube slowly decreases in density (concentration), thus forming a continuous density gradient in the centrifuge tube.

requirements for the medium vary from application to application, different density gradient media have been developed and are available from manufacturers.

ii. RATE ZONAL CENTRIFUGATION

There are two methods of density centrifugation: rate zonal and isopycnic (also known as buoyant density centrifugation). **Rate zonal (size) centrifugation** *is a time-dependent method in which particles are separated as they sediment through a density gradient.* In the rate zonal method, the sample solution is layered on top of a preformed, usually continuous, density gradient. During centrifugation the particles sediment through the gradient. The rate of movement of the particles depends primarily on their size, less so on their density, so different particles move at different rates. As the particles move at different speeds through the gradient, they separate into bands, or zones (Figure 32.8).

In rate zonal centrifugation, the particles are more dense than the most dense part of the gradient. This means that if centrifugation is continued too long, the particles will all move to the bottom of the tube, mixing again into a pellet. Rate zonal centrifugation, therefore, is time dependent and is halted once the sample components have separated into discrete bands.

iii. ISOPYCNIC CENTRIFUGATION

Isopycnic centrifugation (also called buoyant density centrifugation) *is a method in which particles are separated based on their density alone.* As a simple example, suppose that it is necessary to separate two types of particles, one of which has a density of 1.50 g/mL and the other a density of 1.35 g/mL. This can be accomplished by layering the sample on top of a liquid medium that has a density of, for example, 1.40 g/mL. After centrifugation, the less-dense particles will form a floating band, whereas the more dense particles will be in a pellet in the bottom of the tube.

In the simple previous example, there is not actually a density gradient; rather, there is a single density interface, called a *density barrier.* More complex isopycnic separations involve a density gradient. The gradient may be a preformed step gradient, a preformed continuous gradient, or a gradient that forms spontaneously during centrifugation. However the gradient is formed, the density of the medium at the bottom of the gradient must be greater than the density of any of the particles to be separated.

During isopycnic centrifugation, particles move away from the center of rotation if they are more dense than the medium in which they are located. Particles move toward the center of rotation if they are less dense than the medium at that point. Equilibrium is eventually reached when all the particles are banded into the portion of the density gradient where their density equals that of the medium around them.

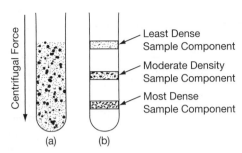

Figure 32.10. Isopycnic Centrifugation. a. A uniform mixture of gradient medium and sample is placed in a tube. **b.** During centrifugation, particles separate into bands such that at equilibrium their density equals that of the medium.

Although the size of a particle does affect its *rate* of migration, buoyant density separations are ultimately based on density, not size. Isopycnic centrifugation thus differs from rate zonal and differential centrifugation where particle size affects the final separation*.

Observe that in isopycnic centrifugation, the sample can be loaded anywhere in the tube. If the sample is placed on top of the gradient, the various particles will migrate down the tube to reach their equilibrium positions. If the sample is loaded in the bottom of the tube with the density gradient on top, the sample components will float up to reach their equilibrium positions. The sample can also be homogeneously mixed throughout a medium that forms a spontaneous gradient. During centrifugation, the gradient forms and particles move till they are in equilibrium with the surrounding medium, see Figure 32.10. Isopycnic centrifugation is therefore not time dependent.

After a density gradient separation is completed, there are various methods to collect the banded sample components. Figure 32.11 shows a technique in which the bottom of the tube is punctured and drops are collected in successive test tubes. Another method of retrieving the sample is to puncture the side of the tube with a needle at the site of a specific band.

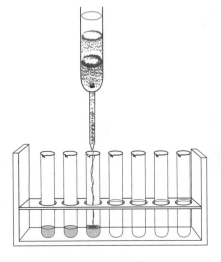

Figure 32.11. Collecting the Fractions after Density Gradient Centrifugation. Collecting fractions from the bottom of a tube.

*There are many techniques that can separate materials on the basis of size, including rate zonal and differential centrifugation, electrophoresis, and chromatography. Isopycnic centrifugation is the only common method that separates materials based solely on differences in their densities.

Table 32.4. MODES OF PREPARATIVE CENTRIFUGATION

Mode	Description
1. Differential Pelleting	Results in two phases: a pellet of sedimented material and a supernatant consisting of liquid and unsedimented particles.
2. Density Gradient Centrifugation	The liquid medium increases in density from the top to the bottom of the centrifuge tube or bottle.
a. Rate Zonal	Sample components form bands due to differing rates of migration through the medium. Run is timed; equilibrium is not reached. Separation is based on size (primarily) and mass.
b. Isopycnic	Sample components form bands where their density equals that of the surrounding medium. Equilibrium is achieved. Separation is based only on density.
3. Continous Centrifugation	For large-volume samples. Sample enters centrifuge and supernatant exits continuously as rotor spins.

C. Continuous Centrifugation

Continuous centrifugation is a mode of centrifugation intended for large-volume samples, such as when cells must be separated from a large-volume fermentation broth. In continuous centrifugation, the sample is continuously fed into a rotor that is rotating at its operating speed. Particles sediment in the rotor while the supernatant is conveyed out of the rotor continuously. Special rotors, centrifuge equipment, and methods are required for continuous centrifugation.

The various modes of preparative centrifugation are summarized in Table 32.4.

III. INSTRUMENTATION: ROTORS, TUBES, AND ADAPTERS

A. Types of Rotors

i. HORIZONTAL ROTORS

Samples are centrifuged inside centrifuge tubes (or bottles) that are placed in compartments in rotors. Individuals who use centrifuges must be able to choose rotors, tubes, and accessories for each application. The user must also be able to properly clean, sterilize (when appropriate), and store these centrifuge components. These practical aspects of centrifuge operation are discussed in this section.

There are three commonly used styles of rotor:

- **Horizontal, also called swinging bucket**
- **Fixed Angle**
- **Vertical**

Examples of these three rotor types are shown in Figure 32.12.

Because of differences in the way samples experience the force of centrifugation in these rotor types, the rotors vary in their relative suitability for specific centrifugal applications, as shown in Table 32.5.

The horizontal (swinging bucket) rotor allows tubes to swing up into a horizontal position during centrifugation, see Figure 32.13. As centrifugation proceeds, particles move toward the bottom of the tube. Particles beginning at the top of the tube move the entire length of the tube to form a pellet. This maximizes the path length through which particles move. (As we will see, the path length is shorter in other types of rotors.) The long path length results in good separations of similar particles by rate zonal centrifugation. Swinging bucket rotors are also useful for differential centrifugation because they give well-formed pellets. A disadvantage to swinging bucket rotors is that, because of their design, they must usually be run at slower speeds than fixed angle or vertical rotors. Because the path length is longer and the speed is slower in horizontal rotors than in other rotors, separations take longer.

(a)

(b)

(c)

Figure 32.12. Common Rotor Types for Centrifugation. a. Horizontal (swinging bucket) rotor. **b.** Fixed angle rotor. **c.** Vertical rotor. (Photos courtesy of Beckman Coulter.)

Table 32.5.   GENERAL SUITABILITY OF ROTORS FOR DIFFERENT CENTRIFUGAL APPLICATIONS

Rotor Type	Applications		
	Differential Pelleting	Rate Zonal Centrifugation	Isopycnic Centrifugation
Horizontal (Swinging Bucket)	+	+++	+++ for cells + for macromolecules
Fixed Angle	++++	+	+++ for macromolecules + for cells
Vertical	---	+++	++++

Observe in Figure 32.13a that the radius of rotation (r) for a particle varies depending on the particle's location in the tube. A particle at the top of the tube, closest to the center of rotation, is at the point where the radius of rotation, called r_{min} (minimum radius), is shortest. In the middle of the tube the radius is called r_{ave} (average radius) and at the bottom of the tube the radius is called r_{max} (maximum radius). When calculating the g-force experienced by a particle, it is necessary to consider its position in the tube. If the position is not known, use the average radius.

ii. FIXED ANGLE ROTORS

Fixed angle rotors *hold centrifuge tubes or bottles in compartments at a particular angle, usually between 15 and 40 degrees*, see Figure 32.14a. The force of

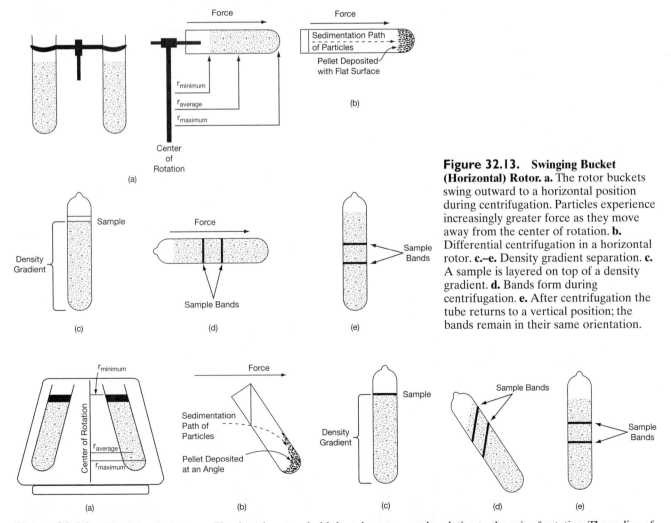

Figure 32.13.   Swinging Bucket (Horizontal) Rotor. a. The rotor buckets swing outward to a horizontal position during centrifugation. Particles experience increasingly greater force as they move away from the center of rotation. **b.** Differential centrifugation in a horizontal rotor. **c.–e.** Density gradient separation. **c.** A sample is layered on top of a density gradient. **d.** Bands form during centrifugation. **e.** After centrifugation the tube returns to a vertical position; the bands remain in their same orientation.

Figure 32.14.   Fixed Angle Rotor. a. Fixed angle rotors hold the tubes at an angle relative to the axis of rotation. The radius of rotation varies depending on the location of a particle in the tube. **b.** Differential centrifugation in a fixed angle rotor. **c.–e. Density gradient separation in a fixed angle rotor. c.** A sample is layered on top of a density gradient. **d.** Particles form bands as centrifugation proceeds. **e.** After centrifugation, when the tube is carefully removed from the rotor, the contents of the tube reorient.

centrifugation causes particles to move outward toward the sides of the tubes. During differential pelleting, particles accumulate at the outer tube wall and then slide downward until they pellet along the side at the bottom of the tube, see Figure 32.14b on p. 609. The path length traveled by particles across the tube in a fixed angle rotor is relatively short. Fixed angle rotors, therefore, result in rapid pelleting (compared to horizontal rotors). A disadvantage to fixed angle rotors for pelleting is that fragile particles, such as intact cells, may be damaged as they are forced against the walls of the tube. Horizontal rotors are gentler.

As in a swinging bucket rotor, the radius of rotation varies for a particle depending on its location in the tube, see Figure 32.14a. At the top of the tube the radius is r_{min}; at the bottom of the tube the radius is r_{max}. Particles at the bottom of a centrifuge tube in a fixed angle rotor therefore experience a higher g force than particles at the top of the tube.

Fixed angle rotors are used for density gradient separations as well as for differential pelleting, see Figure 32.14c–e. Note that the density gradient formed in a fixed angle rotor reorients horizontally when centrifugation ends and the tubes are removed from the rotor.

iii. NEAR VERTICAL AND VERTICAL TUBE ROTORS

Near vertical tube rotors *are a modification of fixed angle rotors that have a shallow angle of about 8 to 10 degrees.* These rotors are recommended for isopycnic separations where short run times are desired.

Vertical rotors hold tubes straight up and down resulting in a very short path length and fast separations, see Figure 32.15. Their main use is for isopycnic separations and they can also be used for rate zonal separations.

iv. κ FACTORS

You may come across the term *k factor* (clearing factor) in reference to a rotor, particularly when working with ultracentrifugation. *The* **k** **factor** *provides an estimate of the time required to pellet a particle of a known sedimentation coefficient.* The k factor varies for different rotors, depending on their minimum and maximum radii and on the speed of rotation. (Manufacturers typically report the k factor at the maximum speed that the rotor can attain.) The smaller the k factor for a rotor, the faster the separations using that rotor. One use of k factors, therefore, is in comparing the efficiency of various rotors.

The *k* factor is calculated by the manufacturer so that it is possible to approximate a particle's sedimentation time, in hours, using the following equation[*]:

$$t = k/S$$

where
t = the predicted time (in hours) to move particles from the top of a tube, at r_{min}, to the bottom of a tube, at r_{max}
k = the clearing factor
S = the sedimentation coefficient in Svedberg units

[*]This formula assumes that the liquid medium is water at 20°C. If the medium is denser or more viscous than water, the sedimentation time will be longer.

Beckman Coulter, Inc. provides a rotor calculation resource at http://www.beckmancoulter.com/resourcecenter/labresources/centrifuges/rotorcalc.asp. This can be used to calculate RPM, RCF (average and maximum), and *k* factor, based on available information about any rotor. They also have a linked resource for run-time conversions with *k* factors.

EXAMPLE

The manufacturer reports that the *k* factor for a rotor is 100. How long will it take ribosomes to sediment through water to the bottom of a tube?

Ribosomes have a sedimentation coefficient of 80 (Table 32.1 on p. 603). Therefore:

$$t = \frac{100}{80} = 1.25 \text{ hours}$$

B. Balancing a Rotor

One of the most important tasks performed by the operator of a centrifuge is making sure that the rotor is used properly and that the tubes or bottles are balanced in the rotor. The purpose of balancing the load is to ensure that the rotor spins evenly on the drive shaft. This is essential for minimizing rotor stress and preventing damage to the centrifuge, rotor, and tubes. In worst-case accidents, improperly used, unbalanced, or damaged rotors have detached from the spindle and flown out of the centrifuge chamber at high velocity, wreaking havoc along their path, see Figure 32.16. Fortunately, modern centrifuges are well armored, which helps prevent rotors from exiting the centrifuge in an accident. However, if a rotor detaches from the spindle at high speed, even if it does not exit the cen-

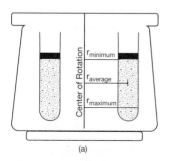

(a)

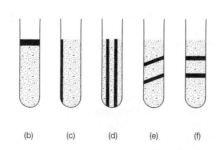

(b)   (c)   (d)   (e)   (f)

Figure 32.15.   Vertical Rotor. a. Tubes are vertical in the rotor compartments. **b–f.** Density gradient separation in a vertical rotor. **b.** A sample is layered on top of a density gradient. **c.** When centrifugation begins, the tube contents reorient under the influence of the relative centrifugal field. **d.** The sample sediments into vertical bands. **e–f.** When the rotor comes to a stop and the tubes are carefully removed from the rotor, the tube contents reorient forming horizontal bands.

Figure 32.16.   Scene Following an Accident in Which an Ultracentrifuge Rotor Failed during Operation. The subsequent "explosion" destroyed the centrifuge, ruined a nearby refrigerator and freezer, and made holes in the walls and ceiling. The centrifuge itself was propelled sideways and cabinets containing hundreds of chemicals were damaged. Fortunately, cabinet doors prevented chemical containers from falling to the floor and breaking. A shock wave from the explosion shattered all four windows in the room and destroyed the control system for an incubator. Luckily, no one was injured. The cause of the accident was the use of a model of rotor that was not approved by the manufacturer for the particular model of centrifuge. (Courtesy of Cornell Office of Environmental Health and Safety, Cornell University, Ithaca, NY.)

trifuge chamber, it will crash into the centrifuge chamber walls with great force, destroying the centrifuge chamber and smashing the rotor into small fragments. The damage to the centrifuge may be irreparable and will at best be extremely costly. A centrifuge accident will also spread the sample throughout the laboratory, usually as an aerosol—a serious problem if the sample is hazardous.

A rotor is balanced by placing samples and their containers symmetrically in the rotor, see Figure 32.17.

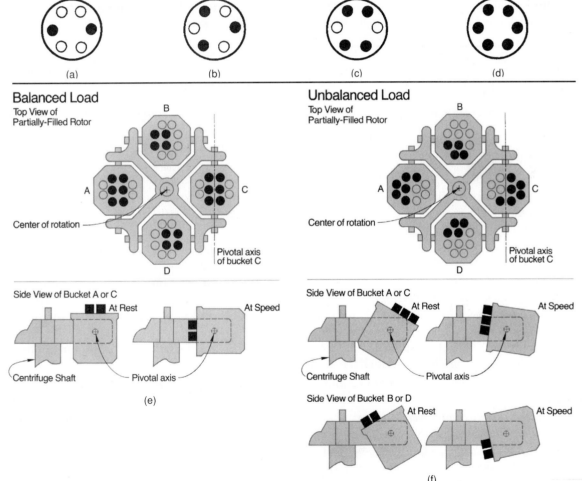

Figure 32.17.   Balancing a Rotor Load. a–d. Rotors containing two, three, four, and six tubes, respectively. Opposing tubes have the same weight (and density profile for ultracentrifuges). **e–f.** The situation is more complex in buckets holding multiple tubes. Each bucket in this rotor holds 10 tubes. **e.** The opposing bucket sets A–C and B–D are loaded with an equal number of tubes, each filled with an equal weight of sample, and are balanced across the center of rotation. Each bucket is also balanced with respect to its pivotal axis. **f.** This rotor is not balanced. None of the four buckets is balanced with respect to its pivotal axis. During centrifugation, buckets A and C will not reach a horizontal position, whereas buckets B and D will pivot past horizontal. Note also that the tubes in B and D are not arranged symmetrically across the center of rotation. (Reprinted by permission of Beckman Coulter, Inc.)

Tubes or bottles across from one another should be of nearly equal weight. The manufacturer specifies how close in weight the tubes that are opposite from one another must be. For example, for a low-speed centrifuge the manufacturer might specify that opposing tubes should be balanced so that there is no more than 0.5 g difference in their weight. For an ultracentrifuge, the manufacturer might specify that opposing tubes must be within 0.1 g of one another. Always adhere to the manufacturer's instructions for balancing tubes. Note also that for ultracentrifuges, tubes or bottles that are across from one another must contain the same density profile as well as match in weight. Guidelines regarding proper balancing of a rotor and operation of a centrifuge are in Box 32.1.

C. Care of Centrifuge Rotors

i. OVERVIEW

Although rotors appear to be massive and sturdy, they are subjected to tremendous forces during centrifugation and can easily be damaged. A damaged or improperly used rotor can fail during centrifugation with catastrophic results. Always read the instructions that come with a rotor. General guidelines for care and cleaning that are relevant to most brands of rotor are discussed in this section and are summarized in Box 32.2.

Recall that the force generated in centrifugation is:

$$RCF = 11.2 \times r \left(\frac{RPM}{1000} \right)^2$$

This equation tells us that the RCF generated is proportional to the square of the speed. For example, if the RPMs are doubled, then the rotor is subjected to four times the original force. The stress on ultracentrifuge rotors, therefore, is considerably greater than that on lower speed rotors. Small imperfections in an ultracentrifuge rotor can cause the rotor to fail when it is centrifuged at high speeds. For this reason, rotors, particularly those for the ultracentrifuge, are made to exacting specifications and must be correctly handled and maintained.

Each rotor is specified by the manufacturer for a maximum allowable speed above which it should never be run. Most high-speed and ultracentrifuge rotors have a protective mechanism that prevents them from being rotated at excessive speeds. This mechanism often

---

Box 32.1. OPERATING A CENTRIFUGE

BE CERTAIN TO CONSULT THE MANUFACTURER'S INSTRUCTIONS WHEN USING A CENTRIFUGE. Guidelines that are generally applicable are shown in this box.

1. *For low- and high-speed centrifuges, opposing tubes can be balanced according to weight without considering density.*
   a. The simplest way to balance tubes or bottles is to place them on opposite sides of a two-pan balance and bring them to equal weight.
   b. Remember to include caps and seals when balancing tubes.
   c. The manufacturer specifies how close in weight the tubes must be. For example, for a low-speed centrifuge the manufacturer may specify that the tubes should be balanced so that there is no more than 0.5 g difference in their weight.
   d. Opposing tubes may both contain sample, or one of the tubes may be a "blank" containing only a liquid, such as water.
2. *In an ultracentrifuge, the tubes that are across from one another must be of the same type with respect to density, as well as be of the same weight.*
   a. Fill opposing tubes with materials of the same density profile and make certain that their weights also match.
   b. Opposing tubes may both contain sample, or one of the tubes may be a "blank" containing only a liquid medium or a density gradient to match the sample tubes.
3. *Tubes that are not directly across from one another can be of different weights and/or have different density profiles.*
4. *Do not exceed the maximum weight in any sample compartment, as specified by the rotor manufacturer.* Be sure to include the weight of any seals, plugs, or caps.
5. *All buckets must be in place in a swinging bucket rotor, even if some contain no samples.* If the buckets are numbered and have caps, they must be run in the correct positions with their caps.
6. *Make sure the rotor cover is present and is properly tightened.*
7. *Do not operate the centrifuge at the critical speed.* At the **critical speed** *any slight imbalance in the rotor will cause vibration.*
8. *Pay attention to the brake setting.* The brake slows the rotor after a run is completed. On many centrifuges it is possible to select a brake setting. If braking is too abrupt, some types of pellets will become resuspended.

Box 32.2.   SUMMARY OF GUIDELINES FOR CLEANING, MAINTAINING, AND HANDLING ROTORS, TUBES, AND ADAPTERS

1. *Regularly clean rotor and adapters.*
   a. Use a mild, nonalkaline detergent and warm water.
   b. Thoroughly remove all disinfectants and detergents using distilled water.
   c. With swinging bucket rotors, clean only the buckets themselves. Do not immerse the rotor portion because its pins cannot readily be dried and will rust if wetted.
   d. Air dry and store rotors upside down to allow liquids to drain. Make sure rotors remain dry during storage. Do not store rotors in a cold room due to the humidity. If a chilled rotor is needed, cool immediately before use.
   e. Disinfect and sterilize rotors according to manufacturer's instructions.
2. *Protect rotors from corrosive agents, including moisture, chemicals, alkaline solutions, and chloride salts.* Aluminum is particularly sensitive to corrosion. The cover and the compartments of titanium rotors may be aluminum; therefore, they are more vulnerable than the rest of the rotor.
   a. Immediately wash the rotor if it is exposed to corrosive materials.
   b. Routinely check the rotor visually for hairline cracks, dents, white spots, and pitting. If damage is detected, do not use the rotor without consulting a service technician.
3. *Do not scratch a rotor.*
   a. Use soft bristle brushes for cleaning.
   b. Do not place metal forceps or other implements into rotor compartments.
   c. Do not use metal implements to remove stuck tubes or to open tight lids, except as provided by the manufacturer.
4. *Check rotor O-rings before every run.* Many rotors have O-rings that maintain a seal during a run. These O-rings and the surfaces they contact must be kept clean and in good condition. Some manufacturers recommend lightly greasing the O-rings with silicone vacuum grease. Check O-rings for cracks, tears, or an uneven surface; replace if necessary.
5. *Consult the ultracentrifuge manufacturer before using a rotor made by a different manufacturer.*
6. *Find out from the manufacturer how long, or for how many runs, a rotor is guaranteed.* Derate or retire a rotor as directed by the manufacturer.
7. *Never exceed the maximum speed of a rotor.*
   a. Never remove overspeed protective devices.
   b. Pay attention to the density of solutions in ultracentrifugation. It may be necessary to run the centrifuge more slowly than usual if a high-density solution is used.
   c. Uncapped tubes, partially filled tubes, and heavy tubes and/or samples may require that the rotor be used at less than its maximum rated speed.
8. *Clean centrifuge tubes and bottles according to the manufacturer's directions.* General guidelines include:
   a. Reusable tubes, bottles, and adapters should be cleaned by hand with a mild detergent using soft, non-scratching brushes. Commercial dishwashers are likely to be too harsh.
   b. Bottles and tubes should not be dried in an oven; rather, they should be air dried. Stainless steel tubes must be thoroughly dried.
   c. Some tubes and bottles can be autoclaved; however, note that this will reduce their life span.
9. <u>*Never use a cracked tube or bottle or one that has become yellowed or brittle with age.*</u>
10. *Make sure that tubes and bottles used for centrifugation are chemically compatible with the sample.*
11. *Always make sure that the outer surfaces of sample tubes are completely dry.* This not only avoids rotor corrosion, but also can reduce the possibility of tubes becoming stuck in the rotor.
12. *Do not spin tubes or bottles at speeds above those for which they are rated.* It may be necessary to reduce the speed when using plastic adapters or metal tubes.
13. *Use adapters when necessary to ensure that tubes or bottles fit snugly in rotor compartments.* Tubes that do not fit snugly in the rotor compartments may shift during centrifugation, resulting in breakage or a poorly formed pellet.
14. *Maintain a rotor log and a centrifuge log.* Document major cleaning and decontamination processes as well as all centrifuge runs.

involves an **overspeed disk** *that has black and white alternating sectors, or lines, and is affixed to the bottom of the rotor.* The number of sectors on an overspeed disk varies depending on the maximum allowable speed for the rotor. In the centrifuge there is a photoelectric device that detects the lines on the overspeed disk during centrifugation. If the lines pass by the photoelectric device faster than an allowable limit, then the centrifuge automatically decelerates.

ii. CHEMICAL CORROSION AND METAL FATIGUE

Chemical corrosion *is a chemical reaction that causes a metal surface to become rusted or pitted.* Common salt solutions, such as NaCl, are particularly likely to cause corrosion and should be removed thoroughly if spilled on a rotor. Other corrosive agents include strong acids and bases, alkaline laboratory detergents, and salts of heavy metals like cesium, lead, silver, or mercury. Moisture that stands in the rotor compartments can also cause corrosion over time, see Figure 32.18.

Even in the absence of chemical corrosion, metals fatigue, or become weakened, due to the repeated stress of centrifugation. This damage occurs inside the rotor, so it is not visible. Ultracentrifuge and high-speed rotors must therefore be derated and/or retired as time goes by and as they are used. *Derating means to run a rotor at a speed less than its originally rated maximum speed.* **Retiring** *a rotor means to discard it or return it to the manufacturer.* The manufacturer specifies the conditions under which a rotor should be derated or retired. Because a rotor has a finite lifetime, it is essential to document its use, typically in a rotor log.

Low-speed rotors can be made of a variety of materials including aluminum, bronze, steel, and even high-impact plastic. High-speed rotors are frequently made of aluminum, which is light, strong, and relatively inexpensive. Aluminum rotors, however, are especially susceptible to corrosion. For ultracentrifugation, more expensive titanium rotors are preferred. Titanium is much more resistant to corrosion than aluminum, so titanium rotors are less likely to fail when repeatedly centrifuged. There are also newer-style, lightweight rotors that are not made of metal, but of a composite material that is composed of spun or woven carbon fibers embedded in epoxy resin. Composite materials tend to be resistant to chemical corrosion and to fatigue. Note that titanium and carbon composite rotors usually have some aluminum or iron components, such as tube compartments, that are susceptible to chemical damage.

iii. SITUATIONS THAT REQUIRE USING ROTORS AT LESS THAN THEIR MAXIMUM RATED SPEED

A rotor is sometimes derated due to its age or the number of times it has been spun. There are other situations that also call for reducing the speed of rotation to avoid overstressing the rotor. One of these situations relates to the density of the samples being centrifuged. Every rotor is designed to centrifuge materials whose densities are below a certain maximum cutoff. This maximum density is often 1.2 g/mL. Some fixed angle and vertical tube rotors are designed to hold a density of 1.7 g/mL, which is the maximum density of CsCl solutions used for DNA separations.

It is sometimes necessary to run a sample that exceeds the density limit for the rotor. In these cases, the rotor cannot be run at its maximum rated speed. The rotor manual will usually provide information about centrifuging samples whose density exceeds the rotor design limits. If the manual does not have this information, then there is an equation to estimate the maximum safe speed when centrifuging dense solutions:

$$\text{Derated speed} =$$
$$\text{Maximum rotor speed in RPM} \sqrt{(RD/SD)}$$

where
RD = the design limit of the rotor, as specified by the manufacturer
SD = the sample density

EXAMPLE PROBLEM

A rotor is specified to have a maximum speed of 25,000 RPM and is designed for densities less than or equal to 1.2 g/mL. A density gradient is being centrifuged with a maximum density at the bottom of the tube of 1.5 g/mL. What is the maximum speed at which this gradient can be centrifuged?

ANSWER

Derated speed =

$$\text{Maximum rotor speed in RPM} \sqrt{(RD/SD)}$$
$$= 25,000 \sqrt{(1.2)/1.5}$$
$$= 22,360 \text{ RPM}$$

This is the derated maximum rotor speed.

Figure 32.18. Examples of Severely Corroded Centrifuge Buckets. (Photo courtesy of DJB Labcare Ltd.)

It is possible that the combination of rotor speed and temperature can be such that a gradient material precipitates out of solution. CsCl gradients in particular can form crystals in the bottom of the centrifuge tube at low temperatures. These crystals have a very high density (4 g/mL) that will produce stress that far exceeds the design limits of most rotors. Consult the manufacturer regarding the use of CsCl gradients in a particular type of rotor.

There are other situations, in addition to the spinning of high-density samples, that require that a rotor be used at less than its normal maximum speed. For example, samples or tubes are sometimes used that exceed the maximum weight for which the rotor was designed. The rotor must then be used at less than its maximum speed, according to the manufacturer's directions.

iv. ROTOR CLEANING

Rotors must be kept clean. Rotors that receive heavy use should be cleaned on a routine basis, and every rotor should be thoroughly washed in the event of a spill. Detergents used to clean rotors must be mild and nonalkaline. Brushes used for cleaning must be soft so as not to scratch the rotor. Manufacturers sell mild detergents and brushes for use with rotors. Detergents must be thoroughly rinsed from rotors with distilled water. Manufacturers recommend air-drying rotors and storing them upside-down in a dry environment with their lids removed to avoid having moisture collect in the compartments.

Some rotors can be safely autoclaved, although repeated autoclaving may weaken ultracentrifuge rotors. Seventy percent alcohol is compatible with all rotor materials and therefore can be used for disinfection. Some manufacturers recommend sterilizing rotors with ethylene oxide, 2% glutaraldehyde, or UV light. Rotors contaminated with pathogenic or radioactive materials should be cleaned with agents that are compatible with the rotor. Note that alkaline detergents designed to remove radioactivity are not compatible with most rotors.

Rotors generally have O-rings and gaskets to ensure tight seals. These O-rings and gaskets can typically be lubricated with a thin layer of silicone vacuum grease. Worn or cracked O-rings and gaskets should be replaced.

Aluminum rotors should frequently be inspected for signs of corrosion, including rough spots, pitting, white powder deposits (which may be aluminum oxide), and heavy discoloration. If problems are observed, consult the manufacturer before using the rotor. Titanium rotors are much more corrosion resistant and require less-frequent inspection.

D. Centrifuge Tubes, Bottles, and Adapters

i. GENERAL CONSIDERATIONS

Centrifuge tubes are containers for smaller volume samples; centrifuge bottles are containers for larger volume samples. Centrifuge tubes and bottles are available in varying styles and sizes, and made of different materials. Some tubes and bottles are used with caps or seals, others are centrifuged open. Some tubes and bottles can be centrifuged when partially filled, whereas others must be completely filled or they will collapse under the force of centrifugation. Some tubes can be washed and used repeatedly, but others must be disposed of after one use. Because there are many types of sample containers for centrifugation, it is necessary to select those that match your samples and rotors.

Capped centrifuge tubes should always be used if a sample is potentially hazardous or volatile. Although screw caps are simple to use and appear to be leakproof when handled on the laboratory bench, they distort under the pressure of centrifugation. This distortion can allow samples to leak from the tubes. Conventional screw caps can only be used with open rotors and adapters. Manufacturers provide cap assemblies that in some way compress onto tubes or otherwise seal firmly.

At low and moderate RCFs, most tubes and bottles can be centrifuged when only partially filled. At higher RCFs, only thick-walled tubes can be run partially filled; thin-walled tubes collapse during centrifugation.

A single rotor can cost thousands of dollars. To maximize their versatility, manufacturers therefore design rotors so that they can flexibly accommodate different styles and sizes of tubes and bottles. This is accomplished with **adapters,** *plastic or rubber inserts that reduce the size of the sample compartment so that smaller*

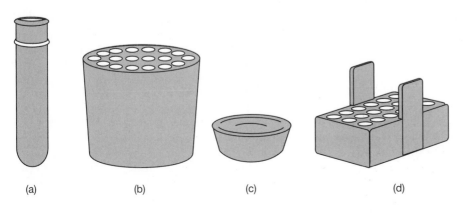

(a)   (b)   (c)   (d)

Figure 32.19.   Adapters. Adapters allow varying sizes and shapes of tubes to be used in a single rotor. **a.** An adapter that lines the rotor compartment and holds one tube. **b.** An adapter that holds eighteen tubes at once. **c.** A cushion to support a glass or plastic bottle in a horizontal low-speed rotor. **d.** An adapter to hold 18 tubes in a horizontal low-speed rotor.

tubes can be run in the same rotor. There are many types of adapters, see Figure 32.19 on p. 615. Some adapters are tube shaped and line the rotor compartment. Other adapters support the tops of tubes and still others the bottoms. There are also adapters to hold 96-well plates. It is important to purchase the correct configurations of tubes, caps, and adapters for each rotor and each type of sample that is to be used in that rotor. It is essential that sample containers fit snugly in the rotor compartments. Be aware that it may be necessary to run a rotor at less than its maximum speed when using adapters.

ii. MATERIALS USED TO MAKE CENTRIFUGE TUBES AND BOTTLES

Materials used to make centrifuge tubes and bottles include glass, stainless steel, and various types of plastics, see Table 32.6. These materials must be able to withstand the high forces of centrifugation. Depending on its composition and its design, every tube and bottle is rated for a maximum g force.

Plastics are relatively light and strong, so they are the preferred material for making tubes and bottles that will experience high centrifugal forces; however, plastics have limitations. Most plastic centrifuge tubes and bottles will soften and deform if they are centrifuged at temperatures above 25°C. Completely filled, thick-wall polyallomer (a type of plastic) tubes can be used at higher temperatures, but lower speeds and shorter run times are preferred. Stainless steel can be used safely at high temperatures, but these tubes are heavy and limit the speed of centrifugation. It is good practice to pretest tubes or bottles that will be used at higher temperatures without sample to determine whether the material can withstand the conditions of centrifugation.

Another limitation is that some plastics are incompatible with various solvents (e.g., phenol, acetone, and alcohols) and with acids and bases. For example, most plastics will soften or dissolve when exposed to phenol. To determine whether a particular tube material is compatible with your sample, consult a chemical compatibility chart from the manufacturer. Even if the sample and centrifuge tube material appear to be compatible based on the chart, it may still be a good idea to perform a test run.

Table 32.6.   MATERIALS USED TO MAKE CENTRIFUGE TUBES AND BOTTLES

1. *Glass. Glass is largely inert, which makes it chemically compatible with almost all samples.* Normal soda glass is unable to withstand forces much above $3000 \times g$, whereas borosilicate glass is able to withstand at least $10,000–12,000 \times g$. The disadvantage to glass is the possibility of breakage. Glass therefore cannot be used at high RCFs. It is best to support glass tubes using a rubber pad or adapter, thus avoiding pressure between glass and metal. It is always good practice to do a test run with glass tubes or bottles before running them with potentially hazardous materials.

2. *Stainless steel. Stainless steel is resistant to organic solvents and to heat.* Stainless steel tubes can be reused many times as long as they are undamaged and uncorroded. These tubes are strong enough to be run partially filled. The disadvantage to metal tubes is that they are heavy and so require that a rotor be spun at less than its maximum speed.

PLASTICS

3. *Polycarbonate. Polycarbonate is strong, transparent, and autoclavable.* Polycarbonate can be constructed with thick walls so that tubes can be only partially filled and do not collapse when centrifuged. Polycarbonate is commonly used for high speed and ultracentrifuges. Polycarbonate, however, is not resistant to many solvents, including phenol and ethanol, both of which are commonly used in biology laboratories. Polycarbonate is also attacked by alkaline solutions, including many common laboratory detergents. Polycarbonate can become brittle and should be visually inspected before use. Discard cracked or brittle tubes.

4. *Polysulphone. This plastic has many of the advantages of polycarbonate, but it is also resistant to alkaline solutions.* It is, however, attacked by phenol.

5. *Polypropylene and polyallomer. Polyallomer is commonly used for high speed and ultracentrifuge tubes.* Both polyallomer and polypropylene can be autoclaved. Both can be made into thin-wall tubes that can be punctured readily to remove bands from gradients. Both plastics tend to be resistant to most organic solvents. Polyallomer is recommended for separating DNA in CsCl gradients because DNA does not adhere to the tube walls.

6. *Cellulose esters. These tubes are transparent, easy to pierce, and strong, but they cannot be autoclaved and are attacked by strong acids and bases and by some organic solvents.*

7. *Other plastics. Many companies make specialized plastic tubes and bottles with various properties and under their own trade names.* Check with the manufacturer before using these with acids, bases, and solvents, and before autoclaving. Note also the maximum speed at which they can be used.

EXAMPLE PROBLEM

A page from a rotor catalog is shown. Answer the following questions about the rotor described.

a. What is the maximum speed that this rotor can withstand?

b. What is the maximum force that this rotor can withstand?

c. How many different types of tubes can be used with this rotor?

d. What is the maximum volume that can be centrifuged in this rotor?

e. Which types of tubes can withstand being centrifuged at 80,000 RPM?

f. What is the function of the adapters? Which tubes require adapters?

g. What happens to the maximum speed that can be used with this rotor if adapters are used?

h. What is the maximum speed at which a density gradient with a maximum density of 1.4 g/mL can be run in this rotor?

i. Is this rotor for a low speed, high speed, or ultra-centrifuge?

FA-80 ROTOR

Description

25-degree fixed angle   Titanium
10 places, 3.5 mL maximum volume
Maximum Speed 80,000 RPM   k-Factor 25.8
Designed for samples with a density less than or equal to 1.2 g/mL

	RCF (× g)	Radius (cm)
maximum	460,000	6.55
average	350,000	5.00
minimum	250,000	3.41

TUBES

Capacity (mL)	Description	Maximum Speed	Cap Assembly	Adapters
3.5	Polyallomer, Thin Walled Tube	80,000	Multipiece	None
2.7	Polycarbonate Thick Walled Tube	45,000	Screw Cap	None
3.5	Polyallomer Thin Walled	80,000	Crimp Seal	None
2.0	Polycarbonate Thick Walled Tube	55,000	None	I per compartment required (Delrin)
2.0	Cellulose Acetate Butyrate Thick Walled Tube	40,000	None	I per compartment required (Delrin)
0.4	Cellulose Acetate Butyrate Thick Walled Tube	40,000	None	I per compartment required (Delrin)

ANSWER

a. The maximum speed for the rotor is 80,000 RPM.

b. The maximum RCF generated by this rotor is 460,000 × g at r_{max}.

c. The manufacturer specifies 6 types of tubes for this rotor.

d. There are 10 compartments each of which can hold a maximum of 3.5 mL, so the maximum volume is 35 mL.

e. Only the polyallomer thin walled tubes can be used at the rotor's maximum speed.

f. Adapters allow tubes smaller than the sample compartment to be used. The tubes which hold a lower volume require adapters.

g. The use of adapters reduces the maximum speed at which the rotor can be used.

h. The rotor is designed for a density ≤ 1.2 g/mL. The square root equation can be used to determine the maximum rate given the higher density as follows:

$$\text{Derated speed} =$$

$$\text{Maximum rotor speed in RPM} \sqrt{(RD/SD)}$$

$$= 80,000 \sqrt{(1.2/1.4)}$$

$$\approx 74,066 = \text{maximum speed for this density}$$

i. This is an ultracentrifuge rotor.

E. Centrifuge Maintenance and Trouble-Shooting

i. MAINTENANCE AND PERFORMANCE VERIFICATION

The details of operation vary from centrifuge to centrifuge, so it is necessary to read the manufacturer's instructions. It is especially important to read the manual when operating an ultracentrifuge because these instruments can easily be damaged with dangerous and costly results.

Centrifuges require periodic maintenance and performance verification that are usually performed by trained service technicians. The technicians make sure that the RPMs that are displayed correspond to the actual speed of the rotor, that the temperature in the chamber is maintained properly, and that timers and other controls work correctly. The vacuum and refrigeration systems must also be checked. Older centrifuges may have brushes that require replacement. All service and maintenance procedures should be documented.

ii. TROUBLE-SHOOTING COMMON PROBLEMS

Some common centrifugation problems are described in Box 32.3.

F. Safety

Various safety issues relating to rotors and centrifuge tubes have already been addressed in this chapter. Another risk that is less dramatic than a catapulting rotor, but is equally dangerous, relates to the centrifugation of hazardous materials such as viruses, or radioactive or carcinogenic compounds. Protection from hazardous materials is of great importance in biological centrifugation. In a worst case, a centrifuge tube that breaks during centrifugation can disperse hazardous aerosols throughout the laboratory and throughout the air-handling system in a building. There are more subtle concerns as well. A poorly sealed centrifuge tube cap can allow aerosols to escape. Screw caps on centrifuge tubes or bottles are likely to distort under the high force of centrifugation, allowing liquids and aerosols to be released. Simply opening a tube, particularly a tube with a "snap-cap," creates aerosols. Pouring and decanting fluids from centrifuge tubes and bottles creates aerosols. Spills can occur while introducing samples into tubes or removing samples from tubes.

Because centrifugation provides many opportunities for contamination, it is necessary to use special precautions when spinning pathogenic or toxic materials. Biological safety cabinets can be used when filling and emptying centrifuge tubes or bottles. There are centrifuges, rotors, and centrifuge tubes that are specifically designed for hazardous materials. Some rotors, for example, are designed to safely contain spilled materials in the event of tube breakage.

Radioactive materials can readily contaminate rotors and centrifuges. The commonly used radioisotope, ^{32}P, tends to bind tightly to metal surfaces, making it difficult to remove. This problem is exacerbated by the fact that alkaline detergents intended to remove radioactive contaminants are damaging to rotors.

Thus, failure to follow proper practices when operating centrifuges can endanger laboratory staff, result in damage to the centrifuge or its accessories, and cause contamination of the centrifuge and the laboratory. Some safe handling guidelines are summarized in Box 32.4.

Box 32.3. *TROUBLE-SHOOTING TIPS*

Symptom 1: *VIBRATION DURING OPERATION*

Vibration is due to rotor imbalance
1. Turn off the centrifuge.
2. Check that all containers are properly balanced and arranged properly.
3. Check that cushions, adapters, and caps match in all sample compartments.
4. Check that the centrifuge is on a level surface.
5. Increase speed gradually.

Symptom 2: *TUBE BREAKAGE*
1. Check that the type of tube in use is rated for the speed and force being used.
2. Reinspect all tubes and discard those with any cracks, pitting, or yellowing.
3. Check that the tubes are properly balanced.
4. Make sure that the adapters and cushions match the type of tubes.
5. Make sure the contents of the tubes are compatible with the materials used in constructing the tubes.
6. If using partially filled tubes, check that sufficient volume is being used to meet manufacturer's specifications.

Symptom 3: *FINE GRAY OR WHITE POWDER OR DUST IN CHAMBER*

This is caused by particles of glass sandblasting the inside of the centrifuge
1. Clean centrifuge thoroughly.
2. Run the centrifuge without samples several times and clean between each operation.

Box 32.4. *GUIDELINES FOR SAFE HANDLING OF CENTRIFUGES AND THEIR ACCESSORIES*

1. *Carefully read the manufacturer's instructions.* Pay attention to warnings provided. Note limitations of rotors. Find out how to use tubes and adapters properly. Careful attention to the manufacturer's directions is especially important when operating high speed and ultracentrifuges.

Box 32.4. (continued)

2. ***Use special precautions when centrifuging pathogenic, toxic, or radioactive materials.***

   a. Make sure that tubes are tightly capped and sealed.

   b. Avoid caps that snap open.

   c. Open tubes containing dangerous materials in a suitable hood or other protected enclosure.

   d. Pour and decant hazardous samples in a hood or protected enclosure.

   e. Clean spills promptly.

   f. Use special containment instrumentation as appropriate. Sealable tubes and rotors are available to contain hazardous materials.

3. ***In laboratories where toxic materials, pathogens, or radioisotopes are used, avoid touching any rotors with bare hands.*** Assume that rotors and the insides of centrifuges are contaminated, but at the same time, try to avoid contamination. Assume that gloves are contaminated after touching a rotor and remove them before touching anything else. Never place bare fingers in any rotor compartment, both because of the possibility of contaminants, and because there may be broken glass.

4. ***Protect service technicians from pathogens and radioactivity by careful cleaning of equipment before any repairs are performed.*** Document the decontamination process.

5. ***Protect yourself from the spinning rotor.*** Although many centrifuges have locking covers that prevent access to a spinning rotor, some centrifuges lack this safety feature. There are also modes of centrifuge operation where the user must have access to the rotor as it spins.

   a. Do not touch a rotor that is in motion and do not slow a rotor with your hand.

   b. If you must work with a spinning rotor, then:
   - remove jewelry, such as necklaces and bracelets
   - remove neckties and scarves
   - tie back long hair
   - roll up and secure shirt sleeves

6. ***Make sure the contents of a rotor are properly balanced before centrifugation.***

7. ***Do not centrifuge large volumes of explosive or flammable materials, such as ethyl alcohol, chloroform, and toluene.*** (Note that in practice, small volumes of these materials are often centrifuged in low-speed centrifuges.)

8. ***Always remain at the centrifuge as it begins to accelerate.*** An unbalanced rotor or other problems will become apparent when the centrifuge reaches critical speed.

PRACTICE PROBLEMS

1. A portion of a table from a manufacturer's rotor manual is shown. Calculate the *RCF*s to fill in the blanks. Refer to the diagram to determine the values for r_{max}, r_{min}, and $r_{average}$.

Speed (in RPMs)	RCFs Generated		
	r_{max}	r_{min}	$r_{average}$
20,000	40,800	17,200	29,000
30,000	———	———	———
40,000	———	———	———

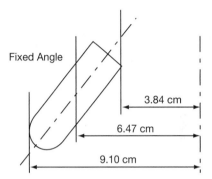

2. The following calculation is incorrect. Why?

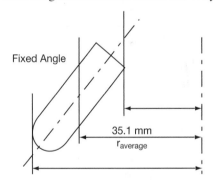

Speed is 45,000 RPM

$$RCF = 11.2 \times r \left(\frac{RPM}{1000} \right)^2$$

$$RCF = 11.2 \times 35.1 \left(\frac{45,000}{1000} \right)^2$$

$$RCF = 796,068$$

Incorrect answer!

3. A journal article includes the statement below. If you have a similar rotor with an r_{ave} of 2.6 cm, how many RPMs will you need to use to duplicate the force used by these authors?

"The extract was centrifuged in a fixed angle rotor at 140,000 × g for three hours."

4. Specifications for three rotors are shown. Answer the following questions about these rotors.

a. Which rotor would be best for separating large volumes of bacterial cells after fermentation? Explain.

b. Which would be used to isolate small, subcellular organelles and for buoyant density separations of macromolecules like DNA and RNA? Explain.

c. Which rotor would be useful for centrifuging 1 mL volumes of precipitated DNA in a molecular biology laboratory? Explain.

	Rotor A	Rotor B	Rotor C
Type	Fixed Angle	Fixed Angle	Fixed Angle
Maximum Speed (RPM)	60,000	16,000	10,000
Maximum Force ($\times$ g)	400,000	40,000	6000
Capacity (# $\times$ mL)	6 $\times$ 25	6 $\times$ 500	12 $\times$ 2

5. A portion from a rotor catalog is shown below. Answer the following questions about the rotor described.

a. What is the maximum speed that this rotor can withstand?

b. What is the maximum force that can be generated with this rotor?

c. What is the maximum volume that can be centrifuged at one time in this rotor?

d. What is the maximum number of tubes that can be centrifuged at once in this rotor? What is the maximum volume they can contain?

e. Is this rotor for a low speed, high speed, or ultracentrifuge?

FA13

Description

28° fixed angle aluminum
Six places, 315 mL maximum volume
Maximum Speed 13,000 RPM   *k*-Factor 2026
Designed for samples with a density less than or equal to 1.2 g/mL
Maximum *RCF*   27,500
Maximum Radius 14.60 cm

Tubes

Capacity (mL)	Description	Maximum Speed	Cap Assembly	Adapters
315	Stainless Steel Bottle	13,000	Stainless Steel	None
280	Polysulfone Bottle	13,000	Polypropylene Seal	None
150	Polycarbonate Thick Walled	13,000	None	1 per compartment required (Delrin)
30	Borosilicate Glass Tube	10,000	None	1 per compartment required (polypropylene) Each adapter holds three tubes
4	Polypropylene	13,000	Polypropylene	1 per compartment required (polypropylene) Each adapter holds 12 tubes

---

QUESTIONS FOR DISCUSSION

High speed and ultracentrifuges are expensive pieces of laboratory equipment that must often be shared by a number of people. Moreover, centrifuge runs may be long; it is not unusual for a procedure to require that samples spin overnight or even longer. Planning, organization, and "centrifuge etiquette" are therefore required to ensure that everyone has a turn to use the centrifuge and that a centrifuge is available at critical points in a procedure. Most laboratories require individuals to schedule their centrifuge use and to sign up in advance to avoid conflicts. **Scenario 19 in the "Skill Standard for the Biosciences Industries" document (Educational Development Corporation, 1995) asks the following question:**

One week ago you reserved time for a 12-hour spin to coincide with the completion of your assay. When you bring your samples to the centrifuge, you discover that it is currently being used. There is no indication of who may be using the centrifuge. What would you do?

CHAPTER 33

Introduction to Bioseparations

I. BIOSEPARATIONS IN THE BIOTECHNOLOGY LABORATORY

A. Introduction

We have so far discussed two types of bioseparation methods: filtration and centrifugation. In this chapter, we discuss how biotechnology laboratories purify specific molecules from complex mixtures. In these cases, it is necessary to devise purification strategies that involve a series of bioseparation steps. **Bioseparation methods** *are the techniques used to separate and purify specific biological products, or biomolecules (such as a particular protein).* These methods play a central role in biotechnology. **Purification,** or isolation, *is the separation of a specific material of interest from contaminants, in a manner that provides a useful end product.*

The ability to isolate specific biomolecules is essential in both research and production settings. In a research setting, a purified biological material may be used for **molecular analysis,** *which is the study of the specific chemical properties of the molecule.* Research into the chemical properties of proteins and other biomolecules can provide basic information about cellular processes and lead, for example, to the development of new drug treatments for human illness. In a production setting, safe and reliable purification processes are required to provide a previously analyzed product in sufficient quantities for large-scale testing or for commercial sale.

In order to purify a particular biological product completely, it is necessary to devise a purification strategy that involves a series of separation steps. Each separation step removes certain contaminants from the product of interest, resulting in a progressively purer preparation. This chapter provides an overview of how purification strategies are developed, with an emphasis on general guidelines.

B. Sources of Biotechnology Products

Biomolecules *are compounds that are produced by some type of biological source, such as a plant, animal, microorganism, or cultured cell.* This source is the starting point for purification of the desired product. The optimal strategy to purify a particular material of interest will depend to some extent on the source of starting material. This is because the best strategy to purify a product depends partly on the nature of the biomolecule being purified and partly on the nature of the contaminants that must be removed. For example, the strategies used to purify insulin differ depending on whether its source is animal pancreas or recombinant *E. coli* that have been genetically engineered to produce insulin.

In an industrial setting, production methods are categorized as upstream or downstream processes. **Upstream processes,** such as fermentation or cell culture, *produce the starting material of interest.* **Downstream processes** *are the separation procedures that result in a purified product.* Whereas downstream processes represent much of the cost of production, they are based on the choices made in the upstream system. Studies show that changes in upstream sources of biomolecules (e.g., switching from an animal source to a genetically engineered microorganism) can have a much greater effect on final production costs than changes in downstream processing.

There are a variety of biological sources that are commonly used for the production of biomolecules, see Table 33.1. Natural sources such as plants, animal tissues, blood, and other biological fluids have long been used as sources of useful products. These natural sources, however, usually provide complex and relatively uncharacterized molecular mixtures, which require complex purification strategies to isolate the biomolecule of interest. Moreover, the amount of a single biomolecule of interest is often quite limited in natural sources. In some cases, as when a biomolecule is found in human tissue or an endangered plant, the natural source cannot be used for mass production.

Table 33.1. SOURCES OF BIOTECHNOLOGY PRODUCTS

Natural Sources	Cell Cultures	Fermentation
Plant tissues	Animal cells	Bacteria
Animal tissues	Plant cells	Yeast
Microorganisms isolated from their natural environment		
Blood and other fluids		

Table 33.2. CONSIDERATIONS IN CHOOSING A BIOLOGICAL SOURCE FOR MOLECULES

There are many variables that can affect the final cost and product yield in bioseparations. These are some of the major considerations when choosing an upstream source for molecules:
- the availability of the source
- the total amount of product present
- the volume of material that will need to be processed
- the stability of the product under upstream conditions
- the amount of product relative to the amount of impurities present
- the nature of the impurities
- the potential pathogenicity of the source
- the cost of media, food, and equipment to grow the source

Due to the complexity of isolating pure products from natural sources and the limited amount of the desired biomolecule that may be available, biotechnologists often use as sources cultured plant or animal cells, or microorganisms grown using fermentation technology. In many cases, microorganisms or cultured cells can be genetically engineered to better produce the molecule of interest, as is the case with insulin. Microbial fermentation is a common source of commercially available biotechnology products, because of the relative ease of developing large-scale upstream operations.

The subcellular localization of a product is significant. As we will discuss shortly, it is much easier to harvest a product that is secreted from cells than one that is sequestered within the cell. Many genetically engineered sources of biomolecules have been designed to secrete proteins that normally would be found in an intracellular location.

In general, the biological source of a product should be chosen or engineered to produce the greatest amount of stable, usable product as possible, as economically as possible. Table 33.2 summarizes some of the major variables in upstream production of biomolecules.

C. Goals and Strategies for Bioseparations

Strategies to purify a specific biomolecule involve a series of fractionation steps. The first steps **extract,** or *release,* the molecule of interest from its biological source. Later steps successively remove various contaminants from the product of interest.

There is no "standard" series of specific steps that can be used to purify all molecules. For example, proteins, the cellular products most likely to be purified in biotechnology facilities, differ greatly in their molecular properties. The specific series of steps used to purify Protein A, therefore, is unlikely to be effective for Protein B, unless the two proteins are closely related in structure. Many separation techniques can be performed under almost

Table 33.3. DESIGNING A PRODUCT PURIFICATION STRATEGY

1. Prioritize the goals of the purification process.
2. Select a source for the product.
3. Develop a specific assay for the product of interest. (see Chapter 24).
4. Determine optimal methods to:
    a. release the product from its source in a soluble form.
    b. reduce the volume of the product.
    c. separate the product from impurities.
    d. purify the product of interest.
5. Verify the identity and purity of the product.

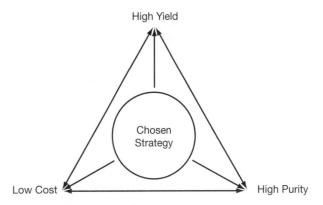

Figure 33.1. Goals for Bioseparation Strategies in the Biotechnology Laboratory.

infinitely variable conditions (such as pH, temperature, and ionic strength). The conditions under which a technique is performed must therefore be individually optimized for every product.

Even though there are no universal procedures applicable to every biomolecule, there are guidelines for the design of appropriate strategies. Table 33.3 provides an outline of these guidelines, which will be discussed throughout this chapter.

The key step in developing a bioseparation strategy is to determine the goals of the process. Questions to ask include: How pure does the product need to be to work effectively? How much of the product is required? How important is low cost? The priorities of these goals vary in research and production facilities and may also vary depending on the nature of the product. For example, a drug product requires the utmost purity, even if the cost is high. In contrast, a product not intended for use in humans may be acceptable in a less purified, cheaper form. In production facilities, the cost of materials and equipment are a major concern when designing a large-scale purification strategy. In research laboratories, minimizing required personnel time may be a primary concern. Another consideration that may be important in a biotechnology company is developing a purification strategy quickly to get a product to market as soon as possible. In an academic research laboratory, speed to publication may be a primary goal.

All purification strategies require compromises. For example, the highest product purities are usually obtained only at high cost and with reduced yield. **Yield** *is defined as the % **recovery**, or percent of the starting amount of the product of interest that can be recovered in purified form using a specific strategy.* Yield is intimately related to cost-effectiveness and maintaining product quality, and will be discussed in greater detail. Optimizing either yield or purity will usually raise the cost of a purification process.

It is important to prioritize goals before developing a strategy. Figure 33.1 shows the three primary goals of a

purification strategy: optimizing purity, optimizing yield, and reducing costs. These three goals are illustrated as a triangle to emphasize that all purification strategies require compromises. Table 33.4 summarizes some of the general concerns that should be considered when planning a product purification scheme.

II. ANALYSIS OF PRODUCT PURITY AND YIELD

A. Assaying the Product of Interest

This section will discuss how product purity and yield are measured and reported. Because biotechnology companies frequently purify proteins for commercial sale, this discussion will focus primarily on proteins.

Devising a purification strategy requires the development of a specific, quantitative assay for the biomolecule of interest. Such an assay enables analysts to detect, quantitate, and determine the degree of purity of the product. The assay is initially used as researchers develop an optimum separation strategy. Later, workers rely on the assay to routinely monitor the separation

Table 33.4. GOALS FOR PRODUCT PURIFICATION STRATEGIES

All purification schemes share three goals:
- maximum product purity, with
- minimum loss of product material or activity, at the
- lowest possible cost.

Each of these goals, however, varies in importance compared with the others, depending on the nature of the final product desired. The following issues must also be considered when formulating bioseparation strategies:
- safety requirements
- simplicity of the procedures
- reliability of the methods
- volumes suitable for processing
- speed to market
- requirement for product activity

Table 33.5. REQUIREMENTS FOR A PRODUCT ANALYSIS METHOD

Product analysis methods for monitoring purification processes vary widely according to the molecule of interest. The assay chosen, however, should have the following characteristics:

- sensitivity to small amounts of the material of interest
- precision and accuracy
- relatively small sample volumes required
- easily applied to large numbers of samples

See Chapter 24 for more information on the general properties of a biological assay.

process. The desired qualities of an analytical assay are summarized in Table 33.5.

In rare cases, the product of interest is distinctive enough to detect without a complex assay. For example, the protein myoglobin can be detected by its red color, which is derived from a heme prosthetic group. Most biological materials are not so obliging, and the detection assay can be complex and potentially destructive to the sample.

During a multistep purification process, the product of interest is successively isolated from multiple impurities. After each separation step, an aliquot of the product mixture is set aside for analysis. Assuming that the molecule to be purified is a protein, the items that are measured are:

1. the amount of the desired protein present, and

2. the amount of total protein present (this includes the protein of interest plus protein contaminants).

Based on these two measurements, it is possible to calculate the specific activity of the protein of interest, and the product yield after each step in the purification process. **Specific activity** *is the amount (or units) of the protein of interest, divided by the total amount of protein in a sample.* As protein purification proceeds, the product of interest should become more pure at each step, so the specific activity of the preparation should increase. Each purification step unfortunately results both in the removal of contaminants and in some unavoidable loss of the biomolecule of interest. As protein purification proceeds, the specific activity therefore increases, but the total amount of the desired product (calculated as the percent yield or recovery) decreases.

B. Protein Assays

In order to determine the purity of a protein product, it is necessary to know both how much of the desired protein is present as well as how much contaminating protein is present. When proteins are purified, the amount of total protein (desired protein plus contaminating protein) is measured after every step. In a purification procedure, the method used to measure total protein must

be reliable, fast, sensitive to low protein concentrations, and relatively insensitive to potential interfering agents (e.g., buffers or nucleic acids). To ensure consistency, a single protein assay method should be chosen and then used throughout the purification process.

C. Specific Activity and Yield Determinations

Biologically active molecules are generally assayed by their activity. Here we will use the example of an enzyme that is assayed by its ability to catalyze a reaction. Enzymatic reactions involve the conversion of a substrate(s) into a product(s):

$$\text{SUBSTRATE(S)} \xrightarrow{\text{enzyme}} \text{PRODUCT(S)}$$

Enzyme assays generally involve the measurement of either the disappearance of substrate or the appearance of product. Enzyme activity is typically expressed as units, based on the rate of the reaction they catalyze. There are two standard expressions of enzyme activity. The **International Unit (IU) of enzyme activity** *is defined as the amount of enzyme necessary to catalyze transformation of 1.0 μmole of substrate to product per minute under optimal measurement conditions.* (Conditions, such as pH and temperature, may affect the activity of an enzyme and so must be optimized.) The **SI unit of enzyme activity** *is defined as amount of enzyme necessary to catalyze transformation of 1.0 mole of substrate to product per second under optimal measurement conditions (such as optimal pH and temperature). This is defined as a* **katal** *or* **kat** *unit.* The ability to measure enzyme activity is dependent on knowing the ideal reaction conditions and also having a quantitative assay that measures the amount of substrate or product. Some enzymes, such as proteases, which digest substrates of variable or unknown molecular weight, cannot be quantitated in these standard units. For these proteins, units are defined according to the assay used, providing a relative measure of enzyme activity. Once the amount of the protein of interest and the amount of total protein have been measured, it is then possible to calculate the specific activity of the protein.

EXAMPLE PROBLEM

An assay for enzyme Q measures the disappearance of substrate. In 15 minutes, 12 mmoles of substrate are converted to product. What is the activity of the enzyme preparation in International Units?

ANSWER

12 mmoles = 12,000 μmoles. The rate of conversion of substrate to product is 12,000 μmoles in 15 minutes = 800 μmoles per minute.

The preparation therefore contains 800 IU of enzyme.

The following example problem shows a sample purification scheme and illustrates how specific activity is calculated and used.

EXAMPLE PROBLEM

The enzyme, β-galactosidase was purified from a culture of *E. coli*. The purification steps were:

1. Ten liters of cells were grown in a fermenter.

2. The 10 L of cells plus broth were centrifuged to separate the cells from the broth. The result was a pellet containing the *E. coli* cells.

3. The cells were suspended in buffer and were broken apart by sonication (a method that uses high-pitched sound). The resulting cell suspension was centrifuged again. The pellet and the supernatant were assayed for β-galactosidase enzymatic activity. β-galactosidase was detected only in the supernatant. This supernatant is called a crude extract.

4. The supernatant was treated with salt to precipitate the β-galactosidase. The precipitated enzyme was collected by centrifugation and was resuspended in buffer. The resulting β-galactosidase preparation was assayed for β-galactosidase enzymatic activity and for total protein.

5. Salt was removed from the β-galactosidase solution using dialysis. The resulting β-galactosidase preparation was assayed for β-galactosidase activity and for total protein.

6. The resulting solution, after dialysis, was passed through a chromatography column. This step separated the β-galactosidase from a number of impurities. The resulting β-galactosidase preparation was assayed for β-galactosidase activity and for total protein.

Observe that the β-galactosidase was extracted from *E. coli* cells and was then purified using a series of steps. After each step, the activity of the β-galactosidase and the amount of total protein were tested. It would be possible to continue with further purification steps at this point, or to stop, depending on the final purity required. The results are shown in the following table.

a. Fill in the blanks in the table.

b. What happened to yield as the purification steps proceeded?

c. What happened to specific activity as the purification proceeded?

PURIFICATION OF β-GALACTOSIDASE FROM E. COLI

Purification Step	Total Protein (mg)	Total Activity (IU)	Specific Activity (IU/mg)	Yield (%)
Crude extract	2140	1360	0.64	100
Salt precipitation	760	1350	___	99
Dialysis	740	___	1.80	___
Chromatography	390	1240	___	___

ANSWER

a. Specific activity equals the amount of activity of the protein of interest divided by the total amount of protein present. Yield is the percent of the amount of starting activity that is still present after each purification step. In this example, 1360 IU represents 100% yield.

PURIFICATION OF β-GALACTOSIDASE FROM E. COLI

Purification Step	Total Protein (mg)	Total Activity (IU)	Specific Activity (IU/mg)	Yield (%)
Crude extract	2140	1360	0.64	100
Salt precipitation	760	1350	<u>1.78</u>	99
Dialysis	740	<u>1332</u>	1.80	<u>98</u>
Chromatography	390	1240	<u>3.18</u>	<u>91</u>

b. As we would predict, some of the β-galactosidase was lost at each step and the percent of total activity remaining decreased as the purification proceeded.

c. As the purification proceeded, the β-galactosidase became more pure; therefore, the specific activity value increased.

Information about specific activity is useful when purchasing enzymes or purified proteins. You will frequently be offered various grades of enzyme, based on either total activity or the amount of protein present. In each case, a specific activity level will be provided by the manufacturer. This will allow you to choose among purity levels depending on your application. In general, less purified enzyme grades are less expensive because less effort is required and greater product yield is possible.

EXAMPLE PROBLEM

You need to purchase enough Enzyme Z to perform 100 reactions. You determine that this will require approximately 8000 U of enzyme. The catalog from your favorite enzyme supplier offers you the following choices:

Enzyme Z

One unit of enzyme activity catalyzes the conversion of 1 μmole of A to B in 1 minute at 25°C at pH = 7.5.

Grade 1	From aardvark liver	100 mg	$ 50
	Activity: 10,000 U/g	500 mg	$ 200
		1 g	$ 350
Grade 2	From aardvark liver	100 mg	$ 200
	Activity: 50,000 U/g	500 mg	$ 700
		1 g	$1500
Grade 3	From aardvark liver	1 g	$ 30
	Activity: 2000 U/g	5 g	$ 110
		10 g	$ 250

a. What enzyme grade would be your best choice if your main priority was to find the least expensive option?

b. What would be your best choice if you needed the purest preparation possible?

ANSWER

First you want to consider the specific activities of each enzyme grade and determine what quantity you need to purchase:

Grade 1: At a specific activity of 10,000 U/g, you will need 0.8 g to provide the 8000 activity units you need. This will require 8 × 100 mg for a total of $400, or 1 g for $350.

Grade 2: At 50,000 U/g, you will need 160 mg, or 2 × 100 mg for $400.

Grade 3: At 2000 U/g, you will need 4 g of protein, at 4 × 1 g for $120, or 5 g at $110.

a. The least expensive option is Grade 3, which is the least pure.

b. The highest specific activity, and therefore the most concentrated activity, is found in Grade 2.

D. Requirements of the Biotechnology Industry

As discussed previously, the required purity of the end product in a production setting will vary according to the intended use of the product. In all cases, undesired biological activities or effects must be eliminated. Products taken from sources containing potentially harmful contaminants will generally require higher standards of purification than products isolated from innocuous sources.

Purification strategies for regulated commercial products must meet federal and state regulations and receive approval from appropriate agencies (see Chapter 5). Once the product and its preparation procedure have been approved, the procedures for product purification must be rigorously followed and documented. In order to change these procedures, the company may again be required to prove the safety and effectiveness of the final product to regulatory agencies, which is a process that can require years of additional testing. There is considerable incentive, therefore, to develop an optimal purification strategy early in the process of product development.

Any product that is administered to humans as a drug must meet especially strict quality specifications for potency and homogeneity. It must be sterile and free of any contaminating substances, especially those such as proteins or nucleic acids that could induce adverse immune responses. Regulatory agencies require a yield analysis for all pharmaceuticals. All starting materials must be accounted for because there cannot be additional product generated by the purification process (this serves as an internal control for the methods), nor should there be significant levels of product breakdown (as opposed to product loss). Because yield obviously affects the production cost of a product, companies use yield analysis as part of their quality control tests.

When drugs are produced using recombinant DNA production systems, product consistency and thorough characterization of the DNA involved are essential.

Table 33.6. CONCERNS FOR RECOMBINANT DNA PRODUCTS

When manufacturing pharmaceutical products using recombinant DNA systems, biotechnology companies must demonstrate that:

- the DNA sequence that was inserted into the host organism is fully characterized
- the amino acid sequence of the produced protein is identical to the predicted product
- products are consistent among production batches, with no variation due to the recombinant DNA process
- any contaminating substances due to the genetic engineering process are removed

Because recombinant DNA products are being manufactured outside their usual environments (e.g., by *E. coli* rather than by pancreatic cells in the case of insulin), the possibility of incorrectly folded, processed, or aggregated proteins must be considered. Table 33.6 summarizes specific concerns for recombinant DNA products (see also Chapter 5).

III. CHOOSING BIOSEPARATION METHODS

A. Parameters for Method Selection

A purification strategy consists of a series of purification steps, each of which removes certain impurities from the product of interest. Each purification step involves a different technique. Most individual purification techniques select one or sometimes two specific molecular properties as the basis for bioseparation. There are many molecular properties that can be exploited for separations. They include:

- Size or molecular weight
- Molecular charge
- Relative solubility of the molecule in water or other solvents
- Relative solubility of the molecule in the presence of salts
- Sensitivity to the effects of pH, light, oxygen, or heat
- **Affinity,** *which is the ability to bind specifically to other biomolecules*

The **selectivity** of a technique *is its ability to separate a specific component from a heterogeneous mixture, based on molecular properties.* By using a sequence of steps that separates molecules based on three or more different properties, it is generally possible to purify any biomolecule, see Figure 33.2.

One of the basic principles of bioseparations is never to purify the product more than necessary. As indicated earlier, the highest product yields will be obtained with the fewest steps. Four to six sequential methods are typically needed to purify and concentrate biomolecules. It is essential to work quickly between steps in order to maintain product activity (if desired), and to coordinate techniques so that the end product of one step provides

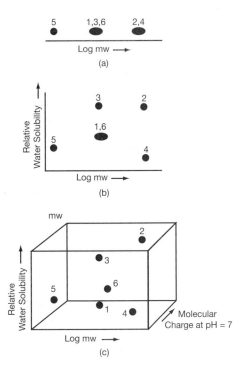

Figure 33.2.   Isolation of Specific Molecules by Their Characteristics. This separation starts with a mixture of six molecules. **a.** Separation on the basis of molecular weight gives clear separation of only Molecule 5. **b.** Separation of these fractions by both molecular weight and relative water solubility purifies all of the components except Molecules 1 and 6. **c.** Adding a third separation parameter, charge, allows all components of the mixture to be purified.

a suitable substrate for the next. For example, after a centrifugation step, the pelleted product should be dissolved in a volume of a buffer that is compatible with the next technique.

Because there are many purification techniques available, it is necessary to choose the techniques to be used for a particular product, and the order in which those techniques are performed. The decisions as to which techniques to use depend partly on the basic biochemical nature of the biomolecule of interest, partly on the probable nature of the impurities, and partly on the goals of the purification process. For example, certain purification techniques are costly and would only be used where the highest purity is required. Certain techniques similarly are most useful for low-volume samples; others are suitable for higher volumes. This is a problem that requires systematic strategy development. The simplest starting point for a bioseparation strategy is to follow the procedure for a similar molecule. This technique varies widely in its success. An example of a flowchart for a purification strategy is shown in Figure 33.3.

All bioseparations start with a relatively crude mixture of materials from the biological source. One of the first priorities is to remove any live organisms from the mixture. Any known biohazardous contaminants should be removed as early in the separation process as possible. Typically the simplest techniques are used early in

the separation process, when the source mixture is complex and represents a relatively large volume. Early steps will generally require the highest **capacity,** *which is the volume of sample that can be processed simultaneously by a technique.* Capacity also refers to the amount of the product of interest that can be separated by the technique. The first definition is usually more significant early in the separation process, when volumes may be high (such as a supernatant from a 1000 L fermentation vessel). Because it is usually more expensive to process large-volume samples, volume reduction is also an early concern. These early steps in a purification strategy also usually take advantage of the major differences between the desired product and the expected impurities. For example, if the product is a large molecule contaminated by many low molecular weight materials, an early separation step based on size would be appropriate.

Later steps in a purification process usually involve techniques that remove impurities similar in nature to the product itself. Later steps are often more expensive and have a lower capacity than earlier steps. Analysts use the terms **low-resolution** and **high-resolution** to distinguish different types of separation techniques.

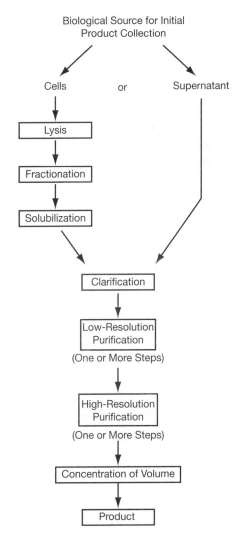

Figure 33.3.   A General Flowchart for Product Purification.

Resolution, or **resolving power,** *is the relative ability of a technique to distinguish between the product of interest and its contaminants.* Earlier steps in the purification process are usually low-resolution; later ones are of higher resolution. Figure 33.3 and Table 33.7 provide an overview of the usual stages of bioseparation strategies.

B. Fractionation and Clarification Methods

As shown in Table 33.7, bioseparation strategies begin with solubilization of the product of interest from cellular and other particulate matter. In the case of secreted molecules, liquid supernatants can simply be separated from solid matter by centrifugation or filtration. In cases where the product is found inside the cell, the starting point is the preparation of a cell homogenate. A **cell homogenate** *is a suspension of cell contents in liquid, produced by disrupting the outer cell membrane, the cell wall if present, and some of the internal structure of the cell.* Cell homogenates generally contain intact organelles, such as mitochondria.

There are many factors to be aware of when preparing a cell homogenate. For example, molecules inside the cell may be soluble in the cytoplasm or may be associated with cellular organelles. Depending on these properties, different buffers and techniques might be used for homogenization. The interior of an intact cell is buffered and protected from oxidizing agents that can

---

Table 33.7. THE STAGES OF A BIOSEPARATION STRATEGY

The sequential stages of a purification process should follow a logical pattern. Although there are occasional exceptions, the series of steps selected will address the following goals. Examples of appropriate techniques are provided for each goal.

1. Solubilization of the product of interest
   • disruption of cells, and/or
   • collection of supernatants containing soluble product
2. Separation of product from contaminating solid material (clarification)
   • centrifugation
   • filtration
3. Separation of product from divergent impurities
   • low-resolution purification techniques, such as:
      salt fractionation
      organic or polymer extractions
4. Purification of product from similar impurities
   • high-resolution purification techniques, such as:
      polyacrylamide gel electrophoresis
      high-performance liquid chromatography
5. Product polishing, concentration, and preparation for end use
   • dialysis to reduce final volume
   • gel filtration to remove salts

Table 33.8. DESTRUCTIVE FACTORS DURING PURIFICATION

Biomolecules removed from their intracellular location can lose biological activity and/or structural integrity due to inhospitable in vitro environments. It is important to consider potential destructive factors that may adversely affect your final product. These factors are present not only during homogenization procedures, but also during other purification steps.
   • Mechanical stress of cell homogenization
   • Dilution and removal of intracellular stabilizing factors
   • Changes in buffering, pH, and temperature
   • Presence of oxidizing agents
   • Product destruction by cellular enzymes

---

destroy molecules or their biological activity. Solutions used for homogenization, therefore, usually maintain a constant pH and include antioxidizing agents. Cell homogenization should be performed as gently as possible, to avoid loss of product due to the factors indicated in Table 33.8.

Many techniques can be used to **lyse** *(break open)* cells, see Table 33.9. Cells with relatively fragile membranes, such as liver cells, will require less strenuous methods than plant cells, which have a tough outer cell wall. Most lysis techniques can be adjusted to be more or less disruptive. Roughness and length of treatment can be varied according to need. It is important to remember that mechanical disruption methods, such as grinding or sonication (the use of high-pitched sound), can generate considerable amounts of heat within the sample. These procedures should be performed under chilled conditions when possible.

After cell lysis, the resulting homogenate should be clarified. **Clarification** *is the removal of unwanted solid matter, usually by centrifugation or filtration.* If the product of interest is still located within a cellular organelle, such as mitochondria, a differential centrifugation can be performed to isolate the organelle of interest (described in Chapter 32).

Table 33.9. COMMON CELL HOMOGENIZATION METHODS

The cell homogenization method used in a specific bioseparation scheme depends on the intracellular location of the molecule of interest, the cell type involved, and other factors, as listed in Table 33.8. Some of the most commonly applied techniques are:
   • Grinding
   • Sonication
   • Freezing and thawing
   • Application of high pressure using a French press
   • Osmotic lysis
   • Enzymatic treatment
   • Chemical treatment

C. Low-Resolution Purification Methods

There is a wide variety of relatively simple, low-resolution purification methods available. Earlier we defined *resolution* as the relative ability of a technique to distinguish between the product of interest and its contaminants. The resolving power of a method is directly dependent on its selectivity. **Low-resolution purification methods** *are those that generally have high capacities and can be performed quickly, but also have relatively low selectivity.* They eliminate contaminants that are highly dissimilar to the product of interest, but not those that have relatively similar molecular properties. These methods provide an efficient starting point for a bioseparation strategy, when the molecule of interest represents only a small portion of source material.

Many of these low-resolution methods exploit the solubility characteristics of the desired product. The most common types of methods are extractions and precipitations. **Extraction methods** *are based on the premise that a specific molecule will exhibit higher solubility in one medium compared with another.* For example, if you need to isolate a **hydrophobic** *(water-fearing)* compound from a cell homogenate, you would add an organic solvent such as ethyl acetate to the aqueous homogenate. By mixing the buffer and ethyl acetate well (usually by shaking), the hydrophobic compound and other hydrophobic contaminants will be extracted into the organic phase. By centrifuging the resulting mixture, you can collect the organic phase containing virtually all of the compound of interest, while removing and discarding water-soluble contaminants in the aqueous phase, see Figure 33.4. Application of this method requires knowledge of the relative solubility of the molecule of interest in nonmiscible liquids.

Other common and simple bioseparation steps include the use of precipitation techniques. **Precipitation methods** *are based on differences between molecules in their tendency to precipitate in particular liquids.* Manipulation of specific characteristics of the liquid medium of the sample will usually result in precipitation of certain molecules, which can then be separated from the liquid. In some cases, the desired product is precipitated; in other strategies, impurities are precipitated and thereby removed. The characteristics of the liquid that are most commonly adjusted are:

- salt concentration
- pH
- temperature
- presence of alcohols, such as ethanol
- presence of organic polymers, such as polyethylene glycol (PEG)

These factors are sometimes combined, as in the precipitation of DNA with ethanol and salt. (Examples of these techniques are also discussed in Chapter 29.)

One of the most common precipitation methods for proteins is to add ammonium sulfate to a cell

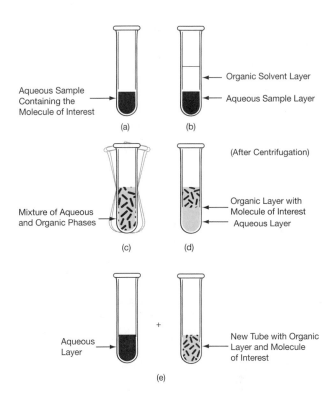

Figure 33.4. An Example of a Simple Organic Extraction. **a.** Start with an aqueous cell homogenate containing the product of interest. **b.** Add an equal volume of ethyl acetate, an organic solvent. **c.** Mix the two liquids thoroughly. **d.** Centrifuge the mixture and separate the organic and aqueous phases. **e.** Collect and assay the organic layer, which contains the product of interest. Note that in this example, the organic layer is of lower density than the aqueous layer. (Relative densities of extraction layers can be determined by consulting O'Neil, Maryadele J., ed. *The Merck Index.* 14th ed. Whitehouse Station, NJ: Merck and Co., 2006.)

homogenate in order to "salt out" (precipitate) the protein of interest. Increasing amounts of the ammonium salt are added to the protein solution. After each addition, the homogenate is centrifuged and the supernatant assayed for the protein of interest. When the appropriate concentration of salt is reached, the desired protein will be (almost) entirely precipitated and can be collected for the next purification step.

D. High-Resolution Purification Methods

i. ELECTROPHORESIS

High-resolution purification methods *are those that have relatively high selectivities, allowing separation of product from similar impurities.* These methods are generally applied after low-resolution techniques have removed the bulk of impurities that are dissimilar to the product of interest. High-resolution methods frequently (but not always) have higher costs and lower capacities relative to the methods discussed in the previous section. The most commonly applied high-resolution purification methods are gel electrophoresis and chromatographic techniques. Gel electrophoresis techniques are low capacity and generally are not used for production systems. In contrast, many chromatographic techniques can be adapted to high-capacity systems.

Electrophoresis *is the separation of charged molecules in an electric field.* In addition to separating specific proteins and DNA fragments of different sizes, electrophoresis techniques are frequently used to determine:

- the length of DNA fragments, in base pairs
- the molecular weight of specific proteins
- the **isoelectric point** of a protein, *the pH value at which a protein exhibits an overall neutral charge*
- the purity of an isolated protein

Here we will briefly discuss two of the most common electrophoretic techniques: agarose gel electrophoresis, most commonly used to separate DNA fragments on the basis of base pair length, and SDS-polyacrylamide gel electrophoresis (PAGE), most commonly used to separate proteins on the basis of molecular weight. Both of these techniques use polymer gels as a separation matrix.

Digested or sheared DNA fragments are usually separated in agarose gels. **Agarose** *is a natural polysaccharide derived from agar, a substance found in some seaweeds,* see Figure 33.5a. When agarose powder is heated at low concentrations (0.8% is typical) in an appropriate buffer, the agarose dissolves into a viscous solution that gels when cooled. This gelling is reversible, so that agarose solutions can be cooled and heated repeatedly. The cooled gel is chemically stable, with a relatively high gel strength for thin sheets of agarose. These sheets can be gently manipulated for staining and transfer of DNA to other media. Because agarose gels generally cannot support their own weights, electrophoresis is performed in a horizontal gel box. An agarose gel electrophoresis apparatus and gel example are shown in Figure 18.9 on p. 321.

Because of the regular occurrence of charged phosphate groups along DNA strands, all DNA molecules have a uniform negative charge. When DNA samples are applied to a gel and exposed to a directional electric field, all DNA fragments will therefore migrate toward the positive electrode. Because the agarose gel acts as a molecular sieve, smaller DNA fragments will move to the positively charged end of the gel faster than larger fragments, which will become entangled in the pores of the gel. As long as the electric field is applied, DNA fragments will migrate through the gel at a rate that is roughly related to the inverse log of the number of base pairs in the fragment. The bands of DNA are most commonly stained with ethidium bromide, allowing visualization of the bands under ultraviolet light. If DNA fragments of known base pair length are applied to the same gel as controls, a reasonable approximation of

$$CH_2 = CHCONH_2 + CH_2(NHCOCH = CH_2)_2$$

Acrylamide | N,N' Methylenebisacrylamide + Activator (Ammonium Persulfate) + Catalyst (TEMED)

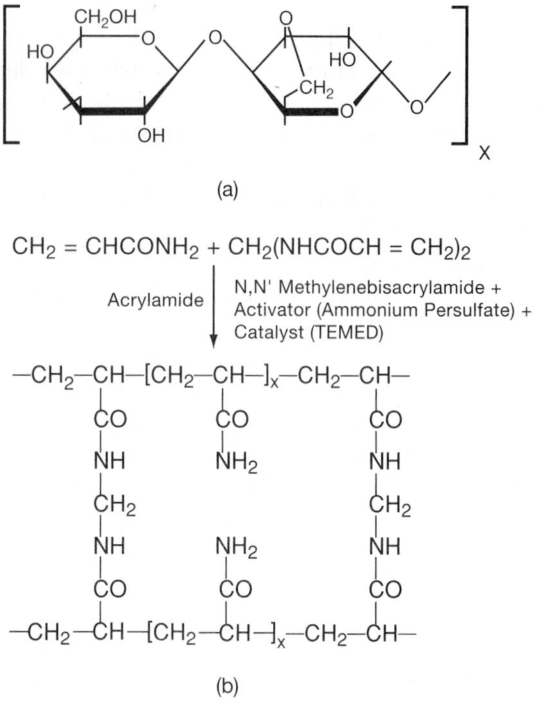

Figure 33.5. Subunit Structures for (a) Agarose and (b) Polyacrylamide Gels.

base pair length for unknown DNA bands can be made. Note that ethidium bromide is a known mutagen, whose handling is discussed in Chapter 11.

Proteins are generally separated on **polyacrylamide gels,** *which are made from acrylamide* (a neurotoxin—see Chapter 11) *and bisacrylamide polymerized in the presence of an appropriate initiator (ammonium persulfate) and catalyst (N,N,N',N'-tetramethylethylenediamine, commonly called TEMED)* (see Figure 33.5b). The pore size of a gel can be adjusted by combining different amounts of acrylamide and bisacrylamide. Polyacrylamide gels are stronger than agarose gels of comparable thickness, so they can be used for vertical gel electrophoresis (PAGE), see Figure 33.6.

Unlike DNA molecules, all proteins are not uniformly charged. The presence of both positively and negatively charged amino acids at a given pH allows the PAGE technique to separate proteins on the basis of charge. A more common adaptation of PAGE, however, is SDS-PAGE. **Sodium dodecyl sulfate-PAGE (SDS-PAGE)** *is a technique that separates proteins on a polyacrylamide gel on the basis of molecular weight.* Protein samples are treated with the negatively charged detergent SDS. SDS binds to proteins at a ratio of approximately one SDS molecule for every two amino acids in the protein. The resulting negative charge conferred by the SDS overshadows any native protein charge so that all proteins will migrate toward the positive electrode in an electric field (as shown in Figure 33.6). In this case, sieving of the proteins occurs in the

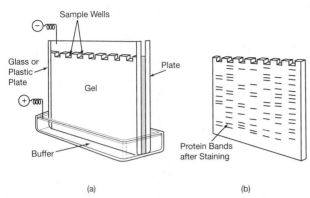

Figure 33.6. Polyacrylamide Gel Electrophoresis (PAGE). a. The gel is poured vertically between two glass plates. Sample is applied to wells at the top of the gel. **b.** Protein bands are separated on the basis of relative molecular weight and visualized with stains. *Note:* This is an example of SDS-PAGE.

gel according to the molecular weight of the protein. The detergent also denatures the protein, minimizing the differential effect of protein shape on migration rate. Higher concentrations of acrylamide and higher levels of cross-linking provide a gel with smaller pore sizes. Adjusting pore size controls the effective range of protein sizes that are separated on an individual gel.

SDS-PAGE is frequently used both for determination of the molecular weight of proteins, and also for the determination of the purity of protein preparations. Figure 33.7 illustrates how this technique can be used to

check product purity after each step in a purification strategy. Each sequential method ideally reduces the number and amount of various contaminants, with the final step resulting in a single protein band on the gel. The presence of a single band on an SDS-PAGE gel is considered one measure of protein purity. However, the protein band may still contain contaminants of the same molecular weight and properties as the protein of interest, and so at least one additional method must be used to confirm the purity of the product.

Electrophoresis requires the same or even greater safety precautions as any other use of electricity. These techniques require extreme care because an electric current is being deliberately applied to a liquid. Electrophoresis generates heat, which may result in the evaporation of system buffer and allow short circuits, equipment damage, or fires. Additional safety suggestions are provided in Chapter 10.

ii. CHROMATOGRAPHY

The term *chromatography* describes the most widely used set of methods applied to the purification of biomolecules. Some type of chromatography is used in most laboratories, for either purification or analytical purposes (or both). This section will begin by discussing some of the most common forms of chromatography practiced in biotechnology laboratories, and then will briefly introduce more specialized techniques.

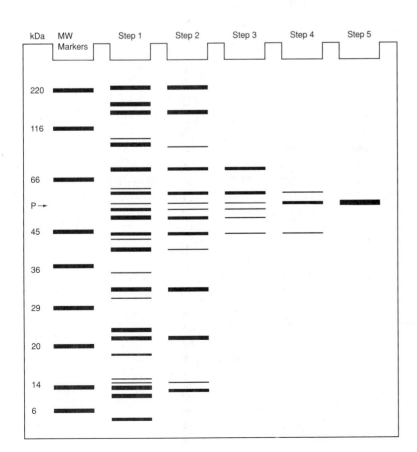

Figure 33.7. SDS-PAGE Analysis of Sequential Purification of a Protein. Lane 1 on the left shows the bands from a protein molecular weight marker standard, used to estimate the size of protein bands in the samples. In this case, the protein of interest (the band labeled P) is known to be approximately 59 kDa (kiloDaltons). The next lane is the sample from Step 1, which is cell homogenization. This sample contains the protein of interest and all protein contaminants initially present. Step 2 in this purification scheme is an ammonium sulfate precipitation. Several contaminants are removed in this low-resolution technique. Gel permeation chromatography in Step 3 removes proteins with a molecular weight significantly different from the desired product. Step 4, partition chromatography, eliminates most other proteins from the sample. The final step is affinity chromatography, which is highly specific for the protein of interest and results in a single protein band. More details about these techniques are provided in the text.

Chromatography refers to a family of bioseparation techniques based on the differential interaction of molecules between a stationary and a mobile phase. As suggested by their names, the **stationary phase** *is an immobile matrix* and the **mobile phase** *is a liquid or gas that moves past the stationary phase.* When performing chromatography, the sample, which contains a mixture of molecules, is introduced into the flowing mobile phase. Sample molecules move along with the mobile phase past the stationary phase material. Different molecules have different physical properties that determine their relative attraction to the mobile phase versus the stationary phase. For example, one molecule might, under certain circumstances, tightly bind the stationary phase matrix while other molecules with different properties remain in the mobile phase and flow right through the chromatography system. Yet another molecule might be weakly attracted by the stationary phase and would therefore be slightly slowed in its flow. Different molecules thus move past the stationary phase at differing rates and hence are separated from one another.

The mobile phase used in chromatography can either be a liquid, in which case the method is termed **liquid chromatography, LC,** or *a gas,* in which case the method is termed **gas chromatography, GC.** A gas mobile phase is typically paired with a liquid stationary phase (sometimes abbreviated as GLC, gas-liquid chromatography) and a liquid mobile phase is most commonly used with a solid stationary matrix.

One of the advantages of chromatography is its versatility of techniques available. Molecules differ in various physical properties, such as their size, shape, charge, and solubility in water. Chromatographic techniques can separate molecules based on differences in any of these molecular properties. Another advantage of chromatography is that it can often separate molecules based on quite subtle differences between them. Another advantage of chromatography is that recovery of separated products after chromatography is generally simple. For industrial applications, chromatographic techniques have the advantage that they are frequently amenable to scale-up from laboratory to large-scale operations.

The first step in choosing a chromatographic method is therefore to determine which molecular properties differentiate the molecule of interest from its contaminants. Table 33.10 discusses some of the molecular properties that are most commonly used as the basis for chromatographic separations.

Once an underlying property is chosen, a specific chromatographic method can be selected. The most commonly used method is **partition chromatography,** *in which a specific molecule will exhibit higher solubility in one chromatographic phase compared to another, and distribute itself accordingly.* This is the same principle applied in extraction techniques (see p. 629). In partition chromatography, the stationary phase is liquid while the mobile phase can be gas or liquid. Partition chromatography is closely related to **adsorption chromatography,** *where specific molecules differ in their tendency to adsorb to a solid stationary phase.* In practice the distinction can be subtle, since in many partition chromatography applications, the "liquid" stationary phase is bound to a solid support matrix and adsorption also occurs. However, partitioning is an essentially nonspecific process based on relative solubility between the two chromatographic phases, whereas adsorption is a relatively specific interaction between the sample and the stationary phase, as in ion exchange chromatography, described below. Adsorption techniques are usually applied to larger molecules such as proteins, whereas partition chromatography is best suited for relatively small molecules.

Chromatographic methods vary in how much instrumentation is required. Early methods used planar (flat) stationary phases, such as paper, that require little specialized equipment. In paper chromatography, small spots of a sample are applied to special chromatography paper and allowed to dry. The end of the paper is then dipped into the mobile phase, which wicks through the paper, carrying along the various components of the sample. **Thin layer chromatography (TLC)** is another planar technique that does not require expensive equipment or materials. In TLC, *a thin layer of stationary phase material, usually silica, is spread on a glass or plastic plate, and the liquid mobile phase passes through the stationary phase by either capillary action or gravity.* For some applications, flexible plastic backing material can provide custom sizes and ease of handling.

Nowadays, planar chromatography is less common, and most chromatography involves columns. **Column chromatography** *is performed with the stationary phase packed into a cylindrical container, the column.* The mobile phase, or **eluent,** passes through the column, propelled either by gravity or mechanically with a pump. **Elution** of molecules from the column *occurs as the molecules in the sample distribute themselves between the*

Table 33.10.   MOLECULAR CHARACTERISTICS COMMONLY USED AS A BASIS FOR CHROMATOGRAPHY

Molecular Property	Chromatography Method
• Polarity, or relative water solubility	Partition chromatography
• Charge	Ion exchange chromatography
• Molecular weight	Gel permeation chromatography
• Hydrophobic properties	Hydrophobic interaction chromatography
• Specific molecular binding properties	Affinity chromatography

two phases according to their affinities for each phase, as shown in Figure 33.8. Fractions of the mobile phase can be collected and assayed as they leave the column.

An LC system can be as simple as the column shown in Figure 33.8, where solvent is added to the top of the column and moves past the stationary phase with the force of gravity. However, a more typical system is shown in Figure 33.9a. In this case, a low-pressure pump is used to maintain a steady flow of mobile phase through the column. The components of the system are connected to one another with chemically inert tubing. The sample is injected into the flowing mobile phase through an injection port located at one end of the column. A detection system that detects molecules as they exit, is located at the other end of the column. The detector is connected to a data recording device, which can be as simple as a strip-chart recorder, or a computer with software to analyze the data. The recorded output is usually in the form of a chromatogram, as shown in Figure 33.8b. Purified sample can be recovered after the detector by collecting sequential fractions (tubes) of the mobile phase.

Small columns, usually made of glass or plastic, can be filled with stationary phase by the user, although these columns are generally used only once and only for low-pressure applications. Prefilled columns can be purchased for convenience, or to provide more delicate stationary phase materials or larger columns. With proper method development, large columns can provide very high product capacities together with high resolution. This accounts for the common use of chromatographic techniques in production settings. Capacity increases with column volume (up to a point), although this increased capacity comes at increased monetary cost as well. Large columns

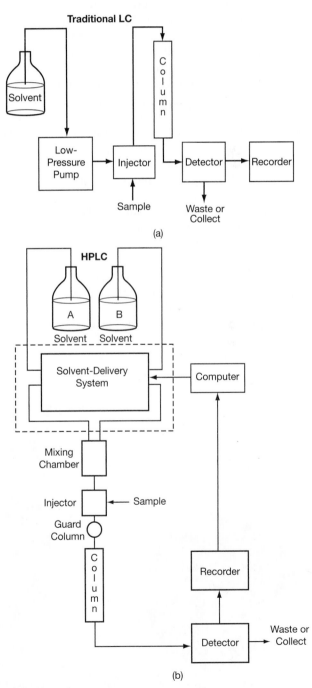

Figure 33.9.   Typical Equipment Systems. a. Traditional Liquid Chromatography. **b.** High-Performance Liquid Chromatography. See text for detailed explanation.

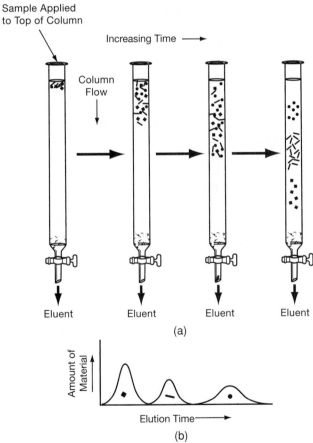

Figure 33.8.   Elution of a Sample Mixture from a Chromatographic Column. a. Separation of Sample Components. A sample containing three different molecules is loaded onto a column. As the mobile phase carries the sample constituents through the column, movement rate of the molecules is affected by the relative tendency of the molecules to adsorb to the stationary phase, slowing their elution. **b.** Sample Chromatogram. The three molecules are graphed according to their elution times from the column.

are usually constructed with steel walls and require pumps to elute the mobile phase.

Selection of the stationary and mobile phases is critical in developing a successful chromatographic technique. The earliest chromatographic techniques used a polar, hydrophilic (water-loving) stationary phase and a mobile phase that was relatively nonpolar (hydrophobic). This is sometimes referred to as *normal-phase* chromatography. However, most modern partition chromatography uses **reversed phase,** *where the stationary phase is nonpolar and the mobile phase is polar relative to the stationary phase*. In practice, reversed-phase chromatography provides higher molecular selectivity and reproducibility than normal-phase systems.

Solid stationary phases for column chromatography are usually based on small microporous silica particles, as shown in Figure 33.10. These particles must be nonreactive and uniform in their physical characteristics, and provide a large surface area for potential interaction with molecules in the mobile phase. For reversed-phase partition applications, a hydrophobic liquid is chemically bonded to the particles. Columns packed with these particles are then paired with aqueous solutions of organic solvent, frequently acetonitrile or methanol. Partition chromatography can be optimized to separate and quantitate a wide variety of molecules by choosing a column with a particular liquid bonded to the stationary silica particles and by selecting a particular concentration of organic solvent.

High-performance liquid chromatography (HPLC) *is a set of chromatographic techniques that involve specially packed columns and high-pressure pumps.* HPLC columns provide stationary phases made of smaller and harder particles than traditional column chromatography. These particles allow the use of faster mobile phase flow rates. Use of HPLC speeds up sample elution ten-

fold or more relative to low-pressure methods. This is convenient, and also allows greater recovery of sample activity. HPLC can be adapted to any of the chromatographic principles shown in Table 33.10 on p. 632.

HPLC columns cannot be packed effectively without special equipment. High-pressure pumps are required to move the mobile phase through the tight packing of the stationary phase. Smaller packing particles provide the highest separation efficiencies, although large-scale separations use larger particle sizes to increase column capacity.

Mobile phase solvents for HPLC must be of the highest quality, generally designated "HPLC grade" by vendors, to avoid introduction of impurities into the system. While occasionally an HPLC separation can be achieved using a constant-solvent elution, most applications require **gradient elution,** *in which the concentration of components in the mobile phase changes during the separation process to progressively elute more components from the stationary phase*. Gradients, which can be performed in steps or as a continuous process, allow separation of molecules with a wider range of properties than single-solvent systems, and also allow faster separations. While several solvents can be mixed in a variety of concentrations, most applications use two solvents, such as water and acetonitrile, to form a continuous gradient.

A simple equipment setup for HPLC is shown in Figure 33.9b on p. 633. While this system is more complex than the low-pressure LC system shown in Figure 33.9a on p. 633, the same principles apply. Solvent from two or more reservoirs moves to a solvent-delivery system, which includes a programmable gradient mixer and high-pressure pump. The sample is injected through a port leading to a **guard column,** *which is an inexpensive, disposable chromatography column that protects the expensive HPLC column from contaminants in the sample.* The

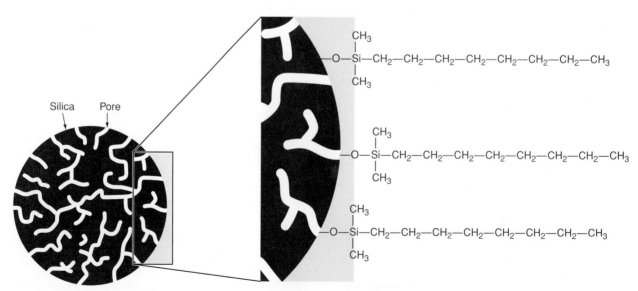

Figure 33.10. Column Packing Material for Reversed Phase Chromatography. a. Microporous beads of silica are used as a base for the stationary phase. The beads are typically 1 to 10 μm in diameter with pores ~0.01 μm wide (for separating small molecules). **b.** The surfaces of the beads and pores are coated with the chemically bonded liquid stationary phase. In this case, the chemical group is a C8 hydrocarbon.

guard column contains a relatively coarse packing material that retains particulate and sometimes specific chemical contaminants from fouling the separation column. After passing through the guard column, the sample proceeds through the HPLC separation column. Because the solvent flows under high pressure, HPLC components are connected with very fine-bore steel tubing and special high-pressure fittings. Two of the essential skills for a chromatographer are thus an understanding of basic plumbing and the ability to use a wrench to maintain proper connections between system components.

So far our discussion has concentrated on partition chromatography separations, but the other methods listed in Table 33.10 on p. 632 are also frequently used in purification and analysis strategies. For example, **hydrophobic interaction chromatography** *uses gentle reversed-phase conditions to separate large biomolecules based on their hydrophobicity.* This method is generally applied to protein purification when it is essential to retain the biological activity of the product. By using a less hydrophobic stationary phase than most reversed-phase applications, purified molecules can be eluted using salt solutions rather than organic solvents capable of disrupting protein folding and function.

Ion exchange chromatography *is an adsorption chromatography technique that separates biomolecules based on their net molecular charges. Inert stationary phase particles are coated with either positively or negatively*

charged molecules to create materials called **ion exchangers.** If the stationary phase particles are coated with positively charged molecules, then they will adsorb negatively charged molecules (anions) that flow by in the mobile phase. In this case, the column is termed an **anion exchange column.** While anions in the sample are adsorbed by the exchanger, impurities that do not bind to the column are washed away with the mobile phase. The anions from the sample can be released from the stationary phase and then recovered from the column by changing the pH and/or salt concentration of the mobile phase. This process is shown in Figure 33.11.

In a similar fashion, the stationary phase may be prepared by binding negatively charged molecules to the stationary phase particles. In this case, the column will adsorb positively charged molecules (cations) that flow by in the mobile phase and the column is termed a **cation exchanger.** Ion exchange chromatography is commonly used in protein purification strategies and with careful development of mobile phase conditions and selection of ion exchange matrices this technique can provide reproducible, high-yield results.

Affinity chromatography *is used to purify biomolecules by exploiting their individual binding properties.* Inert column packing material is coated with a **ligand,** *a molecule that binds specifically to another molecule of interest.* When the sample is passed through the column, only the desired product binds to its ligand, while

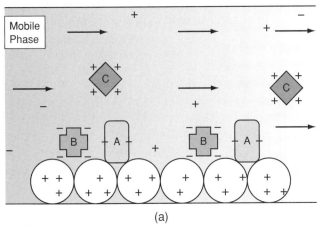

(a)

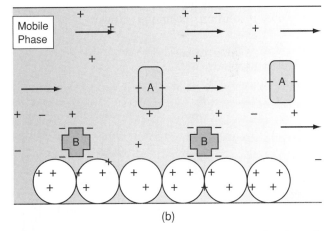

(b)

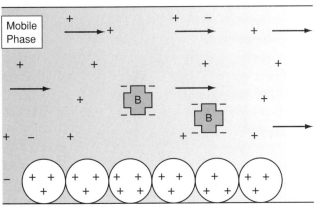

(c)

Figure 33.11. Ion Exchange Chromatography. The example here is an anion exchange column. **a.** A sample with three proteins differing in their relative surface charges passes over a positively charged anion exchange resin. Protein A has a relatively weak negative charge, Protein B has a stronger negative net charge, and Protein C has a net positive charge. The negatively charged proteins are adsorbed by the stationary matrix, while the positively charged protein is washed out with the mobile phase. **b.** The mobile phase is modified to provide a relatively more positive charge than the previous step, and successfully competes with the anion exchanger to wash out some of the negatively charged proteins. **c.** The proteins with the strongest relative negative charge are finally removed from the column with another modification in the mobile phase.

contaminants are eluted with the mobile phase. The desired product can be washed from the column by changing the mobile phase composition. If a suitable ligand exists, this is a very effective method of purifying a desired product away from contaminants. Examples of ligand/biomolecule pairings include enzyme/substrate and antibody/antigen binding.

An LC method that merits separate discussion is gel filtration, or gel permeation chromatography. **Gel permeation chromatography (GPC)** *is a technique for separating molecules by relative size and molecular weight using a column filled with porous gel particles.* GPC is also known as **molecular sieving** or **size exclusion chromatography.** The gel particles are usually highly cross-linked polysaccharides (one typical material is Sephadex, manufactured by GE Healthcare Bio-Sciences) with relatively large pore sizes compared to other chromatographic media. Gel permeation is based on molecular sieving rather than chemical interaction between the sample and stationary phase. In a gel permeation system, small molecules enter the pores of the stationary phase and are retarded on the column. Large molecules are excluded from the pores and elute from the column relatively quickly, see Figure 33.12. In other words, the larger the molecule, the faster it passes through the column. This is the opposite result of the sieving found in electrophoretic techniques, although each of these methods can be used to determine the approximate molecular weight of materials. Unlike other chromatography methods, gel permeation is a relatively low-resolution technique. However, it is frequently used as a final purification step to remove salts from final products.

The multitude and flexibility of chromatographic techniques make the design of purification methods complex. As with other techniques, the properties of the desired product and its impurities must be considered. Because chromatographic methods (with the exception of gel permeation) frequently separate molecules based on subtle differences between them, small changes in stationary or mobile phase materials can make major differences in purification results. Chromatographic methods frequently require extensive fine-tuning to provide optimal results, but can provide extremely powerful bioseparation tools.

E. Common Problems with Purification Strategies

The development of an optimized purification strategy often does not proceed as smoothly or quickly as desired. Perseverance is necessary to ensure that the final strategy will be appropriate on a long-term basis. The most common problem encountered is the apparent loss of product during a specific separation step. If this occurs during initial strategy development, it may suggest that an inappropriate separation method was chosen. For example, if you are performing an affinity chromatography technique and your product will not elute from the column, you may have inadvertently chosen a column matrix that binds irreversibly with the product material. Always develop your separation strategy using an aliquot of a larger sample so that you can return to the

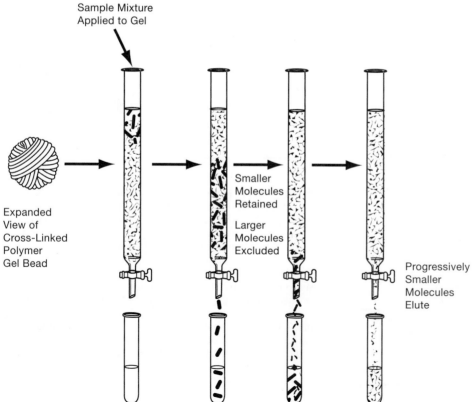

Sample Mixture Applied to Gel

Expanded View of Cross-Linked Polymer Gel Bead

Smaller Molecules Retained

Larger Molecules Excluded

Progressively Smaller Molecules Elute

Figure 33.12.   Diagram of Gel Permeation Chromatography.

results of a previous step and try another strategy.

Product is sometimes unexpectedly lost during a separation step that was previously proven to be effective. In these cases, there is usually a technical difficulty that can be identified. These same difficulties can occur during strategy development, leading to the incorrect assumption that a method is not effective for purifying your product. Whenever large amounts of product disappear during a separation step, consider whether the factors outlined in Table 33.11 may be involved.

Note that some of the problems indicated in Table 33.11 are the result of technical difficulties (e.g., using the wrong buffer) and others are the relatively unpredictable result of source contamination (e.g., the presence of enzyme inhibitors). In the latter case, either the interfering substance must be removed before the product is assayed, an alternative source of product must be used, or a different product assay technique is required.

To summarize the most basic precautions against losing product during a bioseparation strategy:

- In general, the fewer purification steps used, the higher the ultimate product yield.

- With many biomolecules, stability is a major concern. Maintain a constant temperature (usually chilled) and pH for the sample and proceed from one step to the next as quickly as feasible.

- Never discard any fraction during a separation procedure. Save everything until the end of the process, when the product has been recovered.

F. Large-Scale Operations

All purification strategies are developed at the laboratory scale before moving to a production setting, and so the ability to convert a strategy to a large-scale operation is essential. The purification of agricultural products, for example, can require handling tons of source and final materials. **Scaling up** *is the process of converting a small-scale laboratory procedure to one that will be appropriate for large-scale product purification.* Scale-up must be considered from the start of product development. Some laboratory purification procedures may not be economically sound or physically possible in an industrial setting. For example, using sonication to lyse cells may work well in a laboratory, but it becomes impractical when hundreds of liters of cellular material are present. Breaking cells open with grinding or high-pressure techniques is more practical on a large scale.

When testing purification strategies for commercial purposes, the research and development team selects and optimizes methods on a small scale. Development scientists then need to increase the scale of operations and test the robustness of the strategy. The working conditions (e.g., pH, temperature, ion concentrations) must be achievable under large-scale conditions. Analysts need to consider the possibility of variations in biological source materials and must design tech-

Table 33.11. *FACTORS THAT MAY RESULT IN PRODUCT LOSS DURING PURIFICATION PROCEDURES*

There are many factors that can contribute to a major loss of product during a bioseparation procedure. Some of these are relatively easy to identify (at least in retrospect), and others are difficult to predict. Common problems and examples include:

Problems leading to product loss

- Discarding the fraction that contains the product of interest (e.g., discarding a supernatant containing product while keeping a pellet of debris)
- Performing separations using poor-quality reagents (e.g., ion exchange medium may be inadequately prepared)
- Poor product stability (may be due to enzymatic degradation by impurities in the mixture, or inherent instability of product)
- Unexpected precipitation of product (e.g., sample may be suspended in an inadequate amount of liquid)

Problems with the product assay

- Performing an inappropriate product assay (lacking adequate selectivity or specificity)
- Using an incorrect buffer for the product (e.g., certain salts may interfere with enzyme assays)
- Presence of product inhibitors (if presence of product is measured by activity)
- Loss of a required cofactor (e.g., an enzyme may require the presence of magnesium ions for activity)

niques that will be relatively insensitive to the most likely variations.

Some of the key factors to be considered during scale-up are important during methods development, but assume an even greater significance under production conditions. As indicated earlier, reproducibility of the techniques in the production setting is essential. Having a simple assay method and documentation process for each purification step will contribute to cost-effectiveness and quality control. Of course, safety issues must be considered during scale-up. Strategies must be designed to eliminate any potential biohazards early in the process, and appropriate containment facilities and procedures are essential.

In reality, many potentially useful biotechnology products never reach the consumer market due to the high costs of downstream processes. The scale of required equipment and labor costs are major considerations. The ability to automate bioseparation procedures can increase cost-effectiveness to some extent, but not all procedures are easily automated. The ability to reduce product source volumes quickly, apply less expensive low-resolution methods, and achieve high product yields are key factors in reducing purification costs for biotechnology products.

PRACTICE PROBLEMS

1. You have received a vial of enzyme that is labeled as containing 4 kat (SI units) of enzyme. How many IU of enzyme are in the vial?

2. You are presented with the following data from a series of purification steps for the enzyme "comatase." Fill in the blanks in the table:

PURIFICATION OF COMATASE FROM E. coli

Purification Step	Volume (ml)	Total Protein (mg)	Total Activity (IU)	Specific Activity (IU/mg)	Yield (%)
I. Homogenization	100,000	12,350	9000	____	100
II. Dialysis	5000	10,233	8289	0.81	92.1
III. Organic extraction	80	3860	____	1.65	____
IV. Ion exchange chromatography	10	1140	5625	____	62.5
V. PAGE	2	386	4688	12.15	____

3. Given the purification scheme data shown in Problem 2, which purification step gave the greatest:

   a. relative increase in product purity?

   b. loss of product?

   c. absolute reduction of product volume (the greatest decrease in actual volume)?

   d. relative reduction of product volume (the greatest % decrease from the previous step)?

4. Suppose you are given the assignment of isolating an enzyme, "enzylase," which is secreted into the fermentation broth of the bacterium *B. techi*. The enzyme has an approximate molecular weight of 60,000. You grow 3 L of *B. techi*, which you then spin down in a centrifuge. You discard the broth and weigh the cells. You break open the cells using a grinding method and centrifuge the resulting paste. You discard the cellular debris and assay the supernatant for the enzylase. You are disappointed to find no activity. You test your assay with "store-bought" enzylase and find that the detection assay works well. What important mistake did you make?

5. You are interested in an enzyme produced by a particular bacterium. You grow 10 L of the microorganism in a fermenter. You centrifuge the 10 L to obtain a pellet of cells and set aside the supernatant. You resuspend the cell pellet in 20 mL of buffer, break open the cells by sonication, and remove the cellular debris by centrifugation. The enzyme of interest is in the intracellular supernatant. (Assume there is still 20 mL of supernatant.) The supernatant is called the crude extract. You remove 0.5 mL of the crude extract and perform an enzyme assay. You find there are 500 units of enzyme activity in the 0.5 mL

sample of crude extract. You take another 0.5 mL sample of crude extract and perform a protein assay. You find there are 25 mg of protein present in the 0.5 mL of crude extract.

   a. Draw a flow chart of this procedure.

   b. What is the specific activity of enzyme in the crude extract?

   c. How many units of enzyme activity were in the original 10 L?

6. Choose the most probable answer to the following questions:

   a. During purification of an enzyme, the specific activity should:

      i. Increase as the purification proceeds

      ii. Decrease as the purification proceeds

      iii. Remain the same throughout the purification

   b. During purification of an enzyme, the amount of total protein present should:

      i. Increase as the purification proceeds

      ii. Decrease as the purification proceeds

      iii. Remain the same throughout the purification

   c. During purification of an enzyme, the yield is likely to:

      i. Increase as the purification proceeds

      ii. Decrease as the purification proceeds

      iii. Remain the same throughout the purification

7. Before developing a strategy to purify a protein, it is important to learn about the biochemical characteristics of that protein. List at least three features of the protein that are important to know when developing a strategy to purify a protein.

QUESTIONS FOR DISCUSSION

Development and trouble-shooting of purification strategies and procedures are an integral part of work in a biotechnology setting. Quality control and its documentation are critical aspects of product preparation. **Scenario 13 in the Skill Standards for the Bioscience Industry (Education Development Center, Inc., 1995) presents the following topic and question:**

You are responsible for following the protocol for purifying your company's product. Demonstrate the steps you take in product purification.

The 2 L of the crude product has a calculated maximum yield of 10 g/L. You expect an 80% yield. After running the column, you calculate the purified total sample yield as 22 g/L. Show how you would handle this result.

UNIT IX

Computers in the Laboratory

Computers have become not only an essential piece of laboratory equipment, but also an integral part of everyday life. This is rather ironic, considering that in 1943, Thomas Watson, the Chairman of IBM, predicted that there would be "a world market for maybe five computers." Clearly, computers have provided useful functions far beyond those originally envisioned!

One reason for the explosion of computer use in the world is that these machines can perform certain types of functions faster and more reliably than humans. Some of these functions include:

- information storage and organization

- general data processing

- creation of illustrations and graphics

- access to vast stores of information

- local and global communication

The ability of computers to perform these tasks relies on human input. Computer systems will continue to become faster, smaller, and less expensive, but for the foreseeable future, they will require human operators. **Computer literacy** is no longer a specialized skill, but one that is required for effective job performance in a wide variety of fields. Becoming computer literate *means developing a general understanding of computers and their uses.* You need to feel comfortable using a computer to solve problems, without necessarily learning a lot about the technical workings of these machines. The essence of computer literacy is the ability to make the computer work for you.

It would be difficult to overestimate the importance of the Internet in biotechnology research and in our daily lives. In the laboratory and office, the Internet can provide access to levels of information that were only a dream a short time ago. For example, researchers now have access to the complete human genome, as well as those of many other species. The capability to search and analyze these data has exponentially increased the speed of new discoveries in gene structure and function. Many research journals are now available online, including archives. We can consult and download information about chemical safety, government regulations, scientific supplies, and techniques, to name just a few applications. E-mail allows us to be in constant contact with colleagues and friends around the world. We are still a long way from a paperless society, but computers and the Internet will continue to offer efficient alternatives to many manual tasks.

This short unit cannot provide comprehensive information about computers, so it is designed to provide a general overview of how computers function, and how you can use computers and the Internet for data handling. We concentrate on the familiar personal computers, which are the units that you are most likely to encounter, because these machines provide enough power for many of the complex tasks you will encounter in the laboratory. Because these are usually linked together into **networks,** additional computing capacity is available through more powerful computers when needed. We concentrate on PCs running the Windows operating system and Microsoft applications. This is not necessarily an endorsement, but an acknowledgement that these are the most common computing tools found in laboratories, and represent the general capabilities of other systems and software on the market. Because studies show that most individuals who use a computer at work will purchase their own computer for home, this unit provides information that will be applicable outside of work as well as in the laboratory.

There are many Internet resources cited in this unit (and throughout the book). We have chosen sites that have a relatively long history and are likely to be maintained in the future. However, because fluidity is one of the strengths (and weaknesses) of the Internet, it is possible that sites will move, disappear, or be abandoned by their creators. Chapter 36 on p. 691 discusses how to evaluate websites for reliability.

> Chapter 34 introduces the basic hardware and operations of computers. It summarizes the varied means of data input and output, and outlines the major considerations when upgrading to a new PC for work or home.
>
> Chapter 35 discusses the ways to handle data using computer software. It provides an overview of the types of software likely to be found on any computer, and those that are common in the laboratory. The increasingly essential role of bioinformatics in biotechnology research is discussed.
>
> Chapter 36 discusses the Internet and its essential role in the research environment, as well as our daily lives. It provides guidelines for evaluating websites and doing research on the Internet.

References for Unit IX

There are many references available that provide detailed information about computers, software, and the Internet, for both beginners and experts. Unfortunately, the technology is evolving so rapidly that all but the most basic information goes out of date within a year or less. Therefore, when consulting a computer reference, look for the most recent publication date. Many computer books keep current with frequent new editions. The list below provides basic references that should be useful over time.

General Computer Information

Any of the computer topics *For Dummies* books, written by various authors and published by John Wiley, Indianapolis. These books are enjoyable to read, informative, and updated regularly in many cases.

Brookshear, J. Glenn. *Computer Science: An Overview*. Boston: Pearson, 2009.

The Computer Desktop Encyclopedia, published by The Computer Language Company, Inc., http://www.computerlanguage.com/. This is an extremely useful reference on all aspects of computer science. It is available on disc and updated quarterly (purchase includes a one-year subscription).

How Stuff Works has an extensive computer section: http://computer.howstuffworks.com/.

White, R., and Downs, T. *How Computers Work*. 9th ed. London: Que, 2007.

An excellent site for computer ergonomics information is http://www.healthycomputing.com/.

The Internet

Hock, R. *The Extreme Searcher's Internet Handbook: A Guide for the Serious Searcher*. 2nd ed. Medford, NJ: CyberAge, 2007.

Lehnert, W.G., and Kopec, R.L. *Web 101*. 3rd ed. Boston: Pearson, 2008.

Web addresses for popular search engines:
Google: http://www.google.com/
Yahoo!: http://www.yahoo.com/
Ask.com: http://www.ask.com/
MSN: http://www.msn.com/

Web addresses for biology research:
PubMed: http://www.ncbi.nlm.nih.gov/PubMed/
Google Scholar: http://scholar.google.com/
Biology Browser: http://www.biologybrowser.com/

Bioinformatics

Claverie, J.M., and Notredame, C. *Bioinformatics for Dummies*. 2nd ed. Indianapolis: Wiley Publishing, 2007. A clear summary of the basic applications of bioinformatics (and definitely not for dummies).

Harvard University Department of Molecular and Cellular Biology has a directory of biomolecular and biochemical databases: http://mcb.harvard.edu/BioLinks/Sequences.html.

Krane, D.E., and Raymer, M.L. *Fundamental Concepts of Bioinformatics*. San Francisco: Benjamin Cummings, 2003. Contains both basic and more advanced information.

The National Center for Biotechnology Information is probably the best place to start for bioinformatics: http://www.ncbi.nlm.nih.gov/.

CHAPTER 34

Computers: An Overview

I. INTRODUCTION

One of the major functions of laboratories is the production of data. (The word *data* is the plural form of the seldom seen *datum,* which refers to a single number, letter, or other symbol. For this reason, the word *data* is generally used with a plural verb.) **Data** *are the unprocessed facts from which we derive information.* **Information** *is processed data which have been collected, manipulated, and organized into an understandable form.*

Computers have become essential to laboratory operations because they can convert data into information and store that information at rates that are impossible without their use. While their capacities for accurately performing repetitive data manipulations are impressive, humans provide the instructions, through programming, to perform the manipulations. Your goal, as a computer user, is to determine which functions can best be performed by a computer, and to provide appropriate instructions and data.

A computer system includes two basic elements: hardware and software. As the names suggest, **hardware** *includes the solid objects needed for computer functions.* **Software,** on the other hand, *includes the programs that provide the instructions that make computers useful to us.* Specific software **programs,** *which can be considered instruction sets,* allow us to "communicate" with computers, and accomplish specific data manipulation tasks.

The components of a computer system allow five general functions. These are:

1. data input
2. access to network information
3. data and information processing
4. information output
5. data and information storage

This chapter focuses on the computer hardware components that facilitate these functions, while software and the Internet will be discussed in the following chapters.

II. HOW COMPUTERS WORK—HARDWARE

A. Digital Technology—Bits and Bytes

The power of computer operations stems from the digital processing system. Like other digital equipment, the computer understands only two electrical states: off and on. Not all digital technologies are new; more than a century ago, telegraphs were able to provide long-distance communication through digital data in the form of clicks, which were then reconstructed into interpretable human language. Just as you did not personally need to know Morse code in order to receive a telegram, you also do not need to understand computer code in order to use a computer. However, a basic understanding of how computers "see" data can be helpful.

In computer language, "off" and "on" are represented as 0 or 1. *The 0 or 1 represents the smallest available information unit, also called a* **bit** (*bi*nary dig*it*). In contrast, analog equipment measures change as a continuous pattern. As a result, analog signals are subject to distortion and noise (see Chapter 18). A computer data bit can only equal 0 or 1, eliminating ambiguity. One significant advantage of digital data is that it can be reproduced indefinitely without loss of quality.

Since a single bit has only two possible values, multiple bits are required to represent the possibilities for real world data, which include letters of the alphabet, numbers, and other symbols. A computer "word," or **byte**, *is a group of continuous bits* (usually eight, providing 256—2^8—possible combinations) *that define a single understandable symbol or character.* As users, we rarely deal with single bytes of information, and for convenience, we commonly measure the data in computers in a minimum of **kilobytes (K or KB),** *which represent ~1000 bytes* (actually 1024, representing 2^{10}, but the approximation is adequate for ordinary purposes) and **megabytes (MB,** *~1 million bytes,* 2^{20}). As a general approximation, about one half of a plain (not word-processed) text page requires about 1 KB of data, see Figure 34.1. Desktop computers have data storage capacities measured in **gigabytes (GB,** *~1 billion bytes,* 2^{30}) or **terabytes (TB,** *~1 trillion bytes,* 2^{40}).

B. Central Processing Unit

The *working core of a computer* is called the **CPU,** or **central processing unit.** This is the main **microprocessor** that determines the relative power of the computer. A microprocessor *is a small but complex calculator that also controls data input and output.* As shown in Figure 34.2, the microprocessor contains the electronic circuits that carry out the instructions found in the computer's software. Microprocessors are also found in a variety of laboratory and home appliances, such as microwave ovens, but these are dedicated to specific functions. The CPU of a computer is highly versatile because of the diversity of instructions that can be provided by other hardware as well as software.

The "thinking" power of a computer is defined by the type of microprocessor that it contains. Chip manufacturers assign names and model numbers to specific microprocessors, to distinguish the rapidly evolving technologies; however, the main specification that a user can readily interpret is the system **clock speed,** *the speed with which the microprocessor can carry out an instruction.* Clock speed is measured and reported in **gigahertz (GHz),** *representing one billion electrical cycles (hertz) per second.* This means that a 1 GHz

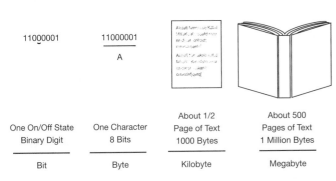

Bit	Byte	Kilobyte	Megabyte	Gigabyte
One On/Off State Binary Digit	One Character 8 Bits	About 1/2 Page of Text 1000 Bytes	About 500 Pages of Text 1 Million Bytes	About 1000 500 Page Volumes 1 Billion Bytes

Figure 34.1. **Comparison of Computer Data Size Units.** (Adapted from Szymanski, R.A., and Szymanski, D.P., and Pulschen, D.M. *Introduction to Computers & Software.* Saddle River, NJ: Prentice Hall, 1996, 74.)

microprocessor can execute one billion instructions per second! Higher clock speeds generally mean faster computers, although the relationship between clock speed and performance is not necessarily linear (in other words, a 2 GHz processor does not usually increase computing speed a full 2-fold over a 1 GHz processor). New microprocessor models usually provide significant increases in computer speed, as shown in Figure 34.3. Improvements in computation speed in recent years have resulted from the use of **dual core** and **quad core technology,** *where the CPU chip contains two or four separate microprocessors that can compute simultaneously.*

C. Memory

Another major performance feature of a computer is its memory capacity. **Memory** *is the ability of a computer to electronically keep track of data.* The amount of memory in a new PC is currently measured in gigabytes. Computer memory exists in two general forms, **ROM (read-only memory)** and **RAM (random access memory).** ROM contains the **BIOS (basic input/output system),** *which is the essential set of routines that sets up the hardware in a PC and boots the operating system when the computer is turned on.* The BIOS is built into a flash

memory chip (discussed below) and contains the instructions necessary to start the computer and perform diagnostic and other basic functions. As read-only implies, these instructions cannot be changed without a major upgrade. The computer user is more interested in **RAM,** *which represents the electronic memory available to work with.* When you turn on the power to your computer, RAM is empty. With prompting from the BIOS, the computer loads part of the operating system (explained below) into RAM. Once the computer and operating system are ready, you choose additional software and data to load into RAM, see Figure 34.4 on p. 644. RAM is the working area of the computer, and instructions (software) and data must be added to RAM before you can use them.

RAM requires electricity to hold its contents. Data found only in RAM are called **volatile,** *because they will disappear when you turn off the computer or a power outage occurs.* In order for data to become a permanent record within your computer, they must be stored. There is a critical distinction between memory and storage in a computer. **Storage** *is the recording of data or information onto hardware such as disks.* Stored information can be retrieved later. Memory is electronic and therefore temporary. One of the most important things that all computer users learn (usually the hard way!) is

Figure 34.2. **A Central Processing Unit for a Personal Computer.** This is the circuitry and packaging for an Intel R Core™ 2 Duo Processor.

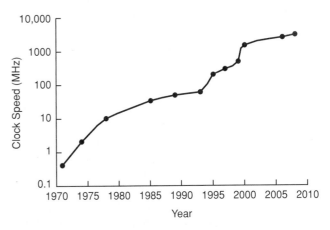

Figure 34.3. **Increasing Speed of Intel's Microprocessors.** This graph shows the progress made since the 1970s in improving the speed of the Intel family of processors that are found in approximately 90% of the personal computers in current use. Note the logarithmic scale on the Y axis.

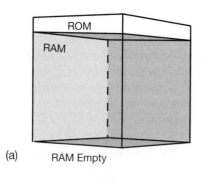

(a)   RAM Empty

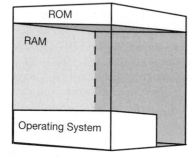

(b)   Operating System Loaded
Into RAM at Startup

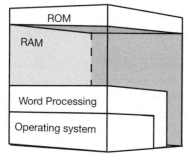

(c) Word-Processing Program
Loaded into RAM

Figure 34.4.   The Use of RAM and ROM within a Computer. a. Computer memory is divided into ROM and RAM sections. **b.** When operating, computers load the operating system into RAM. **c.** Additional programs are added to RAM as needed. (Adapted from Szymanski, R.A., Szymanski, D.P., and Pulschen, D.M. *Introduction to Computers & Software.* Saddle River, NJ: Prentice Hall, 1996, 78.)

to frequently **save** their data, *which means to transfer them from volatile memory to permanent storage.* If you are writing a 20-page report on a computer and even a brief power outage occurs, all of your work in RAM will be lost. Only the portion of the paper that is saved to storage can be recovered. For this reason, many software packages, such as Microsoft Office, provide the option to **Autosave** data at regular intervals, which means *to transfer data to a storage device without specific instructions from the user.* That way, if you set the program to Autosave every 5 minutes, you will not lose more than 5 minutes worth of data in case of a loss of power or other unforeseen problem.

III. HOW COMPUTERS WORK—SOFTWARE

A. System Software

Although hardware provides the tangible part of a computer, the hardware cannot accomplish anything without software. **Software** or **programs** are *instruction sets written in computer language or code.* **Programmers** are those *who write coded instructions to make computer use possible for those of us who have not learned computer language.* Programs provide instructions that enable hardware to perform all necessary functions.

There are two general categories of software: applications and system software. An **application** *is software that addresses specific problems, such as data organization or word processing.* These will be discussed in the next chapter. **System software** performs a series of more basic functions. *These are the programs that operate the system hardware and provide the foundation for application programs.* System software does not manipulate user data or create documents.

The components of system software most familiar to users are the **operating system (OS)** and the user **interface.** The OS *provides four general functions: it operates the hardware, runs software applications, manages file storage, and monitors system operations.* It also supports the user **interface,** *which is the part of the system software that the user sees and interacts with.* All communication between the user and the computer hardware occurs through this interface, allowing the computer to carry out the user's commands, see Figure 34.5. Some software applications provide their own interface as well.

There are a limited number of common OS systems. The oldest system still in occasional use in personal computers is **MS-DOS** (for Microsoft disk-operating sys-

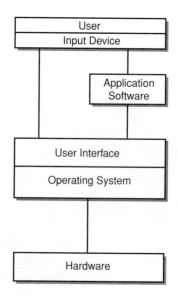

Figure 34.5.   Interactions among Computer Hardware, Software, and User. (Adapted from Szymanski, R.A., Szymanski, D.P., and Pulschen, D.M. *Introduction to Computers & Software.* Saddle River, NJ: Prentice Hall, 1996, 115.)

tem). DOS provides a text-based interface (defined below), and is capable of running only one program at a time. DOS cannot be used in networks, and has been replaced by **Windows,** a **multitasking** OS *that allows the user to open and run more than one program at a time.* The availability of RAM determines how many programs can successfully run simultaneously.

The operating systems above are found in the personal computers widely referred to as PCs. The other OS in common use in personal computers is Mac OS X (ten), found in the Macintosh computers (Macs) manufactured by Apple, Inc. These two families of personal computers are discussed in the Case Study below. Because PCs greatly outnumber Macs, the discussions in this unit use examples from the Windows OS. However, Mac OS X has very similar capabilities.

CASE STUDY

PC versus Mac

At present, there are two well-known types of personal computers available: the PC and Apple Macintosh (Mac) families. While PCs dominate the market, Mac users are a loyal and enthusiastic minority. The PC family of personal computers originated with DOS, a text-based interface, as an operating system. Then in 1984, Apple introduced a user-friendly personal computer with a different OS and a graphical user interface (GUI). These could perform the same tasks as the DOS systems, but required little or no computer expertise. Eventually, Microsoft developed the Windows operating system to provide a GUI for DOS machines. Virtually all PCs now use a version of the Windows operating system.

From the beginning, Apple chose to maintain a proprietary operating system, so that only Apple and licensed companies could produce peripherals and software for Macs. Microsoft, on the other hand, made their operating system available for other companies to develop compatible programs and hardware. Many business analysts believe that this business choice ultimately led to the current market dominance of PCs. Because it was originally intended as an ideal system for graphic design work, Macintosh still remains the standard in this field. At present, though, PCs account for approximately 96% of the personal computers sold in this country.

Many personal computer users describe themselves as either Mac or PC people. In fact, the interfaces of these systems are very similar in function, and Macs now use Intel chips and have the ability to run the Windows OS and its software, so compatibility is no longer a significant issue. In a similar fashion, Macs no longer require their own unique peripheral devices such as keyboards and printers, so many hardware elements have become interchangeable as well. Macs come well-equipped for multimedia applications, and Mac OS X has the significant advantage of being relatively computer virus–free. Given their advantages, along with the innovative technologies for which Apple is known (iPods, iPhones, etc.), Macs remain an attractive choice for computer users.

B. User Interface

Every OS must have a user interface, which is usually based on either text or graphics. In a **command-line,** or **text-based, interface,** *the user types the instructions, or commands, for the computer one line at a time.* MS-DOS and UNIX are examples of text-based operating systems. **UNIX** *is a widely used operating system for large network computers,* which are generally operated by individuals with special training. **Linux,** *a derivative of UNIX,* is used in network servers (see below), a small number of personal computers, and consumer products such as gaming systems. Linux can be used with either a command-line or graphical user interface.

A command-line prompt in MS-DOS looks like:

C:\Documents and Settings\Cynthia Moore>

In order to proceed, the user must know and respond with the specific commands and formatting used in DOS. For this reason, text-based interfaces are currently used almost exclusively by computer specialists.

All personal computers provide a **graphical user interface,** or **GUI** (pronounced "gooey"). Instead of remembering complex commands and grammar, the user can "point and click" using a computer mouse. GUIs operate with the familiar **windows, icons, menus,** and **pointers** (sometimes referred to as WIMPs by those who prefer command-line interfaces). The screen where these elements appear is called a **desktop,** representing the area where your work can be laid out, shown in Figure 34.6.

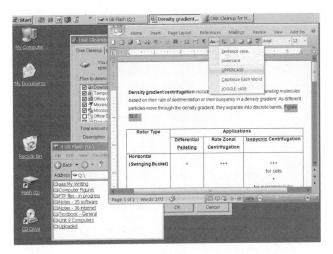

Figure 34.6. Elements of a Graphical User Interface. Three windows are open on the desktop. An open folder and the interface for the Disk Cleanup utility are in the background, with a Microsoft Word document in the active window on top. Note that all three windows have been specifically sized and shaped. Desktop icons are visible on the left of the screen. In Microsoft Word 2007, the menus and toolbars of previous versions have been replaced with the Ribbon and a Quick Access toolbar, which serve the same function. The command "Change Case" has been selected from the toolbar, calling up a menu where Uppercase has been selected to format the highlighted text in the document.

A GUI **window** *is a rectangular viewing and work area.* Each individual document or information display appears in its own window, and many windows can be present on the desktop simultaneously. You can change the shape and size of windows according to your needs. Only one window is active at a time, but the contents of the others are visible if the windows do not overlap.

Icons *are graphic representations of specific objects and computer functions.* In some cases, such as file folders and the Recycle Bin on the Windows desktop, the function may be obvious. You can select and move items into the folders for storage or into the trash can for disposal. Application programs and the files that they create generally have specific icons, as illustrated in Figure 34.7.

Menus come in many forms, but *all have the common feature of offering a series of text options or icons for your selection.* These can be essential in complex applications, because they eliminate the need to memorize commands. Clicking on a main menu frequently causes a drop-down menu or submenu to appear. Many applications also have "floating" menus (called **toolbars** if they contain icons) which can be moved (float) to any convenient position on the desktop while you are working in that program.

The final component of a GUI is the pointer, which acts as a **cursor** *to indicate your location on the desktop or in a window.* The pointer may appear as an arrow, an I-bar, or a blinking rectangle in various applications or locations, or the user can specify a unique appearance. The pointer is controlled with a mouse, although it can also be moved with arrow keys on a keyboard, or a joystick.

C. Application Software

Application programs are designed to provide computer solutions to human problems. A **program** is simply *a series of instructions, written in one of many computer languages, which allows the computer to perform specific functions.* The program that controls a digital wristwatch may only need a thousand instructions, while complex computer applications may require tens of millions of instructions to work properly. This reflects the difference between the limited operations of a watch compared to the versatility of a computer. For most applications, the user provides initial data, which is placed into a selected area within a window supplied by the software. The application will (with proper instructions) process and save the data in a variety of ways. Specific software applications are discussed in Chapter 35.

Utilities *are "housekeeping" programs that enhance computer operations.* Utilities generally do not create new documents or files. Examples include virus-protection software, screensavers, and accessories such as desktop calculators. **Screensavers** *are programs that blank out or create a moving graphic display on the monitor screen after a specified time without user input.* These programs were originally designed to prevent damage to monitors from etching of images on the screen. Monitors no longer have this problem, but most users enjoy these programs anyway.

IV. INFORMATION STORAGE

A. Computer Files

A computer **file** *represents a set of digital data that is grouped to form a single storage unit.* Each file stored on your computer has a name, and represents either a program or a data document. Program files have assigned names which should not be changed. Data files can be named almost anything, although names that clearly reflect the contents of the file make manual file retrieval much easier.

When the computer stores a data file, it uses the language and conventions of the application software that created the document. The resulting **file format** *allows the appropriate application to recognize its own files;* these files may not be readable by other applications. Data layouts and organization are generally program-specific. One example is a **PDF** *(portable document format)* file, which is read with the free Adobe Reader program, see Figure 34.8. These produce complex and attractive text documents that are commonly found on the Internet and that cannot be fully interpreted by most other programs.

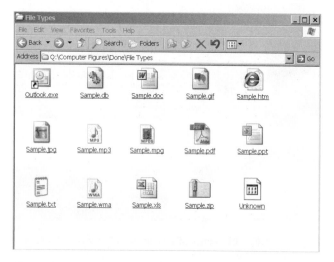

Figure 34.7. Examples of Common File Icons. Windows assigns specific icons to document formats, making them easily identifiable. Note that the Microsoft Outlook icon at the top left has a small arrow in the corner, indicating that this is a Shortcut. The actual program file is located elsewhere on the computer, but clicking on the Shortcut will open that file. The .exe suffix indicates that this is an executable program; these files should never be moved from their installed locations. See Table 34.1 for a description of the file formats shown here.

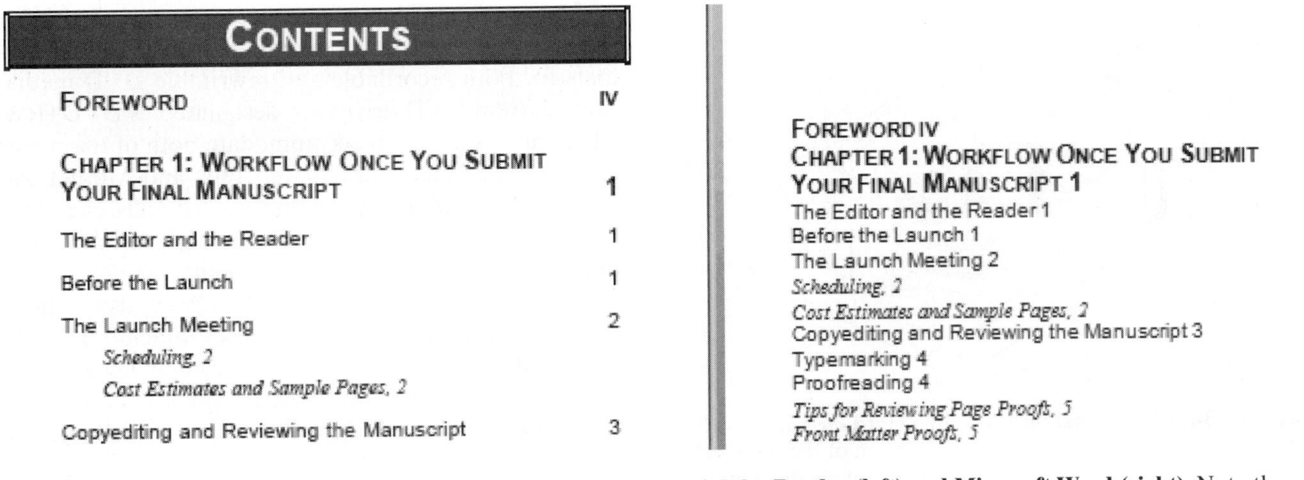

CONTENTS

Figure 34.8.   Comparison of the Contents of a PDF File Read by Adobe Reader (left) and Microsoft Word (right). Note the formatting in the PDF and the loss of the header and spacing in the Word file.

You can determine the format of a file by identifying the icon representing the file, or by the three-letter suffix extension (name.xxx) in the complete file name, as shown in Figure 34.7. These extensions may not be displayed, but format can also be determined by holding the mouse over the file name. Table 34.1 provides the file extensions and applications for the common icons seen in Figure 34.7.

When you **open** a file, *the computer transfers a copy of the contents of that file from storage into memory (RAM).* The original file still exists in storage, and any changes that you make to the file contents remain volatile in RAM until you save them. The changes can be saved under the same file name, which usually writes over the previous version, or the new version can be saved with a different name, so that both versions are stored. Opening a document file (by clicking on the icon) will usually start the application that created the file, although closing a file may not automatically shut down the application program.

B. Storage Media

When comparing computer storage and memory, it may be helpful to remember that computers deal with information much the way humans do: They either keep an idea in memory or write it down for future reference. One of the advantages of information storage on a computer is that digital data can be copied an infinite number of times without changing the data. If you have ever copied a videotape (an analog technology), you are aware that copies lose some of the quality of the original, and that each succeeding copy of a copy is worse than the previous one. Digital recording and storage offer the ability to make large numbers of perfect copies.

Computers can store files on various storage media, representing three common storage mechanisms:

- magnetic storage; a **hard disk** *is a piece of hardware that stores software as magnetic signals.*

- optical storage; **CDs** and **DVDs** *are etched discs that contain computer data in a series of pits that are read by a laser.*

- electronic storage; a **flash drive** *is a small, portable memory card that acts like a hard drive and stores data electronically instead of magnetically.*

Personal computers contain at least one built-in hard disk and the ability to read and write to various types of portable storage media.

A **hard drive** *is a well-protected, sealed unit containing multiple rigid magnetic storage disks (hard disks) and a drive mechanism,* as shown in Figure 34.9 on p. 648. A typical hard drive for a PC has at least 120 GB to several TB of storage space and is ideal for storage of large files and programs. In addition to internal drives, there are external, portable hard drives for long-term storage

Table 34.1.   COMMON FILE FORMATS

File Extension	File Type
.db	Database file
.doc	Word document
.exe	Executable program
.gif	GIF graphics file
.htm	HTML file
.jpg	JPEG graphics file
.mpg	MPEG video file
.mp3	MPEG-1 Audio Layer 3 audio file
.pdf	Portable document format file
.ppt	PowerPoint presentation file
.sys	System file (do not disturb!)
.txt	Text file
.wma	Windows Media Audio file
.xls	Excel spreadsheet
.zip	Compressed file

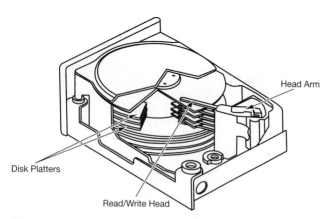

Figure 34.9. Internal Structure of a Hard Disk Drive. The drive casing has been removed and a section of the disks has been cut away to show the series of eight storage disks and read/write heads.

and backup, as shown in Figure 34.20b on p. 657. Because a single dust particle within the case can completely disable a hard drive, these components are usually assembled in cleanrooms (described in Chapter 12).

Optical storage takes the form of CDs and DVDs. Blank CDs and DVDs are inexpensive and durable for long-term storage. With proper care (discussed on p. 662), the lifespan of these discs is estimated to be a hundred years. A **CD-ROM (compact disc read-only memory)** *is an optical disc used to store computer programs and files.* They are also called simply CDs, because music CDs store audio files in the same manner. Like magnetic disks, they contain stored digital information that is reconstructed into the analog signals of sound, text, or video. Note that optical media are called *discs,* while magnetic media are called *disks.* Pre-recorded CDs are made of a thin disk of polycarbonate backed with a reflective layer of aluminum, as shown in Figure 34.10. To record data (music files, for example), pits are stamped into the polycarbonate to encode the binary data, spiraling outward from the center. For playback, these pits are then detected by a laser and translated into data output.

For consumer-recorded CDs, **CD-R (CD-record-able)** media have an additional dye layer adjacent to the polycarbonate, which encodes the data patterns. CD-R media have a maximum capacity of 700 MB of data or 80 minutes of music; either or both capacities may be listed on the label, but all CDs can record both music and other types of data files. There are also **CD-RW (CD-rerewritable)** media, where data can be added or deleted in multiple sessions using a slightly different technology. Most CD drives are CD-RW, which means they can record on either type of CD.

DVDs (digital versatile discs) *are optical discs similar to CDs, but with greater data storage capacity.* They are replacing CDs as storage media because a single layer DVD can store 4.7 GB of data (almost seven times the capacity of a CD), and DVDs can also be recorded in double layers and on both sides (4 possible recording

surfaces total). Their use for data storage has been somewhat hampered by the existence of competing formats for both recordable and rewritable DVD media. Most current DVD drives are designated as DVD±RW, which means they can accommodate both of the major competing formats (designated Plus and Minus). All DVD drives can also read and record CDs. Standard DVDs are slowly being replaced by **high-definition (hi-def) DVDs,** *which are DVDs with a much higher storage capacity than regular DVDs.* Hi-def DVDs also come in competing formats, but Blu-ray discs, which hold up to 50 GB of data, are becoming increasingly common. In order to record and retrieve data (including movies), these discs require a Blu-ray DVD drive, which will also read CDs and standard DVDs.

The third major storage medium is the flash drive, also known as thumb drives, pen drives, USB drives, memory sticks, and a variety of other names. Unlike true drives, these devices have no moving parts, which makes them extremely durable. Flash drives operate with **flash memory,** *which stores data electronically instead of magnetically,* and consist of a thumb-sized circuit board surrounded by a hard case, as shown in Figure 34.11. Flash memory is faster, smaller, and lighter than magnetic storage, and is used in many portable applications such as game consoles and digital cameras. While flash drives can connect to any computer using the common USB port (discussed below), flash memory cards for digital cameras (Figure 34.11) are proprietary and require special reading devices, which are frequently built into new computer systems.

Despite its advantages, flash memory storage is unlikely to replace hard drives in the near future. Hard drives provide much greater storage capacity for the same price. Also, flash memory boards have a limited number of re-writes relative to hard drives. Although it is unlikely a user will wear out a portable

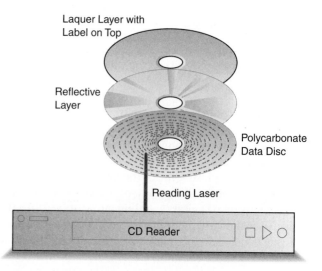

Figure 34.10. Data Storage on a Pre-Recorded Optical Disc. Data are stamped into the bottom polycarbonate layer and the resulting pits are detected with a laser.

Figure 34.11.   Data Storage Using Flash Memory.
Thumb-sized flash drives (left) have become a popular means to transfer files between computers. Digital cameras use flash memory cards (right) to store photos for easy transfer to a computer or printer.

flash drive, flash memory would have a limited lifespan if constantly accessed inside a computer as the main storage device.

V. PERIPHERAL DEVICES

A. Introduction

A **peripheral device,** or simply peripheral, *is any hardware that you plug into your computer.* The computer itself is simply the box which contains the CPU—everything else is a peripheral. Peripherals are necessary for all user input and output. In order to function, these devices require connection ports—cables or wireless transmitters—and driver software.

Cables *are the wiring that connects peripherals to the computer unit.* These connect through specific types of sockets, called **I/O (input/output) ports,** or just **ports,** for each device. *These ports allow communication between computer components.* Historically, most peripherals had their own specific type of port found on either PCs or Macs. While there are still a few specialized ports on personal computers, the current standard for most peripherals is a **USB (universal serial bus)** connection. USB *is a universal type of port that provides high speed, stable connections between up to 127 devices from a single port.* USB allows peripherals to be easily added and removed from computer systems, and interchanged between PCs and Macs.

Another common port type is **FireWire,** *which is Apple's name for a connection particularly well-suited for high speed applications,* such as downloading video from digital camcorders. Multiple USB ports are standard on personal computers, which usually provide at least one FireWire port. These and additional ports are shown in Figure 34.12a.

Both USB and FireWire ports support **daisy chaining,** *which refers to the connection of a series of peripheral devices through a single port,* as shown in

Figure 34.12b. This is accomplished using inexpensive hardware hubs, which provide multiple connection ports (usually 4 to 10). Hubs also provide the convenience of providing USB ports where they are most convenient for the user.

Like all computer hardware, peripheral devices require software in order to be useful. The required program is called a **driver,** *which is the software that allows the computer to control the device.* It establishes compatibility between the CPU and peripheral device by translating commands from applications and the CPU into specific signals recognized by the peripheral, and vice versa. Drivers are OS- and hardware-specific. They generally come either pre-installed in the OS or are added with the peripheral. It is sometimes necessary to obtain current drivers from the Internet.

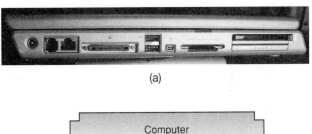

(a)

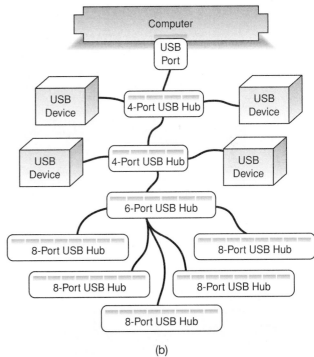

(b)

Figure 34.12.   Computer Ports. a. All computers provide a variety of ports for networking and peripherals. This laptop has (from left to right) 1. power adapter port; 2. telephone connection for dial-up modem; 3. Ethernet port for networking; 4. Expansion port; 5. two vertically stacked USB ports; 6. FireWire port; 7. slot for flash memory cards; 8. two vertically stacked card slots that can be used for many purposes, including network connections, TV tuners, or digital certificates. See text for additional explanations. There are additional ports on the other side of the computer for connecting an external monitor, microphone, and headphones. **b.** USB ports can daisy chain up to a total of 127 devices with the use of multi-port hubs.

B. Input Devices

Computers require data input into RAM before any processing can occur. All necessary programs and data must be either copied from a storage device, such as a DVD, or entered into the system through a peripheral, such as a keyboard. All operations take place in the RAM and the results are then transferred to one or more output devices and (usually) to a storage device. A wide variety of peripherals are available for data input and output; the most common are described here.

i. KEYBOARD

The keyboard is the most familiar input device for the computer (Figure 34.13). A standard desktop keyboard provides not only the basic typing keys, but also:

- a numeric keypad for fast numerical data entry
- arrow keys for moving the cursor
- special editing keys
- a set of **function keys** which are usually defined within specific applications

Both data and commands can be entered through the keyboard, although many commands are more easily entered with a mouse.

Laptop computers *(portable computer systems that can be operated on batteries)* usually have an abbreviated keyboard to minimize size and weight, as shown in Figure 34.13b. There are many distinctive keyboards available for special purposes, such as gaming, mathematics, or multimedia, as well as keyboards adapted for accessibility or ergonomics (Figure 34.13c).

ii. MOUSE

The **mouse** *is a pointing device that controls the on-screen cursor in a graphical user interface.* A standard PC mouse design is a smoothed rectangular device with two or more buttons (a Mac mouse requires only one button). The use of a mouse is called **"point and click"** input, *because the user slides the mouse to point the cursor at an on-screen object, and then presses one of the mouse buttons to select or activate the object.* Because of

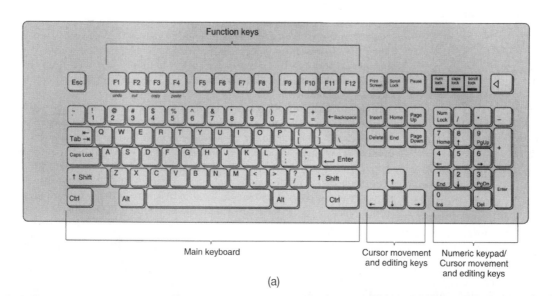

(a)

(b)

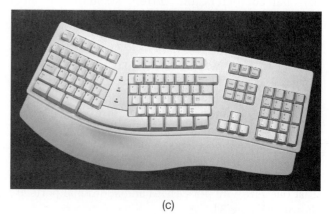

(c)

Figure 34.13. PC Keyboards. Typical layouts are shown for **a.** a desktop computer and **b.** a laptop computer. Specialized keyboards are also available, such as **c.** this ergonomic model. Figure 34.16a on p. 652 shows another keyboard designed for high visibility.

the speed of the cursor movement, using a mouse can provide much faster command entry than the keyboard.

The most common type of mouse currently is the **optical mouse,** *which uses light emitted on the underside of the mouse to track its movement.* While most people still use a mouse pad as the sliding surface for their mouse, it is generally unnecessary with an optical mouse, which can track over almost any surface that is not shiny or transparent.

There are many variants on the standard mouse, illustrated in Figure 34.14. A **trackball** *functions as a stationary, upside-down mechanical mouse.* Instead of rolling a mouse, you roll the ball directly with your fingers or thumb. These are frequently used for graphics and gaming. Most laptop computers have **trackpads,** which have no moving parts. A trackpad (shown in Figure 34.13b) *is a flat rectangular area of the keyboard that controls the cursor.* The user moves a finger on the touch-sensitive pad, which causes the on-screen cursor to move in a similar direction, and then the user taps the pad or clicks a nearby button. Some users feel that these devices give finer cursor control than a standard mouse.

iii. SCANNER

A **scanner** *is an optical device that converts graphic images to digital form and sends them to the computer for manipulation and storage.* The most common types are flatbed or hand-held scanners. Flatbed scanners are similar in appearance and performance to photocopiers, except that they do not directly produce printed copies. Hand-held scanners are manipulated to scan pictures or text in successive passes and are best suited for small images. They are relatively inexpensive but rely on the user to move the scanner in straight lines.

Scanners treat blocks of text as **objects** instead of characters. An **object,** in computer language, *is a graphic image that is treated as a single item, described by a mathematical equation.* In order to recognize and edit scanned text, the computer requires **OCR (optical character recognition)** software. **OCR** technology *allows text to be scanned and recognized as text to be manipulated in a word processing program without retyping.* There are also specialized scanners for photos, slides, and other types of media. Despite the ease of using a scanner, the user always needs to consider copyright issues before reproducing artwork.

iv. SPEECH RECOGNITION

For many users, the fastest and simplest way to input information and commands into a computer would be speaking. **Speech recognition** *refers to the ability of computers to convert human speech into written text or commands.* While the capabilities of this technology are still developing, the software is constantly improving in consistency and usefulness.

Simple applications already exist in many areas. We are all familiar with automated phone systems that rec-

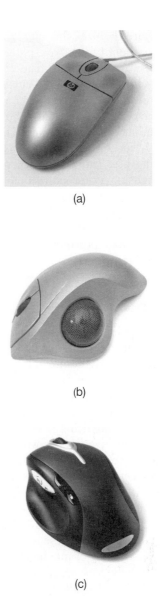

(a)

(b)

(c)

Figure 34.14.   Examples of Three Common Mouse Designs. a. A standard mouse, **b.** a trackball mouse, and **c.** an ergonomically designed mouse. (Photo **a.** Elbaliz Mendez/ Pearson Learning Group; **b.** iStockPhoto; **c.** Natural Wireless Laser Mouse 6000, © 2008 Microsoft Corporation.)

ognize specific words, such as yes and no, and numbers. There are excellent systems for using voice commands to execute computer operations, but again, these are applications with limited vocabularies. What most users want is software that would allow computers to accurately recognize continuous speech.

True speech recognition software provides a computer with the capability to accept voice input by recognizing the speech of an individual user. These systems are already used by scientific and medical professionals who need to keep their hands free for other work, as well as by individuals with physical challenges. They generally work well for individual words, but they require "programming" by each user for continuous speech. To train the system to recognize specific words, the user pronounces words and sentences into a microphone. The software digitizes the analog sound waves and then matches them to the wave patterns of words found in the software dictionary. The computer will then either obey the command or display the spoken text on the monitor. The user corrects any mistakes, which enables the computer to "learn" how each word or phrase is pronounced. While it can be time consuming to program a speech recognition system, time is saved later during use. The amount of programming necessary is decreasing as speech recognition technology becomes more common and versatile.

Computers can also provide voice output, which is relatively simple compared to speech recognition. Through the OS, most personal computers can be directed to "read" text from a file, using a variety of synthetic voices.

v. OTHER INPUT DEVICES

There are many other input devices available for specialized uses. For example, in the biotechnology laboratory, computers can be directly connected to spectrophotometers, radioactivity detectors, and other instruments that output data to the computer.

Artists and other users who input very precise graphic data can use a **digitizing tablet and pen.** These tablets have touch-sensitive screens with electronic circuitry underneath. The X–Y coordinates of the tablet screen exactly correspond to those of the monitor display screen. This allows not only precise drawing but also handwritten input. There are also tablet PCs, which combine the functions of a digitizing tablet with those of a laptop, see Figure 34.15. These are convenient for taking notes while standing or moving around a lab.

Many additional devices are designed for use by individuals with physical challenges. There are head

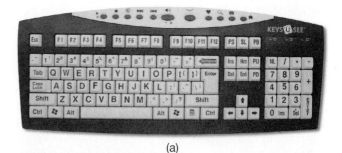
(a)

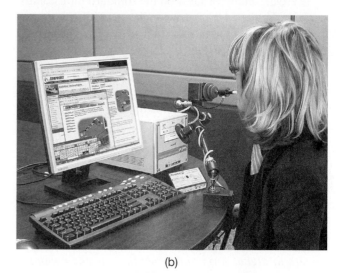
(b)

Figure 34.16.   Accessibility Devices for Data Input.
a. A high-visibility keyboard, with large, high-contrast labels. **b.** A mouth-operated joystick/mouse. (Photo a. Keys-U-See® keyboard/© 2005–2008 Genesis Worldwide Enterprises, Inc. All Rights Reserved. b. Jouse2 by Compusult.)

position trackers that substitute a headset for the mouse, positioning the cursor with head movements. These trackers can have attached mouth tubes to allow "clicking" by puffing air into the tube. There are also eye trackers that use a mounted video camera to track the position of the eye pupils of the user. Examples of accessible input devices are shown in Figure 34.16. Computers, particularly coupled with communications peripherals and software, provide global access to many individuals who might otherwise be isolated from the outside world.

C. Output Devices

After data are manipulated by the computer, the user generally wants to see the results of the processing. Information output is available in two forms; soft and hard copy. **Soft copy** *refers to the information displayed on the monitor and saved in computer files.* **Hard copy** *is a printed paper copy of the information* (the term originated in newspaper offices for the paper translation of Morse code messages). Virtually all computer systems have both a monitor and at least one hard copy output device attached.

Figure 34.15.   Tablet PC. These can be similar in appearance to a laptop computer, with a display that swivels and folds flat like a clipboard so that the touch-sensitive screen can be "written" on with a special pen. (Portégé M700 photo courtesy of Toshiba.)

i. MONITOR

It would be theoretically possible but very difficult to use a computer without a monitor. The **monitor** *is the main soft copy display where you can see your data input and choose commands.* It is important to remember that data shown on the monitor are in RAM and are therefore volatile unless saved.

Bulky CRT (cathode ray tube) monitors have been replaced almost entirely by flat screen **LCD (liquid crystal display) monitors,** *which use LCD technology to provide slim, energy-efficient, digital display screens,* as shown in Figure 34.17. Newer models have uniformly high-quality displays and high **resolution,** *which determines the sharpness of the screen image.* Monitor resolution is described by the number of horizontal **pixels** (*picture elements*) in a line that are separately shown on the display times the number of vertical lines (described as in a "1366 × 768" display, for example; typical for a widescreen laptop). Software applications with a major graphics component frequently require a specific resolution for proper display. If desired, the user can set a lower monitor resolution to increase the visibility of display items such as text and icons.

Monitor screen size is measured diagonally. Most new monitors have widescreen displays, to enhance movie viewing. However, this is also useful in the work setting, where larger amounts of data can be accommodated on screen.

The high quality of modern LCD monitors has lessened, but not eliminated, the problem of eyestrain from staring at a display screen for extended periods. Eyestrain can be minimized by working in a well-lit room and keeping the screen clean. It is helpful to give your eyes a rest by using the 20:20:20 rule: every 20 minutes, shift your focus away from the screen to something 20 feet away for 20 seconds. LCD screens are usually plastic with an anti-glare, anti-scratch coating that can be damaged by rough handling or improper cleaning technique. Tips for cleaning monitor screens are provided on p. 659.

Figure 34.17.   LCD Monitors. LCD monitors are thin and lightweight. (Photos courtesy Shutterstock.)

ii. PRINTER

Printers are by far the most common and familiar hard copy output devices. The main types are laser and inkjet. **Laser printers** work somewhat like photocopiers, with a laser, drum, and toner. They are fast and offer excellent resolution, although many do black-and-white printing only. Color laser printers are becoming more common as prices decrease. **Inkjet printers** *create images by squirting tiny sprays of ink onto paper,* which can produce sharp print. They are quieter and less expensive, but generally slower than laser printers. **Multifunction printers,** *which combine printing, copying, and scanning,* are good for low-volume applications and are available in both laser and inkjet models.

The two measurable output features that vary among printers are resolution and speed. Printer resolution is measured in **dpi** (*dots per inch*), with more dpi indicating greater resolution and a sharper image. Even **low-end** (*relatively inexpensive*) inkjet printers have color outputs up to 4800 × 1200 dpi, higher than most laser printers. Printer speed is measured in text pages printed per minute and is usually higher for laser printers. Keep in mind that these are provided as maximum specifications; in practice, most printers are slower than their advertised speeds.

While printer hardware can be relatively inexpensive, it is important to consider maintenance costs. Ink cartridges have a much higher per-page cost than toner for a laser printer. Similarly, while both types of printer can use standard photocopier paper, better resolution and durability can be obtained by using better-quality paper, especially for color inkjet printing.

VI. COMPUTER NETWORKING

A. Introduction to Computer Networks

Many of the computers that you encounter in laboratories will be part of a network. A **network** *is simply two or more computers that are electronically connected for the purpose of sharing programs, files, information, and/or other resources.* A network can be local or global. The telephone system and the Internet are two examples of global computer networks. Networks provide the resources for **telecommunications,** *the electronic transport of data in the forms of text, audio, and video over significant distances.* By electronic transmission of data, you can send data quickly, as well as avoid the need to physically exchange paper or storage media.

Probably because of the rapidly changing technologies involved, terminology for networks can be confusing. Two terms that are frequently used are intranet and internet. An **intranet** *is an internal, usually private* *network, perhaps within a company or university.* An **internet** *is two or more connected networks.* The **Internet** (with a capital "I") *is a specific internet connected with special sets of communication protocols,* to be discussed in Chapter 36.

In order to be part of a network, computer hardware must be connected in some way to the other computers in the network. There are several common ways to achieve this, including:

- Ethernet (defined below), which is frequently used to connect within an intranet.

- Dial-up modem, which provides low-speed intermittent internet connections through telephone lines, usually in homes.

- DSL (digital subscriber line) and cable "modems," which provide high-speed continuous internet connections.

- Wireless, which use radio frequencies to communicate through either intranets or internets.

Each of these connection methods is discussed below.

B. Local Area Networks

Local networks provide an economical means for relatively small groups to share computer resources. These networks provide a central system for group-licensed software, storage of large data files, group access to expensive peripheral devices, and a central link to outside networks. This eliminates not only the expenses of buying high-end equipment for every laboratory or office in the group, but also the need to store large application programs on multiple computers.

A **LAN** *is a local area network (intranet), generally operating within a building, school, or company.* LANs are private and usually limited to a five-mile radius. Individual computers, also called **workstations,** may be physically connected by network or fiber optic cables, or part of a wireless system. To use this type of network, all participating computers must contain a compatible **network interface card (NIC),** *which is computer hardware that provides an appropriate interface between the computer and a network.* NICs for specific network types, such as wireless, can be built into the computer or added through existing ports. **Ethernet** *is one standard type of LAN interface that is built into all network-ready computers.* Figure 34.18 shows the arrangement of a simple LAN that connects to the Internet and possibly additional smaller networks.

In this type of LAN, all participating computers are wired through an **Ethernet switch or hub,** *which regulates communications between workstations within the network and distributes access to other networks.* Internet access is provided through a **router,** *hardware that sorts and directs incoming and outgoing information between networks.* In a secure network, the router also provides a **firewall,** *which filters incoming and outgoing communica-*

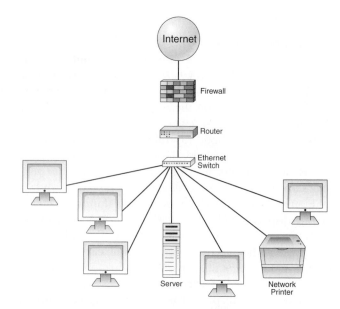

Figure 34.18. Basic Structure of a LAN with Internet Access. See text for an explanation of the various components. Note that the firewall might be within the router.

tions to block known threats to the network. Individual computers can (and should) also maintain the firewalls that are included in Windows and anti-virus software.

Usually a LAN system will have one or more **servers,** *computers that act as a common resource for data storage, program access, and networking.* The server includes both hardware and software to provide user access to the network. Since local networks are private, they usually require users to **log in,** *by entering their name and password into the computer.* Additional identification may also be necessary. Once inside the network, you have access to other computers, programs, and equipment, which are part of the local network. This access is controlled and monitored by the **system administrator**(s), *the individual(s) who serve(s) as managers of the network.* Only administrators can modify the OS or install network software, and assign shared resources for network workstations, although individual users maintain control over the settings on their own computers. On a home computer, the registered user becomes the computer administrator.

Another common type of intranet is the **VPN (virtual private network),** *which uses internet resources (frequently the Internet) to connect members of a private network across longer distances.* These are increasingly useful as more workers shift to mobile computing. Branch offices can thus be connected to the main company network through external resources.

C. Modems

A modem serves as both an input and output device for network computers (as well as individual computers). A true **modem** (short for **mo**dulator/**dem**odulator) *changes (modulates) the digital signal from a computer to an analog signal that can be carried over phone lines or television cables.* These signals must be received by another modem, which changes the signals back to digital form. Modem speed is measured in **Kbps (1000 bits per second) or Mbps (1,000,000 bits per second),** *referring to the speed of data transmission.* A modem can be a separate peripheral device, which is usually the case for DSL and cable modems (which are not technically modems, although they serve the same networking function). Virtually all computers have a built-in low-speed modem, which allows the user to dial up a network connection over a telephone line. Dial-up connections are limited to a maximum speed of 56.6 Kbps and are no longer suitable for the ways most people use networks.

DSL "modems" *connect directly to a phone line, but use high-frequency transmission signals and allow normal telephone use with a proper sound filter installed.* These connections are constantly on and allow much higher connection speeds (3 to 10 Mbps) than regular modems. Similarly, there are even higher-speed network connections (up to 10 times faster than DSL) available from some cable television companies, which use the TV cables instead of phone lines. Unlike Ethernet, where the LAN has its own server to provide high-speed connections to the Internet, the telephone, DSL, and cable systems require the user to subscribe to an **ISP (Internet Service Provider),** *who provides the necessary server connections to the Internet for a fee, as well as other services in many cases.*

D. Wireless Communications

Wireless networks *are networks that use radio frequencies or other means to establish communication between computers without the use of physical connections.* In computers, this usually means Wi-Fi systems. **Wi-Fi (wireless fidelity)** *is a high-speed wireless networking standard that is sometimes referred to as a "wireless Ethernet,"* because it shares the same purposes and characteristics of Ethernet connections. Wi-Fi is used to establish wide-area network access through a web of **hot-spots,** *which are wireless access points.* One hot-spot provides wireless connections over an approximately 300-foot radius (sometimes greater), so a series of regularly spaced hot-spots can cover a large continuous area.

Large businesses and other organizations can establish a wireless LAN with one or more wireless access points connected to an Ethernet system. A wireless LAN can also be set up in homes or small businesses by connecting a wireless router to a DSL or cable modem. A simple wireless access system is illustrated in Figure 34.19.

In order for a computer to connect to a wireless network, it must have either a built-in Wi-Fi connection chip, found in virtually all new portable computers, or an external Wi-Fi connectivity card. These are sometimes referred to as **802.11a/b/g wireless LAN** systems, referencing the official standard for Wi-Fi. Wi-Fi access requires computer configuration, and public hot spots usually require a **PIN** (*personal identification number*) to access the network. It is important to remember that Wi-Fi signals travel as radio waves and it is relatively easy to "eavesdrop" on other computers connected to the same hot-spot. The security section discusses some of the ways you can increase the security of wireless transmissions.

Wi-Fi is only one type of wireless network; the other widely used standard is Bluetooth (named after a 10th century Danish king). **Bluetooth** *is a high-speed wireless networking standard used for short distances.* It is routinely used by cell phones and **PDAs** (*personal digital assistants*) to communicate with computers or wireless headsets. It is also used to replace cables from computers to peripherals such as printers and keyboards, and

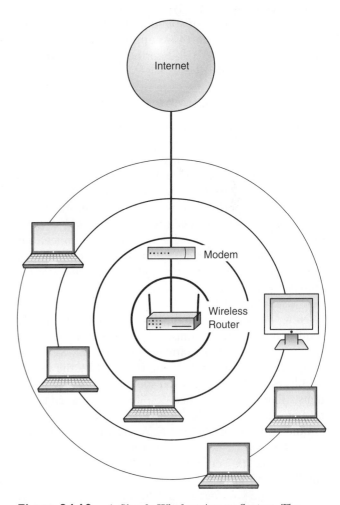

Figure 34.19. A Simple Wireless Access System. The modem could be DSL, cable, or other connection, and could be combined with the router equipment. A firewall might be installed in the router/modem, but in a public network it is essential that individual computers also have their own firewall.

in a variety of testing equipment for field work and medical applications where cable connections are inconvenient. Bluetooth has a much smaller range than Wi-Fi (~30 feet compared to Wi-Fi's ~300 feet), but is cheaper and more energy-efficient. It has the added advantage of detecting and networking with any other Bluetooth-enabled hardware without the need for configuration.

VII. PURCHASING A COMPUTER

A. Introduction

The first consideration when upgrading an "old" computer system or buying a new system is deciding what you need the computer to do. This will determine the major types of software programs you will need. Specific software requirements will then dictate the hardware capabilities that your computer must have, especially speed, the amount of necessary memory, and storage capacity. Some types of programs will also require specific types of sound and graphics cards. We will discuss specific software applications in the next chapter, but here we will describe general considerations for choosing hardware.

B. Upgrading a Computer System

Each "generation" of new computers (and a computer generation is no more than 3 to 4 months long) provides faster and less-expensive machines than the earlier models. This does not mean, however, that you need a new system every four months. In most situations, a computer can perform so much faster than its human operator that adding speed with a new CPU may be unnecessary except for memory-intensive applications such as multimedia or **data mining** (*'digging' through very large amounts of data, such as multiple genome sequences, to discover relationships and other information, discussed in Chapter 35*).

New versions of complex software frequently require much larger amounts of memory and disk space for storage. For example, the Windows XP operating system is currently being replaced by the newer Windows Vista OS. However, some new computer systems still have XP installed despite the upgraded capabilities of Vista. Table 34.2 shows one reason why people still use the older OS. Aside from the greatly increased use of system resources, many computers that predate 2007 do not have the power to effectively run Windows Vista.

If you find your needs outgrowing the capabilities of your computer, consider whether a system **upgrade** will solve the problem; *adding additional memory and storage to the old system*. Many computer problems, such as sluggish performance and full hard disks, can be addressed relatively inexpensively. One of the least

Table 34.2. SYSTEM REQUIREMENTS FOR WINDOWS XP AND WINDOWS VISTA

Specification	Minimum/Recommended Requirements	
	Windows XP	Windows Vista
Processor speed	233 MHz/300 MHz	800 MHz/1 GHz
System memory (RAM)	64 MB/128 MB	512 MB/1 GB
Hard disk space	1.5 GB	15 GB/60 GB free on main disk; must have at least 40 GB/120 GB total capacity

The numbers shown represent minimum system requirements and recommended parameters for better performance. These data are taken from Microsoft's website, http://www.microsoft.com.

expensive and most useful upgrades is the addition of RAM. Memory upgrade modules, which are small circuit boards containing a series of integrated circuits ("chips") that provide RAM (Figure 34.20a), come in standard sizes starting at 512 MB. These are easily installed and can significantly enhance computer performance. The terminology describing different types of memory modules (SoDIMM, DDR2, DDRAM, for example) can be confusing; because each computer system can only work with certain types and combinations of modules, it is important to check the documentation for your computer to determine the proper hardware for a memory upgrade.

If computer performance is adequate but the installed hard disk is full, you can add an external hard drive to the current system for additional storage (Figure 34.20b). These drives can be connected through a USB port at any time to provide the necessary storage capacity. These are an ideal medium for data backup, since they provide a large-volume, portable unit that can be stored at a different location from the original computer for safety.

C. Selecting a New System

With current rates of progress in technology, it usually becomes more cost-effective to replace an inadequate computer rather than upgrade after two to three years. However, choosing a new computer system can be a challenging assignment. Most major computer manufacturers allow you to "build" your own computer online, telling them exactly what components you want. However, there are many reputable computer stores and online dealers that are able to provide an excellent price on a high-quality system with mixed components; this usually requires more knowledge of computer hardware than buying direct from a manufacturer. It is

(a)

(b)

Figure 34.20.   Computer Upgrade Options. a. Extra RAM can be added to your computer in the form of additional memory modules. This is an example of a DIMM (double in-line memory module), with 8 memory chips on a card. **b.** Portable hard drives for extra storage. External hard drives come in many shapes and sizes. They can be stacked vertically or horizontally on site, or transported carefully to other locations. These three drives represent an additional 2 TB of data storage.

easy to become confused about all of the specifications of a computer system, but there are three that are of greatest interest when buying a new system:

- microprocessor
- memory
- hard drive space

Box 34.1 provides a discussion of some of the considerations for evaluating a potential purchase.

The speed of a computer system is determined by the type of microprocessor and the clock speed, in GHz, of the system. Remember that while higher clock speeds indicate a faster computer, many users do not need the fastest system available. In reality, current microprocessors are so fast that the average user will not notice much difference except in very demanding applications.

Usually, the performance you get from your system is most dependent on the size of the RAM, the working area of the computer; therefore, you want a system with as much RAM as possible. Many inexpensive computers are sold with a minimal amount of RAM that may not be adequate for large applications. It is easiest to have the vendor install (and guarantee) all the additional memory you need at the time of purchase.

Other system choices will depend on your intended uses of the computer. You will probably want as large a hard drive as practical, although you can always add additional external hard drives for extra storage. You also need to consider the level of graphics and video support that you need. Some computer systems are designed specifically for high-quality graphics, including gaming systems, but others may need added sound cards and video boards for multimedia applications. These are most easily installed at time of purchase. At least one optical drive is needed for a complete system, since boxed software will be stored on CD or DVD. A DVD±RW drive has the capability to read and write DVDs in different formats, as well as read and write CDs. There are many additional items that can be built into a system, such as webcams, flash memory card readers, and extra ports.

Box 34.1.   COMPUTER PURCHASE CONSIDERATIONS

**Great Buy!   Zoom H27/1.73 GHz
2 GB DDR2/160 GB/DVD±RW**

You may find an online description for a desktop computer that looks like this. It is easy to figure out what the "Great Buy!" part is supposed to mean, but what about the rest? People do not generally buy computers based on their appearance but rather on their performance, which is indicated by the above information. Here is an interpretation of this catalog description.

The first part—**Zoom H27/1.73 GHz**—shows the most basic specifications:

- **Zoom H27:** type and model of microprocessor
- **1.73 GHz:** clock speed

The second part indicates the specific features of this system:

- **2 GB DDR2:** the amount and type of RAM
- **160 GB:** storage capacity of the hard drive
- **DVD±RW:** the type of optical drive provided

Together, these specifications provide enough information to determine the power of a particular system, and whether it can supply the basic needs of the laboratory.

Unless you are buying a laptop, essential peripherals such as the monitor, keyboard, and mouse are usually chosen separately. Monitor size is the criterion that most users consider; bigger is indeed better, up to a point. If you will be sitting close to the monitor, a screen size above 20 to 22 inches can be uncomfortably large, but that is a matter of personal preference. Be sure the monitor is height- and angle-adjustable, so that an ergonomically appropriate arrangement is possible.

As discussed earlier, the keyboard and mouse come in many designs, so it is best to try them out and see what is most comfortable. Many wireless models are available for convenience as well. Adding an adjustable keyboard shelf to the computer desk is generally a worthwhile investment, because it allows you to place the keyboard and mouse at the right height and angle to minimize repetitive stress injuries.

D. Laptop Computers

Laptop computers are now common due to the attraction of mobile computing, as well as the development of lighter and more powerful models. The term *laptop* generally refers to any self-contained, portable computer weighing 4 to 8 pounds. *Notebook* computers once referred to smaller models, but are now essentially the same as *laptops*. Some laptops have become extraordinarily small, although there is a trade-off in ease of use and built-in resources. The considerations when buying a laptop are much the same as those for a desktop. Get as much RAM as possible, because laptops tend to have less room for major upgrades. Hard drives are generally smaller on laptops, but externals drives can be used when needed.

Some people choose to have a laptop as their primary computer rather than a desktop model. Most laptops have ports for connection of external monitors, full-size keyboards and a mouse, and the laptop itself provides the CPU for the system. However, you probably won't be carrying those peripherals with you on the road. Because laptops have built-in keyboards, monitors, and pointing devices, it is important to try out the computer before buying. As shown in Figure 34.13 on p. 650, laptop keyboards are relatively compressed, and smaller models have smaller keyboards. Be sure you can type comfortably on a specific system. Likewise, the laptop will have a built-in mouse substitute, most commonly a trackpad. Some people find trackpads difficult to use, so many laptop users carry a USB mouse. Laptops generally have widescreen displays, but make sure the screen is large enough to use without eyestrain. You will certainly want a built-in wireless adapter and a LAN (Ethernet) port. If laptop size and weight are major concerns, keep in mind that power adapters are never included in the advertised weight, and you will probably be carrying one with you.

VIII. MAINTAINING A HEALTHY COMPUTER

A. Taking Care of Hardware

For the most part, computer hardware is robust, but you can maximize the life of your system (and its data contents) with proper maintenance. Computer components have three major enemies: heat, dust, and electrical spikes. These hazards can be minimized with good practices. Your main goal is to keep your computer in a cool, clean environment, powered through a surge protector.

Computers generate significant heat through the high-speed activity of the CPU, so it is important to install the machine away from heat sources or sunny windows. Make sure there is good air circulation around the computer to help with heat dissipation. Computers have installed fans that cool the CPU and power supply areas. These fans can be the weakest link in the hardware chain; pay attention to any changes in the noise level emanating from your computer. Many fans provide audible warnings in response to mechanical problems. If the fan stops completely, the internal temperature of the unit will rise rapidly and the delicate microprocessors will be "cooked" unless the machine is turned off. Most computers have a mechanism to shut down automatically if the temperature rises dangerously high. Heat generation is a major consideration for laptops, given that their fans are relatively small. The heat vents on laptops are located on the bottom, so always use them on a solid surface, with room underneath for air circulation. They are guaranteed to overheat if operated on a bed or lap with the vents blocked.

Computers do not require much regular maintenance work, but they do need a clean environment to operate efficiently. Fans will pull dust into the computer through the vents, which can damage internal components. Dust can also block the vents, as shown in Figure 34.21, and

Figure 34.21. Dust Adversely Affects Computer Function. Dust can block the air vents of a computer console and build up on internal components.

allow heat build-up. One useful and inexpensive tool for keeping components dust-free is a can of compressed air. If used carefully, the high-pressure air stream can blow dust from small components and crevices that would be impossible to wipe clean. Experts recommend that the interior of a desktop unit be cleaned at least once a year to remove dust, especially if the processor unit is kept on the floor. If the unit is not very dirty, this is done by removing a side panel from the computer and gently using a low-suction hand vacuum or condensed air to remove as much dust as possible without touching the interior components. Do not use a floor vacuum, and be careful not to blow dust farther into the components. If the dust is difficult to remove, it is probably best to have a computer professional do a thorough cleaning.

It is a good idea to clean the exterior of the computer on a regular basis, especially around the heat vents. The plastic components of the system can be dusted and lightly wiped down with an appropriate mild cleaning solution, as recommended by the manufacturer. **Never spray any type of cleaner directly onto the computer!** Instead, spray liquids onto a dust-free rag and then wipe down the computer surface.

It is essential to carefully follow the manufacturer's instructions for cleaning monitor screens. Computer design and materials have changed significantly in recent years. You do not want to follow someone else's suggestions for cleaning your monitor and then find out that they meant a CRT monitor with a glass screen and you have an LCD monitor—with a now-degraded screen. Never use paper products to wipe down the screen; these will leave lint traces. A small cloth for cleaning eyeglasses is usually appropriate for keeping the screen dust-free.

It is best not to eat or drink while working with a computer. The classic mishap is spilling coffee or other beverages on the keyboard, where the liquid can seep under the keys. Small amounts of water are usually not a problem, but sodas and other beverages containing high amounts of sugar can create a sticky residue that renders the keyboard unusable. It is more cost- and time-effective to replace a malfunctioning desktop keyboard than try to fix it. If the problem occurs with a laptop, it is best to have a professional clean the system, since there are essential computer components under the keyboard.

Food crumbs and dust will also fall into the keyboard. It is a good practice to regularly turn your keyboard upside down and gently tap it so that particulates will fall out between the keys, which can then be wiped down gently. Compressed air can be used to blow out any remaining dust behind the keys.

Computers are highly sensitive to variations in electricity. An essential accessory for every computer is a surge protector. These are relatively inexpensive and provide significant system protection in case of a power

Figure 34.22. Main Features in a Surge Protector. The important features of this simple surge protector are: **1.** multiple outlets, including two for bulky transformers; **2.** protected ports for a telephone modem; **3.** central on/off switch; **4.** indicator lights to show that the circuit breaker is working. (Photo courtesy of TrippLite.)

outage. A **surge protector** *acts as a power regulator between the computer and the power outlet.* They are designed to smooth out momentary spikes and fluctuations in power levels. If a computer is operating when a power outage occurs, the contents of RAM are lost, but usually no damage occurs to the system. However, when power is restored, there is frequently an electrical spike which can potentially damage or destroy the CPU. When an outage occurs, it is a good idea to disconnect the computer system from its power source immediately; power strips that act as surge protectors usually have a switch for turning the electricity off without having to unplug anything. Restart the system after the outage has been resolved.

The main features of a surge protector are shown in Figure 34.22. It is important to remember that some inexpensive power strips are not surge protectors; they merely provide extra electrical outlets. Others will be labeled as **surge suppressors,** *which even out only minor fluctuations in power.* Generally you will pay more for a strip with a higher level of electrical spike protection, which will be noted on the product label. Better surge protectors also have circuit breakers that will burn out in the event of a major power spike, stopping the flow of power into the equipment.

Even a surge protector will not guard against a direct lightning strike to the building, which will burn out the internal components of a computer and other electrical equipment. If lightning or power outages are a frequent problem in your area, you might want to invest in an **Uninterruptible Power Supply (UPS),** *which will deliver constant power to the computer for a brief period during a power outage.* A battery-powered UPS may last ten minutes or less, but you will have enough time to save your work and shut down your computer properly. Most UPS units also contain a circuit breaker, which will burn out in case of a lightning strike and protect the computer. Institutions with local networks supporting many computers usually install high-capacity central UPS units with built-in generators; individual surge protectors are still a good idea.

Be careful to avoid mechanical shocks to computer equipment. It is easy for chips and other internal components to disconnect from their sockets. Be sure the computer is turned off when moving the unit and handle it gently. Also keep in mind that while flash drives are tough, portable hard drives are not. These drives have hard casings that resist small bumps during transportation, but dropping a unit will probably destroy the hard drive mechanism and make data recovery difficult. For this reason, it is an excellent idea to use a padded carrying case for drives and laptops during transportation.

Never open or unplug a storage drive while it is operating. Drives almost always have a light that indicates activity; never disrupt the drive while this light is on. This is mechanically stressful and will (obviously) interrupt drive function. If the drive is in the middle of saving a file, it is possible to lose those data entirely, even the previously saved portions, because the file is not properly closed. When in doubt, use the Safely Remove Hardware function of your OS, which will ensure that the component is not in the middle of an operation when disconnected.

Table 34.3 summarizes these tips for maintaining your computer hardware.

B. Taking Care of Data

Damaged or malfunctioning hardware is easy to replace and inexpensive compared to the consequences of data loss. Every computer user dreads a hard drive crash, when all data stored on the hard disk becomes at least temporarily, and sometimes permanently, inaccessible. **The first lesson of computer use is to be sure that your computer files are backed up regularly.** You should have two backup copies of every important data file, with at least one stored on an independent storage device in a safe location away from the main computer. Remember to back up e-mail regularly as well. Most large businesses and institutions with LANs have a **backup server,** *a network computer that can automatically back up data from individual workstations,* but it is good practice to have your own copy as well.

Every OS includes a utility program that will back up system hard drives, although many users purchase an additional program with extra features. **Backup programs** *typically perform automatic backup operations, compress files, and verify data.* These utilities can compress unused files as well as the backup copies. **File compression** *recodes repetitive data within a file to reduce data storage size and save disk space.* It is a good idea to use the copy command to regularly save files onto separate storage media, but backup programs can automatically identify and copy important program and data files from various locations on the hard drive. The

---

Table 34.3. **Hardware Maintenance Guidelines**

The following guidelines will help maximize the life of your computer hardware.

- Avoid any extreme temperatures and direct sunlight.
- Keep the computer area as dust-free as possible.
- Keep liquids away from computer vents, the keyboard, and the mouse.
- Never block the vents in the computer console or place a dust cover over a computer while it is running. This will cause overheating of the CPU.
- Do not ignore changes in the sound of the computer fan(s).
- Buy and use a surge protector.
- Check the power cords and connecting cables on the system regularly. Make sure circuits are not overloaded.
- Never open an optical drive or unplug a computer component while its "active" light is on.

Windows Backup utility is found in the System Tools folder in the Programs menu, as shown in Figure 34.23 on p. 661.

Most backup programs can perform several types of backup that would be tedious for the user. The options usually include:

- Full backup, with copies of all specified file types.
- Differential backup, only saving new versions of files that have been modified since the last backup.
- Incremental backup, similar to a differential backup, but adding new versions of modified files to the old backup copies instead of replacing them.
- System backup, which includes all system files necessary to restore your computer to working order after a system crash or reset.

A system backup creates a **bootable backup,** also called **recovery media,** *that can be used to restore your computer's OS and programs* in case of a major disaster that damages the OS. Most new computers do not include a copy of the OS on disc because the software is preinstalled. Therefore, the first task that should be performed with a new computer system is to create recovery media. Computer manufacturers provide built-in software to perform this task and the system manual will explain the simple (if lengthy) automated procedure.

The Windows OS creates a series of automatic **system restore points,** which *are profiles of the state of a computer at specific times.* These can be used in case of a minor system crash or virus infection, and essentially reset (restore) the computer system to its state before the problem. A system restoration does not remove or change document files that have been saved since the last restore point. In addition to the automatic restore

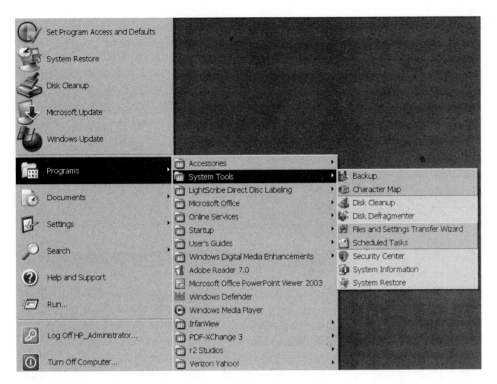

Figure 34.23. System and File Maintenance Utilities in the Windows OS. The Windows OS has an essential folder called System Tools (frequently found within the Accessories folder in the Programs menu). This folder contains important utilities such as the Backup, Disk Cleanup, Disk Defragmenter, and a list of Scheduled Tasks such as automatic updates. System Restore can be accessed here as well. These utilities are further described in the text.

points, you can instruct the computer to create additional points. Always set a new system restore point immediately before installing new software. This can be done easily using the System Restore utility in the System Tools folder.

It is not uncommon for a computer user to delete a file in error. This is why GUIs provide a "trash can" or "recycle bin" icon for storage of discarded files—so the user can still retrieve the file before it is erased. Once you issue the command to empty the recycle bin folder, these files disappear from both the folder and the disk. However, because of the way disk storage works, the file contents are not truly removed. What actually happens is that the markers needed to identify the file are eliminated. The disk space occupied by the file contents is then made available for future storage. If you immediately realize that you have discarded an important file, it may be possible to recover it using a **file recovery utility.** *These programs find the file contents and create a new disk directory entry that allows you to open the file again.* This will work only if you have not stored other information on top of the old file, so the recovery utility must be applied as soon as possible after the file is discarded. It is a good practice to have one of these utilities (there are many free versions available on the Internet) installed on your computer. Otherwise, if you need to download the utility after a file has been erased, be sure to copy the program to an external drive, to avoid overwriting the file you want to save!

There is another consequence of the way computers discard files; confidential information is not removed from the disk immediately after a file is deleted, which

means that some or all of these data could be recovered by outside sources. If you need to reliably eradicate confidential information, you can use a **disk sanitizer,** *a utility that writes over sensitive files immediately.* Never discard an old hard drive without using one of these utilities to destroy your stored files.

CDs and DVDs require care in order to provide reliable long-term function and data storage. Keep discs in a closed container to prevent dust contamination or breakage when not in use. For large quantities of discs (for example, a full system backup set) that will not be handled frequently, you can use a spindle and cover. Individual discs should be stored in plastic jewel cases or individual paper sleeves. The type of plastic sleeves usually found in inexpensive storage notebooks or textbooks can sometimes adhere to the discs, which may be damaged during removal. Discs should be stored away from light or heat sources.

Contamination of the disc storage surface with fingerprints, dust, or liquids can lead to data loss. Handle discs only by the edges with clean hands. If it is necessary to clean a disc, use a lint-free cloth and always wipe straight from the center to the outer edge. Do not wipe in a spiral, because this is the direction of data storage and is more likely to affect data retrieval. Solvents should not be used, unless they are labeled for use with optical discs.

To avoid mechanical damage, do not flex discs or write on the label with a sharp implement such as a ballpoint pen. Use felt-tip markers, or purchase an optical drive that can etch permanent labels onto the disc. Table 34.4 summarizes the guidelines for handling optical discs.

Table 34.4. Optical Disc Maintenance Guidelines

- Keep discs stored away from dust or other contaminants.
- Use plastic jewel cases or high-quality paper sleeves for storage.
- Avoid extreme temperatures, including direct sunlight.
- Handle only the edges of a disc.
- Keep discs away from liquids and foods (probably the most frequently violated guideline). A disc may not survive a coffee spill.
- Never write on a disc with a sharp pen or pencil.
- Do not bend discs or attempt to remove paper labels.

C. Taking Care of System Operations

This section discusses ways that you can guard against general system failures and optimize the operations and speed of your computer system. Although modern OS and commercial application software tend to be stable, every user dreads the occasional **system crash,** *when your computer freezes up and stops working.* Most local system crashes are related to memory problems rather than user error; usually this is the result of too many programs and/or operations competing for inadequate memory space. Software bugs (*functional defects*) can also cause these problems. It is always wise to consider the possibility of a computer virus or other software intruder in your system. These issues are discussed in detail in Chapter 36.

The most common occurrence is a program that does not complete a command, followed by a message that the program is "not responding." In Windows, you can call up the Task Manager by simultaneously pressing the Ctrl+Alt+delete keys. If the problem is only within the nonresponsive program, the Task Manager will appear with a list of open applications and you can tell it to close the specific application and leave the others open and functional. However, if the problem is systemic, the Task Manager may not appear. In that case, you probably need to perform a warm boot by pressing Ctrl+Alt+delete a second time. A **warm boot** *closes down computer operations then restarts the system without turning off the power.* This will take care of almost all system crashes. Unfortunately, after a reset, any unsaved data that was in RAM will be permanently lost, which is good motivation to save files frequently while you work.

In some cases, even this procedure will not work. PCs and Macs have different reset procedures and it is a good idea to keep the instructions posted near the computer in case you need them. Turning off system power should always be a last resort, because this does not shut down the OS properly and could result in further problems when you restart the system.

The best way to keep your computer operating at its best is to perform regular system file clean-up and maintenance. Make sure that your OS and especially your antivirus software are updated regularly. Both of these can be set for automatic updates. Updates generally make your system safer from viruses and other threats, as discussed in Chapter 36. Remember that updates must be both downloaded and installed in order to function. Complete installation frequently requires that the computer be restarted.

Run the disk cleanup software supplied with your OS on a weekly basis. This will remove temporary files, empty the recycle bin, compress unused files, and recover additional storage space. In Windows, the Disk Cleanup utility is located in the System Tools folder, as shown in Figure 34.23 on p. 661. It is also a good idea for the user to delete any unnecessary files and occasionally get rid of unused software. Many programs come with their own uninstall program that you can run yourself, or you can use the Windows Add/Remove Software function that will cleanly uninstall programs. An **uninstall utility** *deletes not only the program folder, but also additional program files in various locations on your hard drive.* This frees up disk space and eliminates "orphan" files, which no longer have an associated application or function.

The other step you should take to optimize your hard drive performance is to use Disk Defragmenter (in Windows), as shown in Figure 34.24. When large files are stored on a hard disk, they may end up divided into fragments to fit the spaces that are open on the disk. This is normal and does not affect the function of the files. However, if enough files are fragmented across the disk, system performance can be slowed (this is less

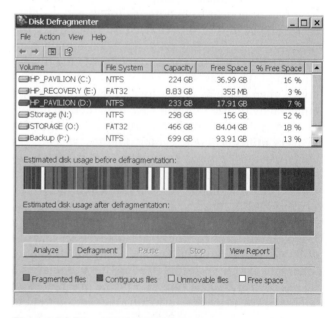

Figure 34.24. Windows Disk Defragmenter. This program analyzes the pattern of file fragmentation on the chosen drive, indicates whether the program can improve drive usage, and if so, reassembles the files in a more organized pattern.

The Hardware

Monitor location (A). The monitor should be located directly in front of you at arm's length with the top at forehand level. Outside windows should be to the side of the monitor to reduce glare. *Monitor features.* The monitor should be high-resolution with anti-glare screens. *Monitor maintenance.* The monitor should be free of smudges or dust buildup. *Keyboard location (B).* The keyboard should be located such that the upper arm and forearms are at a 90-degree angle. *Keyboard features.* The keyboard should be ergonomically designed to accommodate better the movements of the fingers, hands, and arms.

The Chair

The chair should be fully adjustable to the size and contour of the body. Features should include: *Pneumatic seat height adjustment (C); Seat and back angle adjustment (D); Back-rest height adjustment (E); Recessed armrests with height adjustment (F); Lumbar support adjustment (for lower back support) (G); Five-leg pedestal on casters (H).*

The Desk

The swing space. Use wrap-around work space to keep the PC, important office materials, and files within 18 inches of the chair. *Adjustable tray for keyboard and mouse (I):* The tray should have height and swivel adjustments.

The Room

Freedom of movement. The work area should permit freedom of movement and ample leg room. *Lighting.* Lighting should be positioned to minimize glare on the monitor and printed materials.

Other Equipment

Wrist rest (J). The wrist rest is used in conjunction with adjustable armrests to keep arms in a neutral straight position at the keyboard. *Footrest (K).* The adjustable footrest takes pressure off the lower back while encouraging proper posture.

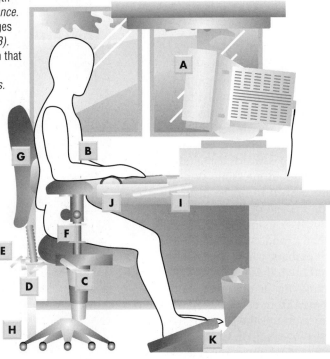

Figure 34.25. **Ergonomic Design of the Computer Workplace.** (From Long, Larry, and Long, Nancy. *Computers.* 5th ed. Saddle River NJ: Prentice Hall, 1998, I12.)

obvious with faster systems). **Defragmentation** *recombines file fragments and optimizes disk usage.* This is an automated process that may run overnight if the program has not been used recently. If used regularly, the process is fairly quick.

Finally, do not forget proper maintenance of the computer user (keeping in mind that eating and drinking at the computer should be avoided). Ergonomic safety is a significant concern for computer users, as discussed in Chapter 10, because the repetitive motions, particularly for the mouse hand, can result in carpal tunnel syndrome. Eyestrain and back and neck problems are also common. The best prevention techniques are to take frequent breaks while working at the computer and to ensure that the computer environment is designed to be both comfortable and efficient. Figure 34.25 summarizes some of the main considerations for setting up an optimal computer workspace.

IX. INTRODUCTION TO INFORMATION SECURITY

Laboratory computers are routinely used to store data that are unique and potentially confidential. For this reason, it is essential to consider the safety of these data. Hardware can be replaced easily (if not cheaply), but that is not usually true of lost data. It is also necessary to consider the security of proprietary information

and employ safeguards against unauthorized access. The increasing use of wireless networks and mobile computers creates additional concerns that must be addressed.

The least complex threat to your data is the possibility of computer theft or physical vandalism. Basic precautions to minimize theft of desktop systems include limiting access to the computer area and to the computers themselves. Install any expensive computer components in nonpublic, locked areas when possible. Be sure to record the serial numbers of all computer components and keep the list in a separate location, which is also important in case of hardware failures.

Laptops are an obvious concern because of their portable nature and the fact that they are frequently used in and transported through public places. They usually have a mechanism for securing a special lock and cable to the computer. Losing a laptop is expensive, but losing the contents of the hard drive is usually much worse. Keep your laptop data backed up at all times and consider the potential consequences of strangers having access to everything stored on the computer.

Unauthorized data access is a more complex threat to consider. Companies that work with proprietary or confidential information will have policies and practices for safeguarding data. This frequently includes user **authentication** (*requiring unambiguous personal* *identification for data access*) procedures and data **encryption** *(storage and transmission of data in coded form)*. Network structure should be designed to prevent unauthorized access as well. It is the responsibility of every worker to learn and apply employer policies to their own computing practices.

The first step in protecting confidential data is to restrict access to specific individuals. Preventing physical access to data storage components is one simple precaution. Another basic step is to require individual passwords to obtain access to computer operations and data files. Passwords vary tremendously in their ability to restrict access. There are password theft programs available that attempt to duplicate passwords using targeted information or simply random guessing. When setting a password, never use a word from the dictionary (theft software checks those) and of course, avoid anything obvious, such as your user name, pet's name, or birthday. Experts recommend a random combination of at least eight letters and numbers, which would be almost impossible to guess, although these are also difficult to remember. Keep in mind that if you forget your password, you may be prevented from using the computer too! Avoid the temptation to use the same passwords for multiple log-in systems and never use a work-related password for personal purposes. Passwords should be changed regularly. If you have too many passwords to remember, write them down by hand and keep the list apart from the computer.

Passwords provide only a first line of defense against intruders; most secure systems require more than one authentication method. **Digital certificates,** *which are physical items such as identification cards or USB plug-in items that confirm your identity to the computer,* are frequently used. Because these can be stolen, many systems now apply biometric identification methods (see Chapter 35). Incorporation of fingerprint readers into a flash drive or mouse is becoming relatively common for this purpose, as shown in Figure 34.26a. More complex biometric systems may include iris or retinal eye scans (Figure 34.26b), or voice analysis. Digitizing tablets can be used to capture biometric signatures, with software to not only match the signature itself, but also the writing mechanics of the individual.

If your computer contains proprietary or sensitive information, you or your employer will want to consider additional levels of data protection, including data encryption. For stored files, the Windows OS and a variety of security programs can lock individual folders, make specified files invisible, and/or provide file encryption on a hard drive or portable media. The additional ability to encrypt information sent out over a network is required to protect the confidentiality of this information. This is especially critical if sensitive data are handled on a wireless network. There are many available forms of **sniffer software** *that allow*

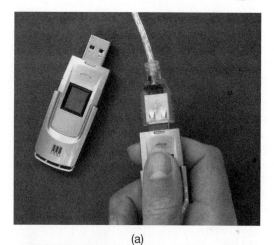

(a)

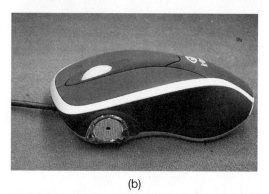

(b)

Figure 34.26. Biometric Identification for Computer Access. a. Flash drive incorporating a fingerprint reader. **b.** Mouse incorporating an iris scanner. (**a.** 2GB BioFlash 2.0 from Imagenix Technologies [Imagenix.com]. **b.** IRIBIO mouse, courtesy of Qritek Japan Co. Ltd [www.qritek. co.jp].)

unauthorized individuals to intercept wireless transmissions from other computers. This is a relatively simple operation at public hot-spots such as airports, where unrelated users are in close proximity and sending and receiving communications on the same radio frequency. Communication with secure Internet sites is safe, but nonsecure sites, e-mails, and **instant messages** (**IMs,** *private messages exchanged in real time, allowing text conversations*) are vulnerable without security precautions. If information security (or just plain privacy) is a concern, it is best to assume that any unencrypted data sent over a wireless network can be intercepted and read by others. For this reason, using encryption software can ensure that even if transmissions are intercepted, they will not be interpretable. **Encryption software** *applies mathematical keys to encode (sender) and decode (recipient) encrypted e-mails and files*, and can be used for both stored files and folders, e-mails, and transmitted files. An employer may supply a specific program for employees, and if so, this should be used for all data transfers and any computer with work-related contents.

Chapter 36 extends the discussion of computer security to general threats encountered through the Internet and other networks.

PRACTICE PROBLEMS

1. Which of the following items are hardware and which are software?

    a. keyboard

    b. GUI

    c. printer cable

    d. printer driver

    e. DVD

    f. database program

2. How many bytes are actually in a megabyte?

3. If you need to make a complete backup of an 80 GB hard disk:

    a. How many 700 MB CD-ROMs are required?

    b. How many 4.7 GB DVDs are required?

    c. How many 50 GB Blu-Ray discs are required?

QUESTION FOR DISCUSSION

Think about the computer system in your laboratory or home. How would you optimize or upgrade the available features?

CHAPTER 35

Data Handling with Computers

I. INTRODUCTION

As discussed in Chapter 34, the scientific process involves gathering data and processing them into useful information. Computers are essential tools for laboratory professionals because they can convert data into information and store that information at rates that are impossible without their use. Data can be input into a computer in various ways, for example, directly from a laboratory instrument, or manually using a keyboard. Once input into the computer, data are processed by software. Data processing functions include:

- recording, storing, and retrieving data
- organizing and preparing data for presentation
- mathematically manipulating numerical data
- sorting and organizing data into compact and easily accessed forms
- discovering patterns within data

This chapter focuses on the common categories of software that are used by laboratory professionals for a variety of tasks. We will explore the basics of how different types of computer programs process data for specific purposes, how to select the software that best matches a particular application, and how to ensure the quality of the program output.

II. BASIC SOFTWARE APPLICATIONS: COMMUNICATION

A. Word Processing Software

Communication is an integral part of the science professional's work and computers greatly facilitate scientific communication in many ways. We begin by discussing three categories of computer software whose major purpose is to enhance communication: word processing software, graphics software, and presentation software.

Word processing is a familiar software function that assists in communication that involves text. In word processing, the data are letters and symbols (i.e., text) that are usually input from a keyboard, but can also be input from scanners, voice recognition systems, or other devices. The computer takes the text that has been input and processes it to provide an organized, formatted document as output.

Word processing programs eliminate most of the mechanical concerns of writing by hand and allow the writer to concentrate on content. These programs also allow written materials to be easily edited, formatted, and saved in multiple versions. Editing options include cutting and pasting blocks of text, finding and replacing words or phrases, and performing spelling and grammar checks. Formatting determines how a document looks, for instance, on the final printed page. General formatting functions of word processing programs include setting the page margins and tabs, line spacing, and fonts to create an easily interpreted document.

In the laboratory, word processing programs facilitate the creation of quality documentation systems. For example, word processors are used to create templates for laboratory standard operating procedures (SOPs), helping to ensure that these documents always have proper revision numbers, that they include all required sections, and that they are prepared and formatted consistently. SOPs stored on computers can be controlled to ensure that only authorized individuals have access and that only the most current revision is used.

There are three types of text handling programs: text editors, true word processors, and desktop publishing programs. A **text editor** *allows the user to enter, edit, save, and print plain text.* All computer systems provide a basic text editor, represented by Notepad in PCs. True **word processing programs** *allow text editing plus more complex actions such as adding superscripts, performing spell checking, creating tables, and making outlines.* Microsoft (MS) Word is probably the most familiar example.

Desktop publishing (DTP) programs *provide complete page layout capabilities for manipulating graphics and text and preparing documents for professional publication.* These programs are designed to create complex page layouts combining blocks of text and graphics, which are treated as objects and positioned on-screen for printing. In computer language, an **object** *is any file or portion of a file that is treated as a unified whole for purposes of movement and processing.* A picture inserted into a word processing document is an example of an object; individual elements within the picture cannot be manipulated within the document. For DTP, text is usually created and edited in a word processing program and then blocks of text are manipulated as objects in the DTP program, which can also perform some basic text manipulation.

Sample output from each of these types of text handling program is shown in Figure 35.1 on p. 668.

B. Graphics Software

i. BITMAPPED GRAPHICS

Graphic images are an essential part of scientific communication; the creation, manipulation, and printing of images is routine in biotechnology laboratories. Examples of scientific images include photographs of electrophoresis gels, immunofluorescent images taken using a microscope, diagrams that illustrate cellular processes, and videos of living cells. **Computer graphics** *are images created or modified through the application of graphics software.* Computers have greatly facilitated the preparation of effective graphics, allowing laboratory personnel to prepare professional-quality results without extensive training.

There are two approaches to creating, manipulating, and storing images: *bitmapped graphics* and *vector graphics.* Each approach has strengths and weaknesses. Most people work with both graphics types without being aware of their features. If you have, for example,

(a)

(b)

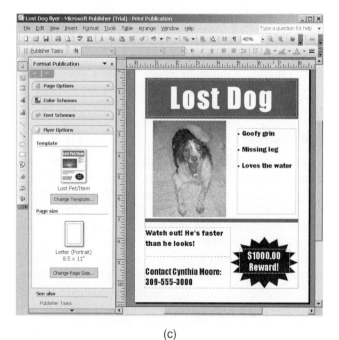

(c)

Figure 35.1. Word Handling Programs. Sample output from: **a.** a text editor (Notepad), **b.** a word processor (Microsoft Word) and **c.** a desktop publishing program (Microsoft Publisher). Note the progressively more sophisticated formatting.

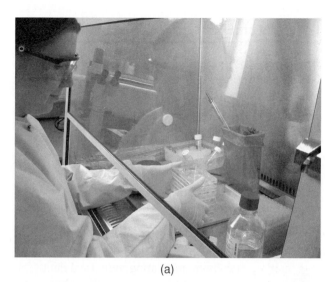

(a)

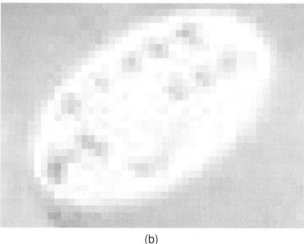

(b)

Figure 35.2. Photographs Are Formed with Pixels. a. The original photo; pixels are too small to see. **b.** The top of one of the tubes is enlarged to such an extent that the individual pixels that compose the image are visible.

ever cropped or enhanced the contrast of a digital photo, you have worked with bitmapped graphics.

Computer monitor, digital camera, and digital video images are examples of bitmapped graphics. In **bitmapped graphics** (also called **raster graphics),** *images are handled as pixels.* **Pixels** *are small, closely spaced dots,* as shown in Figure 35.2. Bitmapped graphics excel in representing complex, detailed images; photographs are always handled in this format.

Image editing programs *are bitmapped graphics programs that provide the capability to modify previously existing images, such as the output of a digital camera or a scanned image.* Image editing manipulations can range from simply adjusting the brightness of the image to major changes in photographic content. This software makes it relatively easy to manufacture digital images of nonexistent events. For this reason, photographic lab data, such as electrophoresis gel images, must never be edited in a way that can lead to inaccurate conclusions. Many scientific journals have policies restricting the use of editing software with images for publication. They generally require that authors provide a copy of the unedited image as well as the publication version.

EXAMPLE

Editing a Photograph for a Presentation

A scientist is preparing a photograph of an electrophoresis gel for a presentation at a professional conference. The presenter uses basic image editing functions, included in standard presentation and word processing software, to make the image easier for the audience to interpret quickly. **a.** The original photo includes not only the electrophoresis gel, but also part of a computer monitor and an interactive computer screen on the right. **b.** The presenter crops the photo in order to eliminate the computer monitor and anything else that might distract the audience from the gel data. **c.** The presenter also increases the brightness and contrast of the image in order to make the gel bands easier to see. This enhancement does not obscure the original gel features or alter data features, and is therefore an acceptable modification. **d.** Finally, the presenter adds labels to the photo by inserting text boxes. The labels assist the audience in quickly interpreting the electrophoresis gel data.

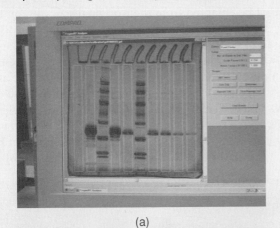

(a)

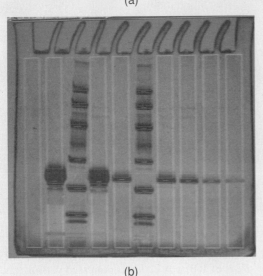

(b)

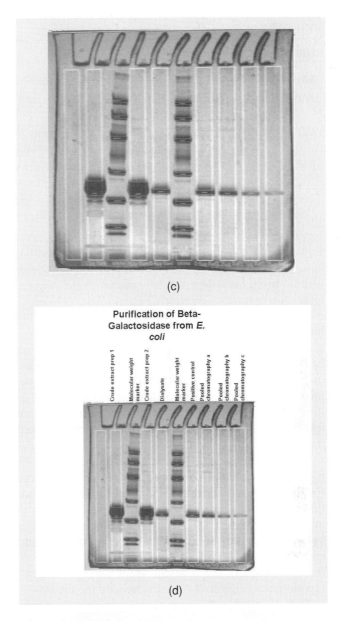

(c)

(d)

Completely new bitmapped/raster graphics are created and edited in **paint programs,** *which provide the ability to create and modify complex images using computerized tools that are similar to those used by an artist on paper.* The resulting graphics are "painted" into a file that contains information about the graphic content of each screen pixel. There are many simple paint programs available, but full-feature software, such as Adobe Photoshop, generally requires training and/or practice to use effectively.

Bitmapped graphics create relatively large files because information about every pixel is individually stored in the image file. Detailed color images such as photographs are input in **32-bit color,** *meaning that every pixel requires 32 binary bits of information for accurate reproduction.* For this reason, bitmapped images are almost always converted to compressed file formats in order to conserve disk space. **Graphics file compression** *abbreviates the image data according to a specified formula for the file format.* Usually compres-

Table 35.1. COMMON GRAPHICS FILE FORMATS

File Format	Translation	Uses	Compression Levels
BMP	Bitmap	Standard bitmap file format.	None
JPEG	Joint Photographic Experts Group	Commonly used for photographs or other detailed images; common for web images.	Multiple compression levels; lossy
GIF	Graphics Interchange Format	Commonly used for drawings; used to create simple animations.	Good compression; not lossy
PNG	Portable Network Graphics	Replacement for the GIF format; wider color palette available but no animations.	Good compression; not lossy
TIFF	Tagged Image File Format	Current standard for scanned images; high quality.	Some compression; not lossy

sion involves combining data for adjacent pixels that have the same color values. File compression can therefore be **lossy,** *because some information in the original image has been lost (eliminated) for storage.*

There are various file formats available for storing bitmapped graphics; common formats and their applications are shown in Table 35.1. A format must be chosen carefully to retain the image quality needed in the final product while also minimizing file size. For example, images posted on the Internet must be relatively small or they will take too long to download. In contrast, printed images from microscopes require high resolution, and therefore large files, to retain subtle structural details.

The standard storage format for photographs, JPEG, is lossy, which means that some information is lost from the image every time the image is saved. This information cannot be restored without going back to the original uncompressed image file. For this reason, it is better to edit photos in a format that is not lossy and convert them to JPEGs as the last step. When saving an image as a JPEG, the user can select the level of compression according to the quality needed for the image.

Figure 35.3 shows two JPEG versions of the same photograph; Figure 35.3a was stored with high image quality but low compression. Figure 35.3b was stored with higher compression, which has combined blocks of pixels to save storage space but has permanently lost too much information to be usable.

Box 35.1 contains guidance for working with bitmapped graphics programs.

ii. VECTOR GRAPHICS

Bitmapped graphics are not suitable for all graphic needs. One major drawback is that bitmapped graphics cannot be enlarged from their original size without losing image quality. There are a fixed number of pixels stored in a file, and the viewing area can become larger only by enlarging the individual pixels, allowing them to become visible. Bitmapped graphics also consume a lot of storage space. Vector graphics do not have these disadvantages. **Vector graphics** *are images based on a set of lines, which can be described and stored by the software as mathematical equations.* An example is shown in Figure 35.4. Another term for this process is **object-**

(a)

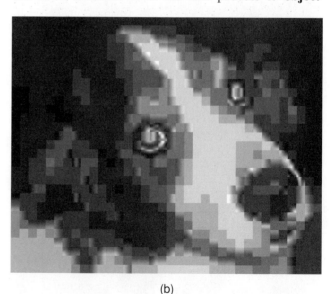

(b)

Figure 35.3. Results of Lossy File Compression. A bitmapped photographic image was saved in two versions of the lossy JPEG file format; **a.** A low-compression version that is suitable for printing. File size is 61 KB. **b.** A highly compressed version that removes additional image data. Blocks of pixels have been combined as single shades that obscure image details. File size is 7.27 KB.

Box 35.1. USING BITMAPPED GRAPHICS SOFTWARE

1. *Decide on the final image quality needed before choosing a program.*
2. *Consider the file size of your final product—storage, e-mail, and transport can present difficulties for larger files.* Photographic prints require high-quality, relatively large files.
3. *When choosing a file format, keep in mind that higher compression levels can mean poorer image quality.*
4. *Because file compression results in permanent loss of graphic information, it is a good idea to keep a copy of the uncompressed image for later recovery.*
5. *Do not repeatedly edit and save an image in JPEG format because each save reduces image quality.*
6. *Consider copyright issues when importing images from the Internet or scanning images from printed materials. Ask for permission before presenting other people's images and give credit to the original creator.*

---

oriented graphics, *images that are created as a set of separate objects, each described separately within the image file.* A number of figures in this textbook were originally created with vector graphics, including Figures 33.11 on p. 635 and 33.12 on p. 636.

Draw programs *are used to create vector graphics* and are ideal for detailed images containing clean lines with precise dimensions, such as biochemical diagrams, blueprints, and maps. The most common programs for vector graphics are Adobe Illustrator and CorelDraw. Many commonly used word processing and data presentation programs, such as MS Word and PowerPoint, contain basic draw functions. These have the advantage of being relatively easy to use, thus allowing individuals with little artistic or technical training to prepare simple graphical images, as demonstrated in the example, at right.

Many full-feature graphics programs, such as Adobe Photoshop, can work with both bitmapped and vector graphics, separately or together. Vector graphics can easily be added to bitmapped images (for example, placing a text label within a photograph), although the entire image will then be stored in a bitmapped format.

A major advantage of vector graphics is that the images can be easily enlarged, because they are described by equations that will maintain the appropriate dimensions and clarity of the image at all sizes. It is also relatively easy to make large-scale changes in vector graphic images because individual objects within the picture can be resized, moved, deleted, and manipulated. Vector graphics are frequently used for images such as company logos, which may be needed for everything from business cards to room-size banners. Professional fonts are also created as vector graphics for easy resizing.

EXAMPLE

Using Vector Graphics to Create a Basic Illustration

Word processing programs provide basic vector graphics tools that can be used to illustrate an experiment for publication. **a.** Using the drawing tools in the program, the scientist draws a test tube by selecting a basic cylindrical shape. **b.** The top of the cylinder does not look right for a test tube, so the scientist creates a second cylinder and places it on top of the first, as indicated by the arrow. **c.** and **d.** Because draw objects are opaque by default, the results are different depending which cylinder is on top (usually referred to as "order" or "placement" by the programs). In this example, the second cylinder belongs on top, as in part d. **e.** Finally, the scientist formats the first cylinder to add color, simulating a liquid.

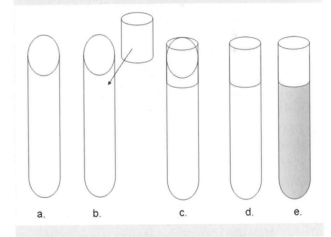

Box 35.2 on p. 672 provides guidelines for choosing between bitmapped and vector graphics.

C. Presentation Software

Verbal communication at meetings, conferences, classes, and other venues is an essential part of the scientific endeavor. Presentation software has greatly enhanced the ability of scientific professionals to effectively display data and convey information. **Presentation software** *provides capabilities for inputting text, graphic images, audio, and*

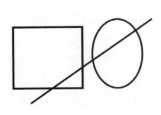

Figure 35.4. Vector Graphics: How a Draw Program Works. For vector graphics, the user creates separate geometric objects, and then combines them to form a picture. In a vector graphics file, this image requires only three equations.

BOX 35.2. WHEN TO USE BITMAPPED VERSUS VECTOR GRAPHICS

Use bitmapped graphics for:

- editing photographs
- creating complex pictures with flowing lines
- maintaining control over subtle differences in color and texture
- taking advantage of artistic techniques such as air brushes and filters
- avoiding clean "cartoon" lines

Use vector graphics to create:

- simple geometric shapes, such as lines, boxes, and circles
- basic diagrams within word processing and presentation documents
- images with precise dimensions; diagrams, charts, graphs, maps, blueprints
- layouts of multiple objects
- images that will need to be enlarged
- smaller image storage files

---

video, and then processing them into "slide" presentations that can be projected onto a screen and printed as hand-outs. This is a form of **multimedia authoring**; *the creation of presentations that use more than one medium to convey information.* One of the major advantages of these programs is the ability to set and maintain slide format to create an attractive and consistent presentation. It is simple to import graphic images, such as photographs,

tables, charts, and graphs, to create visual interest. Video and audio information clips can also be incorporated.

The purpose of this software is to aid in the preparation of effective presentation materials. Presentation software usually includes a large number of background patterns, colors, fonts, animated graphics, and other embellishments for slides, many of which are not particularly useful if your goal is communication—you want the audience to focus on the content of your presentation, not on the slides themselves. Box 35.3 provides guidelines for using presentation programs effectively in a scientific setting.

III. BASIC SOFTWARE APPLICATIONS: MANIPULATING NUMERICAL DATA

A. Spreadsheets

We have so far explored programs that are used primarily for communicating textual and graphical information. We now turn to programs that are primarily designed to facilitate the processing of numerical data. These programs include spreadsheets, graphing software, and statistical packages.

Spreadsheet programs *provide a blank grid or table of* **cells,** *each of which is uniquely designated by a row and column coordinate; calculations can be performed on selected data in the cells.* Computer spreadsheet programs are modeled after traditional paper spreadsheets (familiar to accountants) that were prepared on gridded paper. Electronic spreadsheet programs eliminate the need for repetitive use of a hand-held calculator; they provide checks of input data, perform

---

BOX 35.3. GUIDELINES FOR USING PRESENTATION SOFTWARE EFFECTIVELY

1. *For each presentation, choose a single template and/or theme.* Select one slide background, a single font (or two, at most), and the same color scheme throughout the slides. Using a consistent format allows the audience to focus on content.

2. *Use simple backgrounds and a high-contrast text color.* Fonts should be as large as possible for easy reading.

3. *Use transitions or sound effects sparingly.* These can be effective if used occasionally; otherwise they are distracting.

4. *Minimize the amount of text on each slide.* The visual materials should not include everything you are going to say. This eliminates the temptation to read from your slides, which is guaranteed to bore the audience. Use just enough text to introduce and reinforce the concept you are discussing.

5. *Use high-quality graphic images whenever they enhance understanding of the topic.* Images should be relevant to the discussion to avoid distracting the audience. When in doubt, choose a graphic element over text.

6. *Each slide should contain only one major concept; additional unrelated information is confusing.* If there are multiple supporting elements, they should not all be crowded onto the same slide.

7. *Limit the number of slides in your presentation in order to focus on the key ideas.* Eliminate slides with content you do not intend to discuss or use as a backdrop for discussion.

8. *Avoid showing graphs with irrelevant lines, tables with huge amounts of data, tiny video clips, and any other slide elements that the audience will not be able to see or fully understand.* Adapt your data figures for different audiences.

9. *Use the spell check function in the software and always go through the slide show ahead of time to avoid surprises, such as media clips that won't start or unreadable fonts.*

virtually instantaneous calculations, and can be programmed to automatically generate graphs and other formatted output.

To use an electronic spreadsheet, the user enters labels or other text, numerical values, and mathematical formulas into the cells of the spreadsheet (which is also called a **worksheet**), see Figure 35.5. Once formulas are entered, calculations are automatically performed on selected cells, columns, or rows of numbers and the results are shown in specified cells.

Full-feature spreadsheet programs allow the creation of multiple worksheets within a single file (called a workbook), facilitating input of multiple data sets on separate worksheets that can be cross-referenced. For example, a 2010 budget workbook might consist of 12 monthly worksheets.

Spreadsheet programs generate a variety of output formats. Most can automatically graph numerical data in multiple ways, as discussed in the next section of this chapter. They can produce both numerical and text tables easily, as shown in Figure 35.6. One of the most powerful aspects of spreadsheet software is that changes in data entries can produce automatic recalculation of results in other parts of the spreadsheet.

There are many scientific applications for spreadsheets and their use is becoming routine in biotechnology laboratories. Spreadsheets are used to carry out repetitive calculations, thus saving time and avoiding the errors that inevitably occur when calculations are performed by humans. They are commonly used for graphing data and for basic statistical calculations. Spreadsheets allow analysts to look for patterns in large numerical data sets. Although they are designed for processing numerical data, spreadsheets can also be used for nonmathematical purposes, such as creating lists, calendars, or text tables.

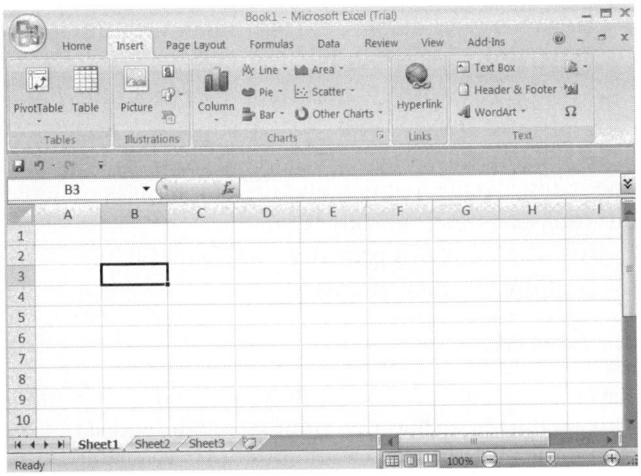

Figure 35.5. A Basic Spreadsheet Screen. Cell B3 is selected, designating column B and row 3. There are three worksheets in the workbook, shown as tabs on the lower left. The Ribbon tab at the top provides access to items that can be inserted into the cell.

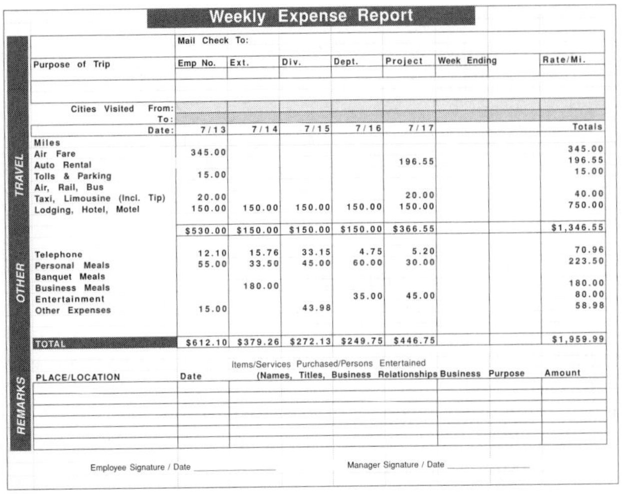

Weekly Expense Report

	Mail Check To:							
Purpose of Trip	Emp No.	Ext.	Div.	Dept.	Project	Week Ending		Rate/Mi.
Cities Visited From:								
To:								
Date:	7/13	7/14	7/15	7/16	7/17			Totals
TRAVEL								
Miles								
Air Fare	345.00							345.00
Auto Rental					196.55			196.55
Tolls & Parking	15.00							15.00
Air, Rail, Bus								
Taxi, Limousine (Incl. Tip)	20.00				20.00			40.00
Lodging, Hotel, Motel	150.00	150.00	150.00	150.00	150.00			750.00
	$530.00	$150.00	$150.00	$150.00	$366.55			$1,346.55
OTHER								
Telephone	12.10	15.76	33.15	4.75	5.20			70.96
Personal Meals	55.00	33.50	45.00	60.00	30.00			223.50
Banquet Meals								
Business Meals		180.00						180.00
Entertainment				35.00	45.00			80.00
Other Expenses	15.00		43.98					58.98
TOTAL	$612.10	$379.26	$272.13	$249.75	$446.75			$1,959.99

REMARKS

PLACE/LOCATION	Date	Items/Services Purchased/Persons Entertained (Names, Titles, Business Relationships	Business Purpose	Amount

Employee Signature / Date ____________ Manager Signature / Date ____________

Figure 35.6. Sample Output from a Spreadsheet Program. A formatted travel expense report. Calculations for subtotals and totals are automated.

As with all software, the results generated by spreadsheet programs will only be valid if the spreadsheet is appropriately used. Box 35.4. provides some guidance for using spreadsheets effectively.

B. Graphing

Graphing numerical data is a standard procedure in nearly every laboratory. Graphs help the viewer to see relationships and patterns among data items. There are specialized computer programs for creating complex graphs and charts, but spreadsheet software is adequate for many situations.

Box 35.4. USING SPREADSHEETS

The general steps for using a data spreadsheet:

- Create the spreadsheet template, with labels and formulas.
- Input the data into the appropriate cells.
- Create the necessary report forms and graphs.
- Modify data and formulas as necessary.

Some general suggestions:

- Check a few random calculations by hand to confirm that the spreadsheet is set up to perform the correct data manipulations.
- Use the data formatting functions of the software to provide validation for data entries.
- Format columns that will contain numbers for the appropriate number of significant figures.
- "#####" as a cell value indicates that the correct number is too large for the cell size (enlarge the cell to see the number).
- Consider the use of pre-existing templates for common tasks.

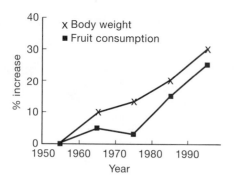

Figure 35.7. Line Graphs Suggest Relationships between Data Sets. Combining these two hypothetical data sets (reflecting real trends in American society) on one graph implies a causal relationship between obesity and fruit intake that is unlikely to be correct.

Most graphing and spreadsheet programs allow the user to rapidly plot the same data in various formats. Creating an effective graph requires selecting a format that best communicates the numerical information. It is thus necessary to understand the purpose of different graphing formats to select the most appropriate one from a menu of options in a graphing or spreadsheet program. The format of a graph can suggest an interpretation of the data, as shown in Figure 35.7, where combining two random data sets on one graph creates the (probably) mistaken impression that the data are causally related.

Line and bar graphs are commonly used to represent scientific data. Line graphs are used to show the relationship between two variables in order to indicate a trend. For example, the line graph in Figure 15.11 on p. 261 illustrates a trend; the production of fruit increases as the level of hormone increases. By manipulating a line graph's characteristics, a scientist can use these graphs to emphasize certain aspects of the data set. Figure 35.8 shows how a single data set can be used to create two different-looking line graphs; one with a log scale y axis and the other with a linear scale, each suggesting a different interpretation of the data.

Bar graphs are frequently used for comparisons; the emphasis is on the specific values shown in the bars, rather than on data trends. For example, bar graphs are effective when directly comparing two or more experimental groups, as shown in Figure 35.9a. Line graphs, however, can also be useful in illustrating these data, as shown in Figure 35.9b.

Histograms are similar in appearance to bar graphs but are used to show the frequency distribution of a variable (refer to Figure 16.9 on p. 281 and Figure 16.10 on p. 282). Bars represent the frequency of data points that fall within a range of values, called an interval, on the x-axis. The width of the intervals, which are usually the same size, must be carefully chosen to accurately reflect the data. If the data intervals are too small, small fluctuations in the data can be confusing. If they are too wide, distribution pattern details will be lost.

In some cases a large set of data points clearly cannot be connected with a line, but still show a general trend.

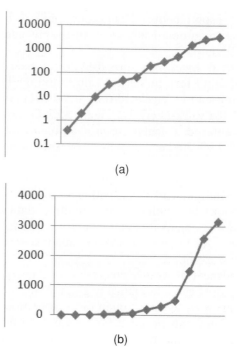

(a)

(b)

Figure 35.8. Line Graphs Show Trends. Line graphs can provide different interpretations based on presentation of the axes. A single data set has been graphed with two different scales on the Y axis (labels have been omitted for this example). **a.** The logarithmic Y axis scale emphasizes a steady exponential rise in the measured variable. **b.** The same data plotted on a linear scale suggests that there was little increase in the measured variable until a threshold point on the X axis. (These data were used to prepare the properly labeled graph in Figure 34.3 on p. 643.)

In these cases, a **scatter plot,** *where all data points are graphed but not grouped or connected,* is appropriate. All data points are plotted and then a statistical analysis is performed to create a **trend line,** *which reflects the general trend in the data values,* if one exists, as shown in Figure 35.10. Data that do not reveal patterns or are not closely related are usually best presented in tables.

Pie charts are used to compare parts, or percents, of a whole. They are inappropriate for many data sets; for example, those shown in Figures 35.8, 35.9, and 35.10. Pie charts are relatively uncommon in scientific literature because they do not clearly show the numerical values of the data. Pie charts work best when there are only a few data categories, and the emphasis is on large differences among the values.

Box 35.5 on p. 676 provides some general guidelines for creating effective graphs.

C. Statistics

Spreadsheet software is usually adequate for simple statistical tests such as determining means and standard deviations. More advanced analyses, however, may require one of the statistical analysis software packages. Statistical programs provide curve-fitting equations, customized calculations, and sophisticated data analysis capabilities. The user needs a solid knowledge of statistics in order to apply these programs effectively.

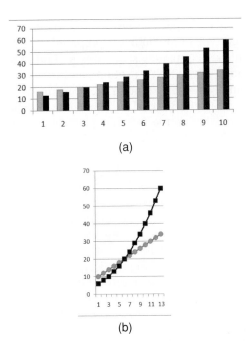

(a)

(b)

Figure 35.9. Bar and Line Graphs Used to Compare the Effectiveness of Two Treatments as the Dose Is Escalated.
a. A bar graph is used to compare the effectiveness of a new treatment (black bars) to the effectiveness of an older treatment (gray bars). The bar graph shows that at lower doses the older treatment is more effective, but as the dose is increased the newer treatment is more effective. **b.** The same data sets are replotted using superimposed line graphs. It is apparent in the line graphs that there is a trend: effectiveness increases as dose increases. For the older treatment (gray line) there is a linear increase in effectiveness as the dose is increased. The trend for the newer treatment is that effectiveness steeply rises as dose is increased.

SAS (Statistical Analysis System) *is a software package designed for statistical analysis.* While there is a menu-driven system for performing some analyses, SAS also operates as a programming package in which the user types text commands and statements that instruct the program how to run the analysis, as shown in Figure 35.11. This provides the user with rigorous documentation about the procedures and data set. Biomedical and pharmaceutical companies use SAS and similar programs to analyze, validate, and document product testing to meet ISO and government compliance standards.

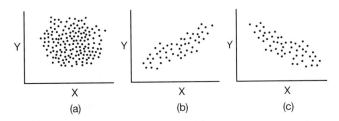

Figure 35.10. Scatter Plots Show the General Distribution of Data Points That Are Not Adequately Defined by a Line. Scatter plots are used for large numbers of data points. **a.** No relationship between *x* and *y* values. **b.** Positive relationship; *y* increases as *x* increases. **c.** Negative relationship; *y* decreases as *x* increases.

```
DATA BUGS;
INPUT POP $ SEX $ WT1 WT2;
CARDS;
RIVER F 5.6 6.3
RIVER M 6.2 7.4
RIVER F 5.1 6.0
RIVER F 4.3 5.2
FIELD M 2.1 4.7
FIELD M 3.0 5.3
FIELD F 2.6 3.1
FIELD F 3.5 5.2
PROC SORT;
BY POP;
RUN;
PROC MEANS;
BY POP;
RUN;
```

Figure 35.11. Example of a Simple SAS Program. This set of instructions tells SAS to create a data set called "Bugs." It then identifies the variables in the data, provides eight sets of data, and instructs the program how to sort and analyze the data. Program courtesy of Dr. Diane Byers.

IV. BASIC SOFTWARE APPLICATIONS: DATA STORAGE AND RETRIEVAL

A. Databases

i. INTRODUCTION

There are many advantages to using a database to organize information:

- Ease of access
- Sorting and manipulation capabilities
- Consistency of format
- Prevention of data duplication
- Data security
- Creation of complex reports and other documents

A **database** *is an organized collection of data that is accessed through database management software.* **Database management (DBM) systems** *allow the user to search for, sort, look for patterns, and report selected data within a database.* The user retrieves information using a **query;** *a text-based set of criteria used to search for and extract a desired subset of data from a database and then present it in a specific format.* A query can be used to specify a database **filter,** *which shows (or hides) records of data with designated characteristics.*

Consider the example of a dictionary. A traditional paper dictionary catalogs words and definitions in alphabetical order. An electronic database dictionary provides a self-organizing and dynamic filing system for information. The database can be queried to quickly identify words with Greek roots, words with more than three syllables, or definitions related to medicine. It would be difficult to extract such information quickly from a paper dictionary. Similarly, given an electronic database containing batch records for cell culture media, a laboratory manager might query to determine which batches were

---

Box 35.5. CREATING INFORMATIVE SCIENTIFIC GRAPHS

All Graphs

- Be sure each part of the graph is easily identified.
- Label the graph clearly with a title, labels, and units on each axis.
- Use appropriate, evenly spaced units on the axes.
- Do not combine an excessive number of colors.
- Do not use graphic tricks, such as unusual geometric shapes or three-dimensional data placement, unless needed to explain the data.
- Limit the number of lines, bars, or segments in a single graph.

Line Graphs and Scatter Plots

- Use line graphs to plot related data points.
- Plot the dependent variable (what was measured) on the Y axis.
- Use large, easily distinguished symbols for data points.
- Do not extrapolate beyond the measured data points.
- Use scatter plots for large numbers of data points that show a general trend.
- Do not connect scattered data points with lines.

Bar Graphs

- Use bar graphs to emphasize individual data points.
- Use bar graphs to directly compare sets of matched data points.
- Limit the number of bars on a single graph.
- Make bars wide enough to be visible.
- Maintain high contrast between adjacent bars.

Histograms

- Use histograms to show frequency distributions.
- Do not leave gaps between bars.
- Choose bar intervals that accurately represent frequency values.

Pie Charts

- Use pie charts only to compare parts, or percents, of a whole.
- Use pie charts when exact numerical values are less important than relative values.

---

formulated by a specific person or the dates on which a particular type of medium was prepared.

Databases containing sequences and other information about DNA, RNA, proteins, and other molecules have rapidly become some of the most important tools of the biotechnologist. Many molecular databases are freely available on the Internet, providing the basis for the rapidly growing field of **bioinformatics,** *which uses computers to analyze molecular data and address biological questions involving large amounts of information.* These huge molecular databases require specialized DBM software, described in more detail below, to search and analyze their contents.

Although DBM programs vary tremendously in complexity, all start with the same database structural elements, as shown in Figure 35.12. A database **table** *contains a set of related information.* Each table is divided into rows of **records,** *which are the basic organizational units of the information.* Each record in turn contains multiple columns of information **fields,** *which provide the individual data items, which can be text, numbers, or other values.* The simplest type of database is a **flat file database,** *which consists of a single data table stored in one computer file.* A familiar example would be a stand-alone computerized address book. Each individual entry is a record, which includes several data fields: names, addresses, and phone numbers.

Many times laboratory workers only need to query and update pre-existing databases, as when accessing a DNA sequence stored on a website or when adding information about a project to an existing database, as shown in Figure 35.13. Most DBM software allows users to develop customized data entry forms (Figure 35.13b), which can speed the process of data input as well as reduce input errors.

Sometimes, however, users need to create their own database. Database creation requires careful consideration of what information will be needed in the future, how that information will be queried, what data will be available for input, and many other considerations. This process is outside the scope of this text; consult the references in the unit bibliography for further information.

ii. RELATIONAL DATABASES

It is relatively easy to design and use a flat file database. All updating and data manipulation take place in one table, so the information relationships are clear. Querying and sorting are straightforward and there are many standard templates that can be modified for specific needs.

Laboratories often deal with complex, interrelated data sets. For example, a laboratory manager might need to record the purchase information for each component used in the preparation of multiple batches of cell culture medium. Entering all this information into one data table would be unwieldy and create redundancy, since multiple components might be purchased from the same vendor.

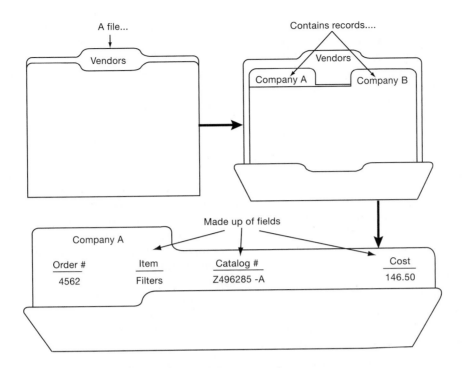

Figure 35.12. Organization of Data in a Flat File Database. The basic units of database organization are tables, records, and fields. In the case of flat file databases, a file represents one table. (Adapted from Szymanski, R.A., Szymanski, D.P., and Pulschen, D.M. *Introduction to Computers & Software*, Saddle River, NJ: Prentice Hall, 1996, 172.)

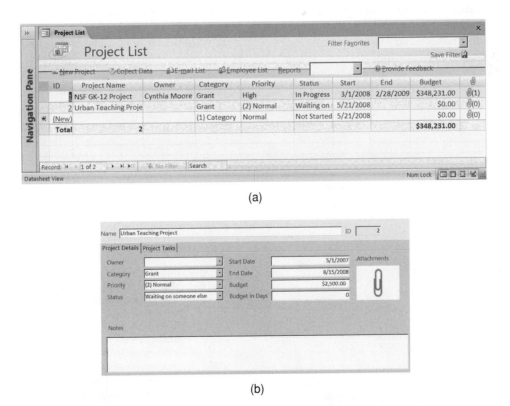

(a)

(b)

Figure 35.13. Entering New Data into an Existing Database. Data can be entered into a database using: **a.** the basic table grid, or **b.** a form designed to maximize ease and accuracy of data input.

In this case, the manager might create a **relational database,** *where multiple data tables are linked by specific relationships, allowing them to work together.* Relational databases are useful when the same information, such as names and addresses, is needed in multiple databases. This information need only be entered or modified in one data table, and it becomes available to all related data tables. Relational DBM software can be more complicated to use than flat file programs, although most common programs provide guidance for basic tasks such as designing forms and reports.

The power of a well-designed relational database to organize and coordinate information is suggested in Figure 35.14, which illustrates the interrelationships of tables within a projects database.

iii. DATABASES VERSUS SPREADSHEETS

It is not unusual to collect a data set and then be uncertain of the optimal computer application for storing, manipulating, and reporting the data. A database table looks very similar to a spreadsheet and one of the most likely sources of confusion is choosing between database and spreadsheet software for analyzing small- to moderate-size data sets. Because of the expanding capabilities built into each generation of software, there are situations where either can be used successfully and other situations where one type of software is much more suitable. Since large-scale data transfer procedures are potential sources of errors as well as extra work, it is better to choose the proper program from the start.

While both databases and spreadsheets have strengths, spreadsheets are frequently used when a database would be superior, simply because spreadsheets are faster and easier to design. However, databases are better suited for long-term data storage.

They provide better data security and validation, and require less RAM than spreadsheets per data unit, since only the required fields and tables are loaded into memory. If a data set is small, a spreadsheet is more convenient than a database, but it is harder to coordinate multiple worksheets than the multiple tables in a well-designed relational database. Databases are superior for sorting, filtering, and storing data; spreadsheets work best when multiple mathematical calculations are required. For example, if you are collecting information about vendors, catalog numbers, dates of purchase, and prices for lab supplies, entering the information into a database will allow you to easily determine which items are purchased from a single vendor and when you last purchased a particular item. Using a spreadsheet would make it easier to use the price information for budgeting purposes. Box 35.6 summarizes issues to consider when choosing between a database and a spreadsheet to record and manipulate data sets.

B. Data Mining

Data mining, *the process of discovering patterns within large quantities of data,* has become an essential tool for biological research. It is a key technique in the field of bioinformatics, which focuses on computer-driven sorting and analysis of molecular data. Modern DNA sequencing techniques have generated vast databases of genomic information, usually available on the Internet. These databases have in turn contributed to **data warehouses,** *which are long-term storage resources for historical database contents.* Data mining allows researchers to sift through these data repositories and find meaningful subsets of information. This process is also referred to as **knowledge discovery in data (KDD).**

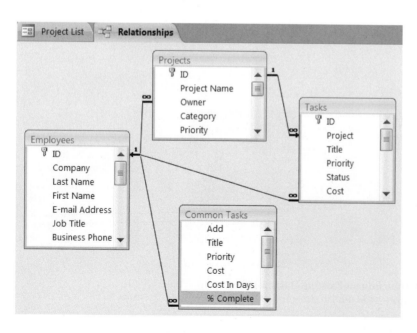

Figure 35.14. Table Relationships within a Relational Database. Connecting lines indicate how four tables are related in a project database. Each project has a unique ID number, designated by the key symbol, used to define data relationships. In this database, for example, each project will have multiple personnel, whose contact information is recorded in the Employees table. Each project also contains multiple tasks, recorded in the Tasks table. Tasks are then assigned to employees. Data for every aspect of the project, including assignments, budgets, and schedules, are thus coordinated.

Box 35.6. DATABASES VERSUS SPREADSHEETS

Both types of software will:

- store and organize information
- allow editing and updating of data sets
- provide basic data validation
- perform simple data sorting (alphabetical or numerical)
- perform simple calculations
- create simple charts

Spreadsheet software excels at:

- performing complex calculations
- revising numerical data and automatically updating calculations
- making graphs and charts with sophisticated formatting
- manipulating data to create "what-if" scenarios (e.g., if gas prices double in the next six months, how will that affect our lab travel budget?)

Database management software excels at:

- handling very large data sets
- customizing data input forms to avoid data entry errors
- performing complex data sorting and filtering based on multiple fields
- discovering relationships among multiple data sets
- allowing complex data searches
- providing stringent data validation
- eliminating redundancy
- installing many levels of data security
- creating specialized report formats

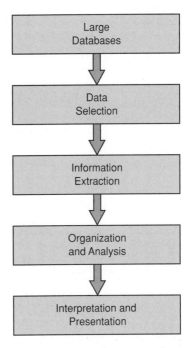

Figure 35.15. The Data Mining Process Requires Several Steps.

a subdivision of NIH that acts as a public resource for molecular data and other biomedical information.

Biomolecular data mining concentrates on many types of knowledge discovery relating to genes, proteins, and other biological molecules. This type of data mining is used for such purposes as determining the function of newly reported gene sequences, and discovering which genes are similar across species and which sets of genes are expressed together in various organisms.

There are many nucleic acid and protein data resources available to scientists, a few of which are shown in Table 35.2 on p. 680. One well-known example is GenBank, which contains virtually all reported DNA sequences. GenBank is maintained by NCBI, along with dozens of additional molecular databases, and is coordinated with several international databases, allowing it to provide a comprehensive data resource. Figure 35.16 on p. 680 shows the number of sequence entries into this database since 1982. GenBank's success in collecting sequences (approximately 84 million in May 2008) lies in the decision by major molecular research journals to require that all published DNA sequences be submitted to a major sequence database. Each sequence is assigned an **accession number,** *a unique identifier for the sequence*, designed to facilitate database searches with that sequence.

The size and complexity of the data repositories involved is what separates data mining from a simple search. Ordinary DBM software is not suited for simultaneously sorting through vast amounts of information in multiple databases on noncentralized computers. Specialized data mining programs analyze and identify data patterns based on open-ended queries from users,

The overall process of data mining is summarized in Figure 35.15. Data mining is not a new idea, but advances in computer technology have greatly enhanced the ability of researchers to separate small subsets of relevant data from massive data sets. For example, current models of global climate change are based on data mining of historical weather records. Many businesses develop marketing strategies based on data mining of consumer practices.

Any large repository of information can be data mined. One major application is **text mining,** *searching databases for topics of interest using keywords*. A familiar form of text mining is using a computer to look for references when writing a paper. PubMed Central, a free online archive of medical and biological journal contents, is an exceptional resource for researching biomedical topics using queries. PubMed Central is maintained by the **National Center for Biotechnology Information (NCBI),**

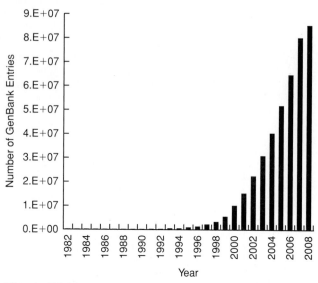

Figure 35.16. **DNA Sequence Entries in GenBank.** The number of entries is shown on a log scale. Notice the relative number of sequences in the database when the first edition of this textbook was published in 2000, compared to 2008.

frequently written in **SQL (structured query language)**, *a computer language specifically designed to find and retrieve data from databases*. Most popular search tools (including search engines such as Google as well as the data mining programs discussed below) offer user-friendly query forms that automatically translate user requests into SQL. However, many serious researchers prefer to write their own SQL queries.

Commercial data mining programs are available, but a variety of free biological data mining software is also available, usually associated with database websites. NCBI offers two well-known mining tools, Entrez and BLAST. **Entrez** *is a data retrieval tool that searches across several dozen molecular databases with data encompassing DNA, RNA, and protein sequences; genomes; taxonomy; diseases; books; and journal articles*.

BLAST (Basic Local Alignment Search Tool) *is a pattern recognition tool used to search for similarities between nucleotide sequences or protein sequences*. This program is one of the foundations of bioinformatics and is routinely used by those who work with nucleic acid or protein sequences. It is easy to do a simple BLAST search that compares a sequence of interest to a wide range of sequence databases. The user either adds a new DNA sequence or provides the accession number for a known sequence, and specifies which data sets to search. BLAST compares the entered nucleotide or protein sequences to database sequences and finds matches or partial matches. BLAST then displays which base pairs or amino acids match in the two sequences.

The results of a simple BLAST search are shown in Figure 35.17. This type of search can be used to help determine the function of an unknown nucleotide or amino acid sequence by finding similar sequences of known function. It can be used to suggest molecular relationships among groups of genes or proteins, or evolutionary relationships among organisms.

V. BASIC SOFTWARE APPLICATIONS: EQUIPMENT AND LABORATORY MANAGEMENT

A. Laboratory Instruments

Laboratory informatics *is the application of computers to collect, analyze, and manage laboratory data and information*. A common example is the association of computers with laboratory instruments, such as chromatography systems, spectrophotometers, DNA sequencers, gel scanners, and microscopes.

Associating a computer with an instrument has many advantages. It eliminates the need for human analysts to take the instrument's data output and manually transfer them to a computer. Computers associated with instruments can record and store data relating to equipment settings and operation and provide automatic time-stamps for documentation purposes. Computers can perform automatic performance checks of the instrument. They can process instrument data in

Table 35.2. **MOLECULAR DATABASES ON THE INTERNET**

Database	Contents	Location
GenBank	DNA sequences	http://www.ncbi.nlm.nih.gov/Genbank
Entrez Gene	Genes	http://www.ncbi.nlm.nih.gov
Ensembl	Genomes (mostly vertebrate), including the human genome	http://www.ensembl.org
UniProt Knowledge Base	Protein sequences; part of ExPASy (Expert Protein Analysis System)	http://www.expasy.ch/
PubMed and PubMed Central	Bibliographic databases; PubMed contains citations and abstracts; PubMed Central has full text articles	http://www.ncbi.nlm.nih.gov
OMIM	Online Mendelian Inheritance in Man; text descriptions and references for human genes and genetic disorders	http://www.ncbi.nlm.nih.gov

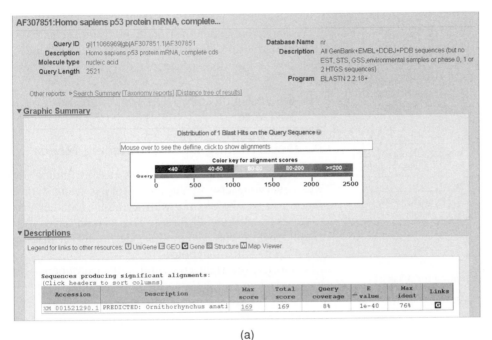

AF307851:Homo sapiens p53 protein mRNA, complete...

Query ID	gi|11066969|gb|AF307851.1|AF307851
Description	Homo sapiens p53 protein mRNA, complete cds
Molecule type	nucleic acid
Query Length	2521

Database Name	nr
Description	All GenBank+EMBL+DDBJ+PDB sequences (but no EST, STS, GSS,environmental samples or phase 0, 1 or 2 HTGS sequences)
Program	BLASTN 2.2.18+

Other reports: ▶Search Summary [Taxonomy reports] [Distance tree of results]

▼ **Graphic Summary**

Distribution of 1 Blast Hits on the Query Sequence ⓦ

Mouse over to see the defline, click to show alignments

Color key for alignment scores

<40	40-50	50-80	80-200	>=200

Query   0   500   1000   1500   2000   2500

▼ **Descriptions**

Legend for links to other resources: Ⓤ UniGene Ⓔ GEO Ⓖ Gene Ⓢ Structure Ⓜ Map Viewer

Sequences producing significant alignments:
(Click headers to sort columns)

Accession	Description	Max score	Total score	Query coverage	E value	Max ident	Links
XM 001521290.1	PREDICTED: Ornithorhynchus anati	169	169	8%	1e-40	76%	Ⓖ

(a)

>❑ref\|XM 001521290.1\| Ⓖ PREDICTED: Ornithorhynchus anatinus similar to p53 protein (LOC100092821),
mRNA
Length=588

 GENE ID: 100092821 LOC100092821 | similar to p53 protein
[Ornithorhynchus anatinus]

 Score = 169 bits (186), Expect = 1e-40
 Identities = 173/226 (76%), Gaps = 0/226 (0%)
 Strand=Plus/Plus

```
Query  509  CGTACTCCCCTGCCCTCAACAAGATGTTTTGCCAACTGGCCAAGACCTGCCCTGTGCAGC  568
            |||||||||| |||||||||| |||||||| ||||| ||||||||||||| ||  ||||
Sbjct  179  CGTACTCCCCATTGCTCAACAAGCTGTTCTGCCAGTTGGCCCGGACCTGCCCCGTTCCAGC  238

Query  569  TGTGGGTTGATTCCACACCCCCGCCCGGCACCCGCGTCCGCGCCATGGCCATCTACAAGC  628
            |||||||| ||  ||| | ||||||||  |||| |||| ||||||| |||||||||||||
Sbjct  239  TGTGGGTCGACTCCCCGCCCCCGCCGGGGGCCCGGGTCGGGGCCATGGCCGTCTACAAGA  298

Query  629  AGTCACAGCACATGACGGAGGTTGTGAGGCGCTGCCCCCACCATGAGCGCTGCTCAGATA  688
            || | |||| ||||| |||| |||| | ||||||||||||||||||| |||||| || |
Sbjct  299  AGACCGACCACAGGGCCGAGGTGGTGAAGAGGTGCCCCCACCACGAGCGCTCTTCCGACG  358

Query  689  GCGATGGTCTGGCCCCTCCTCAGCATCTTATCCGAGTGGAAGGAAA  734
            | ||| ||    ||| || |||||||| |||| ||||| |||||||
Sbjct  359  GTGACGGAGCGGCGCCCGCCCAGCATCTGATCCGCGTGGAGGGGAA  404
```

(b)

Figure 35.17.   BLAST Search Results. These are the results from a BLAST sequence search starting with the human p53 tumor suppressor gene coding sequence and querying if any similar sequences have been reported in the platypus (*Ornithorhynchus anatinus*). **a.** BLAST has found one platypus sequence similar to a small portion of the p53 coding sequence. The table provides the accession number for the discovered sequence, which represents only 8% of the p53 sequence, and shows a base pair match of 76%. The calculated E value of 1×10^{-40} indicates that this match is extremely unlikely to have occurred by chance, which means the two sequences are probably related. **b.** This portion of the results shows the exact alignment of the two sequences.

valuable ways; for example, computers associated with spectrophotometers can generate standard curve graphs, calculate the concentration of an analyte based on a standard curve, estimate the purity of a DNA sample based on A_{260}/A_{280} readings, and perform other common calculation and display functions. Computers associated with microscopes assist the user in obtaining high-quality graphical images. Because of their many advantages, most modern laboratory equipment is either connected to a computer or controlled by internal microprocessors.

B. Laboratory Information Management Systems (LIMS)

An increasing number of companies are adopting comprehensive **LIMS (Laboratory Information Management Systems),** *computerized systems that provide integrated information and resource management for a variety of laboratory activities.* LIMS are especially

valuable in settings where documentation of individual samples and products is essential. In addition to interfacing with laboratory instruments, as described above, LIMS can also track inventories, generate project and auditing reports, provide chain of custody reports for samples, and perform additional activities, as shown in Figure 35.18 on p. 682.

LIMS provide many advantages over manual data collecting and reporting. The software can assist in securing data against loss or unauthorized change. It facilitates sharing of information among company divisions or with collaborators in distant locations. When interfaced with lab equipment, LIMS can be programmed to notify personnel of noncompliance events (e.g., undesired temperature or pH changes in a fermenter). LIMS can be programmed to automatically check data input for mistakes. For example, LIMS can prompt users to input dates so they do not forget to record the day, month, and year, or it can flag temperature values that are outside a specified range. They

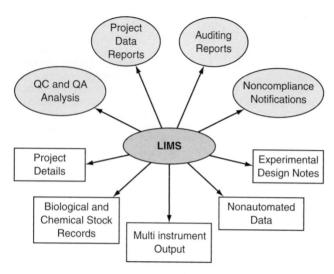

Figure 35.18. LIMS Can Be Programmed for a Variety of Input and Output Functions.

incorporate multiple databases that facilitate searches for information about samples, batches, processes, personnel, analyses, equipment, and materials.

LIMS have the disadvantage that they are often complex and custom designed, making them expensive and time consuming to introduce into a facility. Also, as previously discussed in Chapter 6, these extensive computer systems require major validation efforts to ensure that they are trustworthy and secure. For these reasons, it is more common to see LIMS in a larger company laboratory than in a small research laboratory.

C. Electronic Laboratory Notebooks

Electronic laboratory notebooks (ELNs) are slowly being adopted by researchers. ELNs use software that allows a computer to take the place of conventional paper laboratory notebooks. This software is much more sophisticated than a simple word processor, because there are built-in validation procedures that help ensure reliable documentation. They have many additional advantages, including:

- facilitating the recording of repeated lab procedures
- prompting the user to record information in a timely manner
- providing automatic time stamps
- recognizing common errors in data entry
- allowing automatic data recording from instruments
- allowing individuals to share data easily
- automatically performing routine calculations
- automatically transferring data to other programs, such as searchable databases or graphing programs
- providing password protection for security
- ensuring legibility of entries
- allowing easy back-up with read-only copies

As the trustworthiness of electronic documentation is increasingly being demonstrated, the trend seems to be toward increasing use of ELNs in technical companies, with slower adoption in academic research laboratories.

VI. GENERAL CONSIDERATIONS FOR CHOOSING SOFTWARE

A. Determining Data Processing Needs

Sometimes choosing software is as easy as finding out what everyone else in the laboratory is using. Other times you will be dealing with a unique data set that requires something different. The remainder of this chapter will discuss general considerations for making appropriate software choices.

Choosing the right computer software depends on understanding and defining the data-handling problem you want the software to address. You will choose different types of programs to deal with numerical data subject to multiple calculations than you will to organize and sort text entries. Some of the questions you need to ask are:

- Do I need to manipulate text?
- How important is the format of the text output?
- Do I need to perform numerical calculations and, if so, how complex are these calculations?
- Do I need to store large amounts of data?
- How will I need to sort and retrieve these data?
- Do I need to prepare formal presentations and reports from my data?
- Do I need to create or manipulate artwork, graphs, or photographs?
- Do I need to organize, budget, or manage laboratory projects?

Most of the time you will be able to use standard commercial programs, but major projects involving large amounts of complex or unusual data may need customized software. Starting out with the best software is much easier than transferring data to a new system, although this can happen when a project outgrows its original boundaries.

B. Choosing Software

Once you have determined the type of application software you need (e.g., spreadsheet or database), you then need to choose a specific program. The major factors to consider are overall program reliability, ease of data input, and quality of program output.

There are many potential software sources. These include:

- Full-feature commercial software
- Basic commercial software
- Shareware and freeware
- Web-based software

Most laboratories require complex data-manipulation capabilities that are only found in commercial software. These are programs that are copyrighted and licensed to users for (sometimes) substantial fees. When you buy software, you are really only purchasing a license to use the software on a single computer, although most recent licenses allow installation on both a desktop and laptop computer, with the understanding that they will not be used simultaneously. If you need to install the program on multiple computers, you are legally required to purchase a **site license,** *which grants you the right to a specific number of program copies.* While some software companies use **copy protection** devices within their software or files, preventing you from making more than one or two copies, copy protection also prevents you from making multiple backup copies and can sometimes interfere with the use of legally installed software. Because of consumer complaints, most companies do not copy protect their software, but this does not mean that you are allowed to make extra copies for distribution.

Full-feature commercial software usually comes with a wide variety of basic and advanced features, as well as a premium price. For simple applications, you may not need many of the extra capabilities that use up RAM and slow the program; in this case it is worthwhile to see if there is a more basic version of the same program. If selecting a lesser-known software package, consider the programs that are in use by others who might need copies of the data, or might provide you with a computerized data set. All programs have some type of import/export capabilities to allow data sharing with other programs. For example, it would be unusual to find a spreadsheet program that could not accept data from market leader MS Excel, or allow you to move data to an Excel spreadsheet; this capability might not extend to data from less-common programs.

For simple applications and also for certain specialized applications, you might want to consider shareware or freeware. **Shareware** *is copyrighted software developed by individuals or small companies that is easily available on the Internet for free download and trial use.* Users who keep and use the program are expected to pay a relatively small specified fee to the software author, which frequently entitles the registered user to upgrades and access to additional program features. **Freeware** *is software that is offered to the public free for personal use.* **Public domain programs** *are a subcategory of freeware; noncopyrighted software that the author has chosen to provide for free to the public for any type of use or alteration.* The rapidly evolving field of bioinformatics has nurtured many creative programmers who often share their ideas via shareware and freeware.

These programs are available on the Internet, through networks and computer groups, or from individuals. As with any software, it is essential to know your sources. The best and safest way to find worthwhile programs on the Internet is to read reviews and download the software at reputable websites[1] that try out the products and scan them for viruses and other security threats (discussed in Chapter 36). Another excellent source is colleagues who have developed specialized programs for laboratory use; these programs are frequently described in the bioinformatics literature.

Web-based software *refers to programs that are located on the Internet, allowing users to work with and store data online.* BLAST is an example; while it is possible to **download** the program (*copy it from the Internet to your computer*), running BLAST locally requires high-end computers and is not practical for most laboratories. There are also web-based programs for basic functions; for example, Google Docs currently provides free word processing, spreadsheet, and presentation capabilities and has become fairly popular in the business community. The major advantage of these programs is the ease of sharing documents among multiple authorized individuals, who can work with the document simultaneously if needed. Web-based documents are easy to download and **upload** (*copy from your computer to the Internet*). Many web-based software sites provide a limited amount of secure online storage space and charge a small fee for additional storage.

Once you have identified potential programs, convenience is the next consideration. All software is not equally easy to use. The program interface that one user finds quick and intuitive can be a constant source of frustration for another; individual preferences are a significant factor in software choices. For this reason, it is an excellent policy to try out a program yourself, as well as to search out software reviews in computer magazines and talk to people who have used the program.

Finally, it is essential to examine the types of output provided by a program and determine whether these will be adequate for your needs. If you need specific types of reports, for example, can the program provide them? If not, can you export the data to another program that will? The second option may be acceptable if the software you purchase can organize the data properly. High-end programs usually have the widest assortment of presentation and report options.

New versions of software illustrate another potential concern: **backwards compatibility** *is the ability of a new software version to work with files created in previous program versions.* Sometimes, older versions of software cannot read files created by newer versions, but newer versions can usually save files in earlier formats

---

[1]Reliable sites for general use software include http://www.download.com, http://www.tucows.com, and http://www.zdnet.com.

to avoid this problem. Check on the ease of data transfer to new software versions, or different programs, before purchasing new software. Databases can be especially difficult to transfer because of their unique designs.

When you install new software, you will be asked to accept a license agreement before you can actually use the software. While most of us prefer to simply check the agreement box, it is a good idea to read through the license. In some cases, you may be agreeing to the installation of additional unspecified software, some of which may act as **spyware,** *software that sends information about your computer activities to outside parties* (discussed in Chapter 36). Similarly, during the installation process, look for check boxes that indicate your agreement to install additional items (frequently toolbars, music and video players or services, and web browsers).

The default value is almost always agreement, but reputable software companies will allow you to opt out of the extra downloads by unchecking the box.

Making the best software selections is key to optimizing computer productivity. Box 35.7 offers additional tips for choosing software.

C. Installing and Using Software

Most software must be formally installed, not simply copied to the hard disk. This requires the installation program included with the software, which places program files at various locations on the hard disk so that the application and the system software can find them. If you need to remove a program from your computer, some software packages include an **uninstall program,** *which finds the scattered files and eliminates them.* If

---

Box 35.7. TIPS FOR CHOOSING AN APPLICATION PROGRAM

Memory Requirements

a. How much RAM does the program require?

b. How much RAM does the company recommend?

Remember that most local computer crashes are the result of RAM problems. Can your computer easily provide the required RAM, with memory to spare for other operations? Many companies advertise the minimum system requirements for their software, along with a recommended configuration, as shown in Table 34.2 on p. 656. It is generally advisable to have the recommended configuration. During the trial period, try the program with multiple other operations running. If it seems sluggish in the beginning, it will seem even worse when you begin to put more demands on the software.

Ease of use

a. Does the software come with a manual, either printed, in a document file, or online, and if so, can you use it effectively?

b. Does the program include tutorials?

c. Does the program provide extensive on-screen help and can you access that help without an Internet connection?

d. How long will it take you to become comfortable with the program?

Many programs no longer come with a hard copy manual, which is not necessarily a problem. An electronic manual should be provided in the software or be easily available online. Introductory tutorials and help screens may provide all the assistance you need while using the program. A well-designed program should provide an intuitive interface, which makes it easy to operate.

Compatibility with other programs

a. Can the program read files from other applications?

b. Can the program save files in a variety of formats?

c. Can the program easily import and export data?

The import/export capabilities of a program can be a significant consideration. If you have already developed spreadsheets using different software, can these data be easily transferred to your new program? What if you want to send your database file to someone who uses a different program? Sometimes you are better off choosing a less-than-optimal application program if that program is standard among your colleagues and compatibility is a key issue.

this option is not available, use the *Remove Programs* feature in the operating system. Never change the location of installed program files, although you can place program access shortcuts anywhere convenient.

In many cases, the easiest way to obtain software is to download it from the Internet. In this case, you will usually receive one or more compressed files packaged with an executable program (noted by the .exe suffix in the name). You will have a choice of running this program to install the software, or simply saving the file for later installation. It is a good practice to save downloaded software to disk before installation and to use your anti-virus software to scan it carefully. If software is purchased on disc, keep your original copies in a location away from your main computer.

Most commercial software provides an **activation key,** *a user-specific code that permits the program to function.* This key is almost always required for technical support. Record the key for every program as you install it and keep a copy of the list separate from the computer.

D. Summary

As a laboratory professional, you are likely to routinely work with all the categories of software described in this chapter: communication, calculation, data sorting and mining, and laboratory management. You will sometimes input data into a computer manually and you will almost certainly work with instruments that do so automatically. You will interact with software to select the desired types of data processing. You will also interact with programs to properly format their output.

Nearly all modern software is designed to be used by individuals with minimal training and with minimal computer background. This is the purpose of menus, help screens, prompts, and other cues. However, as with all laboratory equipment, acquiring more understanding permits you to use the advanced tools provided by computers effectively. With increased knowledge you can better select the proper software tool for a particular application and then better apply that tool to obtain trustworthy and useful information. Properly used, computers greatly contribute to the output of a laboratory.

PRACTICE PROBLEMS

1. Which of the following data would best be entered into a database and which should be placed into a spreadsheet?
   a. budget projections for the next fiscal year
   b. vendor addresses
   c. batch records
   d. records of the use of medium batches
   e. time sheets
   f. laboratory maintenance assignments

2. What type of graph would most clearly present the following types of data?
   a. percent of your work time spent among four different projects
   b. number of mice developing tumors over time after treatment with a carcinogenic chemical
   c. comparison of the growth rate of cells in two types of medium
   d. decrease in the effectiveness of a continuous filtration system with increasing use

3. What are the differences between bitmapped and vector graphics?

QUESTIONS FOR DISCUSSION

1. Think about your current laboratory work. What are two specific applications of the following software that could (or do) enhance your productivity?
   a. Spreadsheet
   b. Database
   c. Presentation software

2. Design a simple database that would contain the information required to create an inventory of laboratory equipment. Specify the characteristics of the fields where data will be recorded.

3. Why is it inappropriate to write your lab notebook in a word processing program?

CHAPTER 36

Internet Resources for the Laboratory

I. INTRODUCTION TO THE INTERNET

A. What is the Internet?

The computer has become essential for scientific and other types of communication largely through the technology of the Internet. The **Internet** *consists of hundreds of thousands of loosely connected computer networks that use a common set of guidelines called protocols to send and receive information.* There is no official group that controls and organizes this system. All specifications needed to create an electronic site on the Internet are publicly available and therefore anyone with a minimum of training can create a pres-

ence on the Internet. There are pitfalls to this structure, which will be discussed shortly.

The Internet provides a massive communication medium, both personally and professionally. Its information resources allow contacts and discussion with scientists around the world, as well as data exchange and other collaborative processes. As shown in Table 36.1, the Internet can be applied to many professional situations.

B. Information Transfer on the Internet

Computers cannot communicate over a network unless they share communication **protocols**, *the common languages used by computers for information exchange.* We

Table 36.1. *INTERNET APPLICATIONS*
FOR SCIENTISTS

The Internet has many professional applications. These include:

1. *Exchanging information with other laboratories.*
2. *Performing online research using databases.*
3. *Obtaining bibliographic information.*
4. *Reading online journals.*
5. *Checking vendor catalogs and ordering.*
6. *Joining newsgroups and discussions of research topics.*
7. *Obtaining information about companies and schools.*
8. *Finding contact information for scientists.*
9. *Performing job hunts and hiring.*

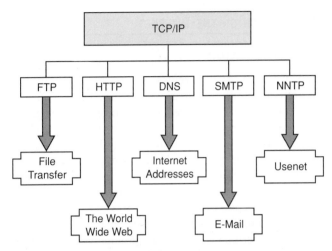

Figure 36.1.   **Internet Services Are Accessed Using TCP/IP Application Protocols.**

do not need to understand how these protocols work in order to use the Internet. However, a basic knowledge of the processes involved helps when a problem arises.

By definition, the Internet operates with **IP (Internet Protocol),** *the standard protocol for packaging, addressing, and sending data through the Internet.* Each computer directly connected to the Internet has a unique IP address, consisting of four sets of numbers separated by periods; for example, 123.231.12.1. These numbers are used within the IP system to address and send information to specific computers. This information is divided into addressed packets, which are sent individually. One drawback to IP software is that it can only send small amounts of data in each packet. Since most data transfers are larger than one packet, Internet users also need a protocol called **TCP (Transmission Control Protocol),** *which works with IP software to make sure that data packets arrive and are reassembled in the correct order, and that the data are intact.* TCP acts as a system checker, without which data transfer on the Internet would be unreliable. All computers come equipped with **TCP/IP software,** *which provides the ability to send and receive information over the Internet.*

Application protocols associated with TCP/IP provide the specific means of Internet data exchange and communication. The applications you are most likely to use are shown in Figure 36.1. The protocol most familiar to Internet users is **HTTP (hypertext transfer protocol),** *which is the protocol used to connect computers on the World Wide Web.*

Many people refer to the Internet and the World Wide Web (or simply the Web) interchangeably, but they are not the same. The Internet is a network of networks, using a set of common protocols to communicate between various Internet services. The **World Wide Web** *is a user-friendly service provided through the Internet using HTTP.* While the Web represents the facet of the Internet seen by most users, there are many additional

services. E-mail is familiar to virtually all computer users. **FTP (file transfer protocol)** *is a service that provides an efficient means to transfer data files between computers.* Like ftp sites, the **Usenet** service, *which provides more than 25,000 public discussion groups or newsgroups, each generally focused on a single topic of interest,* is older than the Internet itself. While other social discussion forums have become more popular, many science newsgroups are still active.

Given the diffuse nature of the Web, navigation requires that each web page be identified by location. **URLs (uniform resource locators)** *provide a unique address for every web page.* As discussed above, each computer connected to the Internet has a numerical IP address. Since these are difficult for users to remember, the **DNS (Domain Name System),** another component of TCP/IP, *allows the assignment of text names to web pages.* These names can then be converted back to IP addresses within the computer system. The composition of a URL is shown in Figure 36.2.

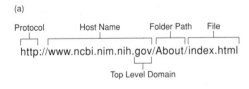

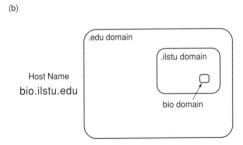

Figure 36.2.   Composition of a URL. a. URL structure.
b. Hierarchy of domain names.

II. RESEARCH ON THE INTERNET

A. Introduction

One of the Internet's greatest assets is the astounding amount of online information that may not be readily available elsewhere. However, this brings its own problems. As these resources increase, it becomes more challenging to narrow down resources to find what you need. This is also a situation where the democratic nature of the Internet can be a problem. Anyone can post on any topic they choose, given the ease of creating and publishing web pages and the lack of any official moderation of Internet content.[1] For example, you will find hundreds of websites describing the dangers of genetically modified foods. Some have accurate and thoughtful information and discussion, some are impassioned but inaccurate, and some are deliberately misleading. The Web is full of parody sites that grossly exaggerate topics for humor; unfortunately this type of site is rarely labeled as a parody (at least directly). Figure 36.3 shows the home page for a satirical campaign to ban dihydrogen monoxide (more commonly known as water).

The remainder of this chapter section discusses basic strategies for locating web pages related to scientific topics, and guidelines for evaluating the credibility of these pages.

B. Internet Search Tools

i. SEARCH ENGINES

Most Internet users are familiar with basic **search engines,** *which are programs that find web pages, create topic indices for these pages, and then identify the ones that match search criteria you provide.* However, there are other search tools available, including metasearch engines and directories (sometimes referred to as subject gateways or portals). Each of these has different ways of finding and identifying information, and choosing the best tool can greatly shorten research time.

Search engines find web pages using **bots** (*programs that perform automated repetitive functions; short for robot*) called **web crawlers** or **web spiders,** *which are browsers that independently move through the Web following every link they find.* Keywords on these pages are then indexed for easy retrieval. Indexing is a com-

Figure 36.3. Ban Dihydrogen Monoxide! This site (http://www.dhmo.org), available in sixteen languages, provides a number of convincing details about the horrors of DHMO, including actual links to government agencies. However, careful readers will notice the disclaimer at the bottom of the page: "Note: content veracity not implied."

plex process, not only identifying keywords on pages, but also assigning relevancy levels according to topic. Some engines also **cache** pages, *keeping copies of web page contents*.

Search engines allow the user to provide one or more keywords describing a topic, and then provide a list of web pages matching the keywords. *The combination of keywords and conditions for a search is called a* **query** (the same term used for other database searches). The engine selects pages that appear to best match the query, generally listing what they consider to be the most relevant links first in their search results.

The default search index for most engines covers the entire Web. However, there are frequently additional, smaller indices for images, maps, and other services. An image index can be particularly helpful, because pages with good images for the keywords often have associated content on the topic. It provides a quick visual way to sort through large numbers of web pages.

In addition to the general search engines, there are many specialized engines available, optimized to search in specific topic areas. One way to find these search resources is to visit sites, such as Easy Searcher (http://www.easysearcher.com), that provide a listing of specialized search sites. Current web browsers provide ready access to multiple search engines in their search bar, as shown in Figure 36.4, so it can be useful to identify several specialized engines and use them when appropriate.

Figure 36.4. Browsers Can Provide Immediate Access to a Variety of Search Engines. Most toolbar search utilities (shown on the right) provides a drop-down menu to access a variety of user-selected search engines.

[1]Some countries such as the United States have legal restrictions in a few areas, such as child pornography and terrorist threats, but these restrictions are not universal. Because the Internet is global, sites that post material that is illegal in our country can (and frequently do) originate in other parts of the world.

Table 36.2.   *WHEN TO USE A SEARCH ENGINE*

All search tools provide better results for some types of queries compared to others.

Search engines excel at:

1. *providing large indices and relatively large numbers of links.*
2. *allowing complex keyword queries.*
3. *indicating the relevance of sites to the query.*

Potential drawbacks are:

1. *providing excessive numbers of links to sort through.*
2. *not providing indications of link quality.*
3. *providing inconsistent search criteria and results among different engines.*

Search engines are optimal when the topic is very specific and you need an exhaustive list of results.

Table 36.3.   *WHEN TO USE A METASEARCH ENGINE*

Metasearch engines excel at:

1. *combining the results of several different search indices.*
2. *narrowing down the number of suggested links to the most relevant.*
3. *eliminating multiple page links from the same website.*

Potential drawbacks are:

1. *discouraging the use of complex queries that may not work in all search engines.*
2. *relying on the relevance rankings of other engines.*
3. *limiting the number of links provided.*

Metasearch engines are optimal when the keywords are unusual and you want an overview of search results from multiple sources.

Table 36.2 summarizes some of the overall pros and cons of using search engines.

While search engines can be an excellent starting point for a general search, they operate with proprietary methods and so each engine may give you a different subset of web pages that match your criteria. This is usually due to differences in the way these engines create indices for web pages. One solution when you can't find what you need in one place is to query several search engines with the same keywords. A quicker option is to use a **metasearch engine,** *a specialized search engine that searches across the indices of other search engines and compiles relevant query results.* Metasearches work best for broader topics, and usually they return a limited number of the most relevant hits from each search engine, although the user can request more. Each link indicates the source engine, which can be useful for determining the most useful search tool for specific queries. Links that have been identified by multiple search engines are frequently good starting points for research. Table 36.3 summarizes some of the pros and cons of using metasearch engines.

ii. WEB DIRECTORIES

A **web directory** *is a collection of links organized by topic.* The hard copy analogy would be the card catalog at a library. Unlike search engines, directories are compiled by humans. This means that they include fewer sites than engines, but these sites are usually better characterized and chosen for quality. Yahoo! provides the best known general directory, with more than a million web pages selected and categorized by a team of editors. Links are arranged in hierarchical fashion, so that individual topics can be followed through increasingly selective categories until reaching the links. For example, if you want to learn about current gene therapy research, you can move through the following Yahoo! hierarchy:

Science > Biology > Genetics > Human Genetics > Gene Therapy > Research

It is also possible to run searches within the directory, which usually matches keywords to page titles or descriptions written by the editors or page authors.

In addition to general directories, there are also specialized directories (sometimes called clearinghouses) that provide access to information that may not be available through search engines. These smaller directories are usually dedicated to specific topics. You can find these resources by adding the keywords "directory" or "clearinghouse" to the topic keywords in an engine search.

Table 36.4 summarizes some of the pros and cons of using subject directories.

iii. ADDITIONAL WEB RESOURCES

Standard search tools are useful for general web searches, but most people are surprised to learn that search engines catalog less than 10% of all web pages. The billions of remaining web pages are referred to as the **Invisible Web**, or the **Deep Web**, *pages that are publicly available but invisible to search engines.* These are frequently dynamic database-driven pages that exist only for a short time; current news pages are an exam-

Table 36.4.   *WHEN TO USE A SUBJECT DIRECTORY*

Directories excel at:

1. *organizing links into clear topic areas.*
2. *providing only appropriate links.*
3. *finding materials that are not identified by search engines.*

Potential drawbacks are:

1. *limiting the number of links available.*
2. *omitting the most recent results due to human limitations.*

Directories are optimal when you want a topic overview or when you want to identify reference or database materials not indexed by search engines.

ple. Feeds are a good way to capture some of this information. A **feed** *consists of continually updated content from a web site.* You will also see these referred to as RSS feeds, XML feeds, syndicated feeds, channels, or web feeds. **RSS (Really Simple Syndication)** *is the most common format used to create feeds.* RSS format is based on **XML (Extensible Markup Language)**, *the computer language (similar to HTML) used to create feeds.* This service is a convenient way to receive automatic updates from web sites that change frequently, such as news sites.

Increasing numbers of websites are including feeds for their changing content. Finding feeds is fairly simple. When you are browsing a site that updates frequently, you can look for icons or text indicating that one or more feeds are available. Feed icons are usually orange; the most common (orange) symbol for RSS feeds is shown in Figure 36.4 on p. 688 on the right side of the address field. You can also search for feeds by topic with a standard Internet search engine or directory or by visiting a site such as http://www.syndic8.com, which provides access to more than half a million feeds, searchable by subject or source. Feeds exploit one of the best features of the Internet: the fact that its content can be updated virtually instantaneously.

There are many reference resources that are not included in standard searches. These include encyclopedias (with the major exception of Wikipedia), almanacs, dictionaries, and catalogs. Specialized directories can help you find these resources. When searching for specific facts or the type of information traditionally found in reference books, try a virtual library such as the Internet Public Library (http://www.ipl.org).

C. Search Strategies

i. GENERAL TOPIC SEARCHES

Many people searching for information on the Web become frustrated when their search engine offers millions of page hits of varying relevance to their purpose. If you are planning to use the Internet for serious research, a single query is rarely adequate to find everything that you need. Most users only touch the surface of what a search engine can do, but there are some basic strategies for performing productive searches.

Before you start a query, think about the information you want to find. Unless the topic is highly specific or your interest is very general, a single keyword is usually inadequate for useable results. For example, if you want to learn more about how viruses infect cells, typing the word "virus" into the Google search engine results in approximately 253 million links (called **hits**). Clearly you need to **refine** the query, *adjusting keywords and conditions to narrow down and focus the search results.*

The results of some simple query refinements are illustrated in Table 36.5. Adding keywords tells the search engine to find links with each of the words, although they are not necessarily found together on the page. Placing quotation marks around a combination of keywords instructs the engine to find that exact phrase.

Search engines list links in perceived order of relevance. If a search is effective, the first 10 to 20 links will include the appropriate information; if you don't find what you want immediately, you need to refine your search. If one site looks like exactly what you want, you can use the links on that page to widen your search. If you tell the search engine that you want to see similar pages, it will add the additional indexed keywords for the specified page to the query.

You can refine any search using **Boolean operators,** *which are terms that determine the logical relationship between keywords.* The Boolean operators commonly used in database queries are AND, OR, and NOT (always capitalized). These operators can be typed directly into the query line with the keywords, or you

Table 36.5. *SEARCHING FOR INFORMATION ABOUT HOW COLD VIRUSES INFECT CELLS*

Search Query	Google Hits (approximate)	Comments on Resulting Links
Virus	253,000,000	Too broad, no focus
Cold virus	1,100,000	Better, but still unfocused
Rhinovirus	429,000	More academic sites
Rhinovirus infection process	33,900	Includes many sites explaining how to avoid colds
Rhinovirus "infection process"	1,310	Many scientific sites in the top twenty, good starting points
Rhinovirus "infection process" molecular	1,160	Little reduction in hits with additional refinement, but Google now recognizes this as an academic search and suggests a few scientific journal articles along with the general results

can use an Advanced Search form, which defines the operators for you, as shown in Figure 36.5.

Box 36.1 summarizes these basic guidelines for performing effective Internet searches.

ii. BIBLIOGRAPHIC SEARCHES

Bibliographic searches are very common in science. An exhaustive study of the literature in your field is time-consuming, but significantly easier using the Internet than working with hard-copy directories. Finding just the right journal reference is best done with a specialized search engine. For example, biomedical researchers frequently use NCBI's Entrez engine to search the PubMed database. Increasing numbers of journals are making their content available through the Internet. However, for most commercially published journals, you will not be able to access the complete text of very new articles online without a subscription. If you locate a citation to a current article that requires a subscription or charges for file access, find out whether your employer's library or research department provides access.

An increasing number of journals are posting archives for previous years, although articles more than 10 to 15 years old may not be available. PubMed Central addresses that issue with its full-text copies of older articles as well as many recent articles. This database has the added advantage of having fully searchable text, while many individual journal archives provide articles only as scanned images.

Keep in mind that web pages are not the only resource for information. Sometimes you need facts that are too specific to find with Web search tools. An excellent way to find answers on your topic is to ask a human expert directly. The Web makes it relatively easy to find contact information for specific individuals. You can run a people search through most general search engines, or find them through company or university websites. Another solution is finding an appropriate mailing list or newsgroup with members who might be able to answer your question. You can find mailing lists on many topics using a list directory such as http://www.tile.net. Usenet newsgroups can be located through Google Groups, which provides

Figure 36.5. Using Boolean Operators to Refine a Search. Google's Advanced Search page provides a set of search options, which it converts to the specific operators, as shown in the bar above the search criteria. There are many additional search conditions not shown here.

access and search capabilities for almost thirty years worth of archives; more than a billion messages. For example, one active biotechnology newsgroup is sci.bio.technology, with more than 851 subscribers. People who participate in these groups tend to be passionate about their topics. If you post a question to the appropriate group, you are likely to get an answer.

D. Assessing Web Page Reliability

While the Web provides access to a wealth of knowledge, it also harbors many questionable sites with inaccurate information, see Figure 36.6 on p. 692. In order to find appropriate research material, it is essential to consider the reliability of the source. There are general guidelines you can follow to assess the quality of the information on a website.

The first thing you should examine is the site's URL. Does the site originate in a .gov or .edu domain? Institutional pages are generally reliable, but look for the tilde (~) that designates personal web pages. Be cautious if the

---

Box 36.1.   *GENERAL STRATEGIES FOR SUCCESSFUL INTERNET SEARCHES*

1. Use popular search engines for general searches, but consider specialized engines for specialized information.
2. Choose multiple keywords and phrases.
3. Use quotation marks to indicate exact phrases.
4. Use the "similar pages" or "more like this" option to mimic the indexing on a preferred link.
5. Use links on the preferred page to identify more resources.
6. Add Boolean operators in Advanced Search options to define relationships among terms.
7. Remember that no single search engine or directory can provide every possible resource.
8. Evaluate the reliability of web page information, as described in the next section.
9. Consider Internet resources beyond web pages, including newsgroups and the experts themselves.

**Figure 36.6.
Don't Believe
Everything You
Read on the
Internet.**
(© CartoonStock)

name domain comes from a free service such as geoci-ties.com; these are almost certainly personal sites, which are only as reliable as the person involved. Compare the site title or content to the domain. Does it seem appropriate? In the early years of the Web, the domains "whitehouse.com" and "nasa.com" linked to "adult" sites (no longer true). Unsuspecting individuals who wanted to contact the White House or the National Aeronautics and Space Administration (which are actually located within the .gov domain) were surprised to find themselves at very different locations.

Determining the purpose of a site can indicate whether the information will be appropriate. Good websites should provide this information on their home page. Who created the page? This may also indicate the site's purpose. For example, the websites for major drug companies usually have excellent information resources for patients with certain disorders. It is a safe bet that the company sells a product to treat these illnesses. Some companies completely separate the drug information from the general education material; others do not. Any information from these sources about specific treatments should be validated with independent references.

If there is a single author, what are their qualifications for providing the information on the page? Providing a name and contact information is frequently a good sign, particularly if the address or URL indicates a reputable institution or organization. If the author claims to be an authority on the topic, check the name in a search engine. If any directories list the web page, this can be a sign of validity.

Note the tone of the page. Is there a clear sales pitch contained in the information? Does the author sound biased or angry? This can be a sign that only data supporting the author's viewpoint will be present. Remember too that propaganda can be subtle.

Does the author provide links or a detailed reference list? If so, do these seem credible and appropriate?

For example, a nutrition site that links to sites that sell expensive vitamins and supplements may be questionable. If the associated sites do appear valid, do they actually support the original page? The parody DHMO site shown in Figure 36.3 on p. 688 contains legitimate links to the Environmental Protection Agency and the American Cancer Society; these sites have no information about DHMO, for obvious reasons.

Are the facts that you are already familiar with correct? If not, there may be more incorrect information on the page. Look for major omissions in the subject matter as well. Try to find out when the site was last updated, or how old the material is. Science pages that have not been updated recently might have obsolete information, and many of the links might no longer be valid.

Table 36.6 summarizes the items that should be examined on any informational website. The Case Study "Anatomy of a Website," along with Figure 36.7, offers an examination of the NCBI website.

CASE STUDY

Anatomy of a Website

The home page for NCBI provides access to a major biotechnology research portal. It exhibits every desirable characteristic described in the text for a reliable site, with many additional features. Excerpts from this page are shown in Figure 36.7. This site is revised continuously and will undoubtedly look different when you visit, but the basic elements should remain the same.

*The URL (http://www.ncbi.nlm.nih.gov) tells you that this is a government site; many visitors will recognize the ".nih" address component for the National Institutes of Health. As you can see in Figure 36.7a, the opening statement on the home page describes the agency and its purpose, along with a link for more information. The text is well-written, and the page is organized to assist the visitor in finding specific contents. There are additional features that improve the usability of the site. There is a prominent search function, as well as links for the most popular page destinations. One of the most useful features for a complex website is a **site map,** which is an organized set of links for every resource and function within the website.*

At the bottom of the home page (Figure 36.7b), the revision date is prominently displayed, and the frequent visitor will note that the page is updated weekly. The contact address for the agency is provided, along with copyright information and the site's privacy statement. There are links to more site resources, including an ftp site, a news display, and RSS feeds. Every feature is designed to assist visitors and give them confidence that this is a reliable website for serious researchers.

Table 36.6. IDENTIFYING CREDIBLE WEBSITES

When evaluating the credibility of web resources, examine:

1. *The URL.*
2. *The purpose of the site.*
3. *The identity of the site author or sponsoring entity.*
4. *The credentials of the author(s).*
5. *The quality of the writing, in terms of spelling and grammar.*
6. *The originality of the material.*
7. *The quality and appropriateness of the provided links and references.*
8. *Whether other credible sources cite this page.*
9. *The accuracy of well-known facts or data.*
10. *The neutrality of the discussion or viewpoint.*
11. *The age of the information.*

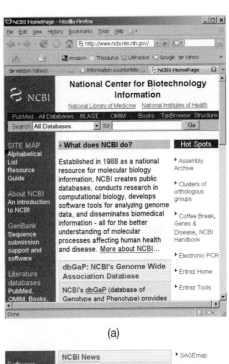

(a)

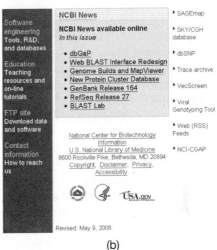

(b)

Figure 36.7. Elements in the NCBI Website. a. Top of the NCBI home page at http://www.ncbi.nlm.nih.gov. **b.** Bottom of the NCBI home page. These page excerpts are discussed in the Case Study on p. 692.

One of the great frustrations of the Web is a URL or link that sounds great but doesn't work. If you receive a Server Not Found message and the URL is correct, the problem is probably with the site's server, which may be temporarily offline. In this case, simply wait and try the URL later. A **404 error** message *indicates that the requested page cannot be found.* Always check the URL for spelling, but in many cases it means that the page is no longer there. Government and educational sites have the best longevity overall, followed by major corporations, professional organizations, and nonprofit groups. Websites run by individuals are the most likely to disappear.

If an unavailable web page seems essential, there are a couple of options to view it. You can take the URL and shorten it section by section, to see if you can open previous folders or find a home page that can point you to the page of interest at a new location. If this strategy doesn't work, remember that many search engines cache page copies. If the web page was active for more than six months, you might be able to access the contents through The Wayback Machine at http://www.archive.org, an Internet archive with more than 60 billion copies of web pages dating back to 1996.

E. Using Wikipedia

No discussion of current online research resources would be complete without acknowledging Wikipedia, one of the most popular sites on the Web. A **wiki** (derived from a Hawaiian expression for "fast") *is a set of collaborative web pages that anyone can create or modify over time.* **Wikipedia** *is a free, open content encyclopedia wiki,* with millions of articles in 253 languages. The basic principles of Wikipedia are simple.

Anyone is free to create new encyclopedia articles or edit existing entries. All articles are required to be neutral in tone and to contain only verifiable information supported by appropriate references. No original research is allowed and all information is heavily cross-linked to other Wikipedia entries when possible.

Wikipedia entries can vary greatly in quality. The project represents a democratic ideal, in which everyone in the online world can contribute information, regardless of their identities in the "real" world. Because of their openness, though, all wikis are subject to inaccuracies and vandalism. These problems are counterbalanced by the ability of everyone else to fix the problems. For this reason, older Wikipedia articles that have been revised multiple times tend to be more reliable than very new entries. Because editing is a continuous process, it is essential to record the exact date when referring to a Wikipedia article.

Using Wikipedia as a reference source carries difficulties in any case. Remember that this is an encyclopedia; Wikipedia does not allow the inclusion of original research. Secondary research sources of this type are not considered legitimate academic or journal references. Certainly many Wikipedia entries are high-quality subject introductions, making them a reasonable place to start learning about an unfamiliar topic. However, all articles are works-in-progress and none are officially peer-reviewed before publication. It is important to use caution and confirm any information found at the Wikipedia site. Many of the evaluation strategies described above can be applied to the individual articles in this website.

III. USING E-MAIL PROFESSIONALLY

E-mail is the predominant form of communication among computer users; however, there is a big difference between personal and professional e-mail styles. With professional e-mail, you are representing an organization or business, and therefore content and presentation are important.

E-mail has an informal feel that sometimes encourages users to apply less common sense to their messages than they might in conversation. Dissemination of potentially offensive humor occurs frequently, and can be grounds for a lawsuit. Virtually all networks and e-mail providers have an **acceptable use policy (AUP)**, *a statement that outlines behaviors that are not allowed within their accounts.* These are considered legal documents and violations can be grounds for losing your job, or at least your e-mail account. Remember that e-mail messages can be saved for a very long time.

It is hard to imagine that when the first edition of this textbook was written, spam was not a major problem for most e-mail users. **Spamming**, *sending copies of unsolicited e-mails (spam[2]) to large numbers of recipients*, has become so widespread that recent studies indicate at least 85 to 90% of all e-mail messages are spam. Do not open spam e-mail or respond to it in any way. All e-mail programs and services have some type of spam filter to remove many of these messages before they even enter your inbox; use these capabilities.

Box 36.2 offers some general tips for using e-mail professionally.

IV. INTERNET SECURITY

A. General Security Threats

This unit has already discussed some common computer security threats, such as electronic eavesdropping and information theft. However, Internet and especially e-mail security are also major concerns for employers, because all that is needed to take down an entire network is one employee who slips up and downloads a virus or reveals a password.

Ensuring the safety of confidential information exchange over the Internet is essential. Do not share private data with non-secure websites. Secure sites use the HTTPS (Hypertext Transport Protocol Secure) protocol, which you can identify from the URL (https://). The other visible sign of a secure site is the lock icon you should see next to the browser address bar. If the lock is missing or looks "open," you do not have a secure connection. Every secure site has a digital certificate for the browser to examine, and when this is recognized, encryption keys are activated for each end of the connection. Because the communication is strongly encrypted, it is safe to share information with these sites.

B. Malware

There are many forms of malware, with varying degrees of destructive power. The most likely delivery method is e-mail attachments, followed by file downloads. If your computer is not part of a network, contains only legally purchased commercial software, and does not come into contact with portable storage devices used with other computers, your system is safe from these unwanted programs. However, the rest of us need to take precautions.

Computer **viruses,** *programs designed to infect any computer that executes the program, then replicate and spread itself to other computers,* are the most familiar form of malware. Some viruses are designed to merely cause annoyance (for example, by displaying strange messages on your monitor screen at random intervals). Others can cause serious damage to data files, up to and including erasing the complete contents of your hard disk. Business losses due to computer viruses have been estimated at tens of billions of dollars each year. The Case Study on the next page demonstrates how this can happen.

There are many variations of the computer virus concept. **Worms** *are self-replicating viruses.* Once they

---

[2]The term *spam* was inspired by a Monty Python comedy sketch about a restaurant that only serves dishes including SPAM, even when the customer does not want it. SPAM is yet another registered trademark that has been co-opted by Internet culture. In this case, however, Hormel Foods Corporation seems to have embraced the situation, resulting in http://spam.com, a distinctly tongue-in-cheek web site celebrating SPAM products with, among other things, a Monty Python-inspired game and a prominent offer to subscribe visitors to the SPAM mailing list.

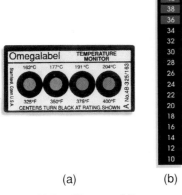

(a)   (b)

Figure 21.7. **Change-of-State Indicators. a.** A label that changes colors at specific temperatures. **b.** Liquid crystal display. (Photos © copyright Omega Engineering, Inc. All rights reserved. Reproduced with the permission of Omega Engineering, Inc., Stamford, CT 06907.)

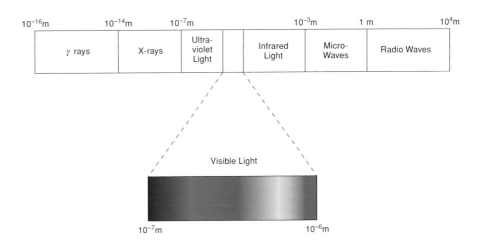

Figure 23.2. **Types of Electromagnetic Radiation and their Wavelengths.** Visible light occupies a small portion of the electromagnetic spectrum. Within the visible portion of the spectrum, different wavelengths are associated with different colors.

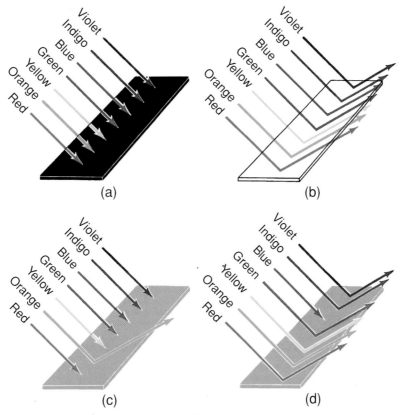

Figure 23.4. **The Colors of Solid Objects. a.** An object appears black if it absorbs all colors of light. **b.** An object appears white if it reflects all colors. **c.** An object appears orange if it reflects only this color and absorbs all others. **d.** An object also appears orange if it reflects all colors except blue, which is complementary to orange.

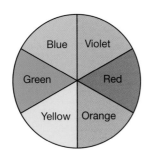

Figure 23.5. A Color Wheel Showing Complementary Colors.

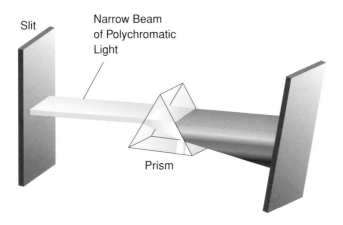

Figure 23.12. The Dispersion of Light into Its Component Wavelengths by a Prism.

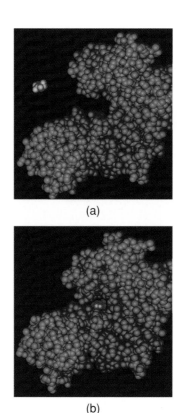

(a)

(b)

Figure 29.1. Example of the Relationship between Protein Structure and Function. An enzyme and its substrate. The smaller substrate molecule **a.** binds within a groove in the larger enzyme that is complementary in shape and ionic properties **b.** Many proteins bind other molecules at sites whose shapes are complementary.

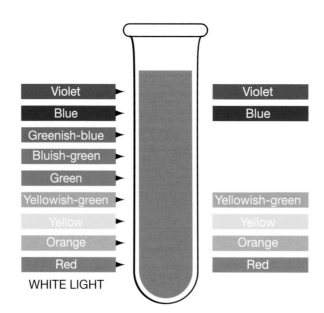

Figure 23.6. The Absorption of Light by a Solution of Red Food Coloring. When white light is shined on a tube containing red food coloring, blue to green colored light is absorbed.

Box 36.2.   GUIDELINES FOR CREATING PROFESSIONAL E-MAIL MESSAGES

Following these guidelines will increase the effectiveness of your messages and protect them from spam filters.

Message header:

1. *Always include a subject line.*
2. *When exchanging e-mails, especially with a group, change the subject line if the topic of the e-mail has changed.*
3. *Be very selective about sending mass e-mails.*
4. *Always check the "To:" line to avoid an unintended "Reply to All."*

Message body:

1. *Keep it as short as possible and get to the point immediately.*
2. *Use a spell checker if available.*
3. *During repeated message exchanges, delete unnecessary or confidential portions of the old messages.*
4. *Never use online language, phrase abbreviations, or other slang in professional e-mail.*

Message attachments:

1. *Avoid sending attachments if possible.*
2. *Only send files that you know are virus-free.*
3. *Always include a personal message with attachments so the recipient knows you sent it yourself.*

E-mail etiquette

1. *Check your e-mail accounts regularly.*
2. *Do not include anything in an e-mail that you might want to retract. Remember that e-mail is not private and can be saved indefinitely.*
3. *DON'T SHOUT! The use of all capital letters is considered rude, at best.*
4. *Be careful when using sarcasm in e-mail. Remember that the recipient of your message cannot necessarily guess your intentions.*
5. *When you write an e-mail about an important or emotional matter, wait ten minutes and reread the message before you send it. You may decide to change the wording at that point.*
6. *Consider whether an old-fashioned telephone call might be the fastest way to get the information you need.*

Keep in mind that e-mails are not private communications. If you need to send any type of sensitive or confidential information by e-mail, always use encryption software. Remember to encrypt any confidential attachments as well.

---

infect a system, they multiply without any additional user action, attempting to escape into any connected network and eventually overrunning the infected computers. LoveLetter, described in the Case Study, was actually a worm. Another type of attacker is the Trojan horse, or simply **Trojan**. *These are malicious programs disguised as legitimate programs.* While their functions

- Reading the Outlook address book and sending virus copies to every address.

- Overwriting a wide variety of program and data files, including standard graphics and video files, Word and Excel documents, and html files, permanently destroying the contents.

- Attempting to download a program for gathering passwords.

LoveLetter replicated and spread itself until entire e-mail systems had to be shut down that day by major businesses and government agencies around the world, including the CIA, Pentagon, and U.S. Senate. Damages were estimated to approach ten billion dollars.

While any computer could receive the e-mail, only those with the Windows OS could be infected. At the time, many casual computer users were unfamiliar with basic precautions associated with e-mail attachments; the LoveLetter virus was a public prototype for learning to protect networked computers from malware.

CASE STUDY

Love in the Workplace

One of the most infamous computer viruses is LoveLetter, sometimes called Lovebug; it was the first virus to cause true global chaos. It was released in the Phillipines on May 4, 2000, and within hours infected millions of computers around the world. LoveLetter was disseminated as an e-mail attachment. The infamous subject line was "ILOVEYOU" and the attached file was named LOVE-LETTER-FOR-YOU.TXT.vbs. Once opened, the virus performed a variety of malicious tasks, including:

can be similar to viruses, Trojans do not spread themselves to other computers. Anytime you find bootlegged software or free programs such as utilities or screensavers, you could end up downloading a Trojan. Ironically, many Trojans spread by claiming to be free anti-malware programs.

Some word processing and spreadsheet programs allow the creation of **macros**, *short sets of commands within a file that create a program shortcut*. While macros can be very useful, they are also potential delivery vehicles for viruses. Fortunately, many programs provide the ability to automatically disable all macros in its files; always use this option unless you specifically need a macro.

There are types of malware that are not virus derivatives. **Bots** *can be programmed to independently roam the Internet looking for unprotected computers that they can control*. They are a major reason why computers face security attacks within minutes of connecting to the Internet. If bots find a vulnerable computer (without a firewall, for example), they move into the system, hide themselves, and then allow the person controlling the bot to perform certain automated functions, such as sending spam or collecting information, without the owner's knowledge.

C. E-Mail Security Issues

Most computer viruses are spread through e-mail attachments, which means you need to be extremely cautious with these files. *Never* open an attachment from an unknown source, and never open attachments you weren't expecting from people that you do know until you confirm that they intended to send it. Because viruses frequently spread by invading the address book on infected computers and using it to send copies to every e-mail address, you are much more likely to receive a virus from someone you know than from a stranger.

While most spam is simply annoying, there are more sinister forms. Avoiding malware attachments will remove one major risk. However, another threat to businesses as well as individuals are phishing e-mails. **Phishing** (pronounced "fishing") *is sending out mass e-mails designed to trick the recipients into revealing private information*, such as passwords and credit card numbers. The most notorious recurring e-mail of this type is the "Nigerian lawyer," "widow," or "good samaritan" who wants to share millions of dollars with you if you simply provide your bank account number. Sadly, this type of scam is successful enough to constantly reappear in various forms for more than a decade. Another type of phishing e-mail supposedly originates from official sources, such as major banks or retailers, urgently requesting that you reply with personal information or visit a specified website. Phishing e-mails have several things in common:

- Most are sent by unfamiliar institutions and are poorly written.
- They are not addressed to you personally.
- They warn of dire consequences unless you respond immediately.
- The e-mail address or website link they provide for your response is disguised.

No legitimate financial institution or e-mail provider will ever ask for personal information in this manner. While the more polished efforts may provide a legitimate-looking link to click, mousing over the link (without clicking) will often reveal a very different website address. This trick is called **spoofing**, *disguising an undesirable website with an apparently legitimate link*. Most e-mail providers have phishing filters, which will **bounce** (*send back*) e-mail that comes from or links to known phishing addresses.

D. Dealing with Malware

How do you know if your computer is infected with malware? You might not; by definition, the files themselves are invisible to the user. The computer might slow down or occasionally flash odd messages. The hard drive might seem to be working overtime while the computer is supposed to be idle. Malware provides yet another reason to back up files at regular intervals, and to keep several previous versions of your hard drive contents in an independent location. In this way, you can replace damaged files that cannot be restored on an infected drive. Removing malware from your computer can be difficult and frequently requires professional assistance. By far the best strategy is avoiding infection entirely.

While it may sound like there are extraordinary numbers of computer security threats associated with the Internet, in most cases, these hazards can be avoided with a few simple precautions. First, take advantage of every security measure available in your OS, browser, and e-mail program. Once you have the programs configured, be diligent about downloading and installing all updates regularly. These are frequently created in response to newly identified security threats.

Then, install high-quality anti-virus software. Despite the name, these programs can protect your computer from most types of attacks. However, anti-virus programs can only protect against known threats. The developers of these programs work continuously to stay informed about new malware, and provide constant software updates. It is essential to update anti-virus programs frequently, preferably automatically. Configure the software to scan all downloads, files, and e-mail attachments automatically as well. Only one anti-virus program should be installed on a computer at any time, because these programs can interfere with each other.

Any individual who is part of a local network has the ability to unintentionally incapacitate that network if

---

Box 36.3.   BASIC INTERNET SECURITY GUIDELINES

You can avoid most security threats by following a few simple guidelines.

The Internet

1. *Configure the security settings of your OS and web browser to provide the appropriate level of protection.*
2. *Download and install updates or patches for your OS, browser, and other software frequently; set up automatic updates if possible.*
3. *Install high-quality anti-virus software and configure it for as much automated security as possible.*
4. *Update your anti-virus software as frequently as feasible; many experts recommend once a day.*
5. *Back up your data regularly and often.*
6. *Avoid downloading questionable files or programs; know your sources.*
7. *Closely examine any web page that requests personal information to confirm its identity; look for the https:// URL and the lock icon.*
8. *Ignore links for freebies, "prizes," or free "virus scans."*

E-Mail

1. *Delete all suspicious e-mails without opening.*
2. *Do not click on links provided by unknown senders.*
3. *Never open attachments from unknown sources.*
4. *Do not open e-mail from acquaintances if the greeting, message, or attachments look suspicious. Confirm the origin first.*
5. *Scan all e-mail attachments with anti-virus software before opening.*

---

they are not diligent about guarding passwords or avoiding virus infections. Do not become part of the problem; execute any security policies that your employer has in place and follow basic safety guidelines when using e-mail and the Internet, as summarized in Box 36.3.

V. THE ETHICS OF THE INTERNET

As in all forms of human expression, there are limits to the types of behavior that are appropriate on the Internet. The lack of central governance, issues of free speech, and a general sense of anonymity create a true sense of freedom on the Internet. However, there are still standards of ethical behavior and good manners. **Netiquette (Internet etiquette)** *is the uncodified but generally accepted standards for behavior while online.* Always remember that the Internet is made up of people. If you would not behave in a certain way if people were present, you should not behave that way on the Internet.

An essential part of ethical behavior when using a network is to refrain from illegally copying commercial software, music, video, and other materials that belong to someone else. Malware is frequently disguised as bootlegged software or pirated media files,

under the premise that if you are willing to download illegal files, you deserve to be targeted.

Assume that everything on the Web is under copyright unless specifically informed otherwise. If you are going to use text or other materials from the Web as references, you must cite them in your bibliography. Currently there is no standard format for electronic citations, but most journals indicate their requirements.

Learn about your employer's AUP and policies regarding personal as well as professional computer activities. In the absence of applicable laws, U.S. courts have generally upheld the rights of employers to monitor and regulate all activities performed using company-owned computers, even when the computers are offsite or the activities take place outside of work hours. This is in accordance with the legal practice of holding employers responsible for illegal or discriminatory activities performed with their computers. Business surveys have shown that approximately half of all companies perform some type of e-mail surveillance, and even more businesses track where their employees visit on the Web. Do not use work computers for personal business if it can be avoided.

Most books on the Internet as well as a number of websites discuss proper behavior and ethical standards in the digital age. Table 36.7 on p. 698 summarizes basic guidelines for ethical use of computers.

Table 36.7. *SOME BASIC PRINCIPLES OF COMPUTER AND INTERNET ETHICS*

1. Do not copy or disseminate materials without permission. The material you see on the Web is copyrighted unless otherwise indicated.
2. Do not use illegal copies of commercial software or download illegal media files.
3. Do not use your employer's computers for web surfing or personal e-mail.
4. In case you are ever tempted to do something even marginally unethical on the Internet, remember that, unless you are a talented hacker, nothing you do is truly anonymous.
5. Remember that you are dealing with human beings on the Internet.

PRACTICE PROBLEMS

1. What is the difference between a search engine and a subject directory?

2. What is the main difference between a virus and spyware?

3. Name three things you should look for on a web page to confirm its reliability.

4. When you are asked for personal information on a commercial website, how can you tell if the site is secure?

5. When responding to a message from a mailing list, when is it appropriate to use Reply, and when is it appropriate to use Reply All?

6. What is the main characteristic of a wiki?

QUESTION FOR DISCUSSION

1. Find four examples of websites that would be useful in your current laboratory.

2. You find an image on another website that you would like to use on your own. How would you handle this?

APPENDIX

Answers to Practice Problems

1. Cloning refers to the production of one or more copies of an original item. In this case, it is a genetic sequence, or gene of interest, that is copied. Observe that the original transformed bacteria are placed in a fermenter where they reproduce many times to form many new bacteria. As the bacteria reproduce, so too do the plasmids within them. Reproduction results in many new plasmids and many new copies of the original genetic sequence of interest.

2. See Glossary.

3. Remember that transfection refers to the introduction of "foreign" DNA into eukaryotic cells and transformation refers to the introduction of "foreign" DNA into prokaryotic cells.

   a. Producing crop plants that are resistant to the herbicide glyphosate involves **transfecting** the cells with a gene that confers resistance.

   b. Growing cultured human skin cells to use for tissue engineering requires growing large numbers of cells. This process does not require the introduction of new genetic material (although one can imagine scenarios in which transfection might be used to improve the procedure).

   c. Making wine does not require the introduction of DNA into cells.

   d. Bacteria do not naturally make human proteins and therefore must be **transformed** with the gene in order to produce human epidermal growth factor.

   e. Harvesting stem cells from an embryo does not require transfection or transformation.

4. **a and b.** Antibodies normally made by a person are proteins that generally recognize and bind foreign agents (e.g., bacteria, viruses). This binding signals other cells in the body to engulf and destroy the invading agent. Mab drugs, like naturally occurring antibodies, recognize and bind specific targets. You can see in Table 1.2 that the targets of Mab drugs, however, are usually not foreign agents, like bacteria or viruses. The targets rather are proteins normally found on the surfaces of cells that for some reason have become involved in a disease state. The binding of the Mab blocks the activity of the target protein. As shown in Table 1.2, many of the targets of currently available Mab drugs are related to cancer. Some are related to autoimmune diseases (such as Remicade) and a few to various other disorders (such as Lucentis and Soliris). At present, few attack bacterial or viral pathogens (Synagis is an exception). (In the future, more Mab drugs are likely to be developed against viral and bacterial pathogens.)

5. **a.** Synagis *recognizes* and *binds* the F protein on the surface of the viral particles. In RSV infection, the F protein binds to surface proteins on lung cells. This binding leads to the internalization of the virus into lung cells, hence, infection. When the Synagis antibody binds the F protein, the virus is *blocked* from binding to surface proteins on the lung cells and therefore is not internalized.

   b. Conventional vaccines use a weakened, killed, or less pathogenic form of the infectious agent to stimulate the body to form *its own antibodies* against the infectious agent. Synagis is itself an antibody and does not stimulate the patient to form his/her own antibodies against RSV.

6. Answers may vary. Some main ideas:
   - The major difference is that the process that was used to create Dolly is intended to generate a whole, adult animal that is a clone of a single parent. Therapeutic cloning is intended to generate stem cells.
   - Both processes involve the introduction of genetic information from only one parent organism into an enucleated egg cell (sexual reproduction is not involved). The egg cell divides to form an embryo.
   - For therapeutic cloning, stem cells are harvested from the blastocyst stage of the embryo. In whole animal cloning the embryo is allowed to develop.
   - To create Dolly, the egg was implanted into a surrogate mother. For therapeutic cloning the egg is cultured in the laboratory until the stem cells are harvested.

7. Biopharmaceuticals are usually proteins. Like the proteins that we consume in food, biopharmaceuticals would be digested if they were taken orally. (Researchers are looking for ways to get around this problem to make oral administration of biopharmaceuticals possible.)

8. Bone marrow stem cells were targeted in the X-SCID experiment. Bone marrow stem cells divide to make blood cells for the lifetime of a person. The cells that were targeted in Ashanti De Silva were not stem cells.

9. You may have various answers. Researchers are presently working on this problem from many angles.
   - Some are trying to fine-tune the cellulase enzymes, for example, by using genetic engineering so that cells (e.g., bacterial cells) make more effective cellulase enzymes, make more abundant enzymes, or make enzymes more cheaply.
   - Researchers are testing different strains of bacteria and yeast to find those that most effectively ferment sugar.
   - Some scientists are studying the biochemical pathways used by microbes naturally found in the guts of cattle, deer, and termites because these organisms readily digest wood into fermentable sugars.
   - Some researchers are working to optimize the fermentation process. It is possible, for example, that bacteria or yeast cells could be genetically engineered so they

can both break down cellulose and ferment it into ethanol in one fermenter.

- Some researchers are developing better sources of cellulose, for example, faster growing plants, plants that grow in marginal habitats where food crops are not grown, and plants whose fiber is easier to break down.

CHAPTER 2

1. There may be more than one possible match to some of these items because there are differences between companies in the division of tasks between functional units. Reasonable answers include:

Research and development: a–c, f, h, k

Production: e, i, j

Quality control: g, m

Quality assurance: d, l

2. You may come up with many ideas; for example, your description of an ideal location, availability of wireless internet, and so on. Be sure to think about the laboratory needs of your company, since start-up companies usually are heavily involved in the development of their products. The majority of your space is likely to be devoted to the laboratory. A good laboratory will probably begin with such features as ample space for equipment and personnel, ample electrical power, access to water, good ventilation systems, laboratory benches, and so on.

CHAPTER 3

1. **a.** iii; **b.** i; **c.** ii; **d.** ii; **e.** ii; **f.** i; **g.** iv; **h.** ii

2. Similarities include:

- Both drugs treat cancer by slowing uncontrolled cell division.
- Both drugs recognize and bind to a specific target in the body that is associated with cancer cells.
- The development of both drugs relied heavily on basic research into the mechanisms of cancer and cell division.

Differences include:

- Gleevec is a small molecule drug; Herceptin is a large molecule (antibody) drug.
- Gleevec treats CML; Herceptin treats breast cancer.
- Gleevec targets the site where ATP binds to an enzyme; Herceptin targets receptors on the surface of breast cells.

You can probably think of other similarities and differences as well.

3. **a.** "Accelerated development/review . . . is a highly specialized mechanism for speeding the development of drugs that promise significant benefit over existing therapy for serious or life-threatening illnesses for which no therapy exists." See the website for more information.

b. "Treatment Investigational New Drugs . . . are used to make promising new drugs available to desperately ill patients as early in the drug development process as possible." See the website for more information.

c. "Another mechanism to permit wider availability of experimental agents is the 'parallel track' policy . . . developed

by the U.S. Public Health Service in response to AIDS. Under this policy, patients with AIDS whose condition prevents them from participating in controlled clinical trials can receive investigational drugs shown in preliminary studies to be promising."

4. **a.** Key concerns include:

- Cells might be contaminated with agents that could cause disease in patients; these agents might be bacteria, fungi, viruses, or prions.
- Host cells might mutate (change genetically), thereby possibly altering the product.
- The genetic construct inserted into the cells might change or be lost over time, thereby changing the product.
- Host cells might induce cancer in patients.

b. End of production cells are tested to see if changes in the cells or the genetic construct have occurred over time. Any changes in the cells or the genetic construct could mean that the biopharmaceutical product is altered. The cells are also tested for contaminating agents.

5. a, b, f, g relate to using genetically modified cells to make a product. These tasks therefore relate specifically to biopharmaceutical production. The other tasks are common to all drugs.

6. There is no single answer to this question. Be sure that your answer discusses the roles of various types of laboratory personnel. Key points include:

- The role of laboratory personnel in performing **basic research that elucidates the workings of cells and organisms** and in uncovering **the basic mechanisms of health and disease.**

  Example: The scientific work of Erwin Chargaff, Rosalind Franklin, Maurice Wilkins, James Watson, and Francis Crick that elucidated the structure of DNA.

- The role of laboratory personnel in the **discovery of potential targets** for drugs and in the **discovery of new drug candidates.**

  Example: The discovery of the Her2 gene and its role in some breast cancers.

- The role of laboratory personnel in the **development of potential products.**

  Example: Dawn Rabbach's work to develop a DNA fingerprinting method.

  Biopharmaceutical development provides many examples including:

  - Testing drugs in animals and humans
  - Optimizing methods for producing drugs in cells
  - Optimizing methods of purifying drugs produced by cells
  - Developing methods to test the activity, potency, and purity of drugs, raw materials, and in-process samples
  - Scaling-up production

- The role of **quality-control analysts who test environmental samples, raw materials, in-process samples, and final products.**

  Example: At the end of Chapter 2 there is an example of how quality-control analysts test a product manufactured in *E. coli* to be sure it meets its specifications.

CHAPTER 4

1. There are a variety of good business reasons to comply with ISO 9000 standards. Companies voluntarily comply with standards to help ensure that they have a quality product. The requirements of ISO 9000 guide the company in setting up organizational systems that promote quality. Companies also comply with these standards to demonstrate to their customers and other interested parties that their product is of good quality. Remember that there are costs associated with a poor-quality product ranging from loss of customers to law suits and legal proceedings.

2. This is a discussion question with various answers. It is particularly important to note that a change in the manufacturing process for a biopharmaceutical might change the resulting drug product. In the event of such a change, the resulting substance is not the one tested in phase I–III clinical trials and therefore is not an approved drug. An unapproved drug cannot be distributed for use in patients. Moreover, because biopharmaceuticals are complex molecules, such changes might be subtle and not be detected through normal laboratory tests.

3. You can think of various answers to this question. Standard procedures help ensure that results from different individuals and from the same individual over time can be compared to one another. Standard procedures that are written by a knowledgeable individual ensure that everyone performs routine tasks correctly. Standard procedures help avoid breakage of equipment due to mistakes. Standard procedures are used to teach individuals how to do the work of that laboratory. When researchers do not follow consistent, correct procedures, their results may be incorrect or inconsistent. At best, this results in wasted time and supplies. At worst, the researchers may reach erroneous conclusions and publish erroneous results.

4. **a.** A product must be tested to see if it meets its specifications before it can be released for sale to the public. This testing is performed by the QC unit. If a test of a product is performed and the product fails to meet the requirements of that test, the result is termed *OOS*. An OOS result means either that the product is defective in some way or that the laboratory test was invalid for some reason. When an OOS result is obtained, there must be an investigation to find its cause. In this particular case study, 70% of all OOS results turned out to be problems with the testing procedure and not the product. The erroneous results were invalidated. This means there were problems in the testing laboratory that were costing the company time and money and leading to unpredictability in how long it would take the laboratory to complete its work before releasing products.

   b. In all the situations in the table, the analysts failed to follow the correct procedure to perform a test. Their failures to perform the tests correctly led to erroneous results. Initially, they tried retraining each analyst in how to perform the tests.

   c. The root cause is the underlying cause of problems. The superficial cause of problems in this case study is that the laboratory analysts did not follow the procedures for performing laboratory tests. The root cause analysis looks at why the analysts failed to follow the procedures. The four hypotheses examined were: The analysts were not sufficiently trained. A second hypothesis was that the procedures were poorly written and therefore the analysts were not able to follow them properly. A third possibility was that the analysts lacked support and supervision. A fourth possibility was that the analysts were rushing through the tests, presumably because they were overworked.

   d. This is not explicitly stated, but we can infer that the root cause investigation showed that the analysts did have deficiencies. Some of the analysts were inexperienced and some were not completely trained on all the methods used in the laboratory. The analysts also were not aware of, or were not convinced of, the importance of complying strictly with the procedures as written and were deviating from them. It appears therefore that the analysts were not adequately trained and that there was insufficient supervision.

   e. To prevent further problems, the company provided more training to the analysts not only in how to perform the test procedures, but also the importance of adhering to the procedures. Group training was particularly effective. The company also provided more supervision to employees in the form of a QA person who supported the laboratory supervisor.

5. **a.** The company apparently made several errors. First, they did not perform a sufficiently rigorous investigation of a problem. Second, they did not initiate aggressive *corrective actions*. Third, they did not take any *preventive actions*. If you read the text carefully you will see that the problem is that product that was placed by QA in quarantine and was not supposed to be shipped was, in fact, shipped to five hospitals. This is a serious error in that these hospitals received "nonconforming" material, material that had not been shown to meet its requirements. There was apparently a form filed relating to a CAPA investigation of this incident(s). The investigation report was incomplete however, since it lacked "the dates of these serious occurrences, the employees involved, he number of instances that product was actually either removed from quarantine or overridden in the computer system (SAP). The CAPA also did not list the number of units that were actually shipped, or the number of hospitals that actually received nonconforming product."

   b. This is a discussion question with various answers. As in the Case Study "Analyst Errors," training of personnel might be part of a prevention plan. Other ideas may occur to you, such as increased security limiting access to the quarantine area, better computer systems to catch mistakes, a system involving a supervisor's review, and so on.

   c. The consequences so far include the shipment of potentially faulty products, which could conceivably adversely affect patients, lead to lawsuits, and damage the company's reputation. This warning letter, like all FDA warning letters, is publicly posted and therefore might negatively impact the company's reputation. The company could lose business and suffer financially. The consequences to a company that consistently fails to comply with FDA can be severe. These consequences include seizing the product from the company, preventing the company from marketing and selling the product, and fining the company.

CHAPTER 5

1. Some points to include in your discussion: In order to have a regulatory process, such as the one outlined in Figure 3.7, the government had to accept responsibility for protecting the consumer. This occurred in 1906 with the passage of the first Food, Drug, and Cosmetic Act. You might also mention the beginnings of the FDA in 1927, because the process outlined in Figure 3.7 includes critical gates controlled by the FDA. Animal and human testing are major parts of the development process shown in Figure 3.7. The 1938 Food, Drug, and Cosmetic Act that was enacted in response to the sulfanilamide incident was of critical importance. This Act mandated that new drugs be tested for safety in animal and clinical studies. Other incidents led to tighter control over the animal and clinical testing stages. In particular, the thalidomide tragedy led to tighter control of human testing, and inspections of animal testing facilities in the 1970s led to tighter control of animal testing.

2. **a.** Congress enacts laws to protect consumers. These laws mandate overall processes for controlling the development and production of these products, and give enforcement powers to government agencies.

   b. The laws passed by Congress are broadly written. The FDA provides specifics in the form of the GLPs, GCPs, cGMPs and also various guidance documents. The FDA controls critical gates during development through an approval process, see Figure 3.7 (see also question #1). The FDA also oversees and inspects the manufacture of products.

   c. Companies must comply with numerous requirements, including those outlined in GLP, GCP, and cGMP. Companies must comply with directives from the FDA. Companies must design their products to be safe and effective, and must put in place systems to ensure the quality of their products.

   d. Consumers must take responsibility for their own health. This includes complying with the guidance shown on product labels (e.g., taking the proper dose of a drug) and storing and preparing food properly.

3. **a.** This section is shown below. It outlines the responsibilities and authorities of the quality-control unit in a pharmaceutical/biopharmaceutical company. Use your own words to explain its provisions.

   b. A separate QC unit is required by law. This means that the production unit cannot take responsibility for these tests.

Subpart B—Organization and Personnel

Sec. 211.22 Responsibilities of quality control unit.

   (a) There shall be a quality-control unit that shall have the responsibility and authority to approve or reject all components, drug product containers, closures, in-process materials, packaging material, labeling, and drug products, and the authority to review production records to assure that no errors have occurred or, if errors have occurred, that they have been fully investigated. The quality-control unit shall be responsible for approving or rejecting drug products manufactured, processed, packed, or held under contract by another company.

   (b) Adequate laboratory facilities for the testing and approval (or rejection) of components, drug product containers, closures, packaging materials, in-process materials, and drug products shall be available to the quality control unit.

   (c) The quality-control unit shall have the responsibility for approving or rejecting all procedures or specifications impacting on the identity, strength, quality, and purity of the drug product.

   (d) The responsibilities and procedures applicable to the quality-control unit shall be in writing; such written procedures shall be followed.

CHAPTER 6

1.

FORMULATION OF XYZ COMPOUND

Clean Gene, Inc.

3550 Anderson St.
MADISON, WI 54909
Revision 01
Master Batch Record # 133
Approved by _____Aaron Reid_____   _____Anna Gold_____   _____Sam Rothstein_____
    Date 2/14/08   Date 2/14/08   Date 2/16/08

Issued by: _Erin Jane_   Date 12/3/08   Lot # 15.987
Product Name: **Very Good Product**
Strength: **10 Units/mL**   Vial Size **10 mL**   Batch quantity: **350 L**

Reference: Refer to separate Formulation SOP Q75 for quantities of each component.
Refer to separate instrument/equipment SOP 76 for ID information

NOTE: *PRODUCT IS TO BE STIRRED CONTINUOUSLY DURING COMPOUNDING AND FILLING*

A. COLLECTION OF WATER FOR INJECTION (WFI), USP.   **wrong temperature**

A1. Collect approximately 370 L of WFI, USP, in a clean, calibrated vessel and cool to 24°C–28°C.
Vessel # __7__   Amount collected __365ℓ__   Initial Temperature (23°C)   Time __8:00__ (am)/pm
Final Temperature __26°C__   Time __08:15__ am/pm
    ↑ **neither is circled**   _LP_ / _GM_
    12/10/08 12/10/08

A2. Close the water for injection valves.   _LP_ / _GM_
    12/10/08 12/10/08

A3. Remove about 20L of the cooled WFI from step 1 and place in a clean, calibrated vessel.
(This water will be used to bring the final formulation to the proper volume.)
Vessel # __2__   _LP_ / _GM_
    12/10/08 12/10/08

A4. Remove about 5L of the cooled WFI from step 1 into a clean, calibrated vessel. (This water will be used to prepare the solutions used to adjust the pH.)

B. PREPARATION OF pH ADJUSTING SOLUTIONS

B1. Collect 750 mL cool WFI from the vessel in step A4 and place into a 1000 mL volumetric flask and add 100 g of NaOH (Sodium Hydroxide # 875) USP and dissolve.   **wrong amount**
Amount of WFI collected __750mL__   Amount NaOH added (115g) (Lot # _____) ← **missing**
Time step completed __09:15__ am/pm   _GM_   **no initials—why did supervisor initial this?**
    ↑ **not circled**   12/10/08

B2. Using a water bath containing cold WFI, cool the solution prepared in B1 to 25°C ± 5°C.
Final Temperature (31°C) ← **wrong temperature**   _LP_ / _GM_
    12/10/08 12/10/08

B3. Bring the solution to 1000 mL with cool WFI from the vessel in step A4.
Approximate volume of WFI added (300mL)   Time completed (09:00) (am)/pm
    ↑   ↑   _LP_ / _GM_
 and so on   **wrong volume**   **time does not make sense**   12/10/08 12/10/08

2. Some ideas: Commitment and directive documents state the intentions of the organization and direct personnel in how to accomplish those objectives. These documents "say what the organization does." Individuals are required to adhere to these written documents. Individuals provide evidence that they have done so by completing data collection documents, such as forms and laboratory notebooks. Data collection documents are the means by which the individual "says what he/she has done." A documentation system provides accountability on the part of the organization and the individuals within it.

3. You can imagine many documents that might be required. A few examples include:

- A batch record might be used to direct production, based on the entrepreneur's original recipe(s) and processes. This document would detail raw materials and equipment, how personnel must prepare, package, and store the cookies, and might provide blanks for them to fill in during the process.

- Standard operating procedures for operating ovens, checking their temperatures, cleaning the bakery, and other tasks would likely be prepared.

- Documents might describe requirements for incoming raw materials.

- Documents might be developed for shipping and receiving, indicating what materials come into the bakery, what products are shipped, and to where they are shipped.

- Other documents might be involved in proving to sanitation inspectors that the bakery operates according to local regulations.

- Documents relating to personnel would be required, such as training records, and time sheets.

- You can imagine that accounting and other business documents (outside the scope of this text) would be required.

4. Some ideas for your discussion:

- The first company did not have a well-written, complete master production (batch) record. This might result in operators making mistakes during manufacturing and labeling of the product, which, in a worst case, could result in a patient being harmed by an improperly manufactured product. This company similarly did not have written procedures to prevent contamination, again potentially putting patients at risk. This company also appears to have maintained incomplete records of testing by their QC department, meaning that it is difficult or impossible to ascertain whether products approved for release were manufactured properly.

- The second company appears to have failed to keep their operating procedures current. This company therefore was not "doing what they say." It is essential that all procedures in a company are up to date.

- The FDA inspection occurred in April, 1997. The company appears to have "lost" all the records pertaining to what happened to their blood products prior to November, 1996; they had, in fact, "lost" all but a few months of records. This is why companies are required to prove

that their electronic documentation systems are secure and reliable *before* putting them into routine use, as discussed previously in this chapter. This company's "loss" of its records could have severe consequences if there is a problem with any blood product that the company had shipped. They will have difficulty investigating the problem or issuing a recall.

CHAPTER 7

1. No. The expiration date must appear on the immediate container. If the expiration date were to be placed on the bottle cap, there would be a greater chance of mix-up at the pharmacy or by the consumer.

2. Yes. In initial studies the company must establish the effectiveness of the preservative and how much is required to protect the product. Then, each time a batch of product is tested, it is evaluated to see whether the proper amount of preservative is present.

3. This is a specification of a process. The setting of specifications is the company's responsibility. In order to determine an acceptable range for detergent residues, the company must determine how residues affect the safety, efficacy, and stability of the product. The ability of the company's cleaning process to eliminate detergent residues is demonstrated during the validation of the process.

4. No. The design of the equipment is a major component of its cleanability; therefore, companies should demonstrate that a cleaning procedure works on each piece of equipment.

CHAPTER 8

1. In the K-Dur extended release coating process example, the analysts essentially tried to "test the sample into compliance." They kept testing it over and over again until they eventually obtained a passing result. They then disregarded the previous eight OOS results. There was no documented basis for disregarding the OOS results. Furthermore no corrective action was taken. They essentially did the same thing in the blend uniformity test, rejecting without basis results that they did not like, and retesting until they obtained a result they did like. They did not document a clear reason for rejecting the OOS results in either situation and they should have conducted a formal investigation. Remember that in cases like this, the product may, in fact, be defective and should not have been released for patient use.

2. This is a general discussion question with many avenues to explore. You might mention overall organizational concerns. For example, in any laboratory, personnel must have the proper training and education to do their jobs; there should be a process to ensure that instruments are routinely calibrated and maintained; a documentation system should be in place; variability should be avoided in routine procedures (e.g., preparation of reagents, operation of instruments). You might also discuss specific issues, such as the types of documents required (e.g., standard operating procedures, laboratory notebooks) and specific educational requirements for analysts and scientists. You might also explore differences between research laboratories and testing laboratories. For example, the processes for change tend to be flexible and informal in research laboratories. Testing laboratories do not change test procedures without careful evaluation and a formal change process.

Various ideas are reasonable with regard to the cases in this chapter. In the incident with the baby bottle, the analysts appeared to have been at fault. Perhaps better hiring and supervision practices would have helped avoid this situation. Perhaps a documented, consistent method of comparing experimental results to controls would help avoid future incidents (see also Chapter 24). In the case of the tissue sample mix-up, a standard and consistent method of labeling samples would have helped avoid problems. An overall quality-management system would help ensure that laboratory workers followed accepted laboratory practices, including labeling. Setting a quality system in place is the responsibility of senior personnel; following quality requirements is the responsibility of all personnel.

3. Various ideas are reasonable. Some topics to consider:

- **Documentation.**

   Standard operating procedures can direct individuals in proper practices that avoid contamination.

   Rigorous recording of all activities can help identify the causes of problems and exposures.

   Chain of custody forms and documents to track samples can ensure that dangerous materials are not misplaced.

   Proper labeling can help ensure that pathogens are not accidentally handled improperly.

- **Training.** All personnel must be adequately trained and their mastery of safety practices should be tested. No untrained personnel should have access to hazardous materials.

- **Regular inspections and audits** can ensure that all safety practices are followed.

- **Regular medical surveillance of staff** can help identify exposures to pathogens.

- **Access and security.** Access to the facility should be controlled by a security system.

- **Individual responsibility.** No quality system will work unless individuals take responsibility for compliance. Senior personnel are responsible for setting a good example and emphasizing the necessity for compliance and the dangers of becoming complacent over time.

CHAPTER 10

1. You should replace the cylinder cap, mark the cylinder as potentially damaged, and return it to the supplier. Do not attempt to use a lever to force the valve open.

2. Canvas shoes can absorb hazardous chemicals or become contaminated with biological materials. They do not provide a solid protective barrier for the feet.

3. Disposable latex gloves are too thin to provide adequate insulation from cryogenic materials. Thick gloves would be more appropriate.

4. A minimum of 15 minutes.

CHAPTER 11

1. The TWA means that the 50 ppm level is a time-weighted average over 8 hours. Actual air concentrations may vary from this average value.

2. **a.** Keeping the floor dry.
   - Being careful walking around water sources.
   - Using a secondary container to transport the solvent bottle.
   - Storing the solvent in a nonglass container.

   b. Notify other personnel of the problem.
   - Consider the toxic properties of the spilled liquid and order an evacuation if appropriate.
   - Be certain that any personal contamination with the solvent is cleaned up first.
   - If safe, proceed with proper spill cleanup techniques using a spill kit.

3. No! The air concentrations of a chemical that can be detected by the human nose are unrelated to toxic concentrations.

4. **a.** NaCl (least toxic, no TLV reported because inhalation is not a hazard); ethyl alcohol, acetone, phenol (most toxic) based on TLV values.

   b. The STEL value is the limit to which a person can be exposed for only 15 continuous minutes, up to four times during an 8-hour day. It is therefore a higher limit than the TLV value, which is the limit to which a person can be exposed 8 hours/day, 40 hours/week.

   c. Acetone. Extinguish using water spray, dry chemical, carbon dioxide, or alcohol-resistant foam.

5. Some of the potential combinations here are dangerously incompatible. It is always essential to consult the MSDS and any manufacturer's recommendations when storing unfamiliar chemicals. Note that this is only excerpted information, and there may be additional considerations in an actual laboratory.

   a. Nitric acid is a strong oxidizer and should be stored separately.

   b. Sodium chlorate is a strong oxidizer and should be stored separately.

   c. 2-amino-2-(hydroxymethyl)-1,3-propanediol (TRIS base), agarose, nicotinic acid (niacin), and D(+)-glactose are all relatively nonhazardous solids and can be stored together.

   d. Ethyl acetate, and 2-propanol are both flammable liquid solvents and can be stored together.

   e. Sodium hydroxide is a strong base and should be stored separately.

   f. Acrylamide is a neurotoxin and should be stored separately.

   g. Sodium is a flammable solid and should be stored separately.

CHAPTER 12

1. **a.** Some basic precautions:
   - Do not use glass containers for biohazards.
   - Use a sealed plastic container.
   - Use a secondary container for the flask.
   - Try to avoid using liquid cultures of pathogens.
   - Wear a lab coat to protect clothes.

   b. This individual's first priority is personal safety (assuming that no one else is involved in the accident). He should remove all contaminated items of clothing and immediately leave the room where the accident occurred. The doors should be closed and no one should reenter the room for at least 10 minutes, to allow any aerosols to settle. The scientist should then wash any skin areas that have been exposed to the pathogen. He should use an antiseptic soap and vigorously scrub these areas without abrading the skin for at least 10 minutes. Following the emergency plan for his institution, he or a colleague should notify the safety office, or set up the materials to decontaminate the laboratory. Cleanup should follow the procedures outlined in Table 12.25.

2. Approximately 864.

3. *Similarities:* Both refer to the importance of washing hands, decontaminating work surfaces, wearing PPE, avoiding mouth-pipetting, proper disposal of waste materials.
Differences: Universal Precautions are OSHA regulations. They were developed for handling human blood and blood-related products. Universal Precautions are more detailed, but all of the requirements of each are compatible.

CHAPTER 13

Manipulation Practice Problems: Exponents and Scientific Notation (from p. 222)

1. **a.** $2^2 = 4$ **b.** $3^3 = 27$
   c. $2^{-2} = 1/4 = 0.25$ **d.** $3^{-3} = 1/27 \approx 0.037$
   e. $10^2 = 100$ **f.** $10^4 = 10,000$
   g. $10^{-2} = 1/100 = 0.01$
   h. $10^{-4} = 1/10,000 = 0.0001$
   i. $5^0 = 1$

2. **a.** $2^2 \times 3^3 = 4 \times 27 = 108$
   b. $(14^3)(3^6) = (2744)(729) = 2,000,376$
   c. $5^5 - 2^3 = (3125) - (8) = 3117$
   d. $\dfrac{5^7}{8^4} = \dfrac{78125}{4096} \approx 19.07$
   e. $(6^{-2})(3^2) = (1/36)(9) = 9/36 = 0.25$
   f. $(-0.4)^3 + (9.6)^2 = (-0.064) + (92.16) \approx 92.1$
   g. $a^2 \times a^3 = a^5$
   h. $\dfrac{c^3}{c^{-6}} = c^9$
   i. $(3^4)^2 = 3^8 = 6561$ **j.** $(c^{-3})^{-5} = c^{15}$
   k. $\dfrac{13}{(43)^2 + (13)^3} = \dfrac{13}{1849 + 2197}$
   $= \dfrac{13}{4046} \approx 0.0032$
   l. $\dfrac{10^2}{10^3} = \dfrac{1}{10} = 10^{-1}$

3. *Larger numbers are **underlined:***
   a. 5×10^{-3} cm, $\underline{500 \times 10^{-1}}$ cm
   b. 300×10^{-3} μL, $\underline{3000 \times 10^{-2}}$ μL
   c. 3.200×10^{-6} m, $\underline{3200 \times 10^{-4}}$ m
   d. $\underline{0.001 \times 10^1}$ cm, 1×10^{-3} cm
   e. $\underline{0.008 \times 10^{-3}}$ L, 0.0008×10^{-4} L

4. **a.** $54.0 = 5.40 \times 10^1$
   ↰
   b. $4567 = 4.567 \times 10^3$
   ↰↰↰
   c. $0.345000 = 3.45000 \times 10^{-1}$
   ↱
   d. $10,000,000 = 1 \times 10^7$
   ↰↰↰↰↰↰↰
   e. $0.009078 = 9.078 \times 10^{-3}$
   ↱↱↱
   f. $540 = 5.40 \times 10^2$
   ↰↰
   g. $0.003040 = 3.040 \times 10^{-3}$
   ↱↱↱
   h. $200,567,987 = 2.00567987 \times 10^8$
   ↰↰↰↰↰↰↰↰

5. **a.** $12.3 \times 10^3 = 12,300$ **b.** $4.56 \times 10^4 = 45,600$
   c. $4.456 \times 10^{-5} = 0.00004456$
   d. $2.300 \times 10^{-3} = 0.002300$
   e. $0.56 \times 10^6 = 560,000$ **f.** $0.45 \times 10^{-2} = 0.0045$

6. **a.** $\dfrac{(4.725 \times 10^8)(0.0200)}{(3700)(0.770)} =$

   $\dfrac{(4.725 \times 10^8)(2.00 \times 10^{-2})}{(3.7 \times 10^3)(7.7 \times 10^{-1})} =$

   $\dfrac{9.45 \times 10^6}{28.49 \times 10^2} \approx 3.32 \times 10^3$

   b. $\dfrac{(1.93 \times 10^3)(4.22 \times 10^{-2})}{(8.8 \times 10^8)(6.0 \times 10^{-6})} =$

   $\dfrac{8.1446 \times 10}{52.8 \times 10^2} = 0.154 \times 10^{-1} \approx 1.5 \times 10^{-2}$

   c. $(4.5 \times 10^3) + (2.7 \times 10^{-2}) =$
   $4500 + 0.027 = 4500.027 = 4.5 \times 10^3$

   d. $(35.6 \times 10^4) - (54.6 \times 10^6) =$
   $(0.356 \times 10^6) - (54.6 \times 10^6) = -54.244 \times 10^6 \approx -5.42 \times 10^7$

   e. $(5.4 \times 10^{24}) + (3.4 \times 10^{26}) = (0.054 \times 10^{26}) + (3.4 \times 10^{26}) = 3.454 \times 10^{26} \approx 3.5 \times 10^{26}$

   f. $(5.7 \times 10^{-3}) - (3.4 \times 10^{-6}) = (5700 \times 10^{-6}) - (3.4 \times 10^{-6}) = 5696.6 \times 10^{-6} \approx 5.7 \times 10^{-3}$

7. **a.** $0.0050 \times 10^{-4} = 0.050 \times 10^{-5} = 0.50 \times 10^{-6} = 5.0 \times 10^{-7}$
   b. $15.0 \times 10^{-3} = 1.50 \times 10^{-2} = 0.150 \times 10^{-1} = 0.00150 \times 10^1$
   c. $5.45 \times 10^{-3} = 54.5 \times 10^{-4}$
   d. $100.00 \times 10^1 = 1.0000 \times 10^3$
   e. $6.78 \times 10^2 = 0.678 \times 10^3$
   f. $54.6 \times 10^2 = 0.00546 \times 10^6$
   g. $45.6 \times 10^8 = 4560 \times 10^6$
   h. $4.5 \times 10^{-3} = 450 \times 10^{-5}$
   i. $356.98 \times 10^{-3} = 0.035698 \times 10^1$
   j. $0.0098 \times 10^{-2} = 0.98 \times 10^{-4}$

Manipulation Practice Problems: Logarithms
(from p. 225)

1. **a.** 2   **b.** 3

2. **a.** $\log 100 = 2$ **b.** $\log 10,000 = 4$
   c. $\log 1,000,000 = 6$ **d.** $\log 0.0001 = -4$
   e. $\log 0.001 = -3$

3. **a.** $\log 7$ between 0 and 1
   b. $\log 65.9$ between 1 and 2
   c. $\log 89.0$ between 1 and 2
   d. $\log 0.45$ between 0 and -1
   e. $\log 0.0078$ between -2 and -3

4. **a.** 4.18   **b.** 2.54   **c.** -2.0
   d. -4.53 **e.** 3.082   **f.** -0.462

5. **a.** antilog $4.8990 \approx 7.925 \times 10^4$
   b. antilog $3.9900 \approx 9.772 \times 10^3$
   c. antilog $-0.5600 \approx 0.2754$
   d. antilog $-0.0089 \approx 0.9797$
   e. antilog $9.8999 \approx 7.941 \times 10^9$
   f. antilog $1.0000 = 10.00$
   g. antilog $8.9000 \approx 7.943 \times 10^8$

6. **a.** 0.35   **b.** 1.35   **c.** 2.35   **d.** 6.49

7. **a.** antilog $-4.56 \approx 2.75 \times 10^{-5}$ M
   b. antilog $-5.67 \approx 2.14 \times 10^{-6}$ M
   c. antilog $-7.00 \approx 1.00 \times 10^{-7}$ M
   d. antilog $-1.09 \approx 8.13 \times 10^{-2}$ M
   e. antilog $-10.1 \approx 7.94 \times 10^{-11}$ M

Manipulation Practice Problems: Measurements
(from p. 228)

1. Larger:

   a. 1 μm   **b.** 1 cm   **c.** 1000 mm
   d. 10 m   **e.** 1000 g = 1 kg   **f.** 1000 μm
   g. 1 mg   **h.** 1 nm   **i.** 10 nm
   j. 1 g   **k.** 1000 μL = 1 mL   **l.** 1 m
   m. 1 L   **n.** 500 cm   **o.** 1 pL = 0.001 nL
   p. 0.1 μm   **q.** 100 cm   **r.** 0.0001 m

2. 100 g = 0.1 kg

3. 12 oz ≈ 355 mL

Manipulation Practice Problems: Equations
(from p. 229)

1. **a.** A is equal to the value obtained if C is doubled.
   b. C is equal to the value obtained if A divided by D.
   c. Y is equal to the value obtained if X is multiplied by 2 and then 1 is added.

2. **a.** RCF increases if RPM increases.
   b. RCF decreases if r decreases.
   c. 11.18, 1000

3. **a.** $A = 2C$ $4 = 2(2)$
   b. $C = \dfrac{A}{D}$ $12 = \dfrac{36}{3}$
   c. $Y = 2(x) + 1$ $15 = 2(7) + 1$

4. **a.** $3X = 15$ $X = \dfrac{15}{3} = 5$
   b. $X = 25(5-4)$ $X = 25(1) = 25$
   c. $-X = 3X - 1$ mg
      1 mg $= 4X$
      $X = \dfrac{1}{4}$ mg
   d. $5X = 3X - 5$ mL $+ 34$ mL
      $2X = (34 - 5)$ mL
      $2X = 29$ mL
      $X = 14.5$ mL
   e. $\dfrac{X}{2} = 25(3)\text{cm}^2$
      $\dfrac{X}{2} = 75 \text{ cm}^2$
      $X = 2(75)\text{cm}^2$
      $X = 150 \text{ cm}^2$
   f. $X = 3.0$ cm $(2.0$ mg/mL$) (2.0)$
      $X = 3.0$ cm $(4.0$ mg/mL$)$
      $X = \dfrac{12.0 \text{ cm mg}}{\text{mL}}$
   g. $X = \dfrac{25 \text{ mg}}{\text{mL}}(4.0 \text{ mL})\dfrac{3.0 \text{ oz}}{\text{mg}}$
      $= \dfrac{300 \text{ mg mL oz}}{\text{mg mL}} = 300$ oz

5. **a.** 798 lb²
   b. $\dfrac{15.2 \text{ g}}{3.1 \text{ g}} \approx 4.9$
   c. $\dfrac{25.2 \text{ cm} \times 34.5 \text{ cm}}{3.00} \approx 290 \text{ cm}^2$
   d. $\dfrac{5 \text{ mL} \times 3 \text{ mL} \times 2 \text{ cm}}{2 \text{ cm}} = 15 \text{ mL}^2$

CHAPTER 14

Manipulation Practice Problems: Proportions
(from pp. 232–233)

1. **a.** $\dfrac{?}{5} = \dfrac{2}{10}$ $? = \dfrac{2 \times 5}{10} = 1$
   b. $\dfrac{?}{1 \text{ mL}} = \dfrac{10 \text{ cm}}{5 \text{ mL}}$
      $? = \dfrac{10 \text{ cm}(1 \text{ mL})}{5 \text{ mL}} = \dfrac{10 \text{ cm}}{5} = 2 \text{ cm}$
   c. $\dfrac{0.5 \text{ mg}}{10 \text{ mL}} = \dfrac{30 \text{ mg}}{?}$
      $? = \dfrac{(30 \text{ mg})(10 \text{ mL})}{0.5 \text{ mg}} = 600 \text{ mL}$
   d. $\dfrac{50}{?} = \dfrac{100}{100}$ $? = \dfrac{(50)100}{100} = 50$
   e. $\dfrac{?}{30 \text{ in}^2} = \dfrac{15 \text{ lb}}{100 \text{ in}^2}$
      $? = \dfrac{15 \text{ lb}(30 \text{ in}^2)}{100 \text{ in}^2} = 4.5 \text{ lb}$

f. $\dfrac{?}{15} = \dfrac{30}{90}$ $? = \dfrac{30(15)}{90} = 5$

g. $\dfrac{100}{10} = \dfrac{50}{?}$ $? = \dfrac{50(10)}{100} = 5$

2. a. $? = $ time to drive to Denver and back

$$\dfrac{?}{2 \text{ way}} = \dfrac{50 \text{ min}}{1 \text{ way}}$$

$$? = \dfrac{(50 \text{ min})(2 \text{ way})}{1 \text{ way}} \qquad ? = 100 \text{ min}$$

b. $? = $ amount of baking soda for 33 loaves

$$\dfrac{1 \text{ t}}{1 \text{ loaf}} = \dfrac{?}{33 \text{ loaf}}$$

$$? = \dfrac{1 \text{ t}(33 \text{ loaf})}{1 \text{ loaf}} \qquad ? = 33 \text{ t}$$

c. $? = $ cost of 100 magazines

$$\dfrac{1.50}{1 \text{ magazine}} = \dfrac{?}{100 \text{ magazine}}$$

$$? = \dfrac{1.50 (100 \text{ magazines})}{\text{magazine}} \qquad ? = 150.00$$

d. $? = $ margarine for 5 batches

$$\dfrac{1/4 \text{ cup}}{1 \text{ batch}} = \dfrac{?}{5 \text{ batch}}$$

$$? = \dfrac{1/4 \text{ cup} (5 \text{ batch})}{1 \text{ batch}} \qquad ? = 1.25 \text{ cups}$$

e. $? = $ baking powder for a half batch

$$\dfrac{1/8 \text{ t}}{1 \text{ batch}} = \dfrac{?}{1/2 \text{ batch}}$$

$$? = \dfrac{1/8 \, t(1/2 \text{ batch})}{1 \text{ batch}} = (1/8 \text{ t})(1/2) \; ? = 1/16 \text{ t}$$

f. $? = $ oz in 4.5 bags of chips

$$\dfrac{1 \text{ bag}}{3 \text{ oz}} = \dfrac{4.5 \text{ bag}}{?}$$

$$\dfrac{(3 \text{ oz})(4.5 \text{ bag})}{1 \text{ bag}} = ? = 13.5 \text{ oz}$$

Application Practice Problems: Proportions (from pp. 232–233)

1. $\dfrac{10^2 \text{ cell}}{10^{-2} \text{ mL}} = \dfrac{?}{1 \text{ mL}}$ $? = 10^4 \text{ cell}$

(10^2 cell in 10^{-2} mL $= 10^4$ cell/mL)

2. $\dfrac{10 \text{ mL}}{1 \text{ tube}} = \dfrac{?}{37 \text{ tubes}}$ $? = 370 \text{ mL}$

3. $\dfrac{5 \times 10^1 \text{ paramecia}}{20 \text{ mL}} = \dfrac{?}{10^5 \text{ mL}}$

$? = 2.5 \times 10^5$ paramecia

4. 5×10^4 g $= 50$ kg

$$\dfrac{315 \text{ larvae}}{1 \times 10^{-1} \text{ kg}} = \dfrac{?}{5 \times 10^1 \text{ kg}}$$

$? = 1.6 \times 10^5$ larvae

5. $\dfrac{1 \times 10^9 \text{ bacteria}}{1 \text{ mL}} = \dfrac{?}{1000 \times 10^3 \text{ mL}}$

$? = 1 \times 10^{15}$ bacteria

6. $\dfrac{30 \text{ g}}{1000 \text{ mL}} = \dfrac{?}{250 \text{ mL}}$ $? = 7.5 \text{ g}$

7. $\dfrac{100 \text{ mL ethanol}}{1000 \text{ mL}} = \dfrac{?}{10^5 \text{ mL}}$

$? = 10,000$ mL ethanol $= 10$ L

8. 10^4 μL $= 10$ mL

$\dfrac{50 \text{ mg}}{1 \text{ mL}} = \dfrac{?}{10 \text{ mL}}$ $? = 500 \text{ mg}$

9. $\dfrac{60 \text{ amino acids}}{\text{min}} = \dfrac{?}{10 \text{ min}}$ $? = 600$ amino acids

1000 amino acids $-$ 600 amino acids

$= 400$ amino acids left after 10 minutes

10.

	100 mL	1500 mL
NaCl	20.0 g	300 g
Na azide	0.001 g	0.015 g
Mg sulfate	1.0 g	15 g
Tris	15.0 g	225 g

11. a. $\dfrac{12 \text{ g}}{1 \text{ mol}} = \dfrac{?}{2 \text{ mol}}$ $? = 24 \text{ g}$

b. $\dfrac{12 \text{ g}}{1 \text{ mol}} = \dfrac{?}{0.5 \text{ mol}}$ $? = 6 \text{ g}$

12. a. $\dfrac{1 \text{ mole}}{58.5 \text{ g}} = \dfrac{1.5 \text{ mole}}{?}$ $? = 87.8 \text{ g}$

Alternatively, 58.5 g/mole $\times$ 1.5 mole $= 87.8$ g

b. $\dfrac{1 \text{ mole}}{58.5 \text{ g}} = \dfrac{0.75 \text{ mole}}{?}$ $? = 43.9 \text{ g}$

Manipulation Practice Problems: Percents (from p. 234)

1. a. $98/100 = 98\%$

b. $110/10,004 \approx \dfrac{1}{100} = 1\%$

c. $\dfrac{3}{15} = \dfrac{1}{5} = \dfrac{20}{100} = 20\%$

d. $\dfrac{45}{45002} \approx \dfrac{1}{1000} = 0.1\%$

2. a. $34\% = \dfrac{34}{100} = 0.34$

b. $89.5\% = \dfrac{89.5}{100} = 0.895$

c. $100\% = \dfrac{100}{100} = 1$

d. $250\% = \dfrac{250}{100} = \dfrac{2.5}{1} = 2.5$

e. $0.45\% = \dfrac{0.45}{100} = 0.0045$

f. $0.001\% = \dfrac{0.001}{100} = 0.00001$

3. a. 15% of 450 = 0.15 × 450 = 67.5

   b. 25% of 700 = 0.25 × 700 = 175

   c. 0.01% of 1000 = 0.0001 × 1000 = 0.1

   d. 10% of 100 = 0.1 × 100 = 10

   e. 12% of 500 = 0.12 × 500 = 60

   f. 150% of 1000 = 1.5 × 1000 = 1500

4. a. $\dfrac{15}{45} = 0.33 = 33\%$

   b. $\dfrac{2}{2} = 1 = 100\%$

   c. $\dfrac{10}{100} = 0.1 = 10\%$

   d. $\dfrac{1}{100} = 0.01 = 1\%$

   e. $\dfrac{1}{1000} = 0.001 = 0.1\%$

   f. $\dfrac{6}{40} = 0.15 = 15\%$

   g. $\dfrac{0.1}{0.5} = 0.2 = 20\%$

   h. $\dfrac{0.003}{89} \approx 0.0000337 \approx 0.0034\%$

   i. $\dfrac{5}{10} = 0.5 = 50\%$

   j. 0.05 = 5%   **k.** 0.0034 = 0.34%

   l. 0.25 = 25%   **m.** 0.01 = 1%

   n. 0.10 = 10%   **o.** 0.0001 = 0.01%

   p. 0.0078 = 0.78%   **q.** 0.50 = 50%

5. a. 20% of 100 = 20   **b.** 20% of 1000 = 200

   c. 20% of 10,000 = 2000   **d.** 10% of 567 = 56.7

   e. 15% of 1000 = 150   **f.** 50% of 950 = 475

   g. 5% of 100 = 5   **h.** 1% of 876 = 8.76

   i. 30% of 900 = 270

6. a. $\dfrac{1 \text{ part}}{100 \text{ parts}} = 1\%$

   b. $\dfrac{3 \text{ parts}}{50 \text{ parts}} = 6\%$

   c. $\dfrac{15 \text{ parts}}{(15 + 45) \text{ parts}} = \dfrac{15 \text{ parts}}{60 \text{ parts}} = 25\%$

   d. $\dfrac{0.05 \text{ parts}}{1 \text{ part}} = 0.05 = 5\%$

   e. $\dfrac{1 \text{ part}}{25 \text{ parts}} = 0.04 = 4\%$

   f. $\dfrac{2.35 \text{ parts}}{(2.35 + 6.50) \text{ parts}} = \dfrac{2.35 \text{ parts}}{8.85 \text{ parts}}$

                     $\approx 0.266 = 26.6\%$

Application Practice Problems: Percents (from pp. 235–236)

1. $\dfrac{25}{55}$ work 20+ hours $\approx 0.4545 \approx 45\%$

   $\dfrac{30}{55}$ work 19− hours $\approx 0.545 \approx 55\%$

2. 50% = 0.5 of total; total = 100 mL

   100 × 0.5 = 50 mL ethanol

3. 30% solution of ethylene glycol

   30% of total (250 mL)

   0.30 × 250 = 75 mL ethylene glycol

   Measure out the ethylene glycol and add water until the volume is 250 mL

4. 10% acetonitrile for 100 mL: 10 mL

   25% MeOH (methanol):   25 mL

   Measure out acetonitrile and MeOH, combine, add water until the volume is 100 mL

5. $\dfrac{5 \text{ mL}}{100 \text{ mL}} = 0.05 = 5\%$ propanol solution

6. $\dfrac{15 \text{ mL EtOH}}{700 \text{ mL}} = 0.021 = 2.1\%$ ethanol solution

7. $\dfrac{10 \text{ μL}}{1 \text{ mL}} = \dfrac{10 \text{ μL}}{1000 \text{ μL}} = \dfrac{1}{100} = 0.01$

   = 1% methanol solution

8. $\dfrac{15 \text{ mL acetone}}{1 \text{ liter}} = \dfrac{15 \text{ mL}}{1000 \text{ mL}}$

   = 0.015 or 1.5% acetone solution

9. 1000 insects, 5% survive

   1000 × 0.05 = 50 insects survive first winter

   50 insects × 100 = 5000 insects after first summer

   5000 × 0.05 = 250 insects after second winter

10. 250,000 people × 0.15 = 37,500 people die

11. Since thymine is 24%, adenine must also be 24% for a total of 48%. This leaves 52% that must be split equally between guanine and cytosine, at 26% each.

12. Of the subjects fed regular chips, 88 experienced gastrointestinal problems compared with 79 subjects fed Olestra chips.

Manipulation Practice Problems: Density (from p. 237)

1. olive oil

2. $\dfrac{57.9 \text{ g}}{3.00 \text{ cm}^3} = \dfrac{?}{1 \text{ cm}^3}$ $? = \dfrac{19.3 \text{ g}}{\text{cm}^3}$

3. density = 1.26 g/mL

   $\dfrac{1.26 \text{ g}}{1 \text{ mL}} = \dfrac{20.0 \text{ g}}{?}$ $? \approx 15.9 \text{ mL}$

Manipulation Practice Problems: Unit Conversions (from p. 240)

1. a. 3.00 ft to cm (proportion method):

    $\dfrac{?}{3.00 \text{ ft}} = \dfrac{12 \text{ in}}{1 \text{ ft}}$

$$? = \frac{(12 \text{ in})(3.00 \text{ ft})}{1 \text{ ft}} \qquad ? = 36 \text{ in}$$

$$\frac{?}{36 \text{ in}} = \frac{2.54 \text{ cm}}{1 \text{ in}}$$

$$? = \frac{(2.54 \text{ cm})(36 \text{ in})}{1 \text{ in}} \qquad ? \approx 91.4 \text{ cm}$$

3.00 ft to cm (conversion factor method)

$$3.00 \text{ ft} \times \frac{12 \text{ in}}{1 \text{ ft}} \times \frac{2.54 \text{ cm}}{1 \text{ in}} \approx 91.4 \text{ cm}$$

b. 100 mg to g (proportion method):

$$\frac{?}{100 \text{ mg}} = \frac{1 \text{ g}}{1000 \text{ mg}}$$

$$? = \frac{(1 \text{ g})(100 \text{ mg})}{1000 \text{ mg}} = 0.1 \text{ g}$$

100 mg to g (conversion factor method)

$$100 \text{ mg} \times \frac{1 \text{ g}}{1000 \text{ mg}} = 0.1 \text{ g}$$

c. 12.0 in to miles (proportion method):

$$\frac{?}{12.0 \text{ in}} = \frac{1 \text{ ft}}{12.0 \text{ in}}$$

$$? = \frac{(1 \text{ ft})(12.0 \text{ in})}{12.0 \text{ in}} = 1 \text{ ft}$$

$$\frac{?}{1 \text{ ft}} = \frac{1 \text{ mi}}{5280 \text{ ft}}$$

$$? = \frac{(1 \text{ mi})(1 \text{ ft})}{5280 \text{ ft}} = \frac{1 \text{ mi}}{5280} \approx 0.000189 \text{ mi}$$

$$? = 1.89 \times 10^{-4} \text{ mile}$$

12.0 in to mile (conversion factor method)

$$12.0 \text{ in} \times \frac{1 \text{ ft}}{12.0 \text{ in}} \times \frac{1 \text{ mi}}{5280 \text{ ft}} \approx 1.89 \times 10^{-4} \text{ mi}$$

d. 100 in to km (proportion method):

$$\frac{?}{100 \text{ in}} = \frac{2.54 \text{ cm}}{1 \text{ in}}$$

$$? = \frac{(2.54 \text{ cm})(100 \text{ in})}{1 \text{ in}} \qquad ? = 254 \text{ cm}$$

$$\frac{?}{254 \text{ cm}} = \frac{1 \text{ m}}{100 \text{ cm}}$$

$$? = \frac{(254 \text{ cm})(1 \text{ m})}{100 \text{ cm}} \qquad ? = 2.54 \text{ m}$$

$$\frac{?}{2.54 \text{ m}} = \frac{1 \text{ km}}{1000 \text{ m}}$$

$$? = \frac{(2.54 \text{ m})(1 \text{ km})}{1000 \text{ m}} \qquad ? = 2.54 \times 10^{-3} \text{ km}$$

100 in to km (conversion factor method):

$$100 \text{ in} \times \frac{2.54 \text{ cm}}{1 \text{ in}} \times \frac{1 \text{ m}}{100 \text{ cm}} \times \frac{1 \text{ km}}{1000 \text{ m}}$$

$$= 2.54 \times 10^{-3} \text{ km}$$

e. 10.0555 lb to oz (proportion method):

$$\frac{?}{10.0555 \text{ lb}} = \frac{16 \text{ oz}}{1 \text{ lb}}$$

$$? = \frac{(16 \text{ oz})(10.0555 \text{ lb})}{1 \text{ lb}} \qquad ? = 160.888 \text{ oz}$$

10.0555 lb to oz (conversion factor method):

$$10.0555 \text{ lb} \times \frac{16 \text{ oz}}{1 \text{ lb}} = 160.888 \text{ oz}$$

f. 18.989 lb to g (proportion method):

$$\frac{?}{18.989 \text{ lb}} = \frac{453.6 \text{ g}}{1 \text{ lb}}$$

$$? = \frac{(18.989 \text{ lb})(453.6 \text{ g})}{1 \text{ lb}} \qquad ? \approx 8613.4 \text{ g}$$

18.989 lb to g (conversion factor method):

$$18.989 \text{ lb} \times \frac{453.6 \text{ g}}{1 \text{ lb}} \approx 8613.4 \text{ g}$$

g. 13 mi to km (proportion method):

$$\frac{?}{13 \text{ mi}} = \frac{1.609 \text{ km}}{1 \text{ mi}}$$

$$? = \frac{(1.609 \text{ km})(13 \text{ mi})}{1 \text{ mi}} \qquad ? \approx 21 \text{ km}$$

13 mi to km (conversion factor method):

$$13 \text{ mi} \times \frac{1.609 \text{ km}}{1 \text{ mi}} \approx 21 \text{ km}$$

h. 150 mL to L (proportion method):

$$\frac{?}{150 \text{ mL}} = \frac{1 \text{ L}}{1000 \text{ mL}}$$

$$? = \frac{(150 \text{ mL})(1 \text{ L})}{1000 \text{ mL}} \qquad ? = 0.150 \text{ L}$$

150 mL to L (conversion factor method):

$$150 \text{ mL} \times \frac{1 \text{ L}}{1000 \text{ mL}} = 0.150 \text{ L}$$

i. 56.7009 cm to nm (proportion method):

$$\frac{?}{56.7009 \text{ cm}} = \frac{1 \text{ m}}{100 \text{ cm}}$$

$$? = 0.567009 \text{ m}$$

$$\frac{?}{0.567009 \text{ m}} = \frac{10^9 \text{ nm}}{1 \text{ m}}$$

$$? = 5.67009 \times 10^8 \text{ nm}$$

56.7009 cm to nm (conversion factor method):

$$56.7009 \text{ cm} \times \frac{1 \text{ m}}{10^2 \text{ cm}} \times \frac{10^9 \text{ nm}}{1 \text{ m}}$$

$$= 5.67009 \times 10^8 \text{ nm}$$

j. 500 nm to μm (proportion method):

$$\frac{?}{500 \text{ nm}} = \frac{1 \text{ μm}}{1000 \text{ nm}}$$

$$? = \frac{(1 \text{ μm})(500 \text{ nm})}{1000 \text{ nm}} \qquad ? = 0.500 \text{ μm}$$

500 nm to μm (conversion factor method):

$$500 \text{ nm} \times \frac{1 \text{ m}}{10^9 \text{ nm}} \times \frac{10^6 \text{ μm}}{1 \text{ m}} = 0.500 \text{ μm}$$

k. 10.0 nm to inches (proportion method):

$$\frac{?}{10 \text{ nm}} = \frac{1 \text{ cm}}{10^7 \text{ nm}} \qquad ? = 1 \times 10^{-6} \text{ cm}$$

$$\frac{?}{1 \times 10^{-6} \text{ cm}} = \frac{1 \text{ in}}{2.54 \text{ cm}} \qquad ? = 3.94 \times 10^{-7} \text{ in}$$

10.0 nm to inches (conversion factor method):

$$10.0 \text{ nm} \times \frac{1 \text{ m}}{10^9 \text{ nm}} \times \frac{100 \text{ cm}}{1 \text{ m}} \times \frac{1 \text{ in}}{2.54 \text{ cm}}$$

$$= 3.94 \times 10^{-7} \text{ in}$$

2. 10 km to miles (proportion method):

$$\frac{?}{10 \text{ km}} = \frac{1 \text{ mi}}{1.609 \text{ km}}$$

$$? = \frac{(1 \text{ mi})(10 \text{ km})}{1.609 \text{ km}} \qquad ? \approx 6.2 \text{ mi}$$

10 km to miles (conversion factor method):

$$10 \text{ km} \times \frac{1 \text{ mi}}{1.609 \text{ km}} \approx 6.2 \text{ mi}$$

3. $26.2 \text{ mi} \times \dfrac{1.609 \text{ km}}{1 \text{ mi}} \approx 42.2 \text{ km}$

4. 5 ft 4 in to meters (proportion method):

$$\frac{?}{5 \text{ ft}} = \frac{12 \text{ in}}{\text{ft}} \qquad ? = \frac{(12 \text{ in})(5 \text{ ft})}{\text{ft}}$$

$? = 60 \text{ in, plus } 4 \text{ in} = 64 \text{ in}$

$$\frac{?}{64 \text{ in}} = \frac{1 \text{ m}}{39.37 \text{ in}}$$

$$? = \frac{(1 \text{ m})(64 \text{ in})}{39.37 \text{ in}} \qquad\qquad ? \approx 1.6 \text{ m}$$

5 ft 4 in to meters (conversion factor method):

$$5 \text{ ft} \times \frac{12 \text{ in}}{1 \text{ ft}} = 60 \text{ in}$$

$60 \text{ in} + 4 \text{ in} = 64 \text{ in}$

$$64 \text{ in} \times \frac{1 \text{ m}}{39.37 \text{ in}} \approx 1.6 \text{ m}$$

5. 45 mi into kilometers (proportion method):

$$\frac{?}{45 \text{ mi}} = \frac{1.609 \text{ km}}{1 \text{ mi}}$$

$$? = \frac{(1.609 \text{ km})(45 \text{ mi})}{1 \text{ mi}} \qquad ? \approx 72 \text{ km}$$

45 mi into km (conversion factor method):

$$45 \text{ mi} \times \frac{1 \text{ km}}{0.6214 \text{ mi}} \approx 72 \text{ km}$$

6. 55 mph to kilometers per hour
(proportion method):

$1 \text{ mph} = 1.609 \text{ kph}$

$$\frac{1 \text{ mph}}{1.609 \text{ kph}} = \frac{55 \text{ mph}}{?}$$

$? \approx 88.5 \text{ kph}$

55 mph to kilometers per hour (conversion
factor method):

$$55 \text{ mph} \times \frac{1 \text{ kph}}{0.6214 \text{ mph}} \approx 88.5 \text{ kph}$$

7. 3.0 ton to kilograms (proportion method):

$$\frac{?}{3.0 \text{ ton}} = \frac{2000 \text{ lb}}{\text{ton}}$$

$$? = \frac{(2000 \text{ lb})(3.0 \text{ ton})}{\text{ton}} \qquad ? = 6000 \text{ lb}$$

$$\frac{?}{6000 \text{ lb}} = \frac{1 \text{ kg}}{2.205 \text{ lb}}$$

$$? = \frac{(1 \text{ kg})(6000 \text{ lb})}{2.205 \text{ lb}}$$

$? \approx 2721.1 \text{ kg}$ (Round to 2700 kg)

3.0 ton to kg (conversion factor method):

$$3.0 \text{ ton} \times \frac{2000 \text{ lb}}{1 \text{ ton}} \times \frac{1 \text{ kg}}{2.205 \text{ lb}} \approx 2721.1 \text{ kg}$$

(Round to 2700 kg)

8. a. \$2.50 for 12 oz (proportion method):

$$\frac{?}{12 \text{ oz}} = \frac{1 \text{ lb}}{16 \text{ oz}}$$

$$? = \frac{(1 \text{ lb})(12 \text{ oz})}{16 \text{ oz}} \qquad ? = 0.75 \text{ lb}$$

$$\frac{?}{0.75 \text{ lb}} = \frac{453.6 \text{ g}}{1 \text{ lb}}$$

$$? = \frac{(453.6 \text{ g})(0.75 \text{ lb})}{1 \text{ lb}} \qquad ? = 340.2 \text{ g}$$

$340.2 \text{ g}/\$2.50 \approx 136.1 \text{ g}/\$1.00$

b. \$3.67 for 250 g (proportion method):

$$\frac{250 \text{ g}}{3.67} \approx 68.12 \text{ g}/ 1.00$$

c. \$4.50 for 0.300 kg:

$$\frac{?}{0.300 \text{ kg}} = \frac{1000 \text{ g}}{\text{kg}}$$

$$? = \frac{(1000 \text{ g})(0.300 \text{ kg})}{\text{kg}} \qquad ? = 300 \text{ g}$$

$300 \text{ g}/\$4.50 \approx 66.7 \text{ g}/\$1.00$

d. \$2.35 for 0.75 lb:

$$\frac{?}{0.75 \text{ lb}} = \frac{453.6 \text{ g}}{\text{lb}}$$

$$? = \frac{(453.6 \text{ g})(0.75 \text{ lb})}{\text{lb}} \qquad ? = 340.2 \text{ g}$$

$340.2 \text{ g}/\$2.35 \approx 144.8 \text{ g}/\$1.00 = \text{least expensive}$

9. (proportion method)

$$\frac{?}{1 \text{ week}} = \frac{7 \text{ days}}{1 \text{ week}} \qquad ? = 7 \text{ days}$$

$$\frac{?}{7 \text{ days}} = \frac{24 \text{ h}}{1 \text{ day}} \qquad ? = 24 \times 7 = 168 \text{ h}$$

$$\frac{?}{168 \text{ h}} = \frac{60 \text{ min}}{1 \text{ h}}$$

$? = 168 \times 60 = 10{,}080 \text{ minutes}$

$$\frac{?}{10080 \text{ min}} = \frac{60 \text{ s}}{1 \text{ min}}$$

$? = 10080 \times 60 = 604,800$ s
(conversion factor method)

$$1 \text{ week} \times \frac{7 \text{ days}}{1 \text{ week}} = 7 \text{ days}$$

$$7 \text{ days} \times \frac{24 \text{ h}}{1 \text{ day}} = 168 \text{ h}$$

$$168 \text{ h} \times \frac{60 \text{ min}}{1 \text{ h}} = 10,080 \text{ min}$$

$$10,080 \text{ min} \times \frac{60 \text{ s}}{\text{min}} = 604,800 \text{ s}$$

10. 1/52 year = 1 week = 168 hours = 10,080 minutes
= 604,800 seconds

11. 1 year = 52 weeks = 364 days = 524,160 minutes
= 31,449,600 seconds

12. 1/5280 mi = 1/3 yds = 1 ft = 12 in
or 0.000189 mi ≈ 0.333 yds ≈ 1 ft ≈ 12 in

13. 0.0156 gal ≈ 0.125 pt = 2 oz

14. $1 \text{ km} = 10^3 \text{ m} = 10^5 \text{ cm} = 10^6 \text{ mm} = 10^9 \text{ μm} = 10^{12} \text{ nm}$

15. $10^{-12} \text{ km} = 10^{-9} \text{ m} = 10^{-7} \text{ cm} = 10^{-6} \text{ mm}$
$= 10^{-3} \text{ μm} = 1 \text{ nm}$

16. $2.5 \times 10^{-5} \text{ km} = 2.5 \times 10^{-2} \text{ m} = 2.5 \text{ cm} = 25 \text{ mm}$
$= 2.5 \times 10^4 \text{ μm} = 2.5 \times 10^7 \text{ nm}$

17. **a.** $6.25 \text{ mm} = 6.25 \times 10^3 \text{ μm}$

b. $0.00896 \text{ m} = 8.96 \text{ mm}$

c. $9876000 \text{ nm} = 9.876 \text{ mm}$

18. $3 \times 10^{-3} \text{ km} = 3 \text{ m} = 3 \times 10^2 \text{ cm} = 1.2 \times 10^2 \text{ in}$

19. $5 \text{ kg} = 5 \times 10^3 \text{ g} = 5 \times 10^6 \text{ mg} = 5 \times 10^9 \text{ μg}$

20. $8.9 \times 10^{-6} \text{ kg} = 8.9 \times 10^{-3} \text{ g} = 8.9 \text{ mg} = 8.9 \times 10^3 \text{ μg}$

21. $2 \times 10^{-17} \text{ kg} = 2 \times 10^{-14} \text{ g} = 2 \times 10^{-11} \text{ mg}$
$= 2 \times 10^{-8} \text{ μg}$

22. **a.** $0.8657 \text{ g} = 8.657 \times 10^{-1} \text{ g} = 8.657 \times 10^2 \text{ mg}$

b. $526 \text{ kg} = 5.26 \times 10^2 \text{ kg} = 5.26 \times 10^8 \text{ mg}$

c. $63 \text{ g} = 6.3 \times 10^1 \text{ g} = 6.3 \times 10^7 \text{ μg}$

d. $2.63 \times 10^{-6} \text{ μg} = 2.63 \times 10^{-15} \text{ kg}$

23. $1 \text{ Ci} = 3.7 \times 10^{10} \text{ dps} = 2.2 \times 10^{12} \text{ dpm}$

24. $1 \text{ Ci} = 1000 \text{ mCi} = 10^6 \text{ μCi} = 3.7 \times 10^{10} \text{ dps}$

25. $2.7 \times 10^{-6} \text{ Ci} = 2.7 \times 10^{-3} \text{ mCi} = 2.7 \text{ μCi} = 10^5 \text{ dps}$

26. $1 \times 10^{-4} \text{ Ci} = 0.1 \text{ mCi} = 100 \text{ μCi} = 3.7 \times 10^6 \text{ dps} \approx 2.2 \times 10^8 \text{ dpm}$

27. $1 \text{ Ci} = 10^3 \text{ mCi} = 10^6 \text{ μCi} = 3.7 \times 10^{10} \text{ dps}$
$= 3.7 \times 10^{10} \text{ Bq}$

28. $2.5 \times 10^{-4} \text{ Ci} = 2.5 \times 10^{-1} \text{ mCi} = 250 \text{ μCi}$
$= 9.25 \times 10^6 \text{ dps} = 9.25 \times 10^6 \text{ Bq}$

Application Practice Problems: Unit Conversions (from p. 241)

1. If the medium requires 5 g/1 L then 125 g of glucose are required for 25 L. The technician added:

$$\frac{?}{0.24 \text{ lb}} = \frac{453.6 \text{ g}}{1 \text{ lb}}$$

$$? = \frac{(453.6 \text{ g})(0.24 \text{ lb})}{\text{lb}} \qquad ? \approx 108.86 \text{ g}$$

The technician, therefore, added too little glucose.

2.

	For 1 L	For 1 mL
NaCl	20.00 g	20 mg
Na azide	0.001 g	0.001 mg
Mg sulfate	1.000 g	1.000 mg
Tris	15.00 g	15.00 mg

3. 680 mg = 0.680 g

so there are 0.680 g enzyme/1 g powder

need $\dfrac{10.0 \text{ oz}}{100.0 \text{ L}} = \dfrac{?}{500.0 \text{ L}}$ $\qquad ? = 50.0 \text{ oz}$

$$50.0 \text{ oz} \times \frac{1 \text{ lb}}{16 \text{ oz}} \times \frac{453.6 \text{ g}}{1 \text{ lb}} = 1417.5 \text{ g}$$

$$\frac{0.680 \text{ g enzyme}}{1 \text{ g powder}} = \frac{1417.5 \text{ g enzyme}}{?}$$

$? \approx 2084.56$ g powder → Need 2.08 kg powder

Manipulation/Application Practice Problems: Concentration (from pp. 241–242)

1. $\dfrac{3 \text{ g}}{250 \text{ mL}} = \dfrac{?}{1000 \text{ mL}}$

$? = \dfrac{(3 \text{ g})(1000 \text{ mL})}{250 \text{ mL}}$ $\quad ? = 12 \text{ g}$

2. $\dfrac{25 \text{ g}}{1000 \text{ mL}} = \dfrac{?}{100 \text{ mL}}$

$? = \dfrac{25 \text{ g} \times 100 \text{ mL}}{1000 \text{ mL}}$ $\quad ? = 2.5 \text{ g}$

3. $\dfrac{1 \text{ mg}}{1 \text{ mL}} = \dfrac{1 \text{ g}}{1 \text{ L}}$ so $\dfrac{1 \text{ g}}{1 \text{ L}} = \dfrac{?}{15 \text{ L}}$ $\qquad ? = 15 \text{ g}$

4. $\dfrac{0.005 \text{ g Tris base}}{\text{L}} = \dfrac{?}{10^{-3} \text{L}}$

$? = \dfrac{(0.005 \text{ g})10^{-3} \text{ L}}{\text{L}}$

$? = 0.005 \text{ g} \times 10^{-3} \text{ g}$ $\quad ? = 5 \times 10^{-6} \text{ g}$

5. $\dfrac{300 \text{ ng dioxin}}{100 \text{ g diapers}} = \dfrac{?}{1000 \text{ g}}$

$? = 3000 \text{ ng} = 3 \text{ μg dioxin}$

6. $\dfrac{?}{5 \text{ mL}} = \dfrac{5 \times 10^{-3} \text{ moles}}{1000 \text{ mL}}$

$? = \dfrac{(5 \times 10^{-3} \text{ moles}) 5 \text{ mL}}{10^3 \text{ mL}}$

$? = 25 \times 10^{-6} \text{ moles} = 2.5 \times 10^{-5} \text{ moles}$

7. $(1 \text{ L} = 10^6 \text{ μL})$

$\dfrac{?}{1 \text{ μL}} = \dfrac{0.1 \text{ moles}}{10^6 \text{ μL}}$ $\quad ? = \dfrac{(0.1 \text{ moles}) \text{ μL}}{10^6 \text{ μL}}$

$? = 0.1 \times 10^{-6} = 1 \times 10^{-7} \text{ moles}$

8. $(1 \text{ L} = 10^3 \text{ mL})$ $\qquad \dfrac{?}{78 \text{ mL}} = \dfrac{10^{-2} \text{ g}}{10^3 \text{ mL}}$

$? = \dfrac{(10^{-2} \text{ g})78 \text{ mL}}{10^3 \text{ mL}}$

$? = 78 \times 10^{-5} \text{ g} = 7.8 \times 10^{-4} \text{ g}$

9. $\dfrac{500 \text{ U}}{\text{mg}} = \dfrac{?}{3 \text{ mg}}$ $\qquad ? = \dfrac{(500 \text{ U})3 \text{ mg}}{\text{mg}}$ $\quad ? = 1500 \text{ U}$

10. $\dfrac{2500 \text{ U}}{\text{mg}} = \dfrac{?}{5 \text{ mg}}$

$? = \dfrac{(2500 \text{ U})5 \text{ mg}}{1 \text{ mg}}$ $\qquad ? = 1.25 \times 10^4 \text{ U}$

11. $\dfrac{?}{10^3 \text{ mL}} = \dfrac{3 \text{ g}}{\text{mL}}$

$? = \dfrac{(3 \text{ g}) \times 10^3 \text{ mL}}{\text{mL}}$ $\qquad ? = 3 \times 10^3 \text{ g}$

12. $\dfrac{5 \text{ µg}}{\text{L}} = \dfrac{5 \text{ µg}}{10^3 \text{ mL}}$

a. $\dfrac{5 \text{ µg}}{10^3 \text{ mL}} = \dfrac{?}{50 \text{ mL}}$

$? = \dfrac{(5 \text{ µg})50 \text{ mL}}{10^3 \text{ mL}}$

$? = \dfrac{250 \text{ µg}}{10^3} = 0.25 \text{ µg}$

b. $\dfrac{5 \text{ µg}}{10^3 \text{ mL}} = \dfrac{?}{500 \text{ mL}}$

$? = \dfrac{(5 \text{ µg})500 \text{ mL}}{10^3 \text{ mL}}$

$? = \dfrac{2500 \text{ µg}}{10^3} = 2.5 \text{ µg}$

c. $\dfrac{5 \text{ µg}}{10^3 \text{ mL}} = \dfrac{?}{100 \text{ mL}}$

$? = \dfrac{(5 \text{ µg})100 \text{ mL}}{10^3 \text{ mL}}$

$? = \dfrac{500 \text{ µg}}{10^3} = 0.5 \text{ µg}$

d. $\dfrac{5 \text{ µg}}{10^6 \text{ µL}} = \dfrac{?}{100 \text{ µL}}$

$? = \dfrac{(5 \text{ µg})100 \text{ µL}}{10^6 \text{ µL}} = \dfrac{500 \text{ µg}}{10^6} = 5 \times 10^{-4} \text{ µg}$

13. a. $\dfrac{0.5 \text{ mg}}{1 \text{ mL}} = \dfrac{?}{5 \text{ mL}}$

$? = \dfrac{(0.5 \text{ mg})5 \text{ mL}}{1 \text{ mL}}$ $\qquad ? = 2.5 \text{ mg}$

b. $\dfrac{0.5 \text{ mg}}{1 \text{ mL}} = \dfrac{?}{0.5 \text{ mL}}$

$? = \dfrac{(0.5 \text{ mg})0.5 \text{ mL}}{1 \text{ mL}}$ $\qquad ? = 0.25 \text{ mg}$

c. $1 \text{ mL} = 10^3 \text{ µL}$ $\qquad \dfrac{0.5 \text{ mg}}{10^3 \text{ µL}} = \dfrac{?}{100 \text{ µL}}$

$? = \dfrac{(0.5 \text{ mg})100 \text{ µL}}{10^3 \text{ µL}}$ $\qquad ? = 0.05 \text{ mg}$

d. $1000 \text{ µL} = 1 \text{ mL}$

$\dfrac{0.5 \text{ mg}}{\text{mL}} = \dfrac{?}{\text{mL}}$ $\qquad ? = 0.5 \text{ mg}$

14. $\dfrac{100 \text{ molecules}}{10^9 \text{ molecules}} = \dfrac{1 \times 10^2}{10^9} = \dfrac{1}{10^7} < \dfrac{1}{10^6}$

Cannot be detected

15. $1 \text{ g}/10^6 \text{ kg} = \dfrac{1 \text{ g}}{10^9 \text{ g}}$ (more pure)

$\dfrac{10^{-2} \text{ mg}}{10^{-3} \text{ kg}} = \dfrac{10^{-5} \text{ g}}{\text{g}} = \dfrac{1 \text{ g}}{10^5 \text{ g}}$

Manipulation Practice Problems: Dilutions (Part A) (from pp. 243–244)

1. a. 1 part orange juice in 4 parts total = 1/4

  b. 1 part orange juice to 3 parts water = 1:3

  c. 1:3 or 1/4

2. a. 1 mL/10 mL $\qquad$ **b.** 1 mL/11 mL

  c. 3 mL/30 mL $\qquad$ **d.** 3 mL/30 mL

  e. 0.5 mL/11.5 mL

3. a. 1/10 $\qquad$ **b.** 1/11 $\qquad$ **c.** 1/4

  d. 1/2 $\qquad$ **e.** 1/5

4. a. 1:9 $\qquad$ **b.** 1:10 $\qquad$ **c.** 1:9

  d. 1:9 $\qquad$ **e.** 1:22

5. $\quad$ 0.5 mL blood
  1.0 mL H$_2$O
  + 3.0 mL reagent
  $\overline{\quad 4.5 \text{ mL} \quad}$
  0.5/4.5 dilution = 1/9

Manipulation Practice Problems: Dilutions (Part B) (from pp. 244–245)

1. $\dfrac{?}{10 \text{ mL}} = \dfrac{1}{10}$ $\qquad ? = \dfrac{(1)10 \text{ mL}}{10}$

$? = 1 \text{ mL blood}$

1 mL + 9 mL will give 1/10 dilution

2. $\dfrac{?}{250 \text{ mL}} = \dfrac{1}{300}$ $\qquad ? = \dfrac{1(250 \text{ mL})}{300}$

$? \approx 0.833 \text{ mL blood}$
  0.833 mL + 249.167 mL = 250 mL

3. $\dfrac{?}{1 \text{ mL}} = \dfrac{1}{50}$ $\qquad ? = \dfrac{1(1 \text{ mL})}{50}$

$? = 0.02 \text{ mL blood}$
  0.02 mL + 0.98 mL = 1 mL

4. $\dfrac{?}{1000 \text{ µL}} = \dfrac{1}{100}$ $\qquad ? = \dfrac{1(1000 \text{ µL})}{100}$

$? = 10 \text{ µL food coloring}$
  10 µL + 990 µL = 1000 µL

5. $\dfrac{?}{23\text{ mL}} = \dfrac{3}{5}$ $\qquad$ $? = \dfrac{3(23\text{ mL})}{5}$

$? = 13.8\text{ mL of Q}$
$\quad 13.8\text{ mL} + 9.2\text{ mL} = 23\text{ mL}$

6. $\dfrac{?}{10\text{ mL}} = \dfrac{1}{10}$ $\qquad$ $? = \dfrac{1(10\text{ mL})}{10}$

$? = 1\text{ mL stock}$
$\quad 1\text{ mL} + 9\text{ mL} = 10\text{ mL}$

7. $\dfrac{?}{15\text{ mL}} = \dfrac{1}{5}$ $\qquad$ $? = \dfrac{1(15\text{ mL})}{5}$

$? = 3\text{ mL stock}$
$\quad 3\text{ mL} + 12\text{ mL} = 15\text{ mL}$

8. $\dfrac{?}{10^3\ \mu\text{L}} = \dfrac{1}{100}$ $\qquad$ $? = \dfrac{1(10^3\ \mu\text{L})}{100}$

$? = 10\ \mu\text{L stock}$
$\quad 10\ \mu\text{L} + 990\ \mu\text{L} = 1000\ \mu\text{L}$

9. $0.01 = 1/100$

$\dfrac{?}{50\text{ mL}} = \dfrac{1}{100}$

$? = 0.5\text{ mL buffer}$
$\quad 0.5\text{ mL} + 49.5\text{ mL}$

Manipulation Practice Problems: Dilutions (Part C) (from p. 246)

1. $\dfrac{(50\%)\,1}{40} = 1.25\%$

2. $\dfrac{10\text{ mg}}{\text{mL}} \times \dfrac{1}{10} = \dfrac{10\text{ mg}}{10\text{ mL}} = 1\text{ mg/mL}$

3. $1{:}1 = \dfrac{1}{2}$ dilution

so: $\dfrac{10\text{ mg}}{\text{mL}} \times \dfrac{1}{2} = \dfrac{10\text{ mg}}{2\text{ mL}} = \dfrac{5\text{ mg}}{\text{mL}}$

4. $1\text{ mL} + 4\text{ mL} = 1/5$ dilution.
1 mL can make 5 mL diluted solution.

5. $1000\text{ mL} \times \dfrac{1}{100} = \dfrac{1000\text{ mL}}{100} = 10\text{ mL original}$

or $\dfrac{?}{1000\text{ mL}} = \dfrac{1}{100}$ $\quad ? = 10\text{ mL}$

Manipulation Practice Problems: Dilutions (Part D*) (from p. 248)

1. $\dfrac{1}{10}$ dilution;

- take 0.1 mL food coloring $+ 0.9\text{ mL}$ diluent

- $\dfrac{1}{10} \times \dfrac{1}{25} = \dfrac{1}{250}$ take 1 mL of above $+ 24\text{ mL}$ diluent

- $\dfrac{1}{250} \times \dfrac{1}{4} = \dfrac{1}{1000}$ take 1 mL of above $+ 3\text{ mL}$ diluent

2. a. $\dfrac{1}{10^6}$ dilution —use three dilutions of 1/100, 1/100, 1/100

- $0.1\text{ mL of culture} + 9.9\text{ mL diluent} = 1/100$
- take $0.1\text{ mL of above} + 9.9\text{ mL}$
  $= 1/100 \times 1/100 = 1/10^4$
- take $0.1\text{ mL of above} + 9.9\text{ mL}$
  $= 1/100 \times 1/10^4 = 1/10^6$

b. $\dfrac{1}{10} \times \dfrac{1}{10} \times \dfrac{1}{10} \times \dfrac{1}{10} \times \dfrac{1}{10} \times \dfrac{1}{10}$

- $0.1\text{ mL culture} + 0.9\text{ mL diluent} = 1/10$
- take $0.1\text{ mL of above} + 0.9\text{ mL}$
  $= 1/10 \times 1/10 = 1/10^2$
- take $0.1\text{ mL of above} + 0.9\text{ mL}$
  $= 1/10 \times 1/10^2 = 1/10^3$
- take $0.1\text{ mL of above} + 0.9\text{ mL}$
  $= 1/10 \times 1/10^3 = 1/10^4$
- take $0.1\text{ mL of above} + 0.9\text{ mL}$
  $= 1/10 \times 1/10^4 = 1/10^5$
- take $0.1\text{ mL of above} + 0.9\text{ mL}$
  $= 1/10 \times 1/10^5 = 1/10^6$

3. • $1\text{ mL blood} + 4\text{ mL diluent} = 1/5$
- take $1\text{ mL of above} + 9\text{ mL} = 1/10 \times 1/5 = 1/50$
- take $1\text{ mL of above} + 4\text{ mL} = 1/5 \times 1/50 = 1/250$

Application Practice Problems: Dilutions (from pp. 249–250)

1. Each tube requires 5 U.

The enzyme concentration is:

$\dfrac{1000\text{ U}}{1\text{ mL}} = \dfrac{1000\text{ U}}{1000\ \mu\text{L}} = \dfrac{1\text{ U}}{1\ \mu\text{L}}$

The $20\ \mu\text{L}$ contains 20 units of enzyme. You can, therefore, make four tubes with 5 U per tube.

2. Each tube requires $5 \times 0.01\text{ U} = 0.05\text{ U}$.

The enzyme concentration is $1\text{ U}/\mu\text{L}$.

$\dfrac{?}{0.05\text{ U}} = \dfrac{1\ \mu\text{L}}{1\text{ U}}$

$? = 0.05\ \mu\text{L} =$ amount needed, but this volume is too low to measure.

Therefore, make 1/100 dilution, $10\ \mu\text{L}$ enzyme $+ 990\ \mu\text{L}$ $= 1/100$. After dilution, there are:

$\dfrac{1\text{ U}}{1\ \mu\text{L}} \times \dfrac{1}{100} = \dfrac{1\text{ U}}{100\ \mu\text{L}} = \dfrac{10^{-2}\text{ U}}{1\ \mu\text{L}}$

Need 0.05 U/tube so use $5\ \mu\text{L}$ of dilution for each tube to 5 mL total volume.

3. Dilute $\dfrac{1}{500,000} = \dfrac{1}{50} \times \dfrac{1}{100} \times \dfrac{1}{100}$

- $0.1\text{ mL antibody} + 4.9\text{ mL diluent} = 1/50$
- take $0.1\text{ mL of above} + 9.9\text{ mL}$
  $\rightarrow 1/100 \times 1/50 = 1/5000$
- take $0.1\text{ mL of above} + 9.9\text{ mL} \rightarrow 1/100 \times 1/5000 = 1/500,000$

4. Have $\dfrac{1 \times 10^9\text{ cells}}{1\text{ mL}}$ want $\dfrac{2 \times 10^3\text{ cells}}{1\text{ mL}}$

$= \dfrac{2 \times 10^2\text{ cells}}{0.1\text{ mL}}$

*Note: For dilution problems, there are often several workable strategies. Only one strategy is shown in the answer key.

$$\frac{1 \times 10^9}{2 \times 10^3} = \frac{1}{2} \times 10^6 \text{ or } 5 \times 10^5 \text{ dilution}$$

$$\frac{1}{100} \times \frac{1}{100} \times \frac{1}{50} = \frac{1}{5 \times 10^5}$$

- 0.1 mL culture + 9.9 mL diluent = 1/100
  → (1/100) × (1 × 10⁹) = 10⁷ cells/mL
- 0.1 mL of above + 9.9 mL → 1/100 × 1/100
  → (1/10⁴) × (1 × 10⁹) → 10⁵ cells/mL
- 0.1 mL of above + 4.9 mL → 1/50 × 1/10⁴

$$\rightarrow \frac{1}{(5 \times 10^5)} \times (1 \times 10^9) \rightarrow 2 \times 10^3 \text{ cells/mL}$$

= 200 cells/0.1 mL

5. The first plate has too many colonies to count and the third plate has fewer colonies than are desirable for a plate count. The middle plate has 21 colonies, so it is the one we will use for calculations. The concentration of cells added to this plate was about 21 cells/0.1 mL = 210 cells/mL. The dilution tube from which these cells were drawn had been diluted 1/10⁷. There were, therefore, originally about 210 × 10⁷ = 2.10 × 10⁹ bacterial cells/mL in the original broth.

6. The first plate can be used for calculations. There are about 43 colonies on the first plate so the concentration of bacteria applied to that plate was about 430 cells/mL. The dilution tube from which these cells were drawn had been diluted 1/10⁶. There were, therefore, about 430 × 10⁶ = 4.30 × 10⁸ cells/mL in the original broth.

7. 10⁷ cells/mL, dilute to $\dfrac{100 \text{ cells}}{0.1 \text{ mL}}$

= 1000 cells/mL final concentration

$$\frac{10^7}{10^3} = 1 \times 10^4 = \text{dilution needed}$$

- 0.1 mL culture + 9.9 diluent = 1/100 = 1/10²
  → 10⁵ colonies/mL
- 0.1 mL culture + 9.9 diluent = 1/100 = 1/10²
  → 10³ colonies/mL

8. $\dfrac{1}{100}$ dilution had $\dfrac{50 \text{ mg}}{\text{mL}}$

$$\frac{50 \text{ mg}}{1 \text{ mL}} \times 100 = \frac{5000 \text{ mg}}{1 \text{ mL}} = \frac{5 \text{ g}}{1 \text{ mL}}$$

= concentration protein in undiluted sample.

$$\frac{5 \text{ g}}{1 \text{ mL}} = \frac{?}{100 \text{ mL}} \qquad ? = 500 \text{ g}$$

9. $\dfrac{1}{50}$ dilution had

$$\frac{87 \text{ mg}}{1 \text{ mL}} \times 50 = \frac{4350 \text{ mg}}{1 \text{ mL}} = \frac{4.35 \text{ g}}{1 \text{ mL}}$$

= concentration protein in undiluted sample.

10. The sample was diluted 1/5 and then 5/25 → (1/5) (5/25) = 5/125 total dilution. There were 3 mg protein in 5 mL = 0.6 mg protein/mL. Multiplying times the dilution: 0.6 mg/mL × 125/5 = 15 mg/mL.

11. The sample was diluted 1/4 and then 10/30 → 10/120 total dilution.

There were 10 mg protein/5 mL = 2 mg/mL protein in the solution. Multiplying times the dilution: 2 mg/mL × 120/10 = 24 mg/mL in original sample.

CHAPTER 15

Manipulation Practice Problems: Graphing (from pp. 255–257)

1. **a.** The amount of product increases in a linear fashion over time.

   b. The amount of product decreases in a nonlinear fashion over time.

   c. The reaction rate is constant, regardless of the amount of compound X present.

2. **a.** (7,5) **c.** (−5,5)
   b. (−4,−4) **d.** (7,−4)

3. change in $X = -11$, change in $Y = -9$,
   change in $X = -1$, change in $Y = 9$

4.

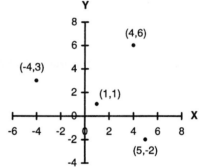

5.

12	61
15	76
20	101

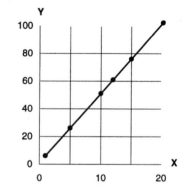

6. $A = 3Q - 4$

Q	A
−1	−7
0	−4
1	−1
2	2
3	5

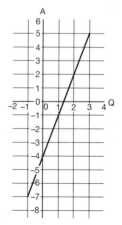

7. a. slope 3, intercept 2
   b. slope 0.2, intercept −1
   c. slope 0.005, intercept 0

8. a. slope 1, intercept (0,2)
   b. slope −1, intercept (0,7.5)
   c. slope 1.25, intercept (0,4)
   d. slope −1, intercept (0,5)

9. a. not linear   **b.** linear
   c. linear   **d.** not linear

10. All will form a line.

11. a., b. *i.* slope = 10 cm/min, Y intercept = (0 min, 1 cm)

   ii. slope = −1, Y intercept = (0,3)

   iii. slope = 7 mg/cm, Y intercept = (0 cm, 12 mg)

c.

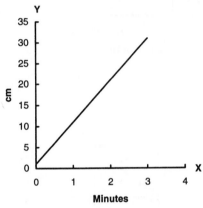

i

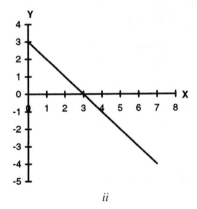

ii

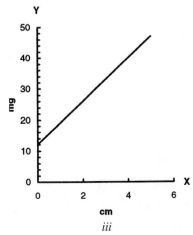

iii

12.

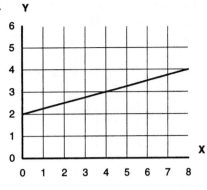

(a)

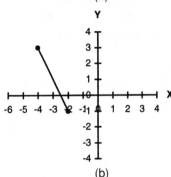

(b)

13. a. $Y = -1X + 2$ **b.** $Y = 2.25X - 9$
   c. $Y = 3$ **d.** $Y = -1X + 5.5$

14. Line A: $Y = \dfrac{3}{4}X$

   Line B: $Y = 2X + 2$

   Line C: $Y = \dfrac{1}{2}X + 2$

15. $Y = 1X + 0$

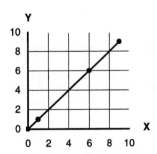

16. a. 9/5 and 32   **b.** 32
   c. 9/5
   d.

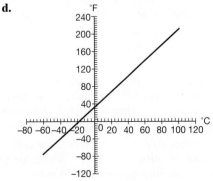

Manipulation and Application Practice Problems: Quantitative Analysis (from p. 260)

1. a.

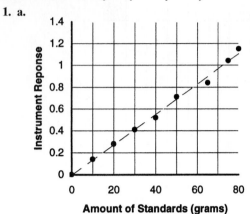

b. A 0.78 instrument response corresponds to 57 g. The sample was diluted 100× so the amount in the undiluted sample was: 57 g × 100 = 5700 g.

2. Stock solution at 100 mg/mL. First, decide what volume to make. For example, a strategy to make 10 mL of each standard:

Standard	Dilution	Stock		Diluent
1 mg/mL	1/100	0.1 mL	+	9.9 mL
5 mg/mL	5/100	0.5 mL	+	9.5 mL
15 mg/mL	15/100	1.5 mL	+	8.5 mL
25 mg/mL	25/100	2.5 mL	+	7.5 mL
50 mg/mL	50/100	5.0 mL	+	5.0 mL
75 mg/mL	75/100	7.5 mL	+	2.5 mL
100 mg/mL		10.0 mL		0 mL

(Another strategy to solve this type of problem is discussed in Chapter 27.)

Application Practice Problems: Graphing (from pp. 262–263)

1. a. Yes. The graph shows a linear relationship between cancer incidence and exposure level, even with the lowest exposures to the agent.

b. No, the graph has a threshold. At low levels of exposure, below the threshold, no change in cancer incidence occurs.

c. Possible explanations include: (1) individuals are able to detoxify and handle low levels of the agent and (2) multiple receptors must be activated before an effect occurs.

d. A background level of cancer is found in the population even in the absence of exposure to the agent.

e. Regulatory agencies need to know whether low level exposure is safe in order to decide what levels (if any) of compound can be allowed. For example, this is important to know when deciding whether a particular pesticide can be used on crops, and if so, what levels are "safe."

2. a. Graph **b** indicates a relationship between the mass of the parent and the offspring where a higher parental mass is associated with a higher offspring mass.

b. Environmental factors may play a major role in determining mass. One experiment would be to take genetically identical clones of the plant and measure their mass under different, controlled environmental conditions. There are other experiments you might devise as well.

3. a.

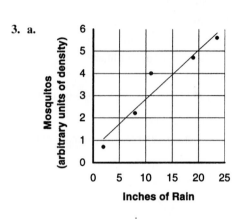

i

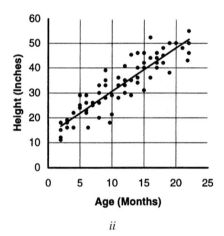

ii

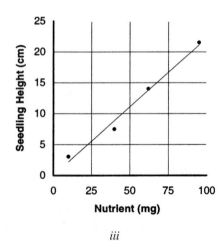

iii

b. *i* slope = 0.22 mosquitos/in rain

ii slope = 1.8 in/month

iii slope = 0.24 cm/mg

c. *i* Y intercept = (0 in rain, 0.6 mosquitos)

   ii Y intercept = (0 months, 13 in)

   iii Y intercept = (0 mg, 0 cm)

d. *i* $Y = (0.22 \text{ mosquitos}/in \text{ rain}) \, X + 0.6 \text{ mosquitos}$

   ii $Y = (1.8 \text{ in/month}) \, X + 13 \text{ in}$

   iii $Y = (0.24 \text{ cm/mg}) \, X + 0 \text{ cm}$

e. 1.7 mosquitos (approximate)

f. $Y = (0.22 \text{ mosquitos/in}) \, (5 \text{ in}) + 0.6 \text{ mosquitos} = 1.7$ mosquitos

g. $Y = (1.8 \text{ in/month}) \, (20 \text{ months}) + 13 \text{ in} = 49 \text{ in}$

h. $Y = (0.24 \text{ cm/mg})(50 \text{ mg}) + 0 \text{ cm} = 12 \text{ cm}$

4. a. 80   **b.** 80

c. The slope is close to 1, so the test scores were roughly equal.

d. The relationship is roughly linear, which means that there is a general relationship between the two scores. Students who did poorly on the midterm tended to do poorly on the final. Midterm scores appear therefore to be fairly predictive, although the next year's class could differ.

5. a. 10^6

b. 10^3

c. Using semilog paper results in a straight line on the graph that is easier to work with than an exponential curve.

6.

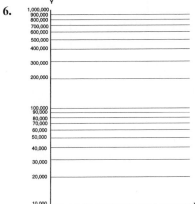

Application Practice Problems: Exponential Equations (from pp. 269–270)

1. a.

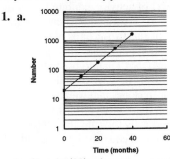

b. $Y = 3^x \, (20)$ where x = number of generations

---

2. $N = (1/2)^t(N_o)$ $N_o = 300 \ \mu\text{Ci}$ half life = 14 days

a. $t = 28 \text{ days}/14 \text{ days} = 2$

   $N_2 = (\tfrac{1}{2})^2 \, (300 \ \mu\text{Ci}) = 75 \ \mu\text{Ci}$

b. $t = 140 \text{ days}/14 \text{ days} = 10$

   $N_{10} = (\tfrac{1}{2})^{10} \, (300 \ \mu\text{Ci}) = 0.293 \ \mu\text{Ci}$

c. $t = 200 \text{ days}/14 \text{ days} = 14.3$

   $N_{14.3} = (\tfrac{1}{2})^{14.3} \, (300 \ \mu\text{Ci}) = 1.49 \times 10^{-2} \ \mu\text{Ci}$

d. $t = 365 \text{ days}/14 \text{ days} = 26.1$

   $N_{26.1} = (\tfrac{1}{2})^{26.1} \, (300 \ \mu\text{Ci}) = 4.17 \times 10^{-6} \ \mu\text{Ci}$

3. a. $N = (\tfrac{1}{2})^{0.38} \, (600 \ \mu\text{Ci}) = 461 \ \mu\text{Ci}$

b. $N = (\tfrac{1}{2})^{0.77} \, (600 \ \mu\text{Ci}) = 352 \ \mu\text{Ci}$

c. $N = (\tfrac{1}{2})^{1.2} \, (600 \ \mu\text{Ci}) = 261 \ \mu\text{Ci}$

4.

RADIOACTIVE MATERIAL RECORD

USE INFORMATION					DISPOSAL INFORMATION					
Activity (μCi or mCi)		Volume (μL or mL)		Date Initials	WASTE TYPES AND ACTIVITY (μCi or mCi)					Date & Initials
removed	remaining	removed	remaining		Solid	Liquid	Animal	Decay	Total	
0	500	0	1.0		12/3/01 NW					
100	400	0.2	0.8		12/8/01 NW					
100	300	0.2	0.6		12/12/01 NW					
								200	200	12/17/01 NW
150	150	0	0.6		12/17/01 NW			150	150	12/17/01 NW

CHAPTER 16

1. a. Sample: The ten vials. Population: All the vials in the batch.

b. Sample: The blood sample tested. Population: All the blood in that individual.

c. The sample is the first graders who were tested. The population depends on the purpose of the study. If the study was intended to test the performance of all first graders in Franklin Elementary School, then the entire population was studied. If the sample was meant to represent all first graders in a city or in a larger group, then the population is that larger group.

2. a. $\bar{x} = 10.000 \text{ g}$ SD ≈ 0.001 g   CV $\approx 0.012\%$

b. $\bar{x} = 84\%$ SD $\approx 9.49\%$ CV $\approx 11.29\%$

c. $\bar{x} = 235 \text{ sec}$ SD ≈ 33.56 sec   CV $\approx 14.28\%$

d. $\bar{x} \approx 237 \text{ sec}$ SD ≈ 12.26 sec   CV $\approx 5.18\%$

The means for the two methods are similar; however, the second method is more consistent. Based on this information, the second method seems better.

3. a. $n = 33$ $\bar{x} \approx 136.76$ mg
SD ≈ 32.90 mg   CV $\approx 24.06\%$

b. $n = 25$ $\bar{x} \approx 8.15$ g
SD ≈ 1.47 g   CV $\approx 18.09\%$

c. $n = 19$ $\bar{x} \approx 1112.32$ mL
SD ≈ 71.55 mL   CV $\approx 6.43\%$

4. a. $n = 20$ $\bar{x} \approx 33.05$ blossoms
median = 32.5 blossoms
range = $21 - 44 = 23$ blossoms
SD ≈ 5.57 blossoms

5. $n = 12$ $\bar{x} \approx 30.39$ kg   median = 28.6 kg
range = $18.2 - 52.2 = 34$ kg   SD ≈ 9.90 kg

6. $\dfrac{\sum X}{30} = 75\%$ $\sum X = 2250\%$

$2250\% + 100\% =$ new total

$\dfrac{2350\%}{31} \approx 75.8\%$

7. a. mean = 4, mode = 4

b. mean is approximately 5, mode is approximately 4

c. mean is approximately 3, mode is approximately 4

8. a. normal

b. bimodal

c. skewed

9. b is less dispersed

10. a is less dispersed

11. about the same

12. a is less dispersed

13. a. $n = 45$ $\bar{x} \approx 9.55$ activity units
SD ≈ 3.60 activity units
range = $0.4 - 15.1$ activity units
     = 14.7 activity units
median = 10.1 activity units

b.

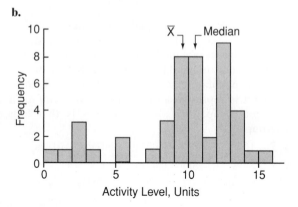

Frequency Table	
Interval	**Frequency**
0–0.9	1
1.0–1.9	1
2.0–2.9	3
3.0–3.9	1
4.0–4.9	0
5.0–5.9	2
6.0–6.9	0
7.0–7.9	1
8.0–8.9	3
9.0–9.9	8
10.0–10.9	8
11.0–11.9	2
12.0–12.9	9
13.0–13.9	4
14.0–14.9	1
15.0–15.9	1

c. Based on just this information, we cannot tell whether the DNA fragment was taken up or not because we do not know what level of enzyme activity exists in cells that have not been treated with the fragment.

14. a. $n = 20$ $\bar{x} \approx 1.29$ activity units
SD ≈ 0.93 activity units

b.

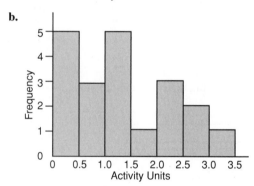

Interval	**Frequency**
0–0.4	5
0.5–0.9	3
1.0–1.4	5
1.5–1.9	1
2.0–2.4	3
2.5–2.9	2
3.0–3.4	1

c. There is variation both in the group treated with the DNA fragment and in the group not treated. Using this assay, there is a low level of activity exhibited even in cells not treated with the DNA fragment; however, the mean enzyme activity of the two groups is different. There is also the suggestion graphically that the clones that were treated with the DNA fell into two categories: (1) those with higher levels of activity and (2) those with lower levels, similar to clones that were not treated with the DNA. Based on these observations, it appears that some of the treated cells took up the DNA and some did not. This hypothesis could be investigated by further study.

15. $n = 100$ $\bar{x} \approx 152.2$ cm range $= 103 - 197$ cm $= 94$ cm
Frequency Distribution Table with data divided into 11 intervals:

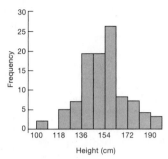

Height (cm)

Interval (cm)	Frequency
100–108	2
109–117	0
118–126	5
127–135	7
136–144	19
145–153	19
154–162	26
163–171	8
172–180	7
181–189	4
190–198	3

16.

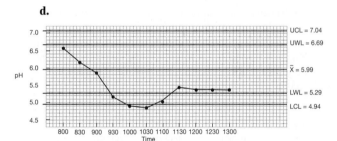

Interval (mg)	Frequency
19–20	2
21–22	4
23–24	7
25–26	3
27–28	1

17. a, b. Very certain. Common sense is sufficient to know this.

    c. Based on the standard deviation of the sample, we would expect 68% of customers to be in this range; however, a sample of five is too small to draw firm conclusions.

    d. Not very certain, a sample size of five is too small to draw firm conclusions.

18. a. 2.25%   **b.** 16%

19. a. 99% (see Table 16.7)   **b.** 2.25%

    c. 68%   **d.** 0.5%

20. a. The average appears to be in the midtwenties and hovers at around ± 5; therefore, 18.1 mg appears a little low.

    b. Mean ≈ 27.16 mg, SD ≈ 3.87 mg. The mean -2 SD is 19.4; therefore, 18.1 mg is outside the range of 2 SD and probably should be investigated.

21. a. The values appear to hover around 65 leaves give or take about 10. It is difficult to tell whether 79 is a cause for concern.

    b. Mean ≈ 64.7 leaves, SD ≈ 9.7 leaves. The mean $+ 2$ SD $= 84.1$; therefore, 79 does not appear to be unreasonable.

22. On several occasions, such as on March 22, the points lie outside the warning range; however, because later points are within the expected range, this is probably due to normal variation. The point on April 15 is out of the control range and therefore the process should be stopped and checked for problems. The points between May 31 and June 23 display an upward trend. A trend in one direction or another is also cause for concern.

23. a. Mean $= 5.99$ pH units, $SD \approx 0.35$ pH units

    b.

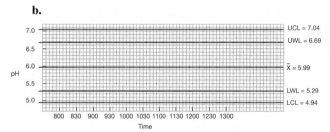

    c. The pH values tend to decline with time.

    d.

    e. This process tended downward from the beginning, eventually reaching the lower control limit.

CHAPTER 17

1. a. To calibrate and check the accuracy of a balance.

    b. Recalibration involves comparison with a primary or secondary standard.

    Recalibration is necessary because working standards may become damaged, dirty, dusty, or may otherwise be altered with time and use.

2. a. Readings will be too high. For example, a 100.0000 g object will be read as more than 100 g.

    b. Subsequent temperature readings will be recorded as higher than they really are.

3. We would expect all the readings made with this balance to be incorrect but their consistency is unlikely to be affected by the improper calibration (adjustment) of the instrument. The accuracy of the instrument, therefore, would be adversely affected but not its precision.

4. See Answer #3.

5. A pH meter needs to stabilize to give accurate results; therefore, this technician will sometimes have inaccurate results. In addition, precision requires that the operator perform a task in a consistent fashion; therefore, the precision achieved by this technician will be poor. Failure to use an instrument properly, in this case, failure to allow for stabilization, is a systematic error.

6. Precision. See text for explanation.

7. The standard deviation is 0.15 mg/mL; therefore, it is reasonable to expect the next measurement to be 5.00 mg/mL ± 0.15 mg/mL.

8. The most accurate measurements have the mean value at the "True Value" (d). The most precise measurements have the narrowest curve with the smallest standard deviation (c).

9.

	Lab 1	Lab 2	Lab 3	Lab 4
mean	37.5 mg/mL	38.1 mg/mL	38.0 mg/mL	38.4 mg/mL
SD	0.32 mg/mL	0.22 mg/mL	1.06 mg/mL	0.69 mg/mL

Laboratories 2 and 3 have good accuracy, but Laboratory 2 has the best precision.

10. mean   37.5 mg/mL
SD   0.46 mg/mL

The results from Laboratory 2 are more precise when the assays are completed at one time rather than over four months. (Many factors might lead to variability over time. Reagents or samples might slowly degrade, instruments may lose calibration or be replaced, personnel might change, and so on.)

11.

	Juan	Chris	Ilana	Mel
mean	5.0002 g	5.0021 g	4.9999 g	5.0247 g
SD	0.0015 g	0.0001 g	0.0001 g	0.0516 g
% error	−0.004%	−0.042%	0.002%	−0.494%

12. mean   3.13
SD   0.019

The number of binding sites for an antibody is a whole number, in this case, presumably 3. The % error (a measure of accuracy) is −4.33%. The SD of the assay (a measure of precision) is ± 0.019 sites.

13. a. Subject 1: SD ≈ 0.15 mg/L
Subject 2: SD ≈ 0.14 mg/L
Subject 3: SD ≈ 0.10 mg/L

b. Pooled mean ≈ 96.6 mg/L

14. a. $\bar{x}$ = 7.16
SD = 0.03

b. Given the consistently high values and the fact that the instrument used to test the buffer was known to be properly working, we can be reasonably confident that

the pH of this lot of buffer is about 7.16. This lot of buffer would be rejected.

15. a.

	Sample 1	Sample 2	Sample 3
$\bar{x}$	1.3 μg/m^3	2.2 μg/m^3	1.6 μg/m^3
SD	0.10 μg/m^3	0.12 μg/m^3	0.10 μg/m^3

b. Pooled sample $\bar{x}$ 1.7 μg/m^3

SD   0.42 μg/m^3

c. As we might expect, the standard deviation is higher for the pooled data because we would expect more variation among samples taken on different occasions than when an individual sample is tested three times.

16. a.

x	10 μg/L	101 μg/L
SD	1.6 μg/L	2.93 μg/L
RSD	16.3%	2.91%
% error	0 %	−1.00%

b. The accuracy and the precision of the assay may change at two different concentrations of the analyte.

c. These analyses evaluate and document the precision and accuracy of the method. Because the assay components may change over time, this verification may be required numerous times.

d. The % error is an indication of the accuracy.

17. a. 56   2 significant figures
b. 62   2 significant figures
c. 35.9865   6 significant figures
d. 8.25   3 significant figures
e. 28.4 mL   3 significant figures
f. 28.44 mL   4 significant figures

18.   ???
a. 2000
    ??? ???
b. 1,000,000
c. 0.~~00~~677
d. 134,908,098

19. a. 5   **b.** 3   **c.** 5   **d.** 3

20. a. 0.0   **b.** 1.0   **c.** 0.6   **d.** 0.2

21. lot a. rounded value = 10 mg/vial; yes, meets specification

lot b. rounded value = 10 mg/vial; yes, meets specification

lot c. rounded value = 8 mg/vial; no, does not meet specification

lot d. rounded value = 9 mg/vial; no, does not meet specification

22. lot a. rounded value = 0.03%, no, does not meet specification

lot b. rounded value = 0.02%, yes, meets specification

lot c. rounded value = 0.03%, no, does not meet specification

lot d. rounded value = 0.02%, yes, does meet specification

23. a, b. The RSD is 1.552% which, when rounded, is 2%. This is higher than the 1% specification; however, this does not necessarily indicate any problem with the spectrophotometer. Biological samples are often not homogeneous and they are often not stable. With biological materials, therefore, variability is more likely to be due to the sample than to the instrument.

CHAPTER 18

1. a. Voltage source—batteries
Resistance—bulb

b. Voltage source—power company (wall outlet)
Resistance—motor in the machine

c. Voltage source—power company via power supply
Resistance—the cells and the buffer in which they are suspended

2. The total standby current is 33.3 W
33.3 W $\times$ 24 hrs/day $\times$ 365 days = 291708 W-hours
291708 W-hours/1000 = 291.708 kW-hours
(291.708 kW-hrs) ($0.09/kW-hr) $\approx$ $26.25 cost
for standby current per year
Multiplied by 50,000, it comes to about $1.3 million.

3. V = I R
0.120 V = 0.012 A (?)
? = 10 ohms

4. P = V I
? = (120 V) (20 A)
? = 2400 W

5. V = I R
6 V = (0.400 A)(?)
? = 15 ohms

6. P = V I
maximum power = 0.400 A $\times$ 500 V = 200 Watts

7. The total carrying capacity of the line in the laboratory is:

$$P = I V$$
$$= (20\ A)(120\ V) = 2400\ W$$

The total power consumed by the fifteen units if they are run at their maximum power is:

$$200\ W\ (15) = 3000\ W$$

The circuit, therefore, cannot handle all the units if they all are using 200 W.

8. P = V I   V = I R
500 W = 120 V (?)   120 V = 4.2 A (?)
? $\approx$ 4.2 A   ? $\approx$ 28.6 ohms

CHAPTER 19

1. Weigh out as close to 15 mg as possible and then calculate the amount of buffer required to dilute the enzyme to a concentration of 15 mg/mL.

2. a. 101.7 mL   **b.** 1.43 mL
  c. 31.134 L   **d.** 10.14 mL

3. Assuming the technician has been careful to avoid temperature effects and drafts, the preparation is probably losing moisture because it was not dry to begin with.

4. Balance (b) has a greater response for a given weight, so it is more sensitive.

5. The mean value = 9.999983333 g. The absolute error is 0.00002 g.
The SD = 0.000116905. Rounded, this is 0.0001.
The balance meets its performance specifications.

6. a. The user first calibrates the balance and then checks the weight of a second standard.

  b. Forms are used to document the result of the performance verification.

7. All subsequent readings with that balance will be a little too high. Dropping the standard is a gross error. The fact that subsequent readings will be a bit high is a systematic error.

8. Midpoint
Full scale = 500.001 g
Weight sum = 500.003 g
Difference = 0.002 g
Linearity error = 0.001 g
25%
Full scale = 500.001 g
Weight sum = 500.001 g
No error
75%
Full scale = 500.001 $\times$ 3 = 1500.003 g
Weight sum = 1499.996 g
Difference = 0.007 g
Linearity error = 0.00175 g

9. a. Because 1000 cm^3 of air is displaced, the weight of air displaced is:
1000 cm^3 $\times$ 1.2 mg/cm^3 = 1200 mg = 1.20 g.

  b. The weight of a sample varies depending on its location—such as whether it is in air or a vacuum. Its mass does not change.

CHAPTER 20

1. a. 4.2 mL

  b. 2.4 mL

  c. 5.2 mL

  d. 6.4 mL

2. a. a volumetric flask or a graduated cylinder.

  b. a graduated cylinder

  c. a pipette or a micropipettor

  d. a micropipettor

  e. a micropipettor

  f. a micropipettor

3. There are many possible answers. Errors in use of a graduated cylinder or volumetric flask include the use of dirty glassware, improper reading of the meniscus, and use of a TC flask where TD is appropriate and vice versa. For micropipettor errors, see Boxes 20.3 and 20.4 and Table 20.3.

4. a. Graduated cylinder, $\pm$ 0.6 mL, volumetric flask, Class A 0.08 mL, Class B 0.16 mL.

   b. Graduated cylinder, $\pm$ 0.6 mL.

   c. A serological pipette $\pm$ 0.02 mL

   d–f From Figure 20.15, mean error in μL, as low as:

   d. $\pm$ 0.8

   e. $\pm$ 0.075

   f. $\pm$ 4

5. Brand B has the best precision and Brand C has the best accuracy, but they are basically similar and either brand is probably acceptable based on these specifications.

6. a. The expected value is ? /0.9982 mg/μL = 100 μL.
? = 99.82 mg

   b. The mean volume for the water in μL is 99.40 mg/0.9982 mg/μL $\approx$ 99.58 μL

$$\% \text{ Error} = \frac{99.58\ \mu L - 100.0\ \mu L}{100\ \mu L} \times 100\%$$

$$= -0.42\%$$

   c. Yes, it is within the specification for accuracy.

7.

Clean Gene, Inc.

VERIFICATION OF PERFORMANCE OF MICROPIPETTOR REPORT
FORM 232

CALIBRATION TECHNICIAN   *J.E.S.*

CALIBRATION DATE   *9/22/07*   NEW DUE DATE FOR NEXT CALIBRATION   *3/22/2008*

MICROPIPETTOR ID NUMBER   *2127*   MANUFACTURER   *Finestt Pipettes Inc.*

MODEL NUMBER   *2127*   LOCATION   *Lab #27*   TEMPERATURE   *20°C*

RANGE   *100–1000μL*   PRIMARY USER   *R.E.S.*

SUMMARY:

PREVERIFICATION CLEANING AND ADJUSTMENTS   *Changed Seals & O-rings*

      PASS/FAIL   *Pass*

STATUS   *Returned to lab for use*

Volume 1

NOMINAL VOLUME (μl)	TUBE NUMBER	INITIAL WEIGHT OF TUBE	WEIGHT AFTER H₂O DISPENSED	NET WEIGHT OF DISPENSED H₂O	H₂O VOLUME
500	1	1.9355 g	2.4352 g	0.4997 g	500.60 μl
500	2	1.9877 g	2.4876 g	0.4999 g	500.80 μl
500	3	1.9787 g	2.4789 g	0.5002 g	501.10 μl
500	4	1.9850 g	2.4873 g	0.5023 g	503.20 μl
500	5	1.9755 g	2.4763 g	0.5008 g	501.70 μl
500	6	1.9387 g	2.4387 g	0.5000 g	500.90 μl

Mean Water Volume   *501.38 μl*   % INACCURACY   *0.28%*   SD   *0.97 μl*   CV   *0.19%*

CHAPTER 21

1. $94°C \approx 201.2°F$
$37°C \approx 98.6°F$
$72°C \approx 161.6°F$

2. $81°C = 177.8°F$

3. $103.2°F \approx 39.6°C$

4. a. The relationship between resistance and temperature appears to be linear for platinum; therefore, a two-point calibration will give accurate results.

   b. This is not the case for nickel.

5. 80°F. Lowest temperature is about 67°F. Highest temperature is about 100°F. Analog.

CHAPTER 22

1. $pH = -\log [H^+]$
$9 = -\log [H^+]$
$[H^+] = 10^{-9}$ moles/liter

2. $pH\ 7.3 = -\log [H^+]$
$= 5.0 \times 10^{-8}$ moles/liter

3. $[H^+] = 1.2 \times 10^{-3}$ moles/liter
$pH = -\log [1.2 \times 10^{-3}] \approx 2.9$

4. $pH = -\log [0.01] = 2$

5. $[1 \times 10^{-4}] [H^+] = 1 \times 10^{-14}$
$[H^+] = 1 \times 10^{-10}$
$pH = 10$

6. a. 59.16 mV/pH unit.

   b. 59.16 mV

   c. 54.2 mV/pH unit

   d. 54.2 mV

7. a. The pH goes down as the bacteria actively metabolize nutrients and their population grows.

   b. pH probes for fermentation are difficult to produce. Difficulties include probe sterilization, which usually involves exposure to pressurized steam. Probes are immersed in a turbid solution that contains numerous materials that can contaminate the electrode surface. They must be reliable because it is difficult to substitute a new probe during a fermentation run.

   c. The pH of a culture is an important indicator of its condition and therefore must be monitored. It may be necessary to adjust the pH during a fermentation run. An unexpected value may indicate a problem. It is important to document that the pH values were those expected during a "good" run. The print recording provides a record of the pH.

CHAPTER 23

1. $100\ \mu m = 100 \times 10^3\ nm = 100,000\ nm$
$30\ cm = 30 \times 10^7\ nm = 300,000,000\ nm$

2. X-rays have the most energy; green light the least.

3. The spectra are probably of the same compound. The heights of the peaks vary depending on the concentration of compound present.

4. Yes.

5. From Table 23.3: **a.** Safranin O red absorbs light at 525 nm, appears red.

   b. Brilliant green absorbs light at 620 nm, appears greenish blue.

   c. Methyl orange absorbs light at 465 nm, appears orange.

6. a. α-carotene absorbs, and makes available to the plant, the energy of light in the violet-blue region of the spectrum, between about 420 and 480 nm.

   b. A yellow-orange color.

7. a. Orange   **b.** greenish blue.   **c.** Blue-green algae

8. a.

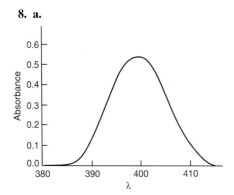

   b. yellow   **c.** band width about 15 nm

9. a.

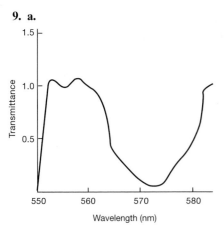

   b. violet to red-violet

10. $0.876 = 87.6\%$ T, $0.776 = 77.6\%$ T,
$0.45 = 45\%$ T, $1.00 = 100\%$ T

11.

	A		A
$t = 0.876$	$\approx 0.057\%$	T = 25%	≈ 0.60
$t = 0.776$	$\approx 0.110\%$	T = 15%	≈ 0.82
$t = 0.45$	$\approx 0.35\%$	T = 95%	≈ 0.02
$t = 1.00$	$\approx 0.00\%$	T = 45%	≈ 0.35

12.

Absorbance	Transmittance (t)	% Transmittance (T)
0.01	0.98	98%
0.25	0.56	56%
2.0	0.01	1.0%

13. $1.24 \approx 0.058\, t$ $0.95 \approx 0.11\, t$
$1.10 \approx 0.079\, t$ $2.25 \approx 0.00562\, t$

14. a. Instrument B. **b.** Instrument B (less stray light). **c.** Instrument B has gel scanning accessories. **d.** Instrument A would be expected to be less costly and is probably suitable for routine work in many laboratories. **e.** Instrument B (smaller spectral slit width). **f.** Instrument B because it has better resolution.

15. Path length, cuvettes, photometer accuracy, wavelength accuracy.

16. Less than 1.95.

CHAPTER 24

1. It depends. If the first vial in each tray tends to be different in some way from the other vials, then this selection method would introduce a systematic error.

2. Specificity (selectivity).

3. a. Robustness.   **b.** Limit of detection and Range.

4. b

5. e

6. This is a positive control and should come up positive, whether or not the person is pregnant. If it does not, then there is likely a problem in the reagents or in the way the test was performed. The test kit instructions inform the user that if the positive control line did not appear, then the test did not work properly.

7. a. No, 10 is not within the range for ill patients.

   b. No.

   c. Maybe. 25 is in the range of both healthy and ill patients.

   d. Yes.

8. a, b. Any patient with a score of 20 or lower is healthy and therefore should have a negative result. Therefore, there is a 0% level of false negatives. However, some patients with scores above 20 are healthy and some are ill. Some people who are healthy, therefore, will get a false positive result if a cutoff score of 20 is chosen. The false positive rate will be about 15%.

9. a, b. Some people with a score of 25 are healthy and some are ill. There will therefore be both false positives and false negative results. The rates of each will be equal, at about 2.5%.

10. a, b. Anyone with a score of 30 or higher is ill, therefore, the level of false positives will be 0. Some people, however, with a score of less than 30 are sick. There will therefore be about 15% false negatives.

11. 20

12. a. Apple juice (experimental sample)

   b. Orange juice (experimental sample)

   c. Pineapple juice (experimental sample)

   d. A solution of glucose (positive control)

   e. A solution of sucrose (negative control)

   f. Pure water would be another negative control, probably in addition to a sucrose solution

13. Onion extract (experimental sample)

   a. Potato extract (experimental sample)

   b. Green pepper extract (experimental sample)

   c. Starch solution (positive control)

   d. Water (negative control)

   e. Glucose solution (negative control)

14 a–c. There should be colonies on Plate A derived from bacteria that successfully took up the ampicillin-containing plasmid. These are the bacteria that would be used to produce the protein product of interest.

Plate B is a control plate that confirms that the cells used for transformation were viable to begin with. Why is this important? Suppose, for example, that no colonies appear on Plate A. There are many explanations for this result. Perhaps none of the bacteria successfully took up the plasmid. Perhaps all the bacteria were killed by the transformation procedure. Perhaps there was a problem with the plasmid. Perhaps the bacteria were not viable to begin with. Plate B allows the investigators to be sure that at least the bacterial cells were viable to begin with. Note also that if the procedure works properly, then Plate B should have far more colonies than Plate A because transformation is a relatively rare event.

Plate C is a control plate. The nontransformed cells should not grow on this plate. If they do, then there is likely a problem with the activity of the ampicillin used to make the plates. (Or, perhaps the cells already had antibiotic resistance.) If colonies appear on this plate, then the procedure is likely invalid, regardless of whether or not there are colonies on Plate A.

Plate D is a control to prove that the nontransformed cells are viable. There should be abundant growth on this plate. If not, perhaps the nutrient medium was improperly prepared, or the cells were not viable to begin with.

Plates B and D may be called "positive controls" in that the researchers expect bacteria to grow on them. Plate C may be considered a "negative control" in that no growth is expected on this plate.

CHAPTER 25

1. **a,b.** A is an absorbance spectrum. X axis, wavelength; Y axis, absorbance. B is a standard curve. X axis, concentration (or amount of analyte); Y axis, absorbance.

   c. An absorbance spectrum is associated with qualitative analysis. A standard curve is associated with quantitative analysis.

2. **a.** quantitative **b.** qualitative **c.** quantitative

   d. quantitative **e.** qualitative **f.** qualitative

   g. quantitative

3. **a.** Compound A is orange. Compound B is red. Compound C is green.

   b. Compound B has the greatest absorptivity constant at 540 nm.

   c. Compound B will be detectable at 540 nm at a lower concentration than the other compounds because it has the largest signal at that wavelength.

4. Graph A: The linear range is 0–250 mg/mL.

   Graph B: The linear range is 0–3.5 mM.

   Graph C: The linear range is 0–15 ppm.

5. **a.**

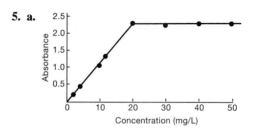

   b. The graph is linear from 2 to 20 mg/L concentrations.

   $$\text{Slope} = m = \frac{Y_2 - Y_1}{X_2 - X_1} = 1.1 \big/ 10 \text{ mg/L} = \frac{0.11 \text{ L}}{\text{mg}}$$

   $$A = \frac{0.11 \text{ L}}{\text{mg}} C + O$$

   c. Absorptivity constant $= \alpha = \dfrac{\text{slope}}{\text{path length}}$

   $$= \frac{0.11 \text{ L}}{(1 \text{ cm}) \text{ mg}}$$

   d. This graph has an upper threshold where the graph plateaus. This is an indication of the limitation of the spectrophotometer at low light transmittances due to stray light.

6. **a.**

   b. $\alpha = \dfrac{\text{slope}}{\text{pathlength}}$

   slope $= 0.3/100$ mM $= 0.003$/mM

   $$\alpha = \frac{0.003}{(\text{mM}) (1 \text{ cm})}$$

   c. $A = \alpha\, b\, C$

   $$C = \frac{A \text{ (mM)}}{0.003}$$

	Absorbance	Concentration (in mM)
Sample A	0.18	60
Sample B	0.31	103.3
Sample C	0.96	320
Sample D	1.0	333.3

7. a.

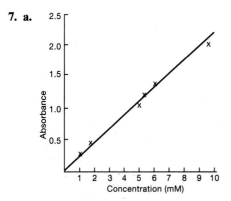

b. $\alpha = \dfrac{\text{slope}}{\text{path length}}$

$\text{slope} = \dfrac{(1.0 - 0.5)}{(4.5 - 2.1)\ \text{mM}}$

$\alpha \approx \dfrac{0.21}{\text{mM (cm)}}$

c. $C = \dfrac{0.30\ (\text{cm})(\text{mM})}{0.21\ (1\ \text{cm})}$

$C \approx 1.43\ \text{mM diluted}$

$C = 1.43\ \text{mM} \times 5$

$\quad = 7.15\ \text{mM undiluted}$

8. a. When the path length decreases the absorbance decreases.

b. 0.8

9. $\dfrac{25\ \text{ppm}}{0.87\ \text{AU}} = \dfrac{40\ \text{ppm}}{?}$ $? \approx 1.39\ \text{AU}$

The absorbance is a linear function of concentration; therefore, this is a proportion problem.

10. $t = 0.680$

$A = -\log(t)$

a. $A = -\log(0.680) \approx 0.167$

b. Because there is a linear relationship between absorbance and concentration:

$\dfrac{A}{16\ \text{mg/mL}} = \dfrac{0.167}{10\ \text{mg/mL}}$

$A = 0.267$

c. $A = -\log(t)$

$0.267 = -\log(t)$

$t \approx 0.540$

11. $A = -\log(0.75) \approx 0.12$

a,b. $\dfrac{A}{60\ \text{mg/mL}} = \dfrac{0.12}{30\ \text{mg/mL}}$

Absorbance = 0.24   transmittance ≈ 0.58

c,d. $\dfrac{A}{90\ \text{mg/mL}} = \dfrac{0.12}{30\ \text{mg/mL}}$

Absorbance = 0.36   transmittance ≈ 0.44

12. $A_{280} = 0.65$ of 1 mg/mL

a. $\dfrac{?}{0.15} = \dfrac{1\ \text{mg/mL}}{0.65}$

$? \approx 0.23\ \text{mg/mL}$

b. The proteins in the preparation may not have the same aromatic amino acid composition as the BSA. There might be nucleic acids in the partially purified preparation that interfere at 280 nm. There are other possible answers as well.

13. $\alpha_{ATP} = 15,400$ L/mole–cm at 260 nm

$A = \alpha\, b\, C$

$1.6 = \dfrac{15,400\ \text{L}\ (1\ \text{cm})\ C}{(\text{mole–cm})}$

$C = \dfrac{1.6\ \text{mole}}{15,400\ \text{L}} \approx 1.04 \times 10^{-4}\ \text{mole/L}$

14. $\alpha_{NADH} = \dfrac{15,000\ \text{L}}{\text{mole–cm}}$ at 260 nm

$A = \alpha\, b\, C$

$0.98 = \dfrac{15,000\ \text{L}}{\text{mole–cm}}\ (1.2\ \text{cm})\ C$

$C = \dfrac{0.98\ \text{mole–cm}}{15,000\ (1.2)\ \text{cm–L}}$

$C \approx 5.44 \times 10^{-5}\ \text{mole/L}$

15. Although the highest absorbance peak for NADH is at 260 nm, it may be practical to use the second peak at 340 nm to determine the concentration because there is no interference from ATP at this wavelength.

16. $A = \alpha\, b\, C$

For lower limit: $0.1 = \dfrac{10,000\ \text{L}\ (1\ \text{cm})(C)}{\text{mole (cm)}}$

$C = \dfrac{0.1\ \text{mole}}{10,000\ \text{L}}$

$C = 1 \times 10^{-5}\ \text{mole/L}$

For upper limit:

$1.8 = \dfrac{10,000\ \text{L}\ (1\ \text{cm})\ (C)}{\text{mole–cm}}$

$C = \dfrac{1.8\ \text{mole}}{10,000\ \text{L}}$

$C = 1.8 \times 10^{-4}\ \text{mole/L}$

The range of concentrations that can be measured for this compound, therefore, is from 1.8×10^{-4} mole/L to 1×10^{-5} mole/L.

17. a. $C = 50\ \mu\text{g/mL}\ (1.25)$

$\quad = 62.5\ \mu\text{g/mL}$

Because the solution was diluted, however, it is necessary to multiply the answer by 10, so 625 μg/mL = concentration of DNA in the original sample

b. $C = (0.63)\ (40\ \mu\text{g/mL})$

$\quad = 25.2\ \mu\text{g/mL}$

$(25.2)\ \mu\text{g/mL} \times 5$

$\quad = 126\ \mu\text{g/mL} =$ concentration of RNA in the original sample

18. a. Substituting into equation 1 from Box 25.2:

Concentration = 50 μg/mL (0.85) = 42.5 μg/mL

b. Substituting into equation 3 from Box 25.2:

40 μg/mL (0.69) = 27.6 μg/mL

27.6 μg/mL ×10 = 276 μg/mL

19. a. No. The results of an assay using a standard curve should be accurate because the values determined are compared with those on the standard curve.

    b. If the single standard was properly prepared and the instrument was properly adjusted with a blank, and if the standard and sample are in the linear range of the assay, then the results should be accurate.

    c. Yes, the value for the absorptivity constant was presumably based on a correctly calibrated instrument and therefore will not be relevant to this spectrophotometer.

20. This problem may be evident when you examine your standard curve if it causes the results to be nonlinear or low. If you prepare a standard curve alongside your samples, therefore, it may be evident that there is a problem.

21. Assuming you prepare a standard curve each time you do the assay, and assuming the reagent still acts in such a way that color is proportional to the amount of analyte, then the standards should be affected by the reagent in the same way as the samples. The results, therefore, will still be accurate. You may observe, however, that the range for the assay is different than usual. Low concentrations of analyte that are normally in the range of the assay may not cause a detectable color change.

22. Yes. If there are substances in the blood that affect absorbance at the analytical wavelength, then the buffer blank will not compensate and the standard curve will not be correct for the samples.

23. a. Concentration of double-stranded
DNA $\approx 50\ \mu g/mL\ (A_{260})$
$\approx 50\ \mu g/mL\ (0.497) \approx 24.9\ \mu g/mL$

    Thus, based on what she weighed, she had a solution of 60 $\mu g/mL$ DNA; based on the absorbance, the solution contained about 25 $\mu g/mL$ DNA. This is not particularly "close," the UV method estimated the concentration of DNA at less than half of what it should have been.

    b. Concentration of protein $\approx 0.276/0.7 \approx 0.394\ mg/mL$

    Thus, based on weight, she had a 0.50 mg/mL solution of BSA. Based on its absorbance, she had a solution with about 0.40 mg/mL BSA. The UV method was thus gave a closer estimate for the protein concentration than for the DNA concentration.

    c. $A_{260} = 0.176 + 0.497 = 0.673$
$A_{280} = 0.276 + 0.260 = 0.536$

    The A_{260} value for the mixture was 0.685 and the A_{280} value for the mixture was 0.547.

    The values for the mixture are very close to the summed values for the pure samples. This means that the absorbances at both wavelengths were additive, as predicted.

    d. For pure DNA: $A_{260}/A_{280} = 0.497/0.260 \approx 1.91$
For pure BSA: $A_{260}/A_{280} = 0.176/0.276 \approx 0.638$
For the mixture: $A_{260}/A_{280} = 0.685/0.547 \approx 1.25$

    All three ratios are consistent with the predicted ratios.

    e. [nucleic acid] $\approx 62.9(0.685) - 36.0\ (0.547)$
$\approx 23.4\ \mu g/mL$

[protein] $\approx 1.55\ (0.547) - 0.757\ (0.685)$
$\approx 0.329\ mg/mL$

Thus, based on what she weighed, the DNA should have been 60 $\mu g/mL$ rather than 23.4 $\mu g/mL$. The protein should have been 0.5 mg/mL rather than 0.329 g/mL.

    f. The UV methods did not give results that exactly matched the expected results for the standards. Many possible explanations of the differences are discussed in the text. Whether or not the estimates provided by the UV methods are "close enough" depends on the situation and the decisions that will be made based on the results. For example, if several DNA purification methods are being compared, then these UV estimates would be adequate to compare the efficacy of the purification methods. If a pharmaceutical company is trying to determine whether a recombinant protein is pure enough for drug use, however, these UV methods are not sufficiently accurate.

CHAPTER 26

Practice Problems: Weight per Volume (from p. 496)

1. $\dfrac{0.1\ g}{1\ mL} = \dfrac{?}{100\ mL}$ $? = 10\ g$

Dissolve 10 g of $AgNO_3$ in water (distilled or purified) and bring to 100 mL total solution in a graduated cylinder or volumetric flask.

2. $\dfrac{2\ mg}{1\ mL} = \dfrac{?}{50\ mL}$ $? = 100\ mg$

3. $\dfrac{100\ \mu g}{1\ mL} = \dfrac{?}{5\ mL}$ $? = 500\ \mu g = 0.0005\ g$

Dissolve 0.0005 g of proteinase K in purified water, bring to a volume of 5 mL.

Practice Problems: Molarity (from p. 499)

1. If you have a 3 L of a solution of potassium chloride at a concentration of 2 M, what is the solute? <u>potassium chloride</u> What is the solvent? <u>water (purified)</u> What is the volume of the solution? <u>3 L</u>
Express 2 M as a fraction.

$$\dfrac{2\ moles\ of\ potassium\ chloride}{liter\ of\ solution}$$

2. Atomic weight of K = 39.10, atomic weight of Cl = 35.45. The FW of KCl, therefore, is: 39.10 + 35.45 = 74.55

74.55 g/mole $\times$ 1 mole/L $\times$ 0.250 L
≈ 18.64 g of solute is required.

3. FW of $KH_2PO_4 = 136.09$.

136.09 g/mole $\times$ 0.3 mole/L $\times$ 10 L = 408.27 g = solute required.
Dissolve the solute in less than 10 L of purified water and BTV.

4. FW of K_2HPO_4 = 174.18.

174.18 g/mole $\times$ 0.1 M $\times$ 0.450 L $\approx$ 7.84 g = solute required.

Dissolve the solute in less than 450 mL and BTV.

5. If 1 L of 1 M Tris buffer requires 121.10 g of solute, then 600 mL of 1 M solute requires:

$$\frac{?}{600 \text{ mL}} = \frac{121.10 \text{ g}}{1000 \text{ mL}}$$

? = 72.66 g

If 600 mL of 1 M Tris buffer requires 72.66 g of solute, then 0.4 M requires:

$$\frac{?}{0.4 \text{ M}} = \frac{72.66 \text{ g}}{1 \text{ M}}$$

? $\approx$ 29.06 g = solute required.

6. You can use the anhydrous form, but you will need to add less than 25 g. The molecular weight of the hydrated form is 270.3 and the molecular weight of the anhydrous form is 162.2. To calculate how much of the anhydrous form to use, set up a proportion:

$$\frac{25 \text{ g}}{270.3 \text{ g}} = \frac{?}{162.2 \text{ g}}$$

? $\approx$ 15.0 g

Use 15.0 g of the anhydrous form instead of 25 g of the hydrated form.

7. 150 mg = 0.150 g. Using a proportion:

$$\frac{1 \text{ mole}}{230.11 \text{ g}} = \frac{?}{0.150 \text{ g}}$$

? $\approx$ 6.52 $\times$ 10^{-4} moles = 652 μmoles

Practice Problems: Percents (from p. 501)

1. 35 mL $\times$ 0.95 = 33.25 mL

Place 33.25 mL of 100% ethanol in a graduated cylinder and BTV 35 mL.

2. 200 g $\times$ 0.75 = 150 g

200 g $-$ 150 g = 50 g

Combine 150 g of resin and 50 g of acetone.

3. 600 mL $\times$ $\dfrac{15 \text{ g}}{100 \text{ mL}}$ = 90 g

Dissolve 90 g of NaCl in water and BTV of 600 mL.

4. $\dfrac{50 \text{ g}}{500 \text{ mL}} = \dfrac{?}{100 \text{ mL}}$

? = 10 g, **a.** so this is a 10% solution **b.** w/v.

5. 100 mg = 0.1 g and 1 L = 1000 mL

$$\frac{0.1 \text{ g}}{1000 \text{ mL}} = \frac{?}{100 \text{ mL}}$$

? = 0.01 g, **a.** so this is a 0.01% solution **b.** w/v.

6. 25% = 25 g/100 mL of NaCl = 250 g/1000 mL

1 M NaCl = 58.44 g/1000 mL

Using a proportion, if 58.44 μg/L is 1 M, then 250 g/L is 4.28 M.

$$\frac{58.44 \text{ g/L}}{1 \text{ M}} = \frac{250 \text{ g/L}}{?} \qquad ? \approx 4.28 \text{ M}$$

A second strategy:

Using the equation in the Chapter Summary: M = w/v % $\times$ 10/FW, you get,

M = 25 $\times$ 10/58.44 $\approx$ 4.28 M

so, 25% NaCl is the same as a 4.28 M solution.

Practice Problems: Parts (from pp. 502–503)

1. 1 part + 3 parts = 4 parts total

$$\frac{1}{4} = \frac{?}{10 \text{ mL}} \qquad ? = 2.5 \text{ mL}$$

$$\frac{3}{4} = \frac{?}{10 \text{ mL}} \qquad ? = 7.5 \text{ mL}$$

Combine 2.5 mL of salt solution with 7.5 mL of water.

2. 1 part + 3.5 parts + 0.6 parts = 5.1 parts total

$$\frac{1}{5.1} = \frac{?}{200 \text{ mL}} \qquad ? \approx 39.2 \text{ mL chloroform}$$

$$\frac{3.5}{5.1} = \frac{?}{200 \text{ mL}} \qquad ? \approx 137.3 \text{ mL phenol}$$

$$\frac{0.6}{5.1} = \frac{?}{200 \text{ mL}} \qquad ? \approx 23.5 \text{ mL isoamyl alcohol}$$

Combine the preceding volumes of each component.

3. 5 parts + 3 parts + 0.1 parts = 8.1 parts total

$$\frac{5}{8.1} = \frac{?}{45 \text{ } \mu\text{L}} \qquad ? \approx 27.8 \text{ } \mu\text{L Solution A}$$

$$\frac{3}{8.1} = \frac{?}{45 \text{ } \mu\text{L}} \qquad ? \approx 16.7 \text{ } \mu\text{L Solution B}$$

$$\frac{0.1}{8.1} = \frac{?}{45 \text{ } \mu\text{L}} \qquad ? \approx 0.56 \text{ } \mu\text{L Solution C}$$

Note that it would be very difficult to accurately pipette 0.56 μL of a solution. You would probably mix a larger volume and remove 45 μL from it.

4. **a.** 3 ppm = $\dfrac{3 \text{ g}}{1 \times 10^6 \text{ g}}$ = $\dfrac{3 \text{ g}}{1 \times 10^6 \text{ mL}}$

$$= \frac{3000 \text{ mg}}{1 \times 10^6 \text{ mL}} = \frac{3 \times 10^{-3} \text{ mg}}{\text{mL}}$$

Another way to think about this is to recall that

1 ppm in water = $\dfrac{1 \text{ } \mu\text{g}}{\text{mL}}$

so 3 ppm = $\dfrac{3 \text{ } \mu\text{g}}{\text{mL}} = \dfrac{3 \times 10^{-3} \text{ mg}}{\text{mL}}$

b. 3 ppm = 3 mg/L

5. $10 \text{ ppb} = \dfrac{10 \text{ g}}{1 \times 10^9 \text{ g}} = \dfrac{10 \text{ g}}{1 \times 10^9 \text{ mL}}$

$= \dfrac{10 \times 10^3 \text{ mg}}{1 \times 10^9 \text{ mL}} = \dfrac{10 \times 10^{-3} \text{ mg}}{1000 \text{ mL}} = \dfrac{0.01 \text{ mg}}{L}$

6. $100 \text{ ppm} = \dfrac{100 \text{ g}}{1 \times 10^6 \text{ mL}} = \dfrac{0.1 \text{ g}}{L}$

You could therefore dissolve 0.1 g of cadmium in water and BTV 1 L.

CHAPTER 27

1. $C_1 V_1 = C_2 V_2$

$5 \text{ X } (?) = 1 \text{ X } (10 \text{ }\mu L)$

$? = 2 \text{ }\mu L$

You need 2 μL of solution A.

2. $C_1 V_1 = C_2 V_2$

$100\% \text{ } ? = 95\% \text{ } (75 \text{ mL})$

$? = 71.25 \text{ mL}$

Take 71.25 mL of 100% ethanol and BTV 75 mL. (The number 250 in the problem is extraneous.)

3. You cannot make a more concentrated solution from a less concentrated one.

4. $C_1 V_1 = C_2 V_2$

$(0.3 \text{ M}) (65 \text{ mL}) = (0.0001 \text{ M}) (?)$

$? = 195,000 \text{ mL} = 195 \text{ L}$

You can make quite a lot—195 L!

5. a. Decide how much stock solution to make. For example, 1 L.

Using the formula given previously:

$(121.1 \text{ g/mole}) (0.1 \text{ mole/L}) (1 \text{ L}) = 12.11 \text{ g}$

Weigh out 12.11 g of Tris and dissolve in about 900 mL of water. Bring the solution to the proper pH using HCl. Then, bring the solution to 1 L. (Assume the solution is to be prepared and used at room temperature, since no temperature was specified.)

b. Decide how much solution to make, for example, 100 mL.

$C_1 \qquad V_1 = \quad C_2 \quad V_2$

$0.1 \text{ M } (?) = 0.01 \text{ M } 100 \text{ mL}$

$? = 10 \text{ mL}$

So, use 10 mL of Tris stock and BTV 100 mL. The pH may change slightly because of dilution, but should be acceptable for most applications. Record the final pH.

6. The pK_a of the buffer should be between 7.5 and 9.5. Any of the buffers in that range (e.g., HEPES or Tris) are acceptable on the basis of pK_a.

7. The pH of a Tris solution will decrease by 0.028 pH units for every degree increase in temperature. In this example the temperature increases 40°C. The pH therefore decreases 1.12 pH units, to 6.4.

8. Stocks

1 M Tris (pH 7.6 at 4°C).

To make 1 L of stock, dissolve 121.1 g of Tris base in water. After the Tris is dissolved, bring its temperature to 4°C by submerging it in an ice bath or placing it in the refrigerator. While the Tris is at this temperature, bring the pH to 7.6 with concentrated HCl. (Tris buffer will change pH as the temperature changes. As the temperature rises, the pH falls. The buffer will therefore require less HCl at 4°C than at room temperature.) Then BTV 1 liter.

1 M Mg acetate. To make 1 L of stock, dissolve 214.40 g of magnesium acetate (tetrahydrate) in water. Bring to volume.

1 M NaCl. To make 1 L of stock, dissolve 58.44 g in water. Bring to volume.

Breaking Buffer to make 200 mL:

0.2 M TRIS	40 mL of 1 M TRIS (pH 7.6 at 4°C.)
0.2 M NaCl	40 mL 1 M NaCl
0.01 M Mg acetate	2 mL 1 M Mg acetate
0.01 M β-mercaptoethanol	142 μL (see note)
5% Glycerol	10 mL glycerol
Bring to volume of 200 mL with water.	

Note: You want 200 mL of a 0.01 M solution of β-mercaptoethanol, so you need 0.156 g. You can use a proportion to figure out how many milliliters will contain 0.156 g:

$\dfrac{1.100 \text{ g}}{1 \text{ mL}} = \dfrac{0.156 \text{ g}}{?}$

$? \approx 0.142 \text{ mL} = 142 \text{ }\mu L$

You therefore need 142 μL of β-mercaptoethanol.

9. Step 1. Each time a reaction is performed, 10 μL of the dNTP mixture is needed. You want to perform the reaction 100 times, therefore, you will need 10 μL × 100 = 1000 μL.

Step 2. Consider only a single dinucleotide, for example, how much of the dATP stock do you need? This is a $C_1 V_1 = C_2 V_2$ problem.

C_1	$V_1 = C_2$		V_2
100 mM	?	1.25 mM	1000 μL

You need 12.5 μL of the dATP. You will similarly need 12.5 μL of the other three dNTPs.

Step 3. Combining 12.5 μL of each dNTP gives a volume of 50 μL. The remaining solution will be water:

12.5 μL 100 mM dATP
12.5 μL 100 mM dCTP
12.5 μL 100 mM dGTP
12.5 μL 100 mM dTTP
 950.0 μL water
1000 μL total solution

Note that it is difficult to bring a solution to a volume of 1000 μL. In this case it is correct to calculate that you need 950 μL of water.

Step 4. How many vials of each dNTP will you need to purchase? The volume of each vial is not listed in the catalogue.

Rather, it tells you that the concentration of dNTP in each vial is 100 mM and there are 40 μmoles (0.040 millimoles) in each vial. So what is the volume per vial?

A concentration of 100 mM means that there are 100 millimoles of the dNTP per liter. So, what volume contains 0.040 millimoles?

$$\frac{100 \text{ mmoles}}{1000 \text{ mL}} = \frac{0.040 \text{ mmoles}}{?}$$

$? = 0.400 \text{ mL} = 400 \text{ }\mu\text{L}$

Each vial contains 400 μL of a given dNTP at a concentration of 100 mM. (This is a good time to review the difference between the words "amount" and "concentration.") One vial has more than enough dNTP to make the solution desired; purchase one vial of each dNTP.

CHAPTER 28

1. Pyrogens are removed by ultrafiltration and reverse osmosis (also by anion exchange resins).

2. Viruses are removed by ultrafilters and reverse osmosis.

3. Table salt is smaller (*FW* = 58.44) than the cutoff for ultrafilters and so passes through them. NaCl is ionized in solution and therefore is removed by deionization.

4. Purified water is unstable. It absorbs gases from the air and leaches materials from the walls of the vessels in which it is contained. Continuous repurification is therefore used for Type I water.

5. 3.7.

CHAPTER 29

Practice Problems: Proteins (from p. 550)

1. a. Storage Buffer
  - Low temperature. Limits bacterial growth and degradation of product, inhibits enzyme activity
  - Tris-HCl. Buffer, maintains pH
  - NaCl. Salt, maintains ionic strength at moderate level
  - EDTA. Chelates Mg^{++}, inhibits enzyme activity during storage
  - DTT. Reducing agent, prevents unwanted disulfide bond formation
  - Triton X-100. Detergent, reduces adsorption of proteins to tube walls
  - BSA. Added protein, reduces loss of protein due to adsorption and protease activity
  - Glycerol. Prevents freezing of protein, stabilizes enzyme at low temperature

    Activity Buffer
  - Tris. Buffer, maintains proper pH for activity
  - NaCl. Salt, maintains ionic strength
  - $MgCl_2$. Salt, required cofactor for enzyme

  b. The two buffers have different functions. Storage requires protection of enzyme from degradation and

adsorption. Activity is not required during storage. Activity buffer optimizes activity.

2. a. Sample Buffer Used When Loading Proteins into Electrophoresis Gel
  - Tris. Buffer, maintains pH
  - Glycerol. Causes proteins to sink into gel wells
  - SDS. Detergent, denatures proteins, confers consistent charge on proteins
  - β-mercaptoethanol. Reducing agent, inhibits disulfide bond formation
  - Bromophenol blue. Dye, for visualization

  b. Restriction Enzyme Buffer (Low Salt)
  - Tris. Buffer, maintains pH
  - $MgCl_2$. Salt, maintains ionic strength, provides Mg^{++} ions as cofactors
  - DTT. Reducing agent, protects against oxidation

3. See Table 29.6 on p. 549.

Practice Problems: DNA (from p. 560)

1. a. TBE
  - Tris-borate. Buffer, maintains pH
  - Boric acid. Bring Tris to proper pH
  - EDTA. Chelates Mg^{++}, inhibits nucleases

  b. TE Buffer
  - Tris. Buffer, maintains pH
  - EDTA. Chelates Mg^{++}, inhibits nucleases

  c. Hybridization Buffer
  - PIPES. Buffer, maintains pH
  - EDTA. Chelates Mg^{++}, inhibits nucleases
  - NaCl. Salt, maintains ionic strength
  - Formamide. Lowers the melting temperature for DNA

2. In 0.4 M salt, DNA tends to form duplexes readily; stringency is relatively low.

3. a. One explanation is that DNA and RNA polymerase can bind one another *both* by nonspecific electrostatic interactions and by another kind of interaction that is specific to a particular site on the DNA. The electrostatic interactions decrease with increasing salt concentration.

  b. This means that when working with RNA polymerase, the salt concentration will have an effect on the products made in the laboratory. One would choose a particular salt concentration based on the specificity required.

4. a. The buffer has a relatively high concentration of salt that facilitates the hybridization of DNA to primer and the synthesis of new strands at 72°C. The melting temperature is raised because of the high salt concentration.

  b. Mg^{++} is required by the *Taq 1* polymerase for its activity.

  c. If *Taq 1* were sensitive to high temperature, it would be destroyed every time the temperature is raised to denature the DNA strands. In that case, more polymerase would have to be added each cycle.

5. Most household detergents and shampoos contain SDS, which solubilizes membranes, denatures proteins, and releases DNA into solution.

6. DNA precipitation requires the presence of higher salt concentrations—bring the solution to 0.25 M Na acetate.

CHAPTER 30

1. A 0.9% solution contains 0.9 g of solute/100 mL or 9 g/L.

The molecular weight of NaCl is 58.44 so a 1 M solution contains 58.44 g/L.

$$\frac{58.44 \text{ g}/1000 \text{ mL}}{1 \text{ M}} = \frac{9 \text{ g}/1000 \text{ mL}}{?}$$

$$? \approx 0.154 \text{ M} = 154 \text{ mM}$$

NaCl dissociates in water to form 2 ions, both of which have an osmotic effect. Therefore multiply by 2 to get milliosmoles:

$2 \times 154 = 308$. So 0.9% NaCl is about 308 mOsm/L.

2. a. 5% (w/w) = 50 g solute/kg.

The molecular weight of glucose-monohydrate is 198.

$$\frac{198 \text{ g}}{1 \text{ mole}} = \frac{50 \text{ g}}{?} \ ? \approx 0.252 \text{ mole}$$

$$= 252 \text{ mmole}$$

Glucose does not dissociate in solution, so 252 mmoles of glucose = 252 milliosmoles.

NaCl dissociates in water into sodium and chloride ions, both of which have osmotic effects. Therefore, 20 millimoles of NaCl = 40 milliosmoles.

Total: 252 milliosmoles + 40 milliosmoles = 292 milliosmoles.

b. 292 mOsm/kg.

3. The $C_1V_1 = C_2V_2$ applies since concentrated stocks are being used.

a. Water:

$C_1V_1 = C_2V_2$

5X (?) = 1X (1000 mL)

? = 200 mL of 5X concentrate is required

b. Concentration of glucose in final solution:

$C_1V_1 = C_2V_2$

20% (10 mL) = ? (1000 mL)   ? = 0.2%

c. Concentration of MgSO4 · 7H2O in final solution:

$C_1V_1 = C_2V_2$

1 M (1 mL) = ? (1000 mL)

? = 1 mM

d. Concentration of thiamine in final solution:

$C_1V_1 = C_2V_2$? = 0.00005%

0.5% (0.1 mL) = (?) (1000 mL)

4. **Table 30.7**

Antibiotic	Amount Stock Required (per L)
Ampicillin (sodium salt)	2 mL
Chloramphenicol	5 mL
Kanamycin	5 mL
Streptomycin	5 mL
Tetracycline HCl	10 mL

5. a. Sodium acetate is the carbon source.

b. Monoammonium phosphate is the nitrogen source.

c. NaCl is used to maintain osmotic balance.

d. Provide potassium, nitrogen, and phosphorus and buffers the medium.

e. This is a selective medium so the likely purpose of the bromthymol blue is to change color to indicate the selective growth of one of the types of bacteria.

(*E. coli* can utilize acetate as a carbon source but *Shigella* cannot. *E. coli* produce alkaline byproducts when they use acetate for growth. Bromthymol blue is a pH indicator that turns blue at higher pHs but is yellowish at lower pHs. A blue color in this selective medium therefore is indicative of *E. coli* growth.)

f. The purpose of agar is to produce a solid surface on which bacteria can grow. The agar is present at a concentration of 2%.

6. 20% of 500 mL = 100 mL, so 100 mL of serum is required. 400 mL of medium is required to bring the volume to 500 mL total.

7. The $C_1V_1 = C_2V_2$ applies since concentrated stocks are being used.

$C_1V_1 = C_2V_2$

100X (?) = 1X (500 mL)   ? = 5 mL

100 mL of serum

5 mL of glutamine

<u>395 mL of medium</u>

500 mL

8. The $C_1V_1 = C_2V_2$ applies since concentrated stocks are being used.

a. 10X cell culture medium concentrate required:

$C_1V_1 = C_2V_2$

10X (?) = 1X (1000 mL)   ? = 100 mL

b. Glutamine stock required:

$C_1V_1 = C_2V_2$

200 mM (?) = 2 mM (1000 mL)   ? = 10 mL

c. Serum required:

1000 mL (0.10) = 100 mL

or

$C_1V_1 = C_2V_2$

100% (?) = 10% (1000 mL)   ? = 100 mL

d. $NaHCO_3$ required:

$C_1V_1 = C_2V_2$

0.89 M (?) = 0.026 M (1000 mL)   ? ≈ 29.21 mL

e. Water required = 1000 mL − 239.21 mL = 760.79 mL

9. You may have slightly different answers due to rounding.

Table 30.8 RECOMMENDED ADDITIONS OF SODIUM BICARBONATE

Medium	mL of NaHCO₃ required, 7.5% stock (per L)	g of solid NaHCO₃ required (per L)	Final concentration NaHCO₃ (mg/L)
DME	49.3	3.70	3700
DME/ Ham's F12	32.5	2.438	2438
Ham's F12	15.7	1.176	1176
MEM Earle's salts	29.3	2.20	2200
MEM Hank's salts	4.7	0.35	350
RPMI 1640	26.7	2.00	2000
McCoy's 5A	29.3	2.20	2200
MEM Alpha	29.3	2.20	2200

Table 30.9 RECOMMENDED ADDITIONS OF L-GLUTAMINE

Medium	Mg/L	mM	mL/L
AMEM	292.3	2.0	10
BME	292.3	2.0	10
DME	584.6	4.0	20
F12K	292.3	2.0	10
Ham's F10	146.2	1.0	5
Ham's F12	146.2	1.0	5
Iscove's DME	584.6	4.0	20
EMEM	292.3	2.0	10
RPMI 1640	300.0	2.05	10.25

11. a. Medium + 10% Serum

500 mL medium = $15.00   500 mL medium = $15.00

50 mL of FBS = $25.00   10 mL of FBS = $5.00

$40.00 $20.00

Reducing the use of serum results in substantial cost savings (50% in this case).

12. Key Similarities

Both are aqueous solutions with solutes that support the growth and reproduction of living cells.

Both must be sterilized and must be handled using aseptic technique.

Both are osmotically balanced.

Both include one or more sources of energy, carbon, vitamins, and trace elements to support the growth and reproduction of cells.

Both may be defined or undefined.

Both may or may not contain hydrolysates and extracts from yeasts, plants, and animals.

Key Differences

Mammalian cell culture tends to be more exacting and more sensitive to contamination.

Mammalian culture media are usually more complex than bacterial.

Agar is used when culturing bacteria, not mammalian cells.

Bacterial media are usually sterilized by autoclaving; mammalian cell media by filtration.

CHAPTER 31

1. NaCl ions, albumin, polio virus, *Serratia marcescens*, red blood cells, pollen grains, sand grains

2. **a.–c.** ultrafiltration   **d.** macrofiltration

   e. microfiltration   **f.** ultrafiltration or ultafiltration

   g. microfiltration   **h.** microfiltration

3. **a.** XYZ   **b.** ABC

   c. XYZ   **d.** "Ready Separation"

CHAPTER 32

1. $r_{max} = 9.10$ cm $r_{min} = 3.84$ cm $r_{average} = 6.47$ cm

SPEED (in RPMs)	RCFs GENERATED r_{max}	r_{min}	$r_{average}$
20,000	40,800	17,200	29,000
30,000	91,700	38,700	65,200
40,000	163,100	68,800	115,900

2. The radius must be in centimeters to use this equation.

3. $1000\sqrt{\dfrac{RCF}{11.2\,r}} = RPM$

$= 1000\sqrt{\dfrac{140,000}{11.2\,(2.6)}}$

$\approx 69,338$ RPM

4. **a.** For separating large volumes of cells from fermentation, a high-capacity rotor is more important than a rotor that can spin at high RCFs. Rotor B is therefore preferred.

   b. The separation of very small particles requires a rotor, like rotor A, which is capable of running at high RCFs.

   c. A microfuge is used for small volume samples like this, rotor C.

5. a. The maximum speed for this rotor is 13,000 RPM.

b. The maximum force that can be generated with this rotor is 27,500 × g.

c. The maximum volume that can be centrifuged in this rotor is 6 × 315 mL = 1890 mL.

d. This rotor can hold 72 tubes each with a volume of 4 mL = 288 mL.

e. High speed.

CHAPTER 33

1. 2.4×10^8 IU

2.

PURIFICATION OF COMATASE FROM E. coli

Purification Step	Volume (ml)	Total Protein (mg)	Total Activity (IU)	Specific Activity (IU/mg)	Yield (%)
I. Homogenization	100,000	12,350	9000	<u>0.73</u>	100
II. Dialysis	5000	10,233	8289	0.81	92.1
III. Organic extraction	80	3860	<u>6369</u>	1.65	<u>70.7</u>
IV. Ion exchange chromatography	10	1140	5625	<u>4.93</u>	62.5
V. PAGE	2	386	4688	12.15	<u>52.1</u>

3. In evaluating the relative merits of specific purification steps, consider the main purpose of each step. In the example given, Step II provided a major reduction in product volume, but little purification (a resulting specific activity of 0.81 vs. 0.73 for the starting material).

a. Step IV provides the greatest relative increase in product purity, with approximately a threefold increase in the specific activity of the product. Step V gave a 2.5-fold purification.

b. The loss of product material in each step can be calculated by dividing the total product activity for each step by the activity present in the preceding step. Step III, therefore, gave the greatest loss of product activity:

(6,369 IU/8,289 IU) × 100%
= 76.8% product recovery in this step.

c. Absolute reduction in product volume can be determined by looking at the final product volume for each step. In this example, Step II gave the greatest volume reduction, from 100,000 mL to 5000 mL. In a production setting, this volume reduction would significantly ease the cost of operations.

d. Relative reduction in product volume is calculated in a similar manner to 3b. The volume after each step is divided by the volume at the beginning of the step. Step III, therefore, gave the greatest *relative* reduction in volume size:

(80 mL/5,000 mL) × 100%
= 1.6% of the starting volume for this step
vs. 5% for Step I

4. The enzyme was secreted into the broth, which was discarded after the first centrifugation step. It is good practice to keep all supernatants and pellets until you are certain you have the product of interest.

5. a. See Figure 33.3 for an example of a flow chart.

b. 20 U/mg protein.

c. If there are 500 units in 0.5 mL of the supernatant, then there are 20,000 units in the entire 20 mL.

6. a. i. Specific activity should increase.

b. ii. Amount of total protein should decrease.

c. ii. Yield will decrease.

7. Some possible answers:

Whether it is found inside cells or is secreted

If it is intracellular, its intracellular location

Molecular weight

Solubility characteristics in different solvents

Molecular charge

Stability at room temperature

CHAPTER 34

1. a. hardware   **d.** software

b. software   **e.** hardware

c. hardware   **f.** software

2. One megabyte = 2 to the twentieth power

$= 2 \times 2 \times 2 \times 2 \times 2 \times 2 \times 2 \times 2 \times 2 \times 2 \times 2 \times 2 \times 2 \times 2 \times 2 \times 2 \times 2 \times 2 \times 2 \times 2$

$= 1,048,576$

3. a. 115

b. 18

c. 2

CHAPTER 35

1. a. spreadsheet   **d.** spreadsheet

b. database   **e.** spreadsheet

c. database (or a spreadsheet, if calculations are needed)   **f.** database

2. a. pie chart   **c.** bar graph or line

b. line graph   **d.** line graph

3. This is the first major version of MegaWhiz, and there have been up to two significant revisions of this version and four minor patches, probably to remove bugs and security problems from the program.

4. Bitmapped graphics are based on individual pixel information, while vector graphics are based on mathematical equations. Bitmapped graphics are used for complex images such as photographs, create larger storage files, and cannot be enlarged without loss of image quality.

CHAPTER 36

1. A search engine catalogs web pages using a web crawler or bot, then indexes the web pages and retrieves them in response to a keyword search. Directories are maintained and indexed by humans. Links are available through subject hierarchies, although they may also be searchable. Search engines generally retrieve more pages than directories.

2. Viruses are designed to spread themselves to multiple computers. Spyware infects only the computer to which it is downloaded.

3. There are many things to look for, including the author's name and credentials, when the page was updated, the purpose of the page, the tone of the writing, the presence and quality of links and references, spelling and grammatical errors, etc.

4. The URL protocol should be https://. There will be a lock symbol on the navigation bar and/or the bottom right of the screen. Clicking on the symbol will bring up the authentication information for the site, as well as a copy of the SSL certificate.

5. You should only Reply to All if the response is intended for the entire mailing list. Otherwise, reply only to the intended recipient.

6. A wiki is open to the public for contributions and editing.

Glossary

Acronyms

ACGIH American Conference of Governmental Industrial Hygienists

ACS American Chemical Society

ADME Absorption, distribution, metabolism, and excretion

ANSI American National Standards Institute

APHIS Animal and Plant Health Inspection Service

API Active pharmaceutical ingredient

ASTM The American Society for Testing and Materials

ATC Automatic temperature compensating (probe)

ATCC American Type Culture Collection

AUP Acceptable use policy

BIOS Basic input/output system

BLA Biological License Application

BLAST Basic local alignment search tool

BMP Bit-Mapped Paint

BPS Bits per second

BSE Bovine spongiform encephalopathy

BSL Biosafety level

BTV Bring to volume

CAPA Corrective and preventive actions

CAS Chemical Abstracts Service

CBER Center for Biologics Evaluation and Research

CD Compact disc

CDC Centers for Disease Control and Prevention

CDER Center for Drug Evaluation and Research

CD-R Compact disc-recordable

CDRH Center for Devices and Radiological Health

CD-ROM Compact disc read-only memory

CD-RW Compact disc-rerewritable

CFR Code of Federal Regulations

CFSAN Center for Food Safety and Applied Nutrition

cGMP current Good Manufacturing Practices

CGPM Conférence Générale des Poids et Mesures/General Conference on Weights and Measures

CHO Chinese hamster ovary

CHP Chemical Hygiene Plan

CIP Clean in place

CIPM Comité International des Poids et Mesures/International Committee of Weights and Measures

CNS Central nervous system

CPU Central processing unit

CRO Contract research organization

CLSI Clinical and Laboratory Standards Institute

DBM Database management

DIN Deutsche Industrial Norms

DNA Deoxyribonucleic acid

DNS Domain name system

DOS Disk operating system

DOT Department of Transportation

DP Drug product

DPI Dots per inch

DS Drug substance

DSL Digital subscriber line

DTP Desktop publishing

DVD Digital versatile disc

ELN Electronic laboratory notebook

EMEA European Agency for the Evaluation of Medicinal Products

EPA Environmental Protection Agency

EtBr Ethidium bromide

ExPASy Expert Protein Analysis System

FAQ Frequently asked questions

FDA Food and Drug Administration

FDCA Food, Drug, and Cosmetics Act

FIFRA Federal Insecticide, Fungicide, and Rodenticide Act

FTP File transfer protocol

GB Gigabyte

GC, GLC Gas-liquid chromatography

GCP Good Clinical Practices

GFI Ground fault interrupt

GHz Gigahertz

GIF Graphics interchange format

GLP Good Laboratory Practices

GMLP Good Microbiological Laboratory Practices

GMO Genetically modified organism

GPC Gel permeation chromatography

GUI Graphical user interface

HACCP Hazard analysis critical control points

HBV Hepatitis B virus

HCS Federal Hazard Communication Standard

HEPA High efficiency particulate air (filter)

HGP Human Genome Project

HIC Hydrophobic interaction chromatography

HIV Human immunodeficiency virus

HPLC High performance liquid chromatography

HTML Hypertext markup language

HTS High throughput screening

HTTP Hypertext transfer protocol

HTTPS Hypertext transfer protocol secure

I/O Input/output

IACUC Institutional Animal Care and Use Committee

IARC International Agency for Research on Cancer

ICH International Conference on the Harmonisation of Technical Requirements for the Registration of Pharmaceuticals for Human Use

IDLH Immediate danger to life and health

ILAR Institute for Laboratory Animal Research

IM Instant messaging

IND Investigational New Drug Application

IP Intellectual property

IP Internet Protocol

IRB Institutional review board

ISE Ion selective electrode

ISO International Organization for Standardization

ISP Internet service provider

IU International unit of enzyme activity

JPEG Joint Photographics Expert Group

JSA Job Safety Analysis

KB Kilobyte

Kbps Kilobits per second

KDD Knowledge discovery in data

LAI Laboratory-acquired infection

LAL *Limulus* amoebocyte test

LAN Local area network

LC Liquid chromatography

LCD Liquid crystal display

LC$_{Lo}$ Lethal dose low

LD$_{50}$ Lethal dose 50%

LIMS Laboratory information management system

Mab Monoclonal antibody

MB Megabyte

Mbps Megabits per second

MCB   Master cell bank

MHz   Megahertz

MPEG   Motion Picture Experts Group

mRNA   Messenger ribonucleic acid

MS-DOS   Microsoft disk operating system

MSDS   Material Safety Data Sheet

MWCO   Molecular weight cutoff

NCBI   National Center for Biotechnology Information

NCCLS   The National Committee of Clinical Laboratory Standards

NCI   National Cancer Institute

NDA   New Drug Application

NFPA   National Fire Protection Association

NIC   Network interface card

NIH   National Institutes of Health

NIOSH   National Institute for Occupational Safety and Health

OCR   Optical character recognition

OD   Optical density

OIML   Organisation Internationale de Métrologie Légale/International Organization of Legal Metrology

OLAW   Office of Laboratory Animal Welfare

OMIM   Online Mendelian Inheritance in Man

OOS   Out of specification

OS   Operating system

OSHA   Occupational Safety and Health Administration

PAGE   Polyacrylamide gel electrophoresis

PDA   Personal digital assistant

PDB   Protein Data Bank

PDF   Portable document format

PEL   Permissible exposure limit

PHMSA   Pipeline and Hazardous Materials Safety Administration

PIN   Personal identification number

PK   Pharmacokinetics

PMT   Photomultiplier tube

PNG   Portable network graphics

PPE   Personal protective equipment

PPM   Parts per million

QA   Quality assurance

QC   Quality control

QS   Quantum Sufficit

R&D   Research and development

RAC   Recombinant DNA Advisory Committee

RAM   Random access memory

RCF   Relative centrifugal field

RCRA   Resource Conservation and Recovery Act of 1976

rDNA   Recombinant DNA

RFLP   Restriction fragment length polymorphism

RNA   Ribonucleic acid

RNAi   RNA interference

RO   Reverse osmosis

ROM   Read-only memory

RPC   Reversed-phase chromatography

RPM   Revolutions per minute

RSI   Repetitive stress injury

RSS   Really simple syndication

SAS   Statistical analysis system

SDS-PAGE   Sodium dodecyl sulfate–polyacrylamide gel electrophoresis

SFM   Serum-free medium

SI   Système International d'Unités

SNP   Single nucleotide polymorphism

SOP   Standard operating procedure

SQL   Structured query language

SRM   Standard reference material

SSL   Secure socket layer

STEL   Short-term exposure limit

STR   Short tandem repeat

TB   Terabyte

TCP   Transmission Control Protocol

TEMED   N,N,N′, N′-tetramethylethylenediamine

TIFF   Tagged image file format

TLC   Thin layer chromatography

TLV   Threshold limit value

TLV-C   Threshold value limit-ceiling

TLV-STEL   Threshold value limit–short-term exposure limit

TLV-TWA   Threshold limit value–time-weighted average

tRNA   Transfer RNA

TSCA   Toxic Substance Control Act

TWA   Time-weighted average

UL   Underwriters Laboratories

UPS   Uninterruptible power supply

URL   Uniform resource locator

USB   Universal serial bus

USDA   United States Department of Agriculture

USP   United States Pharmacopeia

UV   Ultraviolet

VIS   Visible

VPN   Virtual private network

WCB   Working cell bank

WFI   Water for injection

Wi-Fi   Wireless fidelity

WWW   World Wide Web

WYSIWYG   What you see is what you get

XML   Extensible Markup Language

GLOSSARY TERMS

% recovery   See **yield**.

32-bit color   Computer setting where interpretation of each pixel in an image uses 32 binary bits of information for accurate reproduction.

404 error   Browser message indicating that the requested page cannot be found.

Abscissa   In graphing, the X coordinate of a point; the distance of a point along the X axis.

Absolute error   The difference between the true value and the measured value. The plus (+) or minus (−) sign indicates whether the true value is above or below the measured value. *Absolute Error = true value − measured value.*

Absolute zero   The temperature at which thermal energy is virtually nonexistent; defined as 0 kelvin or −273.15°C.

Absorbance scale accuracy   See **photometric accuracy.**

Absorbance spectrum   The plot or graphic representation of absorbance of a particular sample when exposed to light of various wavelengths, typically plotted with wavelength on the X axis versus absorbance on the Y axis.

Absorbance (A) also called **optical density**   A measure of the amount of light absorbed by a sample defined as: $A = \log_{10}(1/t) = -\log_{10} t$ where t is transmittance.

Absorption   Process in which one substance is taken up by another, as when a cotton ball absorbs water or liquid is taken up into the depth of a filtering material.

Absorption (in spectrophotometry)   The loss of light of specific wavelengths as the light passes through a material and is converted to heat energy. *Note:* Spectrophotometers actually detect and measure transmittance, not absorbance. Absorbance is a calculated value.

Absorption (in the context of drug testing)   A process by which a drug substance moves from the site of administration to the blood.

Absorptivity   The inherent tendency of a material to absorb light of a certain wavelength. (Note that absorbance is a measured value that depends on the instrument used to measure it whereas absorptivity is an intrinsic property of a material.) See also **Beer's Law.**

Absorptivity constant (a or α)   A value that indicates how much light is absorbed by a particular substance at a particular wavelength under specific conditions (such as temperature and solvent). It is sometimes calculated as: $\alpha = A/bC$, where A = absorbance, b = path length, and C = concentration, and α has units that vary depending on the units of concentration and path length. See also **Beer's Law.**

Absorptivity constant, molar (ε)   The absorptivity constant when the concentration of analyte is expressed in units of moles per liter.

Acceptable use policy (AUP) A statement issued by networks that outlines behaviors that are not allowed within their accounts.

Acceptance (control chart usage) A decision that a process is operating in control. See also **control chart limits.**

Acceptance criteria (according to GMP) The minimum specifications for a product and the criteria for accepting or rejecting that product, together with a plan for sampling it.

Accepted reference value * "A value that serves as an agreed-upon reference for comparison, and which is derived as: (1) a theoretical or established value, based on scientific principles, (2) an assigned value, based on experimental work of some national or international organization [e.g., see definition of **standard reference material (SRM)**], or (3) a consensus or certified value, based on collaborative or experimental work under the auspices of a scientific or engineering group."

Accession number A unique identifier for a DNA sequence, supplied by GenBank and designed to facilitate database searches with that sequence.

Accuracy Closeness of agreement between a measurement or test result and the true value or the accepted reference value for that measurement or test.** *Note:* The term accuracy, when applied to a set of observed values, is affected both by random error and systematic error (bias). Because random components and bias components cannot be completely separated in routine use, the reported "accuracy" must be interpreted as a combination of these two elements. The use of the term "imprecision" to describe random errors and "bias" to describe systematic errors will emphasize these distinct elements of variation. When assessing the systematic error (bias) of test methods or operations, use of the term "bias" will avoid confusion.

Acid A compound that dissociates in water to release a hydrogen ion and reacts with base to form neutral salts and water; can be corrosive and reactive.

Acid–base error A cause of inaccuracy when using pH indicator dyes. Indicators are themselves acids or bases and when added to unbuffered or weakly buffered solutions will cause a change in the pH of a solution.

Acidic solution An aqueous solution with a pH less than 7.

Acrylamide A small neurotoxic monomer that can be polymerized to relatively nontoxic polyacrylamide; see **polyacrylamide gels.**

Activated carbon Material that adsorbs and removes organic contaminants from water. Derived from wood and other sources that are charred at a high temperature to convert them to carbon. The carbon is "activated" by oxidation from exposure to high temperature steam.

Activation key A user-specific code that permits a program to be installed or activated.

Active pharmaceutical ingredient, (API) The substances or mixture of substances used in the manufacture of a drug product that have effect in the patient.

Activity (in pH measurement) The effective concentration of a solute that accounts for interactions among solutes, temperature, and other effects.

Acute Having a rapid onset.

Acute exposure A short-term contact or single dose of a substance or chemical.

Acute toxicity The harmful effects of any single dose or short-term exposure to a chemical or other substance.

Adapter (in centrifugation) An insert placed in a rotor compartment that permits the use of a smaller-sized tube than could otherwise be accommodated.

Adenine (A) One of the four types of nucleotide subunit that comprise DNA and RNA.

Adherent cells (in mammalian cell culture) Cells that grow in a single layer attached to the surface of their culture vessel.

ADME (absorption, distribution, metabolism, and excretion) Testing performed during the development of a drug product that evaluates where the substance goes in the body, how it is metabolically altered, and how it is excreted or broken down.

Adsorption Process in which a material sticks to the surface of a container, membrane, bead, filter membrane, or other solid.

Adsorption chromatography A chromatographic technique for separating molecules based on their relative affinities for a solid stationary phase and a gas or liquid mobile phase.

Adulterated (according to the FDCA) A food or drug that is produced by methods that do not conform to current Good Manufacturing Practices, or that is made under unsanitary conditions, or that contains unacceptable contaminants.

Adventitious contaminant A contaminant that is accidentally and sporadically introduced from a source connected to the production system, for example, from a contaminated piece of equipment or from a person.

Adverse events Undesired effects or toxicity due to exposure to a pharmaceutical or medical product.

Aerosol Very small particles suspended in the air.

Affinity The ability of a material to bind specifically to other biomolecules.

Affinity chromatography A chromatographic technique for separating molecules by their ability to bind specifically to other molecules that are incorporated onto the surface of a stationary phase.

Agar A hardening agent derived from seaweed; used to make a solid nutrient substrate for bacterial growth.

Agarose A natural polysaccharide derived from agar, a substance found in some seaweeds.

Air displacement micropipettor Device for measuring microliter volumes; designed so that there is an air cushion between the micropipettor and the sample.

Air-purifying respirator A **respirator** that filters room air through canisters of various materials that remove specific contaminants.

Airborne particle retention A measure of the efficiency of an air filtration system.

Airfoils (in fume hoods) Grates located at the bottom and sides of the hood sash that help to reduce air turbulence at the face opening of the hood.

Airlock A small room with interlocked doors between areas with different cleanliness standards.

Alcohol error (in pH measurement) A cause of inaccuracy when using pH indicator dyes. Alcohols may cause indicators to be a different color than they appear to be in aqueous solutions.

Aliquot The portions that result from subdividing a homogenous solution or substance into smaller units.

Alkaline error (in pH measurement) An error in the response of a pH measuring electrode in which the electrode responds to Na^+ and K^+. Alkaline error occurs when the pH values are above 9 or 10.

Alkalinity A measure of the capacity of water to accept H^+ ions; that is, its acid-neutralizing ability. Carbonate, bicarbonate, and hydroxide ions are common contributors to alkalinity.

Allergen A substance that produces an allergic response in some individuals.

Allergy A reaction by the body's immune system to exposure to a specific chemical.

Alpha helix See **secondary structure.**

Alternating current, (AC) Electrical current that cycles between flowing first in one direction and then in the other direction.

Ambient temperature The average temperature of the surrounding air that contacts the instrument or system being studied; often room temperature.

American Association for Accreditation of Laboratory Animal Care (AAALAC) An independent peer-review organization that ensures that companies, universities, hospitals, government agencies and other research institutions surpass minimal animal care standards.

American Chemical Society (ACS) Professional organization that sets standards for purity of chemical reagents.

American Conference of Governmental Industrial Hygienists (ACGIH) An organization of governmental, academic, and industrial professionals who develop and publish recommended threshold exposure limits for chemical and physical agents.

American National Standards Institute (ANSI) A national organization that sets standards related to safety and safety design. The U.S. representative to ISO.

American Society for Testing and Materials (ASTM) An organization that prepares and distributes standards to promote consistent procedures for measurement.

---

*Based on definitions in ASTM Standard E 456-06 "Standard Terminology Relating to Quality and Statistics."

**Based on definitions from American National Standard ANSI/ASQC A8402-1994 "Quality Management and Quality Assurance," American Society for Quality Control, Milwaukee, WI.

American Type Culture Collection (ATCC) a global, nonprofit organization that provides and distributes biological resources, particularly cells and tissues.

Amino acids A class of naturally occurring molecules that are the building blocks of proteins. Every amino acid contains a carbon atom bonded to an amino group ($-NH_2$), a carboxyl group ($-COOH$), a hydrogen atom, and a side chain. Each amino acid has a different side chain that gives it distinctive chemical properties.

Amino group Part of the core structure of amino acids, NH_3.

Ammeter A device to measure electrical current.

Amount How much of a substance is present. For example, 2 grams or 4 cups.

Amperes (A or amp) The units by which current is expressed. When 6.25×10^{18} electrons pass a point in the path of electricity's flow every second, the path is said to be carrying a current of 1 A.

Amplification The boosting of a weak signal.

Amplifier A device that boosts the voltage or current from a detector in proportion to the size of the original signal.

Analog "Smoothly changing." Analog measurement values are continuous, as for example, displayed by a meter with a needle that can point to any value on a scale.

Analog to digital converter (A/D converter) Device that converts an analog signal to a digital signal.

Analysis (in biology) The study of the specific chemical properties of a molecule.

Analyte A substance of interest whose presence and/or level is evaluated in a sample using an instrument, assay or test.

Analytical balance An instrument that can accurately determine the weight of a sample to at least the nearest 0.0001 g.

Analytical method A test used to analyze, identify, or characterize a mixture, compound, chemical, or unknown material.

Analytical ultracentrifuge An ultracentrifuge that is designed to provide information about the sedimentation properties of particles.

Analytical wavelength (in spectrophotometry) The wavelength at which absorbance measurements are made in a particular assay.

Anaphylactic shock A sudden life-threatening reaction to allergen exposure.

Angstrom (Å) Unit of length that is 10^{-10} meters, named after the Swedish physicist A.J. Ångström. A human hair has a diameter of about 500,000 angstroms.

Animal and Plant Health Inspection Service (APHIS) Division of the U.S. Department of Agriculture that oversees protection of animal and plant resources, including overseeing transgenic organisms.

Anion A negatively charged ion.

Anion exchange chromatography Ion exchange chromatography method where the stationary phase particles are coated with positively charged molecules to adsorb negatively charged molecules (anions) that flow by in the mobile phase.

Anion exchange resin A solid matrix material that has negatively charged ions available for exchange with negatively charged ions in a solution.

Anisotropic membrane A membrane in which the pore openings are larger on one side than on the other.

Annotation, genome (according to the Human Genome Project glossary) "Adding pertinent information such as gene coded for, amino acid sequence, or other commentary to the database entry of raw sequence of DNA bases."

Anode A positive electrode.

Anodized surface Thin, protective coating of aluminum oxide deposited electrochemically on aluminum rotors to help protect them from corrosion.

Antibody Protein made by immune system cells that recognizes and binds to substances invading the body and that aids in their destruction.

Antigens Substances that trigger the production of antibodies.

Antilogarithm (antilog) The number corresponding to a given logarithm. For example, $100 = 10^2$. The log of 100 is 2. The antilog of 2 is 100.

Anti-virus software Programs that protect computers from a variety of **malware.**

AOAC International An independent association of scientists devoted to promoting methods validation and quality measurements in the analytical sciences.

Application (in computers) Software that fulfills a specific function, such as data organization or word processing.

Artifact A distortion or error in the data. For example, in electron microscopy an artifact might be a substance that appears to be a component of a cell but, in fact, was acidentally created by the stains used to prepare the sample for visualization.

Aseptic processing A process in which drug or biological products and their containers are sterilized separately, frequently by different methods, and then packaged together under aseptic conditions to create a sterile final product.

Aseptic technique A system of laboratory practices that minimize the risk of biological contamination.

Ash content The percent ash residue remaining when a paper filter is burned.

Asphyxiant A gaseous compound or vapor that can cause unconsciousness or death due to lack of oxygen.

Asphyxiation Interruption of normal breathing, caused by lack of oxygen or from breathing high concentrations of carbon dioxide or other gases.

Assay A test of a sample or system. An assay might measure a characteristic of a sample (e.g., identity, purity, activity), a biological response, or an interaction between molecules. The terms "assay," "test," and "method" are often interchangeably, though the term "assay" is not generally applied to a test of an instrument's performance and generally refers to an analysis of a sample. "Assay" can also be used as a verb, meaning "to determine," for example, "the technician assayed the sample for protein."

Atmospheric pressure Pressure due to the weight of the air that comprises the atmosphere and presses down on every object on earth.

Atomic absorption spectrophotometry A spectrophotometric technique based on the absorption of radiant energy by atoms; used to measure the concentration of metals.

Atomic weight See **gram atomic weight.**

Attenuated vaccine A vaccine prepared from live bacteria or viruses that have been weakened so that they elicit an immune response in the recipient but do not cause disease.

Attenuator An electronic component that reduces the level of the signal from an instrument.

Audit trail A secure, computer-generated, time- and date-stamped record that allows the reconstruction of a course of events relating to the creation, modification, and deletion of an electronic record.

Authentication Unambiguous personal identification for data access.

Autoclave A laboratory pressure cooker that sterilizes materials with pressurized steam.

Automatic temperature compensating (ATC) probe A probe placed alongside pH electrodes that automatically reports the sample temperature to the meter.

Autosave A software function that **saves** data at regular intervals automatically.

Auxiliary scales (in thermometry) Extra scale markings at zero degrees and 100°C to assist in calibration and verification of the performance of the thermometer.

Background (in spectrophotometry) Light absorbance caused by anything other than the analyte.

Backup program Software that performs automatic backup operations, compresses files, and verifies data.

Backup server A network computer that can automatically back up data from individual workstations.

Backwards compatibility The ability of new software versions to work with files created in previous program versions.

Bacterial broth Aqueous mixtures of nutrients to support the growth and reproduction of bacterial cells; prepared without a hardening agent.

Baffles (in fume hoods) Adjustable panels that direct the air flow within a fume hood.

Balance An instrument used to measure the weight of a sample by comparing the effect of gravity on the sample to the effect of gravity on objects of known mass.

Balanced salt solution (BSS) An isotonic mixture of inorganic salts in specific concentrations.

Band pass See **spectral band width.**

Basal liquid medium (mammalian cell culture) Growth medium that contains a defined mixture of nutrients dissolved in a buffered physiological saline solution.

Base (for exponents) A number that is raised to a power. In the expression 10^3, the base is 10 and the exponent is 3.

Base (in the context of DNA and RNA) The molecules that distinguish the nucleotides that comprise DNA and RNA. In DNA there are four types of bases: adenine, cytosine, thymine, and guanine. RNA has uracil instead of thymine.

Base (in the context of pH) 1. A chemical that causes H^+ ions to be removed when dissolved in aqueous solutions. 2. A chemical that releases OH^- ions when dissolved in water.

Base pair (in the context of nucleic acids) Two complementary bases that lie across from each other on opposite strands of DNA. Adenine always pairs with thymine and guanine with cytosine.

Baseline A line that shows an instrument's response in the absence of analyte; a reference point.

Basic properties (in the SI system) The fundamental measured properties for which the SI system defines units. These properties are: length, mass, time, electrical current, thermodynamic temperature, luminous intensity, and amount of substance.

Basic research Research studies that are performed in order to understand nature.

Basic solution (in the context of pH) An aqueous solution with a pH greater than 7.

Batch (according to GMP) "[A] specific quantity of a drug or other material that is intended to have uniform character and quality, within specified limits, and is produced according to a single manufacturing order during the same cycle of manufacture."

Batch record An exact copy of a **master batch record** but with an assigned lot number. The batch record is used to direct the manufacture of a product and is the document in which formulation and manufacturing activities are recorded.

Battery A device that harnesses the potential of electrochemical reactions to do work.

Beam See **lever.**

Beer's Law (also Beer-Lambert or Beer-Bouguer Law) A rule that states that the absorbance of a homogeneous sample is directly proportional to both the concentration (C) of the absorbing substance and to the thickness of the sample in the optical path (b).

Beta-pleated sheet See **secondary structure.**

Bias (relating to measurements) "A systematic error that contributes to the difference between a population mean of the measurements or test results and an accepted or reference value."*

Bimetallic expansion thermometers Type of thermometer made of two different metals, each of which expands and contracts to a different extent as the temperature changes.

Bimodal distribution A frequency distribution with two peaks.

Bioaerosol An aerosol which includes biologically active materials.

Bioanalytical method Sometimes loosely refers to any test of a biological material. The FDA defines the term more narrowly (in "Guidance for Industry, Bioanalytical Method Validation") as a method used for the "quantitative determination of drugs and/or metabolites in biological matrices such as blood, serum, plasma, or urine . . . tissue and skin samples" taken from animal and human subjects.

Bioassay Commonly describes any assay that involves cells, tissues, or organisms as test subjects. Bioassays are used for many purposes, for example, in testing the potency of a drug.

Bioburden The number of contaminating microbes on a material before that material is sterilized.

Biohazard A potentially dangerous biological organism or material.

Bioinformatics The field in biology that uses computers to analyze molecular data and address biological questions involving large amounts of information.

Biologic According to FDA, antitoxins, antivenins, and venoms; blood, blood components, plasma derived products; childhood vaccines, including any future AIDS vaccines; human tissue for transplantation; allergenic extracts used for the diagnosis and treatment of allergic diseases and allergen patch tests; cellular products, including products composed of human, bacterial or animal cells (such as pancreatic islet cells for transplantation); gene therapy products.

Biological activity According to FDA's "Biotechnology Inspection Guide," the level of activity or potency of a product as determined by tests in animals, in cells in culture, or an in vitro biochemical assay.

Biological License Application (BLA) A document submitted by a manufacturer to FDA containing information regarding manufacturing methods, testing results, and clinical trial data for a new biologic product. The application must be approved before the product can be marketed.

Biological macromolecule A large and complex molecule which has biological function; (e.g., a hemoglobin molecule or a strand of RNA).

Biological safety cabinet Enclosure designed for the containment of biological hazards.

Biological solution A laboratory solution that supports the structure and/or function of biological molecules, intact cells, or microorganisms in culture.

Biometrics A method of verifying an individual's identity based on the measurement of physical features or repeatable actions that are unique to that person (e.g., fingerprint, retinal scan). A signature can be considered to be a biometric method.

Biomolecules Compounds produced by some type of biological source, such as a plant, animal, microorganism, or cultured cell.

Biopharmaceutical 1. A drug product that is manufactured using genetically modified organisms as a production system. 2. Most broadly, any drug manufactured by living cells or organisms (whether or not recombinant DNA techniques are involved); whole cells or tissues used therapeutically; any drug that is a large biological molecule, these are usually proteins, but can be RNA or DNA (e.g., DNA for gene therapy).

Bioreactor A specialized growth chamber used for producing a product in cells (usually refers to mammalian cells) in which conditions of temperature, nutrient level, aeration, pH, and mixing are controlled.

BIOS (basic input/output system) The essential set of routines that sets up the hardware in a PC and boots the operating system when the computer is turned on.

Biosafety Level (BSL) Defined by NIH as the combinations of laboratory facilities, equipment, and practices that protect the laboratory, the public and the environment from potentially hazardous organisms.

Biosafety Level 1 (BSL1) The lowest biosafety level; used when working with well-characterized strains of living microorganisms that are not known to cause disease in healthy adult humans.

Biosafety Level 2 (BSL2) The biosafety level used when working with agents that may cause human disease and therefore pose a hazard to personnel.

Biosafety Levels 3 and 4 (BSL3 and BSL4) Biosafety levels generally associated with dangerous agents that are highly infectious.

Bioseparation methods Separation techniques that are used to extract, isolate, and purify specific biological products, or biomolecules (such as a particular protein).

Bit The smallest unit of data storage. Short for binary digit.

Bitmapped graphics Images that are handled as collections of **pixels.** Also called **raster graphics.**

Blank (relating to spectrophotometry) A reference that contains no analyte but does contain the solvent (for a liquid sample) and any reagents that are intentionally added to the sample. The blank is held in a cuvette that is identical to that used for the sample, or the blank is alternately placed in the same cuvette as the sample. A spectrophotometer compares the interaction of light with the sample and with the blank in order to establish the absorbance due to the analyte.

BLAST (basic local alignment search tool) A pattern recognition tool used to search for similarities between nucleotide sequences or protein sequences.

Blood As defined by OSHA, refers to human blood, blood components, and any products made from human blood.

Blood serum The liquid component of blood from which blood cells have been removed.

Bloodborne pathogens As defined by OSHA, pathogenic microorganisms that are present in human blood and can cause disease in humans. These pathogens include, but are not limited to, hepatitis B virus (HBV) and human immunodeficiency virus (HIV).

"Blow out" pipette A type of pipette that is calibrated so that the last drop is to be forcibly expelled from the tip of the pipette.

Blue litmus A pH indicator dye that changes from blue to red, denoting a change from alkaline to acid.

Bluetooth A high-speed wireless networking standard used for short distances, routinely used by cell phones.

Boiling point The temperature at which a substance in the liquid phase transforms to the gaseous phase (under specified conditions of pressure).

Boolean operators (in databases) Mathematical terms that determine the logical relationship

---

*From ASTM E 131-05 "Standard Terminology Relating to Molecular Spectroscopy."

between keywords. The Boolean operators commonly used in database queries are AND, OR, and NOT.

Bootable backup media See **recovery media.**

Borosilicate glass A strong, temperature-resistant glass that does not contain discernable contamination by heavy metals and is used to manufacture general purpose laboratory glassware.

Bot A program that performs automated repetitive functions; short for robot.

Bounce (in computers) To return an e-mail to the sender.

Breakthrough rate (related to gloves) A measurement of the time required for a chemical that is spilled on the outside of a glove to be detected on the inside of the glove.

Brilliant yellow A pH indicator dye that is yellow at pH 6.7 and changes to red at pH 7.9.

Bromocresol green A pH indicator dye that changes from yellow to blue at pH 4.0 to 5.4.

Bromocresol purple A pH indicator dye that changes from yellow at pH 5.2 to purple at pH 6.8.

Browser Software that creates an interface between the user and the Internet.

Brushes (in centrifugation) A component of a centrifuge motor that conducts electrical current. Because brushes require maintenance, brushless motors are found in newer centrifuges.

BSE (bovine spongiform encephalopathy, also known as "mad cow disease") A fatal, neurodegenerative disease of cattle caused by an infectious prion agent. The disease is thought to be transmissible to humans and other animals.

BTV (bring to volume) The procedure in which solvent is added to solute (typically in a volumetric flask or graduated cylinder) until the total volume of the solution is exactly final volume desired.

Buffer A substance or combination of substances that, when in aqueous solution, resists a change in H^+ concentration even if acids or bases are added.

Buffered salt solution Solution that is intended to maintain living cells for short periods (minutes to hours) in an isotonic, pH-balanced environment.

Bug (in computers) A functional defect in a program.

Bulb (of a liquid-in-glass thermometer) A thin glass container at the bottom of a thermometer that is a reservoir for the liquid.

Bunny suit Special protective clothing designed to prevent human introduction of particulates into cleanroom facilities.

Buoyancy (principle of) Any object will experience a loss in weight equal to the weight of the medium it displaces.

Buoyancy error The discrepancy between mass and weight. The displacement of air by an object results in a slight buoyancy effect: Air slightly supports the object. The more air an object displaces, the more buoyant it is and the less it appears to weigh on a balance. Standards for balance calibration are typically made of metal with a density of 8.0 g/cm^3. In contrast, water has a density of only 1.0 g/cm3. Because water is less dense than calibration standards, a 1 kg mass of water takes up more space and displaces more air than a 1 kg mass of metal. Because balances are

calibrated with metal weights, there is a slight buoyancy error when less dense objects are weighed.

Buoyant density gradient centrifugation See **isopycnic centrifugation.**

Burette (also spelled "buret") Long graduated tube with a stopcock at one end that is used to accurately dispense known volumes.

Bypass fume hood Fume hood that has an opening at the top of the hood behind the sash, allowing air to enter the hood and bypass the working face, restricting face velocity when the sash is lowered.

Byte A group of continuous bits (usually eight, providing 256—2^8—possible combinations) that define a single understandable symbol or character.

Cable (in computers) The wiring that connects peripherals to the computer unit.

Caching (in search engines) Storing copies of web page contents.

Calcium A metallic element often found in water, typically as dissolved calcium carbonate; causes water hardness.

Calibration 1. To adjust a measuring system to bring it into accordance with external values. 2. A process that establishes, under specified conditions, the relationship between values indicated by a measuring instrument or measuring system and values of a trustworthy standard. Calibration permits the estimation of the uncertainty of the measuring instrument, or measuring system. The result of a calibration is recorded in a document.

Calibration (of an ion sensitive electrode) The use of standards with known concentrations of the ion of interest to determine the relationship between the voltage response of the electrodes and the ion concentration in the sample.

Calibration (of a pH meter) The use of standards of known pH to determine the relationship between the voltage response of the electrodes and the pH of the sample.

Calibration (of a spectrophotometer) (1) The use of standards with known amounts of analyte to determine the relationship between light absorbance and analyte concentration. (2) Bringing the transmittance and wavelength values of a spectrophotometer into accordance with externally accepted values.

Calibration (of a thermometer) The process by which the display of a thermometer is associated with reference temperature values.

Calibration (of a volume measuring device) 1. Placement of capacity lines, graduations, or other markings on the device so that they correctly indicate volume. 2. Adjustment of a dispensing device so that it dispenses accurate volumes.

Calibration curve See **standard curve.**

Calibration standard (for a balance) Objects whose masses are established and documented.

Calomel A paste consisting of mercury metal and mercurous chloride, Hg/Hg_2Cl_2.

Calomel electrode A type of reference electrode containing calomel.

Cancer A disease that is characterized by the uncontrolled growth of cells in different body organs.

Capacitor A component of electronic circuits that stores electrical energy.

Capacity The relative volume of sample that can be processed simultaneously by a technique; also refers to the amount of the product of interest that can be separated by the technique.

Capacity (for a balance) Maximum load that can be weighed on a particular balance as specified by the manufacturer.

Capacity line A line marked on an item of glassware to indicate the volume if the item is filled to that mark.

Carbon adsorption A method of removing dissolved organic contaminants from water by using activated carbon.

Carboxyl group Part of the core structure of amino acids, $-COOH$.

Carcinogen Compound that is capable or suspected of causing cancer in humans or animals.

Carpal tunnel syndrome A painful medical condition resulting from repetitive wrist and hand movements; caused by pressure on nerves due to deep tissue swelling in the wrist; an example of a **repetitive stress injury.**

Carrier An infected individual who is capable of spreading the infecting agent to other hosts; this individual may or may not show symptoms of disease.

Carryover Material from one sample that is carried to and contaminates another sample.

Cathode A negative electrode.

Cation A positively charged ion.

Cation exchange chromatography Ion exchange chromatography method where the stationary phase particles are coated with negatively charged molecules to adsorb positively charged molecules (cations) that flow by in the mobile phase.

Cation exchange resin A solid matrix material that has positively charged ions available for exchange with positively charged ions in a solution.

CD (compact disc) An etched optical disc that contains computer data in a series of pits that are read by a laser.

CD-R (CD-recordable) An optical disc that can be used to record data.

CD-ROM (compact disc-read only memory) A prerecorded optical disc used to store computer programs and files.

CD-RW (CD-rerewritable) An optical disc where data can be added or deleted in multiple sessions.

Ceiling limit The air concentration of a chemical that is not to be exceeded at any time.

Cell (in spreadsheets) The unit of data entry, arranged in a grid with each uniquely designated by a row and column coordinate.

Cell constant (K) (for conductivity measurements) The ratio of the distance between the electrodes and the area of the electrodes.

Cell culture The process in which living cells derived from the tissue of multicellular animals are maintained in nutrient medium inside petri dishes, flasks, or other vessels.

Cell homogenate A suspension of cell contents in liquid, produced by disrupting the outer cell membrane and wall (if present) and some of the interior structure of the cell.

Cell line   Cells in culture derived from a common ancestor cell that have acquired the ability to multiply indefinitely.

Cell-based assay   Any assay that involves living cells as the test subjects; a type of bioassay. Cell-based assays are used, for example, to test a material for viral contaminants, to look for a response to a drug product, and to look for toxic effects of a compound.

Cell therapy   Clinical methods that treat disease or injury with the use of whole cells or tissues.

Celsius scale (C) (also previously called centigrade scale)   A temperature scale defined so that 0°C is the temperature at which pure water freezes and 100°C is the temperature at which water boils.

Center for Biologics Evaluation and Research (CBER)   FDA division that regulates biologics for human use.

Center for Devices and Radiological Health (CDRH)   FDA division that regulates medical devices and many, but not all, in vitro diagnostic kits.

Center for Drug Evaluation and Research (CDER)   FDA division that regulates drug products (including many, but not all, biotechnology products) such as insulin, heparin, aspirin, and erythropoietin.

Center for Food Safety and Applied Nutrition (CFSAN)   FDA division that promotes and protects the public's health and economic interest by ensuring that food is safe, nutritious, and honestly labeled.

Centers for Disease Control and Prevention (CDC)   An agency of the Department of Health and Human Services whose mission is "to promote health and quality of life by preventing and controlling disease, injury, and disability."

Centi-(c)   A prefix meaning 1/100.

Centimeter (cm)   A unit of length that is 1/100 of a meter.

Central nervous system (CNS)   The biological system that includes the brain, spinal cord, and system of neurons; the target of neurotoxins.

Centrifugal force   The force that pulls a particle away from the center of rotation, in classical mechanics.

Centrifuge   An instrument that generates centrifugal force, commonly used to help separate particles in a liquid medium from one another and from the liquid.

Certification   "Certification* refers to the issuing of written assurance (the certificate) by an independent, external body that has audited an organization's management system and verified that it conforms to the requirements specified in the standard. Registration means that the auditing body then records the certification in its client register. For practical purposes, in the ISO 9001:2000 and ISO 14001:2004 contexts, the difference between the two terms is not significant and both are acceptable for general use."

Certified reference material   Any reference material issued with documentation. See **standard reference material.**

Chain of custody   Refers to the controlled and documented sequence of handling, storage, and disposition of a sample.

Change of state indicators   Products that change color or form when exposed to heat.

Characterization (in the context of drugs)   Process in which a molecular entity's physical, chemical, and functional properties are determined using specific assays. The qualities of the entity are then defined in terms of the results of those assays.

Chelator   An agent that binds to and removes metal ions from solution.

Chemical Abstracts Service (CAS) number   Identification number assigned to every commercial chemical compound.

Chemical corrosion   A chemical reaction that causes a metal surface to become rusted or pitted.

Chemical fume hood   A well-ventilated, enclosed chemical- and fire-resistant work area which provides user access from one side.

Chemical Hygiene Plan (CHP)   A written manual which outlines specific information and procedures necessary to protect workers from hazardous chemicals.

Chemical spill kits   Preassembled materials for controlling and cleaning up small- to medium-size laboratory spills.

Chemically defined growth medium   An aqueous medium to support the growth and reproduction of cells that contains only known ingredients in known quantities.

Chinese hamster ovary (CHO) cells   A cell line derived from the ovary of a Chinese hamster that is commonly used for biopharmaceutical production.

Chip   A thin piece of silicon that contains all the components of an electrical circuit.

Chromatin   The native form of DNA that is composed of DNA bound to several types of proteins.

Chromatography   A group of bioseparation techniques based on the differential interaction of molecules between a stationary and a mobile phase. Because molecules differ in their relative attraction to the mobile phase, they will move past the stationary phase at differing rates.

Chromatophore, chromophore   An atom or group of atoms or electrons in a molecule that absorb light.

Chromosomes   Long DNA macromolecules that contain genes, regulatory sequences, and stretches of DNA of unknown function.

Chronic exposure   Long-term continuous or intermittent contact.

Chronic health hazard   Biological harm resulting from long-term contact, or continuous or intermittent exposure to a hazard.

Circuit   A complete path for current flow.

Circuit board   A card on which components are mounted and connected to one another to form a functional unit.

Circuit breaker   A safety device that limits the current flowing in a circuit.

Clarification (in bioseparations)   The removal of unwanted solid matter after a bioseparation procedure, usually by centrifugation or filtration.

Clarification (in cell culture)   The removal of cells and cellular debris from a culture medium.

Clean bench   A type of laminar flow cabinet designed to provide a sterile work surface but not worker protection.

Clean in place (CIP)   The process of cleaning large, nonmovable items (e.g., fermentation vessels) in their place. CIP often involves pumping and circulating cleaning agents through the items.

Cleanroom   An area with a temperature-, pressure-, and humidity-controlled environment in which the levels of particulate contaminants, including dust, aerosols, vapors, and microorganisms, are significantly reduced.

Clearance studies   1. Studies of a process in which a contaminant (e.g., a virus) is intentionally added to a system in order to test the ability of the process to remove it. 2. In pharmacokinetics, a candidate drug substance is administered to an organism and its clearance from the body is evaluated.

Clearing factor (k) (in centrifugation)   A measure of the time required to sediment a particle in a particular rotor and under specified conditions.

Clearing time (t) (in centrifugation)   The time required to sediment a particle. $t = k/S$, where t = time in hours, k = clearing factor (in hours—Svedbergs), S = the sedimentation coefficient (in Svedberg units, S).

Clinical and Laboratory Standards Institute (CLSI) (formerly NCCLS)   An organization that promotes voluntary consensus standards for clinical laboratory testing of patient samples.

Clinical development   The stages of drug development in which a candidate drug is tested in human volunteers.

Clinical laboratory   Per OSHA, a workplace where diagnostic or other screening procedures are performed on blood or other potentially infectious materials.

Clinical trials   See **clinical development.**

Clock speed   The speed with which the microprocessor can carry out an instruction. Reported in **GHz.**

Clone   A copy. 1. An exact copy of a DNA segment produced using recombinant DNA technology. 2. One or more cells derived from a single ancestral cell. 3. One or more organisms derived by asexual reproduction that is/are genetically identical (or nearly identical) to a parent.

Cloning   Commonly refers to the production of one or more animals that have identical genetic information.

Closed system (in computers)   A computer system in which access is controlled by the people who are responsible for the content of the system's records. For example, a system of computers that is only accessible to the individuals who work in a company is a closed system.

Code of Federal Regulations (CFR)   A numerical system for the classification and identification

---

*Based on definitions in ISO/IEC Guide 25, "General Requirement for the Competence of Testing and Calibration Laboratories," Geneva, Switzerland. August 1996 edition.

of all Federal Regulations; all legally established federal regulations have a CFR number.

Coefficient (as relates to scientific notation) The first part of a number expressed in scientific notation. For example, the number 235 in scientific notation is expressed as 2.35×10^2 where "2.35" is the coefficient.

Coefficient of variation (CV) (relative standard deviation [RSD]) A measure that expresses the standard deviation in terms of the mean.

Coenzyme A complex organic molecule required by an enzyme for activity.

Cofactors A chemical substance required by an enzyme for activity. Cofactors include salt ions, such as Fe^{++}, Mg^{++}, and Zn^{++}.

Colony forming unit (CFU) A unit that measures the number of bacteria in a sample. Measuring CFUs requires incubating a sample and counting the resulting colonies; each colony is assumed to have been derived from one bacterium.

Colorimeter An instrument used to measure the interaction of visible light with a sample.

Colorimetry Technique for measuring color.

Column chromatography Chromatography performed with the stationary phase packed into a cylindrical container that varies in width and length.

Combination electrode A measuring electrode and reference electrode that are combined into one housing.

Combustible Substance that can vigorously and rapidly burn under most conditions.

Comité International des Poids et Mesures/ International Committee of Weights and Measures (CIPM) A body of 18 eminent scientists elected by the CGPM.

Command-line interface An interface that supplies a blank line for the user to type instructions for the computer one line at a time; also called a text-based interface.

Common logarithm (also log or $\log_{10}$) The common log of a number is the power to which 10 must be raised to give that number. For example, $1000 = 10^3$, so the log of 1000 is 3. The log of 5 is approximately 0.6990 which means that $10^{0.6990} \approx 5$.

Common name The brand name, trade name, or name in common use for a chemical compound or mixture.

Compendium (of standard methods) A published collection of accepted methods in a particular field.

Complete immersion thermometer A liquid expansion thermometer that is designed to indicate temperature correctly when the entire thermometer is exposed to the temperature being measured. Compare with **partial immersion thermometer** and **total immersion thermometer**. (Based on definition in ASTM Standard E 344-07 "Terminology Relating to Thermometry and Hydrometry.")

Complete medium (mammalian cell culture) Medium that has all its components, including supplements and components that are added just before use.

Complex growth medium 1. Bacterial cell culture: An aqueous medium to support the growth and reproduction of cells that contains ingredients whose exact composition is unknown (e.g., contains extracts from plant and animal tissues). 2. Mammalian cell culture: Loosely refers to a medium that contains more ingredients than basal media.

Compound A substance composed of atoms of two or more elements that are bonded together.

Concentration A ratio where the numerator is the amount of a material of interest and the denominator is the volume (or sometimes mass) of the entire solution or mixture. For example, if 1 g of table salt is dissolved in water so that the total volume is 1 L, the concentration is 1 g/L. See also **ratio**.

Conception (in the context of patents) The formation, in the mind of the inventor, of the complete invention (as defined in the patent claim).

Condenser (in distillation) Site where cooling water lowers the temperature of water vapor, causing it to condense back to a liquid form.

Conductance A measurement of conductivity; the reciprocal of the resistance in ohms when measured in a 1 cm^3 cube of liquid at a specific temperature. Conductance = 1/resistance. The units are: 1/ohm = 1 mho = 1 Siemen (S). 1 micromho (μmho) = 1mho/1,000,000.

Conductivity The inherent ability if a material to conduct electrical current. Used in water treatment to determine the level of ionized impurities present.

Conductivity meter A measuring instrument that measures the conductivity of a solution.

Conductor A material that offers little resistance to the flow of electricity.

Conférence Générale des Poids et Mesures/ General Conference on Weights and Measures (CGPM) A body consisting of representative of the governments that have subscribed to the Convention of the Metre.

Confidence interval A range of values (calculated based on the mean and standard deviation of a sample) which is expected to include the population mean with a stated level of confidence.

Congo red pH indicator dye that changes from red at pH 3.0 to blue at pH 5.0.

Constant Number in a particular equation that always has the same value.

Constant air volume (CAV) fume hood A fume hood that has a constant air flow through the exhaust duct; in these hoods, raising and lowering the sash changes the face velocity at the sash opening.

Containment The control of hazards and reduction of risk by isolation of the organism from the worker; see **primary containment** and **secondary containment.**

Contaminated Per OSHA, refers to the presence or the reasonably anticipated presence of blood or other potentially infectious materials on an item or surface.

Continuous density gradient A density gradient where there is a smooth increase in density from top to bottom with no sharp boundaries between layers.

Continuous flow centrifugation A mode of centrifugation (requiring special equipment) in which sample flows continuously into and out of the spinning rotor.

Contract research organization (CRO) A company contracted by a sponsor to perform preclinical or clinical drug testing.

Contraction chamber (of a liquid-in-glass thermometer) An enlargement of the capillary bore that holds some of the liquid volume, thereby allowing the overall length of the thermometer to be reduced.

Control 1. *Evaluation:* An evaluation to check, test, or verify; an item used to evaluate or verify a process, method, or experiment. 2. *Authority:* The act of guiding, directing, or managing. 3. *Stability:* A state in which the variability in a process is attributable only to chance (i.e., to the normal variability inherent in the process). (Based on ASTM Standard E 456-06 "Standard Terminology Relating to Quality and Statistics.")

Control chart A graphical display of test results together with limits in which the values are expected to lie if the process is in control. Control charts are useful in distinguishing between the normal variability inherent in a process and variability due to a problem or unusual occurrence.

Control chart limits Boundary lines on a control chart that define the limits of acceptable values based on standard deviations as determined in preliminary studies. The *lower warning limit* (LWL) is typically the mean minus two standard deviations and the *upper warning limit* (UWL) is the mean plus two standard deviations. The warning limits define the range in which 95.5% of all values should lie. The *lower control limit* (LCL) is typically the mean minus three SD and the *upper control limit* (UCL) is the mean plus 3 SD. These control limits define the range in which 99.7% of all points should lie.

Coordinates The address for a point on a graph; values that describe the distances horizontally and vertically for the location of a point.

Copy protection A device incorporated into discs or files to prevent the creation of more than one or two copies.

Corex A type of glass that is resistant to physical stress and is often used to make centrifuge tubes.

Corrective and preventive actions (CAPA) The processes by which a company responds to problems and failures. Corrective action means to fix problems that have already occurred and may happen again. Preventive action involves looking for problems that have not yet occurred and preventing them.

Corrosive Substance that will cause tissue damage or destruction at the site of contact.

Counterbalance A weight used in a mechanical balance that compensates for the weight of the sample.

Covalent bond Strong molecular bond formed by the sharing of electrons between atoms.

Covalent peptide bond A bond that connects adjacent amino acids together thus forming chains; these bonds are relatively strong and are not readily dissociated by environmental changes, such as raised temperature.

CPU (central processing unit) The main **microprocessor** that determines the relative power of the computer.

Crash (in computers) See **system crash.**

Cresol red pH indicator dye that changes from yellow at pH 7.2 to red at pH 8.8.

Creutzfeldt-Jakob disease Fatal, neurodegenerative human disease thought to be caused by misfolded proteins, prions.

Critical speed (in centrifugation) A low speed at which any slight rotor imbalance will cause the rotor to vibrate.

Cross-contamination A situation where cells from one culture accidentally enter another culture.

Cryogenic Extremely cold substances; usually at temperatures below −78°C.

Cubic centimeter (cc, cm³) A unit of volume that is equal to 1 mL.

Culture media See **growth media.**

Current (I) (electrical) The flow of electrical charge. Current may involve the movement of electrons in a conductor or the flow of ions in a solution.

Current Good Manufacturing Practices (cGMP) The currently accepted minimum standards and requirements for the manufacture, testing, and packaging of pharmaceutical products. "Current" means that a practice may be enforced, even if it is not in the published GMP regulations, if the practice has become accepted by industry.

Cursor A pointer used to indicate your location and select options on a computer screen.

Cushion (in centrifugation) Rubber or plastic pads positioned at the bottom of the compartments in a rotor, or placed underneath adapters, which support tubes and minimize breakage by distributing force over a larger area.

Cuvette A sample "test tube" that is designed to fit a spectrophotometer and is made of an optically defined material that is transparent to light of specified wavelengths.

Cytoplasm The substance of a cell outside the nucleus that is a fluid mixture of water, proteins, lipids, carbohydrates, and salts.

Cytosine (C) One of the four types of nucleotide subunit that comprise DNA.

Cytotoxic Damaging to cells.

Daisy chaining The connection of a series of peripheral devices through a single port.

Dalton (D) A unit of mass nearly equal to the mass of a hydrogen atom, 1.0000 on the atomic mass scale. A kilodalton (kd) is a unit of mass equal to 1000 daltons.

Dampening device A balance component that slows the oscillations of the moving parts of a balance so that the weight value can be read more quickly.

Dark current Signal that arises in the photodetector electronic circuits when no light shines on its surface.

Data Observations of a variable (singular, **datum**). The unprocessed facts from which we derive information.

Data mining The process of "digging" through very large amounts of data to discover relationships and other information.

Data warehouse A long-term storage resource for historical database contents.

Database An organized collection of data that is accessed through **database management software.**

Database management (DBM) software The programs that allow the user to search for, sort, look for patterns, and report selected data within a **database.**

Decontamination The use of physical or chemical means to remove, inactivate, or destroy bloodborne pathogens on a surface or item to the point where they are no longer capable of transmitting infectious particles and the surface or item is considered safe for handling, use, or disposal (OSHA).

Deep web See **invisible web.**

Defined medium See **chemically defined growth medium.**

Defragmentation An automated process to gather file fragments and reassemble them in one place on a disk.

Degradation rate (related to gloves) A measurement of the tendency of a chemical to physically change the properties of a glove on contact.

Degree (in thermometry) An incremental value or division in a temperature scale. For example, the Celsius scale is divided so that there are 100 degrees between the freezing point and the boiling point of water, whereas the Fahrenheit scale is divided into 180 degrees between the same two reference points.

Deionization A process in which dissolved ions are removed from a solution by passing the solution through a cartridge containing ion exchange resins. The resins are composed of beads that exchange hydrogen ions for cations in the solution and hydroxyl ions for anions in the solution. The ionic impurities remain associated with the beads wheras the hydrogen and hydroxyl ions combine to form water.

Denaturation (of DNA) Separation of the two complementary strands of DNA from one another.

Denaturation (of proteins) Unfolding of a protein so that its normal three-dimensional structure is lost but its primary structure remains intact.

Denominator The number written below the line in a fraction.

Densitometry A method used to quantify the amount of material on a solid medium (e.g., photographic negative) by measuring its absorbance of light.

Density Mass per unit volume of a substance under specified environmental conditions.

Density gradient centrifugation Techniques for separating molecules based on their rate of sedimentation or their buoyancy in a density gradient. See also **isopycnic centrifugation** and **rate zonal centrifugation.**

Deoxyribonucleic acid (DNA) A linear polymer consisting of four types of molecular subunits,

called nucleotides, connected one after another into long strands. DNA provides the molecular basis for inheritance, which is the passing of traits from parent to offspring.

Department of Transportation (DOT) Agency that regulates the transportation of hazardous materials.

Dependent variable A variable whose value changes depending on the value of the independent variable. For example, if an experimenter measures the growth of seedlings under different light intensities, the growth of the seeds is the dependent variable and the light intensity is the independent variable. See also **variable** and **independent variable.**

Depth filter A filter made of matted fibers or sand that retains particles through its entire depth by entrapment.

Derate (in centrifugation) A situation where a rotor must be run at a speed lower than its originally specified maximum speed.

Dermatitis Redness, inflammation, or irritation of the skin.

Descriptive statistics Statistical methods that are used to describe and summarize data.

Desktop (in computers) The area on a monitor screen where your work can be laid out.

Desktop publishing (DTP) software Programs that provide complete page layout capabilities for manipulating graphics and text and preparing documents for professional publication.

Detection limit (of a detector) The minimum level of the material or property of interest that causes a detectable signal.

Detector Refers to an electronic transducer that generates an electrical signal in response to a physical or chemical property of a sample.

Detector (in a spectrophotometer) Device used to measure the amount of light transmitted through a sample.

Detergent Substances that have both a hydrophobic "tail" and a hydrophilic "head." Detergents therefore have two natures; they can be both water-soluble and lipid-soluble.

Deuterium arc lamp Source that produces light in the UV region, from about 185–375 nm.

Deviation 1. (of a data point) The difference between a data point and the mean. 2. (in a quality context) An unexpected occurrence. 3. (of a measurement)* "The difference between a measurement . . . and its stated value or intended level."

Dewar flask A heavy multiwalled evacuated metal or glass container, used to hold cryogenic liquids.

Dialysis A separation method based on differences in the concentrations of solutes between one side of a membrane and the other.

Differential centrifugation A mode of centrifugation in which the sample is separated into two phases: a pellet consisting of sedimented material and a supernatant.

Differential medium A type of culture medium designed to reveal differences among microorganisms or groups of microorganisms that are growing on the same substrate.

*Based on definitions in ASTM Standard E 456-06 "Standard Terminology Relating to Quality and Statistics."

Differentiation   The process in which an embryonic cell, which originally has the capacity to become any type of cell in the body, matures into a particular cell type with a specialized structure and function (e.g., muscle, nerve, skin).

Diffraction   Bending of light that occurs when light passes an obstacle or narrow aperture.

Diffraction grating   Device consisting of a series of evenly spaced grooves on a surface that is used to separate polychromatic light into its component wavelengths.

Digital   Discontinuous measurement values (in contrast to continuous, analog values), as for example, stored and displayed by a computer.

Digital certificate   A physical item such as an identification cards or USB plug-in that confirms your identity to a computer.

Digital filter   Device used in electronic balances to stabilize their readout in the presence of drafts or vibrations.

Digital microliter pipettor   An instrument used to dispense volumes in the microliter range that can be adjusted to deliver different volumes.

Digital to analog converter   A signal processing device that converts a digitized signal to an analog signal.

Digitizing tablet and pen   An electronic tablet with a touch-sensitive screen with electronic circuitry underneath; the user can "write" on the screen with a special pen-like device.

Diluent   A substance used to dilute another. For example, when concentrated orange juice is diluted, the diluent is water.

Dilution   Addition of one substance (often but not always water) to another to reduce the concentration of the original substance.

Dilution series   A group of solutions that have the same components but at different concentrations.

DIN (Deutsche Industrial Norms)   A German agency that provides engineering and measurement standards.

Diode   An electronic component that allows electricity to flow in only one direction.

Direct current (DC)   Current that always flows in the same direction.

Directory   A search tool with a collection of links organized by topic.

Disc   Refers to optical storage media.

Disinfection   Destruction of most, but not all microorganisms by means of heat, chemicals, or ultraviolet light.

Disk   Refers to magnetic storage media.

Disk sanitizer   A utility that writes over sensitive files repeatedly and erases the data.

Dispersing element (in a spectrophotometer)   A part of a monochromator that separates polychromatic light into its component wavelengths.

Dispersion   The separation of light into its component wavelengths.

Display (in measurement)   Devices, such as meters, strip chart recorders, and computer screens, that display information in a form that is interpretable to a human or a computer.

Dissolved inorganics   Water contaminants, not derived from plants, animals or microorganisms, that usually dissociate in water to form ions.

Dissolved organics   Water contaminants broadly defined to contain carbon and hydrogen. Organic materials in water may be the result of natural vegetative decay processes or may be human-made substances such as pesticides.

Distillation   A water-purification process in which impurities are removed by heating the water until it vaporizes. The water vapor is then cooled to a liquid and collected, leaving nonvolatile impurities behind.

Distribution   The pattern of variation for a given variable.

Disulfide bond   A covalent bond that forms in proteins when sulfurs from two cysteine molecules bind to one another with the loss of two hydrogens.

DNA fingerprinting   A technique for distinguishing individuals based on differences in their DNA sequences.

DNA vaccines   Vaccines that are made from vectors that have been genetically engineered to include the DNA coding for one or two specific proteins from the infectious agent. When injected into the cells of a person (or other animal), the DNA is expressed, leading to the synthesis of proteins from the infectious agent. The recipient's immune system responds to the new proteins by mounting a protective immune response.

DNase   Deoxyribonuclease, a class of enzyme that breaks down the nucleotides in DNA.

DNS (domain name system)   A component of **TCP/IP** that allows the assignment of text names to web pages.

Documentation   Written records that guide activities and that record what has been done.

Dot diagram   Simple graphical technique to represent a data set in which each datum is represented as a dot along an axis of values.

Double-beam spectrophotometer   A type of spectrophotometer in which the sample and blank are placed simultaneously in the instrument so the absorbance of the sample can be continuously and automatically compared to that of the blank.

Double junction   An electrode junction configuration used when the sample is incompatible with the reference electrode filling solution. A double junction separates the reference electrode filling solution from the sample.

Download   To copy a file from the Internet to your computer.

Downstream (in filtration)   The side of a filter facing the filtrate.

Downstream processes   The separation procedures that result in a purified product.

Downstream processing   The stages of processing, including isolation and purification, of a desired product that takes place after the product is made by a fermentation or cell culture process (upstream processing).

DPI (dots per inch)   A measure of printer **resolution,** with more dpi indicating greater resolution and a sharper image.

dpKa/dt   The change of the pH of a buffer in pH units with change in its temperature in degrees Celsius. The larger this value, the more the buffer will change pH with each degree of temperature change. Negative values for dpKa/dt mean that there is a decrease in pH with an increase in temperature and vice versa.

Draw program   Software used to create vector graphics.

Driver (in computers)   Software that allows the computer to control a peripheral device.

Drug   Generally used synonymously with "pharmaceutical" though "drugs" include not only therapeutic compounds but also agents that are used in the body for nontherapeutic (e.g., "recreational") or harmful purposes.

Drug discovery   Methods for identifying new therapeutic agents.

Drug product (DP)   A finished dosage form (e.g., tablet, vial) that contains a drug substance.

Drug substance (DS)   The active ingredient in a drug product. See also **active pharmaceutical ingredient.**

Dry ice   Frozen carbon dioxide in solid form.

DSL (digital subscriber line)   A high-speed Internet connection that uses high-frequency transmission signals over a phone line, allowing normal telephone use with a proper sound filter installed.

Dual core (in microprocessors)   Technology in which a CPU chip contains two separate microprocessors that can compute simultaneously.

DVD (digital versatile disc)   An optical disc similar to a CD but with greater data storage capacity.

Dynamic range (of a detector)   The range of sample concentrations that can be accurately measured by the detector.

Efficacy (in the context of drugs)   The ability of a drug to control or cure an illness or injury.

Electrical ground   A conducting material that provides a pathway for current to the earth.

Electrical potential   The potential energy of charges that are separated from one another and attract or repulse one another; measured in units of volts. Also commonly termed *voltage* and *electromotive force.*

Electrical shock   The sudden stimulation of the body by electricity, when the body becomes part of a electrical circuit.

Electrochemical reaction   Chemical reaction that occurs when metals are in contact with electrolyte solutions.

Electrode   A metal and an electrolyte solution that participate in an electrochemical reaction.

Electrolyte   Substance, such as an acid, base, or salt, that releases ions when dissolved in water.

Electrolyte solution   A solution containing ions that conducts electrical current.

Electromagnetic radiation   A form of energy that travels through space at high speeds. Electromagnetic radiation is classified into types based on wavelength including: gamma rays, X-rays, ultraviolet (UV) light, visible (Vis) light, infrared (IR) light, microwaves, and radio waves.

Electromagnetic spectrum The range of all types of electromagnetic radiation from radiation with the longest wavelengths to those with the shortest.

Electromotive force (EMF) or ε See **electrical potential.**

Electronic balance Instrument that determines the weight of a sample by comparing the effect of the sample on a load cell with the effect of standards.

Electronic laboratory notebook (ELN) A tablet computer with software that allows a computer to take the place of conventional paper laboratory notebooks.

Electronic records Text, graphics, data, audio, pictorial information that is created, modified, maintained, archived, retrieved, or distributed by a computer system.

Electronic signature A computer equivalent to a handwritten signature. In its simplest form, can be a combination of a user ID plus password. It may also include identification based on biometric characteristics.

Electronics Instruments with components such as transistors, integrated circuits, and microprocessors that amplify, generate, or process electrical signals.

Electrophoresis A class of techniques in which molecules are separated from one another based on differences in their mobility when placed in a gel matrix and subjected to an electrical field.

Electroporator An instrument used to introduce drugs, genetic material, and other molecules into living cells suspended in an aqueous medium, by exposing the cells to a pulse of high voltage.

Electrostatic interactions Attractions between positive and negative sites on macromolecules. These weak interactions bring together amino acids and stabilize protein folding.

Eluent The mobile phase in chromatography.

Elution The passage of molecules through a chromatographic column, as the molecules in the sample distribute themselves between the two phases according to their affinities.

Elutriation A specialized centrifugation method in which the sedimentation of particles in a centrifugal field is opposed by the flow of liquid pumped toward the center of rotation. The flow rate and the centrifugation speed can be adjusted in such a way as to wash the cells out with the flowing liquid. This method minimizes the forces experienced by the particles and is useful for isolating fragile, living cells.

Embryotoxin A substance that is harmful to the developing fetus while showing little effect on the mother.

Emergent stem correction (thermometry) A method that corrects for the error introduced when a liquid-in-glass thermometer is immersed to a different depth than that at which it was calibrated.

Emission wavelength In fluorescence, the wavelength of light emitted by the sample as it fluoresces.

Encryption Creating encoded files for storage and transmission of data.

Encryption software Programs that apply mathematical keys to encode (sender) and decode (recipient) encrypted e-mails and files.

Endotoxin Materials derived from lipopolysaccharides that are released from the breakdown of gram negative bacteria and that trigger a dangerous immune response and induce fever in animals. Can negatively affect the growth of cultured cells.

Endotoxin units/mL A measure of the concentration of pyrogens in water.

Energy The ability to do work.

Engineering controls As defined by OSHA, biohazard controls (such as sharps disposal containers) that isolate or remove the bloodborne pathogens hazard from the workplace.

Entrance slit width (in spectrophotometry) The size of the slit through which light enters the monochromator after being emitted by the source.

Entrez A data retrieval tool, maintained by NCBI, that searches across several dozen molecular databases with data encompassing DNA, RNA, and protein sequences; genomes; taxonomy; diseases; books; and journal articles.

Entry routes Method of entry into the body, such as the mouth, lungs, or absorption through the skin.

Environmental Protection Agency (EPA) A U.S. government agency that, among other responsibilities, is involved in the regulation of environmental releases of genetically modified organisms.

Epitope The part of an antigen that an antibody recognizes and binds.

Equation A description of a relationship between two or more entities that uses mathematical symbols.

Equivalent weight (in reference to acids and bases) For an acid, 1 equivalent is equal to the number of grams of that acid that produces 1 mole of H^+ ions. For a base, 1 equivalent is equal to the number of grams of that base that supplies 1 mole of OH^-.

Equivalent weight (in reference to the concentration of solutes in body fluids) The number of grams of the solute that will produce 1 mole of ionic charge in solution.

Error 1. The difference between a measured value and the "true" value (see also **percent error** and **absolute error**). 2. The cause of variability in measurements.

Ethernet One standard type of **LAN** interface which is built into all network-ready computers.

Ethernet switch, or hub Hardware that regulates communications between workstations within the network and distributes access to other networks.

Ethidium bromide (EtBr) A mutagenic dye that reversibly intercalates into DNA molecules, allowing them to be visualized under ultraviolet light.

Ethylene oxide Reactive cyclic ether gas used for sterilization.

Etiological agent An organism that causes a specific disease in an infected host.

Eukaryotic organism Organisms whose cells have nuclei; plants, animals, and yeast are eukaryotic.

European Agency for the Evaluation of Medicinal Products (EMEA) An agency established by the European Union to coordinate scientific resources in member nations, in order to evaluate and supervise medicinal products for human and veterinary use.

Excitation wavelength In fluorescence, the wavelength of light that is absorbed by the molecule of interest and causes the compound to emit fluorescence.

Exit slit (in spectrophotometry) The slit through which light exits the monochromator.

Exit slit width (in spectrophotometry) The size of the slit through which light emerges from the monochromator.

Exothermic A chemical reaction that releases heat.

Expansion chamber (in thermometry) An enlargement of the capillary bore at the top of the thermometer to prevent buildup of excessive pressure.

Expansion thermometers Thermometers that rely on the expansion and contraction of a material in response to temperature.

Explosion A sudden release of large amounts of energy and gas within a confined area.

Explosive Substance that is capable of rapid combustion, causing sudden release of heat, gas, and pressure.

Exponent A number used to show that a value (the base) should be multiplied by itself a certain number of times. The expression 10^3 means: $10 \times 10 \times 10$, which equals 1000. The base is 10, the exponent is 3.

Exponential equation A relationship between two or more entities whose equation includes a variable that is an exponent, for example, $y = 2^x$.

Exponential notation See **scientific notation.**

Exposure incident Defined by OSHA as a specific eye, mouth, mucous membrane, nonintact skin, or parenteral contact with blood or other potentially infectious materials that results from the performance of an employee's duties.

Expression (in the context of genetics) The process in which a cell makes the protein product encoded by a specific gene.

Expression system A host organism that has taken up a vector containing a gene of interest and that produces the protein encoded by the gene.

Extinction coefficient See **absorptivity constant.**

Extractables Contaminants that are leached into water from the materials used to construct filters, storage vessels, filters, tubing, and other items.

Extraction methods Bioseparation techniques based on the fact that molecules differ from one another in their solubility in various liquids.

Face velocity (refers to fume hoods) The rate of air flow into the entrance of the hood, measured in **linear feet per minute (fpm).**

Fahrenheit scale (F) A temperature scale that sets the freezing point of water at 32° and the boiling point at 212°.

FDA Guidance Documents Documents issued by FDA that contain information about new technologies and discuss concerns that should be addressed by pharmaceutical and related companies.

Federal Hazard Communication Standard (HCS) A federal standard which focuses on the availability of information concerning employee hazard exposure and applicable safety measures.

Federal Register A daily publication from the U.S. Government Printing Office that contains major revisions to regulations and announcements of proposed new regulations. It also contains announcements of meetings (e.g., FDA or EPA), seminars, and guidance documents.

Feed (in computers) Electronically transmitted data, usually for the purpose of providing subscribers with updated content from a website. See **RSS.**

Feedwater The source water that enters a treatment process.

Fermentation (in the context of biotechnology) A process in which a product is produced by the mass culture of cells (usually refers to microorganisms) under controlled conditions.

Fermenter A vessel in which cells (usually microorganisms) are grown under controlled conditions of temperature, nutrient levels, aeration, pH, and mixing.

Fibrous junction A type of pH reference electrode junction composed of a fibrous material such as quartz or asbestos commonly used as a general purpose junction.

Field (in databases) Columns of information in a database **table,** which provide the individual data items, which can be text, numbers, or other values.

File (in computers) A set of digital data that are grouped to form a single storage unit.

File compression A system that recodes repetitive data within a file to reduce data storage size and save disk space.

File format The specific data storage characteristics that allow programs to recognize their own files.

File recovery utility These programs find the file contents and create a new disk directory entry that allows you to open the file again.

Filling hole Opening in a reference electrode by which filling solution is introduced into the electrode.

Filling solution A solution of defined composition within an electrode. The filling solution sealed inside a pH measuring electrode is called *internal filling solution,* and it normally consists of a buffered chloride solution. The solution that surrounds the reference electrode metal element is called *reference filling solution.* This electrolyte solution provides contact between the reference electrode metal element and the sample.

Filter A device used to separate components of samples on the basis of size; particles smaller than a certain size pass through a porous filter material; particles larger than a certain size are trapped by the filter.

Filter (in databases) A database command that shows (or hides) data records with designated characteristics.

Filter (in spectrometry) Material that blocks the passage of radiant energy in a particular manner with respect to wavelength.

Filtering (in electronics) Signal processing in which unwanted noise is removed from the signal generated by a detector.

Filter, neutral (in spectrometry) A filter that attenuates radiant energy by the same factor for all wavelengths within a certain spectral region.

Filtrate The fluid and any associated particles that have passed through a filter.

Filtration A separation method in which particles are separated from a liquid or gas by passage through a porous material.

Fire A chemical chain reaction between fuel and oxygen, which requires heat or other ignition source.

Fire triangle The combination of heat, fuel and oxygen required to start a **fire.**

Firewall A combination of hardware and software that filters incoming and outgoing communications to block known threats to the network.

Firewire A common port type that is well-suited for high speed applications, such as downloading video from digital camcorders.

Fixed angle rotor A rotor that holds tubes at a fixed angle.

Fixed reference points (in thermometry) Physical systems whose temperatures are fixed by some physical process and hence are universal and repeatable, such as phase transitions.

Fixed resistor A resistor that provides a single, set level of resistance in a circuit.

Flammable Any substance that will ignite and burn readily in air.

Flash drive A small, portable memory card that acts like a hard drive and stores data electronically instead of magnetically.

Flash memory A type of computer memory chip that stores data electronically instead of magnetically,

Flash point The minimum temperature at which a compound gives off sufficient vapors to be ignited.

Flat file database The simplest type of database, consisting of a single data **table** stored in one computer file.

Flow rate (in filtration) The rate of flow of a gas or liquid through a filter at a given pressure and temperature. The flow rate determines the volume that can be filtered in a given amount of time.

Fluorescence Light emitted by an atom or molecule after it absorbs light with a shorter wavelength.

Fluorescence spectrometer An instrument used to analyze a sample based on its fluorescence.

Fluorophore A molecule that fluoresces when exposed to light.

Food and Drug Administration (FDA) The federal agency empowered by Congress to regulate the production of cosmetics, electronic products (e.g., x-ray machines), food, medical devices, diagnostic devices, and pharmaceuticals for human and veterinary use.

Food, Drug and Cosmetics Act (FDCA) An act passed by the U.S. Congress in 1938 that requires that food and drugs not be "adulterated" and empowers the FDA to enforce food and drug laws.

Formula weight (FW) The weight, in grams, of one mole of a given compound. The FW is calculated by adding the atomic weights of the atoms that compose the compound.

Formulation The form in which a drug is administered to patients including its chemical components, the system of administration (e.g., pill or injection), and the mechanism by which the drug is targeted to its site of action.

Fouling When contaminants, such as packed bacteria, form a crust on a filter, membrane, or other surface, thus blocking further flow.

Freeware Software that is offered to the public free for personal use.

Freezing point The temperature at which a substance goes from the liquid phase to the solid phase.

Freezing point depression A lowering of the freezing point of a liquid due to the addition of impurities.

Freezing point depression osmometer A common device for measuring the osmotic pressure of a solution based on the fact that dissolved particles depress the temperature at which the water freezes. The temperature at which a sample freezes is compared to standards of known osmolality.

Frequency The number of times a particular value or range of values is observed in a data set.

Frequency (in electricity) The rate at which alternating current cycles back and forth measured in cycles/sec or Hertz (Hz).

Frequency (of electromagnetic radiation) The number of waves of electromagnetic radiation that pass a given point per second, expressed in units of Hertz (Hz).

Frequency distribution A listing of the number of times each value for a variable occurs.

Frequency histogram A graphical display of data in which the frequency of each value or range of values is plotted as a bar.

Frequency polygon A graphical display of data in which the frequency of each value or range of values is plotted as a point.

Fritted junction (in pH measurement) A type of relatively slow-flowing pH reference electrode junction composed of a ceramic material consisting of many small particles pressed closely together.

FTP (file transfer protocol) An Internet service that provides an efficient means to transfer data files between computers.

Fume hood See **chemical fume hood.**

Function key The special keys, usually at the top of the keyboard, that have varying functions defined within specific applications.

Functional units (in electronics) Combinations of electronic components connected to one another in various configurations that perform a particular function.

Fuse A safety device that limits the current flowing in a circuit.

Gain The degree to which an input voltage is amplified to become an output voltage.

Galvanometer An instrument that measures small electrical currents.

Gas chromatography (GC) Chromatography methods in which the mobile phase is a gas.

Gas-liquid chromatography (GLC) Chromatography method in which a gas mobile phase is paired with a liquid stationary phase.

Gel filtration See **gel permeation chromatography.**

Gel permeation chromatography (GPC) A chromatographic technique for separating molecules by relative size and molecular weight using a column filled with porous gel particles.

GenBank An online database, maintained by NCBI, that contains virtually all reported DNA sequences.

Gene A sequence of DNA that occupies a specific location on a chromosome and codes for a particular characteristic in an organism.

Gene therapy Replacing a gene that is missing or correcting the function of a faulty gene in order to cure an illness.

Genetic engineering Methods that allow a gene from one organism to be transferred to another. Can also apply to methods used to remove a native gene from an organism.

Genetically modified organism (GMO) An organism into which genetic information from another organism has been transferred.

Genomics Study of the genetic makeup of organisms including the base sequence and map of their DNA.

Germicidal Methods that are capable of killing bacteria or other microorganisms.

GFI (ground fault interrupt) circuits Safety circuits designed to shut off electric flow into the circuit if an unintentional grounding is detected; these are usually installed around sinks and other water sources.

Gigabyte (GB) Approximately 1 billion bytes, 2^{30}.

Gigahertz (GHz) One billion electrical cycles (**hertz**) per second; used to measure clock speed.

Glass electrode See **pH measuring electrode.**

Glycoprotein A protein that has one or more attached sugars.

Glycosylation A common, important type of protein modification in which sometimes complex, branched carbohydrates are attached to a protein after it is assembled from amino acid building blocks.

Good Clinical Practices (GCP) The procedures and practices that govern the performance of clinical trials of potential pharmaceutical products.

Good Laboratory Practices (GLP) 1. *FDA:* The procedures and practices that must be followed when performing laboratory studies in animals in order to investigate the safety and toxicological effects of new drugs. 2. *EPA:* The procedures and practices that must be followed when investigating the effects of pesticides and other agrochemicals. 3. *With small letters, (i.e., good laboratory practices [glp]):* A general term used to refer to all quality practices in any laboratory.

Good Manufacturing Practices (GMPs) See **Current Good Manufacturing Practices.**

Good Microbiological Laboratory Practices (GMLP) The basic practices that should be used when working with all microbiological organisms.

Gradient elution (in chromatography) Elution method in which the concentration of components in the mobile phase changes during the separation process to progressively elute more components from the stationary phase.

Gradient maker An apparatus used to make a continuous density gradient.

Graduated cylinders Cylindrical vessels calibrated to deliver various volumes.

Graduations Lines marked on glassware, plasticware, and pipettes that indicate volume.

Gram (g) The basic metric unit for mass.

Gram atomic weight The weight, in grams, of 6.02×10^{23} (Avogadro's number) atoms of a given element. For example, one mole of the element carbon weighs 12.0 grams.

Graphical user interface (GUI) An operating system interface that uses windows, icons, menus and pointers to interact with the user, instead of a **command line.**

Graphics Images created or modified through the application of graphics software.

Graphics file compression A system that abbreviates graphic image data for storage according to a specified formula for the file format.

Gravimetric method A method that involves the use of a balance; gravimetric methods are used to calibrate volume measuring devices.

Gravimetric volume testing Determination of the "true" volume of a sample based on its weight. The volume of the liquid is calculated from its weight and its density at a particular temperature. This method takes advantage of the high accuracy and precision attainable with modern balances.

Gravity (g) The unit of measure for the rate of acceleration of gravity. The earth's normal gravity is defined as $1 \times g$ (or 1 g force).

Ground wire A separate wire attached to the metal frame of an appliance and connected to the earth through the third prong of the plug. Protects an operator in the event of a short circuit.

Growth factors 1. *Bacterial cell culture:* Often refers to components that cannot be synthesized by bacterial cells and must be obtained from the environment. 2. *Mammalian cell culture:* Refers to a molecular entity released by cells that causes other cells to proliferate.

Growth media Solutions that support the survival and growth of cells in culture. These solutions include nutrients, vitamins, salts, and other required substances.

Guanine (G) One of the four types of nucleotide subunit that comprise DNA and RNA.

Guard column An inexpensive, disposable chromatography column that protects the expensive HPLC column from contaminants in the sample.

Guidelines As defined by the FDA, documents produced by the FDA that establish principles and practices. Guidelines are not legal requirements; rather they are practices that are acceptable to and recommended by the FDA.

Guidelines for Research Involving Recombinant DNA Molecules As defined by NIH, an influential document written by The Recombinant DNA Advisory Committee that covers topics such as: methods of assessing the risk of a recombinant DNA experiment, methods of classifying organisms based on risk, and methods of containing recombinant organisms both on an experimental scale and in production settings.

GXP A term that includes cGMP, GLP, and GCP.

Half-life The time it takes for half of the initial number of radioactive atoms in a given sample to undergo radioactive decay.

Hard copy A printed paper copy of information.

Hard disk A hardware disk that stores software and data as magnetic signals.

Hard drive A well-protected, sealed unit containing multiple rigid magnetic storage disks (**hard disks**) and a drive mechanism.

Hardness (in reference to water) An indication of the concentration of calcium and magnesium salts in water.

Hardware The solid objects needed for computer functions.

Hazard The equipment, chemicals, and conditions that have a potential to cause harm.

Hazard Analysis Critical Control Points (HACCP) A quality program implemented in the food industry that emphasizes reducing hazards throughout the production, slaughter, processing, and distribution of foods.

Hazard diamond system A system developed by the **National Fire Protection Association** to rate chemicals according to their flammability, reactivity, and general health hazards.

Heat The energy associated with the disordered motion of molecules in solids, liquids, and gases; thermal energy.

Heating The transfer of energy from the object with more random internal energy to an object with less.

HEPA filter A high-efficiency filter for particulate matter in air.

Hepatitis B A bloodborne infection that attacks the liver and causes inflammation.

Hepatitis B virus (HBV) The etiological agent for human hepatitis B.

Hertz (Hz) The unit of frequency of alternating current. 1 Hz = 1 cycle per second.

High-definition (hi-def) DVD DVDs with a much higher storage capacity than regular DVDs; Blu-ray is the predominant format.

High-performance liquid chromatography (HPLC) An instrumental technique used to separate, analyze, and sometimes collect the components of a mixture.

High-resolution purification methods Techniques that have relatively high selectivities but relatively low capacities; see also **low-resolution purification methods.**

High-speed centrifuge A centrifuge which spins samples at rates up to about 30,000 RPM and generates forces up to about $100,000 \times g$.

High-throughput screening (HTS) An automated method used by the pharmaceutical industry to quickly test thousands of compounds for possible therapeutic effects.

Hits (in computers) The links returned by a **search engine** in response to a **query.**

Hollow-fiber ultrafilter A cylindrical cartridge packed with ultrafiltration membranes formed into hollow, tubular fibers.

Horizontal laminar flow hood See **clean bench.**

Horizontal rotor (also called swinging bucket rotor) A rotor in which tubes or sample bottles swing into a horizontal position when centrifugal force is applied.

Hot-spot (in computers) A wireless network access point; one hot-spot provides wireless connections over an approximately 300-foot radius (sometimes greater).

Human genome The entire sequence of nucleotide building blocks that make up the DNA in humans; often compared to a "blueprint" that contains the instructions for building a human being.

Human Genome Project A major government-coordinated project that mapped and sequenced the human genome.

Human immunodeficiency virus (HIV) The etiological agent for acquired immunodeficiency syndrome.

Hybrid system (in the context of documentation) A system that uses both electronic and paper records. For example, a laboratory instrument might be attached to a computer that retrieves and processes data from the instrument, and then prints out a result that is signed and dated.

Hybridization The binding of single-stranded DNA or RNA to strands of complementary DNA or RNA.

Hybridoma An immortalized cell line that secretes monoclonal antibodies; results from the fusion of an antibody-producing cell with a cultured myeloma (tumor) cell.

Hydrates Compounds that contain chemically bound water. The weight of the bound water is included in the FW of hydrates. For example, calcium chloride can be purchased either as an anhydrous form with no bound water ($CaCl_2$, FW = 111.0), or as a dihydrate ($CaCl_2 \cdot 2H_2O$, FW = 147.0).

Hydrogen bond A type of weak chemical bond formed when a hydrogen atom bonded to an electronegative atom (e.g., F, O, or N) is shared by another electronegative atom.

Hydrogen ion H^+.

Hydrophilic "Liking" water; a substance that readily dissolves in water.

Hydrophilic (in filtration) A filter's ability to wet with water.

Hydrophobic Molecules that are relatively insoluble in aqueous liquids; water-hating. *In filtration:* A filter's resistance to wetting with water.

Hydrophobic interaction chromatography (HIC) A chromatographic technique for separating molecules based on their **hydrophobic** properties.

Hydroxide ion OH^-.

Hygroscopic compound A compound that absorbs moisture from the air.

Hypertext Transfer Protocol (HTTP) The protocol used to connect computers on the World Wide Web.

Hyperventilation An increased respiration rate that can induce dizziness.

Hypoallergenic A product label indicating that the product is less likely to trigger allergic reactions than similar products; note that this does not mean the product is nonallergenic.

I/O (input/output) port See **port.**

Ice point check (in thermometry) A method of verifying the performance of a thermometer by checking that it reads 0°C (or 32°F) when it is in an ice bath.

Icon (in computers) Graphic representations of specific objects and computer functions.

Image editing program Bit-mapped graphics programs that provide the capability to modify previously existing images, such as the output of a digital camera or a scanned image.

Immediate danger to life and health (IDLH) Designation that indicates environmental conditions requiring respirator use and maximum personal protective equipment.

Immersion (of a liquid-in-glass thermometer) The depth to which a liquid-in-glass thermometer is immersed in the material whose temperature is to be measured.

Immersion line (in thermometry) A line etched onto the stem of a liquid-in-glass thermometer to show how far it should be immersed in the material whose temperature is to be measured.

Immunoassays A group of test methods based on the interaction between antibodies and the antigens to which they attach. Immunoassays can be used to detect the presence of proteins and to quantify how much is present.

Impervious Preventing passage of an organism or material.

Implosion The violent collapsing of a vessel with an internal pressure lower than the outside atmosphere.

In vitro "In glass"; refers to processes occurring in a test tube or other laboratory medium.

In vivo "In life"; refers to processes occurring in a living organism.

Incident light Light that strikes, or shines on, a substance.

Incompatible (in chemical safety) Refers to combinations of chemicals that will react and cause hazardous conditions.

Independence (in sampling) Sampling in such a way that the choice of one member of a sample does not influence the choice of another.

Independent variable A variable whose value is controlled by the experimenter. See also **variable** and **dependent variable.**

Indicator electrode An electrode that responds selectively to a specific type of ion or molecule in solution.

Infection Invasion and multiplication of microorganisms in tissue; may be associated with health effects.

Infectious The ability of an organism to spread to and invade another host organism.

Infectivity assay A method of testing for pathogens in which susceptible cells are grown with the substance being tested to see if the cells become infected.

Inferential statistics Statistical methods that are used to reach conclusions about a population based on a sample.

Inflammable Another term for **flammable** materials.

Information Processed data that have been collected, manipulated, and organized into an understandable form.

Infrared radiation The region of the electromagnetic spectrum from 780–2500 nm.

Infrared thermometer A thermometer that senses the infrared energy emitted by an object, converts the infrared energy into an electrical signal that in turn is converted to a temperature reading.

Infringement (in the context of patents) A situation where one party holds a patent on an invention and another party uses that invention without permission.

Ingestion Entering the body through the mouth and digestive tract.

Inhalation Entering the body through the respiratory system and lungs.

Inkjet printer A printer that creates images by squirting tiny sprays of ink onto paper, which can produce sharp print.

Inorganic In water treatment, matter not derived from plants, animals, or microorganisms and that usually dissociates in water to form ions.

Input connector A connection that enables an instrument to receive electrical signals from another device.

Installation qualification (IQ) A set of activities designed to determine if a piece of equipment is installed correctly.

Instant messaging (IM) Communication method in which private messages are exchanged in real time.

Institute of Laboratory Animal Resources (ILAR) Agency founded under the National Research Council to compile and disseminate information about laboratory animals and their care, and to promote humane care of these animals.

Institutional Animal Care and Use Committee (IACUC) The committee that reviews an institution's programs for humane care and use of animals, inspects institutional animal facilities, and reviews and approves all protocols using live vertebrate animals, among other duties.

Institutional review board (IRB) A panel composed of medical, scientific, and nonscientific community members who are responsible for protecting the rights, safety, and well-being of human experimental subjects.

Instrument response time The time required for an indicating device to undergo a particular displacement following a change in the property being measured.

Insulator (in electricity) Materials in which the outer electrons of atoms are not free to move so electricity does not flow readily.

Integrated circuit (IC) A small electronic circuit with multiple components (such as resistors, capacitors, and diodes) built on a chip.

Integrator A signal processing mode in which the area under a peak is calculated.

Intellectual property (IP) A type of property that encompasses creations of the mind and intellect.

Interface Site where a measuring instrument contacts the material being measured.

Interface (in computers) The part of the system software that the user sees and interacts with.

Interference (in the context of an assay) Any substance in a sample that leads to an incorrect result in an analysis.

International Agency for Research on Cancer (IARC) Agency that determines the relative cancer hazard of materials.

International Conference on the Harmonisation of Technical Requirements for the Registration of Pharmaceuticals for Human Use (ICH) An organization that brings together the regulatory authorities of Europe, Japan, and the United States and experts from the pharmaceutical industry to discuss scientific and technical aspects of pharmaceutical product regulation with the purpose of harmonizing pharmaceutical regulatory requirements internationally.

International Organization of Legal Metrology (OIML) A worldwide intergovernmental organization whose aim is to harmonize regulations and controls relating to metrology in its member nations.

International prototype standard The internationally accepted unit of mass as embodied in a platinum-iridium cylinder housed in France. The mass of this cylinder is, by definition, exactly 1 kg.

International Temperature Scale-90 (ITS-90) An internationally accepted temperature scale established in 1990 by the 18th General Conference on Weights and Measures. The scale uses reference temperatures based on the freezing, boiling, and triple points of various materials including water, hydrogen, and several metals.

Internet Two or more connected networks.

Internet, the (with a capital "I") A specific internet connected with special sets of communication **protocols.**

Internet service provider (ISP) An agency that provides individuals with the necessary server connections to the Internet, as well as other services in many cases.

Intranet An internal, usually private, network, perhaps within a company or university.

Invention The discovery or creation of a new material (e.g., a drug, a protein made by a genetically modified organism), a new process (e.g., a new method of inserting genetic information into cells), a new use for an existing material, or any improvement of any of these.

Investigational New Drug Application (IND) An application filed with the Food and Drug Administration in which a sponsoring organization requests permission to launch clinical trials of an experimental drug.

Invisible web The billions of web pages that are publicly available but invisible to search engines.

Ion An atom or group of atoms with a net charge as a result of having lost or gained electrons. See **anion** and **cation.**

Ion exchange The process in which ions in solution are exchanged with similarly charged ions associated with ion exchange resins.

Ion exchange chromatography A chromatographic technique for separating molecules by molecular charge; based on ionic stationary phases and manipulation of salt concentrations and pH in the mobile phase.

Ion selective electrode (ISE) A class of indicator electrode that responds to a specific ion in solution.

Ionic detergent A **detergent** whose hydrophilic portion is ionized in solution.

Ionic strength A measure of the charges from ions in an aqueous solution.

IP (Internet Protocol) The standard protocol for packaging, addressing, and sending data through the Internet.

Irritant Substance that will cause irritation to the skin, eyes, or respiratory system.

ISO A network of the national standards institutes of 157 countries, on the basis of one member per country, with a Central Secretariat in Geneva, Switzerland, that coordinates the system.

ISO 9000 A set of internationally accepted standards that outline a system for quality management aimed at ensuring the quality of a product or service. ISO 9000 currently includes three quality standards: ISO 9000:2005, ISO 9001:2000, and ISO 9004:2000. ISO 9001:2000 presents *requirements*, while ISO 9000:2005 and ISO 9004:2000 present *guidelines*. All of these are process standards (not product standards).

Isoelectric point The pH value at which a protein exhibits an overall neutral charge.

Isolation See **purification.**

Isopycnic centrifugation A method in which particles are separated based on their density alone; also called buoyant density centrifugation.

Isotonic solution A solution that has the same solute concentration as the interior of the cell.

Isotropic membrane A membrane in which the pore size is the same on both sides of the membrane.

IU (International Unit of enzyme activity) The amount of enzyme necessary to catalyze transformation of 1.0 umol of substrate to product per minute under optimal measurement conditions.

Jack (in the context of electricity) A receptacle for a plug. The plug is inserted into the jack to complete a circuit.

Job Safety Analysis (JSA) A detailed analysis of each step in a procedure, identifying potential hazards and outlining accident prevention strategies.

Junction A part of a reference electrode that allows filling solution to flow into the solution whose pH is being measured. Also called a **salt bridge.**

Kat The designation for **SI units of enzyme activity.**

Kelvin scale A temperature scale based on the Celsius degree with 100 units between the freezing point and boiling point of water; however, the Kelvin scale sets the zero point as absolute zero, rather than the freezing point of water. $0°C = 273.15$ K. (Units are in kelvin, K.)

Kilo- (k) A prefix meaning 1000.

Kilobits per second (Kbps) A measure of the speed of data transmission between computers; 1000 bits per second.

Kilobyte, (K or KB) Approximately 1000 bytes (actually 1024), representing 2^{10}.

Kilogram (kg) A unit of mass equal to 1000 g.

Kilometer (km) A unit of length equal to 1000 m.

Kinetic spectrophotometric assay An assay that measures the changes over time in concentration of reactants or products in a chemical reaction.

Knife edge In mechanical balances, a support on which the beam rests that allows it to swing freely.

Laboratory-acquired infection (LAI) Infections that can be traced directly to laboratory organisms handled or contacted by the infected individuals.

Laboratory informatics The application of computers to collect, analyze, and manage laboratory data and information.

Laboratory information management systems (LIMS) Computerized systems that provide integrated information and resource management for a variety of laboratory activities.

Lachrymator Substance that is an eye irritant and stimulates tear formation.

Lambda (λ) An older term sometimes used to mean 1 μL.

LAN (local area network) An intranet, generally operating within a building, school, or company. LANs are private and usually limited to a five-mile radius.

Laptop computer A portable computer system that can be operated on batteries; usually weighing 4 to 8 pounds.

Large-scale bioproduction Production of biological materials in quantities greater than ten liters.

Laser printer A computer printer that works somewhat like a photocopier, with a laser, drum, and toner.

Latency period Time from the first exposure to a toxic agent to the time when biological effects can be detected.

Latex A natural rubber product that is commonly found in gloves and many pieces of laboratory equipment and household items.

LC_{50} (lethal concentration 50%) Concentration of a compound in air that will kill 50% of test animals.

LC_{Lo} (lethal concentration low) The lowest concentration recorded to cause lethality in test animals.

LCD (liquid crystal display) monitor A flat-panel monitor that uses LCD technology to provide a slim, energy-efficient digital display screen.

LD_{50} (lethal dose 50%) Amount of a toxic compound given in a single dose that will cause death in 50% of test animals.

LD_{Lo} (lethal dose low) The lowest single dosage known to cause lethality in test animals.

Leach (in the context of water purity) A process in which water dissolves substances from materials over which it flows. For example, minerals are leached as water flows over rocks.

Least squares method A statistical method used to calculate the equation for the line of best fit for a series of points.

Leveling Adjusting the balance to a level, horizontal position.

Leveling bubble A bubble that is used to determine whether the balance is level or not. When the balance is level, the bubble is centered in the window.

Leveling screws The screws used to adjust the balance so that it is level.

Lever (or beam) A rigid bar used in mechanical balances on which the sample to be weighed is balanced against objects of known mass.

L-glutamine An essential amino acid required by cultured mammalian and insect cells.

License (in the context of patents) An agreement by a patent holder that it will not enforce the right of exclusion against the licensee (the party wishing to use the patented invention).

Light Electromagnetic radiation with wavelengths from 180–2500 nm; includes the ultraviolet, visible, and infrared regions of the electromagnetic spectrum.

Light Emitting Diode (LED) A diode that emits light when current flows through it.

Light scattering An interaction of light with small particles in which the light is bent away from its initial path.

Light source (in spectrophotometry) The bulb or lamp that emits the light that shines on the sample.

Limit of Detection (LOD) The lowest level of the material of interest that a method or instrument can detect.

Limit of Quantitation (LOQ) The lowest level of the material of interest that a method or instrument can quantify with acceptable accuracy and precision.

***Limulus* amebocyte lysate test (LAL)** Test used to quantify the concentration of pyrogens in a sample that requires an extract of blood from the horseshoe crab, *Limulus polyphemus*. Pyrogens initiate clotting in the presence of this blood extract.

Line of best fit A line connecting a series of data points on a graph in such a way that the points are collectively as close as possible to the line.

Linear dispersion of a monochromator A measure of the ability of the monochromator to separate light into its component wavelengths.

Linear feet per minute (fpm) (in fume hoods) Measurement of **face velocity.**

Linear relationship A relationship between two properties such that when they are plotted on a graph the points form a straight line.

Linearity 1. The characteristic of a direct reading device. If a device is linear, calibration at 2 points (e.g., 0 and full-scale) calibrates the device (2 points determine a straight line). 2. A measure of how well an instrument follows an ideal, linear relationship.

Linearity (for a balance) The ability of a balance to give readings that are directly proportional to the weight of the sample over its entire weighing range.

Linearity check (in pH measurement) A pH meter system is normally calibrated at two pHs, (e.g., pH 7.00 and pH 10.00). A linearity check tests whether the meter's response is linear through its entire range (e.g., also at pH 4.00).

Linux An **operating system** derived from **UNIX;** used in network servers, a small number of personal computers, and consumer products such as gaming systems. Linux can be used with either a command-line or graphical user interface.

Lipopolysaccharide (in reference to water quality) Molecules containing carbohydrate and lipids found in the outer wall of gram negative bacteria. See **pyrogen.**

Liquid chromatography (LC) A general term for any chromatographic technique where the mobile phase is liquid.

Liquid crystal display (LCD) Digital device commonly used to display individual measurement values.

Liquid expansion thermometer See **liquid-in-glass thermometer.**

Liquid nitrogen The liquid form of nitrogen at $-198°C$, supplied in large compressed gas cylinders.

Liquid-in-glass thermometer A thermometer in which liquid, usually mercury or alcohol, expands or contracts with temperature changes and moves up or down the bore within a glass tube.

Liter (L) A metric system unit of volume. $1 L = 1 dm^3$. A liter is slightly more than a quart. See also **volume.**

Load (for a balance) The item(s) exerting force on the balance.

Load (in electricity). The electrical demand of a device expressed as power (watts), current (amps), or resistance (ohms).

Load cell Device that uses an electromagnetic force to compensate for, or counterbalance, the force of an object on the weighing pan.

Log, Logarithm (Log$_{10}$) See **common logarithm.**

Log in Entering a name and password into the computer for identification.

Log to linear conversion Processing in which a signal that has a logarithmic relationship to the property being measured is converted to a signal with a linear relationship to the property being measured.

Lossy Describes graphic file formats that eliminate **pixel** data from an image to reduce storage size.

Lot Defined by GMP: "[A] batch or a specific identified portion of a batch, having uniform character and quality within specified limits . . ."

Lot number (also, control number or batch number) An identifying number that is used to distinguish one lot from another.

Low-resolution purification methods Bioseparation techniques that generally have high capacities and can be performed quickly, but exhibit relatively low selectivity; see also **high-resolution purification methods.**

Low-speed centrifuge A centrifuge that spins samples at rates less than about 10,000 RPM.

Lyophilization Freeze-drying; a preservation method commonly used for proteins, that involves rapid freezing and drying under vacuum.

Lyse To break open cells and release their contents.

Lysis buffer A solution whose primary function is to lyse the cell membrane and/or cell wall.

Macro A short set of commands within a file that create a program shortcut; usually found in high-end word processors.

Macrofiltration The removal of relatively large particles, above about 10 μm, from a liquid by passage through porous materials including paper, glass fibers, and cloth.

Magnetic dampening A device that helps stabilize the weighing pan in mechanical balances when samples are placed on it or weights are added to the beam, thus reducing the time required to reach a stable reading.

Malware Malicious software.

Manual dispenser (for reagent bottle) Device that dispenses set volumes from a bottle.

Mass The amount of matter in an object, expressed in units of grams. Mass is not affected by the location of the object or the effect of gravity.

Master batch record Original step-by-step instructions that detail how to formulate or produce a product, including raw materials required, processing steps, controls, and required testing.

Master cell bank (MCB) Stored cells that are the source of cells used in a process. The stored cells are usually derived from a single cell, are fully characterized, and are aliquoted and stored cryogenically to assure genetic stability. The MCB is the source for the working cell bank.

Material Safety Data Sheet (MSDS) An OSHA-required technical document provided by chemical suppliers, describing the specific properties of a chemical.

Matrix The physical material in which a sample or analyte is located or from which it must be isolated, for example, blood, soil, or fiber.

Maximum tare The maximum container weight that can be automatically subtracted from the weight of a sample.

Mean The average. The sum of all values divided by the number of values. Statisticians distinguish between the true mean of an entire population, represented by μ, and the mean of a sample from that population.

Measurand A quantity that is measured (e.g., the volume of a particular sample under specified conditions, such as temperature and pressure).

Measurement Quantitative observation; numerical description.

Measurement system A related group of units, such as inches, feet, and miles. See also **United States Customary System, metric system,** and **SI System.**

Measures of central tendency Measures of the values about which a data set is centered (e.g., the mean, median, and mode).

Measures of dispersion Measures of the variability in a set of numerical data (e.g., range, variance, and standard deviation).

Measuring pipette A type of pipette calibrated with a series of graduation lines to allow the measurement of more than one volume.

Mechanical balance (laboratory) A balance that uses a lever and that does not generate an electrical signal in response to a sample. The object to be weighed is placed on a pan attached to the lever and is balanced against standards of known mass.

Median A statistic that is the middle value of a data set or the number that is greater than or equal to 50% of the values and less than or equal to 50% of the values.

Medical device Medical items that are not pharmaceuticals and are used in the diagnosis or treatment of disease or injury. Examples include sterilizers, test kits, cell counters, and pacemakers.

Megabyte (MB)   Approximately 1 million bytes, 2^{20}.

Megohm-cm   A unit used to measure the resistivity of water. The fewer dissolved ions in water, the higher its resistivity. The theoretical maximum ionic purity for water is 18.3 megohm-cm at 25°C.

Melting point   The temperature at which a substance transforms from the solid phase to the liquid phase.

Melting temperature (for DNA), T_m   The temperature at which DNA denatures. The bonds between guanines and cytosines are stronger than those between adenines and thymines; therefore, a DNA molecule that is comparatively rich in G-Cs denatures at a higher temperature than one that is rich in A-Ts.

Membrane filter   A filter where particles are retained primarily on the surface of the membrane and in which there are pores or channels of defined size.

Memory (in computers)   The workspace that provides computers with the ability to electronically keep track of data. See **RAM** and **ROM.**

Meniscus   Greek for "crescent moon." The surface of liquids in narrow spaces forms a curve known as a meniscus. The bottom of the meniscus is used as the point of reference in calibrating and using volumetric labware.

Menu (in computers)   A list of command options, either text or icons, for selection with the **mouse.**

Mercury thermometer   A liquid expansion thermometer containing liquid mercury.

Messenger ribonucleic acid (mRNA)   A class of RNA molecule that acts as an intermediary by transferring information from DNA in the nucleus to ribosomes in the cytoplasm of the cell. This information dictates the amino acid sequence of a protein to be manufactured by the cell.

Meta data   Information that describes the content and context of the data. They help to reconstruct the original raw data. For example, a digital camera produces both a picture and also metadata that includes the camera shutter speed, f-stop, and other camera settings when the photo was taken.

Metallic resistance thermometer   Thermometers having metal wires whose resistance increases as the temperature increases.

Metasearch engine   A specialized search engine that searches across the indices of other search engines and compiles relevant query results.

Meter (m)   The basic metric unit for length.

Method   The means of performing an analysis. A method describes the steps necessary to perform an analysis and related details, such as how the sample should be obtained and prepared, the reagents that are required, the set-up and use of instruments, comparisons with reference materials, calculations, and so on.

Method validation   See **validation of an analytical method.**

Methyl red   pH indicator dye that changes from red at pH 4.2 to yellow at pH 6.2.

Metric system   A measurement system used in most laboratories and in much of the world

whose basic units include meters, grams, and liters. These basic units are modified by the addition of prefixes that designate powers of 10.

Metrologist   A person who studies and works with measurements.

Metrology   The study of measurements.

Micro- (μ)   A prefix meaning 1/1,000,000 or 10^{-6}.

Microampere (μA)   One millionth of an ampere.

Microarray   A tool used by scientists to study thousands of genetic sequences simultaneously.

Microbalance   An extremely sensitive balance that can accurately weigh samples to the nearest 0.000001 g.

Microchip   Computer chip; an integrated circuit.

Microelectronics   Electronics with very small components arranged into circuits.

Microfiltration   The filtration of particles whose sizes are in the range from about 0.01 μm to 10 μm using membrane filters.

Microfiltration membrane filter   Plastic polymeric filter that prevents the passage of microorganisms and particles physically larger than the filter's pore size.

Microfuge   A small centrifuge intended for small volume samples in the microliter to 1 or 2 mL range.

Microgram (μg)   A unit of mass that is 1/1,000,000 g, or 10^{-6} g.

Microliter pipette or micropipette   A term used in this book to refer to various styles of device that measure volumes in the 1 to 1000 μL range.

Microliter (μL)   A unit of volume that is 1/1,000,000 L, or 10^{-6} L.

Micrometer (μm)   A unit of length that is 1/1,000,000 m or 10^{-6} m.

Micromolar (μM)   A concentration expression that is 1/1,000,000 M. For example 1 M NaCl is 58.44 g of NaCl in 1 L total volume of solution and 1 μM NaCl is 0.00005844 g of NaCl in 1 L of solution.

Micromole (μmol)   An amount that is 1/1,000,000 of a mole. For example, 1 mole of NaCl is 58.44 g and 1 μmole NaCl is 0.00005844 g of NaCl.

Micron   A micrometer.

Microorganism   A microscopic organism; includes bacteria, fungi, and viruses.

Micropipettor   A term used in this book to refer to an instrument that measures volumes typically in the 1–1000 μL range. This term is not generally used to refer to a simple hollow tube device, but rather refers to a more complex instrument.

Microprocessor   An integrated circuit that can store and process information.

Mil (in gloves)   A measurement of glove thickness, where 1 mil = 0.001 inch.

Milli- (m)   A prefix meaning 1/1000 or 10^{-3}.

Milliampere (mA)   One thousandth of an ampere.

Millibits per second (Mbps)   A measure of the speed of data transmission between computers; 1,000,000 bits per second.

Milligram (mg)   A unit of mass that is 1/1000 g.

Milligrams per cubic meter, mg/m³, of air   A measure of air concentrations of chemicals.

Milliliter (mL)   A unit of volume that is 1/1000 L.

Millimeter (mm)   A unit of length that is 1/1000 m.

Millimolar (mM)   A concentration expression that is 1/1000 molar. For example 1 M NaCl is 58.44 g of NaCl in 1 L total volume of solution and 1 mM NaCl is 0.05844 g in 1 L total volume.

Millimole (mmol)   An amount that is 1/1000 of a mole. For example, 1 mole of NaCl is 58.44 g and 1 mmole of NaCl is 0.05844 g.

Milliosmolality   A concentration expression that is millosmoles/kg of solvent.

Millivolt (mV)   One thousandth of a volt.

Minimal medium   Growth medium that contains only the minimum nutrients required for a particular microbe to survive and reproduce.

Mixed bed ion exchanger   A design in which both cation and anion exchange resins are housed in the same cartridge.

Mobile phase (in chromatography)   The liquid or gas that moves past the **stationary phase** in a chromatography system.

Mode   The value that is most frequently observed in a set of data.

Modem, modulator/demodulator   Hardware that changes (modulates) the digital signal from a computer to an analog signal that can be carried over phone lines.

Mohr pipette   A type of serological pipette that is calibrated so that the liquid in the tip is not part of the measurement.

Molality (m)   An expression of the concentration of a solute in a solution that is the number of moles of solute per kilogram of solvent.

Molarity (M)   An expression of concentration of a solute in a solution that is the number of moles of solute dissolved in 1 L of total solution.

Mole   A mole of any element contains 6.02×10^{23} (Avogadro's number) atoms. Because some atoms are heavier than others, a mole of one element weighs a different amount than a mole of another element. A mole of a compound contains 6.02×10^{23} molecules of that compound.

Molecular sieving   See **gel permeation chromatography.**

Molecular weight (MW) (of a compound)   See **formula weight.**

Molecular weight cutoff (MWCO)   In ultrafiltration, the lowest molecular weight solute that is retained by the membrane.

Monitor (in computers)   The main soft copy display where you can see your data input and choose commands.

Monochromatic   Light that is of one wavelength. In practice, "monochromatic" light is composed of a narrow range of wavelengths.

Monochromator   Device used to separate polychromatic light into its component wavelengths and to select light of a certain wavelength (or narrow range of wavelengths).

Monoclonal antibodies (Mab) Exceptionally homogeneous populations of antibodies directed against a specific target and produced by cells that are derived from a single antibody-producing cell.

Mouse (in computers) A pointing device that controls the on-screen cursor in a graphical user interface.

MS-DOS (Microsoft disk operating system) A command-line operating system that is capable of running only one program at a time; MS-DOS cannot be used across networks.

MSDS See **Material Safety Data Sheet.**

MT Empty; used to mark empty gas cylinders.

Multicomponent spectrophotometric analysis Methods that allow simultaneous quantitation of more than one analyte in a sample.

Multifunction printer A printer that combines printing, copying, and scanning capabilities. Available in both **laser** and **inkjet** models.

Multimedia authoring The creation of presentations that use more than one medium to convey information; may include text, images, audio, and video.

Multitasking (in computers) The ability to open and run more than one program at a time.

Mutagen A substance that can cause changes in DNA, resulting in genetic alterations.

Mycoplasma Simple bacteria that lack a cell wall and are small enough to pass through regular cell culture sterilization filters.

National Center for Biotechnology Information (NCBI) A subdivision of NIH that acts as a public resource for molecular data and other biomedical information.

National Committee of Clinical Laboratory Standards (NCCLS). See **Clinical and Laboratory Standards Institute.**

National Fire Protection Association (NFPA) Organization that developed the visual labeling and rating system for heath, flammability, reactivity and related hazards.

National Institute for Occupational Safety and Health (NIOSH) Public health service that tests and recommends chemical exposure limits.

National Institute for Standards and Technology (NIST) The national standard laboratory for the United States, formerly known as NBS, the National Bureau of Standards. NIST is a federal agency that works with industry and government to advance measurement science and develop standards.

National Institutes of Health (NIH) A part of the U.S. Department of Health and Human Services that is the primary federal agency for conducting and supporting medical research.

Natural band width The width of the peak generated for a specific material when its absorbance is plotted versus wavelength; an intrinsic characteristic of the substance. Natural band width is measured as the width of the peak at half its height.

Natural log (Log$_e$, ln) Logarithms whose base is the number ≈2.7183, called "e."

Near vertical tube rotor A modification of a fixed angle rotor having a shallow angle of about 8–10 degrees.

Netiquette, Internet etiquette The uncodified but generally accepted standards for behavior while online.

Network (in computers) Two or more computers that are electronically connected for the purpose of sharing programs, files, information, and/or other resources.

Network interface card (NIC) Computer hardware that provides an interface between the computer and a network.

Neurotoxin Compound that can cause damage to the **central nervous system.**

Neutral litmus pH indicator dye that is red in acid conditions and blue in alkaline conditions.

Neutral red pH indicator dye that changes from red at pH 6.8 to orange at pH 8.0.

Neutral solution An aqueous solution with an equal number of hydrogen and hydroxide ions and a pH of 7.

Neutralize To chemically react acids and bases to form neutral salts and water; to bring the pH of a material or mixture to 7.0.

New Drug Application (NDA) An application filed with the Food and Drug Administration requesting permission for marketing and sale of a new drug product.

Noise (electrical) Unwanted electrical interference that creates a signal unrelated to the property being measured. Electrical noise may have two components:

> **Short term** Random, rapid spikes in the electronics of an instrument.

> **Long term, or drift** A relatively long-term increase or decrease in readings due to changes in the instrument and electronics.

Nominal pore size (in filtration) Rated pore size; a large percent (often 99.9%) of particles above the nominal pore size should be retained by the filter.

Nominal volume The desired volume for which a pipetting device is set.

Nomogram, nomograph A chart that can be used to determine either RCF, rpm, or rotor radius values (rarely), if the other two values are known.

Non-bypass fume hood A fume hood design where air enters the hood only at the bottom of the sash.

Nonconformance Situation in which a product or raw material does not meet its specifications.

Nondisclosure agreement A contract in which the parties promise to protect the confidentiality of secret information that is disclosed during employment or another type of business transaction.

Nonelectrolyte Substance that does not ionize when dissolved in solution.

Nonfiber-releasing filter A filter that is treated in such a way that it will not release fibers into the filtrate.

Nonflammable Not easily ignited and burned.

Nonionic detergent A **detergent** whose **hydrophilic** portion is not ionized in solution.

Normal distribution The frequency distribution of data that have a bell shape when graphed. The peak of a perfect normal distribution is the mean, median, and mode, and values are equally spread out on either side of that central high point.

Normality An expression of the concentration of a solute in a solution that is the number of equivalent weights of a solute per liter of solution. See **equivalent weight.**

Notebook computer See **laptop computer.**

Nuclease Any member of a class of enzymes that break the covalent bonds of DNA and RNA.

Nucleotide Subunit of RNA and DNA that consists of a phosphate group, a five carbon sugar (ribose in RNA or deoxyribose in DNA), and one of four nitrogenous bases (adenine, guanine, cytosine, or uracil in RNA; adenine, guanine, cytosine, or thymine in DNA).

Numerator The number written above the line in a fraction.

Object (in computers) A graphic image or block of text that is treated as a single item, described by a mathematical equation.

Object-oriented graphics See **vector graphics.**

Occupational Exposure to Hazardous Chemicals in Laboratories Standards (29 CFR Part 1910) A set of federal standards which adapt and expand the **HCS** to apply to academic, industrial, and clinical laboratories.

Occupational exposure As defined by OSHA, reasonably anticipated skin, eye, mucous membrane, or parenteral contact with blood or other potentially infectious materials that may result from the performance of an employee's duties.

Occupational Safety and Health Administration (OSHA) The main federal agency responsible for monitoring workplace safety.

OCR (optical character recognition) software Technology that allows text to be scanned and recognized as text to be manipulated in a word processing program without retyping.

Off-center errors Differences in indicated weight when a sample is shifted to various positions on the weighing area of the weighing pan.

Ohm (Ω) Unit of resistance. One ohm is the value of resistance through which 1 V maintains a current of 1 A.

Ohm's law An equation that relates voltage, current, and resistance in a circuit: Voltage = (current) · (resistance).

Ohmmeter An instrument used to measure electrical resistance in a circuit.

Open (in computers) To transfer a copy of the contents of a file from **storage** into **memory.**

Open system (in the context of computers) A computer system that is not controlled by the persons who are responsible for the content of the system. For example, if a contract laboratory sends data to a company via the Internet, the system is open. Additional security must be in place for open systems as compared to closed systems.

Operating system (OS) The software foundation that operates the computer hardware, runs software applications, manages file storage, and monitors system operations.

Operational qualification (OQ) A set of activities designed to establish that a piece of equipment performs within acceptable limits.

Optical density (OD) See **absorbance.**

Optical mouse A **mouse** that uses light emitted on the underside of the mouse to track its movement.

Order of magnitude One order of magnitude is 10^1. For example, 10^2 is said to be "two orders of magnitude" less than 10^4.

Ordinate In graphing, the Y coordinate for a point; the distance of a point along the Y axis.

Organelle A structurally distinct component of a cell that performs a particular job in the cell (e.g., ribosomes are an organelle that produces proteins).

Organic A broad category that refers to materials containing carbon and hydrogen.

Origin The intersection of the X and Y axes on a two-dimensional graph whose coordinates are (0,0).

O-ring (for a micropipettor) Rubber ring used as a seal to prevent the leakage of air.

Orphan Drug Act Enacted in 1983 in the United States, provides tax relief and some marketing exclusivity for companies that develop drugs for less-common diseases.

Osmolality The number of osmoles of solute per kg of water.

Osmolarity The number of osmoles of solute per liter of solution.

Osmole An osmole of any substance is equal to 1 gram molecular weight of that substance divided by the number of particles formed when the substance dissolves. For example, KCl ionizes when dissolved to form one K^+ and one Cl^- ion; therefore, 1 osmole of potassium chloride is equal to its molecular weight, 74.6 g divided by 2 = 37.3 g. For a substance that does not ionize when dissolved, 1 mole = 1 osmole.

Osmosis The net movement of water through a semipermeable membrane from a region of lesser solute concentration to a region of greater solute concentration.

Osmotic equilibrium A condition where the rate of water flow into and out of a cell are equal.

Osmotic pressure The amount of pressure that would need to be exerted to halt water's movement by osmosis.

Osmotic support The cells of plants, yeasts, and microbes are sensitive to changes in osmotic pressure when their cell walls are removed. Materials such as sucrose, sorbitol, mannitol, and glucose may be added to solutions containing these cells to protect the cell's integrity when the cell wall is removed.

Other potentially infectious materials OSHA-defined term referring to 1. Human body fluids such as semen, vaginal secretions, cerebrospinal fluid, synovial fluid, pleural fluid, pericardial fluid, peritoneal fluid, amniotic fluid, saliva in dental procedures, any body fluid that is visibly contaminated with blood, and all body fluids in situations where it is difficult or impossible to differentiate between body fluids. 2. Any unfixed tissue or organ (other than intact skin) from a human (living or dead). 3. HIV-containing cell or tissue cultures, organ cultures, and HIV- or HBV-containing culture medium or other solutions; and blood. Organs, or other tissues from experimental animals infected with HIV or HBV.

Outlier A data point that lies far outside the range of all the other data points.

Out-of-specification (OOS) result A finding that indicates a product fails to meet its requirements.

Output connector A connection that enables an instrument to send an electrical signal to another device.

Overspeed disc (in centrifugation) A striped black disc attached to the bottom of rotors which, in conjunction with a photoelectric device in the centrifuge, prevents a rotor from being spun at rates above its maximum safe speed.

Oxidation The removal of electrons from a substance.

Oxidizing agent A substance that gains electrons in a reaction or can oxidize another substance.

Ozone sterilization A method of killing bacteria by exposing them to ozone.

PAGE (Polyacrylamide gel electrophoresis) A protein separation technique that involves applying an electric field to a **polyacrylamide gel.**

Paint program A program that provides the ability to create and modify bit-mapped images using computerized tools that are similar to those used by an artist on paper.

Parallel circuit A circuit in which the components are connected so that there are two or more paths for current.

Parameter A numerical statement about a population, comparable to a sample statistic (e.g., the true mean and standard deviation of a population are parameters).

Parametric statistical methods Statistical methods that assume the variables of interest are normally distributed in the population.

Parenteral OSHA defines as referring to piercing mucous membranes or the skin barrier through events such as needle sticks, human bites, cuts, and abrasions.

Partial immersion thermometer A liquid expansion thermometer designed to indicate temperature correctly when the bulb and a specified part of the stem are exposed to the temperature being measured. Compare with **complete immersion thermometer** and **total immersion thermometer.** (Based on definition from ASTM Standard E 344-07 "Terminology Relating to Thermometry and Hydrometry.")

Particle retention A term used in reference to depth filters to indicate the smallest particle size (in micrometers) that is retained by the filter. Particle retention ratings are nominal (i.e., a small percent of particles of the rated size will penetrate the filter).

Particulate A general term used to refer to a solid particle large enough to be removed by filtration.

Partition chromatography A chromatographic technique for separating molecules between two liquid phases based on their relative solubility in each phase.

Parts per billion (ppb) An expression of concentration of solute in a solution that is the number of parts of solute per billion parts of solution. Any units may be used but must be the same for the solute and total solution.

Parts per million (ppm) An expression of concentration of solute in a solution that is the number of parts of solute per 1 million parts of total solution. Any units may be used but must be the same for the solute and total solution.

Pasteur pipette A type of pipette used to transfer liquids from one place to another, but not to measure volume.

Patent A type of intellectual property protection that is an agreement between the government, represented by the Patent Office, and an inventor whereby the government gives the inventor the right to exclude others from using the invention in certain ways.

Path length The distance light passes through the sample; typically measured as the length of the cuvette in centimeters or millimeters.

Pathogen Any biological organism that can cause disease.

Pathogenicity The relative capability of an organism to cause disease in humans or other living organisms.

PDF (portable document format) file A computer file format that preserves the layout of complex text documents, commonly found on the Internet; can be read with Adobe Reader.

Pellet Components of a sample that have settled to the bottom of a container after centrifugation.

Pen drive See **flash drive.**

Peptones Hydrolyzed (cleaved) proteins formed by enzymatic digestion or acid hydrolysis of natural substances including milk, meats, and vegetables.

Percent (%) A fraction whose denominator is 100.

Percent Error

$$\frac{\text{True Value} - \text{Measured Value} (100\%)}{\text{True Value}}$$

where the true value may be the value of an accepted reference material.

Percent solution An expression of the concentration of solute in a solution where the numerator is the amount of solute (in grams or milliliters) and the denominator is 100 units (usually milliliters) of total solution. There are three types of percent expressions that vary in their units:

> **Weight per volume percent, w/v** The grams of solute per 100 mL of solution.

> **Volume percent, v/v** The milliliters of solute per 100 mL of solution.

> **Weight percent, w/w** The grams of solute per 100 g of solution.

Performance qualification (PQ) A set of activities that evaluate the performance of a piece of equipment under the conditions of actual use.

Performance verification A process of checking that an instrument is performing properly.

Peripheral device, peripheral Any hardware that you plug into your computer.

Permeate The fluid and particles that pass through a membrane filter

Permeation rate (in gloves) A measurement of the tendency of a chemical to penetrate glove material.

Permissible exposure limit (PEL) Limit set by OSHA for the allowable concentration of a substance in air.

Peroxide former A chemical that produces peroxides or hydroperoxides with age or air contact.

Personal containment Standard practices used to reduce the spread of the microorganism; procedures used for handling and disposal of the hazard along with the practices of proper laboratory hygiene.

Personal protective equipment (PPE)   Specialized clothing or equipment worn for protection against a hazard.

pH   A measure representing the relative acidity or alkalinity of a solution; expressed as the negative log of the H^+ concentration when concentration is expressed in moles per liter; $pH = -log [H^+]$.

pH indicator dyes   Dyes whose color is pH dependent. Indicators can be directly dissolved in a solution or can be impregnated into strips of paper that are then dipped into the solution to be tested.

pH measuring electrode   An electrode whose voltage depends on the concentration of H^+ ions in the solution in which it is immersed, also called a **glass electrode.**

pH meter   A term that is commonly used to refer to a specialized voltage meter and its accompanying electrodes used to measure pH.

Pharmaceuticals   Chemical agents with therapeutic activity in the body that are used to treat, correct, or prevent the symptoms of illnesses, injuries, and disorders in humans and other animals.

Pharmacokinetics (PK)   The study of how the body acts on a drug; the absorption, distribution, metabolism, and excretion of the drug and its metabolites in the body over time.

Phase I clinical trials   Studies of candidate pharmaceuticals typically performed on healthy human volunteers that focus on the safety of potential products.

Phase II clinical trials   Studies of potential pharmaceuticals performed on a small number of diseased patients to determine the drug's clinical efficacy, dose, and potential side effects.

Phase III clinical trials   Studies of potential pharmaceuticals performed on a population of several hundred to several thousand patients to establish therapeutic efficacy, side effects, longer-term safety, and recommended dose levels.

Phase transition   Condition where a liquid turns to a solid or a vapor, or a vapor turns to liquid.

Phenol red   pH indicator dye that changes from yellow at pH 6.8 to red at pH 8.2.

Phenolphthalein   pH indicator dye that changes from colorless at pH 8.0 to red at pH 10.0.

Philadelphia chromosome   A chromosomal rearrangement in which pieces of two different chromosomes break off and each piece reattaches to the opposite chromosome. This rearrangement fuses part of a specific gene from chromosome 22 (the bcr gene) with part of another gene from chromosome 9 (the abl gene). The resulting chromosome 22 is termed the Philadelphia chromosome and is associated with a type of leukemia.

Phishing   Sending out mass e-mails designed to trick the recipients into revealing confidential information.

Phosphodiester bond   A covalent bond that connects the nucleotides that compose DNA or RNA. A phosphodiester bond is formed when the 5′-phosphate group of one nucleoside is joined to the 3′-hydroxyl group of the next nucleoside through a phosphate group bridge.

Photodetector   A device that responds to light by producing an electrical signal that is proportional to the amount of incident light.

Photodiode array detector (PDA)   A type of detector that can simultaneously determine the absorbance of a sample at a wide range of wavelengths.

Photoemissive surface   A surface coated with a material that gives off electrons when bombarded by light.

Photometer   An instrument that measures light absorbance consisting of a light source, a filter to select the wavelength range, a sample holder, a detector, and a readout device.

Photometric accuracy (also called **absorbance scale accuracy**)   The extent to which a measured transmittance (or absorbance) value agrees with the nationally or internationally accepted value.

Photometric linearity   The ability of a spectrophotometer to give a linear relationship between the intensity of light hitting the detector and the transmittance display of the instrument.

Photomultiplier tube (PMT)   A common type of detector used in spectrophotometers that consists of a series of metal plates coated with a thin layer of a photoemissive material. Light transmitted through the sample "knocks" electrons from the surface, ultimately resulting in an electrical signal.

Photons   Packages of energy of electromagnetic radiation.

Physical containment   Containment strategies that include laboratory design and the physical barriers that workers use.

Physiological saline   0.9% NaCl.

Physiological solution   An isotonic solution that is used to support intact cells or microorganisms in the laboratory.

Pilot plant   A moderate size production facility where development of efficient production methods occurs. Used to produce clinical material until a larger facility is needed for commercial production.

Pipeline and Hazardous Materials Safety Administration (PHMSA)   The agency within the Department of Transportation responsible for regulating and monitoring all transport of hazardous materials within the United States.

Pipette (also spelled "pipet")   Hollow tube that allows liquids to be drawn in and dispensed from one end; generally used to measure volumes in the 0.1–25 mL range.

Pipette-aid   A device used to draw liquid into and expel it from pipettes.

Pixel   The smallest image unit found on a computer monitor screen and output devices; short for picture element.

pKa   The pH at which a buffer stabilizes and is least sensitive to additions of acids or bases.

Plasmid   A circular molecule of DNA found most often in bacteria, which exists separately from the bacterial chromosome and can replicate independently. Plasmids are often used as vectors to transport genetic information into a cell.

Platinum resistance thermometer   A metallic resistance thermometer based on the fact that the resistance in a platinum wire increases as the temperature increases.

Plunger   Part of a manual micropipettor that is compressed by the operator as liquids are taken up and expelled.

Point and click   The technique of using a mouse to point the cursor at an on-screen object, and then press one of the mouse buttons to select or activate the object.

Poise   In a mechanical balance, a sliding weight mounted on a lever used to balance against a sample.

Polarity   The characteristic of having a positive or negative charge.

Polarization layer (in filtration)   A build-up of retained particles or solutes on a membrane.

Polarized   An electrical device that has negative and positive identifications on its terminals. When connecting such devices to a source of voltage the negative terminal of the device is connected to the negative terminal of the voltage source.

Polarized plug   Plugs with different sized prongs to ensure that the plug is oriented correctly in the outlet.

Polished water   Refers to high-quality water after it has gone through a final phase of treatment that removes "all" impurities.

Polyacrylamide gel   A separation gel made from polymerized acrylamide and bisacrylamide in the presence of an appropriate initiator (ammonium persulfate) and catalyst (N,N,N′, N′-tetramethyl-ethylenediamine, also called TEMED).

Polyacrylamide gel electrophoresis   See **PAGE.**

Polychromatic light   A combination of light of many wavelengths.

Polyclonal antibodies   Antibodies derived from different B-cell lines that are produced in response to a particular antigen but potentially recognize a different epitope.

Polymodal distribution   A frequency distribution with more than two peaks.

Population   A group of events, objects, or individuals where each member of the group has some unifying characteristic(s). Examples of populations are all of a person's red blood cells and all the enzyme molecules in a test tube.

Pore size, absolute (in filtration)   The size of particles that are retained with 100% efficiency by a microfilter under specified conditions.

Pore size, nominal (in filtration)   The size of particles that are retained with an efficiency below 100% (typically 90–98%) by a microfilter under specified conditions.

Porosity   The percentage of a filter area that is porous.

Port, I/O (input/output)   The connection outlets that allow communication between computer components.

Positive displacement micropipettor   Volume-measuring device, such as a syringe, where the sample comes in contact with the plunger and the walls of the pipetting instrument.

Positively charged   A material in which electrons are depleted.

Potable water   Water that is suitable for drinking.

Potential energy   Energy is the ability to do work; potential energy is stored energy.

Potentiometer   A type of variable resistor that can be adjusted to control the voltage or current in a circuit.

Power (P) The rate of using energy measured in units of watts: power = (volts) · (current).

Power supply A device that produces varying levels of AC and DC voltage, converts AC current to DC, and regulates the voltage level in the circuit.

Precipitation methods Bioseparation techniques that are based on differences between molecules in their tendency to precipitate from various liquids.

Precision 1. The consistency of a series of measurements or tests obtained under stipulated conditions. (See also **Reproducibility** and **Repeatability**.) 2. The fineness of increments of a measuring device; the smaller the increments, the better the precision.

Precision (for a balance) A measure of the consistency of a series of repeated weighings of the same object.

Preclinical development The stages of drug development preceding the first trials in humans. Encompasses laboratory (in vitro) and animal testing.

Predicate Rules (in the context of 21CFR11) The GMP, GCP, GLP, and other regulations (as contrasted with the 21CFR Part 11 regulations).

Prefilter A filter placed upstream from a membrane filter or from an instrument to protect the membrane or instrument from particulates that would cause clogging.

Preparative centrifugation Centrifugation for the purpose of obtaining material for further use.

Preparative method (in contrast to an analytical method) Method that produces a material or product for further use (perhaps for commercial use, or perhaps for further experimentation). Chromatography, for example, is used to separate the components of a sample from one another. If the components are separated in order to purify an enzyme product that will later be sold commercially, then chromatography is being used as a preparative method. Chromatography is also frequently used to help identify the components present in a mixture, in which case, chromatography is used as an analytical method.

Presentation software Software that provides capabilities for inputting text, graphic images, audio, and video, and then processing them into "slide" presentations that can be projected onto a screen and printed as handouts.

Pretreatment (of water) Initial steps in a water treatment process that remove a large portion of contaminants and prolong the life of filters and cartridges used in later, more expensive steps.

Preventive maintenance A program of scheduled inspections of laboratory instruments and equipment that leads to minor adjustments or repairs and ensures that items are functioning properly.

Primary chemical container Containers supplied by the manufacturer; these should immediately be labeled with a date and user name when they arrive in the laboratory.

Primary containment Protection of the worker and laboratory environment through the use of good microbiological technique and safety equipment.

Primary culture (mammalian cell culture) Cultured cells that are derived directly from excised tissue.

Primary standard In the United States the physical items housed at NIST to which measurements are referenced (traced).

Primary structure (proteins) The linear sequence of amino acids that comprise a protein and are connected by covalent peptide bonds.

Printed circuit board A thin piece of insulating material onto which copper wires are printed using a chemical process. Electronic components are connected to one another on the board.

Prion A type of infectious agent, thought to be composed of misfolded proteins, which causes a variety of neurodegenerative diseases including BSE in cattle and **Creutzfeldt-Jakob disease** in humans.

Probability The likelihood of a particular outcome.

Probability distribution A theoretical distribution of values based on the calculated probabilities of values occurring.

Probe (temperature) A generic term for many types of temperature sensor.

Procedure "[S]pecified way to perform an activity."*

Process Set of interrelated resources (such as personnel, facilities, equipment, techniques, and methods) that transforms inputs into outputs.*

Process development The process in which R&D scientists develop the methods by which a product will be made, especially through increasingly larger scales of production.

Process filtration Filtration of large volumes of liquids, as are found in biotechnology production facilities, pharmaceutical companies, and food-production facilities.

Process validation Activities that prove that a manufacturing process meets its requirements so that the final product will do what it is supposed to do safely and effectively. See **validation.**

Product Results of activities or processes. A product may be a service, a processed material, or knowledge.*

Product development The process in which R&D scientists explore and optimize the features of a potential product.

Product water Water produced by a purification process.

Program A series of instructions, written in one of many computer languages, that allows the computer to perform a specific function.

Programmer An individual who writes the coded instructions for software.

Prokaryotic organism Organisms whose cells have only a few organelles and lack nuclei; includes bacteria and cyanobacteria.

Proportion Two ratios that have the same value but different numbers. For example: 1/2 = 5/10.

Proportionality constant A value that expresses the relationship between the numerators and the denominators in a proportional relationship.

Protease An enzyme which breaks peptide bonds, thus degrading proteins.

Protease inhibitor An agent that inhibits protease activity.

Protein Diverse biological molecules composed of amino acid subunits that control the structure, function, and regulation of cells.

Protein assay A test of the amount or concentration of protein in a sample.

Protein error A cause of error when using pH indicator dyes, where proteins react with indicators and affect their colors.

Proteolysis The breaking apart of the primary structure of a protein by the action of enzymes (proteases) that digest the peptide bonds connecting the amino acids.

Proteomics The study of how proteins interact with one another and work in living systems.

Protocol A step-by-step outline that tells an operator how to perform an experiment that is intended to answer a question.

Protocol (in computers) The common languages used by computers for information exchange.

Public domain program Noncopyrighted software that the author has chosen to provide for free to the public for any type of use or alteration.

Purification The separation of a specific material of interest from contaminants, in a manner that provides a useful end product.

Purified water (USP) Water that meets USP requirements and is used in a number of pharmaceutical and cosmetic applications. **Purified water** is defined by its chemical specifications and may be produced by a variety of techniques.

Pyrogen Materials derived from lipopolysaccharides that are released from gram-negative bacteria and that trigger a dangerous immune response and induce fever, hence their name pyrogen (heat producing). The term *endotoxin* is sometimes used as a synonym because these substances are toxic to animals.

Pyrophoric Chemicals that will ignite on contact with air.

Q.S Latin abbreviation, *quantum sufficit,* "as much as sufficient." See **BTV, bring to volume.**

Quad core (in microprocessors) Technology in which a CPU chip contains four separate microprocessors that can compute simultaneously.

Qualification process "[P]rocess of demonstrating whether an entity is capable of fulfilling specified requirements."*

Qualitative analysis The analysis of the identity of substance(s) present in a sample.

Qualitative filter paper A paper filter that leaves an appreciable percent of ash when burned.

Quality According to ISO, all the features of a product or service that are required by the customer or user.

Quality Assurance (QA) An organizational unit in a company that provides confidence that product quality requirements are fulfilled, in part

---

*Based on definitions from American National Standard ANSI/ASQC A8402-1994 "Quality Management and Quality Assurance," American Society for Quality Control, Milwaukee, WI.

through the effective deployment and management of documentation.

Quality Control (QC) A department responsible for monitoring processes and performing laboratory testing to ensure that products are of suitable quality.

Quality system The organization, structure, responsibilities, procedures, processes, and resources for ensuring the quality of a product or service. *

Quantitation limit The lowest amount of analyte in a sample which can be quantitatively determined with suitable precision and accuracy.

Quantitative analysis The measurement of the concentration or amount of a substance of interest in a sample.

Quantitative filter paper A paper filter that leaves little ash when burned.

Quaternary protein structure A protein complex formed when two or more protein chains associate with one another.

Query (in databases) A text-based set of criteria used to search for and extract a desired subset of data from a database and then present it in a specific format.

R group (of proteins) Amino acids are distinguished from one another by a side chain, or *R group*, which has a particular structure and chemical properties.

Radiant energy Energy of electromagnetic waves.

Radius of rotation The distance from the center of rotation to the material being centrifuged.

RAM (random access memory) Memory chips that provide the portion of electronic memory available to work with.

Random error Error of unknown origin that causes a loss of precision.

Random phenomenon A phenomenon where the outcome of a single repetition is uncertain, but the results of many repetitions have a known, predictable pattern. For example, the outcome of a single flip of a coin might be a head or a tail; the outcome of many flips is predicted to be half heads and half tails.

Random sample A sample drawn in such a way that every member of a population has an equal chance of being chosen.

Random variability (random error) The situation where a group of observations or measurements vary by chance in such a way that they are sometimes a bit higher than expected, sometimes a bit lower, but overall average the "correct" or expected value. See also the contrasting term, **Systematic Variability.**

Range 1. A range of values, from the lowest to the highest, that a method or instrument can measure with acceptable results. 2. A statistical measure that is the difference between the lowest and the highest values in a set of data.

Rankine temperature scale A temperature scale that places 0 at absolute zero (like the Kelvin scale) and whose units,° R, are the same size as a Fahrenheit degree.

Raster graphics See **bitmapped graphics.**

Rate zonal density gradient centrifugation A method of density gradient separation in which particles move through a gradient at different rates based on their size and density.

Ratio The relationship between two quantities, for example, "25 miles per gallon" and "10 mg of NaCl per L."

Raw data The first record of an original observation. Depending on the situation, raw data may be recorded with pen by the operator, may be a paper output from an instrument, or may be recorded directly into a computer medium.

Reactivity The tendency of a chemical to undergo chemical reactions.

Readability (for a balance) Value of the smallest unit of weight that can be read. This may include the estimation of some fraction of a scale division or, in the case of a digital display, will represent the minimum value of the least significant digit. (According to ASTM document E 319-85 [reapproved 1993].)

Readout See **display.**

Reagent-grade water Water suitable to be a solvent for laboratory reagents or for use in analytical or biological procedures.

Receptacle The half of a connector that is mounted on a wall, instrument panel, or other support and into which electrical devices are plugged.

Reciprocal A number related to another so that when multiplied together their product equals 1. To calculate the reciprocal of a number, divide 1 by that number. For example, the reciprocal of 5 is: 1/5 or 0.2.

Recirculation (of water) A process in which purified water is continuously recirculated and repurified. Recirculation helps prevent microbial growth and leaching of materials into the water from storage containers.

Recombinant DNA (rDNA) DNA that contains sequences of DNA from different sources that were brought together by techniques of molecular biology and not by traditional breeding methods.

Recombinant DNA Advisory Committee (RAC) An expert committee assembled by The National Institutes of Health that has reviewed and interpreted what is known about the risks associated with genetic manipulation methods and has published guidelines for safely working with recombinant DNA.

Record (in databases) The basic data unit in a database, represented by the rows in a data **table;** each record has multiple **fields.**

Recovery media Storage media that can be used to restore your computer's OS and programs in case of a major disaster that damages the OS. Also called bootable backup media.

Rectification Conversion of alternating current into direct current.

Red litmus pH indicator dye that changes from red to blue denoting a change from acid to alkaline.

Reducing agent A substance that donates electrons in a chemical reaction.

Reducing agents (in biological solutions) A chemical added to a solution to simulate the reduced intracellular environment and prevent unwanted oxidation reactions.

Reduction to practice (in the context of patents) Constructing a prototype of an invention or performing a method or process (as described in the patent claim).

Reference (in spectrophotometry) See **blank.**

Reference electrode An electrode that maintains a stable voltage for comparison with the pH measuring electrode or an ion sensitive electrode.

Reference junction The cold junction in a thermocouple circuit that is held at a stable known temperature.

Reference material "A material or substance one or more properties of which are sufficiently well established to be used for the calibration of an apparatus, the assessment of a measuring method, or for assigning values to materials."**

Reference point A temperature at which the reading of a thermometer is calibrated or verified.

Reference standards (for weights) Individual or combinations of weights whose mass values and uncertainties are known sufficiently well to allow them to be used to calibrate other weights, objects, or balances.

Refine (in computers) Adjusting a query with additional keywords and conditions to narrow down and focus the search results.

Reflection (of light) An interaction of light with matter in which the light strikes a surface, causing a change in the light's direction.

Regeneration (of ion exchange resins) A process in which the capacity of used ion exchange resins is restored. An acid rinse is used to restore cation resins and a sodium hydroxide rinse is used to restore anion resins.

Regulated waste Liquid or semi-liquid blood or other potentially infectious materials; any contaminated items that would release blood or other potentially infectious materials in a liquid or semi-liquid state if compressed; any items that are caked with dried blood or other potentially infectious materials and are capable of releasing these materials during handling; contaminated sharps; and pathological and microbiological wastes containing blood or other potentially infectious materials (OSHA).

Regulations Requirements imposed by the government and having the force of law.

Regulator (in gas cylinders) A device which attaches to the valve of a compressed gas cylinder and decreases and modulates the pressure of the gas leaving the cylinder.

Regulatory affairs unit A functional unit in a company whose personnel interpret the rules and guidelines of regulatory agencies and ensure that the company complies with these requirements.

Regulatory agency A government body that is responsible for enforcing and interpreting legislative acts. The FDA, for example, is a regulatory agency responsible for ensuring the safety, effectiveness, and reliability of medical products, foods, and cosmetics as set forth in federal legislation.

---

*Based on definitions from American National Standard ANSI/ASQC A8402-1994 "Quality Management and Quality Assurance," American Society for Quality Control, Milwaukee, WI.

**Based on definitions in ISO/IEC Guide 25, "General Requirement for the Competence of Testing and Calibration Laboratories," Geneva, Switzerland. August 1996 edition.

Regulatory DNA   Segments of DNA that act as "switches" to control which genes are turned on at a given time in a given cell.

Regulatory submissions   Documents that are completed to meet the requirements of a government regulatory agency.

Relational database   A complex database in which multiple data **tables** are linked by specific user-defined relationships, allowing them to work together.

Relative centrifugal field (RCF)   The ratio of a centrifugal field to the earth's force of gravity; the units of RCF are in × g. The relative centrifugal field developed in a centrifuge depends on the speed of rotation and the distance to the center of rotation according to the formula: RCF = 11.2 × r (RPM/1000)^2, where RCF is relative centrifugal field, RPM is revolutions per minute, and r is the radius of rotation in cm.

Relative standard deviation   See **coefficient of variation.**

Renaturation (of DNA)   Process in which complementary base pairs reestablish hydrogen bonds, bringing together two strands of DNA.

Repeatability   "Precision under repeatability conditions."[*]

Repeatability conditions   "Conditions where independent test results are obtained with the same test method on identical test items in the same laboratory by the same operator using the same equipment within short intervals of time."[*]

Repeatability standard deviation   "The standard deviation of test results obtained under repeatability conditions."[*]

Repetitive stress injury (RSI)   Physical injury caused by performing constant identical motions over long periods of time.

Representative sample   Per GMP: "[A] sample that consists of a number of units that are drawn based on rational criteria such as random sampling and intended to assure that the sample accurately portrays the material being sampled."

Reproducibility   "Precision under reproducibility conditions."[*]

Reproducibility conditions   "Conditions where test results are obtained with the same test method on identical test items in different laboratories with different operators using different equipment."[*]

Reproducibility standard deviation   "The standard deviation of test results obtained under reproducibility conditions."[*]

Research and development (R&D)   The organizational unit in a company that finds ideas for products, performs research and testing to see if the ideas are feasible, and develops promising ideas into actual products.

Residue analysis   Method in which particles of interest are separated and retained on a filter for further analysis.

Resin (ion exchange)   Beads of synthetic material that have an affinity for certain ions.

Resistance (R, in the context of electricity)   A measure of the difficulty electrons encounter when moving through a material, measured in

units of ohms (Ω) or megohms ($M\Omega$) where 1 megohm = 1,000,000Ω.

Resistant (in safety)   Indicates a relative inability to react with another material.

Resistivity   1/conductivity. The units are ohm-cm.

Resistor   An electronic component that resists the flow of current.

Resolution   The smallest detectable increment of measurement.

Resolution (in bioseparations)   The relative ability of a technique to distinguish between the product of interest and its contaminants. The resolving power of a method is directly dependent on its **selectivity.**

Resolution (in computers)   A measure of the sharpness of a monitor screen image; described by the number of horizontal and vertical pixels.

Resolution (in spectrophotometry)   The separation (in nanometers) of two absorbance peaks that can just be distinguished.

Resource Conservation and Recovery Act of 1976 (RCRA)   Legislation which provides a system for tracking hazardous waste, including toxic or reactive chemicals, from generation to disposal.

Respirator   Breathing devices designed to reduce airborne hazards by manipulating the quality of the air supply.

Restriction endonucleases   Enzymes that cleave DNA at sites with specific base pair sequences.

Restriction fragment length polymorphism (RFLP) analysis   A method of analyzing DNA that can be used to identify individuals based on slight differences in the sequence of their DNA in specific regions of the genome. The method is also used to look for mutations associated with specific genetic diseases. The term *polymorphism* refers to the slight differences that exist between individuals in base pair sequences. The method involves incubating the test subject's DNA with restriction enzymes that recognize and cut the DNA at specific sequences. The resulting DNA fragments are separated from one another by electrophoresis and stained. The DNA fragments form a pattern of bands that differs among individuals depending on their sequence of DNA in the analyzed region.

Retentate   The materials trapped by a filter.

Retiring (a rotor)   Taking a rotor out of use due to age or amount of use it has received.

Reverse osmosis   A process in which liquid is forced through a semi-permeable membrane, leaving behind impurities; can remove very low molecular weight materials, including salts, from a liquid.

Reverse osmosis membranes   Very restrictive filters that prevent the passage or materials as small as dissolved ions.

Reversed phase, (RP, in chromatography)   Any chromatography method where the stationary phase is non-polar and the mobile phase is polar relative to the stationary phase.

Revolutions per minute (RPM)   A measure of the speed of rotation in a centrifuge.

Rheostat   A type of variable resistor that can be adjusted to control the voltage and/or current in a circuit.

Ribonucleic acid (RNA)   A single-stranded molecule comprised of nucleic acids that is found in the nucleus and cytoplasm of cells where it performs various roles, particularly those relating to protein synthesis. See also **mRNA** and **RNAi.**

Ribosomes   The cellular organelle responsible for manufacturing proteins.

Rise (in graphing)   An expression that describes the amount by which the Y coordinate changes on a two-dimensional graph.

Risk   The probability that a **hazard** will cause harm.

Risk assessment   Estimation of the potential for human injury or property damage from an activity.

RNAi, RNA interference   Short stretches of RNA molecules that can turn off the activity of specific genes.

RNase   Ribonuclease. A class of enzymes that catalyze the cleavage of nucleotides in RNA.

Robustness (in the context of an assay method)   The capacity of a method to give acceptable results when there are deliberate variations in method parameters.

ROM (read-only memory)   Memory chips that store permanent information, such as the computer **BIOS.**

Root mean square noise (RMS)   A statistical calculation that "averages" the noise present over a period of time; the lower the value, the less noise is present.

Rotor   The device that rotates in the centrifuge and holds the sample tubes or other containers.

Router   Hardware that sorts and directs incoming and outgoing information between networks.

RSS (Really simple syndication)   The most common format used to create Web feeds.

Ruggedness   Similar to robustness. The ruggedness of an analytical procedure is its ability to tolerate small variations in procedural conditions, which may include variation in volumes, temperatures, concentrations, pH, and instrument settings, without affecting the analytical result. It provides an indication of the applicability of the method in a variety of laboratory conditions.

Run (in graphing)   An expression that describes the amount by which the X coordinate changes on a two-dimensional graph.

Safety   The elimination of hazards and risk, to the extent possible.

Safety rules   Procedures which are designed to prevent accidents, by controlling the risk of hazards in situations where the hazards cannot be eliminated entirely.

Salt   A compound formed by replacing hydrogen in an acid by a metal (or a radical that acts like a metal). (When an acid and base combine, their ionic components dissociate. In solution, the H$^+$ and OH$^-$ ions combine to form water. The other two ions combine to create a salt. For example, the salt, NaCl, is formed by the combination of solutions of NaOH and HCl.)

Salt bridge   See **junction.**

---

[*]*Based on definitions in ASTM Standard E 456-06 "Standard Terminology Relating to Quality and Statistics."*

Salt error   A cause or error when using pH indicator dyes. Salts at concentrations above about 0.2 M can affect the color of pH indicators.

Sample   A subset of the whole that represents the whole (e.g., a blood sample represents all the blood in an individual).

Sample chamber   The location in which the sample is placed.

Sample statistic   A numerical statement about a sample (e.g., the sample mean is a statistic).

SAS (statistical analysis system)   A complex software package designed for statistical analysis.

Save (in computers)   Transfer data from **volatile** memory to permanent storage.

Scale   Deposits of calcium carbonate that, for example, coat the inside surface of boilers or the surfaces of membranes.

Scale (of a thermometer)   Graduations that indicate degrees, fractions of degrees, or multiples of degrees.

Scale-up   1. The process of converting a small-scale laboratory procedure to one that will be appropriate for large-scale product purification. 2. The transition between producing small amounts of a product to the manufacture of large quantities for clinical testing and ultimately commercial sale.

Scanner   An optical device that converts graphic images to digital form and sends them to the computer for manipulation and storage.

Scanning (in spectrophotometry)   Process of determining the absorbance of a sample at a series of wavelengths.

Scatter plot   A plot where all data points are graphed but not grouped or connected.

Scattering (of light)   Redirection of light in many directions by small particles.

Scientific notation (also called exponential notation)   The use of exponents to simplify handling numbers that are very large or very small. For example, 4500 in scientific notation is 4.5×10^3.

Screensaver   A program that blanks out or creates a moving graphic display on the monitor screen after a specified time without user input.

SDS (sodium dodecyl sulfate)   A negatively charged detergent.

SDS-PAGE (sodium dodecyl sulfate-PAGE)   A technique that separates proteins on the basis of molecular size, using **polyacrylamide gel electrophoresis.**

Search engine   A program that finds web pages, create topic indices for these pages, and then identifies the ones that match specific search criteria.

Secondary containment   Protection of the environment outside the laboratory by the use of good laboratory design and safe practices.

Secondary standard   A standard whose value is based on comparison with a primary standard.

Secondary structure (of proteins)   Regularly repeating patterns of twists or kinks of an amino acid chain held together by hydrogen bonds. Two common types of secondary structure are the alpha-helix and β-pleated sheet. Regions of proteins without regularly repeating structures are said to have a "random coil" secondary structure.

Sedimentation   The settling out of particles from a liquid suspension.

Sedimentation coefficient (S)   A measure of the sedimentation velocity of a particle. In practice, the sedimentation coefficient is a function of the size of a particle; larger particles have larger sedimentation coefficients.

Selective medium   A type of growth medium for cells that inhibits the reproduction of unwanted organisms and/or encourages the reproduction of specific organisms.

Selectivity   The ability of a technique to separate a specific component from a heterogeneous mixture based on molecular properties; see **specificity.**

Self-contained breathing apparatus   A **respirator** that contains its own air supply and is appropriate for situations where the user is exposed to highly toxic gases.

Semiconductor   A material that is more resistant to electron flow than a conductor but is less resistant than an insulator. Semiconductors are important in the manufacture of electronic devices.

Semilog paper   A type of graph paper that has a log scale on one axis and a linear scale on the other axis.

Semipermeable membrane   A membrane that allows water and some small molecules to flow through unimpeded, whereas other molecules are restricted; the membrane that surrounds every living cell is semipermeable.

Sensitivity (for a laboratory balance)   The smallest value of weight that will cause a change in the response of the balance that can be observed by the operator.

Sensitivity (of a detector or instrument)   Response per amount of sample.

Sensitizer   A substance or agent that may trigger an allergy directly, or cause an individual to develop an allergic reaction to an accompanying chemical.

Sensor   See **transducer.**

Sephadex   A **gel permeation** medium manufactured by GE Healthcare Life Sciences.

Serial dilutions   Dilutions made in series (each one derived from the one before) and that all have the same dilution factor. For example, a series of 1/10 dilutions of an original sample would be: 1, 1/10, 1/100, 1/1000, etc.

Series circuit   A circuit configuration in which the components are connected in a "string," end to end, so that current can only flow in one pathway.

Serological pipette   Term that usually refers to a calibrated glass or plastic pipette that measures in the 0.1 mL to 25 mL range. These pipettes are calibrated so that the last drop is in the tip. This drop needs to be "blown out" to deliver the full volume of the pipette.

Serum (mammalian cell culture)   The liquid component of blood from which blood cells and most clotting factors have been removed.

Serum-free medium (mammalian cell culture) (SFM)   A culture medium that does not contain serum. These media are supplemented with proteins, growth factors, vitamins, hormones, and other constituents to take the place of the serum.

Server (in computers)   A computer that acts as a common resource for data storage, program access, and networking.

Shareware   Copyrighted software developed by individuals or small companies that is easily available on the internet for free download and trial use.

Sharps   A term to describe laboratory items, such as razor blades and needles, that can cause cuts and lacerations.

Short circuit   An unintended path for current flow that by passes the resistance or load.

Short tandem repeat (STR)   A defined region of DNA containing multiple copies of short sequences of bases (1 to 5 base pairs long) that are repeated a number of times; the number of repeats varies among individuals.

Short-term exposure limit (STEL)   See **TLV-STEL.**

SI System (Système International d'Unités)   A standardized system of units of measurement, adopted in 1960 by a number of international organizations, that is derived from the metric system.

SI unit of enzyme activity   The amount of enzyme necessary to catalyze transformation of 1.0 mol of substrate to product per second under optimal measurement conditions; expressed in units of kat.

Siemen, S   A unit of conductance.

Signal (in electricity)   An electrical change that conveys information. The signal may be a change in voltage, current, or resistance.

Signal processing unit (signal processor)   A device that modifies an electrical signal, for example, to amplify it.

Signal-to-noise ratio   The instrument response due to the sample divided by the electronic noise in the system.

Significant figure   A digit in a number that is a reliable indicator of value.

Silicon   An abundant element that is used in the manufacture of electronic components.

Silver/silver chloride (Ag/AgCl) electrode   A type of reference electrode that contains a strip of silver coated with silver chloride and immersed in an electrolyte solution of KCl and silver chloride.

Single-beam, double-pan balance   A simple balance that compares the weight of the sample to the weight of a standard that is placed on a pan across a beam.

Single nucleotide polymorphisms (SNPs)   Places in DNA where a single nucleotide differs from person to person.

Site license   A software license that grants you the right to use a specific number of program copies.

Site map   A map and organized set of links for every resource and function within a website.

Size exclusion chromatography   See **gel permeation chromatography.**

Skewed distribution   A distribution in which the values tend to be clustered either above or below the mean.

Sleeve junction   A type of reference electrode junction made by placing a hole in the side of the electrode housing and covering the hole with a glass or plastic sleeve. The sleeve junction is relatively fast flowing and is unlikely to become clogged.

Slope (of a line)   Given a straight line plotted on a graph, the slope of the line is how steeply the line rises or falls. It is the ratio of vertical change to horizontal change between any two points on the line. Slope can be calculated by choosing any two points on the line and dividing the change in their Y values by the change in their X values.

Smoke generators   Small tubes of chemicals, frequently including titanium tetrachloride, which generate highly visible white smoke from a chemical reaction; used to check the general function of fume hoods.

Sniffer software   Programs that allow unauthorized individuals to intercept wireless transmissions from other computers.

Soft copy   Information displayed on a computer monitor and saved in computer files.

Softened water   Water that has cations replaced with sodium ions.

Software   The **programs** that provide the instructions that make computers useful.

Solid or semisolid medium (bacterial culture)   Aqueous-based mixtures of nutrients that contain agar as a hardening agent to provide a solid substrate on which or in which microbes may be cultured.

Solid state   An electrical circuit that uses semiconductor diodes and transistors instead of vacuum tubes. (The term "solid" is used because signal flows through a solid semi-conductor material instead of through a vacuum.)

Solid state electrode   Newer type of electrode that relies on a small electronic chip to detect ions.

Solute   A substance that is dissolved in some other material. For example, if table salt is dissolved in water, salt is the solute and water is the solvent.

Solution   A homogeneous mixture in which one or more substances is (are) dissolved in another.

Solvent   A substance that dissolves another. For example, if salt is dissolved in water, then water is the solvent.

Solvent cutoff (in spectrophotometry)   The wavelength at which a particular solvent absorbs a significant amount of light. Solvent absorbance interferes with the analysis of the analyte.

Somatic cell nuclear transfer   See **therapeutic cloning.**

Sonication device (sonicator)   Laboratory equipment that disrupts cells with high frequency sound waves.

Source (in spectrophotometry)   The lamp or bulb used to provide light in a spectrophotometer.

Spam   Unsolicited e-mail sent to large numbers of recipients.

Specific activity   The amount (or units) of the protein of interest, divided by the total amount of protein in a sample.

Specification   The defined limits within which physical, chemical, biological, and microbiological test results for a product should lie to ensure its quality.

Specificity (also called selectivity)   A measure of the extent to which a method or instrument

can determine the presence of a particular compound in a sample without interference from other materials present. Typically these might include impurities, degradants, matrix, and so on.

Spectral band width (in spectrophotometry)   Spectral band width is a measure of the range of wavelengths emerging from the monochromator when a particular wavelength is selected.

Spectral slit width   The physical width of the monochromator exit slit multiplied by the linear dispersion of the diffraction grating.

Spectrometer   Any instrument used to measure the interaction of electromagnetic radiation with matter.

Spectrophotometer   An instrument that measures the effect of a sample on the incident light beam. The instrument consists of a source of light, entrance slit, monochromator, exit slit, sample holder, detector, and readout device.

Spectroscopy   The study and measurement of interactions of electromagnetic radiation with matter.

Spectrum   Electromagnetic radiation separated or distinguished according to wavelength.

Speech recognition   The ability of computers to convert human speech into written text or commands.

Spirit thermometer   A liquid expansion thermometer containing an alcohol-based liquid.

Sponsor   Individual, company, institution, or organization that initiates or pays for testing a potential drug product and submits regulatory documents to the FDA. The sponsor may or may not perform drug development activities itself.

Spontaneous density gradient   A density gradient made of a medium that forms a gradient spontaneously under the influence of a centrifugal field.

Spoofing   Disguising an undesirable website with an apparently legitimate link.

Spore   Dormant microorganism that is resistant to heat.

Spore strips   Dried pieces of paper to which large numbers of nonpathogenic spores are adhered; used to test the success of sterilization.

Spreadsheet software   Programs for storing and manipulating numerical data in a grid of **cells.**

Spyware   Software that sends information about your computer activities to outside parties

SQL (structured query language)   A computer language specifically designed to find and retrieve data from databases.

Stability (in safety)   The chemical characteristic of remaining unchanged over time.

Stability (in the context of pharmaceuticals)   The capacity of a product to remain within its specifications over time.

Stability testing   According to FDA, "The purpose of stability testing is to provide evidence on how the quality of a drug substance or drug product varies with time under the influence of a variety of environmental factors, such as temperature, humidity, and light, and to establish a retest period for the drug substance or a shelf life for the drug product and recommended storage conditions."

Standard   Operating principle or requirement.

Standard (in spectrophotometry)   A mixture including the analyte of interest dissolved in solvent and used to determine the relationship between absorbance and concentration for that analyte.

Standard (relating to measurement)   1. Broadly, any concept, method, or object that has been established by authority, custom, or agreement to serve as a model in the measurement of any property. 2. A physical object, the properties of which are known with sufficient accuracy to be used to evaluate another item; a physical embodiment of a unit. For example, a metal object whose mass is accurately known can be used by comparison to determine the mass of a sample. 3. In chemical or biological measurements, a standard often describes a substance or a solution that is used to establish the response of an instrument or an assay method to the analyte. 4. A document established by consensus and approved by a recognized body that establishes rules or guidelines to make a procedure consistent among various people.

Standard curve (also called a calibration curve)   A graph that shows the relationship between a response (e.g., of an instrument) and a property that the experimenter controls (e.g., the concentration of a substance).

Standard curve (in spectrophotometry)   A graph that shows the relationship between absorbance (on the Y axis) and the amount or concentration of a substance (on the X axis).

Standard deviation   A measure of the dispersion of observed values or results expressed as the positive square root of the variance. A statistical measure of variability.

Standard error of the mean (SEM)   An estimate of the standard deviation of a hypothetical distribution of sample means. Standard Error of the Mean =

$$\frac{\text{Standard deviation of a sample}}{\sqrt{N}}$$

where N = the sample size.

Standard method   A technique to perform a measurement or an assay that is specified by an external organization to ensure consistency in a field. For example, ASTM specifies standard methods to calibrate volumetric glassware; the U.S. Pharmacopeia specifies methods to perform tests of pharmaceutical products.

Standard operating procedure (SOP)   A set of instructions for performing a routine method, manufacturing operation, administrative process, or maintenance operation.

Standard reference material (SRM)   A well-characterized material available from NIST produced in quantity and certified for one or more physical or chemical properties.

Standard weight   Any weight whose mass is known with a given uncertainty.

Star activity   When a restriction enzyme "mistakenly" cleaves DNA at sequences that are not its proper target.

Stationary phase (in chromatography)   The immobile matrix in a chromatography system.

Statistical process control   The desired situation in which a manufacturing or measurement process behaves as expected. A certain amount of variability in the process is expected, but when the process is in control the variability does not exceed previously determined limits.

Stem (of a liquid-in-glass thermometer) A glass capillary tube through which the mercury or organic liquid moves as temperature changes.

Stem cells, adult Undifferentiated cells found among differentiated cells in a tissue or organ that can differentiate when needed to form the specialized cell types found in that tissue.

Stem cells, embryonic Cells from an embryo that have the potential to differentiate into any cell in the body.

Stem enlargement (of a liquid-in-glass thermometer) A thickening of the stem that assists in the proper placement of the thermometer in a device (e.g., in an oven).

Step gradient A density gradient in which there are layers of medium with different densities, each of which is clearly demarcated from the layer above and below.

Sterile technique See **aseptic technique.**

Sterilization Destruction of all life forms on an object.

Sterilize Per OSHA, using a physical or chemical procedure to destroy all microbial life, including highly resistant bacterial spores.

Sterilizing filter A nonfiber releasing filter that produces a filtrate containing no demonstrable microorganisms when tested as specified in the United States Pharmacopeia.

Stock solution A concentrated solution that is diluted to a working concentration.

Storage (in computers) The recording of data or information onto hardware such as hard drives or optical discs.

Stray light Radiation that reaches the detector without interacting with the sample.

Stringency The reaction conditions used when single-stranded complementary nucleic acids are allowed to hybridize. At high stringency, binding occurs only between strands with perfect complementarity, whereas at lower stringency, some mismatches of base pairs are tolerated.

Strip chart recorder An analog display device that records the output of a detector continuously over time.

Strong acid A chemical that completely dissociates in water to release hydrogen ions.

Strong base A chemical that completely dissociates in water to release hydroxide ions.

Strong electrolyte A substance that ionizes completely in solution.

Sublimate Change directly from a solid to a gas form (as in dry ice).

Sum of squares For a set of data, the sum of the squared differences between each data point and the mean.

Superheated Refers to liquids that have been heated past their boiling point without the release of the gaseous phase; superheated liquids can boil over, sometimes violently, if jarred.

Supernatant The liquid medium above a pellet after centrifugation.

Supplements (in cell culture) Loosely defined as materials that are sometimes added to basal media after the medium is prepared and/or that are added in varying concentrations.

Surface filter See **membrane filter.**

Surge protector A multioutlet power regulator between the computer and the main power outlet, designed to smooth out momentary electrical spikes and fluctuations.

Surge suppressors A multioutlet power regulator that evens out only minor fluctuations in power; see **surge protector.**

Suspended cells (mammalian cell culture) Cells that grow suspended in culture medium, either as individual cells or as small clumps of cells.

Suspended solids Undissolved solids that can be removed by filtration.

Svedberg, T. A pioneer in centrifugation who worked on the mathematics, methods, and instrumentation for centrifugation.

Swinging bucket rotor See **horizontal rotor.**

Switch A device that controls current flow in a circuit.

System administrator The individual(s) who serves as manager of a network.

System crash A situation where a computer freezes up and stops working; usually related to memory problems.

System restore point In the Windows operating system, automatic profiles of the state of a computer at specific times; used to reset (restore) the computer system to its state before a problem such as a system crash or virus infection.

System software The programs that operate the system hardware and provide the foundation for application programs.

System suitability A part of method validation based on the concept that the instruments used, the sample, and procedure must together provide acceptable results.

Systematic error An error that causes measurement results to be biased. See **systematic variability.**

Systematic variability (systematic error) The situation where a group of observations or measurements vary in such a way that they tend to be higher or lower than the true value or the expected value; this is also called bias. The cause of systematic error may be known or unknown. See also the contrasting term, **random variability.**

Table (in databases) A set of related information.

Tangential flow filtration A filtration mode used mainly in industry where the fluid to be filtered flows across the filter. The sweeping motion of the fluid clears the surface of the membrane, thus reducing clogging.

Tare A feature that allows a balance to automatically subtract the weight of the weighing vessel from the total weight of the sample plus container.

Target organ The body part or organ most likely to be affected by exposure to a chemical or hazard.

TCP (Transmission Control Protocol) An Internet **protocol** that works with **IP** software to make sure that transmitted data packets arrive and are reassembled in the correct order, and that the data are intact.

TCP/IP software The programs that coordinate the use of **TCP** and **IP** protocols to provide the ability to send and receive information over the Internet.

Technology The application of knowledge to make products useful to humans.

Telecommunications The electronic transport of data in the forms of text, audio, and video over significant distances.

TEMED (N,N,N′,N′-tetramethylethylenediamine) A chemical catalyst for the polymerization of **acrylamide.**

Temperature A measure of the average energy of the randomly moving molecules that make up a substance. The tendency of a substance to lose or gain heat.

Temperature scale A scale derived by choosing two or more fixed reference points.

Terabyte (TB) Approximately 1 trillion bytes, 2^{40}.

Teratogen, teratogenic Compound that can cause defects in a fetus when administered to the mother.

Tertiary structure The three-dimensional globular structure formed by bending and twisting of a protein.

Test According to ISO Guide 17025: a technical operation that consists of the determination of one or more characteristics of performance of a given product, material, equipment, organism, physical phenomenon, process, or service according to a specified procedure. The result of a test is normally recorded in a document sometimes called a test report or a test certificate. Note that a "test" may or may not involve a "sample."

Testing laboratory A place where analysts test samples.

Text-based interface See **command-line interface.**

Text editor A program that allows the user to enter, edit, save, and print plain text.

Text mining Searching databases for topics of interest using key words; see **data mining.**

Therapeutic cloning A technology, not yet accomplished in humans, in which stem cells would be created by removing the genetic material from the cell of a patient and transferring it to an enucleated human egg from a donor woman. The resulting embryo would be allowed to divide a few times, and the embryonic stem cells harvested to treat the patient.

Thermal conductivity The ability of a material to conduct heat.

Thermal equilibrium The condition where two objects are at the same temperature, so that there is no net transfer of internal energy from one to the other.

Thermal expansion An increase in the size of a material due to an increase in temperature.

Thermal expansion coefficient The amount that a particular material expands with a given rise in temperature.

Thermistor A thermometer consisting of a semiconductor material that has a large change in resistance when exposed to a small change in temperature.

Thermocouple A temperature-sensing device that consists of two dissimilar metals joined together. The thermocouple has a voltage output proportional to the difference in temperature between the junction whose temperature is being measured and the reference junction.

Thermocycler An instrument that holds multiple tubes and that repeatedly heats and cools the tubes to specified temperatures.

Thin layer chromatography (TLC) A chromatographic technique where a thin layer of stationary phase material is spread on a glass plate, and the mobile phase passes through the stationary phase by either capillary action or gravity.

Threshold (for a graph) A change in a relationship plotted on a graph.

Throughput (in filtration) The length of time a liquid can flow through a filter before the membrane clogs.

Thumb drive See **flash drive.**

Thymine (T) One of the four types of nucleotide subunit that compose DNA.

Tissue culture The in vitro propagation of cells derived from tissue of higher organisms.

TLV (threshold limit value) The air concentration of a chemical that will not pose a health threat to most normal healthy workers; determined by the ACGIH.

TLV-C (threshold value limit–ceiling) The maximum allowable concentration of a material in air; concentrations should never exceed this value.

TLV-STEL (threshold limit value–short-term exposure limit) The concentration of a toxic material in air that limits exposure of a worker to 15 minutes; determined by the ACGIH.

TLV-TWA (threshold limit value–time-weighted average) The acceptable air concentration of a substance averaged over an 8-hour day; determined by the ACGIH.

To contain (TC) A method of calibrating glassware so that it contains the specified amount when exactly filled to the capacity line. The device will not deliver that amount if the liquid is poured out because some of the liquid will adhere to the sides of the container.

To deliver (TD) A method of calibrating glassware so that it delivers the specified amount when poured.

Tolerance The amount of error allowed in the calibration of a measuring item. For example, the tolerance for a "500 g" Class 1 mass standard is ± 1.2 mg, which means the standard must have a true mass between 500.0012 g and 499.9988 g.

Toolbar (in computers) The "floating" icon **menus** found in many applications.

Total bacteria count An estimation of the total number of bacteria in a solution based on an assay that involves incubating a sample and counting the number of colony-forming units (CFU).

Total dissolved solids (TDS) A value representing all the solids dissolved in a solution.

Total immersion thermometer A liquid-expansion thermometer designed to indicate temperature correctly when that portion of the thermometer containing the liquid is exposed to the temperature being measured. *Compare with **complete immersion thermometer** and **partial immersion thermometer.**

Total organic carbon (TOC) A measure of the concentration of organic contaminants in water.

Total solids (TS) In water treatment, a measure of the concentration of both the dissolved (TDS) and suspended solids (TSS) in a solution.

Total squared deviation See **sum of squares.**

Toxic Poisonous; a substance's ability to cause harm to biological organisms or tissue or to cause adverse health effects.

Toxic materials Substances that are poisonous.

Toxic Substances Control Act Authorizes EPA to, among other things, review new chemicals before they are introduced into commerce, including the examination of microorganisms that have been genetically modified.

Toxicology The study of poisonous materials and their effects on living organisms and tissue; the study of adverse effects of a drug and its metabolites on the body.

Trace analysis Analysis of a material that is present in very low (e.g., ppm or ppb) concentrations.

Traceability The ability to trace the history, application, or location of an **entity** . . . by means of recorded identifications. The term *traceability* may have one of three main meanings:

1. in a product . . . sense it may relate to the origin of materials and parts, the product processing history, the distribution and location of the product after delivery;

2. in a calibration sense, it relates measuring equipment to national or international standards, basic physical constants or properties, or reference materials;

3. in a data collection sense, it relates calculations and data generated throughout the quality loop . . . sometimes back to the requirements for quality . . . for an entity.**

Traceability (for weight standards) A documented genealogy that links a mass standard to the international prototype in Sevres, France.

Traceability (with reference to solutions) The process by which it is ensured that every component of a product can be identified and documented.

Trackball Hardware that functions as a stationary, upside-down mechanical mouse; instead of rolling a mouse, you roll the ball directly with your fingers or thumb.

Tracking error (in spectrophotometry) A problem that can occur when an absorbance spectrum is scanned too quickly, resulting in absorbance peaks that are slightly shifted from their true locations.

Trackpad A flat rectangular area next to a keyboard that controls the cursor instead of a **mouse.**

Trade secrets Private information or physical materials that give a competitive advantage to the owner.

Transcription The cellular process in which information encoded in a DNA sequence is used to synthesize a corresponding mRNA molecule.

Transducer A device that senses one form of energy and converts it to another form.

Transfection Introduction of foreign DNA into host cells; usually refers to eukaryotic host cells.

Transformation 1. A process in which cells undergo genetic changes that alter the normal mechanisms controlling cell growth and reproduction. 2. The introduction of foreign genetic information into bacteria.

Transformer (in electronics) Device used to vary the input voltage entering an instrument.

Transgenic A plant or animal that is genetically modified by the introduction of foreign DNA using the techniques of biotechnology.

Transistor A semiconductor component used primarily for amplification and sometimes to switch a circuit on or off.

Translation (in the context of cell biology) The cellular process in which the information encoded in mRNA is used to manufacture a protein.

Translational medicine Applied medical research as contrasted with basic research.

Transmittance (t) The ratio of the amount of light transmitted through the sample to that transmitted through the blank.

Transmitted light Light that passes through an object.

Trend line In scatter plots, a statistically derived line that reflects the general trend in the data values, if one exists.

Triple point of water The temperature and pressure at which solid, liquid, and gas phases of a given substance are simultaneously present in equilibrium. The triple point for water is 0.01°C.

Tris (tris(hydroxymethyl)aminomethane) One of the most common buffers in biotechnology laboratories. Tris buffers over the normal biological range (pH 7 to pH 9), is nontoxic to cells, and is relatively inexpensive.

Tris base Unconjugated **Tris**; has a basic pH when dissolved in water.

Tris-HCl **Tris** buffer that is conjugated to HCl.

tRNA A small type of RNA that transfers a specific amino acid to a growing polypeptide chain at the ribosome during protein synthesis.

Trojan horse, or **trojan (in computers)** Malware disguised as legitimate programs; while their functions can be similar to viruses, Trojans do not spread themselves to other computers.

True value (for a measurement) The actual value for a measurement that would be obtained in the absence of any error.

Trueness "The closeness of agreement between the population mean of the measurements or test results and the accepted reference value."****

Tungsten filament A thin metal wire that emits light when heated and provides visible light in visible spectrophotometry.

Turbid solution One that contains numerous small particles that both absorb and scatter light.

TWA (time weighted average) The concentration of a substance in air that is allowed when

---

*Based on definition in ASTM Standard E 344-07 "Terminology Relating to Thermometry and Hydrometry."

**Based on definitions from American National Standard ANSI/ASQC A8402-1994 "Quality Management and Quality Assurance," American Society for Quality Control, Milwaukee, WI.

***Based on definitions in ASTM Standard E 456-06 "Standard Terminology Relating to Quality and Statistics."

averaged over an eight-hour day; during the day the actual concentrations will be higher and lower than the daily average concentration.

Two-point calibration If the response of a device is linear, calibration at two points (e.g., 0 and full-scale) calibrates the device.

Two-stage gas regulator A gas regulator with a pair of valves that 1. greatly reduce the pressure of gas leaving the cylinder, and 2. provide the operator with the ability to fine-tune the gas pressure reaching its destination.

U.S. National Prototype Standards (for mass) Platinum-iridium standards with mass values traceable to the International Prototype mass standard.

Ultrafiltration A membrane separation technique that is used to separate macromolecules on the basis of their molecular weight.

Ultrafiltration membrane A filter with pores small enough to remove large molecules. These membranes are rated in terms of their molecular weight cutoff.

Ultramicrobalance An extremely sensitive analytical balance that can accurately weigh samples to the nearest 0.0000001 g.

Ultrapure water Type I or better water.

Ultrasonic washing A method of cleaning items by exposing them to high pitch sound waves that effectively penetrate and clean crevices, narrow spaces, and other difficult-to-reach spaces.

Ultraspeed centrifuge, or ultracentrifuge A centrifuge that rotates at speeds up to about 120,000 RPM and can generate forces up to $700,000 \times g$.

Ultraviolet (UV) light A form of nonionizing radiation which makes up the light spectrum between visible light and x-rays at 180–380 nm.

Uncertainty "An indication of the variability associated with a measured value that takes into account two major components of error: (1) bias, and (2) the random error attributed to the imprecision of the measurement process. . . . Quantitative measurements of uncertainty generally require descriptive statements of explanation because of differing traditions of usage and because of differing circumstances . . ."**

Undefined growth medium See **complex growth medium.**

Underwriters Laboratories (UL) An organization which has developed codes for safe electrical devices.

Uninstall program Software that deletes not only a main program folder, but also additional program files in various locations on the hard drive.

Uninterruptible power supply (UPS) A battery that will deliver constant power to the computer for a brief period during a power outage.

Unit (of measure) A precisely defined amount of a property.

United States Customary System (USCS) The measurement system common in the United States that includes miles, pounds, gallons, inches, and feet.

United States Department of Agriculture (USDA) A U.S. government agency that enforces requirements for purity and quality of meat, poultry, and eggs, and that is involved in nutrition research and education. The USDA is one of the government agencies that plays a role in the regulation of genetically engineered food plants.

United States Pharmacopeia (USP) 1. An organization that promotes public health by establishing and disseminating officially recognized standards for the use of medicines and other health care technologies. 2. The compendium containing drug descriptions, specifications, and standard test methods for such parameters as drug identity, strength, quality, and purity. This compendium is recognized as a legal authority by the FDA.

Universal absorbent Polypropylene, expanded silicates, or other materials that can absorb virtually any liquid safely, including some corrosives; can be purchased in loose or pillow form.

Universal precautions An approach to infection control, where all human blood and certain human body fluids are treated as if known to be infectious for **HIV, HBV,** and other **bloodborne pathogens** (OSHA).

UNIX A widely used operating system for large network computers.

Upgrade (in computers) Improving the performance of a computer by adding additional memory and storage.

Upload To copy a file from your computer to the Internet.

Upstream (in filtration) The side of a filter facing the incoming liquid or gas.

Upstream processes Biological processes, such as fermentation or cell culture, that produce the biomolecule of interest.

Upstream processing In biopharmaceutical manufacturing, the process of growing cells and their production of a desired product.

Uracil (U) One of the four types of nucleotide subunit that comprise RNA.

URL (uniform resource locator) A unique text address for every web page.

USB (universal serial bus) A universal type of port that provides high speed, stable connections between up to 127 devices from a single port.

USB drive See **flash drive.**

Usenet An Internet service that provides more than 25,000 public discussion groups or newsgroups, each generally focused on a single topic of interest.

Utility program A "housekeeping" program that enhances computer operations.

UV oxidation A process in which ultraviolet light breaks down organic impurities.

VAC (in electricity) An abbreviation for voltage when the current is alternating.

Validation A process or a set of activities that ensure that an individual piece of equipment, a process, or an analytical method reliably and effectively performs the function for which it is intended.

> **Process validation** Activities that prove that a manufacturing process meets its requirements so that the final product will do what it is supposed to do safely and effectively.

Equipment validation:

> **Installation qualification (IQ)** A set of activities designed to determine if a piece of equipment is installed correctly.

> **Operational qualification (OQ)** A set of activities designed to establish that a piece of equipment performs within acceptable limits.

> **Performance qualification (PQ)** A set of activities that evaluates the performance of a piece of equipment under the conditions of actual use.

Validation (of an analytical procedure) A process used by the scientific community to evaluate a test method or assay. This involves determining the ability of the method to reliably obtain a desired result, the conditions under which such results can be obtained, and the limitations of the method.

Value (in math) An assigned or calculated numerical quantity.

Variable 1. A property or a characteristic that can have various values (in contrast to a constant that always has the same value). See also **dependent variable** and **independent variable.** 2. A quality of a population that can be measured. For example, for the population of all six-year-old children, one could measure height, eye color, or favorite book. Variables may be classified as: 1. **Discrete variables** are those that can be counted, such as litter size or number of bacterial colonies on a petri dish. 2. **Continuous variables** can be measured and can be whole numbers or any fraction of a whole number, such as weight or temperature. 3. **Qualitative variables** are attributes that are not numeric, such as color or flavor.

Variable air volume (VAV) fume hoods Fume hoods that maintain a relatively constant **face velocity** by changing the amount of air exhausted from the hood.

Variable resistor (in electronics) Device that can be adjusted to provide differing amounts of resistance to electron flow. (See also **rheostat** and **Potentiometer.**)

Variance "A measure of the squared dispersion of observed values or measurements expressed as a function of the sum of the squared deviations from the population mean or sample average." **

VDC (in electricity) An abbreviation for voltage when the current is direct.

Vector An entity, such as a plasmid or a modified virus, into which a DNA fragment of interest is integrated, and that carries the DNA of interest into a host cell.

Vector graphics Images based on a set of lines, which can be described and stored by the software as mathematical equations; also called object-oriented graphics.

Velometer, Velocity meter An instrument used to measure the **face velocity** of a fume hood.

Verification. "Confirmation by examination and provision of objective evidence that specified requirements have been fulfilled . . ."* 1. In connection with the management of measuring equipment, verification provides a means of checking that the deviations between values indicated by a measuring instrument and corresponding known

---

*Based on definitions in ISO/IEC Guide 25, "General Requirement for the Competence of Testing and Calibration Laboratories," Geneva, Switzerland. August 1996 edition.

**Based on definitions in ASTM Standard E 456-06 "Standard Terminology Relating to Quality and Statistics."

values are consistently smaller than the limits of permissible error defined in a standard regulation or specification relevant to the management of the measuring equipment. 2. In connection with measurement traceability, the meaning of the terms *verification* and *verified* is that of a simplified calibration, giving evidence that specified metrological requirements including the compliance with given limits of error are met."

Vertical rotor A rotor in which tubes are held upright in the sample compartments.

Viable organism A living organism that can reproduce.

Virtual private network (VPN) A network that uses internet resources (frequently the Internet) to connect members of a private network across longer distances.

Virulence The capacity of a biological organism or material to cause harm.

Virus Infectious particle containing either DNA or RNA surrounded by a protein coat.

Virus (in computers) Malware designed to infect any computer that executes the program, then replicate and spread itself to other computers.

Visible radiation The region of the electromagnetic spectrum from about 380–780 nm. Light of different wavelengths in this range is perceived as different colors.

Volatile (in computers) Describes data contained in **RAM** that will disappear when you turn off the computer or a power outage occurs.

Volatile (in safety) Refers to chemicals that evaporate quickly at room temperature.

Voltage The potential energy of charges that are separated from one another and attract or repulse one another.

Voltmeter An instrument used to measure voltage.

Volts Units of potential difference between two points in an electrical circuit.

Volume The amount of space a substance occupies, defined in the SI system as length × length × length = meters^3. More commonly, the liter (dm^3) is used as the basic unit of volume. See also **liter.**

Volumetric (transfer) pipette A pipette made of borosilicate glass and calibrated "to deliver" a single volume when filled to its capacity line at 20°C.

Volumetric glassware A term used generally to refer to accurately calibrated glassware intended for applications where high accuracy volume measurements are required.

Warm boot Closing down computer operations then restarting the system without turning off the power.

Waste Any laboratory material which has completed its original purpose and is being disposed of permanently.

Water aspirator A laboratory set-up that creates a vacuum through a side arm to a faucet with flowing water.

Water for injection (WFI) Water used to make drugs for injection. The USP specifies the quality of WFI.

Water softener An ion exchange device that exchanges positive ions that make water hard, primarily calcium and magnesium, with sodium ions.

Watts (W) (in electricity) Units of power.

Wavelength (λ) The distance from the crest of one wave to the crest of the next wave.

Wavelength accuracy The agreement between the wavelength the operator selects and the actual wavelength that exits the monochromator and shines on the sample.

Wayback Machine, The (http://www.archive.org) An Internet archive with more than 60 billion copies of web pages dating back to 1996.

Weak acid A chemical that partially dissociates in water with the release of hydrogen ions.

Weak base A chemical that partially dissociates in water with the release of hydroxide ions.

Weak electrolyte A substance that partially ionizes when dissolved.

Weak molecular interactions Attractive interactions between atoms that are more easily disrupted than covalent bonds.

Web crawler Bots, or browsers, that independently move through the Web following every link they find, identifying web pages for search engines.

Web spider See **Web crawler.**

Web-based software Programs that are located on the Internet, allowing users to work with and store data online.

Weigh boat A plastic or metal container designed to hold a sample for weighing.

Weighing paper Glassine coated paper used to hold small samples for weighing.

Weight The force of gravity on an object.

Wi-Fi (wireless fidelity) A high-speed wireless networking standard conforming to IEEE 802.11 standards.

Wiki A set of collaborative web pages that anyone can create or modify over time; derived from a Hawaiian expression for "fast."

Wikipedia A free, open content encyclopedia wiki, with millions of articles in 253 languages.

Window (in computers) A rectangular viewing and work area.

Windows (in computers) The most popular **operating system** for personal computers.

Wireless network A network that uses radio frequencies or other means to establish communication between computers without the use of physical connections.

Word processor Software that combines text editing with complex formatting and other functions such as spell checking, creating tables, and making outlines.

Workbook (in spreadsheets) A set of **worksheets** stored in a single file.

Workers' Compensation A no-fault state insurance system designed to pay for the medical expenses of workers who are injured on the job, or develop work-related medical problems.

Working Cell Bank (WCB) Cells derived from the master cell bank that are expanded and used for production of a product.

Working standard A physical standard that is used to make measurements in the laboratory and that is calibrated against a primary or secondary standard.

Worksheet (in spreadsheets) A single **spreadsheet** page.

Workstation (in computers) An individual computer connected to a network.

World Wide Web, The A user-friendly service provided through the Internet using HTTP.

Worm (in computers) A self-replicating **virus.**

X axis The main horizontal line on a graph.

"X Solution" A stock solution where X means how many times more concentrated the stock is than normal. A 10 X solution is 10 times more concentrated than the solution is normally prepared.

Xenotransplantation Transplantation of organs, tissues, or cells from one species to another.

XML (Extensible Markup Language) The computer language (similar to HTML) used to create feeds.

Y axis The main vertical line on a graph.

Y intercept The point at which a line passes through the Y axis, where X = 0, on a two-dimensional graph.

Yeast extract The water soluble portion of yeast cells that have been allowed to die so that the yeasts' digestive enzymes break down their proteins into simpler compounds while preserving the vitamins from the yeast.

Yield The **% recovery,** or percent of the starting amount of the product of interest that can be recovered in purified form using a specific strategy.

Z factor A conversion factor that incorporates the buoyancy correction for air at a particular temperature and barometric pressure.

Zonal rotor Rotors that accommodate large volume samples by containing the sample in a large cylindrical cavity rather than in individual tubes or bottles. Zonal rotors are used for both rate zonal and buoyant density gradient separations, but are seldom used for pelleting.

Zoonotic diseases, zoonoses Diseases that can be passed between different species.

Note: *Italicized* page numbers indicate entries in the Glossary section.